# Methods in Enzymology

## Volume 393
## CIRCADIAN RHYTHMS

# METHODS IN ENZYMOLOGY

## EDITORS-IN-CHIEF

### John N. Abelson     Melvin I. Simon

DIVISION OF BIOLOGY
CALIFORNIA INSTITUTE OF TECHNOLOGY
PASADENA, CALIFORNIA

## FOUNDING EDITORS

### Sidney P. Colowick and Nathan O. Kaplan

Methods in Enzymology

Volume 393

# Circadian Rhythms

EDITED BY

## Michael W. Young

LABORATORY OF GENETICS
THE ROCKEFELLER UNIVERSITY
NEW YORK, NEW YORK

ELSEVIER
ACADEMIC
PRESS

AMSTERDAM • BOSTON • HEIDELBERG • LONDON
NEW YORK • OXFORD • PARIS • SAN DIEGO
SAN FRANCISCO • SINGAPORE • SYDNEY • TOKYO

Elsevier Academic Press
525 B Street, Suite 1900, San Diego, California 92101-4495, USA
84 Theobald's Road, London WC1X 8RR, UK

For all information on all Academic Press publications
visit our Web site at www.books.elsevier.com

ISBN: 0-12-182798-4

PRINTED IN THE UNITED STATES OF AMERICA
05  06  07  08  9  8  7  6  5  4  3  2  1

# Table of Contents

## Section I. Genetic Approaches to Circadian Clocks

v

# Section V. Mosaic Circadian Systems

# Section VI. Peripheral Circadian Clocks

# Section VII. Cell and Tissue Culture System

# Section VIII. Intercellular Signaling

# Section IX. Photoresponsive Clocks

# Section X. Sleeping Flies

# Section XI. Circadian Biology of Populations

# Section XII. Circadian Clocks Affecting Noncircadian Biology

# Contributors to Volume 393

Article numbers are in parentheses and following the names of contributors.
Affiliations listed are current.

BIKEM AKTEN (35), *Department of Neuroscience, Sackler Graduate Program of Neuroscience, Tufts University School of Medicine, Boston, Massachusetts 02111*

ROZI ANDRETIC (40), *The Neurosciences Institute, San Diego, California 92121*

JUSTIN BLAU (36), *Department of Biology, New York University, New York, New York 10003*

GENE D. BLOCK (33), *Center for Biological Timing, Department of Biology, University of Virginia, Charlottesville, Virginia 22903*

STEVEN A. BROWN (28), *Department of Molecular Biology, Sciences III, University of Geneva, CH-1211 Geneva-4, Switzerland*

JAY BURNETTE (44), *Department of Biology, University of Virginia, Charlottesville, Virginia 22903*

GREGORY CAHILL (5), *Department of Biology and Biochemistry, University of Houston, Houston, Texas 77204*

LUCA CARDONE (5), *Institut de Génétique et de Biologie Moléculaire et Cellulaire, 67404 Illkirch-Strasbourg, France*

VINCENT M. CASSONE (29), *Center for Research on Biological Clocks, Department of Biology, Texas A&M University, College Station, Texas 77843*

YAN CHENG (27), *Center for Experimental Therapeutics, University of Pennsylvania School of Medicine, Philadelphia, Pennsylvania 19104*

SEHYUNG CHO (23), *Institut de Génétique et de Biologie Moléculaire et Cellulaire, Centre National de la Recherche Scientifique, 67404 Illkirch-Strasbourg, France*

STEPHANIE CRAPO (19), *Huntsman Cancer Institute, University of Utah School of Medicine, Salt Lake City, Utah 84112*

ANNE M. CURTIS (27), *Center for Experimental Therapeutics, University of Pennsylvania School of Medicine, Philadelphia, Pennsylvania 19104*

ALEC J. DAVIDSON (26), *Department of Biology, University of Virginia, Charlottesville, Virginia 22903*

CHARNA DIBNER (28), *Department of Molecular Biology, Sciences III, University of Geneva, CH-1211 Geneva-4, Switzerland*

MASAO DOI (23), *Institut de Génétique et de Biologie Moléculaire et Cellulaire, Centre National de la Recherche Scientifique, 67404 Illkirch-Strasbourg, France*

STUART E. DRYER (25), *Department of Biology and Biochemistry, University of Houston, Houston, Texas 77204*

GILES DUFFIELD (14), *Department of Genetics, Dartmouth Medical School, Hanover, New Hampshire 03755*

JAY C. DUNLAP (1, 14), *Department of Genetics, Dartmouth Medical School, Hanover, New Hampshire 03755*

DAVID J. EARNEST (29), *Center for Research on Biological Clocks and Department of Human Anatomy and Medical Neurobiology, Texas A&M University Health Sciences Center, College Station, Texas 77843*

ISAAC EDERY (18), *Department of Molecular Biology and Biochemistry, Rutgers University, Center for Advanced Biotechnology and Medicine, Piscataway, New Jersey 08854*

ERIK J. EIDE (19), *Huntsman Cancer Institute, University of Utah, Salt Lake City, Utah 84112*

GARRET FITZGERALD (27), *Center for Experimental Therapeutics, University of Pennsylvania School of Medicine, Philadelphia, Pennsylvania 19104*

RUSSELL G. FOSTER (37), *Department of Visual Neuroscience, Imperial College, Charing Cross Hospital, London W6 8RF, United Kingdom*

YING-HUI FU (9), *Howard Hughes Medical Institute, Department of Neurology, University of California, San Francisco, California 94143*

MONICA GALLEGO (19), *Huntsman Cancer Institute, University of Utah School of Medicine, Salt Lake City, Utah 84112*

GINKA K. GENOVA (35), *Department of Neuroscience, Tufts University School of Medicine, Boston, Massachusetts 02111*

REBECCA GEORGE (44), *Department of Biology, University of Virginia, Charlottesville, Virginia 22903*

MARTHA U. GILLETTE (31), *Department of Cell & Structural Biology and Neuroscience Program and Department of Molecular & Integrative Physiology, University of Illinois at Urbana-Champaign, Urbana, Illinois 60801*

CARLA B. GREEN (6), *Department of Biology, University of Virginia, Charlottesville, Virginia 22903*

JEFFREY C. HALL (4), *Department of Biology, Brandeis University, Waltham, Massachusetts 02454*

MARK W. HANKINS (37), *Department of Visual Neuroscience, Imperial College, Charing Cross Hospital, London W6 8RF, United Kingdom*

PAUL E. HARDIN (25), *Department of Biology and Biochemistry, University of Houston, Houston, Texas 77204*

MICHAEL H. HASTINGS (30), *MRC Laboratory of Molecular Biology, Division of Neurobiology, Cambridge CB2 2QH, United Kingdom*

NAOTO HAYASAKA (6), *Department of Biology, University of Virginia, Charlottesville, Virginia 22903 ***

CHARLOTTE HELFRICH-FÖRSTER (21), *University of Regensburg, Institute of Zoology, Universitätsstraße 31, 93040 Regensburg, Germany*

JUN HIRAYAMA (5), *Institut de Génétique et de Biologie Moléculaire et Cellulaire, 67404 Illkirch-Strasbourg, France*

JAY HIRSH (44), *Department of Biology, University of Virginia, Charlottesville, Virginia 22903*

KAREN S. HO (41), *Howard Hughes Medical Institute, Department of Neuroscience, University of Pennsylvania Medical School, Philadelphia, Pennsylvania 19104*

*Present address: Department of Anatomy and Neurobiology, Kinki University School of Medicine, Osaka-Sayama, Osaka, 589-8511 Japan*

JOHN B. HOGENESCH (16), *The Genomics Institute of the Novartis Research Foundation, San Diego, California 92121 and Department of Neuropharmacology, The Scripps Research Institute, La Jolla, California 92037*

TODD C. HOLMES (36), *Department of Biology, New York University, New York, New York 10003*

YANMEI HUANG (35), *Department of Neuroscience, Tufts University School of Medicine, Boston, Massachusetts 02111*

KANAE IIJIMA-ANDO (13), *Cold Spring Harbor Laboratory, Cold Spring Harbor, New York, 11724*

TAKATO IMAIZUMI, (11), *Department of Cell Biology, The Scripps Research Institute, La Jolla, California 92037*

F. ROB JACKSON (35), *Department of Neuroscience, Tufts University School of Medicine, Boston, Massachusetts 02111*

CARL HIRSCHIE JOHNSON (43), *Department of Biological Sciences, Vanderbilt University, Nashville, Tennessee 37235*

CHRISTOPHER R. JONES (9), *Department of Neurology, University of Utah, Salt Lake City, Utah 84132*

MAKI KANEKO (5), *Department of Biology and Biochemistry, University of Houston, Houston, Texas 77204*

HEESEOG KANG (19), *Huntsman Cancer Institute, University of Utah School of Medicine, Salt Lake City, Utah 84112*

STEVE A. KAY (11), *Department of Cell Biology, The Scripps Research Institute, La Jolla, California 92037; Department of Psychiatry, San Diego VA Medical Center, University of California, San Diego, La Jolla, California 92093*

YELENA KLEYNER (35), *Department of Neuroscience, Sackler Graduate Program of Neuroscience, Tufts University School of Medicine, Boston, Massachusetts 02111*

HYUK WAN KO (18), *Graduate Program in Neuroscience, Rutgers University, Center for Advanced Biotechnology and Medicine, Piscataway, New Jersey 08854*

BENOÎT KORNMANN (28), *Department of Molecular Biology, Sciences III, University of Geneva, CH-1211 Geneva-4, Switzerland*

ACHIM KRAMER (34), *Institute of Medical Immunology, Laboratory of Chronobiology, Charité-Universitaets Medizin Berlin, 10115 Berlin, Germany*

SEBASTIAN KRAVES (34), *Department of Neurobiology, Harvard Medical School, Boston, Massachusetts 02115*

PARTHASARATHY KRISHNAN (25), *Department of Biology and Biochemistry, University of Houston, Houston, Texas 77204*

CHARALAMBOS P. KYRIACOU (42), *Department of Genetics, University of Leicester, Leicester LE1 7RH, United Kingdom*

SILVIA I. LA RUE (6), *Department of Biology, University of Virginia, Charlottesville, Virginia 22903*

KEVIN LEASE (44), *Department of Biology, University of Virginia, Charlottesville, Virginia 22903*

CHENG CHI LEE (45), *Department of Biochemistry and Molecular Biology, University of Texas Health Science Center, Houston, Texas 77030*

YI LIU (17), *Department of Physiology, University of Texas Southwestern Medical Center, Dallas, Texas 75390*

JENNIFER J. LOROS (1, 14), *Department of Genetics and Department of Biochemistry, Dartmouth Medical School, Hanover, New Hampshire 03755*

SHARON S. LOW-ZEDDIES (24), *MusWorks, Inc., Rockville, Maryland 20855*

GABRIELLA B. LUNDKVIST (33), *Center for Biological Timing, Department of Biology, University of Virginia, Charlottesville, Virginia 22903*

ELISABETH S. MAYWOOD (30), *MRC Laboratory of Molecular Biology, Division of Neurobiology, Cambridge CB2 2QH, United Kingdom*

DOUGLAS G. MCMAHON (30), *Department of Biological Sciences, Vanderbilt University, Vashville, Tennessee 37235*

MICHAEL MENAKER (26), *Department of Biology, University of Virginia, Charlottesville, Virginia 22903*

JEROME S. MENET (32), *Howard Hughes Medical Institute, Department of Biology, Brandeis University, Waltham, Massachusetts 02454*

MARTHA MERROW (10), *Institute for Medical Psychology, University of Munich, 80336 Munich, Germany*

ANDREW J. MILLAR (2), *Department of Biological Sciences, University of Warwick, Coventry CV4 7AL, United Kingdom*

FELIX NAEF (15), *ISREC, CH-1066 Epalinges, Switzerland*

EMI NAGOSHI (28), *Department of Molecular Biology, Sciences III, University of Geneva, CH-1211 Geneva-4, Switzerland* [†]

PIPAT NAWATHEAN, (32), *Howard Hughes Medical Institute, Department of Biology, Brandeis University, Waltham, Massachusetts 02454*

MICHAEL N. NITABACH (36), *Department of Cellular and Molecular Physiology, Yale University School of Medicine, New Haven, Connecticut 06520*

HITOSHI OKAMURA (20), *Division of Molecular Brain Science, Department of Brain Sciences, Kobe University Graduate School of Medicine, Kobe 650-0017, Japan*

CARRIE L. PARTCH (38), *Department of Biochemistry and Biophysics, University of North Carolina School of Medicine, Chapel Hill, North Carolina 27599*

STUART N. PEIRSON (37), *Department of Visual Neuroscience, Imperial College, Charing Cross Hospital, London W6 8RF, United Kingdom*

JEFFREY L. PRICE (3), *University of Missouri – Kansas City, School of Biological Sciences, Kansas City, Missouri 64110*

LOUIS J. PTÁCEK (9), *Howard Hughes Medical Institute, Department of Neurology, University of California, San Francisco, California 94143*

AKHILESH B. REDDY (30), *MRC Laboratory of Molecular Biology, Division of Neurobiology, Cambridge CB2 2QH, United Kingdom*

MARY A. ROBERTS (35), *Department of Neuroscience, Tufts University School of Medicine, Boston, Massachusetts 02111*

[†]*Present address: Department of Biology – HHMI, Brandeis University, Waltham, Massachusetts 02454*

TILL ROENNEBERG (10), *Institute for Medical Psychology, University of Munich, 80336 Munich, Germany*

MICHAEL ROSBASH (32), *Howard Hughes Medical Institute, Department of Biology, Brandeis University, Waltham, Massachusetts 02454*

R. DANIEL RUDIC (27), *Center for Experimental Therapeutics, University of Pennsylvania School of Medicine, Philadelphia, Pennsylvania 19104*

AZIZ SANCAR (38), *Department of Biochemistry and Biophysics, University of North Carolina School of Medicine, Chapel Hill, North Carolina 27599*

PAOLO SASSONE-CORSI (5, 23), *Institut de Génétique et de Biologie Moléculaire et Cellulaire, Centre National de la Recherche Scientifique, 67404 Illkirch-Strasbourg, France*

UELI SCHIBLER (28), *Department of Molecular Biology, Sciences III, University of Geneva, CH-1211 Geneva-4, Switzerland*

WILLIAM J. SCHWARTZ (22), *Department of Neurology, University of Massachusetts Medical School, Worcester, Massachusetts 01655*

AMITA SEHGAL (41), *Howard Hughes Medical Institute, Department of Neuroscience, University of Pennsylvania Medical School, Philadelphia, Pennsylvania 19104*

PAUL J. SHAW (40), *The Neurosciences Institute, San Diego, California 92121, and Department of Anatomy and Neurobiology, Washington University Medical School, St. Louis, Missouri 63110*

SANDRA M. SIEPKA (7, 8), *Center for Functional Genomics, Northwestern University, Evanston, Illinois 60208*

RAE SILVER (22), *Departments of Psychology, Barnard College and Columbia University and Department of Anatomy and Cell Biology, Health Sciences, Columbia University, New York, New York 10027*

MEGAN M. SOUTHERN (2), *Department of Biological Sciences, University of Warwick, Coventry CV4 7AL, United Kingdom*

JOOWON SUH (35), *Department of Neuroscience, Sackler Graduate Program of Neuroscience, Tufts University School of Medicine, Boston, Massachusetts 02111*

VASUDHA SUNDRAM (35), *Department of Neuroscience, Tufts University School of Medicine, Boston, Massachusetts 02111*

JOSEPH S. TAKAHASHI (7, 8, 12, 24), *Howard Hughes Medical Institute, Department of Neurobiology & Physiology, Northwestern University, Evanston, Illinois 60208*

FILIPPO TAMANINI (20), *Department of Cell Biology & Genetics, Erasmus University, 3000DR Rotterdam, The Netherlands*

ÖZGUR TATAROGLU (26), *Department of Biology, University of Virginia, Charlottesville, Virginia 22903*

ERAN TAUBER (42), *Department of Genetics, University of Leicester, Leicester LE1 7RH, United Kingdom*

STEWART THOMPSON (37), *Department of Visual Neuroscience, Imperial College, Charing Cross Hospital, London W6 8RF, United Kingdom*

SHELLEY A. TISCHKAU (31), *Department of Cell & Structural Biology and Neuroscience Program, University of Illinois at Urbana-Champaign, Urbana, Illinois 60801* [‡]

‡*Present Address: Department of Veterinary Biosciences, University of Illinois at Urbana-Champaign, Urbana, Illinois 60801*

GIJSBERTUS T. J. VAN DER HORST (20), *Department of Cell Biology & Genetics, Erasmus University, 3000DR Rotterdam, The Netherlands*

RUSSELL N. VAN GELDER (39), *Department of Ophthalmology and Visual Sciences, Washington University Medical School, St. Louis, Missouri 63110*

DAVID M. VIRSHUP (19), *Huntsman Cancer Institute, University of Utah School of Medicine, Salt Lake City, Utah 84112*

JOHN R. WALKER (16), *The Genomics Institute of the Novartis Research Foundation, San Diego, California 92121*

CHARLES J. WEITZ (34), *Department of Neurobiology, Harvard Medical School, Boston, Massachusetts 02115*

DAVID K. WELSH, (11), *Department of Cell Biology, The Scripps Research Institute, La Jolla, California 92037; Department of Psychiatry, San Diego VA Medical Center, University of California, San Diego, La Jolla, California 92093*

HERMAN WIJNEN (15), *Laboratory of Genetics, The Rockefeller University, New York, New York 10021*

KAZUHIRO YAGITA (20), *Unit of Circadian Systems, Department of Biological Science, Nagoya University Graduate School of Science, Furo-cho, Chikusa-ku, Nagoya 464-8602, Japan*

SHIN YAMAZAKI (12), *Department of Biological Sciences, Vanderbilt University, Nashville, Tennessee 37235*

FU-CHIA YANG (34), *Department of Neurobiology, Harvard Medical School, Boston, Massachusetts 02115*

JERRY C. P. YIN (13), *Cold Spring Laboratory, Cold Spring, New York 11724*[§]

MICHAEL W. YOUNG (15), *Laboratory of Genetics, The Rockefeller University, New York, New York 10021*

IRENE YUJNOVSKY (23), *Institut de Génétique et de Biologie Moléculaire et Cellulaire, Centre National de la Recherche Scientifique, 67404 Illkirch-Strasbourg, France*

[§]*Present Address: University of Wisconsin, Madison, Genetics Department, Madison, Wisconsin 53705*

# Preface

We've just passed the 20th anniversary of the physical cloning of *period* (*per*), the first circadian rhythm or "clock" gene. Although this can hardly be called a ripe old age, there was much ground-breaking associated with this early fusion of molecular and circadian biology. Among the handful of genes physically isolated in the late 1970s and early 1980s, *per* was uniquely magnetic. Genes encoding a few enzymes and structural proteins of known function were collected in this time-frame, and beyond these, the first genes affecting development and morphology, but *per* stood out as the only gene in hand that controlled animal behavior. *per* was isolated with the specific hope that its anatomy and function would reveal an underlying chemistry of behavioral cycles.

This first phase of this new work, which focused on an insect, *Drosophila*, proved that *per*'s contributions to rhythms were connected to the production of a single protein. Arrhythmia occurred when this protein could not be generated, and transplantations of cloned versions of *per* that restored this protein brought robust behavioral rhythmicity back to clock-less flies. More importantly, fast- and slow-running clocks in mutant flies were found to be connected to subtle changes in this protein's structure. A generation of "chronobiologists" came to believe that the problem of the clock might be reduced to defining the activity of such a protein. A second phase of the work showed that *per* was rhythmically expressed and that such cycles required the collaboration of genes and proteins within a small oscillatory network. Rhythmic appearance of PER and other network proteins came from (1) transcriptional feedback loops (many of the proteins were auto-regulatory transcription factors), and (2) physical interactions among network proteins that determined stability and sub-cellular location.

Today we know that similar molecular rules govern the daily cycles of bacteria, fungi, plants, and "higher" animals, including humans. In fact, orthologs of *per* and its closest partners *timeless, double-time, Clock, cycle,* and *cryptochrome* appear to have been carried over from the fly to play important, and sometimes identical, roles in the mammalian circadian clock.

Research in circadian biology, though centuries old, has been intensively ticking away for more than 50 years now. Much of the work preceding the introduction of molecular biology was essential to the framing of any question focused on genes and proteins. This early work clarified the endogenous nature of circadian clocks, and showed that environmental stimuli could reset them.

Carefully extracted rules indicated that ordered transitions in a chemical mechanism must register internal time of day or night, and strongly indicated that rhythmicity would emerge at the level of single cells. As clock genes and proteins became available, experiments were designed to understand the physical bases for these classical rules.

The many chapters assembled in this edition of Methods in Enzymology show a pattern of rapid growth through assimilation. Today, everyone interested in biological clocks seems to study genes and proteins, but work in this field continues to be carried out by individual investigators who are likely to be conducting experiments involving animal behavior in one part of the laboratory, while electrophysiology, cell biology, protein chemistry, or molecular biology are briskly running a few benches (or walls) away. I believe the range of approaches routinely applied in this sphere is exceptionally broad, and an easy willingness to step across disciplinary boundaries to share tools and special knowledge has been a trademark of this field.

I am deeply grateful to our authors, all of whom have continued in this tradition. They have given us a unique assembly of their expertise. I also thank Minsun Kwon. Minsun kept everything on track, collecting, and formatting chapters as they arrived and coaxing everyone along so that this volume could be published in a timely fashion.

MICHAEL W. YOUNG

# METHODS IN ENZYMOLOGY

# Section I

# Genetic Approaches to Circadian Clocks

# [1]  Analysis of Circadian Rhythms in *Neurospora:* Overview of Assays and Genetic and Molecular Biological Manipulation

*By* JAY C. DUNLAP and JENNIFER J. LOROS

## Abstract

The eukaryotic filamentous fungus *Neurospora crassa* is a tractable model system that has provided numerous insights into the molecular basis of circadian rhythms. In the core circadian clock feedback loop, WC-1 and WC-2 interact via PAS domains to heterodimerize, and this complex acts both as the circadian photoreceptor and, in the dark, as a transcription factor that promotes the expression of the *frq* gene. In the negative step of the loop, dimers of FRQ feed back to block the activity of the WC-1/WC-2 complex (WCC) and, in a positive step, to promote the synthesis of WC-1. Several kinases phosphorylate FRQ, leading to its ubiquitination and turnover, releasing the WC-1/WC-2 dimer to reactivate *frq* expression and restart the circadian cycle. Light and temperature entrainment of the clock arise from rapid light induction of *frq* expression and from the effect of elevated temperatures in driving higher levels of FRQ. Noncircadian candidate slave oscillators, termed FRQ-less oscillators (FLOs), have been described, each of which appears to regulate aspects of *Neurospora* growth or development. Overall, the core FRQ/WCC feedback loop coordinates the circadian system by regulating downstream clock-controlled genes either directly or via regulation of driven FLOs. This article provides a brief synopsis of the system and describes current assays for the *Neurospora* clock. Methods for genetic and molecular manipulation of the core clock are summarized, and accompanying chapters address more specifically aspects of photobiology and output.

## Introduction

*Neurospora* is a filamentous fungus, a widely studied model organism for the kingdom within the eukaryotes having the closest evolutionary relatedness to animals and plants (Stechmann and Cavalier-Smith, 2003). In its life cycle, *Neurospora* employs just a few distinct cell types, some cellular and some syncytial. In its vegetative state it grows as a syncytium, a growth habit more similar to that of muscle cells than of fibroblasts. Because, unlike metazoans or higher plants, there are not a lot of different cell types simultaneously expressing rhythms or making up the whole adult organism, all the power of genetics, genomics, and molecular biology can be brought to bear on how the circadian system works at the level of a cell, from light and temperature input, to the core circadian oscillator, to various heirarchical slave oscillators and outputs. The insights gained from *Neurospora* have resulted from this ability to define and refine chronobiologically meaningful questions in a molecularly tractable way; in answering these questions, we have increased our insight into clocks in all organisms.

This overview, after introducing the organism and its clock, focuses on the tools available for the *Neurospora* system and how they have been used to dissect the clockworks. Separate chapters deal independently with the analysis of output, photobiology, and biochemical examination of phosphorylation and turnover.

Rhythms in *Neurospora*

Fungi, including *Neurospora*, express a variety of different rhythms having widely disparate period lengths and characteristics. These include rhythms in hyphal branching and/or in growth rate or morphology, the period lengths of which are a function of the temperature, or nutritional conditions under which the fungus is growing (Bünning, 1973; Ingold, 1971). Such noncircadian rhythms are often exposed under environmental stresses such as growth on sorbose (Feldman and Hoyle, 1974) or when the circadian clock is disabled through mutation. They can generally be driven by external cycles, especially temperature cycles (Merrow *et al.*, 1999), but often lack the ability to self-sustain or to appear without an external stimulus (Loros and Feldman, 1986). Although important to the biology of the organisms, such rhythms are not characterized as true circadian rhythms, as they lack one or more of the characteristics defining rhythms as "circadian" (Dunlap *et al.*, 2003; Sweeney, 1976). One rhythm, however, in the developmental patterning of *Neurospora*, asexual sporulation, seen when cultures grew over an agar surface, was shown by Pittendrigh and colleagues (1959) to be circadianly controlled. *Neurospora* has since became a favored organism for biological clock studies because of its ease of culture, excellent genetics, and ease of molecular manipulation. Today it represents a salient model system for the analysis and molecular dissection of circadian oscillatory systems.

The importance of the *Neurospora* system initially derived from its tractability to genetic analyses; the first *Neurospora* clock-mutant strains were identified by Feldman and colleagues (1971, 1973) coincident with those found in *Drosophila* by Konopka and Benzer (1971). Subsequently, molecular analysis of the *Neurospora* circadian system was initiated shortly after that of *Drosophila* in the mid-1980s and resulted in cloning of the clock gene *frequency* (*frq*) (McClung *et al.*, 1989) and completion of the first global screens for clock-controlled genes (ccgs) in any organism (Loros *et al.*, 1989). Because of the tractability of the system and the ease of genetic transformation, work in *Neurospora* pioneered studies in which suspected molecular components were manipulated to determine their role in the clock (Aronson *et al.*, 1994b). This work eventually proved *frq* to encode a central component of the core oscillator itself and that the PAS proteins WC-1 and WC-2, acting as a heterodimeric complex, activate the expression of *frq* (Crosthwaite *et al.*, 1997). With this genetic and molecular framework in place, steady progress has been made in understanding the molecular basis for sustainability of the rhythm, period length, resetting of the circadian system by light and temperature cues, and gating of input cues. Because the formal properties and molecular mechanisms underlying the clock appear to be quite similar in all organisms with circadian rhythms, *Neurospora* and other microbial systems are likely to remain favored model systems for genetic and molecular analysis of these clocks.

To provide a framework for the methods described here and in later articles and to place the various genes and their products in a biological context, this article begins with a brief overview of the known molecular components of the *Neurospora* clock and the expected layout of the circadian system (Fig. 1).

## Overview of Molecular Events in the *Neurospora* Circadian Cycle

Transcription of the *frq* gene begins late at night because its activators [WC-1 and WC-2, acting as the white collar complex (WCC)] are present and its repressor, FRQ, is at low levels. The WCC binds to the promoter of the *frq* gene (Froehlich *et al.*, 2003), and the *frq* transcript is produced and begins to appear by late subjective night, even before real or subjective dawn. Primary transcripts are spliced in a complex manner (H. U. Colot, J. J. Loros, and J. C. Dunlap, unpublished results) that determines the ratio of long to short FRQ that is translated (Liu *et al.*, 1997). FRQ proteins appear by early morning (Garceau *et al.*, 1997), dimerize (Cheng *et al.*, 2001a), and soon enter the nucleus (Garceau *et al.*, 1997; Luo *et al.*, 1998) where they fulfill all their known roles in the clock. FRQ participates in three separate essential actions within the clock cycle. First, FRQ interacts physically with the WCC in the nucleus (Cheng *et al.*, 2001a; Denault *et al.*, 2001; Merrow *et al.*, 2001) and blocks the ability of the WCC to activate transcription (Froehlich *et al.*, 2003). This seems to be its dominant role because when FRQ is present, less *frq* is produced and, by midday, WCC activity is declining to its lowest level as FRQ levels rise (Lee *et al.*, 2000).

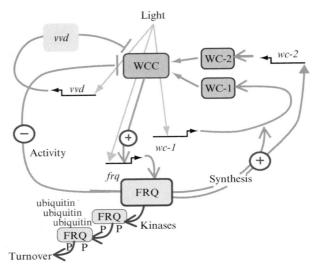

FIG. 1. Known molecular components in the *Neurospora* circadian clock. Genes and proteins with known roles in the clock are shown. Lines ending in bars connote negative regulation or repression, and arrows connote positive regulation (activation). P on FRQ denotes phosphorylation. See text for details. Modified from Lee *et al.* (2000) with permission.

Second, as soon as FRQ appears, it begins to be phosphorylated by casein kinase Ia, casein kinase II, and calcium/calmodulin-dependent kinase, actions that govern the stability of FRQ (Gorl *et al.*, 2001; Liu, 2005; Liu *et al.*, 2000; Yang *et al.*, 2001, 2002, 2003). FRQ phosphorylation may also influence FRQ/WCC interactions (Yang *et al.*, 2002). Third, FRQ promotes the expression of WC-1 from a more or less constant pool of *wc-1* RNA to yield a rhythm in WC-1 and WCC that seems to promote robustness of the overt rhythm (Cheng *et al.*, 2001b, 2002; Lee *et al.*, 2000; Merrow *et al.*, 2001). Although regulated by FRQ and by WC-1 (Cheng *et al.*, 2003), WC-2 levels do not cycle but instead remain high and relatively constant (Denault *et al.*, 2001), probably due to stability of the protein. By midday, FRQ levels are high and rising, and FRQ has entered the nucleus (Luo *et al.*, 1998) to bind to the WCC and is becoming phosphorylated. FRQ interferes with the ability of the WCC to bind DNA (Froehlich *et al.*, 2003), thereby depressing *frq* expression. As a result of inhibited expression, *frq* transcript levels begin to decline, although continued translation causes FRQ protein levels to continue to rise, yielding a lag between peaks in RNA and protein (Aronson *et al.*, 1994a; Crosthwaite *et al.*, 1995; Garceau *et al.*, 1997).

Finally the processive phosphorylation of FRQ promotes its recognition by an SCF-type ubiquitin ligase, which ubiquitinates FRQ (He *et al.*, 2003), leading to its turnover (Garceau *et al.*, 1997; Liu *et al.*, 2000). FRQ-promoted synthesis of WC-1 is balanced by WC-1 degradation, and WC-1 levels peak in the night near to when FRQ levels drop to their lowest point (Lee *et al.*, 2000). The release of WCC following FRQ turnover creates a sharp transition, with high WCC activity initiating the next cycle (Froehlich *et al.*, 2003).

The special virtues of the *Neurospora* system that have fostered this analysis are its excellent genetics and more recently genomics tools, the ability to produce large, biochemically manipulable amounts of tissue that can retain their rhythmic nature, the ease of genetic transformation, and the availability of molecular biological tools that allow the investigator to manipulate the timing and level of expression of trans-genes as a means of determining their functions. These approaches all rely to some degree on the ability to monitor the rhythm.

## Assays for Rhythmicity in *Neurospora*

*Neurospora* is a haploid fungus that can grow either asexually or via a sexual cycle. Both the asexual developmental cycle, which yields asexual spores (conidia), and the sexual cycle, which yields ascospores, are regulated circadianly. In the former, a developmental switch leading to the production of conidia is activated in the late subjective night of the circadian cycle. Later on, early in the subjective day when the switch is reversed, mycelia continue to grow but remain undifferentiated (Sargent and Kaltenborn, 1972). Because of this, the most obvious and prominent circadianly controlled rhythm in *Neurospora* is that of conidiation. Interestingly, the clock controls the potential to develop rather than the developmental process itself, and once the switch is thrown, the actual process of development can take a long time or a short time depending on the temperature and nutritional state of the organism. This switch occurs once a day, yielding a characteristic conidial banding pattern of developed or developing conidiophores that can be assayed readily on plates or in hollow glass culture tubes called race tubes (Ryan *et al.*, 1943) (Fig. 2). As vegetative growth proceeds from the point of inoculation, $CO_2$ levels become elevated and can suppress conidiation and therefore mask the rhythmic banding pattern. For this reason, all commonly used laboratory stocks for circadian rhythm studies carry a mutation in the *band* gene (*bd*) (Sargent *et al.*, 1966), which renders strains about 200-fold less sensitive to the $CO_2$ masking effect. In wild isolates or laboratory strains lacking the *bd* allele, the formation of conidial bands can sometimes be strengthened by turning

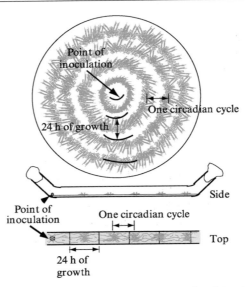

FIG. 2. Monitoring rhythms in *Neurospora* on solid media. (Top) A schematic diagram of a petri dish in which *Neurospora* has been inoculated in the middle. The clock-regulated rhythm in conidiation is manifest as concentric rings. (Middle and bottom) Schematic top and side views of a race tube. Cultures are inoculated on the left, grown for a day in light, and transferred to darkness, which the organisms' clocks interpret as dusk. On either medium, the growing front is marked at 24-h intervals, and because the growth rate is roughly constant, distance along the tube can be interpreted as time passed and rhythm parameters of period and phase can be calculated. Modified from Froehlich (2002) with permission.

the race tubes upside down or by adding rubidium salts to the growth medium (Gall and Lysek, 1981). Too much or too little biotin and an aberrant pH may also interfere with expression of the conidiation rhythm (West, 1975). Finally, on petri dish cultures, once conidia form on the first day, they can break off, float, and germinate at a different point on the plate, thereby obscuring later days of the rhythm. This problem is overcome by including an allele of the *csp-1* or *csp-2* (*conidial separation*) gene in the genetic background of strains being cultures on petri dishes (Mattern and Brody, 1979). Mutation in this gene prevents conidial spores from breaking off after formation.

Detailed instructions for the measurement of circadian rhythms in growth habit, including media recipes and hints, are available on the Web (http://www.fgsc.net/teaching/circad.htm). This source also includes a description of *Neurospora* chronobiology experiments that have been adapted for classroom use. Once growth is complete, plates or race tubes are scanned to create an electronic image that is used for the calculation of period and phase information. The Web site includes directions for using public domain image analysis programs for calculating period and phase information, although many laboratories also use the Chrono program (Roenneberg and Taylor, 2000) for this purpose. Race tubes are typically used in circadian rhythm laboratories, but they must be fashioned by glassblowers and tend to be expensive (>$4 each). For this reason, alternative methods have been developed, including growth on petri dishes or in 15-ml test tubes. Although the rhythms on petri dishes and test tubes are not expressed for as many cycles as on race tubes, nor are they as clear, these alternative techniques are quite sufficient for scoring presence or absence of rhythmicity and estimates of period length, phase, and light or temperature resetting.

### Problems Associated with Assessment of Circadian Characteristics Based on Growth on Solid Media

There are some caveats to the assay of the clock based on growth habit on solid medium. First, the assay for rhythmicity on solid medium assumes a linear rate of growth such that if the growth front is marked once per day, then distance along the surface can be equated with elapsed time in order to calculate period and phase. However, it is known that there is a daily rhythm in the rate of growth (Gooch *et al.*, 2004; Sargent *et al.*, 1966). The instantaneous rate of linear extension can change by as much as 100%, thereby potentially yielding errors of as much as 1.5 h in the calculated phase and 0.75 h in the calculated period.

Second, as noted earlier, all fungi, including *Neurospora*, can exhibit rhythms that are not circadian, but rather whose period lengths and characteristics are dependent on the composition of the growth medium and the genetic background of the strain. Asexual development (conidiation) is a complicated process; a large number of genes can affect it and a large number of environmental factors promote it. Exclusive of the circadian clock, these include but are not limited to various aspects of nutritional starvation, temperature, humidity, nutritional supplements, dessication, and light. One can imagine then that mutations in many genes might influence conidiation, or even noncircadian rhythms in conidiation, without being particularly informative as to circadian regulation. In much of the older literature, no distinction was made between one rhythm and another and no effort was made to be rigorous regarding "circadian." In fact, the "clock" mutant in *Neurospora* describes a gene affecting a noncircadian rhythm (Sussman *et al.*, 1965). It follows that a variety of mutations— morphological mutations, mutations in genes associated with nutrition, or with light or development—might influence the pattern of growth on an agar surface, but not all of these would exert their effect by altering the circadian clock. Some strains, such as *poky*, exhibit a "stopper" phenotype where the linear rate of hyphal extension changes periodically from slow to fast, giving the appearance of a rhythm (Bertrand *et al.*, 1980). Similarly, when the *chol-1* strain is starved for choline it exhibits cyclical changes in growth rate of as much as eightfold and has a marked periodicity, depending on the degree of starvation, of up to 100 h (Lakin-Thomas, 1996).

Third, although the waveform of the conidiation rhythm is generally very consistent from day to day under constant conditions or regular entraining cues, it can vary substantially when these conditions are altered. For instance, the width of a conidial band at half maximum, in hours, can vary from as few as 6 h to as many as 12 h depending on temperature and growth medium. Despite these differences in when conidiation begins and stops, the center of the conidial band occurs dependably in the late subjective evening under constant conditions (Sargent *et al.*, 1966, 1967, 1972) and is taken as the most standard and reliable phase reference point for comparative studies. Although other phase reference points can be used so long as the conditions are held strictly constant, they should not be used to compare phases between conditions (different media, temperatures, or different entraining conditions), as these can all alter the shape of the curve and therefore the reliability of the phase estimate.

## Effects of Medium Composition

Sargent and Kaltenborn (1972) spent considerable time evaluating the effects of different media, carbon, and nitrogen sources on the expression of rhythm. In general, neither the amount nor the type of carbon source has an effect on the clock itself, although high sugar concentrations promote profuse conidiation, thereby obscuring the rhythm. In comparing race tubes run with medium containing different amounts of glucose, for instance, it is clear that the daily bands of conidiation become wider as the amount of glucose increases, although the time of peak conidiation, the center of the band, remains relatively constant between CT 23-2 (CT connotes circadian time). The implication of this is that phase reference points other than the peak change their relationship with the underlying oscillator with changing environmental conditions. Stated differently, the clock does not show phase compensation for the onset of development, although it does show phase compensation for the maximum time of development as well as period compensation. On standard media such as 1.2% sodium acetate 0.5% casamino acids or 0.1% glucose/0.17% L-arginine, the period length of a wild-type strain is about 22 h. The clock runs normally in the absence of the *bd* allele, but expression of the conidial banding phenotype can be obscured, as described earlier. The nitrogen composition of the growth medium can affect the period length to a surprisingly large degree. For instance, supplementation with L-histidine in place of arginine results in an increase in the period length up to ~24 h (Sargent and Kaltenborn, 1972), and other media affects are known. There is no or very little effect of humidity on the period of the wild-type clock, although again this can have large effects on the total amount of development. In particular, high levels of sugar in the medium can result in profuse conidiation so that bands get wider and wider until the overt rhythm is lost.

## Analysis via Genetic and Molecular Genetic Techniques

Both molecular and physiological rhythms in *Neurospora* can also be assayed using liquid cultures (Aronson *et al.*, 1994a; Garceau *et al.*, 1997; Loros *et al.*, 1989; Nakashima, 1981; Perlman *et al.*, 1981). The clock will run normally in mycelial fragments or disks cut from syncytial mats. The individual pieces will remaining synchronous, retaining their endogenous rhythmicity and phase when transferred from liquid to solid growth media (Nakashima, 1981; Perlman *et al.*, 1981). Such cultures remain rhythmic for up to 96 h and retain their sensitivity to light and chemical stimuli and also their compensation against changes in the growth medium; for instance, addition of glucose to the starving cultures does not reset the clock (Nakashima, 1981). So long as mycelia comprising the disks are not physically torn or abused, the disks can be transferred from liquid to solid media and will retain their phase information (Nakashima, 1981; Perlman *et al.*, 1981), thereby allowing chemical or light phase response curves to be performed with complete washout (e.g., Dunlap and Feldman, 1988). The harvesting of individual disks from liquid culture requires some finesse, and gentle shaking prevents the fusion of mycelial disks. An advantage of such mycelial cultures is that they can be grown and harvested at times throughout their circadian cycles and the tissue used for molecular and biochemical assays, including the monitoring of mRNAs and proteins within a cell (e.g., Loros *et al.*, 1989) (Fig. 3). Within about 24 h, if fresh medium is not introduced, such cultures are starving so the spectrum of genes being expressed is substantially curtailed compared with well-fed vegetative mycelia (Zhu *et al.*, 2001), but the clock runs well. Operation of the clock can be verified now by following any of a number of molecular markers (the level of *frq* transcript, FRQ protein, or levels of various ccg's such as *ccg-1* or *ccg-2*).

A caveat of this method, of course, is that growth in liquid is physiologically different from growth on solid substrates: Although the clock itself

Liquid culture method to produce timed tissue

Inoculate $2 \times 10^6$ conidia into 30 ml rich medium (0.3% glucose, 0.5% arginine) in a petri dish.

Grow mycelial mat for 30–40 h without shaking.

Cut disks (about 9 mm) from mat, and transfer to limiting medium (0.03% glucose, 0.05% arginine) in a flask; culture with gentle shaking.

The time of transfer of the flask from light to darkness sets the clock.

Harvest under a red safe light at the appropriate circadian time; process immediately or flash freeze.

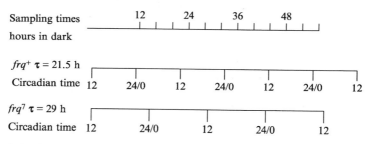

FIG. 3. A schematic showing the method for growing rhythmic tissue in liquid culture.

apparently runs in much the same way under either condition (conserving period and phase), there is little reason to believe that output pathways will be the same. Real-time analysis of rhythmic clock-controlled or core clock gene activity can also be assayed *in vivo* in liquid or on solid media using firefly luciferase as a reporter for gene expression (Mehra *et al.*, 2002; Morgan *et al.*, 2003). Whenever tissue is collected for biochemical or molecular biological analysis, it is essential to bear in mind that the physiology of a culture will reflect the subjective circadian time (directly reflecting the number of hours since the last entraining cue), as well as the developmental state (directly reflecting the nutritional and environmental conditions for growth). These, in turn, will change over time in culture as carbon and nitrogen sources are used up and the medium becomes "conditioned." To be certain that only circadian changes are being tracked, care must be taken to stagger the inoculation times and culture conditions appropriately (Dunlap and Feldman, 1988; Loros *et al.*, 1989). Protocols to achieve this are described more fully in the accompanying chapter on the analysis of circadian output in *Neurospora* and mammalian cell culture. Another method of obtaining tissue that is the same developmental age but at a different clock time is to use period length mutants of the clock. As seen in Fig. 3, after 40 h in the dark, cultures of clock wild type and the long period (29 h) mutant $frq^7$ will have cycled 180° out of phase in circadian time although remaining the same age developmentally (Bell-Pedersen *et al.*, 1996).

## Genetic Analyses

Using these assays, a number of rhythm-affecting alleles have been identified (reviewed in Dunlap *et al.*, 2004; Loros and Dunlap, 2001). This collection of clock mutants includes genes that affect the period length, such as $frq^1$ and $frq^2$ (Feldman, 1967; Feldman *et al.*, 1973, 1979; Gardner and Feldman, 1980), whereas other mutations, e.g., *chr, prd-1, prd-2, prd-3, prd-4, wc-2*, and other *frq* alleles, affect both the period of the rhythm and temperature compensation (Collett *et al.*, 2002; Feldman *et al.*, 1978, 1979). The *frequency* gene has been identified at least eight separate times (Chang and Nakashima, 1998; Feldman and Hoyle 1973; Gardner and Feldman, 1980) (M. Collett, J. Dunlap, and J. Loros, unpublished results) and can give rise to both short (16 h) and long (29 h) period length mutants, as well as mutants lacking circadian rhythmicity (although they do retain noncircadian rhythms). Compilations such as those cited typically include all genes having reported effects on periodicity of the growth habit, but a word of caution is appropriate here. A variety of mutations—morphological mutations, mutations in genes associated with nutrition or with light or development—might influence the pattern of growth on an agar surface as noted earlier, but not all of these would exert their effect by altering the circadian clock.

*Neurospora* has seven chromosomes and a nuclear organization very typical of eukaryotes, except that genes appear closer together and introns are smaller, making the overall size of genes smaller. Genes appear on average every 3.7 kb along the chromosome. The average length of a gene is 1.3 kb, compared with about 1.1 kb for *Saccharomyces* and *Schizosaccharomyces*. Introns are found in nearly all genes and are distributed evenly. Due to whole genome scanning mechanisms, such as repeat-induced point mutation (RIP) (see Selker, 1997) that detect and eliminate duplications, there is a dearth of repeats and gene families (Galagan *et al.*, 2003). Many genes use alternative promoters and many primary transcripts are alternatively spliced. It is common for genes to be controlled in a combinatorial manner by a variety of different factors.

The beauty of genetic analyses, of course, is that they can be used to distinguish between mutations in the same gene that give different period lengths and mutations in different genes that result in the same period lengths. Additionally, genetic analyses can be used to determine whether an allele is dominant or recessive to its wild-type counterpart by observing whether the phenotype (the characteristics of the strain bearing the mutation) of a strain bearing one copy of each allele is like the wild type (a recessive mutant gene) or like the mutant (a dominant mutant gene). The means by which these analyses are achieved in a haploid syncytial organism such as *Neurospora* are slightly different from how they are done in a bacterium or a diploid animal or plant and are worth summarizing here.

*Classical Transmission Genetics.* Genetic analyses in *Neurospora* developed at about the same time as those of *Drosophila* in the 1930s and are now well advanced. To begin to understand this, start from the realization that *Neurospora* has a much more enlightened approach to sex and gender than people. There are two different mating types, A and a, and crosses can happen only between individuals of different mating type. Gender, however, is determined environmentally, so strains of either mating type can act as males or females. When a strain grows on a medium that is limiting for carbon and especially nitrogen, sexuality is induced and proceeds through the elaboration of novel cell types that comprise the female body, a protoperithecium. The perithecia send out specialized hyphae called trichogynes that use pheromone cues to search their environment for conidia or mycelial fragments of the opposite mating type, and when one is found, a nucleus is captured and escorted back to the perithecium where it joins its counterpart of the opposite mating type. Within the perithecium, then, the two nuclei grow and divide in synchrony for period of time and then all the pairs fuse in unison—the only diploid stage in the life cycle—and immediately enter meiosis followed by one mitotic cycle, yielding an ordered ascus containing eight haploid spores. The entire sexual cycle requires about 3 weeks. Detailed methods for carrying out classical genetic analyses are available on the Web (http://www.fgsc.net/methods/genmthds.html) or in publications (e.g., Davis amd deSerres, 1970) and need not be reiterated here.

Genomics efforts (described in greater detail later) have allowed the population of a single nucleotide polymorphism (SNP) map based on comparisons between the laboratory wild-type strain (OR74A) and an exotic *N. crassa* isolated in Mauriceville, Texas. A cross between a mutant gene isolated in the laboratory strain and the Mauriceville strain will segregate the mutation of interest, as well as the thousands of SNPs in the Mauriceville parent. By following the cosegregation of various SNPs (analyzed in parallel via molecular means) and the gene of interest, it is possible to refine the genetic location of most novel mutations to within tens of kilobases within a few days' or weeks' time.

*Heterokaryons.* Hyphae of *Neurospora* that are compatible will fuse and comingle their nuclei and cytoplasm, giving rise to a strain that can have variable proportions of genetically distinct nuclei. Such strains are called heterokaryons, and they allow the manipulation of nuclear ratios beyond the simple 1:1 dictated by diploidy. For instance, a heterokaryon bearing both $frq^+$, and a $frq$-null ($frq^9$) nuclei becomes rhythmic with only 20% of the nuclei being $frq^+$, whereas in strains bearing mixtures of $frq^1$ (period 16 h), $frq^2$ (period 19 h), $frq^+$ (period 22 h), $frq^3$ (period 24 h), or $frq^7$ (period 29 h, or 7 h longer than wild type), the period length of the heterokaryon simply reflects the numerical average of the nuclear composition: A strain 75% $frq^7$ and 25% $frq^+$ has a period length reflecting its nuclear composition (Fig. 4). The ability of strains to fuse is governed by a series of loci known as heterokaryon incompatibility loci (e.g., Sarkar *et al.*, 2002); strains must be identical at all loci, including mating type, so this is a means of self–nonself recognition. Thus, strains can fuse with themselves under vegetative conditions, but can not mate with themselves.

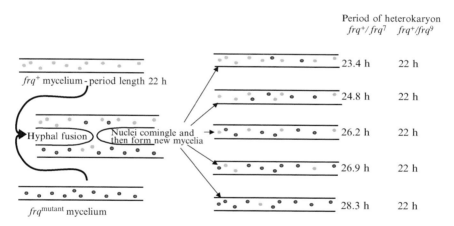

|  | Period of heterokaryon |  |
|---|---|---|
|  | $frq^+/frq^7$ | $frq^+/frq^9$ |
| $frq^+$ mycelium - period length 22 h | 23.4 h | 22 h |
|  | 24.8 h | 22 h |
| Hyphal fusion — Nuclei comingle and then form new mycelia | 26.2 h | 22 h |
|  | 26.9 h | 22 h |
| $frq^{mutant}$ mycelium | 28.3 h | 22 h |

FIG. 4. Use of heterokaryons to assess dominance/recessivity. Light gray circles represent $frq^+$ nuclei, and dark gray circles represent mutant nuclei, either $frq^7$ with a period length of 29 h or $frq^9$, which is arrhythmic on acetate casamino acids medium and has a noncompensated period length on glucose–arginine media. When hyphae fuse (middle left), nuclei comingle in a common cytoplasm. When mycelia fragments are later isolated from this fused culture, they may contain different ratios of wild type to mutant nuclei, and the phenotypes displayed by these heterokaryotic mycelia can be determined. The period lengths observed reflect as a first approximation the numerical average of the period lengths for alleles that are codominant, such as $frq^7$, but reflect the period of the dominant allele ($frq^+$) in the case of $frq^9$. Data follow the trends seen in Gardner and Feldman (1980), Loros (1986), and Loros *et al.* (1986) but are not taken from those studies.

*Molecular Genetic Analyses and Manipulation*

Major advances in the analysis of circadian rhythmicity within the past two decades have all derived from the varied tools of molecular genetics: the ability to clone genes corresponding to informative clock mutations; to determine the sequence of these genes and thereby make educated guesses as to the function of a gene; and, most importantly, to reinsert the cloned gene itself, or structural or regulatory variants of the gene, back into the genome in a controlled manner so that the effects on the clock can be observed. In many respects, work in *Neurospora* pioneered these analyses in the context of rhythms and, conversely, work on rhythms led to the establishment of these tools in the *Neurospora* system. There are three general areas into which such tools naturally fall, each of which is considered separately.

*Transformation.* Transformation refers to the ability to insert foreign DNA back into the genome of an organism, to select the transformants from among the much larger pool of nontransformed individuals, and to have the foreign DNA expressed. The technology allowing transformation of *Neurospora* was developed in the late 1970s, not long after *Saccharomyces* (reviewed in Giles *et al.*, 1985). Parenthetically, the early availability of this technology, several years before it was developed for *Drosophila*, had much to do with the authors of this work choosing this system as a platform for the dissection of a circadian clock. As noted earlier, the molecular characteristics of the *Neurospora* genome are more similar to those of animals than yeasts, and in line with this, initially transformation was achieved only by random ectopic insertion of the DNA(s) into the genome. Because the chromatin context could not be determined, this led to uncontrolled chromatin position effects on the expression of the inserted gene. A breakthrough came later with the introduction of vectors and strains that facilitated the targeted insertion of trans-genes into a neutral chromosomal site, the *his-3* gene (Ebbole and Sachs, 1990). Although insertion at additional alternative sites is now possible, *his-3* remains the most commonly used, and a variety of useful vectors listed at the Fungal Genetics Stock Center Web site (http://www.fgsc.net/clones.html) are available for construction of cassettes that will target integrated genes.

*Regulatable Promoters.* Another key set of tools needed for dissection of a process are promoters that can be used to drive the expression of transgenes at the desired place, time, and level. This was initially achieved in *Neurospora* using the *qa-2* promoter, which can be activated in a regulatable and graded manner by the addition of the gratuitous carbon source quinic acid and repressed (through inducer exclusion) by the addition of the preferred carbon source glucose (reviewed in Geever *et al.*, 1989). The first use of the *qa-2* promoter to regulate heterologous gene expression in *Neurospora* was in the context of chronobiology, where it was used to drive elevated and constant expression of *frq* in studies that proved the existence of a negative feedback loop, containing *frq* and its products, at the core of the circadian clock (Aronson *et al.*, 1994a). Additional promoters useful for constitutive or on/off control of heterologous gene expression have been developed that will respond to different stimuli; these include carbon source (Collins *et al.*, 1991; Temporini *et al.*, 2004), heavy metals (Schilling *et al.*, 1992), light, oxygen, time of day (Wang *et al.*, 1994), or constitutive expression (Orbach, 1994).

*Targeted Gene Knockouts and Creation of Mutations.* A final key set of tools concerns the ability to create null mutations, or altered function mutations of genes of interest. In *Neurospora*, there are several ways in which this can be achieved. The creation of partial loss-of-function strains through transformation with gene fragments or use of small interfering RNAs was described (Cogoni and Macino, 1994) prior to or concurrent with the first observations in *Caenorhabditis elegans*, and in *Neurospora* the phenomenon is known as quelling (Cogoni and Macino, 1997). The *Neurospora* genome contains all of the cognate genes required for siRNA production (RNA-dependent RNA polymerase, DICER, etc.) and the manner in which they act appears to be quite similar to that being determined in animal cells.

A second, and in fact the original, method of gene inactivation in *Neurospora* is based on a bit of physiology that is peculiar to *Neurospora* and a few of its allies within the ascomycetes; this is repeat-induced point mutation (reviewed in Selker, 1997). *Neurospora* possesses the ability to scan its genome just prior to meiosis and to detect any sequences that are duplicated. When duplications are found, both copies of the duplication are methylated and, at high frequency, mutated through the transversion of cytosines to thymines. Depending on the size of the duplicated region and its location within a gene, either the promoter or the coding region of the gene can be mutated, and the severity of the mutation can vary from slight to complete loss of function. There exist protected regions in the genome, such as the repeat containing the ribosomal genes, that are not subject to RIP; however, for unknown reasons it is not possible to target inserted transgenes to this region.

Although one can choose the region of a gene to be RIPed, one cannot choose the degree of RIP, and the only way to establish this is to sequence the RIPed region after the fact. Because of this, the development of methods to achieve complete gene disruptions was essential. This was developed along the lines that are successful in mammalian cells, by creating a knockout cassette in which a selectable marker is flanked by long

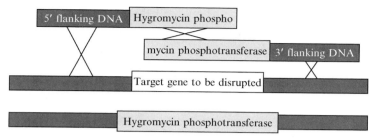

Fɪɢ. 5. Disruption of genes in *Neurospora* through the use of split markers. (Top) Two linear DNA fragments. One contains the DNA on the 5′ flank of the gene to be disrupted, in addition to DNA encoding the N-terminal two-thirds of the selectable marker hygromycin phospho-transferase, which confers dominant selectable resistance to hygromycin. The other fragment encodes the C-terminal two-thirds of the hygromycin phosphotransferase gene followed by DNA on the 3′ flank of the gene to be disrupted. Because neither fragment contains the entire selectable marker, ectopic integration via nonhomologous end joining of either one by itself will not confer resistance. However, if the two cross over within the shared sequences within the marker, then the marker will be flanked by *Neurospora* DNA and will be able to integrate via double homologous crossing over. (Bottom) The chromosome structure of the deleted gene.

(several kilobase) regions of DNA corresponding to the sequences surrounding the gene to be disrupted; the first contexts in which this was used was again the context of circadian clocks, in the disruption a *ccg* (Aronson *et al.*, 1994c) and the *frq* gene in studies that established a noncircadian FRQ-less oscillation (FLO) as the null phenotype (Aronson *et al.*, 1994b). Initially the frequency of disruption by homologous gene replacement was quite low, on the order of a few percent, but steady development of this technique, including the implementation of split mark-er technologies (e.g., Catlett *et al.*, 2003), has yielded a method in which targeted gene disruptions can dependably be achieved for any gene within a few weeks or, if done in groups, with an average throughput of a knockout per day (Fig. 5). This method is the basis of a *Neurospora* genome project in which most of the genes in the genome are being disrupted (http://www.dartmouth.edu/%7Eneurosporagenome/). The bank of knockouts being created in this way is available to the community at large.

Concluding Remarks

This article covered or cited references to the essentials of the care and feeding of *Neurospora* for use in the analysis and dissection of the circadian oscillator and its input and output. The strengths of *Neurospora* lie in the homogeneity of the cell type required to express a rhythm, the ability to produce biochemically useful amounts of material that are synchronous with respect to circadian time and developmental state, the variety of assays of rhythmicity, and the ability to manipulate the system in a state-of-the-art manner through classical or molecular genetic techniques. With ongoing genome projects and the support of a vibrant and growing research community, *Neurospora* will remain a useful model for the cellular analysis of circadian rhythms for the foreseeable future.

## References

Aronson, B., Johnson, K., Loros, J. J., and Dunlap, J. C. (1994a). Negative feedback defining a circadian clock: Autoregulation in the clock gene *frequency*. *Science* **263,** 1578–1584.

Aronson, B. D., Johnson, K. A., and Dunlap, J. C. (1994b). The circadian clock locus *frequency*: A single ORF defines period length and temperature compensation. *Proc. Natl. Acad. Sci. USA* **91,** 7683–7687.

Aronson, B. D., Lindgren, K. M., Dunlap, J. C., and Loros, J. J. (1994c). An efficient method for gene disruption in *Neurospora crassa*. *Mol. Gen. Genet* **242,** 490–494.

Bell-Pedersen, D., Shinohara, M., Loros, J., and Dunlap, J. C. (1996). Circadian clock-controlled genes isolated from *Neurospora crassa* are late night to early morning specific. *Proc. Natl. Acad. Sci. USA* **93,** 13096–13101.

Bertrand, H., Collins, R. A., Stohl, L. L., Goewert, R. R., and Lambowitz, A. M. (1980). Deletion mutants of *Neurospora crassa* mitochondrial DNA and their relationship to the "stop-start" growth phenotype. *Proc. Natl. Acad. Sci. USA* **77,** 6032–6036.

Bünning, E. (1973). "The Physiological Clock," revised 3rd Ed. Springer-Verlag, New York.

Catlett, N. L., Lee, B., Yoder, O. C., and Turgeon, B. G. (2003). Split-marker recombination for efficient targeted deletion of fungal genes. *Fungal Genet News* **50,** 9–11.

Chang, B., and Nakashima, H. (1998). Isolation of temperature sensitive rhythm mutant in *Neurospora crassa. Genes Genet. Syst.* **73,** 71–73.

Cheng, P., Yang, Y., Gardner, K. H., and Liu, Y. (2002). PAS domain-mediated WC-1/WC-2 interaction is essential for maintaining the steady-state level of WC-1 and the function of both proteins in clock and light responses of Neurospora. *Mol. Cell. Biol.* **22,** 517–524.

Cheng, P., Yang, Y., Heintzen, C., and Liu, Y. (2001a). Coiled coil mediated FRQ-FRQ interaction is essential for circadian clock function in *Neurospora. EMBO J.* **20,** 101–108.

Cheng, P., Yang, Y., and Liu, Y. (2001b). Interlocked feedback loops contribute to the robustness of the *Neurospora* circadian clock. *Proc. Natl. Acad. Sci. USA* **98,** 7408–7413.

Cheng, P., Yang, Y., Wang, L., He, Q., and Liu, Y. (2003). WHITE COLLAR-1, a multifunctional *Neurospora* protein involved in the circadian feedback loops, light sensing, and transcription repression of *wc-2. J. Biol. Chem.* **278,** 3801–3808.

Cogoni, C., and Macino, G. (1994). Suppression of gene expression by homologous transgenes. *Antonie Van Leeuwenhoek* **65,** 205–209.

Cogoni, C., and Macino, G. (1997). Isolation of quelling-defective (qde) mutants impaired in posttranscriptional transgene-induced gene silencing in *Neurospora crassa. Proc. Natl. Acad. Sci. USA* **94,** 10233–10238.

Collett, M. A., Garceau, N., Dunlap, J. C., and Loros, J. J. (2002). Light and clock expression of the *Neurospora* clock gene frequency is differentially driven by but dependent on WHITE COLLAR-2. *Genetics* **160,** 149–158.

Collins, M. E., Briggs, G., Sawyer, C., Sheffield, P., and Connerton, I. F. (1991). An inducible gene expression system for *Neurospora crassa. Enzyme Microb. Technol.* **13,** 400–403.

Crosthwaite, S. C., Dunlap, J. C., and Loros, J. J. (1997). *Neurospora wc-1* and *wc-2*: Transcription, photoresponses, and the origins of circadian rhythmicity. *Science* **276,** 763–769.

Crosthwaite, S. C., Loros, J. J., and Dunlap, J. C. (1995). Light-induced resetting of a circadian clock is mediated by a rapid increase in *frequency* transcript. *Cell* **81**, 1003–1012.

Davis, R. L., and deSerres, D. (1970). Genetic and microbial research techniques for *Neurospora crassa*. *Methods Enzymol.* **27A,** 79–143.

Denault, D. L., Loros, J. J., and Dunlap, J. C. (2001). WC-2 mediates WC-1-FRQ interaction within the PAS protein-linked circadian feedback loop of *Neurospora crassa*. *EMBO J.* **20,** 109–117.

Dunlap, J. C., and Feldman, J. F. (1988). On the role of protein synthesis in the circadian clock of *Neurospora crassa*. *Proc. Natl. Acad. Sci. USA* **85,** 1096–1100.

Dunlap, J. C., Loros, J. J., and Decoursey, P., eds. (2003). "Chronobiology: Biological Timekeeping" Sinauer, Sunderland, MA.

Dunlap, J. C., Loros, J. J., Denault, D., Lee, K., Froehlich, A. F., Colot, H., Shi, M., and Pregueiro, A. (2004). Genetics and molecular biology of circadian rhythms. *In* "The Mycota III" (R. Brambl and G. A. Marzluf, eds.), pp. 209–229. Springer-Verlag, Berlin.

Ebbole, D., and Sachs, M. (1990). Targeting chromosomal insertion to the *his-3* locus. *Fung. Genet. News* **37,** 15–16.

Feldman, J., and Hoyle, M. N. (1974). A direct comparison between circadian and noncircadian rhythms in *Neurospora crassa*. *Plant Physiol.* **53,** 928–930.

Feldman, J. F. (1967). Lengthening the period of a biological clock in *Euglena* by cycloheximide, an inhibitor of protein synthesis. *Proc. Natl. Acad. Sci. USA* **57,** 1080–1087.

Feldman, J. F., and Atkinson, C. A. (1978). Genetic and physiological characterization of a slow growing circadian clock mutant of *Neurospora crassa*. *Genetics* **88,** 255–265.

Feldman, J. F., Gardner, G. F., and Dennison, R. A. (1979). Genetic analysis of the circadian clock of *Neurospora*. *In* "Biological Rhythms and their Central Mechanism" (M. Suda, ed.), pp. 57–66. Elsevier, Amsterdam.

Feldman, J. F., and Hoyle, M. (1973). Isolation of circadian clock mutants of *Neurospora crassa*. *Genetics* **75,** 605–613.

Feldman, J. F., and Waser, N. (1971). New mutations affecting circadian rhythmicity in *Neurospora*. *In* "Biochronometry" (M. Menaker, ed.), pp. 652–656. National Academy of Sciences, Washington.

Froehlich, A. C. (2002). "Light and Circadian Regulation in *Neurospora crassa*," Ph.D. thesis. Dartmouth.

Froehlich, A. C., Loros, J. J., and Dunlap, J. C. (2003). Rhythmic binding of a WHITE COLLAR containing complex to the *frequency* promoter is inhibited by FREQUENCY. *Proc. Nat. Acad. Sci. USA* **100,** 5914–5919.

Gall, A., and Lysek, G. (1981). Induction of circadian conidiation by rubidium chloride. *Neurospora Newsl.* **28,** 13.

Galagan, J., Calvo, S., Borkovich, K., *et al.* (2003). The genome sequence of the filamentous fungus *Neurospora crassa*. *Nature* **422,** 859–868.

Garceau, N., Liu, Y., Loros, J. J., and Dunlap, J. C. (1997). Alternative initiation of translation and time-specific phosphorylation yield multiple forms of the essential clock protein FREQUENCY. *Cell* **89,** 469–476.

Gardner, G. F., and Feldman, J. F. (1980). The *frq* locus in *Neurospora crassa*: A key element in circadian clock organization. *Genetics* **96,** 877–886.

Geever, R. F., Huiet, L., Baum, J., Tyler, B., Patel, V., Rutledge, B., Case, M., and H., G. N. (1989). DNA sequence, organization and regulation of the *qa* gene cluster of *Neurospora crassa*. *J. Mol. Biol.* **207,** 15–34.

Giles, N. H., Case, M. E., Baum, J., Geever, R., Huiet, L., Patel, V., and Tyler, B. (1985). Gene organization and regulation in the *qa* (quinic acid) gene cluster of *Neurospora crassa*. *Microbio. Rev.* **49,** 338–358.

Gooch, V., Freeman, L., and Lakin-Thomas, P. (2004). *J. Biol. Rhythms* **19**, 493–503.

Gorl, M., Merrow, M., Huttner, B., Johnson, J., Roenneberg, T., and Brunner, M. (2001). A PEST-like element in FREQUENCY determines the length of the circadian period in *Neurospora crassa. EMBO J.* **20**, 7074–7084.

He, Q., Cheng, P., Yang, Y., He, Q., Yu, Q., and Liu, Y. (2003). FWD-1 mediated degradation of FREQUENCY in *Neurospora* establishes a conserved mechanism for circadian clock regulation. *EMBO J.* **22**, 4421–4430.

Ingold, C. T. (1971). "Fungal Spores." Clarendon Press, Oxford.

Konopka, R. J., and Benzer, S. (1971). Clock mutants of *Drosophila melanogaster. Proc. Natl. Acad. Sci. USA* **68**, 2112–2116.

Lakin-Thomas, P. (1996). Effects of choline depletion on the circadian rhythm in *Neurospora crassa. Biol. Rhythm Res.* **27**, 12–30.

Lee, K., Loros, J. J., and Dunlap, J. C. (2000). Interconnected feedback loops in the Neurospora circadian system. *Science* **289**, 107–110.

Liu, Y. (2005). Analysis of posttranslational regulations in the *Neurospora* circadian clock. *Methods Enzymol.* **393** [17] 2005 (this volume).

Liu, Y., Garceau, N., Loros, J. J., and Dunlap, J. C. (1997). Thermally regulated translational control mediates an aspect of temperature compensation in the *Neurospora* circadian clock. *Cell* **89**, 477–486.

Liu, Y., Loros, J., and Dunlap, J. C. (2000). Phosphorylation of the *Neurospora* clock protein FREQUENCY determines its degradation rate and strongly influences the period length of the circadian clock. *Proc. Nat. Acad. Sci. USA* **97**, 234–239.

Loros, J. J. (1986). "Studies on *frq-9*, a Recessive Circadian Clock Mutant of *Neurospora crassa*," Ph.D. thesis. U.C. Santa Cruz.

Loros, J. J., Denome, S. A., and Dunlap, J. C. (1989). Molecular cloning of genes under the control of the circadian clock in *Neurospora. Science* **243**, 385–388.

Loros, J. J., and Dunlap, J. C. (2001). Genetic and molecular analysis of circadian rhythms in *Neurospora. Annu. Rev. Physiol.* **63**, 757–794.

Loros, J. J., and Feldman, J. F. (1986). Loss of temperature compensation of circadian period length in the *frq-9* mutant of *Neurospora crassa. J. Biol. Rhythms* **1**, 187–198.

Loros, J. J., Richman, A., and Feldman, J. F. (1986). A recessive circadian clock mutant at the *frq* locus in *Neurospora crassa. Genetics* **114**, 1095–1110.

Luo, C., Loros, J. J., and Dunlap, J. C. (1998). Nuclear localization is required for function of the essential clock protein FREQUENCY. *EMBO J.* **17**, 1228–1235.

Mattern, D., and Brody, S. (1979). Circadian rhythms in *Neurospora crassa*: Effects of unsaturated fatty acids. *J. Bacteriol.* **139**, 977–988.

McClung, C. R., Fox, B. A., and Dunlap, J. C. (1989). The *Neurospora* clock gene *frequency* shares a sequence element with the *Drosophila* clock gene *period. Nature* **339**, 558–562.

Mehra, A., Morgan, L., Bell-Pedersen, D., Loros, J., and Dunlap, J. C. (2002). Watching the *Neurospora* Clock Tick. *In* "Society for Research on Biological Rhythms." Amelia Island, FL.

Merrow, M., Bruner, M., and Roenneberg, T. (1999). Assignment of circadian function for the *Neurospora* clock gene *frequency. Nature* **399**, 584–586.

Merrow, M., Franchi, L., Dragovic, Z., Gorl, M., Johnson, J., Brunner, M., Macino, G., and Roenneberg, T. (2001). Circadian regulation of the light input pathway in *Neurospora crassa. EMBO J.* **20**, 307–315.

Morgan, L. W., Greene, A. V., and Bell-Pedersen, D. (2003). Circadian and light-induced expression of luciferase in *Neurospora crassa. Fung. Genet. Biol.* **38**, 327–332.

Nakashima, H. (1981). A liquid culture system for the biochemical analysis of the circadian clock of *Neurospora. Plant Cell Physiol.* **22**, 231–238.

Orbach, M. J. (1994). A cosmid with a HyR marker for fungal library construction and screening. *Genes Dev.* **150,** 159–162.

Perlman, J., Nakashima, H., and Feldman, J. (1981). Assay and characteristics of circadian rhythmicity in liquid cultures of *Neurospora crassa. Plant Physiol.* **67,** 404–407.

Pittendrigh, C. S., Bruce, V. G., Rosenzweig, N. S., and Rubin, M. L. (1959). A biological clock in *Neurospora. Nature* **184,** 169–170.

Roenneberg, T., and Taylor, W. (2000). Automated recordings of bioluminescence with special reference to the analysis of circadian rhythms. *Methods Enzymol.* **305,** 104–119.

Ryan, F. J., Beadle, G. W., and Tatum, E. L. (1943). The tube method for measuring the growth rate of Neurospora. *Am. J. Bot.* **30,** 784–799.

Sargent, M. L., and Briggs, W. R. (1967). The effect of light on a circadian rhythm of conidiation in *Neurospora. Plant Physiol.* **42,** 1504–1510.

Sargent, M. L., Briggs, W. R., and Woodward, D. O. (1966). The circadian nature of a rhythm expressed by an invertaseless strain of *Neurospora crassa. Plant Physiol.* **41,** 1343–1349.

Sargent, M. L., and Kaltenborn, S. H. (1972). Effects of medium composition and carbon dioxide on circadian conidiation in *Neurospora. Plant Physiol.* **50,** 171–175.

Sarkar, S., Iyer, G., Wu, J., and Glass, N. L. (2002). Nonself recognition is mediated by HET-C heterocomplex formation during vegetative incompatibility. *EMBO J.* **21,** 4841–4850.

Schilling, B., Linden, R. M., Kupper, U., and Lerch, K. (1992). Expression of *Neurospora crassa* laccase under the control of the copper-inducible metallothionein-promoter. *Curr. Genet* **22,** 197–203.

Selker, E. U. (1997). Epigenetic phenomena in filamentous fungi. *Trends. Genet.* **13,** 296–301.

Stechmann, A., and Cavalier-Smith, T. (2003). The root of the eukaryote tree pinpointed. *Curr. Biol.* **13,** R665–R666.

Sussman, A. S., Durkee, T., and Lowrey, R. J. (1965). A model for rhythmic and temperature-independent growth in the "clock" mutants of *Neurospora. Mycopathol. Mycol. Appl.* **25,** 381–396.

Sweeney, B. M. (1976). Circadian rhythms, definition and general characterization. *In* "The Molecular Basis of Circadian Rhythms" (J. W. Hastings and H.-G. Schweiger, eds.), pp. 77–83. Dahlem Konferenzen, Berlin.

Temporini, E. D., Alvarez, M. E., Mautino, M. R., Folco, H. D., and Rosa, A. L. (2004). The *Neurospora crassa cfp* promoter drives a carbon source-dependent expression of transgenes in filamentous fungi. *J. Appl. Microbiol.* **96,** 1256–1264.

Wang, Z., Deak, M., and Free, S. J. (1994). A cis-acting region required for the regulated expression of grg-1, a *Neurospora glucose-repressible gene.* Two regulatory sites (CRE and NRS) are required to repress *grg-1* expression. *J. Mol. Biol.* **237,** 65–74.

West, D. J. (1975). Effects of pH and biotin on a circadian rhythm of conidiation in *Neurospora crassa. J. Bacteriol.* **123,** 387–389.

Yang, Y., Cheng, P., and Liu, Y. (2002). Regulation of the *Neurospora* circadian clock by casein kinase II. *Genes Dev.* **16,** 994–1006.

Yang, Y., Cheng, P., Qiyang, Q., He, Q., Wang, L., and Liu, Y. (2003). Phosphorylation of FREQUENCY protein by casein kinase II is necessary for the function of the *Neurospora* circadian clock. *Mol. Cell. Biol.* **23,** 6221–6228.

Yang, Y., Cheng, P., Zhi, G., and Liu, Y. (2001). Identification of a calcium/calmodulin-dependent protein kinase that phosphorylates the *Neurospora* circadian clock protein FREQUENCY. *J. Biol. Chem.* **276,** 41064–41072.

Zhu, H., Nowrousian, M., Kupfer, D., Colot, H., Berrocal-Tito, G., Lai, H., Bell-Pedersen, D., Roe, B., Loros, J. J., and Dunlap, J. C. (2001). Analysis of expressed sequence tags from two starvation, time-of-day-specific libraries of *Neurospora crassa* reveals novel clock-controlled genes. *Genetics* **157,** 1057–1065.

# [2]   Circadian Genetics in the Model Higher Plant, *Arabidopsis thaliana*

*By* MEGAN M. SOUTHERN and ANDREW J. MILLAR

## Abstract

In recent decades, most research on the circadian rhythms of higher plants has been driven by molecular genetics. A wide variety of experimental approaches have discovered mutants in the plant circadian clock, yet the screens are far from saturated and there must still be important clock-related genes to identify. Direct methods to screen for circadian mutants include the original assay of rhythmic luminescence from promoter:luciferase constructs *in planta* or a recently developed assay based on stomatal rhythms. Mutants found through simpler screens of processes only partially controlled by the clock are still identifying novel and interesting circadian phenotypes when their rhythms are tested, while the sequenced genome and the large range of mutant stocks available have made reverse genetics increasingly powerful.

## Introduction

### Arabidopsis *Genetics*

Genetics is an extremely effective tool for unravelling biological processes at the molecular level. The starting point is to find mutations disrupting the process of interest from which to understand the function of the cognate wild-type genes. The ease of this process depends on prior knowledge of the genome of the organism. The small size yet high seed production, short generation time of 6–8 weeks, and small genome of the higher plant *Arabidopsis thaliana* have made it popular among plant geneticists, and the growing base of knowledge and tools developed have made it ever increasingly the model of choice. Screens are usually performed in a wild-type or single mutant background; chromosomal genetic tools have not been developed. *Arabidopsis* is a self-fertilizing hermaphrodite that can be outcrossed manually. It is usually mutagenized chemically using ethyl methanesulfonate (EMS) or by insertional mutagenesis using the T-DNA of an engineered Ti plasmid transferred from *Agrobacterium tumefaciens*. Other possible mutagens include γ radiation and transposons introduced from other plant species.

*Circadian Screens in* Arabidopsis

Several approaches to screening for *Arabidopsis* clock mutants have been successful. All have screened for changes to outputs from the clock (Fig. 1), although screens for changes to oscillator components are now equally possible. Many circadian-related genes were discovered in screens for mutants in processes only partially controlled by the clock, particularly flowering time and hypocotyl length. Others have been found in direct screens for circadian mutants and still more by reverse genetics. So far all screens have been done in wild-type backgrounds; suppressor or enhancer screens in the background of the mutants discovered are now possible.

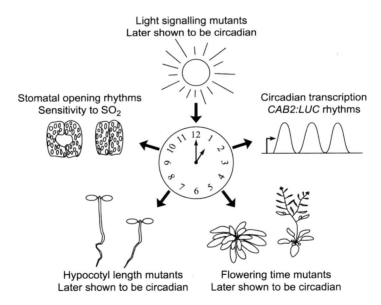

Fig. 1. Phenotypes used to identify circadian clock mutants in *Arabidopsis*. Clock-affecting mutants have been found from screens using many phenotypes. Most primary screens have been for changes to the transcriptional regulation of *CAB2:LUC*; however, mutants have also been identified in a screen for the phase of stomatal opening and thus $SO_2$ sensitivity. Many circadian mutants were identified in secondary screens of mutants originally identified in screens for processes controlled only partially by the clock, such as flowering time or hypocotyl length. Secondary screens of light signal transduction mutants have identified light input pathways to the clock.

Circadian Screens Using Luciferase

A direct screen of the circadian system must monitor a circadian rhythm over at least 24 h; to have a high throughput this needs to be automated. Because *Arabidopsis* lacks a clock-controlled trait suitable for a genetic screen, one was engineered by inserting the firefly luciferase gene under the control of the *CHLOROPHYLL A/B BINDING PROTEIN 2 (CAB2*, also known as *LHCB1\*1*) promoter and has been the basis for most circadian screens. *CAB2* was chosen as it was among the first circadian-controlled promoters discovered in *Arabidopsis* (Millar and Kay, 1991). Many luciferase fusion constructs are now available, controlled by the promoters of clock-controlled output genes or of genes encoding oscillator components.

*Luciferase*

Firefly luciferase is the ideal reporter gene for monitoring circadian rhythms in eukaryotes; bioluminescence imaging is noninvasive, allowing a long time course to be analyzed from one individual, and luciferase activity has a short half-life of about 2–3 h, allowing both increases and decreases in promoter activity to be detected. Luciferase emits a photon at 560 nm upon the reaction of luciferin and $O_2$ to form dehydroluciferin and $CO_2$ using ATP and $Mg^{2+}$. The luminescence levels produced in plants are not bright enough to be detected by anything but an exquisitely sensitive photon detector. The luciferase that we currently use is *LUC+* (Promega, Madison, WI), a modified, cytoplasmic form of firefly luciferase that gives 5- to 20-fold brighter luminescence in transgenic *Arabidopsis* than the native, peroxisomal form. *CAB2:LUC+* in the Ws accession (genetic background) is available from the Nottingham Arabidopsis Stock Centre (NASC; Nottingham, UK), stock number N9352 (Hall *et al.*, 2001). Several clock mutants were identified in screens using *CAB2:LUC* in the C24 accession (NASC stock number N3755; Millar *et al.*, 1992a).

*Manual Luciferase Screens*

The first screens for circadian mutants using *CAB2:LUC* were performed in Nam-Hai Chua's laboratory (Rockefeller University) and in Steve Kay's laboratory (University of Virginia). Luminescence was measured by video imaging of the progeny (M2 generation) of EMS-mutagenized seed (M1 generation). Luminescence measurements must be taken in absolute darkness, although the samples can be illuminated between measurements. Clock mutants should, in theory, have been revealed by their higher luminescence levels when wild-type seedlings were at the trough of luminescence. A single image captured at this time might then have sufficed as a screen. This approach was invalid in practice because the trough level of luminescence varied significantly among wild-type seedlings and an unexpected class of "supernova" mutants gave greatly increased mean luminescence, although the timing of their rhythms was not affected. A successful approach was to capture the second peak of luminescence under constant light; three time points were selected as the imaging was not automated (Fig. 2). Because wild-type seedlings gave similar luminescence levels in the first and third images, the changing luminescence of a mutant with altered timing could be detected visually when the first and third images were alternated rapidly on a video screen. Any M2 seedlings that appeared to have a mistimed peak were selected, reentrained to light–dark cycles, and rescreened directly with higher time resolution (see Fig. 2). From ~8000 M2 seedlings screened, 26 heritable circadian mutations were isolated, although some of these subsequently proved difficult to maintain. The cognate genes for three of these mutations, *timing of cab expression1-1* (*toc1-1*; Millar *et al.*, 1995), *zeitlupe-1* (*ztl-1*; Somers *et al.*, 2000), and *tej* (Panda *et al.*, 2002), have since been cloned in Steve Kay's laboratory; the wild-type gene affected in the *time for coffee* (*tic*; Hall *et al.*, 2003) mutant remains to be identified. Their circadian roles are understood to varying degrees.

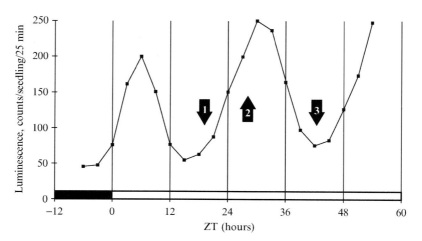

FIG. 2. A minimal screen for circadian mutants using *CAB2:LUC*. Seedlings of the M2 generation were transferred from light:dark cycles to constant light (open box). Arrows indicate times at which three images were collected, to identify plants with a mistimed peak of luminescence. The selected M2 seedlings were reentrained and assayed again over the full time course shown. Data shown are from a control seedling of the parental *CAB2:LUC* line (Miller, PhD Thesis).

## Automated Luciferase Screens Using a TopCount

The three time point screen was most sensitive to mutants of 2–3 h shorter period or >2 h longer period than the wild type. There are of course many other classes of circadian mutant we would like to find, in particular arrhythmic lines that might identify essential components of the circadian clock, but for these the screen must have a higher time resolution. Luminescence can be measured automatically in living plants over a long time course using a luminometer or low-light charge-coupled device camera. We use Packard TopCount scintillation counters (PerkinElmer Life Sciences, UK) and Princeton Instruments (New Jersey) and Hamamatsu Orca cameras (Hamamatsu City, Japan). The TopCount is the most appropriate for a screen, as it has a much higher throughput; our 12-detector instrument measures up to 1920 seedlings in constant darkness with a time resolution of 45 min. If light is present, the time resolution is lower. The seedlings can be monitored for many days, although after 10 days' growth in a 96-well plate, space and nutrients are very limited. In comparison, a camera in our configuration measures up to 484 seedlings with a time resolution of 45 min to 2 h.

## Setting up a Circadian Screen Using a TopCount

The seedlings should be grown in a sterile environment and set up in a flow hood with a low air flow rate to avoid the seedlings wilting and the agar drying out. The seeds are sterilized and then plate in tissue culture dishes containing nutrient agar medium [Murashige and Skoog basal salt mixture (MS; Sigma Aldrich, UK) with 3% sucrose and 1.5% agar] and incubated for 5–7 days in 12-h light:12-h dark cycles at a constant temperature of 22°. The medium can contain an antibiotic to select for the luciferase transgene. The seedlings should be transferred to 96-well microtitre plates for analysis in the TopCount. The plates (Fluorolux HB, DYNEX Technologies Limited, UK) must be black to avoid signal cross-talk between wells, must have solid short edges to hold a barcode label, and not jam in the TopCount. Each well of the plate should be filled with 0.3 ml of nutrient agar and a seedling should then be transferred with forceps, ensuring that the roots contact the agar. This time-consuming process is the rate-limiting step of a TopCount screen. Individual organs can be transferred to test gene expression in, for example, just the roots or cotyledons. The substrate is added to each well. Fifteen microliters of 5 m$M$ luciferin (Biosynth AG, Switzerland) supports luminescence for at least 10 days. The plate is covered with a clear seal (Packard Topseal, PerkinElmer Life Sciences, UK). The seedlings can stick to the underside of the seal, making recovery of mutants very difficult. To avoid this, we cut $\sim\frac{1}{2}$ cm from each side of a seal, attach it to the middle of a complete seal (sticky side to sticky side), and cover the plate. Because both the plants and the luciferase need oxygen, a small hole is made above each well using a sterile needle. Finally, a barcode is attached to the right-hand side of each plate for the TopCount to track, the edges are wiped with WD-40 to help prevent plate jams, and the plate is returned to the 12-h light:12-h dark cycle for another day. Early treatment with luciferin is essential to destroy the active luciferase that has built up during the life of the seedlings (Millar *et al.*, 1992b). Luciferase is deactivated during its reaction, possibly by end-product inhibition, so the luminescence measured during the screen will predominantly be from newly synthesised luciferase.

The TopCount should be set to measure luminescence and cycle through the plates indefinitely, allowing unattended operation. Delays can be included in the assay settings if the time resolution would otherwise be too high. Place the plates in the stacker at the front of the machine, and a "stop" plate on top to signal the end of the cycle to the TopCount (Fig. 3). A weight is needed on top of each stack, and we add extra weights (old plates) to reduce jams caused by plates dropping unevenly. Motorized stackers for the TopCount are available to further reduce jams caused by plates failing to drop under gravity alone.

The TopCount can be kept in a light-tight room to measure seedlings in constant darkness or modified to assay the plants in light (Fig. 3). To provide light, LED arrays or fluorescent bulbs are attached to the sides of the stackers; LEDs are available with a variety of emission wavelengths. A light:dark cycle can be set using a commercial timer. Lights will raise the temperature of the seedlings by a few degrees, although this can be reduced with fans. Reflector plates are needed between each experimental plate to reflect the light from the side of the stack to the seedlings. These are clear plates with mirrors (for design, see http://www.scripps.edu/cb/kay/ianda/spacer_plate.htm). Lighting reduces the time resolution of a TopCount assay because of the time taken to move the reflector plates.

The plant circadian clock maintains rhythms in both constant light and constant darkness, but while the *CAB2* promoter is rhythmic for many cycles in constant light, in constant darkness the rhythms dampen and are very irregular after one or two cycles. We are able to screen for circadian clock mutants in constant darkness using *CAB2:LUC+* because of the high time resolution of the TopCount. Phase differences of 2 h or more are distinct from wild type by the first peak of *CAB2* expression, reducing the duration of data required to 24 h (Fig. 4). This short duration allows high throughput in the first stage of screening, which is the limiting step, but results in a high rate of false positives. A longer screen in constant light would improve the accuracy of the initial screen, as the consistency and

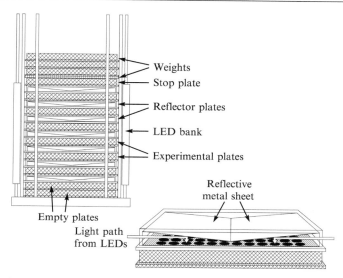

FIG. 3. Arrangement of the TopCount stack with LED banks. Banks of LEDs are attached to the sides of both stackers. We have alternate rows of red and blue LEDs with separate power sources to allow experiments in red, blue, or mixed light qualities. Empty plates are needed at the bottom of the stack to lift the experimental plates to the level of the LEDs. Each experimental plate must be between two reflector plates, which reflect the light to the seedlings. There must be a reflector plate below the bottom experimental plate, because when the plates are in the back stacker, their order is reversed. The reflector plates are clear with a reflective metal sheet.

cumulative effects of the phenotype can be monitored over several successive days. A potential compromise would be to collect just 2 or 3 days of data after 2 or 3 days constant light.

### Selecting Mutants

The TopCount produces a file of tab delimited values for each reading of a plate with the photon count per second from each well and the time it was taken. Andrew Millar's laboratory has developed a Microsoft Excel workbook, TopTempII, to import TopCount files, allowing easy analysis of circadian screens and other circadian data. It converts the file to a time series of photon counts for each well with time points averaged across the whole plate for each reading. It calculates zeitgeber time (ZT; h since the last dawn), allows easy labeling of genotypes, and graphs luminescence over time in groups of four mutant seedlings against the wild type. For this, the wild types should be in wells A1, A2, A6, A7, A11, and A12. It can also

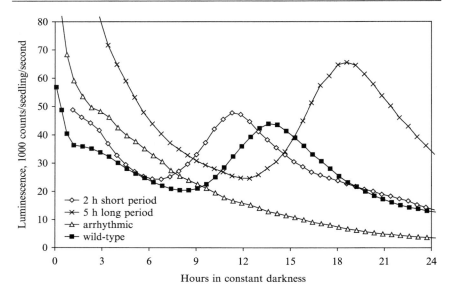

FIG. 4. A screen for circadian mutants in 24-h constant darkness assaying *CAB2:LUC+* in a TopCount. Seedlings of the M2 generation were transferred from light:dark cycles to constant darkness and their luminescence measured in a TopCount. The wild-type control is the average 1000 counts per second from six seedlings. The other traces are from individual M2 seedlings; they were selected from these data and their circadian phenotypes were subsequently shown to be 2-h short period, 5-h long period, and arrhythmic.

graph traces of groups selected by the user. Top Templl and instructions are available on request.

We select mutants by visual analysis, although mathematical analyses are also possible. Because *CAB2:LUC+* luminescence traces from wild-type plants are very variable, it is important to be familiar with the normal range before selecting mutants. Any screen should test some known mutants as positive control subjects and as standards to find or avoid alleles of the known genes.

To map an EMS mutation using molecular markers, the mutant must be crossed to another accession. The substantial genetic variation between the circadian clocks of different *Arabidopsis* accessions makes it difficult to map mutations of small phenotypic effect (less than a 2-h period difference to the wild type, for example). As the periods of each phenotypic class are normally distributed, the classes can overlap so substantially that few mutants can be scored reliably in the F2 generation. We therefore exclude such mutants unless we can find conditions that reveal a stronger phenotype.

## Confirming Candidate Mutants

Circadian screens carried out so far using luciferase have had a large percentage of false positives (up to 95% of putative mutants selected in the M2 generation can be discarded in the rescreen of the M3 progeny). This is partly because sick plants, including those that carry detrimental but non-circadian-related mutations, often have less stable rhythms. The putative mutants can be rescreened as M2 seedlings (see Fig. 2), although this cannot be done easily in the TopCount: Once the seedling has been screened it needs to recover from the limited light and nutrients of the TopCount plate, after which it would be too large for another TopCount experiment. We therefore grow the putative mutants to seed on soil and screen 15 M3 seedlings per line in the TopCount, which also ascertains the dominance of segregating lines. Alternatively, the M2 plant could be grown in an entraining environmental cycle and severed leaves rescreened in the TopCount.

Our screen in 24-h constant darkness found long- and short-period, early- and late-phase, arrhythmic and nondampening mutants (the latter maintain high-amplitude *CAB2:LUC*+ rhythms in darkness, in contrast to the wild type). The most common class was long-period mutants. All screens incorporate some bias and it is worth adapting the rescreen to eliminate the less interesting mutants. Arrhythmic mutants were particularly rare, as many arrhythmic M2 seedlings died and in most others the apparent arrhythmia was not heritable. Such screens are far from saturated—clock-related genes are still being discovered, along with interesting alleles of known genes.

## Circadian Characterization

Once a mutant has been shown to have a reproducible phenotype, the cognate gene can be cloned by map-based cloning for EMS-induced mutants and by inverse polymerase chain reaction for T-DNA insertion mutants. Additionally, the circadian phenotypes can be characterized in more detail. A mutant isolated for *CAB2:LUC*+ rhythms in constant darkness would be assayed in constant light to give period estimates and in different fluence qualities and rates to reveal any light dependence. It is difficult to estimate period accurately without mathematical analysis. Our usual method is fast Fourier transformation nonlinear least squares (FFT-NLLS; Plautz *et al.*, 1997), which finds the closest fitting cosine wave to a raw data trace and estimates the period, amplitude, and associated uncertainty values for the curve. A number of Microsoft Excel worksheets are available to manipulate data from circadian experimental techniques, apply FFT-NLLS, and perform statistical analyses of the results. The most comprehensive, BRASS, is available from http://www.amillar.org/downloads.html. BRASS is designed to take data from TopCount, leaf movement, and luminescence measured in cameras controlled by NightOwl and Metamorph software and can analyze data from all other methods, as the user can specify the content of the data columns. Further characterization of a mutant depends on the results of this initial characterization and is beyond the scope of this article, but it typically includes measuring the effect on known circadian components, flowering time, hypocotyl length, and yeast two-hybrid experiments.

## Circadian Screen Measuring Stomatal Rhythms

Direct screens can of course measure rhythms other than luciferase transgenes. C. Robertson McClung's laboratory (Dartmouth College) developed a screen based on stomatal conductance rhythms (Salomé et al., 2002). Sensitivity to the toxic gas sulfur dioxide ($SO_2$) was shown to have a circadian rhythm. $SO_2$ enters the leaf via the stomata, and the stomatal aperture is clock controlled, so the sensitivity of the plant to $SO_2$ is greater at phases when the stomata are open than when they are closed. Plants entrained to 14-h light:10-h dark cycles then transferred to constant light were sensitive to $SO_2$ for most of the circadian cycle, but resistant for a 1-h window, 2 h before subjective dawn. McClung and colleagues therefore screened 3- to 4-week-old M2 plants derived from EMS mutagenesis to select those whose leaves developed necrotic lesions from $SO_2$ exposure at this time but did not develop necrotic lesions without exposure to $SO_2$. They isolated 65 putative mutants from 6500 plants screened and tested their circadian phenotypes further using leaf movement, $CO_2$ assimilation, and *CAB2:LUC* rhythms. Twelve had altered circadian properties, with long or short periods or altered phase. The phase mutants are the first mutants to have been identified in *Arabidopsis* that affect the phase but not the period of the clock, one has been shown to be an allele of *phyB* (Salomé et al., 2002).

## Testing Circadian Rhythms of Mutants Identified in Other Screens

Many circadian mutants were initially isolated in screens for processes partially controlled by the circadian clock, such as flowering time and hypocotyl length. Their circadian phenotype was later assayed and shown to be aberrant. The original screens did not use luciferase transgenic lines, and it is time-consuming to cross or transform the reporter transgenes into all potential clock mutants. A faster approach has been to assay leaf movement rhythms. The cotyledons and leaves of *Arabidopsis* seedlings move up and down once per day in free-running conditions under the control of the circadian clock. They can be tracked automatically (Millar *et al.*, 1995) and used to measure the circadian rhythm of the mutant. The leaf movement assay has a lower throughput than luciferase assays, but is ideal for a secondary screen of lines enriched in clock mutants. Leaf movement can only be measured in constant light, as light is necessary for the cameras to take the images, although infrared imaging is technically possible. The leaf angle also responds acutely to dark-to-light transitions. Morphological phenotypes of the mutant can affect the ease of measuring the rhythm, e.g., the small, bunched leaves of some mutants mean that individual leaves cannot be tracked easily. Physiological phenotypes can also determine the age of the seedling to measure; leaves of a slowly developing mutant will emerge and be measured later than the wild type.

## Mutants Identified in Noncircadian Screens

The first circadian mutant to be isolated was from a screen for "supervital" mutants by György P. Rédei in the 1950s (Rédei, 1962). He selected *gigantea* (*gi*) alleles for their vigor and characterized their late flowering phenotype and ability to outcompete the wild type in controlled conditions. *gi* has since been linked to the circadian clock; all alleles shorten the period of leaf movement rhythms, whereas some shorten and others lengthen the period of *CAB2:LUC* rhythms (Park *et al.*, 1999). More recent examples include *late elongated hypocotyl* (*lhy*), which was identified as a late-flowering and long hypocotyl mutant, but testing of its circadian phenotypes showed it to be arrhythmic (Schaffer *et al.*, 1998), and *LHY* is now one of the strongest candidate oscillator genes in *Arabidopsis*. Screens of early flowering mutants have been equally fruitful. *early flowering3* (*elf3*) was isolated as an early flowering mutant, for example, and shown to have circadian phenotypes: Its leaf movement and *CAB2::LUC* rhythms are arrhythmic in constant light (Hicks *et al.*, 1996).

A further interesting, although poorly studied, class of candidate clock-affecting mutants alter light input from known photoreceptors to the clock components. Circadian screens of light signal transduction mutants are beginning to uncover their circadian phenotypes (e.g., Staiger *et al.*, 2003).

## Reverse Genetics

The immense worth of a sequenced genome becomes most apparent in reverse genetic approaches. Reverse genetics identifies a candidate gene from its sequence, usually due to homology to other genes in the process of interest. A mutant is then isolated and the phenotype is studied. Mutants cannot be created by homologous recombination in *Arabidopsis*, but enormous databases of sequence-indexed T-DNA insertion mutants are available from the stock centers (e.g., SALK lines; Alonso *et al.*, 2003) with mutations in most genes. In the circadian clock, this approach has been used to address the roles of the *PSEUDO RESPONSE REGULATOR* (*PRR*) family (Eriksson *et al.*, 2003; Michael *et al.*, 2003). T-DNA mutants are usually null alleles, caused by the insertion of a large stretch of DNA in the gene. Novel weak alleles can be created by targeting induced local lesions in genomes (TILLING) (Till *et al.*, 2003). This selects up to 12 EMS mutants from a requested region of a gene. It is particularly suitable if the null alleles are interesting but an allelic series is wanted to give a greater indication of the function of the gene or if the null is lethal. Links to these and other resources can be found at the online databases for *Arabidopsis* (http://www.arabidopsis.info and http://www.arabidopsis.org).

The wide variety of genetic approaches has successfully found many components of the *Arabidopsis* circadian system and we expect that there are still many to discover. The connections between them that make a functioning clock and those linking the clock to the wider regulatory network are slowly being elucidated, and this will be simpler with a more complete set of components.

## References

Alonso, J. M., Stepanova, A. N., Leisse, T. J., Kim, C. J., Chen, H., Shinn, P., Stevenson, D. K., Zimmerman, J., Barajas, P., Cheuk, R., Gadrinab, C., Heller, C., Jeske, A., Koesema, E., Meyers, C. C., Parker, H., Prednis, L., Ansari, Y., Choy, N., Deen, H., Geralt, M., Hazari, N., Hom, E., Karnes, M., Mulholland, C., Ndubaku, R., Schmidt, I., Guzman, P., Aguilar-Henonin, L., Schmid, M., Weigel, D., Carter, D. E., Marchand, T., Risseeuw, E., Brogden, D., Zeko, A., Crosby, W. L., Berry, C. C., and Ecker, J. R. (2003). Genome-wide insertional mutagenesis of *Arabidopsis thaliana*. *Science* **301**, 653–657.

Eriksson, M. E., Hanano, S., Southern, M. M., Hall, A., and Millar, A. J. (2003). Response regulator homologues have complementary, light-dependent functions in the *Arabidopsis* circadian clock. *Planta* **218**, 159–162.

Hall, A., Bastow, R. M., Davis, S. J., Hanano, S., McWatters, H. G., Hibberd, V., Doyle, M. R., Sung, S., Halliday, K. J., Amasino, R. M., and Millar, A. J. (2003). The *TIME FOR COFFEE (TIC)* gene maintains the amplitude and timing of *Arabidopsis* circadian clocks. *Plant Cell* **15**, 2719–2729.

Hall, A., Kozma-Bognár, L., Tóth, R., Nagy, F., and Millar, A. J. (2001). Conditional circadian regulation of *PHYTOCHROME A* gene expression. *Plant Physiol.* **127**, 1808–1818.

Hicks, K. A., Millar, A. J., Carré, I. A., Somers, D. E., Straume, M., Meeks-Wagner, D. R., and Kay, S. A. (1996). Conditional circadian dysfunction of the *Arabidopsis* early-flowering 3 mutant. *Science* **274**, 790–792.

Michael, T. P., Salome, P. A., Yu, H. J., Spencer, T. R., Sharp, E. L., McPeek, M. A., Alonso, J. M., Ecker, J. R., and McClung, C. R. (2003). Enhanced fitness conferred by naturally occurring variation in the circadian clock. *Science* **302**, 1049–1053.

Millar, A. J., Carré, I. A., Strayer, C. A., Chua, N.-H., and Kay, S. A. (1995). Circadian clock mutants in *Arabidopsis* identified by luciferase imaging. *Science* **267**, 1161–1163.

Millar, A. J., and Kay, S. A. (1991). Circadian control of cab gene transcription and mRNA accumulation in *Arabidopsis*. *Plant Cell* **3,** 541–550.

Millar, A. J., Short, S. R., Chua, N.-H., and Kay, S. A. (1992a). A novel circadian phenotype based on firefly luciferase expression in transgenic plants. *Plant Cell* **4,** 1075–1087.

Millar, A. J., Short, S. R., Hiratsuka, K., Chua, N.-H., and Kay, S. A. (1992b). Firefly luciferase as a reporter of regulated gene expression in plants. *Plant Mol. Biol. Rep.* **10,** 324–337.

Panda, S., Poirier, G. G., and Kay, S. A. (2002). *tej* defines a role for poly(ADP-ribosyl)ation in establishing period length of the *Arabidopsis* circadian oscillator. *Dev. Cell* **3,** 51–61.

Park, D. H., Somers, D. E., Kim, Y. S., Choy, Y. H., Lim, H. K., Soh, M. S., Kim, H. J., Kay, S. A., and Nam, H. G. (1999). Control of circadian rhythms and photoperiodic flowering by the *Arabidopsis* GIGANTEA gene. *Science* **285,** 1579–1582.

Plautz, J. D., Straume, M., Stanewsky, R., Jamison, C. F., Brandes, C., Dowse, H. B., Hall, J. C., and Kay, S. A. (1997). Quantitative analysis of *Drosophila* period gene transcription in living animals. *J. Biol. Rhythms* **12,** 204–217.

Rédei, G. P. (1962). Supervital mutants of *Arabidopsis*. *Genetics* **47,** 443–460.

Salomé, P. A., Michael, T. P., Kearns, E. V., Fett-Neto, A. G., Sharrock, R. A., and McClung, C. R. (2002). The *out of phase 1* mutant defines a role for PHYB in circadian phase control in *Arabidopsis*. *Plant Physiol.* **129,** 1674–1685.

Schaffer, R., Ramsay, N., Samach, A., Corden, S., Putterill, J., Carré, I., and Coupland, G. (1998). The *late elongated hypocotyl* mutation of *Arabidopsis* disrupts circadian rhythms and the photoperiodic control of flowering. *Cell* **93,** 1219–1229.

Somers, D. E., Schultz, T. F., Minamow, M., and Kay, S. A. (2000). *ZEITLUPE* encodes a novel clock-associated PAS protein from *Arabidopsis*. *Cell* **101,** 319–329.

Staiger, D., Allenbach, L., Salathia, N., Flechter, V., Davis, S. J., Millar, A. J., Chory, J., and Fankhauser, C. (2003). The *Arabidopsis* SRR1 gene mediates phyB signalling and is required for normal circadian clock function. *Genes Dev.* **17,** 256–268.

Till, B. J., Reynolds, S. H., Greene, E. A., Codomo, C. A., Enns, L. C., Johnson, J. E., Burtner, C., Odden, A. R., Young, K., Taylor, N. E., Henikoff, J. G., Comai, L., and Henikoff, S. (2003). Large-scale discovery of induced point mutations with high-throughput TILLING. *Genome Res.* **13,** 524–530.

# [3]   Genetic Screens for Clock Mutants in *Drosophila*

### By JEFFREY L. PRICE

## Abstract

The isolation and analysis of mutant flies (*Drosophila melanogaster*) with altered circadian rhythms have led to an understanding of circadian rhythms at the molecular level. This molecular mechanism elucidated in fruit flies is similar to the mechanism of the human circadian clock, which confers 24-h rhythmicity to our sleep/wake behavior, as well as to many other aspects of our cellular and organismal physiology. In fruit flies, genes can be mutated to abolish circadian rhythms (i.e., produce arrhythmia) or

METHODS IN ENZYMOLOGY, VOL. 393

alter the period of the circadian rhythm; these genes encode key components of the circadian oscillator mechanism. Other mutations have identified components of the input pathways (by which light and temperature synchronize the circadian clock to environmental cycles) or output pathways (which connect the circadian oscillator to the physiological response). Mutations in genes are typically generated by chemical mutagenesis or mutagenesis with transposable elements. Flies with mutagenized chromosomes are processed in a series of genetic crosses, which allow specific chromosomes to be screened for semidominant mutations, recessive mutations, enhancer/suppressor mutations, or genes that can be overexpressed to alter circadian rhythms. Circadian phenotypes, which are assayed to identify mutants, include eclosion (emergence of the adult from the pupal case), locomotor activity (similar to human sleep/wake behavior), and circadian oscillations of gene expression. It is argued that screens for new circadian genes will continue to reveal novel components of the circadian mechanism.

Introduction

There are few areas in which our knowledge of molecular mechanisms owes as much to genetic studies as in the area of circadian rhythm research. In the pregenetic era, pioneers such as Colin Pittendrigh and Jurgen Aschoff established the formal properties of circadian rhythms and contributed to theoretical models that have been applied successfully to the current molecular models (Pittendrigh, 1974). It was postulated that circadian rhythms were produced by a biological circadian oscillator—a clock-like entity that sustained circadian rhythms with a period of $\sim$24 h in the absence of any environmental driving force. This circadian clock was clearly synchronized (or entrained) by input pathways responding to environmental light and temperature cycles, and communicated timing information to diverse cellular, biochemical, and physiological processes through multiple output pathways. However, for many years the identities of the molecules that comprise the circadian mechanism were largely unknown, as was the mechanism in which they participate. It is precisely such a "black box" that a genetic analysis is ideally suited to address, because a genetic analysis requires no initial knowledge about the nature of the mechanism. Instead, the isolation of mutations that affect the mechanism can identify the molecules, and genetic analysis of the mutations can establish the general nature of the mechanism.

It was essential to have well-designed screens and well-thought-out criteria to identify mutations likely to affect different parts of the circadian mechanism. Null mutations, which by definition eliminate all function of a gene, are predicted to lead to arrhythmicity if they affect a component of the circadian oscillator mechanism. However, mutations that uncouple output pathways from the oscillator are also predicted to cause arrhythmicity in the affected outputs. For instance, mutations that prevent the synthesis of a neurotransmitter conveying temporal information to a particular circadian output are predicted to cause an arrhythmic output, even though the underlying clock mechanism is still functional. Moreover, mutations that uncouple input pathways from the circadian oscillator could also produce overt arrhythmicity, e.g., by introducing asynchrony between multiple oscillators in an organism or population. If a mutation produces arrhythmia in all known output pathways, it is clearly not specific to one output pathway and is therefore more likely to be affecting a central clock mechanism. However, the mutation could be affecting a common output pathway or an input pathway that lies upstream from the circadian oscillator. Therefore, an arrhythmic mutation does not by itself identify a central clock component.

Importantly, mutant alleles for many of the clock genes produce changes in the period of all circadian rhythms exhibited by the organism, rather than lack of overt circadian rhythms (Konopka and Benzer, 1971; Price *et al.*, 1998; Rothenfluh *et al.*, 2000a,c; Rutila *et al.*, 1996). It is unlikely that a circadian period change can be produced solely at the level of an output or input pathway. The first mutant alleles isolated in *Drosophila* all affected the same gene (*per*) and produced the full range of possibilities outlined here—arrhythmicity, short period rhythms, and long period rhythms in all known circadian outputs (Konopka and Benzer, 1971). This seminal work strongly argued that the genetic analysis of single-gene mutations in the fruit fly would probe the nature of the circadian oscillator mechanism.

A brief overview of the circadian mechanism, which is not the focus of this chapter, is offered at the outset to provide some context for the genes mentioned repeatedly throughout the chapter. The circadian oscillator of *Drosophila* is driven by the oscillations of several gene products, which regulate their own synthesis to produce the oscillations (Price, 2004). The *period* (PER) and *timeless* (TIM) proteins are thought to interact with a transcription factor heterodimer, composed of the *dClock* (dCLK) and *cycle* (CYC) proteins, to abrogate the transcriptional activation by dCLK/CYC. When not repressed by PER and TIM proteins, the dCLK/CYC transcription factor activates transcription of the PER and TIM repressors, the *vrille* (VRI) repressor (which binds directly to DNA in the *dClk* promoter to negatively regulate *dClk* mRNA), and a host of clock-controlled genes, which ultimately produce the various circadian outputs. Several kinases, including *doubletime* (DBT), *shaggy* (SGG),

*Time-keeper/Andante*, (TIK/AND), and a phosphatase (PP2A) collaborate to delay the negative feedback afforded by PER and TIM so that levels of dCLK/CYC-dependent mRNAs can increase before repression is accomplished; this delay, coupled with an eventual turnover of PER and TIM that relieves the negative feedback, produces the molecular oscillations that underlie the oscillator mechanism.

Genetic screens have also identified a cryptochrome-like photoreceptor (CRY) involved in entrainment and several proteins dedicated to specific output pathways rather than the circadian oscillator. Mutants in genes dedicated to output pathways typically dampen the rhythmicity of selected circadian responses or alter the phase of these responses, without strong effects on circadian period or the molecular oscillations of core clock genes such as *per* and *tim*.

## Strategies for Mutagenesis

### Chemical Mutagenesis

Only one *Drosophila melanogaster* circadian mutation has been identified in nature (the *tim^{rit}* mutation; Matsumoto *et al.*, 1999); the rest have been isolated from laboratory stocks after mutagenesis. The most commonly used mutagen has been ethylmethane sulfonate (EMS), an alkylating agent that ethylates DNA bases and thereby can introduce errors in DNA replication, with a consequent mutation of the DNA. The mutations are typically single base pair changes, which alter the type of amino acid encoded at a specific position in a polypeptide chain (a missense mutation) or produce a nonsense mutation (if an amino acid codon is replaced with a translational stop codon).

The procedure for EMS mutagenesis of *Drosophila* was specified by Lewis and Bacher (1968). Dehydrate 2- to 3-day-old males at room temperature for 30 min in bottles with Whatmann 3MM paper. Then treat them at room temperature for 16–20 h in a bottle with Whatmann paper soaked in 0.025 *M* ethylmethane sulfonate in a 1% sucrose solution. Then cross the mutagenized males to females at 25°. Because EMS is a mutagen, it should be handled in a chemical fume hood with protective clothing to minimize personal exposure. After use, wash EMS solutions and all containers with deactivation solution (4 g NaOH + 0.5 ml thioglycolic acid in 100 ml) to detoxify them.

### P Element Mutagenesis

P elements are a class of *Drosophila* transposable elements that produce mutations when they insert in the vicinity of a gene. Their insertion can directly disrupt the reading frame of the gene or (more commonly) interfere with transcription or splicing of the gene. P elements can be transposed into new chromosomal locations and then stabilized at that location with exquisite genetic control (Fig. 1) (Sehgal *et al.*, 1991). There are strains carrying P elements that can transpose only in the presence of functional transposase from another source, and other strains that carry a P element that produces high levels of transposase but that cannot transpose itself. The two strains are crossed to produce progeny containing both elements and in which transposition of the transposition-competent element is driven by the transposase-encoding element. Mating of these "transposition" flies and selection of the appropriate progeny ensure that the two P elements are no longer carried in the same progeny so that the transposed P element will be stably inserted into a new location. The mobilized P element carries an eye color marker, the presence of which in the indicated progeny ("? or" in Fig. 1) demonstrates that the P element is no longer present in its original location (the X chromosome in the example of Fig. 1) because it is found in progeny (male in this case) that did not inherit the original chromosome bearing the P element.

The strategy shown in Fig. 1 was used to produce the original *tim$^o$* mutation (Sehgal *et al.*, 1994). The *lark* mutation was produced by a similar P element mutagenesis strategy (Newby and Jackson, 1993), and one of the original *dbt* mutations was found in an existing P element stock (Price *et al.*, 1998). The *pdf$^o$* mutation was produced by an excision of a P element (Renn *et al.*, 1999).

From the standpoint of a forward genetic screen, the advantage of P element mutagenesis is that it tags the gene with an identifiable sequence that can immediately be used to clone the gene. For instance, genomic libraries can be constructed from the mutant strain, and clones encoding the mutant gene can be identified because they will hybridize to P element sequences (as was done for cloning of *lark;* Newby and Jackson, 1993). Most modern P element constructs encode plasmid origins of replication, which allow them to be transformed directly into *Escherichia coli* and replicated as plasmids containing a piece of the adjacent *Drosophila* DNA (as was done for the cloning of *dbt;* Kloss *et al.*, 1998). However, as was demonstrated with the *tim$^o$* mutation, P element-induced mutations are not always associated with a P element insertion, and in these cases cloning is not facilitated by the P element, which is instead a distraction (Myers *et al.*, 1995). Moreover, P element mutagenesis is not as random as EMS mutagenesis. With the single nucleotide resolution afforded by the completed *Drosophila* genome sequence and the extensive collection of ordered clones and P element strains offered by the Berkley *Drosophila*

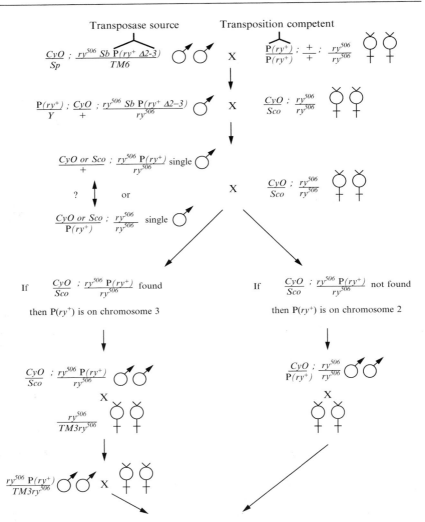

Screen homozygotes (where possible) or heterozygotes for altered circadian rhythms

FIG. 1. Screen of P element insertion stocks for clock mutants.

genome project, a large-scale and unbiased screen employing P element insertional mutagenesis is no longer advisable, although enhancer trap and overexpression screens involving P elements are still extremely useful (see later).

## Genetic Crosses to Generate Genotypes for a Screen

### Crosses for a Screen of the X Chromosome

Screens of the X chromosome for recessive mutations are easy because recessive mutations can be uncovered in the hemizygous condition in males, which only carry one X in *Drosophila*. Figure 2 outlines the series of crosses used by Konopka and Benzer (1971) to produce lines of flies for screens of the X chromosome. EMS-treated males were mated to $X^\wedge X/Y$ females ($X^\wedge X$ is an "attached X:" two X chromosomes held together so that they always segregate together). Note that two X chromosomes produce a female in *Drosophila*, despite the presence of a Y chromosome. This cross produces two kinds of progeny: Y chromosome/X* males in which the mutagenized X (denoted by an asterisk) is inherited from the father and the Y chromosome from the mother, and $X^\wedge X/Y$ females in which the Y chromosome is inherited from the father and the attached X chromosome from the mother. Other possible genotypes (Y/Y and $X/X^\wedge X$) are lethal. Single F1 Y/X* male progeny were collected from this cross and mated to $X^\wedge X/Y$ females; it was important to start this cross with single males because different F1 progeny males from the same mutagenized parent male could carry different mutations, and the goal was to produce

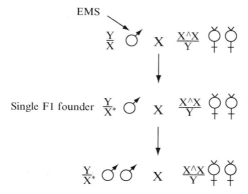

Self-propagating stock in which all males carry a mutagenized X, and all females have a nonmutagenized $X^\wedge X$.

Screen for lines in which the males have an altered circadian rhythm and the females have wild-type rhythms.

FIG. 2. Screen of EMS-mutagenized X chromosomes for clock mutants.

a line in which all males carry the same mutation. This cross produces Y/X^X females and Y/X* males as F2 progeny. All of the males will carry the circadian mutation, if one has been induced, whereas all of the females should exhibit wild-type behavior. After the cross of the F1 male, the line is self-propagating. The original *per* alleles (long, short, and arrhythmic) and the *Andante* mutation (Konopka *et al.*, 1991) arose in lines generated like these.

### Crosses to Screen for Autosomal Recessive Mutations

A line with a homozygous autosome is required in order to screen for recessive mutations, which by definition only produce a phenotype when both of the two alleles of the gene are mutant. The parts of the cross scheme shown after the isolation of the P element transposition in Fig. 1 are designed to identify the autosome (either chromosome 2 or 3; the X chromosome is chromosome 1, and chromosome 4 is so small that it has been neglected in circadian screens) on which the P element has landed, and to produce flies homozygous for that chromosome. The P element carries a gene (the wild-type allele of *ry*), which restores normal eye color to the flies, thereby allowing determination of the chromosome carrying the novel P element insertion during the cross scheme. The final steps of the cross are modified as necessary to produce second or third chromosome lines.

Likewise, the cross scheme depicted in Fig. 3 was employed by Price *et al.* (1998) and Rothenfluh *et al.* (2000a) to isolate mutations of *dbt* and *tim*. It is designed to produce lines of flies that are homozygous for single mutagenized second chromosomes (the *dbt*^s^ mutation, which is found on the third chromosome, was found serendipitously in one of these lines). The cross scheme employs balancers (in this case, *CyO* or *CyO, Gla*), which are multiply rearranged second chromosomes that do not recombine with more normal (but nevertheless mutagenized - *) second chromosomes. Hence, the presence of a balancer ensures identical descendants of the original mutagenized chromosome from the single founder F1 male. The balancers are marked with dominant mutations (such as *Cy* for curly wings or *Gla* for glazed eyes) that allow the presence of a single copy of the balancer to be detected. Because the homozygous condition for the balancer is lethal, balancer chromosomes are maintained in stocks opposite homologous chromosomes that carry lethal mutations or deletions [e.g., *Sco* or Df(2R)XTE11] so that the only viable progeny are heterozygotes (e.g., *Sco/Cyo*).

In this cross scheme, the mutagenized chromosome carries two eye color mutations (*cn* and *bw*), which together produce white eye color so

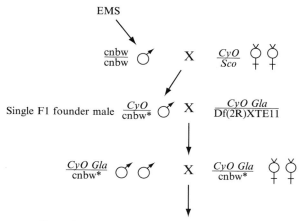

Screen homozygous cnbw flies (where possible) or heterozygotes for altered circadian rhythms.

FIG. 3. Screen of EMS-mutagenized second chromosomes for clock mutants.

that homozygotes are easily scored by visual inspection. They are also scored by the absence of the dominant mutations carried by the balancer. Entirely analogous screens have been conducted on the third chromosome by using third chromosome marker mutations and balancers (e.g., TM3 and TM6).

This type of screen has produced numerous clock mutant alleles: $dbt^S$, $dbt^G$, and $dbt^H$ alleles (Price *et al.*, 1998; Suri *et al.*, 2000), the $dClk^{Jrk}$ allele (Allada *et al.*, 1998), and $tim^{o1}$, $tim^{L1}$, and $tim^{L2}$ alleles (Rothenfluh *et al.*, 2000a; Sehgal *et al.*, 1994).

### Crosses to Screen for Semidominant Autosomal Mutations

The screens for recessive clock mutants described earlier turned up genes that could generally be mutated to produce semidominant alleles, i.e., alleles that detectably altered circadian behavior in the heterozygous condition (mutant gene on one chromosome/wild-type allele on the other). In particular, *per, tim,* and *dbt* alleles, which retained some function, altered period length significantly, but even some null alleles, which lead to arrhythmicity when homozygous, caused circadian period alteration when heterozygous (e.g., $per^o$, which causes an $\sim$0.5- to 1-h circadian period lengthening in $per^+/per^o$ flies; Smith and Konopka, 1982). The cross schemes depicted in Figs. 1 and 3, while designed to generate homozygous mutant flies, also generated many lines that could only be screened in the heterozygous condition because the mutagenized chromosome carried a lethal mutation that precluded screening of homozygous flies. For instance, progeny of homozygous lethal lines generated by the cross scheme outlined in Fig. 3 would be screened as CyO, Gla/cnbw* individuals. In some cases, the lethality on the chromosome arises from the same mutation that causes the circadian effect. In such a case, the only way to assess the circadian phenotype is to score it in the heterozygous condition [such was the case for *lark* (Newby and Jackson, 1993) and *Tik* (Lin *et al.*, 2002a)], although there are ways to work around lethality (see discussion of *sgg* and *slimb* mutants later).

It is possible to design a screen that only detects semidominant mutations, thereby increasing the throughput of the screen. Rothenfluh *et al.* (2000a) devised such a strategy, as depicted in Fig. 4, and Newby and Jackson (1993) employed a P element-based semidominant screen to identify *lark*. The enhanced throughput is possible because the laborious crosses to produce homozygous lines are not undertaken in the vast majority of cases. Male flies are screened in the F1 generation and only mated if they display interesting circadian rhythms; most do not and are discarded. Because individual flies with a wild-type circadian genotype can nevertheless display apparent rhythmicity (see discussion of this in the section entitled "Validation of Candidate Lines Identified in Genetic Screens"), only F1 males with period phenotypes are pursued, i.e., flies with periods less than 23.5 h or greater than 24.5 h. Recall that some alleles that produced arrhythmia when homozygous produce period changes when heterozygous, so the strategy does not preclude the isolation of mutations that produce arrhythmia when homozygous.

Many of the existing alleles of various clock mutants were produced in this way, including the $dbt^{AR}$ and $dbt^L$ alleles (Price *et al.*, 1998; Rothenfluh *et al.*, 2000b), *lark* (Newby and Jackson, 1993), and the $tim^{L3}$, $tim^{L4}$, $tim^{S1}$, $tim^{S2}$, and $tim^{UL}$ alleles (Rothenfluh *et al.*, 2000a,c).

## Crosses to Screen for Suppressors and Enhancers of per Mutations

Rutila *et al.* (1996) undertook a screen for enhancers and suppressors of the $per^L$ and $per^S$ mutations. Enhancers are defined as mutations that increase the severity of a phenotype produced by another mutation (in this case, the $per^L$ or $per^S$ mutation). Suppressors are defined as mutations that reduce the severity of another mutation. Screens for enhancers or suppressors may uncover mutations that would not have been detected in an otherwise wild-type background because the mutant background may sensitize the process to effects that would be too subtle in a wild-type background. The specific strategy employed in this case was to mutagenize $per^L$

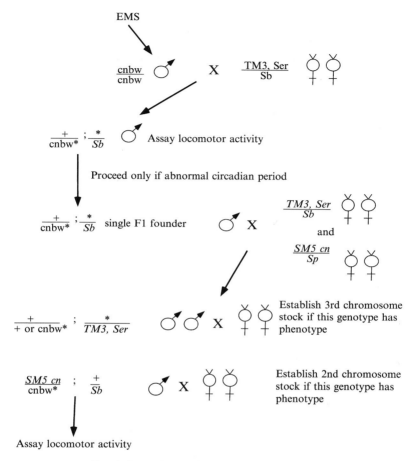

Fig. 4. Screen for semidominant clock mutations.

(or *per^S*); *ry^{506}* (a third chromosome mutation) flies with EMS and cross them in a scheme that resembled the one outlined in Fig. 3, except that all flies carried a *per^S* or *per^L* genotype, and third chromosome balancers were employed instead of second chromosome balancers.

One line was identified in which there was a semidominant suppression of the *per^L* phenotype (ironically, the suppressor mutation mapped to the second chromosome and had to be reextracted from the stock). This mutation (*tim^{sL}*, for suppressor of *per^L*) shortened the *per^L* circadian period dramatically and virtually eliminated the lack of temperature compensation, while having almost no effect on the rhythms of flies carrying a wild-type *per* genotype (Rutila *et al.*, 1996). Such allele-specific interactions between two genes can arise from direct interactions between the gene products; in fact, a large body of molecular data argue that PER and TIM can form a heterodimer, which is required for their nuclear accumulation and negative feedback on dCLK/CYC. The *tim*$^{sL}$ mutation appears to allow the TIM/PER heterodimer to "bypass" the period lengthening caused by the *per*$^L$ mutation, perhaps by altering the time course of TIM phosphorylation.

The majority of mutants produced by this screen have mutant phenotypes that are as obvious or even more obvious in a *per*$^+$ genetic background. These mutants include the *cyc* mutation (Rutila *et al.*, 1998), the recessive *Clk*$^{ar}$ mutation (Allada *et al.*, 2003), and the *Tik* mutation (Lin *et al.*, 2002a) so the enhancer/suppressor strategy does not preclude the detection of mutants that could have been detected in a wild-type background as well.

### Screens for Genes that Alter Circadian Rhythms when Overexpressed

It is possible to use a P element specifically designed to overexpress adjacent genes for this purpose. Such a screen can detect genes that are homozygous recessive lethal mutations and that have little or no effect on circadian rhythms as heterozygotes; these genes would not be detected by analysis of the flies produced in any of the aforementioned cross schemes.

The P elements that drive expression of adjacent genes are termed EP elements (Rorth *et al.*, 1998). Rorth and collaborators produced a collection of 2300 such lines by mobilizing the progenitor EP element and deriving lines with different insertion sites. The mobilized P element contains a binding site for the yeast transcription factor GAL4. If GAL4 is supplied by another P element transgene (the driver) in the EP-containing fly, the promoter sequence in the EP element will transcribe an mRNA that is elongated into the adjacent fly genome. To screen for insertions that could drive expression of adjacent genes affecting circadian rhythms, Martinek *et al.* (2001) crossed a *timeless* (UAS)-GAL4 driver to each of these 2300 EP lines, and the doubly transgenic progeny (driver + EP) were screened for effects on circadian rhythms. The *timeless* promoter in the driver was modified with a GAL4-binding site (termed a "UAS"). This driver becomes active in clock cells, which activate the *timeless* promoter. Then, the GAL4 transcription factor produced by the driver autoactivates its own expression through the UAS sequence, thereby maintaining constitutively high levels of GAL4 only in clock cells (ubiquitous expression is to be avoided, as it is more likely to be lethal). In turn, transcription from the EP element is also activated. This screen identified an EP element that activated expression of the *Drosophila sgg* gene, thereby shortening the period of the circadian clock.

*A priori*, phenotypes produced in such a way are suspect, as the over-expressed or even misexpressed protein may participate in processes it would not normally affect so it is important to support an initial finding of this sort with other genetic evidence. Because loss-of-function *sgg* mutations are lethal, it was not possible to assess loss-of-function effects on behavior directly. Instead, the lethality was complemented by overexpressing SGG from a heat shock promoter, induced with high-temperature pulses in a *sgg* mutant background; then, the temperature was lowered in the adult to reduce the amount of SGG protein. Such a protocol allowed the flies to survive to adulthood and produced flies with a long circadian period, indicative of a loss-of-function phenotype (Martinek *et al.*, 2001).

## Detection of Mutant Circadian Phenotypes

### The Eclosion Assay

The simplest assay, and perhaps the most elegant, is still the original one devised by Konopka and Benzer (1971). It is based on the circadian rhythm of eclosion, which is behavior that produces emergence of the adult from the pupal case. Circadian rhythms of eclosion are revealed in a population of flies with a range of developmental ages. The number of adults that eclose in consecutive intervals (e.g., intervals of 1 h) can be measured the hard way (by day shift and night shift research teams) or with automated fly collection systems. In the latter case, cultures of flies are shaken automatically at regular intervals to cause the adults to fall into collecting vials or photodetectors. [The original "bang" boxes devised by Pittendrigh have been updated by the Trikinetics company (Waltham, MA), which produces versions that interface directly with a computer for direct recording of data.] Because fly eclosion is gated, it occurs in a window of time during the morning following developmental competence for eclosion. If a fly becomes developmentally ready to eclose outside this window, eclosion is delayed until the next window. Therefore, for a population of flies with a range of developmental ages, a plot of the number of eclosing flies at each collection time as a function of collection time demonstrates an eclosion rhythm, with peak eclosion occurring during the morning in wild-type flies. The rhythm persists in constant darkness, with flies eclosing at the time of morning in the previous light:dark cycle (Pittendrigh, 1974).

Even with automated collection of flies, a full-scale eclosion assay would be very ponderous to undertake for a large number of lines.

However, simple observation of a line reared in a 12-h:12-h light:dark cycle can detect aberrant eclosion behavior. Flies can be collected from lines bearing a mutagenized chromosome twice a day—1 h before lights on and 1 h before lights off. In wild-type flies, at least twice as many emerge in the second collection as in the first collection under these conditions. Flies with circadian mutations can produce high levels of flies in the first collection (i.e., during the dark period) (Sehgal *et al.*, 1991). Arrhythmic mutants produce relatively constant eclosion activity during both day and night, mutants with short-period circadian rhythms eclose early to produce high levels of night emergers, and mutants with long-period circadian rhythms eclose late to produce high levels of night emergers. Such a screen identi-fied the original *per* alleles of Konopka and Benzer (1971) (*per^S*, *per^L*, *per^{o1}*), the *tim^{o1}* mutation (Sehgal *et al.*, 1994), and the *lark* mutation (Newby and Jackson, 1993). The *lark* mutation has no discernible effect on locomotor activity and does not affect the circadian period of eclosion in constant darkness. Therefore, it is thought to be an output-specific muta-tion; it affects the phase of eclosion coupling to the central clock without affecting the central clock.

These stripped-down eclosion assays are best conducted in a light:dark cycle rather than constant darkness because the light:dark cycle imposes a consistent phase on period-altering mutants and generally leads to more robust eclosion rhythms (circadian rhythms are damped in constant dark-ness). The consistent phase derives from the entrainment to the 24-h cycle time of the light:dark cycle. In constant darkness, the time of the eclosion peak relative to the collection time would be constantly drifting as the rhythm free runs with its own period. Hence, the imposition of a light:dark cycle ensures clear-cut results in the screen.

Despite the inherent difficulty in an eclosion screen conducted in con-stant darkness, Konopka and collaborators (1991) isolated the first *Andante* mutation in such a screen, which was also designed to detect mutations affecting phase resetting by a temperature pulse. At the end of entrainment to light:dark cycles, fly lines were transferred to constant darkness and given a 12-h, 29° temperature pulse beginning 12 h after the transition to constant darkness. They were then maintained in constant darkness at 22°, and the relative number of males (containing a mutagen-ized X) and females (containing a nonmutagenized X^X; see Fig. 2) emerging in consecutive 12-h periods were compared. *Andante* caused a 1.5-h lengthening of the circadian period, thereby causing the eclosion of male flies to drift out of phase with that of females. Much later, the identity of the *Andante* gene was determined when P element excision mutations were isolated, and it was shown to encode a regulatory subunit that interacts with the TIK protein kinase catalytic subunit (Akten *et al.*, 2003).

## Locomotor Activity Assays

Locomotor assays directly measure the rest/activity cycles of individual flies. This assay has become by far the most common behavioral assay in flies, as it has for mammalian studies. The Trikinetics company (Waltham, MA) manufactures activity monitors, an interface device and the software for computerized data collections. Individual flies are placed in a tube that contains food on one end and a porous plug (e.g., rayon or cotton) on the other end. Infrared light (to which the circadian clock is not sensitive) is passed through the tube and is detected by a photodetector on the other side. Movements of the fly deflect the beam and generate a signal that is recorded by a computer and binned (e.g., into a series of consecutive half-hour bins). Activity per bin, plotted as a function of time of collection, typically demonstrates a circadian rhythm under conditions of constant temperature and darkness. The active period occurs during the times lights were on in the previous entrainment regimen, while the inactive period occurs at the time lights were turned off. $\chi^2$ periodogram analysis is the algorithm used most commonly to assess periodicity and arrhythmicity (Sokolove and Bushell, 1978), but other more sophisticated methods can also be applied (Dowse *et al.*, 1989). The author's laboratory is currently using Clocklab from Actimetrics (Evanston, IL) to analyze our locomotor assays, which typically run for at least a week. In the Canton S wild-type strain, the average period extracted by $\chi^2$ periodogram analysis is 24 h (Sehgal *et al.*, 1992).

There are both advantages and disadvantages of the locomotor assays in comparison with the simple eclosion assay described previously. Automation relieves the researcher from daily involvement with the assay except at the beginning and the end. Whereas the eclosion assay requires a large population of flies, high-resolution records can be obtained from a single individual with locomotor assays (the high-through-put screen described in Fig. 4 could not be conducted with an eclosion-based assay for this reason). Because of the high temporal resolution (typically 0.5-h bins), screens can be performed easily in constant darkness, and the gradual drift of period-altering mutants is discerned readily. In theory, assays performed in constant darkness can reveal mutant phenotypes that might be masked in light:dark cycles, although in practice, most circadian mutants affect circadian behavior under both conditions. Another advantage is that arrhythmicity is revealed more convincingly in individual flies, as overt arrhythmicity in eclosion assays can arise from rhythmic but asynchronous individuals. In fact, eclosion-based assays of dark-reared flies were originally thought to show that light was required to start the *Drosophila* rhythms (Brett, 1955), whereas subsequent work showed that most flies reared in constant darkness displayed locomotor rhythms that were not synchronized (Sehgal *et al.*, 1992).

A disadvantage is that many locomotor records of wild-type flies (defined as such because they will not pass on any rhythm mutations) are too messy to demonstrate clear rhythmicity. A line of flies putatively arrhythmic by this assay requires further analysis to validate the arrhythmicity (see section at the chapter's end on validation). Nevertheless, this assay has produced most of the *Drosophila* circadian mutants, including alleles of *per* (all except the original three isolated by Konopka) (Hamblen-Coyle *et al.*, 1989), of *tim* (all except the original $tim^{o1}$ and $tim^{o2}$ alleles) (Rothenfluh *et al.*, 2000a,c; Stanewsky *et al.*, 1997), of *dbt* (Price *et al.*, 1998; Rothenfluh *et al.*, 2000b; Suri *et al.*, 2000), of *dClk* (Allada *et al.*, 1998, 2003), of *cyc* (Rutila *et al.*, 1998), of *sgg* (Martinek *et al.*, 2001), and of *Tik* (Lin *et al.*, 2002a).

### Real-Time Assays of Molecular Rhythms with a Luciferase Reporter Gene

Brandes *et al.* (1996) developed this cutting-edge methodology. Transcription of a transgenic firefly luciferase gene is under the control of a circadian promoter. When fed luciferin (the substrate for luciferase), wild-type flies produce a rhythmic emission of light that results from the rhythmic expression of luciferase. The light emission can be detected by a sophisticated "luminometer" (actually a Packard Topcount multiplate scintillation counter), which samples immobilized flies at regular intervals to determine how much light they are emitting at that particular time. The resulting real-time "luminescence" record is obtained from a single fly and is therefore the molecular analog of the locomotor activity record.

One screening strategy afforded by this methodology entailed a screen for mutations that affect the molecular oscillations of *per* mRNA. Flies carrying a reporter transgene consisting of the *per* promoter and 5' part of the *per* gene (encoding the first two-thirds of the PER protein), as well as EMS-mutagenized chromosomes, were screened for alterations in the rhythm of luminescence. The reporter transgene (BG construct) reproduces the phase of normal *per* mRNA because it has elements required for both transcriptional and posttranscriptional regulation, so in theory the screen could have detected mutations affecting either of these regulatory processes. The $cry^b$ mutation was identified as a mutation that renders the normal cycles of BG expression arrhythmic in 12 h:12 h LD (Stanewsky *et al.*, 1998). Further analysis demonstrated that $cry^b$ causes PER and TIM to be constitutively expressed in most tissues, in part because TIM is no longer rhythmically degraded in response to light. CRY was shown to mediate light-dependent degradation of TIM by forming an association with it (Ceriani *et al.*, 1999); in short, it is an intracellular photoreceptor. Nevertheless, locomotor activity rhythms persist in this mutant because several neurons that control locomotor activity are entrained redundantly by other photoreceptor pathways (Stanewsky *et al.*, 1998). Hence, the *cry*[b] mutation would not have been identified in a standard locomotor activity screen.

Another application for luciferase reporter technology has been to screen for genes that are transcribed rhythmically with an "enhancer trap" approach (Stempfl *et al.*, 2002). For this screen, the gene for firefly luciferase was placed on a P element under the control of a weak promoter (the transposase promoter), and the P element was moved around the genome to generate insertion lines, as described in Fig. 1. The weak promoter does not produce detectable levels of luciferase unless the P element inserts into a region with enhancers, which can augment the now adjacent weak promoter. If the enhancers produce circadian control of transcription, the rhythmic expression of luciferase can be detected with the same real-time sampling methodology outlined earlier. The sequence of the flanking genomic DNA immediately identifies the nearby genes, as the sequence can be matched to the corresponding region of the completely sequenced *Drosophila* genome. These genes can then be assessed for circadian oscillations, and the P element itself caused a mutation in one case (a *numb* mutation). If the P element does not cause a mutation, imprecise excision of the P element may produce a mutation. In an initial screen, a total of 71 lines showed rhythmic expression of luciferase (6% of insertions). Most required *per* and *tim* for this molecular rhythmicity (Stempfl *et al.*, 2002).

## Assays to Identify Genes Whose Transcript Levels Oscillate with a Circadian Rhythm

Because several known clock genes oscillate in a circadian manner, identification of other genes that exhibit circadian oscillations may identify other genes with circadian oscillator function. Alternatively, these genes may be clock-controlled genes that do not contribute to the central oscillator, but which are driven by it—perhaps to mediate output processes. The first such clock-controlled gene to be isolated strictly by molecular screens were *Drosophila* rhythmically expressed genes (*Dregs*) (Van Gelder *et al.*, 1995) and *Crg*1 (Rouyer *et al.*, 1997). *Dregs* were identified by individually testing several hundred cDNAs known to be expressed in the adult fly head but not the early embryo. Approximately 20 showed diurnal rhythms, which did not in all cases persist under constant darkness or require functional PER for expression; i.e., the oscillations of many were not circadian. *Dreg* 5 mRNA does require PER to oscillate and oscillates in phase with *per* mRNA oscillation. *Crg1* was identified in a subtractive screen, in which reverse transcription of mRNA from ZT3 (3 h after lights are illuminated) and ZT15 (3 h after darkness begins) was followed by subtractive hybridization of ZT3 cDNA (the driver) from ZT15 cDNA (the tracer). Three rounds of subtractive hybridization were followed by cloning enriched ZT15-specific cDNA (e.g., *per* cDNA was 25X enriched). The enriched DNA was further screened with radiolabeled ZT15–ZT3 subtracted cDNA and nonsubtracted ZT15 cDNA. *Crg1* was identified as a clone enriched in the subtracted cDNA, and its oscillation, which requires PER, was independently demonstrated with other techniques. The functional significance of *Dreg* and *Crg1* oscillations is not known.

The *takeout* (*to*) gene was identified in a polymerase chain reaction (PCR)-based cDNA subtraction, in which cDNAs from head mRNA of the null *cyc* mutant (*cyc$^o$*) and of wild-type flies collected at ZT15 were synthesized, and the cDNA products from the *cyc$^o$* flies were subtracted from the wild-type products. The PCR products were cloned and screened to identify clones encoding transcripts that were CYC dependent (So *et al.*, 2000). The *to* mRNA was shown to be underexpressed in *cyc$^o$* flies and to be induced by starvation in a manner that required several of the clock gene products (Sarov-Blat *et al.*, 2000). While this gene was originally implicated in resistance to starvation, it has been implicated more recently in male courtship behavior (Dauwalder *et al.*, 2002).

*vrille* (*vri*) was identified as an mRNA with a circadian oscillation by differential display (Blau and Young, 1999). With this technique, PCR products from hundreds of mRNAs are derived by RT-PCR and resolved on a gel. Bands that are differentially amplified from two different mRNA samples are excised and cloned. *vri* was preferentially amplified from ZT14 in wild-type flies relative to ZT20 flies and was amplified to equal extents from *per$^o$* mRNA at both of these time points. Subsequent molecular analysis showed that *vri* is more than just an oscillating gene, it is a component of the oscillator mechanism (Cyran *et al.*, 2003; Glossop *et al.*, 2003), as described earlier.

Microarray analyses in *Drosophila* have undertaken genome-scale screens for oscillating transcripts (Ceriani *et al.*, 2002; Claridge-Chang *et al.*, 2001; Lin *et al.*, 2002b; McDonald and Rosbash, 2001; Ueda *et al.*, 2002). In these experiments, oligonucleotides complementary to all of the suspected transcripts in *Drosophila* are synthesized on glass slides, which are then hybridized with fluorescently labeled cDNAs from different time points. The amounts of hybridization signals for all of these genes are quantified and analyzed for evidence of rhythmicity. Several hundred candidate rhythmic genes have been identified. Analysis of these candidates will be ongoing for many years.

This approach does not generate mutants directly. However, in several cases (e.g., for *vri* and *to*), existing mutants were already available. For others, recent advances have facilitated reverse genetics (the production of mutants for genes identified by molecular approaches). One of the most promising approaches is to produce transgenic flies that express double-stranded RNAs complementary to a unique sequence for the gene of interest (e.g., RNAi approaches). Expression of these double-stranded RNAi can reduce the level of the endogenous mRNA and produce a loss-of-function phenotype. The utility of such an approach was demonstrated for the *per* gene, as transgenically expressed RNAi produced long-period rhythms (Martinek and Young, 2000). Another approach is to overexpress wild-type protein in flies or to express a dominant-negative protein (Ko *et al.*, 2002), which can interact with other clock proteins but not in a fully functional way—a molecular "wrench in the works." These approaches can also be used to assess the roles of candidate genes, which are covered in the next section.

## Screens of Candidate Genes for Circadian Function

As analysis of the genes initially identified in unbiased forward genetic screens began to elucidate the mechanism of the circadian clock, it became possible to more accurately frame hypotheses about the involvement of other genes. Therefore, many clock-involved genes have been identified by testing mutants initially isolated for other reasons than for analysis of circadian rhythms. For instance, an extensive screen of mutations affecting visual system function and physiology pinpointed the *disconnected* mutation (*disco*) as the only mutation with a strong effect on locomotor rhythmicity (Dushay *et al.*, 1989). The mutation eliminates most of the *per* expression in the region of the brain that is critical for rhythmic locomotor activity. Based on evidence that cAMP-dependent pathways were involved in vertebrate rhythms, mutations affecting this pathway were tested for effects on circadian rhythmicity. As a result, mutations in the cAMP response element-binding protein (CREB) (Belvin *et al.*, 1999), protein kinase A subunits (both catalytic and regulatory) (Majercak *et al.*, 1997; Park *et al.*, 2000), a cAMP-specific phosphodiesterase (*dnc*) (Levine *et al.*, 1994), and neurofibromatosis-1 (*Nf1*) (Williams *et al.*, 2001) have all been shown to affect circadian rhythms of locomotor activity. For the most part, these effects seem to be specific for the locomotor output pathway, although the evidence is somewhat conflicting regarding effects on the central clock mechanism. Despite initial expectations, NF1 mutants appear to affect circadian locomotor behavior by upregulation of the RAS/MAPK pathway in cells that respond to the circadian neuropeptide pigment dispersing factor (PDF) rather than through the cAMP pathway (Williams et al., 2001).

PDF was fingered as an important circadian neuropeptide when antibodies to the crustacean ortholog detected its presence in the clock neurons that control circadian behavior (Helfrich-Forster, 1995). A P element-induced mutation in *pdf* was then used to show that lack of PDF leads to rapid loss of circadian locomotor rhythms (Renn et al., 1999). PDF has been implicated in synchrony of different clock cells or even in intercellular communication necessary for sustained intracellular circadian oscillation (Peng et al., 2003). The cAMP-signaling pathway and MAPK/RAS pathway may be downstream of PDF in postsynaptic cells.

Other tests of candidate genes have identified factors that collaborate with or antagonize the kinases operating on PER. For instance, the F-box/WD-40 repeat protein SLIMB has been overexpressed, expressed as a dominant negative, and monitored as loss-of-function mutations for effects on circadian locomotor activity and PER degradation (Grima et al., 2002; Ko et al., 2002). It was shown to affect circadian rhythms, most likely by interacting preferentially with phosphorylated PER and targeting it for degradation. Because *dbt* and *sgg* are both involved in the *wingless* signal transduction pathway, mutants and overexpression constructs for the PP2A phosphatase, which is also involved in the *wingless* pathway, were assessed for effects on circadian rhythmicity and shown to antagonize clock kinase function (Sathyanarayanan et al., 2004). Finally, because fragile X mental retardation in humans is associated with sleep dysfunction, mutants for the *dfmr1* gene in *Drosophila* were assessed for effects on circadian rhythmicity and were shown to have effects (Dockendorff et al., 2002; Inoue et al., 2002; Morales et al., 2002).

## Validation of Candidate Lines Identified in Genetic Screens

Most of the lines identified initially as potential mutants turn out to be false positives upon further testing. The false-positive problem is particularly relevant to lines that display arrhythmic locomotor activity. It seems that the robustness of the locomotor activity rhythm is easily reduced by many factors in the genetic background and may be produced by reduced viability or activity levels rather than by effects on the clock mechanism.

The first step in validating the mutants is to show that the mutation maps to a single locus (polygenic inheritance of the phenotype is much more difficult to study and has not been pursued extensively in *Drosophila*). The mutant line is outcrossed to a different balancer stock from the ones used in the mutagenesis, and the candidate mutant chromosomes are recovered as homologs of the balancer in the outcrossed line. The circadian rhythm defect must be recovered in the new line, if it does arise from the candidate chromosome. If the mutation is thought to arise from a P element insertion, the P element can be remobilized by a procedure like the one outlined in Fig. 1. If the mutation in fact arises from the insertion of the P element, mobilization of the P element should revert the mutation with high frequency (assuming the P element is not inserted directly in the reading frame) (Sehgal *et al.*, 1994).

Recombination mapping is initiated by crossing the candidate line to a line that has wild-type circadian rhythms but contains a chromosome with multiple visible mutations along its length (markers). Females that are heterozygous for the candidate circadian mutant chromosome and the marker chromosome (mutant?/marker) are recovered and mated to a male containing a balancer chromosome; female heterozygotes are used for this cross because meiotic recombination does not normally occur in male *Drosophila*. Single male progeny heterozygous for the balancer chromosome and a potential recombinant chromosome are again mated to the balancer stock to establish a line that is isogenous for any recombinant chromosome. These lines are tested for the presence of the candidate circadian mutation and are scored for the presence of the visible markers; given the variable penetrance of rhythmic behavior, it is usually preferable to test multiple individuals from each line than to test single individuals that each represent distinct recombination events (Sehgal *et al.*, 1994). If lines are established, it is not necessary for arrhythmia to be 100% penetrant to demonstrate that a mutation is present in a line. In fact, some mutants (e.g., *disco*) do produce a small percentage of rhythmic flies (in the case of the *disco* mutant, these rhythmic individuals typically retain some *per* expression in the brain locomoter centers, unlike their arrhythmic siblings) (Helfrich-Forster, 1998).

Most candidate arrhythmic mutants are no longer positive after the outcross and initial recombination tests. In other words, it is not possible to map the arrhythmic mutation to a specific chromosomal locus, as the phenotype is not consistently recovered in the outcross or recombination test. Sometimes, only part of the original phenotype is recovered. For instance, the original $tim^{01}$-containing line was night active in light–dark cycles (as opposed to the day activity exhibited by wild type) and arrhythmic in constant darkness (Sehgal *et al.*, 1991). The arrhythmia mapped to the left arm of the second chromosome (Sehgal *et al.*, 1994), but the night activity did not map to any particular locus (i.e., it was generated by the general genetic background; J. L. Price and M. W. Young, unpublished data).

Further mapping of the mutation entails recombination tests with additional markers in the region and complementation tests with chromosomes containing deletions (termed *Deficiencies*) or other clock mutations in the region. Failure of Deficiencies or another clock mutation to complement the candidate mutation (i.e., generation of a strongly mutation phenotype with a heterozygous combination of two recessive alleles) suggests that the new mutation is contained within the deleted region or is an allele of the established clock mutation, respectively. Mutations that alter the circadian period can be difficult to assess in complementation tests because they have strong semidominant effects opposite a wild-type allele and it can be difficult to determine whether the mutant phenotype in the complementation test is a lack of complementation or just additive effects of the two different mutations. For circadian period-altering mutations, complementation tests with null alleles are typically most instructive because, with the absence of any competition from functional gene product from the null allele, the period-altering phenotype can be as strong as the homozygous phenotype for the period-altering mutant, thereby distinguishing the failure to complement from the weaker semidominant phenotype when the null and period-altering alleles do complement (Konopka and Benzer, 1971; Price *et al.*, 1998; Rothenfluh *et al.*, 2000a).

The final confirmation of the mutant line comes after molecular analysis has produced a candidate gene. The gene can be sequenced in the mutant line to confirm that its structure is disrupted or its expression can be assessed to determine whether the mutation alters the expression (Kloss *et al.*, 1998). Transgenic flies can be produced carrying a wild-type transgene in a mutant background; the wild-type transgene should restore the phenotype toward the wild-type state (a rescue) if the mutation has affected that gene (Bargiello *et al.*, 1984; Yu *et al.*, 1987).

Concluding Comments

Screens for new clock mutations have provided fodder for a genetic analysis that has been the primary force in molecular chronobiology. One wonders whether additional screens will produce additional clock genes; in other words, have the screens reached saturation? Operationally, saturation is defined as a state in which novel screens only recover additional alleles of previously recovered mutations. While multiple alleles exist for several important clock genes (e.g., *per, tim, dbt*), several important clock genes exist as only single mutant alleles (e.g., *cyc* and *cry*). Altered screening strategies (e.g., luciferase reporter-based screens, EP screens) have produced mutations that would not have been obtained in earlier screens, and the list of candidate genes derived from molecular screens is quite large. The author's sense is that the genetic screens are still incomplete and that genetic analysis will continue to make significant contributions to molecular chronobiology for many years.

## References

Akten, B., Jauch, E., Genova, G. K., Kim, E. Y., Edery, I., Raabe, T., and Jackson, F. R. (2003). A role for CK2 in the Drosophila circadian oscillator. *Nature Neurosci.* **6,** 251–257.

Allada, R., Kadener, S., Nandakumar, N., and Rosbash, M. (2003). A recessive mutant of Drosophila *Clock* reveals a role in circadian rhythm amplitude. *EMBO J.* **22,** 3367–3375.

Allada, R., White, N. E., So, W. V., Hall, J. C., and Rosbash, M. (1998). A mutant Drosophila homolog of mammalian *Clock* disrupts circadian rhythms and transcription of *period* and *timeless*. *Cell* **93,** 791–804.

Bargiello, T. A., Jackson, F. R., and Young, M. W. (1984). Restoration of circadian behavioural rhythms by gene transfer in Drosophila. *Nature* **312,** 752–754.

Belvin, M. P., Zhou, H., and Yin, J. C. (1999). The Drosophila dCREB2 gene affects the circadian clock. *Neuron* **22,** 777–787.

Blau, J., and Young, M. W. (1999). Cycling vrille expression is required for a functional Drosophila clock. *Cell* **99,** 661–671.

Brandes, C., Plautz, J. D., Stanewsky, R., Jamison, C. F., Straume, M., Wood, K. V., Kay, S. A., and Hall, J. C. (1996). Novel features of Drosophila *period* transcription revealed by real-time luciferase reporting. *Neuron* **16,** 687–692.

Brett, W. J. (1955). Persistent diurnal rhythmicity in Drosophila emergence. *Ann. Ent. Soc. Am.* **48,** 119–131.

Ceriani, M. F., Darlington, T. K., Staknis, D., Mas, P., Petti, A. A., Weitz, C. J., and Kay, S. A. (1999). Light-dependent sequestration of TIMELESS by CRYPTOCHROME. *Science* **285,** 553–556.

Ceriani, M. F., Hogenesch, J. B., Yanovsky, M., Panda, S., Straume, M., and Kay, S. A. (2002). Genome-wide expression analysis in Drosophila reveals genes controlling circadian behavior. *J. Neurosci.* **22,** 9305–9319.

Claridge-Chang, A., Wijnen, H., Naef, F., Boothroyd, C., Rajewsky, N., and Young, M. W. (2001). Circadian regulation of gene expression systems in the Drosophila head. *Neuron* **32,** 657–671.

Cyran, S. A., Buchsbaum, A. M., Reddy, K. L., Lin, M. C., Glossop, N. R., Hardin, P. E., Young, M. W., Storti, R. V., and Blau, J. (2003). *vrille, Pdp1*, and *dClock* form a second feedback loop in the Drosophila circadian clock. *Cell* **112,** 329–341.

Dauwalder, B., Tsujimoto, S., Moss, J., and Mattox, W. (2002). The Drosophila *takeout* gene is regulated by the somatic sex-determination pathway and affects male courtship behavior. *Genes Dev.* **16,** 2879–2892.

Dockendorff, T. C., Su, H. S., McBride, S. M., Yang, Z., Choi, C. H., Siwicki, K. K., Sehgal, A., and Jongens, T. A. (2002). Drosophila lacking *dfmr1* activity show defects in circadian output and fail to maintain courtship interest. *Neuron* **34,** 973–984.

Dowse, H. B., Dushay, M. S., Hall, J. C., and Ringo, J. M. (1989). High-resolution analysis of locomotor activity rhythms in *disconnected*, a visual-system mutant of *Drosophila melanogaster*. *Behav. Genet.* **19,** 529–542.

Dushay, M. S., Rosbash, M., and Hall, J. C. (1989). The *disconnected* visual system mutations in *Drosophila melanogaster* drastically disrupt circadian rhythms. *J. Biol. Rhythms* **4,** 1–27.

Glossop, N. R., Houl, J. H., Zheng, H., Ng, F. S., Dudek, S. M., and Hardin, P. E. (2003). VRILLE feeds back to control circadian transcription of *Clock* in the Drosophila circadian oscillator. *Neuron* **37,** 249–261.

Grima, B., Lamouroux, A., Chelot, E., Papin, C., Limbourg-Bouchon, B., and Rouyer, F. (2002). The F-box protein *slimb* controls the levels of clock proteins *period* and *timeless*. *Nature* **420,** 178–182.

Hamblen-Coyle, M., Konopka, R. J., Zwiebel, L. J., Colot, H. V., Dowse, H. B., Rosbash, M., and Hall, J. C. (1989). A new mutation at the *period* locus of *Drosophila melanogaster* with some novel effects on circadian rhythms. *J. Neurogenet.* **5,** 229–256.

Helfrich-Forster, C. (1995). The *period* clock gene is expressed in central nervous system neurons which also produce a neuropeptide that reveals the projections of circadian pacemaker cells within the brain of *Drosophila melanogaster. Proc. Natl. Acad. Sci. USA* **92,** 612–616.

Helfrich-Forster, C. (1998). Robust circadian rhythmicity of *Drosophila melanogaster* requires the presence of lateral neurons: A brain-behavioral study of *disconnected* mutants. *J. Comp. Physiol. A* **182,** 435–453.

Inoue, S., Shimoda, M., Nishinokubi, I., Siomi, M. C., Okamura, M., Nakamura, A., Kobayashi, S., Ishida, N., and Siomi, H. (2002). A role for the Drosophila fragile X-related gene in circadian output. *Curr. Biol.* **12,** 1331–1335.

Kloss, B., Price, J. L., Saez, L., Blau, J., Rothenfluh, A., Wesley, C. S., and Young, M. W. (1998). The Drosophila clock gene *double-time* encodes a protein closely related to human casein kinase Iε. *Cell* **94,** 97–107.

Ko, H. W., Jiang, J., and Edery, I. (2002). Role for *Slimb* in the degradation of Drosophila *Period* protein phosphorylated by *Doubletime. Nature* **420,** 673–678.

Konopka, R. J., and Benzer, S. (1971). Clock mutants of *Drosophila melanogaster. Proc. Natl. Acad. Sci. USA* **68,** 2112–2116.

Konopka, R. J., Smith, R. F., and Orr, D. (1991). Characterization of *Andante*, a new Drosophila clock mutant, and its interactions with other clock mutants. *J. Neurogenet.* **7,** 103–114.

Levine, J. D., Casey, C. I., Kalderon, D. D., and Jackson, F. R. (1994). Altered circadian pacemaker functions and cyclic AMP rhythms in the Drosophila learning mutant *dunce. Neuron* **13,** 967–974.

Lewis, E. B., and Bacher, F. (1968). Method of feeding ethyl methanesulfonate (EMS) to Drosophila males. *Drosophila Inform. Serv.* **43,** 193.

Lin, J. M., Kilman, V. L., Keegan, K., Paddock, B., Emery-Le, M., Rosbash, M., and Allada, R. (2002a). A role for casein kinase 2alpha in the Drosophila circadian clock. *Nature* **420,** 816–820.

Lin, Y., Han, M., Shimada, B., Wang, L., Gibler, T. M., Amarakone, A., Awad, T. A., Stormo, G. D., Van Gelder, R. N., and Taghert, P. H. (2002b). Influence of the *period*-dependent circadian clock on diurnal, circadian, and aperiodic gene expression in *Drosophila melanogaster. Proc. Natl. Acad. Sci. USA* **99,** 9562–9567.

Majercak, J., Kalderon, D., and Edery, I. (1997). *Drosophila melanogaster* deficient in protein kinase A manifests behavior-specific arrhythmia but normal clock function. *Mol. Cell. Biol.* **17,** 5915–5922.

Martinek, S., Inonog, S., Manoukian, A. S., and Young, M. W. (2001). A role for the segment polarity gene shaggy/GSK-3 in the Drosophila circadian clock. *Cell* **105,** 769–779.

Martinek, S., and Young, M. W. (2000). Specific genetic interference with behavioral rhythms in Drosophila by expression of inverted repeats. *Genetics* **156,** 1717–1725.

Matsumoto, A., Tomioka, K., Chiba, Y., and Tanimura, T. (1999). *tim^{rit}* lengthens circadian period in a temperature-dependent manner through suppression of PERIOD protein cycling and nuclear localization. *Mol. Cell. Biol.* **19**, 4343–4354.

McDonald, M. J., and Rosbash, M. (2001). Microarray analysis and organization of circadian gene expression in Drosophila. *Cell* **107**, 567–578.

Morales, J., Hiesinger, P. R., Schroeder, A. J., Kume, K., Verstreken, P., Jackson, F. R., Nelson, D. L., and Hassan, B. A. (2002). Drosophila fragile X protein, DFXR, regulates neuronal morphology and function in the brain. *Neuron* **34**, 961–972.

Myers, M. P., Wager-Smith, K., Wesley, C. S., Young, M. W., and Sehgal, A. (1995). Positional cloning and sequence analysis of the Drosophila clock gene, *timeless*. *Science* **270**, 805–808.

Newby, L. M., and Jackson, F. R. (1993). A new biological rhythm mutant of *Drosophila melanogaster* that identifies a gene with an essential embryonic function. *Genetics* **135**, 1077–1090.

Park, S. K., Sedore, S. A., Cronmiller, C., and Hirsh, J. (2000). Type II cAMP-dependent protein kinase-deficient Drosophila are viable but show developmental, circadian, and drug response phenotypes. *J. Biol. Chem.* **275**, 20588–20596.

Peng, Y., Stoleru, D., Levine, J. D., Hall, J. C., and Rosbash, M. (2003). Drosophila free-running rhythms require intercellular communication. *PLoS Biol.* **1**, 32–40.

Pittendrigh, C. S. (1974). Circadian oscillations in cells and the circadian organization of multicellular systems. *In* "The Neurosciences: Third Program Study" (F.O. Schmitt and F.G. Worden, eds.), pp. 437–458. MIT Press, Cambridge, MA.

Price, J. L. (2004). *Drosophila melanogaster*: A model system for molecular chronobiology. *In* "Molecular Biology of Circadian Rhythms" (A. Sehgal, ed.), pp. 33–74. Wiley, Philadelphia, PA.

Price, J. L., Blau, J., Rothenfluh, A., Abodeeley, M., Kloss, B., and Young, M. W. (1998). *Double-time* is a novel Drosophila clock gene that regulates PERIOD protein accumulation. *Cell* **94**, 83–95.

Renn, S. C., Park, J. H., Rosbash, M., Hall, J. C., and Taghert, P. H. (1999). A *pdf* neuropeptide gene mutation and ablation of PDF neurons each cause severe abnormalities of behavioral circadian rhythms in Drosophila. *Cell* **99**, 791–802.

Rorth, P., Szabo, K., Bailey, A., Laverty, T., Rehm, J., Rubin, G. M., Weigmann, K., Milan, M., Benes, V., Ansorge, W., and Cohen, S. M. (1998). Systematic gain-of-function genetics in Drosophila. *Development* **125**, 1049–1057.

Rothenfluh, A., Abodeely, M., Price, J. L., and Young, M. W. (2000a). Isolation and analysis of six *timeless* alleles that cause short- or long-period circadian rhythms in Drosophila. *Genetics* **156**, 665–675.

Rothenfluh, A., Abodeely, M., and Young, M. W. (2000b). Short-period mutations of *per* affect a double-time-dependent step in the Drosophila circadian clock. *Curr. Biol.* **10**, 1399–1402.

Rothenfluh, A., Young, M. W., and Saez, L. (2000c). A TIMELESS-independent function for PERIOD proteins in the Drosophila clock. *Neuron* **26**, 505–514.

Rouyer, F., Rachidi, M., Pikielny, C., and Rosbash, M. (1997). A new gene encoding a putative transcription factor regulated by the Drosophila circadian clock. *EMBO J.* **16**, 3944–3954.

Rutila, J. E., Suri, V., Le, M., So, W. V., Rosbash, M., and Hall, J. C. (1998). CYCLE is a second bHLH-PAS clock protein essential for circadian rhythmicity and transcription of Drosophila *period* and *timeless*. *Cell* **93**, 805–814.

Rutila, J. E., Zeng, H., Le, M., Curtin, K. D., Hall, J. C., and Rosbash, M. (1996). The *tim^{SL}* mutant of the Drosophila rhythm gene *timeless* manifests allele-specific interactions with *period* gene mutants. *Neuron* **17**, 921–929.

Sarov-Blat, L., So, W. V., Liu, L., and Rosbash, M. (2000). The Drosophila *takeout* gene, is a novel molecular link between circadian rhythms and feeding behavior. *Cell* **101,** 647–656.

Sathyanarayanan, S., Zheng, X., Xiao, R., and Sehgal, A. (2004). Posttranslational regulation of Drosophila PERIOD protein by protein phosphatase 2A. *Cell* **116,** 603–615.

Sehgal, A., Man, B., Price, J. L., Vosshall, L. B., and Young, M. W. (1991). New clock mutations in Drosophila. *Ann. N. Y. Acad. Sci.* **618,** 1–10.

Sehgal, A., Price, J., and Young, M. W. (1992). Ontogeny of a biological clock in *Drosophila melanogaster. Proc. Natl. Acad. Sci. USA* **89,** 1423–1427.

Sehgal, A., Price, J. L., Man, B., and Young, M. W. (1994). Loss of circadian behavioral rhythms and *per* RNA oscillations in the Drosophila mutant *timeless. Science* **263,** 1603–1606.

Smith, R. F., and Konopka, R. J. (1982). Effects of dosage alteration at the *per* locus on the period of the circadian clock of Drosophila. *Mol. Gen. Genet.* **185,** 30–36.

So, W. V., Sarov-Blat, L., Kotarski, C. K., McDonald, M. J., Allada, R., and Rosbash, M. (2000). *takeout,* a novel Drosophila gene under circadian clock transcriptional regulation. *Mol. Cell. Biol.* **20,** 6935–6944.

Sokolove, P. G., and Bushell, W. N. (1978). The chi square periodogram: Its utility for analysis of circadian rhythms. *J. Theoret. Biol.* **72,** 131–160.

Stanewsky, R., Frisch, B., Brandes, C., Hamblen-Coyle, M. J., Rosbash, M., and Hall, J. C. (1997). Temporal and spatial expression patterns of transgenes containing increasing amounts of the Drosophila clock gene *period* and a lacZ reporter: Mapping elements of the PER protein involved in circadian cycling. *J. Neurosci.* **17,** 676–696.

Stanewsky, R., Kaneko, M., Emery, P., Beretta, B., Wagner-Smith, K., Kay, S. A., Rosbash, M., and Hall, J. C. (1998). The *cry^b* mutation identifies cryptochrome as a circadian photoreceptor in Drosophila. *Cell* **95,** 681–692.

Stempfl, T., Vogel, M., Szabo, G., Wulbeck, C., Liu, J., Hall, J. C., and Stanewsky, R. (2002). Identification of circadian-clock-regulated enhancers and genes of *Drosophila melanogaster* by transposon mobilization and luciferase reporting of cyclical gene expression. *Genetics* **160,** 571–593.

Suri, V., Hall, J. C., and Rosbash, M. (2000). Two novel *doubletime* mutants alter circadian properties and eliminate the delay between RNA and protein in Drosophila. *J. Neurosci.* **20,** 7547–7555.

Ueda, H. R., Matsumoto, A., Kawamura, M., Iino, M., Tanimura, T., and Hashimoto, S. (2002). Genome-wide transcriptional orchestration of circadian rhythms in Drosophila. *J. Biol. Chem.* **277,** 14048–14052.

Van Gelder, R. N., Bae, H., Palazzolo, M. J., and Krasnow, M. A. (1995). Extent and character of circadian gene expression in *Drosophila melanogaster:* Identification of twenty oscillating mRNAs in the fly head. *Curr. Biol.* **5,** 1424–1436.

Williams, J. A., Su, H. S., Bernards, A., Field, J., and Sehgal, A. (2001). A circadian output in Drosophila mediated by neurofibromatosis-1 and Ras/MAPK. *Science* **293,** 2251–2256.

Yu, Q., Jacquier, A. C., Citri, Y., Hamblen, M., Hall, J. C., and Rosbash, M. (1987). Molecular mapping of point mutations in the *period* gene that stop or speed up biological clocks in *Drosophila melanogaster. Proc. Natl. Acad. Sci. USA* **84,** 784–788.

# [4]    Systems Approaches to Biological Rhythms in *Drosophila*

*By* JEFFREY C. HALL

The only book that is worth writing is one in a field which is developing so fast that the book will be out of date before you can get it printed.

T. H. Morgan (paraphrased in Sturtevant, 2001)

Abstract

The chronobiological system of *Drosophila* is considered from the perspective of rhythm-regulated genes. These factors are enumerated and discussed not so much in terms of how the gene products are thought to act on behalf of circadian-clock mechanisms, but with special emphasis on where these molecules are manufactured within the organism. Therefore, with respect to several such cell and tissue types in the fly head, what is the "systems meaning" of a given structure's function insofar as regulation of rest-activity cycles is concerned? (Systematic oscillation of daily behavior is the principal overt phenotype analyzed in studies of *Drosophila* chronobiology). In turn, how do the several separate sets of clock-gene-expressing cells interact—or in some cases act in parallel— such that intricacies of the fly's sleep-wake cycles are mediated? Studying *Drosophila* chrono-genetics as a system-based endeavor also encompasses the fact that rhythm-related genes generate their products in many tissues beyond neural ones and during all stages of the life cycle. What, then, is the meaning of these widespread gene-expression patterns? This question is addressed with regard to circadian rhythms outside the behavioral arena, by considering other kinds of temporally based behaviors, and by contemplating how broadly systemic expression of rhythm-related genes connects with even more pleiotropic features of *Drosophila* biology. Thus, chronobiologically connected factors functioning within this insect comprise an increasingly salient example of gene versatility— multi-faceted usages of, and complex interactions among, entities that set up an organism's overall wherewithal to form and function. A corollary is that studying *Drosophila* development and adult-fly actions, even when limited to analysis of rhythm-systems phenomena, involves many of the animal's tissues and phenotypic capacities. It follows that such chronobiological experiments are technically demanding, including the necessity for investigators to possess wide-ranging expertise. Therefore, this chapter includes several different kinds of Methods set-asides. These

METHODS IN ENZYMOLOGY, VOL. 393

techniques primers necessarily lack comprehensiveness, but they include certain discursive passages about why a given method can or should be applied and concerning real-world applicability of the pertinent rhythm-related technologies.

## Introduction

Dealing with rhythmicity in *Drosophila* as a "system" will involve two levels of subject treatment. These are previewed in the next four paragraphs. However, the author warns the reader that this article is somewhat discursive, as opposed to dwelling solely on the nuts and bolts of experiments. That kind of approach would be warranted, given the title of this volume. It obviously includes several chapters about *Drosophila* rhythms, including in turn some reviews that are slanted toward methodological matters. However, the "Methods" component of this article is set aside in boxes, if only to keep the mind alive[1] or at least focused on the discussion of chronobiological issues, conundrums, and other interesting matters that have been shoved under the rug or fallen by the wayside.

With regard to the other noun within the volume's title, the manner by which rhythmicity of fruit flies is regulated could indeed include a fair amount of "enzymology." Coincidentally, perhaps, a fair fraction of the more recently identified rhythm mutants, plus an old one whose gene finally got cloned, involve enzymatic functions. However, information about these catalytic phenomena falls somewhat short of precise determinations of the manner by which products of these "chrono-enzyme genes" act on their known or presumed substrates.

As was just referred to, and as would be expected for any area of biological inquiry involving *Drosophila*, its genes and their products are heavily in play. Thus, the first and perhaps most obvious level at which this particular topic will be handled is to consider rhythmic attributes of the biology of the animal in a broad manner that involves a multistep "biomolecular pathway." To the extent that the various stages of such a sequence, and certain pathway branches, are understood involves definition of these conceptual entities by gene discoveries and elucidation of the functions of their products.

---

[1]In fact, within these boxes, the Methods being considered at the moment will mostly be dealt with by referrals to other articles, including several contained within this volume. In this spirit, several passages within the current Methods boxes will incorporate certain pieces kibitzing about certain procedural inadequacies and antics.

Take, for example, the rhythm of rest versus locomotor activity of the adult fly and contemplate the analysis of this phenotype that subsumes central brain functions, including (a) "molecular pacemaker" ones that are mediated intracellularly by the products of "clock genes" (an ostentatious term that discriminates such core controllers from other genes that are merely rhythm related); (b) the first-stage "outputs" that emanate from such central pacemaking functions, one of whose jobs is to mediate the control of clock-regulated genes; (c) second-stage outputs that would include intercellular communications and do involve knowledge about the roles played by at least one "neurochemical gene," as well as a few others that encode products involved in signal transduction (i.e., mediating communication onto a cell, followed by processing the received information intracellularly); and, finally, (d) interneuronal message transfers that eventually end up at the "distal" motorneuronal control of overt behavioral rhythmicity, about which almost nothing is known from either chronogenetic or neuromuscular perspectives.

Nonetheless, it is arguable that *Drosophila* chronobiology as a subfield should be approaching this level of analysis, compared with the earlier days when investigators seemed primarily concerned with identifying intracellular clock factors and figuring out how they act and interact. In this respect, however, the rhythmic behavior of *Drosophila* was crucial: it created an extremely convenient bioassay for genetic variants, several of which defined bona-fide clock factors and most of which were isolated with respect to circadian locomotor rhythms (see Method 1). This phenotype speaks to one investigatory inadequacy, apart from the desirability of expanding one's horizons *beyond* what is going on *within* a pacemaker neuron: Other features of the circadian biology of the fly have been underanalyzed or even unexplored, either because certain known rhythms are more difficult to assay or because it has been far less than obvious *what* to evaluate with regard to some hypothetical piece of "tissue physiology," for example.

This brings us to the second level of subject treatment: The spatial expression of many rhythm-related genes in *D. melanogaster* has turned out to extend far beyond the nervous system. For example, certain key clock genes generate their products in nearly all organ systems and tissues of the animal, although usually not in a uniform manner within a given anatomical structure (e.g., Hege *et al.*, 1997). Moreover, what has become known as "circadian cycling" of such gene products is widespread spatially—occurring in many peripheral and internally located posterior structures of the organism, although not in all such tissues, in that a clock-gene's products may be present in a certain location but do not exhibit any systematic daily

METHOD 1. *Locomotor behavior* (often cryptically called "activity") of adult *Drosophila* is assessed most powerfully in long-term monitorings (e.g., over the course of a few light:dark cycles, proceeding into a week or more of constant darkness) by automated methods. Almost universally, these rely on a series of elongated, cylindrical glass tubes, which are typically plugged at one end by some fly food and at the other by a cotton stopper. The tubes are snapped into receptacles distributed over a given "monitor board" on which each tube location is flanked by infrared emitter/detector pairs. The standard visual system of the fly and its circadian photoreception are insensitive to light in that wavelength range. A popular source of these boards is Trikinetics, Inc. (Waltham, MA: http://www.trikinetics.com/). Each such device accommodates 32 fly-containing tubes. A simpler version of a board of this kind is depicted in Hamblen *et al.* (1986), although this more primitive device (custom-made at a research institution) contains spaces for only eight tubes. An additional publication that concentrates on this first-stage feature of the monitoring materials and methods, along with subsequent procedural steps, is Klarsfeld *et al.* (2003). The entire setup for automated locomotor recordings (again, in the commercially available version introduced earlier) is called the "TriKinetics Drosophila activity monitoring system." Crucially, it includes hardware and software necessary to store the behavioral events—infrared beam breakages—on a computer; a low-grade such machine can be fed into by a large number of boards, permitting hundreds of flies' worth of behavioral to be monitored and data stored simultaneously. Once a behavior file for each fly is generated by operating the elementary data-crunching subroutines, a wide variety of higher-level processings can be applied to a given behavioral record: displays of intraday and day-by-day locomotor fluctuations, merging of several such records for superposed display of a whole "genotype's worth" of behavior; assessments of whether a given record is defined by a periodic component (accompanied by a best estimate of the cycle duration for a "significantly" rhythmic record); and other metrics (e.g., resulting from "phase" analyses, which are discussed later). How best to effect high-level formal analyses of such records can draw on a variety of analytical principles (and the different kinds of software existing for most such methods): e.g., periodograms or those based on fast Fourier transforms (FFTs), maximum entropy spectral analysis (which in part encompasses FFT), or autocorrelation. The "how-best" concern is up in the air at present, even controversial. For descriptions of (only) some of the analytical strategies tactics, outcomes, and issues revolving round them, see Dowse and Ringo (1991), Levine *et al.* (2002b), Klarsfeld *et al.* (2003), and Price (2004, 2005).

fluctuations there.[2] Another aspect of the manner by which rhythm genes in *Drosophila* are temporally expressed involves timescales much longer than a given day, in the sense that the gene products tend to be found at most or all life cycle stages. However, they could have been present "only late," given the adult behavioral anomalies that were involved in identifying such genes or, in certain cases, the adult emergence phenotype (eclosion) that was used as a circadian bioassay to isolate other rhythm mutants (Method 2). In the latter case, the gene product would have to be functioning during late metamorphosis, at least, and could have persisted into the next stage (adulthood) if the mutation turned out also to affect behavioral rhythmicity of the mature fly. Such dual phenotypic defects were found for several rhythm mutants (e.g., Allada *et al.*, 1998; Konopka and Benzer, 1971; Rutila *et al.*, 1998b; Sehgal *et al.*, 1994), but not all (e.g., Newby and Jackson, 1991, 1993; Wülbeck *et al.*, 2005). In other words, an adult behavioral mutant, an eclosion one, or one that turned out to be doubly defective might have defined a gene that does not make its products during embryonic and larval stages. However, such rhythm-related genes are rare, perhaps nonexistent.

Therefore, *Drosophila* chronobiology as a system involves genetic factors that in and of themselves have taken investigators toward broad considerations of the developmental and organ system biology of the organism. However, it seems that chronobiologists did not always want to be taken there. In this regard, it was probably surprising when expression patterns of the seminal clock gene in *D. melanogaster—period* (*per*)—first began to be assessed and were found to involve multiple adult tissues (Liu *et al.*, 1988; Saez and Young, 1988; Siwicki *et al.*, 1988), as well as various life cycle stages (e.g., Bargiello and Young, 1984; Bargiello *et al.*, 1987; James *et al.*, 1986; Reddy *et al.*, 1984). These (unpleasant?) surprises may be correlated with the fact that dealing with this breadth is still in its embryonic stages (so to speak). Why, for example, are genes such as *per* and *timeless* (*tim*) generating their products during the first day of the animal's life cycle? Hall (2003a) furnished detailed tabulations of these temporal expression patterns, which will be reviewed in a briefer form within a subsequent section of this article.

Yet it is fair to say that some clock genes are expressed at many life cycle stages *not* because their functions necessarily connect with anything

---

[2]Incidentally, it seems as if knowledge about the first clock gene cloned, which occurred in *D. melanogaster*, being so pleiotropically expressed stimulated analogous assessments for the mammalian clock genes. These, too, were found to generate their products widely, and cyclically, in adult rodents (as reviewed, for example, by Allada *et al.*, 2001). Such relatively early findings (late 1990s) catalyzed considerations of mammalian chronogenetics and molecular biology as a "system" that extends well beyond the brain (as reviewed, for example, by Brandstaetter, 2004; Okamura, 2003; Schibler and Sassone-Corsi, 2002).

METHOD 2. *Eclosion*, if it includes periodic daily peaks of adult emergence, must involve developing cultures in which the animals are synchronized chronobiologically (albeit not in terms of developmental stages). This usually involves exposing cultures to 12-h:12-h light:dark cycles (12:12 LD) from their earliest stages onward and then harvesting pupae (even though a one-time light exposure delivered to larvae or pupae is sufficient to induce synchronization). The Trikinetics company (see Method 1) also sells an eclosion-monitoring device. This was used by Konopka *et al.* (1994, who included a photo of the device in their report) and continues to be applied in a few studies (e.g., Mealey-Ferrara *et al.*, 2003; Myers *et al.*, 2003; Wülbeck *et al.*, 2005); "few" refers partly to the fact that eclosion profiles are rarely determined anymore (see later). In brief, harvested pupae are adhered to a disk that gets placed atop an upside-down funnel; emerging adults slide down the funnel an break infrared light beams flanking the stem; each such event is counted and stored as referred to in Method 1. This is followed by operation of data-crunching and -plotting subroutines, along with one or more of the formal analytical programs noted in Method 1 (also see Method 14). Konopka *et al.* (1994) and Mealey-Ferrara *et al.* (2003) each give detailed verbal accounts of these procedures. Back when eclosion monitoring of fly emergence was more in vogue and this phenotype almost exclusively studied in *D. pseudoobscura* (1950s–1970s), a large electromechanical device was often applied, which periodically banged a rotating series of cultures into counting receptacles (e.g., Zimmerman *et al.*, 1968). For an example of how these "bang boxes" were used to monitor eclosion in *D. melanogaster* see, for example, Konopka and Benzer (1971), who drew upon this circadian character to isolate the first clock mutants. However, their first stage assessment of potentially anomalous eclosion involved direct by-eye observation of adult emergence during the light *vs* dark halves of 12:12 cycles. This leads to a mention of "by-hand" eclosion monitoring, whereby a group of investigators laboriously dumps out however many flies emerged within (usually) the previous 2 h, in a "24/7" manner, and simply counts the numbers of young adults for each such time bin. This brute-force approach has been taken several times down the years (e.g., Bargiello *et al.*, 1984; Dushay *et al.*, 1989; Hall and Kyriacou, 1990; Park *et al.*, 2003). It works, however, in the sense that roughly morning peaks of adult-emergence maxima result from these culture-clearing and counting agonies. The adverb just entered was used judiciously because finding any kind of "peaky" emergence in tests can be a hit-and-miss proposition, as some eclosion "runs" do *not* work when expected (even when applying the automated method), e.g., for wild-type control cultures. This problem may be related to the fact that this circadian character in *D. melanogaster* seems rather sloppy: adult-emergence peaks are observed as broad maxima, which can be spread over the late nighttime into the middle of the day. In contrast, eclosion rhythmicity of the aforementioned *D. pseudoobscura* cultures involves appreciably sharper daily recurring peaks. This box now shifts to *genetic methods* for manipulating the eclosion system. As noted within the main text, not much is known about clock variants and how their intracellular functions are "output toward" eclosion control (e.g., do all the factors diagrammed within Fig. 1 play a role at this earlier life cycle stage?). Nonetheless, more and more of the genes that involve such pacemaker-to-eclosion pathways are being isolated and then manipulated to elucidate relatively downstream features of the complex manner by which adult-emergence rhythmicity is directed. Some of the relevant transgene-based tools (Clark *et al.*, 2004; McNabb *et al.*, 1997; Park *et al.*, 2003) are brought up later (also see Baker *et al.*, 1999). With reference to a molecular method that involves a tool in the narrower sense [compared with the neuropeptide-gene(NP)-*gfp* fusions just referred to], please register the existence of a UAS-NP-*gfp* transgene (Husain and Ewer, 2004). This construct (and further forms of it involving a variety of NPs) allows for more than merely marking elements of the eclosion-controlling neuroendocrine axis; also scrutiny of GFP-emitted fluorescence modulations that occur within a pupa (for example) in conjunction with its ramp up to eclosion (e.g., not only identification of pertinent peptidergic neurons, but also how and from where their secretory contents are released in live specimens).

rhythmic during development, but rather that certain such genes play at least dual roles: one of which has nothing to do with circadian phenomena, the other involving sustenance of gene product generation at late life cycle stages when the "second" purpose kicks in on behalf of periodic eclosion, behavior, or both. It follows that these particular rhythm-related factors could be *vital genes*. Several of them are. Among the rhythm factors that turned out to cause lethality in certain of their mutant forms (e.g., Price *et al.*, 1998) are genes independently discovered on this basis by investigators who had no apparent knowledge of or interest in mature animal rhythmicity (e.g., George and Terracol, 1997; Jauch *et al.*, 2002; Shannon *et al.*, 1972; Zilian *et al.*, 1999); or involved dominant rhythm-affecting mutations, which, after their behaviorally based isolations, were found to be recessive lethal variants (Lin *et al.*, 2002; Newby and Jackson, 1993). It must be that any such essential gene is broadly expressed; at least its products would not be generated solely after development is completed.

Intriguingly, however, the spatial manifestations for the products of two vital rhythm genes have been claimed to be rather limited within the brain during adulthood [reviewed by Blau (2003), although this author sidesteps the fact that comprehensive organ system assessments of the expression patterns of these genes in mature flies have not been performed].

## The Circadian Clock of the Fly: Genes and Their Products, Considered in Part from Systems Perspectives

This brings us to the matter of *what* molecularly appreciable functions are defined by genes that got tapped into by elementary chemical mutagenesis (leading to intragenic nucleotide substitutions) and transgene mobilizations and/or manipulations (e.g., Martinek *et al.*, 2001; Newby and Jackson, 1993; also see the review by Phelps and Brand, 1998). The tactics devoted to, and the outcomes from, these forward genetic approaches are reviewed with respect to the rhythm variants of *Drosophila* by Price (2005).

Other rhythm mutants were arrived at by molecular screening, followed by adult rhythm testing of existing variants or engineered ones involving genes originally identified in their normal form from one gene expression perspective or the other [e.g., Blau and Young (1999); also McDonald and Rosbash (2001) and Ueda *et al.* (2002) (for example), as compared with Cyran *et al.* (2003); plus Stempfl *et al.* (2002), as compared with Sathyanarayanan *et al.* (2004)]. It should be reiterated that a fair number of such genes indeed encode essential developmental functions, such as *Andante, lark, double-time, vrille, shaggy, Timekeeper,* and *twins,* but others do not, such as *per, tim, Clock, cycle,* and *Pdp-1.* See Fig. 1, various other chapters within this volume, and Shirasu *et al.* (2003) and Price (2004, 2005) for the basics of what most of these genes are. This article is somewhat slanted toward "extra" functions associated with these factors, including whether lethal mutations have been identified at a given one of these genetic loci. The *inessential* gene category *might* involve molecular functions that are rather dedicated to rhythm control. However, as we will see, this is not necessarily so. Certain of these nonvital genes exhibit broad temporal and spatial expression patterns, not all features of which involve molecular cycling, as previewed earlier, and some components of which seem to have little or nothing to do with rhythmicity, as discussed in terms of mutational effects ostensibly unrelated to cyclical biology.

For the moment, however, some lip service should be given to the manner by which most of the genes mentioned earlier function on behalf of the circadian clock, as well as, for putative nonclock rhythm-related genes, to mediate some feature of molecular or cellular rhythmicity (e.g., an early stage output from the clock, as introduced at the beginning of this

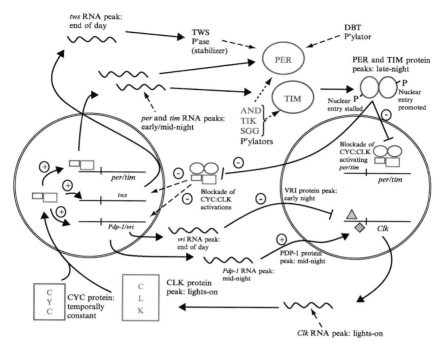

FIG. 1. Expansions of the basic pacemaker mechanism subserving daily rhythmicity in *Drosophila*: two interlocked feedback loops and their posttranslational regulators. This diagram is based loosely on one presented by Glossop *et al.* (1999), the investigators whose experiments prompted them to invoke "extra" gene functions and feedback phenomenettes. These components act beyond the elementary feedback events that are rooted in the negative effects of PER and TIM on their own genes' transcribability, as diagramed here within parts of the left half of the diagram (which has been modified twice from the Glossop *et al.* scheme, i.e., as it was tweaked within a figure appearing in Hall, 2003a). Genes functioning within the schemes ($n = 2$, really, as described later) are represented by italicized symbols, their protein products by the corresponding abbreviations in all uppercase; and activating/ stimulating *vs* inhibitory/repressive functions and events are designated by "plus" and "minus" symbols, respectively. The left half of this picture and the right-hand portion each represent the nucleus within one circadian pacemaker cell. The mechanistic components implied to be operating within and hovering around the two circles are roughly separated in terms of gene product dynamics that occur for *per, tim, Clk,* and *cyc* (left) and occurring in a pretty much temporally distinct manner with regard to *Clk* product dynamics and those of their gene expression regulators (*vri, Pdp-1*) (right). Thus factors entering the nucleus on the right and acting within it are oscillating in a largely out-of-phase fashion as compared with the "interlocked" loop components functioning within the left-hand circle and feeding into that nucleus (in terms of the *per, tim, Pdp-1, vri,* and *tws* genes chosen to be depicted within it). As to how these genes and their products are currently believed to act and interact: In brief, and focusing for the moment only on the transcriptional regulators PER:TIM binding to CYC:CLK, the former two proteins (at least PER) poison the actions of CYC:CLK, which

article). Thus, the core clock genes in *Drosophila* encode, in the main, transcription factors or kinases whose substrates are the former factors. Those regulators of gene expression control, in part, their own expression such that two interlocking feedback loops are formed by the actions of and interactions among the five transcription factors and the three phosphorylating enzymes known so far (Fig. 1, in which four "clock kinase" genes are indicated because two of them encode separate subunits of one such holoenzyme).

To apprehend what goes on within the functioning of each loop, the key mnemonic is that what goes up—in terms of the rising concentration of a given transcription factor, occurring over some hours of time—must come down if oscillations are to be mediated. Thus, the feedback effects are

---

results in not only in depression of *per/tim* activation, but also releases CYC:CLK-dependent repression of *Clk*. The negative function just implied is mediated by a transcription factor encoded by the *vri* gene, but *Clk* begins to be transcriptionally activated at a later cycle time by the action of the transcription factor of *Pdp-1* [specified as PDP-1ε by Cyran *et al.* (2003) and thus symbolically abbreviated here]. Therefore, CLK production eventually goes into its rising phase, resulting from action of the PDP-1 activator. As CLK accumulates, it stimulates increases in *per* and *tim* transcription rates, but inasmuch as the cooperative action of CYC:CLK has also promoted VRI production, whose result is to repress *Clk* transcription, PDP-1-regulated activation of *Clk* is stalled in terms of the latter's primary and secondary products beginning to rise toward their lights-on maximum (*Clk* mRNA and CLK protein rise and fall in concert within an LD cycle). The posttranslationally acting elements of these mechanisms involve an array of catalytic activities, the specific actions and eventual effects of which are verbally noted within portions of the bipartite diagrams. These schemes include symbols for kinase subunits encoded by the genes *dbt*, *And*, *Tik*, and *sgg* (color codings for the latter three gene products are meant to indicate the blue DBT enzyme as using TIM for its rhythm-relate substrate, whereas the bicolored AND and TIK designations stand for the fact that these two enzyme subunits participate in phosphorylating both PER and TIM). Another kind of enzymatic function schematized here involves the *tws* gene, which specifies a protein phosphatase subunit; it is believed to be activated transcriptionally by CLK:CYC, because the cycling of *tws* mRNA shown is flattened by the effects of a *cyc*-null mutation. However, the contribution of TWS/phosphatase to clock gene dynamics—apparently meaning in the main stabilization effects on PER levels, including those within the nucleus—is more complicated, because this polypeptide forms only one (regulatory) subunit of a phosphatase holoenzyme called PP2A. Another such subunit emanates from the *widerborst* (*wdb*) gene whose mRNA seems to cycle, but the relevant peak was observed at the end of the night, less than two times above the apparent transcript level 12 h earlier (or later), whereas *tws* mRNA (albeit for only one of the transcript isoforms emanating from this PP2A subunit gene) maxes out at the end of the day compared with the trough level that is *ca.* five times lower at the end of the night. For simplicity, and because this recently emerged phosphatase contribution to the overall mechanism is difficult to appreciate fully, only TWS is included with the diagram. (For example, particulars of the chrono roles that seem to be played by not only *wdb*-encoded polypeptide, but also by the *catalytic* PP2A subunit encoded by a gene known as *mutagenic star* are left out.) (See color insert.)

negative. For example, the PERIOD (PER) protein is a negatively acting transcription factor not because it binds to DNA but instead by virtue of inhibiting actions of the CLOCK (CLK) and CYCLE (CYC) proteins, one of whose (combined) effects is to activate transcription of the *period* (*per*) gene. Clearly, the three sample factors being considered at the moment must act within the nucleus of a rhythm-regulating cell (which may be a neuron). This truism is valuable to bear in mind because of the oscillatory time course involved. Thus, what if activated *per* were followed quickly by rising PER, which would be able to speedily enter the nucleus and then inhibit its own mRNA and protein productions? This half of the feedback scheme (the left-hand portion of the diagram in Fig. 1) would rapidly collapse or perhaps define a cycle duration much less than *circa* 24 h. It follows that there should be a considerable *delay* between the appearance or *per* RNA, which then increases in abundance during the late day and into the first third of the night, and of the negatively acting protein it encodes. A time delay of this kind is indeed observed: The *per* mRNA maximum leads the PER late-night peak by a few hours so that the protein is going up as the transcript level is going down.

This delay component of the implied mechanism (PER on the rise being causally connected to the fall of mRNA) is at least in part regulated by elements of the kinase controllers, such as the polypeptide encoded by *double-time* (*dbt*). Action of the DBT enzyme (reviewed by Harms *et al.*, 2003; Price, 2004) disallows an immediate rise in PER abundance—subsequent to *per* gene activation occurring late in the day—at least in terms of that protein approaching its peak levels in the nucleus (Fig. 1).[3] Thus, the feedback loops must be understood with respect to transcription-al activation/deactivation as coordinated by posttranslational modifications of the gene regulators. These interactions among gene products (e.g., PER inhibiting CLK, and DBT using PER as one of its substrates) permit an appreciation of slowly rising and falling levels of the transcription factors, such that the time spans between gene product peaks or troughs are each about one day.

Against better advice (author to himself), a few additional details and conceptual problems revolving round the clock model just presented in outline form will now be pointed out. First, do we understand enough

---

[3]This is not all that the action of *double-time* manages: Mutations in this gene change other aspects of feedback-loop function (monitored temporally in terms of *per* and *tim* transcript dynamics); also, when DBT enters the nucleus, it is bound to PER and participates in controlling the timecourse of this protein's decline within that subcellular compartment during the late-night and early morning phases of a daily cycle (reviewed by Harms *et al.*, 2003; Price, 2004).

about how the arguably key clock proteins, PER and TIM, accumulate toward their peak levels after the mRNAs encoding them have risen to or near their maximal intracycle levels? It seems as if DBT, acting in the cytoplasm, phosphorylates PER as it begins to be manufactured in that subcellular compartment during the nighttime, rendering the latter protein "unstable." This implies (a) that PER begins to be translated immediately as its mRNA starts to accumulate, such that the time-delay mechanism would not involve a temporary impingement of the translatability of the *per* mRNA and (b) that phosphorylated forms of clock proteins are more susceptible to degradation than versions of the molecules untouched by posttranslational modification. There are precedences for kinase actions exerting, as one of their myriad effects, instabilities on certain protein substrates, but there is no direct evidence that phospho-PER is per se extradegradable. One is left with information about a clock degradative role for the proteosome (Naidoo *et al.*, 1999) and a chronobiological task carried out by a *Drosophila* protein called SLIMB (Grima *et al.*, 2002; Ko *et al.*, 2002), which targets phosphoproteins to the proteosome. (For more about clock protein proteolysis, see Ko and Edery, 2005.) In any case, when TIM begins to accumulate (but why is its rise time delayed?), it would partner with PER, thus somehow protecting the latter from degradation (see, for example, Ko *et al.*, 2002). This would allow PER—actually PER: TIM dimers—to increase in concentration, including within the nucleus (by now, we have arrived at a late-night cycle time).

The state of TIM during this segment of a daily cycle speaks to how much we do not know about phospho-whatevers being subjected to that modification, potentially to increase the turnover rate of chrono-proteins. Take, for example, the *sgg*-encoded kinase. This enzyme uses TIM as its principal substrate (in terms of the rhythm role played by this versatile gene); such posttranslational modification promotes nuclear translocation of TIM (reviewed by Harms *et al.*, 2003), which is in the direction of an effect *opposite* of destabilization. In this regard, the kinase holoenzyme identified as rhythm relevant by *And* and *Tik* mutations counts both PER and TIM among its substrates, and the results of such phosphorylation events also promote nuclear entry of those proteins (Akten *et al.*, 2003; Lin *et al.*, 2002; Nawathean and Rosbash, 2004).[4] A final point about "how much..." is that next to nothing is understood about *substrate details*, such as the numbers and qualities of amino acids that come into play as PER or TIM get "progressively" phosphorylated (e.g., Edery *et al.*, 1994; Zeng *et al.*, 1996). By the way, CLOCK is another pacemaker protein that is a substrate for one or more (unknown) kinases (Lee *et al.*, 1998).

More about delay mechanisms: There are separate cycle times, which may have nothing to do with phosphorylation of clock factors, when such processes have been inferred to contribute to overall operation of the interlocked feedback loops: (a) If CLK can more or less simultaneously promote the transcription of an activator (PDP-1) and a repressor (VRI) of its own transcription (Fig. 1), no "coherent" rhythmic expression of *Clk* would seem to be possible (as discussed by Allada, 2003). The (descriptive) answer to this conceptual problem is that abundance of the VRI transcription factor rises earlier than the level of the PDP-1 such factor (Cyran *et al.*, 2003), which temporally separates repression and activation of *Clk* transcription (Fig. 1). (b) Focusing on the onset of dawn during a light:dark (LD) cycle, TIM is degraded rather rapidly as a result of that environmental transition (note that photic inputs to the clockworks are dealt with in detail later, but mainly in terms of neuroanatomy); yet PER persists and cycles away for a while (Marrus *et al.*, 1996; Rothenfluh *et al.*, 2000c; Weber and Kay, 2003).[5] Therefore, PER (acting alone during the morning hours) can form an "inhibitor buffer" that effects a stall to "restarting" the next cycle of transcription, as activated by CLK and CYC (Weber and Kay, 2003).

Some conundrums remain attached to this scenario, however (again, as diagrammed in Fig. 1). For example, the notion that PER:TIM can enter nuclei of pacemaker cells only as an obligate heterodimer relies almost exclusively on the behavior of such polypeptides (or engineered fragments thereof) in transfected, cultured cells (Saez and Young, 1996). This heavily used assay system has nothing to do with *Drosophila* neurons and involves little in the way of clockiness (discussed within Method 3). Within cells of actual fly tissues, there is mounting evidence that PER may be the main player insofar as inhibiting CLK:CYC dimers is concerned: (a) Application

---

[4]Also, it is thought that action of the *And/Tik*-specified CK2 enzyme on TIM exerts a "priming" function, such that SGG-mediated phosphorylation is facilitated or enhanced; stated another way, these two separate kinases may act synergistically on behalf of promoting the nuclear translocation of TIM (reviewed by Blau, 2003). The AND/TIK actions, presumably as augmented by that of SGG on its TIM substrate, ultimately result in hyperphosphorylated forms of the latter, which helps target TIM for degradation during the late night and early morning (so the actions of the three kinases just named would not be limited to intranight promotion of the nuclear entry of TIM). The wheels for such proteolysis could be greased by infusion of the activities of SLIMB (discussed by Ko *et al.*, 2002).

[5]Thus, from another angle, in addition to those discussed in the ensuing main-text paragraph, one appreciates that TIM is not needed for PER putatively to act. In the particular context being discussed at the moment, one might have inferred the presence of TIM to be obligatory for that of PER because very little of the latter is detectable in a *tim*-null mutant (Price *et al.*, 1995) and keeping flies in constant light for a long time gets rid of not only TIM, but also most of PER (Price *et al.*, 1995; Zerr *et al.*, 1990). However, these situations are materially different from a natural one in which the animals are progressing through light:dark cycles (again, see Marrus *et al.*, 1996; Weber and Kay, 2003).

of a certain long-period *tim* mutant revealed that the product of this gene could be gotten rid of (by light exposure), leaving PER alone as sufficient to mediate pacemaking function (Rothenfluh *et al.*, 2000c); (b) PER naturally enters the nucleus of certain brain neurons (see later) at least 3 h before any TIM can be detected there (Shafer *et al.*, 2002); (c) PER alone, but not TIM, is able to repress CLK:CYC-mediated transcriptional activation in transfected cells (Ashmore *et al.*, 2003; Nawathean and Rosbash, 2004; Weber and Kay, 2003); (d) When the amounts of daylight *vs* dark were varied systematically, the dynamics of the rises and falls of PER "tracked" the altered LD cycles better than TIM, including that no TIM cycling was observable in a 6-h L:18-h D cycle, within a certain subset of brain neurons that normally express this protein (along with PER) and are believed to be important for regulation of rhythmic behavior (see later); however, the cyclical locomotion of flies in this condition almost certainly still reflected underlying clock activity (Rieger *et al.*, 2003; Shafer *et al.*, 2004).

---

METHOD 3. *Cultured, transfected cells* used to study the cell biochemistry of rhythm-related genes and their products make use of a "Schneider 2" cell line (S2); it was derived long ago from anonymous embryonic cells of *D. melanogaster*. The tactics involved transfecting certain fragments of, for example, a clock gene and/or DNA constructs designed to generate the complete product of some other rhythm gene. Thus, an exemplary type of question would ask whether supplying S2 cells with CLOCK protein leads to activation of an otherwise silent *period*-derived construct. Answer: yes, as monitored by measuring the luciferase activity (also see Method 16) that is put out by a *per-luc* construct cotransfected into such cells (this *luc* ploy is the one typically applied to report whatever gene-activation may be in question). By the way, the affirmative answer just given was interpretable because of the remarkable fact that the *cycle* gene of *D. melanogaster* is constitutively expressed by S2 cells, obviating the need for cotransfecting a quasi-complete $cyc^+$ gene. See Saez and Young (1996), Darlington *et al.* (1998), and Ceriani *et al.* (1999) for early applications of transfected cell principles, techniques, and outcomes; these involved, variously, assessments of interactions between pacemaker proteins or between them and the regulatory regions of companion clock genes. Certain review articles provide entrées to other such studies and describe more about the methods (Ko and Edery, 2005; Price, 2004; Sehgal and Price, 2004).

---

It has "always" been the case that *either* PER or TIM *should* be sufficient to inhibit CLK, CYC, or both: The latter two proteins contain dimerization domains known as PAS—an intrapolypeptide motif that was first identified (unwittingly) within the PERIOD polypeptide itself.[6] PER and TIM can associate via the former's PAS (Saez and Young, 1996). Thus, even though "mapping" the region within PER, which is necessary and sufficient for CLK:CYC inhibition (in the usual cell transfection experiments involving constructs designed to produce various protein fragments), led to the uncovering of an inherently featureless stretch of amino acids that is *well downstream of the PAS domain*; this "CLK:CYC Inhibition" domain also contains a previously unidentified nuclear localization signal within PER (Chang and Reppert, 2003).

Last, we get carried away by the primacy of PER vis-à-vis TIM, let us go nack a few years worth of investigatory time: It seemed during the late 1990s that all PER is doing on behalf of clock functions would be to help get TMI into the nuclear. Once within this subcellular compartment, the latter protein would effect the crucial inhibition of CLK:CYC dimers. This supposition arose, for example, from studies involving homogenates of heads taken from *per*-null flies. This material was used in immunoprecipitation experiments to show that TIM alone is sufficient to "pull down" CLK and CYC (Lee *et al.*, 1998) (Method 4). However, such an artificial situation was probably just as misleading as were certain inferences from the experiments performed in transfected cell cultures. Nevertheless, we need to stand back from all of these protein mechanistic concerns to move back toward the fly's entire system of rhythmicity. What will not quit in this regard is that *tim*-null *Drosophila* are severely arrhythmic and behave just as badly in this regard as do flies suffering from the effects of lack-of-function mutations within the classical companion gene, i.e., *per* mutations or homozygosity for a deletion of the gene (e.g., Hamblen-Coyle *et al.*, 1989). [However, a given set of *tim*-null individuals, compared with *per* nulls, does not behave "as arrhythmically" as do the latter types (discussed within the legend to Fig. 3). Other *tim* mutations cause some of the most dramatic alterations in circadian periodicity known (e.g., Rothenfluh *et al.*,

---

[6]Additional factors were suggested to carry out clock functions in part because they are PAS proteins. For example, (a) the first clock gene cloned in mouse, literally *Clock* and the reason *Drosophila*'s homologous gene was given this name, was honed in on by positional cloning of a mutationally defined locus and identification of a nucleotide sequence that would encode a PAS-containing transcription factor (reviewed by Allada *et al.*, 2001), and (b) searching for the molecular correlates of two arrhythmic mutants of *D. melanogaster* that cause *per* and *tim* products to be constitutively low led to the identification of sequences at or near the respective loci on chromosome *3*, which were good candidates for the genes in question by virtue of encoding, on paper, PAS transcription factors (Allada *et al.*, 1998; Rutila *et al.*, 1998b).

METHOD 4. *Clock gene products in tissue homogenates* are assessed and manipulated in experiments that usually start with the removal of heads from a given large group of flies, e.g., 50 or 300 such, sacrificed at a given time over the course of (almost always) *one* 24-h period; in turn, these operations usually involve 12:12 LD cycles. Some such determinations, however, have involved homogenizing whole flies (e.g., Reddy *et al.*, 1984) or heads *vs* "bodies" (everything posterior to the neck) separately (e.g., Emery *et al.*, 1998; Hardin, 1994). Incidentally, frozen adult animals can readily have their heads broken off from the remainder of the fly and the different body regions separated by use of sieves. It must be mentioned that the separated heads have almost always included the compound eyes of the fly as a major contributor to whatever molecules are going to be monitored (exception: Zeng *et al.*, 1994). For this, the body parts are homogenized and then various extraction procedures are carried out, depending on whether one wishes to track RNA abundances or those of extracted proteins, respectively, by Northern blottings, RNase protections, reverse-transcriptase PCR (the latter not used much yet for fruit fly chrono-molecular biology), or Western blottings, and immunoprecipitations (IPs). Additionally, transcription rates per se have been monitored for certain clock gene products by performing nuclear run-on experiments (e.g., So and Rosbash, 1997), accompanied by the more usual assessments of steady-state mRNA levels (e.g., via RNase protections). A huge number of these chrono-biochemical experiments have been performed—far too many to cite beyond, for example, the case of Lee *et al.* (1998). Thus, see Price (2004) for a review that goes into a bit of detail about the several relevant methods. This author also describes (and cites the usage of) other tactics, in addition to application of IP protocols, for assessing "protein–protein" interactions involving things like dimerization of rhythm-related polypeptides, such as principles and practices of the "yeast two-hybrid" system, which has frequently been used to ask questions about the chrono-proteins of *Drosophila* and even to discover certain of them (Gekakis *et al.*, 1995).

2000a,c) when considering rhythm-altering variants in all the organisms studied chronogenetically. The following additional notion about the importance of *tim* may seem silly, but keep in mind that screening for newly induced rhythm mutations on chromosome 2 of *D. melanogaster*, which has been occurring off and on for 20 years, resulted mainly in the isolation of *timeless* variants. In other words, independently induced "repeat hits" of *tim* kept surfacing again and again and they were worth saving, plus pursuing, because of their solid and usually severe rhythm deficits or abnormalities.[7] No additional, serious rhythm mutants turned out to be inducible or otherwise isolable regarding this 40% of the genome of the organism (reviewed by Hall, 2003a; Price, 2005; Stanewsky, 2003) either in terms of biological rhythm phenotypes or the molecular one noted within footnote 7 (also see Method 16). Further such variants, mutated at other ill-defined 2nd-chromosomal loci (dolefully tabulated by Hall, 2003a), exhibited such mild or dubious defects that they were never followed up and probably no longer even exist.

So *timeless* is one of *the* prime players in the clock mechanism of *Drosophila*. Therefore, what TIM is doing must be figured out fully. This implies that we do not understand the "real" or total functions of TIM well enough (as was detailed in earlier passages within this section). The manner by which such complete understanding will be gained is likely to require interrogating the actions of the *timeless*-encoded protein by experimenting upon the intact system (whole, unhomogenized flies or perhaps intact brains analyzed cellularly and intracellularly in organ culture).

Turning to the kinases diagrammed in Fig. 1, their function on behalf of circadian pacemaking was given short shrift in the foregoing diversional discussion, even though it is notable that elements of the "DBT mechanism" have been deduced from assaying this enzyme molecule and the results of its apparent catalytic action in fly tissues and cells (reviewed by Price, 2004). However, it is unwarranted to pause further for descriptions of the current clock enzymatic picture, which would go on to describe what is known about actions of the *And*, *sgg*, and *Tik* gene products and include a nod to the inevitable involvement of a phosphatase, which also lurks within Fig. 1. [See Sathyanarayanan *et al.* (2004) and please register that this enzyme is encoded by a *Drosophila* gene called *twins*, which has long been known to exert pleiotropic influence on *Drosophila* development and other phenotypes (e.g., Perrimon *et al.*, 1996; Philip, 1998).]

Instead, let us muse about the possibility that these kinds of proteins function only on the periphery of the core pacemaking mechanism. Surely none of these five enzyme-encoding genes (including *dbt*) encodes dedicated clock factors, because all of them have been mutated to cause

---

[7]In particular, eight *tim* mutants were recovered in screens for behavioral or eclosion variants, starting with mutagenesis of rhythm-normal flies (Rothenfluh *et al.*, 2000a; Sehgal *et al.*, 1994); one *tim* mutant was found after "taking 2nd chromosomes" from natural populations (Matsumoto *et al.*, 1999; Murata *et al.*, 1995); another was induced by mutating flies carrying a long period *per* mutation (Rutila *et al.*, 1996), whereby the new *tim* allele largely suppressed the effects of *per^L* (was it a mere coincidence that no such suppressors at any other loci were found in his manner?); and three additional *tim* mutations were isolated by virtue of their effects on luciferase-reported *per* product cycling (Stempfl *et al.*, 2002; Wülbeck *et al.*, 2004).

developmental lethality. Some such mutations—notably the classic *shaggy* ones—were induced and analyzed long before any rhythm involvement could have been even dimly suspected (e.g., Judd *et al.*, 1972; Shannon *et al.*, 1972; also, ironically, Young and Judd, 1978). However, both *And* and *dbt* were originally discovered by non-null, nonlethal mutations that might have identified genes devoted solely to the regulation of rhythmicity.[8]

What if a mutant such as *Andante*, induced during the early days of hunting for rhythm variants in *D. melanogaster*, as documented in the Ph.D. dissertation of Orr (1982), nine years in advance of the seminal *And* publication, had been the first putative clock gene cloned? Mapping the *And* locus was accomplished with enough precision (Orr, 1982) to allow for "early years" positional cloning—the same kind of cytogenetic information available for the *period* gene as of the 1970s into the early 1980s (Konopka and Benzer, 1971; Smith and Konopka, 1981, 1982; Young and Judd, 1978). What, instead or worse, if the *per* locus had never been mutated back in those early days? Molecular identification of the one mutationally defined gene available, *Andante*, would have revealed that "a kinase is involved in the rhythmicity of *Drosophila*." In this regard, it should be mentioned that the essential developmental role of *And* (and those played by all the other kinase encoders in Fig. 1) could mean that the one viable mutant involving this gene is an *anatomical* variant, which exhibits rhythm anomalies because it is slightly brain damaged (although not severely or overtly so, as documented for *And* by van Swinderen and Hall, 1995). However, let us leave this possibility aside and assume that the "moderately slow" clock exhibited by *And* mutant flies and eclosing cultures is caused by a defect in catalytic functioning per se (i.e., during the acute executions of locomotor or adult-emergence control).

In this light, *And* as the initial rhythm-related gene molecularly on the table would have been a conceptual disaster compared with historical reality: *per* was positionally cloned first (Bargiello and Young, 1984; Bargiello *et al.*, 1984; Reddy *et al.*, 1984; Zehring *et al.*, 1984) and then inferred by elementary sequence guesses to encode either a mystery protein (e.g., Shin *et al.*, 1986) or a "proteoglycan" (Jackson *et al.*, 1986; Reddy *et al.*, 1986). The latter presumption still provided little clue as to what the gene product is doing (Bargiello *et al.*, 1987, notwithstanding), whether or not PER as a proteoglycan

---

[8]The *And* mutant sat in this state for many years after it was isolated (Konopka *et al.*, 1991, as compared with Akten *et al.*, 2003; DiBartolomeis *et al.*, 2002; Jauch *et al.*, 2002). The initial *doubletime* short period variant (Price *et al.*, 1998) was found to be altered in a vital gene "moments" after it was induced and identified as a short period mutant (by definition) and soon afterward was identified independently by a set of developmentally dead variants (Zilian *et al.*, 1999).

turned out to be magisterially misconstrued (Edery *et al.*, 1994; Flint *et al.*, 1993; Saez *et al.*, 1992; Siwicki *et al.*, 1992). Turning round this misconception finally to demystify PER as putatively involvednot only in gene regulation, but also in controlling "itself" (*per*) provided the entry-level evidence toward cracking the entire case (Hardin *et al.*, 1990; Siwicki *et al.*, 1988; Zerr *et al.*, 1990). If, however, *Drosophila* chronobiologists had been stuck with a "clock kinase" (the *And* gene product), almost no idea would have suggested itself as to how a circadian pacemaker might work, because phosphorylation of "something" at the level of cellular biochemistry could mean anything and everything, therefore nothing.

However, it is most valuable that we now appreciate the biochemical roles played by rhythm-related kinases. This means more than understanding how DBT contributes to the state and action of PER during different circadian cycle times (see earlier discussion). Just as important is the fact that, as in the case of *tim* mutations, several of the chrono-kinase variants change the pace of the clock so dramatically—or eliminate rhythmicity, depending on the mutant allele—that these factors must continue to be analyzed for their pacemaking contributions. This is because the altered or ruined rhythms just referred to occur at the systems level: The kinase mutants that exhibit dramatic period changes or arrhythmia are quintessentially appreciated in terms of whole animal misbehavior and anomalously noncircadian eclosion (e.g., Price *et al.*, 1998; Rothenfluh *et al.*, 2000b; Suri *et al.*, 2000).

### Clock-Controlled Genes: Tentative Steps That Move the Subject from Central Pacemaking Concerns out into the System as a Whole

Having said all this, let us refocus attention on the regulatees of clock kinases—the transcriptional regulators that are substrates of these enzymes. If the products of certain rhythm-regulating genes can control the transcription of what encodes them, then factors such as *period* and *Clock* should also be able to participate in mediating the cyclical expression of other genes or their products. The latter factors would include those functioning at early steps along the aforementioned output pathways whose end stages involve effectors of actual biological rhythmicity (reviewed by Hall, 2003a; Jackson *et al.*, 2001, 2005; Park, 2002). Indeed, there are many mRNAs now known to cycle in their daily abundances. Most of these gene products were identified in microarray screens involving probe sources taken at different times during the day or night or at different "free-running" cycle times for *Drosophila* sacrificed in constant darkness (reviewed by Etter and Ramaswami, 2002; Jackson and Schroeder, 2001; Jackson *et al.*, 2005; Park, 2002). Because this article claims to be about circadian system biology for this organism, further discussion of these "chip screen" outcomes is unwarranted: too little is known about why (or whether) a given mRNA cycles on behalf of the rhythmic biology of the fly. There were several other putative clock output factors identified by microarrays, not via RNA cycling but instead effects of clock mutations on their gross abundances; but the great majority of these factors, too, are at present unappreciable as to their chronobiological relevance.

In this regard, there is no way that a rhythm-related factor has to cycle. For example, *And* and *dbt* mRNAs stay at the same gross levels over the course of a given 24 h (Akten *et al.*, 2003; Kloss *et al.*, 2001), even though the DBT protein cycles in terms of its movements between the cytoplasm and the nucleus (Kloss *et al.*, 2001). Also note that a positively acting clock transcription factor such as CLK need not cycle either; daily oscillations of a CLK-inhibiting entity, such as PER, would be sufficient. That *Clk*-encoded mRNA and protein nevertheless go up and down each day— and do so in temporal concert, as opposed to the case of PER or the *timeless*-encoded protein (Fig. 1)—is conceptually important in terms of CLK functions forming the key interface between the two out-of-phase feedback loops (again as depicted in Fig. 1 and explicated in its legend). Incidentally, the existence of both loops could cause mere annoyance on the part of fly rhythm fans (or those who try to cope with feedback complexities in other systems, e.g., Merrow and Roenneberg, 2005; Tischkau and Gillette, 2005). Instead, consider that mediation of a variety of peak and trough times for gene regulators (Fig. 1) may be just what is needed for *maximally versatile* pacemaker outputting. This means that different aspects of *Drosophila* rhythmicity involve biological maxima and mimima (with regard to late development or postdevelopmental behavior and physiology) that are semi-scattered around the clock face.

To delve into cyclical features of certain chronobiological output factors, it is perhaps ironic that they were identified either serendipitously or by what might seem to be rather primitive molecular screening—compared with microarray tactics, which have been chronobiologically applied many times to *Drosophila*, as well as to several other organisms, during the early years of this century (reviewed by Duffield, 2003; Sato *et al.*, 2003; Walker and Hogenesch, 2005). Thus, we now welcome a couple of circadian system components to the article. They are chosen from among a host of putative gene-defined candidates (Jackson *et al.*, 2005) for three reasons: (1) These components of the system are tied to some real biology, thanks to *mutant phenotypes* as opposed to gene expression patterns alone. (2) The two apparent output factors are literally that, or least one surmises that the gene products *"escape" from certain cells* in which the clockworks operate (outputs from the clock are sometimes subsumed under the metaphorical term "escapement" by those who wallow in chronobiological jargon).[9] (3) The *phases of these separate subsystem features* (implied by the tissue expression patterns of the two genes to be discussed) are different. This is in line with how the animal should wish its rhythms, taken as a whole, to operate (see previous paragraph).

Take the gene called *takeout*. This factor (*to*) was originally identified in its normal form by old-fashioned "molecular subtraction" methods (Method 5) with a genetic twist: searching for RNAs whose abundance is subnormal in the $cyc^{01}$ mutant (So *et al.*, 2000). Recall that that mutation causes *per* and *tim* mRNAs to be not only low but also not to exhibit any appreciable cycling (Rutila *et al.*, 1998b). In this regard, *to* mRNA was found to exhibit an abundance of oscillations in homogenates of wild-type flies, with peak time late at night (Sarov-Blat *et al.*, 2000; So *et al.*, 2000). Moreover, this gene was reidentified in one of the chip screens alluded to earlier, as were cycling molecular relatives of *takeout* (McDonald and Rosbash, 2001; for more about the "*takeout* gene family," see Dauwalder *et al.*, 2002). To demonstrate the clock control of *to* further, null mutations of *per*, *tim*, and *Clk* were shown to cause much lower than normal transcript levels emanating from the newly discovered gene; $cyc^{01}$ had the same effect, confirming the "isolation phenotype" (Sarov-Blot *et al.*, 2000). The TO protein, whose sequence implies it to be a secreted factor that binds small lipophilic molecules, also exhibits a daily oscillation (same peak time as *to* mRNA); this cycling was found to be blunted in $cyc^{01}$ or $Clk^{Jrk}$ mutants.

Further studies of *takeout* indeed take us out into the elements of the rhythm system of the fly as a whole—meaning in this case, beyond the brain (Sarov-Blat *et al.*, 2000). Thus, *to* gene products are found in peripheral nervous system (PNS) structures such as the eye and antennae and also in the thoracic alimentary system (cardia and crop). Gut cells of *Drosophila* do contain PER (this tissue type being one of many that were found in the early days to express this gene, as inventoried by Hall, 1995). The immunoreactivity of this protein, along with that of TIM, implies daily cycles of abundance for these molecules within alimentary structures (Giebultowicz *et al.*, 2001). This aspect of the *takeout* expression pattern seemed to prompt the specific genetic test that was carried out on behalf of the biological meaning of the gene. Luck permitted such an experiment: Lurking within a marker strain of *D. melanogaster* was a genotypically severe *to* mutant, a

---

[9]In contrast, another kind of conceptually apprehendable output would be *intra*cellular and involve neuronal excitability—chronobiological studies of which have a long history (see Lundkvist and Block, 2005). In *Drosophila* there are chromosomal or engineered variants associated with the excitability substory (e.g., Nash *et al.*, 2002; Nitabach *et al.*, 2002), but these and other features of it are covered by Nitabach *et al.* (2005) (also see Jaramillo *et al.*, 2004).

METHOD 5. *Molecular subtraction methods* have a long history in gene product studies. With regard to the rhythm system of *Drosophila*, these tactics have typically been applied with respect to RNAs taken from different daily cycle times (e.g., Rouyer *et al.*, 1997) to ask whether some newly discoverable gene product can be found by virtue of (for example) "subtracting away" transcript types that are the same levels at two times (e.g., 12 h apart or perhaps fewer, but more than something like 10% of a day). Other RNA subtractions have been based on the possibility that differential abundances of certain transcripts could be determined in extracts of wild-type tissues *vs* those of a rhythm mutant (e.g., So *et al.*, 2000). In such cases the nucleic acids are cloned, as well as being viewed as differentially expressed. Sehgal and Price (2004) reviewed applications of these principles and procedures (preparations of RNA populations and of molecular probes) with respect to the organism at hand. It should also be mentioned that "differential display" has also been applied to tap into *Drosophila* RNAs whose abundances cycle (Blau and Young, 1999). This tactic and those on which molecular subtractions are based are falling out of favor, however. The main text mentioned this matter in terms of the rise of microarrays: How many genes in *Drosophila* generate their products in a time-differential manner? How many will respond to the effects of clock mutations by exhibiting unusual concentrations? Which specific genes are expressed or have their expression altered in these ways? "Gene chip" technologies can lead to this kind of information and identification of materials [see Walker and Hogenesch (2005)].

partial intralocus deletion that results in lower than normal mRNA and TO protein levels; the latter does not cycle (Sarov-Blat *et al.*, 2000). When the locomotion of this mutant was monitored in a situation of food deprivation, such flies died (even) earlier than normal adults do (Sarov-Blat *et al.*, 2000). (Inferentially, there was no particular necessity to uncover this *to*-mutant phenotype in the context of behavioral rhythm monitoring.)

A further connection of *takeout* to nutrition is that starvation was shown to induce enhanced levels of products of the gene, which were observed in a spatially broader pattern than usual (Sarov-Blat *et al.*, 2000). Here is a sour note, however, which may uncouple *to* from having anything to do with rhythmic behavior (also see the parenthetical sentence given earlier). Expression of mRNA of the gene within the brain (Sarov-Blat *et al.*, 2000) proved to be irreproducible when Dauwalder *et al.* (2002) showed this intrahead material to be detectable only within fat body tissue. Moreover, the latter investigators rediscovered *takeout* molecularly from a perspective having nothing to do with rhythms; instead, with regard to *Drosophila* genes that are expressed sex specifically. In particular, *to* products are found only in males, except that its mRNA is detectable in the antenna of both sexes (Dauwalder *et al.*, 2002). Connections of rhythms to sex, including behavioral aspects of the latter, will be taken up in a later section.[10]

The second clock output entity to be judged in considerable detail is a neuropeptide called pigment-dispersing factor (PDF, named after the discovery in crustacea of a highly homologous 18-mer that functions according to its name). This molecule is found only within the central nervous system (CNS) of *Drosophila* and not, as described commonly, solely in certain "clock neurons" within the brain (which are dealt with more generally and in much detail later)—also, in a posterior region of the ventral nerve cord (VNC), and transiently in other brain neurons during late development and in very young adults (reviewed by Hall, 2003a; Helfrich-Förster, 2003). The potentially risky supposition that PDF could be coexpressed with PER, over a small part of the expression domain of *per* within the head of the fly, was taken to heart (Helfrich-Förster, 1995). This test panned out, and the findings prompted cloning the *Pdf* gene of *Drosophila* (Park and Hall, 1998). The mRNA it encodes stays steady during the day and night of a given cycle (Park and Hall, 1998; Park *et al.*, 2000a).

However, *Pdf* and its gene product are clock controlled in two ways. Within a subset of the CNS subset of PER cells (*ca.* 150 out of $10^5$ total neurons), neither *Pdf* mRNA nor PDF itself is detectable in animals homozygous for the original *Clock* mutation ($Clk^{Jrk}$) or in those homozygous for either of the two *cyc* mutations induced *in vivo*.[11] In particular, Park *et al.* (2000a) found $Pdf^+$ to be unexpressible within a cluster of small ventrolateral PER neurons (*s*-LN$_v$s), one of two equally sized LN$_v$ cell groups (Fig. 2), in the adult brains of *Clock* or *cycle* mutants. [The larval precursors of such neurons (discussed in a later section) are also PDF-less in these mutants (Blau and Young, 1999; Park *et al.*, 2000a).] This implies

---

[10]Sarov-Blat *et al.* (2000) made no mention of sex specificity of *to*. It seems as if their assessments of the gene products' abundances involved homogenizing mixtures of males and females, which would of course allow for *to* mRNA and TO to be detected, and that their determinations of the spatial expression of *takeout* patterns would have been performed solely with individual males for some reason.

[11]All three of these clock variants are apparently "null" mutants in terms of the intra-ORF nonsense mutations that were induced by a chemical mutagen then identified by the adult locomotor arrhythmicity they cause (Allada *et al.*, 1998; Park *et al.*, 2000a; Rutila *et al.*, 1998b). Subsequently, a further *Clk* mutant was isolated independently by the same behavioral phenotype and found to be a near null molecularly (Allada *et al.*, 2003).

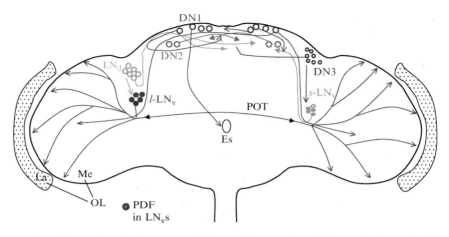

FIG. 2. Neuronal projection patterns of CNS cells expressing clock genes in the brain of a *Drosophila* adult. The three clusters of lateral neurons (LNs) and the three groups of dorsal ones (DNs) are depicted in somewhat of a cartoon fashion—not so much in terms of positions of these cell groups, but, for example, because space limitation forced many fewer of the dorsally located DN3s to be diagrammed here than the actual number: *ca.* 40 bilaterally symmetrical pairs of such cells. For the other cell groups, numbers of colored dots are more representative, in that a given hemibrain (left or right side) contains approximately 4–5 small (*s*)-LN$_v$ cell bodies ("small" referring to *relative* sizes of such perikarya). Five for the large (*l*)-LN$_v$ group nearby, six for the more dorsally located LN$_d$s, and for the two additional groups of DN cells: 15 for DN1s (such that the real number of these cells, as for the DN3s, is diagrammatically underdepicted) and two for DN2s. Totals (including the DN3 number noted earlier) are approximately 70–75 pairs of brain neurons expressing the *period* gene, the *timeless* one, or both (implying that not all the neurons indicated could be observed to coexpress both of the clock gene products, as discussed briefly within the text). These cell counts were originally presented by Ewer *et al.* (1992) and Frisch *et al.* (1994); updated by Kaneko and Hall (2000); and have appeared in various reports of Helfrich-Förster, as summarized in 2003. Within that review is a diagram that includes a large number of small circles (compared with the colored dots within the present picture), representing the *ca.* 1800 *per*-expressing glial cells that are located in several cortical and neuropile regions of the central brain proper and those of the optic lobes (see Fig. 3 for what these glia may be about). The neurites projecting in various directions and potentially to various target neurons are depicted by thin colored lines, accompanied by arrowheads when the so-called targets (referring to distal extents of certain axon terminals) could not be observed in definitive fashion. This intrabrain "wiring diagram" is based on stainings for the PDH immunoreactivity that is present in all of the *l*-LN$_v$s and in of most *s*-LN$_v$s (e.g., Helfrich-Förster, 2003) as well as by application of transgene-driven neurite markers (Kaneko and Hall, 2000; Park *et al.*, 2000a; Renn *et al.*, 1999; Stoleru *et al.*, 2004; Veleri *et al.*, 2003). Focusing on elements of these projection patterns, *l*-LN$_v$ perikarya send axons across the brain midline [via the posterior optic track (POT)] to contralateral LN regions and also project into the optic lobes (OL), but only as far as the medulla (Me), which is underneath the distal-most lamina (La); it is unknown whether a given *l*-LN$_v$ cell body might elaborate both POT neurites as well as centrifugal fibers. *s*-LN$_v$ cells project fibers into a dorsomedial brain region, with these nerve

that PDF is still present in the large (*l*) LN$_v$s within such clock-mutated brains (true).

Thus, there would seem to be exquisitely neuron-specific effects of the *Clk* and *cyc* genes on *Pdf* transcription. This is not the whole substory, however: Either kind of clock mutation just indicated, which leaves the presence of PDF unperturbed within the *l*-LN$_v$s, causes the anatomy of projections from those larger neurons to be anomalous, as was revealed by PDF itself serving a marker for both cell bodies and neurites (Park *et al.*, 2000a). Therefore, the *Clk* and *cyc* mutants may be developmental as well as adult-behavioral variants. This is a harbinger of things to come, when discussion of *Drosophila* chronobiology as a broad system includes consideration of several life cycle stages. [The author refers to chronobiological matters per se—other than those revolving round the aforementioned kinase-encoding genes, along with *vrille* and *twins* (Fig. 1), and the nonrhythm-dedicated roles that all of these genes play in the biology of the animal.]

Does *Pdf* have anything to do with cyclicity at the cellular or whole organismal level? The affirmative answers (×2) revealed first that the gene *product* is clock controlled in a certain way. Thus, PDF immunoreactivity fluctuates in a daily manner with peak time early in the morning (Helfrich-Förster *et al.*, 2000; Park *et al.*, 2000a). This "staining cycle" was observed only at the terminals of axons that project from the aforementioned *s*-LN$_v$ neurons into a dorsal region of the adult brain (Fig. 2). Such neurochemical cycling is eliminated by the effects of either *per*-null or *tim*-null mutations (Park *et al.*, 2000a). However, PDF levels were, in general, undisturbed by the latter two clock variants. Remember that *Clk* or *cyc* nulls knock *per* and *tim* product levels way down (in gross adult head homogenates). Therefore, the cell-specific wipeouts of *Pdf* mRNA and PDF, caused by mutationally eliminating either of these positively acting

---

terminals being near the mushroom body (MB) calyces (see Fig. 3). The integrity of the MB structure may be necessary for thorough normality of the behavioral rhythmicity of *Drosophila* (Helfrich-Förster *et al.*, 2002b). The PDF-nonexpressing LN$_d$ perikarya send their (transgenically marked) axons mainly into the dorsal brain (Kaneko and Hall, 2000), but there is also a minor projection that courses toward the LN region (Stoleru *et al.*, 2004). Many of the DN cells also, and in the main, project locally to regions relatively near the locations of *s*-LN$_v$ and LN$_d$ axon terminals. However, certain DN1 cells also (or instead) project fibers to the vicinity of LN$_v$ perikarya and to that of the esophagus (Es). An analogous situation pertains to DN3 neurites, some of which project rather "dorsolocally" (Kaneko and Hall, 2000), with others coursing toward putative LN$_v$ targets (Veleri *et al.*, 2003). This diagram is based largely on one that appeared in Hall (2003a), as augmented by the LN$_d$-to-LN$_v$ and by the DNs-to-LN-region projections that were uncovered more recently. These six clusters worth of clock gene-expressing neurons (including the two groups that are also PDF-containing) reappear with less anatomical detail in Figs. 3 and 4 in terms of functional meanings associated with several of the cell groups. (See color insert.)

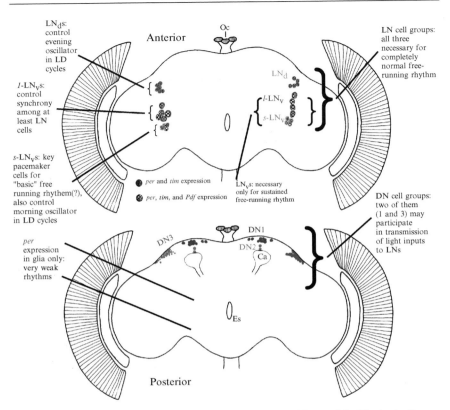

FIG. 3. Neural substrates for the daily locomotor rhythms of *Drosophila*. The basic diagram is from Hall (1998), as augmented by Hall (2003a), and as then modified substantially here. The brain-behavioral responsibilities of the various cell types and neuronal cluster locations diagrammed within the current version of this scheme have been deduced from various points summarized later, with the caveat that this recitation of relevant functions is not comprehensive and that certain features of these illustrative conclusions must be thought of as tentative. In any case, essentially all LN cells are "removed" in a given *disco* mutant individual (Helfrich-Förster, 1998; Zerr *et al.*, 1990); such neurons also have their synaptic-output chrono-functioning ruined by combining the pan-*per* or pan-*tim*-(gal4) drivers with UAS-*tetanus-toxin* (Kaneko *et al.*, 2000b). The three kinds of variants just noted largely genocopy the effects of *per*-or *tim*-null mutations on locomotor rhythmicity in DD (Dowse *et al.*, 1989; Dushay *et al.*, 1989; Hardin *et al.*, 1992; Helfrich-Förster, 1998; Kaneko *et al.*, 2000b; Renn *et al.*, 1999, but see Blanchardon *et al.*, 2001). Proviso: *tim*[0] mutations are from time to time noted as allowing for a bit more locomotor rhythmicity (expressed as the proportion of animals tested that slip into the significantly periodic range), compared with the behavior that is observed routinely for *per*-null mutants (see, e.g., Sehgal *et al.*, 1994; Yang and Sehgal, 2001). PER presence within only a subset of the LNs is mediated by one of the "promoterless" *per* transgenics, discussed in several sections of the text (Frisch *et al.*, 1994), focused attention on LN/rhythmic behavior connections from the opposite perspective

transcription factors (Blau and Young, 1999; Park *et al.*, 2000a), do not occur via $Clk^{Jrk}$- or $cyc^0$-induced lowering of PER and TIM within $s$-LN$_v$ neurons and their larval precursors (*cf.* Allada *et al.*, 1998; Rutila *et al.*, 1998b).

In summary, the *Pdf* gene is clock regulated by two positively acting core transcription factors (but not in terms of mRNA cycling) and PDF is clock controlled by the indirect action of other clock genes (PER and TIM negative transcriptional regulators no doubt indirectly influencing nerve terminal fluctuations of this neuropeptide). The latter effect could have to do with the antigenicity of PDF per se (thus, posttranslational modification of the peptide within the nerve terminals in question mediated as a downstream effect of PER and TIM functions). However, the more conventional hypothesis is that PDF is released cyclically from the dorsal brain $s$-LN$_v$ nerve terminals (as speculated by Park *et al.*, 2000a). But to what effect?—a question that once again is systems related in that we want to know how the neurons within which clock genes and their regulates "feed outward" toward behavioral modulation.

---

(quasi-restoration of normal behavior as opposed to disruption of it). Loosely similar inferences resulted from applying a transgenic type carrying *glass*-gene regulatory sequences fused to PER-coding ones (Vosshall and Young, 1995). Specific elimination of $s$-LN$_v$ and $l$-LN$_v$ cells, mediated by *Pdf*-promoter or *per*-like enhancer-trap driving of cell-killing factors (Blanchardon *et al.*, 2001; Renn *et al.*, 1999), resulted in degrees of arrhythmicity that were not as bad as those caused by the clock-null mutations noted earlier; moreover, LN$_v$-ablated brains caused the flies containing them progressively to lose rhythmicity in DD as opposed to becoming immediately arrhythmic upon proceeding from LD-cycling conditions into constant ones. The behavior of several *disco*-mutant individuals is similar to that of the doubly transgenic type just implied: quasi-rhythmic locomotion early in DD→thoroughgoing arrhythmicity by day 2 or 3 (Wheeler *et al.*, 1993), although the overall penetrance of rhythmic behavior is considerably worse when both LN$_d$s and LN$_v$s are gone (i.e., as caused by either of the two *disco* mutations tested for their effects on locomotor rhythmicity). With respect to the small subset of the latter cluster pair, $s$-LN$_v$s keep getting harped on as "the SCN of the fly" (*cf.* Tischkau and Gillette, 2005), in part because these neurons *may* be able to sustain oscillations of clock gene products in DD better than do the other LN cell types (see text). Concentrating on diel behavior in LD cycles, transgenic ablations of the LN$_d$s or the LN$_v$s, dovetailing with the LN$_v$-specific effects of a *Pdf*-null mutation, led to the supposition that there is differential (neuronal) regulation of "evening" *vs* "morning" oscillators that support locomotor anticipations of lights off *vs* lights on, respectively (Grima *et al.*, 2004; Stoleru *et al.*, 2004). The diagram at the bottom deals with chrono-behavioral functions of DN cell groups, as are mainly explicated within Fig. 4. Also suggested (within this figure) is the possibility of "PER-glial" involvement in behavioral rhythm regulation due to rare *per*$^+$//*per*-null genetic-mosaic individuals (Ewer *et al.*, 1992) in which no LN expression of the normal allele was detectable (also see Method 12); but these flies exhibited weak, long-period rhythms, correlated with the presence of *per*$^+$ in various glia only (depending on the brain region where that allele was retained within the mosaic individual in question). However, *disco* flies exhibit robust and normally cyclical expression of PER in brain glia (Zerr *et al.*, 1990), which therefore is insufficient for routine behavioral rhythmicity in the almost-certain *anatomical* absence of the LN's in this mutant (see earlier discussion). (See color insert.)

In this regard, a further supposition is that certain neuronal targets within the dorsal brain possess PDF receptor molecules, whose daily variations of PDF activation are then transduced into functional modification of the target cells. The latter have not been specifically identifiable, but within these neurons the fly analog of neurofibromatosis-1 (i.e., an NF1-like protein) and interacting intracellular-signaling functions functioning within the Ras/MAPK pathway are inferred to be active (Williams et al., 2001). The clock gene expressing $LN_v$s are manifestly interneurons, as must be the hypothetical dorsal brain target cells just discussed. Therefore, the chronobiological functions mediated or affected by these two sets of brain cells leave us a long anatomical distance from behavioral control, which in this case is affected in the most distal sense by VNC motor neurons and their regulation of walking vs standing still. (This could be an insect equivalent of "sleep vs wake," as discussed later.)

The matter of rest–activity cycles of Drosophila brings us to the second aspect of the involvement of PDF in rhythmicity, which was revealed by another piece of serendipity: the existence of a Pdf-null mutant in a variety of D. melanogaster laboratory strains (Renn et al., 1999). Before its behavior is described, two things: (1) This mutant is relentlessly underdescribed in (20th-century) studies that apply it because the nonsense mutation contained with the Pdf transcription unit (Park and Hall, 1998) knocks out not only PDF itself, but also an inherently featureless and much larger peptide that is encoded upstream of the PDF 18-mer (Renn et al., 1999); the former oligopeptide is coexpressed with PDF in $LN_v$ cells of the adult brain (Renn et al., 1999). (2) Certain dorsal brain neurons stained by application of anti-PDH made against the crab form of this peptide (known as PDH) are completely spurious insofar as bona fide Pdf expression is concerned within the CNS of the fly: the $Pdf^{01}$ mutation wipes out anti-PDH immunoreactivity in all neuronal types mentioned except in the dorsal brain, within which, as well, no Pdf mRNA can be found. Further results confirming this point are that anti-(fly)PDF does not lead to staining of the dorsal neurons in question nor does a certain Pdf-derived transgene drive marker expression in these dorsally located PDH-immunoreactive neurons (Park et al., 2000a; Renn et al., 1999). This side point is not straying too far off track because the transgenic type just mentioned is relevant to the locomotor effects of $Pdf^{01}$ and which PDF-containing neurons would be responsible for rhythmic behavior. In this regard, the implication is that no such responsibilities are necessarily among the job descriptions of the PDH-immunoreactive, dorsally located cells harboring some mysterious material that cross-reacts with PDF.

With this background information and qualifications in mind, it is notable that the PDF-null mutant exhibits quasi-arrhythmicity in constant darkness

(DD): A proportion of the tested adults is aperiodic in this behavior, but other individuals are weakly rhythmic, often in terms of degrading from ephemerally cyclical behavior after proceeding from light–dark cycles into constant darkness, followed by disappearance of periodic locomotion (Renn *et al.*, 1999). It is most important to certify, in terms of core clock functions, that null mutants involving pacemaking functions ($per^0$, $tim^0$, *Clk*-null, or near-null; $cyc^0$) are significantly worse in terms of free-running arrhythmicity than is observed for $Pdf^{01}$. Therefore, other brain neurons aside from those in which PDF is expressed could contribute to overall locomotor control (as discussed in considerable detail later when various neuroanatomical variants and transgene combinations will enter the story), or perhaps factors in addition to PDF are contained in, even released by, the $LN_v$ neurons. The latter notion was undermined by one usage of a *Pdf–gal4* transgene (Method 6), which introduces this molecular–genetic tactic to the story. Thus, the putative regulatory region of the *Pdf* gene (5' to the transcription start site) was fused to DNA encoding the yeast transcription factor GAL4. Another way to effect such local expression is to try and find an "enhancer-trap" strain in which a *gal4*-containing transgene would be produced in only a subset of, for example, the usual *per* pattern (Method 7). Such a strain exists, involving a transposon inserted near some anonymous gene, which leads to expression of markers "mostly in the $LN_v$s" (Blanchardon *et al.*, 2001; see details within Method 7). Shortly thereafter, a description of the utility of this transgenic escalated to the canard that it is an "LNvs-specific *gal4*... driver" (Grima *et al.*, 2002).

Despite these under- and overstatements about the "*per*-like" enhancer trap, these two related molecular–genetic tactics allowed clock neuron-expressing transgenes to be combined with any of a vast number of preexisting (or readily makeable) constructs in which GAL4 targets called upstream activating sequences are fused upstream of useful factors (see Method 6; for an early review of this molecular–genetic ploy, see also Brand and Dormand, 1995). For example, where can *Pdf* expression be inferred to occur in a *Pdf-gal4*/UAS marker combination[12] [the general features of which mixtures are reviewed by Brand (1999) and as was specifically hinted at with regard to what is driven by the aforementioned

---

[12]A note about genetic nomenclature is that doubly or even triply variant *Drosophila* enter into several of the experiments described in this article. When both factors contained, for example, within a double variant type (mutations or transgenes) are located on the same chromosome, the genetic symbols are simply entered one after the other. When separate factors are located on, or inserted into, separate chromosomes, formal rules for the genetics of this organism ask that the two symbols be separated by a semicolon. Because this is ambiguous in terms of gleaning the nature of the text passage in question, this article adapts the mode of using a slash mark to indicate separate chromosomal locations for the genetic factors being described at that moment.

METHOD 6. *Two-component transgene tactics* for monitoring and manipulating gene expression take advantage of the fact that yeast-derived factors are nicely functional in *Drosophila* cells. A review of the early days of applying this "*gal4*/UAS system" was first authored by one of its inventors (Brand and Dorman, 1995) in terms of applying these molecular–genetic methods in neurobiological experiments (therefore on point for the current article). As to molecular constructs in which a transgene of interest can be "driven" by dedicated regulatory sequences, DNA located 5' to the transcription start site and coding region of a gene is subcloned and ligated to the *gal4* gene of *Saccharomyces cerevisiae*; this factor has long been known to participate in activating transcription by binding to upstream activating sequences (UAS) located 5' to the normal forms of GAL4 regulatees in this fungus. It follows that a 5'-flanking-*gal4* fusion transgene should drive the expression of another such factor in which several UAS cassettes (in tandem) are located upstream of a marker-encoding sequence (exemplified in Method 11) or one designed to express some other gene of interest. The latter could include the normal form of, for example, a *Drosophila* clock gene (usually meaning a cDNA that defines its entire open reading frame) or any number of disrupting entities (which would inactivate some cellular function or even ablate the entire cell, as dealt with later in Methods 8 and 9). If this form of the *gal4*/UAS system (compared with the "enhancer-trap" methods described in Method 7) is to work, the 5'-flanking region in question might well have been bioassyed previous to fusing it upstream of *gal4*. For example, potentially complete forms of a clock gene were transformed into the germline of correspondingly clockless host animals and found to "rescue" the effects of the arrhythmia-inducing mutation just implied. In these ways, the 5'-flanking portion of the gene as a whole that is sufficient for normal or near-normal rhythmicity is tacitly identified (e.g., Citri *et al.*, 1987) or a similarly sufficient 5'-flanking region can be identified by fusing such regulatory material to the coding region (only) of that same gene (e.g., Ousley *et al.*, 1998; Rutila *et al.*, 1998a). To let the cat out of the bag, it was *period* and *timeless* regulatory sequences, respectively, that were just referred to. These mutant-rescue findings led to the construction of the requisite *per*-5'-flanking and *tim*-5'-flanking *gal4*-fusion transgenes and transformed *Drosophila* strains (Emery *et al.*, 1998; Plautz *et al.*, 1997a). A word of caution about application of such *per-gal4* and *tim-gal4* transgenics: When combined with various UAS markers (e.g., Method 11), expressions of the GAL4-driven sequences did not always match those of the normal *per⁺* or *tim⁺* spatial patterns. In some of the transgenic lines, it appeared as if *per-gal4*, for example, had "trapped an enhancer" (see Method 7), such that the UAS marker pattern represented a combination of some anonymous enhancer effect along with that of the 5'-flanking sequence of *per* (Kaneko and Hall, 2000). However, for a subset of the transgenic strains in question (involving the same *per-gal4* construct inserted in separate genomic locations), the driven marker signals were the same as one another and better reflected the normal spatial distribution for the PER protein. Nonetheless, some "extra" cells were marked (in this case brain neurons), i.e., the same ones among the (more useful) subset of the transgenic lines. This implied not more unwanted enhancer trapping, but instead that the *per* gene may naturally express its mRNA in certain cells within which the potential protein does not normally accumulate (see, for relevant results and discussion of this issue, Kaneko and Hall, 2000). What amounts to the opposite kind of transgene effect can be achieved experimentally, which means that an inadequate subset of some gene's 5'-flanking region can be deliberately subcloned and fused to *gal4*. This could result in some fraction of the normal gene's spatial pattern (via combination with a UAS marker), which would then allow only those cells to be disrupted (via subsequent combination with the UAS-including transgenes exemplified in Methods 8 and 9). Such "promoter bashing" has frequently been performed in a whole host of biogenetic experiments performed on *Drosophila*. A chrono-genetic example is provided by *Pdf-5'*-flanking-*gal4* fusions; the main text describes usage of such a construct for which GAL4 was expressed exclusively in the normal *Pdf* spatial pattern (at least within the larval and adult CNS). However, a bashed form of the 5'-flanking region in which the "spatially sufficient" 2.4 kb of regulatory DNA was whittled down to 0.5 kb left brain staining for a UAS marker normal but eliminated expression within the abdominal ganglion, where *Pdf* normally generates its mRNA and PDF neuropeptide (Park *et al.*, 2000a).

---

METHOD 7. *Enhancer trapping* methods can be covered in a mercifully brief manner (given Method 6). One uses such transgenes nowadays by almost exclusively applying ones that contain *gal4*. In turn, this usually entails drawing on one or more of the vast collections of "transposon-mobilized" strains of *D. melanogaster* that are extant at various institutions or fly stock repositories. Such strains were generated by genetic crosses (do not ask) designed to mobilize a starting transgene such that it gets reinserted in a huge number of separate genomic locations among the substrains derived. In which of them might *gal4* be activated such that some interesting, useful pattern of driven

UAS-marking would occur? Answer: A considerable fraction of the substrains has been inferable to involve insertion of the *gal4*-containing transposon near enough to some (usually) anonymous enhancer sequence, such that marking occurs "somewhere." Screening through a collection of enhancer-trap transgenic strains can lead one to conclude that the spatial expression pattern *overlaps* that of a gene of interest. The pattern might even be inferred to *match* most of the normal expression domain or correspond quasi-exactly to a *subset* of the "+ allele's" spatial realm. An example is provided by Blanchardon *et al.* (2001), who found a *gal4* enhancer trap that mediates UAS marker expression within a subset of the normal array of PER-containing brain neurons. A word of caution: There is almost certainly not one single enhancer-trap strain in which the desired expression pattern is *solely* where one wants it to be. Blanchardon *et al.* (2001) so divulged with regard to their "partially *per*-like" strain and the fact that various extra, non-PER cells are also marked within the CNS in question. Thus, these chronobiologists divulged that this *per*-like enhancer trap drives expression in approximately one "s-LNv-like cell" ("like" meaning intra-CNS location within a given hemibrain) that is *not* PER containing; and in *ca.* three "LNd-like cells" that also express this *gal4*-transposon but again are PER-less; also, certain cells within the fly head are *gal4* active "at the surface of the medulla [optic lobe] in young adults" (Blanchardon *et al.*, 2001). Additional and worse examples of not-great utility and interpretability of enhancer traps are provided by claims of things like "mushroom body lines," advertised initially to be expressed "predominantly" within that dorsal brain structure (e.g., Connolly *et al.*, 1996), but then later revealed to elicit marker signals in a spectacularly broader number of CNS and other tissue locations (Kitamoto, 2002b). Moreover, it is rare that an enhancer trap strain is assessed for where its *gal4*-mediated marking occurs, in and among various tissues, at all life cycle stages. When combining such a transgene with a UAS disruptor (see Methods 8 and 9), it can be important to know when as well as where the perturbation is induced and whether the stages in question might correspond to the developmental expression patterns of a (normal) gene whose actions may be considered in parallel with the transgene manipulations.

LN enhancer trap]? The answer in this case is the same sites where native *Pdf* mRNA and PDF immunoreactivity are present (Park *et al.*, 2000a; Renn *et al.*, 1999). More to the present point is that when *Pdf-gal4* or the LN$_v$ enhancer trap was combined with UAS cell-killer transgenes,

METHOD 8. *Cell ablation transgenics* are referred to repeatedly in the main text by usage of the vague "UAS cell-killer" term. This means that GAL4 drivees each containing a sequence encoding an apoptosis factor prediscovered to mediate some sort of natural cell death phenomena. These factors are well known from studies of many organisms; reviews pertaining to them, some of which are slanted toward methodological matters, have been presented by Bangs *et al.* (2000), Richardson and Kumar (2002), and McCall and Peterson (2004). It may be that ectopic expression of a gene such as *reaper* (featured prominently in the articles just cited) would ablate cells in which *gal4* causes its products to be made artificially. Therefore, when Renn *et al.* (1999) and Blanchardon *et al.* (2001), for example, used a total of three different UAS cell killers, along with *gal4* drivers that are expressed within subsets of the normal *per*[+] pattern, it was necessary to ask whether the relevant (usually) PER-containing neurons were indeed undetectable in the various doubly transgenic brains. The anatomical validations that resulted allowed for interpretable effects of the cellular ablations on rhythmic behavior, as described in the main text.

the usual PDF cells were indeed ablated (Method 8). However, the behavior of these flies was determined (Renn *et al.*, 1999) or seemed to be (Blanchardon *et al.*, 2001) no worse than the locomotor defect exhibited by *Pdf*-null flies.

## The Multicellular System Operating within the Brain of *Drosophila* to Regulate Behavioral Rhythmicity

It follows that the "brain system" subserving locomotor rhythmicity is not comprised solely of neurons within which clock genes and the *Pdf* one are coexpressed. This notion about broader cellular substrates for this brain behavioral system were gelling more generally from certain earlier observations and from contemporary experiments that extended beyond the PDF case. More recently, additional transgenic-based "CNS dissections" have deepened one's appreciation of the complexity of the system (even in terms of that phrase referring only to the brain-behavioral phenomena on which this section dwells).

It should be repeated that we must get unstuck from the silliness (which is minireviewed relentlessly) that *small* ventrolateral clock gene-expressing neurons are *the* pacemaker cells underlying the rhythmic behavior of *Drosophila*. Thus, flies expressing the effects of *disconnected* (*disco*)

mutations are largely arrhythmic, as follows: (i) most *disco* mutant individuals are $per^0$-like (or $tim^0$-, etc.-like) (Dowse *et al.*, 1989; Dushay *et al.*,
1989; Hardin *et al.*, 1992; Helfrich-Förster, 1998); (ii) with the proviso that
one analyzes a given behavioral record starting 2–3 days into constant
darkness (Renn *et al.*, 1999; Wheeler *et al.*, 1993)[13] and (iii) the great
majority of *disco* flies are devoid of PER immunoreactivity (Zerr *et al.*,
1990) within *all three* of the usual lateral-neuronal clusters (Fig. 2). A part
of this anatomical problem, which almost certainly means that these LNs
are not present as opposed to merely PER-less, has been revealed by
Helfrich-Förster's demonstration in 1998 of no PDF immunoreactivity in
almost all *disco* individuals; these data seem to have encouraged the
misconception that only $s$-LN$_v$s and $l$-LN$_v$s are missing in the brains of this
mutant, compared with reality: the naturally PDF-less LN$_d$s ("d" for
dorsal) are also gone (Zerr *et al.*, 1990). Global disturbance of neuronal
pacemaking functions in the brain leads to distinctly more severe arrhythmia (Kaneko *et al.*, 2000b) than the effects of either the $Pdf^{01}$ mutation or
local ablation of cells that normally express this gene (*cf.* Renn *et al.*, 1999;
Blanchardon *et al.*, 2001); the key disruption was effected by combining a
*tim-gal4* driver that is (indeed) expressed "globally" in nearly all TIM
neurons and the great majority of PER ones (*cf.* Kaneko and Hall, 2000)
by applying a UAS-containing transgene that encodes the tetanus toxin
light chain (TeTxLC) (Method 9). The effect on locomotor rhythmcity was
essentially the same as that of a *tim*-null mutation (Kaneko *et al.*, 2000b). In
contrast, *Pdf-gal4* with UAS-*TeTxLC* combinations (which do *not* ablate
the usual PDF-containing neurons) left behavioral rhythmicity essentially
undisturbed (Kaneko *et al.*, 2000b), the same result as that obtained when
the *per*-like enhancer-trapped driver whose expression is (sort of) limited
to LN$_v$ neurons was combined with this toxin-encoding transgene
(Blanchardon *et al.*, 2001). None of the results currently under discussion
here undermine the behavioral significance of LN$_v$ functions, but instead
suggest that PDF release (if any) from these neurons does not depend on
synaptobrevin (the target of TeTxLC). We will see later (according to the
transgenic experiments described later) that the reason for severe $tim^0$- or
$per^0$-like arrhythmicity in a situation where tetanus toxin was engineered to
be expressed well beyond the LN$_v$s (Kaneko *et al.*, 2000b) likely involved
disrupting synaptic transmission emanating from one further cell cluster,
the LN$_d$s. This would fit with the results of pitting the effects of *disco*

---

[13]However, Blanchardon *et al.* (2001) reported a lower percentage of fly-by-fly arrhythmicity
for this *disco* mutant, which yours truly does not believe, because five separate studies of these
variants' behavior (other than the one reported in 2001) have each found well over 90% of
*disco* individuals to behave arrhythmically in constant darkness.

METHOD 9. *Neuron-inactivating transgenics* typically involve application of UAS-containing transgenes designed to block synaptic transmission from neurons in which some interesting *gal4* driver is expressed [exception: see review by Nitabach *et al.* (2005) about neuronal excitability disruptions]. The two kinds of synaptic blockers used are sequences encoding a tetanus toxin light chain (which will so block but cannot spread to other neural regions, i.e., outside the domain of the *gal4* driver) or the mutated form of a *Drosophila* protein called dynamin (which functions in many organisms on behalf of synaptic vesicle recycling). Methods revolving round the TeTxLC tactic are reviewed by Martin *et al.* (2002), along with summaries of many "brain-behavioral" results obtained from application of this transgene type. Similarly, Kitamoto (2002a) reviewed the invention of the *shibire*-mutant ploy aimed at dynamin disruption. This UAS-*shi*$^{TS}$ factor comes into play briefly in one chronobiological experiment mentioned later (footnote 30). It is notable that the temperature-sensitive (*TS*) feature of the *shibire*-based method allows synaptic transmission to be turned off by heat treating the animals at any life cycle stage. This has almost always involved exposing adults to a "nonpermissive" high temperature, followed by assessments of the "acute" behavioral effects (Kitamoto, 2002a). In contrast, a *gal4* driver expressed during more than one stage—as the great majority of such factors almost certainly are—will expose neurons to the damaging effects of TeTxLC "unconditionally." This can lead to neuroanatomical problems during the development of such doubly transgenic *Drosophila*, undermining the interpretability of synaptic transmission blocking that would also be occurring as the behavior of an eventual adult fly is observed. Such interpretational problems are discussed by Martin *et al.* (2002), in part by referrals to those authors' own induction of tetanus toxin upsets within various regions of the *Drosophila* nervous system during developing and adult stages.

mutations against the more limited behavioral ones of *Pdf*-controlled LN$_v$ ablations (see earlier discussion).

These findings—reflecting an ensemble of gene expression determinations (Method 10) and behavioral variant effects—form a platform for considering the next set of observations: Many of them stemmed from the application of rhythm-related transgenic types—either to further assessments of where the clock genes are expressed or with the hope that additional kinds of engineered abnormalities would be behaviorally informative. In these ways, a fuller understanding of the brain-behavioral system

METHOD 10. *Tissue expression assessments* have relied on two methods to determine where clock gene products and related molecules are made: *in situ* hybridization (ISH) and immunohistochemistry (IHC). Early application of ISH methods gave RNA signals within tissues, especially within the cells they contain, that suffered from low signal to noise and poor resolution (e.g., James *et al.*, 1986; Liu *et al.*, 1988; Saez and Young, 1988), although occasionally the chronobiologically relevant images were better interpretable (e.g., Bargiello *et al.*, 1987; Lorenz *et al.*, 1989). As histologists graduated to usage of ISH probes that give enzymatic "color reaction" or fluorescent readouts, the image qualities associated with rhythm-related tissues and cells have improved (e.g., Beaver *et al.*, 2002; Peng *et al.*, 2003; Rachidi *et al.*, 1997; Rouyer *et al.*, 1997; Yang and Sehgal, 2001), although this is not always the case (e.g., So *et al.*, 2000). Some of these protocols allow for double labeling of two gene products expressed within a given chrono-relevant cell type (Zhao *et al.*, 2003). This brings us to IHC methods, starting with an example of double labeling for the presence of both a protein emanating from one rhythm-connected gene and the mRNA from another (Yang and Sehgal, 2001; *cf.* Sauman and Reppert, 1996). Immunogens created to elicit mammal-produced antibodies have entailed synthetic peptides (e.g., Siwicki *et al.*, 1988; Park *et al.*, 2000a), or recombinantly produced fusion proteins (allowing for semipurification of, for example, a clock protein fragment bound covalently to some carrier factor, e.g., Bargiello *et al.*, 1987; Myers *et al.*, 1996; Reddy *et al.*, 1986), or essentially the entirety of a given clock protein as manufactured in transformed or transfected heterologous cells (e.g., Edery *et al.*, 1994; Liu *et al.*, 1992; Stanewsky *et al.*, 1997a). It is impossible to mention all the studies that have "antibody-stained" proteins with rhythm relevance to *Drosophila*; but see Siwicki *et al.* (1988), Zerr *et al.* (1990), Frisch *et al.* (1994), Stanewsky *et al.* (1997a), and Kaneko and Hall (2000) for some examples with which the author is familiar (these studies represent more than one immunogen type and signal read-out method). A couple of additional studies are notable, because one of them involved the only electron microscopic determination of the intracellular localization of a clock protein within various cell types of the fruit fly head (Helfrich-Förster *et al.*, 2002a; Liu *et al.*, 1992; Yasuyama and Meinertzhagen, 1999). The other study (Ewer *et al.*, 1992) exemplifies double labeling for two separate proteins within so-called clock neurons, in this case to demonstrate that putative PER neurons indeed coexpress a neuronal-only protein within various brain cells and within compound eye photoreceptors (see Fig. 2), but that clock protein lacks the neuronal marker within a host of other CNS cell types (see, e.g., Fig. 3). The marker in question is the ELAV-binding protein (Robinow and White, 1991; Yao and White, 1994), the gene for which has also been used to make the "pan-neuronal" *gal4* driver type referred to in certain passages of the main text (see Myers *et al.*, 2003; Stoleru *et al.*, 2004; Yang and Sehgal, 2001). Note, however, that the likely expression of *elav* "in all neurons" and "only" them encompasses PNS as well as CNS locations; also putative neuronally ubiquitous production of GAL4 by an *elav*-derived driver does not mean that the UAS-encoded polypetide will *accumulate* within all central and sensory neurons (Helfrich-Förster *et al.*, 2000). Finally, it should be mentioned methodologically that most of the early tissue expression assessments being microreviewed here involved serial frozen sections of whole animals (e.g., Ewer *et al.*, 1992; Zerr *et al.*, 1990), whereas the more recent trend is to dissect out CNS specimens and assess gene product or marker patterns in whole-mounted material (e.g., Blanchardon *et al.*, 2001; Peng *et al.*, 2003).

underlying *Drosophila* rhythmicity has emerged, embracing information about the roles played by differentially located neurons or even brain cell types, but also including some puzzles or a least a sharper focus on "what we do not know" (Blau, 2001). Thus, the anatomical outputs from brain neurons expressing *per* and *tim* have been revealed, in stages: (i) PDF-immunoreactive axons project from $s$-LN$_v$s into the aforementioned dorsal brain region (Fig. 2); similarly determined fiber pathways project from $l$-LN$_v$s across the brain midline, terminating near the bilaterally symmetrical, contralateral cluster of ventro-LNs, and out into the optic lobes (in other words, it may be that the axons of certain individual $l$-LN$_v$ cells form a minicommissure, whereas others send their neurites in a centrifugal direction only); (ii) *per-gal4* or *tim-gal4* drivers, when combined with UAS marker transgenes designed to fill axons as well as neuronal cell bodies (Method 11), confirmed the patterns just described, as well as showing that LN$_d$ axons project to a dorsal brain region near to (or the same as) the $s$-LN$_v$ "target," along with revealing a semicomplex "wiring diagram" for the dorsal neurons (DNs) that naturally contain PER and TIM; and (iii) more recently uncovered neurite pathways emanating from certain of these neuronal clusters involve axons projecting from a subset of the DN3 cells (Fig. 2) toward the LN$_v$ region (Veleri *et al.*, 2003) and to the latter region from certain LN$_d$ cells (Stoleru *et al.*, 2004).

The DNs have been known for some time to comprise the majority of brain neurons in which these clock genes are expressed (*ca.* 60 pairs of DNs

METHOD 11. *Neurite-marking projections* emanating from circadian pacemaker neurons can be permitted by the fortuitous discovery (Helfrich-Förster and Homberg, 1993) of a factor that is coexpressed with clock factors in certain neurons (Helfrich-Förster, 1995). As was just alluded to and as mentioned within several sections of the main text, immunoreactivity of the PDF neuropeptide is an axonal as well as a cell-body marker for a subset of PER/TIM neurons in the brain (also see Figs. 2 and 3). More general marking of neurites—mainly meaning *axons* in terms of the neuronal process that have been scrutinized and interpreted—has been chrono-neuronally accomplished by combining *gal4* drivers (see Methods 6 and 7) with UAS-containing trans-genes designed to produce "axon fillers." If a driver seems "strong," then combining it with a UAS-*lacZ* factor can lead to expressing of (*Escherichia coli*-derived) β-galactosidase throughout most of the neur-ites' extents. See, for example, Park *et al.* (2000a) who used a UAS-*lacZ* that encodes "cytoplasmic" β-GAL and read out the result of the enzymatic activity of that factor (there are other UAS-*lacZ* constructs engineered to restrict a β-GAL fusion protein to the nucleus; this kind of intracellular marking can be read out by virtue of the activity of the galactosidase or immunoreactivity of that bacterial protein; the latter method has also been applied to cytoplasmic β-GAL). Other kinds of neurite markers are provided by UAS-*tau*-containing transgenes (de-signed to produce a mammal-derived form of this microtubule-associating protein, which fills neuronal processes, by itself or as fused to another polypeptide such as β-GAL). Examples of UAS-*tau* applications in a chronobiological context are salted throughout the report of Kaneko and Hall (2000). Around that time, however, it was discovered that TAU protein can disrupt the normal morphology of at least some *Drosophila*, neurons in which its presence has been affected gratu-itously (Williams *et al.*, 2000), although one must say that the pattern of PDF immunoreactivity (with regard to the LN$_v$ neuronal processes that are diagrammed in Fig. 2) seemed exactly to match the antibody-mediated TAU staining that was observed in *per-gal*/UAS-*tau* and *tim-gal*/UAS-*tau* transgenic brains (Kaneko and Hall, 2000). Nevertheless, investigators in this field have gone over to marking rhythm-related neurites mostly by combining a given *gal4* driver with UAS-*gfp* transgenes. A primitive such type was applied by investigators such as Kaneko and Hall (2000), Helfrich-Förster *et al.* (2002a), and Veleri *et al.* (2003) in the sense that the GFP they set up to be produced under the control of a given rhythm-related driver was plain old green fluorescent (GFP) protein. It did lead to quite good marking of the processes radiating from PER- and TIM-containing neurons. It could be that the newly observed GFP-marked axonal tracts emanating from certain such cell bodies (see Fig. 2) were TAU damaged in conjunction with the earliest assessments of the rhythm wiring diagram of the brain (see earlier and Kaneko and Hall, 2000). A so-called enhanced GFP has also been used in combinations of *gal4* drivers with UAS-*(e)gfp*, which encodes an *in vitro*-mutagenized form of the fluorescent protein designed to emit extra-intense signals (for chronobiological examples, see Emery *et al.*, 2000a, and Zhao *et al.*, 2003). Another engineered form of GFP has come to the fore, thanks to the efforts of Lee and Luo (1999, 2001); these *Drosophila* neurobiologists fused a membrane-targeted (mt) protein to GFP ("mt" is a mouse lymphocyte polypeptide from mouse called CD8), resulting in fine filling of the animal's neuronal processes, perhaps maximally promoting the transport of MT-GFP all the way to the axon terminal. However, to the author's knowledge, only Stoleru *et al.* (2004) have used an UAS-*(mt)gfp* transgene (which had been generated by Lee and Luo, 1999) in CNS markings involving the rhythm system of the fly, in this case to "uncover" a newly appreciated $LD_d$-to-$LN_v$ axonal pathway (see Fig. 2).

compared with *ca.* 15 pairs of LNs), but the word "these" raises the question as to which pacemaking factors are expressed within a given brain cell. In this regard, (i) not all LN or DN cells could be shown to coexpress *per* and *tim* (Kaneko and Hall, 2000); (ii) the *double-time* (kinase) gene may be expressed in all such neurons and in a host of additional CNS cells (Kloss *et al.*, 2001); (iii) both *vri* and *tws* genes make their products (at least as inferred from transgene-reported marking) in rather widespread patterns within the brain other than their colocalization within certain pacemaker neurons (respectively, Blau and Young, 1999; Sathyanarayanan *et al.*, 2004); however, (iv) the *Andante* and *Timekeeper* (kinase) genes exhibit more limited patterns of expression, i.e., within the $LN_v$s and two (Lin *et al.*, 2002) to eight (Akten *et al.*, 2003) ill-defined, more dorsally located brain neurons; (v) information about where *other* clock genes make their products within the adult brain is pathetic or at best stems from crude observations in which one's scrutiny has been limited to the "LN region" (worse-case states of knowledge in this regard: tissue expression patterns for the $Clk^+$ and $cyc^+$ alleles, as reviewed by Hall, 2003a).

In this regard, *Clk*- or *cyc*-null mutations cause minimal, but above-zero, levels of *per* and *tim* products in head homogenates (see earlier discussion) but result in apparently normal levels of the latter two genes' expression within certain brain neurons (Kaneko and Hall, 2000). This makes one wonder whether the $Clock^+$ and $cycle^+$ genes are (naturally) expressed at all within such neurons. A neuropeptide gene newly isolated from *Drosophila*, of which the rhythm system of the animal may take advantage, is on point: Sequences encoding a corazonin oligopeptide are expressed within a variety of CNS regions, although in a low number of cells; these patterns were undiminished spatially or quantitatively in $Clk^{Jrk}$ mutant larvae or in adults expressing that mutation or a *cyc*-null one (Choi *et al.*, 2005). However, *Crz* gene expression resulted in two ectopically labeled cells within the adult brain of either mutant type. Are CLK and CYC normally present within these neurons, such that their removal alleviates normal inhibition of *Crz* activation in these locations? This is unknown. With regard to *per* and *Crz*, coexpression of these two genes was observed in a few dorsal brain cells, involving a small handful of PER DNs. However, neither *per*- nor *tim*-null mutations affected the normal CRZ cell pattern (Choi *et al.*, 2005). CLK and CYC *could* be possessed by these dorsally located cells, in which mutations eliminating either of those transcription factors would presumably bring down *per* and *tim* product levels (*cf.* Allada *et al.*, 1998; Rutila *et al.*, 1998b), but the noneffects of eliminating *Clk* and *cyc* functions directly on *Crz*-encoded material make this a moot point.

DN cells, and clock factors functioning within them, are neither necessary nor sufficient for free-running rhythmicity because (i) a truncated *per* gene (lacking 5′-flanking material) led to expression only within a subset of the LNs, which allowed for rather robust free-running rhythmicity (Frisch *et al.*, 1994); (ii) *disco* brains retain DNs and cyclical clock gene expression within them (Blanchardon *et al.*, 2001; Kaneko and Hall, 2000) but do not support anything like routinely rhythmic locomotion (see earlier discussion); and (iii) a different kind of 5′-truncated *per* construct led to solidly rhythmic expression of the gene within DNs only, but (again) no rhythmic behavior in DD (Veleri *et al.*, 2003).[14]

The *period* gene is expressed within hundreds of glial cells within central brain and optic lobe ganglia (reviewed by Hall, 2003a; Helfrich-Förster,

---

[14]By the way, the DN-limited expression of this particular promoterless-*per* construct makes absolutely no sense against a background of 8-kb subsets of the gene that have been shown to rescue the DD arrhythmicity that would (otherwise) have been caused by *per*-null variants in the genetic background of the relevant transgenics (Hamblen *et al.*, 1986; Zehring *et al.*, 1984). These findings stemmed from the early days when it was unknown that the 5′ end of this DNA fragment is located within the first intron of *per*, upstream of the first coding exon. This (intronic) restriction site is the self-same location of the 5′ end for the promoterless *per-luc* fusion gene of Veleri *et al.* (2003). The behavioral efficacy, in free-running conditions, of the plain-old 8-kb *per*⁺ fragment continued to be demonstrated (Yu *et al.*, 1987), including a study that showed this piece of DNA *not* to be limited in its brain expression to dorsal brain regions (Liu *et al.*, 1991). Let us bear in mind for later that the dorsal brain expression of the *per*-8-kb-*luc* fusion construct of Veleri *et al.* (2003) not only produces almost completely full-length PER protein, but also that its DN expression pattern is correlated with efficacious behavioral control in certain circumstances, although not constant darkness, as is discussed later (*cf.* Fig. 3).

2003) [also see Kaneko and Hall (2000) for similar information about "TIM glia"]. PER immunoreactivity in such glia cycles just as nicely as it does in the relevant neurons (Zerr *et al.*, 1990), and many of these non-neuronal brain cells decorate the axons projecting from PDF-containing LNs (Helfrich-Förster, 1995). Do glia containing clock gene products contribute to the regulation of rhythmic locomotion? These two factors (anatomical and gene expressional) are neither necessary nor sufficient for such behavior, because no PER was detectable within glia in the aforementioned 5'-less but behaviorally rhythmic *per* transgenic (Frisch *et al.*, 1994), and because PER cycling in glial cells was shown to be normal in LN-less *disco* brains (Zerr *et al.*, 1990). However, behavioral analysis of a series of genetic mosaics (Method 12), in which part of each fly was *per*+, the remainder *per*⁰ (Ewer *et al.*, 1992), suggested that functioning of the normal allele in "glia only" could allow for weakly rhythmic behavior. No LN *marking* of the normal allele's expression was detectable in such

---

METHOD 12. *Genetic mosaics* applied in brain-behavior chronobiologica experiments. This phrase usually refers to the creation of *Drosophila* in which parts of the nervous system are normal for the expression of a given rhythm gene, with the remainder of such an animal being rhythm mutated. In turn, most such mosaics are generated nowadays by combining transgenes. Implicit within Methods 6 and 7, and as mentioned in certain main text sections, is the generation of "partially *per*" flies; for example, one can (and has) combined a *Pdf-gal4* driver with a UAS-*per*+ transgene in flies whose genetic background includes a nonfunctional *per* gene. (Therefore, how solid or perhaps subtly defective will the rhythmic behavior of the fly be?) Other mosaics of the more classical type have been generated by "loss" of X chromosomes early in the development of *Drosophila*. Examples in the rhythm area are provided by Konopka *et al.* (1983, 1996) and Ewer *et al.* (1992). Several details of these procedures are reviewed by Price (2004). Absent any full description of the genetic tactics and outcomes, suffice it to say here that if an X chromosome containing a dominant or semidominant allele of a clock factor (such as *per*+, which is essentially dominant to the effects of a *per*-null mutation, or *per*ˢ, which exhibits semidominance vis-à-vis *per*+) is "unstable," it can get lost during mitotic divisions occurring within a very young embryo. If that animal starts out as an XX female—the other of whose X chromosomes contains the recessive allele (e.g., *per*⁰)—then something like half or one-quarter of the eventual fly will have the effects of the mutation exposed (i.e., in all single X cells and tissues derived from the embryonic nucleus that lost the other X). One way to symbolize such a mosaic animal is $per^+/per^+//per^0$, with the double slash distinguishing the genetically normal diplo-X tissues from the $per$-null haplo-X ones. The question is: In which neural locations are the results of $per^+$ action necessary and sufficient for normal loco-motor rhythmicity? Brain-behavioral analysis of many such mosaic in-dividuals is necessary because each one exhibits a unique distribution of normal $vs$ mutant tissue (exemplified by Ewer $et\ al.$, 1992). This is in contrast to the transgenically produced mosaics noted earlier; each doubly transgenic individual possesses the same internal pattern of rhythm-enabling $vs$ clock-mutated cells.

$per^+//per^0$ individuals (Ewer $et\ al.$, 1992); but those neurons were $present$, as opposed to the case of $disco$ (Helfrich-Förster, 1998; Zerr $et\ al.$, 1990).

We must pause for a moment before enumerating further experiments involving these neural substrates, because of various matters that arose earlier in context of such items being inadequately investigated, let alone densely presented in the foregoing paragraphs. For example, known or suspected brain behavioral functions of certain cell types are now on the table, but there is only negative evidence so far for the DNs and the clock gene products they contain. Thus, one residual issue involves additional information obtained to elucidate functional meanings of the separate neuronal groups. A second such concern connects (literally) with the ostensible attachments between and among various of the neuronal groups—the wiring diagram that may coordinate crucial elements of this brain system's chronobiological functioning. This second issue will be taken up first in what follows, because it also speaks to job descriptions for the different LN cell groups. Thus, the axons of $s$-LN$_v$s may terminate on or near dorsal brain cells acting within a key output pathway (Fig. 2), even the major one that eventually feeds into locomotor control. This PDF-containing anatomical conduit could involve pure throughput along the way to thoracic regions of the CNS and thus to the eventual control of walking behavior (as inferred, for example, by Park $et\ al.$, 2000a; Renn $et\ al.$, 1999; Williams $et\ al.$, 2001). However, the neuropeptide just noted also might meaningfully communicate between $l$-LN$_v$s and the contralater-al LNs (Fig. 2). These hypothetical signals could act as a go-between that contributes to intrabrain synchrony between bilateral pacemaker structures. Absent this hypothetical feature of the function of PDF, the flies would slip into arrhythmicity in constant darkness, which they do, as opposed to becoming immediately aperiodic in DD (Renn $et\ al.$, 1999), because those LN structures are causally slipping out of phase with respect to each other. Petri and Stengl (1997, 2001) presented and formally analyzed brain-behavioral findings on this point derived from PDF-related studies of cockroach rhythms. Closer to home, it is interesting that a supposedly eyeless and ocelliless *sine oculis* mutant (see legend to Fig. 4) is actually more pleiotropically defective: many such *so* individuals also lack the posterior optic tract that interconnects LN cell groups (Helfrich-Förster and Homberg, 1993; *cf.* Fig. 2). Concomitantly (it must be), a fair fraction of *so* mutant flies that have been behaviorally monitored in constant darkness exhibited "splitting" of their free-running locomotor activity into two separate free-running components with different circadian period values (Dushay *et al.*, 1989; Helfrich, 1986).

Therefore, the dorsally coursing axons from the smaller ventrolateral neurons may have nothing to do with the locomotor rhythm defect exhibited by $Pdf^{01}$ *Drosophila*. In this regard additional, provocative findings from the analysis of $Pdf^{01}$ brains (*cf.* Renn *et al.*, 1999) were obtained by Peng *et al.* (2003), starting with their demonstration that *tim* mRNA rhythms normally persist for several days in DD within several of the neuronal types normally expressing this gene. They also exhibit persistent free-running rhythmicity within various neuronal clusters that naturally express the products of the *cryptochrome* gene (which is dealt with mostly in a subsequent topic). Parallel histochemical observations of PDF-null brains showed that these RNA oscillations disappeared gradually over the course of several DD days (Peng *et al.*, 2003). Therefore, widely dispersing neurochemical communications among neurons containing chrono-gene products seem to be required for the maintenance of intercellular synchrony or for rhythmicity. These mutationally induced defects were found for neurons that naturally contain this peptide and/or putative PDF-containing axonal inputs to them, as well as for other cell groups (LN$_d$s, DNs) that are descriptively unrelated to the expression of this neuropeptide. Moreover, within the LN$_d$s the *cry* mRNA rhythm amplitude was crushed in the $Pdf^{01}$ mutant, in that levels of this transcript did not cruise along at some intermediate level, compared with the normal cycling, but instead were constitutively low (Peng *et al.*, 2003). At least superficially, this feature of the results is not consistent with analogous findings from Lin *et al.* (2004). Flies suffering from the effects of the same $Pdf^{01}$ mutation were shown to "maintain" free-running PER rhythms of inferred abundance and nuclear translocation within $s$-LN$_v$ cells, although the amplitude of the cycling of this clock protein diminished within the LN$_d$s after a week in DD. [mRNA levels or *per* oscillations were not monitored, as Peng *et al.* (2003) had effected for products of the other rhythm-related genes.] The key result obtained by Lin *et al.* (2004) was that brain cycling of PER displayed progressive dispersion of intercellular phasing of immunohistochemically inferred abundance oscillations in constant darkness, during which no external time giver (Zeitgeber) was operating to keep the separate cells in sync.

Nevertheless, both of these studies indicate that PDF-mediated communication among clock neurons is important for the "brain system" to function normally. In this regard, some of these PDF-mediated interactions among neurons may involve local paracrine messages as opposed to humoral brain bathing. Thus, Peng et al. (2003) used a labeled form of this substance to reveal PDF binding within the s-LN$_v$ cluster, within a potential target of contralaterally extending l-LN$_v$ axons (see earlier discussion) and possibly in the vicinity of DN3 neurons (which could be a target of dorsally projecting s-LN$_v$ neurites; see earlier discussion). An additional and highly specific element of PDF-mediated influence on sustained DD cycling was inferred from the scrutiny of DN1 neurons (Klarsfeld et al., 2004): rapid dampening of PER oscillations in those dorsal cell bodies was observed soon after Pdf mutant flies were transferred from LD to DD. Recall that some of the PDF-containing lateral neurons indeed project neurites to within hailing distance of DN1 cell bodies [as reviewed by Helfrich-Förster (2003) and depicted within Fig. 2]. A bonus from the study of Peng et al. (2003) did not involve effects of the Pdf mutation, but nonetheless further disabuses us of an overly intense focus on the LN$_v$ neurons acting alone. This extra experiment involved restoration of normal cycle gene function solely within these ventrolateral cells, i.e., in Pdf-gal4/UAS-cyc$^+$ transgenic flies whose genetic background included homozygosity for a cyc-null mutation; these flies behaved in an aperiodic manner in DD (Peng et al., 2003).

These results speak to the brain as a broadly functioning coordinated system for behavioral control. This scenario was presaged (i) by the brain behavioral effects of a sine oculis mutation (Helfrich, 1986; Helfrich-Förster and Homberg, 1993): frequent lack of connectivity between contralateral LN cells and splitting of free-running locomotion into two periodic components (not a feature of Peng and co-workers' findings, although the neurochemical mutation they applied should more widely disrupt communication among pacemaker neurons); and (ii) by the fact that intracellular cycling of clock factors can drift out of synchrony within an individual fly brain (Helfrich-Förster et al., 2001), absent in this case the coordinating influence of external stimuli (see later). There are, however, results that undermine the notion that communication among neurons containing clock proteins (or them along with PDF) is essential for all features of free-running molecular cycling within these cells: Mutant disco brains, in which no outputs would be conveyed from lateral neurons, let alone from PDF-containing ventrolateral cells, exhibit oscillations of

PER immunoreactivity within the dorsal neurons that normally contain this protein. These DN staining cycles were observed in this mutant not only in LD cycles (see earlier discussion) but were also sustained over the course of 5 days in constant darkness (Veleri *et al.*, 2003).

Further genetic dissections of the behaviorally significant tasks performed by separate clusters of supposed clock neurons concentrated on locomotion in more natural conditions: laboratory mimics of LD cycles. For this, additional kinds of rhythm-related *gal4* drivers came into play, both in terms of gene regulatory sequences and the nature of the yeast-derived transcription factor (Stoleru *et al.*, 2004). [Quite similar experiments were performed and results obtained by Grima *et al.* (2004), but involving somewhat different molecular-genetic tactics, by Grima *et al.* (2004).] First, consider that a *cry-gal4* driver (against a background of the brief information given earlier about the brain expression of *cryptochrome*) promotes UAS marker expression in all $LN_v$ and $LN_d$ cells, along with two neurons with the DN1 cluster (*cf.* Fig. 2). Combining this *cry-gal4* with a UAS cell killer (see earlier discussion) led to diel behavior (in LD) that mimicked the locomotor anomalies exhibited by clock-null flies such as *per⁰*: no behavioral anticipation of either D-to-L or L-to-D transitions [*cf.* Wheeler *et al.* (1993), but see later for discussion of Helfrich-Förster (2001)]. In contrast, *Pdf-gal4*/UAS cell-killer transgenics, which are solely $LN_v$-less, exhibited "clock-like" anticipatory behavior, especially of dusk, in this condition (Renn *et al.*, 1999). These results were confirmed here by Stoleru *et al.* (2004), who went on to show that both *Pdf-gal4*/UAS cell-killer and *Pdf⁰¹* singly mutant flies exhibit diminished dawn-anticipatory behavior (also implied in certain data displayed by Renn *et al.*, 1999). As would be expected, the *cry-gal4*/UAS cell-killer flies behaved aperiodically in DD as if they were thoroughgoingly clockless anatomically; this would make them behave like *per⁰*, which (stated again) goes arrhythmic immediately entering DD (e.g., Wheeler *et al.*, 1993), in contrast to *Pdf-gal4*/UAS cell-killer or *Pdf⁰¹* variants. We pause for a moment to realize once more that retention of (most) DN neurons in the severely abnormal transgenic type, in which the *cry-gal4* driver-induced neuronal ablations in a rather widespread manner, is insufficient for normal locomotor rhythms.

Really digging into brain behavior dissectional experiments took advantage of a GAL4 inhibitory factor encoded by *gal80* sequences taken from yeast (Method 13). Therefore, *cry-gal4* was combined with a newly constructed *Pdf-gal80* transgene aimed at leaving GAL4 active within $LN_d$s and the two DN2 cells, but blocking its action within $LN_v$s. Triply transgenic flies carrying *cry-gal4/Pdf-gal80*/UAS cell killer were, in the first place, "specifically" ablated of their $LN_d$ cells (and presumably of the

METHOD 13. *GAL4 inhibitory effects of GAL80* were initially introduced into *Drososphila* gene manipulation systems by Lee and Luo
(1999), as reviewed by these investigators in 2001. In the context (originally) of genetic mosaics (*cf.* Method 12) produced by mitotic recombination, an *S. cerevisiae*-derived *gal80* sequence was found efficaciously to
produce its "GAL4-blocking" protein, which antagonizes GAL4 activity
by binding to the activation domain of GAL4; this prevents interaction
between GAL4 and its target regulatory DNA sequences. [See Ma and
Ptashne (1987) and description of UAS in Method 6; the reference just
entered is provided not solely for a yeast historical reason, but also
because the name of this paper's second author happened to be the
inspiration for the shortest *period* mutant known, referring to its free-
running cycle durations (Konopka *et al.*, 1994).] Drawing on these GAL4/
GAL80-based results (yeast findings and those stemming from the constructs of Lee and Luo, 1999), the inhibitory factor can also be introduced
into flies carrying a given *gal4* driver along with a newly designed
regulatory-DNA *gal80* transgene. For example, if the regulatory sequence in the relevant *gal4*-containing factor is expressed in a relatively
broad pattern of brain neurons, but a sequence of this sort, derived from
another gene and about to be fused to *gal80*, had been predetermined to
mediate expression in a known subset of such cells, then the driver
combination at hand will result in activation of whatever UAS-containing
transgene is of interest only in the *gal80*-less cells, i.e., in a configuration
that conceptually subtracts the *gal80* expression pattern from that of the
*gal4* driver (see Stoleru *et al.*, 2004).

smaller number of those pesky DNs); and, second, their behavior indicated
suppressed anticipation of the environmental transition in the evening
(Stoleru *et al.*, 2004). Previous to the performance of these experiments,
behavioral significance of the more dorsally located LN cells was appreciable only from indirect evidence (e.g., "adding" the behaviorally more
severe effects of the elimination of *disco* of all lateral neurons to the milder
effects of LN$_v$ disruptions in *Pdf* variants).

Narrow-sense conclusion: Functions of LN$_v$ cells underlie regulation of
the "morning" component of the diel behavior of *D. melanogaster*, whereas
LN$_d$s independently affect significant contributions to "evening" activity
(Grima *et al.*, 2004; Stoleru *et al.*, 2004). Flies of this species classically
exhibit such bimodally rhythmic patterns of locomotion in LD cycles
(e.g., Hamblen-Coyle *et al.*, 1992). In free-running conditions, however, only
the evening locomotor peak tends to persist (reviewed by Hall, 2003a).
Interestingly, however, the LD/evening-peakless *cry-gal4/Pdf-gal80/*UAS

cell-killer flies not only exhibited solid rhythmicity in DD, but the behavioral patterns also indicated persistence of only the LD morning peak, i.e., as such $LN_d$-less flies proceeded from cyclical to environmentally constant conditions (*cf.* Grima *et al.*, 2004; Stoleru *et al.*, 2004). Furthermore, the $LN_v$-less type (see earlier discussion) sustained only an evening peak of locomotion and only during early stages of its DD monitoring (consistent with the diel behavior of *Pdf-gal4*/UAS cell-killer flies and their inability to remain rhythmic for long in constant conditions).

These transgenic manipulations went on to include intrabrain dissection experiments opposite to the kinds reviewed earlier, i.e., now involving neuronally effected "rescues" of rhythm defects. For this, a UAS-*per$^+$* construct was applied by Stoleru and co-workers (as had been used previously in various *gal*-driver contexts by Blanchardon *et al.*, 2001; Kaneko *et al.*, 2000b; Yang and Sehgal, 2001). Here, a pan-neuronal driver was applied [which had first been used chronobiologically by Helfrich-Förster *et al.* (2000) (see earlier discussion)]; when it was combined with UAS-*per$^+$* in flies that were otherwise *per$^0$*, rhythmicity was restored [confirming Yang and Sehgal (2001), as comes into play later], but blocking *per$^+$* expression within $LN_v$s and $LN_d$s by inclusion of yet another new construct—*cry-gal80*—reversed this behavioral rescue (Stoleru *et al.*, 2004). When the aforementioned *Pdf-gal80* transgene was combined with the pan-neuronal driver—to limit the mutational effect of *per$^0$* to $LN_v$ cells and create a putative genotypic mimic of *Pdf*-mediated ablation of these neurons—the behavioral effects of this doubly transgenic type was indeed the same (Stoleru *et al.*, 2004; *cf.* Renn *et al.*, 1999). However, in LD, pan-neuronal-*gal4* with *Pdf-gal80* flies exhibited locomotor anticipation of both lights on and lights off, unlike the behavior of $LN_v$-ablated flies (see earlier discussion). Stoleru and co-workers thus inferred "recovery" of an $LN_v$ output function, which is possible in pan-neuronal/*Pdf-gal80* flies, inasmuch as they retain those neurons anatomically. This induction of $LN_v$ functionality was further inferred to emanate from the $LN_d$s, which are not only present but also *per$^+$* enabled in the doubly transgenic type just noted. In other words, the inherently clockless $LN_v$s might somehow be "directed" by the $LN_d$s to regulate normal behavior around the time of dawn. This supposition was correlated with an inclusion in this study of a "new" intrabrain anatomical pathway (unobserved in the Kaneko and Hall study that established the first blush clock neuronal wiring diagram for this part of the CNS), i.e., a putative efferent projection from one or more $LN_d$ neurons to the region where $LN_v$ cells are located (Stoleru *et al.*, 2004; *cf.* Fig. 2).

Broader-sense conclusion: There is a "two-oscillator system" functioning within the *Drosophila* brain on behalf of temporally separate features of the rest–activity cycles of the fly. The dissectable anatomical components of this system are regulated by the same intracellular entities (such as *per*, *cf.* Yoshii *et al.*, 2004). However, the discrete structures use different output entities—attention on which is focused by lack of knowledge about inter-cellular communicators that are presumably contained in dorsolateral clock gene-expressing neurons, compared with what we know about PDF in ventrolateral such cells.[15] Whereas the separate oscillator structures can "independently" affect temporally discrete features of daily rhythmicity, another output issue is that these neuronal pacemakers may be meaning-fully "coupled" (Stoleru *et al.*, 2004). This would jibe in general terms with the results of other recent studies, described earlier.

However, is there something more or different about a two-oscillator system for the locomotor rhythm of *Drosophila*? This could be so, as the ensuing passages will outline—briefly—because this substory is largely devoid of brain-behavioral findings (although see Yoshii *et al.*, 2004). Nevertheless, consider that the morning activity peak in LD is minimally altered by period-altering *per* variants (Hamblen-Coyle *et al.*, 1992) (Method 14). Perhaps the normal form of this gene does not affect that feature of the diel behavior of the fly so that other factors would be the primary regulators of dawn activity. [Such gedanken genes might act, on

---

METHOD 14. *Phase determination* for rhythmic characters typically means scrutinizing peak times per day for things like locomotor behav-ior (Method 1) or molecular oscillations (e.g., Method 16). Examples of objectively formal methods so applied are given in Hamblen-Coyle *et al.* (1992), Wheeler *et al.* (1993), Plautz *et al.* (1997b), and Levine *et al.* (2002a,b,c). Also, Klarsfeld *et al.* (2003) reviewed elements of some of the computational approaches that have been applied to *Drosophila* rhythms. One really needs to process formally the time course data (as just implied). For example, the absolute locomotor peak at a given cycle time within a given day may involve a "spiky" peak of activity occurring for a particular 30-min bin worth of data, but the major components of the behavioral rise and fall could imply a different phase maximum. Therefore, it is warranted (probably necessary) to filter raw data digital-ly (e.g., Hamblen-Coyle *et al.*, 1992; Levine *et al.*, 2002b) such that high-frequency components are removed and a "smooth" peak falls out of

---

[15]In this regard, Taghert *et al.* (2001) disrupted the posttranslational processing of neuro-peptides in experiments designed to upset molecules of this type well beyond PDF and in brain cells extending far outside the CNS regions that contain the neurons on which this article focuses. The effects of these transgenic perturbations on behavioral rhythmicity implied that there are broadly distributed chronobiological functions for these substances, although involving unknown numbers and types of neuropeptides.

this first-stage processing of the time course. Moreover, it could be necessary to hone in on two separate windows of time within a given 24-h cycle if plots of the rhythmic character appear bimodal. This is the case for lomocotor cycles of *D. melanogaster* in LD (the morning and evening peaks discussed in the main text), and one report of *Clk* mRNA cycling (Fig. 1) indicated *two* transcript maxima per day (Darlington *et al.*, 1998). For rhythm assessments involving recurring peaks of behavior, eclosion, or, even in some cases molecular cycles (Method 16), one asks the computer to determine phase values for the smoothed maxima each day; these calculations lead to an average phase for that character (e.g., the evening locomotor peak) for that fly. Such means tend to be accompanied by very small standard errors (e.g., Hamblen-Coyle *et al.*, 1992), but lurking within the properly analyzed time courses is the fact that the "average phase" for *Drosophila* of a given genotype involves quoting means of means. With respect to phase values among individual animals, one final point about crunching the numbers in rather heavy fashion is that the statistical significance of rhythm phases can be determined, e.g., how well clustered are the locomotor peaks for several flies of a given genetic type? To answer this question formally, "circular" approaches to phase determinations have been taken; this means that peak values are plotted around the "clock face" (Klarsfeld *et al.*, 2003) and the extent to which they are nicely grouped together or too scattered is analyzed statistically (e.g., Levine *et al.*, 2002a,b).

behalf of this particular behavior, in neurons other than those containing PER, TIM, CRY, etc.; however, see Stoleru *et al.* (2004), who find it unnecessary to invoke roles played by hypothetical "new" genes.]

As for *per*[0] flies, which are by definition arrhythmic in DD, the lack of effect of that mutation on the morning oscillator might be moot because there is naturally little or no free-running locomotor peak at that phase (see earlier discussion but recall the quasi-contrary results of Grima *et al.*, 2004 and Stoleru *et al.*, 2004). In other words, the aperiodic behavior of *per*[0] in constant conditions is able to reveal itself, even if it is unable to affect the cyclically occurring morning maximum of locomotion; because, with reference to *per*[+] flies monitored in DD, there is no such behavioral peak for *per*[0] putatively to eliminate. These elements of the dual-oscillator supposition segué into the experiments of Helfrich-Förster (2001), in which she varied LD cycle durations in the long direction. Previous and analogous experiments varied such cyclical daylengths (called T cycles) downward from 12-h:12-h LD (Wheeler *et al.*, 1993) and concluded that *per*[0] flies "merely respond" to the environmental transitions (e.g., quickly after 6 h of darkness ends or after that amount of daylight is over, in 6:6 LD cycles). The later experiments, however, revealed this would-be arrhythmic mutant to anticipate D-to-L transitions in conditions of cycles longer than 24 h. Such behavioral expectations of the environmental change in question do not alone demand that clock functions underlie such behavior. However, dynamics of the anticipatory phases—when the locomotor peaks occurred vis-à-vis the photic transitions, among the varying long T cycle conditions—were used to argue that $per^0$ retains a morning oscillator that regulates bona fide clock-controlled behavior in such conditions (Helfrich-Förster, 2001).[16] If these conclusions have force—and if Stoleru et al. (2004) might have overinterpreted their results in terms of "per only" and its known companion functions regulating both morning and evening components of the locomotor rhythmcity of Drosophila—then fly people are in for some further gene hunting: Will novel factors be discoverable as devoted mainly to regulating rhythmic behavior at dawn?

We will now grapple with whether the "per and companion functions" just noted must include molecular rhythmicities if periodic behavior is to occur. In this respect, Yang and Sehgal (2001) used a variety of transgenic drivers and drivees to jam per and tim expressions into temporally constitutive modes within brain neurons—at least in terms of transcriptional cycling. Yet the key transgenic types were rhythm-enabled behaviorally.

Before discussing some details of these results and their particular implications, here is an interlude about free-running molecular rhythms within brain neurons, including those observed in conjunction with Yang and Sehgal assessing free-running behavioral rhythmicity of the per/tim-constitutive transgenic types. Initial attempts to infer oscillating levels of a clock protein in DD showed immunohistochemical cycling of PER to be weak (low amplitude) as of a few days into this environmentally constant condition (Zerr et al., 1990). This could have meant that the animals had begun to drift out of synchrony with respect to each other, i.e., at the whole animal level. (Indeed, free-running behavioral cycles of D. melanogaster involve period variations extending at least down to 23 h and up to 25 h.) In other words, taking a histochemical time point based on the temporal parameter alone (local time of day in the laboratory) would involve sacrificing flies that vary for their inherent cycle times. Moreover, Zerr et al. (1990) were monitoring LN cycles of PER staining by considering those neurons as a broad group of laterally located cell bodies (prior to appreciation of the three kinds of LN subgroups; see Ewer et al., 1992; Frisch et al.,

---

[16]These analytical considerations, which are based a tad too much on formalistically assertional pieces of chronobiological dicta, were recycled from an earlier study of Merrow et al. (1999), performed on the conidiation rhythm of Neurospora.

1994). In this regard, Yang and Seghal (2001) reported that the $s$-LN$_v$ cluster exhibits solid free-running staining rhythms of PER and TIM for two days into DD, but the $l$-LN$_v$ cells could not be detected to cycle at all in this manner (these two investigators also showed, via protein homogenates, that TIM oscillations dampen rapidly after the transfer of flies from LD to DD). These findings are essentially congruent with those of Shafer *et al.* (2002) and Lin *et al.* (2004), based on their examinations of temporal dynamics for, respectively, both TIM and PER or of PER during the initial hours of DD. Veleri *et al.* (2003) tracked these kinds of cytochemical time courses during five days of constant darkness and reported a "significant difference(s) in PER staining intensity between the two time points (12 h apart)" within cells of the $s$-LN$_v$, LN$_d$, DN3, and DN2 neuronal groups (as mentioned elsewhere in this article, DN2s cycle out of phase compared with cells in other clusters). No such persistent, free-running PER oscillations were detectable within either $l$-LN$_v$s (à la Yang and Sehgal, 2001) or DN1s. These findings do not jibe fully with elements of results presented by Klarsfeld *et al.* (2004): "robust PER cycling" was said to be sustained for at least 2.5 days in DD, within $s$-LN$_v$s, $l$-LN$_v$s, and LN$_d$s; the same was so for dorsally located DN3 cells, and for at least two days of free-running PER cycles within DN1s. [Also see Lin *et al.* (2004) for similar results obtained by tracking the LN$_d$ dynamics of PER in DD.] Other investigators took a different empirical tack—*in situ* hybridization—and started with the demonstration that *tim* RNA cycling persists over the course of two DD days within essentially all neuronal groups that express this gene (Zhao *et al.*, 2003). As introduced previously, Peng *et al.* (2003) went on to report persistent free-running cycling of *tim* transcripts for more than one week, asserting in parallel that "on the eighth day of DD, the locomotor cycles of the fly population were still in close synchrony."[17] Again, all *tim* neuronal types were said to sustain such cellular oscillations, including the remarkable finding of a constant-dark recovery period associated with the $l$-LN$_v$s; that is, subsequent to the *tim* RNA rhythmicity, which is easily monitorable in LD within these particular LNs, this molecular phenotype was not detectable during the first two days of DD (consistent with Yang and Sehgal, 2001), but it eventually "adapted" (!) such that these large ventrolateral neurons were observed to cycle for this molecular character on day 8 (Peng *et al.*, 2003). This is three days after Veleri *et al.* (2003) were unable to

---

[17]These investigators did not look into the possibility of persistent free-running *per* mRNA cycling, in light of the odd fact that one's wherewithal to assess the presence of this transcript *in situ* has gotten lost in the shuffle (*cf.* Yang and Sehgal, 2001), despite the earlier successes reported by Rachidi *et al.* (1997) and Rouyer *et al.* (1997). See Method 10.

detect $l$-LN$_v$ cycling for the other state variable in play within this paragraph, i.e., immunohistochemically inferred levels of PER protein.

Therefore, we do not have a complete or clear descriptive picture of what cycles within the brain—cellularly and molecularly—and for how long in the environmentally constant condition that allows *Drosophila* to sustain locomotor rhythmicity for at least a month (reviewed by Hall, 2003a). Nonetheless, Yang and Sehgal (2001) plunged ahead, severely to impinge on whatever such gene product cycling there may be and to assess the behavioral corollaries. In brief, they found that two widely expressed *gal4* drivers, applied mainly because they were each expected to mediate constitutive (*const*) expression of GAL4 and thus of a given UAS-encoded product, indeed forced *tim*-encoded mRNA to be temporally constant (*per* probes designed to detect the transcript of that gene by *in situ* hybridization did not work, as discussed in footnote 17). Such *const-gal4* transgenes, when combined with either UAS-*per*$^+$ or UAS-*tim*$^+$, quasi-rescued—let us say—the effects of *per*$^0$ or *tim*$^0$ in the respective clock mutant genetic backgrounds. Furthermore, triply transgenic flies carrying one of the *const-gal4* drivers along with UAS-*per*$^+$ or UAS-*tim*$^+$—fighting the simultaneous presence of *per*$^0$ and *tim*$^0$—exhibited rhythmic locomotion for "forty to fifty percent" of the behaviorally monitored individuals (Yang and Sehgal, 2001). Within the brains of doubly transgenic animals (-*gal4*/UAS-), and notwithstanding the nonoscillating levels of (at least) *tim* mRNA in the supposedly key lateral neurons, PER and TIM proteins were observed to cycle, i.e., by temporally controlled immunohistochemistry and observations of the *s*-LN$_v$ cells. The cyclically varying TIMELESS protein encoded by UAS-*tim*$^+$ in this circumstance (involving the one *const-gal4* type that was applied in those brain stainings) was said to be "significant but low amplitude" (Yang and Sehgal, 2001).

The findings of Yang and Sehgal (2001) have inspired some to wonder whether the feedback loop components that "return to the genome" each day (in this case to the *per* and *tim* regulatory DNAs) are all that interesting from the perspective of actual rhythmicity, as opposed to that which is all the rage molecularly (Fig. 1). It is important to contemplate some specific, counterargumentative phenomena in the context of tacitly posing this disturbing question [whether or not the outcome of the following discourse should be accompanied with the wise words about believing "half of what you see, son, and none of what you hear" (Whitfield and Strong, 1967)]. The relevant experiments have shown that manipulated forms of clock genes, notably *per*, can produce an oscillation of steady-state mRNA levels and lateral-neuronal PER cycling (Frisch *et al.*, 1994) that is not underpinned by a transcription-rate oscillation (So and Rosbash, 1997; also see Vosshall and Young, 1995), but which supports solid free-running rhythms of locomotion, such that, for example, 94 and 96% of the transgenic individuals (from two separate lines of the same promoter-*per* transgenic type) behaved in a significantly periodic manner in DD at 25° (Frisch *et al.*, 1994), compared with the rather weak and anomalously periodic behavior of flies from among the "constitutive *per/tim*" lines reported by Yang and Sehgal (2001) in what were stated to be the exact same conditions. Also, the quality of mRNA transcribed under the control of the 5'-flanking DNA of *per* (Brandes *et al.*, 1996; Stanewsky *et al.*, 1997b) significantly influences the dynamics of steady-state transcript cycling (i.e., the "*per* promoter region" and its transcriptional control features are insufficient for a normal temporal profile of the encoded RNA species). It is even the case that no RNA cycling at all (neither transcriptional nor in terms of steady-state levels) can nonetheless support PER protein cycling in the eye (Cheng and Hardin, 1998), as well as both PER and TIM oscillations within the brain (Yang and Sehgal, 2001).

These kinds of extra complexities, possibly annoyances, should not allow us to get carried away into some sense of mRNA-cycling irrelevance. The counterargument would go that all levels of molecular cycling control—starting from that mediated transcriptionally and ending with the contribution of posttranslational oscillations—may be necessary in sum to mediate the maximally robust biological rhythms exhibited by wild-type flies. In this regard, something like "protein cycling only" is unknown as to whether it could support anything in the realm of normal biological rhythmicity [*cf.* Cheng and Hardin (1998), discussed further in the sections about peripheral tissue oscillators and clock gene pleiotropies]. Moreover, nontranscriptional mRNA oscillations or flat transcript levels—with either such phenomenon proceeding to a crude version of final gene-product (PER) cycling—are correlated not only with less than robust periodic behavior (overall, among their transgenic types), but also with the matter of Yang and Sehgal (2001) tending to observe not fully "penetrant" locomotor rhythmicity. Let us pit these molecular–genetic, brain behavioral results against those obtained by similar analyses of the truncated-*per* transgenic type that reentered the picture shortly earlier: Absent transcriptional control per se of *period* RNA and PER protein cycling in these "nearly rescued" transgenics, the penetrance values of Frisch *et al.* (1994) were nevertheless near unity (in terms of the percentages of rhythmic individuals quoted earlier); yet, these locomotor rhythms were not like those of wild-type *Drosophila*, because the behavior of flies from the two truncated *per* strains gave free-running periods that were 9% and 10% longer than normal. Incidentally, it is odd that Sehgal and colleagues persist in citing Frisch *et al.* with regard to "behavioral rhythms [that can be] restored in . . . *per*-null flies by constitutively expressing *per*" (Sathyanarayanan *et al.*, 2004). The adverb just quoted (with emphasis added) is palpably false in the context of that 1994 paper ("believe half of what you see...").

Some additional aspects of intrabrain cycling of clock gene products are weird or at least unappreciable for their chronobiological relevance. Take a subset of the DN cells that contain *per* and *tim* gene products: Within the larval brain (discussed in more detail in a subsequent section), one cluster of DNs exhibits PER and TIM staining cycles that are completely out of phase with reference to those exhibited by other larval DNs and their brain lobe LN cells (Kaneko *et al.*, 1997). This anomaly is mirrored in the adult brain, for which cells in the DN2 cluster (Fig. 2) show a similar PER peak in constant darkness (Veleri *et al.*, 2003). DN2 maxima within the larva and the fly therefore occur at the *trough* times for immunohistochemically inferred levels of one or more clock proteins, within neurons of the DN1, DN3, $LN_d$, and $LN_v$ clusters; along with the "usual" late-night peaks and troughs exhibited for PER within brain and optic lobe glia (Zerr *et al.*, 1990). Do the out-of-phase DN2s support some aspect of brain-controlled rhythmicity that has nothing to do with the chronobiology of the other cell groups? There is no answer to this question, which is asked in a rhythm-systems spirit. But further studies of DN2 neurons, which took into account an additional component of their cellular contents, were informative in the context of circadian photoreception. This phenomenon is the major subject in the following section.

## Input Systems Subserving Clock Resetting in *Drosophila*

### Light

The most salient environmental input into the clock system of the fly is light. Widening knowledge about how photic signals are mediated in terms of gene functions and tissue structures causes one's view of the mechanisms and pathways that deal with such inputs to be a system to beat the band (or at least to battle several portions of the bandwidth's worth of photic wavelengths). Accordingly, the upcoming treatment of it focuses mostly on the multiple component features of light resetting, along with certain intriguing enigmas associated with some of these phenomena. (For terser descriptions of them, see the review of Helfrich-Förster, 2002.)

The first point is that external photoreceptors are not necessary for light to "get to the clockworks" of *Drosophila*. Such findings jibe with revelations of extraocular circadian photoreception in many animals (e.g., Page, 1982; Truman, 1976), with the notable exception of mammals (van Gelder, 2005; also see footnote 32). In the case of the fly, external eyeless mutants have long been known to be entrainable such that they behave in synchrony with light:dark cycles (reviewed by Helfrich-Förster, 2002). It is useful to register as well that the ocelli could have been the secret external eyes for such photic inputs, compared with the obviously more conspicuous compound eyes. However, this will not wash because double mutants in which both the latter and the former structures are eliminated still entrain (e.g., Stanewsky *et al.*, 1998); also, *no-receptor-potential* (*norpA*) mutations, which cause physiological blindness of both the main eyes and the ocelli (Hu *et al.*, 1978), similarly allow solid entrainment (e.g., Wheeler *et al.*, 1993).[18]

Participation of the simple eyes (Oc within Fig. 4) in circadian photoreception has been teased out genetically. Experiments that expanded the picture in this manner—and on behalf of all the other structures known or suspected to be involved in these processes—involved locomotion occurring in so-called extreme photoperiods (Rieger *et al.*, 2003). For example, the compound eyes and phototransducing functions operating within them were inferred to be especially important for behavioral synchronization to LD cycles containing only 4 h of light or darkness per 24 h. Due to the fact that these inferences and others, including the apparent contribution of the ocelli, were derived from analyses of locomotion occurring in such occult LD cycles (these details of Rieger *et al.*'s findings are described by burying them within the legend of Fig. 4).

Let us then focus on the major features of extraocular photoreception in *Drosophila*. How might photic stimuli arrive at the clockworks in such a manner? One way is that light could go straight into pacemaking neurons. That this is so was revealed by discovery of the aforementioned *cryptochrome* gene in *D. melanogaster* (reviewed by Hall, 2000). *cry* indeed makes its mRNA and "blue-absorbing" protein within various of the LN and DN cells, which were beaten to death within previous sections in terms of the clock-gene products they contain and the rhythmic behavior they control (see earlier discussion and Emery *et al.*, 2000b; Klarsfeld *et al.*, 2004). The probable meaning of CRY within such CNS neurons has been disclosed in two ways:

---

[18]The *norpA* mutation used in this particular study is a loss-of-function variant called *norpA*$^{P24}$ (Pearn *et al.*, 1996; Zhu *et al.*, 1993); *norpA*$^{P41}$ is another allele used in several analogous rhythm experiments (described later), which seems also to be an amorphic mutant based on its elementary visual-response defects, but it has not been proved to be molecularly null. Another visual system mutant, *glass*, enters the story a few pages from now; it was equally desirable to apply a loss-of-function mutation of this kind of variant; and the allele that always has been used, *gl*$^{60j}$, is indeed a null mutant from molecular analysis of the intragenic change (Moses and Rubin, 1991).

1. From a chrono-molecular angle (reviewed by Hall, 2000, 2003a), which began with the discovery of "TIM crashing" after *Drosophila* are exposed to light; this is apparently presaged by light-induced CRY:TIM, along with CRY:PER protein interactions. (The former such light effect would somehow expose TIM to degradation, whereas the latter could mean that CRY:PER helps pull PER away from TIM, potentially making it even more vulnerable to proteolysis.) Clearly, if light can "reach CRY" as it sits within LN brain cells, it would also be arriving at the clockworks by the direct input route just implied. Moreover, the semi-sudden disappearance of a pacemaking state variable (TIM) during its rising phase perforce resets the clock in a delaying direction whereas the same molecular effect of light during the falling phase must result in a phase advance. Indeed, early night light pulses cause circadian clocks (in any and all organisms) to be delayed, whereas late-night pulses induce advances. Here is an irritation, however: PER may be of prime importance for the functioning of the "works" in question (as labored over in a previous section). Should not this protein, too, disappear soon after exposing *Drosophila* to light? That is, if the classical and ubiquitous clock resetting scenario, which was cavalierly outlined earlier, applies to this insect. Well, PER does not respond to light in such a robust manner. Whereas, for example, the onset of experimental dawn light leads to the rapid decline of TIM (as expected from the earlier light pulse/TIM crash results), PER levels (measured grossly) were found to remain rather high for a considerable period afterward (Marrus *et al.*, 1996; Weber and Kay, 2003). This was way too long a time for facile appreciation of the role of this state variable in light-induced clock-resetting. However, perhaps thoroughly functional forms of PER are removed not long after light onset (*cf.* Nawathean and Rosbash, 2004), as would not be revealable by crude histochemical monitoring of PER levels in homogenates that once again rear their ugly experimental heads (*cf.* Method 4). For example, the light-induced PER-with-CRY interaction mentioned previously may block the ability of the former protein to carry out its (usual) negative feedback function (Fig. 1).

2. The behavioral meaning of CRY in the brain has been unveiled mostly by virtue of one way that the gene was tapped into originally: isolation of the *cry*[b] mutation, which leads to very low levels of an immunochemically detectable protein (Stanewsky *et al.*, 1998). Singly mutant *cry*[b] flies are readily entrainable for LD-synchronized behavior; curiously, however, they cannot have their locomotor cycles phase shifted by short light pulses (Stanewsky *et al.*, 1998). They are circadian blind in another way as well in that *cry*[b] eliminates or severely attenuates the arrhythmia-inducing effects of constant light (LL) on adult locomotion (Emery *et al.*, 2000a; Helfrich-Förster *et al.*, 2001; Mealey-Ferrara *et al.*, 2003).

These results are not thoroughly consistent among each other. Thus, does $cry^b$ lead to locomotor periodicities indistinguishable from those of wild-type *Drosophila* in DD, i.e., all-out circadian blindness in LL (Emery *et al.*, 2000a) or is the rhythmicity in the latter condition somewhat long-period (Helfrich-Förster *et al.*, 2001), which would imply that this mutant perceives "medium-level" LL to be very dim light? Why that notion? Wild-type *D. melanogaster* behave rhythmically when the constant light level is very low, but the cycle durations are longer than those measured in DD (Konopka *et al.*, 1989). The further LL experiment of Mealey-Ferrara *et al.* (2003) failed to resolve this micro-issue about the behavior of $cry^b$ in LL as compared with the two earlier studies just cited. However, the effects of certain engineered *cry* variants, from which C-terminal amino acids were removed, are consistent with the dim-light effects on wild-type locomotor rhythmicity. These truncated CRYs have a "constitutively active" property, which leads to lower than normal levels of TIM and PER along with weak amplitudes of their daily oscillations (*cf.* Busza *et al.*, 2004; Dissel *et al.*, 2004). Overexpression of this inappropriately lively cryptochrome molecule led to longer than normal behavioral cycles in DD (Dissel *et al.*, 2004), presumably mimicking the modestly activating effects of dim light on normal CRY and the TIM/PER diminutions that follow.

There is another point of tacit disagreement among behavioral experiments involving the $cry^b$ mutant (ostensibly unrelated to the issues just discussed). When Yoshii *et al.* (2004) monitored locomotion in LL, the mutant exhibited *two* free-running "components": one with a relatively short and the other with a longer circadian cycle duration. This "rhythm dissociation" was not observed by other investigators (e.g., Emery *et al.*, 2000a; Helfrich-Förster *et al.*, 2001; Mealey-Ferrara *et al.*, 2003), apparently because they applied levels of constant light appreciably more intense than those used by Yoshii *et al.* (2004). The latter investigators went on to show that the light input route, which somehow stimulates behavior driven by "two circadian clocks," uses the external photoreceptors, because dissociation into two rhythmic locomotor components was prevented in double mutants for which an eye-removing mutation or a *norpA*-null one was combined with $cry^b$ (Yoshii *et al.*, 2004).

To continue consideration of (near) circadian blindness in $cry^b$ in certain circumstances, note that the entrainability of this singly mutant type to "standard" LD cycles can be rationalized by the retained functions of external photoreceptors. This prompted combining $cry^b$ with a *norpA*-null mutation; the flies were much less sensitive in terms of entraining to (dim)LD cycles (Stanewsky *et al.*, 1998). Also, singly mutant *norpA*-null or eyeless types exhibited mediocre sensitivity (*cf.* Ohata *et al.*, 1998), driving home the point that the external photoreceptors are "circadian involved"

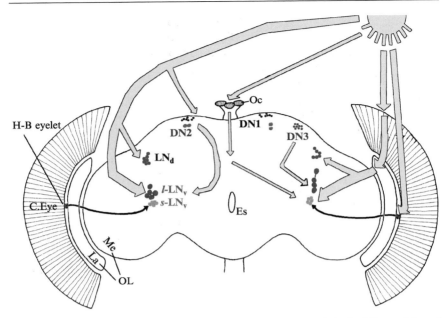

FIG. 4. Input routes for light-mediated resetting of the behavioral clock of *Drosophila* and molecules functioning in these anatomical pathways. This diagram was adapted from Helfrich-Förster *et al.* (2001) as presented in modified form within the review of Hall (2003a); also see two reviews by Helfrich-Förster (2002, 2003). With respect to photic influences on the rhythm system of the fly, these three articles mainly summarize results reported by Ohata *et al.* (1998), Stanewsky *et al.* (1998), Yasuyama and Meinertzhagen (1999), Emery *et al.* (2000a,b), Helfrich-Förster *et al.* (2001, 2002a), and Malpel *et al.* (2002); also see the "postreview" papers of Mealey-Ferrara *et al.* (2003), Rieger *et al.* (2003), Veleri *et al.* (2003), and Klarsfeld *et al.* (2004). Essentially all of the individual studies just cited were based on applications of mutations, transgenics, or dietary treatments (see later) and stemmed from earlier and more basic experiments in which only the external eyes were eliminated or inactivated, which nevertheless allowed *D. melanogaster* adults to entrain to light:dark (LD) cycles (e.g., Dushay *et al.*, 1989; Helfrich, 1986). The current picture shows the input routes inferred to be knocked out or remaining under the influence of a given "treatment" (usually a genetic one). One element of the scheme is based on the entrainability to LD cycles, and reentrainability to shifted ones, of *Drosophila* that express a *norpA*-null mutation; it causes physiological blindness of photoreceptor cells (PRs) within the compound eyes (C.eye), the ocelli (Oc), and probably the H-B eyelet as well; or of flies that express an *eyes-absent* (*eya*) mutation in combination with an *ocelliless* one. The afferents of C.eye PRs may transmit results of their light inputs to dendrites of *l*-LN$_v$ cells (Helfrich-Förster, 2003; Helfrich-Förster *et al.*, 2002a); axon terminals projecting from H-B PRs have been more definitively observed to overlap with those particular LN arborizations as well as with those of the *s*-LN$_v$s (Helfrich-Förster *et al.*, 2002a); however, it is unknown whether Oc afferents connect with LN cells (or with any of the DNs for that matter), either directly or via dorsal-brain interneurons (Helfrich-Förster, 2003). Additional visual system variants have come into play as follows: (A) A *sine oculis* (*so*) mutant, which is missing C.eyes and Oc, although not all flies in this strain are thoroughly

devoid of the former structures (e.g., Helfrich-Förster *et al.*, 2002a) so mutant *Drosophila* retain their H-B eyelets. (B) A *glass* (*gl*-null) mutation that wipes out all external PRs and the eyelet as well (Helfrich-Förster *et al.*, 2001). (C) A cryptochrome mutation (*cry*[b]), which leads to little or no function of CRY, a blue-light absorbing protein present in many *Drosophila* tissues, including various neurons (see later). (D) A *histidine decarboxylase* (*hdc*) mutant, which lacks histamine in both the external eyes and the H-B eyelet such that these structures are unable to mediate neurotransmission to CNS neurons [Rieger *et al.* (2003) and references therein]. Usage of this *hdc* mutant can be regarded as a valuable adjunct to applications of pleiotropic PR-anatomical mutants such as *so* or *gl*, which are known or suspected to suffer some brain damage beyond their external PR and "circadian CNS" problems (see Helfrich-Förster and Homberg, 1993; Klarsfeld *et al.*, 2004). In any case, one can draw on these genetic tools to do things like conceptually subtracting the circadian-PR functions of external structures (*so*) from those of external eyes + H-B eyelet (*gl*), or one can cause flies to retain only the eyelet function (*so*/*cry*[b]). Nutritional rearing provides an additional tool to affect functioning of the (eventual) animals' visual system, in particular by growing flies on vitamin A-deficient medium (Ohata *et al.*, 1998; *cf.* Zimmerman and Goldsmith, 1971); this results in severe knockdowns of the rhodopsins that are naturally contained within the main eyes, the simple ones atop the head, and the eyelet underneath the retina. Thus, with regard to all external photoreceptors and the H-B eyelet (summing to seven such PR structures), sensitivity for LD entrainment is sharply reduced in the mutant types that are, for example, missing their C.eyes or them plus Oc (Helfrich-Förster *et al.*, 2002a; Stanewsky *et al.*, 1998). These results are consistent with the effects of rearing *Drosophila* on vitamin A-less food (Ohata *et al.*, 1998). Because LD entrainability was not eliminated in these situations, a light-to-clock input role of extraocular photoreception was implied (e.g., in the aforementioned studies from the 1980s). Reentraining flies to either advanced or delayed LD cycles (in separate experiments) implied that the eyelet, cooperating with deep brain photoreception (CRY), is involved mainly in mediating behavioral phase delays (Helfrich-Förster *et al.*, 2002a). A further study of this sort applied the mutants and genetic combinations just noted; also *Drosophila* devoid of their C.eyes only (*eya*); or of them plus CRY (*eya*/ *cry*[b]); and even flies that were missing all relevant external structures (*n* = 5 eyes) plus the H-B eyelet, along with communication between these PR organs and second-stage interneurons of the visual system (*so*/*gl*/*hdc*). Most of these experiments, in turn, were based on entrainability of animals partially or massively gutted of putative circadian photoreceptive functions in non-12:12 LD cycles (Rieger *et al.*, 2003). Thus it was deduced, from analysis of behavior in 16:8 or (especially) 20:4 LD cycles, that all overtly eye-like PR types (C.eye, Oc, and H-B) donate functions pertinent to entrainment in such heavily L-enriched conditions; also that the C.eyes and Oc both contribute to (non entrained) period lengthening that can occur for certain mutants (notably *hdc* expressed by itself) in "long photoperiods" (again, as encompassed by LD cycles totaling 24 h in duration). However, most of this period-lengthening effect was deduced to "go through" the H-B eyelet and the light absorption CRY (Rieger *et al.*, 2003). Focusing on deep brain circadian photoreception, the *cry*[b] mutation eliminates behavioral phase shifts that are normally induced by short light pulses in the early or late subjective night (in DD). Within the brain, CRY is found in LN pacemaker neurons and DN cells as well. Transgenically mediated restoration of *cry*[+] function to certain LNs only (in flies homozygous for *cry*[b]) allowed for solid light pulse responsiveness, as if extraocular photoreception is mediated partly by pacemaker cells themselves. That *cry*[+] is expressed within DNs is interesting in light of the DN1- and DN3-to-LN axonal pathways shown in Fig. 3 and that the aforementioned *glass* mutation eliminates at least a substantial subset of the DN1s as well as the H-B eyelet. Anatomical studies of that internally located, eye-like entity revealed an eyelet-to-LN region axonal projection (as originally specified in empirical detail by Yasuyama and Meinertzhangen, 1999), designated by even though they are not required. In LD cycles involving not so dim light levels, the *norpA* with *cry*[b] combination allowed for reasonably good entrainment (Stanewsky *et al.*, 1998). The latter result is key because if this double mutant were intrinsically blind circadianly, behavioral synchronization to these environmental cycles would not be possible no matter what the light level. Perhaps *cry*[b], which is "only" a missense mutant (Stanewsky *et al.*, 1998), retains some low but meaningfully functional level of the CRY protein. In fact, the potential absence of the protein in this mutant is creeping upward in terms of immunodetectability (Busza *et al.*, 2004; Emery *et al.*, 2000a).[19]

Another possibility is that there are additional input routes by which the effects of light reception reach the relevant brain cells. This supposition

---

[19]The latter group of investigators isolated a second mutation at the locus called *cry*[m]. It is accounted for by a stop codon within the ORF, which could create a null mutant. However, the location of this nonsense mutation near the 3′ end of the coding region of the gene would allow >96% of the polypeptide to be translated. Indeed the CRY[m] protein is detectable in extracts of the new mutant at an apparently lower concentration than for CRY[b] (Busza *et al.*, 2004). Yet certain *in vivo* tests (e.g., a "bright-light" PRC) indicated that *cry*[m] flies retain a higher functional level of CRY compared with *cry*[b]. However, this new cryptochrome variant has not yet been subjected to the battery of whole animal tests that have been performed on *cry*[b], often in combination with other visual system mutations, to determine the residual light response capabilities of a genetically inadequate clock input system.

---

the two mildly curving horizontal lines. The external-PR-less and DN-disrupted *gl* mutant could be readily reentrained in LD tests (as implied earlier), albeit with reduced sensitivity. This is similar to the effects of a *norpA*-null mutation (acting by itself) or of carotenoid depletion (see earlier discussion), keeping in mind that those two effects should not alter light reception by any of the relevant CNS neurons. Such resynchronization responses exhibited even lower sensitivity when a *norpA* mutation was combined with *cry*[b], and the responsiveness was worse still (perhaps zero) in *gl cry*[b] double mutants. Both of these doubly mutant types should be anatomically and/or physiologically nonfunctional for external + eyelet + deep-brain photoreception. Inasmuch as the *gl*-null *cry*[b] combination leads to extremely poor nonentrainability, the *glass* mutation (but not *norpA*) is presumed to affect a "fourth pathway" by which photic stimuli get to the deep brain clock. Further mutationally based findings are on point (Rieger *et al.*, 2003): With the external PRs gone and the DNs rendered nonfunctional in *so/cry*[b] flies, only half of these doubly mutant individuals—despite their retention of H-B eyelet functioning—could entrain to 12:12 LD cycles, fewer to 8:16 or 16:8, and none to 4:20 or 20:4. When the *hdc* neurochemical mutation was combined with *cry*[b], presumably to eliminate all five of the separate input routes (C.eyes, Oc, H-B, LNs, and DNs) in a similar manner to the effects of *gl cry*[b], a minority of *hdc/cry*[b] double mutants entrained to 12:12 and none in the other (> or < 12-h) photoperiods. This genetic combination can be inferred to be less severe than that of *gl cry*[b] (Helfrich-Förster *et al.*, 2001) or perhaps roughly equivalent (*cf.* Mealey-Ferrara *et al.*, 2003). In any case the effects of these maximally pleiotropic mutant combinations reinforce the reality of pathway #4 and that it would involve the DNs-to-LNs route noted earlier. (See color insert.)

has been on the table for a while due to the discovery of an eye-like structure lurking in a quasi-internal location: between the proximal region of the compound eye and the distal-most optic lobe (Hofbauer and Buchner, 1989). This entity has become known as the "H-B eyelet" (see Fig. 4). It was later shown that key elements of substucture discerned from electron microscopic observations of this eyelet look photoreceptor-like ultrastructurally and that these cells send their axons to sites near the LN cell bodies within the adult brain (Helfrich-Förster et al., 2002a; Yasuyama and Meinertzhagen, 1999).

Functional meaning of the presence of the eyelet was uncovered— sort of—by further studies of double-mutant behavior. For this, a *glass* mutation was used in combination with $cry^b$. The former variant (a *gl*-null, as cited within footnote 18) ruins the ability of the developing animal to form external photoreceptors *and* seems entirely to eliminate the H-B eyelet (Helfrich-Förster et al., 2001). The *gl* effects, combined with (near?) absence of CRY, caused circadian blindness in certain LD cycle situations (as opposed to light pulse experiments, which do not require inclusion of visual mutations other than $cry^b$ for such a severe resetting defect). Thus, the *gl $cry^b$* mutant type could not be "shifted over" to behave in synchrony with new LD cycles effected experimentally by an abrupt 8-h change to later times of lights-on and -off (Helfrich-Förster et al., 2001).[20]

This same *gl $cry^b$* double mutant was retested in another circumstance whereby flies were subjected to 12-h:12-h LD cycles for a few days before "gently" proceeding into 13:13 cycles.[21] Against a background of the fact that wild-type *Drosophila* can readily entrain to LD cycles that are way off the usual daylength (e.g., Helfrich-Förster, 2001; Wheeler et al., 1993), it was valuable to learn that a fair number of *gl $cry^b$* individuals not infrequently reentrained to the 13:13 cycles, in that they exhibited 26-h locomotor cycle durations even in dim L conditions (Mealey-Ferrara et al., 2003). The same quasi-normal in-synch locomotion was observed for *norpA*-null/$cry^b$ flies that simultaneously were suffering the effects of

[20]An additional complexity involving the two mutations just noted and another eye anatomical variant speaks to the photoreceptive structures that are involved in phase-shift *delays* of behavior compared with *advances* (Helfrich-Förster et al., 2002a). The specifics of this matter are taken up within the legend to Fig. 4.
[21]Consider in this regard how difficult it would be to tell whether a circadian-blind fly is exhibiting entrained behavior in tests that would solely make use of 12:12 cycles. In other words, 24-h locomotor periodicities in that LD condition could reflect the mere manifestation of free-running rhythmicity, in context of this organism's "natural" clock pace being quite near to 24 h (reviewed by Hall, 2003a; Price, 2005).

H-B eyelet ablation (mediated by an eyelet *gal4* driver combined with one of the aforementioned types of UAS cell-killer transgene).[22] It was as if this 12:12 → 13:13 experimental tactic better revealed the residual circadian photoreceptive capacities of eyeless/H-B-less/CRY-less flies, compared with the ostensibly more demanding situation in which such multiply impaired animals had to achieve an 8-h behavioral phase shift (Helfrich-Förster *et al.*, 2001).

To delve into these empirical and interpretational problems one deeper step, let us entertain the possibility of additional photic input routes (instead of recycling the notion that *cry^b* is not a severe enough mutant to wipe out its feature of the circadian photoreceptive system). We start with the fact that the *norpA^+* gene makes its "phototransduction" protein [a phospholipase C (PLC)] within H-B eyelet cells (Malpel *et al.*, 2002). Therefore, the *norpA*-null/*cry^b* flies of Stanewsky *et al.* (1998) should have been pleiotropically defective enough to be just as entrainment impaired as the *gl cry^b* type, but the latter was shown to be worse off (even in the hands of Mealey-Ferrara *et al.*, 2003). Thus, it would seem as if the *glass* mutation by itself causes more pleiotropic problems than elimination of external eyes and the H-B eyelet. In fact, certain clock gene-expressing brain neurons are "gone" in this *gl* mutant, which lacks detectable DN1 cells (Helfrich-Förster *et al.*, 2001, 2002a, *cf.* Fig. 2) or at least is missing an appreciable subset of this particular dorsal neuronal cluster (Klarsfeld *et al.*, 2004). Intriguingly, the GL protein is coexpressed with PER in a proportion of the DN1s—in wild-type adults, after the developmental functions of *gl^+* have ended (Klarsfeld *et al.*, 2004); alternatively, the continued presence of GL might be required to maintain the integrity of these dorsal neurons within the brain of the fly.

Dorsally located, clock gene-expressing neurons could communicate the effects of light reception to the would-be principal pacemaking cells, because (a) all three DN clusters in adult brains contain CRY protein, although not all such cells, e.g., only about one-quarter of the DN3s

[22]This driver took advantage of one of two rhodopsin (*Rh*) genes expressed within the H-B eyelet (Helfrich-Förster *et al.*, 2002a; Malpel *et al.*, 2002; Yasuyama and Meinertzhagen 1999), leading to the production of an *Rh-gal4* tranegene that was applied by Mealey-Ferrara *et al.* (2003). It might be implicit from previous implications, whereby the PLC product of *norpA* functions downstream of light absorption by rhodopsins (Pak and Leung, 2003), that all external photoreceptors exercise their primary light-absorbing capacities via the products of various and sundry *Rh* genes. However, as seen in a later section about chrono-tissue pleiotropy, not all such genes are expressed solely within eyes or eye-oid structures such as the H-B one.

(Klarsfeld *et al.*, 2004),[23] and (b) two types of axonal output from these dorsal neurons include a DN1-to-LN region pathway (Kaneko and Hall, 2000) and a DN3-to-LN one (Veleri *et al.*, 2003). The latter observation added this ventrally extending DN3 projection to the previously known pathway that runs from this particular cluster of dorsal neurons to a more dorso-medial brain region (Kaneko and Hall, 2000).

Veleri *et al.* (2003) were mainly concerned with the expression patterns and behavioral effects of a truncated *per* transgene (in play within previous sections for other reasons). This particular 5′-less construct leads to PER production in all three DN cell clusters but in none of the LNs (Veleri *et al.*, 2003). Accordingly, this "promoterless" transgenic type exhibited *no rescue* of the effects of a $per^0$ mutation on free-running arrhythmicity (see earlier discussion). However, such flies were able to entrain quite well in LD-cycling conditions (Veleri *et al.*, 2003). Therefore, the DNs may be broadly involved in receiving light and sending such results to more centrally located portions of the rhythm system of the CNS. At least the DN1 cells are thought (better) to be involved, thanks to the following recent experiment. The behavioral entrainability of a *gl*-null mutant was compared with that of transgenic flies containing a construct in which GLASS-responsive regulatory DNA had been fused to a cell-killer gene; the former, straight-mutant genotype knocks out all known photoreceptors (PRs) and (as a reminder) some or all of the DN1s, whereas the latter engineered variant leads to PR-lessness but spares these dorsal neurons (Klarsfeld *et al.*, 2004). Pitting the behavioral effects of the *glass* mutation against those of the fusion transgene showed that, in a dim-light condition, the DN1-less mutant resynchronized to 8-h-shifted LD cycles in a relatively poor manner, i.e., needing more days to reentrain than were required for the PR-ablated but DN1-containing transgenic. The latter behaved as well as a *norpA* mutant in terms of behavioral phase shifting (Klarsfeld *et al.*, 2004), which suggests that the sensory transduction enzyme encoded by this gene is not involved in circadian photoreceptive events that (now) seem established to occur within DN1 neurons.

Some nagging questions remain, however.

1. Why are DN3s and the recently uncovered efferent projection from them (Veleri *et al.*, 2003) apparently insufficient for light entrainment (given that *glass* affects only DN1 cells insofar as so-called neurons within

---

[23] Also recall that the aforementioned *cry-gal4* transgene, with regard to DN expression, drives marking solely within a subset of the DN1 cells (Stoleru *et al.*, 2004). This implies that the stretch of 5′ regulatory DNA fused to *gal4* does not contain all the cell-specific enhancers possessed by the normal form of this *cry* gene.

the CNS are concerned)? Are the effects of this *gl* mutation more pleiotropic (in that the DN3 cells bodies would be intact, but their centripetally coursing axons may be gone)? Or, in turn, are the residual entrainment capacities that were teased out in tests of the *gl cry*[b] double mutant by Mealey-Ferrara *et al.* (2003) subserved by a "*DN3*-to-LN-only" input route (absent external photoreceptors, the H-B eyelet, as well as most or all of both DN1 cells and CRY functioning)?

2. Is cycling of PER and other clock factors involved in circadian photoreception? For example, PER naturally cycles within DN neurons (discussed again and again in previous sections), and the cyclical expression of the gene is correlated with the light entrainability of the fly (Veleri *et al.*, 2003). Why is there molecular cycling within these DNs? Not for core behavioral control (free-running rhythmicity); thus, possibly to process properly the light inputs that are received by certain DN cells, then need to be "handled" in some way by cyclical intracellular activities before being sent on to the LN cells.[24]

3. Are all three DN clusters involved in light-to-clock functioning? This is an inquiry that arises in the context of question 2, because the DN2s exhibit PER and TIM oscillations that are so wildly unphased with respect to the cellular cycles operating within the DN1s and DN3s (Klarsfeld *et al.*, 2004; Veleri *et al.*, 2003); mercifully, perhaps, no neurites have been observed that could connect DN2 cells to LNs.

4. However many DN cell types may contribute to photic entrainment, does this require an interplay with the more ventrally located pacemaker neurons? The effects of externally blinding, arrhythmia-inducing *disconnected* mutations are interesting in this regard: Most *disco* individuals lack optic lobes (hence are eye–brain disconnected) and are devoid of almost or all LN cells (on a fly-by-fly basis). The latter defect, but not the former, would account for the arrhythmic locomotion in DD exhibited by this mutant type. However, such flies are able to entrain to LD (e.g., Wheeler *et al.*, 1993), even though they are LN-less. This point was made most tellingly in context of the following assumption: If *disco* flies in LD cycles behave in an anatomically clockless manner, such mutations should be epistatic to the effects of a period-altering variant involving another gene. Quite the opposite occurred when a *disconnected* mutation was combined with the *per*[S] one, which moved the usual

---

[24]This supposition cannot be dismissed out of hand because it seems as if "receptor–structure rhythmicity" is involved in mediating another kind of behaviorally entraining stimulus (discussed later).

"*disco*-only" peak of evening locomotion to a distinctly earlier phase (Hardin *et al.*, 1992).

This could mean that DN regulation of behavior in environmentally cycling conditions involves not so much messages transmitted from PER/TIM-containing dorsal neurons, which are retained in *disco* brains, to the lateral neurons, but instead stand-alone control of diel behavior by the DNs themselves. In contrast, studies by Veleri *et al.* (2003) of their transgenic type with PER in DNs only might imply that the entrainability of these flies would involve DN-to-LN axonal communications (*cf.* Fig. 2), in the sense that these transgenic brains retain LN cells (although they are PER-less). This is a similar supposition to that made by Stoleru *et al.* (2004) regarding the modulation of entrained diel behavior by $LN_d$-to-$LN_v$ communiqués (again, involving absence of PER in the later neuron types). However, a scenario for these brain neuronal interplays will not scour for the *disco* case because there are no LNs available to be signaled by output functions of the more dorsally located neurons. Yet, here is an alternative to the "DN stand-alone" hypothesis about the LD-synchronized behavior of *disco*: Intrabrain functions that may support the entrainment capacities of this mutant could involve the clock capacities of *glial* cells. Thus, within such LN-less flies, glia might, by some intrabrain miracle, regulate entrained locomotion in LD cycles (once again, all by themselves). Whereas *disco* mutant heads of course lack the many PER-containing glial cells normally present within the optic lobes (diagrammed by Helfrich-Förster, 2003), central brain cells of this type are still present *and* exhibit robust cyclical fluctuations in the apparent abundance of this clock protein (Zerr *et al.*, 1990).

Finally, and perhaps most troublesome, why on earth is the $cry^b$ mutant circadian blind in two circumstances—responses to short light pulses and constant illumination? External photoreceptors and the H-B eyelet, which do not use CRY but opsins for their primary photoreception, are sufficient to feed photic inputs to the clockworks in LD entrainment situations. Therefore, it seems as if these structures would allow the effects of light pulses or of constant light to make *cry*-mutated flies exhibit behavioral phase shifts or arrhythmicity, respectively. However, the compound eyes and the eyelet are rendered effectively functionless in these photic circumstances, for which the primacy of CRY actions was revealed (Emery *et al.*, 2000a; Stanewsky *et al.*, 1998). There is only a minimally satisfactory way out of this dilemma. If we are on the cusp of understanding elements of the paradox, the refined photomanipulative experiments performed by Rieger *et al.* (2003) are germane. Thus, CRY elimination was shown to be associated with period lengthening in LD cycles involving long photoperiods; these conditions (lots of "L" per 24-h cycle) approach a constant-light situation in which cryptochrome's control of light-to-clock input takes over.[25] CRY was also demonstrated to be more important than other light input routes (Fig. 4) for entrainment to cycles involving short photoperiods (Rieger *et al.*, 2003)—an experimental situation that approaches short light pulse testing the phase shiftability of $cry^b$, which was found to be nearly nil (Stanewsky *et al.*, 1998).

In summary, we know more and more about the several factors and entities that take light into *Drosophila*—not for its effects on moment-to-moment behavior, but instead to form a complex system that functions on behalf of slowly changing locomotor fluctuations (Fig. 4). Why this complicated set of substances and structures? As has been discussed several times (e.g., Hall, 2000), the circadian photoreceptive system may require some extra versatility, compared with the generic "white light" setup that subserves the acute responses observed when flies are stimulated to exhibit phototactic and optomotor behaviors. Thus, says *Drosophila*, "I will use CRY during the dim phases of dawn and dusk, when light that reaches me is enriched in the blue; I will use my opsins during the brighter phases of morning and evening because those molecules are most sensitive to UV light and that which is mainly in the green range [and even the red one (Helfrich-Förster *et al.*, 2002a; Zordan *et al.*, 2001)]; furthermore I require several different kinds of cellular structures to mediate this versatility, because my tissue differentiational capacities have not allowed all kinds of photoreceptive substances to be contained within one cell type." Having said all this, the fly might wonder why there is in fact lots of CRY protein within its compound eye. This puzzle is taken up in the penultimate section of the article, which focuses on the chronobiological functions of tissues outside the central nervous system.

*Temperature*

Daily fluctuations of temperature have been known, forever, to entrain circadian rhythms of things like cold-blooded animals and microbes (e.g., Merrow *et al.*, 1999; Zimmerman *et al.*, 1968). For example, adult *Drosophila* behave in synchrony to temperature cycles, as long as their amplitudes are ~3° or more (Tomioka *et al.*, 1998; Wheeler *et al.*, 1993). That the

---

[25]To reclarify, $cry^b$ *Drosphila* are either insensitive to arrhythmia induction by LL (Emery *et al.*, 2000a), or this singly mutant type is very poorly sensitive in that condition in tests that have shown the locomotor cycles of this mutant to be anomalously long period (Helfrich-Förster *et al.*, 2001). Either such result implies the "takeover" in question, in the sense that the *cry*-unaffected inputs from structures, such as external photoreceptors, are unable meaningfully to "flood" the system with LL effects and induce behavioral arrhythmia.

locomotor cycles in this condition reflect bona fide entrainment is signified (but not proved) by the fact that recurring evening lowerings of temperature were *preceded* by activity maxima—an even earlier anticipation than is observed during the evening of LD cycles (Wheeler *et al.*, 1993).

Temperature fluctuations do not provide all that reliable an indicator of "what time of day or night?" This kind of environmental change might better be detected with respect to an animal altering its daily biology over the course of different *seasons*. In this regard, the pioneering studies that probed circadian-system meanings of temperature changes were performed squarely in a seasonal context (Majercak *et al.*, 1999; Sidote *et al.*, 1998). These molecular phenomena are not rereviewed here very much (see Hall, 2003a), because too little is known about temperature changes from the perspective of systems chronobiology (cells, tissues). Suffice it to say that the temperature level could be perceived directly by the clock neurons of *Drosophila* alone, just as well or better than the manner by which such cells can respond to photic inputs.

Based in part on the inchoate belief just stated, let us focus solely on two recent reports, which stemmed from analysis of the interplay between temperature changes and light (Collins *et al.*, 2004; Majercak *et al.*, 2004). Remembering that alternative splicing of the 3′-untranslated region (UTR) of *per* is thermosensitive (Majercak *et al.*, 1999), these further studies showed that a combination of low temperature and short day conditions is especially stimulating for splicing out the "extra" sequences within this UTR. Who cares? Well, increasing proportions of this spliced-out isoform of *per* RNA are correlated with earlier rises of the abundance and advances of the transcript in the flies, evening peaks of locomotion (Majercak *et al.*, 1999). Mnemonic: during the winter it is OK or even advantageous to be active during the later phases of daytime, whereas in summer, being out and about during that cycle time could cause debilitating desiccation of the little flies. Consistent with this supposition, daytime splicing becomes extra-attenuated as the temperature rises, and this feature of the interacting environmental conditions was found to require normal *per* and *tim* functioning (Collins *et al.*, 2004; Majercak *et al.*, 2004). Further mnemonic: warm daytimes, dessication-avoiding activity pushed later into the early night. Now the so-called phototransduction factor specified by *no-receptor-potential-A* comes into play. It was shown long ago that *Drosophila* suffering from the blinding effects of *norpA* mutations exhibit an earlier than normal evening peak of activity in LD (Wheeler *et al.*, 1993). The corollary that *norpA*-null flies run slightly fast in DD may have been assumed to connect with the fact that this gene is expressed not only in the peripheral nervous system, but also in the CNS (reviewed by Hall, 2003b). Did these *norpA*-mutant phenotypes imply some unknowable epiphenomenological effect on central brain pacemaking? Probably not, thanks to clarification of these sensory/central issues by Collins *et al.* (2004) and especially by Majercak *et al.* (2004). In head extracts of a *norpA*-null mutant, the proportions of *per* transcripts in which the 3'UTR sequence is spliced out were found to be abnormally high and exhibited an immodest daytime decrease, even when the temperature was relatively high. Therefore, irrespective of the thermal conditions, RNA splicing regulation in this phospholipase C mutant exhibited characteristics normally observed only at relatively low temperatures. That molecular phenotype connects with the anomalous diel behavior of *norpA* mutants, whose evening activity is advanced even on warm days. This genetic effect could not be explained by the visual blindness of *norpA* (Collins *et al.*, 2004; Majercak *et al.*, 2004). These findings and conclusions almost certainly fit with the fact that the PLC in question is not only found in many tissues, *norpA* mutants are also more than visual-response variants. For example, this signal transduction enzyme is likely to be present in the olfactory system of *Drosophila*, because mutations at the locus cause odor response impairments (reviewed by Hall, 2003b). Thus, *norpA*$^+$ could be broadly concerned with regulating sensory inputs. Now the actions of its PLC encompass the sensation of temperature changes. Inputting such stimuli could occur partly within PNS structures; one place that the results of such "thermal reception" would be sent is to the CNS clockworks.

*Chemosensory Cues*

Among "nonphysical" stimuli that can reset animal rhythms are social cues. Certain such phenomena have been reviewed by Levine (2004), primarily in the context of summarizing social resetting of the clock of *Drosophila*. Thus, one set of flies can cause locomotor phase shifts of another fly group in its midst (Levine *et al.*, 2002a). The presence alone of resetters, as opposed to their behavior as well, may provide sufficient phase-shifting stimuli. This mini-conclusion stemmed from the demonstration that behavioral resets could be induced by an airstream moving from one group of flies to a separated set of potentially resettable "targets." Mutants also came into play in these experiments. For example, grouping wild-type target flies with arrhythmic *per*$^0$ ones caused "phase dispersal" of individuals within the former group. Prominent among the genetic variants usefully applied were (a) *smellblind* mutants (*para*$^{sbl}$, altered in a sodium channel gene), which—when they were the targets of incoming cues from a potential resetter group—exhibited little or no behavioral response, and (b) a truncated *per* transgenic type, the one that expresses this gene product only in LN neurons, enabling such flies behaviorally to entrain in LD cycling conditions and to free run (Frisch *et al.*, 1994). If these promoterless *per* transgenics were the targets of incoming social cues, their inherent capacity for locomotor rhythmicity would allow them to be phase shifted, but the behavior of this variant type was found not to be influenced by mixing such flies with putative resetter ones (Levine *et al.*, 2002a). Was it the absence of PER as such in peripheral tissues that caused this nonresponsiveness, or was it because this transgenic type cannot mediate the rhythmicity of PNS structures? What does that mean? First, register that one such sensory entity is the antenna; one of the PNS functions of this appendage is to act at the first stage of an input route for odors. Second, PER-lessness of the antenna of the transgenic is correlated with absence of an odor sensitivity rhythm (recorded from the appendage, as discussed in a later section), even though these flies exhibited their basic responsiveness to odorants, i.e., as applied to the antenna of a promoterless *per* individual (Krishnan *et al.*, 1999; again, see later).

This is why the notion of a rhythmically functioning input structure was surmised to be relevant for proper processing of social resetting stimuli. These may also be received rhythmically, in the sense that flies putting out the signals might release them in a periodic manner (recall that *per*-null flies desynchronized the behavior of individuals in a target group). In any case, the incoming cues were further concluded to be chemicals coming in mainly via antennae (Levine *et al.*, 2002a). However, it is important to keep in mind, on the one hand, that *Drosophila* antennae are also the "ear" of this insect (reviewed by Tauber and Eberl, 2003) and, on the other, that $per^+$ is normally expressed in essentially *all* peripheral structures of *Drosophila* (reviewed later). Thus, the promoterless *per* transgenic would have been rhythm disabled throughout its PNS. The major social cue in these experiments was nevertheless hypothesized to be olfactory (as opposed to contact chemosensory via mouthparts and legs or mechanosensory via bristles all over the animal). This presumption was derived not only from the *smellblind* and odor-blow effects (Levine *et al.*, 2002a), but also because surgical removal of antennae from wild-type target flies caused them to be as unresponsive as nonphase-shiftable $para^{sbl}$ individuals or those carrying the truncated transgene (unpublished data mentioned in Levine, 2004).

Food has come back into vogue as a stimulus studied for its entrainment effects on mammalian rhythms (reviewed by Schibler *et al.*, 2003). A fly-food chrono connection was discussed earlier with regard to the *takeout* gene, its expression in the alimentary system, and effects of a *to* mutation on the survival of nutritionally deprived flies. The phenomena just reiterated do not, however, presage any relationship of gustatory cues to clock resetting in *Drosophila*. Indeed, a study performed by Oishi and co-workers (2004) was unable to detect any food-anticipatory locomotion (*cf.* Schibler *et al.*, 2003) nor any effect of restricted feeding schedules on gene-expression profiles of the $per^+$ or $tim^+$ genes.

Therefore, we are not faced with tracing anatomical pathways originating in the mouthparts or legs of the fly (see earlier discussion), whose destination would be one or more groups of clock gene-expressing neurons. However, the clock-resetting effects of visual inputs initiated at the periphery, and analogous kinds of olfactory effects, indicate that not enough is known about PNS-to-CNS connections that contribute to the operation of the rhythm system of *Drosophila*. For example, a remarkable number of findings are emerging about where antennal afferents "go," including where the odor inputs in question end up deep within the brain (e.g., Jefferis *et al.*, 2002), but these data include no information about the potential for an antenna-to-LN pathway (whether direct or via interneurons that would eventually target those laterally located brain pacemakers). With regard to compound eye into CNS pathways—forget it, little hard knowledge has been obtained that goes deeper than the optic lobes, even from a general brain-behavioral perspective, let alone a specifically chronobiological one.[26]

### Actions of Rhythm-Related Genes During *Drosophila* Development

This subtopic should start with a disclaimer about how little is known about clock genes expressed among life cycle stages, except for the cases of *period* and *timeless*. Reasonably intensive assessments have been executed as to when the products of these two genes are detectable during development. Indirectly, in a sense, we also know about the stage-by-stage expression of certain additional rhythm-related factors, referring to genes with distinct development functions as such; this of course speaks to the proteins that were classically or semiclassically uncovered initially from *non*-chrono standpoints in relatively recent times.[27] However, there is a hapless paucity

---

[26]With regard to external eye to CNS with the rhythm system, there are only some brief speculations about possible pathways: Helfrich-Föster *et al.* (2000a) stated their idea that centripetals from certain compound eye photoreceptors, called R7 and R8, which are centrally located within a given facet, may connect with *l*-LN$_v$ dendrites; the latter neurons are located near the surface of the medulla and their "receiving neurites" may ramify into that optic lobe where the R7,8 afferents terminate. Helfich-Förster (2003) went on to conjecture that, in addition to or instead of R7,8 involvement, peripherally located intra-ommatidia photoreceptors called R1–6 could have their light inputs "reach" the same *l*-LN$_v$s indirectly through certain lamina interneurons that are innervated by R1–6 axons projecting inward to this distally located optic lobe.

[27]Such as what is encoded at the *dbt* = *dco* locus (Price *et al.*, 1998; Zilian *et al.*, 1999) and that of *tws*, along with two other genes known to specify additional subunits of the "TWS phosphatase" [see Sathyanarayanan *et al.* (2004) and the legend to Fig. 1], or as began to be recognized in the mid-1970s for *sgg* (the classically lethal alleles for which are cited elsewhere).

of information about when nonvital chronogenes in addition to *per* and *tim* become active and then sustain their expressions, except with regard to transcript or protein production during adulthood. Also, these determinations tend to have been performed at low spatial resolution. For instance, *Clock* makes its mRNA in adult "bodies," i.e., tissues posterior to the head, where the presence of this transcript has been monitored in gross homogenates (Darlington *et al.*, 1998), in parallel with a report of low-magnification, low-resolution images showing where *Clk* (also *cyc*) mRNAs are found *in situ* within the adult head (So *et al.*, 2000). However, one would like to know when these particular genes kick into action at an earlier stage, given the neuroanatomical problems caused by mutations in these genes (see earlier discussion and re-refer to Park *et al.*, 2000a). One is therefore forced to limit the upcoming descriptions and discussions to these kinds of data as collected for *per* and *tim* during the three major stages of development.

### Embryos

The *period* gene starts to make its RNA (James *et al.*, 1986) and protein (Siwicki *et al.*, 1988) about halfway through the embryonic stage [buttressed by the report of Liu *et al.* (1988) of embryonic staining for the β-galactosidase activity that was encoded by a *per-lacZ* fusion transgene]. Resolving such *per* expression showed it to occur only within the CNS and in every such neural ganglion, including observations of ventrally located signals within the thoracic and abdominal ganglia of the VNC. Are the products of this gene appearing in such an early manner merely to gear up for larval functions (see next subsection), or could this gene exercise some chonobiological task during relatively late embryonic life? The first such question needs to dovetail with the fact that the embryonic pattern (spatially) is neurally broader than that which comes down within the larval CNS, although the precursors of certain lateral neurons, and adult ones as well (see later), are born or at least begin to express *per* at the end of embryogenesis (Malpel *et al.*, 2002). The second question suggests that it would be interesting to ask whether *per* and PER exhibit intraday cycling in embryos. Inasmuch as the duration of this stage is only 1 day at ~25°, only a "half-cycle" would be determinable in that thermal condition; so intraembryonic time points should be collected for animals progressing through this stage much more slowly, at temperatures less than 20°—but why would a clock gene bother to make its products cycle in embryos? The rhythm system of *Drosophila* does not include circadian gating of larval hatching, whereas other insects with elongated embryonic periods do operate a day-by-day clock during that stage for the purpose of regulating rhythmicity of the embryo-to-larval hatch [e.g., Sauman and Reppert (1998); although see Sauman et al. (1996) vs Sauman et al. (2000) for how *period* gene involvement in this event, associated with the early development of a moth, was discounted].

Just for the hell of it, let us entertain one particular temporal-type function for PER in embryos. For this, one resurrects long-forgotten observations that revealed the total developmental time to be changed under the influence of *per* mutations (Kyriacou et al., 1990). That these findings demonstrated squeezing down the durations of more than one major stage in developing $per^{Short}$ animals or lengthening them in $per^{Long}$ ones defocuses attention on embryos. However, it could be that the stage is set for this gene's temporal regulation of total development according to its products' appearances during the first stage. The chronobiological function so surmised would not necessarily have anything to do with a circadian function, if only because the overall timescale is approximately 10 days; also the extent to which $per^{S}$ and $per^{L}$ change free-running periodicities (5 h in appositive directions) did not correlate with the magnitudes of these mutations' effects on development. Incidentally, the "gating" for end-of-metamorphosis eclosion times in these *per*-mutant cultures was decoupled from effects on developmental durations, because these mutations retained their influence on embryo-to-adult time spans even in LL, a condition in which eclosion is completely aperiodic (Kyriacou et al., 1990).

The *timeless* gene might also contribute to the control of developmental timing, in that this gene also makes its first expressional appearance within the CNS of embryos (J. Blau and M. W. Young, unpublished observations cited in Hall, 2003a). Both *tim* and *per* continue to churn out their mRNAs and proteins during all subsequent stages, naturally, as will begin to be considered in the immediately ensuing passages.

## Larvae

The *per* gene, and increasingly the *tim* plus *cry* ones, are increasingly being dealt with in terms of "where expressed" and "what do the products mean" during larval stages. Thus, the *per* pattern within the brain consists of lateral and dorsal neuronal groups; stainings of PER antigenicity and of β-GAL-reported *per* gene expression showed such cellular configurations to arise during the first larval instar (L1) and then to "mature" during L2 and L3. Most of the focus was devoted to the latter (prepupal) stage and included the following sorts of details (Helfrich-Förster, 1997; Helfrich-Förster et al., 2002a; Kaneko and Hall, 2000; Kaneko et al., 1997; Klarsfeld *et al.*, 2004; Malpel *et al.*, 2002), which subsume larval CNS assessments of not only *per* and *tim* expressions, but also those of the *Pdf* gene and those encoding photopigments: (a) There are fewer larval CNS LNs and DNs compared with the numbers that emerge during metamorphosis and are sustained into adulthood. (b) Most, but not quite all, of the LNs coexpress PDF, beginning in early L1. (c) These larval LNs are precursors of the heavily discussed $s$-LN$_v$ cells within the adult brain, monitored by L3 → pupal monitorings of both *per* and *Pdf* expression within such developing neurons; it follows that the $l$-LN$_v$ cells of adults are born late or they are present earlier but undetectable because they lack PER and PDF until the end of larval life. (d) Cycling of PER and TIM within all LNs plus DNs is robust within L3 brains, both in LD and in DD; however, larval DN2s (which may be precursors of the neuronal cluster given the same name for adult brains) cycle 180° out of phase compared with staining cycles quantified for the other cell groups; this means, for example, that TIM is "high" during the daytime within DN2 cell bodies, which knocks down the wooden soldier that might have imagined this protein to be inherently light sensitive in terms of its rapid proteolysis in other circumstances (see earlier discussion). (e) The LNs and two DN cell groups (1 and 3) have been shown to contain *cry* products by *in situ* observations (Emery *et al.*, 2000b; Klarsfeld *et al.*, 2004) during L3 [contrary to the false-negative results of Ishikawa *et al.* (1999) from Northern blotting assessments], whereas DN2s do not. Recall that CRY:TIM interactions are believed to trigger degradation of the latter when cells containing both proteins are exposed to light (by the way, CRY cannot accumulate either during the L phase of an LD cycle, e.g., Emery *et al.*, 1998). (e) Introduction of a UAS-*cry*$^+$ construct into this cluster of L3 dorsal neurons (via a *tim-gal4* driver) led to complete reversal of the (usually) anomalous PER cycling with DN2s; this CRY function, artificial as it is at least within DN2 cells, *might* speak to an actual pacemaking function for this protein within larval brain neurons, an issue that should be kept in mind when reading the next major section (following the subsection about pupae).[28]

What are the actions of these clock genes and what are the larval brain neurons in which they are expressed doing? No locomotor rhythmicity of L3 animals has been detectable (Sawin *et al.*, 1994). However, certain features of rhythm-system functionality have been inferred to be running during larval life.

---

[28]Incidentally, the DN2s of adults naturally contain CRY, as do neurons within the other cell groups diagrammed within Fig. 2, such that the out-of-phase clock protein cycling for DN2s in mature flies was a candidate for the "CRY phase reversal" experiment (Klarsfeld *et al.*, 2004).

First, consider circadian photoreception at this life cycle stage: Input routes to, and functions subserving, the relevant neurons, light responsiveness seem simpler for larval brains, in that combination of a *norpA*-null mutation with *cry*[b] knocked out the usual LD synchronization of PER and TIM cyclings within the L3 stage LNs containing these clock proteins (Kaneko *et al.*, 2000a). (That the *cry* mutation has this profound effect, provided that it is combined with the *norpA* one, indicates further that larvae *do* produce cryptochrome and thus that probing a homogenate for *cry* mRNA is an insufficiently sensitive measure.) The phototransduction factor defined by the mutation in the *norpA* gene and the rhodopsins acting upstream of the PLC it encodes almost certainly exert their chronobiological functions within the anteriorly located larval photoreceptor organ; it contains this PLC and sends an axonal projection to the LN region within L3 brain lobes (Helfrich-Förster *et al.*, 2002a; Kaneko *et al.*, 1997; Malpel *et al.*, 2002). To jump the gun somewhat in terms of life cycle stages, it should be first specified that the structure in question is called Bolwig's organ (BO), an anteriorly located part of the anatomy that is classically viewed as the main input route for light into an insect larva. That the BO is likely to be involved in the rhythm system of *Drosophila* operating during these stages was first pointed out by Kaneko *et al.* (1997), who observed the centripetal projection of Bolwig's nerve to terminate near the Lateral Neurons of L3 larvae. Subsequently, rhodopsin gene expression was observed in the BO during this stage (which jibes with the *norpA* effect noted earlier). This persists through metamorphosis (Helfrich-Förster *et al.*, 2002a; Malpel *et al.*, 2002), allowing one to discern that the BO turns into an "eyelet" structure, which functions within the adult system of circadian photoreception (see later).

The second subtopic within this subject engages the behavioral meaning of larval photoreception. We start by noting that light pulses delivered to larvae can entrain not only periodic eclosion, occurring days later, as taken up in the next subsection, but also cause phase shifts of adult behavior, occurring even longer afterward (e.g., Brett, 1955; Sehgal *et al.*, 1992). This implies operation of a time-memory clock within the larval CNS, and it was found to be *per* controlled by Kaneko *et al.* (2000a). Light pulse pairs delivered to developing *per*[S] or *per*[+] larvae demanded the conclusion that the time memory clock runs faster than normal in this mutant, as revealed by the eventual locomotor phases of pretreated L3 animals. Malpel and co-workers (2004) delved into the later (behavioral) consequences of the L3 photoreceptive system in some further genetic detail by showing that subjecting *glass*-null larvae to LD cycles led to arrhythmic behavior for fully half of the eventual mutant adults, whereas extension of such an entrainment regimen into adulthood restored near-normal locomotor cycles (consistent with the *gl*-mutant results obtained by Helfrich-Förster *et al.*, 2001; Vosshall and Young, 1995). It would seem, however, that the *gl*-null effects on the flies' eventual locomotor rhythmicity (following larval stage light treatments) could be covered by direct CRY-mediated photoreception within L3 LN cells (see earlier discussion). However, impinging on afferent inputs to more centrally located neurons can, in general, lead to malformation of such targets. In this regard, *glass* mutations are among the visual system variants of *Drosophila* that befoul development of the BO, discussed early on by Vosshall and Young (1995) and Kaneko *et al.* (2000a) in a rhythm context (there is more about this mutant later); and differentiation of larval LNs is defective in animals lacking the BO-to-LN axonal pathway [Malpel *et al.* (2002), within which this particular experiment took advantage of the previously described cell-killer transgene fused downstream of GL-responsive regulatory sequences]. Malpel *et al.* (2004) extended their lit-up larva experiments to monitor the eclosion of *glass*-mutant cultures: Flies emerging from them eclosed aperiodically after being put through LD cycles, or given a one-time light exposure, as larvae (harking back to the findings of Brett, 1955), whereas these mutant cultures emerged rhythmically when reared in LD through late metamorphosis. This brings us to the final "clock developmental" subtopic.

*Pupae*

Expression of our two workhorse clock genes *per* and *tim*, along with the *Pdf*-encoded marker for some of the relevant neurons, provides almost all we know about descriptions of metamorphic development occurring within the rhythm system of *Drosophila*'s CNS (Helfrich-Förster, 1997; Kaneko and Hall, 2000; Malpel *et al.*, 2002). Suffice it to say that the larval-to-adult CNS pattern basically matures during the pupal period, obviously becoming more adultlike as metamorphosis progresses. Certain details, or residual questions, are worth noting, however: The extent to which the neural patterns of larvae are retained as the developing animal moves toward adulthood is not fully understood, except insofar as the early born, then relentlessly persisting, precursors and mature brain cells called $s$-LN$_v$s are concerned (see earlier discussion). Those PDF-containing neurons (from early larvae onward) are added to by similarly differentiated $l$-LN$_v$s in the middle portion of the pupal period. However, it is unknown whether the later developmental appearance of the latter cell group, which in adults also contain PDF but send axons to very different locations compared with $s$-LN$_v$ projection (Figs. 2 and 3), involves late birthdays of novel cells or the (at last) production in preexisting ones of PER, TIM, and PDF proteins. It is intriguing perhaps that certain PER-continuing DN cells within larvae either disappear during metamorphosis, to be replaced metamorphically by analogously positioned neurons that turn on their *per* gene soon after they are born, or such cells could be retained anatomically, turn off the expression of that gene, and then reinvigorate PER production later. The latter events seem more likely because *tim* expression in these DN cells is sustained before, during, and after the *per* lull (Kaneko and Hall, 2000).

Another aspect of DNs in development and in flies was addressed by Klarsfeld *et al.* (2004). Recall that those investigators detected GLASS protein within a subset of DN1 neurons, in context of their belief that a *gl*-null mutation results in the eventual elimination of only a portion of such cells with the adult brain (contra Helfrich-Förster *et al.*, 2001, who detected no such cells within the imaginal brain of the same mutant). Klarsfeld and co-workers went on to deduce that the $gl^+$-expressing DN1 cells in mature flies represent new neurons of this type; i.e., cells that did not contain GL during larval life are unaffected by elimination of this protein during that stage but take on the expression of this gene (along with that of $per^+$) as they "appear during metamorphosis." Alternatively, such cells were always there but could not be marked by either anti-GL or anti-PER.

Who cares about new birthing of neurons that then make clock-gene products, compared with cell differentiations that are the only late-occurring events? Well, it is worth musing about the extent to which the larval rhythm system, such as it is for actual chronobiology during this stage, is fundamentally similar anatomically to the late-stage systems (although the unmarkable early progenitors would not possess all the chrono-gene products present in the mature forms of these cells), as opposed to the possibility that altogether separate early *vs* late neurons are rhythm regulators only in vaguely analogous ways (with many of the relevant early *vs* late neurons containing clock factors and related ones, but bearing no cellular relationship to each other).

One final and heretofore unmentioned item about *per* expression compared with that of *tim*, as *Drosophila* converts itself from a maggot to an imago: VNC expression is rather prominent neuronally for the former gene during larval life; much so for the latter, including that *tim* products are quite spatially restricted in this ventroposterior portion of the developing CNS (Kaneko and Hall, 2000). As the pupal period proceeds, VNC expression of these genes takes on a new character because at least the *per* products are detectable only in glia (Ewer *et al.*, 1992). This matter resurfaces in a later section about "temporal-phenotype" pleiotropy and the influence of clock factors on certain noncircadian behaviors.

Chronobiological significance of outputs from the pupal system of rhythm factors that are expressed in the nervous system (whether in adult-homologous neurons or not) must be rather special. Pupae do not behave until mature such animals are ready to emerge. These eclosion-related actions involve a given individual insect exercising its one-time behavioral subroutine, such that the circadian feature of this phenomenon is appreciable only at the population (whole culture) level. This is of course in distinct contrast to what a mature fly does as it cycles in and out of robust locomotion for days on end. Therefore, the output system for behavioral rhythmicity might involve factors that minimally overlap with those acting downstream of clock gene actions carried out by pupae. This brings us to deliberations about gene products and other factors functioning during metamorphosis to regulate rhythmic eclosion.

It must be that "hands" of the pupal clocks are extant and have meaning for an overt feature of *Drosophila* rhythmicity—eclosion—compared with the inferences and formalisms revolving round the larval clock and that it possesses pacemaking "cogs" only (*cf.* Shirasu *et al.*, 2003), with no known hands for outputting these core functions into any heretofore appreciable piece of biology. First, let us remember that most clock mutants in *Drosophila* affect periodic eclosion by either exhibiting none of it or causing appreciable alterations in time spans between adult emergence peaks in constant darkness. (Incidentally, there are way too few studies in which eclosion of *D. melanogaster* cultures has been monitored in LD cycles.)

Therefore, this subsection is, in principle, connected to major themes running through nearly all previous sections. Having said this, eclosion in *Drosophila*, and even as it of course occurs in other better-studied insects is a pain to deal with verbally and otherwise. One such fly-o-centric problem relates to the small size of *D. melanogaster*, which has militated against performing the direct animal interventions that can tease out information about the manner by which the "neuroendocrine axis" of this organism functions to regulate transitional stages of the life cycle (not only pupal to adult, but also prepupal molts). The so-called axis just noted also speaks to the fact that the regulatory factors, both cells and substances, mostly have to do with phenomena not yet tied into the clockworks of *Drosophila* or any other insect. Even when eclosion-related genes are involved inferentially, such as those encoding the peptides mentioned later, or even when certain of these factors have been cloned and manipulated to allow the performance of genetic experiments on eclosion and other developmental transitions, what these studies reflect is a fair amount of neuroendocrine wheel spinning: (1) Softly apprehendable phenomena, involving a variety of empirically based and conceptual ambiguities (see later), and (2) little in the way of firm relationships to substantive elements of the pupal clockworks (with a couple of exceptions, which are dwelled upon briefly at the end of this subsection).

With reference to item 1 just stated, let us start by referring interested readers to a handful of reviews that variously cover eclosion in *Drosophila* and other insects, prepupal ecdyses in both (but mainly summarizing non-*Drosophila* studies) and the relevant neurohormones (as directly encoded by genes or not).[29] Also, here are some samples of what is going on in this area, as reflected in a handful of primary papers. Studies of the expression of ecdysis- and/or eclosion-controlling hormones, along with manipulation of certain genes related to actions of these factors, have to led to some conclusions about the functional hierarchies by which these regulatory factors operate. For instance, cloning in *Drosophila* of the *Eclosion-Hormone* (*EH*) gene and that which encodes a homolog of crustacean cardio-active peptide (CCAP) created the wherewithal to ablate neurons that usually contain these peptides (e.g., Clark *et al.*, 2004; McNabb *et al.*, 1997; Park *et al.*, 2003). It is worth remarking on the misnomer that characterizes the *EH* gene product: ablation of all cells that normally contain this hormone leave *Drosophila* cultures eclosion enabled, although dynamics of the adult emergence profiles and of certain microevents associated with eclosion biology and behavior were anomalous (McNabb *et al.*, 1997). Additional and more recently obtained results of this sort suggested that EH and ecdysis triggering hormone (ETH) affect their mutual ecdysis-supporting release by positive feedback effects (in *Drosophila*, too, but as previously established from neurohumoral experiments performed on larger insects, notably moths). However, ETH could be released in the absence of EH; both substances were found to contribute to ecdysis-associated filling of trachea; yet EH was inferred to be the major regulator for this (Clark *et al.*, 2004). This factor is "classically" appreciated to act upstream of the role played by CCAP (*cf.* Park *et al.*, 2003), but current results suggested that EH can act independently of the latter (in terms of how the axis eventually "gets to" functions of CCAP within the VNC, as exemplified in the next paragraph).[30] The ensemble of these

---

[29]See Truman and Morton (1990), Truman (1992), Jackson *et al.* (2001), Ewer and Reynolds (2002), and Hall (2003a), the third and fifth of which such reviews are slanted toward reviewing eclosion phenomena in *Drosophila*.

[30]For the moment, make note of the *CCAP-gal4* driver applied in these (Clark *et al.*, 2004) and previous experiments (Park *et al.*, 2003). In order to disrupt the "CCAP subsystem," this transgene was combined with a UAS cell killer, naturally or in other experiments. The latter focused on and led to transgene-induced defects in eclosion-associated wing expansion, whereby *CCAP-gal4* was combined with a UAS drivee that encodes a conditional blocker of synaptic transmission (*cf.* Kitamoto, 2002a). The latter (known as UAS-*shi^{TS}*) is a better tool than the aforementioned, analogously acting UAS-*TeTxLC* transgene, which cannot have its effects turned on or off during a given life cycle stage (Martin *et al.*, 2002). These matters are discussed within Method 9.

findings, and those resulting from additional accompanying experiments, forced Clark and co-workers to worry that they are "not easily explained by currently available hypotheses" (*cf.* Ewer and Reynolds, 2002). Stay tuned for more neuroendocrine grinding, although such studies will potentially be given further boosts by the gene-manipulation opportunities just exemplified with respect to stage transition studies involving *D. melanogaster* development and emergence into adulthood.

What about the rhythmic feature of eclosion (item 2, described earlier) and its regulation by factors other than the "far downstream" ones implied in the foregoing paragraph? It seemed as if a value-free screen for eclosion variants might have tapped investigators into genes, some of which could function closer to the core control of these processes. In this regard, the *lark* mutant was isolated as a transposon-induced, early enclosing variant (Newby and Jackson, 1993). The mutation was quickly shown not to affect adult behavior rhyhmicity but *lark* is most pleiotropic otherwise, because only *lark*/+ flies could be eclosion monitored, i.e., the mutation is a recessive embryonic lethal (Newby and Jackson, 1993). A wealth of further information about the *lark* gene product, its tissue expression patterns, and eclosion involvement has been generated (see Hall, 2003a; Jackson *et al.*, 2001). Only a few highlights are mentioned here (also see the legend to Fig. 5): *lark* expression is not rhythmic molecularly, although the encoded protein—an RNA-binding factor—does cycle within certain neural cells of late pupae (not in larvae, however). The nadir that LARK reaches within the last day before eclosion led to the notion that this naturally occurring decrement could alleviate a repressive effect on adult emergence. Implicationally—and whereas LARK seems not to be a *core* clock factor because it only influences one known rhythm—the hypothetical "alleviation" would involve a *regulatory* process as opposed to this protein being a distal effector of eclosion-associated events (the neuropeptides described earlier). The hypothesis just stated received support from some formal genetics (results that were obtained more or less in the same spirit of how *lark* was identified in a "forward genetic" screen, compared with cloning a priori known genes such as *EH* or *CCAP*). Thus, extra doses of the *lark*+ allele were found to cause later than normal eclosion peaks, the opposite effects of the decrement of function genotype that led to the "early bird" isolation phenotype. Finally, a nod to the expression patterns of *lark*, which are spatially rather vastly; the same for life cycle stages (see reviews cited earlier).

Here is one curiosity about a microcomponent of where LARK finds itself within the CNS of *Drosophila*: This RNA-binding protein is coexpressed with CCAP in pupae (it is unknown what such cocalization means in terms of any findings involving or ideas about the cellular biochemistry). LARK exhibits a histochemically observable staining cycle within such neurons—most conspicuously those distributed down the length of the VNC (e.g., Park *et al.*, 2003), which parallels results obtained from temporally controlled homogenates of pupae (see earlier discussion). Interestingly, when late metamorphosing animals carrying a *per*-null mutation were sacrificed and stained for LARK at different times throughout a 24-h period, the profile for this protein stayed flat. However, recall that *per*[+] has been concluded to be expressed only within glia of the VNC (Ewer *et al.*, 1992), so one wonders whether regulation by the *period* gene of oscillating LARK levels within CCAP neurons is mediated intercellularly.

Do any of the neural cells alluded to or mentioned in the foregoing discussions—the ones introduced by specifying items 1 and 2 near the top of this subsection—have *any* connection with the core clockworks of *Drosophila*? "Connection" in this sense could refer to certain components of clock gene actions, which might influence the production or release of *EH* gene products and the like (leaving aside for the moment the micromatter of the relationship of *per* to *lark* in rather downstream anatomical regions of the animal); refer to cellular interactions that might underlie influences of pacemaker neurons with those located elsewhere (even within the brain) and contribute to circadian-gated eclosion by way of releasing "distal effector" molecules (once again, let us forget about LARK matters and the soft possibility that PER glia might interact with CCAP neurons); or refer to both. There is but one set of experiments that has begun to cope with the historical inadequacies just whined about.

Myers *et al.* (2003) began this study with manifest knowledge about *per* and *tim* actions, including their regulation of rhythmic adult emergence (both genes were in fact discovered via aperiodic eclosion mutants); the less well-remembered phenomenon of prothoracic gland rhythmicity in *Drosophila* (Emery *et al.*, 1997), which has been demonstrated in this insect only in terms of intra-tissue PER oscillations (this was the seminal case of a not strictly neural rhythm in *D. melanogaster* shown to operate autonomously with a relatively peripheral organ); and the reasonably well-appreciated fact that the *Pdf* gene begins to make its product way before the only known chronobiological function of PDF is mediated (in adults). Myers and co-workers (2003) thus expressed normal *timeless* function only in neurons (by combining the pan-neuronal-*gal4* driver described earlier with UAS-*tim*[+]); eclosion "remained" arrhythmic, i.e., according to the background genotype of these late pupae, which included homozygosity for a *tim*-null mutation. Among the many tissues that were TIM-less in these animals was the PG. It was known to be eclosion involved because a drop in ecdysteroid release from PG cells just prior to adult emergence is necessary for that event in other insects (Ewer and Richardson, 2002). Might the PG require clockwork functioning for eclosion to be robustly gated? Myers *et al.* (2003) thus looked for PER and TIM oscillations within the PG by taking late pupal time points and found staining cycles for both (*cf.* Emery *et al.*, 1997). In the PG experiment that ensued, temporally constitutive expression of *tim* was affected by a driver that produces GAL4 in this organ; the eclosion profiles of such doubly transgenic cultures were essentially arrhythmic. Bringing the possibility of neuronal involvement back into the picture, these investigators inferred from evidence reported elsewhere that projections from LN cells may innervate the PG "indirect-ly." Thus, our old friend the *Pdf-ga4*/UAS cell-killer combination was applied to eliminate pupal LNs; eclosion was "gated" for only a couple of DD days in these transgenic cultures. Therefore, normal LN clockworks (intracellularly) are insufficient for sustained, free-running, periodic eclo-sion (i.e., if the PG cells are $tim^0$ in neuronally $tim^+$ pupae), but the presence of LN cells is necessary for such rhythmicity.

PDF is of course present within pupal LNs. Myers *et al.* (2003) tested the effects of eliminating this peptide and found that $Pdf^{01}$ cultures ex-hibited aperiodic eclosion, which was correlated with a lack of the usual intra-PG cycling of PER or TIM in late pupal specimens. This led Myers *et al.* (2003) to conclude that the coordinated action of pupal LNs and the PG involves LN-to-PG communication via PDF-mediated humoral or paracrine influences of the former neuronal cells on clock functioning within cells of the latter endocrine organ. Is the "LN PDF release" just implied cyclical, as it is inferred to be within the dorsal brain of adult flies? Recall Park *et al.* (2000a) in this regard, but also see Helfrich-Förster *et al.* (2000), who contemporaneously reported a PDF staining cycle in the region of centripetally projecting $s$-LN$_v$ axon terminals. The latter such study is the key one to recite here, because it included eclosion monitorings of cultures in which a UAS-$Pdf^+$ construct was ectopically and also perhaps overexpressed (by, e.g., combining the transgene just noted with a set of brain "enhancer traps," like the strain of that sort that was introduced into this brain-behavioral story many pages ago). Eclosion profiles were severely disrupted (at least being erratic if not flat-out ar-rhythmic) under the influence of these transgene combinations (Helfrich-Förster *et al.*, 2000). This provided the first tentative evidence suggesting the presence of PDF during late development to have a clock output meaning.

The principal *input* to the eclosion clock of an insect is light, as usual (although see Zimmerman *et al.*, 1968). Long ago, it was demonstrated by phase-shifting adult emergence rhythms that the external eyes of meta-morphosing *D. pseudoobscura* are not only not needed as a conduit for light stimuli, but they are also irrelevant (Zimmerman and Goldsmith, 1971). Correlative findings from action spectra determinations indicated (although not very definitively) that the maximal sensitivity for photic phase shifts of eclosion peaks was "in the blue" (Frank and Zimmerman, 1969; Klemm and Ninnemann, 1976).[31] It was as if a flavoprotein might be involved as opposed to rhodopsin (*cf.* the vitamin A deprivation experiment of Zimmerman and Goldsmith, 1971). CRYs are in the former category. Therefore, a façile prediction suggests itself, which is that *cry*[b] cultures of *D. melanogaster* would not be eclosion entrainable. This formed part of the paper by Myers *et al.* (2003). However, the result (singular, an aperiodic profile from one mutant eclosion "run") flies completely in the face of the *cry*[b] eclosion testing that was performed contemporaneously by Mealey-Ferrara *et al.* (2003). Developing mutant animals were quite solidly entrained to eclose rhythmically even, in most of the companion experiments when a *norpA*-null mutation was present simultaneously. One might gingerly favor the latter results because several independently performed eclosion runs had their profiles depicted, their adult emergence data formally analyzed and tabulated, or both. In Myers *et al.*'s favor, however, *cry*[b] was *also* shown to eliminate the aforementioned prothoracic gland rhythm, which usually can be detected by temporally controlled histochemical experiments performed with pupae. Who knows how to resolve these discrepancies? Given the last author of Mealey-Ferrara *et al.* (2003), the current author proposes gingerly that it would not beg the imagination for more than one photic route into the circadian pacemaker for eclosion to be operating; in other words, a *cry* (and even *norpA*-)-independent one in addition to that which was knocked out to no effect on periodic adult emergence in one of these studies.

---

[31]These kinds of action spectral data for adult locomotor rhythmicity are far more complex, as inferred by the diagram in Fig. 4, especially the dense description of it in the legend. In brief, the partial (Zordan *et al.*, 2001), more complete (Ohata *et al.*, 1998; Suri *et al.*, 1998), or quite comprehensive findings (Helfrich-Förster *et al.*, 2002a) pertinent to this mature fly point are consistent with photopigment contributions, to the entrainment of daily rhythms in LD cycles, of both rhodopsins functioning within external light-receiving structures and of cryptochrome functioning more centrally as a circadian photoreceptor molecule. Moreover, the detailed light *level* sensitivity measurements of Helfrich-Förster *et al.* (2002a) dovetail nicely with preliminary determinations of such behaviorally associated values by Ohata *et al.* (1998); e.g., that nonfunctionality or absence of external eyes causes locomotor-rhythm entrainability of such deprived flies to exhibit poor sensitivity (*cf.* Stanewsky *et al.*, 1998). See Fig. 4, especially its legend, for much more about these matters, primarily involving all sorts of visual mutant applications.

The Rhythm System of *Drosophila* as Defined Genetically Influences
Weakly Appreciable Circadian Phenotypes, as Well as
Noncircadian but Temporally Based Characters

The findings to be reviewed within this section may be comprehended
as cases of errant pleiotropy. That is, these effects of rhythm-related
variants may seem to stray into random areas of *Drosophila* biology,
by analogy to the fact that vital clock genes in their null state cause an
assortment of defects that are uninteresting from any chronobiological
standpoint. However, the current discussion of (in the main) *inessential*
rhythm factors and associated phenotypes still leaves us masters of various
temporal domains. (Facts brought to light in the subsequent, penultimate
section of this article will suggest that we have finally lost our way
chronobiologically.)

*Sleep*

Do fruit flies sleep during the former phase of their rest–activity cycles?
It is believed that they and other insects do (e.g., Sauer *et al.*, 2003; Tobler,
1983), according to findings that began to appear early in this decade.
However, reviews about the *Drosophila* side of this subtopic are beginning
to outstrip the weight of empirical evidence that keeps getting summarized
(e.g., Greenspan *et al.*, 2001; Hendricks, 2003; Hendricks and Sehgal, 2004;
Ho and Sehgal, 2005; Shaw, 2003). Nonetheless, the "sleep-rebound" effects
of rest deprivations, to which *Drosophila* have been subjected in several of
the experiments cited in these articles, is one if not the best piece of support
for the resting of this insect being a "sleep-like state," by analogy to the
effects of sleep-depriving mammals (e.g., Kilduff, 2000). It should be noted,
however, that not very many of the sleep-related studies performed on
*D. melanogaster* make an explicit connection to circadian locomotor
rhythms as such. A more general concern (brought to the author's attention
by D. M. Edgar, personal communication) is discussed between here and the
end of the paragraph; this quickly brings us back to the rest-rebound side of
the fly sleep story. Thus, we ask whether compensatory inactivity after
forced sustained activity constitutes bona fide, compensatory sleep in an
insect (for studies pertinent to the impending wet-blanket discussion, see
Levenbook, 1950; Tobler, 1983). Such animals have difficulty managing acid
buildup in the hemocoel, which must be "ventilated away" through the
trachea excreted via the Malpighian tubules (which happen to enter our
chronological discussion in a different context later on). Sustained motor
activity, particularly as mediated by the low-efficiency musculature of insect
legs, causes the hemolymph pH to drop (as is directly measurable in larger
insects). Such buildup causes a neurological cessation in motor activity until

the acid is sufficiently cleared. That this can be ameliorated and affect rest *vs* activity has been revealed by sampling the pH of the hemolymph of an insect as a function of forced sustained motor activity, followed by introduction of a buffer into the hemocoel to achieve long durations of sustained motor activity or accelerated rates of recovery from sustained activity.

Have these pieces of rather "low-level" physiology and blood chemistry been addressed appropriately in the context of rest in *Drosophila* being one of the ol' "model systems" for studying sleep-like processes and thus potentially to reflect upward toward higher animal systems whose potential for genetic analysis is less forceful? The answer is no. However, it is for the reader to decide, mainly by perusing the reviews cited here, as to whether fly people are into sleep research as part of their overall chrono-enterprise.

Two additional points will be made, which were tacitly previewed within the foregoing paragraph: Rhythm genetics of *Drosophila*, and certain elements of its biochemical genetics, have been brought to bear in experiments that purport to monitor sleep-like phenotypes. Only one of these studies will be mentioned, wherein Shaw *et al.* (2002) started by following up the earlier applications of *per*- and *tim*-null mutations in sleep-rebound tests. The former was normal for this character; but the latter was reported to be reboundless after relatively brief rest deprivations, yet exhibited robust rebounds after *ca.* one-third to one-half day deprivations. The most interesting clock mutational effect was observed in analogous tests of *cyc*-null flies: After deprivation times of 10 h or more, this mutant type exhibited substantial mortality; around one-third of the individuals died over the rather short time spans involved, whereas none of the other clockless mutants were killed to any noticeable extent (Shaw *et al.*, 2002). Given the nature of the gene product of *cyc* and how it affects day-to-day locomotor rhythmicity, what does this mortality mean for sleep control? Nothing, for now. This dying-while-rebounding character, however, does drive home the point that there is a good deal of "noncircadian pleiotropy" associated with the clock genes of *D. melanogaster*. In fact, if we wanted we could call the *cycle* gene a vital one, in that effects of one of its null mutations can kill, however baroque the circumstances just described may be, and that *cyc* mutants exhibit neuroanatomical abnormalities (Park *et al.*, 2000a). Is it a wayward further coincidence that extension of the *cycle* gene beyond the rhythm realm was suggested in an auxiliary manner by the fact that it is the only known clock factor to be expressed "promiscuously" within the cultured, nonneuronal cells referred to about a zillion pages earlier? (See Method 3.)

The second matter arising—in the context of "sleep or not" in the fly—is that if "yes," real brain functions should be on the table. They are, investigatorily, as revealed by the titles of certain primary papers that include phrase such "electrophysiological correlates of rest and activity" (Nitz *et al.*,

2002) and "uncoupling of brain activity from movement" (van Swinderen *et al.*, 2004). These studies come under the heading of asking whether "arousal" processes are operating within the brain of *Drosophila* [reviewed by van Swinderen and Andretic (2003), but harking back to some much older studies in other insects, e.g., Kaiser and Steiner-Kaiser (1983); Tobler (1983)]. Certain such connections between electrode readouts and behavioral monitorings are an explicit feature of the primary papers just cited. In addition, these findings move this subsystem into obviously deeper levels of analyses compared with those that cope solely with whether the animal is moving around or not, displaying a rest rebound, or whatever whole animal activities may be grossly observable. Let us contemplate, however, the peroration that appeared at the end of one such "arousal state" study (van Swinderen *et al.*, 2004). This attribute of *Drosophila* brain activity— as inferred by these authors from recordings they performed in conjunction with supervising whether the appendages of the fly were moving—is, "like consciousness in humans . . . unlikely to be localized to a unique set of cells in the brain . . . [but] probably recruits dynamic networks extending throughout" the entire anterior portion of the CNS of the fly. This is heady stuff. So let us scrutinize momentarily what "unique set of cells" or "dynamic [neuronal] networks" may mean. The author is not previewing a generic cheap shot: "nothing yet," instead pointing out that it would seem most interesting to determine eventually whether any of the rhythm-related neurons, about which much is known morphologically and functionally, bear any relationships to the quasi-anatomical terms just quoted from within this publication. Connections in this respect could even mean that certain cells in question (within the brain regions recorded from by Nitz *et al.*, 2002; van Swinderen *et al.*, 2004) involve some of the so-called clock neurons—even explicit communication with LN cells and the like (Fig. 2) by physical contacts feeding from or into wherever and whatever the arousal centers may be.

## Learning and Memory

These phenomena involve "la recherché du temps." The mechanisms underlying it have ranged far, wide, and deep for decades. Take, for example, cyclic AMP. Metabolism of that small molecule and the knockon consequences of modulating its levels have long been appreciated to regulate any number of simple- as well as higher-learning phenomena (reviewed, for instance, by Dubnau, 2004; see the next subsection). The signaling functions swirling round cAMP prompted heavy analyses of related learning variants in *Drosophila*, such as those involving the *dunce* gene (cAMP phosphodiesterase), one that encodes the catalytic subunit of a cAMP-dependent kinase, another specifying a regulatory subunit for this enzyme, and also CREB factors (transcriptional regulators whose acronym decodes as cyclic AMP response element binders). Could it conceivably be that these four factors are *uninvolved* also in the rhythmicity of the fly (*cf.* Li *et al.*, 1998; Tischkau and Gillette, 2005; Whitmore and Block, 1996; Zatz, 1996)? The answers were "no" (Levine *et al.*, 1994), "no" (Majercak *et al.*, 1997), "no" (Park *et al.*, 2000b), and "no" (Belvin *et al.*, 1999). However, the effects of such chromosomal or molecularly engineered variants have never been dug into, for example, which brain cells might share these cAMP-controlling or -responding entities along with the products of rhythm genes, or at least how putatively separate sets of CNS cells (which may not simultaneously contain something like *dunce*'s cA-PDE and the PER protein) interact synaptically or otherwise.

The investigatory inadequacies just discussed speak to an additional short story in this area, which recounted the possibility of even tighter connectivity between circadian pacemaking functions and those involved in learning. This dreary tale started with the report of defects in experience-dependent *Drosophila* behavior observed for a subset of the clock variants—notably long-period *per* mutants and *Andante*, but also including some additional ne're-do-well rhythm mutants. Therefore, conditioned courtship (which is reviewed more generally and more recently by Siwicki and Ladewski, 2003) was said to be subnormal in these mutants that are, if you will, slothful in terms of how they "handle time" in an internally slow-moving manner (Jackson *et al.*, 1983). However, these genetic effects proved to be irreproducible (Gailey *et al.*, 1991; van Swinderen and Hall, 1995) and also nonextendable in tests of all three *period* mutant types for associative "classical conditioning" (again, *cf.* Dubnau, 2004).

The author will not bother the reader with citing the initially positive outcome of applying rhythm variants to learning tests (from >21 years ago), followed by registering the reversals (early to mid-1990s), if it were not for the fact that a possibility for chrono/mnemo connections will not seem to quit (*cf.* Arvanitogiannis *et al.*, 2000; Daan, 2000; de Groot and Rusak, 2000). Sakai *et al.* (2004) are now saying they find *Drosophila* males to exhibit substantial memory defects in certain *period* mutants. This recent resurgence of the potential for clock gene effects on experience-dependent courtship involves, if not apples and oranges with regard to the earlier studies, then at least Macintosh *vs* Golden Delicious. Thus, Sakai *et al.* (2004) were principally analyzing long-term memory after subjecting (mainly) $per^0$ males to lengthy training sessions with mated females (*cf.* Siwicki and Ladewski, 2003), followed by testing courtship-suppression a few hours later. (If any of the seminal experiments of this sort had validity, the positive results stemmed solely from courtship tests of long-period rhythm variants; $per^0$ ones were found to learn normally.) Males lacking *per* function were shown by Sakai and co-workers to be defective in such suppression or at least in terms of the robust and lengthy retention of it that was exhibited by wild-type controls (called long-term Memory in this reproductive-behavioral context). These *per*[0]–derived findings are belied by one of the courtship-learning experiments reported by McBride *et al.* (1999). Nonetheless, Sakai and co-workers went on to demonstrate that transgenically induced overexpression of *per*[+] enhanced LTM in their paradigm. However, circadian clock disturbances brought to bear in other ways led to no decrements in this experience-dependent behavior in that neither constant-light exposure nor tests of *Clk*-null and *cyc*-null males impinged on memory retention. One conclusion, therefore, is that "the … mechanism of LTM formation is independent of the circadian clock"—the *per*-null effects notwithstanding. However, these particular findings of Sakai and co-workers harmonize with the same kind of experiments performed by others on another feature of the reproductive behavior of *Drosophila* [see Beaver and Giebultowicz (2004), discussed two subsections hence].

## Habituation and Sensitization

The two experience-modulated characters of habituation and sensitization are examples of simple learning, as alluded to earlier. Experimental manipulations of animals or parts thereof, and the behavioral or physiological readouts of these *nonassociative* happenings, of course involve a time base. So perhaps it is not shocking that electrical stimulation of the so-called giant fiber pathway of *Drosophila*, which originates in the visual system and brain, *habituates*. Anterior CNS stimulations led to response decrements, measured at one of the pathway's end points in thoracic flight muscles (Megighian *et al.*, 2001). Curiously, flies expressing either of two *per*[0] mutations exhibited an earlier onset of such habituation when maintained in constant light compared with light:dark cycling conditions (Megighian *et al.*, 2001). The habituation values for wild type tended to be the same as baseline metrics determined for these mutants in LD, but exposing *per*[+] flies to constant light did not lead to shortened time courses for habituation to physiological stimuli.

In experiments involving the opposite kind of nonassociative learning, a rather wide array of clock mutants in *Drosophila* (five genes' worth) was found to exhibit subnormalities or anomalies of *sensitization* to drug applications (reviews: George *et al.*, 2005; Hirsh, 2001). Correlated with these findings were those with a neurochemical component, but against a background of some ancient epiphenomena associated with the *period* gene. Thus, accumulation of newly synthesized octopamine and tyramine was found by Livingstone and Tempel (1983) to be reduced threefold in brains dissected out of $per^0$ adults. Concomitantly, tryosine decarboxylase activity (TDC, which is required for production of the octopamine precursor tyramine) was reduced to around one-third normal by the effects of that $per$-null mutation, and TDC was claimed to be at about half-normal levels in $per^S$ and $per^L$ (Livingstone and Tempel, 1983). Elements of these seemingly meddlesome results were unwittingly examined further in another context by Hirsh and co-workers (see the reviews cited earlier). For this, Andretic et al. (1999) showed that TDC is induced by repeated exposure to cocaine, i.e., in experiments involving application of the drug to intact flies (see later) at 6-h intervals. In this context, tyramine was shown to be required for behavioral sensitization to such cocaine administrations (McClung and Hirsh, 1999). Mutant $per^0$, $Clk^{Jrk}$, and $cyc^0$ flies did not exhibit drug-induced increases in TDC activity, whereas a $tim$-null mutant showed normal induction after exposure to cocaine (Andretic et al., 1999). These results provided a platform for testing these clock mutations (and others) for effects on the behavioral-sensitization component of responsiveness to this drug (see earlier discussion).

The only experiments in this ballpark (or perhaps bullpen) with a flagrant connection to temporal control per se were based on induction by a "D2-like" dopamine receptor of locomotor-like actions; this occurred in a daily rhythmic manner (Andretic and Hirsh, 2000). The flies in question were only partly that, as they were decapitated, facilitating quinpirole application to the resulting hole at the neck (Method 15). Responsiveness peaked during the middle of the night in an LD cycle and exhibited a mild free-running rhythm during 1 day of constant light. The curve for the latter

METHOD 15. *Behavioral effects of drug applications* have conspicuously involved testing rhythm variants for possible anomalies of these quasi-behavioral aftereffects. Such sensitizations (see Fig. 5, top) are quantified after exposure of whole flies to volatized forms of the drug (reviewed by Hirsh, 2001). George et al. (2005) update this story methodologically, focusing on the effects of cocaine administration and in terms of the reflex-type behaviors (jumping, "twirling") that can be induced by this pharmacological agent. Another method for administration of pharmacological agents (e.g., dopamine receptor agonists) is based on applying a given substance to the "neck hole" that results from decapitating *Drosophila* adults, as reviewed by Hirsh (1998). Fruit flies and other insects sustain survival of sorts in this horrifying circumstance and can exhibit modest degrees of locomotion (implied in Fig. 5, middle), which are quantified by counting numbers of grid line crossings.

time course was flattened in tests of headless $per^0$ flies, although this mutant showed an early nighttime peaklet in LD. A *per*-influenced "body oscillator" was invoked to explain the circadian modulation of quinpirole-induced locomotion (Andretic and Hirsh, 2000), which should be kept in mind when discussion of the neural substrate for a short-term behavioral rhythm enters into this section (see later).

## A Peripheral Tissue Rhythm

Meanwhile, we seemed to have wandered back into the circadian realm. Let us therefore recall the several mentions of apparent circadian oscillations running all over the periphery of *Drosophila*. This means one solid set of descriptive phenomena: the *period, timeless, Clock*, and *cryptochrome* genes oscillate in all body regions, conspicuously including those posterior to the head, thus well outside the brain and implicationally beyond the ventro-posterior CNS.

Most such findings involve such far-flung cyclings of *per* and *tim*, whereby explicit organ-by-organ monitorings of non-CNS and peripheral appendage rhythmicities at the molecular level are known for these two genes (most of this information, in turn, will be taken up in the next section). However, it is only within the antenna of the fly that a biological rhythm is correlated with *period* and *timeless* product cyclings therein. It is, for example, well established that isolated antennae exhibit daily fluctuations of reported expression of these genes. For this, *period-luciferese* and *tim-luc* transgenic constructs were generated, transformed into the *Drosophila* germline (Brandes *et al.*, 1996; Stanewsky *et al.*, 1998), and then used in part for real-time reporting of the glow rhythms that are nicely detectable in cultured antennae (Krishnan *et al.*, 2001; Levine *et al.*, 2002c; Plautz *et al.*, 1997a) (Method 16). Such isolated appendage rhythmicity indicates *autonomy* of clockwork operations as such within the antenna, and that independent attribute goes on to include stand-alone *light sensitivity* of the structure, meaning in particular LD entrainability (e.g., Plautz *et al.*, 1997a). This could be mediated by the light-receptive capacities of CRY, which perhaps can occur "anywhere" on and in the fly, given that $cry^+$ is indeed expressed from stem to stern of the adult animal (e.g., Emery *et al.*, 1998), and the pleiotropic presence of its transcript includes readily detectable *cry* RNA in the antenna (Okano *et al.*, 1999).

Is antennal expression of these three rhythm-related genes merely recreational in that they are expressed in that appendage but would not doing anything meaningful—only spinning their wheels to make fluctuating levels of gene products? The answer to this is apparently "no" because

METHOD 16. *Real-time monitoring of molecular rhythms* is a powerful method for assessing the daily oscillations of clock gene products. The principles, if you will, are concerned with the undesirability of sacrificing one, up to some huge number, of flies at a given time point (which would be followed by tissue homogenization, and so forth, as outlined in Method 4). This approach "loses" each such animal in terms of the molecular rhythmicity it may sustain during later times. Also, it could be that flies within a group sacrificed at a given time are not all in phase with regard to their daily rhythmicities. (No one has done something like monitor the locomotor cycles of individual flies and then group a bunch of them, which could be determined "on line" to be at the same apparent behavioral phase, before grinding up their heads.) Therefore, a sequence taken from a species of beetle (firefly) that encodes luciferase (*luc*) has been fused to certain clock-related pieces of DNA. Most of the ensuing experiments have involved a variety of *per-luc* transgenes (Brandes *et al.*, 1996; Stanewsky *et al.*, 1997b, 2002) and one form of a *tim-luc* fusion (Stanewsky *et al.*, 1998). Nearly all such constructs have drawn upon upstream regulatory sequences associated with these two clock genes (but see Veleri *et al.*, 2003). Hazelrigg (2000), Stanewsky (2005), and Yu and Hardin (2005) described the procedures devoted to monitoring whole flies or parts thereof, in and from which clock-controlled luciferase activity can report fluctuating levels of the gene products or of "promoter" activity alone (i.e., when only 5'-flanking sequences of the clock factor are fused to *luc*, as opposed to that regulatory region and part or even most of the coding region of the *Drosophila* gene). Due to the heavy methodological features of the articles just cited, along with the contents of a companion chapter in this volume (Walsh *et al.*, 2005), only a few chronobiological comments will be made about LUC in flies: *luc*-mediated bioluminescence is measured indirectly but (and this is crucial) automatically so, using a machine called TopCount (originally manufactured by Packard, now PerkinElmer:http://las.perkinclmer.com/catalog/Category.aspx?Category Name=TopCount&is Family=Truc). It is a finicky device whose operation requires a shallow learning curve, much patience, and benzodiazapene self-medication (*cf.* Turek and Van Reeth, 1988). *Drosophila* do not make the LUC substrate, so one supplies luciferin to a bit of fly food within a given well of a plate, which contains an individual animal underneath a tiny immobilizing plastic dome, and is repeatedly cycled in and out of the TopCount for light emission measurements, or by supplementing the tissue culture medium with luciferin when body parts torn off or dissected out of an animal are being LUC monitored. *Comment*:

The high-throughput features of these LUC-monitoring methods facilitate application of *luc*-containing transgenics and the various associated procedures in screens for novel rhythm variants, such as those that would cause abnormalities of *per-luc* cycling (see Stanewsky *et al.*, 1998; Stempfl *et al.*, 2002; Wülbeck *et al.*, 2005). Moreover, a modification of the transposon mobilization tactics for identifying arrays of enhancer-trap strains (see Method 7) led to the recovery of novel transgenic types in which a *luc*-containing construct that is intrinsically expressed at a low and temporally constitutive level "landed" near putative enhancers, which kicked the construct into cycling mode; this also provided immediately applicable genetic variants at the loci for these presumed clock-regulated genes (Stempfl *et al.*, 2002). One of the downsides of transgenically effected LUC monitoring involves substrate depletion. This is a problem in live fly monitorings, causing counts-per-second ordinate values in the x–y plotted time courses to slide inexorably downward over the course of a week. However, one can "detrend" these values (e.g., Levine *et al.*, 2002a) better to reveal the daily ups and downs of luciferase-reported rhythmicity. This speaks to wider concerns about how best to analyze these molecular timecourses formally; this matter is somewhat up in the air, contentious even [try pitting the innards of Plautz *et al.* (1997b) against those of Levine *et al.* (2002a)]. Again with regard to real-time reporting of rhythms running in live flies, it seems almost certain that most of the LUC signals result from relatively peripheral tissues (which, however, are replete with clock gene products—one of the principal themes of this chapter). One problem with this matter—actually, representing a set of substantive findings—is that molecular rhythms emanating from *per-luc* or *tim-luc* transgenics poop out soon after transferring individuals from LD cycling to constant-dark conditions. If free-running oscillations of clock factors are in operation within some fraction of the internal tissues of a fly, this rhythmicity is very likely swamped by the nonsustenance of peripheral undulations. Therefore, one must dissect out a *luc*-containing internal organ and put it in the aforementioned culture medium (placed within a given plate well for TopCounting). Alternatively, a manipulated form of a clock gene suspected to allow for spatial expression only within certain internally located cells could be fused to *luc*—aimed at determining whether sustained free-running reportage of the molecular rhythm might be able to "escape" from inside the animal and allow for live-fly monitoring of the local rhythmicity in question: this approach is beginning to work, as reported literally and figuratively by Veleri *et al.* (2003).

of the odor sensitivity rhythm that has been monitored by time-based recordings of electroantennograms (EAGs) distributed over the course of a given daily cycle (Method 17). In particular, the sensitivity maxes out during the nighttime of an LD time course and roughly during the middle of the second 12 h of a DD cycle. Null mutations in either *per* or *tim* genes led to flat odorant sensitivities over the course of a day and night (Krishnan *et al.*, 1999), even though marker-reported expression of neither gene could be detected within the sensory *neurons* of *per-gal4* or *tim-gal4* antennae (Kaneko and Hall, 2000).

---

METHOD 17. *Rhythmicity of antennal sensitivity* has been recorded, so far, by relatively crude though nevertheless skill-requiring recordings of odor-elicited responses of this adult appendage. Krishnan *et al.* (2005) deals with the various procedural details. Some highlights: It is much tougher to "stick an electrode" into the relevant antennal segment (a proximal one), compared with, for example, what was done almost exclusively in the early days of studying sensory inputs to fruit flies (recordings of light-elicited electroretinograms into the relatively beach ball-sized compound eye). The electrode so stuck is designed, in studies of fly antennal physiology, to record extracellularly the (very small) voltage changes caused by odorant applications. Moreover, it is warranted nowadays—even in terms of the chronophysiology of antenna—to record from particular subsets of the organ's sensillae [see Tanoue *et al.* (2004) for why and how]. The manner by which such odorants are applied is also a ticklish proposition (as it were): What concentration ranges of the solutions should be applied (speaking to the matter of narrow-sense sensitivity, irrespective of daily fluctuations for these parameters)? How to deliver such substances, via some controlled distance from the receptor cells? How reproducible are the EAG signals? The last question alludes to the fact that the "nighttime peak" of olfactory sensitivity is (1) not easily compared with a putative trough value (EAGs taken at other intra-night times or throughout the day give values that are somewhat flat with respect to each other), and (2) that the peak-to-plateau amplitude is not large. Therefore, an EAG timecourse for flies of a given genotype requires multiple recordings at a given cycle time; only then can one be convinced that, for example the temporally based plot of values recorded from *per*-null antennae are thoroughly flat compared with the wild type (see Krishnan *et al.*, 1999). Finally, and as was just alluded to, it has not been possible to generate a timecourse for a given individual fly in an online fashion, which would necessitate repeated recordings from the self-same appendage.

Tanoue *et al.* (2004) presented additional evidence that is very much on point. First, they extended clock mutational effects on the peripheral rhythm (beyond what happens in *per*- and *tim*-null mutants) by constructing UAS-*Clk* and UAS-*cyc* transgenes in which the transcription factor coding sequences were deleted internally (in part), aimed at achieving dominant negativity (*dom-neg*) for the usual CLK and CYC functions. Driving either UAS-containing construct with *tim-gal4* reduced *tim* mRNA abundance and its usual cycling, including that a solid oscillation for this particular clock transcript was (newly) detected in unperturbed antenna. Concomitantly, the *tim-gal4*/UAS-*Clk*$^{dom-neg}$ or *tim-gal4*/UAS-*cyc*$^{dom-neg}$ combinations abolished EAG rhythmicity (Tanoue *et al.*, 2004). These investigators next turned their attention to effects on clock factors within olfactory receptor neurons (ORNs) by applying ORN-*gal4* drivers whose expression pattern includes antennal sensillae known to respond to acetates (one of the odorant types used in the experiments that established antenna physiological rhythmicity in the first place) or subsumes expression of only a subset of such sensillae. Either driver type, combined with either UAS-*Clk*$^{dom-neg}$ or UAS-*cyc*$^{dom-neg}$, knocked out the odor response rhythm (Tanoue *et al.*, 2004). Are the *Clk*$^+$ and *cyc*$^+$ alleles actually expressed within this appendage and within the relevant "basiconic" sensillae alluded to earlier? That is unknown. However, these descriptive assessments really must be performed, because one never knows whether the expected clock factors will be detectable within a given tissue (*cf.* Kaneko and Hall's inability to infer *per*$^+$ or *tim*$^+$ expression within antennal sensory neurons).

These findings are reviewed by Hardin *et al.* (2003). Included in that summary (also see Krishnan *et al.*, 2005) are mentions of the usage of whole flies in these experiments, i.e., tethered animals into whose antennae electrodes get inserted. Sensory responsiveness so recorded does not reveal any autonomy for the antennal *bio* rhythm (*cf.* Levine *et al.*, 2002c; Plautz *et al.*, 1997a).[32] However, the systematic fluctuation of

---

[32]Some recently reported comparative chronobiology is interesting here and not merely as a backup to the fly work (*cf.* Stengl *et al.*, 1992). Thus Page and Koelling (2003) analyzed an analogous olfactory rhythm in the antenna of *Leucophaea medera*. It persisted for 2 weeks in constant darkness and exhibited a much greater amplitude than in the case of *Drosophila* (*cf.* Hardin *et al.*, 2003). How autonomous is this cockroach rhythm? It might not have that property because it has long been known that this insect must use its eyes for circadian photoreception (a rare situation among invertebrates, e.g., Helfrich-Förster *et al.*, 1998; Page, 1982; Truman, 1976), and the inputs for these photic stimuli travel a relatively short distance to "the" circadian pacemaking neurons located within an optic lobe (reviewed by Homberg *et al.*, 2003). It might follow in a way, and did do experimentally (twice), that the odor-response rhythm should be abolished by ablating the compound eyes of the cockroach or by severing the centripetally coursing optic tracts (Page and Koelling, 2003).

sensitivity to odorants appears to run "by itself" within this appendage, as indicated by its elimination in recordings from a certain *period*-derived transgenic. Once more, we infuse into the discussion a truncated, 5′-flanking-less *per* construct (*cf.* Frisch *et al.*, 1994)—here to note that its retention of CNS PER, but absence of that product anywhere else, was correlated with EAG nonrhythmicity of the transgenic antenna. This indicates that central clock control cannot feed out toward the appendage to regulate this physiological cycling (Krishnan *et al.*, 1999). Hardin and co-workers also performed the opposite kind of experiment to assess autonomy of the physiological rhythmicity of the antenna: they ablated LN neurons by *Pdf-gal4*/UAS cell-killer combinations (as usual) and found the fluctuating responsiveness of the appendage to be essentially undisturbed (Tanoue *et al.*, 2004).

With reference to the inherent light responsiveness of the antenna as putatively mediated by cryptochrome, here is a twist to this part of the peripheral clock tale. Remember one way that the CRY-encoding gene of *Drosophila* was discovered. In this regard, the $cry^b$ mutant was not detected via any light response phenotype as such—instead, because the mutation eliminated luciferase-reported cycling in a *per-luc* transgenic background (Stanewsky *et al.*, 1998). At first blush, therefore, this mutant is a clockless variant. Further real-time monitorings of *per-luc* in whole flies suggested that most, perhaps all, of the luminescent signals originate from peripheral tissues (Stanewsky *et al.*, 1997b). That the normal such glow cycling is flattened by $cry^b$ might mask an underlying retention of clock protein oscillations within the brain [*cf.* Veleri *et al.* (2003), who reported—literally and figuratively—the little bit of glow that can eke out of the brain when *luc* expression is limited to that internal location]. Indeed, it is the case that PER and TIM cycles within $s$-LN$_v$ neurons keep going within the brain of $cry^b$ flies (Helfrich-Förster *et al.*, 2001; Stanewsky *et al.*, 1998) and that of mutant larvae (Ivanchenko *et al.*, 2001; Kaneko *et al.*, 2000a). These results focused one's attention on the possibility that CRY is a circadian pacemaking factor in peripheral tissues, and perhaps a circadian-photoreceptive molecule as well, whereas it may solely mediate the latter function within "deep brain" locations (*cf.* Emery *et al.*, 2000b). Support for CRY playing a core clock role at least in the antenna was gathered by the experiments of Krishnan *et al.* (2001), as augmented empirically and further analyzed by Levine *et al.* (2002c): $cry^b$ caused *per-* or *tim-luc* glow rhythms to be squashed in LD plus DD monitorings of cultured appendages (naturally, *cf.* Stanewsky *et al.*, 1998), and the odor sensitivity time courses were similarly flat. However, these mutational effects were not solely a matter of nonentrainability to a photic Zeitgeber, because the molecular and physiological rhythms could neither occur in nor be entrained by warm:cool temperature cycles (Krishnan *et al.*, 2001). Interpretation of these findings in terms of "CRY within the clockworks" are not airtight, but one must entertain a scenario in which this molecule carries out different kinds of rhythm-related functions in central *vs* peripheral tissues. Analogous suppositions are on the table regarding the manners by which the two *Cry* genes possessed by the genome of a given mammal are functioning rhythmwise, i.e., in separate ways within the eyes compared to the chronobiological tasks carried by mCRYs in the CNS of a mouse [reviewed by Peirson *et al.* (2005) and van Gelder (2005), presumably larded over by discussions of the controversies that continue to swirl around mammalian photoreception].

*Reproductive Behavior*

This last subtopic sends us spinning out of the circadian realm again and portends the description of miscellaneous attributes of rhythm-related genes in the next section (the last one that contains descriptions of any chronogenetic results). For the moment, however, here is a brief summary of courtship song rhythmicity in *Drosophila*. This periodically fluctuating character involves modulation of the rate by which "pulses" of tone are produced by the wing vibrations of a courting male, and the cycle durations for these InterPulseInterval oscillations are in the range of around 0.6 to 1.2 min, depending on the species (reviewed by Hall and Kyriacou, 1990; also see Demetriades *et al.*, 1999). *period* mutants have long been known to affect this short-term (ultradian) rhythm; at least the first report of these mutational effects goes way back. [See Hall and Kyriacou (1990) for a review of the song cycle controversy that surfaced in the late 1980s and Alt *et al.* (1998) plus Ritchie *et al.* (1999) for confirmation of the original and second-stage publications by Kyriacou and Hall (1980, 1982).] Moreover, transfer of the $per^+$ gene from *D. simulans* (whose males display 35- to 40-s song rhythms) into $per^0$ *D. melanogaster* (whose wild-type rhythms are defined by 55- to 60-s periodicities) caused host males to take on *simulans*-like cyclical singing (Wheeler *et al.*, 1991).

This long-ago finding would not be worth mentioning if it were not for the next two items: (1) Vaguely analogous experiments have been performed involving interspecific *per* gene transfers and the manner by which they affect daily cycles of both generic locomotion as well as a provocative rhythm of sexual behavior and mating (Peixoto *et al.*, 1998; cf. Petersen *et al.*, 1988; Sakai and Ishida, 2001; Tauber *et al.*, 2003). (2) A further feature of the courtship song studies dipped into some biosystem biology. Thus, it was useful to reveal that the control of rhythmic singing is not exerted by any kind of ultradian pacemaker in the brain of the fly, which might even have been running within the same neurons that regulate circadian behavior; instead, $per^+//per^0$ mosaics showed the VNC to contain the song clock (Konopka *et al.*, 1996). The main reason for pointing out *this* finding (which is getting long in the tooth) is as a mnemonic device: no PER-containing *neurons* have been observable within thoracic regions of the CNS. This refocuses a bit of attention on "PER glia" within the VNC (Ewer *et al.*, 1992) and makes one wonder how that cell type contributes to the functioning of one or more such thoracic ganglia in order that a male's rate of tone-pulse production fluctuates systematically within a given minute of his progression through the courtship sequence.

At the end of a successful such performance, the male and female initiate copulation. Duration of such mating is in the range of 15–20 min for wild-type *D. melanogaster*, implying that this species-specific character is under genetic control [see Lee *et al.* (2001) and references therein]. It was therefore intriguing to learn that $per^0$ or $tim^0$ males mate for durations up to 70% longer than wild-type male–female pairs (Beaver and Giebultowicz, 2004). These mutant types have fertility problems (see later), but there was no correlation between mating durations and progeny production. The broader point is that the *tim* and *per* mutants in question are unable to mediate proper time-based behaviors in other circumstances; but they still displayed tight control of mating durations, even though the values are well outside the normal range. Other aspects of "standard" circadian phenomena, including additional components of the clockworks, were found to be uninvolved in this temporally controlled reproductive character: Exposing males to constant light for a few days prior to pairing them with females did not affect copulation time, and there were no effects on this phenotype of null mutations in the *Clk* or *cyc* genes (Beaver and Giebultowicz, 2004).

### The Rhythm System, as Defined Genetically, Influences Elements of *Drosophila* Biology with No Particular Temporal Components

We have arrived for better or, more likely, worse at the doorstep of chronogenetics that has little or nothing to do with the *periodic* biology of *Drosophila*. However, these effects of rhythm variants, or the expression of engineered molecular entities, do involve all of our old gene friends. These cases of further pleiotropies will, however, not go so astray that you are on the verge of reading about morphological foul-ups caused during development by lethal alleles of genes such as *double-time* (*discs overgrown*) or the "segment polarity" factor encoded by *shaggy*. Instead, here are some miscellaneous behavioral findings that connect rhythm gene actions to phenotypes devoid of sharply defined temporal aspects, and findings of another sort that may eventually be known to marry the action of these genes with rhythmic non-neural physiology. Even if that does not come to pass, we will not be surprised that so-called clock factors can spread their influences beyond the time-related realm—again, forgetting for the moment about the enzymes encoded within genes that are easily mutated to cause lethality.

Several of the other pacemaking controllers are not mythical proteins such as "clockin" or "timin", but instead are transcriptional regulators. So why would they necessarily be dedicated only to temporal features of the lifestyle of the fly? It is true that most of these factors are expressed in daily-cycling manners (Fig. 1), at least in part. This means that data are incomplete for the more recently discovered gene regulators. Such a concern is best highlighted by the fact that the first of these factors to be discovered, the *period* gene, was eventually shown to produce *per* mRNA and PER in a fluctuating fashion in most, but not all, regions of its expression domain (Hardin, 1994); nor does this gene carry out any known time-related function within the ovarian tissue where its products conspicuously fail to oscillate. The phenotype just alluded to will be taken up as the current section unfolds. In it, nowhere near all claims about chronogenetic pleiotropies will be rereviewed, given that they were agonizingly tabulated within Hall (2003a); and few items on that list involve reports of clock mutational effects that were analyzed further. Thus, let us pick and choose and in fact pick up where we left off in the previous section with some more sex.

Four of the core clock factors had null mutations in genes that encode them tested for progeny production. Thus, matings performed by $per^0$, $tim^0$, $Clk^{Jrk}$, and $cyc^0$ males all were found to result in *ca.* 40% fewer offspring than normal, due to decreased numbers of eggs laid by their wild-type female partners; the latter also put out anomalously high percentages of unfertilized eggs (Beaver *et al.*, 2002). Within the reproductive organs of these clockless male types, aberrantly low numbers of sperm were observed to be released from testis to seminal vesicle (SV). The *period* gene has long been known to make its products in the reproductive system of *Drosophila* males (Hall, 1995). Beaver *et al.* (2002) expanded this picture (metaphorically) by showing that both *per* and *tim* are expressed strongly in the lower testis and SV and more weakly in the upper testis and ejaculatory duct. With reference to these particular male-specific structures, a "sperm release rhythm" was discovered long ago in moth testis (reviewed by Giebultowicz, 1999, 2000, 2001). Is it a coincidence that Beaver *et al.* (2002) observed molecular rhythmicity in these tissues? First, they found cycling of PER and TIM immunoreactivities, which were limited to the lower testis and SV organ. These investigators went on to provide the only *in situ* demonstration of *Clock* expression outside the fly head (*cf.* Emery *et al.*, 1998; So *et al.*, 2000). Imagining that *Clk* mRNA would be present within male reproductive organs (given the $Clk^{Jrk}$ effect) and that it could cycle out of phase with *per* transcript levels (*cf.* Fig. 1), Beaver and co-workers (2002) indeed detected *Clk* expression in the lower testis and SV by *in situ* hybridization. Furthermore, the inferred level of *per* mRNA was low at a cycle time (4 h after lights on) when the apparent amount of *Clk* transcript level was high (vice versa for 4 h after lights off).

Periodic sperm release occurs autonomously in moth using a rhythmic subsystem that includes photoreceptive capacities (Giebultowicz, 1999, 2000, 2001). It is therefore on point that testis-SV complexes ran luciferase-reported rhythms of *per* and *tim* expression for the majority of specimens dissected out of male abdomens and cultured (Beaver *et al.*, 2002). In a subsequent experiment, inherent photoreception of these organs was inferred indirectly by demonstrating weak circadian cycling of *per-luc* or *tim-luc* testis-SV explants after transfer to constant darkness, followed by restimulation of higher-amplitude molecular rhythmicity upon return to LD. Light-absorbing candidates, and related molecules, that *could* mediate the ability of these male organs to see involve distally located representatives of the opsin system of *Drosophila* in male reproductive organs (Alvarez *et al.*, 1996). The possible presence of cryptochrome in these structures has not been looked for. More important, it is unknown whether a sperm release rhythm is running in these reproductive structures of *Drosophila* males. Such a demonstration would close the case in a way, i.e., bring it chronobiologically in line with the moth phenomenon.

The *period* gene is also expressed within the female reproductive system, naturally (Liu *et al.*, 1988, 1992; Saez and Young, 1988). However, the relevant structures within such organs are molecularly distinct because they do not operate cycles of *per* mRNA abundance (Hardin, 1994). In particular, within the follicle cells that surround young oocytes, PER immunoreactivity cruises along at an apparently constant level, and signals were never observed in nuclei (Beaver *et al.*, 2003; *cf.* Liu *et al.*, 1992). Beaver *et al.* (2003) found the same temporally constitutive cytoplasmic expression for TIM protein within these ovarian cells. Such descriptive aspects of this posterior component of the fly's system of clock-gene expression suggest that the internal reproductive biology of *Drosophila* females does not possess pacemaking features. Consistent with this supposition, neither $Clk^{Jrk}$ nor $cyc^0$ mutations affected PER or TIM levels in the ovarian follicle cells, nor did exposure of females to constant light (Beaver *et al.*, 2003)—the same kind of negative results that led Beaver and Giebultowicz (2004) and Sakai *et al.* (2004) to infer "noncircadian" effects of clock mutations on, respectively, male-controlled mating durations and the conditioned courtship behavior that can be exhibited by such flies. Beaver *et al.* (2003) gave their study of female reproduction in *Drosophila* a loosely analogous biological twist by showing that *per*-null and *tim*-null females generated only half-number numbers of progeny after mating, and either mutation caused decrements in production of mature eggs within virgin females.

We might retain a dim memory of the fact that the chrono-molecularly discovered *takeout* gene (So *et al.*, 2000) makes its products almost exclusively in one sex (Dauwalder *et al.*, 2002). Subsequent to the original identification of this factor, a *to* mutant was stumbled upon but was found to exhibit no particularly rhythm-related defect (Sarov-Blat *et al.*, 2000). Indeed, most of what is known currently about the lipophilic molecule-binding capacities putatively possessed by the TO protein take its functions outside the rhythm realm. From the standpoint of molecular sex, *to* was shown to be activated in males only by the combined actions of "sex determination/sex differentiation" factors encoded by the *doublesex* (*dsx*) and *fruitless* (*fru*) genes (whose general properties are reviewed by Baker *et al.*, 2001). *dsx* generates a female (F)-specific form of its transcriptional activator in that kind of *Drosophila*, and $DSX^F$ was shown to repress *to* expression in females (Dauwalder *et al.*, 2002), with no requirement for *fru* involvement (a hallmark of the latter gene is to produce certain male-specific transcripts and a $FRU^M$ form of this hypothetical transcription factor). One can feminize portions of a *Drosophila* male by transgenic manipulation of *transformer* gene expression (*tra* acts "upstream" of *dsx* and *fru* in the sex determination "hierarchy," again as reviewed by Baker *et al.*, 2001). Doing so with regard to *takeout* expression in terms of where that gene normally makes its products in males led to a severe reduction of male courtship performance (Dauwalder *et al.*, 2002).[33]

---

[33]The *to-gal4* fusion transgenes newly constructed by Dauwalder *et al.* (2002) and combined with a UAS marker essentially reproduced the spatial patterns of *to* expression, which conspicuously includes fat body tissue near the brain; also less intense expression in fat cells distributed throughout the thorax and abdomen, within the alimentary system, and in the antenna [*cf.* Sarov-Blat *et al.* (2000) for information about the latter two tissue types, and see Wolfner (2003) for further cases of connections between fat bodies, and *Drosophila* genes with sex-biased expression]. Recall that clock mutations cause markedly decreased levels of *to*⁺ gene products (Sarov-Blat *et al.*, 2000); this is a principal reason for viewing *takeout* to be rhythm-related as opposed to concerned solely with sex. Could such mutations (at the *per*, *tim*, *Clk*, or *cyc* loci) decrement *to*⁺ expression by direct effects within fat bodies, in the sense that these clock genes are normally expressed within that tissue? There is little documented evidence on this point—only a line item tabulated by Kaneko and Hall (2000), indicating that *per-gal4* and *tim-gal4* each drive marker expression within *abdominal* fat bodies. Two papers show pictures of what look like PER immunoreactivity in head fat bodies (Ewer *et al.*, 1992; Stanewsky *et al.*, 1997a), although these signals were not controlled as to whether a *per*-null mutation eliminated the antibody-mediated stainings. Furthermore, no data have been collected as to whether the *cycle*⁺ gene is expressed within the head's version of that tissue type; this is pointed out, because *takeout* was originally discovered via an mRNA present at anomalously low abundance in head extracts of a *cyc*-null mutant (So *et al.*, 2000).

Concomitantly, the *to* mutant was found to court subnormally (but only in one genetic background), and there were "synergistic" such defects when this mutation was combined with a heterozygosity for each of two separate *fruitless* mutations. [Several of the *fru*-mutant alleles, when homozygous by themselves, caused decrements in male actions directed at females, although a given *fru*/+ male courts normally (Villella *et al.*, 1997).]

The author would not have gone semi-deep into the case of *takeout* were it not for the fact that the gene was discovered as a clock gene regulatee (So *et al.*, 2000) then found to make its products according to a daily rhythm, in the seminal studies and as reconfirmed by taking a different molecular tack (reviewed by Etter and Ramaswami, 2002). Furthermore, there is by now a reasonably sized handful of additional relationships between sex control and that of *Drosophila* rhythmicity, which we have encountered even before dealing with these matters in the current section (also see later).

Geotaxis has a modest connection to sex differences, believe it or not. This vignette starts with referral to a molecular screen for genes whose expression levels apparently differed in lines of *D. melanogaster* selected (long ago) for relatively high *vs* low geotaxis scores (Toma *et al.*, 2002). The *cryptochrome* and *Pdf* genes were among those implicated by carrying out the usual kind of microarray protocols (as cited in other contexts near the beginning of the article). These features of the molecular results prompted geotaxis testing of the $cry^b$ mutant and of *Pdf* variants, including transgenics involving the latter. Toma *et al.* (2002) thus found that the cryptochrome mutant gave elevated geotactic scores over various baseline values, although not reaching the level of the high geotaxis line; *cry* was tapped into in this study because its "microarrayed" mRNA level was *lower* than expected in the "high" line, correlated with a similar behavioral phenotype caused by the $cry^b$ mutation. Similarly, *Pdf* mRNA was observed by Toma *et al.* (2002) to be down in the line that gives high geotactic scores; these authors attempted to describe the effects on this behavior of the *Pdf*-null mutation and/or of gene dosage alterations involving a normal form of this gene (*cf.* Renn *et al.*, 1999); an additional genetic complexity involved the manner by which these *Pdf* variants interacted with the effects on geotaxis of one *vs* two X chromosomes. Perhaps the latter element of this behavioral phenogenetics has something to do with the previous demonstration by Park and Hall (1998) that *Pdf* mRNA is more abundant in head extracts of males compared with females (involving *D. melanogaster* and three closely related species). These molecular phenotypes, by the way, might also contribute to sex differences in adult locomotor rhythmicity (Helfrich-Förster, 2000). As for geotaxis, most of the behavioral results presented by Toma *et al.* (2002) are more or less comprehensible in the narrow sense less so for possible relationships among antigravity locomotion, sexual dimorphisms thereof, and PDF. However, they make little or no overall sense in light of the "main effects" of these rhythm factors.

The following seems, in contrast, as it if *will* be chronobiologically more sensible, but this may not occur in a fulsome fashion until some dark night in the future. Nevertheless, we once more shift our gaze caudally, toward the abdominally located Malpighian tubules (MTs). In *Drosophila*, this excretory organ runs rhythms of clock gene expression, as monitored in part (and of course) by explanting the structure and measuring *per-* or *tim*-promoted luciferase cycling. These tests provided one way for this posterior tissue rhythm to be apprehended as autonomous (most recently, Giebultowicz *et al.*, 2000). Such independence of MT molecular rhythmicity includes the inherent photoreceptive capacities of the organ. Cryptochrome would thus be inferred to operate in the MTs. Indeed, Ivanchenko *et al.* (2001) could readily detect *cry* mRNA in extracts of this *Drosophila* organ and found that the *cry$^b$* mutation attenuated the MT light responsiveness in terms of molecular rhythm entrainment. Curiously, these investigators found *norpA* mRNA also to be present within MTs, although a null mutation in that gene (with or without the presence of *cry$^b$*) had no effect on light resettability of the clock gene cycling of the organ. The only known effect of the PLC of *norpA* on MT function is that one of its mutations was found to attenuate the response of the organ (previously known) to diuretic peptides (Pollock *et al.*, 2003). Once again, we appreciate the phospholipase emanating from *norpA$^+$* to be involved in biologically broad aspects of signal transduction. Further analysis of the effects of the *cry* mutation on MT clock-gene rhythmicity in constant darkness seemed to reflect actual *pacemaking* problems (Ivanchenko *et al.*, 2001). These could reflect a "more core clock" role for CRY in this internal organ, by analogy to inferences about how that protein functions within the antenna (reviewed by Hardin *et al.*, 2003).

However, what does the MT clock as a whole do? Nothing is known about putative physiological rhythms associated with the functioning of this organ [although more and more is being uncovered about details of its "static" physiology, as reviewed by Dow and Davies (2003) and exemplified by several of the additional findings in Pollock *et al.* (2003)—let alone how genes such as *per* or *tim* (and *cry*?) may feed into whatever chrono-renal regulation there may be.

Clock genes in *Drosophila* are expressed nonneurally in organs and other structures ranging way beyond the Malpighian tubules just discussed; also farther than the alimentary and reproductive systems dealt with earlier. However, next to nothing is known about the biological meanings of these expression patterns, including what their temporal components might mean for organs such as the MT and gut, whose abdominal functions are closely associated. Is it that the fly is "regular"?[34] Instead of reiterating what we do not know about findings obtained with regard to *period* and *timeless* expressions (this being the state of affairs for most of the clock-gene tissue pattern data), let us make one last referral to the descriptive inventories (Hall, 1995, 2003a; Kaneko and Hall, 2000) and then go on to make some remarks about organs on (not in) the periphery of the fly.

The eye is most mentionable in this respect, perhaps because fluctuations of PER immunoreactivity in the photoreceptors of this structure provided the first findings about clock gene product cyclings in organisms (Siwicki *et al.*, 1988). The many subsequent investigations that were triggered by these results unwittingly determined oscillating molecular timecourses mostly in homogenized eyes. That is, the investigators would grind up *Drosophila* heads and apply Northern blotting, RNase protection, or Western blotting protocols to the extracted macromolecules (see Method 4). *per, tim,* and other clock gene products are present within approximately 13,000 eye cells; this would swamp the molecular signals stemming from co-homogenizing brain tissue within fly heads.

Let us dwell for a moment on expression of one other gene expressed within this anterior region of the fly: *cryptochrome*. We saw earlier that the functions of CRY could be teased out from experiments involving only a tiny subset of the head (a few neurons in the brain, also the antenna). What the hell is CRY doing in the eye? That photoreceptors (PRs) of the structure are rhodopsin ridden. Maybe cryptochrome supplies a "blue backup" within such cells, with the proviso that the protein would be found within compound eye PRs. This implies that *cry*+ expression has not yet been detectable *in situ*, which is the case (Egan *et al.*, 1999; Emery *et al.*, 2000b). However, eye tissue must be rich with CRY protein for the following reason: The second-stage molecular phenotype found for the *cry*[b] mutant (Stanewsky *et al.*, 1998) was inappropriately high levels of TIM protein throughout the day and night of an LD cycle, determined (you guessed it) by extracting proteins from fly heads. In wild-type extracts, TIM is driven to superlow levels during the daytime (irrespective of the dynamics of *tim* mRNA cycling). Given that most of the *timeless* gene products in an adult head are within the eye, according to the original and all subsequent assessments of spatial expression patterns for *tim in situ*

---

[34]Perhaps the enigmatic TAKEOUT factor, which is in part a gut one, would be correlated with erratic IPIs (that is, InterPoopIntervals, in this case) in males within which this protein is mutationally eliminated (as it were)—encouraging investigators to rename the gene *colonblowout.*

(Hunter-Ensor *et al.*, 1996; Myers *et al.*, 1996; Sehgal *et al.*, 1995), here is the inescapable conclusion: The "TIM high" phenotype observed in whole head protein extracts of the $cry^b$ mutant is primarily the result of disallowing normal light-induced degradation of TIMELESS polypeptides within the compound eye. Assuming that CRY could be coexpressed with rhodopsins in the PRs of that structure, it is intriguing to wonder whether the manifest molecular clock operating within such cells is contributed to by a *cry*-specified function. This would make the control of eye rhythmicity, such as it is (see later), analogous to what is surmised for regulation of the clockworks within another sensory structure on the head—the antenna (Hardin *et al.*, 2003).

Back to specific issues revolving around extraction, overwhelmingly, of not only eye CRY, but also of PER, TIM, CLK, and so forth, from that tissue in head homogenates. This would be okay if eye rhythms progress in lock step with those running within the remainder of the head. Such a salutary conclusion was presented by Zeng *et al.* (1994) via Western blottings resulting from tearing off eyes and measuring PER cycles in that material versus what was left; but it is difficult to believe that the "eye peak" or trough could really be resolved from the "head minus eye" ones (especially in a one-time profile), i.e., imagining that the tissue-specific time courses are in reality different. Indeed, internally controlled monitorings of PER cycles within all head locations where the protein was immunohisto-chemically detectable revealed a distinctly earlier late-day trough time in compound-eye photoreceptors compared with the nadirs observed for both lateral neuronal and glial oscillations (Zerr *et al.*, 1990). This kind of finding fits loosely with detailed determinations of *phase values* that resulted from monitoring *per-luc* and *tim-luc* rhythms in the culturings of various torn-off appendages (Levine *et al.*, 2002c). These molecular timecourses defined their own peak times for a given structure (although the separate values were not far off), belying an earlier, briefly stated claim about all these phases being the same among separately cultured specimen types (Plautz *et al.*, 1997a). What might these different peak and trough times mean biologically? We do not know, but nonetheless bear in mind that an animal really does not want *all* aspects of its oscillating biology to max- or-bottom out at the same times. Recall the dawn and dusk peaks for locomotor rhythmicity of *D. melanogaster* compared with the midnight one for odor sensitivity. Therefore, why not set up antenna, leg, and wing sensillae (which are distributed all over these structures) to be maximally sensitive to stimuli in their own good time?

This brings us to the second item concerning molecular rhythms in the eye of the fly. Why there? No clue: Microspectrophotomical and electro-physiological attempts failed to connect a daily rhythm of rhabdomere turnover within photoreceptors or fluctuating visual sensitivity (*cf.* Bennett, 1983) with varying *per* functions (Chen *et al.*, 1992). Other kinds of eye rhythms are known in insects (e.g., Sakura *et al.*, 2003), but these parameters have not been examined in clock-gene contexts for *Drosophila*.

An object lesson suggests itself in this opaque framework for the meaning of molecular rhythmicity in *Drosophila* eyes. Skulking in the fine print of any early section (footnote 3) was mention of *double-time* mutational effects on the dynamics of macromolecules encoded by other pacemaking factors. Regarding one feature of such findings, much was made by Suri *et al.* (2000) of these particular observations. Homogenates of a certain *dbt* mutant, one of two novel such variants isolated by these investigators, led to molecular time courses in light:dark cycles whose results included elimination of the famed delay between *per* or *tim* transcript peaks and those of their encoded proteins (*cf.* Fig. 1). These findings have been ignored ever since. This permits one to speculate with no compunction that the only tissues within which the extracted mRNAs and polypeptides peaked at the same time were those of the compound eyes. Thus the unknown biological rhythmicity, struggling to operate within the external photoreceptors of this *dbt* variant (allele name *dbt$^h$*), would be severely disturbed. However, flies manifesting the effects of Suri *et al.*'s mutation *behaved* in a periodic manner (although free-running locomotor cycles were an hour longer than normal in *dbt$^h$*/+ flies and 5 h so in mutant homozygotes). The author therefore believes that such *biological* rhythmicity is underpinned by a retention of temporal discordance between the transcript and protein maxima within pertinent cells of the central brain (Fig. 3).

What were the levels of such neuronal macromolecules at a given cycle time for this *double-time* mutant? Such experimental information was of course lost in the whole head soup. Therefore, work the system! If flies must be sacrificed to take a temporal time point, acquire brain specimens and simultaneously monitor the apparent abundances of both kinds of clock gene products within neurons known to possess such RNAs and proteins. This can be achieved by marrying methods for *in situ* hybridization with those that lead to immunohistochemical signals. The wherewithal to affect such double labelings for the contents of pacemaker neurons has been around for a while (Sauman and Reppert, 1996; Yang and Sehgal, 2001). Even better, for a truly intensive focus on what is going on within the head of the *double-time* mutant just discussed, section the entirety of that adult tissue and then ask whether, for example, *per* and PER exhibit nicely phase-discordant oscillations inside the brain but cycle in temporal concert within the eye. Thus, these gene-product dynamics would be spatially analyzed by a double-labeling experiment in a directly comparable fashion.

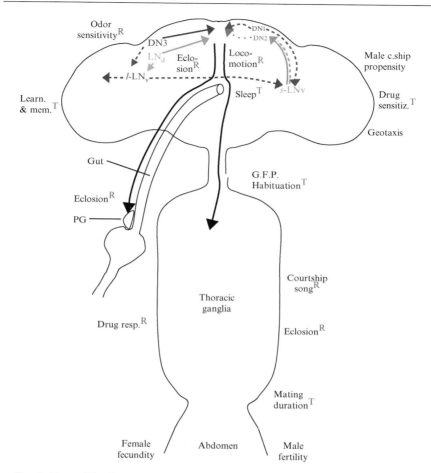

FIG. 5. *Drosophila* phenotypes caused by rhythm mutations and tissue expressions of the corresponding genes. This diagram summarizes many of the biological correlates of such genetic effects and gene product locations. The former are indicated by uppercase words or phrases. These are placed within or next to a body region from which one or more rhythm-related mutations are known or presumed to regulate the phenotype or at least in some way to influence it. Abbreviations associated with certain such attributes of the behavior or physiology of the fly: C.SHIP, courtship; G.F.P., giant fiber pathway. Most of the phenotypes referred to have a temporal component, e.g., daily rhythmicity (**R**) of locomotion, eclosion, odor sensitivity, drug resp. (responsiveness), or *ca.* 1-min rhythmicity of courtship song. Other biological parameters, such as Sensitiz. (sensitization) to DRUG exposures, are associated with some sort of "time base" (**T**) not explicitly connected to rhythmicity; but certain additional phenotypes, notably female fecundity and male fertility, are not. The expression patterns of genes connected to *Drosophila* chronobiology are only alluded to for most of the relevant tissues, as opposed to being diagrammed in detail with regard to tissues and organs

However, how could a clock mutation such as *dbt* have such differential tissue effects on gene products encoded by other such genes? Here is a possible answer: The basic features of brain *vs* eye cyclings of PER start out with temporally varying control. Recall that these spatially separable dynamics have distinct phases, as revealed by temporally controlled sectionings and stainings of wild-type heads (Zerr *et al.*, 1990). This suggests that the overall pacemaking milieu is by no means equivalent among all the separate cell types in which clock-gene products find themselves, augmented by the distinct possibility that not all such products (*cf.* Fig. 1) are even coexpressed within every part of the fly. Therefore, it might be

---

within a given body region and within which particular genes are known to make their products in these locations. Inside the head, colored brain neurons that contain "clock proteins" are shown along with most features of their neurite "wiring diagram" (*cf.* Fig. 2). Additional substances are colocalized within certain such cells, axons, or both; e.g., the PDF neuropeptide within ventro(v)-Lateral Neurons of the brain; cryptochrome in most of the six neuronal (N) cell groups. Clock genes are also expressed in many peripherally located sensory structures, as exemplified by the pictorial allusion to an antennal rhythm of odor sensitivity. However, certain rhythm-related genes are probably not neurally expressed within the head; e.g., *takeout*, which makes its products in fat-body tissue juxtaposed to the brain and is surmised to influence male c.ship thereby (via unknown molecular and cellular actions). An eclosion-regulating endocrine structure within the anterior throrax—the prothoracic gland (PG)—is pulled off to the side for diagrammatic purposes (same for the gut, within which certain clock genes make their products but for unidentified biological reasons). Other thoracic structures are not specifically depicted. In this regard it should be pointed out that the *period* clock gene, the one that has been assessed most extensively for tissue expression patterns, makes its thoracic ganglia products only within glial cells. These portions of the *Drosophila* CNS are the three anterior-most such ganglia within the ventral nerve cord (VNC). Elements of this figure depict eclosion control as originating (if you will) in anterior structures (LNs, PG). This regulation is ultimately "read out" via operation of the VNC and the locomotor events associated with emergence of an adult from the pupal case. The same kind of anterior → posterior pathway naturally underpins overall features of the daily behavioral cycles of the adult. However, the locomotion label is not reiterated with regard to this VNC involvement, because no gene-defined chronobiological factors are known to regulate the adult activity rhythm of the fly from within this body region. In contrast, an eclosion indicator is tacitly placed in association with the VNC because of expression of a specific gene in this portion of the CNS; its products are coexpressed with an eclosion-involved neuropeptide called CCAP within the VNC, in context of the relevant *lark* genetic variants influencing the timing of periodic adult emergence. (LARK protein is also contained within CCAP neurons in the brain, which project axons through the neck into the VNC.) The abdominal ganglion (not designated) is actually within the thorax and is the most posteriorly located portion of the VNC. As the diagram vaguely suggests, control of mating duration may occur via expression of clock genes in abdominal ganglionic cells (again, these are mostly or exclusively glia). The abdomen per se is barely included within the diagram, even though three clock genes are known to be expressed in various posterior organs. Among them are the female and male reproductive systems, whose local functions could be abnormal in the clock mutants that exhibit the aforementioned deficits in fecundity or fertility. (See color insert.)

"easy" for mutationally altered structure and function of one such product, in this case the DBT kinase, differentially to affect how its own substrate (PER) and companion clock factors cycle in the brain compared with the eye.

Whether the experimental call to action just trumpeted has any force, many further appraisals of the state in which clock-gene products find themselves at a given time will necessitate biochemical inspections. For this, one should not continue to grind along (literally) and monitor molecular cyclings primarily in a tissue that has no known chronobiological meaning. Why not, instead, take advantage of the *D. melanogaster* tool kit and use—from now on—heads removed from eyeless mutant flies? What gets ground up during the next stage of a given procedure will at least be enriched for clock factors, in whatever intra-extract state they find themselves, within the more or less chrono-interpretable brain of the animal.

Conclusion

With this soft (±) claim and tacit hope, the author is therefore tempted merely to give a précis of one subsection located roughly in the middle of the article: "Three blind flies . . . did you ever see such a sight in your life?" The latter of the phrases just quoted perhaps provides a sufficient summary passage for this last sectionette of the piece. However, the author augments these final remarks a bissel by saying, first of all, that studying biological rhythms in *Drosophila* continues to plumb the mechanistic *depths* (as partly elaborated earlier); much of this is soporific stuff by now (irrespective of however inadequately the author may have summarized the clock-gene side of the story). Second and more salutary, the additional types of investigations described and evaluated in nearly all other sections, which comprise a wide array of relatively stealth studies, are *broadening* one's viewpoints about what the rhythm *system* of the fly means in a more holistic manner. However, this system's side of the story still relies on *molecular genetics*; identification of *novel* genes, along with *when* and especially *where* the chronobiologically functioning molecules are found within the animal; and, finally, *what* they are doing in these far-flung biolocations. A synopsis of this wide-ranging information base, along with various presumptions stemming therefrom, is presented in Fig. 5.

That diagram may sufficiently summarize the genetics and biology of this system, such that the article can end with this (real) précis: An appreciation of recently identified chronobiological factors, along with expanding knowledge about how these genes and the pacemaking molecules are connected to *Drosophila* rhythmicity, are arguably of more present-day interest than matters revolving tightly around the heart of the clock.

However, this enlarging picture is still only a sketch. A representation of "nonclock" factors and the increasingly multidimensional attributes of the core regulators guide our gaze back to the quote at the beginning of the article. Such a reminder dovetails with the way that a contemporary practitioner of *Drosophila* rhythmology judiciously views this enterprise: from the standpoint of "what we know and what we do not know" (Blau, 2001).

## Acknowledgments

I am most grateful for the collaborative work that led to certain of the studies reviewed in this chapter; these cooperative efforts have involved the Hall laboratory along with those of Michael Rosbash, Steve A. Kay, Ralf Stanewsky, and Charlotte Helfrich-Förster. I appreciate discussions with, and references supplied by, Dale M. Edgar, Jae H. Park, Kathleen K. Siwicki, and Monika Stengl. Charalambos P. Kyriacou and Ralf Stanewsky furnished valuable comments on the manuscript. Several of the studies referred to have been supported in the Hall laboratory by grants from the U.S. NIH: a pair that have run their course (GM-33205, MH-51573) and two that are extant (NS-44232, GM-66778).

## References

Akten, B., Jauch, E., Genova, G. K., Kim, E. Y., Edery, I., Raabe, T., and Jackson, F. R. (2003). A role for CK2 in the *Drosophila* circadian oscillator. *Nature Neurosci.* **6,** 251–257.

Allada, R. (2003). Circadian clocks: A tale of two feedback loops. *Cell* **112,** 284–286.

Allada, R., Emery, P., Takahashi, J. S., and Rosbash, M. (2001). Stopping time: The genetics of fly and mouse circadian clocks. *Annu. Rev. Neurosci.* **24,** 1091–1119.

Allada, R., Kadener, S., Nandakumar, N., and Rosbash, M. (2003). A recessive mutant of *Drosophila Clock* reveals a role in circadian rhythm amplitude. *EMBO J.* **22,** 3367–3375.

Allada, R., White, N. E., So, W. V., Hall, J. C., and Rosbash, M. (1998). A mutant *Drosophila* homolog of mammalian *Clock* disrupts circadian rhythms and transcription of *period* and *timeless*. *Cell* **93,** 791–804.

Alt, S., Ringo, J., Talyn, B., Bray, W., and Dowse, H. (1998). The *period* gene controls courtship song cycles in *Drosophila melanogaster. Anim. Behav.* **56,** 87–97.

Alvarez, C. E., Robison, K., and Gilbert, W. (1996). Novel $G_q\alpha$ isoform is a candidate transducer of rhodopsin signaling in a *Drosophila* testes-autonomous pacemaker. *Proc. Natl. Acad. Sci. USA* **93,** 12278–12282.

Andretic, R., Chaney, S., and Hirsh, J. (1999). Requirement of circadian genes for cocaine sensitization in *Drosophila. Science* **285,** 1066–1068.

Andretic, R., and Hirsh, J. (2000). Circadian modulation of dopamine receptor responsiveness in *Drosophila melanogaster. Proc. Natl. Acad. Sci. USA* **97,** 1873–1878.

Arvanitogiannis, A., Stewart, J., and Amir, S. (2000). Conditioned stimulus control in the circadian system: Two tales tell one story. *J. Biol. Rhythms* **15,** 292–293.

Ashmore, L. J., Sathyanarayanan, S., Silvestre, D. W., Emerson, M. M., Schotland, P., and Sehgal, A. (2003). Novel insights into the regulation of the timeless protein. *J. Neurosci.* **23,** 7810–7819.

Baker, B. S., Taylor, B. J., and Hall, J. C. (2001). Are complex behaviors specified by dedicated regulatory genes? Reasoning from *Drosophila*. *Cell* **105,** 13–24.

Baker, J. D., McNabb, S. L., and Truman, J. W. (1999). The hormonal coordination of behavior and physiology at adult ecdysis in *Drosophila melanogaster*. *J. Exp. Biol.* **202,** 3037–3048.

Bangs, P., Franc, N., and White, K. (2000). Molecular mechanisms of cell death and phagocytosis in *Drosophila*. *Cell Death Differ.* **7,** 1027–1034.

Bargiello, T. A., Jackson, F. R., and Young, M. W. (1984). Restoration of circadian behavioural rhythms by gene transfer in *Drosophila*. *Nature* **312,** 752–754.

Bargiello, T. A., Saez, L., Baylies, M. K., Gasic, G., Young, M. W., and Spray, D. C. (1987). The *Drosophila* clock gene *per* affects intercellular junctional communication. *Nature* **328,** 686–691.

Bargiello, T. A., and Young, M. W. (1984). Molecular genetics of a biological clock in *Drosophila*. *Proc. Natl. Acad. Sci. USA* **81,** 2142–2146.

Beaver, L. M., and Giebultowicz, J. M. (2004). Regulation of copulation duration by *period* and *timeless* in *Drosophila melanogaster*. *Curr. Biol.* **14,** 1492–1497.

Beaver, L. M., Gvakharia, B. O., Vollintine, T. S., Hege, D. M., Stanewsky, R., and Giebultowicz, J. M. (2002). Loss of circadian clock function decreases reproductive fitness in males of *Drosophila melanogaster*. *Proc. Natl. Acad. Sci. USA* **99,** 2134–2139.

Beaver, L. M., Rush, B. L., Gvakharia, B. O., and Giebultowicz, J. M. (2003). Noncircadian regulation and function of clock genes period and timeless in oogenesis of *Drosophila melanogaster*. *J. Biol. Rhythms* **18,** 463–472.

Belvin, M. P., Zhou, H., and Yin, J. C. P. (1999). The *Drosophila* dCREB2 gene affects the circadian clock. *Neuron* **22,** 777–787.

Bennett, R. (1983). Circadian rhythm of visual sensitivity in *Manduca sexta* and its development from an ultradian rhythm. *J. Comp. Physiol. A* **150,** 164–174.

Blanchardon, E., Grima, B., Klarsfeld, A., Chélot, E., Hardin, P. E., Préat, T., and Rouyer, F. (2001). Defining the role of *Drosophila* lateral neurons in the control of activity and eclosion rhythms by targeted genetic ablation and PERIOD overexpression. *Eur. J. Neurosci.* **13,** 871–888.

Blau, J. (2001). The *Drosophila* circadian clock: What we know and what we don't know. *Semin. Cell Dev. Biol.* **12,** 287–293.

Blau, J. (2003). A new role for an old kinase: CK2 and the circadian clock. *Nature Neurosci.* **6,** 208–210.

Blau, J., and Young, M. W. (1999). Cycling *vrille* expression is required for a functional *Drosophila* clock. *Cell* **99,** 661–671.

Brand, A. (1999). GFP as a cell and developmental marker in the *Drosophila* nervous system. *Methods Cell Biol.* **58,** 165–181.

Brand, A. H., and Dormand, E. L. (1995). The GAL4 system as a tool for unravelling the mysteries of the *Drosophila* nervous system. *Curr. Opin. Neurobiol.* **5,** 572–578.

Brandes, C., Plautz, J. D., Stanewsky, R., Jamison, C. F., Straume, M., Wood, K. V., Kay, S. A., and Hall, J. C. (1996). Novel features of *Drosophila period* transcription revealed by real-time luciferase reporting. *Neuron* **16,** 687–692.

Brandstaetter, R. (2004). Circadian lessons from peripheral clocks: Is the time of the mammalian pacemaker up? *Proc. Natl. Acad. Sci. USA* **101,** 5699–5700.

Brett, W. J. (1955). Persistent diurnal rhythmicity in *Drosophila* emergence. *Ann. Entomol. Soc. Am.* **48,** 119–131.

Busza, A., Emery-Le, M., Rosbash, M., and Emery, P. (2004). Roles of the two *Drosophila* CRYPTOCHROME structural domains in circadian photoreception. *Science* **304,** 1503–1506.

Ceriani, M. F., Darlington, T. K., Staknis, D., Más, P., Petti, A. A., Weitz, C. J., and Kay, S. A. (1999). Light-dependent sequestration of TIMELESS by CRYPTOCHROME. *Science* **285**, 553–556.

Chang, D. C., and Reppert, S. M. (2003). A novel C-terminal domain of *Drosophila* PERIOD inhibits dCLOCK:CYCLE-mediated transcription. *Curr. Biol.* **13**, 758–762.

Chen, D.-M., Christianson, J. S., Sapp, R. J., and Stark, W. S. (1992). Visual receptor cycle in normal and *period* mutant *Drosophila*: Microspectrophotometry, electrophysiology, and ultrastructural morphometry. *Vis. Neurosci.* **9**, 125–135.

Cheng, Y., and Hardin, P. E. (1998). *Drosophila* photoreceptors contain an autonomous circadian oscillator that can function without *period* mRNA cycling. *J. Neurosci.* **18**, 741–750.

Choi, Y. J., Lee, G., Hall, J. C., and Park, J. H. (2005). Comparative analysis of corazonin-encoding genes (*Crz*'s) in four *Drosophila* species and functional insight into *Crz*-expressing neurons. *J. Comp. Neurol.* In press.

Citri, Y., Colot, H. V., Jacquier, A. C., Yu, Q., Hall, J. C., Baltimore, D., and Rosbash, M. (1987). A family of unusually spliced biologically active transcripts encoded by a *Drosophila* clock gene. *Nature* **326**, 42–47.

Clark, A. C., del Campo, M. L., and Ewer, J. (2004). Neuroendocrine control of larval ecdysis behavior in *Drosophila*: Complex regulation by partially redundant neuropeptides. *J. Neurosci.* **24**, 4283–4292.

Collins, B. H., Rosato, E., and Kyriacou, C. P. (2004). Seasonal behavior in *Drosophila melanogaster* requires the photoreceptors, the circadian clock, and phospholipase C. *Proc. Natl. Acad. Sci. USA* **101**, 1945–1950.

Connolly, J. B., Roberts, I. J., Armstrong, J. D., Kaiser, K., Forte, M., Tully, T., and O'Kane, C. J. (1996). Associative learning disrupted by impaired Gs signaling in *Drosophila* mushroom bodies. *Science* **274**, 2104–2107.

Cyran, S. A., Buchsbaum, A. M., Reddy, K. L., Lin, M. C., Glossop, N. R., Hardin, P. E., Young, M. W., Storti, R. V., and Blau, J. (2003). *vrille, Pdp1*, and *dClock* form a second feedback loop in the *Drosophila* circadian clock. *Cell* **112**, 329–341.

Daan, S. (2000). Learning and circadian behavior. *J. Biol. Rhythms* **15**, 296–299.

Darlington, T., Wager-Smith, K., Ceriani, M. F., Staknis, D., Gekakis, N., Steeves, T., Weitz, C., Takahashi, J. S., and Kay, S. A. (1998). Closing the circadian loop: CLOCK-induced transcription of its own inhibitors, *per* and *tim*. *Science* **280**, 1599–1603.

Dauwalder, B., Tsujimoto, S., Moss, J., and Mattox, W. (2002). The *Drosophila takeout* gene is regulated by the somatic sex-determination pathway and affects male courtship behavior. *Genes Dev.* **16**, 2879–2892.

de Groot, M. H., and Rusak, B. (2000). Responses of the circadian system of rats to conditioned and unconditioned stimuli. *J. Biol. Rhythms* **15**, 277–291.

Demetriades, M. C., Thackeray, J. R., and Kyriacou, C. P. (1999). Courtship song rhythms in *Drosophila yakuba*. *Anim. Behav.* **57**, 379–386.

DiBartolomeis, S. M., Akten, B., Genova, G., Roberts, M. A., and Jackson, F. R. (2002). Molecular analysis of the *Drosophila miniature-dusky* (*m-dy*) gene complex: *m-dy* mRNAs encode transmembrane proteins with similarity to *C. elegans* cuticulin. *Mol. Genet. Genom.* **267**, 564–576.

Dissel, S., Codd, V., Fedic, R., Garner, K. J., Costa, R., Kyriacou, C. P., and Rosato, E. (2004). A constitutively active cryptochrome in *Drosophila melanogaster*. *Nature Neurosci.* **7**, 834–840.

Dow, J. T., and Davies, S. A. (2003). Integrative physiology and functional genomics of epithelial function in a genetic model organism. *Physiol. Rev.* **83**, 687–729.

Dowse, H. B., Dushay, M. S., Hall, J. C., and Ringo, J. M. (1989). High-resolution analysis of locomotor activity rhythms in *disconnected*, a visual-system mutant of *Drosophila melanogaster*. *Behav. Genet.* **19**, 529–542.

Dowse, H. B., and Ringo, J. M. (1991). Comparisons between "periodograms" and spectral analysis: Apples are apples after all. *J. Theor. Biol.* **148**, 139–144.

Dubnau, J. (2004). Neurogenetic dissection of conditioned behavior: Evolution by analogy or homology? *J. Neurogenet.* **17**, 295–326.

Duffield, G. E. (2003). DNA microarray analyses of circadian timing: The genomic basis of biological time. *J. Neuroendocrinol.* **15**, 991–1002.

Dushay, M. S., Rosbash, M., and Hall, J. C. (1989). The *disconnected* visual system mutations in *Drosophila* drastically disrupt circadian rhythms. *J. Biol. Rhythms* **4**, 1–27.

Edery, I., Zwiebel, L. J., Dembinska, M. E., and Rosbash, M. (1994). Temporal phosphorylation of the *Drosophila period* protein. *Proc. Natl. Acad. Sci. USA* **91**, 2260–2264.

Egan, E. S., Franklin, T. M., Hildebrand-Chae, M. J., McNeil, G. P., Roberts, M. A., Schroeder, A. J., Zhang, X., and Jackson, F. R. (1999). An extraretinally expressed insect cryptochrome with similarity to the blue light photoreceptors of mammals and plants. *J. Neurosci.* **19**, 3665–3673.

Emery, I. F., Noveral, J. M., Jamison, C. F., and Siwicki, K. K. (1997). Rhythms of *Drosophila period* gene expression in culture. *Proc. Natl. Acad. Sci. USA* **94**, 4092–4096.

Emery, P., So, W. V., Kaneko, M., Hall, J. C., and Rosbash, M. (1998). CRY, a *Drosophila* clock and light-regulated cryptochrome, is a major contributor to circadian rhythm resetting and photosensitivity. *Cell* **95**, 669–679.

Emery, P., Stanewsky, R., Hall, J. C., and Rosbash, M. (2000a). *Drosophila* cryptochrome: A unique circadian-rhythm photoreceptor. *Nature* **404**, 456–457.

Emery, P., Stanewsky, R., Helfrich-Forster, C., Emery-Le, M., Hall, J. C., and Rosbash, M. (2000b). *Drosophila* CRY is a deep brain circadian photoreceptor. *Neuron* **26**, 493–504.

Etter, P. D., and Ramaswami, M. (2002). The ups and downs of daily life: Profiling circadian gene expression in *Drosophila*. *BioEssays* **24**, 494–498.

Ewer, J., Frisch, B., Hamblen-Coyle, M. J., Rosbash, M., and Hall, J. C. (1992). Expression of the *period* clock gene within different cells types in the brain of *Drosophila* adults and mosaic analysis of these cells' influence on circadian behavioral rhythms. *J. Neurosci.* **12**, 3321–3349.

Ewer, J., and Reynolds, S. (2002). Neuropeptide control of molting in insects. *In* "Hormones, Brain and Behavior" (D. W. Pfaff, A. P. Arnold, S. E. Fahrbach, A. M. Etgen, and R. T. Rubin, eds.), vol. 3, pp. 1–92. Academic Press, San Diego.

Flint, K. K., Rosbash, M., and Hall, J. C. (1993). Transfer of dye among salivary gland cells is not affected by genetic variations of the *period* clock gene in *Drosophila melanogaster*. *J. Membr. Biol.* **136**, 333–342.

Frank, K. D., and Zimmerman, W. F. (1969). Action spectra for phase shifts of a circadian rhythm in *Drosophila*. *Science* **163**, 688–689.

Frisch, B., Hardin, P. E., Hamblen-Coyle, M. J., Rosbash, M., and Hall, J. C. (1994). A promoterless *period* gene mediates behavioral rhythmicity and cyclical per expression in a restricted subset of the *Drosophila* nervous system. *Neuron* **12**, 555–570.

Gailey, D. A., Villella, A., and Tully, T. (1991). Reassessment of the effects of biological rhythm mutations on learning in *Drosophila melanogaster*. *J. Comp. Physiol. A.* **169**, 685–697.

Gekakis, N., Saez, L., Delahaye-Brown, A.-M., Myers, M. P., Sehgal, A., Young, M. W., and Weitz, C. J. (1995). Isolation of timeless by PER protein interaction: Defective interaction between TIMELESS protein and long-period mutant PER$^L$. *Science* **270**, 811–815.

George, R., Lease, K., Burnette, J., and Hirsh, J. (2005). A "bottom-counting" video system for measuring cocaine-induced behaviors in *Drosophila. Methods Enzymol.* **393**(44), 837–847 (this volume).

George, H., and Terracol, R. (1997). The *vrille* gene of *Drosophila* is a maternal enhancer of *decapentaplegic* and encodes a new member of the bZIP family of transcription factors. *Genetics* **146**, 1345–1363.

Giebultowicz, J. M. (1999). Insect circadian rhythms: Is it all in their heads? *J. Insect Physiol.* **45**, 791–800.

Giebultowicz, J. M. (2000). Molecular mechanism and cellular distribution of insect circadian clocks. *Annu. Rev. Entomol.* **45**, 767–791.

Giebultowicz, J. M. (2001). Peripheral clocks and circadian timing: Insights from insects. *Phil. Trans. Roy. Soc. Lond. B* **356**, 1791–1799.

Giebultowicz, J. M., Ivanchenko, M., and Vollintine, T. (2001). Organization of the insect circadian system: Spatial and developmental of clock genes in peripheral tissues of *Drosophila melanogaster. In* "Insect Timing: Circadian Rhythmicity and Seasonality" (D. L. Denlinger, J. Giebultowicz, and D. S. Saunders, eds.), pp. 31–42. Elsevier Science B. V., Amsterdam.

Giebultowicz, J. M., Stanewsky, R., Hall, J. C., and Hege, D. M. (2000). Transplanted *Drosophila* excretory tubules maintain circadian clock cycling out of phase with the host. *Curr. Biol.* **10**, 107–110.

Greenspan, R. J., Tononi, G., Cirelli, C., and Shaw, P. J. (2001). Sleep and the fruit fly. *Trends Neurosci.* **24**, 142–145.

Grima, B., Chélot, E., Xia, R., and Rouyer, F. (2004). Morning and evening peaks of activity rely on different clock neurons of the *Drosophila* brain. *Nature* **431**, 869–873.

Grima, B., Lamouroux, A., Chélot, E., Papin, C., Limbourg-Bouchon, B., and Rouyer, F. (2002). The F-box protein slimb controls the levels of clock proteins period and timeless. *Nature* **420**, 178–182.

Hall, J. C. (1995). Tripping along the trail to the molecular mechanisms of biological clocks. *Trends Neurosci.* **18**, 230–240.

Hall, J. C. (1998). Molecular neurogenetics of biological rhythms. *J. Neurogenet.* **12**, 115–181.

Hall, J. C. (2000). Cryptochrome: Sensory reception, transduction and clock functions subserving circadian systems. *Curr. Opin. Neurobiol.* **10**, 456–466.

Hall, J. C. (2003a). Genetics and molecular biology of rhythms in *Drosophila* and other insects. *Adv. Genet.* **48**, 1–286.

Hall, J. C. (2003b). A neurogeneticist's manifesto. *J. Neurogenet.* **17**, 1–90.

Hall, J. C., and Kyriacou, C. P. (1990). Genetics of biological rhythms in *Drosophila. Adv. Insect Physiol.* **22**, 221–298.

Hamblen, M., Zehring, A. A., Kyriacou, C. P., Reddy, P., Yu, Q., Wheeler, D. A., Zwiebel, L. J., Konopka, R. J., Rosbash, M., and Hall, J. C. (1986). Germ-line transformation involving DNA from the *period* locus in *Drosophila melanogaster:* overlapping genomic fragments that restore circadian and ultradian rhythmicity to *per^o* and *per^-* mutants. *J. Neurogenet.* **3**, 249–291.

Hamblen-Coyle, M. J., Wheeler, D. A., Rutila, J. E., Rosbash, M., and Hall, J. C. (1992). Behavior of period-altered circadian rhythm mutants of *Drosophila* in light:dark cycles. *J. Insect Behav.* **5**, 417–446.

Hardin, P. E. (1994). Analysis of *period* mRNA cycling in *Drosophila* head and body tissues indicates that body oscillators behave differently from head oscillators. *Mol. Cell. Biol.* **4**, 7211–7218.

Hardin, P. E., Hall, J. C., and Rosbash, M. (1990). Feedback of the *Drosophila period* gene product on circadian cycling of its messenger RNA levels. *Nature* **343**, 536–540.

Hardin, P. E., Hall, J. C., and Rosbash, M. (1992). Behavioral and molecular analyses suggest that circadian output is disrupted by *disconnected* mutants in *D. melanogaster. EMBO J.* **11**, 1–6.

Hardin, P. E., Krishnan, B., Houl, J. H., Zheng, H., Ng, F. S., Dryer, S. E., and Glossop, N. R. (2003). Central and peripheral oscillators in *Drosophila. Novartis Found. Symp.* **253**, 150–160.

Harms, E., Young, M. W., and Saez, L. (2003). CK1 and GSK3 in the *Drosophila* and mammalian circadian clock. *Novartis Found. Symp.* **253**, 267–277.

Hazelrigg, T. (2000). GFP and other reporters. In "*Drosophila* Protocols" (W. Sullivan, M. Ashburner, and R. S. Hawley, eds.), pp. 313–343. Cold Spring Harbor Laboratory Press, Cold Spring Harbor, NY.

Hege, D. M., Stanewsky, R., Hall, J. C., and Giebultowicz, J. M. (1997). Rhythmic expression of a PER-reporter in the Malpighian tubules of decapitated *Drosophila*: Evidence for a brain-independent circadian clock. *J. Biol. Rhythms* **12**, 300–308.

Helfrich, C. (1986). Role of the optic lobes in the regulation of the locomotor activity rhythms in *Drosophila melanogaster*: Behavioral analysis of neural mutants. *J. Neurogenet.* **3**, 321–343.

Helfrich-Förster, C. (1995). The period clock gene is expressed in CNS neurons which also produce a neuropeptide that reveals the projections of circadian pacemaker cells within the brain of *Drosophila melanogaster. Proc. Natl. Acad. Sci. USA* **92**, 612–616.

Helfrich-Förster, C. (1997). Development of pigment-dispersing hormone-immuno-reactive neurons in the nervous system of *Drosophila melanogaster. J. Comp. Neurol.* **380**, 335–354.

Helfrich-Förster, C. (1998). Robust circadian rhythmicity of *Drosophila melanogaster* requires the presence of lateral neurons: A brain-behavioral study of *disconnected* mutants. *J. Comp. Physiol. A* **182**, 435–453.

Helfrich-Förster, C. (2000). Differential control of morning and evening components in the activity rhythm of *Drosophila melanogaster*-sex-specific differences suggest a different quality of activity. *J. Biol. Rhythms* **15**, 135–154.

Helfrich-Förster, C. (2001). The activity rhythm of *Drosophila melanogaster* is controlled by a dual oscillator system. *J. Insect Physiol.* **47**, 877–887.

Helfrich-Förster, C. (2002). The circadian system of *Drosophila melanogaster* and its light input pathways. *Zoology* **105**, 297–312.

Helfrich-Förster, C. (2003). The neuroarchitecture of the circadian clock in the brain of *Drosophila melanogaster. Microsc. Res. Tech.* **62**, 94–102.

Helfrich-Förster, C., Edwards, T., Yasuyama, K., Wisotzki, B., Schneuwly, S., Stanewsky, R., Meinertzhagen, I. A., and Hofbauer, A. (2002a). The extraretinal eyelet of *Drosophila*: Development, ultrastructure, and putative circadian function. *J. Neurosci.* **22**, 9255–9266.

Helfrich-Förster, C., and Homberg, U. (1993). Pigment-dispersing hormone-immunoreactive neurons in the nervous system of wild-type *Drosophila melanogaster* and of several mutants with altered circadian rhythmicity. *J. Comp. Neurol.* **337**, 177–190.

Helfrich-Förster, C., Stengl, M., and Homberg, U. (1998). Organization of the circadian system in insects. *Chronobiol. Int.* **15**, 567–594.

Helfrich-Förster, C., Täuber, M., Park, J. H., Mühlig-Versen, M., Schneuwly, S., and Hofbauer, A. (2000). Ectopic expression of the neuropeptide pigment-dispersing factor alters behavioral rhythms in *Drosophila melanogaster. J. Neurosci.* **20**, 3339–3353.

Helfrich-Förster, C., Winter, C., Hofbauer, A., Hall, J. C., and Stanewsky, R. (2001). The circadian clock of fruit flies is blind after elimination of all known photoreceptors. *Neuron* **30**, 249–261.

Helfrich-Förster, C., Wülf, J., and de Belle, J. S. (2002b). Mushroom body influence on locomotor activity and circadian rhythms in *Drosophila melanogaster. J. Neurogenet.* **16**, 73–109.

Hendricks, J. C. (2003). Sleeping flies don't lie: The use of *Drosophila melanogaster* to study sleep and circadian rhythms. *J. Appl. Physiol.* **94,** 1660–1672.

Hendricks, J. C., and Sehgal, A. (2004). Why a fly? Using *Drosophila* to understand the genetics of circadian rhythms and sleep. *Sleep* **27,** 334–342.

Hirsh, J. (1998). Decapitated *Drosophila*: A novel system for the study of biogenic amines. *Adv. Pharmacol.* **42,** 945–948.

Hirsh, J. (2001). Time flies like an arrow: Fruit flies like crack? *Pharmacogeno. J.* **1,** 97–100.

Ho, K. S., and Sehgal, A. (2005). *Drosophila melanogaster*: An insect model for fundamental studies of sleep. *Methods Enzymol.* **393**(41), 770–791 (this volume).

Hofbauer, A., and Buchner, E. (1989). Does *Drosophila* have seven eyes? *Naturwissenschaften* **76,** 335–336.

Homberg, U., Reischig, T., and Stengl, M. (2003). Neural organization of the circadian system of the cockroach *Leucophaea maderae. Chronobiol. Int.* **20,** 577–591.

Hu, K. G., Reichert, H., and Stark, W. S. (1978). Electrophysiological characterization of *Drosophila* ocelli. *J. Comp. Physiol. A* **126,** 15–24.

Hunter-Ensor, M., Ousley, A., and Sehgal, A. (1996). Regulation of the *Drosophila* protein timeless suggests a mechanism for resetting the circadian clock by light. *Cell* **84,** 677–685.

Husain, Q. M., and Ewer, J. (2004). Use of targetable gfp-tagged neuropeptide for visualizing neuropeptide release following execution of a behavior. *J. Neurobiol.* **59,** 181–191.

Ishikawa, T., Matsumoto, A., Kato, T., Jr., Togashi, S., Ryo, H., Ikenaga, M., Todo, T., Ueda, R., and Tanimura, T. (1999). DCRY is a *Drosophila* photoreceptor protein implicated in light entrainment of circadian rhythm. *Genes to Cells* **4,** 57–65.

Ivanchenko, M., Stanewsky, R., and Giebultowicz, J. M. (2001). Circadian photoreception in *Drosophila*: Functions of cryptochrome in peripheral and central clocks. *J. Biol. Rhythms* **16,** 205–215.

Jackson, F. R., Bargiello, T. A., Yun, S. H., and Young, M. W. (1986). Product of *per* locus of *Drosophila* shares homology with proteoglycans. *Nature* **320,** 185–188.

Jackson, F. R., Gailey, D. A., and Siegel, R. W. (1983). Biological rhythm mutations affect an experience-dependent modification of male courtship behavior in *Drosophila melanogaster. J. Comp. Physiol. A* **151,** 545–552.

Jackson, F. R., Genova, G. K., Huang, Y., Kleyner, Y., Suh, J., Roberts, M. A., Sundram, V., and Akten, B. (2005). Genetic and biochemical strategies for identifying *Drosophila* genes that function in circadian control. *Methods Enzymol.* **393**(35), 661–680 (this volume).

Jackson, F. R., and Schroeder, A. J. (2001). A timely expression profile. *Dev. Cell* **1,** 730–731.

Jackson, F. R., Schroeder, A. J., Roberts, M. A., McNeill, G. P., Kume, K., and Akten, B. (2001). Cellular and molecular mechanisms of circadian control in insects. *J. Insect Physiol.* **47,** 833–842.

James, A. A., Ewer, J., Reddy, P., Hall, J. C., and Rosbash, M. (1986). Embryonic expression of the *period* clock gene in the central nervous system of *Drosophila melanogaster. EMBO J.* **5,** 2313–2320.

Jaramillo, A. M., Zheng, X., Zhou, Y., Amado, D. A., Sheldon, A., Sehgal, A., and Levitan, I. B. (2004). Pattern of distribution and cycling of SLOB, Slowpoke channel binding protein, in *Drosophila. BMC Neurosci.* **5**(1), 3.

Jauch, E., Melzig, J., Brkulj, M., and Raabe, T. (2002). *In vivo* functional analysis of *Drosophila* protein kinase CK2 $\beta$ subunit. *Gene* **298,** 29–39.

Jefferis, G. S., Marin, E. C., Watts, R. J., and Luo, L. (2002). Development of neuronal connectivity in *Drosophila* antennal lobes and mushroom bodies. *Curr. Opin. Neurobiol.* **12,** 80–86.

Judd, B. H., Shen, M. W., and Kaufman, T. C. (1972). The anatomy and function of a segment of the X chromosome of *Drosophila melanogaster. Genetics* **71,** 139–156.

Kaiser, W., and Steiner-Kaiser, J. (1983). Neuronal correlates of sleep, wakefulness and arousal in a diurnal insect. *Nature* **301,** 707–709.

Kaneko, M., Helfrich-Förster, C., and Hall, J. C. (1997). Spatial and temporal expression of the *period* and *timeless* genes in the developing nervous system of *Drosophila*: Newly identified pacemaker candidates and novel features of clock gene product cycling. *J. Neurosci.* **17,** 6745–6760.

Kaneko, M., Hamblen, M. J., and Hall, J. C. (2000a). Involvement of the *period* gene in developmental time-memory: Effect of the *per$^{Short}$* mutation on phase shifts induced by light pulses delivered to *Drosophila* larvae. *J. Biol. Rhythms* **15,** 13–30.

Kaneko, M., and Hall, J. C. (2000). Neuroanatomy of cells expressing clock genes in *Drosophila*: Transgenic manipulation of the *period* and *timeless* genes to mark the perikarya of circadian pacemaker neurons and their projections. *J. Comp. Neurol.* **422,** 66–94.

Kaneko, M., Park, J. H., Cheng, Y., Hardin, P. E., and Hall, J. C. (2000b). Disruption of synaptic transmission or clock-gene-product oscillations in circadian pacemaker cells of *Drosophila* cause abnormal behavioral rhythms. *J. Neurobiol.* **43,** 207–233.

Kilduff, T. S. (2000). What rest in flies can tell us about sleep in mammals. *Neuron* **26,** 295–298.

Kitamoto, T. (2002a). Targeted expression of temperature-sensitive dynamin to study neural mechanisms of complex behavior in *Drosophila. J. Neurogenet.* **16,** 205–228.

Kitamoto, T. (2002b). Conditional disruption of synaptic transmission induces male-male courtship behavior in *Drosophila. Proc. Natl. Acad. Sci. USA* **99,** 13232–13237.

Klarsfeld, A., Leloup, J. C., and Rouyer, F. (2003). Circadian rhythms of locomotor activity in *Drosophila. Behav. Processes* **64,** 161–175.

Klarsfeld, A., Malpel, S., Michard-Vanhee, C., Picot, M., Chelot, E., and Rouyer, F. (2004). Novel features of cryptochrome-mediated photoreception in the brain circadian clock of *Drosophila. J. Neurosci.* **24,** 1468–1477.

Klemm, E., and Ninnemann, H. (1976). Detailed action spectra for the delay phase shift in pupal emergence of *Drosophila pseudoobscura. Photochem. Photobiol.* **24,** 364–371.

Kloss, B., Rothenfluh, A., Young, M. W., and Saez, L. (2001). Phosphorylation of PERIOD is influenced by cycling physical associations of DOUBLE-TIME, PERIOD, and TIME-LESS in the *Drosophila* clock. *Neuron* **30,** 699–706.

Ko, H. W., and Edery, I. (2005). Analyzing the degradation of PERIOD protein by the ubiquitin-proteosome pathway in cultured *Drosophila* cells. *Method Enzymol.* **393**(18), 392–406 (this volume).

Ko, H. W., Jiang, J., and Edery, I. (2002). Role for Slimb in the degradation of *Drosophila* Period protein phosphorylated by Doubletime. *Nature* **420,** 673–678.

Konopka, R. J., and Benzer, S. (1971). Clock mutants of *Drosophila melanogaster. Proc. Natl. Acad. Sci. USA* **68,** 2112–2116.

Konopka, R. J., Hamblen-Coyle, M. J., Jamison, C. F., and Hall, J. C. (1994). An ultrashort clock mutation at the *period* locus of *Drosophila melanogaster* that reveals some new features of the fly's circadian system. *J. Biol. Rhythms* **9,** 189–216.

Konopka, R. J., Kyriacou, C. P., and Hall, J. C. (1996). Mosaic analysis in the *Drosophila* CNS of circadian and courtship-song rhythms affected by a *period* clock mutation. *J. Neurogenet.* **11,** 117–139.

Konopka, R. J., Pittendrigh, C., and Orr, D. (1989). Reciprocal behaviour associated with altered homeostasis and photosensitivity of *Drosophila* clock mutants. *J. Neurogenet.* **6,** 1–10.

Konopka, R., Wells, S., and Lee, T. (1983). Mosaic analysis of a *Drosophila* clock mutant. *Mol. Gen. Genet.* **190,** 284–288.

Konopka, R. J., Smith, R. F., and Orr, D. (1991). Characterization of *Andante*, a new *Drosophila* clock mutant, and its interactions with other clock mutants. *J. Neurogenet.* **7**, 103–114.

Krishnan, B., Dryer, S. E., and Hardin, P. E. (1999). Circadian rhythms in olfactory responses of *Drosophila melanogaster*. *Nature* **400**, 375–378.

Krishnan, P., Dryer, S. E., and Hardin, P. E. (2005). Measuring circadian rhythms in olfactory using electroantennograms. *Methods Enzymol.* **393**(25), 493–506 (this volume).

Krishnan, B., Levine, J. D., Lynch, K. S., Dowse, H. B., Funes, P., Hall, J. C., Hardin, P. E., and Dryer, S. E. (2001). A novel role for cryptochrome in a *Drosophila* circadian oscillator. *Nature* **411**, 313–317.

Kyriacou, C. P., and Hall, J. C. (1980). Circadian rhythm mutations in *Drosophila melanogaster* affect short-term fluctuations in the male's courtship song. *Proc. Natl. Acad. Sci. USA* **77**, 6929–6933.

Kyriacou, C. P., and Hall, J. C. (1982). The function of courtship song rhythms in *Drosophila*. *Anim. Behav.* **30**, 794–801.

Kyriacou, C. P., Oldroyd, M., Wood, J., Sharp, M., and Hill, M. (1990). Clock mutations alter developmental timing in *Drosophila*. *Heredity* **64**, 395–401.

Lee, C., Bae, K., and Edery, I. (1998). The *Drosophila* CLOCK protein undergoes daily rhythms in abundance, phosphorylation and interactions with the PER-TIM complex. *Neuron* **21**, 857–867.

Lee, T., and Luo, L. (1999). Mosaic analysis with a repressible cell marker for studies of gene function in neuronal morphogenesis. *Neuron* **22**, 451–461.

Lee, T., and Luo, L. (2001). Mosaic analysis with a repressible cell marker (MARCM) for *Drosophila* neural development. *Trends Neurosci.* **24**, 251–254. Review. (Erratum: *Trends Neurosci.* **24**, 385.)

Lee, G., Villella, A., Taylor, B. J., and Hall, J. C. (2001). New reproductive anomalies in *fruitless*-mutant *Drosophila* males: Extreme lengthening of mating durations and infertility correlated with defective serotonergic innervation of reproductive organs. *J. Neurobiol.* **47**, 121–149.

Levenbook, L. (1950). The physiology of carbon dioxide transport in insect blood. III. The buffering capacity of *Gastrophilus* blood. *J. Exp. Biol.* **27**, 184–191.

Levine, J. D. (2004). Sharing time on the fly. *Curr. Opin. Cell Biol.* **16**, 210–216.

Levine, J. D., Casey, C. I., Kalderon, D. D., and Jackson, F. R. (1994). Altered circadian pacemaker functions and cyclic AMP rhythms in the *Drosophila* learning mtuant *dunce*. *Neuron* **13**, 967–974.

Levine, J. D., Funes, P., Dowse, H. B., and Hall, J. C. (2002a). Resetting the circadian clock by social experience in *Drosophila melanogaster*. *Science* **298**, 2010–2012.

Levine, J. D., Funes, P., Dowse, H. B., and Hall, J. C. (2002b). Signal analysis of behavioral and molecular cycles. *BMC Neurosci.* **3**(1), 1.

Levine, J. D., Funes, P., Dowse, H. B., and Hall, J. C. (2002c). Advanced analysis of a cryptochrome mutation's effects on the robustness and phase of molecular cycles in isolated peripheral tissues of *Drosophila*. *BMC Neurosci.* **3**(1), 5.

Li, X., Borjigin, J., and Snyder, S. H. (1998). Molecular rhythms in the pineal gland. *Curr. Opin. Neurobiol.* **8**, 648–651.

Lin, J.-M., Kilman, V., Keegan, K., Paddock, B., Emery-Le, M., Rosbash, M., and Allada, R. (2002). A role for casein kinase $2\alpha$ in the *Drosophila* circadian clock. *Nature* **420**, 816–820.

Lin, Y., Stormo, G. D., and Taghert, P. H. (2004). The neuropeptide pigment-dispersing factor coordinates pacemaker interactions in the *Drosophila* circadian system. *J. Neurosci.* **24**, 7951–7957.

Liu, X., Lorenz, L., Yu, Q., Hall, J. C., and Rosbash, M. (1988). Spatial and temporal expression of the *period* gene in *Drosophila melanogaster*. *Genes Dev.* **2,** 228–238.

Liu, X., Yu, Q., Huang, Z., Zwiebel, L. J., Hall, J. C., and Rosbash, M. (1991). The strength and periodicity of *D. melanogaster* circadian rhythms are differentially affected by alterations in *period* gene expression. *Neuron* **6,** 753–766.

Liu, X., Zwiebel, L. J., Hinton, D., Benzer, S., Hall, J. C., and Rosbash, M. (1992). The *period* gene encodes a predominantly nuclear protein in adult *Drosophila*. *J. Neurosci.* **12,** 2735–2744.

Livingstone, M. S., and Tempel, B. L. (1983). Genetic dissection of monoamine neurotransmitter synthesis in *Drosophila*. *Nature* **303,** 67–70.

Lorenz, L. J., Hall, J. C., and Rosbash, M. (1989). Expression of a *Drosophila* mRNA is under circadian clock control during pupation. *Development* **107,** 869–880.

Lundkvist, G. B., and Block, G. D. (2005). Role of neuronal membrane events in circadian rhythm generation. *Methods Enzymol.* **393**(33), 621–640 (this volume).

Ma, J., and Ptashne, M. (1987). The carboxy-terminal 30 amino acids of GAL4 are recognized by GAL80. *Cell* **50,** 137–142.

Majercak, J., Chen, W. F., and Edery, I. (2004). Splicing of the *period* gene 3′-terminal intron is regulated by light, circadian clock factors, and phospholipase C. *Mol. Cell. Biol.* **24,** 3359–3372.

Majercak, J., Kalderon, D., and Edery, I. (1997). *Drosophila melanogaster* deficient in protein kinase a manifests behavior-specific arrhythmia but normal clock function. *Mol. Cell. Biol.* **17,** 5915–5922.

Majercak, J., Sidote, D., Hardin, P. E., and Edery, I. (1999). How a circadian clock adapts to seasonal decreases in temperature and day length. *Neuron* **24,** 219–230.

Malpel, S., Klarsfeld, A., and Rouyer, F. (2002). Larval optic nerve and adult extra-retinal photoreceptors sequentially associate with clock neurons during *Drosophila* brain development. *Development* **129,** 1443–1453.

Malpel, S., Klarsfeld, A., and Rouyer, F. (2004). Circadian synchronization and rhythmicity in larval photoperception-defective mutants of *Drosophila*. *J. Biol. Rhythms* **19,** 10–21.

Marrus, S. B., Zeng, H., and Rosbash, M. (1996). Effect of constant light and circadian entrainment of *per*^S flies: Evidence for light-mediated delay of the negative feedback loop in *Drosophila*. *EMBO J.* **15,** 6877–6886.

Martin, J.-R., Keller, A., and Sweeney, S. T. (2002). Targeted expression of tetanus toxin: A new tool to study the neurobiology of behavior. *Adv. Genet.* **47,** 1–47.

Martinek, S., Inonog, S., Manoukian, A. S., and Young, M. W. (2001). A role for the segment polarity gene *shaggy*/GSK-3 in the *Drosophila* circadian clock. *Cell* **105,** 769–779.

Matsumoto, A., Tomioka, K., Chiba, Y., and Tanimura, T. (1999). *tim*^rit lengthens circadian period in a temperature-dependent manner through suppression of PERIOD protein cycling and nuclear localization. *Mol. Cell. Biol.* **19,** 4343–4354.

McBride, S. M., Giuliani, G., Choi, C., Krause, P., Correale, D., Watson, K., Baker, G., and Siwicki, K. K. (1999). Mushroom body ablation impairs short-term memory and long-term memory of courtship conditioning in *Drosophila melanogaster*. *Neuron* **24,** 967–977.

McCall, K., and Peterson, J. S. (2004). Detection of apoptosis in *Drosophila*. *Methods Mol. Biol.* **282,** 191–206.

McClung, C., and Hirsh, J. (1999). The trace amine tyramine is essential for sensitization to cocaine in *Drosophila*. *Curr. Biol.* **9,** 853–860.

McDonald, M. J., and Rosbash, M. (2001). Microarray analysis and organization of circadian gene expression in *Drosophila*. *Cell* **107,** 567–578.

McNabb, S. L., Baker, J. D., Agapite, J., Steller, H., Riddiford, L. M., and Truman, J. W. (1997). Disruption of behavioral sequence by targeted death of pepidergic neurons in *Drosophila*. *Neuron* **19**, 813–823.

Mealey-Ferrara, M. L., Montalvo, A. G., and Hall, J. C. (2003). Effects of combining a cryptochrome mutation with other visual-system variants on entrainment of locomotor and adult-emergence rhythms in *Drosophila*. *J. Neurogenet.* **17**, 171–221.

Megighian, A., Zordan, M., and Costa, R. (2001). Giant neuron pathway neurophysiological activity in $per^0$ mutants of *Drosophila melanogaster*. *J. Neurogenet.* **15**, 221–231.

Merrow, M., Brunner, M., and Roenneberg, T. (1999). Assignment of circadian function for the *Neurospora* clock gene *frequency*. *Nature* **399**, 584–586.

Merrow, M., and Roenneberg, T. (2005). Enhanced phenotyping of complex traits with a circadian clock model. *Methods Enzymol.* **393**(10), 248–263 (this volume).

Moses, K., and Rubin, G. M. (1991). *glass* encodes a site-specific DNA-binding protein that is regulated in response to positional signals in the developing *Drosophila* eye. *Genes Dev.* **5**, 583–593.

Murata, T., Matsumoto, A., Tomioka, K., and Chiba, Y. (1995). *Ritsu*: A rhythm mutant from a natural population of *Drosophila melanogaster*. *J. Neurogenet.* **9**, 239–249.

Myers, E. M., Yu, J., and Sehgal, A. (2003). Circadian control of eclosion: Interaction between a central and peripheral clock in *Drosophila melanogaster*. *Curr. Biol.* **13**, 526–533.

Myers, M. P., Wager-Smith, K., Rothenfluh-Hilfiker, A., and Young, M. W. (1996). Light-induced degradation of TIMELESS and entrainment of the *Drosophila* circadian clock. *Science* **271**, 1736–1740.

Naidoo, N., Song, W., Hunter-Ensor, M., and Sehgal, A. (1999). A role for the proteosome in the light response of the timeless clock protein. *Science* **285**, 1737–1741.

Nash, H. A., Scott, R. L., Lear, B. C., and Allada, R. (2002). An unusual cation channel mediates photic control of locomotion in *Drosophila*. *Curr. Biol.* **12**, 2152–2158.

Nawathean, P., and Rosbash, M. (2004). The doubletime and CKII kinases collaborate to potentiate *Drosophila* PER transcriptional repressor activity. *Mol. Cell* **13**, 213–223.

Newby, L. M., and Jackson, F. R. (1991). *Drosophila ebony* mutants have altered circadian activity rhythms but normal eclosion rhythms. *J. Neurogenet.* **7**, 85–101.

Newby, L. M., and Jackson, F. R. (1993). A new biological rhythm mutant of *Drosophila melanogaster* that identifies a gene with an essential embryonic function. *Genetics* **135**, 1077–1090.

Nitabach, M. N., Blau, J., and Holmes, T. C. (2002). Electrical silencing of *Drosophila* pacemaker neurons stops the free-running circadian clock. *Cell* **109**, 485–495.

Nitabach, M. N., Holmes, T. C., and Blau, J. (2005). Membranes, ions, and clocks: Testing the Njus–Salzman–Hastings model of the circadian oscillator. *Methods Enzymol.* **393**(34), 643–661 (this volume).

Nitz, D. A., van Swinderen, B., Tononi, G., and Greenspan, R. J. (2002). Electrophysiological correlates of rest and activity in *Drosophila melanogaster*. *Curr. Biol.* **12**, 1934–1940.

Ohata, K., Nishiyama, H., and Tsukahara, Y. (1998). Action spectrum of the circadian clock photoreceptor in *Drosophila melanogaster*. *In* "Biological Clocks, Mechanisms and Applications" (Y. Touitou, ed.), pp. 167–170. Elsevier Science B. V., Amsterdam.

Oishi, K., Shiota, M., Sakamoto, K., Kasamatsu, M., and Ishida, N. (2004). Feeding is not a more potent Zeitgeber than the light-dark cycle in *Drosophila*. *Neuroreport* **15**, 739–743.

Okamura, H. (2003). Integration of molecular rhythms in the mammalian circadian system. *Novartis Found Symp.* **253**, 161–170.

Okano, S., Kanno, S., Takao, M., Eker, A. P. M., Isono, K., Tsukahara, Y., and Yasui, A. (1999). A putative blue-light receptor from *Drosophila melanogaster*. *Photochem. Photobiol.* **69,** 108–113.

Orr, D. P.-Y. (1982). "Genetic Analysis of the Circadian Clock System of *Drosophila melanogaster*." Ph.D. thesis, California Institute of Technology, Pasadena, CA.

Ousley, A., Zafarulluh, K., Chen, Y., Emerson, M., Hickman, L., and Sehgal, A. (1998). Conserved regions of the *timeless* (*tim*) clock gene in *Drosophila* analyzed through phylogenetic and functional studies. *Genetics* **148,** 815–825.

Page, T. L. (1982). Extraretinal photoreceptors in entrainment and photoperiodism in invertebrates. *Experientia* **38,** 1001–1013.

Page, T. L., and Koelling, E. (2003). Circadian rhythm in olfactory response in the antennae controlled by the optic lobe in the cockroach. *J. Insect Physiol.* **49,** 697–707.

Pak, W. L., and Leung, H. T. (2003). Genetic approaches to visual transduction in *Drosophila melanogaster*. *Receptors Channels* **9**(3), 149–167.

Park, J. H. (2002). Downloading central clock information in *Drosophila*. *Mol. Neurobiol.* **26,** 217–233.

Park, J. H., and Hall, J. C. (1998). Isolation and chronobiological analysis of a neuropeptide pigment-dispersing factor gene in *Drosophila melanogaster*. *J. Biol. Rhythms* **13,** 219–228.

Park, J. H., Helfrich-Förster, C., Lee, G., Liu, L., Rosbash, M., and Hall, J. C. (2000a). Differential regulation of circadian pacemaker output by separate clock genes in *Drosophila*. *Proc. Natl. Acad. Sci. USA* **97,** 3608–3613.

Park, S. K., Sedore, S. A., Cronmiller, C., and Hirsh, J. (2000b). Type II cAMP-dependent protein kinase-deficient *Drosophila* are viable but show developmental, circadian, and drug response phenotypes. *J. Biol. Chem.* **275,** 20588–20596.

Park, J. H., Schroeder, A. J., Helfrich-Förster, C., Jackson, F. R., and Ewer, J. (2003). Targeted ablation of CCAP neuropeptide-containing neurons of *Drosophila* causes specific defects in execution and circadian timing of ecdysis behavior. *Development* **130,** 2645–2656.

Pearn, M. T., Randall, L. L., Shortridge, R. D., Burg, M. G., and Pak, W. L. (1996). Molecular, biochemical, and electrophysiological characterization of *Drosophila norpA* mutants. *J. Biol. Chem.* **271,** 4937–4945.

Peirson, S. N., Thompson, S., Hankins, M. W., and Foster, R. G. (2005). Mammalian photoentrainment: Results, methods, and approaches. *Methods Enzymol.* **393**(37), 695–724 (this volume).

Peixoto, A. A., Hennessy, J. M., Townson, I., Hasan, G., Rosbash, M., Costa, R., and Kyriacou, C. P. (1998). Molecular coevolution within a *Drosophila* clock gene. *Proc. Natl. Acad. Sci. USA* **95,** 4475–4480.

Peng, Y., Stoleru, D., Levine, J. D., Hall, J. C., and Rosbash, M. (2003). *Drosophila* free-running rhytms require intercellular communication. *Pub. Lib. Sci. Biol.* **1,** 32–40.

Perrimon, N., Lanjuin, A., Arnold, C., and Noll, E. (1996). Zygotic lethal mutations with maternal effect phenotypes in *Drosophila melanogaster*. *Genetics* **144,** 1681–1692.

Petersen, G., Hall, J. C., and Rosbash, M. (1988). The *period* gene of *Drosophila* carries species-specific behavioral instructions. *EMBO J.* **7,** 3939–3947.

Petri, B., and Stengl, M. (1997). Pigment dispersing hormone shifts the phase of the circadian pacemaker of the cockroach *Leucophaea maderae*. *J. Neurosci.* **17,** 4087–4093.

Petri, B., and Stengl, M. (2001). Phase response curves of a molecular model oscillator: Implications for mutual coupling of paired oscillators. *J. Biol. Rhythms* **16,** 125–141.

Phelps, C. B., and Brand, A. H. (1998). Ectopic gene expression in *Drosophila* using GAL4 system. *Methods* **14,** 367–379.

Philip, A. V. (1998). Mitotic sister-chromatid separation: What *Drosophila* mutants can tell us. *Trends Cell Biol.* **8,** 150.

Plautz, J. D., Kaneko, M., Hall, J. C., and Kay, S. A. (1997a). Independent photoreceptive circadian clocks throughout *Drosophila. Science* **278,** 1632–1635.

Plautz, J. D., Straume, M., Stanewsky, R., Jamison, C. F., Brandes, C., Dowse, H. B., Hall, J. C., and Kay, S. A. (1997b). Quantitative analysis of *Drosophila period* gene transcription in living animals. *J. Biol. Rhythms* **12,** 204–217.

Pollock, V. P., Radford, J. C., Pyne, S., Hasan, G., Dow, J. A., and Davies, S. A. (2003). *NorpA* and *itpr* mutants reveal roles for phospholipase C and inositol (1,4,5)-trisphosphate receptor in *Drosophila melanogaster* renal function. *J. Exp. Biol.* **206,** 901–911.

Price, J. L. (2004). *Drosophila melanogaster:* A model system for molecular chronobiology. *In* "Molecular Biology of Circadian Rhythms" (A. Sehgal, ed.), pp. 33–74. Wiley, New York.

Price, J. L. (2005). Genetic screen for clock mutants in *Drosophila. Methods Enzymol.* **393**(3), 35–60 (this volume).

Price, J. L., Blau, J., Rothenfluh, A., Abodeely, M., Kloss, B., and Young, M. W. (1998). *double-time* is a new *Drosophila* clock gene that regulates PERIOD protein accumulation. *Cell* **94,** 83–95.

Price, J. L., Dembinska, M. E., Young, M. W., and Rosbash, M. (1995). Suppression of PERIOD protein abundance and circadian cycling by the *Drosophila* clock mutation *timeless. EMBO J.* **14,** 4044–4049.

Rachidi, M., Lopes, C., Benichou, J.-C., and Rouyer, F. (1997). Analysis of *period* circadian expression in the *Drosophila* head by *in situ* hybridization. *J. Neurogenet.* **11,** 255–263.

Reddy, P., Zehring, W. A., Wheeler, D. A., Pirrotta, V., Hadfield, C., Hall, J. C., and Rosbash, M. (1984). Molecular analysis of the *period* locus in *Drosophila melanogaster* and identification of a transcript involved in biological rhythms. *Cell* **38,** 701–710.

Reddy, P., Jacquier, A. C., Abovich, N., Petersen, G., and Rosbash, M. (1986). The *period* clock locus of *D. melanogaster* codes for a proteoglycan. *Cell* **46,** 53–61.

Renn, S. C. P., Park, J. H., Rosbash, M., Hall, J. C., and Taghert, P. H. (1999). A *pdf* neuropeptide gene mutation and ablation of PDF neurons each cause severe abnormalities of behavioral circadian rhythms in *Drosophila. Cell* **99,** 791–802.

Richardson, H., and Kumar, S. (2002). Death to flies: *Drosophila* as a model system to study programmed cell death. *J. Immunol. Methods* **265,** 21–38.

Rieger, D., Stanewsky, R., and Helfrich-Förster, C. (2003). Cryptochrome, compound eyes, Hofbauer-Buchner eyelets, and ocelli play different roles in the entrainment and masking pathway of the locomotor activity rhythm in the fruit fly *Drosophila melanogaster. J. Biol. Rhythms* **18,** 377–391.

Ritchie, M. G., Halsey, E. J., and Gleason, J. M. (1999). *Drosophila* song as a species-specific mating signal and the behavioural importance of Kyriacou & Hall cycles in *D. melanogaster* song. *Anim. Behav.* **58,** 649–657.

Robinow, S., and White, K. (1991). Characterization and spatial distribution of the ELAV protein during *Drosophila melanogaster* development. *J. Neurobiol.* **22,** 443–461.

Rothenfluh, A., Abodeely, M., Price, J. L., and Young, M. W. (2000a). Isolation and analysis of six *timeless* alleles that cause short- or long-period circadian rhythms in *Drosophila. Genetics* **156,** 665–675.

Rothenfluh, A., Abodeely, M., and Young, M. W. (2000b). Short-period mutations of *per* affect a *double-time*-dependent step in the *Drosophila* circadian clock. *Curr. Biol.* **10,** 1399–1402.

Rothenfluh, A., Young, M. W., and Saez, L. (2000c). A TIMELESS-independent function for PERIOD proteins in the *Drosophila* clock. *Neuron* **26,** 505–514.

Rouyer, F., Rachidi, M., Pikielny, C., and Rosbash, M. (1997). A new gene encoding a putative transcription factor regulated by the *Drosophila* circadian clock. *EMBO J.* **16,** 3944–3954.

Rutila, J. E., Maltseva, O., and Rosbash, M. (1998a). The *tim^SL^* mutant affects a restricted portion of the *Drosophila melanogaster* circadian cycle. *J. Biol. Rhythms* **13,** 380–392.

Rutila, J. E., Suri, V., Le, M., So, W. V., Rosbash, M., and Hall, J. C. (1998b). CYCLE is a second bHLH-PAS clock protein essential for circadian rhythmicity and transcription of *Drosophila period* and *timeless. Cell* **93,** 805–814.

Rutila, J. E., Zeng, H., Le, M., Curtin, K. D., Hall, J. C., and Rosbash, M. (1996). The *tim^SL^* mutant of the *Drosophila* rhythm gene *timeless* manifests allele-specific interactions with *period* gene mutants. *Neuron* **17,** 921–929.

Saez, L., and Young, M. W. (1988). *In situ* localization of the per clock protein during development of *Drosophila melanogaster. Mol. Cell. Biol.* **8,** 5378–5385.

Saez, L., and Young, M. W. (1996). Regulated nuclear localization of the *Drosophila* clock proteins PERIOD and TIMELESS. *Neuron* **17,** 911–920.

Saez, L., Young, M. W., Baylies, M. K., Gasic, G., Bargiello, T. A., and Spray, D. C. (1992). *Per*: No link to gap junctions. *Nature* **360,** 542.

Sakai, T., Tamura, T., Kitamoto, T., and Kidokoro, Y. (2004). A clock gene, *period*, plays a key role in long-term memory in *Drosophila. Proc. Natl. Acad. Sci. USA* **101,** 16058–16063.

Sakai, T., and Ishida, N. (2001). Circadian rhythms of female mating activity governed by clock genes in *Drosophila. Proc. Natl. Acad. Sci. USA* **98,** 9221–9225.

Sakura, M., Takasuga, K., Watanabe, M., and Eguchi, E. (2003). Diurnal and circadian rhythm in compound eye of cricket (*Gryllus bimaculatus*): Changes in structure and photon capture efficiency. *Zool. Sci.* **20,** 833–840.

Sarov-Blat, L., So, W. V., Liu, L., and Rosbash, M. (2000). The *Drosophila takeout* gene is a novel link between circadian rhythms and feeding behavior. *Cell* **101,** 647–656.

Sathyanarayanan, S., Zheng, X., Xiao, R., and Sehgal, A. (2004). Posttranslational regulation of *Drosophila* PERIOD protein by protein phosphatase 2A. *Cell* **116,** 603–615.

Sato, T. K., Panda, S., Kay, S. A., and Hogenesch, J. B. (2003). DNA arrays: Applications and implications for circadian biology. *J. Biol. Rhythms* **18,** 96–105.

Sauer, S., Kinkelin, M., Herrmann, E., and Kaiser, W. (2003). The dynamics of sleep-like behaviour in honey bees. *J. Comp. Physiol.* A **189,** 599–607.

Sauman, I., and Reppert, S. M. (1996). Circadian clock neurons in the silkmoth *Antheraea pernyi*: Novel mechanism of Period protein regulation. *Neuron* **17,** 889–900.

Sauman, I., and Reppert, S. M. (1998). Brain control of embryonic circadian rhythms in the silkmoth *Antheraea pernyi. Neuron* **20,** 741–748.

Sauman, I., Tasi, T., Roca, A. L., and Reppert, S. M. (1996). Period protein is necessary for circadian control of egg hatching behavior in the silkmoth *Antheraea pernyi. Neuron* **20,** 901–909.

Sauman, I., Tsai, T., Roca, A. L., and Reppert, S. M. (2000). Erratum. *Neuron* **27**(1), [last page of issue].

Sawin, E. P., Dowse, H. B., Hamblen-Coyle, M. J., Hall, J. C., and Sokolowski, M. B. (1994). A search for locomotor activity rhythms in *Drosophila melanogaster* larvae. *J. Insect Behav.* **7,** 249–262.

Schibler, U., Ripperger, J., and Brown, S. A. (2003). Peripheral circadian oscillators in mammals: Time and food. *J. Biol. Rhythms* **18,** 250–260.

Schibler, U., and Sassone-Corsi, P. (2002). A web of circadian pacemakers. *Cell* **111,** 119–122.

Sehgal, A., Price, J. L., Man, B., and Young, M. W. (1994). Loss of circadian behavioral rhythms and *per* RNA oscillations in the *Drosophila* mutant *timeless. Science* **263,** 1603–1606.

Sehgal, A., Price, J., and Young, M. W. (1992). Ontogeny of a biological clock in *Drosophila*. *Proc. Natl. Acad. Sci. USA* **89,** 1423–1427.

Sehgal, A., Rothenfluh-Hilfiker, A., Hunter-Ensor, M., Chen, Y., Myers, M. P., and Young, M. W. (1995). Rhythmic expression of *timeless*: A basis for promoting circadian cycles in *period* gene autoregulation. *Science* **270,** 808–810.

Sehgal, A., and Price, J. L. (2004). Genetic and molecular approaches used to analyze rhythms. *In* "Molecular Biology of Circadian Rhythms" (A. Sehgal, ed.), pp. 17–29. Wiley, New York.

Shafer, O. T., Levine, J. D., Truman, J. W., and Hall, J. C. (2004). Flies by night: Effects of changing day length on *Drosophila*'s circadian clock. *Curr. Biol.* **14,** 424–432.

Shafer, O. T., Rosbash, M., and Truman, J. W. (2002). Sequential nuclear accumulation of the clock proteins Period and Timeless in the pacemaker neurons of *Drosophila melanogaster*. *J. Neurosci.* **22,** 5946–5954.

Shannon, M. P., Kaufman, T. C., Shen, M. W., and Judd, B. H. (1972). Lethality patterns and morphology of selected lethal and semi-lethal mutations in the *zeste-white* region of *Drosophila melanogaster*. *Genetics* **72,** 615–638.

Shaw, P. (2003). Awakening to the behavioral analysis of sleep in *Drosophila*. *J. Biol. Rhythms* **18,** 4–11.

Shaw, P. J., Tononi, G., Greenspan, R. J., and Robinson, D. F. (2002). Stress response genes protect against lethal effects of sleep deprivation in *Drosophila*. *Nature* **417,** 287–291.

Shin, H.-S., Bargiello, T. A., Clark, B. R., Jackson, F. R., and Young, M. W. (1985). An unusual coding sequence form a *Drosophila* clock gene is conserved in vertebrates. *Nature* **317,** 445–448.

Shirasu, N., Shimohigashi, Y., Tominaga, Y., and Shimohigashi, M. (2003). Molecular cogs of the insect circadian clock. *Zool. Sci.* **20,** 947–955.

Sidote, D., Majercak, J., Parikh, V., and Edery, I. (1998). Differential effects of light and heat on the *Drosophila* circadian clock proteins PER and TIM. *Mol. Cell. Biol.* **18,** 2004–2013.

Siwicki, K. K., Eastman, C., Petersen, G., Rosbash, M., and Hall, J. C. (1988). Antibodies to the *period* gene product of *Drosophila* reveal diverse tissue distribution and rhythm changes in the visual system. *Neuron* **1,** 141–150.

Siwicki, K. K., Flint, K. K., Hall, J. C., Rosbash, M., and Spray, D. C. (1992). The *Drosophila period* gene and dye coupling in larval salivary glands: A re-evaluation. *Biol. Bull.* **183,** 340–341.

Siwicki, K. K., and Ladewski, L. (2003). Associative learning and memory in *Drosophila*: Beyond olfactory conditioning. *Behav. Processes* **64,** 225–238.

Smith, R. F., and Konopka, R. J. (1981). Circadian clock phenotypes of chromosome aberrations with a breakpoint at the *per* locus. *Mol. Gen. Genet.* **183,** 243–251.

Smith, R. F., and Konopka, R. J. (1982). Effects of dosage alterations at the *per* locus on the circadian clock of *Drosophila*. *Mol. Gen. Genet.* **185,** 30–36.

So, W. V., and Rosbash, M. (1997). Post-transcriptional regulation contributes to *Drosophila* clock gene mRNA cycling. *EMBO J.* **16,** 7146–7155.

So, W. V., Sarov-Blat, L., Kotarski, C. K., McDonald, M. J., Allada, R., and Rosbash, M. (2000). *takeout*, a novel *Drosophila* gene under circadian clock transcriptional regulation. *Mol. Cell. Biol.* **20,** 6935–6944.

Stanewsky, R. (2003). Genetic analysis of the circadian system in *Drosophila melanogaster* and mammals. *J. Neurobiol.* **54,** 111–147.

Stanewsky, R. (2005). Analysis of rhythmic gene expression in adult *Drosophila* using the firefly-luciferase reporter gene. *In* "Circadian Rhythms: Reviews and Protocols" (E. Rosato, ed.). Humana Press, Totowa, NJ. In Press.

Stanewsky, R., Frisch, B., Brandes, C., Hamblen-Coyle, M. J., Rosbash, M., and Hall, J. C. (1997a). Temporal and spatial expression patterns of transgenes containing increasing amounts of the *Drosophila* clock gene *period* and a *lacZ* reporter: Mapping elements of the PER protein involved in circadian cycling. *J. Neurosci.* **17,** 676–696.

Stanewsky, R., Jamison, C. F., Plautz, J. D., Kay, S. A., and Hall, J. C. (1997b). Multiple circadian-regulated elements contribute to cycling *period* gene expression in *Drosophila*. *EMBO J.* **16,** 5006–5018.

Stanewsky, R., Kaneko, M., Emery, P., Beretta, B., Wager-Smith, K., Kay, S. A., Rosbash, M., and Hall, J. C. (1998). The *cry*[b] mutation identifies cryptochrome as a circadian photoreceptor in *Drosophila*. *Cell* **95,** 681–692.

Stanewsky, R., Lynch, K. S., Brandes, C., and Hall, J. C. (2002). Mapping of elements involved in regulating normal *period* and *timeless* RNA expression patterns in *Drosophila melanogaster*. *J. Biol. Rhythms* **17,** 293–306.

Stempfl, T., Vogel, M., Szabo, G., Wülbeck, C., Liu, J., Hall, J. C., and Stanewsky, R. (2002). Identification of circadian-clock regulated enhancers and genes of *Drosophila melanogaster* by transposon mobilization and luciferase reporting of cyclical gene expression. *Genetics* **160,** 571–593.

Stengl, M., Hatt, H., and Breer, H. (1992). Peripheral processes in insect olfaction. *Annu. Rev. Physiol.* **54,** 665–681.

Stoleru, D., Peng, Y., Agosto, J., and Rosbash, M. (2004). Coupled oscillators control morning and evening locomotor behavior in *Drosophila*. *Nature* **431,** 862–868.

Sturtevant, A. H. (2001). Reminiscences of T. H. Morgan (published in the year cited, but based on notes taken of a lecture delivered by Sturtevant at Woods Hole, MA, 1967). *Genetics* **159,** 1–5.

Suri, V., Hall, J. C., and Rosbash, M. (2000). Two novel *doubletime* mutants alter circadian properties and eliminate the delay between RNA and protein in *Drosophila*. *J. Neurosci.* **20,** 7547–7555.

Suri, V., Qian, Z., Hall, J. C., and Rosbash, M. (1998). Evidence that the TIM light response is relevant to light-induced phase shifts in *Drosophila melanogaster*. *Neuron* **21,** 225–234.

Taghert, P. H., Hewes, R. S., Park, J. H., O'Brien, M. A., Han, M., and Peck, M. E. (2001). Multiple amidated neuropeptides are required for normal circadian rhythmicity in *Drosophila*. *J. Neurosci.* **21,** 6673–6686.

Tanoue, S., Krishnan, P., Krishnan, B., Dryer, S. E., and Hardin, P. E. (2004). Circadian clocks in antennal neurons are necessary and sufficient for olfaction rhythms in *Drosophila*. *Curr. Biol.* **14,** 638–649.

Tauber, E., and Eberl, D. F. (2003). Acoustic communication in *Drosophila*. *Behav. Processes* **64,** 197–210.

Tauber, E., Roe, H., Costa, R., Hennessy, J. M., and Kyriacou, C. P. (2003). Temporal mating isolation driven by a behavioral gene in *Drosophila*. *Curr. Biol.* **13,** 140–145.

Tischkau, S. A., and Gillette, M. U. (2005). Oligodeoxynucleotide methods for analyzing the circadian clock in the suprachiasmatic nucleus. *Methods Enzymol.* **393**(31), 591–608 (this volume).

Tobler, I. (1983). Effect of forced locomotion on the rest-activity cycle of the cockroach. *Behav. Brain Res.* **8,** 351–360.

Toma, D. P., White, K. P., Hirsch, J., and Greenspan, R. J. (2002). Identification of genes involved in *Drosophila melanogaster* geotaxis, a complex behavioral trait. *Nature Genet.* **31,** 349–353.

Tomioka, K., Sakamoto, M., Harui, Y., Matsumoto, N., and Matsumoto, A. (1998). Light and temperature cooperate to regulate the circadian locomotor rhythm of wild type and *period* mutants of *Drosophila melanogaster*. *J. Insect Physiol.* **44,** 587–596.

Truman, J. W. (1976). Extraretinal photoreception in insects. *Photochem. Photobiol.* **23,** 215–225.

Truman, J. W. (1992). The eclosion hormone system of insects. *Prog. Brain Res.* **92,** 361–374.

Truman, J. W., and Morton, D. B. (1990). The eclosion hormone system: An example of coordination of endocrine activity during the molting cycle of insects. *Prog. Clin. Biol. Res.* **342,** 300–308.

Turek, F. W., and Van Reeth, O. (1988). Altering the mammalian circadian clock with the short-acting benzodiazepine, triazolam. *Trends Neurosci.* **11,** 535–541.

Ueda, H. R., Matsumoto, A., Kawamura, M., Iino, M., Tanimura, T., and Hashimoto, S. (2002). Genome-wide transcriptional orchestration of circadian rhythms in *Drosophila. J. Biol. Chem.* **277,** 14048–14052.

van Gelder, R. N. (2005). Nonvisual ocular photoreception in the mammal. *Methods Enzymol.* **393**(39), 744–753 (this volume).

van Swinderen, B., and Andretic, R. (2003). Arousal in *Drosophila. Behav. Processes* **64,** 133–144.

van Swinderen, B., and Hall, J. C. (1995). Analysis of conditioned courtship in *dusky-Andante* rhythm mutants of *Drosophila. Learn. Mem.* **2,** 49–61.

van Swinderen, B., Nitz, D. A., and Greenspan, R. J. (2004). Uncoupling of brain activity from movement defines arousal states in *Drosophila. Curr. Biol.* **14,** 81–87.

Veleri, S., Brandes, C., Helfrich-Förster, C., Hall, J. C., and Stanewsky, R. (2003). A self-sustaining, light-entrainable circadian oscillator in the *Drosophila* brain. *Curr. Biol.* **13,** 1758–1767.

Villella, A., Gailey, D. A., Berwald, B., Ohshima, S., Barnes, P. T., and Hall, J. C. (1997). Extended reproductive roles of the *fruitless* gene in *Drosophila melanogaster* revealed by behavioral analysis of new *fru* mutants. *Genetics* **147,** 1107–1130.

Vosshall, L. B., and Young, M. W. (1995). Circadian rhythms in *Drosophila* can be driven by *period* expression in a restricted group of central brain cells. *Neuron* **15,** 345–360.

Walker, J. R., and Hogenesch, J. B. (2005). RNA profiling in circadian biology. *Methods Enzymol.* **393**(16), 364–374 (this volume).

Walsh, D. K., Imalzumi, T., and Kay, S. A. (2005). Real-time reporting of circadian-regulated gene expression by luciferase imaging in plants and mammalian cells. *Methods Enzymol.* **393**(11), 267–286 (this volume).

Weber, F., and Kay, S. A. (2003). A PERIOD inhibitor buffer introduces a delay mechanism for CLK/CYC-activated transcription. *FEBS Lett.* **555,** 341–345.

Wheeler, D. A., Hamblen-Coyle, M. J., Dushay, M. S., and Hall, J. C. (1993). Behavior in light-dark cycles of *Drosophila* mutants that are blind, arrhythmic, or both. *J. Biol. Rhythms* **8,** 67–94.

Wheeler, D. A., Kyriacou, C. P., Greenacre, M. L., Yu, Q., Rutila, J. E., Rosbash, M., and Hall, J. C. (1991). Molecular transfer of a species-specific behavior from *Drosophila simulans* to *Drosophila melanogaster. Science* **251,** 1082–1085.

Whitfield, N., and Strong, B. (1967). "I Heard It through the Grapevine." Motown Music, Detroit, MI.

Whitmore, D., and Block, G. D. (1996). Cellular aspects of molluskan biochronometry. *Semin. Cell Dev. Biol.* **7,** 781–789.

Williams, D. W., Tyrer, M., and Shepherd, D. (2000). Tau and tau reporters disrupt central projections of sensory neurons in *Drosophila. J. Comp. Neurol.* **428,** 630–640.

Williams, J. A., Su, H. S., Bernards, A., Field, J., and Sehgal, A. (2001). A circadian output in *Drosophila* mediated by *Neurofibromatosis-1* and Ras/MAPK. *Science* **293,** 2251–2256.

Wolfner, M. F. (2003). Sex determination: Sex on the brain? *Curr. Biol.* **13**, R101–R103.

Wülbeck, C., Szabo, G., Shafer, O. T., Helfrich-Förster, C., and Stanewsky, R. (2005). The novel *Drosophila tim^blund* mutant affects behavioral rhythms but not periodic eclosion. *Genetics.* In press.

Yang, Z., and Sehgal, A. (2001). Role of molecular oscillations in generating behavioral rhythms in *Drosophila. Neuron* **29**, 453–467.

Yao, K.-M., and White, K. (1994). Neural specificity of *elav* expression: Defining a *Drosophila* promoter for directing expression to the nervous system. *J. Neurochem.* **63**, 41–51.

Yasuyama, K., and Meinertzhagen, I. A. (1999). Extraretinal photoreceptors at the compound eye's posterior margin in *Drosophila melanogaster. J. Comp. Neurol.* **412**, 193–202.

Yoshii, T., Funada, Y., Ibuki-Ishibashi, T., Matsumoto, A., Tanimura, T., and Tomioka, T. (2004). *Drosophila cry^b* muation reveals two circadian clocks that drive locomotor rhythm and have different responsiveness to light. *J. Insect Physiol.* **50**, 478–488.

Young, M. W., and Judd, B. H. (1978). Nonessential sequences, genes and the polytene chromosome bands of *Drosophila melanogaster. Genetics* **88**, 723–742.

Yu, Q., Jacquier, A. C., Citri, Y., Hamblen, M., Hall, J. C., and Rosbash, M. (1987). Molecular mapping of point mutations in the *period* gene that stop or speed up biological clocks in *Drosophila melanogaster. Proc. Natl. Acad. Sci. USA* **84**, 784–788.

Yu, W., and Hardin, P. E. (2005). Use of firefly luciferase activity assays to monitor circadian molecular rhythms *in vivo. In* "Circadian Rhythms: Reviews and Protocols" (E. Rosato, ed.). Humana Press, Totowa, NJ. In press.

Zatz, M. (1996). Melatonin synthesis: Trekking toward the heart of darkness in the chick pineal. *Semin. Cell. Dev. Biol.* **7**, 811–820.

Zehring, W. A., Wheeler, D. A., Reddy, P., Konopka, R. J., Kyriacou, C. P., Rosbash, M., and Hall, J. C. (1984). P-element transformation with *period* locus DNA restores rhythmicity to mutant arrhythmic *Drosophila melanogaster. Cell* **39**, 369–376.

Zeng, H., Hardin, P. E., and Rosbash, M. (1994). Constitutive overexpression of the *Drosophila* period protein inhibits *period* mRNA cycling. *EMBO J.* **13**, 3590–3598.

Zeng, H., Qian, Z., Myers, M. P., and Rosbash, M. (1996). A light-entrainment mechanism for the *Drosophila* circadian clock. *Nature* **380**, 129–135.

Zerr, D. M., Hall, J. C., Rosbash, M., and Siwicki, K. K. (1990). Circadian fluctuations of *period* protein immunoreactivity in the CNS and the visual system of *Drosophila. J. Neurosci.* **10**, 2749–2762.

Zhao, J., Kilman, V. L., Keegan, K. P., Peng, Y., Emery, P., Rosbash, M., and Allada, R. (2003). *Drosophila Clock* can generate ectopic circadian clocks. *Cell* **113**, 755–766.

Zhu, L., McKay, R. R., and Shortridge, R. D. (1993). Tissue-specific expression of phospholipase C encoded by the *norpA* gene of *Drosophila melanogaster. J. Biol. Chem.* **268**, 15994–16001.

Zilian, O., Frei, E., Burke, R., Brentrup, D., Gutjahr, T., Bryant, P. J., and Noll, M. (1999). *double-time* is identical to *discs overgrown*, which is required for cell survival, proliferation and growth arrest in *Drosophila* imaginal discs. *Development* **126**, 5409–5420.

Zimmerman, W. F., and Goldsmith, T. H. (1971). Photosensitivity of the circadian rhythm and of visual receptors in carotenoid-depleted *Drosophila. Science* **171**, 1167–1169.

Zimmerman, W. F., Pittendrigh, C. S., and Pavlidis, T. (1968). Temperature compensation of the circadian oscillation in *Drosophila pseudoobscura* and its entrainment by temperature cycles. *J. Insect Physiol.* **14**, 669–684.

Zordan, M., Osterwalder, N., Rosato, E., and Costa, R. (2001). Extra ocular photic entrainment in *Drosophila melanogaster. J. Neurogenet.* **15**, 97–116.

## [5]  Analysis of Circadian Rhythms in Zebrafish

*By* Jun Hirayama, Maki Kaneko, Luca Cardone,
Gregory Cahill, and Paolo Sassone-Corsi

### Abstract

The zebrafish probably constitutes the best animal system to study the complexity of the circadian clock machinery and the influence that light has on it. The possibilities of producing transgenic fishes, to establish light-responsive cultured cells, and to directly explore light phototransduction on single clock cells are all remarkable features of this circadian system. This article describes some of the most useful methodologies to analyze the behavioral, cellular, and molecular aspects of the zebrafish circadian clock system.

### Introduction

The zebrafish (*Danio rerio*) constitutes an attractive alternative to the mouse in the study of circadian rhythms in vertebrates. In addition, the zebrafish is useful for a comparative analysis of the molecular organization of the circadian clock in various systems, thereby providing essential information into the mechanisms governing rhythmicity (Cahill, 2002; Pando *et al.*, 2002). In the zebrafish, the pineal gland and the retina have been identified as the primary pacemakers that regulate its physiology and behavior (Cahill, 1996), although to date no functional equivalent to the SCN (suprachiasmatic nucleus) has been described in the zebrafish system. Both pineal gland and retina are directly light entrainable and contain circadian oscillators that drive rhythmic melatonin synthesis.

As in mammals, the zebrafish circadian system is composed of both central and peripheral clocks (Schibler and Sassone-Corsi, 2002). Organ and tissue culture explant experiments have demonstrated that peripheral circadian oscillators are present throughout the tissues and organs of the zebrafish and that they display the remarkable feature of being light responsive (Cahill, 1996; Cermakian *et al.*, 2000; Whitmore *et al.*, 1998, 2000). In addition, cultured lines of embryonal zebrafish cells have been established, which display light responsiveness and an intrinsic autonomous clock mechanism (Pando *et al.*, 2001; Whitmore *et al.*, 2000). The Z3 cell line, which recapitulates most features of the zebrafish clock system (Pando *et al.*, 2001), has been instrumental for the dissection of the intracellular signaling pathways implicated in light transduction and clock function (Cermakian *et al.*, 2002).

Characterization of the molecular components of the zebrafish circadian oscillator has revealed duplications for most clock genes. There are three homologs of *Clock* genes (Ishikawa *et al.*, 2002; Whitmore *et al.*, 1998), three *Bmal1* (Cermakian *et al.*, 2000; Ishikawa *et al.*, 2002), four *Per* (Pando *et al.*, 2001; Vallone *et al.*, 2004), and six *Cry* genes (Cermakian *et al.*, 2002; Kobayashi *et al.*, 2000). CLOCK:BMAL heterodimers provide the central transcriptional potential that drives the cell autonomous clocks. In zebrafish, both *Clock* and *Bmal* display rhythmic oscillations in gene expression, which, on average, peak early during the night phase. Both genes are expressed in most tissues of the animal but display differences in the peak, levels, and kinetics of expression. Expression variations are also observed for the same gene when comparing between different tissues (Cermakian *et al.*, 2000). This suggests that the exact composition of CLOCK:BMAL heterodimers changes during time and between tissues. The two *zfBmal* genes are most divergent in their carboxy-terminal transcription activation domains. This is thought to allow the central transcription complex of the circadian oscillator to have precise control over its transcriptional potential, facilitating the proper response to general and tissue-specific entraining stimuli. Similarly, circadian expression profiles, light inducibility, and regulation of the numerous *Per* and *Cry* genes are also differential (Cermakian *et al.*, 2002; Kobayashi *et al.*, 2000; Pando *et al.*, 2001), indicating the high molecular complexity of the zebrafish oscillator and suggesting a differential contribution of the various components in clock regulation for various peripheral tissues.

## General Methods

The following general guidelines are used in various laboratories, although slight variations are common. It is useful to refer to methods described (Westerfield, 1995) and to recommandations presented and updated in the ZFIN (The Zebrafish Information Network) Web site (http://www.grs.nig.ac.jp:6070/).

### Fish

*Food.* Fish are fed twice daily. The food consists either of ground dry trout pellets or of dry flake such as Tetra brand or daphnia, both of which are available at most pet stores.

*Water.* Adult fish are maintained in distilled water, to which a small amount of salts and minerals is added. The water is heated at 26–29°, filtered, and recycled continuously. Embryos and young larvae are raised and maintained in egg water (~60 mg/L Instant ocean in deionized water, pH 7.0, conductivity 100 $\mu$S/cm, aerated at least 12 h).

*Light/Dark Cycle.* Fish are classically maintained in a 14 light:10 dark cycle. Because zebrafish is photoperiodic in breeding, keeping a proper cycle is important for embryo preparation.

*Fish Dissection.* Fish are killed by rapid immersion in chilled water followed by decapitation. After dissection, tissues are immediately frozen on dry ice for subsequent analysis.

## Ex Vivo *Organ Study*

Zebrafish peripheral tissues (e.g., heart, kidney, and spleen) have been shown to be light responsive even when explanted from the animal and isolated in a culture dish. This remarkable feature allows direct entrainment of peripheral circadian clocks, independently from the central clock system (Cermakian *et al.*, 2000; Whitmore *et al.*, 1998, 2000). Therefore, *ex vivo* tissue systems represent a very useful tool to study light-dependent circadian gene induction.

*Materials.* Prepare forceps, six-well plates, anatomical microscope, and culture medium (composition: L15 medium supplemented with 15% fetal calf serum, 2 m$M$ glutamine, gentamycin, streptamycin, and penicillin).

### Procedure

1. Raise and sacrifice fish as described earlier.
2. Isolate tissues with sharp forceps from an individual fish between ZT9 and ZT12 (Zeitgeber time, ZT0 corresponding to light on and ZT14 to light off) under an anatomical microscope.
3. Place the dissected tissues in a six-well plate in L15 medium.
4. Keep the tissues at 25° and atmospheric $CO_2$ concentration.

### Preparation of Embryo

*Breeding.* Adult zebrafish (aged between 7 and 18 months) should be used for breeding. Zebrafish lay eggs every morning, shortly after sunrise. It is not advisable to collect embryos more than 2 days in a row from the same couple of fish. In our experience, zebrafish breed better if fed adult brine shrimp (*Artemia* sp.) once a day.

### Procedure

1. Keep males and females in separate tanks with up to 8 females or 16 males per 10-gal tank. Clean the tank once per day by replacing one-third water.

2. On the breeding day, feed the fish and clean the tank 1 or 2 h before the end of the light period. Transfer the males to the tank with the females at a rate of one male to two females.
3. Add marbles to cover the bottom of the tank.
4. After the beginning of the next light cycle, collect the embryos (see later).
5. Transfer the males back to their tank, and scoop out the marbles with a net and clean them by autoclaving.

## Collecting Embyros

*Material.* Prepare a siphon made of a plastic or glass tube (1 cm i.d. and 30–50 cm long) covered at one end with a piece of tygon tubing.

### Procedure

1. Draw water with a siphon through a medium-mesh nylon net, sweeping the bottom of the tank from side to side.
2. Invert the net over a petri dish filled with egg water to let the embryos fall off into the dish.
3. Culture embryo at 26° under atmospheric $CO_2$ concentration.

### Locomotor Activity Rhythms

Locomotor (swimming) activity is rhythmic in larval (5–20 day old) zebrafish maintained individually in 0.5-ml wells for a week or more in constant conditions. An automated infrared video image analysis system was developed for high-throughput recording of these rhythms. The system described here can monitor simultaneously the activity of up to sixty-three 5- to 12-day-old fish that have never been fed or up to 150 larger, 10- to 20-day-old fish that have been fed *Paramecium ad libitum* from day 5 to days 10–12. The maximum numbers are limited by the size of the image (related to the volume of water required to keep fish healthy in the absence of water changes) and the resolution of the digitized image. This system has been used to screen for clock mutations, to characterize mutant phenotypes, and to characterize the development of circadian rhythmicity in zebrafish (Cahill, 1998, 2002; DeBruyne *et al.*, 2004; Hurd and Cahill, 2002). Strong behavioral rhythmicity can be recorded from larvae of AB and SJD strains; less robust rhythmicity was observed in C32, TU, and some pet store strains.

During activity monitoring, larval zebrafish are maintained in a rectangular array of oval wells drilled in a translucent white polyethylene specimen plate. To avoid disturbance, the fish are not fed and water is

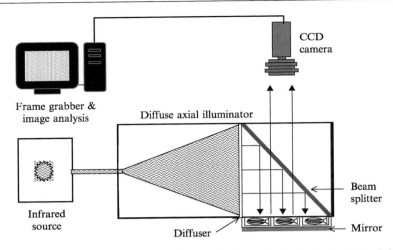

Fig. 1. Video-recording apparatus for monitoring behavioral rhythmicity of larval zebrafish. The diffuse axial illuminator ensures even illumination and eliminates glare from the water surface.

not changed during activity recording. The temperature is lowered to 24° to slow metabolism, and virtually all fish survive and can be recovered after a week. A CCD video camera is focused on the plate from above, and an infrared diffuse axial illuminator is used to produce even illumination without glare from the water surface (Fig. 1). The specimen plate is placed on a mirror to backlight the wells. The key imaging challenge is to ensure that all fish are darker than all background so that a single threshold can be set for automated object identification. Image acquisition and analysis are controlled by customized software running on a desktop computer. For each activity sample, a series of images is captured at a rate of 1/s, digitized, and stored in RAM (Fig. 2). The position of each fish in each image is determined, and the distance moved during the sample period is calculated from the series of coordinates. Data are stored to a text file, the images are erased from memory, and a new cycle of image capture and analysis is initiated. We typically sample 30–60 s of activity every 4 min and then average six samples to produce activity records with 24-min resolution.

*Materials*

Light-tight refrigerated incubator maintained at 24°; monochrome video camera with a 50-mm macro lens, and a 1.7-cm CCD sensor with the IR-blocking filter removed, automatic gain control, shading correction

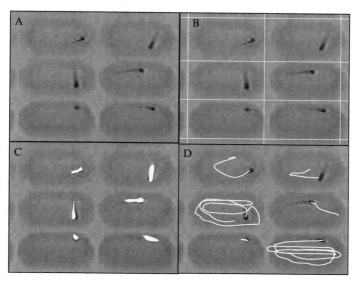

FIG. 2. Measurement of larval zebrafish behavioral rhythms by automated video image analysis. (A) Infrared video image of 6 wells of the 150-well specimen plate with 10-day-old zebrafish. (B) The image is divided into a grid of cells, each containing 1 well. (C) Thresholds are set to detect fish. (D) Simulation of paths swum by fish during a 1-min sampling period.

and horizontal center resolution >750 TVL; a desktop computer with a 640 × 480 pixel, 8-bit gray scale frame capture card (Flashpoint 128), and a Windows operating system; Optimate 6.2 (MediaCybernetics, Silver Spring, MD) image processing software with the Swimming1.1 macro (Meyer Instruments, Houston, TX); a diffuse axial illuminator with infrared light source (custom made, but smaller versions are available commercially from Edmund Industrial Optics, Barrington, NJ; RVSI/NER, Weare, NH; or Advanced Illumination, Rochester, VT), a translucent white polyethylene specimen plate, 18.5 cm × 13.5 cm × 8 mm thick, with oval wells, 12 × 6 × 7 mm, arranged in 10 × 15 array with 1-mm-thick walls; an 18.5 × 13.5-cm flat mirror; egg water; and large-bore transfer pipettes.

*Procedure*

1. Raise zebrafish to desired age (up to 100/1-liter beaker) at 28.5° in a 14:10 LD cycle. Change water daily and feed liberally with *Paramecium* from day 5 to the day before transferring them to the recording apparatus.
2. In the morning, wash larvae repeatedly with fresh egg water and starve the fish for at least 6 h before transferring them to recording wells.

3. Strain fish from the beaker with fine nylon mesh and wash into a 100-mm petri plate with fresh egg water.

4. Use a large-bore pipette to transfer fish to wells filled with fresh egg water.

5. Install the specimen plate on the reflecting surface of the mirror and under the diffuse axial illuminator (Fig. 1). The camera and illumination system are installed in the incubator, which is humidified by bubbling air through a large open reservoir of water.

6. Launch Optimate and the Swimming macro (Fig. 3), and click on *Acquire* to view the live-digitized video. Focus the camera so that the array of occupied wells fills the image field.

7. Set the camera controller to use automatic gain control. Use the four shading correction dials on the camera controller to even out any variation in background shading from center to periphery, side to side, or up and down. Freeze the image and open the *Monochrome Threshold* dialog window; pixels with values between the lower and the upper threshold will be highlighted. Set the lower threshold to 0 and scan the upper threshold down until all fish are below and all background is above the upper threshold.

8. With the image frozen, click and drag to set the *Region of Interest*, a rectangle that encompasses all occupied wells. Set the number of *Rows* and *Columns* of wells. Use the *Test Layout* button to superimpose a grid dividing the region of interest into the specified rows and columns onto the image. Repeat these steps if every cell in the grid does not encompass one and only one well.

9. Set sampling parameters, including *Images/Cycle*, *Delay* between image captures within a series, *Cycle Length*, which must be greater than the time required for acquisition and analysis of an image series, and the number of *Cycles* to *Process* before automatically shutting down.

10. Create an *ASCII Data File* with a .txt extension for the measurements. The first column of this tab-delimited file will contain time stamps, and each subsequent column will have data for one well.

At the end of the recording period, transfer the data file to your favorite time series analysis program. We find Chrono 4.5.1 (Roenneberg and Taylor, 2000) to be a useful and versatile Macintosh program.

## Bioluminescence Rhythms in *per3-luc* Transgenic Zebrafish

In a variety of species, circadian rhythms of bioluminescence have been produced by expressing a *luciferase* transgene under the control of a clock-driven promoter (Brandes *et al.*, 1996; Millar *et al.*, 1992). This provides a

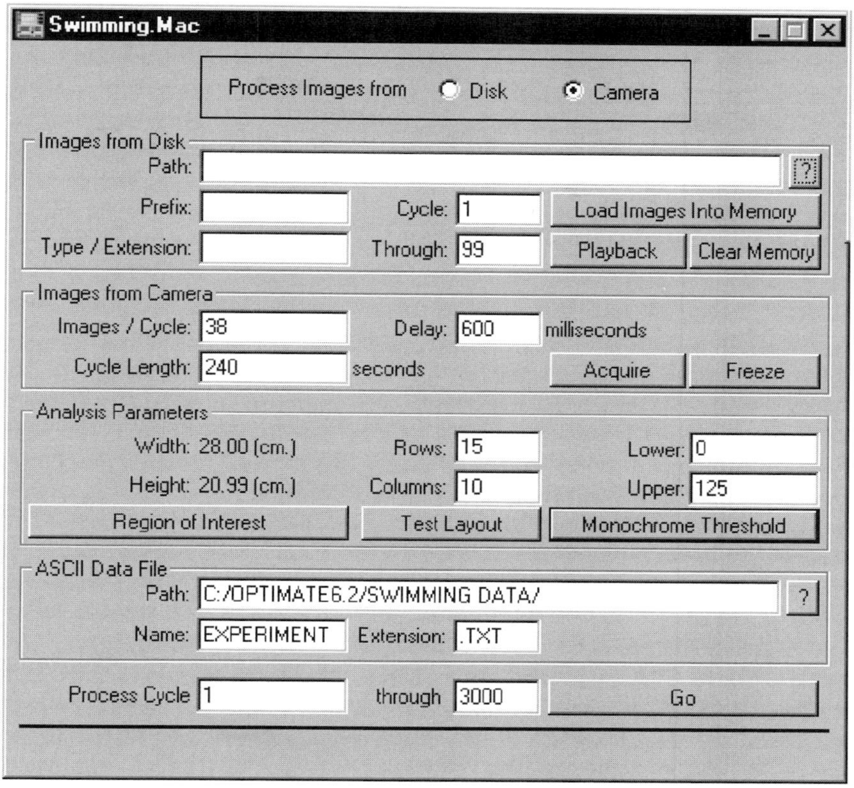

FIG. 3. Control panel for Swimming macro. Copyright Matthew Batchelor.

convenient method for high-throughput assays of molecular circadian rhythmicity. Zebrafish lines carrying a transgene with *luciferase* driven by the zebrafish *period3* promoter(*per3-luc*) were produced. When incubated with luciferin, live larval fish and cultured adult organs, including retina, heart, spleen, and gall bladder from these transgenic fish, glow rhythmically (Kaneko and Cahill, 2004a,b).

The transgene is a modified bacterial artificial chromosome (BAC) with a zebrafish genomic DNA insert that included a 5′ coding sequence of the *per3* gene as well as >8.3 kb of an upstream sequence (clone 8M06). The *per3* sequence from the initiation codon to the end of the first coding exon was replaced with a gene cassette containing *luciferase* and *kanamycin resistance* genes (Muyrers *et al.*, 1999). This construct was linearized by digestion with *Not*I and injected into one to two cell stage embryos

(Higashijima *et al.*, 1997). Five germline transformant lines were recovered; luc expression levels were variable among the lines. *Luc* expression was highest in line 23, and we have focused further studies on this line. These fish will be made available through the Zebrafish International Resource Center (http://zfin.org/zirc).

## Bioluminescence Rhythms in Live Larval Zebrafish

### *Materials*

TopCount multiplate scintillation counter (Perkin-Elmer) in a refrigerated incubator maintained at 22°, white 96-well Optiplates (Perkin-Elmer), TopsealA sheets (Perkin-Elmer), Holtfreter/luciferin solution (7.0 g NaCl, 0.4 g $NaHCO_3$, 0.2 g $CaCl_2$, 0.1 g KCl in 21 ml $ddH_2O$, pH 7.0, with 0.5 m$M$ D-luciferin potassium salt and 0.013% Amquel Instant Water Detoxifier) aerated overnight, and a large-bore pipette.

### *Procedure*

1. Raise zebrafish to 6 days of age at 22° under a 14:10 LD cycle.
2. Pipette fish into every other well of Optiplates.
3. Replace water in each occupied well with 200 $\mu$L Holtfreter/ luciferin solution.
4. Cover each plate with a sheet of TopsealA and perforate over each occupied well twice with a 23-gauge needle.
5. Install loaded plates in the TopCount plate stacker and set the TopCount to monitor each well for 4.8 s every 30 min.
6. At the end of the experiment, transfer data to your favorite time series analysis program.

## Bioluminescence Rhythms in Cultured Organs

### *Materials*

TopCount multiplate scintillation counter (Perkin-Elmer) in a refrigerated incubator maintained at desired temperature (20–32°), white 96-well Optiplates (Perkin-Elmer), TopsealA sheets, dissecting medium [Liebowitz L-15 culture medium (GIBCO) diluted 2:1 in $ddH_2O$, 10% fetal bovine serum (FBS), 10% antibiotic/antimycotic solution (GIBCO)], and culture medium [Liebowitz L-15 culture medium (GIBCO) diluted 2:1 in $ddH_2O$, 10% FBS, 1% penicillin/streptomycin (GIBCO), 0.5 m$M$ luciferin potassium salt].

*Procedure*

1. Maintain adult zebrafish at 28.5° under a 14:10 LD cycle.
2. Dissect organs into dissecting medium.
3. Transfer organs to every other well of Optiplates with 260 $\mu$l of culture medium.
4. Cover each plate with a sheet of TopsealA.
5. Install loaded plates in the TopCount plate stacker and set the TopCount to monitor each well for 15 s every hour.
6. At the end of the experiment, transfer data to your favorite time series analysis program.

## Recapitulating the Zebrafish Clock in Cultured Cells

The most remarkable and unique feature of the zebrafish system is the ability to respond to light. Several clock-related genes show circadian expression in *ex vivo* peripheral tissues like heart, kidney, and spleen (Whitmore *et al.*, 1998, 2000) and in zebrafish-derived cell lines such as the Z3 line (Pando *et al.*, 2001). Remarkably, circadian gene expression in Z3 cells can be synchronized (i.e., entrained) by light, proving that zebrafish cells contain not only a circadian oscillator, but also phototransduction mechanisms sufficient for light entrainment. In particular, we have established a zebrafish cell line, designated Z3, which derives from zebrafish embryos, and have demonstrated that several circadian clock components display distinct and differential light-dependent activation and expression profiles under various light conditions (Pando *et al.*, 2001). Z3 cells, therefore, nicely recapitulate the zebrafish clock system and constitute an invaluable tool for investigating the link between light-dependent gene activation and the signaling pathways responsible for the generation of vertebrate circadian rhythms.

## Establishment of the Z3 Cell Line

*Materials*

Prepare sterile forceps, hood, sterile beaker, 0.25% trypsin, 0.5% bleach, 1× phosphate-buffered saline (PBS), L15 medium [GIBCO/BRL, supplemented with 15% fetal calf serum (FCS), 2 m$M$ glutamine, gentamycin, streptamycin, and penicillin], 25-cm$^2$ flasks, and incubator (26°, atmospheric $CO_2$ concentration).

*Procedure*

1. Rinse the 24-h-old embryos in 0.5% bleach for 2 min and then rinse three times in sterile PBS.

2. After the final rinse in PBS, dechorionate manually the embryos using sterile forceps.
3. Transfer dechorionated embryos to a tissue culture hood and place them in a sterile beaker containing sterile PBS and rinse twice for 5 min.
4. Place the embryos in 0.25% trypsin at a concentration of 50 embryos/ml and incubate them at room temperature to dissociate them. During trypsinization, pipette the embryos 5–10 times through a P1000 (Gilson) every 3–5 min, until mostly single cells are obtained.
5. Rinse the cell suspension twice in 10 ml of culture medium.
6. Spin down the cells at 300$g$ and resuspend in complete L15 medium at a concentration of 20 embryos/ml.
7. Place 5 ml of the resuspended cells in sealed 25-cm$^2$ flasks and place in an incubator maintained at 26° and atmospheric $CO_2$ concentration.
8. Split cells at a 1:2 dilution once reaching confluence. After several passages, the Z3 cellular subpopulation is able to survive under the given culture conditions by growing steadily (Fig. 4).

Culture Conditions for Z3 Cells

Temperature and $CO_2$ concentration: 26° and atmospheric $CO_2$ concentration. Medium: L-15 medium (GIBCO-BRL), containing 10% FCS and gentamycin. Incubator: should be water jacketed, thermostatically controlled, and light sealed (Fig. 5A). Control of light/dark conditions:

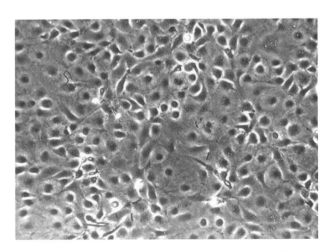

FIG. 4. Confluent Z3 cells in culture. Magnification 100×.

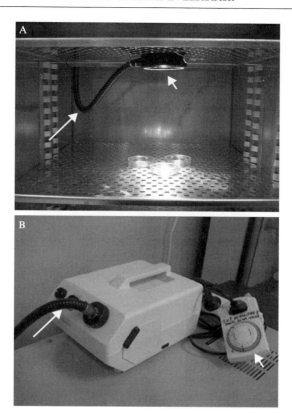

Fig. 5. Incubator for Z3 cell culture. (A) Z3 cells exposed to light. Illumination is achieved by a halogen light (short arrow) fed into the incubator through a fiber-optic line (long arrow). The incubator is water jacketed, thermostatically controlled, and light sealed. (B) A programmable timer, which controls light–dark cycles (short arrow) and the halogen light source (long arrow).

Illumination is achieved by using a halogen light source fed into the incubator through a fiber-optic line (Fig. 5A). A programmable timer connected to the light source controls the light cycles (Fig. 5B).

Passage: Tripsinize and split cells at a 1:3 dilution once reaching confluence. Under the culture conditions described, Z3 cells should attain confluence at 3–4 days after passage. Passages should be done under the dark, using a red lamp.

Other Zebrafish-Derived Cells for Circadian Studies

In addition to Z3 cells, two other zebrafish cell lines have been employed for the study of the circadian clock: PAC2 cells, established by

Nancy Hopkins at the Massachussets Institute of Technology (Cambridge, USA), and BRF41 cells (Ishikawa *et al.*, 2002). These lines are derived from the embryo and caudal fin, respectively, and are cultured under conditions highly similar to Z3 cells.

## Molecular Methods

### RNA Isolation

RNASolv reagent (Omega Bio-tek) can be used for isolating RNA from Z3 cells as well as from zebrafish tissues.

### RNA Analysis

The RNase protection assay (RPA) has been essential for studying circadian clock gene expression at the RNA level in zebrafish because of its robustness, reproducibility, sensitivity, and reliability (Cermakian *et al.*, 2000; Ishikawa *et al.*, 2002; Kobayashi *et al.*, 2000; Pando *et al.*, 2001; Whitmore *et al.*, 1998, 2000). To visualize circadian clock gene expression within fish tissues, *in situ* hybridization has also been applied successfully (Cermakian *et al.*, 2000; Whitmore *et al.*, 2000). These methodologies have been described in detail elsewhere (Macho and Sassone-Corsi 2003; Sassoon and Rosenthal 1993; Tessarollo and Parada 1995). DNA sequences selected for protection or *in situ* analyses are cloned into a suitable vector to obtain the antisense probe by RNA polymerase. A suitable restriction site is required for linearization of the plasmid that would give rise to probes in the range of 100–400 nucleotides. To obtain a reliable riboprobe, several fragments need to be tested. Successfully utilized riboprobes for zebrafish clock genes are listed in Table I.

### Signaling Inhibitors Treatment

To study light-induced signaling pathways leading to the regulation of clock gene expression, a panel of inhibitors can be used. We used this approach on Z3 cells. For example, U0126 and Ro-31-8220, MEK, and PKC inhibitors, respectively, have been shown to block *Per2* light induction (Cermakian *et al.*, 2002). To study *Per2* induction, inhibitors are added directly on confluent Z3 cells. After a 1-h treatment, cells are exposed to light and then harvested at each time point of interest for RNA analysis. Inhibitors are dissolved in dimethyl sulfoxide (DMSO). For control experiments, cells are exposed to DMSO without the inhibitor. It is important to note that the treatment should be done under a dim red light. Similar approaches can be used to analyze the signaling pathways utilized by the clock machinery at different times of the circadian cycle, as well as to explore the transduction routes implicated in the temperature compensation process.

TABLE I
RIBOPROBES FOR EXPRESSION STUDY OF ZEBRAFISH CIRCADIAN CLOCK GENES

| Gene | From–to | Protected fragment | Ref. |
|---|---|---|---|
| *zfClock1* | 639–731 | 92 | Ishikawa *et al.* (2002) |
| *zfClock2* | 1404–1527 | 123 | Ishikawa *et al.* (2002) |
| *zfClock3* | 1301–1462 | 161 | Ishikawa *et al.* (2002) |
| *zfBmal1* | 1328–1759 | 431 | Cermakian *et al.* (2000) |
| *zfBmal2* | 1199–1599 | 400 | Cermakian *et al.* (2000) |
| *zfBmal3* | 1646–1971 | 231 | Ishikawa *et al.* (2002) |
| *zPer1* | 1–223 | 222 | Pando *et al.* (2001) |
| *zPer2* | 1981–2369 | 88 | Pando *et al.* (2001) |
| *zPer3* | 1218–1572 | 354 | Pando *et al.* (2001) |
| *zPer4* | 265–675 | 410 | Vallone *et al.* (2004) |
| *zCry1a* | 1535–1671 | 136 | Cermakian *et al.* (2002) |
| *zCry2a* | 1679–1854 | 175 | Cermakian *et al.* (2002) |
| *zCry1b* | 1661–1821 | 200 | Cermakian *et al.* (2002) |
| | Plus 40 nt after stop codon | | |
| *zCry2b* | 1491–1804 | 313 | Cermakian *et al.* (2002) |
| *zCry3* | 1767–1797 | 353 | Cermakian *et al.* (2002) |
| | Plus 323 nt after stop codon | | |
| *zCry4* | 1621–1677 | 113 | Cermakian *et al.* (2002) |
| | Plus 56 nt after stop codon | | |
| *zfBeta-actin* | 957–1070 | 113 | Kobayashi *et al.* (2000) |

## Action Spectrum Analysis of Z3 Cells

To elucidate phototransduction pathways and the respective photoreceptors, the action spectrum analysis is an essential step. For details on how to perform an action spectrum, please look at specialized literature (Payne and Sancar, 1990). This section provides simple information on the preparation of Z3 cells for this analysis. Indeed, we used the direct light-responsiveness of Z3 cells by scoring for *Per2* induction to perform the action spectrum of Z3 cells (Cermakian *et al.*, 2002).

*Materials.* Integrated monochromator-actinometer and cuvette (Quantacount Photon Technology International).

*Procedure*

1. Trypsinize Z3 cells and resuspend them in 8 ml of the culture medium.
2. Maintain the cells in suspension by rocking for 4 h.
3. Transfer cells to a cuvette and irradiate light with specific wavelengths through an integrated monochromator–actinometer.
4. Maintain cells in darkness for 2.5 h and then isolate RNA for analysis.

Retroviral Infection

Transient transfection efficiency of cultured zebrafish-derived cells is relatively low (Hirayama *et al.*, 2003). To circumvent this problem, we have found that retroviral infection of Z3 cells is a very attractive option.

*Solutions and Materials*

$2 \times$HBS: Dissolve HEPES (5 g) and NaCl (8 g) in 400 ml sterile MilliQ $H_2O$. Adjust pH to 7.1 with 10 $N$ NaOH and bring to 500 ml. Autoclave and store at room temperature.

$100 \times$ phosphate solution: Dissolve 4.97 g $Na_2HPO_4$, 4.2 g $NaH_2PO_4$, and 400 ml sterile MilliQ $H_2O$ and bring to 500 ml. Autoclave and store at room temperature.

2 $M$ $CaCl_2$: Dissolve 58.8 g $CaCl_2$ $2H_2O$ in sterile MilliQ $H_2O$ and bring to 200 ml. Filter sterilize with a 0.2-$\mu$m filter and store at $4°$.

0.1% gelatin (porcine skin, Sigma): Add gelatin (1 g) to 1 liter of sterile MilliQ $H_2O$ and autoclave (gelatin will not dissolve until autoclaved). Store at room temperature.

293gagpol packaging cell line : Gagpol selection can be done by adding 200 $\mu$l of 1 mg/ml blastocidin per 10 ml medium.

Medium: DMEM + 4.5 g/L glucose + 10% FCS + gentamycin.

Retroviral vectors: The RetroMax expression system (IMGENEX) can be used because it provides several kinds of retroviral vectors, such as pCLXSN, pCLNCX, pCLNRX, and pCLNDX, in which the cloned genes are under the control of SV40, CMV, RSV, and DHFR promoters, respectively. In our experience, pCLNCX seems to be one of the most useful vectors because of the high infection efficiency and the significant gene expression levels revealed in infected Z3 cells (unpublished data). As an enveloping vector, pMD.G/vsv-g is recommended because it is available commercially and has given good results in Z3 cells.

## Preparation of Virus Solution

### Gelatin Coat

1. Add 3 ml 0.1% gelatin to a 10-cm plate.
2. Incubate at room temperature for 10 min.
3. Remove gelatin from plate and rinse the plate once with 5 ml PBS.

### Transfection

1. Split 293 gagpol packaging cells onto gelatin-coated 10-cm plate at 40% confluence and let cells recover overnight at 37° and 5% $CO_2$.
2. Prepare solution A by mixing 7.5 $\mu$l of 100× phosphate solution and 375 $\mu$l of 2× HBS.
3. Prepare solution B by mixing 45 $\mu$l of 2 $M$ $CaCl_2$ and 368 $\mu$l of DNA containing 20 $\mu$g of virusvector, 5 $\mu$g of envelope vector, and 5 $\mu$g of carrier vector (pBluescript).
4. Add solution B to solution A dropwise while tapping the tube.
5. Mix well and incubate at room temperature for 15 min.
6. Add mixture to cells dropwise and mix gently.
7. Incubate cells at 37° under 5% $CO_2$ for 24 h. Refresh medium to remove growth-inhibiting factors 24 h after transfection and further incubate cells under the same conditions for an additional 24 h.

### Recovery of Viral Particles

1. Recover all the medium (around 9 ml) from dish and add 1 ml FCS and 10 $\mu$l 4 mg/ml polybrene.
2. Filter the medium through a 0.22-$\mu$m (low protein binding) syringe filter.
3. Store at 4° for short-term storage (2 days maximum) and freeze on dry ice and store at $-20°$ for long-term storage.

## Z3 Cell Infection

1. Split Z3 cells onto a 6-cm plate to 80% confluence and let cells recover overnight at 25° under atmospheric $CO_2$ concentration.
2. Dilute recovered viral particle solution fivefold or more.
3. Remove culture medium from the plate of Z3 cells and add 2.5 ml of diluted virus solution.

4. Incubate for 2–3 h at 25°. Remove virus solution and repeat infection four times.
5. Remove virus solution from the plate and replace with usual complete medium for Z3 cells and grow as usual. Score for expression 24–48 h postinfection.

### Preparation of Nuclear Extracts

Z3 cells represent a valuable tool for the study of general light-dependent gene induction. Using the electrophoretic mobility shift assay (EMSA) with nuclear extract from Z3 cells cultured in either light or dark conditions, light-responsive promoter sequences could be identified. The general method of EMSA (Rimbach et al., 2001) can be applied for the Z3 cell line. We have performed a wide-search analysis for transcription factors whose DNA binding to their respective specific recognition sequence would be induced by light in Z3 cells (J. Hirayama and L. Cardone, unpublished results). This approach proved very successful, and this section describes the methodology used.

#### Solutions

Hypotonic buffer (prepare just before use): 10 m$M$ HEPES–KOH (pH 7.8), 10 m$M$ KCl, 0.1 m$M$ EDTA, 1 m$M$ dithiotheitol (DTT), 0.15% Triton X-100, 1 m$M$ phenylmethylsulfonyl fluoride (PMSF), 1× protease inhibitor cocktail, 50 m$M$ NaF, and 100 $\mu M$ Na$_3$ VO$_4$.

Hypertonic buffer (prepare just before use): 20 m$M$ HEPES–NaOH (pH 7.8), 400 m$M$ NaCl, 1 m$M$ EDTA, 1 m$M$ DTT, 1 m$M$ PMSF, 1× protease inhibitor cocktail, 50 m$M$ NaF, and 100 $\mu M$ Na$_3$ VO$_4$.

#### Procedure

1. Rinse Z3 cells cultured in 75-cm$^2$ flask two times in sterile PBS.
2. Add 500 $\mu$l of hypotonic buffer directly to the cells and scrape and transfer them to a 1.5-ml tube.
3. Leave the tube on ice for 10 min.
4. Centrifuge for 5 min at 700$g$ and 4° and remove the supernatant. The supernatant can be used as the cytosolic fraction.
5. Add 500 $\mu$l of hypotonic buffer to the precipitate and resuspend.
6. Centrifuge at 700$g$ for 5 min at 4° and remove the supernatant.
7. Add 50 $\mu$l of hypertonic buffer to the precipitate and shake at 4° for 30 min.
8. Centrifuge at 10,000$g$ for 30 min at 4°.

9. Transfer the supernatant to a 1.5-ml tube. The supernatant is the nuclear fraction.

10. Freeze the nuclear extract on dry ice and store at $-80°$ for long-term storage. Normally, the expected yield from a 75-cm$^2$ confluent flask is 50–60 $\mu$g of total nuclear protein as determined by the Bio-Rad protein assay kit.

This procedure could be extended to applications that include immunoprecipitation and coimmunoprecipitation of specific proteins and consequent analysis by Western blotting and chromatin-Immuno precipitation, valuable to unravel the transcriptional complex recruitment to specific gene regulatory elements.

## References

Brandes, C., Plautz, J. D., Stanewsky, R., Jamison, C. F., Straume, M., Wood, K. V., Kay, S. A., and Hall, J. C. (1996). Novel features of *Drosophila period* transcription revealed by real-time luciferase reporting. *Neuron* **16,** 687–692.

Cahill, G. M. (1996). Circadian regulation of melatonin production in cultured zebrafish pineal and retina. *Brain Res.* **708,** 177–181.

Cahill, G. M. (1998). Circadian rhythmicity in the locomotor activity of larval zebrafish. *Neuroreport* **9,** 3445–3449.

Cahill, G. M. (2002). Clock mechanisms in zebrafish. *Cell Tissue Res.* **309,** 27–34.

Cermakian, N., Pando, M. P., Thompson, C. L., Pinchak, A. B., Selby, C. P., Gutierrez, L., Wells, D. E., Cahill, G. M., Sancar, A., and Sassone-Corsi, P. (2002). Light induction of a vertebrate clock gene involves signaling through blue-light receptors and MAP kinases. *Curr. Biol.* **12,** 844–848.

Cermakian, N., Whitmore, D., Foulkes, N. S., and Sassone-Corsi, P. (2000). Asynchronous oscillations of two zebrafish CLOCK partners reveal differential clock control and function. *Proc. Natl. Acad. Sci. USA* **97,** 4339–4344.

DeBruyne, J., Hurd, M. W., Gutiérrez, L., Kaneko, M., Tan, Y., Wells, D. E., and Cahill, G. M. (2004). Characterization and mapping of a zebrafish circadian clock mutant. *J. Neurogenet.* In press.

Higashijima, S.-I., Okamoto, H., Ueno, N., Hotta, Y., and Eguchi, G. (1997). High-frequency generation of transgenic zebrafish which reliably express GFP in whole muscles or the whole body by using promoters of zebrafish origin. *Dev. Biol.* **192,** 289–299.

Hirayama, J., Fukuda, I., Ishikawa, T., Kobayashi, Y., and Todo, T. (2003). New role of zCRY and zPER2 as regulators of sub-cellular distributions of zCLOCK and zBMAL proteins. *Nucleic Acids Res.* **31,** 935–943.

Hurd, M. W., and Cahill, G. M. (2002). Entraining signals initiate behavioral circadian rhythmicity in larval zebrafish. *J. Biol. Rhythms* **17,** 307–314.

Ishikawa, T., Hirayama, J., Kobayashi, Y., and Todo, T. (2002). Zebrafish CRY represses transcription mediated by CLOCK-BMAL heterodimer without inhibiting its binding to DNA. *Genes Cells* **7,** 1073–1086.

Kaneko, M., and Cahill, G. M. (2004a). Development of circadian molecular oscillations revealed by bioluminescence in transgenic zebrafish. Manuscript in preparation.

Kaneko, M., and Cahill, G. M. (2004b). Real-time measurement of circadian gene expression in peripheral tissues from transgenic zebrafish. *Soc. Res. Biol. Rhythms Abstr.* **9**, 90.

Kobayashi, Y., Ishikawa, T., Hirayama, J., Daiyasu, H., Kanai, S., Toh, H., Fukuda, I., Tsujimura, T., Terada, N., Kamei, Y., Yuba, S., Iwai, S., and Todo, T. (2000). Molecular analysis of zebrafish photolyase/cryptochrome family: Two types of cryptochromes present in zebrafish. *Genes Cells* **5**, 725–738.

Macho, B., and Sassone-Corsi, P. (2003). Functional analysis of transcription factors CREB and CREM. *Methods Enzymol.* **370**, 396–415.

Millar, A. J., Short, S. R., Chua, N. H., and Kay, S. A. (1992). A novel circadian phenotype based on firefly luciferase expression in transgenic plants. *Plant Cell.* **4**, 1075–1087.

Muyrers, J. P. P., Zhang, Y., Testa, G., and Stewart, A. F. (1999). Rapid modification of bacterial chromosomes by ET-recombination. *Nucleic Acids Res.* **27**, 1555–1557.

Pando, M. P., Pinchak, A. B., Cermakian, N., and Sassone-Corsi, P. (2001). A cell-based system that recapitulates the dynamic light-dependent regulation of the vertebrate clock. *Proc. Natl. Acad. Sci. USA* **98**, 10178–10183.

Pando, M. P., and Sassone-Corsi, P. (2002). Unraveling the mechanisms of the vertebrate circadian clock: Zebrafish may light the way. *Bioessays* **24**, 419–426.

Payne, G., and Sancar, A. (1990). Absolute action spectrum of E-FADH2 and E-FADH2-MTHF forms of *Escherichia coli* DNA photolyase. *Biochemistry* **29**, 7715–7727.

Rimbach, G., Saliou, C., Canali, R., and Virgili, F. (2001). Interaction between cultured endothelial cells and macrophages: *In vitro* model for studying flavonoids in redox-dependent gene expression. *Methods Enzymol.* **335**, 387–397.

Roenneberg, T., and Taylor, W. (2000). Automated recordings of bioluminescence with special reference to the analysis of circadian rhythms. *Methods Enzymol.* **305**, 104–119.

Sassoon, D., and Rosenthal, N. (1993). Detection of messenger RNA by *in situ* hybridization. *Methods Enzymol.* **225**, 384–404.

Schibler, U., and Sassone-Corsi, P. (2002). A web of circadian pacemakers. *Cell* **111**, 919–922.

Tessarollo, L., and Parada, L. F. (1995). *In situ* hybridization. *Methods Enzymol.* **254**, 419–430.

Vallone, D., Gondi, S. B., Whitmore, D., and Foulkes, N. (2004). E-box function in a period gene repressed by light. *Proc. Natl. Acad. Sci. USA* **101**, 4106–4111.

Westerfield, M. (1995). "The Zebrafish Book: A Guide for the Laboratory Use of Zebrafish (*Brachydanio rerio*)." University of Oregon Press.

Whitmore, D., Foulkes, N. S., Strahle, U., and Sassone-Corsi, P. (1998). Zebrafish Clock rhythmic expression reveals independent peripheral circadian oscillators. *Nature Neurosci.* **1**, 701–707.

Whitmore, D., Foulkes, N. S., and Sassone-Corsi, P. (2000). Light acts directly on organs and cells in culture to set the vertebrate circadian clock. *Nature* **404**, 87–91.

## [6]   Genetic Manipulation of Circadian Rhythms in *Xenopus*

*By* NAOTO HAYASAKA, SILVIA I. LARUE, and CARLA B. GREEN

### Abstract

*Xenopus laevis* retina is an important experimental model system for the study of circadian oscillator mechanisms, as light input pathways, central oscillator mechanisms, and multiple output pathways are all contained within this tissue. These retinas continue to exhibit robust circadian rhythms even after being maintained in culture for many days. The usefulness of this system has been improved even further by the development of a technique for simple genetic manipulation of these animals, which is complemented by expanded genomics resources (*Xenopus* genome project, microarray, etc.). By taking advantage of the transgenic technique in *Xenopus* described in this article, many types of analysis can be done on the primary transgenic animals within a couple of weeks after transgenesis. The availability of many cell-type-specific promoters and well-characterized cell types within the *Xenopus* retina provides the advantage of cell-specific modification of clock function using this method; in other words, contributions of different cell types within the circadian system can be analyzed independently by "molecular dissociation" of these cells. This article describes both how this transgenic technique is useful and various considerations that should be taken into account when these types of experiments are planned and interpreted. Application of these new techniques to studies of clock function provide an opportunity to rapidly assess gene expression and/or function in the context of the intact retina.

### Introduction

#### History of Circadian Clock Studies in Xenopus

*Xenopus* retinas have been used for cell biological or biochemical studies of vision for many years, even before circadian pacemakers were found in this tissue. There are several advantages of using *Xenopus* retinas for this kind of research (e.g., ease of culture, large size of the cells, 1:1 distribution of rod and cone cells), and these studies resulted in an accumulation of physiological evidence and also development of different techniques for the retinal studies, some of which are also useful for current circadian rhythm research.

METHODS IN ENZYMOLOGY, VOL. 393

Since a circadian oscillator was first identified in the retina of *Xenopus laevis* (Besharse and Iuvone, 1983), the *Xenopus* retina has been used as a good model to study circadian systems and has become arguably one of the best-studied nonbrain oscillators due to its many experimental advantages, including its robustness in culture. Several decades of retinal clock studies in *Xenopus* have revealed that many aspects of retinal physiology (e.g., disc shedding, light sensitivity, neurohormone synthesis) are under the control of a local circadian clock(s) (Anderson and Green, 2000; Cahill and Besharse, 1995; Green and Besharse, 2004).

*Advantages of Using* Xenopus

As mentioned earlier, *Xenopus* provides many advantages for studies of the eye/retina, including the examination of circadian rhythms. Extensive studies on these retinas have provided much knowledge about the structure and physiology of the vertebrate retina and have also resulted in the development of many assays and methods that continue to be useful. For example, the development of a perfusion culture system for the long-term maintenance of eyecups or retinas allows circadian rhythms of melatonin release to be measured easily (Cahill and Besharse, 1991). A sensitive radioimmunoassay (RIA) for melatonin (Rollag and Niswender, 1976) allows this neuromodulator to be measured even at very low concentrations as found in tadpole eyes (Green *et al.*, 1999).

Despite these experimental advantages, the *Xenopus* system has traditionally been limited with regard to genetic manipulation. However, this limitation has been overcome by the development of the method of restriction enzyme-mediated integration (REMI; Kroll and Amaya, 1996). This transgenic technique for *Xenopus* makes it possible to produce dozens of transgenic *Xenopus* in a single day. Therefore, transgenic animals carrying several different transgenes or a transgene driven by distinct promoters can all be generated at the same time so that their phenotypes can be compared directly. Moreover, because the transgenic embryos generated in this method are not mosaics, the primary transgenics can be analyzed as soon as they develop to the point where they exhibit the tissue and/or physiology of interest. Because circadian clocks are functional in differentiated retinas by about 4–5 days after fertilization, these transgenic animals can be evaluated very rapidly. Compared with transgenic procedures in other vertebrate species, such as mouse or rat, this method is technically more simple and less time-consuming and costly, making *Xenopus* a very advantageous experimental animal for transgenic studies.

The ability to make transgenic *Xenopus* is complemented by the recent advancement in *Xenopus* genome research and development of new genetic tools. Although *X. laevis* is the species that has been predominantly used historically for biological research, *Xenopus tropicalis*, another closely related species, has been adopted as a model system with some significant advantages for genetic studies [e.g., shorter generation time, diploidy versus pseudotetraploidy in *X. laevis* (Amaya *et al.*, 1998; Hirsch *et al.*, 2002)]. The REMI method has already been modified and applied to *X. tropicalis* (Offield *et al.*, 2000), and the short generation time allows the establishment of transgenic lines in this species within a period of a few months.

The *Xenopus* genome project is underway, already resulting in large numbers of expressed sequence tags (ESTs) from both *X. laevis* and *X. tropicalis* and a draft version of the *X. tropicalis* genome sequence (see http://www.xenbase.org/genomics/genomics.html). This information also makes genome-wide analyses possible, as in the development of tools such as the *Xenopus* Affymetrix GeneChip (see http://www.xenbase.org/genomics/microarrays/Xenbase_affy_upd_v5.html). All of these advantages have made *Xenopus* a powerful model system in biological fields such as developmental biology and chronobiology.

## Transgenic Method

### General Considerations

*Design of Transgenes.* Before manipulating gene function *in vivo*, both the design of the transgenes and the choice of promoters need to be considered. First, the design of a transgene obviously depends on the aim of the research, but because the transgenic method in *Xenopus* does not currently allow genetic manipulation by homologous recombination ("knockouts" or "knockins"), the transgene must be something that will work when expressed in an animal along with the normal endogenous genes. The most common examples of transgenes are reporter genes or mutant forms of genes that are expected to have a "dominant-negative" effect. In our studies, we have used both these approaches: a dominant-negative form of the *Clock* gene that perturbs central clock function (discussed here) and a green fluorescent protein (GFP) reporter gene used for promoter analysis of the *nocturnin* gene (discussed briefly later). It is also possible to introduce extra copies of normal genes in order to study gene dosage effects or to introduce RNAi expression constructs to knockdown function of an endogenous gene (Dykxhoorn *et al.*, 2003).

In order to make a transgene that would disrupt normal clock function when overexpressed in clock cells, we designed a dominant-negative CLOCK (a core circadian clock component) that retains its DNA-binding domain and dimerization domains but lacks a transactivation domain. Based on previous data from studies of mutant animals (Allada *et al.*, 1998; Gekakis *et al.*, 1998; King *et al.*, 1997), we expected this form to competitively abolish normal endogenous CLOCK's function as a transcriptional activator. This mutant form of CLOCK (XCLΔQ) was initially tested in transient transfection experiments to verify its dominant-negative action in a well-controlled system before making transgenic animals. This assay showed that the mutant form prevented activation of transcription by normal CLOCK/BMAL1 heterodimer in a dose-dependent manner (Hayasaka *et al.*, 2002).

Another important design consideration is to build the transgene in such a way that it can be easily distinguished from the endogenous gene. In most cases, this can be accomplished by the addition of a "tag" sequence. In our studies, we added an in-frame enhanced GFP tag to the C terminus of the truncated CLOCK. This tag allowed us to examine localization of the expressed transgene, as well as giving us a unique sequence for quantitation of the expression level of the transgene by real-time polymerase chain reaction (PCR) (discussed later). As is the case anytime a tagged protein is used, the fusion protein must be tested to make sure that the addition of the tag does not alter cellular localization (if it is critical), function, or stability.

*Choice of Promoters.* To manipulate transgene expression in a cell-type/tissue-specific manner, an appropriate promoter has to be selected. In our case, the selection of cell-type-specific promoters was relatively straightforward, as the different cell types within the retina have been characterized so extensively. We tried various 5'-flanking regions derived from several photoreceptor-specific genes and initially chose three different promoters to target transgenes to one or both photoreceptor cell types. A promoter from the interphotoreceptor retinol-binding protein (IRBP; Boatright *et al.*, 1997) was used for expression in all photoreceptor cell classes, a rod opsin promoter (XOP; Knox *et al.*, 1998) for rod-specific expression, and a cone arrestin promoter (CAR; Zhu *et al.*, 2002) for cone-specific expression. In our hands, we have found that promoters from other species often give appropriate spatial expression, thereby expanding the number of available promoters that can be tested. For example, our IRBP promoter and our CAR promoter are both from mammals. Regardless of the source of the promoter, the cell-type-specific expression of the transgene must be confirmed. In our case, GFP expression from the XCLΔQ–GFP fusion protein was examined histologically in the transgenic retinas. Our goal was to use different promoters for cell-specific ablation of circadian clock function with the hope that this could provide information on the organization of the clock system within the retina.

Another consideration in selecting promoters is the required expression levels of the transgene. Some promoters drive much higher levels of expression on average than other promoters, which can have a major impact on the success of the transgenic experiment. If a transgene contains a reporter gene, then the expression needs to be strong enough to detect the reporter protein easily. Similarly, if a dominant-negative transgene is being introduced, the expression level must be high enough to compete with the endogenous gene product. In some cases, it may be advantageous to use promoters that do not drive high expression, particularly in cases where the transgene could be deleterious at high levels or in cases where an intermediate phenotype is desired. Even with careful consideration given to promoter strength, there will likely be a high level of variability in the expression level of the transgene in different animals due to variable copy numbers and different integration sites (position effects). This can be an advantage or a disadvantage, depending on the experimental design. This issue is discussed more thoroughly later.

*Modified REMI Method: Protocol*

The REMI transgenic method was originally developed by Kroll and Amaya (1996) and has subsequently been modified by several other laboratories (e.g., Hutcheson and Vetter, 2002; Rollag *et al.*, 2000; Sive *et al.*, 2000). This simple rapid method allows the production of many transgenic embryos in just a few hours' time. The first part of this method is the actual REMI reaction where sperm nuclei are incubated with a decondensating extract (made from eggs), a small amount of restriction enzyme, and the linearized DNA construct that is being introduced. In this short reaction (10–15 min) the DNA is integrated stably into the genome of sperm nuclei at random sites. The second portion of this technique is introduction of transgenic sperm nuclei into unfertilized eggs by microinjection. Eggs that exhibit normal cleavage patterns are selected and allowed to develop until the effect of the transgene can be assayed. Because this technique has been described in great detail in recent publications, we refer the reader to other accounts for this information (Hutcheson and Vetter, 2002; Kroll and Amaya, 1996; Rollag *et al.*, 2000; Sive *et al.*, 2000). In particular, a very informative and detailed description of this modified technique is given by Hutcheson and Vetter (2002). A simple flowchart showing how this technique can be used to study circadian rhythms in *Xenopus* is shown in Fig. 1.

In our laboratory, we routinely inject 500–1000 eggs per REMI reaction (this takes about 1 h). Of these eggs, approximately 50% will cleave normally (this is highly dependent on beginning with very good quality eggs from healthy females). These normally dividing eggs are separated

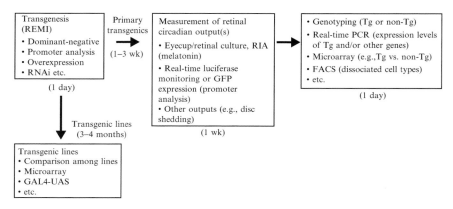

Fig. 1. A flowchart of circadian rhythm studies of *Xenopus* retina using transgenic and other techniques. Primary transgenic animals can be used at the age of 1–3 weeks for short-term analyses, whereas transgenic lines are available for studies that require numbers of animals as a group. One advantage of using primary transgenic animals is to be able to complete a set of experiments within a period of 1 month. Tg, transgenic.

from the abnormal cleavers (and noncleavers), and usually about 50% of these will successfully develop to the tailbud stage. At this point, it is often necessary to determine which of these embryos is actually carrying the transgene. This can be done in several ways. Perhaps the simplest way is to have a fluorescent tag (such as GFP) encoded by the transgene. This allows the identification of transgenic animals by simple observation under a fluorescent microscope. However, in order for this method to be used, the transgene must be expressed at high enough levels to generate a sufficient signal for discrimination. In addition, the region of transgene expression must be visible in the whole embryo. In many of our experiments this is not the case: the promoter being used is fairly weak, the site of expression is inside the pigmented eye, and/or the addition of a fluorescent tag is not recommended for some reason (inactivity of a fusion protein for example). Therefore, for many of our experiments we discriminate the transgenic versus nontransgenic tadpoles by genotyping, as described later. We perform our genotyping on tadpoles that are 1 week of age or older by clipping off a small portion of their tail. Metamorphosed frogs can be genotyped by clipping a small piece of webbing from between their toes. It has been reported that it is possible to select for transgenic tadpoles by including a neomycin resistance marker on the plasmid containing the transgene and then maintaining the developing embryos in G418 (Moritz *et al.*, 2002).

*Genotyping*

Dissect a small portion of the tail of the tadpole (tadpole should be at least 1 week of age) by first anesthetizing the tadpoles in a solution of 0.02% Tricaine (MS222; ethyl 3-aminobenzoate methanesulfonate salt; Sigma #A5040) for a few minutes until they stop moving. Carefully transfer the tadpole to a clean surface and, using a clean sharp razor blade (sterilized in 70% ethanol and wiped dry), remove about one-third to one-half of the tail. (Cutting too much will kill the animal and cutting too little will not yield enough DNA.) Place the tail in a microcentrifuge tube and proceed to DNA isolation or store at $-80°$ for long periods. (*Note*: If collecting tissue from metamorphosed frogs, hold the frog firmly and cut a triangle of webbing from between its toes and continue as just described. No anesthetic is necessary unless this is a terminal dissection.) Isolation of genomic DNA is done by using the Qiagen DNeasy tissue kit and following the manufacturer's instructions except that the final elution of DNA is done in 80 $\mu$l of 1/10 TE buffer. Purified genomic DNA is stored at $-20°$.

The genotyping is done by PCR, using primers that will amplify the transgene but not an endogenous gene. If no such primers exist, then primers should be chosen that flank an exon so that the endogenous gene can be distinguished by size from the transgene. The following protocol is for identifying a transgene that contains a enhanced GFP tag, as in our XCL$\Delta$Q-GFP example (Hayasaka *et al.*, 2002).

*GFP Primers*

The GFP primers that we use have the following sequences: forward primer is 5′-CAAGCTGACCCTGAAGTTCATCTG-3′ and reverse primer is 5′-CGGATCTTGAAGTTCACCTTGATG-3′. [This primer pair amplifies a 383-bp portion of the EGFP cDNA (from 219 to 602).] PCR was done using standard conditions with Amplitaq Gold polymerase, 2.5 m$M$ MgCl$_2$, and 10 pmol of each primer. PCR conditions were as follows: 95° for 10 min, 30 cycles of 94° for 40 s, 55° for 1 min, 72° for 1 min, and 72° for 10 min. A positive control (either a known positive sample or 1 ng of a plasmid containing the transgene) and a negative control (no DNA) are always included. Analyze the results on a 1.5% agarose gel. Positive animals are identified by the presence of a 383-bp band.

## Measurement of Circadian Outputs

Although there are many different circadian rhythms that could be analyzed in the *Xenopus* retina, the measurement of melatonin rhythmicity has many advantages, the greatest of which is that the rhythms can be measured from individual living retinas over time. Another advantage is that melatonin can be detected from tadpole eyes as early as about 1 week postfertilization (Green *et al.*, 1999). This means that individual transgenic retinas can be analyzed for rhythmicity very rapidly after the transgenesis procedure. We modified a method originally developed for adult *Xenopus* eyes by Cahill and Besharse (1991) to apply to tadpole eyes (Green *et al.*, 1999) and investigated an effect of the XCLΔQ transgene expression on circadian melatonin rhythmicity. Although isolated retinas can be used from adult *Xenopus*, when tadpole eyes are used for the analysis, eyecup culture is optimal because the retina is too small to be isolated easily. This is accomplished by dissecting the eye from the tadpole (while anesthetized with MS222) and then removing the anterior part of the eye, including the lens, with a sharp pair of forceps to allow efficient media and oxygen exchange. In this method, the eyecups are perfused continuously with culture medium at a constant rate, and media coming from the eyecups are collected in a fraction collector in 4-h bins (Green *et al.*, 1999; Hayasaka *et al.*, 2002). These fractions are then used for melatonin analysis by RIA (Rollag and Niswender, 1976). Although younger tadpoles can be used, we often use 2- to 3-week-old tadpoles for our eyecup cultures, as the melatonin levels are higher (and the eyes larger) in these older animals. Since in our experience, the REMI procedure results in the production of transgenic tadpoles and nontransgenic siblings at a ratio of about 1:1, we choose tadpoles randomly for phenotypic analysis in flow-through culture, saving their tails (at −80°) for genotyping, which is done after the culture and melatonin analysis are complete. This allows us to determine circadian melatonin profiles from transgenic and nontransgenic eyes in the same conditions and to analyze the results blindly, with no knowledge of the genotype until after the analysis is complete. Because there is some variability between tadpoles (there are no inbred strains of *X. laevis*) and the transgene integration site and copy number is different in each individual, this type of unbiased analysis and the use of nontransgenic siblings as controls are critical for appropriate interpretation of the effect of the transgene.

## Analysis of Transgene Expression Level by Real-Time PCR

As mentioned earlier, because individual primary transgenic frogs/ tadpoles are predicted to carry different copy numbers of a transgene in different integration sites, quantitative analysis of the transgene expression levels is critical, especially when phenotype(s) will be analyzed in the primary transgenic animals. The optimal method for this analysis is real-time PCR because this method can quantitate mRNA levels very precisely from very small tissue samples. In our studies, we performed real-time PCR on RNA samples isolated from individual pairs of transgenic tadpole eyes. Relative levels of mRNA encoding XCLΔQ-GFP were quantified using GFP primers to distinguish the transgene-encoded mRNA from the endogenous *xClock* mRNA. This quantitation was done immediately following the phenotypic analysis. Expression levels of XCLΔQ were compared among each pair of transgenic eyecups and correlated with the melatonin rhythmicity phenotype.

After the phenotypic analysis, such as the flow-through culture, is complete, each pair of eyes is collected and total RNA is extracted using Trizol according to the manufacturer's instruction. GlycoBlue coprecipitant (Ambion, Inc.) is added to the samples just prior to the precipitation step to aid in the visualization of the pellet, as the total amount of RNA isolated from these tadpole eyes is very low. The final RNA pellet is resuspended in 20 $\mu$l of RNase-free water and one-quarter of this (5 $\mu$l) is used in the Superscript II (Invitrogen) reverse transcriptase reaction as described by the manufacturer. We have found that this amount of cDNA is sufficient for 20 real-time PCR reactions [using the Bio-Rad iCycler with IQ SYBR Green Supermix (BioRad)] for the detection of our transgene messages using IRBP, XOP, and CAR promoters. Details on how to determine mRNA levels accurately using real-time PCR are beyond the scope of this article. Results from these analyses demonstrated significant differences in transgene expression between different transgenic eyes (with a range of more than two orders of magnitude). In our experimental paradigm, these differences in expression were useful and allowed us to study dose–response effects. In short, we observed that the abnormalities observed in melatonin rhythms in the transgenic eyecups correlated well with the expression levels of XCLΔQ; the higher the expression, the more severe the abnormality in the rhythm, indicating that XCLΔQ abolishes/alters circadian melatonin rhythmicity in a dose-dependent manner (Hayasaka *et al.*, 2002). Although this variability was a benefit to our studies because it allowed dose analysis, in some experiments this variability could be a problem. For these cases, it is advantageous to raise lines of transgenic animals (see later) and then real-time PCR can be used to characterize the expression levels of the animals from which the lines will be generated.

## Changes in Expression Profiles of Other Genes in Transgenic Eyes

Real-time PCR can also be used for the quantification of other genes in the same samples to investigate the molecular mechanism underlying the phenotype(s) observed in the transgenic eyes. For example, in our XCLΔQ experiments, it is predicted that expression profiles of other genes (e.g., clock/clock-controlled genes, photoreceptor-specific genes) will also be altered. Real-time RT-PCR is a powerful tool to perform this kind of sensitive quantitative analysis using mRNA from a single pair of eyes from a tadpole in which the transgene expression is already characterized. Although microarray analysis could theoretically be used to study genome-wide gene expression profiles in the transgenic animals, the small sample size is a limitation. This issue, however, can be solved by making transgenic lines as mentioned later, or by amplifying cDNAs in a quantitatively reliable way.

## Generation of Lines

The power of generating transgenic lines has been demonstrated in other species, such as *Drosophila*, mouse, and rat, whereas making such lines in *X. laevis* was not so straightforward, primarily because of its long generation time. However, improved husbandry methods have shortened the time to sexual maturity in *X. laevis*, making this a more realistic approach. In our laboratory, we can routinely get males to sexual maturity in 5–6 months. In addition, the use of *X. tropicalis*, with its much shorter generation time (Amaya *et al.*, 1998), has made the generation of lines much more expedient (3–4 months). There are several advantages of using transgenic lines. First of all, once transgenic lines with certain transgene expression levels are established, all the animals in a line can be used as a group. This extends the possibilities of analyses compared with the use of individual animals. For instance, comparison of temporal gene expression profiles in different lines can be achieved by collecting tissue at different time points from each line of transgenic animals and analyzing RNA levels by real-time PCR or microarray. Second, the "dose–effect" of a transgene can be evaluated more fully. If a phenotype(s) in transgenic animals is observed in a dose-dependent manner as demonstrated in our study, it would be possible to elucidate the difference by comparing multiple individuals from different transgenic lines with strong and weak transgene expression levels. Third, the availability of animals from lines makes it possible to use binary systems for transgene expression such as the GAL4-UAS system originally employed in *Drosophila* (Brand and Perrimon, 1993). This has already been tested and verified to work in both *X. laevis* and *X. tropicalis* (Chae *et al.*, 2002; Hartley *et al.*, 2002). It has also been reported that the Cre-loxP system for conditional expression also works in *Xenopus* (Werdien *et al.*, 2001). These techniques will contribute to the use of *Xenopus* as a valuable system for genetic approaches.

## Use of REMI Method for Reporter Gene Expression

In addition to using the REMI method for perturbation of gene function, this method is also a valuable way to introduce reporter genes for analysis of promoters or for labeling of specific cell types or tissues. In our laboratory we used the GFP reporter for analysis of the 5′-flanking region of the *nocturnin* gene (Liu and Green, 2001). Using this approach, we were able to define a novel element that was necessary and sufficient to drive *nocturnin* gene transcription specifically in rod and cone photoreceptors (Fig. 2). Several other promoters have also been characterized using this method (e.g., Knox *et al.*, 1998; Mani *et al.*, 1999; Moritz *et al.*, 1999; Warkman and Atkinson, 2004; Zhu *et al.*, 2002). Another valuable use of reporter gene expression in transgenic *Xenopus* is to allow detection of gene expression in the living animal. Although this has not yet been used in *Xenopus* to follow rhythmic gene expression, it has been used to follow temporal and spatial patterns of expression during development (e.g., Offield *et al.*, 2000).

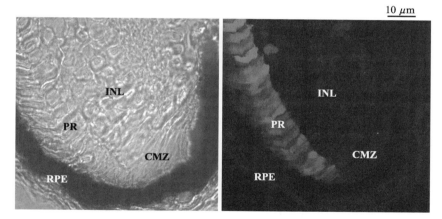

Fig. 2. A small portion of the nocturnin promoter is sufficient to drive GFP reporter gene expression specifically to differentiated photoreceptor cells. A piece of the *Xenopus nocturnin* promoter (−108/+20) was cloned upstream of a GFP reporter and used to generate transgenic *Xenopus*. Phase-contrast (left) and fluorescent (right) images of a portion of a transgenic tadpole eye show GFP expression in the cell bodies of the rod and cone photoreceptors (PR), but not in the other layers of the retina nor in the ciliarly marginal zone (CMZ) where undifferentiated retinal stem cells are located. RPE, retinal pigment epithelium; INL, inner nuclear layer. Reprinted with permission from Liu and Green (2001). (See color insert.)

*Future Directions for the Use of Transgenic* Xenopus *in
Studying Circadian Rhythmicity*

One of the attractive features of using transgenic *Xenopus* for analysis
of circadian oscillators is that in addition to investigations of the molecular
machinery that makes up the clock, this model system also allows one
to address the significance of the clock within the context of the retinas,
which are known to have many aspects of their physiology under circadian
control. Analysis of a variety of rhythms in animals with perturbed circadian clocks in specific cell types (such as our XCLΔQ transgenic animals) will provide other ways to study how oscillators in different cells
contribute to the overall rhythmic physiology of the retina. Like melatonin,
dopamine synthesis is also regulated in a circadian manner, as is retinal
sensitivity and many cellular events, such as rod disc shedding and intracellular signaling pathways. It will be interesting to know whether the same
rod and cone clocks control these other rhythms in the same manner
that they control melatonin rhythmicity. For analysis of these types of
rhythms (which cannot be followed easily from individual living retinas),
it will be important to generate lines of animals so that many tadpoles/frogs
with the same expression levels are available to compare across time
points.

Another possibility for exploring the circadian oscillator in lines of
animals is to analyze the circadian phenotype(s) of each cell type independently by dissociating and sorting cells. As a transgene product can be
designed to be expressed as a GFP fusion protein, and its expression can
be put under the control of a cell-type-specific promoter (such as in our
rod- or cone-specific XCLΔQ transgenics), it would be possible to collect
only cells expressing the transgene by dissociating retinas and collecting
GFP-positive cells using a fluorescence-activated cell sorter (FACS). These
cells could then be used for analysis of gene expression or activation of
signaling pathways or other intracellular rhythms.

Finally, the use of real-time monitoring of rhythmic gene expression
by reporters such as luciferase or destabilized GFP under the control of
rhythmic promoters has been very valuable for studying the clock in
other model systems (as described elsewhere in this volume). Development
of this technique for *Xenopus* eyes has been hampered by the dark pigment that surrounds the tadpole eye, making reporter detection problematic. However, the production of lines of animals bearing these reporter
constructs would allow monitoring gene expression from the larger
tadpole or frog eyes in which the pigmented cells could be removed more
easily. *Xenopus* has an advantage for this approach because it is easy
to introduce more than one transgene into each animal. Therefore, analysis

of rhythmicity could be done in both wild-type animals and animals expressing other transgenes.

## Concluding Remarks

The retinal circadian clock has traditionally been an excellent model system for circadian studies because light input pathways, robust circadian oscillators, and multiple output pathways are contained in this well-characterized and robust tissue. As described in this article, this model system has recently become even more powerful, thanks to the combination of several techniques (e.g., REMI, retinal flow-through culture, melatonin measurement), tools such as cell-type-specific promoters, and information coming from the *Xenopus* genome project. Studies have suggested in mammals that circadian oscillators exist both in brain and in other peripheral tissues (even cultured cells) containing circadian oscillators, and significance of the oscillatory mechanisms in each tissue has been uncovered. Genetic manipulation of *Xenopus* will allow further investigation into both the detailed molecular clock machinery and the organization and physiological significance of the circadian system within this important tissue.

## References

Allada, R., White, N. E., So, W. V., Hall, J. C., and Rosbash, M. (1998). A mutant Drosophila homolog of mammalian Clock disrupts circadian rhythms and transcription of period and timeless. *Cell* **93,** 791–804.

Amaya, E., Offield, M. F., and Grainger, R. M. (1998). Frog genetics: *Xenopus tropicalis* jumps into the future. *Trends Genet.* **14,** 253–255.

Anderson, F. E., and Green, C. B. (2000). Symphony of rhythms in the *Xenopus laevis* retina. *Microsc. Res. Tech.* **50,** 360–372.

Besharse, J. C., and Iuvone, P. M. (1983). Circadian clock in *Xenopus* eye controlling retinal serotonin *N*-acetyltransferase. *Nature* **305,** 133–135.

Boatright, J. H., Buono, R., Bruno, J., Lang, R. K., Si, J. S., Shinohara, T., Peoples, J. W., and Nickerson, J. M. (1997). The 5' flanking regions of IRBP and arrestin have promoter activity in primary embryonic chicken retina cell cultures. *Exp. Eye Res.* **64,** 269–277.

Brand, A. H., and Perrimon, N. (1993). Targeted gene expression as a means of altering cell fates and generating dominant phenotypes. *Development* **118,** 401–415.

Cahill, G. M., and Besharse, J. C. (1991). Resetting the circadian clock in cultured *Xenopus* eyecups: Regulation of retinal melatonin rhythms by light and D2 dopamine receptors. *J. Neurosci.* **11,** 2959–2971.

Cahill, G. M., and Besharse, J. C. (1995). Circadian rhythmicity in vertebrate retinas: Regulation by a photoreceptor oscillator. *Progr. Retinal Eye Res.* **14,** 267–291.

Chae, J., Zimmerman, L. B., and Grainger, R. M. (2002). Inducible control of tissue-specific transgene expression in *Xenopus tropicalis* transgenic lines. *Mech. Dev.* **117,** 235–241.

Dykxhoorn, D. M., Novina, C. D., and Sharp, P. A. (2003). Killing the messenger: Short RNAs that silence gene expression. *Nature Rev. Mol. Cell Biol.* **4,** 457–467.

Gekakis, N., Staknis, D., Nguyen, H. B., Davis, F. C., Wilsbacher, L. D., King, D. P., Takahashi, J. S., and Weitz, C. J. (1998). Role of the CLOCK protein in the mammalian circadian mechanism. *Science* **280,** 1564–1569.

Green, C. B., and Besharse, J. C. (2004). Retinal circadian clocks and control of retinal physiology. *J. Biol. Rhythms* **19,** 91–102.

Green, C. B., Liang, M.-Y., Steenhard, B. M., and Besharse, J. C. (1999). Ontogeny of circadian and light regulation of melatonin release in *Xenopus laevis* embryos. *Dev. Brain Res.* **117,** 109–116.

Hartley, K. O., Nutt, S. L., and Amaya, E. (2002). Targeted gene expression in transgenic *Xenopus* using the binary Gal4-UAS system. *Proc. Natl. Acad. Sci. USA* **99,** 1377–1382.

Hayasaka, N., LaRue, S. I., and Green, C. B. (2002). *In vivo* disruption of *Xenopus* CLOCK in the retinal photoreceptor cells abolishes circadian melatonin rhythmicity without affecting its production levels. *J. Neurosci.* **22,** 1600–1607.

Hirsch, N., Zimmerman, L. B., and Grainger, R. M. (2002). *Xenopus,* the next generation: X *Tropicalis genetics* and genomics. *Dev. Dyn.* **225,** 422–433.

Hutcheson, D. A., and Vetter, M. L. (2002). Transgenic approaches to retinal development and function in *Xenopus laevis. Methods* **28,** 402–410.

King, D. P., Zhao, Y., Sangoram, A. M., Wilsbacher, L. D., Tanaka, M., Antoch, M. P., Steeves, T. D., Vitaterna, M. H., Kornhauser, J. M., Lowrey, P. L., Turek, F. W., and Takahashi, J. S. (1997). Positional cloning of the mouse circadian *Clock* gene. *Cell* **89,** 641–653.

Knox, B. E., Schlueter, C., Sanger, B. M., Green, C. B., and Besharse, J. C. (1998). Transgene expression in *Xenopus* rods. *FEBS Lett.* **423,** 117–121.

Kroll, K. L., and Amaya, E. (1996). Transgenic *Xenopus* embryos from sperm nuclear transplantations reveal FGF signaling requirements during gastrulation. *Development* **122,** 3173–3183.

Liu, X., and Green, C. B. (2001). A novel promoter element, photoreceptor conserved element II, directs photoreceptor-specific expression of nocturnin in *Xenopus laevis. J. Biol. Chem.* **276,** 15146–15154.

Mani, S. S., Besharse, J. C., and Knox, B. E. (1999). Immediate upstream sequence of arrestin directs rod-specific expression in *Xenopus. J. Biol. Chem.* **274,** 15590–15597.

Moritz, O. L., Biddle, K. E., and Tam, B. M. (2002). Selection of transgenic *Xenopus laevis* using antibiotic resistance. *Transgenic. Res.* **11,** 315–319.

Moritz, O. L., Tam, B. M., Knox, B. E., and Papermaster, D. S. (1999). Fluorescent photoreceptors of transgenic *Xenopus laevis* imaged *in vivo* by two microscopy techniques. *Invest. Ophthalmol. Vis. Sci.* **40,** 3276–3280.

Offield, M. F., Hirsch, N., and Grainger, R. M. (2000). The development of *Xenopus tropicalis* transgenic lines and their use in studying lens developmental timing in living embryos. *Development* **127,** 1789–1797.

Rollag, M. D., and Niswender, G. D. (1976). Radioimmunoassay of serum concentrations of melatonin in sheep exposed to different lighting regimens. *Endocrinology* **98,** 482–489.

Rollag, M. D., Provencio, I., Sugden, D., and Green, C. B. (2000). Cultured amphibian melanophores: A model system to study melanopsin photobiology. *Methods Enzymol.* **316,** 291–309.

Sive, H. L., Grainger, R. M., and Harland, R. M. (2000). "Early Development of *Xenopus laevis:* A Course Manual." Cold Spring Harbor Laboratory Press, Cold Spring Harbor, NY.

Warkman, A. S., and Atkinson, B. G. (2004). Amphibian cardiac troponin I gene's organization, developmental expression, and regulatory properties are different from its mammalian homologue. *Dev. Dyn.* **229,** 275–288.

Werdien, D., Peiler, G., and Ryffel, G. U. (2001). FLP and Cre recombinase function in *Xenopus* embryos. *Nucleic Acids Res.* **29,** E53.

Zhu, X., Ma, B., Babu, S., Murage, J., Knox, B. E., and Craft, C. M. (2002). Mouse cone arrestin gene characterization: Promoter targets expression to cone photoreceptors. *FEBS Lett.* **524,** 116–122.

## [7]  Forward Genetic Screens to Identify Circadian Rhythm Mutants in Mice

*By* Sandra M. Siepka and Joseph S. Takahashi

### Abstract

This article describes the methods and techniques used to produce mutagenized mice to conduct high-throughput forward genetic screens for circadian rhythm mutants in the mouse. In particular, we outline methods to safely prepare and administer the chemical mutagen $N$-nitroso-$N$-ethylurea (ENU) to mice. We also discuss the importance of selecting mouse strain and outline breeding strategies, logistics, and throughput to produce these mutant mice. Finally, we discuss the breeding strategies that we use to confirm mutation heritability.

### Introduction

High-throughput forward mutagenesis screens are ongoing at a variety of institutions to identify mouse behavioral mutants using the strategy of forward genetics (from phenotype to gene) (Justice *et al.*, 1999; Moldin *et al.*, 2001). Unlike reverse genetic (from gene to phenotype) approaches, forward genetic approaches make no assumptions concerning the underlying genes involved to illicit a given behavior. Random point mutations are introduced into the mouse genome with the chemical mutagen $N$-nitroso-$N$-ethylurea (ENU), and mutant mice are identified based on their altered phenotype. In forward genetic screens, a deep understanding of phenotype is the essence of successful endeavors.

We use circadian wheel running activity to identify mice expressing abnormal circadian rhythms. Although activity rhythms are robust, quantitative, and automated, other factors, such as mouse strain, ENU preparation and dosage, and colony breeding strategies are integral to the success of the screen. This article describes the methods employed to produce mutagenized mice to conduct forward genetic screens for circadian rhythm mutants in mouse.

Strain Choice

The choice of mouse strain is a major consideration for any mouse mutagenesis production colony. Strain choice dictates the effectiveness of the ENU to introduce mutations, the efficiency of the breeding scheme to produce mutant mice, and ultimately the ability to detect mutants in the phenotypic screens. To avoid inconsistencies and variability in any and all of these factors, we have elected to use only inbred strains for our muta-genesis program. The two strains that we have used, BTBR/J and C57BL/6J, were chosen for their ENU mutation rate, their fertility, and, most importantly, their circadian behavior. Bear in mind, however, that a high mutation rate and/or great fecundity can never compensate for a weak or inconsistent phenotype.

ENU is a very potent and powerful mutagen (Russell et al., 1979). Male mice are treated with the highest dose of ENU that can be tolerated without causing infertility. The primary target of ENU in the germline of mice is the spermatogonia, which can be highly mutagenized and go on to produce mutant gametes. Any animal sired from one of these mutagenized males will harbor multiple mutations in its genome. In behavioral pheno-types such as circadian rhythms, the number of gene targets is relatively low (perhaps 20–30 genes contribute to the behavior). The success of any mutation screen, therefore, is dependent on the forward mutation rate. Some mouse strains, such as BTBR/J, can tolerate a single high dose of ENU to achieve relatively high forward mutation rates ($\sim$1/250 per locus per gamete) (McDonald et al., 1990; Shedlovsky et al., 1993). Other strains, such as C57BL/6J, required multiple weekly injections of lower ENU doses to achieve a somewhat comparable forward mutation rate ($\sim$1/800 per locus per gamete) (Justice et al., 2000).

Second, it is important to select a mouse strain that is a good breeder. Average litter size and the tendency of female mice to eat their first litters will influence the breeding strategy and logistics used to produce mutant mice. In general, we have found that our mice breed better as trios (one male and two females) as opposed to pairs. Whenever possible, we rotated males through cages of females to increase the number of litters sired by each male.

It is possible to use mixed strain F1s to increase mutation frequencies and to increase fecundity. Although these hybrid mice are better breeders overall, it is far better from a phenotyping perspective to use inbred strains for a mutagenesis production colony. Hybrid and outbred strains perform inconsistently in most behavioral assays and their mixed genetic backgrounds can make it difficult to detect true phenotypic outliers. For example, [C57BL/6J × BALB/cJ]F2 hybrid mice display a wider range of circadian behavioral measurements as compared with parental inbred strains (Shimomura et al., 2001). If used in a mutagenesis screen, it would be difficult to determine whether any altered circadian phenotypes were due to a mutation or simply a result of the mixed genetic background.

Finally, the specific inbred strain used in a mutagenesis study can determine whether a mutant phenotype can even be detected. For example, C57BL/6J mice are robust runners with a somewhat invariant free running period. BALB/cJ mice, however, are sloppy runners with a more variable free running period (Schwartz and Zimmerman, 1990; Shimomura et al., 2001). It would be difficult to detect true outliers using BALB/cJ mice. Strain also influences the observed phenotype. For example, on a CF-1 background the epidermal growth factor receptor (EGF-R) knockout is lethal at embryonic day 7.5 (Sibilia and Wagner, 1995; Threadgill et al., 1995). The same EGF-R knockout on a 129/Sv background is lethal at midgestation, lethal perinatally on a CD-1 background, and lethal postnatally on a 129/Sv × C57BL/6J background. Although we cannot predict a priori, which inbred strain is best suited to display a given mutant behavior or phenotype, we can select strains that have consistent and somewhat invariant phenotypic measurements to ensure that we will be able to detect a true phenotypic outlier.

## ENU Safety Procedures

N-Nitroso-N-ethylurea is purchased in ISOPAC containers (Sigma # N3385 ~1 g/bottle) as a wetted solid and stored at −20° until needed. ISOPAC containment minimizes investigator contact with the ENU. We prepare the ENU and inject our mice within a SPF barrier facility. All investigators handling ENU wear a disposable laboratory coat, latex gloves, hair bonnet, disposable shoe covers, and safety goggles. To further minimize contact with the ENU, all procedures are performed within a biological safety cabinet lined with two layers of disposable bench paper. All solutions are introduced into and removed from the ISOPAC bottle through needles inserted through the rubber stopper of the container. All ENU-contaminated materials are placed in a solid waste container, collected, and removed from our facility by our hazardous waste staff. Per recommendation of our office of research safety, ENU-contaminated materials are not decontaminated with an alkaline sodium isothiosulfate solution. Although alkaline treatment neutralizes ENU, EPA guidelines dictate that investigators do not attempt to neutralize and dispose of hazardous materials within the laboratory. The following protocol produces approximately 100 ml of ENU stock solution. We rarely use more than 5 ml of solution per week. All excess ENU stock solution is disposed of immediately after use, and we prepare fresh ENU stocks for each round of injections.

## ENU Preparation

To prepare an ENU solution stock, the powdered ENU is first dissolved in 10 ml of 95% ethanol within the ISOPAC bottle. The ethanol solution is injected into the bottle with an 18-gauge needle attached to a 10-ml syringe, the needle and syringe are removed, and the solution is agitated gently until the ENU is completely dissolved. An 18-gauge needle is inserted through the stopper of the bottle to act as a pressure vent. A second 18-gauge needle attached to a 60-ml syringe is used as an injection port to introduce 70 ml of PC buffer (0.1 $M$ $Na_2HPO_4$, 0.05 $M$ Na citrate pH adjusted to 5.0 and 0.2 $\mu$m filter sterilized). The needles are removed and the solution is swirled gently. A small aliquot of ENU solution is removed ($\sim$300 $\mu$l) using a 1-ml tuberculin syringe, the sample is diluted 1:5 with PC buffer, and the $OD_{398}$ is determined. ENU concentration is calculated where 1 $OD_{398}$ = 0.72 mg/ml. For safety, the spectophotometer is kept in the biological safety cabinet and is used exclusively for ENU concentration determinations. Plastic disposable cuvettes are used in place of quartz cuvettes.

Enough additional PC buffer is added to the ENU solution to create a 10-mg/ml stock solution. Between injections the bottled ENU solution is vented with an 18-gauge needle. If the solution is not vented often, air bubbles accumulate within the injection syringes, preventing accurate ENU dose administration to the mice.

## Injections

ENU treatments are initiated at 6 weeks of age. The mice are individually weighed and injected in the intraperitoneal cavity (ip) using 28-gauge short-needle insulin syringes (Becton-Dickson # 328438) with the proper dose of ENU. For C57BL/6J mice, we typically administer three weekly doses of 100 mg ENU/kg body weight (3 $\times$ 100). We mutagenize BTBR/J mice with a single dose of 250 mg/kg body weight (1 $\times$ 250).

To inject a mouse, we hold it by the scruff of the neck with the thumb and forefinger to splay the front legs back to prevent head movement. The tail is held between the ring and little fingers to extend the rear legs and expose the dorsal surface of the mouse. ENU is injected just anterior to the leg and ~1 cm to the right or left of midline. This ensures that the ENU is not injected either into the bladder or into any major organ. If the ENU is injected into the bladder, the ENU is excreted and little, if any, ENU reaches the spermatogonia. If the ENU is injected into a major organ, the mouse does not survive more than 24 h postinjection. Because the final ENU stock solution contains ~10% ethyl alcohol, all of the mice exhibit some symptoms of acute alcohol intoxication. The effect is short-lived and the mice resume their normal level of activity 2–4 h later.

We selected the 3 × 100-mg treatment as the highest dose of ENU that the C57BL/6J mice could tolerate, regain fertility, and live long enough to produce 50–100 offspring. There is some risk, however, that all of the mice injected with this high dose never regain fertility, develop tumors, and/or die. To ensure against this possibility, we also inject a second set of mice with a slightly lower dose of ENU (3 × 90 mg).

After injection, the mice are placed in a clean cage (five per cage) with fresh bedding, food, and water and are kept overnight within a biological safety cabinet. During this time, the mice excrete nearly all of the injected ENU. The following morning, the mice are placed in clean cages with fresh bedding and are returned to cage racks in the production facility. The ENU-contaminated bedding is disposed of as hazardous waste. The contaminated cages are cleaned by rinsing with soap and hot water and then washed and autoclaved along with the other cages in our animal care facility.

### Breeding Strategies

We have used two different breeding schemes to produce mutant mice—the backcross scheme and the intercross scheme (Fig. 1). Each production scheme, however, starts the same. Male mice (G0) are treated with ENU, pass through a sterile period, and eventually recover their fertility (but are now producing mutant sperm). These G0 mice are mated with wild-type females to produce G1 mice. Each G1 mouse represents one mutagenized gamete and can be phenotyped to conduct a first-generation screen for semi-dominant or dominant circadian mutants. The *Clock* mutant was identified in such a first-generation dominant screen (Vitaterna *et al.*, 1994).

To conduct a recessive screen for circadian rhythm mutants, G1 mice are mated to found three-generation kindreds. The resulting G3 mice are phenotyped to determine whether the G1 founder harbors any circadian mutations. These G3 mice can be produced using either a backcross scheme or an intercross scheme.

In the backcross scheme, G1 males are mated with wild-type females to produce G2 daughters. These G2 daughters are selected at random to

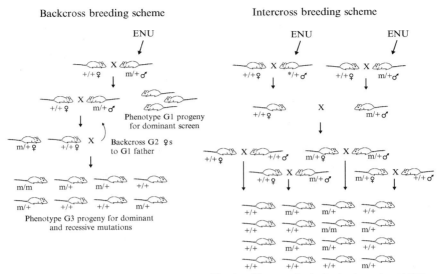

Fig. 1. Backcross and intercross breeding schemes for mutant production. See the text for a full description. Note that the intercross breeding scheme only depicts the inheritance pattern of a mutation contributed by the male G1 mouse. The female G1 mouse, in fact, also contributes mutations to the pedigree, but has not been depicted for clarity.

backcross to their G1 fathers. The resulting G3 mice are phenotyped to identify mutants. One-half of all of the G2 females in the kindred will harbor the same mutation as the G1 kindred founder. In G3 litters produced by these G2 carriers, one-fourth of the pups will be homozygous for the mutation carried by the G1 founder, one-half of the pups in these litters will be heterozygous carriers, and the remaining one-fourth will be homozygous wild type. One-half of the G3 pups produced from the noncarrier G2 daughters will be heterozygous for the G1 founder mutation. Because G3 mutant carriers can be either heterozygous or homozygous for the mutation, a recessive screen is, in fact, both a dominant and a recessive screen. We conducted our BTBR recessive screen using this backcross breeding scheme.

Alternatively, G3 mice can be produced by the intercross scheme. Here each kindred is founded by one G1 male and one nonsibling G1 female. In other words, two mutagenized gametes are used to found the kindred. The G2 mice produced are then intercrossed to produce G3 mice for phenotyping. This breeding strategy is similar to that used in the zebrafish where two G1 kindred founders are used to compensate for a low forward mutagenesis rate (Amsterdam *et al.*, 1999). Because the C57BL/6J strain has a somewhat lower forward mutagenesis rate than that of BTBR/J, but is a well-characterized strain for many behavioral phenotypes (including circadian rhythm behavior), we explored the feasibility of using the intercross breeding scheme to produce mutant mice to conduct a recessive mutant screen.

## The Efficiency of Scanning the Genome to Detect a Recessive Mutant

The efficiency of scanning the genome to detect a recessive mutation in a given kindred using the backcross breeding scheme is determined using the following formula (Shedlovsky *et al.*, 1986):

$$1 - [0.5 + 0.5(0.75)^n]^k$$

or 1 minus the probability of *not* detecting a recessive mutation, where $k$ is the number of G2 daughters backcrossed to the G1 founder and $n$ is the number of progeny/G2 female. If we phenotype five offspring from each of four G2 daughters (or 20 G3 mice), we have a probability of 0.85 of detecting a recessive mutant in any given kindred.

The efficiency of scanning the genome to detect a recessive mutation in a given kindred using the intercross breeding scheme is determined using the following formula:

$$1 - [0.75 + 0.25(0.75)^n]^k$$

or 1 minus the probability of *not* detecting a recessive mutation, where $k$ is the number of G2 intercross pairs per kindred and $n$ is the number of progeny phenotyped/intercross pair. If we phenotype six offspring from each of six G2 intercrosses (or 36 G3 mice), we have a probability of 0.75 of detecting a recessive mutant in any given kindred if the kindred is founded with one G1 parent (or mutagenized gamete). Remember, however, that we initiate each kindred with two G1 mice (or two mutagenized gametes), so we are scanning 1.5 (0.75 × 2) genomes for every set of 36 G3 mice.

The intercross breeding scheme appears to be more attractive from both theoretical and practical perspectives. Theoretically, we can scan more of the genome (1.5) per 36 G3 mice as compared with 0.85 genome per 20 G3 mice in the backcross scheme. From a practical perspective, the reproductive burden placed on the G1 male mouse is less in the intercross scheme than in the backcross scheme. In the intercross scheme the G1 mice need only to produce two to three litters of G2 mice. In the backcross scheme the G1 male must first produce G2 daughters and then later sire the G3 pups once those G2 daughters are old enough to breed. In other words, the G1 males must remain viable and fertile for weeks longer than in the intercross scheme. Second, no male mice are rotated through cages of females in the intercross scheme. We simply set up mating pairs and wait for them to produce G3 litters. This reduces our work flow greatly in the production facility.

Using C57BL/6J mice, however, the intercross scheme proved to be much less effective than the backcross scheme. Fertility of the G1 and G2 females was less than expected (Table I). Matings in the intercross scheme are set up as breeding pairs. Only 45% of all G1 intercross pairs and 55% of all G2 intercross pairs were fertile. In the backcross scheme (where we set up mating trios instead of pairs), 83% of all G1 × wt matings were fertile. Second, the fraction of G2 litters that survived to weaning was greater in the backcross scheme (67%) than in the intercross scheme (44%). Finally, the establishment of G2 breeding pairs was dependent on the production of an equal number of G2 males and G2 females. Because the average litter size for the G2 litters produced by the G1 intercross pairs was five, at best we could only set up two G2 breeding pairs from each litter 33% of the time. One-third of the time we could only set up one breeding pair and the other third of the time we were unable to set up any breeding pairs. In order to compensate for the low number of G2 mating pairs, we kept the G1 breeding pairs in the colony longer to produce more G2s. The net result was a drastic increase in the number of mating cages in our breeding colony with a net decrease in the efficiency of scanning of the genome. This scheme could possibly succeed in another mouse strain with a much higher average litter size, but this theoretical strain must also breed well using mating pairs and not trios. We now use the backcross scheme exclusively to produce our G3 mutant mice.

## Mouse Production Logistics

After the final ENU injection, the G0 mice go through a sterile period. During the first 4–6 weeks postinjection, the ENU treatment kills most of the spermatogonia of each mouse and mature unmutagenized sperm levels fall to zero. Mutagenized spermatogonia begin to replicate, and by the end

TABLE I
MATING EFFICIENCIES—INTERCROSS VS BACKCROSS BREEDING SCHEMES

| Mating type | G1 × G1 | G1 × 2 wt | G2 × G2 | G1 × G2 |
|---|---|---|---|---|
| Breeding scheme | Intercross | Backcross | Intercross | Backcross |
| Total matings | 1048 | 918 | 1067 | 1978 |
| Fertile matings | 475 (45%) | 760 (83%) | 590 (55%) | 886 (45%) |
| Total litters | 2030 | 2560 | 2587 | 3154 |
| Litters that survive | 898 (44%) | 1724 (67%) | 1411 (55%) | 1979 (63%) |
| Litter size at birth | 5.8 | 6.1 | 4.5 | 5.4 |
| Litter size at weaning | 5.1 | 5.4 | 4 | 4.6 |

of the sterile period sperm levels return to normal levels but the sperm produced are now highly mutagenized.

After 6 weeks, at 14-day intervals we begin to rotate the G0 mice through three to four cages of 6- to 8-week-old C57BL/6J female pairs. Because each G1 mouse produced is the product of one mutagenized gamete and the ENU treatment greatly reduces the number of different spermatogonia per G0 mouse, we limit G1 production to 25 G1 males per G0 father to prevent mutant line duplication in our colony.

Approximately 20–30% of our ENU treatment C57BL/6J mice never regain their fertility after ENU treatment. By 14 weeks postinjection, all of the G0 mice that will regain fertility have already done so. Only 7% of the fertile G0 mice reach G1 production limits. Many G0 mice die or develop tumors before they reach production goals. On average, each fertile G0 male sires 10 G1 males. Half of all G1 litters produced by the G0 males do not survive to weaning, and the average litter size at birth is six pups. Nearly all of the viable pups, however, survive to weaning. This is no different than that observed for a typical C57BL/6J wild-type colony. C57B6/6J females are not the best mothers and usually eat their first litters. We have found that production is improved when our mice are mated as trios rather than as pairs. Two mothers per cage increase the chances that at least one of the females is capable and willing to nurse and care for any litter born into that cage.

At 24 days of age, the G1 litters are weaned, G1 female mice are retired, and G1 male mice are kept as kindred founders. At 6 weeks of age, each G1 mouse is set up in a single mating with two C57BL/6J female mice. These mating trios are kept until the G1 father sires four to six G2 females. If the G1 male does not sire any viable litters within 10 weeks, he is removed from the production facility. By this time, he is 16 weeks old and unlikely to ever sire any litters. If he does sire any G2 litters after this time, he will be even older by the time his G2 daughters are old enough to backcross to produce G3 mice.

Once the first G2 daughter is weaned, the G1 male is separated from his wild-type mates and housed alone until his daughters are 6 weeks old and ready to backcross. The G1 male is then rotated at 14-day intervals through two to three cages of G2 daughter pairs. We call this set of mating cages a backcross group. Once each G2 daughter has each produced five pups, the backcross group is retired.

## Production Throughput

The goal of our mutagenesis production facility is to produce 10,000 G3 mice each year for phenotyping. This target is dictated by our phenotyping capacity, in which we can initiate approximately 200 circadian assays per week. We estimate that we can produce 200 G3 mice per week from 10 G1 backcross kindreds. This means that we only need to produce 500 G1 males per year. If every G0 male survived the ENU treatment and produced 25 G1 males, we would only need to inject 20 G0 males per year. In practice, however, 50% of all G0 males do not survive the ENU treatment, many of the G1 males are infertile, G1 males sometimes produce fewer than four G2 females, and several of the G2 females are infertile. To prevent production bottlenecks, therefore, we inject 150–200 G0 males per year (50 males every 3–4 months) to overproduce G1 males. We also establish twice as many (20) G1 backcross groups than theoretically necessary to produce 10,000 G3 mice per year. We phenotype the G1 male mice not used to found kindreds to screen for dominant circadian rhythm mutants.

### Mutant Heritability Tests

All putative mutants (or putants) are mated to determine trait heritability. If any offspring from the putant express the same deviant phenotype (in the expected Mendelian pattern), the putant is considered a true mutant. As a first step in this heritability test, putants are mated to wild-type mice. If the putant originates from a G1 dominant screen, we simply produce and phenotype G2 offspring. In the simplest case we would expect one-half of the G2 offspring to have the same phenotype as the putant parent. If the putant originates from a G3 recessive screen, we intercross its G4 offspring and phenotype the G5 offspring. Here, the expected phenotype of the G5 mice can vary depending on the nature of the G3 mutation.

If the G3 putant is heterozygous for the mutation, then only one-half of its offspring are carriers of the mutation and only some of the G4 intercrosses will be between heterozygous carriers. The phenotypes of G5 mice, therefore, can vary from completely wild type to that of the G3 putant to some extreme and previously undetected phenotype (including homozygous lethal). If the G3 putant is a homozygous carrier, all of the G4 offspring are heterozygous carriers and one-fourth of G5 mice will be wild type, one-half will be heterozygous for the mutation, and one-fourth homozygous of the mutation.

Taken together, the patterns of behavior of all of the G3 mice from a given kindred can hint at the mutation mode of inheritance, and we adjust our heritability mating strategies accordingly. Occasionally a G3 putant will have siblings or cousins with the similar or weaker, but still similar, phenotype. This strongly suggests (but does not prove) that the putant is, in fact, a true mutant. Every one of these siblings or cousins is mated to increase the probability that the putant line is not lost to infertility. Once we have reproduced the behavior in putant line offspring, the putant is upgraded to mutant and we expand the kindred to generate enough mice for genetic mapping and to create a homozygous mutant mouse line.

## Acknowledgments

This work was supported by NIH grant #U01_MH61915. J. S. Takahashi is an Investigator in the Howard Hughes Medical Institute.

## References

Amsterdam, A., Burgess, S., Golling, G., Chen, W., Sun, Z., Townsend, K., Farrington, S., Haldi, M., and Hopkins, N. (1999). A large-scale insertional mutagenesis screen in zebrafish. *Genes Dev.* **13,** 2713–2724.

Justice, M. J., Carpenter, D. A., Favor, J., Neuhauser-Klaus, A., Hrabe de Angelis, M., Soewarto, D., Moser, A., Cordes, S., Miller, D., Chapman, V., Weber, J. S., Rinchik, E. M., Hunsicker, P. R., Russell, W. L., and Bode, V. C. (2000). Effects of ENU dosage on mouse strains. *Mamm. Genome* **11,** 484–488.

Justice, M. J., Noveroske, J. K., Weber, J. S., Zheng, B., and Bradley, A. (1999). Mouse ENU mutagenesis. *Hum. Mol. Genet.* **8,** 1955–1963.

McDonald, J. D., Bode, V. C., Dove, W. F., and Shedlovsky, A. (1990). The use of N-ethyl-N-nitrosourea to produce mouse models for human phenylketonuria and hyperphenyla-laninemia. *Prog. Clin. Biol. Res.* **340C,** 407–413.

Moldin, S. O., Farmer, M. E., Chin, H. R., and Battey, J. F., Jr. (2001). Trans-NIH neuroscience initiatives on mouse phenotyping and mutagenesis. *Mamm. Genome* **12,** 575–581.

Russell, W. L., Kelly, E. M., Hunsicker, P. R., Bangham, J. W., Maddux, S. C., and Phipps, E. L. (1979). Specific-locus test shows ethylnitrosourea to be the most potent mutagen in the mouse. *Proc. Natl. Acad. Sci. USA* **76,** 5818–5819.

Schwartz, W. J., and Zimmerman, P. (1990). Circadian timekeeping in BALB/c and C57BL/6 inbred mouse strains. *J. Neurosci.* **10,** 3685–3694.

Shedlovsky, A., Guenet, J. L., Johnson, L. L., and Dove, W. F. (1986). Induction of recessive lethal mutations in the T/t-H-2 region of the mouse genome by a point mutagen. *Genet. Res.* **47,** 135–142.

Shedlovsky, A., McDonald, J. D., Symula, D., and Dove, W. F. (1993). Mouse models of human phenylketonuria. *Genetics* **134,** 1205–1210.

Shimomura, K., Low-Zeddies, S. S., King, D. P., Steeves, T. D., Whiteley, A., Kushla, J., Zemenides, P. D., Lin, A., Vitaterna, M. H., Churchill, G. A., and Takahashi, J. S. (2001). Genome-wide epistatic interaction analysis reveals complex genetic determinants of circadian behavior in mice. *Genome Res.* **11,** 959–980.

Sibilia, M., and Wagner, E. F. (1995). Strain-dependent epithelial defects in mice lacking the EGF receptor. *Science* **269,** 234–238.

Threadgill, D. W., Dlugosz, A. A., Hansen, L. A., Tennenbaum, T., Lichti, U., Yee, D., LaMantia, C., Mourton, T., Herrup, K., Harris, R. C., Barnard, J. A., Yuspa, S. H., Coffey, R. J., and Magnuson, T. (1995). Targeted disruption of mouse EGF receptor: Effect of genetic background on mutant phenotype. *Science* **269,** 230–234.

Vitaterna, M. H., King, D. P., Chang, A. M., Kornhauser, J. M., Lowrey, P. L., McDonald, J. D., Dove, W. F., Pinto, L. H., Turek, F. W., and Takahashi, J. S. (1994). Mutagenesis and mapping of a mouse gene, *Clock*, essential for circadian behavior. *Science* **264,** 719–725.

## [8]  Methods to Record Circadian Rhythm Wheel Running Activity in Mice

*By* SANDRA M. SIEPKA and JOSEPH S. TAKAHASHI

### Abstract

Forward genetic approaches (phenotype to gene) are powerful methods to identify mouse circadian clock components. The success of these approaches, however, is highly dependent on the quality of the phenotype—specifically, the ability to measure circadian rhythms in individual mice. This article outlines the factors necessary to measure mouse circadian rhythms, including choice of mouse strain, facilities and equipment design and construction, experimental design, high-throughput methods, and finally methods for data analysis.

### Introduction

We have undertaken a forward genetics approach to identify the components governing circadian behavior in the mouse. Benzer and colleagues used forward genetics to identify *Drosophila* genes involved in such behaviors as learning and memory (Dudai *et al.*, 1976), courtship, and circadian rhythms (Konopka and Benzer, 1971). The success of their screens depended on their abilities to design and build equipment to perform high-throughput screens for these robust and quantitative behaviors and to utilize the tools of *Drosophila* genetics.

In *Drosophila*, the screen for circadian rhythm mutants was performed on populations of flies rather than individuals. Daan and Pittendrigh (1976a–e) questioned whether these same methods used to measure circadian rhythms in populations of flies could be applied to individual rodents, designed the equipment to do so, and ultimately published a series of papers in which they describe their methods and detailed descriptions of rodent running-wheel behavior. In the late 1980s the first mammalian circadian rhythm mutant *tau* was identified (Ralph and Menaker, 1988), confirming a genetic basis for circadian behavior in mammals. By the early 1980s, methods were developed to produce highly mutagenized mice (Russell *et al.*, 1979). A highly automated assay, the identification of a genetic basis for the circadian behavior in mammals, and identification of a powerful mutagen for mice coupled with the tools of mouse genetics made a forward genetics screen for circadian rhythms mutants in mice possible. The result was identification of the first mouse circadian mutant, *Clock* (Vitaterna *et al.*, 1994). We have initiated a large-scale recessive screen to identify even more mouse circadian rhythm mutants. The goal is to screen 10,000 mutant mice per year to identify both new genes and mutant alleles of known genes that control circadian behavior. This article describes our refinement of the running wheel equipment designs and the methods we used to record circadian rhythm wheel running activity in mice.

## Strain Choice

The choice of mouse strain is the most important consideration for any mouse circadian rhythm screen and ultimately dictates the ability to identify mutants. Poor runners or variable wheel running activity will muddle the interpretation of results and increase the number of false-positive mutants. For example, BALB/cJ mice would be a poor choice of strain for a mutant screen. These mice are poor runners and have a highly variable circadian free running period (Shimomura *et al.*, 2001). Similarly, mice with mixed genetic backgrounds or outbred strains would also have a highly variable free running period and would complicate mutant identification. To minimize the number of false positives in our screen, we have elected to use inbred strains with robust and relatively invariant wheel running activity for our mutagenesis program. The two strains that we use, BTBR/J and C57BL/6J, were also chosen for their *N*-nitroso- *N*-ethylurea (ENU) mutation rate and their favorable reproductive behavior.

## Facilities and Equipment Needed for a Circadian
   Rhythm Screen

The basic equipment required to conduct a circadian activity assay are as follows: a dedicated room for the assay, one or more light-tight chambers, running wheel cages, and a computer to collect data.

Our rooms are approximately 12 × 15 ft in size. All windows are permanently sealed and the doors are fitted with gaskets and sweeps to keep the room light tight. One of our rooms has a locked anteroom to further minimize light leaks from the door into the room. A dead bolt with an interior latch is installed in the room door to prevent accidental entries while performing procedures in complete darkness. Alternatively, a darkroom revolving door can be used in place of a regular door. Darkroom doors, however, are harder to secure and are too small to allow for carts of wheel cages and mice into the room.

In addition to the regular fluorescent light fixtures in the room, we have installed ceiling light fixtures (Kodak Darkroom lamp #152 1178) fitted with light filters (Kodak #11) to illuminate the room when we use infrared night vision goggles (http://www.nightvisionweb.com).

All cables leading from the light-tight boxes to the data collection computer are housed in 3-in PVC conduit mounted 8 ft above ground to protect the wires from damage during room cleaning and sanitization.

Although we are constantly refining and improving our cabinet designs, all have the same basic features and dimensions—horizontal light-tight wooden cabinets (approximately 6 ft long, 2.5 ft high, and 2.5 ft deep) constructed out of 1/2 × 3/4-in plywood. The exterior is finished with clear varnish and the interior with flat black epoxy paint. The cabinets are stacked four high on rolling platforms and are arranged around the perimeter of the room. Each cabinet contains an interior shelf. Ten to 12 wheel cages are arranged on each shelf for a total of 20–24 cages per box. The doors to the cabinets hinge on the bottom of the door for easy access. The door for the top box of each stack, however, hinges on the top and is held open by automotive hatchback hydraulic pumps.

The box doors are light sealed with heavy-duty piano felt. Three or four equally spaced heavy-duty brass window latches are used to secure the doors closed. The doors are kept closed and latches are locked at all times to prevent the wood doors from warping due to seasonal humidity changes. All holes drilled through the cabinets to accommodate wires are sealed with black caulk.

Each cabinet door is fitted with a baffle to allow air to flow into the cabinet. Baffled air exhaust holes with fans are at the rear of the cabinet and are linked directly to the HVAC system of the room to eliminate odors and minimize particulate material in the room. The exhaust system generates a considerable amount of "white noise," which masks the noises we generate (carts moving, cages banging, human voices, etc.) when we work in the room.

We control the light cycle in each box with panels of LED lights mounted above each shelf. Previously, we used fluorescent light fixtures mounted on the rear wall of the box. We found, however, that the fluorescent fixtures generated so much heat that it was difficult to design an airflow system to control the temperature adequately inside the cabinets. We now use a panel of green LEDs (Newark Electronics Cat. No. 88C0836, Agilent Technologies, Part No. HLMP-AM01-Q0000) arranged in a two-dimensional matrix spaced every 8–10 cm for uniform illumination. Green LEDs burn cooler, use less energy, are longer lived than fluorescent lights, are brighter than white LEDs (100 lux 25 cm from the bottom of the wheel cage), and, most importantly, emit light in the proper part of the spectrum to entrain mice to a light cycle (Takahashi et al., 1984).

Each box is also equipped with light sensors, door sensors (to detect door openings), and temperature and humidity probes (VWR Cat No. 61161-378). All wires within the boxes are contained in metal conduit to prevent escaped mice from damaging the equipment.

After the boxes are assembled, we test for light leaks with photosensitive paper. In darkness, the paper is placed in each box, the boxes are closed, and the room lights are left on for 24 h. The paper is removed in darkness the next day and is developed to detect light leaks in the cabinets. We also test for light leaks by eye. An investigator sits in the dark room for 30 min to adapt to the darkness. The lights in the boxes are turned on in each light-tight box (with the doors latched) and the investigator looks for light leaks by eye.

We construct running wheel cages using polycarbonate cages (Fischer Scientific, Cat. Nos. 01-288-1B and 01-288-21) fitted with 4.75-in stainless-steel diameter running wheels (custom built for us by Lab Products, Inc., Seaford, DE). All metal hardware used in cage assembly is made of stainless steel to prevent rusting. Two holes are drilled into the side of the cage and are fitted with banana plug receptacles (Newark Electronics, Cat. No. 35F713) to mount the microswitches to the side of each wheel cage. Because this equipment must be durable enough to collect data for 1 month and to survive cage washing, we use locking nuts and lock-tight whenever possible to keep the equipment intact.

Wheel revolutions activate microswitches (Newark Electronics, Cat. No. 84F514) modified with two single banana plugs (Newark Electronics, Cat. No. 39F889) to mount the switch to the side of the wheel cage. Heavy-duty lamp wire (Newark Cat. No. 19122-250-1) is soldered to the switch and a double banana plug (Newark Electronics, Cat. No. 34F857) is attached at the end of the wire. The microswitch assembly plugs into a panel of banana plug receptacles. This panel feeds microswitch counts to the data collection computer through a National Instruments data acquisition board (Model PCI-6023e). With time, the physical switch wears down, but it can be easily removed and replaced with a new switch. All other parts in the switch assembly are reusable, and we keep a stock of assembled switches on hand to rapidly replace and repair bad switches. Detailed instructions for cage and microswitch assembly are available by contacting the authors.

Running wheel revolutions are counted using ClockLab software (Actometrics, Evanston, IL) by computers housed in a protective enclosure (http://www.itsenclosures.com). The computer both controls the lights within the light-tight box and collects data from up to three stacks of light control cabinets (or the equivalent of 256 data collection channels). The collection computers are kept on a UPS battery backup in case of power failure and are backed up nightly to an off-site server. The computer clocks are synchronized nightly. Additionally, the computer clocks are adjusted for daylight savings time, but the collection software clock does not change. Time in the collection computer, therefore, is maintained as experimental time and not real time. During daylight savings time, we simply adjust our schedules to account for the 1-h time difference. Data files are copied from the collection computers weekly for analysis.

Experimental Design and Throughput

The goal of our phenotyping facility is to assay 10,000 mice for wheel running behavior per year. Because of the high-throughput nature of the screen, we keep the experimental design fairly straightforward. Circadian behavior is measured in 200 mice each week by continuously monitoring running wheel activity for a period of 1 month. The assay is divided into two distinct data collection phases. During the first phase of the assay, mice are placed in individual running wheel cages and are maintained under the same 12:12 LD cycle to which they entrained since birth. After collecting running wheel activity data for 1 week, the mice are released into constant darkness (DD) and activity data are collected for an additional 21 days. The screen can also be altered to identify mice that have abnormal behaviors under constant light conditions or have altered responses to light pulses. To collect data for either of these conditions, however, requires extended time on the running wheel or, in the case of the light pulse screen, direct intervention of the investigator. Because our goal is to assay as many mice per year using an automated system, we have not included an LL phase or any light pulses as part of our screen.

Circadian rhythm mutants are identified as mice that display abnormal measurements in one or more of the following parameters: free running period, phase angle of entrainment to the LD cycle, period amplitude, and daytime running activity. Each of these measurements is extracted from the wheel running records (or actograms) (Fig. 1) using ClockLab software (http://www.actimetrics.com). In general, mice with circadian measurements more than 3 SD from the mean for any of these parameters are selected as putants (or putative mutants) and mated to determine trait heritability.

Because the final step to identify a mutant mouse involves a heritability test, it is important that the mice finish the assay in a timely manner. In general, both male and female mice are mature to reproduce by 5–6 weeks of age. If not mated for the first time until 12 weeks of age, however, there is a dramatic decrease in female fertility. In general, we place mice

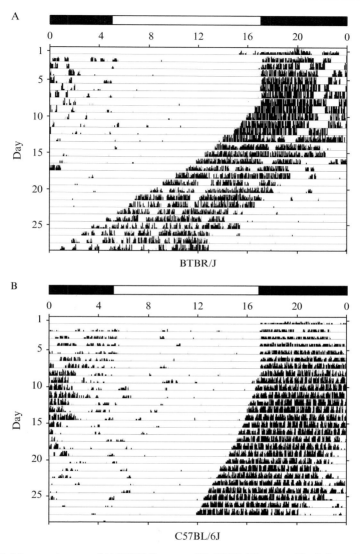

FIG. 1. Mouse actograms. (A) BTBR/J mouse. (B) C57BL/6J mouse. The light cycle for the LD portion of the experiment is represented by a bar at the top of the actogram where lights are on at 5:00 h and off at 17:00 h. Activity bouts are represented as histogram bars (the sum of wheel revolutions per discrete time interval) across a 24-h period. Each horizontal line represents one 24-h period of the experiment. Mice were maintained on a LD cycle for the first 7 days of the experiment and DD for the last 21 days. The difference in free running periods between these two mice is solely due to background strain differences.

on wheels at 6 weeks of age to not compromise running wheel activity and no later than 12 weeks of age to prevent fertility problems at the end of the assay. This age range for optimal wheel running activity and reproductive behavior can and does vary by mouse strain.

One other consideration in the experimental design concerns the way in which the mice are handled before and after the screen. We frequently assay our mice for other behavioral phenotypes, and the order in which these assays are performed can have a dramatic effect on the circadian behavior. Furthermore, the circadian assay screen can be particularly stressful for the mice, as they are singly housed for 1 month during the screen. Mice that have gone through the assay tend to be jumpy, more difficult to handle, and more aggressive if group housed after the assay. For these reasons, we cannot screen postwheel mice for any stress-related phenotypes. Conversely, we try to avoid any invasive assays before placing mice on wheels. Any treatment that alters activity or affects mobility would have a tremendous negative impact on the screen.

We start an experiment by first assembling all materials needed—cages, bedding, food, and water bottles. Each wheel is first lubricated with silicon spray and spun to detect any bent wheels. A handful of bedding is placed in each cage (too much bedding prevents the running wheels from moving freely) and the cage is placed in position in the light-tight box and connected to the appropriate microswitch. The wheel is spun again to ensure that the microswitch is activated by each wheel revolution. One to 2 h before lights out, we bring the mice into the room and place them one by one into individual wheel cages. We initiate the experiment at the end of the day (lights off) for two reasons. First, mice are nocturnal. By initiating the experiment at the end of the light cycle we do not disturb the mice in the middle of their rest period. Second, the wheel cage is novel to the mouse and novel wheel running behavior causes phase shifts in circadian behavior (Janik et al., 1994; Van Reeth and Turek, 1989). To minimize phase shifts due to placement into a novel wheel cage, we initiate the experiment at the end of the light portion of the LD cycle when the mouse begins its normal activity period. Once the mice are in the wheel cages we add water bottles and food and activate the data collection channels on the computer. The mice are monitored daily to look for problems—leaky water bottles, inactivity, high activity (lack of food or water, malocclusion), broken switches, unaligned switches, etc. During the last couple of hours of the light cycle of the LD phase of the experiment we top off the food bins in each cage. As much as possible, we try to detect and resolve any problems with the equipment during the LD phase of the assay.

Data Analysis

We analyze activity data 3.5 weeks into the assay. All activity records (or actograms) are first inspected visually to remove records that will give aberrant results in the batch analysis. In particular, we look for actograms of mice with extraordinarily low activity levels. These actograms will not yield accurate circadian period values. By visual inspection we are also able to detect abnormal patterns of activity not usually detected by analysis program—long rest period, abnormal onset patterns, etc. We extract five different measurements to assess the circadian behavior of the mice—free running period (as measured two ways), phase angle of entrainment, circadian amplitude, and average daily activity levels.

The circadian period can be extracted from the constant darkness phase of the wheel running activity records by either $\chi^2$ periodogram analysis or by linear regression analysis of activity onset. Both of these measurements can be optimized in the ClockLab analysis program to account for actograms with low activity levels and slightly abnormal activity onsets. To date, the single most important measurement to detect circadian rhythms mutants is the free running period. Both hamster *tau* and mouse *Clock* homozygotes' periods differed from those of wild type mice by almost 4 h (Ralph and Menaker, 1988; Vitaterna *et al.*, 1994).

The phase angle of entrainment reflects the difference between the onset of activity and the onset of the dark period in nocturnal animals. In addition to determining this value while the animals are exposed to the LD cycle, the value is also determined by examining the onset of daily activity after the transfer to DD to verify that the exposure to light was not masking the true onset of activity. This is accomplished by determining the linear regression of activity onset for the first 7 days in constant darkness and extrapolating this line to the last day of the LD cycle. Animals with an abnormal phase angle of entrainment could be carrying a mutation affecting a photoreceptor or some component of the input pathway to the circadian pacemaker. A mutation in the period of the circadian clock itself could also lead to a change in the phase angle of entrainment, which would be assessed directly when the animals are transferred to DD.

To assess the dominant circadian component for the activity rhythm, power spectral density of the circadian peak by fast Fourier transformation (FFT) is measured. Circadian amplitude is a measurement of the robustness of the rhythm. A low circadian amplitude could reflect an unstable circadian clock or a mouse with low activity levels. An extremely low circadian amplitude may also indicate that the animal is arrhythmic.

Average daily activity is calculated by the sum of the number of wheel revolutions divided by the total numbers of days. In LD or DD, low activity levels may indicate other physical abnormalities and serve as a general assay for physical robustness. Low activity levels may also correspond with low circadian amplitude. Because circadian measurements can be interrelated, it is important to look at all measurements in addition to examining the actogram itself to characterize the circadian phenotypes of the animals.

Putants are group housed (two per cage, single sex) to resocialize and reentrain the mice to a 12:12 LD cycle. After 1 week in a 12:12 LD cycle, the putants are mated to wild-type mice, initiating the trait heritability test as described in the previous article.

In conclusion, circadian wheel running assays using automated data acquisition systems in a highly parallel fashion (>800 channels) as described here provide an extremely quantitative and sensitive phenotype for genetic screens as well as many other circadian experiments.

## Acknowledgments

This work was supported by NIH grant #U01_MH61915. J. S. Takahashi is an Investigator in the Howard Hughes Medical Institute.

## References

Daan, S., and Pittendrigh, C. S. (1976a). A functional analysis of circadian pacemakers in nocturnal rodents. I. The stability and lability of spontaneous frequency. *J. Comp. Physiol. A* **106**, 223–252.

Daan, S., and Pittendrigh, C. S. (1976b). A functional analysis of circadian pacemakers in nocturnal rodents. II. The variability of phase response curves. *J. Comp. Physiol. A* **106**, 253–266.

Daan, S., and Pittendrigh, C. S. (1976c). A functional analysis of circadian pacemakers in nocturnal rodents. III. Heavy water and constant light: Homeostasis of frequency. *J. Comp. Physiol. A* **106**, 267–290.

Daan, S., and Pittendrigh, C. S. (1976d). A functional analysis of circadian pacemakers in nocturnal rodents. IV. Entrainment: Pacemaker as clock. *J. Comp. Physiol. A* **106**, 291–331.

Daan, S., and Pittendrigh, C. S. (1976e). A functional analysis of circadian pacemakers in nocturnal rodents. V. Pacemaker structure: A clock for all seasons. *J. Comp. Physiol. A* **106**, 333–355.

Dudai, Y., Jan, Y. N., Byers, D., Quinn, W. G., and Benzer, S. (1976). Ounce, a mutant of *Drosophila* deficient in learning. *Proc. Natl. Acad. Sci. USA* **73**, 1684–1688.

Janik, D., Godfrey, M., and Mrosovsky, N. (1994). Phase angle changes of photically entrained circadian rhythms following a single nonphotic stimulus. *Physiol. Behav.* **55**, 103–107.

Konopka, R. J., and Benzer, S. (1971). Clock mutants of *Drosophila melanogaster*. *Proc. Natl. Acad. Sci. USA* **68**, 2112–2116.

Ralph, M. R., and Menaker, M. (1988). A mutation of the circadian system in golden hamsters. *Science* **241**, 1225–1227.

Russell, W. L., Kelly, E. M., Hunsicker, P. R., Bangham, J. W., Maddux, S. C., and Phipps, E. L. (1979). Specific-locus test shows ethylnitrosourea to be the most potent mutagen in the mouse. *Proc. Natl. Acad. Sci. USA* **76**, 5818–5819.

Shimomura, K., Low-Zeddies, S. S., King, D. P., Steeves, T. D., Whiteley, A., Kushla, J., Zemenides, P. D., Lin, A., Vitaterna, M. H., Churchill, G. A., and Takahashi, J. S. (2001).

Genome-wide epistatic interaction analysis reveals complex genetic determinants of circadian behavior in mice. *Genome Res.* **11,** 959–980.

Takahashi, J. S., DeCoursey, P. J., Bauman, L., and Menaker, M. (1984). Spectral sensitivity of a novel photoreceptive system mediating entrainment of mammalian circadian rhythms. *Nature* **308,** 186–188.

Van Reeth, O., and Turek, F. W. (1989). Stimulated activity mediates phase shifts in the hamster circadian clock induced by dark pulses or benzodiazepines. *Nature* **339,** 49–51.

Vitaterna, M. H., King, D. P., Chang, A. M., Kornhauser, J. M., Lowrey, P. L., McDonald, J. D., Dove, W. F., Pinto, L. H., Turek, F. W., and Takahashi, J. S. (1994). Mutagenesis and mapping of a mouse gene, *Clock*, essential for circadian behavior. *Science* **264,** 719–725.

# [9]  Genetic Approaches to Human Behavior

*By* LOUIS J. PTÁCEK, CHRISTOPHER R. JONES, and YING-HUI FU

## Abstract

Tremendous progress in the field of human genetics has made a major impact over the last two decades into our understanding of many Mendelian disorders affecting humans. It is much more difficult to approach the genetics of complex disorders and is particularly challenging in the case of dissecting genetics of the wide range of behavioral variation in the general population. Knowledge of the biology and genetics of Mendelian traits is beginning to inform studies in complex genetics. Furthermore, the convergence of this growing knowledge with increasingly powerful new tools now put complex genetics of behavior within reach. The bottleneck in such studies and the greatest challenge to investigators is the rigorous phenotyping of human research subjects.

## Introduction

The realization that recombination mapping could allow localization and ordering of genes on a chromosome first came in the field of phage genetics. The mathematical algorithms for statistical analysis of mapping data were established over 50 years ago (Haldane, 1934; Morton, 1955). However, it was not until the latter part of the last century that application of these principles to the mapping and cloning of genes causing human diseases was proposed (Botstein *et al.*, 1980).

This effort was led by Ray White, whose group collected a majority of the 60 multigenerational CEPH families that are used worldwide for the mapping of human genetic markers. Ascertainment of four living grandparents, two parents, and a sibship of at least 10 children was required for inclusion. These families are necessary for tracing the segregation of polymorphic alleles in families and recognition of whether alleles from two markers are traveling together or separated by recombination during meiosis. It is critical to "set phase" by determining which contributions of an individual's DNA came from mother and father. While simple in concept, the collection of these families was the major effort that enabled all future mapping and cloning efforts in human genetics. Arguably, these initial gene-mapping efforts marked the beginning of the Human Genome Project.

Following this paradigm, scientists began to make progress late in the 1980s and early 1990s with the identification of genes for diseases such as Duchenne muscular dystrophy (Koenig *et al.*, 1987), neurofibromatosis type 1 (Cawthon *et al.*, 1990; Viskochil *et al.*, 1990), fragile X mental retardation (Fu *et al.*, 1991), and myotonic dystrophy (Fu *et al.*, 1992), to name a few. During the past 15 years, these reagents have led to an explosion of the molecular characterization of many Mendelian disorders. The paradigm is one of phenotyping large families segregating Mendelian disease alleles and then using genetic markers to map their location in the genome. Subsequently, positional cloning in the critical region defined by recombination events in these families can lead to identification of the disease gene.

Thus, a genetic marker that is close to a disease allele in the genome is physically linked and can only be separated in the event of rare recombinational events between the marker and the disease gene itself. Even if recombination has occurred between a marker allele and the disease mutation, linkage can still be seen in large families as one can calculate the probability of a marker allele to segregate with the disease phenotype as a function of being linked *vs* traveling with the disease allele by sheer coincidence.

Human behavior is genetically programmed to an extent. In addition, genes interact with the environment to manifest as the rich spectrum of human behaviors that we recognize in the population. Because of the complexity (presumably many genes contributing to a particular behavior and the interaction of those genes with other genes and with the environment), the genetics of behavior was a topic that could not be addressed until recently.

The first efforts at characterizing genetic aspects of behavior began in fruit flies in the late 1960s by Seymour Benzer and colleagues (Konopka and Benzer, 1971). Benzer and colleagues hypothesized that by devising simple assays of fruit fly behavior (circadian activity rhythms, learning and memory paradigms, courting behavior), genetic screens could be done to identify phenotypic variants caused by single genes. This approach was remarkably successful and continues to develop in modern behavioral genetics. The earliest mutant recognized as associated with a Mendelian behavioral phenotype was the *period* mutation (Konopka and Benzer, 1971). Additional screening identified an allelic series of mutants causing both long and short periods. These investigations culminated a number of years later in identification of the *period* gene and subsequent characterization of the genetic variants in the mutant flies (Bargiello *et al.*, 1984; Zehring *et al.*, 1984). Since that time, additional *Drosophila* circadian rhythm genes have been identified, and such studies have also expanded to include rodent circadian rhythm genes (Dunlap, 1999; King and Takahashi, 2000; Reppert, 1998; Wager-Smith and Kay, 2000).

The challenges of translating this work into the study of genetic factors contributing to human behavioral variation are onerous. It is not ethical to lock humans in a dark room and observe their activity rhythm for weeks at a time without their permission. The modern-day free running experiments in humans are extremely laborious and costly. Furthermore, because of the commitment of time and energy of such study subjects, it is difficult to recruit particular individuals of interest with any high frequency. Finally, it is not possible or desirable to do mutagenesis in humans and screening for mutant phenotypes.

However, extending these kinds of studies into humans is absolutely critical to understanding human circadian function. There is no doubt by experts in the field that many elements of circadian clocks are conserved among mammals and even vertebrates and into invertebrates. However, humans are unique and distinctly different from other organisms, including rodents. While the genomes of humans and rodents are quite homologous, the differences that allow us to function in uniquely human endeavors promise to be an exciting and productive area for research over the coming decades. Finally, because of the ability of human subjects to communicate far more extensively than nonhuman research organisms, it is possible to ask research questions in human populations that go far beyond simple behavioral paradigms used to study other organisms. Thus, it is critically important to translate current knowledge to a study of human behaviors, including circadian behavior.

Historically, it has only been possible to gain insights into genetically programmed phenotypes through the study of genetic variants in an organism that have a causal relationship with the phenotype. This has, until recently, been impossible in human circadian function as there were no recognized Mendelian variants in circadian function. This changed, however, with the description of the first Mendelian circadian rhythm disorder called familial advanced sleep phase syndrome (FASPS) (Jones *et al.*, 1999).

## Identifying Mendelian Circadian Variants

### Identification of Probands

Because genetic screens are not available in humans, one must begin by identifying individuals with phenotypes of interest. For long- and short-period circadian rhythm variants, one would predict that such individuals should have an advanced or delayed sleep phase, respectively. Because the variation in sleep- and wake-time preference is so large in the general population, selecting individuals with the most dramatic phenotypes has the highest probability of identifying a single-gene circadian trait. Then, through examination and phenotyping of first- and second-degree relatives, it is possible to establish whether the trait is segregating in a Mendelian fashion. However, one must be mindful that the wide spectrum of variation in both general population and FASPS families can lead to errors in phenotyping (Fig. 1).

### Enrollment and Sampling of Research Subjects

Research subjects can be enrolled after appropriate informed consent. A huge bureaucracy has evolved around research in humans, particularly in genetics. Rules and regulations vary from institution to institution, and further discussion of this area would be fodder for an entire *Methods in Enzymology* series. Once subjects are enrolled, DNA must be obtained from each, usually by phlebotomy. DNA is purified from leukocytes through very standard methods (Ptáček *et al.*, 1991).

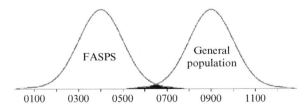

Fig. 1. Normal distributions of sleep time preferences in FASPS families and the general population. The spectrum of sleep and wake time in the general population is quite large and spans from morning larks to night owls. Within FASPS kindreds that have been identified, a large spectrum is also present, although the entire distribution is significantly shifted to earlier times. Note that the right tail of the FASPS distribution overlaps with the left tail of sleep-time preference in the general population. The challenge of the molecular characterization of FASPS is to maximize the sensitivity for this phenotype while maintaining the specificity.

*Phenotyping Subjects*

To date, there is only one Mendelian circadian rhythm trait recognized in humans. FASPS is characterized by onset in infancy, childhood, or young adult life as a tendency to sleep before 8:30 PM and to wake spontaneously before 5:30 AM (in the absence of alarm clocks, light therapy, medications or other chemicals, etc.). These individuals have a normal quality and quantity of sleep, but their sleep bout is "advanced" in the solar day relative to that of more conventional sleepers (Jones *et al.*, 1999). Inherent challenges exist because of the large variation of such traits in the general population and the masking of innate biological sleep/wake tendencies by psychosocial and familiocultural factors to which humans are subject. Thus phenotyping must focus on disentangling the biological tendency from the milieu of factors affecting human behavior.

One approach is to assess the circadian period ($\tau$) in research subjects. $\tau$ can be estimated under constant, so-called "free-running" (Kleitman and Kleitman, 1953) or forced desynchrony conditions (Dijk and Czeisler, 1991). One FASPS subject was shown to have a remarkably short $\tau$ under constant conditions (Fig. 2, Jones *et al.*, 1999). Performing such studies on all research subjects would be ideal but is not possible because not all subjects would submit to such time-consuming studies and because they are prohibitively difficult and expensive. The constant routine (Czeisler and Brown, 1985) is less time-consuming than estimating $\tau$ but still requires 1–2 full days in the laboratory. The dim light melatonin onset (DLMO) is a reliable phase marker of the suprachiasmatic nucleus and can be accomplished in one evening using saliva samples (Lewy and Sack, 1989). Like all measures of circadian phase, the DLMO can theoretically be shifted somewhat by self-imposed light–dark cycles; the magnitude of this effect in research subjects is difficult to predict. The phase of the daily core body temperature nadir can now be measured outside the laboratory by radiotelemetry using swallowed temperature transducers (Hamilos *et al.*, 1998) but seems to be a noisier signal than the DLMO.

The rest–activity rhythm can be measured conveniently and reliably in ambulatory subjects using small, wrist-worn movement detectors known as actigraphs (Ancoli-Israel *et al.*, 2003). Actigraphy correlates well with sleep recorded in the laboratory and is thought to be complemented by sleep logs. Again, many individuals are capable of behaviorally shifting their activity rhythm so as to mask or attenuate the magnitude of their "biological" circadian preference. Self-assessment instruments such as the Horne-Östberg and Munich Chronotype questionnaires have been validated as tools for assessing sleep/wake preferences in humans (Horne and Östberg, 1976; Roenneberg *et al.*, 2003). The Horne-Östberg morningness–eveningness score has been shown to correlate rather well with measured $\tau$ (Duffy *et al.*, 2001).

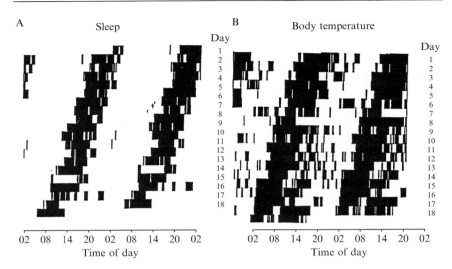

Fɪɢ. 2. FASPS subject with a short circadian period. Sleep–wake (A) and body core temperature (B) rhythms of a 69-year-old female from kindred 2174 studied in time isolation for 18 days. Data are double plotted. (A) Sleep data (filled bars) are derived from polygraphically recorded sleep, scored using standard criteria. (B) Temperature plot shows body temperature below the daily mean (filled bars). $\chi^2$ periodogram analysis showed a free-running period of 23.3 h for both variables during the 18-day recording interval [copyright permission from *Nature Med.* **5**, 1064 (1999), http://www.nature.com/.]

It is also critical to rule out other reasons for early sleep and wake times. Any cause of excessive daytime sleepiness (e.g., obstructive sleep apnea and narcolepsy) could manifest with early sleep times. Major depression causes early-morning awakening in many individuals. Thus, a general history and a depression index are critical to rule out other sleep disorders. Numerous tools are available for these purposes, and there is no clear consensus which would be of highest sensitivity and specificity for clinical circadian studies. At least for now, this must be left to the judgment of individual clinicians evaluating the subjects.

Our tendency to sleep and wake is further complicated by exogenous compounds that are commonly consumed by humans, including alcoholic beverages and caffeinated products. Furthermore, many of us make value judgments about our own or others' tendencies to sleep or wake at various times, and these may affect reporting by research subjects. Phenotyping of such subjects is extremely challenging for the reasons just mentioned and efforts are underway to improve on these assessments as discussed elsewhere in this volume (Chapter 10).

## Mapping and Cloning Human Circadian Genes

Genetic linkage mapping requires polymorphic markers that can be genotyped in DNA samples from research families being studied. This is now usually done using polymerase chain reaction amplification of highly polymorphic repeat markers. The notion that such markers could be used to map genes causing human traits is predicated on the fact that genetic loci will segregate independently if they are on different chromosomes or far apart on the same chromosome. Marker alleles close to the genetic variant causing a trait will segregate with the trait (Fig. 3). The statistical measure of whether a marker allele is segregating with a trait because it is linked (*vs* by chance) has been studied beginning with the work of Haldane (1934) and development of the LOD score method based on sequential testing procedures (Morton, 1955). This method was incorporated in a computer algorithm (LINKAGE) that continues to be used today (Lathrop *et al.*, 1985).

Once a locus has been localized, subsequent mapping with a dense array of markers allows more precise localization of recombinational events that delimit the critical region containing the gene of interest. The

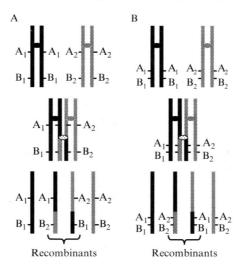

Fig. 3. Genetic linkage mapping and recombination. Crossing over occurs between homologous chromosomes in meiosis and leads to a new combination of maternal and paternal alleles on the recombinant chromosomes. (A) If a recombination (white oval) occurs between two markers (or a disease locus and a marker), then marker allele $A_1$ will be separated from marker allele (or disease allele) $B_1$ (B) If the two loci are close together, recombination between them will be much less common, and $A_1$ and $B_1$ alleles will travel together and so will $A_2$ and $B_2$ alleles.

likelihood of narrowing the critical region increases with the number and size of the families being studied. Other methods can sometimes be used to further narrow the region and include homozygosity mapping and linkage disequilibrium mapping but are beyond the scope of this article (Hastbacka et al., 1992; Lander and Botstein, 1987).

After narrowing the critical interval to the extent possible, identification of the causative gene and mutation requires identification of genetic variants. This can be done using mRNA if the gene is expressed in available tissue, but generally, mutation analysis is performed with genomic DNA; this has the added advantage that splice site mutations in introns can sometimes be seen.

Many techniques for mutation detection have been employed, including single-strand conformation polymorphism (SSCP) (Orita et al., 1989) and denaturing HPLC (O'Donovan et al., 1998). However, decreased cost and increased throughput for sequencing make this the best approach in many cases. Typically, mutation screening is first performed on exons and flanking intronic sequence of coding exons and then on noncoding exons. This technique is not sensitive to gene duplications or deletions and this caveat must be considered if mutations cannot be found. Also, mutations in introns and intragenic regions may contribute to differences in expression levels and would not be detected by sequencing of exons.

Proof that a gene is causative of a Mendelian trait requires that the variant is not found in control samples and that it segregates with the phenotype in the pedigree(s). Causation is further supported (although not proven) if the residue is highly conserved and if the variant results in a dramatic change (dramatic charge or size change, truncation of protein, etc.). The strongest genetic evidence of causation is (1) identification of multiple independent mutations in the same gene in different families with the same phenotype and/ or (2) occurrence of a de novo mutation in a sporadic case of a rare phenotype. These cases represent convergences of multiple rare events, thus decreasing the likelihood that they occurred by chance.

Ultimately, demonstration of functional consequences of the genetic variant in vitro or recapitulation of the phenotype in vivo (generation of animal models) provides very strong evidence for a causal relationship. However, differences between human and rodent physiology lead to some human disease gene mutations that do not cause phenotypes in mice.

Approaches to Complex Genetics

In contrast to genetic approaches for identifying causative genes in Mendelian disorders, characterization of complex genetic contributions to human behavior or complex diseases is not searching for the single causative mutation. Rather, genetic variations are sought that contribute to *risk* of a disorder, such as hypertension or Alzheimer's, for example. With regard to traits that are not diseases themselves, genetic variation contributes to normal variations in the population. This is the case with traits such as sleep/wake time preferences, for example. Variations at multiple loci *summate*, and the resulting phenotype is the result of contributions from multiple loci.

In the field of complex genetics, investigators typically pursue either case control or family-based tests for linkage. In family-based studies, data are generated within a family with unaffected individuals representing controls for their affected relatives. In this case, some of the potentially confounding factors are better controlled, as family members generally share more similar environmental factors, as well as sharing a more significant percentage of their DNA (depending on the relationship between such individuals). Transmission disequilibrium tests have been used to this end and examine the frequencies of alleles transmitted to affected individuals *vs* alleles that are not transmitted (Spielman *et al.*, 1993).

Case control studies investigate the association between a trait and genotypes using data generated from a cohort of unrelated patients and a population of unrelated control individuals. For example, if one allele in a candidate gene occurs at a higher rate in affected subjects versus controls, this would raise the possibility that this variant itself (or another variant in linkage disequilibrium) might contribute to that phenotypic trait. The likelihood of this association being significant can be calculated statistically. The assumption made here is that the matching between patients and controls is appropriate and that these groups do not differ in marker frequencies for reasons other than the locus being correlated with the trait.

This is a relatively new area and the specific tools for statistically analyzing such data are still evolving. There are multiple approaches to searching for associations between genetic variations and quantitative traits. No one of the tools currently available is clearly the best approach for dealing with data in such experiments.

## Challenges of Behavioral Genetics

The single largest limitation to identification of molecular determinants of behavior is the robustness of the observed phenotype. This is simpler in model organisms where it is possible, for example, to measure activity rhythms under conditions of continuous darkness. Given the broad spectrum of variation in behaviors that are considered "normal," it has been very difficult to find strong behavioral phenotypes segregating as highly penetrant Mendelian traits.

Even within a large family segregating an autosomal-dominant trait for FASPS, phenocopies (other causes of the same phenotype) can be present. The ASPS of aging, for example, is separate from FASPS but could lead one to make the diagnosis of FASPS in an aged individual who is not a carrier for an FASPS gene variant. For this reason, it is critical that the early morning awakening of FASPS must be present before the age of 40 (Jones *et al.*, 1999). There are individuals in the population who do not carry a single gene causing early bird phenotype who, because of the constellation of genes in their genome, are in the tail of distribution of sleep time preferences (Fig. 1). This turned out to be the case in a small branch of kindred 2174, the family in which FASPS was first described

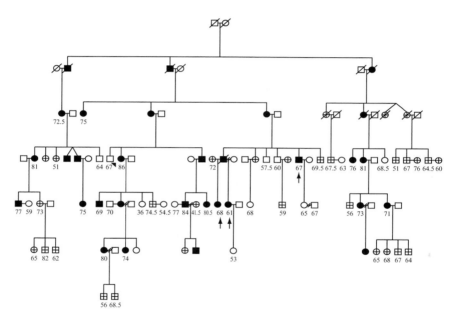

Fig. 4. FASPS kindred 2174. Circles represent women and squares represent men. Filled symbols show individuals with the FASPS phenotype. Empty symbols are those considered unaffected by FASPS, and a cross-hatched symbol marks an individual who cannot be classified as definitely affected or definitely unaffected. Horne-Östberg scores, where available, are shown underneath each individual. The mean Horne-Östberg scores were 75.1 ($n = 18$, SD $\pm$ 6.45) for individuals meeting FASPS criteria, 61.9 ($n = 15$, SD $\pm$ 9.9) for those meeting non-FASPS criteria, and 64.4 ($n = 22$, SD $\pm$ 9.1) for individuals whose phenotype was indeterminate. One branch includes three subjects considered affected in the initial genetic studies (Horne-Östberg scores of 67, 68, and 61). However, these individuals turned out not to carry the allele linked to FASPS nor the genetic variant in the *hPer2* gene. Note that a woman marrying into this branch of the family has a high Horne-Östberg score (72) and presumably carries multiple genetic factors, leading her to manifest an early bird phenotype.

(Jones *et al.*, 1999) and in whom the molecular basis of FASPS was first established (Toh *et al.*, 2001) (Fig. 4).

Thus, there are many challenges to characterizing human behavioral phenotypes. These pose significant challenges for the molecular characterization of human behavior. Still, as shown in FASPS, it is now possible to identify the genetic basis of human behavior and sharpen our focus on the critical issues of precise phenotyping required for such accomplishments.

## Acknowledgments

This work was supported by Grant HL59596 (from NIH). Dr. Ptacek is an Investigator of the Howard Hughes Medical Institute.

## References

Ancoli-Israel, S., Cole, R., Alessi, C., Chambers, M., Moorcroft, W., and Pollak, C. P. (2003). The role of actigraphy in the study of sleep and circadian rhythms. *Sleep* **26**, 342–392.

Bargiello, T. A., Jackson, F. R., and Young, M. W. (1984). Restoration of circadian behavioral rhythms by gene transfer in *Drosophila*. *Nature* **312**, 752–754.

Botstein, D., White, R. L., Skolnick, M., and Davis, R. W. (1980). Construction of a genetic linkage map in man using restriction fragment length polymorphisms. *Am. J. Hum. Genet.* **32**(3), 314–331.

Cawthon, R., Weiss, R., Xu, G., Viskochil, D., Culver, M., Stevens, J., Robertson, M., Dunn, D., Gesteland, R., O'Connell, P., and White, R. (1990). A major segment of the neurofibromatosis type 1 gene: cDNA sequence, genomic structure, and point mutations. *Cell* **62**, 193–201.

Czeisler, C., and Brown, E. N. (1985). *Sleep Res.* **14**, 295.

Dijk, D. J., and Czeisler, C. (1991). *Sleep Res.* **20A**, 531.

Duffy, J. F., Rimmer, D. W., and Czeisler, C. A. (2001). Association of intrinsic circadian period with morningness-eveningness, usual wake time, and circadian phase. *Behav. Neurosci.* **115**, 895–899.

Dunlap, J. C. (1999). Molecular bases for circadian clocks. *Cell* **96**, 271–290.

Fu, Y.-H., Kuhl, D. P., Pizzuti, A., Pieretti, M., Sutcliffe, J. S., Richards, S., Verkerk, A. J., Holden, J. J., Fenwick, R. G., Jr., and Warren, S. T. (1991). Variation of the CGG repeat at the fragile X site results in genetic instability: Resolution of the Sherman paradox. *Cell* **67**(6), 1047–1058.

Fu, Y.-H., Pizzuti, A., Fenwick, R. G., Jr., King, J., Rajnarayan, S., Dunne, P. W., Dubel, J., Nasser, G. A., Ashizawa, T., de Jong, P., Wieringa, B., Korneluk, R., Perryman, M. B., Epstein, H. F., and Caskey, C. T. (1992). An unstable triplet repeat in a gene related to myotonic muscular dystrophy. *Science* **255**(5049), 1256–1258.

Haldane, J. B. S. (1934). *Ann. Eugen. Lond.* **5**, 26.

Hamilos, D. L., Nutter, D., Gershtenson, J., Redmond, D. P., Clementi, J. D., Schmaling, K. B., Make, B. J., and Jones, J. F. (1998). Core body temperature is normal in chronic fatigue syndrome. *Biol. Psychiatry.* **43**, 293–302.

Hastbacka, J., de la Chapelle, A., Kaitila, I., Sistonen, P., Weaver, A., and Lander, E. (1992). Linkage disequilibrium mapping in isolated founder populations: Diastrophic dysplasia in Finland. *Nat. Genet.* **2**, 204–211.

Horne, J. A., and Östberg, O. (1976). A self-assessment questionnaire to determine morningness-eveningness in human circadian rhythms. *Int. J. Chronobiol.* **4**, 97–110.

Jones, C. R., Campbell, S. S., Zone, S. E., Cooper, F., DeSano, A., Murphy, P. J., Jones, B., Czajkowski, L., and Ptáček, L. J. (1999). Familial advanced sleep-phase syndrome: A short-period circadian rhythm variant in humans. *Nature Medicine* **5**, 1062–1065.

King, D. P., and Takahashi, J. S. (2000). Molecular genetics of circadian rhythms in mammals. *Annu. Rev. Neurosci.* **23**, 713–742.

Kleitman, N., and Kleitman, E. (1953). Effect of non-twenty-four-hour routines of living on oral temperature and heart rate. *J. Appl. Physiol.* **6**, 283–291.

Koenig, M., Hoffman, E. P., Bertelson, C. J., Monaco, A. P., Feener, C., and Kunkel, L. M. (1987). Complete cloning of the Duchenne muscular dystrophy (DMD) cDNA and preliminary genomic organization of the DMD gene in normal and affected individuals. *Cell* **50**(3), 509–517.

Konopka, R. J., and Benzer, S. (1971). Clock mutants of *Drosophila melanogaster*. *Proc. Natl. Acad. Sci. USA* **68**, 2112–2116.

Lander, E. S., and Botstein, D. (1987). Homozygosity mapping: A way to map human recessive traits with the DNA of inbred children. *Science* **236**, 1567–1570.

Lathrop, G. M., Lalouel, J. M., Julier, C., and Ott, J. (1985). Multilocus linkage analysis in humans: Detection of linkage and estimation of recombination. *Am. J. Hum. Genet.* **37**, 482–498.

Lewy, A. J., and Sack, R. L. (1989). The dim light melatonin onset as a marker for circadian phase position. *Chronobiol. Int.* **6**, 93–102.

Morten, N. E. (1955). Sequential tests for the detection of linkage. *Am. J. Hum. Genet.* **7**, 277–318.

O'Donovan, M. C., Oefner, P. J., Roberts, S. C., Austin, J., Hoogendoorn, B., Guy, C., Speight, G., Upadhyaya, M., Sommer, S. S., and McGuffin, P. (1998). Blind analysis of denaturing high-performance liquid chromatography as a tool for mutation detection. *Genomics* **52**, 44–49.

Orita, M., Iwahana, H., Kanazawa, H., Hayashi, K., and Sekiya, T. (1989). Detection of polymorphisms of human DNA by gel electrophoresis as single-strand conformation polymorphisms. *Proc. Natl. Acad. Sci. USA* **86**, 2766–2770.

Ptáček, L. J., Goerge, A. L., Griggs, R. C., Tawil, R., Kallen, R. G., Barchi, R. L., Robertson, M., and Leppert, M. F. (1991). Identification of a mutation in the gene causing hyperkalemic periodic paralysis. *Cell* **67**, 1021–1027.

Reppert, S. M. (1998). A clockwork explosion! *Neuron* **21**, 1–4.

Roenneberg, T., Wirz-Justice, A., and Merrow, M. (2003). Life between clocks: Daily temporal patterns of human chronotypes. *J. Biol. Rhythm.* **18**, 80–90.

Spielman, R. S., McGinnis, R. E., and Ewens, W. J. (1993). Transmission test for linkage disequilibrium: The insulin gene region and insulin-dependent diabetes mellitus (IDDM). *Am. J. Hum. Genet.* **52**, 506–516.

Toh, K. L., Jones, C. R., He, Y., Eide, E. J., Hinz, W. A., Virshup, D. M., Ptáček, L. J., and Fu, Y. H. (2001). An hPer2 phosphorylation site mutation in familial advanced sleep phase syndrome. *Science* **291**, 1040–1043.

Viskochil, D., Buchberg, A., Xu, G., Cawthon, R., Stevens, J., Wolff, R., Culver, M., Carey, J., Copeland, N., Jenkins, N., White, R., and O'Connell, P. (1990). Deletions and a translocation interrupt a cloned gene at the neurofibromatosis type 1 locus. *Cell* **62**, 187–192.

Wager-Smith, K., and Kay, S. A. (2000). Circadian rhythm genetics: From flies to mice to humans. *Nature Genet.* **26**, 23–27.

Zehring, W. A., Wheeler, D. A., Reddy, P., Konopka, R. J., Kyriacou, C. P., Rosbash, M., and Hall, J. C. (1984). P-element transformation with period locus DNA restores rhythmicity to mutant, arrhythmic *Drosophila melanogaster*. *Cell* **39**, 369–376.

## [10]   Enhanced Phenotyping of Complex Traits with a Circadian Clock Model

*By* MARTHA MERROW and TILL ROENNEBERG

### Abstract

Models of biological systems are increasingly used to generate and test predictions *in silico*. This article explores the basic workings of a multi-feedback network model of a circadian clock. In a series of *in silico* experiments, we investigated the influence of the number of feedbacks by adding and removing one or more. We further explore the possibilities of testing *in silico* models in classic "circadian" protocols. In addition, we performed an *in silico* mutagenesis screen (by altering parameters throughout the network), creating a library of mutants (based on "phenotype," not "genotype"), and subjected them to a variety of straightforward "circadian" protocols. The results of this mutant "taxonomy" are surprising. While most mutants can be identified (separated) using a limited set of experimental protocols, some resist such a separation, even when "mutations" are at vastly different locations within the complex model. Furthermore, some protocols distinguish similar alleles of the same component, which would be counterproductive. The described taxonomy invites experimental verification, *in vivo*, and may ultimately streamline genotyping of complex traits, which may have been based previously on imprecise phenotypes.

### Introduction

One approach in modeling biology is to mimic cellular processes as close to "reality" as possible, with precise kinetics of known reactions being converted into mathematical algorithms. This approach has been successful in modeling circadian properties, such as free running rhythm in *virtual* constant darkness (DD) and constant light (LL), and temperature compensation (Gonze *et al.*, 2001; Leloup *et al.*, 1997; Rensing *et al.*, 1997). As new genetic components are discovered, models are accordingly amended (Gonze *et al.*, 2001). High and low "levels of abstraction" can also be combined, effectively moving from phenotype, such as behavior, to mathematical principles. Traditional engineering uses models in this way. In the biological sciences, research on chemotaxis works back and forth, between the overt physiology, physical properties, genetics, and mathematics to understand the system as well as modules within it (Iglesias *et al.*,

2002). Description of biological phenomena with physical terms can result in identification of critical parameters that must be described at the upper level (Elowitz *et al.*, 2000; Zak *et al.*, 2003).

An alternative approach is to move up in levels of abstraction and model concepts rather than data. This top-down approach can, for example, be driven by a given phenotype: algorithms of the model are tuned to match the phenotype as closely as possible. The circadian system is ideal for this type of modeling, as a set of common features has been described in organisms from single cells to mammals. Combining high and low levels of abstraction has resulted in creative experiments involving, for example, the construction of artificial transcriptional regulatory networks in bacteria (Elowitz *et al.*, 2000). We have chosen to primarily model concepts or global features rather than modelling the specific kinetics of known components.

So what are some of the common circadian features that a clock model should mimic? This list will depend on one's concept of the clock and will influence how and to what end the model is used. Because circadian systems have evolved in a world of daily changes, systematic entrainment should be considered the single most important function of the clock (Roenneberg *et al.*, 2003a). This means that the phase of entrainment, although ultimately controlled by zeitgeber cycles, is clearly not driven, for example, by the cyclical transitions from light to dark or warm to cold. Depending on the length and/or the strength of the zeitgeber cycle or on its relative proportions between day and night (e.g., photoperiod or thermoperiod), the phase of entrainment will be different and not simply locked to one of the transitions. A good example is entrainment by light/dark cycles of asexual spore formation in *Neurospora*: onsets of this developmental process occur at around midnight in all 24-h photoperiod cycles (Tan *et al.*, 2004). Another example is hamsters that entrain earlier or later in the day depending on zeitgeber strength (Pittendrigh *et al.*, 1976). It is the phase of entrainment that is reflected in the chronotype distribution in humans (Roenneberg *et al.*, 2003c). We have much information regarding the internal phase of entrainment of hormones, such as melatonin in various conditions, but when one looks at individuals, this phase is different according to chronotype (Bailey *et al.*, 1991, 2001; Duffy *et al.*, 1999).

A second feature of circadian systems is a *ca.* 24-h free running period in constant conditions. It should be noted that the ability to free run in constant conditions is, most likely, as much a consequence as a prerequisite of how evolution has solved the "problem" of entrainment (Roenneberg *et al.*, 2002)—a triggered hourglass mechanism would not develop a free running rhythm. Other features, such as Aschoff's rule (the dependence of the free running period on the intensity of the constant light) or temperature compensation (Aschoff, 1979; Pittendrigh, 1960), can also be included in a list of "phenotypic" demands posed on a model.

Finally, when putting pen to paper, some unifying features of circadian physiology might be incorporated into the formal model. The signature of our models (Roenneberg *et al.*, 1998a, 1999, 2002) has been to designate an input feedback, acknowledging the important regulation of sensory input (especially light) pathways by the clock (Fleissner *et al.*, 1988). We call this a *Zeitnehmer* (German for "time taker," Roenneberg *et al.*, 1998b), and it represents the endogenous, clock regulation on the input system to the clock itself.

At its simplest, the original *Zeitnehmer* model looks like two coupled feedback loops—one representing the input pathway and the other the rhythm generator (Roenneberg *et al.*, 1998a) (but both are important for the circadian system to function normally). Simulations show that this model has a free running period of *ca.* 24 h in constant conditions and that it entrains systematically to different zeitgeber cycles (i.e., it has a generic phase–response curve, PRC). The state variables in both oscillators are rhythmically expressed. As constructed, and as a consequence of the interlocked oscillators, the downstream oscillator is arrhythmic in the absence of the rhythmic, clock-controlled input pathway. The components of both oscillators would, in practice, be called "clock genes" or "clock proteins."

What does the *Zeitnehmer* model imply for biological circadian systems? It presents several dilemmas: how would one distinguish between clock genes that are on the input feedback versus the rhythm generator? How is the coupled oscillator model compatible with the simple transcription/translation loop model? Facing these questions, we used traditional circadian protocols, namely T cycles, on *Neurospora* wild-type and clock mutant strains (Merrow *et al.*, 1999). Using entrainment with temperature cycles, a systematic series of phase angles shows the existence of additional *ca.* 24-h oscillatory mechanisms in arrhythmic clock mutant strains. Several clock mutant strains also display rhythmic conidiation in the circadian range, even though the rhythm lacks the precision of the wild type and some of the compensation to changes in temperature or nutrition (Lakin-Thomas *et al.*, 2000; Loros *et al.*, 1986). Furthermore, we and others found that the clock gene *frequency* (*frq*) in *Neurospora* is placed squarely on the light input pathway regulating the sporulation rhythm. When there is no FRQ, there is no light-regulated conidiation (Chang *et al.*, 1997; Lakin-Thomas *et al.*, 2000; Merrow *et al.*, 1999). This clock gene would, therefore, qualify for placement in the input oscillator. Thus, a simple eukaryotic model system experimentally fulfills hypotheses predicted by computational modeling.

In the meantime, the molecular mechanism of the circadian clock has grown increasingly complex (Young *et al.*, 2001). The mechanistic backbone of the molecular loop remains, however, transcription and translation based. Any of these central feedback reactions could be completed in a few hours (Merrow *et al.*, 1997) and, indeed, feedbacks in relatives of clock genes have been timed with very short periods (Hirata *et al.*, 2002)! Furthermore, we still lack a satisfactory explanation for the precision in circadian rhythms. We, therefore, revisited the coupled oscillator model to function as a network of several, interconnected feedback loops, thereby increasing its complexity.

The Network Model

*The Model*

The basic model (Fig. 1) is a network of five feedback loops, all of which oscillate with periods of far under 24 h when uncoupled from each other (Roenneberg *et al.*, 2002). Individual feedbacks are constructed around two state variables ($S_1$ and $S_2$), resembling the transcription/translation loop idea, although they equally could represent any other negative feedback in the metabolism of the cell (e.g., enzyme-product feedbacks). $S_1$ is regulated by production and degradation rates, by negative feedback from $S_2$, and by a coupling factor that connects it to the next feedback in the network. All feedbacks ($FB_1$ through $FB_5$) damp rapidly when not connected within the network. Although basically constructed in the same way, each of the feedbacks has different rates and

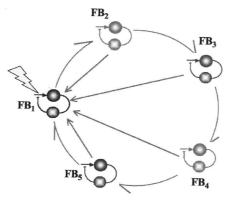

FIG. 1. The network model. Each feedback loop (FB) is a short-period, damping oscillator with two state variables. FBs are coupled in a head-to-tail fashion. In addition, each feeds back onto the input oscillator, $FB_1$, a so-called *Zeitnehmer* effect. The network is entrained through $FB_1$.

coupling factors. Only when the system is connected in a network do the oscillations of *all* components remain self-sustained with a period in the circadian range.

This model was constructed in consideration of the ubiquity of feedbacks within molecular, biochemical systems, the commonly described short periods of their (damped) oscillations, and the lack of explanation to date of the biochemical basis of the *ca.* 24-h period. It is likely that prior to a clock with properties such as entrainability by light, temperature, compensation, robustness, and precision, the metabolism of the cell already contained many such feedbacks that would need to be coordinated so that cellular biochemistry remains controllable and predictable in a cyclic world (Roenneberg *et al.*, 2002). Construction of this model showed us that the most common result of stringing together multiple feedbacks was chaotic behavior, which would be incompatible with efficient cellular function. Recently described oscillations in yeast may be a real example of this sort of precircadian clock metabolic programming (Adams *et al.*, 2003; Klevecz *et al.*, 2004). Although we generally use the model as a representation of the molecular clock, it could just as well be used to explore coupling of numerous peripheral oscillators (e.g., liver and brain) or even interacting circadian systems of ecologically interdependent organisms.

In constructing the model, incorporation of complexity, specifically using sub-24-h feedbacks, and entrainability of the system was all that was demanded. The individual feedbacks were systematically assigned increasing rates (rates 1 through 5 to 100 through 500, respectively, in steps of 100). These values govern both production and destruction of the $S_1$ component of each feedback. Tuning the model involved simply adjusting the coupling factors between the feedbacks. What continues to surprise us is that when the model is probed *in silico* with circadian protocols, it commonly "behaves" like real biological circadian systems (Roenneberg *et al.*, 2002). Originally we also explored the effect of zeitgeber and *Zeitnehmer* strength on the system and its entrainment. The *Zeitnehmer* is necessary for reliable and precise entrainment. The model shows systematic phases of entrainment with different zeitgeber strengths. The impact of coupling within the network was examined, and in this model it is the key to self-sustained rhythms. It also determines how susceptible or resistant the system is to entraining with a given zeitgeber strength (Roenneberg *et al.*, 2002).

### How Many Feedbacks?

The model is based on an assumed evolution of a circadian clock by assembling existing cellular components (feedbacks) in a network rather than creating circadian components *de novo*. The actual molecular mechanism will certainly be very different in configuration and will differ widely from organism to organism or even between different tissues in the same organism (Roenneberg *et al.*, 2003b). The head-to-tail configuration, for instance, should look more like lacework, with coupling between all or at least several feedbacks. In an attempt to understand what model features contribute which circadian properties, or which ones are critical, we explored the effect of changing the number of oscillator components within the network.

A six-feedback model was constructed by duplicating each of the existing oscillators and systematically placing it, in turn, between all feedbacks (yielding, by permutation, 25 individual experiments). The general structure was maintained, with the connections disrupted for the insertion and then repaired to incorporate the new feedback. The results were surprising, in that, except for one constellation, none of the resulting models exhibited a self-sustained rhythm with either a constant "zeitgeber" influx (*in silico* LL) or in the absence of a zeitgeber (*in silico* DD, Table I). Almost half of the new networks entrained in zeitgeber cycles. When the length of the zeitgeber phase was titrated (like changing photoperiods), some of the six-feedback models entrained in fewer cycle protocols, others in more, relative to the wild type. In all cases where entrainment was observed, the entrained phase changed with the zeitgeber cycle, indicating entrainment and not driven synchronization. The loss of the free run was thus not a complete loss of the clock. Rather, in about half of the six-oscillator models, the hallmark of the clock in nature—entrainment—is intact.

We increased the number of feedbacks in the model further, making a module of duplicated $FB_2$ and $FB_4$, coupled as though in tandem. This unit was systematically placed between all of the original five FBs. This time, all configurations showed all of the basic properties that are found in the five oscillator model (Table II). When the model was decreased to three oscillators, similarly, it behaved like a circadian system, whereas a

TABLE I

CHARACTERISTICS OF A SIX-OSCILLATOR NETWORK MODEL

| In silico protocol | Result |
| --- | --- |
| No zeitgeber ("DD") | One is self-sustained[a] |
| Constant zeitgeber ("LL") | None are self-sustained |
| Zeitgeber cycle | About half entrain[b] |
| Changing phase in various zeitgeber cycles | All that entrain |

[a] When $FB_3$ is placed between $FB_4$ and $FB_5$.
[b] When FBs are placed between $FB_3$ and $FB_4$, $FB_4$ and $FB_5$, or $FB_5$ and $FB_1$.

TABLE II
CHARACTERISTICS OF A SEVEN-OSCILLATOR NETWORK MODEL

| In silico protocol | 1, 2[a] | 2, 3[a] | 3, 4[a] | 4, 5[a] | 5, 1[a] | Wild type |
|---|---|---|---|---|---|---|
| No zeitgeber ("DD") | 22.1 | 22.1 | 28.7 | 30.6 | 24 | 18.3 |
| Constant zeitgeber ("LL") | 26.5 | 28.7 | 26.3 | 26.2 | 28.6 | 24 |
| Entrain in zeitgeber cycle (like photoperiods, results given in % "L") | 20–80 | 10–80 | 10–90 | 10–90 | 10–80 | 20–90 |

[a] Indicates that the two-oscillator module (constructed from $FB_2$ and $FB_4$) was inserted between these two FBs to make a seven-oscillator network.

four-oscillator network was like the six-feedback model, arrhythmic but entrainable (data not shown). Thus, for the specifications of this network, an odd number of feedbacks is essential for yielding the full spectrum of experimentally probed circadian properties. We may be able to exploit this observation further to understand dynamic aspects of coupling feedbacks. For instance, if interactions between the feedbacks (in the coupling factor) were manipulated, then circadian properties might be recovered. If coupling were imposed between all feedbacks (as described earlier), the odd number of feedbacks might not be critical. Furthermore, if a model with an even number of feedbacks had initially been tuned to optimally mimic a circadian system, it might have been the odd-number models that showed arhythmicity. This possibility still has to be investigated.

*Additional Properties of the Network Model:*
  In Silico *Clock Experiments*

As published, the five-feedback model shows some basic circadian properties, namely free running rhythm and systematic entrainment. We ran additional simulations to probe the model using some of the many protocols that have described specific clock properties.

Aschoff's rule [the systematic modification of frequency in constant light of increasing fluence (Aschoff, 1979)] was mimicked by running the model in constant (i.e., noncyclical) zeitgeber strengths from 0 (DD) to 50. The wild-type model typically uses a zeitgeber strength of 40 and becomes arrhythmic with a constant zeitgeber input (LL) of 50. However, at all lower fluence rates, period correlates positively with the strength of the input signal (Fig. 2A).

In its entrainment, the model is similar to biological systems—it approaches a steady state over several days rather than locking onto a given phase immediately. Furthermore, in different photoperiods, it has phase

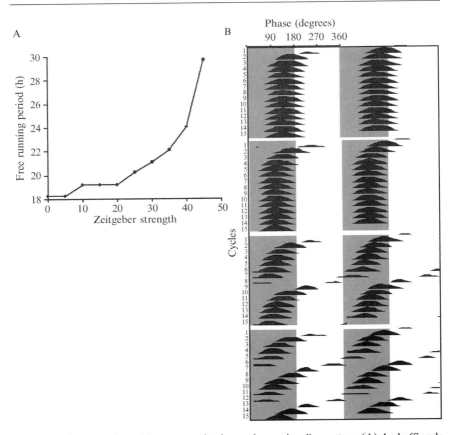

FIG. 2. The network model reacts to zeitgeber as does a circadian system. (A) Aschoff's rule was simulated by holding the zeitgeber strength constant between levels of 0 and 50. The wild-type model specifies a level of 40. (B) The system was entrained in a 50% photoperiod (comparable to a symmetrical 12:12-h LD cycle). As the rate of $FB_1$ was increased from 100 to 108 (top two panels), the entrained phase advanced. Increasing this value further, from 112 to 116 (bottom two panels), created a system that failed to stably entrain, showing relative coordination.

angles that shift relative to the zeitgeber transitions, as has been shown in mammals, flies, and fungi (Aschoff *et al.*, 1978; Pittendrigh *et al.*, 1976; Tan *et al.*, 2004). Thus alterations of the zeitgeber quantity (either strength or duration) result in changes in entrained state. By changing the rate of $FB_1$ (affecting production and destruction of state variable 1), a stable entrainment in a light:dark cycle of 12 h in each condition (LD 12:12) becomes a free run, which occasionally shows relative coordination to zeitgeber cycles (bottom 2 panels, Fig. 2B) (Wever, 1979). In the top two panels of Fig. 2B, the system entrains, showing an advancing phase relationship to the zeitge-ber with increasing rate (top panel, Fig. 2B). The bottom two panels of Fig. 2B show an increasing tendency to free run as the rate is increased further. The amount of increase in the $FB_1$ rate is less than 20%, with 100% (i.e., the value chosen in the original, "wild-type" model) in the top panel and 118% in the bottom one.

Note that the model was initially not tuned for these specifically circa-dian response characteristics. This still does not mean that the network model resembles a circadian system, but it does give us a tool with which to probe the elements of the model that confer or impact various circadian responses.

### In Silico Clock Mutants

#### Evolution of a Circadian System

Circadian systems produce distinct, quantifiable phenotypes (free run-ning periods in constant conditions or phases of entrainment) that reflect the activity of many genes (Young *et al.*, 2001). As our network model is designed, it lends itself to "genetic" and "evolutionary" descriptions that can be compared to, for example, the mammalian circadian system (Roenneberg *et al.*, 2003b). The model is based on the assumption that the intact system, with all of its properties (self-sustainment, compensation, precision, etc.), evolved through successive mutation, which would effec-tively tune the network for optimal (circadian) function (Roenneberg *et al.*, 2002). Some polymorphisms within the network will, therefore, change the properties of the network system so that it may have an altered free running rhythm and/or so that it entrains to zeitgeber cycles with a differ-ent phase relationship. Polymorphisms have been associated with in-dividually different circadian response characteristics (Ebisawa *et al.*, 2001; Ralph *et al.*, 1988; Toh *et al.*, 2001). Figure 2B demonstrates that *in silico* "polymorphisms" (by changing parameters) in our model excel-lently mimic the behavior of real circadian clocks by inducing altered phases of entrainment or even loss of entrainability. We have, therefore, used the model to further probe the consequences of mutations/polymorphisms placed in various locations within the network.

#### Making Mutants

To simulate genetic mutations, we changed the following parameters: the rates of each feedback, their coupling factor, and the strength of the negative feedback on the production of $S_1$. For each "mutation," only a single parameter in a given FB was altered. Values were changed within a range between half and double of the original ("wild type") value by running series of simulations. When no mutant phenotype emerged, the value was changed more. The *Zeitnehmer* function and zeitgeber strengths were also targeted in this *in silico* "mutagenesis." This latter would be comparable to mutations in sensitivity to light (i.e., light signal reception or transduction). It could, however, also simulate nongenetic effects such as shielding clocks from light by living predominantly inside. The resultant polymorphic population of models was "screened" in a LD cycle of 12:12 h in each condition, and the resulting time series were evaluated in CHRONO (Roenneberg *et al.*, 2000) for their phase of entrainment.

The described virtual mutant screen reveals network components that are more or less sensitive or tolerant to changing the phenotype. The input feedback loop is generally the most susceptible in terms of yielding an early or late chronotype upon alteration of its components (perhaps this is not surprising, as the screen was phase of entrainment in 12:12, and this is the input feedback). In contrast, the oscillator symmetrically furthest away from the input feedback loop (FB$_3$, see Fig. 1) proved most resistant to showing a mutant phenotype. Because the system as a whole relies on every one of the oscillators to be part of the network for rhythmicity with normal circadian properties, these observations suggest that the clock genes discovered in mutant screens to date would disproportionately represent the input feedback.

We grouped the mutants (many showed a "wild-type" phenotype) according to phase of entrainment. They included some that entrain early (larks) and some that do so late (owls). In this respect, the *in silico* mutants approximately resemble the chrono-type distribution in a human population (Roenneberg *et al.*, 2003c), an excellent example of a complex genetic trait. However, our understanding of the genetic basis of human chronotypes is limited. Only a minority of early or late chronotypes have the same genetic polymorphisms, and very few have, so far, been identified. This could reflect the relatively crude phenotyping tools that we have for humans. Until recently, a subjective questionnaire was the primary instrument in this regard (Horne *et al.*, 1976). We now prefer one that asks, although still subjectively, for bedtime and wakeup time (Roenneberg *et al.*, 2003c) for both work and free days. With better, more quantitative information, one might design simple protocols or tests to enhance phenotyping. We propose to use the model to try to uncover which of the circadian protocols (or a combination of different circadian protocols) is most powerful to separate apparently similar phenotypes, in this case similar phase of entrainment in LD 12:12.

TABLE III
*In Silico* Mutants with Delayed Phase of Entrainment

| Mutant | Description | Phenotype (phase, h after lights off) |
|---|---|---|
| Wild type | As described in text and in Roenneberg *et al.* (2002) | 14.4 |
| sen1-1 | Zeitgeber at 60; 50% increased sensitivity to light | 17.7 |
| sen1-2 | Zeitgeber at 80; 100% increased sensitivity to light | 18.5 |
| no1-1 | 40% decrease in negative feedback in $FB_1$ | 18.0 |
| no1-2 | No negative feedback in $FB_1$ | 19.8 |
| no5-1 | 40% decrease in negative feedback in $FB_5$ | 17.9 |
| rate1-1 | 50% decrease in rate constant in $FB_1$ | 18.8 |
| rate2-1 | 95% decrease in rate constant in $FB_2$ | 18.1 |
| cf3-1 | 400% increase in coupling factor between $FB_3$ and $FB_4$ | 16.0 |
| cf4-1 | 400% increase in coupling factor between $FB_4$ and $FB_5$ | 17.5 |
| zn1-1 | 60% decreased *Zeitnehmer* strength | 20.9 |
| zn2-2 | 40% decreased *Zeitnehmer* strength | 18.3 |

We now concentrate on mutants with a delayed phase of entrainment in an *in silico* LD 12:12 (Table III). The mutants are named as though they were genetic mutants. The mutants *sensitivity 1-1* and *sensitivity 1-2* (*sen*) are two alleles (i.e., altering the same parameter), in this case, zeitgeber strength. *no1-1* and *1-2* represent changes in the strength of negative feedback in $FB_1$. *no5-1* has a decreased feedback in $FB_5$. Rate mutations (*rate1-1* and *2-1*) represent rate decreases in $FB_1$ and $FB_2$, respectively. Coupling factor changes in $FB_3$ and $FB_4$ are described by *cf3-1* and *cf4-1*. Finally, the *Zeitnehmer* function was decreased to yield *zn1-1* and *zn1-2*.

## A Phenotypic Taxonomy

We then asked the question of how we could use the model in combination with circadian protocols to distinguish the mutants from one another. Specifically, which circadian protocols are capable of identifying subgroups that correspond to mutations in different components? The mutants were first run through a constant condition routine, with no zeitgeber (simulating DD, Fig. 3). Two individuals retained a wild-type free running period, two became arrhythmic, six had long periods, and one was markedly long. Thus, with this first protocol, the arrhythmic mutant can be clearly isolated from the rest. Next, LL was simulated. This separated another two mutants from the group of six that had a long period in DD. In LL, one mutant in this group was arrhythmic and

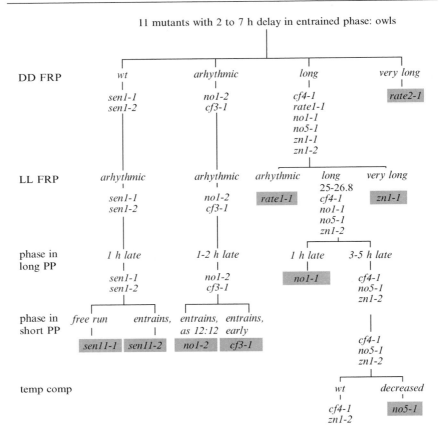

Fig. 3. A phenotypic taxonomy. Eleven mutants with a delayed phase of entrainment were submitted to a series of circadian protocols to distinguish them from one another. Except for a coupling factor and a *Zeitnehmer* mutant, all could be separated, using free running period (FRP) in DD, FRP in LL, entrained phase in a long and a short photoperiod, and temperature compensation.

another showed a very long period compared with the rest within this group. The entrained phase in long photoperiods further distinguished the negative feedback mutants from each other and from a coupling factor and a *Zeitnehmer* mutant. Interestingly, the mutants that were clustered together based on their wild-type period or arrhythmicity in the DD protocol still show no differences in their *in silico* phenotype. They did show clear differences, however, when they were examined in short-photoperiod LD cycles. This protocol also separated two "alleles," a result that would be confounding in the absence of concrete genetic information. (In other words, in this case, enhanced phenotyping would create a cluster within a locus rather than clustering the mutants together.) Finally, a temperature compensation experiment was simulated (by uniformly increasing all rates with increasing "temperature") to separate the remaining "mutants" (*no5-1*, *cf4-1*, and *znl-2*). The negative feedback mutant showed decreased compensation relative to the coupling factor and the *Zeitnehmer* mutants, which were both compensated similarly compared with the wild-type model.

## Conclusions

Our conclusion from this exercise is that enhanced phenotyping of delayed or advanced (data not shown) entrained phase can improve genotyping. Subtle clusters can be defined by exploring additional circadian properties with appropriate circadian protocols. For humans, this might include surveying them throughout the year with sleep logs or actimeters as they experience different photoperiods, a practical, nonclinical and cost-effective approach. For mice, the benefits are potentially greater, as they can indeed be submitted to most of these protocols relatively easily.

A caveat is that some mutations (i.e., *cf4-1* and *znl-2*, as described here) will be difficult, if not impossible, to distinguish phenotypically. Furthermore, in the process of making the mutant population, some mutation sites proved resistant to yielding a mutant phenotype (namely rates 3, 4, and 5). This highlights a critical role for biochemistry in following the path of oscillators, molecule by molecule, from input to core, to output, and back to input again. An additional implication is that the clock components discovered so far are predominantly involved in handling sensory input. Circadianly regulated input pathways are an integral part of the clock mechanism. This is also evident for our *in silico* clock model. The traditional distinction between input pathways and core clock mechanism is, therefore, conceptually unproductive because mutations in all relevant parts of the system are true clock mutants. The bottom line is that clock genes have been discovered for their profound effects on entrained phase or free running rhythm (Feldman *et al.*, 1971; Kondo *et al.*, 1994; Konopka *et al.*, 1971). While the input feedback has a large impact on the precise timing of phase, other loops can yield similar effects, and it is only an accurate timer in combination with the rest of the system to support it.

This exercise also addresses the question "why model?" or, more specifically, "Why model concepts without direct correlation to biochemical reactions?" As clearly shown here, the process allows one to incorporate information collected from many systems and protocols into a cohesive system. Furthermore, the exercise produces many hypotheses that are experimentally testable.

## Acknowledgments

We thank Frank Doyle and Eduardo Mendoza for helpful comments on the manuscript. Our work is supported by the Deutsche Forschungsgemeinschaft, the 5th Framework Programme of the European Union ("Brain Time"), the Dr.-Meyer-Struckmann-Stiftung, and the Eppendorf Company, Hamburg.

## References

Adams, C. A., *et al.* (2003). The Gtsl protein stabilizes the autonomous oscillator in yeast. *Yeast* **20**, 463–470.

Aschoff, J. (1979). Circadian rhythms: Influences of internal and external factors on the period measured under constant conditions. *Z. Tierpsychol.* **49**, 225–249.

Aschoff, J., *et al.* (1978). Phase relations between a circadian rhythm and its *zeitgeber* within the range of entrainment. *Naturwiss* **65**, 80–84.

Bailey, S. L., *et al.* (1991). Morningness-eveningness and early-morning salivary cortisol levels. *Biol. Psychol.* **32**(2–3), 181–192.

Bailey, S. L., *et al.* (2001). Circadian rhythmicity of cortisol and body temperature: Morningness-eveningness effects. *Chronobiol. Int.* **18**(2), 249–261.

Chang, B., *et al.* (1997). Effects of light-dark cycles on the circadian conidiation rhythm in *Neurospora crassa*. *J. Plant Res.* **110**, 449–453.

Duffy, J. F., *et al.* (1999). Relationship of endogenous circadian melatonin and temperature rhythms to self-reported preference for morning or evening activity in young and older people. *J. Invest. Med.* **47**(3), 141–150.

Ebisawa, T., *et al.* (2001). Association of structural polymorphisms in the human *period3* gene with delayed sleep phase syndrome. *EMBO Rep.* **2**, 342–346.

Elowitz, M. B., *et al.* (2000). A synthetic oscillatory network of transcriptional regulators. *Nature* **403**, 335–338.

Feldman, J. F., *et al.* (eds.) (1971). New Mutations Affecting Circadian Rhythmicity in Neurospora. National Academy of Sciences, Washington, DC.

Fleissner, G., *et al.* (1988). Efferent control of visual sensitivity in arthropod eyes: With emphasis on circadian rhythms. *In* "Information Processing in Animals" (M. Lindauer, ed.). Vol. 5, 1–67. Gustav Fischer, Stuttgart.

Gonze, D., *et al.* (2001). A model for a network of phosphorylation-dephosphorylation cycles displaying the dynamics of dominoes and clocks. *J. Theor. Biol.* **210**(2), 167–186.

Hirata, H., *et al.* (2002). Oscillatory expression of the bHLH factor Hes1 regulated by a negative feedback loop. *Science* **298**, 840–843.

Horne, J. A., *et al.* (1976). A self-assessment questionnaire to determine morningness-eveningness in human circadian rhythms. *Int. J. Chronobiol.* **4**(2), 97–110.

Iglesias, P., *et al.* (2002). Modeling the cell's guidance system, Science's STKE.

Klevecz, R. R., *et al.* (2004). A genomewide oscillation in transcription gates DNA replication and cell cycle. *Proc. Natl. Acad. Sci. USA* **101**(5), 1200–1205.

Kondo, T., *et al.* (1994). Circadian clock mutants of cyanobacteria. *Science* **266**, 1233–1236.

Konopka, R., *et al.* (1971). Clock mutants of *Drosophila melanogaster. Proc. Natl. Acad. Sci. USA* **68**, 2112–2116.

Lakin-Thomas, P. L., *et al.* (2000). Circadian rhythms in *Neurospora crassa*: Lipid deficiencies restore robust rhythmicity to null *frequency* and *white-collar* mutants. *Proc. Natl. Acad. Sci. USA* **97**, 256–261.

Leloup, J. C., *et al.* (1997). Temperature compensation of circadian rhythms: Control of the period in a model for circadian oscillations of the per protein in *Drosophila. Choronbiol. Int.* **14**, 511–520.

Loros, J. J., *et al.* (1986). Loss of temperature compensation of circadian period length in the *frq-9* mutant of *Neurospora crassa. J. Biol. Rhythms* **1**, 187–198.

Merrow, M., *et al.* (1997). Dissection of a circadian oscillation into discrete domains. *Proc. Natl. Acad. Sci. USA* **94**(8), 3877–3882.

Merrow, M., *et al.* (1999). Assignment of circadian function for the *Neurospora* clock gene *frequency. Nature* **399**, 584–586.

Pittendrigh, C. S. (1960). Circadian rhythms and the circadian organization of living systems. *Cold Spring Harb. Symp. Quant. Biol.* **25**, 159–184.

Pittendrigh, C. S., *et al.* (1976). A functional analysis of circadian pacemakers in nocturnal rodents. IV. Entrainment: Pacemaker as clock. *J. Comp. Physiol. A* **106**, 291–331.

Ralph, M. R., *et al.* (1988). A mutation of the circadian system in golden hamsters. *Science* **241**, 1225–1227.

Rensing, L., *et al.* (1997). Temperature comensation of the circadian period length: A special case among general homeostatic mechanism of gene expression. *Chronobiol. Int.* **14**(5), 481–498.

Roenneberg, T., *et al.* (1998a). Molecular circadian oscillators: An alternative hypothesis. *J. Biol. Rhythms* **13**, 167–179.

Roenneberg, T., *et al.* (1998b). Cellular mechanisms of circadian systems. *Zoology* **100**, 273–286.

Roenneberg, T., *et al.* (1999). Circadian clocks and metabolism. *J. Biol. Rhythms* **14**(6), 449–459.

Roenneberg, T., *et al.* (2000). Automated recordings of bioluminescence with special reference to the analysis of circadian rhythms. *Methods Enzymol.* **305**, 104–119.

Roenneberg, T., *et al.* (2002). Life before the clock: Modeling circadian evolution. *J. Biol. Rhythms* **17**(6), 495–505.

Roenneberg, T., *et al.* (2003a). The art of entrainment. *J. Biol. Rhythms* **18**(3), 183–194.

Roenneberg, T., *et al.* (2003b). The network of time: Understanding the molecular circadian system. *Curr. Biol.* **13**, R198–R207.

Roenneberg, T., *et al.* (2003c). Life between clocks: Daily temporal patterns of human chronotypes. *J. Biol. Rhythms* **18**(1), 80–90.

Tan, Y., *et al.* (2004). "Photoperiodism in *Neurospora crassa. J. Biol. Rhythms* **192**, 135–143.

Toh, K. L., *et al.* (2001). An *hPer2* phosphorylation site mutation in familial advanced sleep phase syndrome. *Science* **291**(5506), 1040–1043.

Wever, R. (1979). "The Circadian System of Man." Springer, Berlin.

Young, M. W., *et al.* (2001). Time zones: A comparative genetics of circadian clocks. *Nature Rev. Genet.* **2**, 702–715.

Zak, D. E., *et al.* (2003). "Continuous-time identification of gene expression models. *OMICS* **7**(4), 373–386.

# Section II

# Tracking Circadian Control of Gene Activity

# [11]   Real-Time Reporting of Circadian-Regulated Gene Expression by Luciferase Imaging in Plants and Mammalian Cells

*By* DAVID K. WELSH, TAKATO IMAIZUMI, and STEVE A. KAY

## Abstract

Luciferase enzymes have been used as reporters of circadian rhythms in organisms as diverse as cyanobacteria, plants, fruit flies, and mice. This article details methodology for real-time reporting of circadian-regulated gene expression by imaging of luciferase bioluminescence in plants and mammalian cells.

## Introduction

Luciferases are naturally occurring protein enzymes that catalyze emission of light from a substrate (luminescence). Many structurally heterogeneous forms of luciferase exist, in a wide range of species (Greer and Szalay, 2002). The best known is firefly luciferase, which catalyzes the emission of green (∼560 nm) photons from its natural substrate, firefly luciferin, in the presence of oxygen and ATP. Many luciferases have relatively short half-lives of just a few hours, or even ∼1 h in some cases (Leclerc *et al.*, 2000; Yamaguchi *et al.*, 2003), making them good reporters of changes in transcriptional rate on a circadian timescale. Genes for firefly luciferase and a few others have been cloned and are available commercially.

Luciferase luminescence is extremely dim, often too dim to be seen by the unaided eye in experimental applications. Various much brighter fluorescent reporters are available (Zhang *et al.*, 2002), and two of these have been used successfully to monitor circadian rhythms: green fluorescent protein (GFP) (Kuhlman *et al.*, 2000) and "Cameleon," a sensor of cytoplasmic $[Ca^{2+}]$ (Ikeda *et al.*, 2003). Luciferase, however, enjoys two principal advantages over the fluorescent reporters. First, luciferase does not require exogenous illumination: the light detected is emitted from the biological sample itself. Fluorescent reporters, on the other hand, require bombardment with toxic, relatively high-energy photons, so that the reporter can be excited and then re-emit lower-energy photons. Over long-term circadian experiments, phototoxicity is a serious concern with fluorescent reporters but a nonissue with luciferase. Second, background emission of light by biological samples (or by luciferase substrate alone) is generally

METHODS IN ENZYMOLOGY, VOL. 393

extremely low, whereas autofluorescence background levels can be as high as the fluorescence signal of the reporter itself (Billinton and Knight, 2001). GFP fluorescence, for example, spectrally overlaps that of riboflavin (vitamin $B_2$), an essential enzyme cofactor found in biological tissues and culture media (Zylka and Schnapp, 1996). Thus, despite its modest photic output, luciferase is well suited as an optical reporter of circadian function.

The first use of a luciferase to monitor circadian function was in *Gonyaulax*, a marine dinoflagellate exhibiting a natural circadian rhythm of bioluminescence (Hastings, 1989). Subsequently, exogenous luciferase genes, under control of promoters conferring circadian regulation, have been introduced into a wide variety of organisms, including cyanobacteria (*Synechococcus*) (Kondo *et al.*, 1993), plants (*Arabidopsis*) (Millar *et al.*, 1992), insects (*Drosophila*) (Brandes *et al.*, 1996; Plautz *et al.*, 1997), and rodents (Asai *et al.*, 2001; Wilsbacher *et al.*, 2002; Yamaguchi *et al.*, 2003; Yamazaki *et al.*, 2000; Yoo *et al.*, 2004). These transgenic organisms exhibit robust (although quite dim) circadian rhythms of bioluminescence, useful for a wide range of genetic and biochemical studies of circadian clock mechanisms. In mammals, luciferase has been used to report circadian rhythms of gene expression in peripheral tissue explants (Yamazaki *et al.*, 2000), cultured suprachiasmatic nucleus (SCN) slices (Yamaguchi *et al.*, 2003), and even *in vivo* (Yamaguchi *et al.*, 2001). This article focuses on methodologies used in our laboratory for monitoring circadian output from plant seedlings and mammalian cells.

## Plants

### Vectors

For real-time monitoring of circadian rhythms in plants, we use *Arabidopsis thaliana* seedlings transgenic for the firefly luciferase gene, under control of a promoter conferring circadian transcription. The circadian promoter elements we use routinely are derived from the "*chlorophyll a/b binding protein 2*" (*cab2*) gene (Millar *et al.*, 1992) or the "*cold circadian rhythm RNA-binding 2*" (*ccr2*) gene (Strayer *et al.*, 2000). We use a modified luciferase sequence (*luc+*), available from Promega, which is 10–100× brighter than the native luciferase when expressed in plants. Binary vectors developed in our laboratory incorporate *luc+* as well as sequences necessary for delivery and expression of the gene in plants. Our standard *cab2* and *ccr2* reporter lines were created using the respective promoter elements, inserted into a promoterless vector (pATM-DOmega). For analysis of *cis* elements in promoters, we use a vector containing a minimal nopaline synthase (*nos*) promoter (pATM-nos). The vectors, their sequences, and the transgenic plants are available upon request.

## Plant Culture and Transformation

Seeds are surface sterilized with bleach and embedded into agar plates made with Murashige-Skoog (MS) salts (Caisson Labs MSP001), 3% sucrose, pH 5.8, in 8 g/L agar. Plants are grown under sterile conditions in an incubator set at 23° and entrained for 5–10 days to an LD 12:12 lighting cycle (12 h light, 12 h dark, light is 50–60 $\mu$mol/m$^2$/s cool white fluorescent light). Plants are transformed using conventional agrobacterium-mediated methods (Clough and Bent, 1998), and transformants are selected by gentamycin or kanamycin drug resistance in MS agar plates with 75 mg/L gentamycin or 50 mg/L kanamycin.

## Equipment and Reagents

To monitor luciferase activity, we use a "VIM" intensified charge-coupled device (CCD) camera equipped with an ARGUS-50 photon-counting imaging system (Hamamatsu). For some experiments, we have also used a "Night Owl" cooled CCD camera (EG&G Berthold). (See later for a discussion of state-of-the-art CCD cameras.) Depending on the desired field of view, we use wide-angle (17- or 35-mm) or standard (50-mm) photographic lenses mounted on the camera. The light-collecting ability of a photographic lens depends on its maximal aperture and is inversely proportional to the square of its f number (f), so lenses with low f are recommended [e.g., Schneider Kreuznach 17 mm (f/0.95), Nikon Nikkor 35 mm (f/1.4) or 50 mm (f/1.2)]. A light-tight black box is used for imaging, and the camera mount should have a light-tight single or double O-ring seal. Light leaks can be fixed with black RTV silicone rubber cement (Dow Corning) or black electrical tape.

As a substrate of luciferase, we use firefly D-luciferin (potassium salt, Biosynth L-8820). Luciferin is dissolved in dH$_2$O at 100 m$M$ and stored in aliquots at −20° in the dark. A working solution (1–5 m$M$ luciferin in 0.01% Triton X-100) is prepared and filter sterilized prior to the experiment. We use a small spray bottle to apply the working solution to plants.

## Imaging

One day before starting the experiment, plants are pre-sprayed with 5 m$M$ luciferin to reduce the background activity of accumulated luciferase (which is more stable in the absence of substrate). When imaging begins,

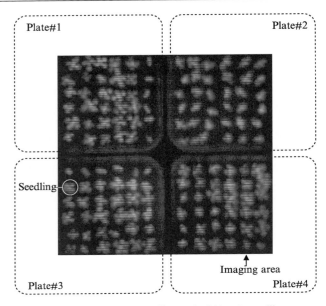

FIG. 1. Bioluminescent *Arabidopsis* seedlings. *Arabidopsis* seedlings transgenic for the bioluminescent reporter gene *cab2::luc* were grown for 2 weeks in square MS-agar plates (9.5 × 9.5 cm) and imaged using a VIM intensified CCD camera with an ARGUS-50 photon-counting imaging system (Hamamatsu). The imaging area (6 × 6 cm) includes portions of four plates and a total of 144 seedlings. This image was produced by accumulating photon counts over a 25-min exposure and then adding together 41 such exposures taken at 2.5-h intervals over a 4-day experiment.

plants are transferred to constant lighting conditions (constant white light, red light, blue light, or dark). At intervals of 2.5 h for 5 days, plants are temporarily moved to the light-tight black box for 25-min imaging sessions. Bioluminescence from up to 144 individual seedlings is captured by the camera in each 25-min exposure (Figs. 1 and 2). To ensure that luciferin substrate is not limiting, the plants are sprayed again before each time point with luciferin (5 m$M$ on the first day and 1 m$M$ thereafter).

## Mammalian Cells

Methods for monitoring circadian rhythms of bioluminescence in mammalian cultures were pioneered by Shin Yamazaki in Michael Menaker's laboratory (Yamazaki *et al.*, 2000), and we follow the methodology he describes in the next article of this volume. Shun Yamaguchi, in Hitoshi Okamura's laboratory, extended these methods to allow imaging of single

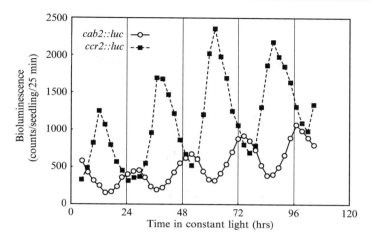

Fig. 2. Circadian rhythms of bioluminescence in *Arabidopsis* seedlings. Bioluminescence from individual *Arabidopsis* seedlings transgenic for either the *cab2::luc* or the *ccr2::luc* transgene was recorded over a 4-day experiment. Each data point represents photon counts from one seedling over a 25-min exposure. Note the higher amplitude and oppositely phased circadian rhythm in the *ccr2::luc* seedling compared with the *cab2::luc* seedling.

cells in SCN slices (Yamaguchi *et al.*, 2003) (see also Figs. 3 and 4). This section focuses on our optimization of these methods for studying dissociated cells (Welsch *et al.*, 2004; see also Figs. 5 and 6).

## Reporter Design

Particularly for single cell studies, it is important to optimize circadian regulation of the luciferase reporter to ensure highly rhythmic expression in cells of interest. Several groups have used promoter sequences from the clock gene *mPer1* to drive expression of luciferase (Asai *et al.*, 2001; Wilsbacher *et al.*, 2002; Yamazaki *et al.*, 2000) or GFP (Kuhlman *et al.*, 2000) in transgenic rats or mice. However, *mPer1* may not be the best choice of promoter, as it appears dispensable for the expression of circadian rhythms at the molecular level (Bae *et al.*, 2001; Zheng *et al.*, 2001), and its expression may be less robust than *mPer2* in some SCN cells (Hamada *et al.*, 2001). Also, while random insertion may produce transgenic animals with multiple copies of the reporter and correspondingly higher levels of expression, there is always the worry that crucial but distant enhancer elements could be omitted or that sequences adjacent to the insertion site may interfere with expression. Indeed, constructs using longer *mPer1* promoter sequences appear to have produced more robust rhythmic expression. An *mPer2* knockin strategy, in which the endogenous rhythmic

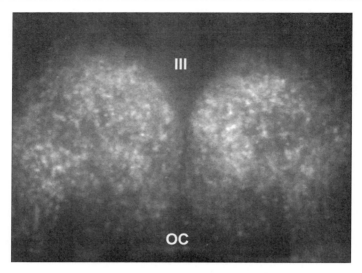

FIG. 3. Bioluminescent SCN slice from *mPer2::luc* knockin mouse. An SCN slice from a newborn *mPer2::luc* knockin mouse was cultured for 6 weeks in medium containing 1 m*M* luciferin. The slice was then imaged on an inverted Olympus IX70 microscope using a UPlanApo 10× objective and the Orca II ER cooled CCD camera (Hamamatsu), with 2 × 2 binning. To eliminate spurious events, this image was constructed by pixel-by-pixel minimization of two consecutive 55-min exposures. Note the clear demarcation of luminescence from individual cells. III, 3rd ventricle. OC, optic chiasm.

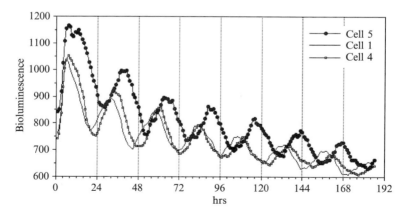

FIG. 4. Circadian rhythms of individual neurons in an SCN slice. Bioluminescence was recorded from individual neurons in an *mPer2::luc* SCN slice over an 8-day experiment. Each data point represents average luminescence intensity (in A/D units) within a single cell region of an image like that in Fig. 3. Note the clear circadian phase differences among cells.

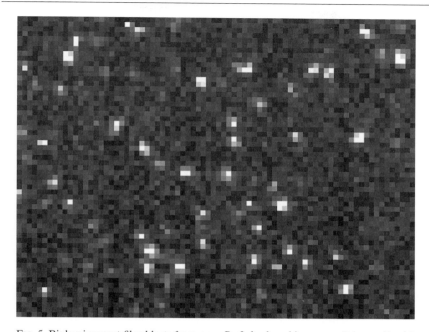

Fig. 5. Bioluminescent fibroblasts from an *mPer2::luc* knockin mouse. Primary fibroblasts were dissociated from the tail of a newborn *mPer2::luc* knockin mouse and cultured for 3 months in medium containing 1 m*M* luciferin. Cells were then imaged using a UPlanApo 4× objective and a Series 800 cooled CCD camera (Special Instruments), with 8 × 8 binning. To eliminate spurious events, this image was constructed by pixel-by-pixel minimization of two consecutive 29.9-min exposures. The bright spots are individual cells. One pixel = 26 $\mu$m.

*mPer2* gene is replaced by a fusion of *mPer2* and *luc+*, obviates these concerns by coopting as much as possible of the endogenous, evolutionarily optimized transcriptional and translational regulation of mPER2. We therefore use the *mPer2::luc* knockin mice developed by Yoo *et al.* (2004).

Further improvements of reporter design will no doubt be made in the future. For instance, it might be possible to boost expression by the addition of exogenous enhancer elements (e.g., SV40). Brighter luciferases from other species (e.g., *Renilla*) or further enhancements of firefly luciferase (beyond *luc+*) may also prove useful. Most circadian reporter studies so far have introduced the reporter gene by germline transformation to create transgenic organisms with stable, uniform expression of the transgene. However, circadian reporters can be introduced into particular tissues or cells directly using adenovirus vectors (Lai *et al.*, 2002; Le Gal La Salle *et al.*, 1993) or other methods (Ikeda *et al.*, 2003), which may be more efficient in some cases.

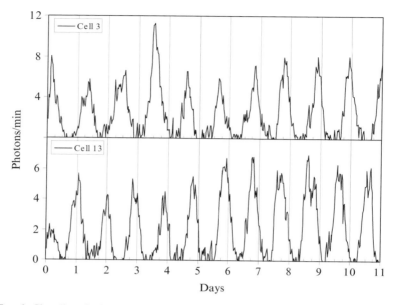

FIG. 6. Circadian rhythms of two individual fibroblasts. Bioluminescence was recorded from individual *mPer2::luc* fibroblasts over an 11-day experiment. Each data point represents total luminescence (in photons/min) within a single cell region of an image like that in Fig. 5. Values were converted from A/D units using the camera's specified gain and QE.

## Tissue Culture

We culture tissue explants or dissociated cells in HEPES-buffered, air-equilibrated Dulbecco's Modified Eagle's Medium (DMEM) (GIBCO 12100-046) supplemented with 1.2 g/L $NaHCO_3$, 10 m$M$ HEPES, 4 m$M$ gln, 25 U/ml penicillin, 25 $\mu$g/ml streptomycin, and 2% B-27 (GIBCO 17504-044). Cells are placed in a 35-mm culture dish covered by a 40-mm circular coverslip (Erie Scientific 40CIR1), which is sealed in place with vacuum grease to prevent evaporation. Brain slices are cultured on Millicell-CM membrane inserts (Fisher PICMORG50). We add 1 m$M$ luciferin (BioSynth L-8220) to the culture medium at the start of the experiment.

## Luminometry

To monitor circadian rhythms of bioluminescence from cultures, we place the 35-mm dishes in a luminometer device constructed for this purpose (LumiCycle, Actimetrics, Inc.), which fits inside a standard tissue culture incubator kept at 36°, 0% $CO_2$. The LumiCycle device is a light-tight box containing four photomultiplier tubes (PMTs; see Table II), hand-selected for low dark counts (10–40 Hz at 36°). Under the downward-gazing PMTs is a turntable with 32 slots for the 35-mm culture dishes. The turntable rotates four dishes at a time into position under the PMTs, and luminescence from each dish is measured for ~70 s at intervals of 10 min. The LumiCycle comes with analysis software that allows convenient plotting of luminescence rhythms, as well as computations of circadian period, phase, amplitude, damping rate, and magnitude of phase shifts (see later).

*Microscopy*

Single PMTs are highly sensitive detectors of photons, but of course they do not provide any spatial information; luminescence is measured from an entire dish at once. Many important questions about cellular organization of circadian clocks require longitudinal monitoring of circadian rhythms from single cells, which requires microscopic imaging.

Long-term luciferase imaging of single cells has been elusive until recently. One reason for this is the difficulty of maintaining fragile cell cultures on the microscope stage for long periods of time. We have solved this problem by use of a lucite chamber (Solent Scientific, UK) custom engineered to fit around the stage of our inverted microscope (Olympus IX70), which rests on an antivibration table (TMC). The environmental chamber keeps the stage at a constant 36°, with no focus drift. The chamber also accommodates gassing with 5% $CO_2$ for pH control of bicarbonate-buffered media, and the gas is thoroughly humidified to minimize evaporative loss. However, many cells do well in the simple configuration used for our luminometer experiments (HEPES-buffered, air-equilibrated DMEM medium, in a 35-mm culture dish sealed to prevent evaporation; see earlier discussion).

*Maximizing the Signal*

With healthy, luminescent cells on the microscope stage, the remaining challenge is to maximize the signal-to-noise ratio (S/N) so that the very dim luminescence of single cells can be detected (Christenson, 2002). This requires optimizing the chemistry of the luciferase reaction, the transmission of bioluminescence from sample to detector, and the sensitivity of the detector itself.

The bioluminescence signal should be maximized by optimizing conditions for the luciferase reaction. In the case of firefly luciferase, the substrate (luciferin) is quite hydrophilic and can be supplied in the culture medium at saturating concentrations (1–5 m$M$). The brand of luciferin may be important, as some brands reportedly contain inhibitory by-products. We use 1 m$M$ luciferin from BioSynth with good results. Temperature, [ATP], and [$O_2$] affect bioluminescence, but these are usually nearly optimal under standard tissue culture conditions.

Transmission of bioluminescence out of the culture dish should be maximized as well. Attenuation of the signal by phenol red in the culture medium may be up to 50% in a luminometer or upright microscope, where the detector is mounted above the culture dish (Erika Hawkins, Promega, personal communication) but is less significant in inverted microscopy, where the detector looks through only a thin film of medium. We use DMEM with phenol red (with GIBCO 12100-046). For inverted microscopy, the underlying culture substrate is more of a concern. Millicell membranes are surprisingly transparent when wet, however. Glass or polystyrene culture vessels are also typically quite transparent, if they are clean.

Collection of photons by the objective lens must also be maximized. It is critical to use a lens with high light gathering power (LGP), which depends on numerical aperture (NA), magnification (Mag), and the mode of illumination:

$$\text{LGP for epifluorescence} = (NA^2/Mag)^2 \times 10^4 \qquad (1)$$

$$\text{LGP for luminescence} = (NA/Mag)^2 \times 10^4 \qquad (2)$$

Numerical aperture of a lens is defined as

$$NA = n \cdot sin\theta, \qquad (3)$$

where $n$ is the refractive index of the medium between the lens and the sample and $\theta$ is the half-angle of the light cone collected by the lens. Note that LGP depends less on NA for luminescence (and for transmitted light) than it does for epifluorescence. This is because light passes through the lens twice for epifluorescence but only once for luminescence or transmitted light. Lower-magnification lenses have higher LGP because light is spread over fewer pixels at the detector; they also provide a greater field of view. Thus, a high-Mag lens with the highest available NA may be best for epifluorescence, but still not as good for luminescence as a lower Mag lens with more modest NA. See Table I for a list of suggested lenses.

The light gathered by an objective lens is still subject to reflective losses at all glass interfaces, both in the lens itself and beyond. The transmission of the lens is defined as the percentage of light gathered by the objective that is actually transmitted through it at a particular wavelength. In the past, reflective losses could be as great as 5% at each interface, and in high-quality

TABLE I
LIGHT-GATHERING POWER (LGP) OF LENSES SUGGESTED FOR LUCIFERASE IMAGING

| Manufacturer | Name | Magnification | NA | NA/magnification | LGP |
|---|---|---|---|---|---|
| Nikon | Plan Apo | 10× | 0.45 | 0.045 | 20 |
| Nikon | Plan Apo | 4× | 0.20 | 0.050 | 25 |
| Olympus | UPlanApo | 10× | 0.40 | 0.040 | 16 |
| Olympus | UPlanApo | 4× | 0.16 | 0.040 | 16 |
| Olympus | XLFLUOR[a] | 4× | 0.28 | 0.070 | 49 |
| Zeiss | FLUAR | 10× | 0.50 | 0.050 | 25 |
| Zeiss | FLUAR | 5× | 0.25 | 0.050 | 25 |

[a] Macro lens requires nonstandard mounting.

objective lenses with many elements for correction of spherical and chromatic aberration, the reflective losses could be as high as 50% overall. Thus, there was some advantage to using inexpensive lenses for luminescence applications because they had fewer lens elements and higher transmission. With recent advances in multilayer antireflection coating technology, however, this is no longer true: modern, high-quality lenses transmit more than 99.9% of the normal incidence light at visible wavelengths (see http://www.microscopyu.com/articles/optics/objectiveproperties.html). Furthermore, more expensive, highly corrected lenses tend to have higher NA (and thus higher LGP) for a given Mag.

Reflective losses beyond the objective should also be minimized. If the detector can be mounted on an inverted microscope with a bottom port (an unusual configuration), light can pass directly from the objective lens in a straight shot to the detector with no intervening mirror or other glass, which otherwise may attenuate the signal by 2–3% (R. Nazar, Olympus, personal communication). Of course, any glass surfaces that must remain in the optical path should be cleaned carefully, as dust specks or fingerprints can scatter considerable light.

*Stray Light*

It is also important to prevent extraneous light from reaching the detector. Even if stray light were reproducible in pattern and intensity so that it could be subtracted out from data images, it would still add greatly to the noise. The microscope should be located in a dark, windowless room, isolated by black curtains or a revolving light-tight door, with black walls and no phosphorescent paint or other materials. Equipment pilot lamps and LEDs should be covered by black electrical tape. Light leaks should be checked by eye after a 20-min dark adaptation. After positioning and focusing the sample using bright-field illumination, the microscope can be turned off. For our inverted microscope (Olympus IX70), we also place a small (bottomless) black lucite box over the sample so that no light enters the lens from above the stage. As it is very difficult to completely eliminate extraneous light (e.g., LEDs inside computers), the microscope itself should be draped with blackout cloth (Thorlabs BK5). Computer monitors should be turned off during data acquisition. As a routine test, one can measure the noise (SD) of an image in the absence of a sample; this will increase if there are new light leaks.

*Cameras*

Perhaps the most critical decision for low-light luminescence microscopy is the choice of a highly sensitive, low-noise detector (Christenson, 2002). The best such devices are digital cameras, known as CCD cameras (see Web sites in Table II). These cameras rely on the photoelectric effect to detect light: incident photons liberate electrons from individual spots on a silicon chip, and the electrons are channeled in a controlled fashion so as to read out the number of electrons in each of these picture elements (pixels). The proportion of incident photons actually detected is known as the quantum efficiency (QE), which depends on wavelength ($\sim$560 nm for firefly luciferase). Due to thermal energy, some electrons are generated in the absence of incident photons, which is known as dark current, the fluctuation of which is dark noise. Dark current is proportional to exposure time, and therefore dark noise tends to predominate at the long exposure times required for very dim samples. There is also some uncertainty in the readout process itself, which is known as read noise. Thus, it is crucial to choose a CCD camera with high QE at 560 nm, low dark current, and low read noise.

Noise of a CCD camera can be measured as follows. First, to estimate total camera noise ($N_{camera}$), take a long exposure with the camera shuttered and calculate the standard deviation (SD) of pixel intensity values across the image. This is the "root-mean-squared" (rms) noise. Then, to

TABLE II
WEB SITES PROVIDING PMT AND DIGITAL IMAGING TUTORIALS

http://www.olympusfluoview.com/theory/pmtintro.html
http://www.microscopyu.com/articles/digitalimaging/digitalintro.html
http://www.olympusmicro.com/primer/digitalimaging/
http://www.emccd.com/tutorial.htm

estimate read noise ($N_{read}$), take a 0-s exposure and again calculate the SD across the image. Finally, the dark noise ($N_{dark}$) can be calculated from

$$N_{camera}^2 = N_{read}^2 + N_{dark}^2 \qquad (4)$$

For comparative purposes, read noise is expressed in electrons/pixel and dark current in electrons/pixel/second. To convert A/D output intensity values to electrons, use the gain value supplied by the manufacturer. Alternatively, if the A/D range and the full well capacity (FWC) of the chip are properly matched, divide by the range of the A/D converter (e.g., 12-bit is $2^{12} = 4096$) and multiply by FWC (e.g., 18,500 electrons /pixel for the Orca II ER). Now calculate dark current (D) from

$$D = N_{dark}^2 / t \qquad (5)$$

where $t$ is the exposure duration used to measure $N_{dark}$. The most critical parameter for comparative purposes is dark current, as single-cell luminescence usually requires long exposure times (i.e., 15–60 min).

Perhaps the best currently available detector for single-cell luminescence is the cooled, back-thinned CCD camera. This device was originally developed for astronomy applications, such as detecting dim stars on a dark background, where very long exposures are practical because the target does not change quickly. In the back-thinned design, the chip is thinned to transparency so that it can be illuminated from behind, and photons do not have to pass through the electron channeling structures on the front of the chip. An antireflection coating is applied to the exposed surface of the chip. This design increases the QE to ~95% at 560 nm. At the slowest readout speed (which minimizes read noise by reducing electronic bandwidth), such chips should have read noise of only ~3 electrons.

Cooling the CCD to temperatures of −80 to −100° greatly reduces thermal dark current ($< 10^{-3}$ electrons/pixel/s). Traditionally, such low temperatures were achieved using liquid nitrogen, but long-term imaging requires daily refilling, and thermoelectric (Peltier) cooling is more practical. Convection, fans, or circulating liquid is used to draw heat from the Peltier elements. It is important that CCD cooling be independent of ambient temperature variations. With one camera, we found a small diurnal variation in dark current that correlated with ambient temperature fluctuations in the building. Warming and recooling the CCD may be advisable after bright-field imaging; this may reduce dark current by clearing residual charge. We have tested cooled, back-thinned CCD cameras made by Andor Technology, Roper Scientific, and Spectral Instruments, and all perform well for single-cell luminescence imaging, with higher QE and lower dark current than the Orca II ER (Hamamatsu) we used initially.

Another option is the photon-counting intensified CCD (iCCD) camera, such as the VIM system we use for plant imaging (see earlier discussion). These cameras were originally developed for military applications requiring rapid imaging of dimly lit moving objects. In the latest designs, a miniature array of photomultipliers, known as a microchannel plate (MCP), or intensifier, preamplifies the input signal before it reaches the CCD, which is now used mainly as a readout device. The QE of the best iCCD intensifiers ($\sim$40–50%) is much lower than that of back-thinned CCDs. However, because the CCD read noise is negligible due to preamplification, there is no penalty for taking more frequent, shorter exposures, and both the photocathode of the intensifier and the CCD itself can be cooled, resulting in very low dark current comparable to conventional cooled CCDs. An additional source of noise in these detectors arises from random fluctuations in intensifier gain, but this can be reduced using two MCPs in series so that their fluctuations tend to cancel. For dim luminescence, iCCD cameras are best operated in a "photon-counting" mode, such that (rare) individual photon events are saturating. In this mode, exposure times must be kept relatively short (e.g., 1 s) so that the chance of overlapping photon events is minimal. However, useful images may require summing many minutes of data, and this obviously presents some data-processing challenges for long-term experiments. Another disadvantage is that iCCDs, like PMTs, are vulnerable to damage if exposed to high light levels. We tested one new photon-counting iCCD camera, the XR/MEGA-10Z (Stanford Photonics), and it compares favorably with conventional cooled CCD cameras for single-cell luminescence imaging.

Several other CCD technologies are potentially suitable for low-light luminescence imaging. The electron bombardment CCD (EB-CCD) is similar to an iCCD, except that preamplification is achieved by placing the CCD inside a single large, high-voltage photomultiplier instead of downstream from an array of tiny ones (the MCP). The electron-multiplying CCD (EM-CCD) is a CCD in which electrons are multiplied on the chip itself before they are read out. This is done by passing electrons through a "gain register" portion of the chip using very high voltages so that, occasionally, a fast-moving electron causes impact ionization as it is transferred, resulting in additional mobile electrons. Thus, like the iCCD, both of these technologies preamplify the signal so that CCD read noise is negligible by comparison, and very fast frame rates are possible at low light levels. However, at the extremely low signal levels of single-cell luminescence imaging, where longer integration times are required, dark noise dominates over read noise. We are unaware of any EB-CCD or EM-CCD cameras with dark current as low as conventional cooled CCD cameras.

## Temporal and Spatial Resolution

For a given sample and a given detector, one can still improve the S/N by measuring more photons, at the expense of either temporal or spatial resolution. One can increase exposure time, sample a greater area per pixel using a lower-power objective, or combine adjacent pixels (binning). When photons are gathered over a longer time or from a wider area, the signal increases proportionally, but most types of noise do not:

$$N_{total}^2 = N_{shot}^2 + N_{dark}^2 + N_{read}^2 \qquad (6)$$

The random statistical fluctuation of the signal ("shot noise") increases only as the square root of the number of photons. Dark current increases in proportion to the signal for longer exposure times or binning, but not for lower magnification. Therefore, dark noise (which is really just shot noise for the dark current) increases only as the square root of the signal for longer exposures or binning, and does not increase at all with lower magnification. Read noise does not increase at all with integration time, decreased magnification, or "on-chip" binning (in which electrons from an array of pixels are read out collectively, as if they were from a single pixel). Fortunately, the inherent temporal dynamics of circadian rhythms are quite slow, such that 15–30 min of resolution is more than adequate, and only very coarse spatial resolution is required to discriminate one cell from another. Thus, even extremely dim circadian reporters may be useful at low temporal and spatial resolution.

## Data Analysis

### Image Correction

Postprocessing of images can correct for certain imperfections in the CCD detector. First, for each readout mode of a CCD camera (i.e., binning, readout rate), there is an associated pattern of current at zero exposure duration, which is known as a bias image. Second, for a given readout mode and nonzero exposure duration in the dark, there is a pattern of dark current (in excess of the bias current), which is known as a dark image. Third, for a given exposure duration and a uniformly illuminated field, there is a nonuniform pattern of current (arising from nonuniform QE), known as a flat field image. For the flat field image, illumination should be perfectly uniform, near half-saturating in intensity, and similar to luciferase luminescence in spectral composition. We use green luminescent liquid from commercial "light sticks" (Extreme Glow) in a 35-mm culture dish. Low-noise bias, dark, and flat field images should be created by averaging multiple exposures of the appropriate type. Then, for each data image: (1) subtract the bias image, (2) subtract a dark image (scaled by exposure time if necessary), and (3) multiply all pixels by the average intensity of the flat field image and then divide (pixelwise) by the flat field image. Any experimental image can be improved by correcting for these three types of CCD imperfections.

## Bright Spots

Bright spot artifacts are common in CCD images. Some of these are consistent from one exposure to the next and may result from light-scattering dust or scratches on glass surfaces in the optical path or from imperfections in the CCD itself ("hot pixels"). Such consistent artifacts can be removed from data images by subtracting a background image or by two-dimensional (2D) interpolation. Larger, extremely bright spots appearing sporadically in long exposures ($\sim$100/h, in random image locations) are "cosmic ray artifacts," or "spurious events." Radioactive laboratory reagents, concrete, or radon gas are possible sources of ionizing radiation producing such artifacts and should be kept away from the CCD camera. Another significant source is K-40, a naturally occurring isotope of potassium. CCD chip anomalies can also cause spurious events. But many of the spots really do result from "cosmic rays": a stream of particles of extraterrestrial origin (e.g., muons, protons), which liberate electrons when they strike the CCD chip. The imaging setup could be protected from some cosmic rays by lead shielding, but this is usually impractical. Instead, cosmic ray artifacts are generally removed by image processing: median filtering or 2D interpolation within a single image or averaging adjacent images from a time series. One sensible approach is to collect images more frequently than the desired final temporal resolution and then average each set of two to three adjacent images while excluding pixels brighter than a threshold value. An even simpler algorithm is to use the minimum value for each pixel among the two to three adjacent images (although this throws away the improved S/N gained by averaging).

## MetaMorph

Images are then analyzed using MetaMorph software (Universal Imaging Corp.). For each image, average luminescence intensity is measured within a region of interest defined manually for each plant or cell. The position of the region is adjusted if necessary to accommodate movements of cells during the experiment, but its size is kept constant across the time series. Our "Import and Analysis" (I and A) macro tool (http://www.scripps.edu/cb/kay/ianda) facilitates import of long time series into a Microsoft Excel file for further analysis.

## $\chi^2$ *Periodogram*

There are several reasonable approaches to analyzing circadian rhythm data (Refinetti, 1993). Perhaps the simplest approach is the $\chi^2$ periodogram (Sokolove and Bushell, 1978). In this method, raw data are simply folded at various candidate periods in the circadian range (i.e., averaging together points 20.0 h apart, 20.1 h apart, etc.), and the period producing the average waveform with the greatest amplitude is selected. Amplitude is defined as the variance or SD of the values in the folded waveform. For circadian phase determinations, data can be smoothed using a 2-h moving average. Circadian phase is then defined by the time of the smoothed peak or trough of the rhythm. The $\chi^2$ periodogram has the advantage of making no assumptions about circadian rhythm waveform and is therefore valid for any type of circadian rhythm data. Fortunately, however, luminescence rhythms are usually nicely approximated by damped sine or cosine curves; methods based on fitting data to such curves can provide more statistically tractable estimates of rhythm parameters.

## *FFT-NLLS*

For plant data, we use a method developed by Marty Straume at University of Virginia, known as fast Fourier transform-nonlinear least-squares analysis (FFT-NLLS) (Plautz *et al.*, 1997; Straume *et al.*, 1991). In this method, data are fitted to a series of linearly damped cosine curves with linear baseline drift:

$$L(t) = (c_0 + c_1 t) + (A_0 + A_1 t) \cdot cos(2\pi t/\tau - \phi) + ... \qquad (7)$$

where $L$ is luminescence intensity, $t$ is time, $c_0$ is luminescence at $t = 0$, $c_1$ is linear rate of change of luminescence with time, $A_0$ is amplitude at $t = 0$, $A_1$ is linear rate of change of amplitude with time, $\tau$ is circadian period, and $\phi$ is circadian phase. Additional terms from the Fourier analysis are added until there is no significant residual amplitude. In this analysis, the amplitude of the residuals relative to the primary fitted amplitude (relative amplitude, or "Rel Amp") is a useful metric of rhythm significance, varying from 0 (perfect cosine fit) to 1 (residual amplitude equals fitted amplitude). With this method, joint confidence limits can be estimated for all parameters at a criterion of 95% probability.

## *LumiCycle Analysis*

For mammalian data, we use the data analysis package accompanying the LumiCycle luminometer developed by David Ferster (Actimetrics). In this method, baseline fluctuations fit to a polynomial curve are first subtracted from raw data, and subtracted data are then fit to a single sine wave with exponentially decaying amplitude:

$$L(t) = A_0 \cdot e^{-t/k} \cdot sin(2\pi t/\tau - \phi) \tag{8}$$

where $L$ is luminescence intensity, $t$ is time, $A_0$ is amplitude at $t = 0$, k is time constant for exponential decay of amplitude, $\tau$ is circadian period, and $\phi$ is circadian phase. This method produces better results for mammalian data, which sometimes show complex baseline fluctuations and rapid damping that is modeled more accurately by exponential decay than by the linear damping assumed in FFT-NLLS.

## Acknowledgments

We thank Elisabeth Gardiner and Kathy Spencer for expert microscopy support. We thank Andrew Liu, Tom Schultz, and Gary Sims for thoughtful comments on the manuscript. Supported in part by K08 MH067657 (DKW) and MH51573 (SAK).

## References

Asai, M., Yamaguchi, S., Isejima, H., Jonouchi, M., Moriya, T., Shibata, S., Kobayashi, M., and Okamura, H. (2001). Visualization of *mPer1* transcription *in vitro. Curr. Biol.* **11,** 1524–1527.

Bae, K., Jin, X., Maywood, E. S., Hastings, M. H., Reppert, S. M., and Weaver, D. R. (2001). Differential functions of *mPer1, mPer2,* and *mPer3* in the SCN circadian clock. *Neuron* **30,** 525–536.

Billinton, N., and Knight, A. W. (2001). Seeing the wood through the trees: A review of techniques for distinguishing green fluorescent protein from endogenous autofluorescence. *Anal. Biochem.* **291,** 175–197.

Brandes, C., Plautz, J. D., Stanewsky, R., Jamison, C. F., Straume, M., Wood, K. V., Kay, S. A., and Hall, J. C. (1996). Novel features of *Drosophila period* transcription revealed by real-time luciferase reporting. *Neuron* **16,** 687–692.

Christenson, M. A. (2002). Detection systems optimized for low-light chemiluminescence imaging. *In* "Luminescence Biotechnology: Instruments and Applications" (K. Van Dyke, C. Van Dyke, and K. Woodfork, eds.), pp. 469–480. CRC Press, Boca Raton, FL.

Clough, S. J., and Bent, A. F. (1998). Floral dip: A simplified method for Agrobacterium-mediated transformation of *Arabidopsis thaliana. Plant J.* **16,** 735–743.

Greer, L. F., 3rd, and Szalay, A. A. (2002). Imaging of light emission from the expression of luciferases in living cells and organisms: A review. *Luminescence* **17,** 43–74.

Hamada, T., LeSauter, J., Venuti, J. M., and Silver, R. (2001). Expression of *Period* genes: Rhythmic and nonrhythmic compartments of the suprachiasmatic nucleus pacemaker. *J. Neurosci.* **21,** 7742–7750.

Hastings, J. W. (1989). Chemistry, clones, and circadian control of the dinoflagellate bioluminescent system. The Marlene DeLuca memorial lecture. *J. Biolumin. Chemilumin.* **4,** 12–19.

Ikeda, M., Sugiyama, T., Wallace, C. S., Gompf, H. S., Yoshioka, T., Miyawaki, A., and Allen, C. N. (2003). Circadian dynamics of cytosolic and nuclear $Ca^{2+}$ in single suprachiasmatic nucleus neurons. *Neuron* **38,** 253–263.

Kondo, T., Strayer, C. A., Kulkarni, R. D., Taylor, W., Ishiura, M., Golden, S. S., and Johnson, C. H. (1993). Circadian rhythms in prokaryotes: Luciferase as a reporter of circadian gene expression in cyanobacteria. *Proc. Natl. Acad. Sci. USA* **90,** 5672–5676.

Kuhlman, S. J., Quintero, J. E., and McMahon, D. G. (2000). GFP fluorescence reports *Period 1* circadian gene regulation in the mammalian biological clock. *Neuroreport* **11,** 1479–1482.

Lai, C. M., Lai, Y. K., and Rakoczy, P. E. (2002). Adenovirus and adeno-associated virus vectors. *DNA Cell Biol.* **21,** 895–913.

Leclerc, G. M., Boockfor, F. R., Faught, W. J., and Frawley, L. S. (2000). Development of a destabilized firefly luciferase enzyme for measurement of gene expression. *Biotechniques* **29,** 590–598.

Le Gal La Salle, G., Robert, J. J., Berrard, S., Ridoux, V., Stratford-Perricaudet, L. D., Perricaudet, M., and Mallet, J. (1993). An adenovirus vector for gene transfer into neurons and glia in the brain. *Science* **259,** 988–990.

Millar, A. J., Short, S. R., Chua, N. H., and Kay, S. A. (1992). A novel circadian phenotype based on firefly luciferase expression in transgenic plants. *Plant Cell* **4,** 1075–1087.

Plautz, J. D., Straume, M., Stanewsky, R., Jamison, C. F., Brandes, C., Dowse, H. B., Hall, J. C., and Kay, S. A. (1997). Quantitative analysis of *Drosophila period* gene transcription in living animals. *J. Biol. Rhythms* **12,** 204–217.

Refinetti, R. (1993). Laboratory instrumentation and computing: Comparison of six methods for the determination of the period of circadian rhythms. *Physiol. Behav.* **54,** 869–875.

Sokolove, P. G., and Bushell, W. N. (1978). The chi square periodogram: Its utility for analysis of circadian rhythms. *J. Theor. Biol.* **72,** 131–160.

Straume, M., Frasier-Cadoret, S. G., and Johnson, M. L. (1991). Least-squares analysis of fluorescence data. "Topics in Fluorescence Spectroscopy" (J. R. Lakowicz, ed.), Vol. 2, pp. 117–240. Plenum, New York.

Strayer, C., Oyama, T., Schultz, T. F., Raman, R., Somers, D. E., Mas, P., Panda, S., Kreps, J. A., and Kay, S. A. (2000). Cloning of the *Arabidopsis* clock gene TOC1, an autoregulatory response regulator homolog. *Science* **289,** 768–771.

Welsch, D. K., Yoo, S.-H., Liu, A. C., Takahashi, J. S., and Kay, S. A. (2004). Bioluminescence imaging of individual fibroblasts reveals persistent, independently phased circadian rhythms of clock gene expression. *Curr. Biol.* **Immediate Early Publication,** November, 2004.

Wilsbacher, L. D., Yamazaki, S., Herzog, E. D., Song, E. J., Radcliffe, L. A., Abe, M., Block, G., Spitznagel, E., Menaker, M., and Takahashi, J. S. (2002). Photic and circadian expression of luciferase in *mPeriod1-luc* transgenic mice *in vivo*. *Proc. Natl. Acad. Sci. USA* **99,** 489–494.

Yamaguchi, S., Isejima, H., Matsuo, T., Okura, R., Yagita, K., Kobayashi, M., and Okamura, H. (2003). Synchronization of cellular clocks in the suprachiasmatic nucleus. *Science* **302,** 1408–1412.

Yamaguchi, S., Kobayashi, M., Mitsui, S., Ishida, Y., van der Horst, G. T., Suzuki, M., Shibata, S., and Okamura, H. (2001). View of a mouse clock gene ticking. *Nature* **409,** 684.

Yamazaki, S., Numano, R., Abe, M., Hida, A., Takahashi, R., Ueda, M., Block, G. D., Sakaki, Y., Menaker, M., and Tei, H. (2000). Resetting central and peripheral circadian oscillators in transgenic rats. *Science* **288,** 682–685.

Yoo, S. H., Yamazaki, S., Lowrey, P. L., Shimomura, K., Ko, C. H., Buhr, E. D., Siepka, S. M., Hong, H. K., Oh, W. J., Yoo, O. J., Menaker, M., and Takahashi, J. S. (2004). PERIOD2::LUCIFERASE real-time reporting of circadian dynamics reveals persistent circadian oscillations in mouse peripheral tissues. *Proc. Natl. Acad. Sci. USA* **101,** 5339–5346.

Zhang, J., Campbell, R. E., Ting, A. Y., and Tsien, R. Y. (2002). Creating new fluorescent probes for cell biology. *Nat. Rev. Mol. Cell Biol.* **3,** 906–918.

Zheng, B., Albrecht, U., Kaasik, K., Sage, M., Lu, W., Vaishnav, S., Li, Q., Sun, Z. S., Eichele, G., Bradley, A., and Lee, C. C. (2001). Nonredundant roles of the mPer1 and mPer2 genes in the mammalian circadian clock. *Cell* **105**, 683–694.
Zylka, M. J., and Schnapp, B. J. (1996). Optimized filter set and viewing conditions for the S65T mutant of GFP in living cells. *BioTechniques* **21**, 220–226.

# [12]  Real-Time Luminescence Reporting of Circadian Gene Expression in Mammals

*By* SHIN YAMAZAKI and JOSEPH S. TAKAHASHI

## Abstract

Luminescence reporters have been used successfully in studies of circadian rhythms. Real-time measurements of circadian variations in gene expression were made in living cells, cultured tissues, and whole organisms. Because this technique is relatively easy and continuous noninvasive measurement from tissue cultures allows for a drastic reduction in the number of experimental animals, we believe this method will become a common technique for studying circadian rhythms. Using a multichannel recording apparatus, it may also become a powerful tool for the discovery of new drugs. In the past, measurements were done using hand-made apparatuses or by modifying commercially available equipment. We, along with other investigators, have developed user-friendly equipment for performing circadian rhythms experiments, and these systems are now available commercially. This article describes the use of luminescence reporters in circadian research and provides detailed methods used in these experiments. One of our goals in this article is to reduce experimental variability in different laboratories by proposing standard protocols.

## Introduction

Ever since luciferase was introduced in real-time luminescence monitoring of gene expression rhythms in plants and cyanobacteria (Kondo *et al.*, 1993; Millar *et al.*, 1992), luminescence reporter techniques have become a powerful tool used in noninvasive assays of circadian oscillations. This method has faithfully monitored the rhythms of circadian genes in the fly (Brandes *et al.*, 1996), mouse (Asai *et al.*, 2001; Geusz *et al.*, 1997; Wilsbacher *et al.*, 2002; Yoo *et al.*, 2004), rat (Yamazaki *et al.*, 2000), and fungi (Morgan *et al.*, 2003), as well as immortalized cell lines driven from the rat (Izumo *et al.*, 2003; Ueda *et al.*, 2002), zebrafish (Vallone *et al.*,

2004), and human (Maronde and Motzkus, 2003). Using this noninvasive assay, we are able to measure real-time expression of circadian and circadian output genes, as well as the protein dynamic of the circadian genes. Serial measurement of rhythms from individual tissue samples greatly reduces the intersample variability seen using conventional sampling methods and, importantly, reduces the number of tissue samples required. In animal experiments, this has the potential to drastically reduce the number of experimental animals. Circadian rhythms can be measured from whole mice (Collaco and Geusz, 2003) and from cultured tissues (Geusz et al., 1997). By detecting the first peaks in cultured tissues, we can estimate the phase of each tissue in vivo (ex vivo experiment, Stokkan et al., 2001; Yamazaki et al., 2000, 2002). Alternatively, using fiber optics implanted into the animal, rhythms can be measured from freely behaving animals (Yamaguchi et al., 2001). This article describes detailed methods for tissue culture and recording apparatuses used in circadian rhythms studies in mammals.

## Animals and Cells

Transgenic animals with a DNA construct in which the promoter of the gene is fused to the luciferease reporter gene were made for monitoring the circadian oscillation of the transcription. c-fos::luc mice, CMV::luc mice, Per1::luc rats, and Per1::luc mice have been examined in circadian rhythm studies (Asai et al., 2001; Geusz et al., 1997; Herzog et al., 2004; Wilsbacher et al., 2002; Yamazaki et al., 2000, referencing only the original publication for each animal). In Per1::luc mice, the phase of Per1 transcription rhythm in the suprachiasmatic nucleus (SCN) in vivo matched the rhythm of luciferase mRNA in vivo, as well as the luminescence rhythm from cultured SCN (Wilsbacher et al., 2002). This indicates that the luminescence reporter can be used in real-time monitoring of Per1 transcription and that the first peak in cultured SCN reflects the peak of in vivo Per1 mRNA rhythm.

The circadian rhythm of the PER2 protein can also be monitored using a luminescence reporter (Yoo et al., 2004). Using homologous recombination in embryonic stem cells, we knocked in the reporter (luciferase) into the C terminus of the Per2 gene to create a PER2::LUCIFERASE fusion protein. Using this knockin mouse line, we were able to record the PER2 protein oscillations from cultured tissues. Luminescence rhythms from cultured SCN matched the PER2 protein rhythm in the SCN in vivo, but not the Per2 mRNA rhythm (there is a time lag between the mRNA and protein rhythms). Because homozygous knockin mice (both copies of Per2 are replaced with the Per2::luc fusion gene) exhibited completely normal activity rhythms and behavioral phase shifts in response to a light pulse, we concluded that the fusion of luciferase to the PER2 protein did not disrupt the normal clock function of the PER2 protein (Yoo et al. 2004). It is known that the loss of functional PER2 protein altered these circadian parameters (Bae et al., 2001; Zheng et al., 1999).

Circadian rhythms of transcription can also be monitored from immortalized cell lines. Both stably and transiently transfected cell lines have been used successfully (Izumo et al., 2003; Maronde and Motzkus, 2003; Ueda et al., 2002). To measure circadian oscillations in cell cultures, a number of stimuli such as forskolin (10 $\mu M$, 30 min), horse serum (50%, 120 min), or other stimuli must be used to induce circadian oscillations (Balsalobre et al., 2000; Tsuchiya et al., 2003).

Tissue Culture

Using sterile technique, tissue explants can be made from adult mice or rats, aged rats (2 years old), or neonates (Asai et al., 2001; Yamaguchi et al., 2003; Yamazaki et al., 2002). Exposing animals to light at night (light pulse) or terminating light during the day (dark pulse) induces both transient and permanent changes in their circadian organization. To avoid this phase shift, we usually perform tissue preparation at just before (within 1 h) lights off. Preparing the samples immediately after lights on is also acceptable. If the animal has been kept in constant darkness, sampling should be done in the dark, using an infrared viewer without exposing the animal to visible light. Because the retina is known as the only circadian photoreceptive tissue in mammals, we anesthetize and enucleate in the dark and perform subsequent procedures in the light. The animals should be anesthetized with $CO_2$ [or anesthesia suggested by Institutional Animal Care and Use Committee (IACUC) in each institution] and euthanized by decapitation (cervical dislocation without anesthesia is also accepted by IACUC at some institutions). Due to both animal welfare and scientific reasons, stress during euthanasia should be kept to a minimum (it is known that stress can alter the phase of circadian rhythms). Brain, eyes, pineal, pituitary, liver, lung, and kidney, as well as any other tissues to be used in the experiment, should be removed quickly and be kept in chilled Hanks' buffered salt solution (HBSS, Table I). For a short period of time, they can be also kept in warm HBSS to minimize temperature-induced phase shifts.

Coronal sections of the brain (200–400 $\mu$m) should be made shortly after removal using a Vibroslicer (horizontal or sagittal slices can be used as well; a tissue chopper is not appropriate for acute recordings because it disrupts rhythms). The suprachiasmatic nucleus (master circadian

TABLE I
CONTENTS FOR HBSS (FINAL VOLUME OF 1 LITER)[a]

1. Hanks' balanced salt solution 10X (H1641, Sigma Chemical Co., St Louis, MO) 100 ml
2. Sodium bicarbonate solution (S8761, 7.5%, Sigma) 4.7 ml
3. 1 $M$ HEPES buffer (#H0887, Sigma) 10 ml
4. Penicillin–streptomycin (#15140-122, 10,000 unit/ml–10,000 μg/ml, Gibco, Invitrogen Co., Grand Island, NY) 10 ml

[a] All contents should be dissolved in autoclaved Milli-Q water; the total volume should be adjusted to 1 liter and kept at 4°.

pacemaker and phase organizer), retrochiasmatic area, arcuate nucleus, or any other brain areas can be dissected and cultured separately on Millicell culture plate inserts (PICMORG50, Millipore, Bedford, MA). The dissected brain area should be smaller than 15 mm$^2$, as the culture will not survive for a long period of time using this static organotypic method. Long-lasting high-amplitude oscillations can be measured by photon-counting devices from the paired SCN, which contains minimal volumes of the non-SCN brain area. Luminescence imaging with image analysis can also be used to measure the SCN rhythmicity from a hypothalamic slice preparation, which contains the SCN and surrounding non-SCN area. Pineal, pituitary, and retina should be cultured on Millicell inserts. Pineals need to be cut halfway through, flattened, and placed on the culture inserts. Pituitaries taken from rats should be hand sliced and reduced to a small piece (about 1–4 mm$^2$; whole pituitary can be used in mouse experiments). Small pieces of retina (about 1 mm$^2$) should be placed on Millicell inserts to obtain better rhythms.

Other peripheral tissues can be cultured without the use of culture inserts. However, when using a carousel unit, these tissues should be placed on a polypropylene mesh sheet (Spectra/Mesh, Medical Industries, Inc., Los Angeles, CA), as movement of the tissue will produce a huge noise on the baseline. Hand-sliced liver, lung, kidney, or other peripheral tissues should be dissected into small pieces (1–9 mm$^2$). Whole corneas dissected from the eyes can also be cultured.

Each tissue should be cultured in a 35-mm petri dish with culture medium (see Table II). The volume of the medium is important when using Millicell culture inserts; we obtain the best results using either 1.2 ml of culture medium in Falcon dishes or 1.0 ml of medium in Corning dishes. Peripheral tissue can be cultured with 1.0–3.0 ml of culture medium. Culture dishes need to be airtight to prevent evaporation of the culture medium. Autoclaved high-vacuum grease (Dow Corning; #14-635-5D, Fisher) should be pasted around the edge of the culture dish using a

TABLE II
CONTENTS FOR THE RECORDING MEDIUM (FINAL VOLUME OF 1 LITER)[a]

1. DMEM powder[b] (13000-021, high glucose, with L-glutamine, with pyrldoxine hydrochloride, without phenol red, without sodium pyruvate, without sodium bicarbonate, Gibco)
2. Sodium bicarbonate solution (S8761, 7.5%, Sigma) 4.7 ml or 0.35 g culture grade sodium bicarbonate powder
3. 1 $M$ HEPES buffer (#H0887, Sigma) 10 ml
4. Penicillin–streptomycin (#15140-122, 10,000 unit/ml–10,000 $\mu$g/ml, Gibco) 2.5 ml
5. B27 supplement (#17504-044, Gibco) 20 ml/or fetal bovine serum 50 ml
6. Beetle luciferin potassium salt (0.1 m$M$ final concentration, E1602, Promega, stock solution should be made with 0.1 $M$ concentration with sterilized water and a small aliquot should be kept $-80°$ with light protection)

[a] All contents (except Nos. 5 and 6) should be dissolved in autoclaved Milli-Q water, and total volume should be adjusted to 1 liter. pH should be stable at around 7 in a few days and osmolality should be around 300. Medium should be kept light protected at $4°$. Luciferin should be added, and necessary amounts of the medium should be warmed up to $37°$.
[b] We used this medium in all experiments published in 2000–2004. However, DMEM powder has been discontinued and is only available through custom orders. The DMEM powder (#13000-021 Gibco) can be replaced with DMEM power (#D-2902, with L-glutamine and 1000 mg glucose, without phenol red and sodium bicarbonate, Sigma) and 3.5 g of D-glucose powder (G7021, Sigma).

3-ml disposable syringe (Becton Dickinson and Co., Franklin Lakes, NJ) and air sealed with a 40-mm microscope glass cover (40 Circle #1, Fisher or VWR).

Recording Medium

In our experience, high-glucose, high-glutamine Dulbeccos Modified Eagles Medium works for most of the tissues. Because most of the equipment we have used (except LM-2400, Hamamatsu, see later) requires the use of airtight static cultures, the concentration of sodium bicarbonate needs to be adjusted to the optimal concentration (350 mg/L), which can be buffered with the $CO_2$ concentration of the air (0.03%). We also add 10 m$M$ of HEPES and antibiotics. B27 supplement or fetal bovine serum (5%) can be used for most cultures. SCN cultures can be maintained without a B27 supplement or fetal bovine serum. Because phenol red reduces light signal penetration, the culture media should not contain phenol red. The medium must be sterilized by filtering. Luciferin (final concentration 0.1 m$M$, beetle luciferin, potassium salt, Promega Co., Madison, WI) should be added just before the luciferase activity assay. The medium should be warmed at $37°$ before being used.

Recording Apparatuses

*Instruction for Custom-Built (Hand-Made) Apparatus*

*Choosing the Optimal Sensor for Photon Counting.* Although many types of photosensitive devices are available, photomultiplier tubes (PMTs) remain the preferred light detectors for many applications. This vacuum tube photodetector has a potential advantage for photon counting. Because luminescence from cultured SCN is extremely weak, we must choose highly sensitive and low-noise (low dark count, nonspecific noise) PMTs. We have been using two different types of photon-counting units (side-on and head-on), and both have enough sensitivity to detect signals from SCN explants. Relatively inexpensive PMTs may be used for peripheral tissue and cell cultures because the luminescence signal can be intensified by increasing the number of cells or the size of cultured tissue.

The H6240 photon-counting head (Hamamatsu, Bridgewater, NJ) consists of a side-on PMT, voltage divider, amplifier, discriminator, and high-voltage power supply, and it works by supplying 5 V without any adjustment. TTL pulses with corresponding photons (at 560 nm, about 8% of the photons reaching the PMT will convert to TTL pulses) can be recorded by personal computer via a TTL counting board (eight-channel counter board, PCI 6602 with SCB-68 connector box; National Instruments, Austin, TX). For SCN explants, we asked the manufacturer to change the standard PMT to R7518P, which is hand selected with dark counts of less than 10 counts per second (cps) at room temperature. Although we have not yet tested the R4220P (selected <10 cps), it has a similar (although slightly less sensitive) performance and can be used in the H6240 photon-counting head.

The HC135-01 (Hamamatsu) is a self-contained head-on photon-counting module with a microprocessor and requires only 5 V power. Counts can be stored on a personal computer using the RS-232C interface. For SCN explants, modifications by the manufacturer (PMT replaced with R3550 < 10 cps, reduce rescale factor to 2) should be made.

In both the H6240 and the HC135-01, there is a linear relationship between the incident number of photons and the count rate of up to 10,000,000 cps. Therefore, no data correction is required in most of luminescence measurements.

*Setting up PMTs.* To ensure stable recordings, it is better to place the PMT about 3–10 mm above the culture sample. Recoding from the bottom of the culture sample is possible; however, heat insulation between the PMT and the sample will be required because a slightly increased temperature from the bottom will produce condensation on top of the culture dish and osmolality changes in the medium will disturb the culture sample. To prevent condensation, the cover of the culture dish should be a little warmer than the culture medium. Because the PMT produces heat, placing it on top of the culture dish will create optimal conditions. Both the PMT and the culture sample need to be placed in *absolute* darkness at 36–37°. Shielding the inside of the incubator with black cardboard or building a light-tight box inside the incubator is necessary (Fig. 1). Because each PMT produces a little heat, the inside of the light-tight box will become warmer than the inside of the incubator. Therefore, when building a light-tight box inside the incubator, the incubator temperature should be set a little below the optimal temperature. A plywood box with black matte paint will be sufficient for a permanent setup. When the power of the PMT is on, exposing the PMT to visible light will immediately damage the PMT. Therefore, the power must be tuned off before loading the samples. Because the HC-135 contains a microprocessor and memory, turning off the power of the HC135 will terminate the recording. The PMT can be inactivated using a command from the personal computer, which will allow for continuous recording. Turning off the power of the H6240 will not terminate the TTL counting because the TTL pulse counting is operated by an independent circuit located within the personal computer.

*Flow-Through System.* For a continuous supply of nutrition and substrate (luciferin) or for pharmacological stimulation, a flow-through system has potential advantages. Because the tubes with the medium can act as fiber-optic conductors, both intake and drain tubes, as well as the pumps, need to be kept in the light-tight boxes. Because the slightest movement of the water surface produces a noise, the culture chamber needs to be an air-free closed system and oxygen supply into the medium is required.

## Commercially Available Circadian Luminescence Recording Apparatuses

*LumiCycle.* The LumiCycle (Actimetrics Inc., Evanston, IL) is a 32-channel carousel unit designed by our group to be used in circadian experiments. Four manufacture-modified H6240 units (see earlier) count photons from 32 cultures. This unit has no thermo-controller and needs to be placed in an incubator. Generally, the incubator temperature should be set 1° below the optimal temperature. However, because temperature uniformity of each incubator is different, temperature inside the Lumi-Cycle should be confirmed by a temperature data logger (HOBO H8 Pro, Onset, Pocasset, MA). One of the unique features of this unit is that it is designed for multiusers. Loading new samples, taking out old samples, and/or pharmacological stimulations can be done without disrupting the

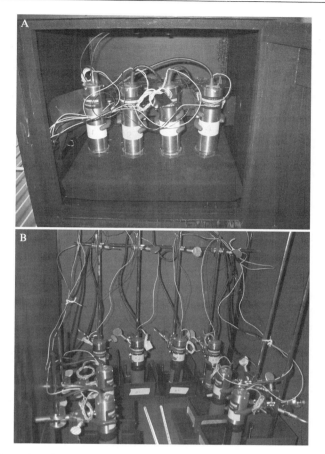

FIG. 1. Examples of custom-built PMT setup for luminescence recordings. (A) Example of the light-tight box for PMT housing. Four HC135 photon-counting modules are located inside the light-tight box. The box is kept in an environmental chamber in which the temperature is set at 35.5°. A small fan with baffles is used for circulating the air inside the light-tight box and temperature inside of the box stays at 36.5°. (B) Example of the use of a light-tight incubator. Eight HC135 photon-counting modules are located inside the incubator. Black cardboard is used for the shielding of light, and black plastic sheet is placed between the inside glass door and the metal outside door to prevent light leaks.

recordings on other channels. One personal computer can handle up to two units. Using data analysis software, circadian parameters (including phase, period, amplitude, and damping rate) can be obtained after the baseline subtraction. To perform pharmacological stimulation at a specific circadian time, the phase of the next cycle can be estimated by the software. Both low and baseline subtracted data can be exported to ClockLab software (Actimetrics Inc.) for detailed analysis. For dual reporting with two different wavelengths of luminescence reporters, optical filters can be installed on the front of two PMTs and the luminescence of 16 samples can be measured by those two PMTs.

*LM-2400.* The weak light detection unit (LM-2400, Hamamatsu) is a 24-channel carousel unit designed by Dr. Tei's group for use in circadian rhythm experiments. Two head-on PMTs (low dark counts selected R1924; similar to R3550) measure the luminescence from 24 cultures. This unit can be used in a regular $CO_2$ incubator. Therefore, $CO_2$-buffered culture medium can be used in the open, in a water-saturated atmosphere. All 24 channels of recordings need to be started and terminated at the same time. Although we have not yet tested this unit, Dr. Tei's group has been using it for *Per 1* :: *luc* measurement from cultured SCN. Temperature inside this unit is less than 1° higher than the environmental temperature.

*LM-300.* The low-light detection unit (LM-300, C8801–01, Hamamatsu) is a one-channel unit, which can be used for SCN culture samples. Because this unit is not a carousel, it is suitable for experiments requiring higher time resolution. One personal computer can handle up to five units. The incubator temperature should be set about 1–2° lower than optimal culture temperature.

*Kronos.* The Kronos (AB-2500, Atto Co., Tokyo, Japan) is an eight-channel carousel unit. Because we have yet not tested this apparatus, we describe the characteristics based on data in the manufacturer's brochure. This unit contains a temperature controller and can be used without an incubator. However, temperature accuracy of $\pm 0.5°$ (per brochure) remains a concern because luciferase activity changes due to small changes in temperature. An automated controller of the optical filters can be used for dual reporters coding different light wavelengths. One personal computer can handle up to five units.

*POLARstar.* The POLARstar Optima (BMG Labtechnologies Inc, Durham, NC) is a plate reader that can be used for circadian rhythms studies. The TopCount (a plate reader, Packard Instrument Co., Meriden, CT) has been used to measure circadian luminescence rhythms of fly and zebrafish at room temperature (Stanewsky *et al.*, 1997; Vallone *et al.*, 2004). However, because of condensation issues (it is very difficult to set the temperature inside the machine to be exactly the same as the temperature outside the machine and because the plates stay outside the machine most of the time except during luminescence reading and small temperature changes produce condensation on top of the culture sample; see earlier discussion), the TopCount is a difficult unit to use in mammalian tissue culture experiments at 36°. The POLARstar has two heaters; one is located on the top and the other at the bottom of the culture plate. Because the top heater is set a little higher than the bottom heater and the culture plate stays between the heaters at all times, no condensation is observed (we noticed a little condensation on the #1 well, home position of the plate). The POLARstar uses H6240 with fiber optics, and the signal can be recorded from either the top or the bottom of the culture sample. To increase the sensitivity (especially in SCN culture recordings) we asked the manufacturer to direct the positioning of the PMT on top of the culture plate without using fiber optics. We have successfully measured luminescence rhythms from rat-1 cells cultured in 24-well plates using this modified apparatus (Izumo *et al.*, 2003).

## Luminescence Imaging

Luminescence Imaging allows us to obtain spatial information that we are not able to obtain using the photon-counting methods described earlier. Although luminescence from luciferase reporter is extremely weak, several CCD cameras can detect the signals. The *Per1* :: *luc* oscillations from single SCN neurons have been measured (Yamaguchi *et al.*, 2003).

We briefly describe the luminescence imaging methods in this chapter. Principles and detailed methods for imaging are described in Welsh *et al.* (2005).

*Cameras.* An intensified CCD camera (ICCD or VIM, C2741-35, Hamamatsu) with a photon-counting mode has been used in living cell imaging expressing luciferase (Brandes *et al.*, 1996; Geusz *et al.*, 1997; Millar *et al.*, 1992). Because a microarray of narrow cylinder-shaped PMTs (photo intensifier) converts each photo to electrons, this camera has high sensitivity to light, but has fewer pixels compared with cooled CCDs.

A back-thinned, back-illuminated CCD camera can be used with cooling devices. Liquid nitrogen is often used for cooling (Yamaguchi *et al.*, 2003). Electrical cooling devices, such as Peltier or Cryotiger, are also used. An electron bombardment (special vacuum chamber with photocathode) CCD camera is also available with a cooling device (Water-cooled EB-CCD camera, C7190-10, Hamamatsu). Single-cell imaging from SCN explants can be captured by this camera using a microscope with high NA, long working-distance objectives (H. Tei *et al.*, personal communication).

*Optics.* The most efficient way to collect a signal from a culture sample is by direct coupling of 35-mm camera lenses to the CCD. We used two coupled 50-mm f1.2 Nikkor lenses (HLN50-12 Nikon) face to face using a connecting ring attached directly to the ICCD camera using a Nikon F/c adaptor. Although the magnification of this setup is 1 (unilateral SCN covers approximately 375 pixels), we were able to obtain an image from the SCN using a 25-min exposure setting (Fig. 2). High NA, long working-distance microscope objectives can also be used for higher-magnification images using an extension barrel attached directly to the camera (Geusz, 2001). A microscope can also possibly be used; however, a longer integration time is required.

*Recording Stages.* Although heating the camera may increase dark currents, putting the whole setup inside the incubator (or environmental chamber) is a simple way to control the temperature of the culture sample.

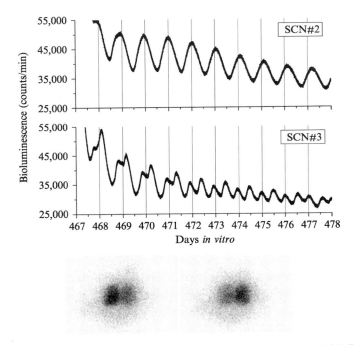

FIG. 2. Circadian rhythms of *Per1 :: luc* activity from long-term cultured rat SCN. Four SCN explants were made and maintained *in vitro* for a long period of time. Two examples of the luminescence rhythms are shown (top). Although SCN#2 showed robust unimordal rhythms all the time, the peak of the rhythms in SCN#3 started breaking down into two components at 468 days in culture and both peaks were phase locked at 180° of the phase (bimodal). At the end of the recording (478 days), SCN#3 explants were transferred to the luminescence imaging setup and images were obtained using an ICCD camera (VIM, photon counting mode with 25-min exposure) at 30-min intervals. We were able to confirm the paired SCN structure in 478-day-old SCN explants. On the first image (bottom left), the left side of the SCN showed a significantly stronger signal than the one on the right side, indicating that the peak, which occurred on the vertical lines (top) came from the left side of the SCN. The image 12 h later (bottom right) shows that the right side of the SCN is brighter than the left side.

A Sykes-Moore type of chamber can also be used to control the temperature with good optical access (Geusz, 2001).

## Data Analysis

### Phase Map

By measuring the peak time between 12 and 36 h in culture, we can estimate the *in vivo* phase of each tissue. We usually use an adjacent averaging method with 2-h running means (moving average) to smooth data. The peak is calculated as the highest point of smoothed data. By plotting peaks of each tissue, we are able to draw the "phase map" and we have been successful using this method to observe the effects of aging and *in vivo* treatments (such as shifting the light and dark cycle or restricting feeding) on circadian organization (Stokkan *et al.*, 2001; Yamazaki *et al.*, 2000, 2002).

### Circadian Parameters, Period, Amplitude, Waveform, Damping Rate

Several circadian analysis programs can be used to analyze luminescence rhythms: *Chrono* (courtesy of T. Roenneberg, University of Munich, Munich, Germany), *LumiCycle* (Actimetrics Inc.), and *ClockLab* (Actimetrics). To analyze the circadian parameters, we first need to remove baseline changes because there are often drastic changes in the baseline in the first few days in culture, and these changes are different in each tissue sample. The baseline drift can be obtained by either adjacent averaging method with 24 or 48 h (*Chrono*) or by fitting a polynomial curve (*LumiCycle*). This can be used for trend correction by subtracting it from data. Because most circadian analysis programs cannot handle negative values, adding a minimum value to detrended data is necessary. The period can be obtained from regression analysis of various circadian markers (peak, trough, half-maximum, half-minimum) or other periodgram analyses. Other circadian parameters, such as amplitude, bandwidth, and damping rate, can also be calculated by those programs.

## Conclusion

We expect that luminescence reporting with an organotypic culture will become a powerful system for circadian studies. SCN explants have been maintained for up to 681 days *in vitro*. The organization in long-term cultured SCN explants was well maintained and the oscillation of *Per1* :: *luc* activity continued (Fig. 2). The initial phase in culture can be used for estimation of the *in vivo* phase. Tissues taken from adult animals or even from aged animal can be cultured. Therefore, tissue cultures shortly after *in vivo* treatments can be used to measure the effect of *in vivo* treatments on circadian organization. Immortalized cells (and primary cell cultures) with a transfected reporter construct can be used for drug screening and molecular analysis. Previously, transfection assays were done only to monitor acute activation. This can now be extended to measure circadian parameters, such as changes in period, phase, and amplitude of circadian oscillations.

## Acknowledgment

We thank Susan J. McMahon for reading of the manuscript.

## References

Asai, M., Yamaguchi, S., Isejima, H., Jonouchi, M., Moriya, T., Shibata, S., Kobayashi, M., and Okamura, H. (2001). Visualization of mPer1 transcription *in vitro*: NMDA induces a rapid phase shift of mPer1 gene in cultured SCN. *Curr. Biol.* **11**, 1524–1527.

Bae, K., Jin, X., Maywood, E. S., Hastings, M. H., Reppert, S. M., and Weaver, D. R. (2001). Differential functions of mPer1, mPer2, and mPer3 in the SCN circadian clock. *Neuron* **30**, 525–536.

Balsalobre, A., Marcacci, L., and Schibler, U. (2000). Multiple signaling pathways elicit circadian gene expression in cultured Rat-1 fibroblasts. *Curr. Biol.* **10**, 1291–1294.

Brandes, C., Plautz, J. D., Stanewsky, R., Jamison, C. F., Straume, M., Wood, K. V., Kay, S. A., and Hall, J. C. (1996). Novel features of drosophila period Transcription revealed by real-time luciferase reporting. *Neuron* **16**, 687–692.

Collaco, A. M., and Geusz, M. E. (2003). Monitoring immediate-early gene expression through firefly luciferase imaging of HRS/J hairless mice. *BMC Physiol.* **19**, 8.

Geusz, M. E. (2001). Bioluminescence imaging of gene expression in living cells and tissues. *In* "Methods in Cellar Imaging," pp. 395–408. Oxford University Press, Oxford.

Geusz, M. E., Fletcher, C., Block, G. D., Straume, M., Copeland, N. G., Jenkins, N. A., Kay, S. A., and Day, R. N. (1997). Long-term monitoring of circadian rhythms in c-fos gene expression from suprachiasmatic nucleus cultures. *Curr. Biol.* **7**, 758–766.

Izumo, M., Johnson, C. H., and Yamazaki, S. (2003). Circadian gene expression in mammalian fibroblasts revealed by real-time luminescence reporting: Temperature compensation and damping. *Proc. Natl. Acad. Sci. USA* **100**, 16089–16094.

Herzog, E. D., Aton, S. J., Numano, R., Sakaki, Y., and Tei, H. (2004). Temporal precision in the mammalian circadian system: A reliable clock from less reliable neurons. *J. Biol. Rhythms* **19**, 35–46.

Kondo, T., Strayer, C. A., Kulkarni, R. D., Taylor, W., Ishiura, M., Golden, S. S., and Johnson, C. H. (1993). Circadian rhythms in prokaryotes: Luciferase as a reporter of circadian gene expression in cyanobacteria. *Proc. Natl. Acad. Sci. USA* **90**, 5672–5676.

Maronde, E., and Motzkus, D. (2003). Oscillation of human period 1 (hPER1) reporter gene activity in human neuroblastoma cells *in vivo*. *Chronobiol. Int.* **20**, 671–681.

Millar, A. J., Short, S. R., Chua, N. H., and Kay, S. A. (1992). A novel circadian phenotype based on firefly luciferase expression in transgenic plants. *Plant Cell* **4**, 1075–1087.

Morgan, L. W., Greene, A. V., and Bell-Pedersen, D. (2003). Circadian and light-induced expression of luciferase in *Neurospora crassa*. *Fung. Genet. Biol.* **38,** 327–332.

Stanewsky, R., Jamison, C. F., Plautz, J. D., Kay, S. A., and Hall, J. C. (1997). Multiple circadian-regulated elements contribute to cycling period gene expression in Drosophila. *EMBO J.* **16,** 5006–5018.

Stokkan, K. A., Yamazaki, S., Tei, H., Sakaki, Y., and Menaker, M. (2001). Entrainment of the circadian clock in the liver by feeding. *Science* **291,** 490–493.

Tsuchiya, Y., Akashi, M., and Nishida, E. (2003). Temperature compensation and temperature resetting of circadian rhythms in mammalian cultured fibroblasts. *Genes Cells* **8,** 713–720.

Ueda, H. R., Chen, W., Adachi, A., Wakamatsu, H., Hayashi, S., Takasugi, T., Nagano, M., Nakahama, K., Suzuki, Y., Sugano, S., Iino, M., Shigeyoshi, Y., and Hashimoto, S. (2002). A transcription factor response element for gene expression during circadian night. *Nature* **418,** 534–539.

Vallone, D., Gondi, S. B., Whitmore, D., and Foulkes, N. S. (2004). E-box function in a period gene repressed by light. *Proc. Natl. Acad. Sci. USA* **101,** 4106–4111.

Welsh, D. K., Imaizumi, T., and Kay, S. A. (2005). Real-time reporting of circadian-regulated gene expression by luciferase imaging in plants and mammalian cells. *Methods Enzymol.* **393** (11), 2005 (this volume).

Wilsbacher, L. D., Yamazaki, S., Herzog, E. D., Song, E. J., Radcliffe, L. A., Abe, M., Block, G. D., Spitznagel, E., Menaker, M., and Takahashi, J. S. (2002). Photic and circadian expression of luciferase in mPeriod1-luc transgenic mice *in vivo. Proc. Natl. Acad. Sci. USA* **99,** 489–494.

Yamaguchi, S., Kobayashi, M., Mitsui, S., Ishida, Y., van der Horst, G. T., Suzuki, M., Shibata, S., and Okamura, H. (2001). View of a mouse clock gene ticking. *Nature* **409,** 684.

Yamaguchi, S., Isejima, H., Matsuo, T., Okura, R., Yagita, K., Kobayashi, M., and Okamura, H. (2003). Synchronization of cellular clocks in the suprachiasmatic nucleus. *Science* **302,** 1408–1412.

Yamazaki, S., Numano, R., Abe, M., Hida, A., Takahashi, R., Ueda, M., Block, G. D., Sakaki, Y., Menaker, M., and Tei, H. (2000). Resetting central and peripheral circadian oscillators in transgenic rats. *Science* **288,** 682–685.

Yamazaki, S., Straume, M., Tei, H., Sakaki, Y., Menaker, M., and Block, G. D. (2002). Effects of aging on central and peripheral mammalian clocks. *Proc. Natl. Acad. Sci. USA* **99,** 1081–1086.

Yoo, S. H., Yamazaki, S., Lowrey, P. L., Shimomura, K., Ko, C. H., Buhr, E. D., Siepka, S. M., Hong, H. K., Oh, W. J., Yoo, O. J., Menaker, M., and Takahashi, J. S. (2004). PERIOD2:: LUCIFERASE real-time reporting of circadian dynamics reveals persistent circadian oscillations in mouse peripheral tissues. *Proc. Natl. Acad. Sci. USA* **101**(15), 5339–5346.

Zheng, B., Larkin, D. W., Albrecht, U., Sun, Z. S., Sage, M., Eichele, G., Lee, C. C., and Bradley, A. (1999). The mPer2 gene encodes a functional component of the mammalian circadian clock. *Nature* **400,** 169–173.

## [13]   Transgenic cAMP Response Element Reporter Flies for Monitoring Circadian Rhythms

*By* Kanae Iijima-Ando and Jerry C. P. Yin

### Abstract

The cAMP response Element (CRE)–binding protein (CREB) is involved in many adaptive behaviors, including circadian rhythms. In order to assess CREB activity *in vivo*, we made transgenic flies carrying a CRE-luciferase reporter and showed that this reporter is CRE and dCREB2 responsive. *dCREB2* is the *Drosophila* homolog of mammalian CREB/CREM. The transgenic luciferase activity cycles with a 24-h periodicity, suggesting that *dCREB2* and *period* are somehow linked. The CRE-luciferase reporter is a useful monitor of circadian activity, and mutations can be found that affect its periodicity, baseline activity, or amplitude. Analysis of such mutations should reveal information about how particular genes affect the molecular machinery of circadian cycling and how different genes affect the activity of dCREB2.

### Introduction

Adaptive behaviors partially result from relevant extracellular signals that alter patterns of gene expression. Stimuli, including growth factors, stress, learning-relevant cues, and circadian rhythms, activate a number of "immediate early" response genes, including the cAMP response element (CRE)–binding protein (CREB) (Lonze and Ginty, 2002). CREB is a member of the bZIP transcription factor superfamily, which contains a subfamily of cAMP and calcium-responsive transcriptional activators. In mammals, this subgroup contains CREB, CREM, and ATF-1. CREB family members contain a C-terminal domain enriched for basic amino acid residues that mediate DNA binding and a leucine zipper that is required for protein hetero- and homodimerization. Dimers bind to *cis*-regulatory consensus sequences, with TGACGTCA having the highest affinity (Mayr and Montminy, 2001).

CREB has two transactivation domains: one that is responsible for constitutive (Q2) activation and one [the kinase-inducible domain (KID)] that is required for inducible gene expression (Brindle *et al.*, 1993; Quinn, 1993). In addition to cAMP, CREB is activated by calcium influx, UV, stress, and growth factors (Mayr and Montminy, 2001). These signals are thought to regulate phosphorylation of at least the Ser133 residue located within the KID, whose phosphorylation is necessary but not sufficient for recruitment of the transcriptional coactivator CBP and its paralogue p300 (Arias *et al.*, 1994; Chrivia *et al.*, 1993). A variety of kinases can phosphorylate this site, including the cAMP-dependent protein kinase (PKA), protein kinase C (PKC), $Ca^{2+}$/calmodulin (CaM)-dependent protein kinases (CaMKs), and mitogen-activated protein kinases (MAPKs) (for review, see Shaywitz and Greenberg, 1999). There are also several other phosphorylation sites in addition to Ser133 in the KID that regulate the transcriptional activity of CREB (Gau *et al.*, 2002; Gonzalez and Montminy, 1989; Kornhauser *et al.*, 2002; Parker *et al.*, 1998; Sun *et al.*, 1994).

Drosophila has two CREB genes; *dCREB1* (or dCREBA) and *dCREB2* (Abel *et al.*, 1992; Smolik *et al.*, 1992; Usui *et al.*, 1993; Yin *et al.*, 1995b). dCREB2 is most similar to mammalian CREB/CREM (Yin *et al.*, 1995b). It contains a bZIP domain, with up to 90% protein identity to mammalian CREB family members, as well as a KID, in which the phosphorylation site corresponding to mammalian Ser133 is conserved as Ser231 (Yin *et al.*, 1995a). The phosphorylation of Ser231 is necessary for the activation of dCREB2, as mutations that change Ser231 to Ala abolish transcriptional activity of dCREB2.

The interactions between molecules that constitute the intrinsic circadian clock and light-induced synchronization work together to produce circadian oscillations (for review, see Stanewsky, 2003). One component of the circadian clock is a transcription feedback loop. In *Drosophila*, Clock (Clk) and Cycle (Cyc) form a heterodimer and mediate transcription of Period (Per) and Timeless (Tim). When the levels of Per and Tim are high enough, they heterodimerize, enter the nucleus, and bind to Clk/Cyc to inhibit transcription. Phosphorylation, mediated by Doubletime (Dbt), Shaggy (Sgg), and Time keeper (Tik), regulate cytoplasmic accumulation, nuclear entry, and degradation of Per and Tim (for review, see Stanewsky, 2003). This basic principle of a transcription feedback loop is conserved in mammals. In mammals, the central circadian clock is located in the suprachiasmatic nucleus (SCN) of the hypothalamus, which is also required for light entrainment, while the lateral neurons contain the *Drosophila* central clock.

The cAMP signal transduction pathway has been reported to be involved in light-mediated resetting of the clock in a number of species, including *Aplysia california* (Eskin and Takahashi, 1983), *Neurospora crassa* (Techel *et al.*, 1990), and the rat (Prosser and Gillette, 1989, 1991). In *Drosophila*, mutation of *dunce*, which encodes cAMP-specific phosphodiesterase, causes increased phase delays after light pulses (Levine *et al.*, 1994). Mutations in *DCO*, which encodes the catalytic subunit of

PKA, cause arrhythmic locomotor activity (Levine *et al.*, 1994; Majercak *et al.*, 1997).

*dCREB2* has been shown to be part of the oscillation loop regulated by *per* (Belvin *et al.*, 1999). A loss-of-function mutant in *dCREB2* diminishes oscillation of the *period* gene transcript and shortens circadian locomotor rhythm to an average of 22.8 h (Belvin *et al.*, 1999). In rodents, CREB has been reported to be involved in the integration of external light stimuli and resetting of the internal clock (Ding *et al.*, 1997; Gau *et al.*, 2002; Ginty *et al.*, 1993). Phosphorylation of CREB residues Ser133 and Ser142 in the KID oscillates and is rapidly induced in response to a light pulse during the dark period in the SCN (Ding *et al.*, 1997; Gau *et al.*, 2002; Ginty *et al.*, 1993). This change in the phosphorylation level of Ser133 is correlated with increased transcriptional activity of CREB monitored using a CRE-lacZ transgenic mouse strain (Obrietan *et al.*, 1999). It has been reported that knockin mice carrying a Ser142-to-alanine point mutation show impairments in behavioral phase shifting, as well as the induction of mPer-1 and c-Fos expression induced by a light pulse. These results demonstrate that CREB is necessary for synchronizing the circadian clock (Gau *et al.*, 2002).

We established transgenic fly lines carrying CRE-responsive reporters to monitor dCREB2 activity *in vivo* (Belvin *et al.*, 1999). This transgene contains an artificial enhancer that consists of three tandem CRE sites driving expression of a luciferase reporter gene. Because there is sufficient luciferase activity produced that can penetrate cellular and tissue barriers, it can be measured quantitatively in the behaving fly, and this measurement can be confirmed *in vitro*. Luciferase activity is short lived and suitable for dynamic measurements. This reporter assay reveals that dCREB2 activity oscillates in a circadian-dependent manner in *Drosophila. dCREB2* and *per* affect each other, suggesting they participate in the same feedback loop. The transgenic reporter can also be used to screen for mutations that affect periodicity or amplitude of cycling transcription.

## Methods

### CRE Reporter Construct

The enhancer-detecting vector pCaSpeR hs43 $\beta$-Gal was modified so that 3X CRE (TGACGTCA) sites were placed in front of the hsp70 TATA box region, and the lacZ gene was replaced with a luciferase reporter gene. These sequences were flanked by the *scs* and *scs'* insulator elements (Kellum and Schedl, 1992; Udvardy *et al.*, 1985; Vazquez and Schedl, 1994) to reduce potential position effects caused by the random insertion site of the transgene (Henikoff, 1994). The entire cassette was subcloned into a P-element vector and injected into *Drosophila* embryos to create transgenic lines. A mutant CRE reporter was also generated in which the consensus CRE sites were mutated to TGAAATCA (mCRE) (see Fig. 1A). Cycling reporter activity was detected whether the reporter gene was flanked by scs and scs' elements or not (data not shown).

### Recording Luciferase Activity from Transgenic Flies Carrying CRE-luc Reporter In Vivo

#### Materials

TopCount microplate scintillation and luminescence counter (Perkin-Elmer, Wellesley, MA)

$CO_2$ for anesthetizing flies

Microfluor 1, black flat-bottom microtiter plates 96 wells (Thermo-Labsystems, Franklin, MA) Clear microtiter dishes for spacing

Polymerase chain reaction (PCR) covers (MicroAmp Caps; Perkin Elmer) for plastic domes

Adhesive plastic sheets (Top-Seal; Perkin Elmer)

18-gauge needles for punching holes

Top layer: 5 m$M$ D-luciferin firefly, potassium salt (synthetic) (Biosynth, Naperville, IL)

1% agar/5% sucrose solution

Bottom layer: 1% agar/5% sucrose solution

Luciferase activity in CRE–luc flies is monitored as described previously (Brandes *et al.*, 1996; Hazelrigg, 2000; Stanewsky *et al.*, 1997). Flies are maintained on a 12-h light:12-h dark cycle at 25° on standard food. Each well of black 96-well microtiter dishes is filled with 250 $\mu$l of a 1% agar/5% sucrose solution (bottom layer), followed by 100 $\mu$l of a 1% agar/ 5% sucrose solution supplemented with 5 m$M$ luciferin (top layer). Flies are anesthetized with $CO_2$ and placed individually into the wells. Alternatively, anesthetized flies are placed individually into the wells of microtiter dishes filled with 100 $\mu$l of top layer agar and covered with plastic domes containing holes are punched by a needle. Plates are covered with adhesive plastic sheets, and air holes are punched over each well with a needle. Plates are loaded into a Perkin Elmer TopCount microplate scintillation and luminescence counter and maintained either on a 12-h light:12-h dark cycle (LD) or constant darkness (DD) according to the experiment. Clear plates are inserted between the black plates alternatively to allow light to penetrate. The expression level of each fly is measured hourly over a period of 6–10 days.

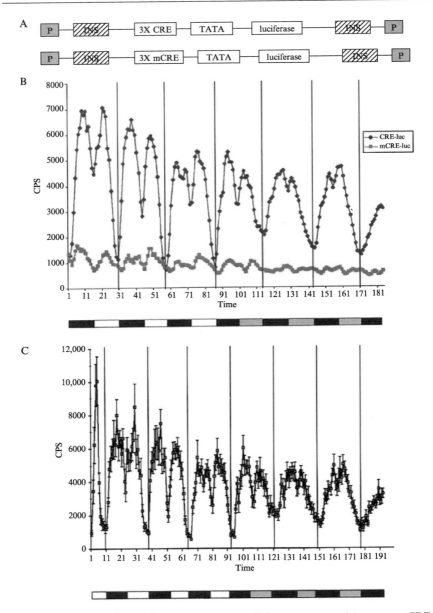

Fig. 1. Cycling of a dCREB2-responsive reporter. (A) Constructs used to generate CRE-luc and mCRE-luc transgenic flies. P, P transposable element inverted repeats; INS, SCS, and SCS' insulator elements; TATA, TATA box sequence from hsp 70 promoter. These constructs were cloned into the pCaSpeR transformation vector. (B) *In vivo* cycling of

*Data Collection and Analysis*

When a run on the TopCount is completed, data are copied to a disk and transferred to Microsoft Excel using the Import & Analysis (I&A) macro set (Carl Strayer, Camilo Orzco, Jeffery D. Plautz, and Steve Kay, available at http://www.scripps.edu/cb/kay). The first 12 h of data are removed, and averages of multiple flies (usually 20–40 flies per experiments) are calculated. The averages are subjected to smoothing, where each data point represents the average of itself plus the two data points on ether side.

*Luciferase Activity from Transgenic Flies Carrying CRE-luc*
   *Reporter* In Vitro

   *Materials*

      Luciferase assay kit (Promega, Madison, WI)
      Laser blade
      $CO_2$ for anesthetizing flies
The expression level of luciferase at a single time point can be measured from dissected fly heads carrying the CRE-luc reporter gene using the luciferase assay kit. Five to 10 fly heads are dissected. Luciferase activity from the head extracts is measured following the manufacturer's instruction. The luciferase signal is normalized by protein level.

Results

*CRE-luc Activity Shows Circadian Cycling and Is dCREB2 and*
   *CRE Responsive*

The transgenic CRE-luciferase activity oscillates with a 24-h rhythm, both in light/dark (L/D) and in constant darkness (Fig. 1). Because the pattern of rhythmic transcription is sustained in constant darkness, it is regulated by the circadian system, rather than simply being a response to light. The main peak of activity occurs right after lights out. Under L/D conditions, a second peak is observed during the daytime. However, this peak gradually blends together under constant darkness and is not detectable if luciferase activity from these flies is measured *in vitro*

CRE-luc and mCRE-luc reporter expression as measured in a Packard TopCount luminometer. All time points represent an average of data points from 30 flies. The bar below the graph indicates light/dark conditions. White box, light period; black box, dark period; gray box, dark period during former light hours (subjective day). Vertical bars in the graph represent lights out and are spaced 24 h apart. (C) Similar graph as in B but with standard error bars added. (See color insert.)

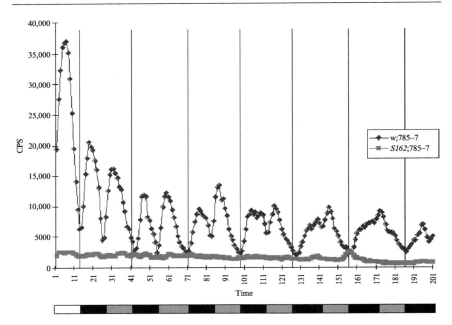

FIG. 2. Expression of the CRE-luc reporter in S162 mutant flies. *S162*/FM7 females were mated to males homozygous for the CRE-luc reporter transgene. Escaper males of the genotype *S162*/Y;CRE-luc/+ were assayed in the luminometer. Traces represent an average of data from 15 flies (*S162* mutants) or 30 flies (wild type). See Fig. 1 legend for a description of the graph. (See color insert.)

(data not shown). Therefore, it is likely that the daytime peak is due to some aspect of the response to lights on, perhaps feeding, rather than a circadian response (Belvin *et al.*, 1999). The cycling reporter activity is CRE dependent, as mutating the consensus CRE to a mutant CRE (mCRE; TGA<u>AA</u>TCA) reduces cycling dramatically (see Fig. 1). The dCREB2 protein binds this mutant CRE site with at least a 20-fold lower affinity in gel shift experiments (Yin *et al.*, 1995b). The CRE-luc signal is also dCREB2 dependent, as it is diminished greatly when a loss-of-function mutation (S162) (Belvin *et al.*, 1999) in dCREB2 is introduced (see Fig. 2).

## CRE-luc Activity and Per Affect One Another

The oscillating CRE-luc activity that occurs in constant darkness is similar to cycling seen in flies containing reporters where the *per* promoter controls luciferase (Belvin *et al.*, 1999; Stanewsky *et al.*, 1997). In S162 escaper flies, *per* promoter-driven reporter flies show diminished amplitudes and a shorter periodicity than in wild-type flies. Western analyses of Per and Tim show abnormalities in the kinetics of protein cycling across a 24-h period (Belvin *et al.*, 1999). Together these results show that dCREB2 affects *per* expression. When the CRE-luc reporter is placed into mutant *per* backgrounds, the periodicity of cycling is a function of the particular *per* mutation (Belvin *et al.*, 1999). In a *per0* null mutant, CRE-luc cycling is abolished, whereas in *perS* (19-h period) and *perL* (29-h period) backgrounds (Konopka and Benzer, 1971), the cycling is altered as expected. Because the two genes mutually affect each other, they must somehow function in a feedback loop (Fig. 3).

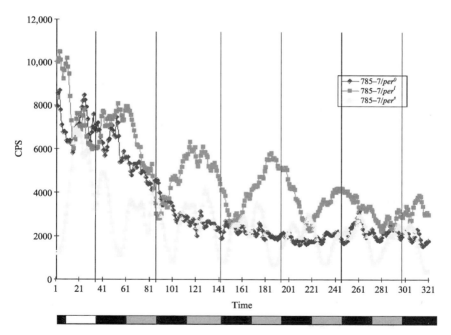

FIG. 3. Expression of the CRE-luc reporter in *per* mutant backgrounds. Females homozygous for one of three *per* mutations (described in the text) were crossed to males homozygous for the CRE-luc reporter, and male progeny of the genotype *per*/Y;CRE-luc/+ were assayed in the luminometer. *per⁰* or *per* null; *perˡ* or *per* long; *perˢ* or *per* short. Flies were entrained on a 12-h light:12-h dark cycle for 4 days before the start of the experiment. Flies were then switched to constant darkness for the duration of the experiment. Each trace represents the average of data from 40 flies. (See color insert.)

## Mutations That Affect CRE-luc Periodicity

In addition to demonstrating the interrelationship between *dCREB2* and *per*, the changes in CRE-luc cycling in the various mutant *per* backgrounds show that the reporter can be used to screen for other mutations that affect periodicity and thus participate in, modulate, or are downstream of the central clock. The stability and nuclear translocation of Per are regulated by phosphorylation. Doubletime (dbt) encodes the fly homolog of CK Iε, which phosphorylates Per and regulates its stability in the nucleus (Kloss *et al.*, 1998; Price *et al.*, 1998). Another kinase, CK2α, has been reported to regulate Per activity (Akten *et al.*, 2003; Lin *et al.*, 2002). A dominant mutation named Tik causes approximately a 1.5-h lengthening in the circadian period. Tik maps to CK2α, and CK2 activity is decreased in these flies. The nuclear entry of Per is delayed in Tik flies, presumably due to a decrease in CK2-mediated phosphorylation of Per (Akten *et al.*, 2003; Lin *et al.*, 2002). When CRE-luc is crossed into a Tik background, there is a delay in the main peak of CRE-luc activity (Fig. 4). It is not known why the delay in molecular cycling (~4 h) is much longer than the delay in behavioral rhythms (1.5 h) in a Tik mutant background.

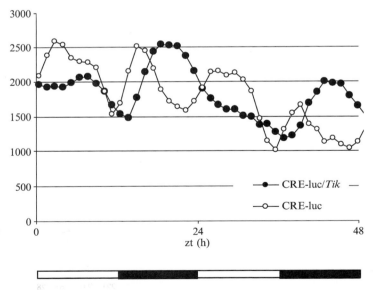

Fig. 4. Expression of the CRE-luc reporter in *Tik* mutant backgrounds. Females homozygous for the CRE-luc reporter were crossed to males heterozygous for the *Tik* mutation, and male progeny of the genotype +/Y;CRE-luc/*Tik* were assayed in the luminometer. Flies were entrained on a 12-h light:12-h dark cycle and kept in this cycle during the experiment. Each trace represents the average of data from 20 flies.

*Mutations That Affect CRE-luc Amplitude*

Neurofibromin, the gene product of neurofibromatosis-1 (Nf1), inactivates the Ras oncogene through hydrolysis of guanosine triphosphate (GTP) and stimulates rutabaga-encoded adenylyl cyclase, thus stimulating PKA activity (Guo *et al.*, 1997; The *et al.*, 1997). In *Drosophila*, loss-of-function mutations in *Nf1* have been reported to cause increased MAPK activity and abnormal circadian locomotor activity. Williams *et al.* (2001) found that Nf1 regulated dCREB2 activity but not its oscillating activity. The CRE-luc signal in *Nf1* mutant flies oscillates in a circadian-dependent manner as in the parental lines. However, baseline luciferase levels in *Nf1* mutant flies is up to three times higher than those of control flies, but the CRE-luc amplitude is reduced (Williams *et al.*, 2001). Oscillation of Per and Tim mRNA, as well as the expression of Per and Tim in LNs, was not affected by *Nf1* mutation. These results indicate that Nf1 is not involved in the cycling of the central clock components but is somehow downstream of the clock when it affects behavioral output (Williams *et al.*, 2001).

Discussion

Transcriptional feedback loops and light-mediated synchronization contribute to the circadian oscillation of molecules. Use of the CRE-luciferase reporter demonstrates that dCREB2 participates in this feedback loop (Belvin *et al.*, 1999). The *per* and Tik mutants, which affect the cycling of Per and the clock, change the periodicity of CRE-luc cycling. In contrast, Nf1 mutants, which do not change Per or Tim cycling, cause arrhythmic behavior and changes in the amplitude of CRE-luc signaling without affecting periodicity. It is possible that changes in the oscillation of CRE-luc reflect changes in a component of the central clock, but that changes in amplitude do not. However, further analysis of both CRE-luc activity and the behavioral phenotype of dCREB2 will be required.

How the circadian clock regulates the cycling of dCREB2 activity is unknown. Because dCREB2 mRNA levels do not change over a 24-h period, it is likely that the central clock modulates an intracellular signaling pathway that is involved in activating dCREB2 (Belvin *et al.*, 1999). During light-synchronized resetting of the clock, glutamate and the subsequent influx of $Ca^{2+}$ activate NMDA receptors in the SCN, ultimately leading to phosphorylation and activation of CREB (Ding *et al.*, 1997; Gau *et al.*, 2002; Schurov *et al.*, 1999). The extracellular signal(s) that triggers dCREB2 oscillation remains unknown, as does the identity of the cells that contribute to reporter activation. It is likely, though, that a considerable part of CRE-luc activity occurs in the visual system of the transgenic fly, as luciferase staining is very prominent in photoreceptors and parts of the brain that process visual input (data not shown).

The mechanism of dCREB2 activation itself is not fully understood. The sequence of the KID of CREB is highly conserved across all species (Yin et al., 1995b). The dCREB2 Ser231 residue is equivalent to the mammalian Ser133 residue, which can be phosphorylated by PKA and is necessary for activity (Yin et al., 1995a). However, unlike mammalian CREB, whose activity is correlated with the phosphorylation level of Ser133, a large portion of dCREB2 contains Ser231 in a phosphorylated state, suggesting that other mechanisms must be important in regulating its activity (Horiuchi et al., 2003). Regulation of CKII phosphorylation and DNA binding may be the dominant regulatory mechanism for dCREB2 (Horiuchi et al., 2003). Currently, it is not known how these phosphorylation events are regulated in vivo. The existence of a transgenic CRE-luciferase reporter system, together with the abundant genetic resources of Drosophila, should provide the necessary tools to unravel the upstream signals regulating CREB activity. Because CREB is involved in many complex behaviors, including memory formation (Silva et al., 1998), sleep (Hendricks et al., 2001), and addiction (Chao and Nestler, 2004), the analysis of dCREB2 activation will shed light on these behaviors as well as the role of dCREB2 in circadian rhythms.

## References

Abel, T., Bhatt, R., and Maniatis, T. (1992). A Drosophila CREB/ATF transcriptional activator binds to both fat body- and liver-specific regulatory elements. Genes Dev. 6, 466–480.

Akten, B., Jauch, E., Genova, G. K., Kim, E. Y., Edery, I., Raabe, T., and Jackson, F. R. (2003). A role for CK2 in the Drosophila circadian oscillator. Nature Neurosci. 6, 251–257.

Arias, J., Alberts, A. S., Brindle, P., Claret, F. X., Smeal, T., Karin, M., Feramisco, J., and Montminy, M. (1994). Activation of cAMP and mitogen responsive genes relies on a common nuclear factor. Nature 370, 226–229.

Belvin, M. P., Zhou, H., and Yin, J. C. (1999). The Drosophila dCREB2 gene affects the circadian clock. Neuron 22, 777–787.

Brandes, C., Plautz, J. D., Stanewsky, R., Jamison, C. F., Straume, M., Wood, K. V., Kay, S. A., and Hall, J. C. (1996). Novel features of drosophila period transcription revealed by real-time luciferase reporting. Neuron 16, 687–692.

Brindle, P., Linke, S., and Montminy, M. (1993). Protein-kinase-A-dependent activator in transcription factor CREB reveals new role for CREM repressors. Nature 364, 821–824.

Chao, J., and Nestler, E. J. (2004). Molecular neurobiology of drug addiction. Annu. Rev. Med. 55, 113–132.

Chrivia, J. C., Kwok, R. P., Lamb, N., Hagiwara, M., Montminy, M. R., and Goodman, R. H. (1993). Phosphorylated CREB binds specifically to the nuclear protein CBP. Nature 365, 855–859.

Ding, J. M., Faiman, L. E., Hurst, W. J., Kuriashkina, L. R., and Gillette, M. U. (1997). Resetting the biological clock: Mediation of nocturnal CREB phosphorylation via light, glutamate, and nitric oxide. *J. Neurosci.* **17**, 667–675.

Eskin, A., and Takahashi, J. S. (1983). Adenylate cyclase activation shifts the phase of a circadian pacemaker. *Science* **220**, 82–84.

Gau, D., Lemberger, T., von Gall, C., Kretz, O., Le Minh, N., Gass, P., Schmid, W., Schibler, U., Korf, H. W., and Schutz, G. (2002). Phosphorylation of CREB Ser142 regulates light-induced phase shifts of the circadian clock. *Neuron* **34**, 245–253.

Ginty, D. D., Kornhauser, J. M., Thompson, M. A., Bading, H., Mayo, K. E., Takahashi, J. S., and Greenberg, M. E. (1993). Regulation of CREB phosphorylation in the suprachiasmatic nucleus by light and a circadian clock. *Science* **260**, 238–241.

Gonzalez, G. A., and Montminy, M. R. (1989). Cyclic AMP stimulates somatostatin gene transcription by phosphorylation of CREB at serine 133. *Cell* **59**, 675–680.

Guo, H. F., The, I., Hannan, F., Bernards, A., and Zhong, Y. (1997). Requirement of Drosophila NF1 for activation of adenylyl cyclase by PACAP38-like neuropeptides. *Science* **276**, 795–798.

Hazelrigg, T. (2000). *In* "*Drosophila* Protocols" (W. Sullivan, M. Ashburner, and R. S. Hawley, eds.), pp. 337–339. Cold Spring Harbor Laboratory Press, Cold Spring Harbor, NY.

Hendricks, J. C., Williams, J. A., Panckeri, K., Kirk, D., Tello, M., Yin, J. C., and Sehgal, A. (2001). A non-circadian role for cAMP signaling and CREB activity in Drosophila rest homeostasis. *Nature Neurosci.* **4**, 1108–1115.

Henikoff, S. (1994). A reconsideration of the mechanism of position effect. *Genetics* **138**, 1–5.

Horiuchi, J., Jiang, W., Zhou, H., Wu, P., and Yin, J. C. P. (2004). Phosphorylation of conserved casein kinase sites regulates CREB DNA binding in Drosophila. *J. Biol. Chem.* **279**, 12117–12125.

Kellum, R., and Schedl, P. (1992). A group of scs elements function as domain boundaries in an enhancer-blocking assay. *Mol. Cell. Biol.* **12**, 2424–2431.

Kloss, B., Price, J. L., Saez, L., Blau, J., Rothenfluh, A., Wesley, C. S., and Young, M. W. (1998). The Drosophila clock gene double-time encodes a protein closely related to human casein kinase I epsilon. *Cell* **94**, 97–107.

Konopka, R. J., and Benzer, S. (1971). Clock mutants of *Drosophila melanogaster*. *Proc. Natl. Acad. Sci. USA* **68**, 2112–2116.

Kornhauser, J. M., Cowan, C. W., Shaywitz, A. J., Dolmetsch, R. E., Griffith, E. C., Hu, L. S., Haddad, C., Xia, Z., and Greenberg, M. E. (2002). CREB transcriptional activity in neurons is regulated by multiple, calcium-specific phosphorylation events. *Neuron* **34**, 221–233.

Levine, J. D., Casey, C. I., Kalderon, D. D., and Jackson, F. R. (1994). Altered circadian pacemaker functions and cyclic AMP rhythms in the Drosophila learning mutant dunce. *Neuron* **13**, 967–974.

Lin, J. M., Kilman, V. L., Keegan, K., Paddock, B., Emery-Le, M., Rosbash, M., and Allada, R. (2002). A role for casein kinase 2alpha in the Drosophila circadian clock. *Nature* **420**, 816–820.

Lonze, B. E., and Ginty, D. D. (2002). Function and regulation of CREB family transcription factors in the nervous system. *Neuron* **35**, 605–623.

Majercak, J., Kalderon, D., and Edery, I. (1997). *Drosophila melanogaster* deficient in protein kinase A manifests behavior-specific arrhythmia but normal clock function. *Mol. Cell. Biol.* **17**, 5915–5922.

Mayr, B., and Montminy, M. (2001). Transcriptional regulation by the phosphorylation-dependent factor CREB. *Nature Rev. Mol. Cell. Biol.* **2**, 599–609.

Obrietan, K., Impey, S., Smith, D., Athos, J., and Storm, D. R. (1999). Circadian regulation of cAMP response element-mediated gene expression in the suprachiasmatic nuclei. *J. Biol. Chem.* **274,** 17748–17756.

Parker, D., Jhala, U. S., Radhakrishnan, I., Yaffe, M. B., Reyes, C., Shulman, A. I., Cantley, L. C., Wright, P. E., and Montminy, M. (1998). Analysis of an activator:coactivator complex reveals an essential role for secondary structure in transcriptional activation. *Mol. Cell* **2,** 353–359.

Price, J. L., Blau, J., Rothenfluh, A., Abodeely, M., Kloss, B., and Young, M. W. (1998). Double-time is a novel Drosophila clock gene that regulates PERIOD protein accumulation. *Cell* **94,** 83–95.

Prosser, R. A., and Gillette, M. U. (1989). The mammalian circadian clock in the suprachiasmatic nuclei is reset *in vitro* by cAMP. *J. Neurosci.* **9,** 1073–1081.

Prosser, R. A., and Gillette, M. U. (1991). Cyclic changes in cAMP concentration and phosphodiesterase activity in a mammalian circadian clock studied *in vitro. Brain Res.* **568,** 185–192.

Quinn, P. G. (1993). Distinct activation domains within cAMP response element-binding protein (CREB) mediate basal and cAMP-stimulated transcription. *J. Biol. Chem.* **268,** 16999–17009.

Schurov, I. L., McNulty, S., Best, J. D., Sloper, P. J., and Hastings, M. H. (1999). Glutamatergic induction of CREB phosphorylation and Fos expression in primary cultures of the suprachiasmatic hypothalamus *in vitro* is mediated by co-ordinate activity of NMDA and non-NMDA receptors. *J. Neuroendocrinol.* **11,** 43–51.

Shaywitz, A. J., and Greenberg, M. E. (1999). CREB: A stimulus-induced transcription factor activated by a diverse array of extracellular signals. *Annu. Rev. Biochem.* **68,** 821–861.

Silva, A. J., Kogan, J. H., Frankland, P. W., and Kida, S. (1998). CREB and memory. *Annu. Rev. Neurosci.* **21,** 127–148.

Smolik, S. M., Rose, R. E., and Goodman, R. H. (1992). A cyclic AMP-responsive element-binding transcriptional activator in *Drosophila melanogaster*, dCREB-A, is a member of the leucine zipper family. *Mol. Cell. Biol.* **12,** 4123–4131.

Stanewsky, R. (2003). Genetic analysis of the circadian system in *Drosophila melanogaster* and mammals. *J. Neurobiol.* **54,** 111–147.

Stanewsky, R., Jamison, C. F., Plautz, J. D., Kay, S. A., and Hall, J. C. (1997). Multiple circadian-regulated elements contribute to cycling period gene expression in Drosophila. *EMBO J.* **16,** 5006–5018.

Sun, P., Enslen, H., Myung, P. S., and Maurer, R. A. (1994). Differential activation of CREB by $Ca^{2+}$/calmodulin-dependent protein kinases type II and type IV involves phosphorylation of a site that negatively regulates activity. *Genes Dev.* **8,** 2527–2539.

Techel, D., Gebauer, G., Kohler, W., Braumann, T., Jastorff, B., and Rensing, L. (1990). On the role of Ca2(+)-calmodulin-dependent and cAMP-dependent protein phosphorylation in the circadian rhythm of *Neurospora crassa. J. Comp. Physiol. B* **159,** 695–706.

The, I., Hannigan, G. E., Cowley, G. S., Reginald, S., Zhong, Y., Gusella, J. F., Hariharan, I. K., and Bernards, A. (1997). Rescue of a Drosophila NF1 mutant phenotype by protein kinase A. *Science* **276,** 791–794.

Udvardy, A., Maine, E., and Schedl, P. (1985). The 87A7 chromomere: Identification of novel chromatin structures flanking the heat shock locus that may define the boundaries of higher order domains. *J. Mol. Biol.* **185,** 341–358.

Usui, T., Smolik, S. M., and Goodman, R. H. (1993). Isolation of Drosophila CREB-B: A novel CRE-binding protein. *DNA Cell Biol.* **12,** 589–595.

Vazquez, J., and Schedl, P. (1994). Sequences required for enhancer blocking activity of scs are located within two nuclease-hypersensitive regions. *EMBO J.* **13,** 5984–5993.

Williams, J. A., Su, H. S., Bernards, A., Field, J., and Sehgal, A. (2001). A circadian output in Drosophila mediated by neurofibromatosis-1 and Ras/MAPK. *Science* **293,** 2251–2256.

Yin, J. C., Del Vecchio, M., Zhou, H., and Tully, T. (1995a). CREB as a memory modulator: Induced expression of a dCREB2 activator isoform enhances long-term memory in Drosophila. *Cell* **81,** 107–115.

Yin, J. C., Wallach, J. S., Wilder, E. L., Klingensmith, J., Dang, D., Perrimon, N., Zhou, H., Tully, T., and Quinn, W. G. (1995b). A Drosophila CREB/CREM homolog encodes multiple isoforms, including a cyclic AMP-dependent protein kinase-responsive transcriptional activator and antagonist. *Mol. Cell. Biol.* **15,** 5123–5130.

# [14]  Analysis of Circadian Output Rhythms of Gene Expression in *Neurospora* and Mammalian Cells in Culture

*By* Giles Duffield, Jennifer J. Loros, and Jay C. Dunlap

## Abstract

The true biology of chronobiology lies in the spectrum of processes that are controlled by the circadian clock. Although this biology plays out at the level of the whole organism, it derives, finally, from clock-driven changes in physiology, and frequently in gene expression, that occur at the level of individual cells. For this reason, analysis of gene expression rhythms measured in cell culture or in organisms that elaborate only a few cell types provides insights not possible in multicellular organisms. In this context we have used mammalian fibroblasts in culture as well as the eukaryotic filamentous fungus *Neurospora crassa* to study circadian output, in particular the output rhythms in gene expression that underlie so much of circadian biology. Each cell type has its own advantages: Data from mammalian cells are obviously immediately pertinent to animal cell rhythms, but the system allows little genetics and only limited amounts of material can be collected. Alternatively, *Neurospora* allows genetic and molecular analyses and is useful for developing concepts and models of output that can be examined in other contexts. This methods article focuses on these two systems for analysis, providing an overview of how control is presently viewed followed by current methods for its analysis.

## Introduction

Although, historically, most interest in the field of chronobiology has been focused on the molecular mechanisms through which circadian time is generated, it is the spectrum of clock-controlled processes that first drew biologists' attention to circadian rhythms. In multicellular organisms, the spectrum of clock-regulated processes is extensive and complex, but all of these complex organismal phenotypes reduce at the end to clock-regulated metabolism and physiology in individual cells (Loros *et al.*, 2003). For this reason, there has been continued interest in the mechanisms by which intracellular circadian oscillators control the physiology and behavior of individual cells.

An early assertion made in *Neurospora* (Loros *et al.*, 1989) that has proven to be universally true in all circadian systems is that daily clock control of gene expression would be a major aspect of output. The first systematic screens for clock-regulated genes were executed in *Neurospora* (Loros *et al.*, 1989), and the genes so identified were called *clock-controlled genes* (*ccgs*). Over 180 ccgs have since been identified in this model eukaryote through a combination of differential hybridization (Bell-Pedersen *et al.*, 1996), cDNA sequencing (Zhu *et al.*, 2001), and cDNA microarrays (Correa *et al.*, 2003; Nowrousian *et al.*, 2003), as discussed more fully later. In multicellular organisms, the identification of ccgs was on a gene-by-gene basis until the advent of microarrays (Duffield, 2003; Harmer *et al.*, 2000). While in the aggregate these studies have demonstrated, in a beautifully graphical manner, the extensive nature of clock control in various tissue and organ systems of plants and animals, it has been generally highly difficult to associate the expression of individual genes with individual cell types. For this reason, analysis of clock-regulated gene expression from cells in culture provides a promising route to examine the molecular avenues leading to clock regulation. Adopting the same paradigms used in whole organisms, we have followed gene expression rhythms in fibroblasts in culture (Duffield *et al.*, 2002). This article considers both the biological context in which such experiments must be designed and interpreted and also the methods with which time series data can be analyzed, initially focusing on *Neurospora* and later covering research on mammalian cells in culture.

## Neurospora

### Biology and Clock Regulation in Neurospora

*Neurospora* grows vegetatively as a syncytium in which cellular compartments are bounded by incomplete septae that allow the passage of nuclei and cytoplasm. A variety of processes within the life cycle of the fungus can be clock regulated (Fig. 1). The best-studied circadian phenotype involves the formation of asexual spores during the vegetative phase

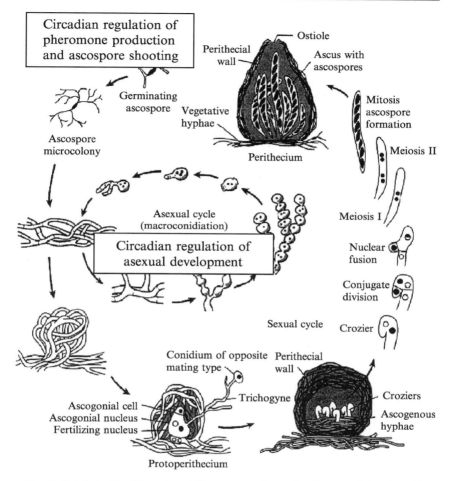

FIG. 1. Points in the life cycle of *Neurospora* where circadian regulation is evident. Modified from Davis (2000), with permission.

of the life cycle, a process known as macroconidiation. Additionally, this can be triggered by environment signals, including blue light, desiccation, and nutrient starvation, as well as by the endogenous circadian clock; in all cases it involves a major morphological change that in turn entails major changes in gene expression. This process begins with the production of aerial hyphae that eventually branch and elaborate hyphae having complete septa and are no longer syncytial, all in all requiring many novel gene products. Other physiological rhythms in the vegetative stage have been described, including the production of $CO_2$ (Sargent and Kaltenborn, 1972), a number of enzymatic activities, lipid and diacylglycerol metabolism (e.g., Ramsdale and Lakin-Thomas, 2000), and even growth rate (V. Gooch and P. Lakin-Thomas, manuscript in preparation; Sargent et al., 1966). Microarray studies executed using circadian liquid cultures (see previous article and later) have shown that all aspects of the biology of the organism are under clock regulation (Table I). Microarrays have also served to highlight the importance of the spectrum of genes on the arrays in relation to the results obtained. One microarray study (Nowrousian et al., 2003) used genes derived from a previously studied cDNA library (Bell-Pedersen et al., 1996) made from RNAs expressed when cultures were slowly starving in the dark. Exhaustive cDNA sequencing (Zhu et al., 2001) yielded a UniGene set of about 1100 genes, about 10% of the genome. Another study (Correa et al., 2003) used 1343 genes (~14% of the genome) derived largely from Expressed Sequence Tag (EST) analysis of growing conidial, mycelial, and sexual tissues where many genes are expressed; these showed relatively little overlap with the first study. Both studies found that most genes are expressed at the time of conidiation (late night to morning), but interestingly, only about 5% of the genes from starving cultures were clock regulated versus nearly 20% from the rapidly growing cultures.

Associated with the sexual cycle, which is initiated under conditions of nutrient starvation, the production of both male and female pheromones is clock controlled (Bobrowicz et al., 2002), as is the time at which the ascospores are shot from the mature female fruiting body, or perithecium

TABLE I
FUNCTIONS OF *Neurospora* ccgs[a]

| Functional category | No. # of ccgs |
| --- | --- |
| Cell division | 1 |
| Signaling/communication | 16 |
| Cell structure/cytoskeleton | 8 |
| Cell defense | 4 |
| Development | 11 |
| Gene regulation | 5 |
| Metabolism | 42 |
| Protein processing | 10 |
| Protein synthesis | 33 |
| Unclassified | 50 |

[a] Expanded from Correa et al. (2003) and Nowrousian et al. (2003).

(Lakin-Thomas *et al.*, 1990). Compared with asexual development, much less is known about the developmental processes associated with the sexual cycle in this organism. It appears, though, that there are several novel cell types elaborated, as well as novel signaling pathways, so it is likely that there will be a large number of directly or indirectly clock-regulated genes associated with sexual development.

Although in the wild-type system all of these processes can be controlled by the circadian clock, that is not the only factor that can control them. First of all, many environmental factors can directly elicit sexual or asexual development. Second and perhaps more important, fungi can, in fact, express a variety of rhythms, many of which are not by definition circadian (Bünning, 1973). A difficulty in the analysis of output pathways has been distinguishing circadian clock regulation from other metabolic or developmental regulations. In general, in *Neurospora*, as in all organisms, it is important to remember that not every rhythm is a circadian rhythm, nor is every rhythmic process controlling conidiation necessarily a part of the circadian clock.

## *The* Neurospora *Circadian System: Circadian and Noncircadian Output Rhythms*

*Neurospora*, like other organisms, contains a circadian system, not just a circadian oscillator. Work over the past two decades has shown that in addition to simple clock-regulated genes and proteins, the system includes other oscillators as well. This is not a new idea in circadian biology: Pittendrigh and Bruce modeled the *Drosophila* circadian system as including a core oscillator whose activity either drove or entrained a slave oscillator tied more directly to output (Pittendrigh and Bruce, 1959) and a similar model may be applicable to *Neurospora* (Iwasaki and Dunlap, 2000; Merrow *et al.*, 2001). Noncircadian putative slave oscillators are generally revealed upon loss of *frq* or *wc* genes or during growth on stressful media (Sussman *et al.*, 1964, 1965) or starvation of choline auxotrophs for choline (Lakin-Thomas, 1996). The term FRQ-less oscillator (FLO) has been used to distinguish these noncircadian oscillators from the FRQ/WC circadian feedback loop (Iwasaki and Dunlap, 2000). Several FLOs have been identified but none have yet been shown either to be inherently circadian or to influence the operation of the FRQ/WCC feedback loop in the circadian clock. A variety of data summarized elsewhere (Dunlap and Loros, 2004) have led to the view of the *Neurospora* system shown in Fig. 2. Here a core circadian feedback loop serves to coordinate a variety of oscillators whose activities, absent input from the core loop, would be neither coordinated nor circadian. Because many or all of these oscillators can influence gene expression, it is useful, with this model in

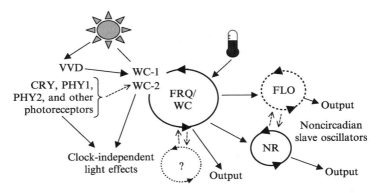

FIG. 2. A schematic view of the inputs, oscillators, and networks controlling and integrating light and temperature influences in the *Neurospora* circadian system. Solid loops, lines, and arrows represent known feedback loops or regulatory relationships, and dotted lines represent those predicted to exist in the wild-type organism. Adapted from National Academies Keck Futures Initiative Signaling National Conference (http://63.251.167.36/nakfi/progressive/GeneticandMolecularDissectionofE/index.htm).

mind, to review briefly what is known about noncircadian oscillators in *Neurospora*.

The original noncircadian oscillation is the rhythm in hyphal branching that is expressed when cultures are grown on sorbose (Sussman *et al.*, 1964, 1965), a sugar that interferes with cell wall biosynthesis. As expected, depending on the composition of the solid medium, the same culture can be made to express either circadian or noncircadian rhythms (Feldman and Hoyle, 1974). A second FLO may be observed when the *frq* gene is deleted (Aronson *et al.*, 1994). The *frq*-null phenotype included, on solid medium, expression of a nontemperature-compensated rhythm of a variable period length in conidial banding (Loros *et al.*, 1986a,b, 1993): The rhythm displays a highly variable period length ranging from 12 to 35 h, appears in only a fraction of all race tube cultures and then only after a few days, cannot be entrained by light cycles, and displays a period strongly dependent on carbon source and temperature. Based on its slow and sporadic appearance, it has been suggested that the FLO may not normally exist in cells at all but rather represents an alternative oscillatory state, seen only in the absence of the clock, that may be formed after the system has passed an "induction period" (e.g., Christensen *et al.*, 2004), by analogy to what is observed in chemical oscillators (Field and Burger, 1985). The rhythm can, however, be driven by temperature cycles. In such a protocol, the waveform of the rhythm changes with the period of the entraining cycle, requiring (as expected) half of the driving cycle period length to transit from trough to peak and vise versa. Peaks always occur in the cold period and troughs always in the warm period, with absolutely no dependency of phase on the driving period, strongly suggesting that the rhythm is driven and not entrained. However, when a phase reference point on the rise of the curve from trough to peak is used for phase estimations instead of peaks or troughs, the derived phase of the banding rhythm can be seen to vary with the period length of the cycle (Merrow *et al.*, 1999), rather than being constant. Although this has been taken as evidence that this oscillator is the circadian "rhythm generator" (Merrow *et al.*, 2001; Roenneberg and Merrow, 1998), a simpler explanation may be just that this reflects the systematic change in the waveform of the rhythm noted earlier rather than being a reflection of a cryptic metabolic oscillator. Given this, temperature cycles might be used to track the changes in gene expression associated with this FLO. This is what Nowrousian *et al.* (2003) attempted, but unfortunately they detected no changes in gene expression in liquid cultures undergoing a 5° temperature cycle in the absence of a functional *frq* gene, the conditions under which FLO is observed. Because such changes in gene expression must be occurring in cultures on solid media, this absence in liquid highlights the different physiologies under the two culture conditions.

Correa and colleagues (2003) have identified what appears to be a second FLO. Examination of the phase of a few novel ccgs identified in a microarray study showed anomalies in wild-type versus the long-period clock mutant strain *frq*[7]; further studies showed expression of these genes cycling in a *frq*-null strain. The phase appears to be set by the LD transfer, making this oscillator distinct from the original FLO described earlier. Although this FLO operates absent the FRQ/WCC feedback loop, the difference in phase between *frq*[+] and *frq*[7] suggests that the FRQ/WCC oscillator influences this FLO.

A separate class of FLO also noted previously is seen in strains bearing the choline-reparable morphological mutant *chol-1* (e.g., Lakin-Thomas and Brody, 2000). Unsupplemented *chol-1* strains show abnormal colony morphology and grow episodically so that the fastest rate of linear extension can be eight times the slowest. The period length of the cycle from fast to slow growth and back again can be varied from as fast as about a day up to over 100 h (Lakin-Thomas, 1998). The rhythm lacks temperature compensation but interestingly is still pH compensated (Ruoff and Slewa, 2002). Unsupplemented *chol-1* strains express a normal rhythm in FRQ expression for several days in liquid culture and are rhythmic by other physiological assays (M. Shi, J. Loros, and J. Dunlap, manuscript in preparation). It may be that the robust rhythms uncovered by choline starvation in the mutant reflect morphological cycles whose appearance masks, or perhaps is gated by, normal circadian regulation of banding.

Still another example of a FLO comes from a study of a rhythm in nitrate reductase (Christensen *et al.*, 2004) that persists in *frq*- and *wc*-null strains. Because nitrate reductase expression is regulated by metabolites arising from assimilated nitrate, there is the basis for a feedback loop oscillator within nitrate metabolism that could constitute this FLO.

Figure 2 provides a plausible view unifying these observations. All the FLOs are seen as slave oscillators that can be coupled to the FRQ/WCC-associated circadian system. Absent the FRQ/WCC loop, they can run on their own in a noncircadian manner. This provides a good context in which to view circadian gene expression, how to study it, and how to understand it. If this model holds, then there will be some directly clock-regulated genes, as well as a variety of genes that are regulated more indirectly by the core clock. Because expression of many genes may be most immediately regulated by a FLO, even absent a circadian oscillator, it may not be easy to distinguish which genes are directly versus indirectly clock-regulated unless they can be observed in a long free run or under different entraining conditions.

*Methods of Analysis of Circadian Gene Expression in* Neurospora

Our last article (Chapter 1) provided a description of methods for generating timed tissue from liquid or solid cultures. The aforementioned discussion established that individual genes can be regulated in a combinatorial manner by both developmental and rhythmic signals and that the rhythmic signals can come directly from the circadian clock or indirectly via a circadian-coordinated slave oscillator. Given this complexity, methods have been developed to try to control for noncircadian effects where possible.

In *Neurospora*, liquid cultures have provided the most widely used avenue for the production of temporally regulated tissues for analysis. Although development can be elicited by a variety of culture conditions in liquid, in general, the developmental and physiological state of the organism will be closely connected with the length of time it has been in a liquid culture. Because of this, if one simply transfers a culture from light to dark and begins to take samples every few hours, the changes in gene expression observed will reflect changes in subjective circadian time as well as changes in developmental and physiological state. Protocols to control for this have been developed, and one of these is shown in Fig. 3. The rationale of this method is that the clock-independent developmental stage will be closely correlated solely with the duration in liquid culture, whereas the subjective circadian time of a culture will be best correlated with the length of time since the light-to-dark transfer. Given this, by balancing the

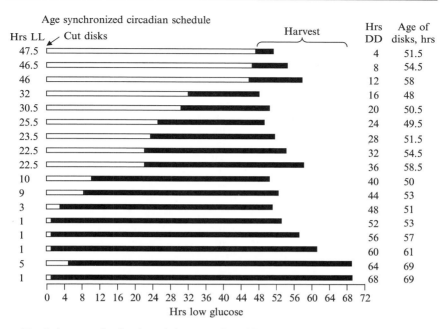

FIG. 3. An example of an inoculation, growth, and harvest schedule designed to yield tissue that is of approximately the same developmental age but represents subjective circadian times spanning several circadian cycles.

time in culture with the time in the dark, one can produce a series of parallel cultures of approximately the same developmental age but having circadian phases scanning more than two circadian cycles (Loros *et al.*, 1989, 1991). Such a sampling protocol is useful for generating tissue to examine the degree of clock-regulated expression of a gene of interest.

The aforementioned discussion concerning noncircadian rhythms, however, raises a useful caveat in that under conditions of stress, the organism may express a rhythm that is not by definition circadian. Table II provides data for a useful cross check under such growth conditions. If a gene is truly regulated by the circadian clock, then the periodicity of its temporal expression should vary in a systematic but different way in strains bearing different alleles of a gene affecting period length. Such a protocol was used by Bell-Pedersen *et al.* (1996) to produce cultures of identical developmental stage but different circadian phase for cross-comparison in a differential hybridization scheme aimed at identifying ccgs (Fig. 4).

TABLE II
CIRCADIAN TIMES IN DIFFERENT STRAINS

| Hours DD | $frq^+$ @ $25°$ $\tau = 21.5$ h | $frq^1$ @ $25°$ $\tau = 16$ h | $frq^7$ @ $25°$ $\tau = 29$ h | $frq^7$ @ $20°$ $\tau = 32$ h |
|---|---|---|---|---|
| 4 | 17 | 18 | 15 | 15 |
| 8 | 21 | 24/0 | 19 | 18 |
| 12 | 1 | 6 | 22 | 21 |
| 16 | 6 | 12 | 1 | 24/0 |
| 20 | 10 | 18 | 5 | 3 |
| 24 | 15 | 24/0 | 8 | 6 |
| 28 | 20 | 6 | 11 | 9 |
| 32 | 24/0 | 12 | 15 | 12 |
| 36 | 4 | 18 | 18 | 15 |
| 40 | 9 | 24/0 | 21 | 18 |
| 44 | 13 | 6 | 24/0 | 21 |
| 48 | 18 | 12 | 4 | 24/0 |
| 52 | 22 | 18 | 7 | 3 |
| 56 | 3 | 24/0 | 10 | 6 |
| 60 | 7 | 6 | 14 | 9 |
| 64 | 11 | 12 | 17 | 12 |
| 68 | 16 | 18 | 20 | 15 |

## Systematic Methods of Identifying Clock-Controlled Genes

The original method used for the systematic identification of ccgs in *Neurospora* was subtractive hybridization (Loros *et al.*, 1989). This method is based on the ability to compare two large and complex mixtures of nucleic acids and to enrich for those sequences present in one but not the other mixture. The method works well, but a decade of experience in many laboratories has shown that, for technical reasons, it tends to select for those sequences that are most highly expressed and tends to underrepresent transcripts from poorly expressed genes. This was also the case in *Neurospora*, where although the two genes identified in the initial subtractive hybridization screen have both proven to be extremely useful as reference standards, they are also known to be among the most highly expressed genes in the organism (Garceau, 1996; Lindgren, 1994; Zhu *et al.*, 2001). A second method that has been used is differential hybridization, in which a cDNA library is plated out and hybridized to labeled probes corresponding to two different times of day (Bell-Pedersen *et al.*, 1996): Spots corresponding to genes more highly expressed at one time of day will collect more of the radioactive probe. The methods again work efficiently, but again tend to identify genes that are among the more highly expressed genes at the time of study (Nelson *et al.*, 1997; Zhu *et al.*, 2001).

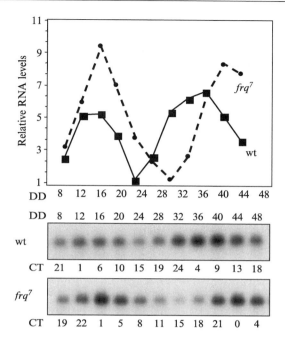

FIG. 4. Strains with different period lengths will cycle out of phase, yielding developmentally synchronous growth conditions with out-of-phase clocks. Wild-type and *frq⁷* strains have period lengths of 22 and 29 h, respectively. Note that at 16 h after the light-to-dark transition both strains are expressing this particular gene, eas(ccg-2), at high levels, but by 32 h in the dark the wild-type strain is expressing the gene at CT 24/0 while the *frq⁷* strain is at CT15, a low point.

A third collection of methods that seeks to avoid this preference for highly expressed genes is to use DNA sequencing to count the number of times an individual gene transcript is seen among a population that is sequenced. One variation of this is SAGE analysis (Velculescu *et al.*, 1995, 1997), in which a short oligonucleotide "tag" is associated with each unique transcript, but the same end result can be achieved simply by exhaustively sequencing a cDNA library (Zhu *et al.*, 2001). This method relies heavily on the means used to generate the library; in particular, experience has shown that amplification of primary libraries serves to enrich for abundant clones at the expense of inabundant ones so that the library can soon become skewed, yielding little useful information in terms of temporal expression (Zhu *et al.*, 2001). However, if the libraries are carefully prepared, this is the method that potentially provides the least bias toward abundantly expressed genes, and it suffers only from cost:

Because genes tags are simply picked at random, sequenced, and then counted, in order to identify rare transcripts, abundant transcripts must be sequenced many times. In *Neurospora* this is achievable, but less so in mammalian cells: We might assume, for the purpose of calculation, ca. 45,000 different transcripts in a mammalian cell and ca. 90,000 to 140,000 actual total transcripts varying in abundance from one to a few hundred copies. Of these, a few will vary in abundance as a function of time but most will not. Assuming random sampling of this population (120,000) in a Monte Carlo simulation, analysis of 200,000 cDNAs would provide an 81% chance of identifying a transcript present at a single copy per cell, 300,000 would give a 92% chance, and 400,000 a 97% chance. At $1.50 per sequencing run that provides the identity of 15 SAGE tags, 92% coverage would cost about $30,000 for a single time point, beyond the scope of most laboratories.

The most recent method of analysis, and one whose utility is still being expanded, is the use of microarrays. Here, instead of the "target" on the solid matrix being limited and the "probe" in solution being unlimited as is the case in classic Northern blots, the situation is reversed: The unlimiting material for hybridization (the "probe") is now in a spot on a microscope slide and the "target" in solution is limited (DeRisi *et al.*, 1997). Since its original application in a circadian context by Harmer *et al.* (2000), microarrays have been used extensively to probe quantitative and qualitative aspects of clock-regulated gene expression (Duffield, 2003). In general, microarrays still suffer some from an inability to detect rare transcripts. In addition, they are not particularly quantitative, with many studies (e.g., Duffield *et al.*, 2002; Nowrousian *et al.*, 2003) showing reduced amplitudes of expression as estimated through microarrays as compared with Northern or quantitative RT-PCR (reverse transcriptase-polymerase chain reaction) analysis. That said, microarray analysis remains the best current method for sampling clock-regulated gene expression at the cellular level. Many references are available, including our own (Duffield *et al.*, 2002; Nowrousian *et al.*, 2003), to describe methods for target collection and labeling and for normalization of one microarray to the next in a series, but a particular issue for circadian experiments concerns data handling—how to identify clock-regulated transcripts from among the noise of highly variable gene expression.

*Methods for Analyzing Circadian Time Series from a*
  *Microarray Experiment*

After normalization of the signal to control for target labeling and hybridization, we have used three different methods to find clones corresponding to clock-regulated genes. The first (Nowrousian *et al.*,

2003) was empirical: Data files were further analyzed in Excel (Microsoft, Redmond, WA), Cluster, and TreeView (Eisen *et al.*, 1998) to find clones that fulfilled the following criteria. (1) Expression patterns were consistent with clock control (one peak and one trough approximately 12 h apart in wild type connected by increasing or decreasing values, respectively, within the observed time span); (2) amplitudes were at least 2-fold in one and at least 1.8- and 1.5-fold in other biological replicate experiments; (3) peaks and troughs of expression for the individual experiments were in phase (no more than 4 h apart); and (4) several putative clock-regulated genes identified by microarray analysis were confirmed by either Northern or quantitative RT-PCR analysis. The latter is particularly important given the squashing of the amplitude of transcript levels in microarray experiments that was noted earlier. Additional details of methods for time series analysis appear in the following section on analysis of output rhythms in mammalian cell culture.

## DNA Microarray Analysis of Clock Control in Mammalian Cell Culture

The basis of the method to generate circadian rhythms in gene expression in nondividing immortalized fibroblasts was originally developed by Balsalobre *et al.* (1998). In this study, 22.5-h rhythms were observed in rat-1 cells of the canonical clock genes *rPer1* and *rPer2* and in the clock associated genes *DBP* and *Rev-erb* $\alpha$. Others working in rat-1 cells (Yagita *et al.*, 2001) and in mouse NIH3T3 cells (Duffield *et al.*, 2000) expanded this set of genes and observed circadian rhythms in a more complete set of clock genes (*per1*, *per2*, *per3*, *bmal1*, *cry1*, *cry2*) and constitutive expression of *CLOCK* as well as *casein kinase Ie/∂*. The rhythmic and constitutive expression patterns of these genes, and the phase relationships of the rhythmic genes, were found to be similar to those observed in the master circadian clock, the suprachiasmatic nucleus (SCN) of the hypothalamic brain. This highlighted quite profoundly that the molecular machinery responsible for generating rhythms in the central nervous system could be produced locally in the individual peripheral tissue, in this case the fibroblast. The ability to grow and manipulate immortalized cells (e.g., using inhibitory RNA technologies and pharmacological treatments) makes this *in vitro* model system of the molecular clock highly useful, in particular in understanding how output from the clock regulates rhythmic cell and tissue physiology through the identification of mammalian ccgs and how the circadian clock can be reset by multiple signaling pathways (Duffield *et al.*, 2002; Grundschober *et al.*, 2001; Hirota *et al.*, 2002). The development of DNA microarray technology is now allowing researchers to profile gene expression on a global scale. Its application to circadian biology has begun and increased with rapid progression, becoming a powerful tool in understanding how the molecular clock orchestrates circadian physiology and behavior via rhythmic transcriptional regulation (Duffield, 2003; Grundschober et al., 2001). These studies (Duffield et al., 2002; Grundschober et al., 2001) reported the use of DNA microarray analysis with this model in vitro tissue culture system to evaluate cellular clock control and to examine potential input pathways to the clock (Duffield et al., 2002; Hirota et al., 2002). This section focuses primarily on the methods employed by our laboratory (Duffield et al., 2002). These methods are applicable to analysis of any immortalized cell types. For example, immortalized SCN neurons and hepatoma cells have been shown to exhibit rhythmic properties of physiology and gene expression, respectively (Balsalobre et al., 1998; Earnest et al., 1999). See Fig. 5 for example experimental design for microarray analysis of ccg expression from immortalized cells.

## Cell Treatment and RNA Collection

Rat-1 or mouse NIH3T3 fibroblasts are seeded into 10-cm culture dishes ($5 \times 10^5$ cells), or scaled as necessary, and allowed to grow to confluence in Dulbecco's modified Eagle medium (DMEM) with 5% fetal calf serum (FBS) and antibiotic penicillin–streptomycin–glutamine (PSG) at 37° in 5–10% $CO_2$. Cells reach confluency after ~4 days. After 7 days, cells are treated with 50% horse serum medium (in DMEM + PSG) for 2 h (a 30-min duration also produces similar results; Duffield, Best, Loros, and Dunlap, unpublished results), which is replaced with serum-free DMEM + PSG for the duration of the experiment. Alternatively, cells can be grown in 10% FBS for 4 days, replaced with serum-free DMEM + PSG for 2 days, and then treated with 50% horse serum media. Other stimuli will equally induce circadian rhythms in clock gene expression, including glucocorticoids, retinoic acid, and glucose, along with a slew of secondary messenger activators such as the MAPK (mitogen activated protein kinase) activator, TPA (12-0-tetradecanoylphorbol-13-acetate) (Akashi and Nishida, 2000; Balsalobre, 2002; Hirota et al., 2002). Moreover, serum and retinoic acid have also been shown to reset the already rhythmic clock in tissue culture (Balsalobre, 2002; Best et al., 2002). The cells are taken at time = 0, 1, and every 4 h for a 48- to 72-h period (start of serum treatment was designated time = 0 h, and RNA was collected over 2–3 circadian days), rinsed with 1X phosphate buffered saline, the total RNA harvested, and purified using 1 ml of Trizol (Invitrogen) or RNeasy (Qiagen, Chatsworth, CA). Approximately 20–40 $\mu$g total RNA is obtained per petri dish.

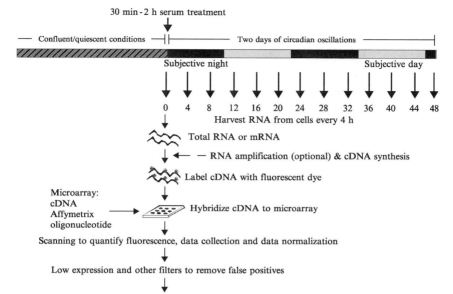

Fig. 5. An example of an experimental design for microarray profiling of clock-controlled gene expression. Rat-1 or mouse NIH3T3 cells are seeded at low density to reach confluency at day 4. Cells remain in low serum until day 6/7, at which point they are treated with an appropriate activating agent (e.g., serum or TPA). RNA is harvested from the tissue at 4-h intervals over a 48- to 72-h period. If derived from a small population of cells, mRNA is amplified by a T7 linear amplification protocol. cDNA is then synthesized and labeled with a fluorescent dye (Cy3 or Cy5; the former may give a better signal). Labeled cDNA is hybridized to microarrays, and fluorescence intensity is measured for each gene. Background level expression and other filters are used to reduce the occurrence of false positives. Data are fitted to 20- to 24-h cosine waves, and a statistical measure of goodness of fit is produced from a randomization of data points. The identified oscillating genes are validated by independent methods that quantify levels of RNA and through the inclusion on the microarray of known genes with expectant profiles (e.g., rhythmic nature of *per2* and constitutive expression of *RORα*). The identified clock-controlled genes are characterized according to phase and amplitude of their oscillation, their functional grouping, and the identification of known (e.g., E-box) and novel motifs in 5′ upstream and intronic regions of their corresponding genomic DNA sequences.

Note that as little as 1 $\mu$g total RNA is sufficient for microarray analysis if an amplification step is used. However, it is useful to have greater yields so as to be able to check the quality of RNA and for validation purposes. It is recommended that several independent time series be collected to identify consistent patterns of gene expression. Examination of different passage of cells shows no obvious influence on subsequent gene expression results (Duffield *et al.*, 2002).

*Microarray Analysis*

  *cDNA Target Preparation.* RNA is treated with DNase I and repurified. To produce acceptable levels of cDNA for array hybridization, either the RNA collection is scaled up (>200 $\mu$g total RNA, depending on the dimensions of the array matrix) or RNA is treated to one round of T7 polymerase-based linear RNA amplification by reverse transcription of RNA with the T7 promoter oligo(dT) primer. This produces ~200-fold amplification in starting mRNA, and the resulting amplified RNA is then used as template in a reverse transcription reaction in the presence of Cy3-cCTP (Amersham), producing the Cy3-dCTP-labeled fluorescent cDNA target. Full details describing the amplification and synthesis of cDNA targets are found elsewhere (Salunga *et al.*, 1999). Amplified RNA template is degraded using RNase H (10 units) and RNase A (10 units), incubated at 37° for 20 min. Targets are purified with a PCR purification kit (e.g., Qiagen, Chatsworth, CA), vacuum dried, and resuspended in 50 $\mu$l hybridization buffer [version 2 hybridization buffer (Amersham Pharmacia Biotech, Piscataway, NJ) with 50% formamide] containing human Cot1 DNA (Invitrogen).

  *cDNA Array Construction.* DNA microarrays come in essentially three forms: cDNA, oligonucleotide, and commercial oligonucleotide. Readers may consult other texts for the merits and methods of the constructing and hybridization of the different array types (Botwell and Sambrook, 2003; Duggan *et al.*, 1999; Schena, 1999). The procedure herein is for the preparation and hybridization of cDNA microarrays using a single fluorescent dye, as used by Duffield *et al.* (2002) and Luo *et al.* (1999). cDNAs as PCR products (0.6–2.4 kb) of mouse, rat, or human clones from libraries of commercial (e.g., Incyte), public (e.g., Integrated Molecular Analysis of Genome and their Expression, IMAGE, consortium), or private origin are spotted robotically onto a matrix. It is very useful to include cDNAs representing a series of control genes to monitor predicted gene expression profiles, such as canonical clock genes (e.g., *bmal1* and *per2*) and immediate early genes (e.g., *c-fos* and *per2*). It is recommended that clones are sequence verified prior to PCR amplification, as "bookkeeping" mistakes whereby sequences for genes are misassigned to incorrect wells in plates are common in EST sequencing projects. cDNAs should be printed in duplicate or up to six times on amino silane-coated slides using a micro-array spotter, such as a Generation III microarray spotter (Molecular Dynamics). cDNAs are PCR amplified and purified with a Qiagen 96 PCR purification kit and then mixed 1:1 with a 10 $M$ NaSCN printing buffer. The spots are ~250 $\mu$m in diameter with a ~280 $\mu$m center-to-center spacing. It is recommended that each microarray includes nonanimal genes for the determination of nonspecific hybridization, e.g., 15–30 plant-specific genes (Duffield *et al.*, 2002; Luo *et al.*, 1999).

*Array Hybridization.* Printed microarrays are incubated in isopropanol at room temperature for 10 min. cDNA targets (5–10 $\mu$g) are heated to 95° for 2 min, to room temperature for 5 min, and applied to the slides. The slides are covered with glass coverslips, sealed with DPX (Fluka, St. Louis, MO), and hybridized at 42° overnight. Each target is hybridized to two duplicate microarrays, each containing at least two spots for each cDNA target.

*Primary Analysis.* Fluorescence intensity for each feature of the array is obtained by using AUTOGENE (Biodiscovery, LA) or similar software (e.g. SCANALYZE, available online at http://rana.lbl.gov/EisenSoftware. htm). A single raw expression value for each gene is derived from the average of the spots representing each gene, and the coefficient of variation (CoV, standard deviation/mean × 100%) and SEM are calculated to control data quality. To determine whether a signal corresponding to a particular cDNA was reproducible between arrays, the coefficient of variation is calculated for each spot of each gene set (e.g., $n$ = 4 to 12). For example, in Duffield *et al.* (2002) the overall average CoV for all arrays within each individual experiment was 12–18%. The intensity level of each microarray is normalized so that the 75th percentile of the expression levels is equal across the microarrays. Before calculating ratios, a threshold is assigned to any gene with an expression level below a particular fluorescence intensity value; this value, derived from the plant cDNAs also spotted on the array, represents the background intensity level. Any gene for which no time point over the 48-h time course had a value greater than the background is discounted from further analysis. In Duffield *et al.* (2002), this value was 30, and the proportion of cDNA targets with less than background fluorescence values for all times across the 48-h time course was 26%, suggesting that these genes were not expressed at significant levels in rat-1 fibroblasts.

## Microarray Data Analysis

Algorithms of hierarchical clustering have been powerful tools in identifying defined patterns of expression. Several array studies of circadian biology have used clustering tools to assist in the identification of both rhythmic and induced gene expression (e.g., Duffield, 2003; Nowrousian *et al.*, 2003). "*Cluster*" and its viewing program "*TreeView*" are available as freeware (http://rana.lbl.gov/EisenSoftware.htm, Eisen *et al.*, 1998), and most commercially available microarray analysis packages include such clustering tools (e.g., *GeneSpring*, Silicon Genetics, Redwood City, CA). It is, however, limited because it requires the subjective judgment on the part of the researcher to visually inspect the full complement of gene expression within the study and select the pattern(s) of interest. Other more objective and often less laborious methods have been specifically developed to identify expression patterns of a rhythmic nature. These include a range of algorithms that have been used to objectively identify rhythmic profiles of gene expression, including autocorrelation analysis (Storch *et al.*, 2002), spectral analysis (Grundschober *et al.*, 2001), and 24-h Fourier component. A recent and very useful methods paper concerning these analyses can be found in Straume (2004).

In Duffield *et al.* (2002), cycling genes with a 20- to 28-h periodicity were identified using the computer program algorithm Circadian CORR-COS [Martin Straume, University of Virginia Center for Biomathematical Technology, Charlottesville, VA; (Harmer *et al.*, 2000; Straume, 2004); http://www.sciencemag.org/cgi/content/full/290/5499/2110/DC1]. Also see the algorithm COSOPT, a modified version of CORRCOS (Panda *et al.*, 2002). CORRCOS tests empirically for statistically significant ($p < 0.05$) cross-correlation between the time series arising from hybridization to each cDNA target and cosine waves of a specific period and phase (Straume, 2004). A significant correlation with a cosine wave of an optimal period between 20 and 28 h is deemed circadian. The analysis is independent of signal strength and amplitude of change, which makes it ideal in identifying low-amplitude rhythms of weakly expressed genes. Note that over 70% of rhythmically expressed genes identified in fibroblasts, liver, and heart have less than a 2.5-fold amplitude (Duffield, 2003). With this in mind, the investigator should be careful about setting arbitrary amplitude filters and instead use repeated measures to provide more robust confidence limits. When subjecting data to such methods, it is wise to remove data that might skew the analysis. For example, in serum-treated confluent fibroblasts, many genes are acutely induced/suppressed, thus potentially interfering with the identification of rhythmic expression of these same genes. It is therefore recommended that data from the first 4 h following stimulation of cells be removed prior to subjecting data to analysis.

A prototype to another cross-correlation-based algorithm, *rhythmic analysis of gene expression* (RAGE) (Langmead *et al.*, 2003), was initially developed using microarray data of Duffield *et al.* (2002) and resulted in the identification of a similar, but not identical, set of rhythmic genes to that from using CORRCOS (Duffield, Langmead, Loros, and Dunlap, unpublished data). Like CORRCOS, this algorithm uses autocorrelations, but uses phase-independent transformations of data rather than examining multiple phases and the Hausdorff distance metric instead of the correlation coefficient to cluster expression profiles.

Deciding criteria to identify ccgs with confidence should be given some cautionary consideration. It is apparent from the collective of array studies of circadian biology thus far that there is no miracle formula, and where one study may have included false negatives, another has provided false positives. It cannot be overstated that repeated measures can reduce error in a gamut of expression changes that are mostly in the 1.3- to 2.5-fold range (Duffield, 2003). Inclusion of positive and negative control genes on the array can assist in this selection process. A strong recommendation is to have a sliding scale of stringencies, thereby producing a table of genes in which the researcher can be most confident, followed by other supplementary lists of genes of sequentially lower confidence. This high stringency issue may explain why published studies employing similar experimental methods and subject species/tissue have resulted in a remarkably low overlap in ccgs identified (Duffield, 2003). This approach would enhance the power of cross-study comparisons.

More comprehensive reviews of microarray studies of circadian rhythms, including details of analysis type, are available (Duffield, 2003; Sato *et al.*, 2003). For more details on analyses used, we refer the reader to the reports of original array studies.

Other patterns of expression can be identified using standard parametric and nonparametric statistical methods, such as the *t* test for a two-sample comparison and analysis of variance (ANOVA). For example, in Duffield *et al.* (2002), the study was interested in analyzing immediate early gene induction following serum treatment. Early response genes were identified using a simple criteria: genes exhibiting a $\geq 1.4$-fold increase in expression at 1-h postserum shock versus time at 0 h using the Student's *t*-test ($p < 0.05$, with a sample size of 5); followed at 4–8 h by a drop to or below time at 0-h levels or to subsequent continuous basal levels; and only genes with maximal expression at time at 1 h versus 4 h.

*Validation of Microarray Data*

It is important to sequence verify the clones used to create the array, as there is often a 1–5% error rate with the fabrication of cDNA microarrays; *Affymetrix* arrays are also not free of annotation errors (Knight, 2001). It is recommended that the identities of the genes of interest are reappraised by performing BLAST searches of the public GenBank database. This and experimental error are reasons why validation using independent techniques is deemed a critical part of a gene-profiling study by microarray analysis. While there are several different methods available, namely *in situ* hybridization, Northern blot, RNase protection assay, quantitative RT-PCR, and real-time quantitative RT-PCR (Brent *et al.*, 2003), real-time quantitative RT-PCR is probably the most sensitive method available to quantify mRNA. This explains its increasing popularity in the validation of microarray data (Duffield, 2003; Duffield *et al.*, 2002; Grundschober *et al.*, 2001). Whatever method is employed, for added confidence it is recommended that a combination of experimental samples be used: (1) identical RNA/cDNA samples as used for the microarray hybridization and (2) samples prepared independently as part of a separate experimental run. In all these methods, another important issue is deciding upon a suitable constitutively expressed gene(s) against which expression of the test genes is to be normalized. GAPDH has been used with success for time-of-day specific samples from SCN and liver (e.g., Ueda *et al.*, 2002), but was found to be unsuitable for rat-1 cells (Duffield *et al.*, 2002) and is known to be a ccg in both microbial and mammalian systems (e.g., Fagan *et al.*, 1999; Iwasaki *et al.*, 2004; Shinohara *et al.*, 1998). Normalization to *casein kinase Iδ, ROR* α, and *tbp* expression are suitable for rat-1 cells (Balsalobre *et al.*, 1998; Duffield, unpublished results; Duffield *et al.*, 2002). *GAPDH* and ROR α have been appropriate for mouse NIH3T3 cells (Akashi and Nishida, 2000; Duffield, unpublished results). Unfortunately, the old favorite, β-*actin*, is not suitable for serum-treated fibroblasts, as it is elevated during the first 12 h following serum treatment. In the case of Northern blots, signals can be normalized to ethidium bromide-stained 28S ribosomal band signals on the membrane.

*Deciphering Data*

Once the researcher has identified a set of *bona fide* rhythmic genes, the next step is making biological sense of their relationship to one another and how their expression might be controlled. Genes may be grouped according to known function and structure and by information derived from their behavior observed from their expression profiles: acrophase and amplitude of rhythm or magnitude of induction. Cellular function can be assigned to the ccgs based on literature searches and classification schemes utilized by databases such as the *expressed gene anatomy database* [*EGAD*, http://www.tigr.org/docs/tigrscripts/egad_scripts/role_report.spl; (White and Kerlavage, 1996)] and *Kyoto Encyclopedia of Genes and Genomes* (*KEGG*) (http://www.kegg.com).

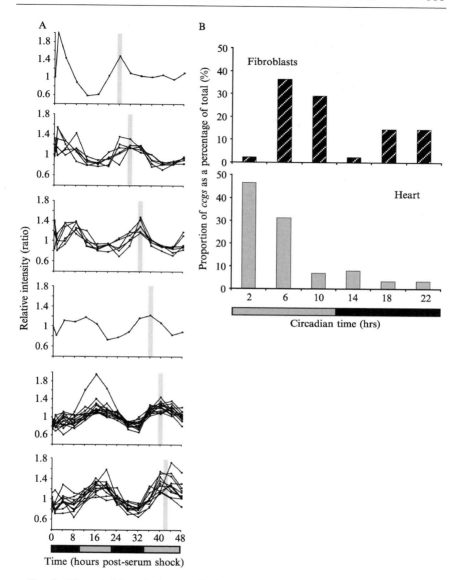

FIG. 6. Characterizing clock control of gene expression by microarray analysis. (A) Different clock-controlled genes (ccgs) peak at different phases of the circadian cycle. Circadian rhythms of gene expression in rat-1 fibroblasts following treatment by serum at time = 0 h to induce rhythmicity, identified by microarray analysis (Duffield *et al.*, 2002). Six profiles of expression were identified clustered according to the phase of the second peak

Genes grouped according to their acrophase can provide information on temporal activity along one or several biochemical or regulatory pathways (see Fig. 6A). For example, in rat-1 cells, *ccgs* associated with protein turnover, such as ubiquitin-conjugating enzymes and components of the proteasome complex, have peak phases of 40-h postserum treatment (predicted midlate day) (Duffield *et al.*, 2002). A similar relationship was seen for members of the Ras/MAPK pathway, highlighting the possibility of phase-specific sensitivity of the fibroblast to extracellular changes at the light-to-dark transition (Duffield *et al.*, 2002). Also associated with the peak in these Ras/MAPK signaling pathway components are components of the actin cytoskeleton, highlighting a possible time-specific gating of fibroblast migration. Grouping ccgs by phase is also helpful when searching for common potential promoter regions as candidate circadian consensus sites within the 5′ upstream regions of the genes of interest (Duffield, 2003; Duffield *et al.*, 2002; Harmer *et al.*, 2000; Sato *et al.*, 2003; Ueda *et al.*, 2002). Histograms of the circadian time of peak expression of the entire population of identified ccgs allows appreciation of general physiological activity of the tissue of interest. For example, rhythmic genes identified in rat-1 cells show peak activity toward the middle end of the subjective day and middle end of subjective night, with little activity at the beginning of day and night (Duffield *et al.*, 2002). This is in stark contrast to other tissues. For example, the heart shows most activity clustered around the beginning of the subjective day (Duffield, 2003; Storch *et al.*, 2002) (see Fig. 6B).

Because the tissue culture system is not directly responsive to natural external time cues, it is only possible to measure endogenous free-running circadian rhythms. However, if we assume that immortalized fibroblasts are behaving in a similar manner to other peripheral tissues *in vivo*, it is possible to estimate circadian time and subjective day and night from the peak and

---

starting at $t = 24$ h (top) through to $t = 44$ h (bottom). Values are the mean ratio of time-specific expression relative to medial expression value from five independent experiments. Circadian day and night were predicted from peak phases of known clock genes (*per2* and *bmal1*), illustrated by horizontal gray:black (subjective day:subjective night) bars. Peak phases of rhythms are highlighted by vertical gray bars. Adapted with permission from Duffield *et al.* (2002). (B) Phase histograms of the proportions of ccgs identified in rat-1 fibroblasts and the murine heart, revealing different phase-specific patterns of ccg expression. ccgs distributed according to the phase of peak expression. Circadian time refers to an internal measure of time of day under constant lighting conditions and where CT12 is defined as activity onset of overt locomotor activity of the mouse, which may be measured or inferred. Circadian day and night are indicated by the horizontal gray:black (subjective day:subjective night) bars below the histograms. Analysis is based on comparing gene expression data from different tissues and studies (Duffield *et al.*, 2002; Storch *et al.*, 2002).

nadir phases of the rhythms of known rhythmic clock genes (e.g., *per2* and *bmal1*) that behave predictably under entrained and constant conditions. Using this approach *in vivo*, it is estimated that 24- to 32-h postserum shock represents night phase and 36- to 44-h postserum shock represents day phase (Duffield *et al.*, 2002). Defining the phase of the identified *ccg* rhythms in such a manner can assist in placing gene expression patterns in a greater physiological context.

## Concluding Remarks

Studies focused on the analysis of circadian output are likely to increase in prominence in future years as the field of circadian rhythms research moves on from determining the mechanism of circadian oscillators to understanding the biology of chronobiology. In efforts to understand how clock regulation works at the level of the nucleus and its surrounding cytoplasm, studies on unicellular models including both *Neurospora* and mammalian cells in culture will remain important. The level of understanding of the *Neurospora* oscillator is advanced with most of its components being known, providing an excellent context in which to study output. Recent efforts are beginning to describe a constellation of noncircadian slave oscillators that surround and are coordinated by the circadian oscillator to comprise a circadian system. At the technical level, dependable protocols exist for the production of time tissue samples and their analysis by Northern or Western analysis or through the use of microarrays. In mammalian cells, techniques such as siRNA, allowing the transient knockdown of genes, and bioluminescence-based reporter systems are enhancing the tractability of the system. There is every reason to expect that work on these systems will continue to inform work on animal and plant as well as fungal clocks, in just the way good models should.

## References

Akashi, M., and Nishida, E. (2000). Involvement of the MAP kinase cascade in resetting of the mammalian circadian clock. *Genes Dev.* **14**, 645–649.

Aronson, B. D., Johnson, K. A., and Dunlap, J. C. (1994). The circadian clock locus *frequency:* A single ORF defines period length and temperature compensation. *Proc. Nat. Acad. Sci. USA* **91**, 7683–7687.

Balsalobre, A. (2002). Clock genes in mammalian peripheral tissues. *Cell Tissue Res.* **309**, 193–199.

Balsalobre, A., Damiola, F., and Schibler, U. (1998). A serum shock induces circadian gene expression in mammalian culture cells. *Cell* **93**, 929–937.

Bell-Pedersen, D., Shinohara, M., Loros, J., and Dunlap, J. C. (1996). Circadian clock-controlled genes isolated from *Neurospora crassa* are late night to early morning specific. *Proc. Nat. Acad. Sci. USA* **93**, 13096–13101.

Best, J. D., Duffield, G. E., Loros, J. J. and Dunlap, J. C. (2002). A molecular circadian clock similar to that of the suprachiasmatic nucleus is present in immortalized mouse NIH3T3 fibroblasts. *In* "3rd Forum of European Neuroscience," [Abstract 24.6].

Bobrowicz, P., Pawlak, R., Correa, A., Bell-Pedersen, D., and Ebbole, D. J. (2002). The *Neurospora crassa* pheromone precursor genes are regulated by the mating type locus and the circadian clock. *Mol. Microbiol.* **45,** 795–804.

Botwell, D., and Sambrook, J. (2003). "DNA Microarrays: A Molecular Cloning Manual." Cold Spring Harbor Laboratory Press, Cold Spring Harbor, NY.

Brent, R., Kingston, R. E., Seidman, J. G., Strukl, K., Ausubel, F. M., Benson Chanda, V., Moore, D. D., Seidman, J. G., and Ausubel, F. M. (2003). "Current Protocols in Molecular Biology." Wiley, New York.

Bünning, E. (1973). "The Physiological Clock" 3rd Ed., Springer-Verlag, New York.

Christensen, M., Falkeid, G., Hauge, I., Loros, J. J., Dunlap, J. C., Lillo, C., and Ruoff, P. (2004). A *frq*-independent nitrate reductase rhythm in *Neurospora crassa. J. Biol. Rhythms* **19,** 280–286.

Correa, A., Lewis, Z. A., Greene, A. V., March, I. J., Gomer, R. H., and Bell-Pedersen, D. (2003). Multiple oscillators regulate circadian gene expression in *Neurospora. Proc. Natl. Acad. Sci. USA* **100,** 13597–13602.

Davis, R. H. (2000). "Neurospora: Contributions of a Model Organism." Univ. Press, Oxford.

DeRisi, J. L., Vishwanath, R. I., and Brown, P. O. (1997). Exploring the metabolic and genetic control of gene expression on a genomic scale. *Science* **278,** 680–686.

Duffield, G. E. (2003). DNA microarray analyses of circadian timing: The genomic basis of biological time. *J. Neuroendocrinol.* **15,** 991–1002.

Duffield, G. E., Best, J. D., Meurers, B. H., Bittner, A., Loros, J. J., and Dunlap, J. C. (2002). Circadian programs of transcriptional activation, signaling, and protein turnover revealed by microarray analysis of mammalian cells. *Curr. Biol.* **12,** 551–557.

Duffield, G. E., Best, J. D., Wahleithner, J. A., Schwartz, W. J., Loros, J. J. and Dunlap, J. C. (2000). Transcriptional profiling of central and peripheral mammalian circadian clocks. *In* "7th Meeting of the Society for Research on Biological Rhythms." [Abstract 37].

Duggan, D. J., Bittner, M., Chen, Y., Meltzer, P., and Trent, J. M. (1999). Expression profiling using cDNA microarrays. *Nature Genet.* **21**(1 Suppl.), 10–14.

Dunlap, J. C., and Loros, J. J. (2004). The Neurospora circadian system. *J. Biol. Rhythms* **19,** 414–424.

Earnest, D. J., Liang, F. Q., Ratcliff, M., and Cassone, V. M. (1999). Immortal time: Circadian clock properties of rat suprachiasmatic cell lines. *Science* **283,** 693–695.

Eisen, M. B., Spellman, P. T., Brown, P. O., and Botstein, D. (1998). Cluster analysis and display of genome-wide expression patterns. *Proc. Natl. Acad. Sci. USA* **95,** 14863–14868.

Fagan, T., Morse, D., and Hastings, J. W. (1999). Circadian synthesis of a nuclear-encoded chloroplast glyceraldehyde-3-phosphate dehydrogenase in the dinoflagellate *Gonyaulax polyedra* is translationally controlled. *Biochem.* **38,** 7689–7695.

Feldman, J., and Hoyle, M. N. (1974). A direct comparison between circadian and noncircadian rhythms in *Neurospora crassa. Plant Physiol.* **53,** 928–930.

Field, R. J., and Burger, M. (1985). "Oscillations and Traveling Waves in Chemical Systems." Wiley, New York.

Garceau, N. (1996). "Molecular and Genetic Studies on the *frq* and *ccg-1* loci of Neurospora." Ph.D., Dartmouth.

Grundschober, C., Delaunay, F., Puhlhofer, A., Triqueneaux, G. V. L., Bartfai, T., and Nef, P. (2001). Circadian regulation of diverse gene products revealed by mRNA expression profiling of synchronized fibroblasts. *J. Biol. Chem.* **276,** 46751–46758.

Harmer, S. L., Hogenesch, J. B., Straume, M., Chang, H.-S., Han, B., Zhu, T., Wang, X., Kreps, J. A., and Kay, S. A. (2000). Orchestrated transcription of key pathways in *Arabidopsis* by the circadian clock. *Science* **290,** 2110–2113.

Hirota, T., Okano, T., Kokame, K., Shirotani-Ikejima, H., Miyata, T., and Fukada, Y. (2002). Glucose down-regulates Per1 and Per2 mRNA levels and induces circadian gene expression in cultured Rat-1 fibroblasts. *J. Biol. Chem.* **277,** 44244–44251.

Iwasaki, H., and Dunlap, J. C. (2000). Microbial circadian oscillatory systems in *Neurospora* and *Synechococcus*: Models for cellular clocks. *Curr. Opin. Microbiol.* **3,** 189–196.

Iwasaki, T., Nakahama, K., Nagano, M., Fujioka, A., Ohyanagi, H., and Shigeyoshi, Y. (2004). A partial hepatectomy results in altered expression of clock-related and cyclic glyceraldehyde 3-phosphate dehydrogenase (GAPDH) genes. *Life Sci.* **74,** 3093–3102.

Knight, J. (2001). When the chips are down. *Nature* **410,** 860–861.

Lakin-Thomas, P. (1996). Effects of choline depletion on the circadian rhythm in *Neurospora crassa*. *Biol. Rhythm Res.* **27,** 12–30.

Lakin-Thomas, P. (1998). Choline depletion, *frq* mutations, and temperature compensation of the circadian rhythm in *Neurospora crassa*. *J. Biol. Rhythms* **13,** 268–277.

Lakin-Thomas, P., Coté, G., and Brody, S. (1990). Circadian rhythms in *Neurospora*. *CRC Crit. Rev. Micro.* **17,** 365–416.

Lakin-Thomas, P. L., and Brody, S. (2000). Circadian rhythms in *Neurospora crassa*. *Proc. Natl. Acad. Sci. USA* **97,** 256–261.

Langmead, C. J., Yan, A. K., McClung, C. R., and Donald, B. R. (2003). Phase-independent rhythmic analysis of genome-wide expression patterns. *J. Comput. Biol.* **10,** 521–536.

Lindgren, K. M. (1994). "Characterization of *ccg-1*, a Clock-Controlled Gene of *Neurospora crassa*," Ph.D, Dartmouth.

Loros, J., and Dunlap, J. C. (1991). *Neurospora crassa* clock-controlled genes are regulated at the level of transcription. *Mol. Cell. Biol.* **11,** 558–563.

Loros, J. J., Denome, S. A., and Dunlap, J. C. (1989). Molecular cloning of genes under the control of the circadian clock in *Neurospora*. *Science* **243,** 385–388.

Loros, J. J., and Feldman, J. F. (1986a). Loss of temperature compensation of circadian period length in the *frq-9* mutant of *Neurospora crassa*. *J. Biol. Rhythms* **1,** 187–198.

Loros, J. J., Hastings, J. W., and Schibler, U. (2003). Adapting to life on a rotating world at the gene expression level. *In* "Chronobiology: Biological Timekeeping" (J. C. Dunlap, J. J. Loros, and P. Decoursey, eds.), pp. 254–289. Sinauer, Sunderland, MA.

Loros, J. J., Lichens-Park, A., Lindgren, K., and Dunlap, J. C. (1993). Molecular genetics of genes under circadian temporal control in *Neurospora*. *In* "Molecular Genetics of Biological Rhythms" (M. W. Young, ed.), pp. 55–72. Dekker, New York.

Loros, J. J., Richman, A., and Feldman, J. F. (1986b). A recessive circadian clock mutant at the *frq* locus in *Neurospora crassa*. *Genetics* **114,** 1095–1110.

Luo, L., Salunga, R., Guo, H., Bittner, A., Joy, K., Galindo, J., Xiao, H., Rogers, K., Wan, J., Jackson, M., and Erlander, M. (1999). Gene expression profiles of laser captured adjacent neuronal subtypes. *Nature Med.* **5,** 117–122.

Merrow, M., Bruner, M., and Roenneberg, T. (1999). Assignment of circadian function for the *Neurospora* clock gene *frequency*. *Nature* **399,** 584–586.

Merrow, M., Roenneberg, T., Macino, G., and Franchi, L. (2001). A fungus among us: The *Neurospora crassa* circadian system. *Semin. Cell Biol. Dev.* **12,** 279–285.

Nelson, M. A., Kang, S., Braun, E., Crawford, M., Dolan, P., Leonard, P., Mitchell, J., Armijo, A., Bean, L., Blueyes, E., and Natvig, D. (1997). Expressed sequences form conidial, mycelial, and sexual stages of *Neurospora*. *Fung. Genet. Biol.* **21,** 348–363.

Nowrousian, M., Duffield, G. E., Loros, J. J., and Dunlap, J. C. (2003). The *frequency* gene is required for temperature-dependent regulation of many clock-controlled genes in *Neurospora crassa. Genetics* **164,** 922–933.

Panda, S., Antoch, M. P., Miller, B., Su, A., Schook, A., Straume, M., Schultz, P., Kay, S., Takahashi, J., and Hogenesch, J. B. (2002). Coordinated transcription of key pathways in the mouse by the circadian clock. *Cell* **109,** 307–320.

Pittendrigh, C., and Bruce, V. (1959). Daily rhythms as coupled oscillator systems and their relation to thermoperiodism and photoperiodism. *In* "Photoperiodism and Related Phenomena in Plants and Animals" (R. B. Withrow, ed.), pp. 475–505. AAAS, Washington, DC.

Ramsdale, M., and Lakin-Thomas, P. L. (2000). sn-1,2-Diacylglycerol levels in the fungus *Neurospora crassa* display circadian rhythmicity. *J. Biol. Chem.* **275,** 27541–27550.

Roenneberg, Y., and Merrow, M. (1998). Molecular circadian oscillators: An alternative hypothesis. *J. Biol. Rhythms* **13,** 167–179.

Ruoff, P., and Slewa, I. (2002). Circadian period lengths of lipid mutants (*cel, chol-1*) of *Neurospora* show defective temperature, but intact pH-compensation. *Chronobiol. Int.* **19,** 517–529.

Salunga, R. C., Hongqing, G., Luo, L., Bittner, A., Joy, K. C., Chambers, J. R., Jackson, S. W., Jackson, M. R., and Erlander, M. G. (1999). Gene expression analysis via cDNA microarrays of laser capture microdissected cells from fixed tissue. *In* "DNA Microarrays: A Practical Approach," pp. 121–137. Oxford Univ. Press, Oxford.

Sargent, M. L., Briggs, W. R., and Woodward, D. O. (1966). The circadian nature of a rhythm expressed by an invertaseless strain of *Neurospora crassa. Plant Physiol.* **41,** 1343–1349.

Sargent, M. L., and Kaltenborn, S. H. (1972). Effects of medium composition and carbon dioxide on circadian conidiation in *Neurospora. Plant Physiol.* **50,** 171–175.

Sato, T., Panda, S., Kay, S., and Hogenesch, J. B. (2003). DNA arrays: Applications and implications for circadian biology. *J. Biol. Rhythms* **18,** 96–105.

Schena, M. (1999). "DNA Microarrays: A Practical Approach." Oxford Univ. Press, Oxford.

Shinohara, M., Loros, J. J., and Dunlap, J. C. (1998). Glyceraldehyde-3-phosphate dehydrogenase is regulated on a daily basis by the circadian clock. *J. Biol. Chem.* **273,** 446–452.

Storch, K. F., Lipan, O., Leykin, I., Viswanathan, N., Davis, F. C., Wong, W., and Weitz, C. J. (2002). Extensive and divergent circadian gene expression in liver and heart. *Nature* **417,** 78–83.

Straume, M. (2004). DNA microarray time series analysis: Automated statistical assessment of circadian rhythms in gene expression patterning. *Methods Enzymol.* **383,** 149–166.

Sussman, A. S., Durkee, T., and Lowrey, R. J. (1965). A model for rhythmic and temperature-independent growth in the "clock" mutants of *Neurospora. Mycopathol. Mycol. Appl.* **25,** 381–396.

Sussman, A. S., Lowrey, R. J., and Durkee, T. (1964). Morphology and genetics of a periodic colonial mutant of *Neurospora crassa. Am. J. Bot.* **51,** 243–252.

Ueda, H. R., Chen, W., Adachi, A., Wakamatsu, H., Hayashi, S., Takasugi, T., Nagano, M., Nakahama, K., Suzuki, Y., Sugano, S., *et al.* (2002). A transcription factor response element for gene expression during circadian night. *Nature* **418,** 534–539.

Velculescu, V., Zhang, L., Vogelstein, B., and Kinzler, K. (1995). Serial analysis of gene expression. *Science* **270,** 484–487.

Velculescu, V., Zhang, L., Zhou, W., Vogelstein, J., Basrai, M., Bassett, D., Hieter, P., Vogelstein, B., and Kinzler, K. (1997). Characterization of the yeast transcriptome. *Cell* **88,** 243–251.

White, O., and Kerlavage, A. R. (1996). TDB: New databases for biological discovery. *Methods Enzymol.* **266**, 27–40.

Yagita, K., Tamanini, F., van Der Horst, G. T., and Okamura, H. (2001). Molecular mechanisms of the biological clock in cultured fibroblasts. *Science* **13**, 278–281.

Zhu, H., Nowrousian, M., Kupfer, D., Colot, H., Berrocal-Tito, G., Lai, H., Bell-Pedersen, D., Roe, B., Loros, J. J., and Dunlap, J. C. (2001). Analysis of expressed sequence tags from two starvation, time-of-day-specific libraries of *Neurospora crassa* reveals novel clock-controlled genes. *Genetics* **157**, 1057–1065.

# [15] Molecular and Statistical Tools for Circadian Transcript Profiling

By Herman Wijnen,[1] Felix Naef,[1] and Michael W. Young

## Abstract

This article describes methods used to evaluate mRNA expression patterns on microarrays and their application in circadian biology. With the intention of complementing rather than duplicating the existing literature, particular emphasis is placed on experimental design, data analysis techniques, and independent verification. Both comparative and temporal study designs are discussed, and their use in circadian research is illustrated with examples. Data analysis methods to assess periodic components in time series data are outlined in detail.

## Array Platforms

A number of different platforms for producing and analyzing nucleic acid microarrays are currently available. Commonly, collections of larger DNA fragments or shorter oligonucleotides are arrayed onto a solid support surface. Commercially available formats include high-density oligonucleotide arrays produced using photolithography and combinatorial chemistry (Affymetrix) or ink-jet technology (Agilent), medium-density arrays containing long oligonucleotides printed onto plastic film or pretreated glass slides (e.g., BD Biosciences), and low-density arrays of cDNA fragments printed on nylon membranes (e.g., BD Biosciences, Schleicher and Schuell Bioscience). Many laboratories print their own medium- or low-density microarrays using either cDNA fragments or oligonucleotides. An elaborate discussion of the technologies associated with the various microarray platforms is beyond the scope of this article. For detailed information

---

[1] Herman Wijnen and Felix Naef contributed equally to this chapter.

concerning commercially available array platforms, we refer the reader to the manufacturers, and extensive instructions for printing microarrays can be found elsewhere (Bowtell and Sambrook, 2003; Brownstein and Khodursky, 2003; Schena, 2003; http://microarrays.org, http://research. nhgri.nih.gov/microarray/, http://cmgm.stanford.edu/pbrown, http://www. protocol-online.org/prot/Genetics_Genomics/Microarray/index.html). The choice of microarray technology has important implications for cost, flexibility of array design, and experimental design. For example, the widely used Affymetrix GeneChip platform is relatively expensive and inflexible and does not offer the option of simultaneous cohybridization of experimental and reference samples. Nevertheless, Affymetrix GeneChips are a popular choice because they offer the advantage of highly standardized arrays and protocols that do not require a great deal of technical skill and experience on the part of the investigator and tend to produce data sets of reliable quality. In contrast, cDNA microarrays can be custom printed at relatively low cost on glass slides and this technology is suited to cohybridizations. The disadvantages associated with this approach are a greater variability in array quality, cross-hybridization between homologous sequences, and greater demands on the time and technical skills of the investigator. Many academic institutions now have their own facilities dedicated to microarray technology and data analysis. Scientists who are planning their first microarray experiments may want to adjust their choice of array platform and experimental design to the preexisting resources available to them locally.

## Experimental Design

This section focuses on general principles of experimental design that are applicable to all types of array technology. If simultaneous cohybridization of two differently labeled targets is used, some additional considerations apply, such as the use of dye-swapping controls and the use of a standard reference sample versus alternative designs. Detailed descriptions of these issues can be found elsewhere (e.g., Kerr, 2003a; Yang and Speed, 2003).

Two basic types of microarray experiments used to study circadian biology are discussed (1) simple comparisons between two or more conditions (treatments, genotypes, etc.) and (2) temporal studies of relatively uniform populations.

### Simple Comparisons

Simple comparison experiments offer a straightforward approach applicable to a wide spectrum of biological questions. It is relatively easy to compare and contrast microarray data sets obtained for two different experimental conditions. It is important to keep in mind, however, that differences in the results reflect not only changes in the biological property of interest, but also unintended experimental and biological variation. The effect of unwanted sources of experimental and biological variability can be reduced by collecting and processing the samples representing the different conditions in parallel using standardized procedures. However, in order to verify that the obtained results are reproducible and robust to slight variations in experimental procedures or biological background, it is necessary to repeat a given experiment several times independently. Thus, a successful experimental design for a simple comparison consists of several independent repetitions of an experiment in which the different conditions are tested in parallel. While it is essential to maintain uniformity in technical procedure and biological background within each experimental repeat, depending on the relative emphasis placed on sensitivity of the assay versus robustness of the interpretation, one could decide to allow limited differences in experimental procedure or biological background between the independent repeats of an experiment. For example, a microarray study aimed at identifying differences in mRNA expression associated with loss-of-function mutations at a particular genetic locus could consist of several independent experiments, each comparing a different but similar mutant allele to a wild-type control in the appropriate genetic background. Studies of the effect of genetic mutations are inherently susceptible to the confounding effect associated with differences between the genetic backgrounds of mutant and wild type. Mutant/wild-type pairs with isogenic backgrounds are not always available. An acceptable alternative may be the inclusion of a "wild-type" control in which the mutation has been rescued transgenically. Aside from genetic background, another complication associated with simple mutant/wild-type comparisons is that they do not distinguish direct from indirect effects. Direct transcriptional responses can be studied with the use of inducible/repressible expression systems or conditional mutations. An important question in experimental design is how many repetitions of a given experiment to perform. In practice, exploratory studies can be performed with a relatively small number, but the predictive value of microarray results can be enhanced greatly by adding a substantial number of experimental iterations. There is a trade-off between the statistical power gained by performing more repetitions and the associated effort and cost. If in doubt, more replications are likely advisable.

As an example of a simple comparison microarray study addressing a problem in circadian biology, we consider an exploratory analysis of transcript profiles in the heads of different arrhythmic mutants of *Drosophila* (Claridge-Chang *et al.*, 2001). Circadian transcripts fail to oscillate in arrhythmic *Drosophila* carrying a *period-0* (*per⁰*), *timeless-01* (*tim⁰¹*), or *Clock-jrk* (*Clk^{jrk}*) mutation (Claridge-Chang *et al.*, 2001; Lin *et al.*, 2002; McDonald and Rosbash, 2001; H. Wijnen, F. Naef, C. Boothroyd, and M. W. Young, unpublished results), but they do sometimes show a mutation-dependent up- or downregulation of average daily expression levels relative to wild type. For example, mRNA abundance of *per, tim*, and *vrille* (*vri*) genes is downregulated in *Clk^{jrk}* flies, but is maintained at intermediate or high levels in *per⁰* or *tim⁰¹* flies, whereas the *Clk* gene is oppositely regulated (Allada *et al.*, 1998; Blau and Young 1999; Hardin *et al.*, 1990; Sehgal *et al.*, 1994, 1995). A comparative microarray study was conducted to detect changes in average daily transcript levels in the three aforementioned arrhythmic mutants. For each mutant, an expanded population of flies was reared in the presence of an environmental 12-h light/12-h dark cycle and samples were collected at 6-h intervals for 1 day. The four samples for each mutant were processed and separately hybridized to microarrays. The transcript profiles for each of the mutants were compared with a larger set of wild-type microarray data that had been collected previously (Claridge-Chang *et al.*, 2001). Since the goal of this study was to identify mutant-specific patterns of constitutively high or low expression throughout the day, different time points were treated as independent replicates. The Wilcoxon rank-sum test was used to search for significant differences between mutant and wild-type expression levels throughout the day. The percentage of genes with detectable expression in wild-type heads that showed a significant prediction ($p \leq 0.05$) for up- or downregulation in the *tim⁰¹*, *per⁰*, or *Clk^{jrk}* mutant was, respectively, 31, 19, and 27%. A naïve interpretation of these observations suggests that a conservative estimate of the genes regulated by TIM, PER, and CLK can be obtained by subtracting the 5% rate of false positives expected from a random data set (at $p \leq 0.05$). However, these estimated numbers of regulated genes are likely to be artificially high due to experimental variability, population-specific effects, and differences in genetic background. Interpretation of the results was initially focused on a set of genes for which a response to clock defects was predicted *a priori* because they showed strong circadian oscillations at the transcript level. Included in this set were a number of genes with confirmed responses in average daily expression levels. The Wilcoxon rank-sum test correctly predicted regulation for the *tim, vri, to*, and *Clk* transcripts (Fig. 1; Claridge-Chang *et al.*, 2001). Based on the fact that all three mutations affect the activity of the transcription factor CLK, a positive correlation between the three Wilcoxon rank-sum test *p* values was anticipated. The Spearman rank correlation coefficients calculated for the three pairwise comparisons of the *p* values indicated that this is indeed the case, both for the set of circadian transcripts and in the broader context

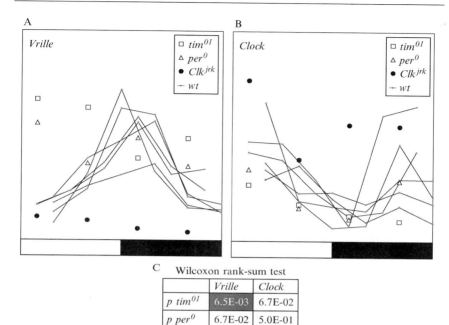

FIG. 1. An example of comparative microarray analysis. (A and B) Microarray signals for *vrille* and *Clock* genes from three arrhythmic mutants ($per^0$, $tim^{01}$, and $Clk^{jrk}$) and wild-type control data. Mutant data are indicated as separate time points, whereas wild-type data are indicated as six line graphs representing time course measurements during 3 LD and 3 DD days. Daily time on the $x$ axis is indicated relative to the daily 12-h light/12-h dark cycle used prior to or during collection of data. Expression is plotted as noise-corrected average difference values (MAS4.0). (C) Wilcoxon rank-sum test probabilities comparing expression for *vrille* and *Clock* in each of the three arrhythmic mutants to wild type. Different time points are treated as replicate observations. Shaded cells indicate $p$ values $\leq 0.05$.

of all genes expressed in the wild-type fly head. The $per^0$ and $tim^{01}$ mutations both produce effects opposite to those of the $Clk^{jrk}$ mutation because they result in defective repression of the CLK transcription factor. Consistent with the similar molecular defects of $per^0$ and $tim^{01}$, none of the genes with strong circadian regulation showed opposing effects in absolute expression levels in response to these two mutations (Claridge-Chang *et al.*, 2001). Confidence in the value of this analysis as an exploratory study was further reinforced by independent confirmation of the observed microarray profiles for the *CG5798* and *Slob* transcripts on Northern blots (data not shown). Although somewhat cumbersome, it would be feasible to conduct independent verification analyses using Northern blots or quantitative polymerase chain reaction (PCR) for the 72 robustly oscillating transcripts whose average expression levels are predicted to respond to $per^0$, $tim^{01}$, or $Clk^{jrk}$. A similar strategy would, however, not be a realistic option for the total set of >3000 transcripts that are both detected at reliable levels in wild type and predicted to show up- or downregulation in one or more of the arrhythmic mutants. The best strategy for distinguishing mutation-dependent effects from population-specific fluctuations or effects due to differences in genetic background in this larger set would be to expand the microarray study with independent repetitions using different populations and isogenic genetic backgrounds.

## Time Course Experiments

The second basic type of microarray experiment discussed here is a temporal study of a relatively uniform population. Examples of such experiments are time course analyses of the effects of a controlled stimulus (e.g., drug treatment, induction of transgenes or conditional phenotypes) or intrinsic biological regulation (e.g., developmental patterns, cell cycle, circadian rhythms). Most of the experimental design considerations discussed earlier for simple comparisons also apply in the case of time course studies. An advantage of the time course design is that samples for a given experiment are all derived from a single relatively homogeneous population, making the results much less sensitive to population-specific effects or slight differences in genetic background. Experimental designs can be tailored to the specific selection of patterns representing the biological regulation of interest by using multiple independent samplings, by incorporating controls to avoid confounding patterns, and by capitalizing on the intrinsic properties of the patterns of interest. For example, targets of a particular gene encoding a transcription factor may be studied using an inducible transgene in a null-mutant context. The first sample of a typical time course would be taken before induction; subsequent samples would then be collected shortly after induction and at various later time points. The influence of confounding effects unrelated to the transgene (e.g., temperature-dependent effects when a heat shock promoter is used) could be reduced by conducting a parallel time course for an isogenic control lacking the inducible transgene. An even more sophisticated design could include separate time courses describing low, intermediate, and full induction of the transgene. Putative direct transcriptional targets would be identified based on an early and quantitative response to induction of the transgene.

Studies that do not require perturbation but simply describe a natural biological process allow for an efficient and unambiguous identification of the patterns of interest. This is certainly true for the analysis of clock-dependent circadian transcript oscillations. Circadian transcript patterns can be distinguished from all other patterns by the fact that even under constant conditions they are faithfully repeated with a period length of ~24 h. In principle, the repeated daily pattern can take on a variety of amplitudes and shapes, but, in practice, most known circadian transcript profiles resemble sine waves with 24-h periods. Circadian transcripts tend to peak once a day; in the context of measurements recorded at multiple-hour intervals from multiple cells and often even multiple individual organisms, it is understandable that the experimental representations of such patterns resemble sine waves. An important question in the design of a microarray time course study is whether to place more emphasis on sampling rate per day or on the extension of time courses over additional days. Most published circadian microarray studies have settled on an empirical consensus of sampling six times per day for 2 or more days (Ceriani *et al.*, 2002; Claridge-Chang *et al.*, 2001; Duffield *et al.*, 2002; Harmer *et al.*, 2000; Lin *et al.*, 2002; Panda *et al.*, 2002; Storch *et al.*, 2002; Ueda *et al.*, 2002a,b). In order to be able to distinguish clock-controlled from environmentally driven signals, it is necessary to record at least 1 day's worth of data representing free running of the clock under constant conditions. Depending on the initial synchrony between different cells and individual organisms, as well as the variation in period length and the expected rate of damping, it may be preferable to limit the monitoring of free-running conditions to the first few days. As an example, we discuss the experimental design of a circadian microarray time course study carried out to determine the set of circadian transcript oscillations in the adult *Drosophila* head (Claridge-Chang *et al.*, 2001). The *Drosophila* head contains various clock-bearing tissues, but the molecular oscillations in most of them show progressive damping in constant conditions (Plautz *et al.*, 1997). Three independent repetitions were generated of a 2-day experiment with 4-h sampling intervals consisting of 1 day (24 h) in the presence of a 12-h light/12-h dark (LD) cycle and a subsequent day in constant darkness (DD). This design takes advantage of the increased oscillatory amplitude and synchrony found during both light/dark entrainment and the first few days of free run. Moreover, the use of experimental time courses spanning both the last day of LD entrainment and the first day of DD free run allowed analysis of the effect of photic entrainment independently from population-specific effects or other sources of experimental variability (see later). Data analysis methods benefited substantially from the inclusion of three independent repetitions of the 2-day time course experiment. For example, global analyses of the full 6-day data set or partial 4-day data sets indicated a specific enrichment for transcript oscillations at the circadian period length (24 h), but this was not the case for any separate analysis of an individual 2-day experiment (see Fig. 5 and Data Analysis Techniques section). Several methods for selecting circadian transcripts were compared, guided by the set of known circadian transcripts from the literature as well as results from an independent verification analyses on Northern blots. It appears that a particularly efficient method for predicting circadian regulation at the transcript level consists of fitting the three normalized time course experiments simultaneously to a sine wave of fixed phase (Fig. 2). Specifically, the normalized $\log_2$ expression ratios for each of the three experimental repetitions were appended and the 24-h spectral power $F_{24}$ was extracted using Fourier analysis (for details, see Data Analysis Techniques section). Then, for each profile, $pF_{24}$ was calculated by observing a 24-h spectral component of equal or higher value from a permuted data set where the time points of each experiment have been reshuffled randomly. A set of transcripts predicted to show especially robust circadian

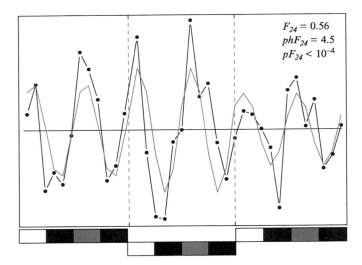

$$F_{24} = 0.56$$
$$phF_{24} = 4.5$$
$$pF_{24} < 10^{-4}$$

FIG. 2. Extraction of the 24-h Fourier component to detect circadian periodicity. Black dots and lines indicate expression data from three 2-day LD/DD time course experiments for the *Ugt35b* gene. Data are presented in the form of $\log_2$ expression ratios normalized per experiment. The 24-h Fourier component shown in red was extracted from these normalized $\log_2$ expression ratios after correction for variations in amplitude between experimental repetitions (see text). The $F_{24}$ score (expressed in range 0–1) indicates the relative strength of the extracted circadian component, and $pF_{24}$ represents its peak phase relative to the onset of light at ZT0. $pF_{24}$ represents the probability of observing an $F_{24}$ score from randomly permuted data that is of equal or greater strength than the extracted Fourier component. Parts of this figure are reprinted from Claridge-Chang *et al.* (2001), with permission from Elsevier. (See color insert.)

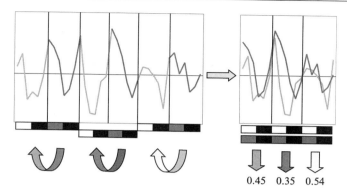

0.45    0.35    0.54

Fig. 3. Calculation of 24-h autocorrelations to detect circadian periodicity. Normalized $\log_2$ expression ratio data for the gene *Ugt35b* (same as in Fig. 2) are used as an example. The 24-h autocorrelation for each 2-day experimental repetition is calculated by fitting the six time points collected during the first day of the time course in the presence of a light/dark cycle (orange lines) to those collected on the second day under conditions of constant darkness (blue lines). For details, see text. (See color insert.)

oscillations was defined by applying a threshold value for $pF_{24}$ in addition to a series of noise filters. An alternative strategy for determining the circadian periodicity of transcript profiles is to measure the correlation between time points that are 24 h apart (see later). One property of this 24-h autocorrelation measure is that it also considers circadian patterns that are not well described by sine functions. Disadvantages are an increased sensitivity to small variations in period length and amplitude and the inability to align experiments that use different sampling times. In our analyses, we applied a 24-h autocorrelation measure ($AC_{24}$) to each experiment separately and used the three $AC_{24}$ values as a secondary filter for circadian patterns (Fig. 3).

## Independent Verification of Microarray Results

Microarray analyses are uniquely suited to study global trends in gene expression, but they also provide experimental information for thousands of genes on an individual basis. It is impractical to reexamine all of the individual expression measurements made on a microarray using conventional analysis methods. Nevertheless, independent verification for a sample of the microarray expression profiles can provide useful information about the predictive value of the microarray results. The question is which transcript profiles to select for further study. Generally, noisy patterns and patterns that remain unchanged throughout the experiment will of course be of less interest than those that suggest some type of meaningful regulation. In selecting a set of microarray profiles that are predicted to show some type of regulation, one has to carefully balance the risk of including false positives with the risk of excluding biologically meaningful responses. In general, such selections are made by ranking microarray expression profiles according to a variable predicted to be correlated with the regulation of interest and applying a threshold value. Independent verification analyses can be helpful in estimating the correlation between the ranking criterion and the regulation of interest and in adjusting the choice of the threshold value to desirable false-positive and false-negative rates. One strategy for determining false-positive and false-negative rates may be to consider several segments of the ranked transcript profiles and for each segment pick a number of profiles at random for independent verification. Often there is *a priori* knowledge of genes that should be included in the selected set of microarray profiles. Such positive controls can also inform the choice of the ranking criterion and threshold value, but because they do not represent an unbiased selection, they do not obviate the need for independent experimental verifications. Information from positive controls and independent experimental verifications may prompt a change of ranking criterion and threshold value. It is important to realize that new follow-up experiments may then be needed for an unbiased analysis of the modified selection. Ultimately, it will be a matter of personal choice how many follow-up experiments are performed and what levels of false positives and false negatives are deemed to be appropriate. The two main techniques used for independent confirmation of microarray results, besides additional microarray analyses, are Northern blot analysis and quantitative PCR. Both of these methods have been used successfully to assess the value of microarray results. If RNA samples and nucleic acid probes of sufficient length are readily available, Northern analysis provides a reliable and cost-efficient method of independent verification. Quantitative PCR requires much smaller quantities of RNA material and does not require separate preparation of nucleic acid probes by the investigator. For a current review of several different quantitative PCR platforms, we refer to Bustin (2000), Wilhelm and Pingoud (2003), Stirling (2003), and Kerr (2003b). Specialized equipment and modified oligonucleotide primers are required for some PCR techniques, and the need to empirically determine amplification conditions for each combination of primers should also be taken into consideration.

As an example of the use of independent verifications, we discuss the use of Northern analysis to address a set of circadian transcript profiles derived from microarray studies in the adult *Drosophila* head (Claridge-Chang *et al.*, 2001). After excluding noisy and unregulated patterns, transcript profiles were ranked based on the probability associated with the

24-h Fourier component ($pF_{24}$) extracted from the full 6-day data set. The reliability of this ranking was illustrated by observations that the two highest ranked genes, *vrille* and *timeless*, encoded known circadian transcripts and that all other known positive controls with one exception were found in the top 60. However, it was important to know to what extent novel circadian profiles could also be experimentally confirmed. Therefore, a series of 20 Northern blots was conducted for genes that were selected without deliberate bias. In 19 of 20 cases a circadian component of the predicted phase was detected. An example is given in Fig. 4. Based on these results, a $pF_{24}$ threshold value of 0.02 was chosen, which defined a circadian set of 158 genes, including 5 positive controls and 16 of the independent verifications. Three remaining independent verifications and one confirmed noncircadian gene fell below this selection threshold.

## Molecular Techniques

For preparation and hybridization of target samples for microarrays, we refer to protocols provided for specific array platforms (Bowtell and Sambrook, 2003; Brownstein and Khodursky, 2003; Schena, 2003; and

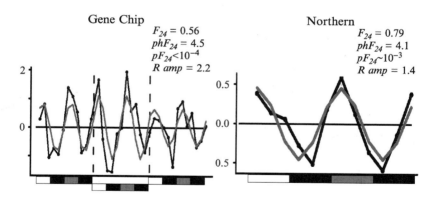

Fig. 4. Independent verification of circadian periodicity. (Left) The circadian pattern detected by microarray analysis for the *Ugt35b* gene (black pattern) along with its extracted 24-h Fourier component (red pattern; see Fig. 2). (Right) Data from a 2-day LD/DD time course Northern analysis for the same gene. The Northern signals for *Ugt35b* were normalized to a loading control and expressed as $\log_2$ ratios relative to the experimental average. The extracted 24-h Fourier component is shown in red. Note the high $F_{24}$ scores and significant $pF_{24}$ values, as well as the similarity in the peak phases ($phF_{24}$) for microarray and Northern data. The relative amplitude (R amp) indicates the estimated fold change at the peak under the assumption that the patterns are perfect sine waves. For a detailed description of R amp calculation, see the on-line supplement for Claridge-Chang *et al.* (2001). Parts of this figure are reprinted from Claridge-Chang *et al.* (2001), with permission from Elsevier. (See color insert.)

information supplied with commercially available reagents). We describe our protocol for purification of RNA from *Drosophila* heads that is suitable both for preparation of microarray target samples and for Northern blot analyses. Northern blots are both useful prior to microarray analysis for assessing the quality of the RNA sample selected for microarray target preparation and after microarray analysis to independently verify the predicted transcript profiles.

## CsCl Preparation of RNA from Adult Drosophila Heads

While shorter and more convenient protocols exist for RNA preparation from *Drosophila* heads, we prefer the use of this method because it yields pure and intact RNA and can accommodate relatively large amounts of tissue. The CsCl-purified samples are compatible with an additional RNA cleanup procedure on RNeasy minicolumns (Qiagen). We have experienced problems with this step when using RNA produced from *Drosophila* heads by simple phenol/guanidine–thiocyanate extractions. This protocol is adjusted to the use of $1/2 \times 2$ in. centrifuge tubes for ultracentrifugation at 45,000 rpm. A single Sw55Ti rotor (Beckman) has space for six such tubes. The volumes used can accommodate at least 500 $\mu$l of frozen heads, but as little as 100 $\mu$l will give sufficient yield for microarray hybridization and several Northerns.

Prior to extraction, flies are harvested on dry ice and stored at $-80°$. Heads are isolated by breaking frozen flies and separating body parts over a set of two sieves (a first sieve with 850- to 1000-$\mu$m pores to retain thoraces and abdomens and a second sieve with 200- to 350-$\mu$m pores to retain heads). It is important to keep the body parts frozen during this process, and we therefore recommend working on dry ice and prechilling the sieves at $-80°$. Heads can be stored at $-80°$.

Prepare for this protocol by prechilling the ultracentrifuge rotor and buckets at 4°.

1. Mix guanidine–thiocyanate/NaOAc/EDTA solution (4 ml/sample; see later) with $\beta$-mercaptoethanol (40 $\mu$l/sample) and a filtered solution of 20% Sarkosyl (40 $\mu$l/sample). Put on ice.
2. For each sample, add 3.7 ml of the mixture to the bottom of a Tenbroeck tissue grinder (Wheaton) (on ice). Save the rest on ice.
3. Transfer the frozen heads to the bottom of the tube with the use of a funnel (chilled on dry ice).
4. Grind the heads thoroughly.
5. Transfer the homogenates to Falcon 2059 tubes and spin in SA-600 rotor (Beckman) with adapters for 10 min at 10,000 rpm at 4°.

6. Meanwhile, label the sides and bottoms of 1/2 × 2 in. ultraclear Beckman centrifuge tubes and add 1.5 ml of CsCl/NaOAc/EDTA (see later).

7. Using a clean long-tip Pasteur pipette, slowly top off the CsCl/NaOAc/EDTA aliquots with the supernatants from the spin.

8. Add guanidine–thiocyanate/NaOAc/EDTA/$\beta$-mercaptoethanol/Sarkosyl solution to achieve equal volumes, leaving the meniscus a few millimeters from the top of each tube.

9. Spin over night at 45,000 rpm, 4°, with slow acceleration and deceleration settings.

10. After centrifugation, remove the tubes with a forceps.

11. Carefully remove most of the supernatant from each tube with a Pasteur pipette.

12. Drain the tubes on clean paper towels.

13. Use a clean razor blade to cut the tubes about 1 cm from the bottom.

14. Resuspend the RNA pellets with nuclease-free water, transfer to 1.5-ml microcentrifuge tubes, and adjust volumes to 400 $\mu$l.

15. Add 40 $\mu$l 2 $M$ NaOAc, pH 5.0, and 1 ml ethanol (stored at $-20°$).

16. Precipitate at $-20°$ for $\geq$ 1 h. Samples can be stored at $-20°$ at this point.

17. Spin for 15 min at 4° in a microcentrifuge at maximum speed.

18. Wash pellets with 1 ml 80% ethanol.

19. Resuspend in nuclease-free water or 1 m$M$ EDTA, pH 7.0 (e.g., use 100 $\mu$l for a preparation of 100 $\mu$l heads). This may require several minutes of incubation in a 65–70° water bath or heat block.

20. Take a 2-$\mu$l sample and add 98 $\mu$l H$_2$O. Measure the optical density at 260, 280, and 320 nm to determine purity and yield.

21. Aliquot the samples for use on microarrays and Northerns (20 $\mu$g per blot) and store at $-80°$ or in ethanol-precipitated form at $-20°$.

*guanidine–Thiocyanate/NaOAc/EDTA*

4 $M$ guanidine–thiocyanate
0.1 $M$ NaOAc (sodium acetate), pH 5.0 (from 10× stock)
5 m$M$ EDTA
Filter and store at 4°

*CsCl/NaOAc/EDTA*

75 g CsCl per 100 ml
0.1 $M$ NaOAc, pH 5.0
5 m$M$ EDTA
Filter and store at 4°

Northern blot analyses can be carried out according to commonly used standard protocols (e.g., Ausubel, 2001; Sambrook and Russel, 2001). We use Nytran SuperCharge membranes (Schleicher and Schuell) and UltraHyb hybridization buffer (Ambion).

## Data Analysis Techniques

This section documents methods for processing raw microarray data with the goal of studying rhythmic and/or time-dependent phenomena. In addition to applications to the circadian clock, many of these methods will be applicable to time series data in general, i.e., cell cycle data.

### Sampling Rate

From an analysis point of view, it is worth considering equal time sampling with no missing points. This makes the analyses simpler and more powerful. For instance, to obtain unbiased results with autocorrelation and Fourier analysis, evenly sampled data are an enormous plus. If absolutely necessary, missing data can be interpolated or more advanced algorithms can be applied (Press *et al.*, 1994), but we recommend the use of sampling at regular intervals whenever possible.

### Signal Estimation

Depending on the array platform used, specific issues arise concerning the extraction of expression signals from the raw output files produced by scanners. For example, Affymetrix GeneChip arrays probe transcripts using 10–20 different 25-base oligonucleotides, together with an equal number of point mutants designed as negative controls. The challenge is to extract a concentration estimate that combines the 20–40 probes, taking into account the variable probe affinities. Although recent methods have improved with respect to rejecting cross-hybridizing signals, optimal signal detection will remain a focus of research. In addition to the standard solutions by Affymetrix, the BioConductor project (http://www.bioconductor.org) has provided a unified platform for alternate methods, with the advantage that new developments are rapidly included and made broadly available to the public. BioConductor is perhaps the most flexible environment now available, but it demands more advanced programming skills than other packages.

Current signal estimation approaches with potentially useful features are (1) the RMA method (Irizarry *et al.*, 2003), which lacks sensitivity at low intensities but appears to minimize false-positive signals, and (2) the GCRMA algorithm (Wu and Irizarry, 2004), which estimates nonspecific background from probe sequence information (Naef and Magnasco, 2003), thereby offering increased sensitivity at low intensity. However, the performance of the latter method has only been explicitly examined with human genome arrays. The normalization of interexperiment variability is accomodated as described earlier.

## Data Transformation

Once the expression signals ($s$) have been extracted, data are organized in a matrix $s_{gt_i} = s_g (t_i)$, such that the row index $g$ represents the different genes and the column index indicates the discrete time $t_i$ (e.g., for regular sampling at 4-h intervals $t_0 = 0$ h, $t_1 = 4$ h, $t_2 = 8$ h, etc.). The next issue is whether to work with the natural or a transformed version of the signal scale. This choice depends on the distributions of signal and noise. For circadian experiments using the Affymetrix GeneChip system, we found that oscillations of known genes are quite accurately sine shaped in logarithmic coordinates. The use of logarithmic transformation requires a truncation for values below a noise threshold to avoid large negative outliers. For example, when using Affymetrix Microarray Suite (MAS) 5.0 signals, a simple procedure is to set all signals below a noise threshold equivalent to the scaled noise factor Q (as described for Affymetrix MAS 4.0 and 5.0) to that threshold before taking logarithms. The original matrix is therefore replaced by $\log_2 [\max(s_{gt_i}, Q)]$. Methods such as RMA return signal estimates directly in log coordinates, and no truncation is required.

## Detrending

Detrending a time series is an attempt to correct for systematic errors that would accumulate during the time of the experiment. For example, trends would occur if steadily decreasing amounts of cRNA were hybridized. Some studies have used linear detrending. This approach can produce effects that differ when applied in natural versus transformed scale. The increase in cell number in a cell culture experiment is an example of a possible source for trends. In the absence of reasons that justify the possibility of trends, we do not recommend systematic detrending. For example, in the time course experiments for populations of adult *Drosophila* described in this article, it is unclear what external cause would translate into a linear trend in log coordinates (equivalent to an exponential trend in the natural scale).

## Treatment of Experimental Repeats

Often microarray experiments are repeated. If the samples used to repeat the same time point are identical (i.e., technical replicates), it is justified to average the measurements to reduce the noise levels, although the benefits from averaging a small number of replicates (e.g., 2–3) is not always obvious. If instead the whole time course is repeated independently using a different population, possibly even in a different genetic background, we found it detrimental to average time points among such biological replicates, even if normalization strategies should, in principle, flatten out interexperimental variability. We have observed that the amplitudes of oscillations of confirmed circadian transcripts tend to show more variation between independently repeated time course experiments than phase or period length. We, therefore, found it advantageous to standardize each individual time course separately, meaning that after transformation the standard Euclidean vector norm equals 1 and the mean vanishes for each gene (row) in the matrix. The following section describes methods to estimate rhythmic behavior from the standardized matrix $\hat{s}^e_{gt_i}$ where the $\wedge$ stands for standard and $e$ is the experiment index.

### Detection of Circadian Periodicity

*Spectral Analysis.* The representation of a time-dependent signal as a weighted sum of sine and cosine functions with different frequencies is called spectral or Fourier analysis. In the context of circadian time course studies, extraction of the 24-h Fourier component is of particular interest. The formal definition of discrete Fourier analysis as a linear decomposition of the signal $(\hat{s}_{t_i})$ into oscillating modes with weights $f_k$ (complex numbers) and period $T_k$ is given as

$$\hat{s}_{t_i} = \frac{1}{\sqrt{N}} \sum_{k=0}^{N-1} f_k e^{-i2\pi k \frac{t_i}{N\Delta t}} \tag{1}$$

where $T_k$ is equivalent to the term $N\Delta t/k$, with $\Delta t$ the sampling rate and $N$ the number of time points sampled (here $N$=12). Although the individual coefficients $f_k$ are complex numbers, the aforementioned summation evaluates to real numbers, as it should since expression values are real numbers. The gene and experiment indices have been omitted for simplicity. The power spectrum $P_k$ is defined for $k = 0, 1, \ldots, \frac{N}{2}$ as $P_k = |f_k|^2$ for $k = 0$, $\frac{N}{2}$ and $P_k = 2|f_k|^2$ for $k = 1, \ldots, \frac{N}{2} - 1$ (Press *et al.*, 1994). The description of the signal in terms of Fourier mode is complete in the sense that there is a sum rule as indicated in Eq. (2):

$$\sum_{k=0}^{\frac{N}{2}} P_k = \sum_{n=0}^{N-1} \hat{s}(t_n)^2 = 1 \tag{2}$$

This implies that a perfectly sine-shaped signal with one of the aforementioned periods $T_{k'}$ has $P_{k'} = 1$ and all others $P_k = 0$. So for an experiment with $N = 12$ points spaced by $\Delta t = 4\ h$, the periods are $\infty$, 48, 24, 16, 12, 9.6, 8. The Nyquist limit $2\Delta t$ is the smallest period that can be probed with a sampling rate $\Delta t$, corresponding to points up, down, up, down, etc. To enforce phase in a multiexperiment analysis, for example, for three 12-time-point time courses ($N_{exp} = 3$), we append the time courses before the decomposition. Then, $N = 12 * N_{exp} = 36$ and thus the periods are given by $T_k = 144/\ k = \infty$, 144, 72, 48, 36, 28.8, 24, etc. As we are interested in patterns that repeat across the different experiments, we usually only consider a subset of the possible periods, namely those that are integer fractions of 48 h (48, 24, 16, 12, 9.6, 8). The ratio between this reduced (fractions of 48 h) spectral power and total spectral power can be used as a measure of reproducibility of the repeats. For reduced components, the procedure just described is strictly equivalent to averaging the standardized matrices $\hat{s}^e_{gt_i}$ over the $N_{exp}$ repetitions. In a screen for circadian pattern, it is sufficient to consider the $T = 24$ component. The phase of a given mode in radians $\phi \in [-\pi, \pi]$ can be extracted from $f_k$ values as

$$\phi_k = \tan^{-1}\left(\frac{\mathrm{Im}f_k}{\mathrm{Re}f_k}\right) \qquad (3)$$

In cases where Re $f_k < 0$, $2\pi$ is added to obtain the phase. Re and Im denote the real and imaginary parts of the complex valued coefficient $f_k$. To convert $\phi_k$ to hours, we multiply by $T_k/(2\pi)$, which for 24-h oscillations is a factor $12/\pi$. Here we use the notation $F_{24}$ to denote the 24-h spectral power $P_{\frac{N\Delta t}{24}}$. This notation has the advantage of being independent from the number of experiments and sampling time. Note that the $F_{24}$ values in the supplemental information associated with Claridge-Chang *et al.* (2001) refer to

$$\sqrt{\frac{N}{2}} P_{\frac{N\Delta t}{24}}$$

There are several alternative methods for spectral estimation, many of which are designed for highly sampled data and so are of limited use in our case. A majority of these estimate continuous (infinite dimensional) spectra, which is a daunting task given 12 measurements. We refer to the Matlab (Mathworks, Inc.) signal processing toolbox for a comprehensive collection of such methods. Given our data and the questions we are addressing, we found no obvious reason to depart from the standard Fourier approach. Because of its relative simplicity, it will most often be applied first. Finally, we emphasize that choices such as natural versus log-transformed signals will affect the number of identified rhythmic genes at least as much as the choice of an alternate spectrum estimation algorithm.

*Sinusoidal Fits.* A similar approach to Fourier analysis is to fit the cosine function of Eq. (4) to data using the least-squares method:

$$s(t) = A\cos(\omega t + \phi) \qquad (4)$$

where frequency $\omega$ can take arbitrary values. For the restricted (but complete) set of frequencies defined in the previous paragraph (when $\omega$ takes a value of $2\pi \frac{k}{N\Delta t}$ for a fixed integer $k$), the procedures are equivalent. In this case, the relationship between fitted amplitude $A$ and the spectral power $P_k$ can be can be expressed as

$$A^2 = P_k \qquad (5)$$

where the phase satisfies the relationship in Eq. (3). Thus, the 24-h spectral power $F_{24}$, defined as $P_k$ for $k = \frac{N\Delta t}{24}$, is also equal to $|A|^2$ at frequency $\omega = \frac{\pi}{12}$. $F_{24}$ can therefore be computed from a least-squares fit to the function in Eq. (4). Sinusoidal fits are typically carried out for a large number of test frequencies in a frequency window. A frequency window could, for example, represent periods $T = \frac{2\pi}{\omega}$ between 20 and 30 h. One important difference with Fourier decomposition is that fits can be done for arbitrary frequencies and therefore allow rhythms of modified periods such as 23 h. However, it is unclear whether 12-point data sets sampled at 4-h intervals allow unambiguous discrimination between 23 and 24 periods given the typical noise levels in microarray experiments.

*Autocorrelations.* Fourier analysis or sinusoidal fitting can be applied to a single circadian period. In this case, high scores will select genes that have accurate sinusoidal shapes, without considering whether the pattern was repeated. In contrast, circadian autocorrelation analysis requires at least 2 full days worth of data. Autocorrelation of lag $j\Delta t$ is defined as

$$a(j\Delta t) = a_j = \sum_{i=1}^{N} \hat{s}_{t_i}\hat{s}_{t_{i+j}} \qquad (6)$$

Intuitively, it measures the overlap between the time course $\hat{s}_{t_i}$ and a version $\hat{s}_{t_{i+j}}$ shifted by $j\Delta t$. In the circadian case, we consider lags of 24 h ($j = 6$ if $\Delta t = 4$ h) and we denote $a_6$ by $AC_{24}$. The maximal score will be obtained if the pattern in the first day exactly repeats itself during the second day, no matter what the initial shape is. For example, autocorrelation will accommodate a sharp *spike-shaped* signal that would not necessarily be identified using Fourier analysis. However, identifying a rhythmic gene on the basis of Fourier analysis requires a certain degree of smoothness in its expression pattern and is therefore less susceptible to artifacts. By definition, good $F_{24}$ patterns will be credited with high $AC_{24}$ scores, but the opposite does not necessarily apply. In our experiments, we found considerable overlap between good $F_{24}$ and $AC_{24}$ patterns.

With repeated time courses, the options are to (i) average the $\hat{s}^e_{gt_i}$ over the $N_{exp}$ repetitions prior to computing $AC_{24}$ and (ii) compute $N_{exp}$ separate $AC_{24}$ values (e.g., Claridge-Chang et al., 2001).

*Testing for Circadian Expression Using the Kruskal–Wallis Statistic.* If data are available from multiple independent time course experiments using the same sampling scheme, then the presence of a significant circadian effect on expression can be detected by grouping the normalized $\log_2$ expression ratios according to the time of day and performing the nonparametric Kruskal–Wallis test. If, for example, two independent 12-point LD/DD time courses have been collected at time points ZT4-8-12-16-20-24-CT4-8-12-16-20-24, then the data points are divided among six groups corresponding to the daily time points ZT/CT4, ZT/CT8, ZT/CT12, ZT/CT16, ZT/CT20, and ZT/CT24. Similar to 24-h autocorrelation measurements, the Kruskal–Wallis statistic indicates whether there is a circadian effect on expression independently from any assumption about the shape of daily expression patterns.

## Empirical Significance Testing

To test whether a 24-h spectral power $F_{24}$ is statistically significant, the simplest solution is to generate a null hypothesis using random permutations of the time points. For each gene, $F_{24}$ is computed for $N$ different realizations of scrambled data, and the score for real-time ordering is ranked in the list of scores from scrambled data. In the situation where we append repeated time courses, permutations are restricted to scrambling of data points within each repeat. The $p$ value is then computed as the rank of the real score divided by $N$. The procedure can be applied identically to other statistics such as $AC_{24}$. $p$ values are denoted by $pF_{24}$, $pAC_{24}$, etc.

## Representation of Distributions

Quantile–quantile plots can be used to study how the circadian oscillator affects transcription globally. In contrast to the approach presented in the previous section, the null model is not computed on a gene-by-gene basis, but considers all transcripts simultaneously. For all transcripts, the 12 points in each experiment are permuted 1000 times, and Fourier scores for the resulting fictitious time series are stored. A number of quantiles equal to the number of probe sets represented in original data is then derived from the 1000 × (number of genes) fictitious Fourier scores. For example, the $x$ value of a probe set X that ranks at the 80th percentile in the list of Fourier scores is given by the 80th percentile of the list of 1000 × (number of genes) permuted scores. An example is shown is Fig. 5.

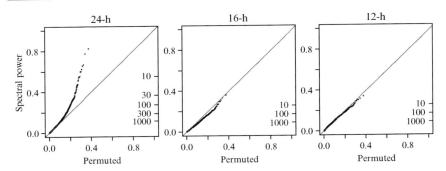

FIG. 5. Global analysis of circadian periodicity using quantile–quantile plots of spectral power. Global trends of rhythmic expression in the heads of wild-type fruit flies are visualized for a microarray data set consisting of four LD/DD time course experiments. Quantile–quantile plots of spectral power (24-h, 16-h, 12-h Fourier components) are shown at circadian (24-h) and noncircadian (16-h, 12-h) periods. The real Fourier score for each probe set is indicated on the $y$ axis, whereas quantiles on $x$ values represent results from 1000 permutations of the time ordering for all probe sets. Quantile–quantile pairs connect identical ranks in the value-ordered sets for real and permuted data. Tick marks on the right show the number of probe sets selected by the indicated lower threshold values. The upward deviation from the diagonal (brown) seen at the 24-h period for highly ranked quantile–quantile pairs indicates enrichment in time traces with a strong circadian component. Fourier components with 16- and 12-h periods do not show similar deviations. (See color insert.)

## Low-Amplitude Filtering

To limit false discoveries, additional filters can be applied. For instance, a threshold on the overall signal intensity can be set. When using MAS 4.0–5.0 data, we required that more than half of all measured expression values for a given gene be larger than four times the scaled noise Q.

Standardized signals no longer contain information about the actual magnitude of the changes. To avoid considering transcripts with a very small range of oscillation, an adequate measure is to express the peak-to-trough amplitude in units of the noise level Q. For selecting circadian transcripts, we have empirically determined an effective threshold value for the peak-to-trough amplitude at 7Q (Claridge-Chang *et al.*, 2001).

## Supervised Pattern Discovery

This section refers to the situation in which more than one mechanism contributes to an expression pattern, as happens for transcripts that are dually regulated by light and the circadian clock in an LD/DD experiment. Given an explicit form for the response to each separate effect, we can fit the full response to a superposition of the individual responses. Inclusion of mutant data with expected phenotypes is also possible. Such approaches are complementary to unsupervised methods (such as standard clustering) and allow us to verify that certain genes behave according to a given model.

For sets of wild-type and arrhythmic mutant LD/DD data, we fit the logarithm-transformed expression levels to the functions in Eqs. (7) and (8)

$$WT(t) = B_{wt} + P_{wt} * f_{onset}(t) + C * \cos\frac{2\pi}{24}(t - \tau) \tag{7}$$

$$MUT(t) = B_{mut} + P_{mut} * f_{onset}(t) \tag{8}$$

assuming independence of the effects. The fits use standard least-squares regression on the combined wild-type $WT(t)$ and mutant $MUT(t)$ time courses. Equal weight is given to all data points; notice that we do not standardize data here. The model posits that the log expression profiles for wild-type $WT(t)$ and clock mutants $MUT(t)$ are sums of a genotype-dependent baseline $B$ and light susceptibility $P$. Only wild-type data can have a circadian component of strength $C$. The peak phase is described by time $\tau$, and the pulse function $f_{onset}(t)$ is a smoothed rectangular pulse of 12 h starting at time $onset$, taking value 1 when the lights are on and 0 otherwise. $f_{onset}(t)$ is the only parameter that is not fit, but is determined by the least residual fit to the other six parameters, when all integer values in the interval $[-2, 6]$ are tested. Thus, $f_{onset}(t)$ allows for a delay in the light response for up to 6 h, which may reflect various kinetic properties of the RNA accumulation process. Examples can be visualized at http://sib-pc27.unil.ch/cgi-bin/felix/FitsME/viewFits2.cgi. All parameters but $\tau$ have units of $\log_2$ ratio.

## Implementation, Use of R

The R programming language (http://www.r-project.org/) for statistical computing provides an optimal environment for implementing the analyses.

We provide a sample code that computes the spectral power for a fictitious gene sampled at 4-h intervals for three repetitions of 2 days. The script computes the 24-h spectral power $F_{24}$, the estimated peak phase, and $AC_{24}$ for a noisy cosine oscillation.

```
# here is an example for computing the 24-hr Fourier
# score
# for the case of three repeated 12 point time course

# can be run as source('example24.r')
# first define a couple of functions
```

```
norm <- function(x) sqrt(sum(x*x) )

#compute fourier components f_k for periods that are
#fractions of 48hr
ff <- function(time, data) {
  pi24 <- pi/24
  f <- double()
  for(n in 0:6)
  {
    f[n+1]  <-  sum(data*exp(- li*pi24*n*time) )  /
      length(time)
  }
  f
}

#autocorrelation
shift <- function(x,n) {
  1 <- length(x)
  if (1 == 1) return(x)
  else {
    s <- c(0:(1-1) )
    s <- (s+n) %%1 +1
    return(x[s])
  }
}

ac <- function (x) {
  1 <- length(x)
  sj <- double(1/2+1)
  for (n in 0:(1/2) ) {
    sj [n+1] <- sum(x * shift (x,n) )
  }
  sj / sj[1]
}

#remove mean and normalize
standardize <- function(dat)
{
  dat=dat-mean(dat)
  s2=mean(dat^2)
  dat=dat/sqrt(s2)
}
```

```
rhythms <- function(times, dat1, dat2, dat3)
{
  nexp=3
  #standardize
  dat1=standardize(dat1)
  dat2=standardize(dat2)
  dat3=standardize(dat3)

  #concatenate
  dat=c(dat1, dat2, dat3)/sqrt(nexp)

  fo <- ff(times, dat)
  fr <- Re(fo)
  fi <- Im(fo)

  F24=2*(fr[3]^2+fi[3]^2) / nexp     #the third
  element corresponds to T=24hr
  phase24=atan2(-fi[3], fr[3])
  if(fr[3]<0) phase24 = phase24 + 2*pi
  phase24=phase24*24/(2*pi)

  c(F24=F24, phase24=phase24)
}

#================
# main start here
#================

#define the times at which the data is sampled, here
#t=0,4,8,...,44

times <- c(0:11)*4
# fictitious data, peak is at 17hr
a=cos(2*pi/24*(times-17) )
d1=a+rnorm(12,0,.2) # add some noise
d2=a+rnorm(12,0,.2)
d3=a+rnorm(12,0,.2)

score=rhythms(times, d1, d2, d3)
print(score)
```

```
plot(rep(times, 3), c(d1,d2,d3), pch=c(rep(1,12),
  rep(2,12), rep(3,12) ),
  xlab='time', ylab='fictitious expression')

#this computes AC24 for a single time course
ac24=ac(standardize(d1) ) [7]

print(c(AC24=ac24) )
```

## Acknowledgments

We thank Catharine Boothroyd, Adam Claridge-Chang, and the staff of the gene array facility at the Rockefeller University for their contributions to the experiments used to illustrate this article and Joanne Edington for help with the creation of Web sites. This work was supported by NIH GM54339 (M.W.Y.) and postdoctoral fellowships from NIMH (PHS MH63579) to H.W. and Bristol-Myers Squibb to F.N.

## References

Allada, R., White, N. E., So, W. V., Hall, J. C., and Rosbash, M. (1998). A mutant Drosophila homolog of mammalian Clock disrupts circadian rhythms and transcription of period and timeless. *Cell* **93,** 791–804.

Ausubel, F. M. (2001). "Current Protocols in Molecular Biology." Wiley, New York.

Blau, J., and Young, M. W. (1999). Cycling vrille expression is required for a functional Drosophila clock. *Cell* **99,** 661–671.

Bowtell, D., and Sambrook, J. (2003). "DNA Microarrays: A Molecular Cloning Manual." Cold Spring Harbor Laboratory Press, Cold Spring Harbor, NY.

Brownstein, M. J., and Khodursky, A. B. (2003). "Functional Genomics: Methods and Protocols." Humana Press, Totowa, NJ.

Bustin, S. A. (2000). Absolute quantification of mRNA using real-time reverse transcription polymerase chain reaction assays. *J. Mol. Endocrinol.* **25,** 169–193.

Ceriani, M. F., Hogenesch, J. B., Yanovsky, M., Panda, S., Straume, M., and Kay, S. A. (2002). Genome-wide expression analysis in Drosophila reveals genes controlling circadian behavior. *J. Neurosci.* **22,** 9305–9319.

Claridge-Chang, A., Wijnen, H., Naef, F., Boothroyd, C., Rajewsky, N., and Young, M. W. (2001). Circadian regulation of gene expression systems in the Drosophila head. *Neuron* **32,** 657–671.

Duffield, G. E., Best, J. D., Meurers, B. H., Bittner, A., Loros, J. J., and Dunlap, J. C. (2002). Circadian programs of transcriptional activation, signaling, and protein turnover revealed by microarray analysis of mammalian cells. *Curr. Biol.* **12,** 551–557.

Hardin, P. E., Hall, J. C., and Rosbash, M. (1990). Feedback of the *Drosophila* period gene product on circadian cycling of its messenger RNA levels. *Nature* **343,** 536–540.

Harmer, S. L., Hogenesch, J. B., Straume, M., Chang, H. S., Han, B., Zhu, T., Wang, X., Kreps, J. A., and Kay, S. A. (2000). Orchestrated transcription of key pathways in Arabidopsis by the circadian clock. *Science* **290,** 2110–2113.

Irizarry, R. A., Bolstad, B. M., Collin, F., Cope, L. M., Hobbs, B., and Speed, T. P. (2003). Summaries of Affymetrix GeneChip probe level data. *Nucleic Acids Res.* **31,** e15.

Kerr, K. (2003a). Experimental design to make the most of microarray studies. *In* "Functional Genomics: Methods and Protocols" (M. J. Brownstein and A. B. Khodursky, eds.), pp. 137–148. Humana Press, Totowa, NJ.

Kerr, R. (2003b). Quantitation of multiple RNA species. *In* "PCR Protocols" (J. M. S. Bartlett and D. Stirling, eds.), 2nd Ed., pp. 211–216. Humana Press, Totowa, NJ.

Lin, Y., Han, M., Shimada, B., Wang, L., Gibler, T. M., Amarakone, A., Awad, T. A., Stormo, G. D., Van Gelder, R. N., and Taghert, P. H. (2002). Influence of the period-dependent circadian clock on diurnal, circadian, and aperiodic gene expression in *Drosophila melanogaster. Proc. Natl. Acad. Sci. USA* **99,** 9562–9567.

McDonald, M. J., and Rosbash, M. (2001). Microarray analysis and organization of circadian gene expression in *Drosophila. Cell* **107,** 567–578.

Naef, F., and Magnasco, M. O. (2003). Solving the riddle of the bright mismatches: Labeling and effective binding in oligonucleotide arrays. *Phys. Rev. E. Stat. Nonlin. Soft Matter Phys.* **011906,** 1–4.

Panda, S., Antoch, M. P., Miller, B. H., Su, A. I., Schook, A. B., Straume, M., Schultz, P. G., Kay, S. A., Takahashi, J. S., and Hogenesch, J. B. (2002). Coordinated transcription of key pathways in the mouse by the circadian clock. *Cell* **109,** 307–320.

Plautz, J. D., Kaneko, M., Hall, J. C., and Kay, S. A. (1997). Independent photoreceptive circadian clocks throughout *Drosophila. Science* **278,** 1632–1635.

Press, W. H., Teukolsky, S. A., Vetterling, W. T., and Flannery, B. P. (1994). "Numerical Recipes in C," 2nd Ed. Cambridge Univ. Press, Cambridge.

Sambrook, J., and Russell, D. W. (2001). "Molecular Cloning: A Laboratory Manual." Cold Spring Harbor Laboratory Press, Cold Spring Harbor, NY.

Schena, M. (2003). "Microarray Analysis." Wiley-Liss, Hoboken, NJ.

Sehgal, A., Price, J. L., Man, B., and Young, M. W. (1994). Loss of circadian behavioral rhythms and per RNA oscillations in the *Drosophila* mutant timeless. *Science* **263,** 1603–1606.

Sehgal, A., Rothenfluh-Hilfiker, A., Hunter-Ensor, M., Chen, Y., Myers, M. P., and Young, M. W. (1995). Rhythmic expression of timeless: A basis for promoting circadian cycles in period gene autoregulation. *Science* **270,** 808–810.

Stirling, D. (2003). Qualitative and quantitative PCR: A technical overview. *In* "PCR Protocols" (J. M. S. Bartlett and D. Stirling, eds.), 2nd Ed., pp. 181–183. Humana Press, Totowa, NJ.

Storch, K. F., Lipan, O., Leykin, I., Viswanathan, N., Davis, F. C., Wong, W. H., and Weitz, C. J. (2002). Extensive and divergent circadian gene expression in liver and heart. *Nature* **417,** 78–83.

Ueda, H. R., Chen, W., Adachi, A., Wakamatsu, H., Hayashi, S., Takasugi, T., Nagano, M., Nakahama, K., Suzuki, Y., Sugano, S., Iino, M., Shigeyoshi, Y., and Hashimoto, S. (2002a). A transcription factor response element for gene expression during circadian night. *Nature* **418,** 534–539.

Ueda, H. R., Matsumoto, A., Kawamura, M., Iino, M., Tanimura, T., and Hashimoto, S. (2002b). Genome-wide transcriptional orchestration of circadian rhythms in Drosophila. *J. Biol. Chem.* **277,** 14048–14052.

Wilhelm, J., and Pingoud, A. (2003). Real-time polymerase chain reaction. *Chembiochemical* **4,** 1120–1128.

Wu, Z., and Irizarry, R. A. (2004). Stochastic models inspired by hybridization theory for short oligonucleotide arrays. *Proceedings of RECOMB,* 98–106.

Yang, Y. H., and Speed, T. P. (2003). Design and analysis of comparative microarray experiments. *In* "Statistical Analysis of Gene Expression Microarray Data" (T. P. Speed, ed.), pp. 35–91. CRC, Boca Raton, FL.

## [16]  RNA Profiling in Circadian Biology

*By* JOHN R. WALKER and JOHN B. HOGENESCH

### Abstract

DNA arrays have become indispensable tools in the study of transcriptional output of the circadian clock and have enabled an increase in the number of outputs from tens to thousands. However, despite their widespread use, many challenges exist in accurately and sensitively identifying all transcriptional targets of the clock. This article discusses aspects of experimental design, including RNA–amplification strategies, data condensation methods, and other aspects that impact the use of these tools for the study of circadian biology.

### Introduction

The past two decades of research have seen insight into the mechanisms of circadian rhythmicity in several organisms and have uncovered a common theme of oscillators as coupled transcriptional/translational feedback loops (Dunlap, 1999; Panda *et al.*, 2002). In that regard, transcriptional output of the clock is an essential function for the maintenance of oscillator function, as well as the central conduit by which the clock coordinates physiology. In mammals, this is evidenced by several animal models in which clock factors have been either deleted via homologous recombination or mutated (reviewed in Reppert and Weaver, 2002). For example, inactivation of the core E-box regulator *Bmal1* by homologous recombination results in complete arrythmicity of both transcriptional and behavioral rhythmicity (Bunger *et al.*, 2000). The *Clock* mutant mouse has also been described (Antoch *et al.*, 1997; King *et al.*, 1997). Homozygous *Clock/Clock* animals display a reduction in transcription for direct targets of the Clock/Bmal1 complex. The consequence of this misregulation for behavior is manifest in a long-period phenotype in constant darkness, followed after a few days by arrythmicity. Similar observations have been made in flies, plants, fungi, and bacteria—clock components tend to be transcription factors (or modify these factors), and disruption of their function produces aberrant effects on circadian control of transcription and, ultimately, physiology (Dunlap, 1999; Panda *et al.*, 2002).

Having established the importance of transcription for the functioning of the clock, it then follows that the global study of transcription regarding clock function is a worthy pursuit. Not surprisingly, this activity has taken many forms over the past decade, from differential display methods to arrays (for examples, see Akhtar *et al.*, 2002; Claridge-Chang *et al.*, 2001; Duffield *et al.*, 2002; Green and Besharse, 1996; Harmer *et al.*, 2000; Kornmann *et al.*, 2001; Lin *et al.*, 2002; McDonald and Rosbash, 2001; Storch *et al.*, 2002; Ueda *et al.*, 2002). Many articles in *Methods in Enzymology* have been devoted to the experimental and informatic tools that are essential to these pursuits (Britton, 2003; Burgess and McParland, 2002; Martin and Rosner, 2003; Roy *et al.*, 2002; Straume, 2004; Thomas *et al.*, 2002; Wittliff and Erlander, 2002). This article focuses on a few specialized aspects of RNA profiling (also known as "chipping") that we have found to impact the study of circadian transcriptional output in plants, flies, and mice. While these issues are largely generic, they are discussed using examples from our work done on the mouse and the technology we have the most experience with—high-density oligonucleotide arrays and the Affymetrix platform.

Experimental Design and Wet Work

Detection of circadian rhythms in transcription is confounded by many of the same sources of error that plague the global study of RNA dynamics in other fields (Fig. 1). Major sources of error can include genetic background of the animal models (Sandberg *et al.*, 2000), the age and condition of the animals (Balasubramaniam and Del Bigio, 2002; Thomas *et al.*, 2002), feeding (Zinke *et al.*, 2002), dissection (Sandberg *et al.*, 2000), and, of course, time of day. For example, many "knockout" models are on a mixed background, and just as background effects can influence locomotor period length estimation, they can influence underlying gene expression.

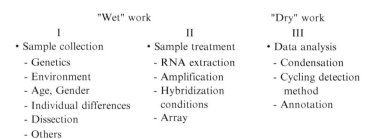

FIG. 1. Biological, environmental, technical, and informatic factors that influence RNA profiling. A schematic of the three principal factors that influence RNA profiling. Under the heading "wet work" comes both steps associated with acquiring a sample for the chipping process, as well as the technical steps involved in producing a scan. "Dry work" encompasses the steps after the image has been generated.

Many genes are responsive at the transcriptional level to environmental variables such as feeding, sickness, hormonal cycle, and even social cues. Furthermore, while dissections of whole liver are relatively straightforward, several structures of particular import to mammalian circadian biologists such as suprachiasmatic nucleus (SCN), retina, and pineal require much practice to master and become consistent. All of these issues increase variability in the experiment and decrease our ability to detect the changes dependent upon the biology of interest. Even worse, many of these sources of error cannot be discerned *a priori* and can only be evaluated comprehensively when many independent replicates can be obtained.

Once samples are obtained, the technical aspects of chipping take over. First, the sample has to be converted to cRNA—a process that can be accomplished using a linear amplification procedure that employs T7 polymerase (the Van Gelder and Eberwine protocol) (Van Gelder *et al.*, 1990). The current Affymetrix *in vitro* transcription kit can be used with as little as 1 $\mu$g of total RNA as starting material; however, for most single round amplifications, we suggest starting with 5 $\mu$g as the failure rate for obtaining the 15–20 $\mu$g of cRNA that is hybridized to the array for this amount is much lower. By modifying this protocol and going through two rounds, double amplification, it is possible to use much less initial starting material. In Panda *et al.* (2002), for example, 10–30 ng of total RNA derived from the SCN was used for each time point (Panda *et al.*, 2002). Despite this, very little difference was seen in the number of cycling transcripts detected compared with several other tissues. A more thorough study examined the effects of single versus double amplification on human kidney biopsies from individual patients (not pools) (Scherer *et al.*, 2003). When RNA from two different patients was either single or double amplified, the correlation coefficient for both methods was almost identical (Fig. 2). However, when one sample was single amplified and the other double amplified, the reproducibility suffered somewhat (Fig. 2). Postamplification, hybridization conditions, washing, and even the arrays themselves can affect performance. We have noted the percentage of transcripts called present (%P) can vary according to the protocol and type of wash station used in the experiment. Although the newer fabrication techniques and algorithms largely reduce batch effects of the arrays, in the past we have also observed the effect of chip lot on results from RNA profiling experiments. Thus, the chipping process itself can have a major impact on the outcome of an experiment—this becomes especially evident when comparing experiments done at different times or at different facilities.

The design that we and others have applied to strike a balance between the aforementioned considerations, cost, and the detection of transcripts with a circadian pattern of variation is as follows. Mice (or other organisms

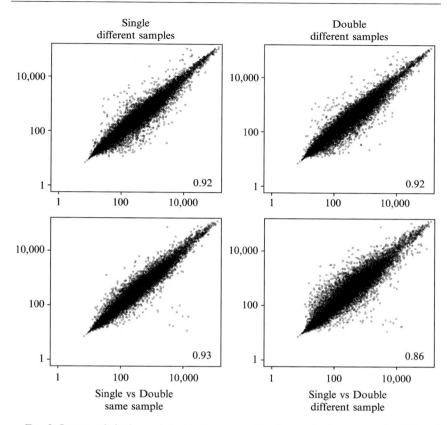

FIG. 2. Impact of single- and double-linear amplification methods on reproducibility of chipping. Log–log plots are shown for single- and double-amplified samples. (Upper left) Single-amplification strategy on two different human tissue samples derived from biopsy. (Upper right) Double-amplification strategy on these same samples. (Lower left) The same sample was both single and double amplified. (Lower right) One sample was single amplified, and the other derived from a separate patient was double amplified. Samples and methods are described in Scherer *et al.* (2003), and were subjected here to condensation with the GCRMA algorithm (Wu *et al.*, 2004) (http://www.bioconductor.org) and plotted using the R-language and environment (Ihaka and Gentlemen, 1996). The coefficient of determination ($R^2$) values is indicated in the lower right-hand corner of the plot.

of interest) are housed in controlled lighting and environmental conditions for at least 2 full weeks or for a period of time that ensures entrainment. Following this period, the mice are released into free-running conditions, usually constant darkness, and samples are collected every 4 h for 2 full days (from CT18 through to CT62). At each time point, two groups of three to five animals of an identical genetic background and similar age (8–12 weeks, C57BL/6J) are sacrificed and dissected. Each group is then subsequently treated as independent replicates for the duration of the experiment. This design allows for the evaluation of cycling over 2 consecutive days and takes into account individual animal and dissection variability, as well as the bias introduced during the chipping process itself. Of course, resources permitting a finer time resolution (every 3 or even 2 h) and/or additional days would be preferable from a statistical standpoint. Finally, like most types of experiments, one size does not fit all—it is assured that other experimental designs and strategies will be used effectively in the detection of cycling transcripts.

Experimental Methodology—Dry Work

Following scanning of the image, there are four *in silico* steps critical to an experiment. First, the image file is converted to a CEL file, essentially a text file that contains the intensity values for each of the probes in a probe set. Second, these values (usually more than 11 perfect match and 11 mismatch probes) are "condensed" to a single intensity value that theoretically approximates the amount of transcript in the sample. This can be accomplished by algorithms included with the Affymetrix system or by third-party tools such as Rosetta Resolver or algorithms penned in academia. Third, statistical tests, filters, and algorithms can be employed to focus on the most relevant genes for the underlying biology. Finally, the lists of transcripts can themselves be explored for interesting pathways and physiologies that change meaningfully in an experiment using informatics. Because several of these informatic steps have themselves been the topic of *Methods in Enzymology* articles (Burgess and McParland, 2002; Straume, 2004; Thomas *et al.*, 2002), the remainder of this section focuses on those aspects that are particularly important for circadian biology.

*Condensation Algorithms*

In our experience, the Affymetrix software does an excellent job of gridding the image files and converting images of probes to intensity values of probes—CEL file creation. However, it is prudent to visually inspect the gridding process, as sometimes a scratch or a speak of dust can result in errors that propagate and compromise downstream analysis. The second step of this process is "condensation" of the multiple perfect match and mismatch probes to a single intensity value that represents the concentration of that transcript. The original algorithm we and others use for circadian studies (Harmer *et al.*, 2000; Lin *et al.*, 2002; Ueda *et al.*, 2002), MAS4 (Affymetrix, Microarray Suite 4), frequently generates intensity values that are negative, as the intensity of the mismatch (MM) probes can exceed that of the perfect match (PM) probes due to cross hybridization. To address this and other problems, Affymetrix and academic groups have devised alternative condensation algorithms—MAS5 (Affymetrix, Microarray Suite 5), PLIER (Affymetrix, Probe Logarithmic Intensity Error Estimation), RMA (Robust Multiarray Average), GCRMA (gc Robust Multiarray Average), dCHIP (algorithm MBEI, Model Based Expression Indexes), and PDNN (Positional Dependent Nearest Neighbor Model)— to solve the tricky problem of accurately and sensitively reporting the intensity value of a single transcript from multiple measures (Irizarry et al., 2003; Li and Hung Wong, 2001; Storch et al., 2002; Wu et al., 2004; Zhang et al., 2003). Unlike MAS4 and MAS5, these other algorithms are model based and rely on reading an entire experiment at a time before emitting an intensity value for any given array.

This condensation step can have a major impact on the detection of transcripts that change meaningfully between samples and on related analysis such as the detection of cycling transcripts. For example, Fig. 3A shows log–log plots of two biological replicates of a single circadian time point condensed with the MAS4, MAS5, Rosetta, dCHIP, RMA, and GCRMA algorithms. While the correlation coefficients for all of these algorithms are similar (~0.99), the model-based dCHIP, RMA, and GCRMA methods introduce less noise at low levels of expression (Fig. 3A). A second important factor in choosing a condensation algorithm is dynamic range, which governs the magnitude of change between samples as well as amplitude of change in a cycling assay. Figure 3A shows that the MAS5 and GCRMA algorithms generate about a log more dynamic range of expression than the other algorithms.

The effect of these different condensation algorithms on detection of cycling is profound. Figure 3B shows the output of a cycling experiment where the same CEL files were first condensed with MAS4, MAS5, or GCRMA and then processed for cycling transcripts with the exact same detection algorithm (Cosopt) (Straume, 2004). Perhaps due to its superior signal/noise performance at the low and high ends of expression, the GCRMA algorithm was able to detect about a third more cycling transcripts than the other algorithms, with MAS4 and MAS5 calling approximately the same number. A comparison of these transcripts reveals that less than a third were called by all three algorithms; these in turn tended to be the most robust cycling transcripts (Fig. 3B). The GCRMA algorithm, in addition, generally reported higher amplitude cycling, followed by MAS5 and then MAS4. Taken in sum, these results show the pervasive effect of the condensation algorithm on downstream-dependent processes such as the detection and amplitude of cycling.

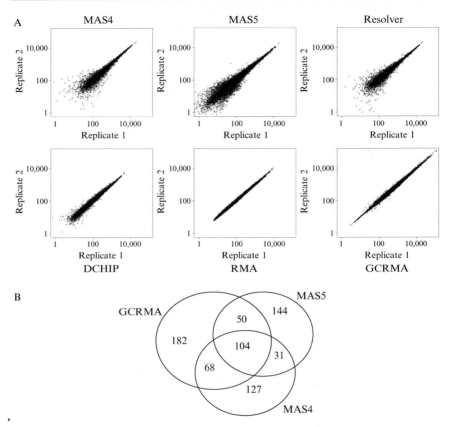

Fig. 3. (A) Impact of condensation algorithms on reproducibility and dynamic range. A time course of samples was collected for 2 days every 4 h from independent pools of three to five animals. These samples were amplified, hybridized, and scanned using the Affymetrix platform [data available at http://www.ncbi.nlm.nih.gov/geo (Edgar *et al.*, 2002), ID=GDS404]. The resultant CEL files were subsequently "condensed" using the MAS4 (Affymetrix), MAS5 (Affymetrix), Resolver (Rosetta Biosoftware, Seattle, WA), DCHIP (Li and Hung Wong, 2001), RMA (Irizarry *et al.*, 2003), and GCRMA (Wu *et al.*, 2004). Data were subsequently scaled to a common median value, and the two replicates were log–log plotted in the R-language and environment (Ihaka and Gentlemen, 1996). (B) A Venn diagram showing overlap of cycling transcripts. The data set was processed with Cosopt to determine cycling transcripts with the GCRMA, MAS4, and MAS5 algorithms. The lists were subsequently compared with a relational database and a Venn diagram was generated.

## Algorithms to Detect Cycling

More intuitive to the circadian biologist is the anticipated effect the algorithm used to detect cycling has on these data sets. As a recent chapter in *Methods in Enzymology* covered this aspect of circadian biology, we provide an example to illustrate its effect relative to the condensation problem described earlier (Straume, 2004). Output from GCRMA or MAS4 procedures was subjected to an implementation of a "moving window" style algorithm (Akhtar *et al.*, 2002) or the latest iteration of Cosopt (Straume, 2004). These were further investigated for overlap. This comparison reveals approximately the same number of transcripts called cycling by both methods; however, only about half are called so by both methods (Fig. 4). Clear examples of robust cycling transcripts called by one algorithm, but not the other, are seen. Of note, differences seen between these two drastically different cycling algorithms, one parametric and one based on curve fitting, are roughly equivalent to those seen between the various condensation algorithms. In addition, these differences are

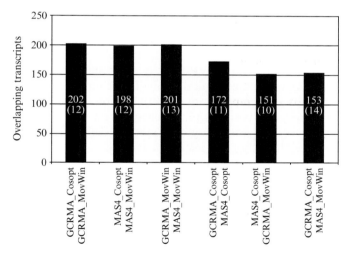

FIG. 4. Effects of condensation and detection of cycling algorithms. The MAS4 and GCRMA data sets from Fig. 3 were analyzed with Cosopt or an implementation of a "moving window" (MovWin) style algorithm (Akhtar *et al.*, 2002). The lists were compared in a relational database and plotted in the R-language and environment. White numbers indicate the number of transcripts in common between two data sets, while numbers in parentheses are numbers expected by chance. The agreement of the two different cycling detection algorithms was generally better than the same algorithm applied to data sets condensed with MAS4 and GCRMA.

independent, using both a different condensation algorithm and a different method of detection produce even fewer common cycling transcripts (Fig. 4).

## Recommendations

It has been more than 6 years since the first chipping papers addressed the cell cycle, and nearly 4 years since RNA profiling using arrays was employed to study circadian transcriptional output (Cho *et al.*, 1998; Harmer *et al.*, 2000). In that time, most of the technical aspects of a chipping experiment have changed—labeling reagents and strategies, the arrays themselves, condensation, and cycling detection algorithms. This observation suggests that researchers interested in applying RNA profiling tools in their circadian studies should keep an eye, lab bench, computer, and checkbook open. Current commercially available RNA-labeling strategies are effective even with very limited starting material (we have tested as low as 1 ng of total). When using these strategies, though, we recommend caution in comparing these double-linear amplification experiments too literally with single-amplification strategies. The latest model-based condensation algorithms, GCRMA and PLIER, are drastic improvements on their predecessors by reducing low end noise while maintaining dynamic range. These offer the opportunity to explore both old and new data sets to more thoroughly wring out biology. Although thoroughly rewarding and challenging, we suspect it is easier to study a cycling transcript than to get universal agreement on how to find them. Try several, and use those that best fit the needs of your experiment. In addition, as the lists of cycling genes grow longer, it is becoming increasingly clear that strategies to take advantage of these data are emerging (Sato *et al.*, 2004; Ueda *et al.*, 2002). Finally, it is likely that chipping itself will eventually be toppled by a more robust technology, such as massively parallel signature sequencing. As these and other approaches mature, additional insights into the circadian clock are sure to follow.

## References

Akhtar, R. A., Reddy, A. B., Maywood, E. S., Clayton, J. D., King, V. M., Smith, A. G., Gant, T. W., Hastings, M. H., and Kyriacou, C. P. (2002). Circadian cycling of the mouse liver transcriptome, as revealed by cDNA microarray, is driven by the suprachiasmatic nucleus. *Curr. Biol.* **12,** 540–550.

Antoch, M. P., Song, E. J., Chang, A. M., Vitaterna, M. H., Zhao, Y., Wilsbacher, L. D., Sangoram, A. M., King, D. P., Pinto, L. H., and Takahashi, J. S. (1997). Functional identification of the mouse circadian Clock gene by transgenic BAC rescue. *Cell* **89,** 655–667.

Balasubramaniam, J., and Del Bigio, M. R. (2002). Analysis of age-dependent alteration in the brain gene expression profile following induction of hydrocephalus in rats. *Exp. Neurol.* **173,** 105–113.

Britton, R. A. (2003). DNA microarrays and bacterial gene expression. *Methods Enzymol.* **370,** 264–278.

Bunger, M. K., Wilsbacher, L. D., Moran, S. M., Clendenin, C., Radcliffe, L. A., Hogenesch, J. B., Simon, M. C., Takahashi, J. S., and Bradfield, C. A. (2000). Mop3 is an essential component of the master circadian pacemaker in mammals. *Cell* **103,** 1009–1017.

Burgess, J. K., and McParland, B. E. (2002). Analysis of gene expression. *Methods Enzymol.* **356,** 259–270.

Cho, R. J., Campbell, M. J., Winzeler, E. A., Steinmetz, L., Conway, A., Wodicka, L., Wolfsberg, T. G., Gabrielian, A. E., Landsman, D., Lockhart, D. J., and Davis, R. W. (1998). A genome-wide transcriptional analysis of the mitotic cell cycle. *Mol. Cell* **2,** 65–73.

Claridge-Chang, A., Wijnen, H., Naef, F., Boothroyd, C., Rajewsky, N., and Young, M. W. (2001). Circadian regulation of gene expression systems in the *Drosophila* head. *Neuron* **32,** 657–671.

Duffield, G. E., Best, J. D., Meurers, B. H., Bittner, A., Loros, J. J., and Dunlap, J. C. (2002). Circadian programs of transcriptional activation, signaling, and protein turnover revealed by microarray analysis of mammalian cells. *Curr. Biol.* **12,** 551–557.

Dunlap, J. C. (1999). Molecular bases for circadian clocks. *Cell* **96,** 271–290.

Edgar, R., Domrachev, M., and Lash, A. E. (2002). Gene Expression Omnibus: NCBI gene expression and hybridization array data repository. *Nucleic Acids Res.* **30,** 207–210.

Green, C. B., and Besharse, J. C. (1996). Identification of a novel vertebrate circadian clock-regulated gene encoding the protein nocturnin. *Proc. Natl. Acad. Sci. USA* **93,** 14884–14888.

Harmer, S. L., Hogenesch, J. B., Straume, M., Chang, H. S., Han, B., Zhu, T., Wang, X., Kreps, J. A., and Kay, S. A. (2000). Orchestrated transcription of key pathways in Arabidopsis by the circadian clock. *Science* **290,** 2110–2113.

Ihaka, R., and Gentlemen, R. (1996). R: A language for data analysis and graphics. *J. Comput. Graph. Stat.* **5,** 299–314.

Irizarry, R. A., Hobbs, B., Collin, F., Beazer-Barclay, Y. D., Antonellis, K. J., Scherf, U., and Speed, T. P. (2003). Exploration, normalization, and summaries of high density oligonucleotide array probe level data. *Biostatistics* **4,** 249–264.

King, D. P., Zhao, Y., Sangoram, A. M., Wilsbacher, L. D., Tanaka, M., Antoch, M. P., Steeves, T. D., Vitaterna, M. H., Kornhauser, J. M., Lowrey, P. L., Turek, F. W., and Takahashi, J. S. (1997). Positional cloning of the mouse circadian clock gene. *Cell* **89,** 641–653.

Kornmann, B., Preitner, N., Rifat, D., Fleury-Olela, F., and Schibler, U. (2001). Analysis of circadian liver gene expression by ADDER, a highly sensitive method for the display of differentially expressed mRNAs. *Nucleic Acids Res.* **29,** E51.

Li, C., and Hung Wong, W. (2001). Model-based analysis of oligonucleotide arrays: Model validation, design issues and standard error application. *Genome Biol.* **2,** RESEARCH0032. Epub, August 2003.

Lin, Y., Han, M., Shimada, B., Wang, L., Gibler, T. M., Amarakone, A., Awad, T. A., Stormo, G. D., Van Gelder, R. N., and Taghert, P. H. (2002). Influence of the period-dependent circadian clock on diurnal, circadian, and aperiodic gene expression in *Drosophila melanogaster*. *Proc. Natl. Acad. Sci. USA* **99,** 9562–9567.

Martin, R. G., and Rosner, J. L. (2003). Analysis of microarray data for the marA, soxS, and rob regulons of *Escherichia coli*. *Methods Enzymol.* **370,** 278–280.

McDonald, M. J., and Rosbash, M. (2001). Microarray analysis and organization of circadian gene expression in *Drosophila*. *Cell* **107,** 567–578.

Panda, S., Antoch, M. P., Miller, B. H., Su, A. I., Schook, A. B., Straume, M., Schultz, P. G., Kay, S. A., Takahashi, J. S., and Hogenesch, J. B. (2002). Coordinated transcription of key pathways in the mouse by the circadian clock. *Cell* **109**, 307–320.

Panda, S., Hogenesch, J. B., and Kay, S. A. (2002). Circadian rhythms from flies to human. *Nature* **417**, 329–335.

Reppert, S. M., and Weaver, D. R. (2002). Coordination of circadian timing in mammals. *Nature* **418**, 935–941.

Roy, S., Khanna, S., Bentley, K., Beffrey, P., and Sen, C. K. (2002). Functional genomics: High-density oligonucleotide arrays. *Methods Enzymol.* **353**, 487–497.

Sandberg, R., Yasuda, R., Pankratz, D. G., Carter, T. A., Del Rio, J. A., Wodicka, L., Mayford, M., Lockhart, D. J., and Barlow, C. (2000). Regional and strain-specific gene expression mapping in the adult mouse brain. *Proc. Natl. Acad. Sci. USA* **97**, 11038–11043.

Sato, T. K., Panda, S., Miraglia, L. J., Reyes, T. M., Rudic, R. D., McNamara, P., Naik, K. A., FitzGerald, G. A., Kay, S. A., and Hogenesch, J. B. (2004). A functional genomics strategy reveals Rora as a component of the mammalian circadian clock. *Neuron* **43**, 527–537.

Scherer, A., Krause, A., Walker, J. R., Sutton, S. E., Seron, D., Raulf, F., and Cooke, M. P. (2003). Optimized protocol for linear RNA amplification and application to gene expression profiling of human renal biopsies. *Biotechniques* **34**, 546–556.

Storch, K. F., Lipan, O., Leykin, I., Viswanathan, N., Davis, F. C., Wong, W. H., and Weitz, C. J. (2002). Extensive and divergent circadian gene expression in liver and heart. *Nature* **417**, 78–83.

Straume, M. (2004). DNA microarray time series analysis: Automated statistical assessment of circadian rhythms in gene expression patterning. *Methods Enzymol.* **383**, 149–166.

Thomas, R. P., Guigneaux, M., Wood, T., and Evers, B. M. (2002). Age-associated changes in gene expression patterns in the liver. *J. Gastrointest. Surg.* **6**, 445–453; discussion 454.

Thomas, R. S., Rank, D. R., Penn, S. G., Craven, M. W., Drinkwater, N. R., and Bradfield, C. A. (2002). Developing toxicologically predictive gene sets using cDNA microarrays and Bayesian classification. *Methods Enzymol.* **357**, 198–205.

Ueda, H. R., Chen, W., Adachi, A., Wakamatsu, H., Hayashi, S., Takasugi, T., Nagano, M., Nakahama, K., Suzuki, Y., Sugano, S., Iino, M., Shigeyoshi, Y., and Hashimoto, S. (2002). A transcription factor response element for gene expression during circadian night. *Nature* **418**, 534–539.

Van Gelder, R. N., von Zastrow, M. E., Yool, A., Dement, W. C., Barchas, J. D., and Eberwine, J. H. (1990). Amplified RNA synthesized from limited quantities of heterogeneous cDNA. *Proc. Natl. Acad. Sci. USA* **87**, 1663–1667.

Wittliff, J. L., and Erlander, M. G. (2002). Laser capture microdissection and its applications in genomics and proteomics. *Methods Enzymol.* **356**, 12–25.

Wu, Z., Irizarry, R. A., Gentlemen, R., Murillo, F. M., and Spencer, F. (2003). A model based background adjustment for oligonucleotide expression arrays. Johns Hopkins University Dept. of Biostatistics Working Papers, 1–29.

Zhang, L., Miles, M. F., and Aldape, K. D. (2003). A model of molecular interactions on short oligonucleotide microarrays. *Nature Biotechnol.* **21**, 818–821.

Zinke, I., Schutz, C. S., Katzenberger, J. D., Bauer, M., and Pankratz, M. J. (2002). Nutrient control of gene expression in *Drosophila*: Microarray analysis of starvation and sugar-dependent response. *EMBO J.* **21**, 6162–6173.

# Section III

# Molecular Cycles: Clock Protein Rhythms

## [17]    Analysis of Posttranslational Regulations in the *Neurospora* Circadian Clock

*By* Yi Liu

### Abstract

Posttranslational modification of circadian clock proteins by phosphorylation is an essential regulatory process in the control of eukaryotic circadian clocks. In the *Neurospora* circadian clock, the key clock protein FREQUENCY (FRQ) is progressively phosphorylated. The phosphorylation of FRQ is regulated by both kinases and phosphatases, and the phosphorylation is important for regulating FRQ stability and its function in the circadian negative feedback loop. The degradation of FRQ is mediated by the ubiquitin/proteasome pathway. This article discusses posttranslational regulations of the *Neurospora* clock and describes the methods used in the studies of FRQ phosphorylation, FRQ kinases and phosphatases, and FRQ degradation.

### Introduction

Posttranslational modification of circadian clock proteins by phosphorylation is an essential regulatory process in the control of eukaryotic circadian clocks (Dunlap, 1999; Young and Kay, 2001). The phosphorylation state of a protein is dynamically regulated by protein kinases and protein phosphatases. Several conserved protein kinases, including casein kinase I (CKI) and casein kinase II (CKII), have been shown to phosphorylate and regulate the key clock components in *Neurospora, Drosophila,* mammals, and *Arabidopsis* (Akten *et al.*, 2003; Gorl *et al.*, 2001; Kloss *et al.*, 1998; Lin *et al.*, 2002; Lowrey *et al.*, 2000; Price *et al.*, 1998; Sugano *et al.*, 1999; Yang *et al.*, 2002, 2003). However, protein phosphatase 1 (PP1) and PP2A have been implicated in the control of the *Neurospora* and *Drosophila* circadian clocks (Sathyanarayanan *et al.*, 2004; Yang *et al.*, 2004). Phosphorylation has been demonstrated to be important for regulating the degradation and activity of the circadian clock proteins (Nawathean and Rosbash, 2004; Yang *et al.*, 2002; Young and Kay, 2001). The degradation process of clock proteins is a critical process in generating circadian rhythmicity, and the phosphorylation-triggered and ubiquitin/proteasome-mediated degradation of clock proteins appears to be another common theme in the eukaryotic clock systems (Akashi *et al.*, 2002; Grima *et al.*, 2002; Kim *et al.*, 2003; Ko *et al.*, 2002; Mas *et al.*, 2003; Naidoo *et al.*, 1999; Yagita *et al.*, 2002). Because of the similarities of the posttranslational regulations in different eukaryotic clock systems, it has been proposed that the molecules mediating the modifications and degradation of clock proteins may be the common foundation that allows the evolution of circadian clocks in eukaryotes (He *et al.*, 2003; Lin *et al.*, 2002).

Posttranslational Regulations in the *Neurospora* Circadian Clock

In *Neurospora*, FREQUENCY (FRQ), WHITE COLLAR-1 (WC-1), and WC-2 proteins are the three essential components in the core circadian negative feedback loop. In constant darkness, a heterodimeric complex formed by WC-1 and WC-2 binds to the *frq* promoter and activates its transcription (Cheng *et al.*, 2001b, 2002, 2003; Crosthwaite *et al.*, 1997; Froehlich *et al.*, 2003; Talora *et al.*, 1999). However, homodimeric FRQ inhibits its own transcription by physically interacting with the WC-1/WC-2 complex (Aronson *et al.*, 1994; Cheng *et al.*, 2001a; Denault *et al.*, 2001; Froehlich *et al.*, 2003). Like the Period proteins in *Drosophila* and mammals, the FRQ protein is progressively phosphorylated after its synthesis and becomes extensively phosphorylated prior to its degradation (Garceau *et al.*, 1997; Liu *et al.*, 2000). In addition to the circadian rhythm in FRQ levels in constant darkness, its phosphorylation status also oscillates robustly so that both the level and the phosphorylation status of FRQ define the time of the clock during a circadian cycle. Both WC proteins are also known to be phosphorylated *in vivo* (Schwerdtfeger and Linden, 2000; Talora *et al.*, 1999), but the roles of their phosphorylation are not known.

An early study of FRQ phosphorylation suggested that it has several functions (Liu *et al.*, 2000). First, phosphorylation of FRQ promotes its degradation. A kinase inhibitor that blocks FRQ phosphorylation *in vivo* also reduces the degradation rate of FRQ and lengthens the period of the clock. In addition, mutation of one FRQ phosphorylation site at Ser-513 leads to the slow-down of FRQ degradation and a very long period length of the clock. Thus, phosphorylation of FRQ appears to be an important determining factor for FRQ stability and for the period length of the clock. This study also suggested that FRQ phosphorylation might have other roles as well, as the mutation of another FRQ phosphorylation site (Ser-519) resulted in a short period and low amplitude of overt rhythm.

To identify the kinases that phosphorylate FRQ and regulate its function, we took a biochemical purification approach and identified two kinases. CAMK-1, a calcium/calmodulin-dependent kinase in *Neurospora*, was the first FRQ kinase identified. CAMK-1 accounts for close to half of the FRQ-kinase activity *in vitro*. Although modest clock phenotypes were found in the *camk-1* null strain, we did not observe significant changes of the FRQ phosphorylation profile in the mutant strains. These data indicate that FRQ is phosphorylated by multiple kinases *in vivo*.

Using the *camk-1* null strain, a second FRQ kinase was purified and identified as casein kinase II (Yang *et al.*, 2002). The *Neurospora* CKII holoenzyme consists of three subunits, one catalytic ($\alpha$) subunit (*cka*) and two different regulatory ($\beta$) subunit genes (*ckb1* and *ckb2*), forming a $\alpha_2\beta_2$ heterotetramer. In a *Neurospora* strain in which the catalytic subunit gene of CKII (*cka*) is disrupted, FRQ levels are high and more stable, and FRQ proteins are hypophosphorylated. The circadian rhythms of *frq* RNA, FRQ protein, and clock-controlled genes are abolished in this mutant. In addition, despite high FRQ levels in the *cka* mutant, *frq* mRNA levels are high, reminiscent of the *frq* level in a *frq⁹* strain (Aronson *et al.*, 1994), suggesting that the circadian negative feedback loop is impaired. When one of the CKII regulatory subunit genes, *ckb1*, was disrupted in *Neurospora*, FRQ proteins are also hypophosphorylated, and long periods but low amplitudes of circadian rhythms are observed in the mutant (Yang *et al.*, 2003). Together, these data suggest that the phosphorylation of FRQ by CKII is essential for the clock function and that phosphorylation may have at least three roles: regulating the stability of FRQ, regulating the interactions between FRQ and the WC proteins, and important for closing of the *Neurospora* circadian negative feedback loop.

CK-1a, one of the two CKIs in *Neurospora*, has also been shown to be a FRQ kinase (Gorl *et al.*, 2001). It interacts physically with FRQ *in vivo* and can phosphorylate one of the FRQ PEST domains *in vitro*. Deletion of the PEST domain leads to increased FRQ stability and a long-period circadian rhythm. The *in vivo* function of CK-1a is still unclear, as disruption of the *ck-1a* gene is lethal. CK-1b is the other CKI in *Neurospora*. Like CK-1a, it can also phosphorylate FRQ *in vitro*. Disruption of the *ck-1b* gene, however, showed that it is not required for FRQ phosphorylation and clock function, despite its important roles in growth and developmental processes (Yang *et al.*, 2003).

Unlike the large number of protein kinases in eukaryotes, there are only a few highly conserved catalytic subunits of protein phosphatases, and protein phosphatase 1 (PP1) and PP2A are two major classes of the eukaryotic protein phosphatases. Protein phosphatases carry out their diverse cellular functions with the help of a large number of regulatory proteins. Studies have shown that both PP1 and PP2A are important regulatory components in the *Neurospora* clock. In a *ppp-1* (the catalytic subunit of PP1) mutant strain, which has significant reduction of PP1 activity due to four missense mutations, FRQ is less stable, resulting in a significant phase advance and a short-period phenotype (Yang *et al.*, 2004).

In contrast, disruption of the *Neurospora rgb-1* gene (encodes for a regulatory subunit for PP2A) does not affect the stability of FRQ. In the *rgb-1* mutant, however, the levels of FRQ protein and *frq* mRNA are low, and the clock oscillates with a low-amplitude and long-period rhythm. These data suggest that PP1 and PP2A have different roles in regulation of the *Neurospora* circadian clock: PP1 influences the clock by regulating the stability of FRQ, whereas PP2A is important for the function of negative feedback loop. Furthermore, it was shown that endogenous FRQ can be dephosphorylated directly by *Neurospora* expressed PP1 and PP2A *in vitro* and that the FRQ phosphorylation profile in the *rgb-1* mutant strain is altered. These data suggest that both PP1 and PP2A may regulate the clock by desphosphorylating FRQ. TWS, the *Drosophila* homolog of RGB-1, has also been shown to be a critical clock component in the fly (Sathyanarayanan *et al.*, 2004).

Like the Period proteins in fly and mammals, the degradation of FRQ following its phosphorylation is mediated by the ubiquitin/proteasome pathway (He *et al.*, 2003). We have shown that FRQ can be ubiquitinated *in vivo*, and its proper degradation requires FWD-1, an F-box/WD-40 repeat-containing protein and the *Neurospora* homolog of the *Drosophila* protein Slimb. In *fwd-1* disruption strains, FRQ degradation is severely impaired, resulting in the accumulation of hyperphosphorylated FRQ, further indicating the role of FRQ phosphorylation in mediating its degradation. In addition, the circadian rhythms of gene expression and the circadian conidiation rhythms are abolished in these mutants. Moreover, FRQ and FWD-1 interact physically *in vivo*, suggesting that FWD-1 is the substrate-recruiting subunit of an SCF-type ubiquitin ligase responsible for FRQ ubiquitination and degradation. The interaction between FRQ and FWD-1 in a wild-type strain is, however, transient, and a stable complex can only be formed when the F-box domain of FWD-1 is deleted, suggesting that the FWD-1-associated FRQ is ubiquitinated and degraded rapidly by the 26S proteasome in a wild-type strain. Together, these data are consistent with a model that the progressive phosphorylation of FRQ at multiple independent sites is a dynamic process that fine-tunes the stability of FRQ and thus the period of the circadian clock (He *et al.*, 2003). In this model, the degree of FRQ phosphorylation determines the affinity between FRQ and FWD-1 and the stability of FRQ. Thus, the progressive nature of FRQ phosphorylation (like those of the Period proteins in animals) provides an explanation for the critical delay needed in the circadian negative feedback loop that allows the clock to cycle with a 24-h periodicity.

The combination of biochemical, genetic, and molecular approaches has proven very powerful in the studies of the *Neurospora* circadian clock and its posttranslational regulations. This article describes some of the methods that have been used successfully in our studies of FRQ phosphorylation, FRQ kinases and phosphatases, and FRQ degradation.

Description of Methods

*Immunoprecipitation (IP) and Phosphatase Treatment to*
  *Demonstrate the Phosphorylation of a* Neurospora *Protein*

This method is used when a protein is suspected to be a phosphoprotein in *Neurospora*. λ phosphatase often does not work well when it is directly added to the cell extracts, but it works efficiently with the immunoprecipitation products (Cheng *et al.*, 2001a; Garceau *et al.*, 1997; Talora *et al.*, 1999).

*Immunoprecipitation.* Incubate 0.5–1 mg of soluble protein extracts in 0.5–1 ml protein extraction buffer (50 m$M$ HEPES, pH 7.4, 150 $M$ NaCl, 10% glycerol, 1 μg/ml pepstatin A, 1 μg/ml leupeptin, 1 m$M$ phenylmethylsulfonyl fluoride) at 4° for 1–2 h with the antiserum (1:200–1000 dilution) against the protein of interest. Then add protein G-agarose beads (10 μl bead volume, Amersham Pharmacia) and incubate for 1–2 h at 4° with gentle mixing. Afterward, wash the protein G-agarose beads four times with ice-cold extraction buffer before suspending the IP products in protein-loading buffer or 50 μl of phosphatase buffer.

*Phosphatase Treatment.* Add 1000 units λ phosphatase (New England Biolabs) to the IP products resuspended in the phosphatase buffer (supplied with the λ phosphatase) and incubate at 30° for 30 min before loading on SDS–PAGE gels for Western blot analysis. To demonstrate the role of the λ phosphatase, incubate the separate IP product with phosphatase buffer alone or with phosphatase plus 20 m$M$ sodium vanadate (a phosphatase inhibitor). If the protein of interest is phosphorylated, the phosphatase treatment will frequently result in a faster gel mobility protein form than the protein from the control treatment (Garceau *et al.*, 1997).

In Vitro *Kinase Assays*

Three kinds of *in vitro* kinase assays have been used successfully to study the phosphorylation of FRQ (Yang *et al.*, 2001, 2002). The first two assays use recombinant GST-FRQ as the kinase substrate, whereas the third assay uses the endogenous FRQ in cell extracts as the substrate.

In Vitro *Solution Kinase Assay Using GST-FRQ as the Substrate*

PURIFICATION OF GST-FRQ. Grow *Escherichia coli* strain BL21 cells (Invitrogen) carrying the GST-FRQ fusion protein expression construct (containing FRQ amino acids 425–683) in 37° to an $A_{600}$ of 0.5–0.7.

Then transfer the cultures to 25° and induce with 1 m$M$ isopropyl-β-D-thiogalactoside. After 3-h induction, harvest cultures resuspend in 1/50 volume of phosphate-buffered saline (PBS), and purify on a glutathione-agarose column.

KINASE ASSAY. Incubate GST-FRQ protein (~5 μg) with *Neurospora* protein extracts (5–10 μg of cell extracts) in kinase assay buffer containing 25 m$M$ HEPES–NaOH (pH 7.9), 10 m$M$ MgCl, 2 m$M$ MnCl$_2$, 25 μ$M$ ATP, and 10 μCi/ml [γ-$^{32}$P] ATP (total reaction volume of 125 μl). When used to monitor the kinase activity of calcium/calmodulin-dependent kinases, add 0.15 m$M$ CaCl$_2$ and 14 μg/ml calmodulin (Sigma) to the reaction buffer. Incubate the reaction mixture at room temperature for 1 h. Then add 0.5 ml of PBS and 10 μl of glutathione-agarose beads. After a 30-min incubation at room temperature with gentle mixing, centrifuge the glutathione-agarose beads down and wash twice with PBS before resuspending and boiling in 1× SDS–PAGE-loading buffer. Then subject the samples to 10% SDS–PAGE. After electrophoresis, dry and subject the gel to autoradiography.

*In-Gel Kinase Assay.* The in-gel kinase assay we used is adopted from previously described methods (Hibi *et al.*, 1993; Liu *et al.*, 1997b; Yang *et al.*, 2001). The advantage of this assay is that it allows for estimation of the molecular weight and the number of the potential kinases involved in phosphorylating a specific substrate. However, one major drawback of this assay is that the kinase of interest must be able to refold correctly after the denaturation process and can function by itself. Thus, it will not identify kinases that cannot refold well under the condition used or require separate regulatory subunits to function.

First, polymerize 8–10% SDS–PAGE gel in the presence of 100 μg/ml of the purified GST (a negative control) or GST-FRQ protein. To save substrates, one gel can be separated into two to three sections (using spacers) containing different substrates. Subject *Neurospora* cell extracts (20–40 μg) prepared from the wild-type strain to electrophoresis using the substrate-containing gel. After electrophoresis, wash the gel twice with 50 m$M$ Tris–HCl (pH 8.0) in 20% isopropanol for 30 min and twice with buffer B (50 m$M$ Tris–HCl, pH 7.5, 5 m$M$ β-mercaptoethanol) for 30 min. Denature the proteins in the gel by incubating the gel twice for 30 min in buffer B containing 6 $M$ guanidium–HCl. Then renature the proteins overnight or longer at 4° in buffer B containing 0.05% Tween 20 with gentle shaking. During the renaturalization process (this step is crucial for the success of this assay), the buffer needs to be changed at least four times. Afterwards, incubate the gel in kinase buffer (25 m$M$ HEPES–NaOH, pH 7.9, 10 m$M$ MgCl$_2$, 2 m$M$ MnCl$_2$, 25 μ$M$ ATP) containing 10 μCi/ml of [γ-$^{32}$P]ATP at room temperature for 1 h. When used, add

0.15 m$M$ CaCl$_2$ and 14 $\mu$g/ml calmodulin (Sigma) to the reaction buffer. Wash the gel five to seven times in 5% trichloroacetic acid and 1% sodium pyrophosphate to remove background radiation. Then dry the gel and subject to autoradiography.

In Vitro *Assay of Phosphorylation of the Endogenous FRQ Protein in* Neurospora *Cell Extracts.* This assay is used to monitor the phosphorylation of endogenous FRQ protein by the endogenous kinases in *Neurospora* cell extracts. It is simple and useful for examining FRQ phosphorylation profile changes *in vitro* and for pharmacological studies.

First, prepare *Neurospora* cell extracts in extraction buffer (50 m$M$ HEPES, pH 7.4, 150 m$M$ NaCl, and 10% glycerol) from cultures harvested at a time point when most of the FRQ proteins are not phosphorylated extensively (e.g., at DD14–16). Adjust the protein concentration of the extracts to 1–2 mg/ml with the extraction buffer. To start the phosphorylation reaction, add 5 m$M$ ATP (or using a ATP regeneration system), 5 m$M$ MgCl$_2$, and 2 m$M$ MnCl$_2$ to the extracts and incubate the reaction mixture for 1–8 h at room temperature before adding an equal volume of 2x SDS–PAGE loading buffer. Subject the samples to SDS–PAGE and Western blot analysis using FRQ antiserum. For pharmacological treatments, drugs can be added to the cell extracts before the start of the phosphorylation reaction.

## Biochemical Purification of Calcium/Calmodulin-Dependent Kinase-1 from Neurospora

CAMK-1 is the first FRQ kinase identified and it is responsible for about half of the FRQ kinase activity *in vitro* (Yang *et al.*, 2001). For the purification of CAMK-1, both in-gel kinase and *in vitro* solution kinase assays (described earlier) in the presence of calcium and calmodulin are used to monitor its kinase activity.

The entire purification procedure should be carried out at 4°. Prepare 100 mg wild-type *Neurospora* cell extracts in 20 ml of 50 m$M$ HEPES, pH 7.4, 137 m$M$ NaCl, and 10% glycerol. Load the extracts onto a DEAE-Sepharose column (50-ml bed volume) that has been equilibrated with buffer A [50 m$M$ Tris–HCl, pH 7.5, 1 m$M$ dithiothretrol (DTT), 1 m$M$ EDTA]. Collect the flow-through portion (containing CAMK-1) and load onto a CM-Sepharose column (15-ml bed volume) preequilibrated with buffer A. Wash the column with a 2-column volume of buffer A before eluting the proteins on the column with buffer A containing 100 m$M$ NaCl. Load the eluted protein fraction (containing CAMK-1) onto a CaM affinity column (1-ml bed volume, Amersham Pharmacia) that has been equilibrated with buffer A containing 5 m$M$ MgCl$_2$ and 1 m$M$ CaCl$_2$. Then wash the column with buffer A containing 5 m$M$ $MgCl_2$, 1 m$M$ $CaCl_2$, and 100 m$M$ NaCl. Elute proteins on the column first by a 5-column volume of buffer A containing 5 m$M$ EGTA and then by buffer A containing 5 m$M$ EGTA and 1 $M$ NaCl. CAMK-1 binds to the CaM column tightly, and the final high salt wash is important for its elution from the column. The final eluted products contain only two proteins: CAMK-1 and the *Neurospora* elongation factor-1.

## Biochemical Purification of Casein Kinase II from Neurospora

CKII is the second FRQ kinase identified by biochemical purification approach and has been shown to be an essential clock component in *Neurospora* (Yang *et al.*, 2002, 2003).

To purify *Neurospora* CKII using GST-FRQ as a substrate, an automatic fast protein liquid chromatography (FPLC) station (Amersham-Phamacia) is needed and all procedures should be carried out at 4°. Buffers and protein samples need to be filtered by 0.2-$\mu$m filters. Use the *in vitro* solution kinase assay described earlier for monitoring CKII activity. Prepare *Neurospora* protein S-100 extracts (1.5 g proteins from 10 to 15 liters of *Neurospora*) in 50 m$M$ HEPES (pH 7.4), 137 m$M$ NaCl, and 10% glycerol from the *camk-1$^{ko}$* strain. Apply the protein extracts onto a Q-Sepharose column (80-ml bed volume, Amersham Pharmacia) equilibrated with buffer A (50 m$M$ Tris–HCl, pH 7.5, 1 m$M$ DTT). After washing with buffer A, elute proteins on the column with a 800-ml linear gradient from 20 to 500 m$M$ NaCl in buffer A. Collect and assay fractions for kinase activity. Pool and load active fractions ($\sim$0.45 $M$ NaCl, CKII binds to the Q-Sepharose column tightly and elutes late in the elution, apart from most of the cellular proteins) on a hydroxyapatite column (7-ml bed volume, Bio-Rad) equilibrated previously with 10 m$M$ $KPO_4$ buffer. Wash the column with 10 m$M$ $KPO_4$ followed by 10 m$M$ $KPO_4$ containing 1 $M$ NaCl. Then elute the proteins with a 15-ml linear gradient from 50 to 350 m$M$ $KPO_4$ (pH 7.5). Pool and concentrate CKII fractions (0.24–0.34 M $KPO_4$, $\sim$2 ml) to $\sim$0.5 ml using a Centricon (Millipore). Then load it on a Superdex 200(10/30) gel filtration column (Amersham Pharmacia) preequilibrated with buffer A containing 150 m$M$ NaCl and elute with the same buffer. Pool and load active fractions (molecular mass around 150 kDa, $\sim$2 ml) onto a Mono Q(5/5) column (Amersham Pharmacia) and elute with a 10-ml linear gradient from 300 to 700 m$M$ NaCl in buffer A. Collect and assay fractions of 0.5 ml for activity. Store active fractions at −80° in the presence of 10% glycerol. SDS–PAGE followed by silver staining can be used to examine the yield and purity of the purified kinase.

## Determination of FRQ Stability *in* Neurospora

Phosphorylation of FRQ regulates its stability and is an important determinant of the period length of the clock (Gorl *et al.*, 2001; He *et al.*, 2003; Liu *et al.*, 2000; Yang *et al.*, 2002, 2003, 2004). There are two convenient and reliable methods to compare FRQ degradation rates among different *Neurospora* strains and both methods are usually used together to complement each other (Liu *et al.*, 1997a, 2000). In the first method, transfer *Neurospora* strains that are grown in constant light (LL) for at least 1 day into constant darkness (DD) and monitor the levels of FRQ proteins by Western blot analysis for the first 12 h in DD. Because the levels of FRQ proteins are high and evenly phosphorylated in LL and the LD transfer triggers rapid degradation of *frq* mRNA, the rate of reduction of FRQ protein levels in the first 12 h after the LD transfer is frequently a good indicator of FRQ stability in various *Neurospora* strains.

In the second method, add the protein synthesis inhibitor cycloheximide (CHX, 10 $\mu$g/ml, stock solution: 10 mg/ml in ethanol) to the cultures grown in LL to block *de novo* synthesis of FRQ, and monitor the degradation rate of FRQ after the addition of the drug by Western analysis. The CHX concentration used has been shown to inhibit greater than 95% of protein synthesis in *Neurospora* (Dunlap and Feldman, 1988; Johnson and Nakashima, 1990), and FRQ phosphorylation and degradation proceed in the presence of the inhibitor. This approach has also been used to monitor the stability of the WC-1 protein (Lee *et al.*, 2000; Schwerdtfeger and Linden, 2001).

## Phosphatase Activity Measurement *in* Neurospora *Extracts*

Protein phosphatase 1 and protein phosphatase 2A are two of the major protein phosphatases in eukaryotic cells. We have shown that both phosphatases are important regulators of the *Neurospora* circadian clock (Yang *et al.*, 2004). PP1 influences the clock by regulating the stability of FRQ while PP2A may prevent the closing of the circadian negative feedback loop. Measurement of specific phosphatase activity in *Neurospora* extracts is more difficult than kinase assays due to following reasons: (1) There are large number of regulatory proteins for each phosphatase catalytic subunit to regulate its activity, specificity, and cellular localization; (2) phosphatase assays typically require measurement of the amount of phosphate released from the substrate rather than examining the inhibition of the substrate phosphorylation so that the substrate needs to be phosphorylated properly before the assay; and (3) both PP1 and PP2A are very fragile in cell extracts (e.g., very sensitive to freeze–thaw cycles). We have successfully used two types of *in vitro* phosphatase assays using *Neurospora* extracts in our studies. The first assay, modified from the traditionally used phosphorylase b-based method (Krebs and Fischer, 1962), is used to measure the general activities of PP1 and PP2A in *Neurospora* extracts. This assay can inform whether the overall activity of the phosphatase of interest is altered in a mutant strain. However, it does not provide information on specific substrates. In the second assay we developed, using a coimmunoprecipitation method to couple the *Neurospora*-expressed phosphatase and substrate together, we can ask whether a specific *Neurospora* phosphoprotein can be a substrate for a phosphatase *in vitro*.

### Protein Phosphatases 1 and 2A Assay In Vitro

PREPARATION OF SUBSTRATE-PHOSPHORYLATION OF PHOSPHORYLASE B. First, resuspend 10 mg of phosphorylase b (Sigma) in 500 $\mu$l reaction mix containing 150 m$M$ Tris–HCl, pH 8.2, 10 m$M$ MgCl2, 1 m$M$ ATP, 0.5 m$M$ CaCl2, 80 m$M$ $\beta$-glycerolphosphate, and 0.5 mCi of $[\gamma\text{-}^{32}\text{P}]$ATP. To start the phosphorylation reaction, add 20 units of phosphorylase kinase (Sigma) to the mix. Vortex well and incubate the mix at 30° for 1 h. Stop the reaction by adding 0.5 ml 90% saturated ammonium sulfate solution. Incubated the mixture on ice for 30 min to allow precipitation of the phosphorylase. Centrifuge at 12,000g for 10 min at 4°. Wash the pellet four times or more with 45% saturated ammonium sulfate solution. Dissolve the pellet in 3 ml of solubilization buffer (50 m$M$ Tris–HCl, pH 7.0, 0.1 m$M$ EDTA, 1 m$M$ DTT). Then purify the $^{32}$P-labeled phosphorylase a by passing through a bio-30 column (Bio-Rad) to further remove free radionucelotide. If the background radiation is high in the phosphatase assay, clean the labeled products with an additional bio-30 column purification. Afterward, the substrates can be kept in 4° for up to 2 weeks. However, incubation of the substrates in 4° for a long period of time does increase the background radiation level.

PHOSPHATASE ASSAY. Determine the activities of protein phosphatase 1 and 2A in *Neurospora* cell extracts by measuring the release of $[^{32}\text{P}]$Pi from $^{32}$P-labeled phosphorylase a. PP1 activity is normally defined as the activity inhibited by inhibitor-2 (Sigma, a specific protein inhibitor of the PP1 catalytic subunit), whereas PP2A-like activities are defined as the activity inhibited by 1–5 n$M$ okadaic acid (Sigma). Okadaic acid can inhibit both PP1 and PP2A, but at a 1–5 n$M$ concentration, it mostly inhibits PP2A. We suggest that only freshly made *Neurospora* extracts be used in the assay. First incubate 20 $\mu$g of *Neurospora* cell extracts incubated in the presence or absence of inhibitor-2 (100 U) or okadaic acid in 40 $\mu$l of phosphatase buffer (20 m$M$ MOPS, pH 7.0, 1 m$M$ DTT) on ice for 15 min. Initiate the reactions with the addition of 10 $\mu$l of the $^{32}$P-labeled phosphorylase a and incubate at 30° for 5 min and then stop by adding

100 $\mu$l of 25% trichloroacetic acid (final concentration of 10%) and 100 $\mu$l 6 mg/ml bovine serum albumin (BSA) (Sigma). BSA works as a carrier to improve precipitation of the substrate and enzyme. Incubate the mixtures on ice for 5 min and centrifuge for 5 min at 12,000g. Count 150 $\mu$l of the resulting supernatant by a scintillation counter.

In Vitro *Dephosphorylation Assay Following Coimmunoprecipitation of Phosphatase and Substrate from* Neurospora *Extracts.* In this assay, the phosphatase and its substrate are immunoprecipitated down together using protein G-agarose beads coupled with antibodies specific for the phosphatase and the substrate before the dephosphorylation reaction is performed. The effects of the phosphatase on the phosphorylation of the substrate are revealed by Western blot analysis of the phosphorylation profile of the substrate. Thus, antiserum specific for the phosphatase and the substrate are required for this assay. In our study of PP1 and PP2A in their roles in dephosphorylating FRQ (Yang *et al.*, 2004), we created *Neurospora* strains in which PP1 or PP2A is c-Myc epitope tagged, and the cell extracts of the strains are mixed with the extracts of a strain expressing c-Myc-tagged FRQ (Cheng *et al.*, 2001a) or a wild-type strain (lacking c-Myc-tagged FRQ and serving as a negative control) before immunopreicpitation using protein G-agarose beads coupled with a c-Myc monoclonal antibody. Because both the phosphatase and the substrate are linked to the beads after immunoprecipitation and are not freely available to each other, significantly more extracts from the strain expressing the Myc-tagged phosphatase than extracts expressing Myc-FRQ are needed in the immunoprecipitation mixture.

First, make fresh cell extracts in extraction buffer (50 m$M$ HEPES, pH 7.4, 150 $M$ NaCl, 10% glycerol, 1 $\mu$g/ml pepstatin A, 1 $\mu$g/ml leupeptin). To perform immunoprecipitation, this 50 $\mu$g of the *Neurospora* cell extracts of *frq^10^*, Myc-FRQ with 1-mg extracts of Myc.PPP-1, Myc.PPH-1, or wild-type strains (total volume of 0.5–1 ml). Then incubate the mixed extracts at 4° for 2 h with the monoclonal c-Myc antibody (2 $\mu$g) (Santa Cruz Biotechnology). Subsequently, add protein G-agarose beads (10 $\mu$l) and incubate the mixture for 1–2 h at 4° with gentle mixing. Afterward, wash the beads four times with ice-cold extraction buffer and once with phosphatase buffer (20 m$M$ MOPS, pH 7.0, 1 m$M$ DTT) before resuspending the IP products in 30 $\mu$l of phosphatase buffer. To carry out the dephosphorylation reaction, incubate the beads in phosphatase buffer at 30° for 1 h before stopping the reaction by adding SDS–PAGE sample buffer and subjecting to Western blot analysis using FRQ antiserum. Changes in the FRQ phosphorylation profile revealed by Western blot analysis will indicate whether FRQ is a substrate for the phosphatase.

*Ophiobolin A Treatment to Examine Whether a Protein Can Be Ubiquitinated in* Neurospora

The ubiquitin/proteasome pathway mediates the degradation of clock components in all eukaryotic circadian systems examined from *Neurospora*, plants, *Drosophila*, and mammals and is important for clock functions (Akashi *et al.*, 2002; Grima *et al.*, 2002; He *et al.*, 2003; Ko *et al.*, 2002; Mas *et al.*, 2003; Naidoo *et al.*, 1999; Yagita *et al.*, 2002). One of the hallmarks of substrates of the ubiquitin/proteasome pathway is that, after the addition of proteasome inhibitors, there is an accumulation of polyubiquitinated forms of these proteins, which migrate as ladders (or smears) of high molecular weight bands on SDS–PAGE followed by immunoblotting. In *Neurospora*, however, all the commonly used proteasome inhibitors, including LLNL, MG115, MG132, and lactacystin, cannot enter the *Neurospora* cells, so the role of the ubiquitin/proteasome system on the degradation of the clock components cannot be studied using these inhibitors. The inability of these drugs to penetrate the *Neurospora* cell wall and/or cell membrane is not unusual, as they also fail to enter the wild-type yeast cells (Lee and Goldberg, 1996).

In our analysis of FRQ degradation by the ubiquitin-proteasome pathway, we identified a known calmodulin inhibitor, ophiobolin A, which, when added to *Neurospora* cultures, triggers a high molecular weight smear of FRQ-specific proteins, reminiscent of the polyubiquitinated proteins. Ophiobolin A treatment also results in a significant increase in the ubiquitinated cellular proteins in *Neurospora*. Such an accumulation of ubiquitinated cellular proteins is very similar to that observed for animal cells following the treatment of proteasome inhibitors. These observations and the identification of FWD-1 as the E3 ligase for FRQ ubiquitination led us to conclude that FRQ can be ubiquitinated *in vivo* and is a substrate for the proteasome system (He *et al.*, 2003). We have also shown that ophiobolin A could induce ubiquitination of cyclin B in *Xenopus* egg extracts. These data suggest that ophiobolin A has a general stimulatory effect on the ubiquitination of eukaryotic proteins. However, unlike proteasome inhibitors, it does not block the degradation of FRQ and, in fact, speeds up its degradation. Therefore, its effect on protein ubiquitination is not due to the inhibition of the activity of the proteasome. We did not observe a similar effect with other calmodulin inhibitors, suggesting that the effects of ophiobolin A on protein ubiquitination are due to an unknown effect of the drug.

For ophiobolin A treatment, treat *Neurospora* cultures that have been grown for 1 day in liquid media with 50–100 $\mu M$ of ophiobolin A (Sigma) for 2–4 h before harvesting. Then subject the samples to Western blot analysis. From our experience, the effects of the drug on old cultures are less prominent.

## Concluding Remarks

*N. crassa*, which has one of the best understood circadian clock systems, offers an excellent experimentally accessible paradigm for understanding the clock mechanism at the molecular level. The methods described here have been instrumental in elucidating the functions of posttranslational control mechanisms in the *Neurospora* clock. The similarities of posttranslational regulations in clock control between *Neurospora* and higher eukaryotes allow the knowledge obtained in *Neurospora* to be used as a guide in other systems. *Neurospora* grows quickly, and large amounts of material can be obtained easily, perfect for performing biochemical experiments needed in posttranslational and other studies. A wide array of molecular, genetic, physiological, and pharmacological approaches, including gene disruption and manipulation of gene expression, can also be applied easily in *Neurospora*. Together with the availability of the complete *Neurospora* genome sequences, they have made *Neurospora* an ideal system for clock studies.

## Acknowledgments

This work was supported by grants from the National Institutes of Health (GM062591 and GM068496) and Welch Foundation. The author is the Louise W. Kahn Scholar in Biomedical Research at University of Texas Southwestern Medical Center.

## References

Akashi, M., Tsuchiya, Y., Yoshino, T., and Nishida, E. (2002). Control of intracellular dynamics of mammalian period proteins by casein kinase I epsilon (CKIepsilon) and CKIdelta in cultured cells. *Mol. Cell. Biol.* **22,** 1693–1703.

Akten, B., Jauch, E., Genova, G. K., Kim, E. Y., Edery, I., Raabe, T., and Jackson, F. R. (2003). A role for CK2 in the *Drosophila* circadian oscillator. *Nature Neurosci.* **6,** 251–257.

Aronson, B., Johnson, K., Loros, J. J., and Dunlap, J. C. (1994). Negative feedback defining a circadian clock: Autoregulation in the clock gene *frequency*. *Science* **263,** 1578–1584.

Cheng, P., Yang, Y., Gardner, K. H., and Liu, Y. (2002). PAS domain-mediated WC-1/WC-2 interaction is essential for maintaining the steady state level of WC-1 and the function of both proteins in circadian clock and light responses of *Neurospora*. *Mol. Cell. Biol.* **22,** 517–524.

Cheng, P., Yang, Y., Heintzen, C., and Liu, Y. (2001a). Coiled-coil domain mediated FRQ-FRQ interaction is essential for its circadian clock function in *Neurospora*. *EMBO J.* **20,** 101–108.

Cheng, P., Yang, Y., and Liu, Y. (2001b). Interlocked feedback loops contribute to the robustness of the *Neurospora* circadian clock. *Proc. Natl. Acad. Sci. USA* **98,** 7408–7413.

Cheng, P., Yang, Y., Wang, L., He, Q., and Liu, Y. (2003). WHITE COLLAR-1, a multifunctional *Neurospora* protein involved in the circadian feedback loops, light sensing, and transcription repression of *wc-2*. *J. Biol. Chem.* **278,** 3801–3808.

Crosthwaite, S. K., Dunlap, J. C., and Loros, J. J. (1997). *Neurospora wc-1* and *wc-2*: Transcription, photoresponses, and the origins of circadian rhythmicity. *Science* **276,** 763–769.

Denault, D. L., Loros, J. J., and Dunlap, J. C. (2001). WC-2 mediates WC-1-FRQ interaction within the PAS protein-linked circadian feedback loop of *Neurospora*. *EMBO J.* **20,** 109–117.

Dunlap, J. C. (1999). Molecular bases for circadian clocks. *Cell* **96,** 271–290.

Dunlap, J. C., and Feldman, J. F. (1988). On the role of protein synthesis in the circadian clock of *Neurospora crassa*. *Proc. Natl. Acad. Sci. USA* **85,** 1096–1100.

Froehlich, A. C., Loros, J. J., and Dunlap, J. C. (2003). Rhythmic binding of a WHITE COLLAR-containing complex to the frequency promoter is inhibited by FREQUENCY. *Proc. Natl. Acad. Sci. USA* **100,** 5914–5919.

Garceau, N., Liu, Y., Loros, J. J., and Dunlap, J. C. (1997). Alternative initiation of translation and time-specific phosphorylation yield multiple forms of the essential clock protein FREQUENCY. *Cell* **89,** 469–476.

Gorl, M., Merrow, M., Huttner, B., Johnson, J., Roenneberg, T., and Brunner, M. (2001). A PEST-like element in FREQUENCY determines the length of the circadian period in *Neurospora crassa*. *EMBO J.* **20,** 7074–7084.

Grima, B., Lamouroux, A., Chelot, E., Papin, C., Limbourg-Bouchon, B., and Rouyer, F. (2002). The F-box protein slimb controls the levels of clock proteins period and timeless. *Nature* **420,** 178–182.

He, Q., Cheng, P., Yang, Y., He, Q., Yu, H., and Liu, Y. (2003). FWD1-mediated degradation of FREQUENCY in *Neurospora* establishes a conserved mechanism for circadian clock regulation. *EMBO J.* **22,** 4421–4430.

Hibi, M., Lin, A., Smeal, T., Minden, A., and Karin, M. (1993). Identification of an oncoprotein and UV-responsive protein kinase that binds and potentiates the c-Jun activation domain. *Genes Dev.* **7,** 2135–2148.

Johnson, C. H., and Nakashima, H. (1990). Cycloheximide inhibits light-induced phase shifting of the circadian clock in *Neurospora*. *J. Biol. Rhythms* **5,** 159–167.

Kim, W. Y., Geng, R., and Somers, D. E. (2003). Circadian phase-specific degradation of the F-box protein ZTL is mediated by the proteasome. *Proc. Natl. Acad. Sci. USA* **100,** 4933–4938.

Kloss, B., Price, J. L., Saez, L., Blau, J., Rothenfluh, A., and Young, M. W. (1998). The *Drosophila* clock gene *double-time* encodes a protein closely related to human casein kinase I$\varepsilon$. *Cell* **94,** 97–107.

Ko, H. W., Jiang, J., and Edery, I. (2002). Role for Slimb in the degradation of *Drosophila* Period protein phosphorylated by Doubletime. *Nature* **420,** 673–678.

Krebs, E. G., and Fischer, E. H. (1962). Phosphorylase b kinase from rabbit skeletal muscle. *Methods Enzymol.* **5,** 373–376.

Lee, D. H., and Goldberg, A. L. (1996). Selective inhibitors of the proteasome-dependent and vacuolar pathways of protein degradation in *Saccharomyces cerevisiae*. *J. Biol. Chem.* **271,** 27280–27284.

Lee, K., Loros, J. J., and Dunlap, J. C. (2000). Interconnected feedback loops in the *Neurospora* circadian system. *Science* **289,** 107–110.

Lin, J. M., Kilman, V. L., Keegan, K., Paddock, B., Emery-Le, M., Rosbash, M., and Allada, R. (2002). A role for casein kinase 2alpha in the *Drosophila* circadian clock. *Nature* **420,** 816–820.

Liu, Y., Garceau, N., Loros, J. J., and Dunlap, J. C. (1997a). Thermally regulated translational control mediates an aspect of temperature compensation in the *Neurospora* circadian clock. *Cell* **89,** 477–486.

Liu, Y., Loros, J., and Dunlap, J. C. (2000). Phosphorylation of the *Neurospora* clock protein FREQUENCY determines its degradation rate and strongly influences the period length of the circadian clock. *Proc. Natl. Acad. Sci. USA* **97,** 234–239.

Liu, Z. P., Galindo, R. L., and Wasserman, S. A. (1997b). A role for CKII phosphorylation of the cactus PEST domain in dorsoventral patterning of the *Drosophila* embryo. *Genes Dev.* **11,** 3413–3422.

Lowrey, P. L., Shimomura, K., Antoch, M. P., Yamazaki, S., Zemenides, P. D., Ralph, M. R., Menaker, M., and Takahashi, J. S. (2000). Positional syntenic cloning and functional characterization of the mammalian circadian mutation tau. *Science* **288,** 483–492.

Mas, P., Kim, W. Y., Somers, D. E., and Kay, S. A. (2003). Targeted degradation of TOC1 by ZTL modulates circadian function in *Arabidopsis thaliana. Nature* **426,** 567–570.

Naidoo, N., Song, W., Hunter-Ensor, M., and Sehgal, A. (1999). A role for the proteasome in the light response of the timeless clock protein. *Science* **285,** 1737–1741.

Nawathean, P., and Rosbash, M. (2004). The doubletime and CKII kinases collaborate to potentiate Drosophila PER transcriptional repressor activity. *Mol. Cell* **13,** 213–223.

Price, J. L., Blau, J., Rothenfluh, A., Adodeely, M., Kloss, B., and Young, M. W. (1998). *double-time* is a new *Drosophila* clock gene that regulates PERIOD protein accumulation. *Cell* **94,** 83–95.

Sathyanarayanan, S., Zheng, X., Xiao, R., and Sehgal, A. (2004). Posttranslational regulation of *Drosophila* PERIOD protein by protein phosphatase 2A. *Cell* **116,** 603–615.

Schwerdtfeger, C., and Linden, H. (2000). Localization and light-dependent phosphorylation of white collar 1 and 2, the two central components of blue light signaling in *Neurospora crassa. Eur. J. Biochem.* **267,** 414–422.

Schwerdtfeger, C., and Linden, H. (2001). Blue light adaptation and desensitization of light signal transduction in *Neurospora crassa. Mol. Microbiol.* **39,** 1080–1087.

Sugano, S., Andronis, C., Ong, M. S., Green, R. M., and Tobin, E. M. (1999). The protein kinase CK2 is involved in regulation of circadian rhythms in *Arabidopsis. Proc. Natl. Acad. Sci. USA* **96,** 12362–12366.

Talora, C., Franchi, L., Linden, H., Ballario, P., and Macino, G. (1999). Role of a white collar-1-white collar-2 complex in blue-light signal transduction. *EMBO J.* **18,** 4961–4968.

Yagita, K., Tamanini, F., Yasuda, M., Hoeijmakers, J. H., van der Horst, G. T., and Okamura, H. (2002). Nucleocytoplasmic shuttling and mCRY-dependent inhibition of ubiquitylation of the mPER2 clock protein. *EMBO J.* **21,** 1301–1314.

Yang, Y., Cheng, P., He, Q., Wang, L., and Liu, Y. (2003). Phosphorylation of FREQUENCY protein by casein kinase II is necessary for the function of the *Neurospora* circadian clock. *Mol. Cell. Biol.* **23,** 6221–6228.

Yang, Y., Cheng, P., and Liu, Y. (2002). Regulation of the *Neurospora* circadian clock by casein kinase II. *Genes Dev.* **16,** 994–1006.

Yang, Y., Cheng, P., Zhi, G., and Liu, Y. (2001). Identification of a calcium/calmodulin-dependent protein kinase that phosphorylates the *Neurospora* circadian clock protein FREQUENCY. *J. Biol. Chem.* **276,** 41064–41072.

Yang, Y., He, Q., Cheng, P., Wrage, P., Yarden, O., and Liu, Y. (2004). Distinct roles for PP1 and PP2A in the *Neurospora* circadian clock. *Genes Dev.* **18,** 255–260.

Young, M. W., and Kay, S. A. (2001). Time zones: A comparative genetics of circadian clocks. *Nature Rev. Genet.* **2,** 702–715.

# [18]  Analyzing the Degradation of PERIOD Protein by the Ubiquitin–Proteasome Pathway in Cultured *Drosophila* Cells

*By* HYUK WAN KO and ISAAC EDERY

## Abstract

Time-of-day specific changes in the levels of key clock proteins are critical for the normal progression of circadian pacemakers. Evidence indicates a major role for the ubiquitin–proteasome pathway (UPP) in the temporal control of clock protein stability. A conserved feature of animal clocks is that PERIOD (PER) proteins undergo daily rhythms in abundance. The stability of PER proteins is regulated by differential phosphorylation, whereby hyperphosphorylated isoforms are selectively degraded by the UPP. The use of transformed stable cell lines has been instrumental in advancing our understanding of the mechanisms underlying the intersection of the UPP and clock protein metabolism. This article describes several standard methodologies used to analyze the UPP-mediated degradation of *Drosophila* PER (dPER) expressed in cultured *Drosophila* cells (Ko *et al.*, 2002). Although this article focuses on dPER as a case study, general issues are discussed that should have broad application to other cell culture-based systems and clock proteins. For example, we discuss (i) advantages/disadvantages of cultured cells, (ii) types of expression vectors and "peptide tags" for recombinant protein production and surveillance, and (iii) standard approaches to determine whether a protein of interest is modified by ubiquitin and degraded by the proteasome. Prior to the discussion on methodologies, the article provides a brief overview of diverse strategies by which clock proteins in a variety of systems are regulated by the UPP.

## Overview of Different Mechanisms Mediating the Degradation of Clock Proteins by the Ubiquitin/Proteasome Pathway

It is now well established that the basic units of circadian pacemakers are cell-autonomous entities based on species or tissue-specific sets of clock genes, whose RNA and protein products participate in interconnected positively and negatively acting transcriptional–translational feedback loops (Dunlap, 1999). As a result of the design principles inherent in these autoregulatory molecular loops, one or more key clock RNA and protein products manifest daily rhythms in abundance. Tightly controlled oscillations in the levels of clock proteins are essential for the normal progression of circadian clocks (e.g., Yang and Sehgal, 2001). Despite the prominent appearance of rhythmic clock RNAs in circadian oscillators, posttranslational regulatory schemes make significant contributions to the temporal regulation of clock protein levels. An emerging theme is that the ubiquitin (Ub)–proteasome pathway (UPP) plays a major role in regulating the stabilities of clock proteins (Akashi et al., 2002; Grima et al., 2002; He et al., 2003; Imaizumi et al., 2003; Kim et al., 2003; Ko et al., 2002; Kondratov et al., 2003; Lin et al., 2001; Mas et al., 2003; Naidoo et al., 1999; Yagita et al., 2002).

In general, proteins targeted to the 26S proteasome for rapid destruction are first modified by the covalent attachment of multiple ubiquitin moieties (Ciechanover et al., 2000; Pickart, 2001; Weissman, 2001). Substrate ubiquitination is catalyzed by a cascade of enzymes termed ubiquitin-activating enzymes (E1s), ubiquitin-conjugating enzymes (E2s or UBCs), and ubiquitin-protein ligases (E3s), which act sequentially to catalyze the covalent attachment of a multiubiquitin chain to the substrate.

How are clock proteins selected for destruction by the UPP? While this is a relatively new area of investigation, growing evidence suggests a prominent role for phosphorylation in regulating the stabilities of clock proteins. One such strategy that appears conserved involves the progressive phosphorylation of de novo-synthesized clock proteins until presumably a highly phosphorylated state is attained many hours later that triggers polyubiquitination and degradation by the 26S proteasome, restarting a new round of protein accumulation. The prototypical clock proteins that define this behavior are the PERIOD (PER) proteins in animal clocks (Akashi et al., 2002; Grima et al., 2002; Ko et al., 2002; Yagita et al., 2002) and FREQUENCY (FRQ) in the Neurospora clock (He et al., 2003). Casein kinase 1ε (CK1ε) [termed DOUBLETIME (DBT) in Drosophila] and CK2 have major roles in regulating the phosphorylation and abundance of PER proteins in animals (Akten et al., 2003; Kloss et al., 1998; Lin et al., 2002; Lowrey et al., 2000; Nawathean and Rosbash, 2004; Price et al., 1998) and FRQ in Neurospora (Gorl et al., 2001; Yang et al., 2002). Although PER proteins and FRQ share little or no sequence homology, they function in the negative limb of circadian transcriptional feedback loops by blocking their own expression (Dunlap, 1999). Thus, it is possible that the slow progressive phosphorylation of these clock proteins acts as a biochemical time constraint delaying turnover and hence enabling them to act as autoinhibitors for many hours despite the lack of de novo synthesis (Edery, 1999; Garceau et al., 1997).

The conserved strategy for degrading PER and FRQ has also highlighted a prominent circadian role for a class of E3 ubiquitin ligases called Skp1/Cullin/F-box (SCF) complexes that recognize substrates in a phosphorylation-dependent manner (Craig and Tyers, 1999; Kipreos and Pagano, 2000; Tyers and Jorgensen, 2000). The main component determining substrate specificity is the F-box protein. Findings indicate that the F-box proteins Slimb (*Drosophila* homolog of $\beta$-TrCP) and FWD1 (F-box and WD40 repeat-containing protein 1; the *Neurospora* homolog of Slimb/$\beta$-TrCP) mediate the degradation of highly phosphorylated *Drosophila* PER (dPER) and *Neurospora* FRQ, respectively (Grima *et al.*, 2002; He *et al.*, 2003; Ko *et al.*, 2002). Also, in *Arabidopsis* the F-box containing proteins FKF1 and ZTL are important in photoperiodism and clockworks, respectively (Imaizumi *et al.*, 2003; Mas *et al.*, 2003). In the plant case there is the added twist that FKF1 and ZTL are under circadian control (Kim *et al.*, 2003; Nelson *et al.*, 2000).

The stabilities of some clock proteins not only change in a circadian manner but also in response to environmental signals, most notably visible light. A role for the UPP has been demonstrated in signal-dependent degradation of clock proteins. This was first shown for the *Drosophila* TIMELESS (dTIM) protein, a key partner of dPER (Naidoo *et al.*, 1999). Again, substrate phosphorylation appears to be an important trigger in the UPP-mediated degradation of dTIM. Presumably, an unknown tyrosine kinase (Naidoo *et al.*, 1999) and the Ser/Thr kinase SHAGGY (SGG, an ortholog of glycogen synthase kinase 3; GSK-3) (Martinek *et al.*, 2001) stimulate the light-enhanced degradation of TIM. Moreover, the blue-light photoreceptor CRYPTOCHROME (dCRY) is important for the photosensitivity of dTIM (Emery *et al.*, 1998; Stanewsky *et al.*, 1998). dCRY itself is degraded by the UPP in a light-mediated fashion; however, it appears that this occurs via a mechanism that does not require phosphorylation of dCRY (Lin *et al.*, 2001).

Another interesting trend that is emerging involves phase-specific heteromeric interactions in the regulation of clock protein stabilities. A role for phosphorylation is also evident in this type of regulation. For example, the interaction of dTIM with dPER presumably slows down the ability of DBT to phosphorylate dPER, hence modulating the timing of dPER turnover and the duration of feedback repression (Kloss *et al.*, 2001; Ko *et al.*, 2002). A striking example in the mammalian circadian system is the binding of BMAL1 to CLOCK, which results in the phosphorylation and subsequent degradation of nuclear localized CLOCK through a proteasome-dependent mechanism (Kondratov *et al.*, 2003).

Analyzing dPER Degradation by the UPP in
  Cultured *Drosophila* Cells

The ability to recapitulate partial or complete clock reactions in tissue culture-based systems has led to remarkable progress in our understanding of the molecular underpinnings governing circadian pacemakers. This article focuses on transfection-based systems using permanently established cell lines to study the degradation of dPER by the UPP (Ko *et al.*, 2002). The overall strategy is to express a recombinant target protein (e.g., dPER) and apply standard methodologies to determine whether it is ubiquitinated and degraded by the UPP.

None of the techniques described in this article are specific for the degradation of clock proteins *per se*, but are well-established methods that are used routinely to study UPP-mediated degradation of proteins in cultured cells. Although this article focuses on dPER as a case study and the techniques described are based on the commonly used and commercially available *Drosophila* Schneider-2 (S2) cells, general issues are discussed that should be relevant to the study of clock protein degradation in other cell culture systems. A comprehensive overview of the basic protocols used in working with S2 cells is outside the scope of this article. There are many good sources that deal with the subject of *Drosophila* cultured cells, from historical accounts to practical methodologies (e.g., Cherbas and Cherbas, 1998; Cherbas *et al.*, 1994; Echalier, 1997). In addition, working with permanent *Drosophila* cell lines has been dramatically simplified and optimized with the availability of numerous commercial sources for obtaining cells, culture media, expression vectors, and transfection procedures. Most of our experience is based on using the *Drosophila* Expression System (DES) from Invitrogen.

### Advantages and Limitations of S2 Cells

While there are many well-established mammalian cell lines, including some that can be induced into exhibiting robust circadian molecular rhythms (Balsalobre *et al.*, 1998), choices for *Drosophila*-derived ones are more limited. Although Ashburner (1989) listed some 100 independent cell lines derived from *Drosophila*, only a small fraction of these are routinely used and readily available. Among the most popular are *Drosophila* S2 cells, which are suitable hosts for high-level expression of functional insect and mammalian recombinant proteins. S2 cells were established from primary embryo cultures (Schneider, 1972) and have several advantages, including ease of growth (e.g., not requiring $CO_2$ incubators), nonlytic expression systems, and adaptability to high-density growth in suspension.

A unique characteristic of S2 cells is that several hundred to thousand copies of the transfected plasmid can stably integrate per cell. Thus, heterologous expression is usually high in the polyclonal stably transfected cells, rendering it unnecessary to select and screen for expression from clonal cell lines (discussed further later). Furthermore, S2 cells are robust and fast growing, doubling about every day. S2 cells are not optimal for all applications, however. For example, Kc but not S2 cells exhibit many of the typical responses to the hormone $\beta$-ecdysone (Echalier, 1997). Nonetheless, given the ubiquitous function of the UPP, most cell lines should provide a reasonable context in which to investigate a potential role for proteasome-mediated degradation of your protein of interest.

Irrespective of the cell lines used, it is important to be aware that endogenously expressed factors might influence the pathway under study, sometimes in unknown ways. For example, contradictory findings have been reported on various aspects concerning the regulation of mammalian PER proteins (mPER1, 2 and 3) in cultured cell systems; most notably how the subcellular distributions of mPERs are regulated (e.g., Kume et al., 1999; Vielhaber et al., 2000; Yagita et al., 2000). Several factors have been suggested that might explain the different results, including cell type-specific variations in the endogenous levels of key players. With regard to studying the UPP-mediated degradation of a recombinant clock protein in cultured cells, a key in evaluating the potential physiological significance of any findings is if one or more known regulatory features can be recapitulated; e.g., progressive phosphorylation by a known kinase (e.g., Camacho et al., 2001; Keesler et al., 2000; Ko et al., 2002) or the requirement for a stimulus such as light (e.g., Lin et al., 2001).

Having simple and efficient strategies to investigate potential endogenous effects by, for example, gene silencing (Worby et al., 2001) or producing trans-dominant negative versions should be considered when deciding on the choice of cell lines. However, while tissue culture-based analysis of clock pathways should be viewed as a part of the circadian toolbox, results obtained need to be examined further in whole animal approaches.

From the perspective of studying functional aspects of clock proteins in S2 cells, several of the known Drosophila clock proteins exhibit little or no endogenous expression in naïve nontransfected cells. We know this to be true for dPER, dTIM, and dCLOCK (dCLK). There is sufficient endogenous CYC levels such that its exogenous production is not required (or apparently limiting) to drive dCLK-CYC-mediated expression of E-box containing reporter genes (Darlington et al., 2000). Depending on the assay, it is possible to obtain measurable effects of endogenous DBT, CK2, Slimb, protein phosphatase 2A (PP2A), and dCRY on clock relevant pathways in S2 cells (Ko et al., 2002; Lin et al., 2001; Nawathean and

Rosbash, 2004; Sathyanarayanan et al., 2004). We are not aware of any information on the endogenous levels of other clock or clock-associated proteins, including the transcription factors VRILLE (VRI) and Par Domain Protein 1 (PDP1) (Cyran et al., 2003). It should not be surprising that some of the clock modulatory activities, such as kinases, phosphatases, and proteasome-associated factors, are present even in clockless cells as they function in many cellular pathways.

Within a first approximation, it is reasonable to suggest that despite inherent limitations, clockless cell lines such as S2 can provide a useful context to investigate specific parts of much larger circadian molecular circuits. As noted earlier, this is especially true for the biochemical analysis of clock protein degradation by the UPP as the basic proteolytic machinery should be quite similar and expressed ubiquitously in all cells. A useful advantage of clockless cells is the enhanced ability to investigate causal relationships in the absence of confounding multiple feedback effects. Yet even in the absence of extensive feedback control circuits, there are still potential pitfalls when trying to establish direct causal relationships. A good example is the finding that although hyperphosphorylated dPER is preferentially located in the nucleus, this was speculated to be a secondary consequence of its increased ability as a transcriptional repressor (Nawathean and Rosbash, 2004).

*Expression Vectors*

We begin with the premise that at least one of the components in question will be expressed heterologously and hence require the design of an expression vector. This may not always be the case if all the factors in a sought-after pathway are functional in naïve cultured cells. However, even under this circumstance, further studies would likely necessitate the design of variant forms of relevant components, the introduction of "affinity tags" or "peptide handles" for simpler surveillance, or the desire to control expression from a heterologous promoter.

Several factors should be considered when deciding on the type of expression vector used to drive the production of target clock proteins. This includes the use of constitutive or inducible promoters (for an overview of different commonly used promoters, see Cherbas et al., 1994; Echalier, 1997). The most popular expression vectors for heterologous protein production in cultured *Drosophila* cells employ either the actin5C (pAct) (Krasnow et al., 1989; Thummel et al., 1988) or the copper-inducible metallothionein (pMT) (Bunch et al., 1988; Johansen et al., 1989) promoters to attain constitutive or inducible expression, respectively. In our experience, the maximal levels of target proteins attained with the actin5C promoter are similar to those obtained with the MT promoter, as reported originally (Angelichio *et al.*, 1991). Approximately 30- to 50-fold induction of chimeric transcript levels can be attained (from 24 to 48 h after treatment) using the pMT-inducible system. Of course, results can vary depending on the stability of the recombinant protein produced. Probably all inducible promoters, such as pMT, are likely to be leaky, so your "time 0" might not accurately reflect the situation in the absence of the induced factor.

Another popular inducible expression system is based on the heat-inducible *hsp70* promoter (Rio and Rubin, 1985; Thummel *et al.*, 1988), but should probably be avoided when studying protein degradation, as heat shock treatment of cells induces *ub* expression and can alter protein stability (Echalier, 1997), including those of dPER and dTIM (Sidote *et al.*, 1998). In contrast, copper induction of the pMT promoter is a more innocuous treatment and does not appear to have a noticeable impact on cellular physiology. We mention the *hsp70* promoter here because it has been used to express *Drosophila* clock proteins in cultured cells (e.g., Saez and Young, 1996).

When to use a constitutive versus an inducible promoter is likely to be ultimately reached empirically depending on the objective and which factors are expressed endogenously in the cultured cell employed. An advantage of placing the expression of at least one clock factor under the control of an inducible promoter is that it offers a more dynamic setting that can better reflect the temporal regulation observed *in vivo*. For example, when we investigated the progressive phosphorylation and degradation of dPER in S2 cells, the expression of *dper* was controlled from the actin5C constitutive promoter (pAct-*per*), whereas the inducible pMT promoter regulated *dbt* expression (pMT-*dbt*) (Ko *et al.*, 2002). Induction of *dbt* led to progressive decreases in the phosphorylation and abundance of dPER that are similar to those observed for native dPER protein in heads from wild-type *Drosophila*. However, when both dPER and DBT were coexpressed under constitutive promoters, we could not observe temporal changes in dPER phosphorylation and stability. Rather, dPER was constantly low in abundance and highly phosphorylated (H. W. Ko and I. Edery, unpublished observations).

Another crucial variable to consider is that results can vary depending on the absolute and relative amounts of plasmids used in the transfection. Increasing the ratio of pMT-*dbt* to pAct-*per* led to progressive decreases in the rates of hyperphosphorylation and degradation of dPER (Ko *et al.*, 2002). When establishing conditions, assaying interactions, or determining the effects of mutations it is critical to vary the amounts of plasmids used. Whether an effect is observed can be highly dependent on the absolute and relative concentrations of the relevant players under study. A notable example where quite different results are obtained depending on the amount of plasmid used is in measuring the dependence of dPER on dTIM for efficient repression of dCLK-CYC-mediated transcription in transiently transfected S2 cells. At low concentrations of dPER, the addition of dTIM increases the magnitude of the transcriptional inhibition but there is no effect of dTIM when large amounts of *dper* expressing plasmids are used (Chang and Reppert, 2003; Nawathean and Rosbash, 2004).

Another issue that arises when generating expression vectors is consideration of the type and position of "tags" for recombinant protein surveillance and purification. Although a comprehensive discussion of the advantages and disadvantages of the myriad of protein tags that have been used successfully is outside the scope of this article, we briefly share some of our experience that could be useful in the analysis of clock protein degradation. (For a comprehensive overview of different peptide tags and their usage, see Hearn and Acosta, 2001.)

In general, regarding recombinant protein surveillance, we have had good experience with many of the commonly used and commercially available small epitope tags (e.g., myc, HA, FLAG, and V5). For maximum flexibility in experimental design, it is best to generate several different tagged versions of each individual recombinant protein. For example, in situations where we want to establish an accurate measurement of the absolute and relative levels of exogenously expressed proteins, we use versions that are modified with the same tag. This helps normalize signal intensities. For this objective, we routinely use a tag that has an associated commercially available antibody, such as the V5 eptitope tag system (Ko *et al.*, 2002).

When purification is desirable, we usually do not rely as much on epitope tags, but instead use the popular histidine tag (His tagged) in combination with immobilized metal affinity chromatography (IMAC) (Gaberc-Porekar and Menart, 2001; Hochuli *et al.*, 1987). Of course, if multiple recombinant proteins are coexpressed, only one target protein can have the tag that will be selected for during affinity purification. The main reasons we use the IMAC method is because the His tag is small and offers the ability to purify large amounts of target proteins under native and denaturing conditions. However, while yields can be high, the complexes obtained are usually quite "dirty" with nonspecific factors copurifying. This is not a problem if the presence of specific proteins in these complexes is assayed using SDS–polyacrylamide gel electrophoresis (PAGE) followed by immunoblotting.

Does tagging change the stability of dPER in S2 cells? Not from what we have observed. Does it matter if the tag is at the amino- or carboxyl-terminal region of the target protein? Again, not from what we have observed. However, it is possible that the position of the tag could affect turnover rates, yields during purification, or interactions with other key components. A final suggestion based on the fact that the phosphorylation of many clock proteins has been assayed by mobility changes following SDS–PAGE and immunoblotting is that a large tag (e.g., GST, MBP) will likely decrease resolution.

## Transient versus Stable Transfections

The main advantage of transient assays is that they are considerably faster; a few days rather than 3–4 weeks are required for stable transformation. Moreover, average levels of target gene expression are measured from the entire population of transfected cells instead of the considerable variability typically observed among individual clones of stably transformed cells. In addition, because the plasmid remains unintegrated during the short assay period, any potential influence of neighboring host chromosomal sequences is avoided.

We sometimes use stable transformant cells when performing experiments that require at least several days of consistent target protein expression, e.g., when testing the effects of endogenous activities on dPER phosphorylation/degradation using the powerful RNAi methodology (Worby et al., 2001). Usually, many days of incubating with double-stranded RNA (dsRNA) are required to efficiently silence endogenous activities by RNAi. It is still possible to perform efficient RNAi-mediated silencing in transient assays by preincubating the cells with dsRNA for several days prior to transfection (e.g., Ko et al., 2002).

## Measuring Protein Stability

When investigating the regulated degradation of a target protein, a common variable to measure is its stability. The typical strategy involves using cycloheximide to block de novo protein synthesis, followed by measuring the decline in the levels of the target protein until an estimate of its half-life ($t_{1/2}$; time required for 50% of the signal to disappear) is obtained. Such an experiment would be quite difficult in a whole fly approach, but more reasonable in explanted tissues and straightforward in S2 cells. However, caution is required because among undesired side effects, apoptosis can be induced in S2 cells treated with cycloheximide at 5–10 µg/ml for more than 8 h (Fraser and Evan, 1997). Commonly used reporter proteins (e.g., luciferase, GFP, and $\beta$-galactosidase) have half-lives typically ranging from 3 to 48 h, offering a wide range of control proteins with different stabilities that can be followed simultaneously with the protein of interest

(Corish and Tyler-Smith, 1999; Thompson *et al.*, 1991). However, the *in vivo* degradation rates of many proteins do not follow simple first-order kinetics (Levy *et al.*, 1996; Suzuki and Varshavsky, 1999). Other more sophisticated techniques, such as UPR (ubiquitin/protein/reference), increase the accuracy of measuring turnover rates by providing a cotranslated reference (Levy *et al.*, 1996).

*Proteasome Inhibitors*

If you suspect that the 26S proteasome is involved in degrading your protein of interest, routine analysis involves measuring the effects of specific inhibitors on target protein stability and testing for the covalent attachment of multi-Ub chains (discussed in the next section).

Several types of low molecular weight inhibitors of the proteasome have been identified that can readily enter cells and selectively inhibit this degradation pathway (Lee and Goldberg, 1996). The most widely used are the peptide aldehydes, such as carbobenzoxy (Cbz)-leu-leu-leucinal (MG132), cbz-leu-leu-norvalinal (MG115), and acetyl-leu-leu-norleucinal (ALLN). These agents are substrate analogs and potent transition-state reversible inhibitors, primarily targeting the chymotrypsin-like activity of the proteasome (Rock *et al.*, 1994). They are highly potent, e.g., the $K_i$ of MG132 is a few nanomolars for the chymotryptic activity of pure 20S proteasomes, and its $IC_{50}$ is a few micromolars for the inhibition of proteolysis in cultured cells. Furthermore, after exposure to these inhibitors, cell viability and growth are not generally affected for 10–20 h (Rock *et al.*, 1994). However, these compounds also inhibit certain lysosomal cysteine proteases and the calpains. Therefore, it is important when using these inhibitors for study of proteasome function in cells to also show that selective inhibitors of lysosomal function or calpains do not have similar effects. Lactacystin and its derivative clasto-lactacystin $\beta$-lactone, which are natural products structurally different from the peptide aldehydes, are much more specific and result in irreversible effects.

*Following Ubiquitination of Target Proteins*

Not all ubiquitinated proteins are targeted for degradation by the proteasome, and not all the proteins that are degraded by the proteasome are ubiquitinated (Pickart, 2001). Despite this disclaimer, the covalent addition of a multi-Ub chain is a characteristic molecular signal used to target proteins for ultimate destruction by the proteasome. Alongside inhibition studies using pharmacological strategies (see earlier discussion), demonstrating that your protein of interest is polyubiquitinated offers strong biochemical evidence that the UPP is involved. There are a variety of experimental tools to show that a protein is ubiquitinated, including the usage of differentially tagged or modified recombinant Ub moieties.

While a summary of different approaches is outside the scope of this article, we have had reasonable success with a standard strategy of using Ub that is modified with a hexahistidine (6His-Ub). The recombinant 6His-Ub can be expressed to high levels and behaves similarly to the wild-type version, meaning that in general cells need to be treated with proteasome inhibitors in order to assay covalent adducts between dPER (or other target proteins) and 6His-Ub. An advantage of His-tagged Ub is that S2 cells can be lysed in strong denaturing conditions to prevent deubiquitination and give you confidence that the association between dPER and Ub is covalent. 6His-Ub containing material is purified using IMAC, and dPER identified by immunoblotting. Other versions of Ub include those that cannot undergo chain growth (e.g., K48R version, or methylated ubiquitin) (Chau et al., 1989; Gregori et al., 1990), and hence ubiquitination can be followed in the absence of targeted degradation. A general word of caution, overexpression itself could result in some fraction of the recombinant protein being targeted in a "nonspecific" manner to the UPP.

Concluding Remarks

A role for the UPP in regulating clock protein levels appears to be a prominent feature of clocks. In many cases, time-of-day or stimulus-dependent phosphorylation events are critical biochemical signals targeting clock proteins for degradation by the proteasome. The advent of tissue culture systems that can mimic numerous aspects of clock biochemistry have been instrumental in revealing how the UPP intersects with the clockworks in a number of model systems. In *Drosophila* the usage of S2 cells offers several advantages, including high-level expression of recombinant proteins, ease of transfection, very efficient RNAi methodologies, and an array of commercially available expression vectors. For example, we used RNAi to demonstrate a role for Slimb in the degradation of phosphorylated dPER by the UPP (Ko et al., 2002). However, it is always important to remember that results obtained for recombinant proteins expressed in cultured systems, even *bona fide* clock cells, need to be probed in whole animal approaches.

Acknowledgments

Our studies are supported by grants from the National Institutes of Health.

## References

Akashi, M., Tsuchiya, Y., Yoshino, T., and Nishida, E. (2002). Control of intracellular dynamics of mammalian period proteins by casein kinase I epsilon (CKIepsilon) and CKIdelta in cultured cells. *Mol. Cell. Biol.* **22**, 1693–1703.

Akten, B., Jauch, E., Genova, G. K., Kim, E. Y., Edery, I., Raabe, T., and Jackson, F. R. (2003). A role for CK2 in the *Drosophila* circadian oscillator. *Nature Neurosci.* **6**, 251–257.

Angelichio, M. L., Beck, J. A., Johansen, H., and Ivey-Hoyle, M. (1991). Comparison of several promoters and polyadenylation signals for use in heterologous gene expression in cultured *Drosophila* cells. *Nucleic Acids Res.* **19**, 5037–5043.

Ashburner, M. (1989). *"Drosophila*: A Laboratory Handbook." Cold Spring Harbor Laboratory Press, Cold Spring Harbor, NY.

Balsalobre, A., Damiola, F., and Schibler, U. (1998). A serum shock induces circadian gene expression in mammalian tissue culture cells. *Cell* **93**, 929–937.

Bunch, T. A., Grinblat, Y., and Goldstein, L. S. (1988). Characterization and use of the Drosophila metallothionein promoter in cultured *Drosophila melanogaster* cells. *Nucleic Acids Res.* **16**, 1043–1061.

Camacho, F., Cilio, M., Guo, Y., Virshup, D. M., Patel, K., Khorkova, O., Styren, S., Morse, B., Yao, Z., and Keesler, G. A. (2001). Human casein kinase Idelta phosphorylation of human circadian clock proteins period 1 and 2. *FEBS Lett.* **489**, 159–165.

Chang, D. C., and Reppert, S. M. (2003). A novel C-terminal domain of drosophila PERIOD inhibits dCLOCK:CYCLE-mediated transcription. *Curr. Biol.* **13**, 758–762.

Chau, V., Tobias, J. W., Bachmair, A., Marriott, D., Ecker, D. J., Gonda, D. K., and Varshavsky, A. (1989). A multiubiquitin chain is confined to specific lysine in a targeted short-lived protein. *Science* **243**, 1576–1583.

Cherbas, L., and Cherbas, P. (1998). Cell culture. *In "Drosophila*: A Practical Approach" (D. B. Roberts, ed.), pp. 319–346. Oxford Univ. Press, Oxford.

Cherbas, L., Moss, R., and Cherbas, P. (1994). Transformation techniques for Drosophila cell lines. *In "Drosophila melanogaster*: Practical Uses in Cell and Molecular Biology" (L. S. B. Goldstein and E. A. Fyrberg, eds.), pp. 161–179. Academic Press, San Diego.

Ciechanover, A., Orian, A., and Schwartz, A. L. (2000). Ubiquitin-mediated proteolysis: Biological regulation via destruction. *Bioessays* **22**, 442–451.

Corish, P., and Tyler-Smith, C. (1999). Attenuation of green fluorescent protein half-life in mammalian cells. *Protein Eng.* **12**, 1035–1040.

Craig, K. L., and Tyers, M. (1999). The F-box: A new motif for ubiquitin dependent proteolysis in cell cycle regulation and signal transduction. *Prog. Biophys. Mol. Biol.* **72**, 299–328.

Cyran, S. A., Buchsbaum, A. M., Reddy, K. L., Lin, M. C., Glossop, N. R., Hardin, P. E., Young, M. W., Storti, R. V., and Blau, J. (2003). vrille, Pdp1, and dClock form a second feedback loop in the *Drosophila* circadian clock. *Cell* **112**, 329–341.

Darlington, T. K., Lyons, L. C., Hardin, P. E., and Kay, S. A. (2000). The period E-box is sufficient to drive circadian oscillation of transcription *in vivo. J. Biol. Rhythms* **15**, 462–471.

Dunlap, J. C. (1999). Molecular bases for circadian clocks. *Cell* **96**, 271–290.

Echalier, G. (1997). *"Drosophila* Cells in Culture." Academic Press, New York.

Edery, I. (1999). Role of posttranscriptional regulation in circadian clocks: Lessons from *Drosophila. Chronobiol. Int.* **16**, 377–414.

Emery, P., So, W. V., Kaneko, M., Hall, J. C., and Rosbash, M. (1998). CRY, a *Drosophila* clock and light-regulated cryptochrome, is a major contributor to circadian rhythm resetting and photosensitivity. *Cell* **95**, 669–679.

Fraser, A. G., and Evan, G. I. (1997). Identification of a *Drosophila melanogaster* ICE/CED-3-related protease, drICE. *EMBO J.* **16**, 2805–2813.

Gaberc-Porekar, V., and Menart, V. (2001). Perspectives of immobilized-metal affinity chromatography. *J. Biochem. Biophys. Methods* **49**, 335–360.

Garceau, N. Y., Liu, Y., Loros, J. J., and Dunlap, J. C. (1997). Alternative initiation of translation and time-specific phosphorylation yield multiple forms of the essential clock protein FREQUENCY. *Cell* **89**, 469–476.

Gorl, M., Merrow, M., Huttner, B., Johnson, J., Roenneberg, T., and Brunner, M. (2001). A PEST-like element in FREQUENCY determines the length of the circadian period in *Neurospora crassa*. *EMBO J.* **20**, 7074–7084.

Gregori, L., Poosch, M. S., Cousins, G., and Chau, V. (1990). A uniform isopeptide-linked multiubiquitin chain is sufficient to target substrate for degradation in ubiquitin-mediated proteolysis. *J. Biol. Chem.* **265**, 8354–8357.

Grima, B., Lamouroux, A., Chelot, E., Papin, C., Limbourg-Bouchon, B., and Rouyer, F. (2002). The F-box protein slimb controls the levels of clock proteins period and timeless. *Nature* **420**, 178–182.

He, Q., Cheng, P., Yang, Y., Yu, H., and Liu, Y. (2003). FWD1-mediated degradation of FREQUENCY in Neurospora establishes a conserved mechanism for circadian clock regulation. *EMBO J.* **22**, 4421–4430.

Hearn, M. T., and Acosta, D. (2001). Applications of novel affinity cassette methods: Use of peptide fusion handles for the purification of recombinant proteins. *J. Mol. Recogn.* **14**, 323–369.

Hochuli, E., Dobeli, H., and Schacher, A. (1987). New metal chelate adsorbent selective for proteins and peptides containing neighbouring histidine residues. *J. Chromatogr.* **411**, 177–184.

Imaizumi, T., Tran, H. G., Swartz, T. E., Briggs, W. R., and Kay, S. A. (2003). FKF1 is essential for photoperiodic-specific light signalling in *Arabidopsis*. *Nature* **426**, 302–306.

Johansen, H., van der Straten, A., Sweet, R., Otto, E., Maroni, G., and Rosenberg, M. (1989). Regulated expression at high copy number allows production of a growth-inhibitory oncogene product in *Drosophila* Schneider cells. *Genes Dev.* **3**, 882–889.

Keesler, G. A., Camacho, F., Guo, Y., Virshup, D., Mondadori, C., and Yao, Z. (2000). Phosphorylation and destabilization of human period I clock protein by human casein kinase I epsilon. *Neuroreport* **11**, 951–955.

Kim, W. Y., Geng, R., and Somers, D. E. (2003). Circadian phase-specific degradation of the F-box protein ZTL is mediated by the proteasome. *Proc. Natl. Acad. Sci. USA* **100**, 4933–4938.

Kipreos, E. T., and Pagano, M. (2000). The F-box protein family. *Genome Biol.* **1** REVIEWS 3002.

Kloss, B., Price, J. L., Saez, L., Blau, J., Rothenfluh, A., Wesley, C. S., and Young, M. W. (1998). The *Drosophila* clock gene double-time encodes a protein closely related to human casein kinase Iε. *Cell* **94**, 97–107.

Kloss, B., Rothenfluh, A., Young, M. W., and Saez, L. (2001). Phosphorylation of period is influenced by cycling physical associations of double-time, period, and timeless in the *Drosophila* clock. *Neuron* **30**, 699–706.

Ko, H. W., Jiang, J., and Edery, I. (2002). Role for Slimb in the degradation of *Drosophila* Period protein phosphorylated by Doubletime. *Nature* **420**, 673–678.

Kondratov, R. V., Chernov, M. V., Kondratova, A. A., Gorbacheva, V. Y., Gudkov, A. V., and Antoch, M. P. (2003). BMAL1-dependent circadian oscillation of nuclear CLOCK: Posttranslational events induced by dimerization of transcriptional activators of the mammalian clock system. *Genes Dev.* **17**, 1921–1932.

Krasnow, M. A., Saffman, E. E., Kornfeld, K., and Hogness, D. S. (1989). Transcriptional activation and repression by Ultrabithorax proteins in cultured *Drosophila* cells. *Cell* **57**, 1031–1043.

Kume, K., Zylka, M. J., Sriram, S., Shearman, L. P., Weaver, D. R., Jin, X., Maywood, E. S., Hastings, M. H., and Reppert, S. M. (1999). mCRY1 and mCRY2 are essential components of the negative limb of the circadian clock feedback loop. *Cell* **98**, 193–205.

Lee, D. H., and Goldberg, A. L. (1996). Selective inhibitors of the proteasome-dependent and vacuolar pathways of protein degradation in *Saccharomyces cerevisiae*. *J. Biol. Chem.* **271**, 27280–27284.

Levy, F., Johnsson, N., Rumenapf, T., and Varshavsky, A. (1996). Using ubiquitin to follow the metabolic fate of a protein. *Proc. Natl. Acad. Sci. USA* **93**, 4907–4912.

Lin, F. J., Song, W., Meyer-Bernstein, E., Naidoo, N., and Sehgal, A. (2001). Photic signaling by cryptochrome in the *Drosophila* circadian system. *Mol. Cell. Biol.* **21**, 7287–7294.

Lin, J. M., Kilman, V. L., Keegan, K., Paddock, B., Emery-Le, M., Rosbash, M., and Allada, R. (2002). A role for casein kinase 2alpha in the *Drosophila* circadian clock. *Nature* **420**, 816–820.

Lowrey, P. L., Shimomura, K., Antoch, M. P., Yamazaki, S., Zemenides, P. D., Ralph, M. R., Menaker, M., and Takahashi, J. S. (2000). Positional syntenic cloning and functional characterization of the mammalian circadian mutation tau. *Science* **288**, 483–492.

Martinek, S., Inonog, S., Manoukian, A. S., and Young, M. W. (2001). A role for the segment polarity gene shaggy/GSK-3 in the *Drosophila* circadian clock. *Cell* **105**, 769–779.

Mas, P., Kim, W. Y., Somers, D. E., and Kay, S. A. (2003). Targeted degradation of TOC1 by ZTL modulates circadian function in *Arabidopsis thaliana*. *Nature* **426**, 567–570.

Naidoo, N., Song, W., Hunter-Ensor, M., and Sehgal, A. (1999). A role for the proteasome in the light response of the timeless clock protein. *Science* **285**, 1737–1741.

Nawathean, P., and Rosbash, M. (2004). The doubletime and CKII kinases collaborate to potentiate *Drosophila* PER transcriptional repressor activity. *Mol. Cell* **13**, 213–223.

Nelson, D. C., Lasswell, J., Rogg, L. E., Cohen, M. A., and Bartel, B. (2000). FKF1, a clock-controlled gene that regulates the transition to flowering in Arabidopsis. *Cell* **101**, 331–340.

Pickart, C. M. (2001). Mechanisms underlying ubiquitination. *Annu. Rev. Biochem.* **70**, 503–533.

Price, J. L., Blau, J., Rothenfluh, A., Abodeely, M., Kloss, B., and Young, M. W. (1998). double-time is a novel *Drosophila* clock gene that regulates PERIOD protein accumulation. *Cell* **94**, 83–95.

Rio, D. C., and Rubin, G. M. (1985). Transformation of cultured *Drosophila melanogaster* cells with a dominant selectable marker. *Mol. Cell Biol.* **5**, 1833–1838.

Rock, K. L., Gramm, C., Rothstein, L., Clark, K., Stein, R., Dick, L., Hwang, D., and Goldberg, A. L. (1994). Inhibitors of the proteasome block the degradation of most cell proteins and the generation of peptides presented on MHC class I molecules. *Cell* **78**, 761–771.

Saez, L., and Young, M. W. (1996). Regulation of nuclear entry of the *Drosophila* clock proteins period and timeless. *Neuron* **17**, 911–920.

Sathyanarayanan, S., Zheng, X., Xiao, R., and Sehgal, A. (2004). Posttranslational regulation of *Drosophila* PERIOD protein by protein phosphatase 2A. *Cell* **116**, 603–615.

Schneider, I. (1972). Cell lines derived from late embryonic stages of *Drosophila melanogaster*. *J. Embryol. Exp. Morphol.* **27**, 353–365.

Sidote, D., Majercak, J., Parikh, V., and Edery, I. (1998). Differential effects of light and heat on the *Drosophila* circadian clock proteins PER and TIM. *Mol. Cell Biol.* **18**, 2004–2013.

Stanewsky, R., Kaneko, M., Emery, P., Beretta, B., Wager-Smith, K., Kay, S. A., Rosbash, M., and Hall, J. C. (1998). The cryb mutation identifies cryptochrome as a circadian photoreceptor in *Drosophila*. *Cell* **95,** 681–692.

Suzuki, T., and Varshavsky, A. (1999). Degradation signals in the lysine-asparagine sequence space. *EMBO J.* **18,** 6017–6026.

Thompson, J. F., Hayes, L. S., and Lloyd, D. B. (1991). Modulation of firefly luciferase stability and impact on studies of gene regulation. *Gene* **103,** 171–177.

Thummel, C. S., Boulet, A. M., and Lipshitz, H. D. (1988). Vectors for *Drosophila* P-element-mediated transformation and tissue culture transfection. *Gene* **74,** 445–456.

Tyers, M., and Jorgensen, P. (2000). Proteolysis and the cell cycle: With this RING I do thee destroy. *Curr. Opin. Genet Dev.* **10,** 54–64.

Vielhaber, E., Eide, E., Rivers, A., Gao, Z. H., and Virshup, D. M. (2000). Nuclear entry of the circadian regulator mPER1 is controlled by mammalian casein kinase I epsilon. *Mol. Cell Biol.* **20,** 4888–4899.

Weissman, A. M. (2001). Themes and variations on ubiquitylation. *Nature Rev. Mol. Cell. Biol.* **2,** 169–178.

Worby, C. A., Simonson-Leff, N., and Dixon, J. E. (2001). RNA interference of gene expression (RNAi) in cultured *Drosophila* cells. *Sci STKE 2001* PL1.

Yagita, K., Tamanini, F., Yasuda, M., Hoeijmakers, J. H., van der Horst, G. T., and Okamura, H. (2002). Nucleocytoplasmic shuttling and mCRY-dependent inhibition of ubiquitylation of the mPER2 clock protein. *EMBO J.* **21,** 1301–1314.

Yagita, K., Yamaguchi, S., Tamanini, F., van der Horst, G. T. J., Hoeijmakers, J. H. J., Yasui, A., Loros, J. J., Dunlap, J. C., and Okamura, H. (2000). Dimerization and nuclear entry of mPER proteins in mammalian cells. *Genes Dev.* **14,** 1353–1363.

Yang, Y., Cheng, P., and Liu, Y. (2002). Regulation of the Neurospora circadian clock by casein kinase II. *Genes Dev.* **16,** 994–1006.

Yang, Z., and Sehgal, A. (2001). Role of molecular oscillations in generating behavioral rhythms in Drosophila. *Neuron* **29,** 453–467.

# [19]   Casein Kinase I in the Mammalian Circadian Clock

*By* ERIK J. EIDE, HEESEOG KANG, STEPHANIE CRAPO, MONICA GALLEGO, and DAVID M. VIRSHUP

## Abstract

The circadian clock is characterized by daily fluctuations in gene expression, protein abundance, and posttranslational modification of regulatory proteins. The *Drosophila* PERIOD (dPER) protein is phosphorylated by the serine/threonine protein kinase, DOUBLETIME (DBT). Similarly, the murine PERIOD proteins, mPER1 and mPER2, are phosphorylated by casein kinase I $\varepsilon$ (CKI$\varepsilon$), the mammalian homolog of DBT. CKI$\varepsilon$ also phosphorylates and partially activates the transcription factor BMAL1. Given the variety of potential targets for CKIε and other cellular kinases, the precise role of phosphorylation is likely to be a complex one. Biochemical analysis of these and other circadian regulatory proteins has proven to be a fruitful approach in determining how they function within the context of the molecular clockworks.

## Introduction

In terms of overall mechanism, molecular clocks share considerable similarity between species. The murine clock is essentially a negative feedback loop in which the heteromeric transcription factor CLOCK/BMAL1 drives transcription of its negative regulators *Per* (*mPer1* and *mPer2*) and *Cryptochrome* (*mCry1* and *mCry2*). The net effect is a daily cycle in which transcript and protein abundance fluctuates (reviewed extensively in Dunlap, 1999; Reppert and Weaver, 2002). A striking feature of the clock is that many of these proteins are phosphorylated at some point during the cycle. Phosphorylation of clock proteins may in fact be an indispensable feature for maintaining circadian rhythmicity. The *Drosophila* PERIOD (dPER) protein, as well as mammalian mPER1 and mPER2, is phosphorylated rhythmically throughout the day (Edery *et al.*, 1994; Lee *et al.*, 2001). The transcriptional activator BMAL1 is phosphorylated as well (Eide *et al.*, 2002). The consequences of protein phosphorylation in circadian regulation include alterations in activity, subcellular localization, protein–protein interactions, and protein stability. The first protein kinase shown to regulate the circadian clock is casein kinase I ε (CKIε). Several alleles of the homologous *doubletime* (*dbt*) gene that conferred aberrant period length were identified in a mutagenesis screen designed to identify novel clock genes (Kloss *et al.*, 1998; Price *et al.*, 1998). *In vitro* and tissue culture studies examining the effect of CKIε on clock proteins such as mPER2 complement genetic and whole animal studies, allowing insights into molecular mechanisms of circadian timing. This article discusses purification of CKIε and mPER2, as well as assays allowing analysis of CKIε function in extracts and *in vivo*.

## Bacterial Expression and Purification of an Active Form of CKIε

The carboxyl terminus autoregulatory domain of CKIε and CKIδ potently autoinhibits the kinase activity after unopposed autophosphorylation (Gietzen and Virshup, 1999; Graves and Roach, 1995). It is therefore critical when performing *in vitro* experiments to use a form of the kinase that does not autophosphorylate. Generally, we use bacterial expression to produce a truncated form of CKIε with a stop codon after residue 319, a form with good *in vitro* activity that does not autoinhibit. Experience has shown that polymerase chain reaction (PCR) amplification and site-directed mutagenesis of the CKIε gene are much more efficient when dimethyl sulfoxide (DMSO) (4% final concentration) is added to the reactions. Alternatively, CKIε can be activated by dephosphorylation (although this can complicate additional assays) or by proteolytic removal of the carboxyl terminus.

The yield of soluble active CKIε using bacterial expression is enhanced markedly by induction at room temperature (20–25°) for 6 h or more, as opposed to short induction periods at 37°.

### Purification of Casein Kinase I

CKIε(Δ319) was cloned by PCR into pET-32 Xa/LIC vector (Novagen), which contains a 105-amino acid thioredoxin tag upstream of the inserted gene, as well as His and S tags. Other constructs without the thioredoxin tag have similarly worked well in our experience.

1. Transform competent BL21(DE3) *Escherichia coli* with the CKIε(Δ319) expression plasmid. Then inoculate 10 ml of LB medium supplemented with ampicillin (100 μg/ml final concentration) with a single colony from a freshly streaked plate.

2. The next day, start a new culture by adding the overnight culture (diluted 1:100) and incubate at 37° with vigorous shaking until the $OD_{600}$ reaches 0.5–0.7.

3. Cool the culture to room temperature by brief immersion in cold water. Next add isopropyl-β-D-thiogalactoside (0.1 mM final concentration) to the culture, followed by incubation at 28° for 5–7 h. Induction conditions, including IPTG concentration, temperature, and time, should be determined by a series of pilot experiments beforehand.

4. Collect the bacteria by centrifugation for 10 min at 4°.

5. Freeze the pellet at −80° for at least 15 min and then resuspend in CelLytic B bacterial cell lysis extraction reagent (Sigma #B3553) supplemented with lysozyme (0.5 μg/ml), 1 mM Phenylmethylsulfonyl fluoride (PMSF), 10 mM $MgCl_2$, and DNase I (10 μg/ml) and incubate on ice for 30 min.

6. Sonicate for three 30-s pulses using output power 20% to shear genomic DNA. Remove cell debris and insoluble recombinant kinase by centrifugation at 30,000g in a JA-17 rotor at 4° for 30 min. This centrifugation step can be repeated once if necessary.

7. Dialyze the supernatant against cation-exchange column buffer [20 mM HEPES, pH 7.5, 10 mM NaCl, 0.02% NP-40, 1 mM EDTA, 1 mM dithiothreitol (DTT), 10% sucrose] overnight with several buffer changes.

8. Use cation-exchange column as a first purification step to reduce contaminants in the final purification step, taking advantage of the basic isoelectric point of Trx-His-CKI$\varepsilon$(D319) (pI~10). Load the dialyzed supernatant from 2 liter of bacterial culture onto a 50-ml bed volume S-Sepharose column preequilibrated with column buffer.

9. Wash the column with 10 column volumes of column buffer.

10. Step elute the bound proteins in a volume of about 200 ml using column buffer supplemented with 300 m$M$ salt.

11. Dialyze the S-Sepharose eluate against Ni-NTA column buffer (50 m$M$ HEPES, pH 8.0, 300 m$M$ NaCl, 10 m$M$ imidazole, pH 7.9). After dialysis, adjust the pH of the column buffer to 7.9.

12. Load the dialyzed protein solution by gravity flow onto a preequilibrated Ni-NTA column (20-ml bed volume) with column buffer (50 m$M$ HEPES, 300 m$M$ NaCl, 10 m$M$ imidazole, pH 7.9).

13. Wash with 10 column volumes of column buffer containing 20 m$M$ imidazole.

14. Elute bound proteins with elution buffer (50 m$M$ HEPES, 300 m$M$ NaCl, 300 m$M$ imidazole, pH 7.0) in 5-ml volume.

15. Concentrate and buffer exchange the Ni-NTA column eluate using ultrafiltration (Millipore, Amicon Ultra centrifugal filter Cat. No. UFC801024) into CKI storage buffer [20 m$M$ HEPES (pH 7.5), 25 m$M$ NaCl, 1 m$M$ EDTA, 1 m$M$ DTT, 0.02% NP-40, 10% sucrose].

16. The protein is stable and can be stored at $-80°$. The protein retains full activity for several freeze–thaw cycles.

*Purification of MBP-mPER2(450–763)*

Attempts to produce full-length mPER1 and mPER2 using several bacterial expression systems and a baculovirus expression system were hampered by minimal solubility of the recombinant protein. However, expression of a fragment of mPER2 in *E. coli* as a maltose-binding protein fusion protein [MBP-PER2(450–763)] was readily accomplished.

The procedure is similar to the CKI purification except cells are grown in LB with 0.2% glucose and lysed in 30 m$M$ HEPES (pH 7.5), 150 m$M$ NaCl, 2 m$M$ DTT, 2 m$M$ EDTA, 0.1% NP-40. Overall, the protocol recommended by New England Biolabs is followed with a slight modification: Binding of the MBP fusion protein to amylose beads is carried out for 2 h (in batch) in column buffer (20 m$M$ HEPES, pH 7.5, 200 m$M$ NaCl, 1 m$M$ EDTA, 1 m$M$ DTT) and washes are also performed in batch with 10 bead volumes of column buffer. The bound MBP fusion protein is eluted with column buffer supplemented with 100 m$M$ maltose and then dialyzed against protein storage buffer

[20 m$M$ HEPES(pH 8.0), 25 m$M$ NaCl] overnight with two to three buffer changes.

Figure 1A shows the purified proteins, and Fig. 1B illustrates the stoichiometry of mPER2 phosphorylation *in vitro*.

## Examination of mPER2 Stability in Tissue Culture Cells

Several effects of phosphorylation on circadian rhythm proteins have been described, including regulation of nucleocytoplasmic shuttling and protein stability (Eide and Virshup, 2001). Examination of protein localization and nuclear export is described elsewhere (Vielhaber *et al.*, 2000, 2001). Several approaches to examining the effect of phosphorylation on protein stability are available. mPER2 stability assays utilize transiently expressed protein in cultured mammalian cells. As circadian rhythms have been observed in multiple tissue culture lines, it is likely that relevant biological pathways can be explored using these systems. To examine the effect of phosphorylation on protein stability, mPER2 and various mutants are expressed, and the cells are then treated with specific inhibitors of cellular kinases, phosphatases, or the 26S proteasome. Cells are then harvested at different times points, and the remaining protein levels are analyzed by immunoblotting (Fig. 2).

The experiment presented here is designed to assess the role of CKI on PER2 stability. Data suggest that CKIε phosphorylates PER2 targeting it for ubiquitin-mediated proteasomal degradation, whereas protein phosphatases dephosphorylate and therefore stabilize PER2. The first part of this experiment describes the assay of mPER2 stability after treatment of transfected HEK 293 cells with the phosphatase inhibitor calyculin A. The second part shows how pretreatment with the CKIε inhibitor IC261 prevents calyculin A-induced mPER2 degradation.

## Induce mPER2 Phosphorylation and Degradation by Phosphatase Inhibition with Calyculin A

### Description

Myc-tagged PER2 (or specific mutants and fragments) is expressed in HEK 293 cells. The day after transfection, cells are treated with cycloheximide (CHX), a protein synthesis inhibitor, in order to block new protein synthesis. Twenty minutes later, calyculin A is added to the

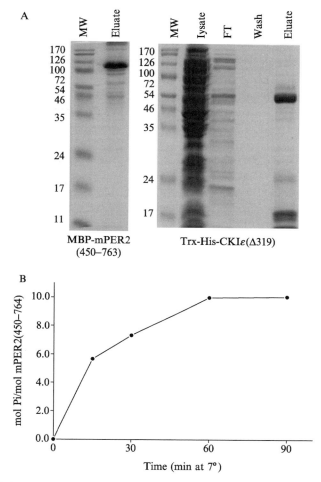

FIG. 1. (A) Purified thioredoxin-His-CKIε(Δ319) and MBP-mPER2(450–763). Proteins were purified as described and 5 μg was analyzed by SDS–PAGE. (B) Recombinant CKIε(Δ319) phosphorylates mPER2 to high stoichiometry. Phosphorylation of mPER2(40 p-pmol) by CKIε(Δ319) (20 pmol) was carried out at 37° in 30 m$M$ HEPES (pH 7.5), 10 m$M$ MgCl$_2$, 1 m$M$ DTT, 1 μg bovine serum albumin, 250 μ$M$ ATP, 200 μCi γ-[γ-ATP]$^{32}$P. At each time point, the reaction was stopped by adding 1X SDS–PAGE sample buffer followed by boiling for 3 min at 95°. Proteins in the reactions were separated on 12% SDS–PAGE. The phosphorylation of mPER2 was quantitated by PhosphorImager and ImageQuant software.

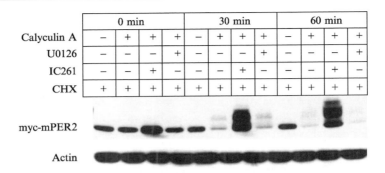

|  | 0 min | | | | 30 min | | | | 60 min | | | |
|---|---|---|---|---|---|---|---|---|---|---|---|---|
| Calyculin A | − | + | + | + | − | + | + | + | − | + | + | + |
| U0126 | − | − | − | + | − | − | − | + | − | − | − | + |
| IC261 | − | − | + | − | − | − | + | − | − | − | + | − |
| CHX | + | + | + | + | + | + | + | + | + | + | + | + |

FIG. 2. Effect of kinase inhibitors on mPER2 stability. HEK 293 cells overexpressing myc-mPER2(450–763) were pretreated with the CKIε inhibitor IC261 (50 $\mu M$), MAPK inhibitor U0126 (30 $\mu M$), or vehicle for 4 h. They were then treated with cycloheximide (25 $\mu$g/ml) for 20 min and with 80 n$M$ calyculin A or DMSO for 0, 30, and 60 min. Cells lysates were analyzed by SDS–PAGE and immunoblotting with antibodies recognizing the myc epitope of PER2 and the actin proteins.

growth medium. Cells are lysed at 0, 30, and 60 min and mPER2 protein abundance is analyzed by SDS–PAGE and Western blotting.

*Protocol*

1. Transfect HEK 293 cells (70–80% confluent) in a six-well dish with a plasmid encoding myc-mPER2(450–763).

2. About 20 h after the end of the transfection, start pretreatment with CHX or vehicle. Aspirate the old media and replace it with 1 ml fresh Dulbecco's Modified Eagles Medium (DMEM) containing CHX (25 $\mu$g/ml). Incubate for 20 min at 37°. The concentration of cycloheximide may have to be adjusted, depending on the sensitivity of specific cell types.

3. To ensure that *de novo* protein synthesis remains inhibited throughout the experiment, the cycloheximide concentration should be altered as little as possible. To this end, a master mix of calyculin A should be prepared first. Dilute calyculin A in DMEM to a concentration sufficient so that only a small volume is required to give a final concentration of 80 n$M$ after adding it to the 1 ml of medium already in the well.

4. For each well, add the appropriate volume of the diluted calyculin A stock to the existing media (80 n$M$ final concentration) and continue to incubate the cells at 37° for 0, 30, or 60 min. The appropriate concentration

of calyculin A may have to be determined empirically, as prolonged exposure is toxic to tissue culture cells.

5. Immediately after the addition of calyculin A, take the 0-min time point. To harvest the cells, aspirate the old media and wash with phosphate-buffered saline (PBS). Then add 200 $\mu$l of lysis buffer (0.1% NP-40, 150 m$M$ NaCl, 20 m$M$ HEPES, pH 7.5, 1 m$M$ EDTA, and freshly added protease inhibitor cocktail and 2 m$M$ DTT) and immediately place the dish on ice.

6. Analyze protein abundance by SDS–PAGE and immunoblotting.

## Testing Degradation or Mobility Shift: Alkaline Phosphatase Treatment

One confounding factor in the quantitation of highly phosphorylated proteins is the presence of a mobility shift. Thus, it is important to determine whether the apparent decrease in protein abundance is due to degradation or diffusion in the gel because of hyperphosphorylation. Treatment of cell lysates with calf alkaline phosphatase (CIP) should convert hyperphosphorylated species to a single unphosphorylated form. Therefore, any previously phosphorylated mPER2 will resolve as a compact band after SDS–PAGE.

For the CIP assay, 100 $\mu$g of extracts from cells treated with calyculin A is mixed with CIP incubation buffer (final concentration 100 m$M$ NaCl, 70 m$M$ Tris–HCl, pH 7.5, and 10 m$M$ MgCl$_2$) in a reaction volume of 30 $\mu$l. Dephosphorylation is achieved by the addition of 5 units of CIP (New England Biolabs) and incubation of the samples for 1 h at 37°. The reaction is stopped by adding 5X SDS sample buffer, and the samples are analyzed by SDS–PAGE and immunoblotting.

## Analyzing mPER2 Protein Degradation in *Xenopus* Egg Extracts

In addition to expression in mammalian cells, protein stability can also be analyzed in other systems. The *Xenopus laevis* egg extract contains all the components necessary for degradation of protein substrates via the ubiquitin–proteasome pathway (Salic *et al.*, 2000). One advantage of this system is that additional components such as inhibitors and recombinant proteins can be added and their effects on degradation can be assessed easily.

In this protocol, *in vitro*-synthesized [35S]methionine labeled mPER2 and luciferase (as a pipetting and gel-loading control) is added to the *Xenopus* egg extract. Okadaic acid is then added, and at various times aliquots are taken and the amount of mPER2 remaining is analyzed by SDS–PAGE and autoradiography.

*Protocol for Preparing Egg Extract*

1. Dejelly *Xenopus* eggs in freshly prepared 2% L-cysteine, pH 8.2.

2. Wash eggs three to four times in 0.2X MMR (dilute from a 10X stock, 100 m$M$ NaCl, 2 m$M$ KCl, 1 m$M$ MgCl$_2$, 2 mM CaCl$_2$).

3. Wash eggs three times in 1X XB (100 m$M$ KCl, 0.1 m$M$ CaCl, 1 m$M$ MgCl$_2$, 10 mM HEPES, pH 7.7, 50 m$M$ sucrose). After the last wash, add a sufficient volume of XB to resuspend the eggs.

4. Using a large-bore transfer pipette, place the eggs in a 1.5-ml microfuge tube and spin at the lowest speed setting for 30 s at room temperature in a microcentrifuge. The eggs are very fragile, and care must be taken to not break them at this point. After spinning, remove the liquid supernatant carefully.

5. To depolymerize the microtubules, pipette 2 $\mu$l of cytochalasin B (10 mg/ml) directly into the packed eggs. Next, spin the eggs at full speed at 4° for 10 min in a microcentrifuge. There should be three distinct layers: An upper yolk layer, an opaque cytoplasm layer, and a dark debris layer at the bottom of the tube.

6. Take a syringe with a 21-gauge needle and poke a hole in the side of the tube, just at the bottom of the cytoplasm layer. Carefully remove the cytoplasm, taking care to avoid the pelleted debris or yolk. Place the cytoplasm in a fresh microfuge tube and repeat the centrifugation steps two more times. After the third spin, the cytoplasm should be nearly clear but some particulate matter may still be present.

7. Supplement the cytoplasm fraction with 1/20 volume energy mix (150 m$M$ creatine phosphate, 20 m$M$ ATP, pH 7.4, 2 m$M$ EGTA, pH 7.7, 20 m$M$ MgCl$_2$) and protease inhibitors, each at a final concentration of 10 $\mu$g/ml (from a stock of leupeptin, pepstatin and chymostatin, 10 mg/ml each).

8. Freeze extracts in liquid nitrogen and store at −80°.

*Protocol for Degradation Assay*

Because many of the components of the assay are present in small amounts, it is important to design the experiment such that master mixes can be used. This will diminish many sources of error, especially pipetting. All reactions should be assembled on ice to prevent premature degradation of the protein of interest. The following protocol is based on previously described methods (Li *et al.*, 2001; Salic *et al.*, 2000).

1. Determine the total volume of egg extract required for all experimental conditions. Although this will depend on the specific experimental design, a typical volume for a single degradation reaction

is 15 $\mu$l. To the total egg extract, add bovine ubiquitin (1 mg/ml stock dissolved in XB buffer) to a final concentration of 40 $\mu$g/ml. Also add 1 $\mu$l of [$^{35}$S]methionine-labeled luciferase to every 25 $\mu$l of egg extract.

2. Split the master mix into siliconized tubes, one for each degradation reaction.

3. Add labeled protein of interest. Generally we add 3 $\mu$l of [$^{35}$S]methionine-labeled protein synthesized from a programmed reticulocyte lysate.

4. Further divide into smaller aliquots, usually a control and experimental tubes.

5. To test the effect of phosphatase inhibition on mPER2 stability, add okadaic acid in DMSO to a final concentration of 1 $\mu$M. Add an equal volume of DMSO to the control reactions. Mix the contents by flicking the side of the tube gently.

6. Incubate at room temperature.

7. At each desired time point, remove 3 $\mu$l of the reaction and add to a tube containing 25 $\mu$l of 1X Laemmli sample preparation buffer and store on ice or freeze at $-20°$ until analysis.

8. Resolve the proteins by SDS–PAGE and analyze by autoradiography or PhosphorImager.

## Concluding Remarks

*In vitro* and transfection-based assays for the effects of protein kinases and phosphatases on circadian rhythm have the advantages of using the well-established methods of biochemistry and signaling transduction fields. These methods allow straightforward structure function analysis of end points such as protein stability, protein–protein interaction, and subcellular localization. The challenge in the circadian rhythm field is to move from these mechanistic findings to determine their role in circadian rhythm regulation in cycling systems.

## References

Dunlap, J. C. (1999). Molecular bases for circadian clocks. *Cell* **96,** 271–290.
Edery, I., Zwiebel, L. J., Dembinska, M. E., and Rosbash, M. (1994). Temporal phosphorylation of the *Drosophila* period protein. *Proc. Natl. Acad. Sci. USA* **91,** 2260–2264.
Eide, E. J., Vielhaber, E. L., Hinz, W. A., and Virshup, D. M. (2002). The circadian regulatory proteins BMAL1 and cryptochromes are substrates of casein kinase Iepsilon. *J. Biol. Chem.* **277,** 17248–17254.
Eide, E. J., and Virshup, D. M. (2001). Casein kinase I: Another cog in the circadian clockworks. *Chronobiol. Int.* **18,** 389–398.

Gietzen, K. F., and Virshup, D. M. (1999). Identification of inhibitory autophosphorylation sites in casein kinase I epsilon. *J. Biol. Chem.* **274,** 32063–32070.

Graves, P. R., and Roach, P. J. (1995). Role of COOH-terminal phosphorylation in the regulation of casein kinase I delta. *J. Biol. Chem.* **270,** 21689–21694.

Kloss, B., Price, J. L., Saez, L., Blau, J., Rothenfluh, A., Wesley, C. S., and Young, M. W. (1998). The *Drosophila* clock gene double-time encodes a protein closely related to human casein kinase Iepsilon. *Cell* **94,** 97–107.

Lee, C., Etchegaray, J. P., Cagampang, F. R., Loudon, A. S., and Reppert, S. M. (2001). Posttranslational mechanisms regulate the mammalian circadian clock. *Cell* **107,** 855–867.

Li, X., Yost, H. J., Virshup, D. M., and Seeling, J. M. (2001). Protein phosphatase 2A and its B56 regulatory subunit inhibit Wnt signaling in *Xenopus. EMBO J.* **20,** 4122–4131.

Price, J. L., Blau, J., Rothenfluh, A., Abodeely, M., Kloss, B., and Young, M. W. (1998). Double-time is a novel *Drosophila* clock gene that regulates PERIOD protein accumulation. *Cell* **94,** 83–95.

Reppert, S. M., and Weaver, D. R. (2002). Coordination of circadian timing in mammals. *Nature* **418,** 935–941.

Salic, A., Lee, E., Mayer, L., and Kirschner, M. W. (2000). Control of beta-catenin stability: Reconstitution of the cytoplasmic steps of the wnt pathway in *Xenopus* egg extracts. *Mol. Cell* **5,** 523–532.

Vielhaber, E., Eide, E., Rivers, A., Gao, Z. H., and Virshup, D. M. (2000). Nuclear entry of the circadian regulator mPER1 is controlled by mammalian casein kinase I epsilon. *Mol. Cell. Biol.* **20,** 4888–4899.

Vielhaber, E. L., Duricka, D., Ullman, K. S., and Virshup, D. M. (2001). Nuclear export of mammalian PERIOD proteins. *J. Biol. Chem.* **276,** 45921–45927.

# [20]  Nucleocytoplasmic Shuttling of Clock Proteins

*By* Filippo Tamanini, Kazuhiro Yagita, Hitoshi Okamura, and Gijsbertus T. J. van der Horst

## Abstract

The mammalian circadian clock in the neurons of suprachiasmatic nuclei (SCN) in the brain and in cells of peripheral tissues is driven by a self-sustained molecular oscillator, which generates rhythmic gene expression with a periodicity of about 24 h (Reppert and Weaver, 2002). This molecular oscillator is composed of interacting positive and negative transcription/translation feedback loops in which the heterodimeric transcription activator CLOCK/BMAL1 promotes the transcription of E-box containing *Cryptochrome* (*Cry1* and *Cry2*) and *Period* (*Per1* and *Per2*) genes, as well as clock-controlled output genes. After being synthesized in the cytoplasm, CRY and PER proteins feedback in the nucleus to inhibit the transactivation mediated by positive regulators. The mPER2 protein acts at the interphase between positive and negative feedback loops by indirectly promoting the circadian transcription of the *Bmal1* gene (through RevErbα) (Preitner *et al.*, 2002; Shearman *et al.*, 2000) and by interacting with mCRY proteins (Kume *et al.*, 1999; Yagita *et al.*, 2002) (for a detailed review, see Reppert and Weaver, 2002). In addition to cyclic transcription of clock genes, immunohistochemical studies on SCN neurons have revealed that mCRY1, mCRY2, mPER1, and mPER2 proteins undergo near synchronous circadian patterns of nuclear abundance (Field *et al.*, 2000). The delay of ~6 h between the peak in clock mRNA production and maximal levels of protein expression in the nucleus is believed to originate from posttranslational modification steps involving phosphorylation, ubiquitination, and proteosomal degradation. Thus, the timing of entry, as well as the residence time of core clock proteins into the nucleus, is a critical step in maintaining the correct pace of the circadian clock. Several clock proteins have been shown to contain nuclear export signal, sequences, on top of nuclear import signals, that facilitate their cellular trafficking (Chopin-Delannoy *et al.*, 2003; Miyazaki *et al.*, 2001; Yagita *et al.*, 2002). This type of dynamic intracellular movement not only regulates protein localization, but also often affects functions by determining interactive partners and protein turnover. Because most of the clock genes have been identified by genetic screening in *Drosophila* and by gene knockdown in mammals, the development of innovative cellular techniques is essential in learning the structure–function and regulation of the corresponding proteins. This article discusses approaches, limitations, and applicable protocols to study the regulation of cellular localization of mammalian clock proteins, with a particular focus on mammalian CRY1 and PER2 proteins.

## Nuclear Localization Signals in Clock Proteins

Whereas small nuclear proteins (<~50 kDa) can spontaneously accumulate in the nucleus, the transport of larger proteins across the nuclear membrane is an active process that requires the presence of nuclear localization signal (NLS) sequences. An NLS is a short amino acid sequence that is not only necessary, but also sufficient for the translocation of a protein from the cytoplasm to the nucleus. An increased number of clock proteins (CRY, PER, TIM, BMAL1, REV-ERBα) have been shown to contain one or more NLS sequences (see Table I), which nicely reflects their predominant localization in the nucleus and function in transcription regulation. Two major classes of NLS sequences have been identified and characterized: (1) the classic monopartite NLS [represented by the simian virus 40 T-antigen NLS (Kalderon *et al.*, 1984)], composed of a single cluster of basic amino acids (P<u>KKKRK</u>V) and (2) the bipartite NLS [represented by

TABLE I

SUMMARY OF SUBCELLULAR LOCALIZATION AND NUCLEAR IMPORT (NLS) AND
EXPORT (NES) SEQUENCES UTILIZED BY MAMMALIAN CLOCK PROTEINS

| | Localization | | NLS | NES |
|---|---|---|---|---|
| | *In vivo* | In transfected cells | | |
| mCRY1 | Nuclear | Nuclear | Bipartite NLS$_C$ (amino acids 585–602) | Unknown |
| hCRY2 | Nuclear | Nuclear | Monopartite NLS$_N$ (amino acids 274–278) Bipartite NLS (amino acids 560–579) | Unknown Unknown |
| rPER2 | Nuclear | Cytoplasmic | Bipartite NLS (amino acids 778–795) | Rev-like (amino acids 109–118, 460–469, 963–990) |
| hREV-ERBα | Nuclear | Nuclear | DBD domain (amino acids 128–191) | Unknown |
| mTIM | Nuclear | Nuclear | Multiple (not characterized) | Multiple (not characterized) |
| mBMAL1 | Nuclear | Nuclear | Multiple (not characterized) | Multiple (not characterized) |
| mCLOCK | Nuclear | Cytoplasmic | Unknown | Unknown |
| CKε | Nuclear | Cytoplasmic | Unknown | Unknown |

the NLS of nucleoplasmin (Robbins *et al.*, 1991)], composed of two clusters of basic amino acids that are separated by a mutational-tolerant space (<u>KRPX$_N$KKKK</u>). Both types of nuclear import sequences are recognized by an NLS receptor (composed of a heterodimeric complex of importin$\alpha$ and $\beta$proteins) that facilitates the nuclear import through the nuclear pore and involves cyclic GTP hydrolysis by the small GTPase Ran (Fried and Kutay, 2003).

To determine the presence and functionality of a clock protein NLS, the first step usually consists of visual examination of the primary amino acid sequence for the presence of potential cellular localization motifs. This is followed by the generation of a cDNA expression construct in which the putative NLS of the protein of interest has been modified. Next, this construct is transiently expressed in mammalian cells (e.g., COS7, HEK293, NIH3T3), after which the subcellular localization of the mutant protein (fragment) is compared to that of its wild-type counterpart by immunocytochemical or (in the case fluorescent protein tags have been added) immunofluorescence detection.

Typical examples of this approach include expression studies with (i) deletion mutant proteins, in which putative NLS-containing regions have been removed from the protein of interest; (ii) chimeric proteins, in which the region of interest is fused in frame to a neutral carrier protein (e.g., GFP, $\beta$-galactosidase); and (iii) full-length mutant proteins in which the critical basic amino acid arginine/lysine of the putative NLS has been changed into the neutral amino acid alanine by site-directed mutagenesis.

Often, a first indication for the presence of a functional NLS is provided by expression constructs in which a putative NLS-containing region has been removed from the protein of interest. However, while differences in subcellular localization between the wild type (nuclear) and the deletion mutant protein (cytoplasmic) indeed may point to the presence of a functional NLS in the deleted region, one should be aware of potential false-positive and negative-results. Even when a functional NLS sequence has been deleted, the protein may still accumulate in the nucleus due to the presence of other (less defined) NLS sequences. Alternatively, a mutant protein may no longer localize to the nucleus because the deletion, rather than removing a functional NLS, interferes with the overall protein structure and nuclear entry. Thus, deletion analysis can be indicative for but certainly does not prove the presence of a functional NLS.

To obtain further proof that a potential NLS sequence is functional, the region of interest can be fused in frame to a neutral carrier protein (e.g., GFP, $\beta$-galactosidase). This approach will not only provide information on whether the putative NLS is necessary, but also on whether it is sufficient to mediate nuclear import. Because proteins with a molecular size smaller than 50 kDa can diffuse freely between cytoplasmic and nuclear compartments through the nuclear pore, it is important that the chimeric protein measures at least 50 kDa in order to reduce the chance for artifacts. However, particularly when the size of the sequence under investigation extends significantly beyond that of the NLS (which often is the case), one should be aware of the presence of other sequences that may drive nuclear localization of the carrier protein. For example, it is possible that a chimeric protein with a nonfunctional NLS can still significantly localize to the nucleus because it retains the capacity to interact with NLS containing binding partners that carry the chimeric protein over the nuclear membrane (piggy-back mechanism). Alternatively, the sequence of interest may contain an unknown "nuclear retention domain" (such as a DNA or histone-binding motif) that allows NLS-independent nuclear accumulation of a chimeric protein with a slight tendency to diffuse freely into the nucleus. As evidence for cross-talk between clock proteins and chromatin is accumulating (Etchegaray et al., 2003), it may be difficult to distinguish between nuclear import and nuclear retention mechanism by only analyzing the subcellular localization of restricted regions of a clock protein fused to GFP.

In our experience and that of many other laboratories, the most reliable test for the functionality of a putative NLS involves site-directed mutagenesis of the critical basic amino acid arginine/lysine of the proposed NLS into the neutral amino acid alanine in the context of the full-length protein. This is well illustrated by our study on the functionality of the NLS sequences of the mouse CRY1 protein (Tamanini et al., 2005). By analyzing the subcellular distribution of CRY1 deletion proteins and CRY1–GFP chimeric proteins, we mapped a potential NLS in the last 30 amino acids of the unique C-terminal tail of mCRY1 (Fig. 1). We named this sequence $NLS_C$ to distinguish it from another proposed NLS that is located in the middle of the mCRY1 protein (called $NLS_N$) (Hirayama et al., 2003). The $NLS_C$ ($KRPX_{11}KVQR$) is located at amino acid positions 585–602 of mCRY1, has similarities with the bipartite NLS of nucleoplasmin, and is preserved among human, chicken, and Xenopus CRY1 homologues, which suggests an evolutionary conserved function. The functionality of the $NLS_C$ has been demonstrated by amino acid mutagenesis in which the two critical basic amino acids of the first cluster have been replaced with alanines ($AAPX_{11}KVQR$) to generate the mCRY1mutNLS_C protein that is localized predominantly in the cytoplasm of COS7 and HEK293 cells (F. Tamanini et al., manuscript submitted). Surprisingly, small in-frame internal deletions, as well as random in-frame insertions of five amino acids in the N-terminal

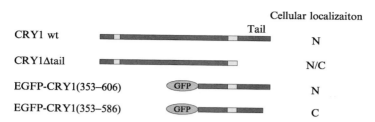

FIG. 1. Examples of deletion, chimeric, and combined deletion/chimeric constructs used to delineate the C-terminal NLS in the mCRY1 protein. Proteins were expressed transiently in COS7 cells, and the subcellular localization of mCRY1 was determined by (immuno)fluorescence. The subcellular localization of the expressed protein is indicated on the right (N, nuclear; N/C, both nuclear and cytoplasmic; C, cytoplasmic). (See color insert.)

$\alpha/\beta$ domain of mCRY1 (between amino acids 1 and 125), render the protein completely cytoplasmic. Because no evident NLS are present in the N terminus of mCRY1, this suggests that the tertiary structure of the protein plays an additional, critical role in the nuclear localization of mCRY1. Therefore, only *in vitro* mutagenesis of the C-terminal NLS in full-length mCRY1 protein allowed us to draw unmistakable conclusions on the functionality of this protein domain.

As discussed earlier, functional analysis of NLS sequences (as well as other subcellular localization signals) involves transient expression of wild-type and mutant proteins in cultured cells. As the amount of the expressed protein in such experiments usually exceeds the physiological range, this to some extent may blur results. For example, whereas the endogenous mouse PER2 protein shows nuclear localization *in vivo* (Field *et al.*, 2000), over-expressed mPER2 is detected more prominently in the cytoplasm than in the nucleus (discussed in a later section), despite the presence of a functional bipartite NLS in the middle region (amino acids residues 778–795) of the mPER2 protein (Miyazaki *et al.*, 2001). To overcome this problem, one ideally should select for stably transfected cell lines that constitutively produce low levels of the protein of interest. Indeed, constitutive low expression of *mPer2* cDNA in NIH3T3 fibroblasts renders fully nuclear mPER2 protein (K. Yagita, unpublished results).

Finally, the type of mammalian cell lines used in transfection studies needs to be carefully considered. Monkey kidney COS7 cells are a good choice because the endogenous expression of many clock proteins is virtually absent in these cells, thus allowing testing the "per se" functionality of subcellular localization motifs in a clock protein. In contrast, all clock proteins are expressed endogenously in NIH3T3 cells, resulting in a functional ticking clock after serum shock. Therefore, cellular trafficking properties of any exogenously expressed clock protein in NIH3T3 cells may, in part, reflect its interaction with endogenous partners (piggy-back mechanism) rather than the activity of cellular localization signals.

## Nuclear Export Signals (NES) in Clock Proteins

Apart from NLS sequences, nuclear proteins may also contain sequence motifs that allow active transport of the protein back to the cytoplasm, resulting in nucleocytoplasmic localization of the protein. Such signal sequences are quite different from those used for the nuclear import and bind a different type of transporter molecule. The most common type of NES is represented by the consensus peptide sequence $Lx_{(1-3)}Lx_{(2-4)}LxL(V/I/M)$, which is rich in closely spaced large hydrophobic amino acids, particularly leucine or isoleucine. This sequence is recognized and bound by the nuclear export receptor CRM1/exportin1, a 115-kDa protein that is localized in the nucleoplasm and particularly enriched at the nuclear envelope (Fornerod *et al.*, 1997; Fukuda *et al.*, 1997; Stade *et al.*, 1997). The association between NES and CRM1/exportin 1 can be abolished by replacing the leucines with the neutral amino acid alanine. In 1995, the first protein found to be exported from the nucleus through the CRM1/exportin 1 pathway was the HIV regulatory protein REV. Since identification of the REV NES, similar sequences have been functionally identified in many proteins, such as p53, MDM2, APC, cyclin B1, and many others.

Analogous to the functional analysis of NLS motifs, the most convincing manner to test the functionality of NES sequences is by probing the subcellular localization of transiently expressed proteins with an *in vitro*-mutagenized NES motif. However, it is important to consider that the primary amino acid sequence of a protein may harbor multiple NES sequences, in which case inactivation of a single NES by mutagenesis may cause only a minor shift from nucleocytoplasmic to nuclear localization. To obtain the full picture, it is therefore necessary to perform an extensive subcellular localization scoring of the mutant protein in different cell lines, as well as with different combinations of mutagenized NES in the full-length protein, as is well illustrated by the study of nuclear export of the mouse PER2 protein (Yagita *et al.*, 2002). Although mPER2 carries a functional NLS, its subcellular localization in transiently transfected cells is mainly cytoplasmic due to the presence of three independent NES sequences (NES1 at residue position 109–118, NES3 at residue position 460–469, and NES2 at residue position 983–990). In comparison with wild-type mPER2, leucine-to-alanine substitutions in NES1 enhanced the percentage of cells showing nuclear localization of the mutant protein.

Similarly, mutations in NES2 and NES3 caused a relatively small but still significant shift from cytoplasmic to nuclear localization. Therefore, all three NES motifs of mPER2 are functional, although NES1 seems more efficiently used by the nuclear export machinery than NES2 and NES3. Importantly, only after combined mutagenesis of all three NES motifs, overexpressed full-length mPER2 is exclusively nuclear, indicating that NES1, NES2, and NES3 act additively in nuclear export of the mPER2 protein. Evidently the presence of NLS as well as NES sequences in one and the same protein now provides mPER2 with the capacity to shuttle between cytoplasm and nucleus.

Nucleocytoplasmic Shuttling of Clock Proteins

Further evidence for nuclear import and subsequent export (nucleocytoplasmic shuttling) of the protein of interest can be provided by subcellular localization studies in transiently transfected cells, grown in the presence of leptomycin B (LMB). LMB is an inhibitor of nuclear export and acts on the NES export factor CRM1/exportin 1 (Kudo et al., 1999). The mammalian CRM1 gene is expressed ubiquitously and, in cultured cells, transcript levels are regulated by the cell cycle control machinery (Kudo et al., 1997). Moreover, in proliferating mammalian cell cultures, LMB has been shown to cause G1 cell cycle arrest when given at nanomolar concentrations (Kudo et al., 1997). LMB was originally discovered during an antifungal antibiotics screen in Schizosaccharomyces pombe and is available commercially from Sigma.

The previous section described the presence of both NLS and NES motifs in the mouse PER2 protein. When overexpressed in COS7 cells, the protein mainly localizes in the cytoplasm. However, after treatment of cells with LMB (10 ng/ml for 1–3 h), mPER2 was shown to exclusively accumulate in the nucleus. This suggests that mPER2, after having entered the nucleus, can be transported back to the cytoplasm via the nuclear export machinery (Yagita et al., 2002). Inhibition of the translation machinery by cyclohexamide (CHX; 100 μg/ml) did not affect the subcellular localization of mPER2 in the absence or presence of LMB. This showed that LMB-mediated nuclear accumulation of mPER2 in the nucleus is not coupled to de novo protein synthesis, but instead involves already existing mPER2. Using the LMB approach, mammalian BMAL1 (Tamaru et al., 2003) and Drosophila dTIM (Ashmore et al., 2003) have also been shown to undergo CRM1/exportin 1-dependent nucleocytoplasmic shuttling (see Table I). However, the responsible cellular localization motifs in these proteins have not been functionally characterized.

Nucleocytoplasmic Shuttling and the Heterokaryon Assay

An alternative method used to study nuclear export of a molecule is the heterokaryon assay. The principle of the assay is that nuclei of cell line X (grown in the presence of LMB) are preloaded with the protein of interest, after which cells are fused with that of cell line Y. In the resulting heterokaryon cells (sharing cytoplasm but having distinct nuclei), the protein under investigation will only start to move from nucleus X to nucleus Y when it is subject to nuclear export, and only after release of the LMB block (see Fig. 2A). This assay is particularly suitable for proteins that are strictly nuclear under steady-state conditions and has been used successfully to demonstrate shuttling properties of p53, MDM2, $\beta$-catenin, and many other proteins. Moreover, the heterokaryon assay also addresses the question whether a protein can only move once from cytoplasm to nucleus and back (for instance, as a result of protein modifications) or whether it can continuously shuttle between these cellular compartments.

An example of the heterokaryon nucleocytoplasmic shuttling assay is shown in Fig. 2B, where mouse NIH3T3 cells are preloaded with the PER2(1–916)-GFP protein (carrying a deletion of the C-terminal 341 amino acids, including NES2) and fused with human HeLa cells. Under steady-state conditions, the PER2(1–916)-GFP protein localizes predominantly in the nucleus, despite the presence of functional NES signals (NES1 and NES3) in this protein (Yagita et al., 2002). Whereas the PER2(1–916)-GFP protein remains in murine nuclei when heterokaryons are grown in LMB-containing medium (Fig. 2B, left), washing out LMB results in the redistribution of the mutant PER2 protein between HeLa nuclei (recognized by antihuman ERCC1 immunoreactivity) and NIH3T3 nuclei (Fig. 2B, right). These findings not only indicate that the remaining NES1 and NES3 domains allow active transport of PER2 from the nucleus to the cytoplasm, but also prove that nuclear import/export is not limited to one cycle and that in fact the protein can shuttle continuously between nucleus and cytoplasm.

*Heterokaryon Nucleocytoplasmic Shuttling Assay*

1. At day 1, transfect cells of mouse origin (NIH3T3) with a plasmid encoding the protein of interest, preferably carrying a (fluorescent protein) tag to facilitate detection. That same day, seed cells of human origin (HeLa) onto glass coverslips at ~40% confluence in a separate dish. Culture cells in Dulbecco's modified Eagle's medium (DMEM), supplemented with 10% fetal calf serum, 1% penicillin, and 1% streptomycin.

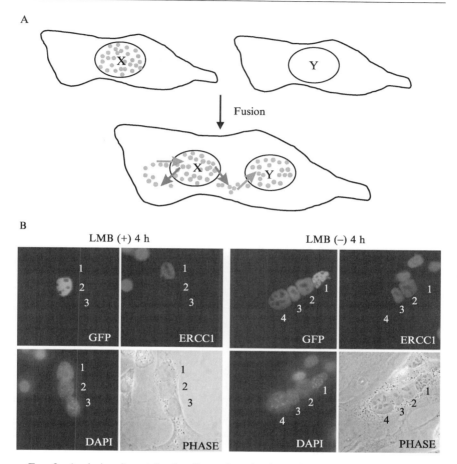

Fig. 2. Analysis of protein shuttling using the heterokaryon assay. (A) Schematic representation of the principle of the heterokaryon assay. For an explanation, see text. (B) Example of the heterokaryon shuttling assay in which transfected NIH3T3 cells that transiently express nuclear mPER2(1–916)-GFP were fused to nontransfected HeLa cells in the presence of cycloheximide (CHX; used to block *de novo* protein synthesis) and in the presence or absence of leptomycin B (LMB; used to inhibit NES-mediated nuclear export). Shown are the subcellular localization of mPER2-GFP(1–916) (top left), as well as phase-contrast photographs of the heterokaryon (bottom right) and their DAPI-stained nuclei (bottom left). HeLa cell-derived nuclei were identified by immunofluorescence using antibodies recognizing the human ERCC1 (top right). Reprinted from Yagita *et al.* (2002), with permission. (See color insert.)

2. At day 2, trypsinize the transfected NIH3T3 culture, after which cells are seeded onto coverslips containing the HeLa cells. Allow the cells to attach and stretch on the glass.

3. When the mixed cell population on the coverslip has reached a confluency of ~80% (optimal condition for heterokaryon formation), start the assay by culturing the cells for 4 h in medium containing 10 ng/ml LMB and 50 $\mu$g/ml CHX. LMB (stock solution at 5 mg/ml in ethanol, stored at $-20°$) blocks nuclear export (and thus establishes preloading of NIH3T3 nuclei with the protein under investigation) prior to formation of the heterokaryon, whereas CHX (stock solution at 30 mg/ml in ethanol stored at $-20°$) blocks *de novo* protein synthesis, allowing visualization of preexisting proteins only.

4. At 30 min before cellular fusion, increase the CHX concentration to 100 $\mu$g/ml. Subsequently, wash cells with phosphate-buffered saline (PBS, 1X), after which excess moisture is removed from the coverslip by tipping the edge with a piece of filter paper. Next, apply a drop of 100–200 $\mu$l of 50% PEG-6000 solution (sterile made in Hanks solution) on cells for 2 min and subsequently wash cells gently three to five times with PBS. After addition of fresh medium containing either CHX (100 $\mu$g/ml), or CHX (100 $\mu$g/ml) + LMB (10 ng/ml), culture cells for 3–4 h at 37° and finally fix in 3% paraformaldehyde for 15 min. Throughout the experiment, prewarm buffers and media at 37°.

5. During (immuno)fluorescence microscopy, the protein of interest can be visualized via specific antibodies against the protein or its tag, whereas (human) HeLa cell-derived and (mouse) NIH3T3 cell-derived nuclei are identified using antibodies that discriminate between human and mouse homologs of a nuclear protein. In absence of such antibodies, counterstaining of cells with 4′,6-diamidino-2-pheylindole (DAPI) will be sufficient to distinguish nuclei of murine origin (with speckles) from those of human origin (no speckles).

## Clock Protein Dynamics: Nucleocytoplasmic Shuttling by FLIP

Evidently the generation of time by the circadian core oscillator is a highly dynamic process, involving timed synthesis, nuclear accumulation, and degradation of clock proteins. Thus far, the behavior of clock proteins over time has been studied predominantly by (semi)quantitative immunohisto- and immunocytochemical approaches in which tissues (SCN, liver, etc.) and cells (following serum shock) have been fixed at 2- to 6-h time intervals. Although this approach unmistakably generated a wealth of information on clock protein expression (including abundance, subcellular distribution, protein–protein interactions, and protein modifications), detailed information on the kinetics of these processes is lacking. With the development of molecular tags for the visualization of proteins in living cells, real-time imaging of the spatiotemporal behavior of clock proteins at the single-cell level is within reach. Particularly, the green fluorescent protein (GFP) from the jellyfish *Aequorea victoria* is a powerful tool, as it can be fused to virtually any protein of interest without affecting its functionality. Furthermore, mutations in the GFP cDNA provided new spectral variants with different nonoverlapping absorbance and emission spectra, such as cyan (CFP) and yellow fluorescent protein (YFP). This allows simultaneous visualization by confocal microscopy of two different fluorescent-tagged proteins over time and at single-cell level. Fluorescent protein technology not only allows time-lapse imaging of the steady-state level and distribution of proteins in the living cell, it also allows further analysis of the kinetic properties of a molecule, such as diffusion rate (indicative for the size of the protein or protein complex) or exchange rate between cellular compartments. Notable among these techniques are (i) fluorescence recovery after photobleaching (FRAP), (ii) fluorescence loss after photobleaching (FLIP), and (iii) fluorescence resonance energy transfer (FRET).

When applying photobleaching techniques, a small three-dimensional area of a cell is exposed to a high-intensity laser pulse, causing GFP-labeled proteins in that particular area to lose the ability to emit a fluorescent signal when exposed to exciting light. Next, the influx of unbleached molecules from neighboring areas into the bleached area (FRAP), as well as the reduction in the level of unbleached molecules in the nonbleached compartment (FLIP), is recorded by time-lapse microscopy. Subsequent analysis of the redistribution of fluorescence provides information on the diffusion rate of the labeled protein (indicative for its molecular mass) and on the fraction of molecules that remains immobile (as is the case when the protein is bound, for instance, to DNA). Importantly, photobleaching only alters the fluorescence state of the tagged protein but does not affect the biological activity of the protein.

FRET is a distance-dependent (10–100 Å) physical process by which energy is transferred from an exited molecular fluorophore (the donor) to another fluorophore (the acceptor). If FRET occurs, the donor channel (CFP) will be quenched and the acceptor channel (YFP) signal will be sensitized or increased (Sekar and Periasamy, 2003). For example, when cells that (transiently) express both CFP-CRY1 and YFP-PER2 proteins are exposed to light that excites CFP, the YFP channel is only expected to sense light when the PER2 and CRY1 proteins interact physically. Thus,

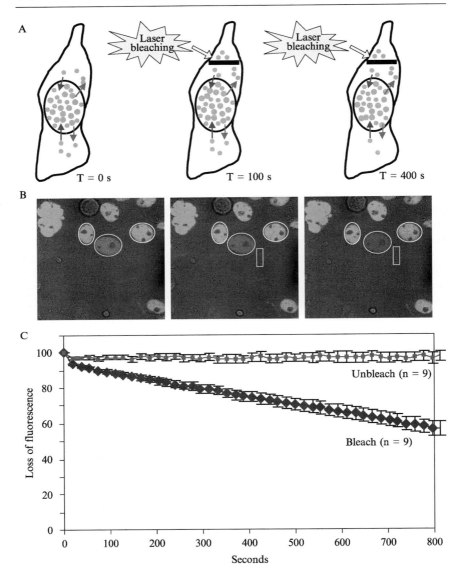

FIG. 3. Analysis of protein shuttling using the fluorescence loss in photobleaching (FLIP) technique. (A) Schematic representation of the principle of FLIP: GFP-tagged proteins are photobleached by laser pulses when crossing a defined area of the cytoplasm. Fluorescence is detected in the nucleus and the loss of signal is calculated. (B) Confocal images of COS7 cells transiently expressing the shuttling mPER2(1–916)-GFP protein taken at various time points after the start of the bleaching experiment. Note the loss of fluorescence in nuclei of cells

FRET allows visualization of protein–protein interactions in living cells with the additional power to simultaneously analyze different cellular compartments (e.g., the nucleus, the nucleolus, the cytoplasm, the cellular membrane).

We envisage that the application of these techniques to, for instance, serum-shocked cultured fibroblasts or tissue explants (containing GFP-tagged cDNAs or transgenes, respectively) will be instrumental in studying the dynamics and kinetics of clock protein interactions and cellular trafficking during the circadian cycle. In our laboratory, we have started to implement FLIP technology in circadian research by studying nucleocytoplasmic shuttling of mPER2. In a pilot study with cultured COS cells that express PER2-GFP transiently, we asked whether application of a series of laser-bleaching pulses to a defined region of the cytoplasm would cause a decrease in nuclear fluorescence. A similar experimental approach has been applied successfully to study adenomatous polyposis coli (APC) protein shuttling (Rosin-Arbesfeld et al., 2003). As shown in Fig. 3A, photobleaching of cytoplasmic PER2-GFP causes a decline in the intensity of fluorescence in the nucleus, which suggests that nuclear PER2-GFP continuously exchanges with the very small population of PER2-GFP molecules in the cytoplasm, and thus shuttles along the nuclear membrane.

*Fluorescence Loss in Photobleaching*

1. Seed COS cells on a coverslip 2 days prior to imaging. The next day, transfect cells with a GFP-tagged expression construct for the protein of interest, using Fugene (Roche). In our experience, optimal quantitative results are obtained with COS7 cells, as they are larger and flatter than other type of cells, such as HEK-293 or HeLa. During imaging, keep cells in DMEM medium at 37° and buffer continuously with $CO_2$. Alternatively, it is possible to supplement culture media with 50 mM HEPES buffer (pH 7.0) and seal the chamber with silicon grease to prevent evaporation.

2. Perform data collection using a Zeiss LSM 510 confocal microscope using the 488-nm line of laser operating at 100% laser power and 0.3% transmission (imaging intensity). Take images with a 1.5-$\mu$m confocal section within the nucleus to minimize noise from cytoplasm above and below the nucleus. Start data collection after 30 min of full system operation to maintain a constant plane of focus relative to the coverslip and to avoid temperature-dependent focus drift during the experiment.

---

subject to photobleaching. (C) Graphic representation of the loss of nuclear fluorescence in time in mPER2(1–916)-GFP expressing COS7 cells that received photobleaching light pulses and their neighboring (unbleached) cells. (See color insert.)

3. Designate a small region of the cytoplasm (excluding the nucleus) for bleaching using the region of interest function and bleach using a FLIP protocol. For each cycle, bleach cells for 100 iterations using the 488-nm line operating at 100% laser power and 100% transmission (bleaching intensity). Record an image at imaging intensity after each round of bleaching (±20 s apart; 30 cycles in total). In another type of protocol we enlarged the region of interest to the full cytoplasm and reduced the number of iterations to 25, obtaining similar results.

4. Select cells with moderate and comparable nuclear fluorescence levels for imaging. For each construct, analyze at least 10 cells.

5. To graph fluorescence lost in photobleaching, set the background-corrected fluorescence at the time before the initial photobleach at 100%. Divide the corrected fluorescence at each time point after bleaching by prebleach intensity and multiply by 100 to give the percentage of initial fluorescence. Include values of nonbleached control cells to control internal monitor bleaching.

## Functionality of Clock Protein Shuttling

The biological significance of nucleocytoplasmic shuttling is that it provides the cell with a mechanism to rapidly reallocate existing protein pools when a rapid response is required, thereby circumventing the need for *de novo* protein synthesis. Indeed, many proteins in signaling pathways have been shown to carry NLS and NES motifs, allowing shuttling and relocalization in response to various stimuli. Evidence is accumulating that the kinetics equilibrium of nucleocytoplasmic shuttling can be changed quickly following posttranslational modifications (i.e., phosphorylation, ubiquitylation, sumoylation) of specific sequences within or in proximity to NLS and NES. For example, the NLS-mediated nuclear import of APC protein is negatively regulated by serine phosphorylation at a putative protein kinase A (PKA) site adjacent to its NLS (Zhang *et al.*, 2000). The NES in cyclin B1 is also inactivated by phosphorylation of the amino-terminal region of the protein, which allows nuclear accumulation of the cyclin B1/Cdc2 complex upon onset of mitosis (Yang *et al.*, 1998). Another example is the p53 protein, in which two serine residues ($S_{15}$ and $S_{20}$) located within the N-terminal NES of p53 become phosphorylated when cells have been exposed to DNA-damaging agents. This event not only blocks p53 nuclear export, but also stabilizes p53, allowing the cell to initiate the p53-dpendent cellular response (Zhang and Xiong, 2001).

How would protein shuttling mechanistically add to the core oscillator mechanism? One of the key features of the self-sustaining circadian feedback loop is the phase delay of several hours between mRNA and protein peaks for genes involved in the circadian core oscillator. This delay is the net result of the kinetics of (i) protein synthesis, (ii) posttranslational modification that affects protein stability (phosphorylation and ubiquitylation), (iii) timed nuclear accumulation, and (iv) protein degradation. It has been shown that the complete nuclear localization of the mPER2–mCRY1 complex requires the NLS of mPER2 (Miyazaki *et al.*, 2001), as well as the presence of full-length mCRY1 (Kume *et al.*, 1999; Yagita *et al.*, 2002), and that mPER2, but not mPER1, is degraded rapidly in the liver and SCN of *mCry1/mCry2* double-deficient mice (Shearman *et al.*, 2000). Moreover, we provided evidence that the mPER2 protein can be ubiquitylated in mammalian cells and forms a target for proteolysis by the 26S proteasomal machinery and that mCRY proteins inhibit ubiquitylation of mPER2, which is expected to extend the half-life of the latter protein (Yagita *et al.*, 2002). This finding, together with the nucleocytoplasmic shuttling behavior of the mPER2 protein, led us to propose a working model for mPER2 nuclear accumulation, in which shuttling mPER2 is ubiquitylated in the cytoplasm and degraded by the proteasome system, unless it is kept in the nucleus by mCRY proteins. Thus, according to this model, mCRY proteins can shift the subcellular localization of mPER2 from cytoplasmic to nuclear, thereby slowing down multiubiquitination of mPER2 in the cytoplasm.

Although most clock proteins are heavily phosphorylated (particularly mPER2) (Eide *et al.*, 2002; Lee *et al.*, 2001), it is yet unknown whether this type of posttranslational modification may modulate nucleocytoplasmic shuttling properties of mammalian clock proteins. The impact of both nucleocytoplasmic shuttling and posttranslational modifications on the function of a *Drosophila* clock protein has been investigated in detail by Nawathean and Rosbash (2004). Using cultured S2 cells, they showed that dPER, like mammalian PER2, is subject to nucleocytoplasmic shuttling and that RNAi-mediated knocking down of Doubletime and CKII kinases (responsible for phosphorylation of dPER) does not cause a shift in the equilibrium between nuclear import and export of dPER. In contrast, evidence shows that phosphorylation of dPER directly improves the capacity of the protein to suppress CLK/CYC-mediated transcription activation. A model is launched in which the cellular dPER pool is composed of nuclear transcriptional "active" molecules (phosphorylated), which are associated with DNA or chromatin, and by nucleocytoplasmic shuttling "inactive" molecules (unphosphorylated), which once into the nucleus can be exchanged with and refill the active pool.

Evidently, in order to obtain a comprehensive and functional view of the clock shuttling and the relevance of this mechanism for mammalian core oscillator functioning, it will be of prime importance to establish cell lines that express fluorescent protein-tagged clock proteins from endogenous promoters. This will allow simultaneous analysis of multiple clock proteins (each carrying a different fluorescent protein tag) in serum-shocked cell in real-time mode. It is expected that this approach will provide a wealth of information on the behavior (e.g., subcellular localization, complex formation, binding to DNA/chromatin) of clock proteins throughout the circadian cycle.

## Acknowledgments

This work was supported in part by grants from the Netherlands Organization for Scientific Research (ZonMW Vici 918.36.619 and NWO-CW 700.51.304) and the European Community (Brain Time QLG3-CT-2002-01829) to GTJvdH.

## References

Ashmore, L. J., Sathyanarayanan, S., Silvestre, D. W., Emerson, M. M., Schotland, P., and Sehgal, A. (2003). Novel insights into the regulation of the timeless protein. *J. Neurosci.* **23,** 7810–7819.

Chopin-Delannoy, S., Thenot, S., Delaunay, F., Buisine, E., Begue, A., Duterque-Coquillaud, M., and Laudet, V. (2003). A specific and unusual nuclear localization signal in the DNA binding domain of the Rev-erb orphan receptors. *J. Mol. Endocrinol.* **30,** 197–211.

Eide, E. J., Vielhaber, E. L., Hinz, W. A., and Virshup, D. M. (2002). The circadian regulatory proteins BMAL1 and cryptochromes are substrates of casein kinase Iepsilon. *J. Biol. Chem.* **277,** 17248–17254.

Etchegaray, J. P., Lee, C., Wade, P. A., and Reppert, S. M. (2003). Rhythmic histone acetylation underlies transcription in the mammalian circadian clock. *Nature* **421,** 177–182.

Field, M. D., Maywood, E. S., O'Brien, J. A., Weaver, D. R., Reppert, S. M., and Hastings, M. H. (2000). Analysis of clock proteins in mouse SCN demonstrates phylogenetic divergence of the circadian clockwork and resetting mechanisms. *Neuron* **25,** 437–447.

Fornerod, M., Ohno, M., Yoshida, M., and Mattaj, I. W. (1997). CRM1 is an export receptor for leucine-rich nuclear export signals. *Cell* **90,** 1051–1060.

Fried, H., and Kutay, U. (2003). Nucleocytoplasmic transport: Taking an inventory. *Cell Mol. Life Sci.* **60,** 1659–1688.

Fukuda, M., Asano, S., Nakamura, T., Adachi, M., Yoshida, M., Yanagida, M., and Nishida, E. (1997). CRM1 is responsible for intracellular transport mediated by the nuclear export signal. *Nature* **390,** 308–311.

Hirayama, J., Nakamura, H., Ishikawa, T., Kobayashi, Y., and Todo, T. (2003). Functional and structural analyses of cryptochrome: Vertebrate CRY regions responsible for interaction with the CLOCK:BMAL1 heterodimer and its nuclear localization. *J. Biol. Chem.* **278,** 35620–35628.

Kalderon, D., Richardson, W. D., Markham, A. F., and Smith, A. E. (1984). Sequence requirements for nuclear location of simian virus 40 large-T antigen. *Nature* **311,** 33–38.

Kudo, N., Khochbin, S., Nishi, K., Kitano, K., Yanagida, M., Yoshida, M., and Horinouchi, S. (1997). Molecular cloning and cell cycle-dependent expression of mammalian CRM1, a protein involved in nuclear export of proteins. *J. Biol. Chem.* **272,** 29742–29751.

Kudo, N., Matsumori, N., Taoka, H., Fujiwara, D., Schreiner, E. P., Wolff, B., Yoshida, M., and Horinouchi, S. (1999). Leptomycin B inactivates CRM1/exportin 1 by covalent modification at a cysteine residue in the central conserved region. *Proc. Natl. Acad. Sci. USA* **96,** 9112–9117.

Kume, K., Zylka, M. J., Sriram, S., Shearman, L. P., Weaver, D. R., Jin, X., Maywood, E. S., Hastings, M. H., and Reppert, S. M. (1999). mCRY1 and mCRY2 are essential components of the negative limb of the circadian clock feedback loop. *Cell* **98,** 193–205.

Lee, C., Etchegaray, J. P., Cagampang, F. R., Loudon, A. S., and Reppert, S. M. (2001). Posttranslational mechanisms regulate the mammalian circadian clock. *Cell* **107,** 855–867.

Miyazaki, K., Mesaki, M., and Ishida, N. (2001). Nuclear entry mechanism of rat PER2 (rPER2): Role of rPER2 in nuclear localization of CRY protein. *Mol. Cell. Biol.* **21,** 6651–6659.

Nawathean, P., and Rosbash, M. (2004). The doubletime and CKII kinases collaborate to potentiate Drosophila PER transcriptional repressor activity. *Mol. Cell* **13,** 213–223.

Preitner, N., Damiola, F., Lopez-Molina, L., Zakany, J., Duboule, D., Albrecht, U., and Schibler, U. (2002). The orphan nuclear receptor REV-ERBalpha controls circadian transcription within the positive limb of the mammalian circadian oscillator. *Cell* **110,** 251–260.

Reppert, S. M., and Weaver, D. R. (2002). Coordination of circadian timing in mammals. *Nature* **418,** 935–941.

Robbins, J., Dilworth, S. M., Laskey, R. A., and Dingwall, C. (1991). Two interdependent basic domains in nucleoplasmin nuclear targeting sequence: Identification of a class of bipartite nuclear targeting sequence. *Cell* **64,** 615–623.

Rosin-Arbesfeld, R., Cliffe, A., Brabletz, T., and Bienz, M. (2003). Nuclear export of the APC tumour suppressor controls beta-catenin function in transcription. *EMBO J.* **22,** 1101–1113.

Sekar, R. B., and Periasamy, A. (2003). Fluorescence resonance energy transfer (FRET) microscopy imaging of live cell protein localization. *J. Cell Biol.* **160,** 629–633.

Shearman, L. P., Sriram, S., Weaver, D. R., Maywood, E. S., Chaves, I., Zheng, B., Kume, K., Lee, C. C., van der Horst, G. T., Hastings, M. H., and Reppert, S. M. (2000). Interacting molecular loops in the mammalian circadian clock. *Science* **288,** 1013–1019.

Stade, K., Ford, C. S., Guthrie, C., and Weis, K. (1997). Exportin 1 (Crm1p) is an essential nuclear export factor. *Cell* **90,** 1041–1050.

Tamanini, F., Chaves, I., Yagita, K., Barnhoorn, S., Okamura, H., and van der Horst, G. T. J. (2005). Multiple pathways for the nuclear import of mammalian CRY1 converge on the C-terminal tail. In press.

Tamaru, T., Isojima, Y., van der Horst, G. T., Takei, K., Nagai, K., and Takamatsu, K. (2003). Nucleocytoplasmic shuttling and phosphorylation of BMAL1 are regulated by circadian clock in cultured fibroblasts. *Genes Cells* **8,** 973–983.

Yagita, K., Tamanini, F., Yasuda, M., Hoeijmakers, J. H., van der Horst, G. T., and Okamura, H. (2002). Nucleocytoplasmic shuttling and mCRY-dependent inhibition of ubiquitylation of the mPER2 clock protein. *EMBO J.* **21,** 1301–1314.

Yang, J., Bardes, E. S., Moore, J. D., Brennan, J., Powers, M. A., and Kornbluth, S. (1998). Control of cyclin B1 localization through regulated binding of the nuclear export factor CRM1. *Genes Dev.* **12,** 2131–2143.

Zhang, F., White, R. L., and Neufeld, K. L. (2000). Phosphorylation near nuclear localization signal regulates nuclear import of adenomatous polyposis coli protein. *Proc. Natl. Acad. Sci. USA* **97,** 12577–12582.

Zhang, Y., and Xiong, Y. (2001). A p53 amino-terminal nuclear export signal inhibited by DNA damage-induced phosphorylation. *Science* **292,** 1910–1915.

# Section IV

# Anatomical Representation of Neural Clocks

# [21]  Techniques that Revealed the Network of the Circadian Clock of *Drosophila*

*By* CHARLOTTE HELFRICH-FÖRSTER

## Abstract

The techniques are reviewed that revealed the neuronal network of the circadian clock in the brain of the fruit fly as well as the function and localization of peripheral oscillators. Three principal techniques helped characterize the circadian clock network of *Drosophila* consisting of pacemaker centers in the brain and oscillators in peripheral tissues: (1) Immunolabeling with antibodies raised against specific clock proteins detected the tissues and cells that express the clock proteins, revealed the subcellular localization of clock molecules, and illuminated their abundance at different time points during the day; (2) reporter genes unraveled the network of clock neurons and reported the circadian cycling of the clock genes *in vivo*; and (3) genetic manipulations of clock gene expression elucidated the function of specific clock genes and clock cells. These techniques and the results gained by them are reviewed briefly.

## Immunocytochemistry

Immunocytochemistry relies on the specific binding of antibodies to the antigens (here clock molecules) they were raised against. Immunolabeling of clock molecules has been performed successfully on whole mounts of the central nervous system and on cryostat sections of the head or body of flies and can be regarded as a standard method established in most circadian laboratories.

Antibodies against the clock molecules PERIOD (PER) and TIMELESS (TIM) revealed a broad distribution of both proteins within and outside the central nervous system, including sensory cells in the compound eyes and ocelli, the antennae and bristles on body and wings, cells of the reproductive system, cells of the Malphigian tubules, the prothoracic glands, the gut, etc. (reviewed by Hall, 1998; Helfrich-Förster, 2002). Except for the ovaries, the clock proteins varied cyclically in their abundance in all cells, demonstrating the molecular cycling of the clock. Immunostaining was highest at the end of the night and lowest at the end of the day. Furthermore, the clock proteins were merely cytoplasmatic in the middle of the night and entirely nuclear at the end of the night, whereby slight differences in the timing were observed between both proteins (Shafer *et al.*, 2002).

In the brain, which is relevant for behavioral rhythmicity, PER and TIM are found in many glia cells and in a few neurons named lateral neurons (LN) and dorsal neurons (DN) according to their position in the brain (Ewer *et al.*, 1992; Frisch *et al.*, 1994; Siwicki *et al.*, 1988; Zerr *et al.*, 1990; Fig. 1A). The LN and DN can each be subdivided into three cell clusters (Fig. 1A): LN cells in the anterior brain consisting of a more dorsally located cluster of 5–8 cells—the $LN_d$—and two ventrally located cell clusters that differ in size [4–6 large $LN_v$ (l-$LN_v$) and 5 small $LN_v$ (s-$LN_v$)]. DN cells consist of $\sim$ 15 $DN_1$, 2 $DN_2$, and $\sim$40 $DN_3$. $DN_1$ and $DN_2$ clusters are composed of middle-sized neurons that are located posterior in the dorsal superior brain, whereas the $DN_3$ cluster contains rather small cells that lie in a very lateral position of the dorsal brain.

Because the clock proteins are located predominantly in nuclei and cell bodies of neurons, immunolabeling with anti-PER and anti-TIM could not reveal the morphology of those neurons. This is, however, necessary to understand the neuronal functioning of clock gene-expressing cells. Partial help came from an antiserum against a neuropeptide—the pigment-dispersing factor (PDF)—that is present in the *l*-$LN_v$ cluster and in four cells of the *s*-$LN_v$ cluster (Helfrich-Förster, 1995; Kaneko *et al.*, 1997). Arborizations of the other neurons remained unknown until reporter genes were used to label their neurites and dendrites.

## Reporter Gene Expression in Clock Neurons

A molecular–genetic technique called the UAS-GAL4 expression system allows the expression of any cloned gene under the control of a specific promoter (Brand and Perrimon, 1993). It uses the yeast transcription factor GAL4, which is put under control of the desired promoter (e.g., the *per* or *tim* promoter), for activating the yeast upstream activating sequence "UAS" fused to the gene wished to be expressed. The gene driven by UAS can be a cell marker gene (reporter) such as green fluorescent protein (GFP) or $\beta$-galactosidase. As a result, the reporter gene is expressed in all cells in which the *per* (or *tim*) promoter is active. Unlike PER or TIM, the reporter protein is not restricted to cell bodies but diffuses into the entire neuron and thus labels its arborizations. Reporter proteins such as GFP and $\beta$-galactosidase have the additional advantage of being very stable. They are thus visible throughout the circadian cycle, making it even possible to detect PER and TIM in weakly expressing cells that cycle out of phase with the others.

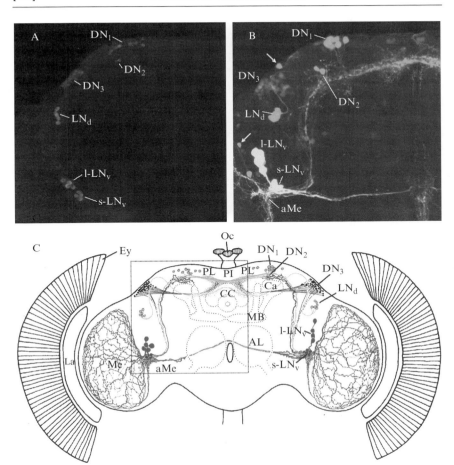

FIG. 1. Clock gene-expressing neurons in the brain of *Drosophila melanogaster*. (A) Confocal image showing a frontal view of the left brain hemisphere of adult flies stained with anti-PER around lights on in a LD cycle. PER immunostaining is visible in nuclei of large and small ventral lateral neurons (s-LN$_v$ and l-LN$_v$), dorsal lateral neurons (LN$_d$), dorsal neuron 1 (DN$_1$), dorsal neuron 2 (DN$_2$), and dorsal neuron 3 (DN$_3$). (B) Confocal image showing arborizations of all clock gene-expressing neurons in a *tim*-GAL4; *UAS*-GFP fly, again in the left brain hemisphere. Arrows point to cells that do not express natural *tim*. (C) Reconstruction of all clock gene-expressing neurons: aMe–accessory medulla, Ca–Calyx of the mushroom body (MB), CC–central complex, Ey–compound eye, La–lamina, Me–medulla, Oc–ocelli, PI–pars intercerebralis, PL–pars lateralis.

The UAS-GAL4 reporter technique has been used successfully to trace all clock neurons in the larval brain and to show the spatial distribution of clock genes in peripheral tissues of larvae and adults (Kaneko and Hall, 2000). In adult brains, the tracing of neurons was more difficult because staining was also found in many cells that were not PER and TIM immunoreactive. Some of this "extra staining" was the result of position effects associated with the transgenes' chromosomal insertion. However, certain extra cells were stained in several of the different transgenic lines generated, indicating real promoter activity of the clock genes in these additional cells. This suggests that sequences downstream the promoter region contribute to the spatial regulation of natural gene expression. Indeed, sequences in the coding region of *per* appear to be necessary for the full spatial expression pattern and for normal kinetics of *per* oscillations (Stanewsky *et al.*, 1997a,b). Under certain circumstances, this information in the *per* coding region alone—without any *per* promoter—is even sufficient to provoke expression in subsets of the clock neurons (Frisch *et al.*, 1994).

Despite the aforementioned difficulties, *per* and *tim* reporter proteins combined with PER and TIM immunostainings allowed visualization of the neuronal processes of the $LN_d$, $DN_1$, $DN_2$, and $DN_3$ clusters, additionally to those of the *l*-$LN_v$ and *s*-$LN_v$ clusters in adults (Kaneko and Hall, 2000). The arborization pattern of all cells is shown in Fig. 1 and was described in detail previously (Helfrich-Förster, 2003; Kaneko and Hall, 2000).

The neurites of all *per/tim*-expressing neurons largely overlap, suggesting that they are functionally connected. All *per/tim*-expressing neurons except l-$LN_v$ cells send their main projections into the dorsal protocerebrum (Fig. 1A and B). The s-$LN_v$, l-$LN_v$, $DN_1$, and $DN_3$ have additional projections toward the accessory medulla—a small neuropil that was shown to house the circadian clock in other insects (for a review, see Helfrich-Förster *et al.*, 1998). l-$LN_v$ cells connect furthermore both accessory medullae via fibers in the posterior optic tract and send a network of fibers onto the surface of the second optic neuropil—the medulla. The arborization pattern of the different clock gene-expressing neurons suggests that all provide a common circadian output signal that is transferred to the dorsal protocerebrum and, via the *l*-$LN_v$ cells, perhaps also into the optic lobe. The dorsal protocerebrum has connections to most sites of the brain and furthermore houses the neurosecretory system of the fly. Thus, circadian signals arising from the entity of the clock neurons may be transferred electrically and/or via humoral pathways to the effector organs.

Reporter gene expression was also used to demonstrate the cycling of *per* and *tim* expression for several days *in vivo*. For that purpose, the promoter region of *per* or *tim* was fused to the luciferase gene of the firefly (Brandes *et al.*, 1996). In contrast to GFP and $\beta$-galactosidase, firefly luciferase has a very short live time and enables real-time reporting of gene expression. When fed with luciferin, the living fly shows a cyclic luminescence, which can be recorded with a sensitive scintillation counter. Notably, this recordable luminescence rhythm stems from the compound eyes and other peripheral oscillators, whereas the cycling of the relative few neurons in the brain is not visible in the luminescence output.

These peripheral oscillators showed a prominent luciferase cycling under light–dark conditions (Brandes *et al.*, 1996). Even cultured isolated parts of the fly did so, indicating that probably each clock gene-expressing cell contains a light-sensitive autonomous oscillator (Plautz *et al.*, 1997). However, the luciferase cycling dampened more or less rapidly after transfer into constant conditions. The latter observation speaks against a role of the peripheral oscillators in driving behavioral rhythmicity that continues for weeks in darkness. The cycling luminescence in the neurons became visible only after restricting the luciferase reporter to these cells. This was possible with the aforementioned technique using just the spatial information for *per* expression present within the *per* gene itself. The promotorless *per* gene was fused to the luciferase gene and inserted into the fly genome. Several transgenic lines were found with luciferase expression restricted to the DN (Veleri *et al.*, 2003). These flies showed a stable and undampened luciferase cycling, which stands in clear contrast to the cycling of the peripheral cells and shows unequivocally that at least some clock neurons have the capability to drive rhythmic behavior for weeks (see later).

In addition to revealing *per* and *tim* cycling *in vivo*, the luciferase technique was also used successfully to screen for new genes involved in the molecular machinery of the circadian clock. Chemical or p-element-based mutagenesis was performed on the *per* luciferase reporter strain, resulting in the isolation of several mutations that altered luciferase cycling (Stanewsky *et al.*, 1998; Stempfl *et al.*, 2002). So far, only mutants could be detected that alter *per* cycling in the peripheral oscillators, but the same techniques could be applied to the strain that expresses the luciferase gene only in specific neurons. Such a screen might identify genes that are specifically involved in rhythm generation in clock neurons.

Genetic Manipulations That Identified Clock Neurons Acting as
  Circadian Pacemakers for Behavioral Rhythmicity

Although perhaps all clock neurons contain undampened circadian
oscillators, not all of them are equally important for maintaining behavioral rhythmicity under constant conditions. This was revealed by elimination
of specific neurons through cell death genes or mutation and genetic
manipulations of the clock genes in the different neurons. All these studies
showed that the three LN groups are more important for driving behavioral
rhythmicity under constant conditions than the three DN groups. The three
DN groups contribute, however, to rhythmic behavior and can mediate
rather normal rhythmicity under light–dark conditions in the absence of
functional LN groups.

One option to find out the role of the different clusters of clock neurons
was the restriction of PER expression to certain cell groups. When the
promotorless *per* luciferase construct (see earlier discussion) was expressed
in the three DN clusters of $per^0$ mutants, activity became normal under LD
conditions but remained arrythmic under DD conditions (Veleri *et al.*,
2003). When a similar construct (just without the luciferase gene) was
expressed solely in the three LN clusters, the activity rhythm of $per^0$
mutants was rescued under LD and DD conditions (Frisch *et al.*, 1994).
Studies with mutants showed the same result: *disco* mutants that retain *per*
cycling in the DN clusters but lack the three LN clusters were more or less
normally rhythmic under LD conditions but arrhythmic under DD conditions (Dushay *et al.*, 1989; Hardin *et al.*, 1992; Helfrich-Förster, 1998).
However, $per^0$/wild-type mosaics with *per* expression in LN clusters
showed robust activity rhythms under both conditions (Ewer *et al.*, 1992).
Nevertheless, DN cells contribute to the control of rhythmic activity under
DD conditions, as *disco* mutants show residual rhythms during the first
days in constant conditions (Blanchardon *et al.*, 2001; Helfrich-Förster,
1998) and transgenic flies with *per* only in the LN cells have rhythms with
lower power and longer period than wild-type flies (Frisch *et al.*, 1994).

Although LN cells appear to be the main circadian pacemakers in the
fruit fly, it is not clear whether all three LN clusters ($LN_d$, $s$-$LN_v$S, $l$-$LN_v$)
are equally essential. An important step toward testing the role of the three
cell clusters in the circadian system was the cloning of the *pdf* gene that is
only expressed in the $l$-$LN_v$ and $s$-$LN_v$ clusters (Park and Hall, 1998). These
two cell clusters could now be manipulated specifically, e.g., by expression
of channel, toxin, cell death, or apoptosis genes under control of the *pdf*
promotor. Killing the cells through cell death genes left the flies normally
rhythmic under light–dark cycles, but rendered them arrhythmic several
days after transfer into constant darkness (Renn *et al.*, 1999). The same

behavior occurred in a *pdf*-null mutant (Renn *et al.*, 1999). Electrical silencing of the cells by expression of a permanently open $K^+$ channel also had little influence on the rhythmicity under light–dark cycles, but again, most flies became arrhythmic under constant darkness (Nitabach *et al.*, 2002). In summary, ablation of the PDF-containing $LN_v$ had milder effects on rhythmic behavior than ablation of the $LN_v$ and $LN_d$ (in *disco* mutants). This shows that the $LN_d$ are as important as the $LN_v$ for the control of rhythmicity under DD conditions.

Another option to investigate the role of different cell clusters is the overexpression of clock genes in specific clock cells in a wild-type background. This should disrupt the molecular feedback loop specifically in overexpressing cells. Indeed, extensive overexpression of *per* in the photoreceptor cells of the compound eyes was shown to stop the clock in this tissue, but not in the other *per*-expressing cells (Cheng and Hardin, 1998); thus, behavioral rhythmicity was undisturbed. However, when clock genes (done for *per*, *tim*, *vri*) were overexpressed in all clock gene-expressing cells (neurons and photoreceptors) using the *per* or *tim* promotor as driver, behavior became arrhythmic under DD conditions (Blau and Young, 1999; Kaneko *et al.*, 2000; Yang and Sehgal, 2001). The behavior under LD conditions was only studied for per overexpression and was changed significantly, although the flies became not totally arrhythmic (Kaneko *et al.*, 2000). When *per* was overexpressed in the LN clusters (strongly in the s-$LN_v$ and l-$LN_v$ and weaker in the $LN_d$), behavior became arrhythmic under DD, but remained normal under LL conditions (Blanchardon *et al.*, 2001), again indicating that LN clusters are most important for behavioral rhythmicity under DD. Restricting per overexpression to the $LN_v$ clusters alone did not result in arrhythmicity for adult behavior (Yang and Sehgal, 2001) but disrupted the eclosion rhythm of the flies out of their pupae (Myers *et al.*, 2003).

In summary, the mentioned results indicate that the circadian clock controlling behavioral rhythmicity is composed of a network of clock gene-expressing neurons that interact in a distinct way. Under the clock neurons a clear hierarchy is visible, with LN groups standing in the first place and DN groups in the second. Of the LN groups, LNd and PDF-expressing LNv appear similarly important in controlling rhythmic activity under DD conditions. The specific roles of the s-$LN_v$ and l-$LN_v$ subclusters, as well as those of the three DN clusters, remain to be revealed. A promising technique might be the combination of the GAL4 system with the GAL80 system (Duffy, 2002). GAL80 represses GAL4-driven gene expression. Thus, two different promoters can be combined: one driving GAL4 in several cell clusters and the other driving GAL80 in certain subclusters. A suited promoter could, for example, drive GAL4 in all three LN clusters and another GAL80 only in the two LN$_v$ clusters of the same fly; then GAL4 will only be activated in the LN$_d$. Consequently, the LN$_d$ could be eliminated specifically by expression of a cell death gene under the control of GAL4.

## Ectopic Expression of Clock and Clock-Related Genes

The UAS-GAL4 system and related techniques were also used to drive clock gene expression in cells that normally do not express any clock gene in order to learn more about specific clock factors.

In 1995, Vosshall and Young expressed the *per* gene under control of the *glass* promoter in a *per$^0$* mutant background. Interestingly, they could rescue the arrhythmic behavior of the flies, suggesting that *glass* is expressed in clock neurons or that ectopic cells can overtake the role of the clock cells. Both turned out to be true: The *glass* gene is expressed in the DN$_1$ cluster and in a group of cells in the lateral brain close to the LN$_d$ cluster (Klarsfeld et al., 2004; Vosshall and Young, 1995). The latter cells even have similar projections to that of the LN$_d$, but were clearly not identical with them (Klarsfeld et al., 2004). Since the DN$_1$ group was recently found to be the only DN cluster without self-sustained PER oscillations and thus not suited to control the activity rhythm under DD conditions (Klarsfeld et al., 2004; Veleri et al., 2003), it is more likely that the LN$_d$-like cluster can overtake the role of the LN$_d$ and drive rhythmic behavior. This is a special case of ectopic clock gene expression where the ectopic cells were similar to the naturally *per*-expressing neurons.

However, in most cases, ectopic clock gene expression is quite different to the natural expression and results in a change or disruption of behavioral rhythmicity. The kind of change and degree of disruption may tell something about the role of the ectopically expressed clock factors, as was shown for *Clk* and *PDF*. *Clk* is a special clock gene because it is also involved in development and its lack (e.g., in the *clk$^{JRK}$* mutant) results in aberrant arborizations of certain clock neurons (Park et al., 2000) in addition to behavioral arrhythmicity. Interestingly, ectopic *Clk* expression could induce molecular clocks in cells that normally do not express any clock gene, demonstrating again the role of Clk in development, perhaps as a kind of master gene (Zhao et al., 2003). The induced ectopic clocks appeared functional and interfered with the natural clock cells, leading to disturbed behavioral rhythms. Unlike Clk, PDF is not involved in the mechanism of the core clock, but acts as a neuropeptide transmitter in the LN$_v$. PDF release is under clock control, as PDF immunoreactivity varies cyclically in the terminals of the s-LN$_v$ cells, the cycle has a short period in *per$^s$* mutants, and is arrhythmic in *per$^0$* and *tim$^0$* flies (Park et al.,

2000). Furthermore, PDF levels were extremely low in $cyc^{01}$, $clk^{JRK}$ mutants, and *vrille*-overexpressing flies (Blau and Young, 1999; Park *et al.*, 2000). When PDF is expressed ectopically in the dorsal brain, the flies are hyperactive, either lengthen period or show splitting in several free-running components, and often become arrhythmic (Helfrich-Förster *et al.*, 2000). This behavior might be caused by elevated PDF levels in the dorsal brain, which interfere with the naturally released PDF from the s-LN$_v$ terminals, putatively producing conflicting signals. Both findings are consistent with a role of PDF as a mediator for circadian signals to downstream neurons. Nevertheless, PDF might also serve as a coupling factor between the different clock neurons (Peng *et al.*, 2003). Then, it may feedback indirectly or even directly on the molecular cycle of these neurons, influencing their period. This would explain the lengthened period of flies that express PDF ectopically. It is furthermore consistent with a short-period rhythm found in $pdf^0$ mutants for several days before the flies become arrhythmic (Renn *et al.*, 1999).

Role of Peripheral Oscillators

From the previously mentioned results, it became quite clear that the peripheral oscillators are not important for behavioral rhythmicity. So what is the physiological function of these peripheral light-entrainable autonomous clocks? Some answers were found for the antennae, compound eyes, and in the male gonads. The antennae show a circadian rhythm in the electroantennogram, which is likely to be responsible for circadian differences in sensitivity to olfactory stimuli (Krishnan *et al.*, 1999). A similar role for the circadian clock has also been proposed in the modulation of the photic sensitivity of the compound eye (Chen *et al.*, 1992). In addition, the male gonads' spermatophore production and sperm mobility was found to occur in a rhythmic manner, and the possession of an intact rhythm in these functions significantly increases the reproductive fitness (Beaver *et al.*, 2002). It is quite imaginable that coordinated rhythms in gut, malpighian tubules, and fat body have similar adaptive advantages for digestion, detoxification, and fat metabolism.

In mammals, analogous peripheral oscillators are also present, showing a remarkable independence from the master clock in the brain (reviewed by Schibler and Sassone-Corsi, 2002). In contrast to flies, these are not light sensitive, but are sensitive to chemical cues or to temperature cycles (Balsalobre *et al.*, 2000; Brown *et al.*, 2002; Schibler *et al.*, 2003). Furthermore, the master clock in the brain of mammals controls the peripheral oscillators hierarchically by humoral and/or neuronal signals. Only if the master clock is lesioned is circadian gene expression dampened in the peripheral tissues

(Reppert and Weaver, 2002). In *Drosophila*, the peripheral clocks dampen even *in vivo* (thus with normal connection to the master clock) as soon as the animals are exposed to constant conditions (Plautz *et al.*, 1997). Thus, the peripheral clocks of *Drosophila* are not controlled by the master clock in the brain, but are governed directly by the external light–dark cycle. Dampening is most likely caused by a desynchrony of the peripheral oscillators due to the lack of a synchronizing light input.

Concluding Remarks

Paired with classical immunocytochemistry, genetic manipulations have been most successful in unraveling the main network of the circadian clock of *Drosophila*. Peripheral oscillators with remarkable independence from the brain were identified using reporter genes. These peripheral oscillators appear to govern rhythms in receptor sensitivity as well as different metabolic and physiological parameters. Behavioral rhythmicity, however, depends on certain pacemaker neurons in the brain, as was revealed by generating genetic mosaics, mutants, and manipulation of clock gene expression. Reporter genes unraveled a neuronal network formed by all clock gene-expressing neurons in the brain, enabling mutual interaction between them. Manipulating certain clusters using the UAS-GAL4 system revealed a clear hierarchy in this network. LN clusters appear more important for behavioral rhythmicity under constant conditions than DN clusters. The specific roles of the three LN and DN subclusters have yet to be revealed. Combination of the UAS-GAL4 system with the GAL80 system might be a promising technique to further dissect the circadian pacemaker system of the fly.

References

Balsalobre, A., Brown, S. A., Marcacci, L., Tronche, F., Kellendonk, C., Reichardt, H. M., Schutz, G., and Schibler, U. (2000). Resetting of circadian time in peripheral tissues by glucocorticoid signaling. *Science* **289**, 2344–2347.

Beaver, L. M., Gvakharia, B. O., Vollintine, T. S., Hege, D. M., Stanewsky, R., and Giebultowicz, J. M. (2002). Loss of circadian clock function decreases reproductive fitness in males of *Drosophila melanogaster. Proc. Natl. Acad. Sci. USA* **99**, 2134–2139.

Blanchardon, E., Grima, B., Klarsfeld, A., Chelot, E., Hardin, P. E., Préat, T., and Rouyer, F. (2001). Defining the role of *Drosophila* lateral neurons in the control of activity and eclosion rhythms by targeted genetic ablation and PERIOD overexpression. *Eur. J. Neurosci.* **13**, 871–888.

Blau, J., and Young, M. W. (1999). Cycling *vrille* expression is required for a functional *Drosophila* clock. *Cell* **99**, 661–671.

Brand, A. H., and Perrimon, N. (1993). Targeted gene expression as a means of altering cell fates and generating dominant phenotypes. *Development* **118**, 401–415.

Brandes, C., Plautz, J. D., Stanewsky, R., Jamison, C. F., Straume, M., Wood, K. V., Kay, S. A., and Hall, J. C. (1996). Novel features of *Drosophila* period transcription revealed by real-time lucifer e reporting. *Neuron* **16**, 687–692.

Brown, S. A., mbrunn, G., Fleury-Olela, F., Preitner, N., and Schibler, U. (2002). Rhythms of mamm: in body temperature can sustain peripheral circadian clocks. *Curr. Biol.* **12**, 1574–158?

Chen, D.-M. hristianson, J. S., Sapp, R. J., and Stark, W. S. (1992). Visual receptor cycle in normal a *period* mutant *Drosophila*: Microspectrophotometry, electrophysiology, and ultrastru iral morphometry. *Vis. Neurosci.* **9**, 125–135.

Cheng, Y., a Hardin, P. E. (1998). *Drosophila* photoreceptors contain an autonomous circadian oscillate hat can function without *period* mRNA cycling. *J. Neurosci.* **18**, 741–750.

Duffy, J. E 2002). GAL4 system in *Drosophila*: A fly geneticist's Swiss army knife. *Genesis* **34**, 1–1

Dushay, N S., Rosbash, M., and Hall, J. C. (1989). The *disconnected* visual system mutations in Dr: phila drastically disrupt circadian rhythms. *J. Biol. Rhythms* **4**, 1–27.

Ewer, J., sch, B., Hamblen-Coyle, M. J., Rosbash, M., and Hall, J. C. (1992). Expression of the *perio* :lock gene within different cell types in the brain of *Drosophila* adults and mosaic anal: s of these cells' influence on circadian behavioral rhythms. *J. Neurosci.* **12**, 3321–3349.

Frisch, , Hardin, P. E., Hamblen-Coyle, M. J., Rosbash, M., and Hall, J. C. (1994). A p moterless *period* gene mediates behavioral rhythmicity and cyclical *per* expression in a r: ricted subset of the *Drosophila* nervous system. *Neuron* **12**, 555–570.

Hall, : C. (1998). Genetics of biological rhythms in *Drosophila*. *Adv. Genet.* **38**, 135–184.

Hardi P. E., Hall, J. C., and Rosbash, M. (1992). Behavioral and molecular analyses suggest th circadian output is disrupted by *disconnected* mutants in *D. melanogaster*. *EMBO J.* 1: 1–6.

Helf h-Förster, C. (1995). The *period* clock gene is expressed in CNS neurons which also p duce a neuropeptide that reveals the projections of circadian pacemaker cells within : brain of *Drosophila melanogaster*. *Proc. Natl. Acad. Sci. USA* **92**, 612–616.

He ich-Förster, C. (1998). Robust circadian rhythmicity of *Drosophila melanogaster* requires e presence of lateral neurons: A brain-behavioral study of *disconnected* mutants. *J. Comp. Physiol. A* **182**, 435–453.

H rich-Förster, C. (2002). The circadian system of *Drosophila melanogaster* and its light nput pathways. *Zoology* **105**, 297–312.

F lfrich-Förster, C. (2003). The neuroarchitecture of the circadian clock in the *Drosophila* brain. *Microsc. Res. Tech.* **62**, 94–102.

:lfrich-Förster, C., Stengl, M., and Homberg, U. (1998). Organization of the circadian system in insects. *Chronobiol. Int.* **15**, 567–594.

elfrich-Förster, C., Täuber, M., Park, J. H., Mühlig-Versen, M., Schneuwly, S., and Hofbauer, A. (2000). Ectopic expression of the neuropeptide pigment-dispersing factor alters behavioral rhythms in *Drosophila melanogaster*. *J. Neurosci.* **20**, 3339–3353.

Kaneko, M., and Hall, J. C. (2000). Neuroanatomy of cells expressing clock genes in *Drosophila*: Transgenic manipulation of the *period* and *timeless* genes to mark the perikarya of circadian pacemaker neurons and their projections. *J. Comp. Neurol.* **422**, 66–94.

Kaneko, M., Helfrich-Förster, C., and Hall, J. C. (1997). Spatial and temporal expression of the *period* and *timeless* genes in the developing nervous system of *Drosophila*: Newly identified pacemaker candidates and novel features of clock gene product cycling. *J. Neurosci.* **17**, 6745–6760.

Kaneko, M., Park, J. H., Cheng, Y., Hardin, P. E., and Hall, J. C. (2000). Disruption of synaptic transmission or clock-gene-product oscillations in circadian pacemaker cells of *Drosophila* cause abnormal behavioral rhythms. *J. Neurobiol.* **43**, 207–233.

Klarsfeld, A., Malpel, S., Michard-Vanheé, C., Picot, M., Chélot, E., and Rouyer, F. (2004). Novel features of cryptochrome-mediated photoreception in the brain circadian clock of *Drosophila. J. Neurosci.* **24,** 1468–1477.

Krishnan, B., Dryer, S. E., and Hardin, P. E. (1999). Circadian rhythms in olfactory responses of *Drosophila melanogaster. Nature* **400,** 375–378.

Myers, E. M., Yu, J., and Sehgal, A. (2003). Circadian control of eclosion: Interaction between a central and peripheral clock in *Drosophila melanogaster. Curr. Biol.* **13,** 526–533.

Nitabach, M. N., Blau, J., and Holmes, T. C. (2002). Electrical silencing of *Drosophila* pacemaker neurons stops the free-running circadian clock. *Cell* **109,** 485–495.

Park, J. H., and Hall, J. C. (1998). Isolation and chronobiological analysis of a neuropeptide pigment-dispersing factor gene in *Drosophila melanogaster. J. Biol. Rhythms* **13,** 219–228.

Park, J. H., Helfrich-Förster, C., Lee, G., Liu, L., Rosbash, M., and Hall, J. C. (2000). Differential regulation of circadian pacemaker output by separate clock genes in *Drosophila. Proc. Natl. Acad. Sci. USA* **97,** 3608–3613.

Peng, Y., Stoleru, D., Levine, J. D., Hall, J. C., and Rosbash, M. (2003). *Drosophila* free-running rhythms require intercellular communication. *PLoS Biol.* **1,** E13.

Plautz, J. D., Kaneko, M., Hall, J. C., and Kay, S. A. (1997). Independent photoreceptive circadian clocks throughout *Drosophila. Science* **278,** 1632–1635.

Renn, S. C. P., Park, J. H., Rosbash, M., Hall, J. C., and Taghert, P. H. (1999). A *pdf* neuropeptide gene mutation and ablation of PDF neurons each cause severe abnormalities of behavioral circadian rhythms in *Drosophila. Cell* **99,** 791–802.

Reppert, S. M., and Weaver, D. R. (2002). Coordination of circadian timing in mammals. *Nature* **418,** 935–941.

Schibler, U., Ripperger, J., and Brown, S. A. (2003). Peripheral circadian oscillators in mammals: Time and food. *J. Biol. Rhythms* **18,** 250–260.

Schibler, U., and Sassone-Corsi, P. (2002). A web of circadian pacemakers. *Cell* **111,** 919–922.

Shafer, O. T., Rosbash, M., and Truman, J. W. (2002). Sequential nuclear accumulation of the clock proteins *period* and *timeless* in the pacemaker neurons of *Drosophila melanogaster. J. Neurosci.* **22,** 5946–5954.

Siwicki, K. K., Eastman, C., Petersen, G., Rosbash, M., and Hall, J. C. (1988). Antibodies to the *period* gene product of *Drosophila* reveal diverse tissue distribution and rhythm changes in the visual system. *Neuron* **1,** 141–150.

Stanewsky, R., Frisch, B., Brandes, C., Hamblen-Coyle, M., Rosbash, M., and Hall, J. C. (1997a). Temporal and spatial expression patterns of transgenes containing increasing amounts of the *Drosophila* clock gene period and a lacZ reporter: Mapping elements of the PER protein involved in circadian cycling. *J. Neurosci.* **17,** 676–696.

Stanewsky, R., Jamison, C. F., Plautz, J. D., Kay, S. A., and Hall, J. C. (1997b). Multiple circadian-regulated elements contribute to cycling *period* gene expression in *Drosophila. EMBO J.* **16,** 5006–5018.

Stanewsky, R., Kaneko, M., Emery, P., Beretta, B., Wager-Smith, K., Kay, S. A., Rosbash, M., and Hall, J. C. (1998). The *cry^b* mutation identifies cryptochrome as a circadian photoreceptor in *Drosophila. Cell* **95,** 681–692.

Stempfl, T., Vogel, M., Szabo, G., Wülbeck, C., Liu, J., Hall, J. C., and Stanewsky, R. (2002). Identification of circadian-clock-regulated enhancers and genes of *Drosophila melanogaster* by transposon mobilization and luciferase reporting of cyclical gene expression. *Genetics* **160,** 571–593.

Veleri, S., Brandes, C., Helfrich-Förster, C., Hall, J. C., and Stanewsky, R. (2003). A self-sustaining, light-entrainable circadian oscillator in the *Drosophila* brain. *Curr. Biol.* **13,** 1758–1767.

Vosshall, L. B., and Young, M. W. (1995). Circadian rhythms in *Drosophila* can be driven by *period* expression in a restricted group of central brain cells. *Neuron* **15,** 345–360.

Yang, Z., and Sehgal, A. (2001). Role of molecular oscillations in generating behavioral rhythms in *Drosophila*. *Neuron* **29,** 453–467.

Zerr, D. M., Hall, J. C., Rosbash, M., and Siwicki, K. K. (1990). Circadian fluctuations of *period* protein immunoreactivity in the CNS and the visual system of *Drosophila*. *J. Neurosci.* **10,** 2749–2762.

Zhao, J., Kilman, V. L., Keegan, K. P., Peng, Y., Emery, P., Rosbash, M., and Allada, R. (2003). *Drosophila* clock can generate ectopic circadian clocks. *Cell* **113,** 755–766.

## [22]  The Suprachiasmatic Nucleus is a Functionally Heterogeneous Timekeeping Organ

*By* RAE SILVER and WILLIAM J. SCHWARTZ

## Abstract

Ever since the locus of the brain clock in the suprachiasmatic nucleus (SCN) was first described, methods available have both enabled and encumbered our understanding of its nature at the level of the cell, the tissue, and the animal. A combination of *in vitro* and *in vivo* approaches has shown that the SCN is a complex heterogeneous neuronal network. The nucleus is composed of cells that are retinorecipient and reset by photic input; those that are reset by nonphotic inputs; slave oscillators that are rhythmic only in the presence of the retinohypothalamic tract; endogenously rhythmic cells, with diverse period, phase, and amplitude responses; and cells that do not oscillate, at least on some measures. Network aspects of SCN organization are currently being revealed, but mapping these properties onto cellular characteristics of electrical responses and patterns of gene expression are in early stages. While previous mathematical models focused on properties of uniform coupled oscillators, newer models of the SCN as a brain clock now incorporate oscillator and gated, nonoscillator elements.

## The Brain's Clock as a Construct

The function of the suprachiasmatic nucleus (SCN) was discovered during the era that identified the hypothalamus as the site of several brain "centers" governing homeostatically regulated behaviors. Ablation of the lateral hypothalamus resulted in a significant reduction of eating behavior

METHODS IN ENZYMOLOGY, VOL. 393

and weight loss, hence an eating center. Destruction of the ventromedial hypothalamus produced obesity, hence a satiety center. The opportunity to electrically self-stimulate the brain was so powerful a reward that all other motivated behaviors were put aside, hence pleasure centers. Lesions of the SCN led to dramatic behavioral and physiological arrhythmicity, hence a timekeeping center.

Today, most of these center constructs have fallen out of favor. For the eating center, it was realized that "specific" hypothalamic lesions damaged fibers of passage, that lesioned animals showed sensory neglect and no longer responded to afferent inputs, and that other, major consequences accompanied weight loss in addition to disrupted feeding. For the pleasure centers, once the neural network and transmitter systems involved in self-stimulation were delineated, it was obvious that no single center controlled the behavior. The notion of brain centers began to lose its heuristic value, and the centrist view was even labeled a "millstone" rather than a "milestone" to progress in neurobiology (Coscina, 1976).

For the SCN, such a revision has not occurred. The body of evidence identifying the nucleus as the master circadian pacemaker in mammals is multidisciplinary in nature, and the strength of this functional localization is unsurpassed by that of any other structure in the vertebrate brain (for review, see Klein et al., 1991). Lesions of the SCN result in a breakdown of the generation or entrainment of a wide array of rhythms and they never recover, no matter how early in development ablation is performed. The circadian oscillation of SCN is seen in vivo and in vitro, using metabolic, electrophysiological, and molecular assays, and electrical or pharmacological stimulation causes predictable phase shifts of these rhythms. Neural grafts of fetal SCN tissue reestablish overt rhythmicity in arrhythmic, SCN-lesioned recipients, and the rhythms restored by the transplants display properties characteristic of the circadian pacemakers of the donors rather than those of the hosts. Thus, among the hypothalamic "centers" of decades ago, the SCN as a timing center has retained its conceptual value.

What we are learning now is that the physical center is not an indivisible homogeneous cellular syncytium but a complex heterogeneous neuronal network with intracenter localization and specialization of function. This article highlights how the locally distributed network properties within the SCN are key to its pacemaker function. It is only through the combined use of advanced morphological, physiological, molecular, and genetic tools that researchers have begun to delineate the functional compartmentalization of this tissue that acts as a circadian clock for the brain.

In the Beginning: Anatomical Heterogeneity but
Functional Homogeneity

The decade of the 1970s is remembered as the time during which the
SCN was implicated as the site of a mammalian circadian pacemaker. In
1972, lesions of the rat SCN were reported to abolish circadian rhythms of
behavioral (wheel running and drinking) (Stephan and Zucker, 1972) and
endocrine (corticosterone) (Moore and Eichler, 1972) activity, and later in
the decade, endogenous rhythms of SCN glucose utilization (measured by
$^{14}$C-labeled deoxyglucose uptake) (Schwartz and Gainer, 1977) and electrical activity (recorded as the overall firing rate of multiple neurons)
(Inouye and Kawamura, 1979) were demonstrated in intact rats. It was
recognized even then, and certainly by 1980, that SCN cells were not a
homogeneous population. Nissl and silver stains, Golgi impregnations, and
electron microscopy of the rat SCN revealed two predominant subdivisions: cells in the dorsomedial part of the nucleus were smaller and more
tightly packed than those in the ventrolateral part (van den Pol, 1980). This
dorsal/ventral distinction corresponded to the ventrolateral segregation of
most SCN inputs (from the retina, raphe, and lateral geniculate) and was
recapitulated by immunohistochemical identification of peptides in SCN
cell bodies [e.g., arginine vasopressin (VP) dorsomedially and vasoactive
intestinal polypeptide (VIP) ventrolaterally] (Inouye and Shibata, 1994;
Moore et al., 2002).

Despite this morphological evidence for regional compartments within
the SCN, physiological data seemed to favor equipotentiality across the
SCN, without a clear localization of function. Small, partial electrolytic
lesions were made in an effort to determine whether specific regions might
govern different rhythms (Pickard and Turek, 1985; van den Pol and
Powley, 1979), but this approach, in the absence of markers for specific
cell phenotypes and assessment of destruction of passing axons, could not
characterize the critical factor in the observed dysrhythmias. What was
generally found was a correlation between the volume of SCN destroyed—
without regard to the unilaterality or regionality of the damage—and a
shortened free-running period of the wheel-running rhythm, prompting the
view that "SCN tissue is homogeneous in its contribution to rhythmicity
even though the anatomical and biochemical heterogeneity of the nucleus
in rats would suggest otherwise" (Davis and Gorski, 1984). Supporting this
view were the discoveries that circadian oscillation of the fetal SCN ante-
dated its regional specialization (Reppert and Schwartz, 1984) and, in the
adult, that the intranuclear distribution of $^{14}$C-labeled deoxyglucose uptake
(Schwartz et al., 1987) and multiunit electrical discharge activity (Bouskila
and Dudek, 1993) showed no obvious dorsal/ventral difference.

## Reducing SCN Tissue to Cells and Slices *In Vitro*

In 1995, individual dissociated SCN neurons were shown to oscillate independently with different circadian periods *in vitro* (Welsh *et al.*, 1995), demonstrating that a circadian clock was localized within individual cells rather than arising as an emergent property of an SCN network. The results were interpreted to suggest that the oscillatory capacity could not be restricted to any particular subset of neuropeptide-containing cells; altogether, immunochemically identified cells accounted for only 23% of all the neurons in culture, whereas 50% of all the neurons were rhythmic. The authors suggested that their data were consistent with the possibility that the nonoscillating cells in culture were actually from outside the SCN and that all SCN neurons were functioning "clock" cells. The cellular homogeneity implied by this interpretation stimulated models for how such cells might be synchronized within a coupled network (Liu *et al.*, 1997).

In order to study SCN cells over multiple circadian cycles *in vitro*, but with a model that would preserve an *in vivo*-like dorsal/ventral architecture, hypothalamic slices containing the SCN have been incubated for weeks (either embedded in plasma clots on coverslips in rotating roller tubes or adhered to filters in stationary cultures). Such "organotypic slice cultures" lost at least 70% of their neurons and flattened to a few cell layers thick, but they survived, expressed immunoreactive AVP and VIP regionally as *in vivo* (Belenky *et al.*, 1996; Tominaga *et al.*, 1994), and exhibited circadian rhythms of the release of both peptides into the medium. When the cultures were treated with antimitotics, the two peptide rhythms appeared to free run separately with different circadian periods (Shinohara *et al.*, 1995), suggesting that they represented a chemical manifestation of comparable, equipotent oscillators in both dorsal and ventral SCN subdivisions. A later study of such rat slice cultures, recording multichannel electrical activity along with rhythms of AVP and VIP release, confirmed oscillations in both dorsal and ventral SCN, although they did report very subtle dorsal/ventral differences, including evidence for a less stable VIP than AVP rhythm and a lower proportion of ventrolateral than dorsomedial neurons with rhythmic firing rates (Nakamura *et al.*, 2001). Most recently, organotypic slices have been made from neonatal transgenic mice expressing a luciferase reporter driven by an oscillating clock gene promoter (*mPer1*) (Yamaguchi *et al.*, 2003). Irrespective of location, virtually all of the luminescent cells (99.2% of a total 1177) exhibited circadian rhythmicity, although the proportion of total SCN cells that were luminescent was not described. Details of slice preparation, thickness, culturing methods, species of origin, and measures used to assay rhythmicity may lead to variable results.

That cultured slices might be an incomplete model of SCN tissue organization is perhaps not too surprising, as they likely undergo some degree of reorganization. For example, although AVP and VIP are present "organotypically" in cultured slices, gastrin-releasing peptide (GRP; a prominent ventrolateral neurotransmitter) expression at the adult level is lost (Wray et al., 1993). Importantly, the apparent functional redundancy in cell and slice cultures contrasts with data obtained from acutely prepared slices. As early as 1984, it was reported that the circadian rhythm of spontaneous single-unit discharge rates in the ventrolateral, but not the dorsomedial, SCN was abolished in slices made from rats housed in constant darkness (DD) or bilaterally enucleated (Shibata et al., 1984). Studies of this kind have generally been performed by sampling the extracellular spike activity of single units for short time intervals at different phases across the circadian cycle, pooling data from a number of slices and phases, and then determining the phase of peak firing rate for the population of units as a whole. More recently, acute slices made from neonatal transgenic mice expressing a short half-life green fluorescent protein (GFP) reporter driven by the *mPer1* promoter (Quintero et al., 2003) indicate that in slices made from mice housed in a light–dark cycle (LD), 11% of the imaged cells exhibited nonrhythmic *Per1*::GFP expression; from mice housed in DD, this proportion increased to 26%. Rhythmic cells in LD slices were more likely ventral than dorsal (64% versus 36%, respectively), whereas the opposite distribution was found in DD slices (43% versus 57%). While these methods suggested functional heterogeneity among SCN cells, they did not associate function with phenotype and left unexplained the differences among experimental results.

## SCN Tissue Organization and Heterogeneous Gene Expression

In addition to cells and slices *in vitro*, SCN tissue must be studied *in vivo* to learn how it actually functions in intact animals. This goal was invigorated in the 1990s by the identification of putative "clock" genes, molecules that appear to lie at the core oscillatory mechanism of the clock as two interacting feedback loops (for review, see Van Gelder et al., 2003). In one loop, transcription of the *Period (Per), Cryptochrome*, and possibly *Timeless* genes is negatively regulated by their protein products, which inhibit (with a time delay) the DNA-binding activity of the positive bHLH transcription factors Clock and Bmal1/Mop3 (in mammals). Within this loop, the essential time delay is provided by a phosphorylation-dependent variation in the stability of the Per protein (mediated by casein kinase 1 epsilon) and its complex with cryptochrome. In the second loop,

*Bmal1/Mop3* mRNA and protein oscillate (in antiphase to *Per*) by driving circadian expression of the *Clock* repressor, Rev-erb-$\alpha$.

In the SCN, the most extensively studied of these clock components have been the *Per* genes (*Per1, Per2,* and *Per3*), encoding mRNAs that both oscillate with a circadian rhythm, expressing high levels during the subjective day and low levels during the subjective night, and are photo-inducible, with a phase dependence similar to that of light-induced phase shifts of behavioral rhythmicity. Initial studies of *Per1* and *Per2* mRNA (Shearman *et al.*, 1997; Sun *et al.*, 1997; Tei *et al.*, 1997) and immunoreactive protein (Field *et al.*, 2000; Hastings *et al.*, 1999) suggested that expression occurred throughout the extent of the mouse SCN. This impression led to the conception of an idealized (linear) signal transduction pathway in which axons of specialized retinal ganglion cells form the retinohypothalamic tract (RHT) in the optic nerves and release glutamate at conventional synapses onto SCN "clock" cells. The resulting membrane depolarization leads to $Ca^{2+}$ influx, CREB phosphorylation, and gene transcription, with the expression of proteins that reset the core auto-regulatory transcription–translation loop of the circadian pacemaker (for review, see Meijer and Schwartz, 2003).

However, anatomical and physiological evidence from animals indicated that this simple concept lacked a critical *inter*cellular dimension. Anatomically, only a subset of all rodent SCN cells are directly retino-recipient (for review, see Lee *et al.*, 2003a). Physiologically, estimates of photoresponsiveness range from about one-fifth (by c-Fos immunoreactivity) (Castel *et al.*, 1997) to about one-third (by electrophysiology) located in the ventrolateral subdivision of the SCN (Meijer *et al.*, 1986, 1998). No more than about 20% of the cells in this subset can be attributed to any one identified peptidergic phenotype (Castel *et al.*, 1997; Romijn *et al.*, 1996), and initial work suggested that no peptidergic cell type uniformly produced photosensitive responses.

Clues to solving this problem have been provided by using new markers for distinct SCN subregions and for individual SCN cells. In the hamster, calbindin-containing cells lying in the caudal core of the SCN were shown to be directly retinorecipient by tract-tracing and double-label electron microscopy (Bryant *et al.*, 2000) and to virtually uniformly express immunoreactive Fos following a light pulse (Silver *et al.*, 1996). Furthermore, light-induced *Per1* and *Per2* mRNAs were concentrated in this calbindin region (Hamada *et al.*, 2001). Remarkably, endogenously rhythmic *Per1, Per2,* and *Per3* expression was not detectable in this region but was observed instead in the dorsomedial SCN region marked by VP containing cells. Of note, electrophysiological recordings of identified calbindin cells in acutely prepared hamster SCN slices have also demonstrated an absence of endogenous rhythmicity (Jobst and Allen, 2002).

A similar regional separation of rhythmic and nonrhythmic (but photo-inducible) gene expression has been found for other genes, e.g., c-*fos* in rats and hamsters (Guido *et al.*, 1999a,b; Schwartz *et al.*, 2000; Sumová *et al.*, 1998), and for *Per* in other animals, e.g., rats (Dardente *et al.*, 2002; Yan and Okamura, 2002; Yan *et al.*, 1999). It had been thought that functional subdivisions in the mouse SCN were not so clearly segregated (King *et al.*, 2003), but it has been shown that *Per1* and *Per2* are light induced but not detectably rhythmic in the GRP-containing cells of the mouse SCN (Karatsoreos *et al.*, 2004). These results provide a clear example of functional segregation by SCN phenotype. Importantly, region-specific SCN rhythmicity extends beyond the genes; in electrophysiological studies *in vivo*, photically responsive neurons did not express the significant circadian rhythm in discharge rate that could be observed in photically insensitive cells (Jiao *et al.*, 1999; Saeb-Parsy and Dyball, 2003). Clearly, SCN neurons are not all functionally equivalent—light responsivity and endogenous rhythmicity of whole SCN tissue are based on a cellular division of labor.

Dissection of the Retinorecipient Subdivision of the SCN

The two major peptidergic phenotypes that receive photic input via the RHT are VIP- and GRPergic neurons. Their mRNA and peptide levels exhibit oppositely phased responses to light, with high levels of VIP during the dark and GRP during the light (Shinohara *et al.*, 1993; Zoeller *et al.*, 1992). In mice, GRP cells express *Per* genes following a light pulse but are not rhythmic in this response (Karatsoreos *et al.*, 2004). In hamsters, calbindin delineates the region of light-induced *Per* expression and approximately 40% VIP and 60% GRP cells contain calbindin (Hamada *et al.*, 2001; LeSauter *et al.*, 2002). In rats, lateral (but not medial) VIP neurons in the ventral subdivision coexpress GRP; the lateral (but not the medial) VIP cells receive retinal innervation and express photoinducible *Per1* (Kawamoto *et al.*, 2003). Taken together, data suggest an important role for GRP in photosensitivity. The intercellular mechanisms for coupling photoreceptive cells to and then resetting endogenously rhythmic cells are unknown, although a role for GRP is suggested. Intracerebroventricular injection of GRP during early night increased *Per* mRNA primarily in the *dorsal* mouse SCN, while the photic induction of *Per* was reduced in GRP receptor-deficient mutant mice, an effect also occurring primarily in the *dorsal* rather than in the ventral part of the nucleus (Aida *et al.*, 2002). *Per1* and *Per2* expression in the dorsal SCN may be crucial to resetting the SCN by phase advances or delays (Yan and Silver, 2002).

The retinorecipient subdivision of the SCN appears to play a critical role in the generation, and not just the entrainment, of circadian rhythmicity. Microlesions that ablated the calbindin region of the hamster SCN resulted in a loss of all measurable circadian rhythms, even though significant portions of the SCN with endogenously rhythmic *Per*-expressing cells survived the lesion (Kriegsfeld *et al.*, 2004b; LeSauter and Silver, 1999). VIP and VPAC2 receptor knockout mice exhibit disrupted behavioral, molecular, and electrophysiological rhythmicity (Colwell *et al.*, 2003; Piggins and Culter, 2003). These kinds of data have stimulated a new model of SCN tissue organization, in which retinorecipient nonrhythmic "gate" cells provide resetting and synchronizing signals to individually rhythmic "clock" cells with different intrinsic periods (Antle *et al.*, 2003). The gate provides daily input to the oscillators and is in turn regulated (directly or indirectly) by the oscillator cells. Individual oscillators with initial random phases can self-assemble so as to maintain cohesive rhythmic output. In this view, SCN circuits are important for self-sustained oscillation, and network properties distinguish the SCN from other tissues that lack resetting signals but rhythmically express clock genes.

### Heterogeneity of Phase at Tissue and Single-Cell Levels

Immunohistochemical detection of phosphorylated ERK/MAP kinase activity has demonstrated two distinct oscillations running in antiphase in different SCN regions (Coogan and Piggins, 2003; Lee *et al.*, 2003b; Nakaya *et al.*, 2003; Obrietan *et al.*, 1998). During the subjective day, pERK expression overlapped (but was not coexpressed) with that of VP (Lee *et al.*, 2003b), while during the night, pERK expression was confined to a small region of "cap" cells (so named because they form a cap over the calbindin cells in the hamster SCN). This latter population of rhythmic cells behaved like a slave oscillator and was dependent upon the eye, even in conditions of constant darkness. Of note, within a central zone of the mouse SCN has been described a cluster of immunoreactive Per-expressing cells, in antiphase to the peak of Per expression at the end of the day (King *et al.*, 2003); the relationship of these cells to the pERK region is unknown.

It is known that environmental lighting can dramatically alter regional phase relationships within SCN tissue. In the rat SCN, a sudden advance or delay of the LD cycle resulted in a transient desynchronization between the ventrolateral and dorsomedial subdivisions; ventrolateral gene expression shifted rapidly, while dorsomedial expression resynchronized only gradually over days to weeks (Nagano *et al.*, 2003). A stable, "forced" desynchronization of ventrolateral and dorsomedial subdivisions (again assessed by patterns of gene expression) has also been achieved by exposing rats to an artificially short 22-h LD cycle (de la Iglesia *et al.*, 2004). In this situation, the SCN was in a unique, reconfigured state; even though intercellular coupling *between* SCN subdivisions was lost, coupling *within* each subdivision was retained, suggesting that inter- and intradivisional synchronizing mechanisms may be different (for review, see Michel and Colwell, 2001). Perhaps the most dramatic example of functionally reconfigured SCN tissue is the phenomenon known as "splitting" in hamsters maintained in constant light, in which the single daily bout of locomotor activity in an animal dissociates into two components that each free run with different periods until they become stably coupled 180° (about 12 h) apart. Splitting appears to be the consequence of a reorganized SCN with left and right halves oscillating in antiphase, as mRNAs characteristic of day and night are simultaneously expressed on opposite sides of the paired SCN (de la Iglesia *et al.*, 2000).

Analysis of the SCN as an integrated tissue is now aided by powerful methods that make possible real-time, simultaneous measurements of oscillatory gene activity over repeated cycles from multiple individual cells. As mentioned previously, bioluminescent rhythms in single neurons have been measured in tissue slices made from *Per1::luc* (Yamaguchi *et al.*, 2003) and *Per1*::GFP (Quintero *et al.*, 2003) transgenic mice. It has been shown that individual SCN cells in slices showed rather large phase differences in the peaks of their bioluminescent rhythms that persisted over repeated cycles (individual cellular periods were similar and stable). The phase order was not a stochastic property of the network because it was restored in *Per1::luc* slices after cycloheximide was applied to first stop and then reset the cellular oscillations to the same initial phase. Moreover, intercellular phase differences were not an artifact of these transgenic preparations because similar differences have been demonstrated by electrophysiological methods in rat SCN slices (Schaap *et al.*, 2003). In general, dorsomedial cells appeared to phase lead (but did not appear to drive) ventrolateral ones in *Per1::luc* slices (Yamaguchi *et al.*, 2003), while a lateral-to-medial gradient was described in the *Per1*::GFP slices (Quintero *et al.*, 2003) or none at all in electrical activity in the rat slices (Schaap *et al.*, 2003). In the SCN harvested from hamsters sacrificed across the circadian cycle, the daily spread of gene expression was from dorsal to ventral (Hamada *et al.*, 2004). What has become clear from all of these studies is that the duration of high molecular and electrophysiological activities of individual SCN cells appears to differ from the composite activities of the tissue as a whole (which generally lasts for most of the subjective day). The functional significance of heterogeneous cellular phases, as well as the mechanisms that keep the cells out of phase and direct their spatial organization, is not known. It is possible that their distribution and clustering can be configured by afferent input (Quintero *et al.*, 2003). It could be that such plasticity of phase differences permits the encoding of a photoperiodic signal (Schaap *et al.*, 2003).

## Building a Global View: From Clock Genes to Circadian Behavior

Of course, it is the neural activity of the SCN in its proper context *in situ*, not gene expression in an isolated SCN *in vitro*, that regulates circadian behavior in whole organisms. Perhaps heterogeneous cellular phases might play a role as part of the temporal programming of SCN outputs (Kalsbeek and Buijs, 2002). Indeed, a recent single-unit electrophysiological study of the rat SCN *in vivo* has demonstrated that antidromically identified SCN neurons innervating arcuate or supraoptic nuclei express a very different firing rate rhythm (with peaks at the light–dark and dark–light transition phases) than cells without such output projections (Saeb-Parsy and Dyball, 2003). Also, anatomical tracing studies have indicated that there is a subset of SCN neurons that are both efferent to other hypothalamic nuclei and responsive to light (de la Iglesia and Schwartz, 2002; Munch *et al.*, 2002); and in hamster, cells in the retinorecipient region delineated by calbindin and those in the rhythmic region delineated by VP both project to all of the same SCN target sites (Kriegsfeld *et al.*, 2004a). These pathways provide a possible direct channel through the SCN for photic inputs to influence neural outputs. Such a route could underlie the immediate effect of light on the nocturnal rhythm of pineal melatonin secretion in which light acts to acutely suppress nighttime melatonin production by an SCN pathway that is physiologically (Nelson and Takahashi, 1991) and pharmacologically (Paul *et al.*, 2003) distinct from that mediating rhythm entrainment. Furthermore, cells from retinorecipient and from rhythmic regions of the SCN both project to all known targets of SCN neurons (Kriegsfeld *et al.*, 2004a), providing another potential substrate for integration of photic and rhythmic information.

## From Center to Network

The view of the SCN as a brain clock composed of 20,000 "clock" cells has been an extremely useful heuristic for advancing our knowledge of the molecular basis of circadian rhythmicity and for modeling formal properties of oscillators. At the same time, it has been remarkable to many students of the mammalian brain that the construct of a timing center has survived the experimental dissection of its component parts. The fact that the clock function of the SCN can be studied "in a dish" in acute slices or long-term cultures, in dispersed, dissociated cells, and in artificial cell lines likely accounts for its continuing heuristic value. It is hardly worth noting that the other once-popular brain centers, pleasure, satiety, and hunger, did not share these easy tools of analysis. Nevertheless, the evidence is clear that specialization of function occurs within the SCN, and its network properties and signaling pathways are only starting to be revealed. Future approaches to understanding these properties—both electrophysiological (Pennartz et al., 1998) and genetic (Low-Zeddies and Takahashi, 2001)—will require regional and cellular levels of resolution.

## Acknowledgments

The work reported here is supported by NINDS Grants RO1 NS37919 (RS) and R01 NS46605 (WJS). The contents of this publication are solely the responsibility of the authors and do not necessarily represent the official views of the NINDS.

## References

Aida, R., Moriya, T., Araki, M., Akiyama, M., Wada, K., Wada, E., and Shibata, S. (2002). Gastrin releasing peptide mediates photic entrainable signals to dorsal subsets of suprachiasmatic nucleus via induction of *Period* gene in mice. *Mol. Pharmacol.* **61**, 26–34.

Antle, M. C., Foley, D. K., Foley, N. C., and Silver, R. (2003). Gates and oscillators: A network model of the brain clock. *J. Biol. Rhythms* **18**, 339–350.

Belenky, M., Wagner, S., Yarom, Y., Matzner, H., Cohen, S., and Castel, M. (1996). The suprachiasmatic nucleus in stationary organotypic culture. *Neuroscience* **70**, 127–143.

Bouskila, Y., and Dudek, F. E. (1993). Neuronal synchronization without calcium-dependent synaptic transmission in the hypothalamus. *Proc. Natl. Acad. Sci. USA* **90**, 3207–3210.

Bryant, D. N., LeSauter, J., Silver, R., and Romero, M.-T. (2000). Retinal innervation of calbindin-$D_{28K}$ cells in the hamster suprachiasmatic nucleus: Ultrastructural characterization. *J. Biol. Rhythms* **15**, 103–111.

Castel, M., Belenky, M., Cohen, S., Wagner, S., and Schwartz, W. J. (1997). Light-induced c-Fos expression in the mouse suprachiasmatic nucleus: Immuno-electron microscopy reveals colocalization in multiple cell types. *Eur. J. Neurosci.* **9**, 1950–1960.

Colwell, C. S., Michel, S., Itri, J., Rodriguez, W., Tam, J., Lelievre, V., Hu, Z., Liu, X., and Waschek, J. A. (2003). Disrupted circadian rhythms in VIP- and PHI-deficient mice. *Am. J. Physiol.* **285**, R939–R949.

Coogan, A. N., and Piggins, H. D. (2003). Circadian and photic regulation of phosphorylation of ERK1/2 and Elk-1 in the suprachiasmatic nuclei of the Syrian hamster. *J. Neurosci.* **23**, 3085–3093.

Coscina, D. V. (1976). "Lateral Hypothalamic Syndrome: Milestone or Millstone, Symposium." Special Interest Group in Physiological Psychology, Society for Neuroscience, Toronto.

Dardente, H., Poirel, V.-J., Klosen, P., Pévet, P., and Masson-Pévet, M. (2002). Per and neuropeptide expression in the rat suprachiasmatic nuclei: Compartmentalization and differential cellular induction by light. *Brain Res.* **958**, 261–271.

Davis, F. C., and Gorski, R. A. (1984). Unilateral lesions of the hamster suprachiasmatic nuclei: Evidence for redundant control of circadian rhythms. *J. Comp. Physiol. A* **154**, 221–232.

de la Iglesia, H. O., Cambras, T., Schwartz, W. J., and Diez-Noguera, A. (2004). Forced desynchronization of dual circadian oscillators within the rat suprachiasmatic nucleus. *Curr. Biol.* **14**, 796–800.

de la Iglesia, H. O., Meyer, J., Carpino, A., Jr., and Schwartz, W. J. (2000). Antiphase oscillation of the left and right suprachiasmatic nuclei. *Science* **290**, 799–801.

de la Iglesia, H. O., and Schwartz, W. J. (2002). A subpopulation of efferent neurons in the mouse suprachiasmatic nucleus is also light-responsive. *NeuroReport* **13**, 857–860.

Field, M. D., Maywood, E. S., O'Brien, J. A., Weaver, D. R., Reppert, S. M., and Hastings, M. H. (2000). Analysis of clock proteins in mouse SCN demonstrates phylogenetic divergence of the circadian clockwork and resetting mechanisms. *Neuron* **25**, 437–447.

Guido, M. E., de Guido, L. B., Goguen, D., Robertson, H. A., and Rusak, B. (1999a). Daily rhythm of spontaneous immediate-early gene expression in the rat suprachiasmatic nucleus. *J. Biol. Rhythms* **14**, 275–280.

Guido, M. E., Goguen, D., de Guido, L., Robertson, H. A., and Rusak, B. (1999b). Circadian and photic regulation of immediate-early gene expression in the hamster suprachiasmatic nucleus. *Neuroscience* **90**, 555–571.

Hamada, T., Antle, M. C., and Silver, R. (2004). Temporal and spatial expression patterns of canonical clock genes and clock-controlled genes in the suprachiasmatic nucleus. *Eur. J. Neurosci.* **19**, 1741–1748.

Hamada, T., LeSauter, J., Venuti, J. M., and Silver, R. (2001). Expression of *Period* genes: Rhythmic and nonrhythmic compartments of the suprachiasmatic nucleus pacemaker. *J. Neurosci.* **21**, 7742–7750.

Hastings, M. H., Field, M. D., Maywood, E. S., Weaver, D. R., and Reppert, S. M. (1999). Differential regulation of mPER1 and mTIM proteins in the mouse suprachiasmatic nuclei: New insights into a core clock mechanism. *J. Neurosci.* **19**, RC11.

Inouye, S. T., and Kawamura, H. (1979). Persistence of circadian rhythmicity in a mammalian hypothalamic "island" containing the suprachiasmatic nucleus. *Proc. Natl. Acad. Sci. USA* **76**, 5962–5966.

Inouye, S. T., and Shibata, S. (1994). Neurochemical organization of circadian rhythm in the suprachiasmatic nucleus. *Neurosci. Res.* **20**, 109–130.

Jiao, Y.-Y., Lee, T. M., and Rusak, B. (1999). Photic responses of suprachiasmatic area neurons in diurnal degus (*Octodon degus*) and nocturnal rats (*Rattus norvegicus*). *Brain Res.* **817**, 93–103.

Jobst, E. E., and Allen, C. N. (2002). Calbindin neurons in the hamster suprachiasmatic nucleus do not exhibit a circadian variation in spontaneous firing rate. *Eur. J. Neurosci.* **16**, 2469–2474.

Kalsbeek, A., and Buijs, R. M. (2002). Output pathways of the mammalian suprachiasmatic nucleus: Coding circadian time by transmitter selection and specific targeting. *Cell Tissue Res.* **309**, 109–118.

Karatsoreos, I. N., Yan, L., LeSauter, J., and Silver, R. (2004). Phenotype matters: Identification of light-responsive cells in the mouse suprachiasmatic nucleus. *J. Neurosci.* **24**, 68–75.

Kawamoto, K., Nagano, M., Kanda, F., Chihara, K., Shigeyoshi, Y., and Okamura, H. (2003). Two types of VIP neuronal components in rat suprachiasmatic nucleus. *J. Neurosci. Res.* **74**, 852–857.

King, V. M., Chahad-Ehlers, S., Shen, S., Harmar, A. J., Maywood, E. S., and Hastings, M. H. (2003). A *hVIPR* transgene as a novel tool for the analysis of circadian function in the mouse suprachiasmatic nucleus. *Eur. J. Neurosci.* **17,** 822–832.

Klein, D., Moore, R. Y. and Reppert, S. M. (eds.) (1991). "Suprachiasmatic Nucleus: The Mind's Clock," Oxford, New York.

Kriegsfeld, L. J., Leak, R. K., Yackulic, C. B., LeSauter, J., and Silver, R. (2004a). Organization of suprachiasmatic nucleus projections in Syrian hamsters (*Mesocricetus auratus*): An anterograde and retrograde analysis. *J. Comp. Neurol.* **468,** 361–379.

Kriegsfeld, L. J., LeSauter, J., and Silver, R. (2004b). Targeted microlesions reveal novel organization of the hamster suprachiasmatic nucleus. *J. Neurosci.* **24,** 2449–2457.

Lee, H. S., Billings, H. J., and Lehman, M. N. (2003a). The suprachiasmatic nucleus: A clock of multiple components. *J. Biol. Rhythms* **18,** 435–449.

Lee, H. S., Nelms, J. L., Nguyen, M., Silver, R., and Lehman, M. N. (2003b). The eye is necessary for a circadian rhythm in the suprachiasmatic nucleus. *Nature Neurosci.* **6,** 111–112.

LeSauter, J., Kriegsfeld, L. J., Hon, J., and Silver, R. (2002). Calbindin-$D_{28K}$ cells selectively contact intra-SCN neurons. *Neuroscience* **111,** 575–585.

LeSauter, J., and Silver, R. (1999). Localization of a suprachiasmatic nucleus subregion regulating locomotor rhythmicity. *J. Neurosci.* **19,** 5574–5585.

Liu, C., Weaver, D. R., Strogatz, S. H., and Reppert, S. M. (1997). Cellular construction of a circadian clock: Period determination in the suprachiasmatic nuclei. *Cell* **91,** 855–860.

Low-Zeddies, S. S., and Takahashi, J. S. (2001). Chimera analysis of the *Clock* mutation in mice shows that complex cellular integration determines circadian behavior. *Cell* **105,** 25–42.

Meijer, J. H., Groos, G. A., and Rusak, B. (1986). Luminance coding in a circadian pacemaker: The suprachiasmatic nucleus of the rat and the hamster. *Brain Res.* **382,** 109–118.

Meijer, J. H., and Schwartz, W. J. (2003). In search of the pathways for light-induced pacemaker resetting in the suprachiasmatic nucleus. *J. Biol. Rhythms* **18,** 235–249.

Meijer, J. H., Watanabe, K., Schaap, J., Albus, H., and Détári, L. (1998). Light responsiveness of the suprachiasmatic nucleus: Long-term multiunit and single-unit recordings in freely moving rats. *J. Neurosci.* **18,** 9078–9087.

Michel, S., and Colwell, C. S. (2001). Cellular communication and coupling within the suprachiasmatic nucleus. *Chronobiol. Int.* **18,** 579–600.

Moore, R. Y., and Eichler, V. B. (1972). Loss of a circadian adrenal corticosterone rhythm following suprachiasmatic lesions in the rat. *Brain Res.* **42,** 201–206.

Moore, R. Y., Speh, J. C., and Leak, R. K. (2002). Suprachiasmatic nucleus organization. *Cell Tissue Res.* **309,** 89–98.

Munch, I. C., Møller, M., Larsen, P. J., and Vrang, N. (2002). Light-induced c-Fos expression in suprachiasmatic nuclei neurons targeting the paraventricular nucleus of the hamster hypothalamus: Phase dependence and immunochemical identification. *J. Comp. Neurol.* **442,** 48–62.

Nagano, M., Adachi, A., Nakahama, K., Nakamura, T., Tamada, M., Meyer-Bernstein, E., Sehgal, A., and Shigeyoshi, Y. (2003). An abrupt shift in the day/night cycle causes desynchrony in the mammalian circadian center. *J. Neurosci.* **23,** 6141–6151.

Nakamura, W., Honma, S., Shirakawa, T., and Honma, K. (2001). Regional pacemakers composed of multiple oscillator neurons in the rat suprachiasmatic nucleus. *Eur. J. Neurosci.* **14,** 666–674.

Nakaya, M., Sanada, K., and Fukada, Y. (2003). Spatial and temporal regulation of mitogen-activated protein kinase phosphorylation in the mouse suprachiasmatic nucleus. *Biochem. Biophys. Res. Commun.* **305,** 494–501.

Nelson, D. E., and Takahashi, J. S. (1991). Comparison of visual sensitivity for suppression of pineal melatonin and circadian phase-shifting in the golden hamster. *Brain Res.* **554**, 272–277.

Obrietan, K., Impey, S., and Storm, D. R. (1998). Light and circadian rhythmicity regulate MAP kinase activation in the suprachiasmatic nuclei. *Nature Neurosci.* **1**, 693–700.

Paul, K. N., Fukuhara, C., Tosini, G., and Albers, H. E. (2003). Transduction of light in the suprachiasmatic nucleus: Evidence for two different neurochemical cascades regulating the levels of *PER1* mRNA and pineal melatonin. *Neuroscience* **119**, 137–144.

Pennartz, C. M. A., De Jeu, M. T. G., Geurtsen, A. M. S., Sluiter, A. A., and Hermes, M. L. H. J. (1998). Electrophysiological and morphological heterogeneity of neurons in slices of rat suprachiasmatic nucleus. *J. Physiol.* **506**(1), 775–793.

Pickard, G. E., and Turek, F. W. (1985). Effects of partial destruction of the suprachiasmatic nuclei on two circadian parameters: Wheel-running activity and short-day induced testicular regression. *J. Comp. Physiol. A* **156**, 803–815.

Piggins, H. D., and Cutler, D. J. (2003). The roles of vasoactive intestinal polypeptide in the mammalian circadian clock. *J. Endocr.* **177**, 7–15.

Quintero, J. E., Kuhlman, S. J., and McMahon, D. G. (2003). The biological clock nucleus: A multiphasic oscillator network regulated by light. *J. Neurosci.* **23**, 8070–8076.

Reppert, S. M., and Schwartz, W. J. (1984). The suprachiasmatic nuclei of the fetal rat: Characterization of a functional circadian clock using $^{14}$C-labeled deoxyglucose. *J. Neurosci.* **4**, 1677–1682.

Romijn, H. J., Sluiter, A. A., Pool, C. W., Wortel, J., and Buijs, R. M. (1996). Differences in colocalization between Fos and PHI, GRP, VIP and VP in neurons of the rat suprachiasmatic nucleus after a light stimulus during the phase delay versus the phase advance period of the night. *J. Comp. Neurol.* **372**, 1–8.

Saeb-Parsy, K., and Dyball, R. E. J. (2003). Defined cell groups in the rat suprachiasmatic nucleus have different day/night rhythms of single-unit activity *in vivo. J. Biol. Rhythms* **18**, 26–42.

Schaap, J., Albus, H., vanderLeest, H. T., Eilers, P. H. C., Détári, L., and Meijer, J. H. (2003). Heterogeneity of rhythmic suprachiasmatic nucleus neurons: Implications for circadian waveform and photoperiodic encoding. *Proc. Natl. Acad. Sci. USA* **100**, 15994–15999.

Schwartz, W. J., Carpino, A., Jr., de la Iglesia, H. O., Baler, R., Klein, D. C., Nakabeppu, Y., and Aronin, N. (2000). Differential regulation of *fos* family genes in the ventrolateral and dorsomedial subdivisions of the rat suprachiasmatic nucleus. *Neuroscience* **98**, 535–547.

Schwartz, W. J., and Gainer, H. (1977). Suprachiasmatic nucleus: Use of $^{14}$C-labeled deoxyglucose uptake as a functional marker. *Science* **197**, 1089–1091.

Schwartz, W. J., Lydic, R., and Moore-Ede, M. C. (1987). *In vivo* metabolic activity of the suprachiasmatic nuclei: Non-uniform intranuclear distribution of $^{14}$C-labeled deoxyglucose uptake. *Brain Res.* **424**, 249–257.

Shearman, L. P., Zylka, M. J., Weaver, D. R., Kolakowski, L. F., Jr., and Reppert, S. M. (1997). Two *period* homologs: Circadian expression and photic regulation in the suprachiasmatic nuclei. *Neuron* **19**, 1261–1269.

Shibata, S., Liou, S. Y., Ueki, S., and Oomura, Y. (1984). Influence of environmental light-dark cycle and enucleation on activity of suprachiasmatic neurons in slice preparations. *Brain Res.* **302**, 75–81.

Shinohara, K., Honma, S., Katsuno, Y., Abe, H., and Honma, K. (1995). Two distinct oscillators in the rat suprachiasmatic nucleus *in vitro. Proc. Natl. Acad. Sci. USA* **92**, 7396–7400.

Shinohara, K., Tominaga, K., Isobe, Y., and Inouye, S. (1993). Photic regulation of peptides located in the ventrolateral subdivision of the suprachiasmatic nucleus of the rat: Daily variations of vasoactive intestinal polypeptide, gastrin-releasing peptide, and neuropeptide Y. *J. Neurosci.* **13,** 793–800.

Silver, R., Romero, M.-T., Besmer, H. R., Leak, R., Nunez, J. M., and LeSauter, J. (1996). Calbindin-D$_{28K}$ cells in the hamster SCN express light-induced Fos. *NeuroReport* **7,** 1224–1228.

Stephan, F. K., and Zucker, I. (1972). Circadian rhythms in drinking behavior and locomotor activity of rats are eliminated by hypothalamic lesions. *Proc. Natl. Acad. Sci. USA* **69,** 1583–1586.

Sumová, A., Trávníčková, Z., Mikkelsen, J. D., and Illnerová, H. (1998). Spontaneous rhythm in c-Fos immunoreactivity in the dorsomedial part of the rat suprachiasmatic nucleus. *Brain Res.* **801,** 254–258.

Sun, Z. S., Albrecht, U., Zhuchenko, O., Bailey, J., Eichele, G., and Lee, C. C. (1997). RIGUI, a putative mammalian ortholog of the *Drosophila period* gene. *Cell* **90,** 1003–1011.

Tei, H., Okamura, H., Shigeyoshi, Y., Fukuhara, C., Ozawa, R., Hirose, M., and Sakaki, Y. (1997). Circadian oscillation of a mammalian homologue of the *Drosophila period* gene. *Nature* **389,** 512–516.

Tominaga, K., Inouye, S. T., and Okamura, H. (1994). Organotypic slice culture of the rat suprachiasmatic nucleus: Sustenance of cellular architecture and circadian rhythm. *Neuroscience* **59,** 1025–1042.

van den Pol, A. N. (1980). The hypothalamic suprachiasmatic nucleus of rat: Instrinsic anatomy. *J. Comp. Neurol.* **191,** 661–702.

van den Pol, A. N., and Powley, T. (1979). A fine-grained anatomical analysis of the role of the rat suprachiasmatic nucleus in circadian rhythms of feeding and drinking. *Brain Res.* **160,** 307–326.

Van Gelder, R. N., Herzog, E. D., Schwartz, W. J., and Taghert, P. H. (2003). Circadian rhythms: In the loop at last. *Science* **300,** 1534–1535.

Welsh, D. K., Logothetis, D. E., Meister, M., and Reppert, S. M. (1995). Individual neurons dissociated from rat suprachiasmatic nucleus express independently phased circadian firing rhythms. *Neuron* **14,** 697–706.

Wray, S., Castel, M., and Gainer, H. (1993). Characterization of the suprachiasmatic nucleus in organotypic slice explant cultures. *Micros. Res. Tech.* **25,** 46–60.

Yamaguchi, S., Isejima, H., Matsuo, T., Okura, R., Yagita, K., Kobayashi, M., and Okamura, H. (2003). Synchronization of cellular clocks in the suprachiasmatic nucleus. *Science* **302,** 1408–1412.

Yan, L., and Okamura, H. (2002). Gradients in the circadian expression of *Per1* and *Per2* genes in the rat suprachiasmatic nucleus. *Eur. J. Neurosci.* **15,** 1153–1162.

Yan, L., and Silver, R. (2002). Differential induction and localization of *mPer1* and *mPer2* during advancing and delaying phase shifts. *Eur. J. Neurosci.* **16,** 1531–1540.

Yan, L., Takekida, S., Shigeyoshi, Y., and Okamura, H. (1999). *Per1* and *Per2* gene expression in the rat suprachiasmatic nucleus: Circadian profile and the compartment-specific response to light. *Neuroscience* **94,** 141–150.

Zoeller, R. T., Broyles, B., Earley, J., Anderson, E. R., and Albers, H. E. (1992). Cellular levels of messenger ribonucleic acids encoding vasoactive intestinal peptide and gastrin-releasing peptide in neurons of the suprachiasmatic nucleus exhibit distinct 24-hour rhythms. *J. Neuroendocrinol.* **4,** 119–124.

# Section V

# Mosaic Circadian Systems

## [23]  Transplantation of Mouse Embryo Fibroblasts: An Approach to Study the Physiological Pathways Linking the Suprachiasmatic Nucleus and Peripheral Clocks

*By* Sehyung Cho, Irene Yujnovsky, Masao Doi, and Paolo Sassone-Corsi

Abstract

One of the unresolved issues in the field of circadian biology is dissection of the communication pathways between central and peripheral oscillators. We have developed an experimental procedure in which an implant of mouse embryo fibroblasts of a specific genotype can be successfully grafted into a host animal of a different genotype. This methodology provides an excellent tool to study how peripheral clocks are entrained under various physiological settings and the contribution of individual signaling effectors in this process.

Signaling Pathways and Peripheral Clock

The understanding of the physiological and functional relationship between central and peripheral clocks is essential in circadian biology. The question of how and which signals emanating from the suprachiasmatic nucleus (SCN) may reach other brain regions and/or peripheral tissues is still unresolved. In addition, SCN-derived signals may operate with a somewhat different timing on diverse target peripheral cells. This concept and additional experimental evidence have inspired the interpretation that the SCN, rather than being the unique pacemaker, acts as an "orchestra's conductor" implicated in the synchronization of a web of peripheral tissues (Schibler and Sassone-Corsi, 2002). Circadian clocks are not only operative in peripheral cells of intact organisms, but even exist in immortalized zebrafish (Pando *et al.*, 2001) and mammalian tissue culture cells (Balsalobre *et al.*, 2000). Remarkably, in zebrafish cells the phase can directly be entrained by light–dark cycles (see Hirayama *et al.*, 2005). The finding of daily oscillators in cells kept *ex vivo* suggests that peripheral tissue clocks can manifest a remarkable independence from the so-called master clocks. For example, in *Drosophila* the daily fluctuations in the sensitivity of olfaction can be recorded in chemosensory cells of severed antennae kept in organ culture (Krishnan *et al.*, 1999). Various zebrafish organs placed in culture maintain the original rhythmicity if kept in constant darkness but can be entrained to any phase by appropriate light–dark cycles (Whitmore *et al.*, 2000).

We have developed an experimental procedure in the mouse that allows the study of the physiological links that exist between the SCN and peripheral clocks (Pando *et al.*, 2002). Taking advantage of the drastically shortened period (20 h) that *Per1*-deficient mouse embryo fibroblasts (MEFs) display when placed in culture, we investigated the functional dependence of peripheral clocks on the central pacemaker. We surgically implanted *Per1*-deficient MEFs encapsulated in a collagen disc into a wild-type host animal. Under normal physiological conditions, the SCN is able to rescue the altered rhythmicity of the implanted *Per1*-deficient MEFs, whose period returned to 24 h within only one circadian cycle. Thereby the SCN seems to be able to compensate for genetic defects affecting the period of peripheral clocks. The adoption by MEFs of the phenotypic characteristics of the host SCN illustrates the striking hierarchical dominance existing between central and peripheral clocks in mammals.

This method was applied to various host mice and MEFs carrying different types of targeted mutations (Pando *et al.*, 2002), stressing its potential use in a variety of settings. Indeed, it is envisageable to use MEFs derived from mice with mutations in any signaling molecule, including membrane receptors, intracellular transduction adaptors, kinases, phosphatases, and nuclear effectors. Similarly, the host mice could also be mutants for components of various signaling pathways. Host mice could be placed under different lighting conditions or given a light pulse during their subjective night.

In addition, the impact of food entrainment on peripheral clocks is an important aspect that can be addressed by employing this technique. It has been demonstrated that restricting feeding to daytime is able to uncouple circadian oscillators in peripheral tissues from the central pacemaker in the SCN (Damiola *et al.*, 2000). Indeed, we applied food restriction to mice that had been implanted with MEFs and found that although food is relatively a poor Zeitgeber for implanted MEFs, it can actually entrain, to some extent, the clock of the MEF (Pando *et al.*, 2002). Thus, it would be possible to feed mice with various diets that could provide the possiblity to study the interactions between metabolism and the circadian clock. This methodology thereby constitutes an attractive opportunity to unravel the physiological connections between the SCN and peripheral clocks.

Outline of the Procedure

To study peripheral clocks *in vivo* and to examine the contribution of individual genes in peripheral clock entrainment, we established a method to transplant MEFs of one genotype into the living animal of another genotype (Pando *et al.*, 2002). As outlined in Fig. 1, MEFs of a specific genotype are obtained from a 13.5-day *postcoitum* (*dpc*) embryo whose genotype is determined either by parents (when both parents are homozygotes for the given allele) or by genotyping the established MEFs (when any of the parents is heterozygote). After propagating MEFs for several passages, condensed collagen matrices embedded with MEFs are generated overnight and grafted subcutaneously into recipient male mice whose genotype was selected on the basis of the biological question to be addressed. The genotype of the host mouse can be different from the grafted MEFs. The recipient males are entrained accordingly before and after transplantation, depending on the experimental design. After circadian manipulations of the graft-harboring animals for several days, MEF–collagen implants are recovered and processed for analyses. These may include scoring for gene expression of circadian genes, levels of specific proteins involved in intracellular signaling, and their possible posttranslational modifications. At the same time, the SCN and other peripheral tissues of the host mouse can be collected and analyzed for comparison with the implant. This will allow one to determine the contribution of individual genes in the process of peripheral clock entrainment.

### Preparation of Mouse Embryo Fibroblasts from Single Mouse Embryo

Fibroblasts probably represent the easiest cell type to grow in culture and have been widely used for various molecular and cellular studies. Because fibroblasts in culture exhibit circadian rhythmicity of gene expression when given proper stimuli (Balsalobre *et al.*, 1998), they constitute an excellent model system to study molecular clocks *in vitro*. The procedure described here constitutes an extension of the use of fibroblasts to the study of peripheral clocks *in vivo*. To analyze the role of individual clock or clock-related genes in peripheral clock entrainment, the use of MEFs derived from wild-type or mutant embryo is quite advantageous. The method described here to obtain primary MEFs from single embryo is simple and reliable. Experimentally, MEFs are generated by mechanical dissociation so that both wild-type and mutant MEFs can be obtained simultaneously from the same littermates by crossing of heterozygotic parents. MEFs can also be obtained by enzymatic digestion as described in detail elsewhere (Spector *et al.*, 1998).

### Animals

Prepare pregnant female mice at 13.5 *dpc* as follows: Set up a couple of mating cages with healthy male and female mice of known genotype as needed and in accordance to the guidelines from your institution. From the

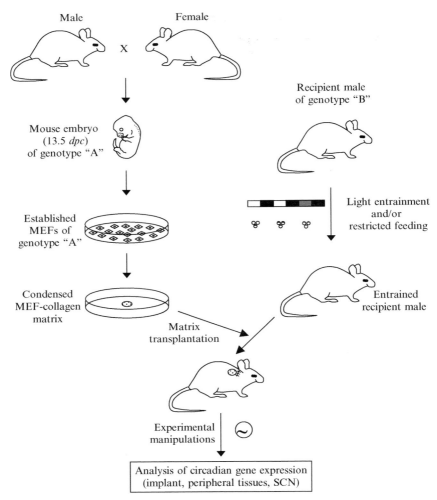

FIG. 1. Schematic outline of the MEF transplantation procedure used to study peripheral clocks in the mouse. Primary MEFs of desired genotype (designated as "A") are obtained from a 13.5-*dpc* embryo and condensed to a tissue-like structure in the presence of concentrated collagen fibers. The resulting disc is grafted into the back of an entrained recipient male mouse of possibly different genotype (designated as "B"). After circadian manipulation of the graft-harboring animal for several days, the collagen disc is recovered and analyzed for the circadian gene expression. Central and peripheral clock gene expression of the recipient animal can be examined simultaneously.

following day on, check for vaginal plug every morning and transfer plug-positive females into new breeding cages. The first day in which the vaginal plug is observed is regarded as 0.5 day *postcoitum*. Breed the plug-positive female mice for 13 more days (13.5 *dpc*).

*Materials*

Prepare sterilized surgical instruments (scissors, forceps, blades, etc.), 70% alcohol swab, phosphate-buffered saline (PBS), sterile 18-gauge needle (1.2 × 40 mm, Terumo Europe), sterile 5-ml syringe (Terumo Europe), culture medium [Dulbecco's modified Eagles medium (DMEM) supplemented with 10% fetal calf serum and gentamicin], 75-cm$^2$ flasks, culture dishes, and pipettes.

*Procedure*

1. Sacrifice a 13.5-*dpc* pregnant female mouse by cervical dislocation, and swab the entire animal immediately with 70% alcohol in a sterile hood.

2. With sterile forceps and scissors, make a horizontal incision across the abdomen under the forelegs and pull the skin down to the back legs with forceps.

3. Using another set of sterile scissors and forceps, cut open the abdomen and remove the uterus containing the embryos to a sterile 10-cm dish.

4. Cut open the membrane surrounding the embryos and transfer the embryos to a fresh 10-cm cell culture dish containing sterile PBS.

5. Rinse the embryos several times with PBS until the solution is clear.

6. Transfer one embryo to a new dish. With fine scissors and forceps, remove head and liver from the embryo. (This step can be omitted, as only fibroblasts can survive the subsequent subculture conditions.)

7. Rinse the embryo in sterile PBS and transfer into a sterile 5-ml syringe with an 18-gauge needle.

8. Add 1 ml of sterile PBS and then carefully replace the piston (the tip of the needle should be pointed toward a tilted 35-mm dish to collect the leaking PBS).

9. Pass the embryonic tissues five times through the needle into the tilted 35-mm dish.

10. Transfer the entire cell suspension to a 75-cm$^2$ flask containing 15 ml of culture medium and incubate at 37° under 5% $CO_2$ tension.

11. One day later, discard the culture medium from the flask and wash twice with sterile PBS to remove unattached cells or floating cell clumps.

Replace with fresh medium and further incubate for 2 more days. Usually at this time point, cells reach confluence. Otherwise, replace with fresh medium and grow cells further until they reach confluence.

12. Split cells evenly into three 75-cm² flasks by conventional trypsinization using 0.25% trypsin. Continue splitting and propagating cells at this dilution (1:3) until enough cells are obtained for the designed experiment.

## Preparation of Condensed MEF–Collagen Matrix

Fibroblasts are known to condense a hydrated collagen lattice to a tissue-like structure (Bell *et al.*, 1979) in a way that resembles *in vivo* wound contraction when cells interact with protein fibers, mainly collagen (Grinnell, 1994). The resulting three-dimensional collagen matrices embedded with fibroblasts are quite rigid, enough for further manipulations, such as transplantation into the back of the mouse (Pando *et al.*, 2002). The fibroblast–collagen matrices have been used successfully to study interactions that occur between cells and surrounding matrix and are relatively well characterized (reviewed in Grinnell, 2003). The rate of collagen matrix

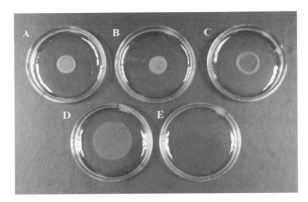

FIG. 2. A representative picture showing MEF–collagen matrix condensation occurring in 60-mm dishes with a different number of cells. MEFs were resuspended to $3.4 \times 10^6$ (A), $1.4 \times 10^6$ (B), $5.6 \times 10^5$ (C), or $2.2 \times 10^5$ (D) cells/ml as described in step 2 of the procedure (see text) and subjected to matrix condensation. Without any cells (E), no condensation was observed. The rate of matrix condensation clearly depends on the number of cells used: with recommended cell density (A), a clear collagen disc (around 1.5 cm in diameter) formed within 12 h of incubation, while collagen matrices were still contracting when smaller numbers of cells were adopted (B–D). This picture was taken after 24 h of incubation for comparison.

condensation is mainly a function of collagen content and cell number (Bell *et al.*, 1979; Fig. 2). The size of the condensed matrix also depends on the size of the dish used. Here we present a protocol suitably modified from the original method (Palmer *et al.*, 1989; St. Louis and Verma, 1988) to be fit for the MEF transplantation experiments. The presented protocol gives condensed MEF–collagen matrix rigid enough for further manipulations within 12 h of incubation.

*Cells, Materials, and Solutions*

Prepare sufficient MEFs of known genotype as needed, 3 mg/ml collagen (type I from rat tail, Sigma-Aldrich Chemie GmbH, C7661, dissolved to 3 mg/ml in filter-sterilized 0.1 $N$ acetic acid overnight at 4°), 0.25% trypsin, complete medium (DMEM supplemented with 10% fetal calf serum), 2× concentrated DMEM without additives, fetal calf serum, 0.1 $N$ NaOH (freshly prepared from 10 $N$ stock and filter sterilized), 74 g/L $KHCO_3$ (filter sterilized), 60-mm dishes, hemocytometer, inverted microscope, and 37° water bath.

*Procedure*

1. Prewarm the solutions to near 37° in a water bath before starting.
2. Trypsinize and resuspend MEFs in a small volume of complete medium. Count cells using a hemocytometer and adjust cell density to $3.4 \times 10^6$ cells/ml with complete medium. (Take the following steps to make one implant. Each implant should be made separately. Keep the addition order.)
3. Into a sterile 50-ml tube, add 1.5 ml of 2× concentrated DMEM.
4. Add 1.5 ml of fetal calf serum to the tube and mix by swirling the tube gently.
5. Add 1.5 ml of 3 mg/ml collagen to the tube and mix immediately by swirling the tube.
6. Add 250 $\mu$l of 0.1 $N$ NaOH and mix immediately by swirling the tube.
7. Add 150 $\mu$l of 74 g/L $KHCO_3$ and mix well as described earlier.
8. Using a 10-ml pipette, take 1.5 ml of MEF suspension prepared in step 2 and add to the tube containing collagen mixture. Flush the entire suspension twice through the 10-ml pipette and pour into a 60-mm dish.
9. Incubate the dish untouched at 37° under 5% $CO_2$ tension overnight.
10. On the following morning, check for the formation of a condensed collagen disc. The resulting disc should be around 1.5 cm in diameter and will be floating in the center of the dish.

Subcutaneous Implantation Procedure

We transplant the resulting MEF collagen disc subcutaneously into the back of the recipient animal because this operation is very simple, easy, and reliable. Depending on the experimental purpose, other places could be considered as transplantation sites.

*Recipient Animals*

For light entrainment, the recipient adult male mice of desired genotype are housed individually on a 12-h light:12-h dark (LD) cycle for at least 2 weeks prior to implantation. Except for experiments in LD, mice are placed in constant darkness (DD) for 3 days after 2 weeks of light entrainment. For the food entrainment study, food availability is restricted to the daytime of a LD cycle for 10 days prior to transplantation. The operation is performed between ZT1 and ZT3 on the day of operation.

*Materials and Solutions*

Prepare MEF collagen discs, ketamine/Rompun working solution (mix 1 ml of ketamine 1000, 0.6 ml of Rompun 2%, and 4 ml of physiological saline before use), suturing strings and needles, complete medium (DMEM supplemented with 10% FCS), PBS, Vetedine gauze, sterilized scissors and forceps, and electronic shaver.

*Procedure*

1. Collect collagen discs in 50-ml tubes containing 30 ml of complete medium. Prepare three tubes of 50 ml PBS.
2. Anesthetize the adult male mouse with ketamine/Rompun working solution (*ip* injection, averaging 0.2 ml of the working solution per 30-g mouse).
3. Shave a small patch below the head on the back. Wipe three times with Vetedine gauze. Use new gauze for each wiping.
4. Using scissors and forceps, make a small incision behind the neck and open a pocket under the skin using sterile scissors.
5. With blunt-ended forceps, pick one implant and rinse it sequentially in three tubes of PBS. Drain the implant gently.
6. Slide the implant into the pocket under the skin and suture the cut with three to four stitches as follows.

    a. Wind the string around the forceps twice in one direction. Take the end of the string with the forceps and pull gently.

b. Wind the string around the forceps only one time in the opposite direction with respect to the first step. Take the end of the string with the forceps and pull gently.

c. Wind the string around the forceps only one time in the same direction as in the first step. Take the end of the string with the fingers and close it gently.

7. Place the mouse under a lamp until it wakes up and perform circadian manipulations as planned.

### Analysis of Circadian Gene Expression by RNase Protection Assay and *In Situ* Hybridization

After circadian manipulations of the graft-harboring animals, the MEF–collagen implants are recovered from the animal at the desired time points. For comparison, other peripheral tissues (such as liver, kidney, or skeletal muscle), as well as whole mouse brain (to examine clock gene expression in the SCN), should be collected simultaneously from the same animal. For RNA isolation, the implant is directly harvested into 3 ml of Trizol (or equivalent) broken up by flushing with a P1000 for a few minutes, and processed further following the standard protocol. Peripheral tissues and whole mouse brain are dissected out and frozen rapidly on dry ice for further investigation. These may involve procedures for a RNase protection assay (Macho and Sassone-Corsi, 2003) and *in situ* hybridization (Crosio *et al.*, 2000), which have been described previously. Additional analyses may include the study of protein levels and related posttranslational modifications using appropriate antibodies.

### References

Balsalobre, A., Damiola, F., and Schibler, U. (1998). A serum shock induces circadian gene expression in mammalian tissue culture cells. *Cell* **93,** 929–937.

Balsalobre, A., Marcacci, L., and Schibler, U. (2000). Multiple signaling pathways elicit circadian gene expression in cultured Rat-1 fibroblasts. *Curr. Biol.* **10,** 1291–1294.

Bell, E., Ivarsson, B., and Merrill, C. (1979). Production of a tissue-like structure by contraction of collagen lattices by human fibroblasts of different proliferative potential *in vitro*. *Proc. Natl. Acad. Sci. USA* **76,** 1274–1278.

Crosio, C., Cermakian, N., Allis, C. D., and Sassone-Corsi, P. (2000). Light induces chromatin modification in cells of the mammalian circadian clock. *Nature Neurosci.* **3,** 1241–1247.

Damiola, F., Le Minh, N., Preitner, N., Kornmann, B., Fleury-Oleda, F., and Schibler, U. (2000). Restricted feeding uncouples circadian oscillators in peripheral tissues from the central pacemaker in the suprachiasmatic nucleus. *Genes Dev.* **14,** 2950–2961.

Grinnell, F. (1994). Fibroblasts, myofibroblasts, and wound contraction. *J. Cell Biol.* **124,** 401–404.

Grinnell, F. (2003). Fibroblast biology in three-dimensional collagen matrices. *Trends Cell Biol.* **13**, 264–269.

Hirayama, J., Kaneko, M., Cardone, L., Cahill, G., and Sassone-Corsi, P. (2005). Analysis of circadian rhythms in the zebrafish. *Methods Enzymol.* **393**, Chapter 5.

Krishnan, B., Dryer, S. E., and Hardin, P. E. (1999). Circadian rhythms in olfactory responses of *Drosophila melanogaster. Nature* **400**, 375–378.

Macho, B., and Sassone-Corsi, P. (2003). Functional analysis of transcription factors CREB and CREM. *Methods Enzymol.* **370**, 396–415.

Palmer, T. D., Thompson, A. R., and Miller, A. D. (1989). Production of human factor IX in animals by genetically modified skin fibroblasts: Potential therapy for hemophilia B. *Blood* **73**, 438–445.

Pando, M. P., Morse, D., Cermakian, N., and Sassone-Corsi, P. (2002). Phenotypic rescue of a peripheral clock genetic defects via SCN hierachical dominance. *Cell* **110**, 107–117.

Pando, M. P., Pinchak, A. B., Cermakian, N., and Sassone-Corsi, P. (2001). A cell-based system that recapitulates the dynamic light-dependent regulation of the vertebrate clock. *Proc. Natl. Acad. Sci. USA* **98**, 10178–10183.

Schibler, U., and Sassone-Corsi, P. (2002). A web of circadian pacemakers. *Cell* **111**, 919–922.

Spector, D. L., Goldman, R. D., and Leinwand, L. A. (1998). "Cells: A Laboratory Manual," Vol. 1, pp. 4.1–4.7. Cold Spring Harbor Laboratory Press, Cold Spring Harbor, NY.

St. Louis, D., and Verma, I. M. (1988). An alternative approach to somatic cell therapy. *Proc. Natl. Acad. Sci. USA* **85**, 3150–3154.

Whitmore, D., Foulkes, N. S., and Sassone-Corsi, P. (2000). Light acts directly on organs and cells in culture to set the vertebrate circadian clock. *Nature* **404**, 87–91.

# [24] Mouse Chimeras and Their Application to Circadian Biology

*By* SHARON S. LOW-ZEDDIES and JOSEPH S. TAKAHASHI

## Abstract

Chimeric mice are versatile model systems for the study of mammalian circadian biology. In chimeras, genetically different cells are combined within single animals, making them useful for assessing how normal cells interact with genetically altered cells in intact biological systems. In particular, the primary circadian pacemaker in the suprachiasmatic nucleus is amenable to analysis using series of chimeras that incorporate cells carrying mutations in circadian genes. The study of chimeras carrying circadian mutations can contribute to a better understanding of the function of the altered genes and of the fundamental physiology of circadian timing. Chimera analysis is a valuable approach for studying network properties in complex, integrated biological systems like that which controls circadian behavior in mammals.

Introduction

The study of aggregation chimeric mice is a historically embryological technique that we have adopted to study circadian rhythmicity in mammals (Low-Zeddies and Takahashi, 2001). A mouse chimera is a genetically composite animal whose cell populations are derived from two different embryos. Aggregation chimeric mice are made by bringing two embryos together *in vitro* and allowing them to fuse together and then allowing the composite embryo to complete development. Chimeras can be made to contain a combination of wild-type and mutant cells (whether chemically induced, spontaneous, or the result of genetic engineering). Mutants are useful resources for studying physiology in that genetic perturbations enable one to better understand normal function (Benzer, 1973). A series of chimeric mice that incorporates a given mutant line enhances its utility and extends the information that can be obtained from the mutation. By studying chimeras in which normal cells are combined with cells carrying a mutation that affects circadian function, one potentially addresses two fundamental questions: what is the function of the altered gene and how does altering the function of the gene affect the physiology of the circadian system? Using chimeric mice, these two basic questions can be addressed at multiple levels of biological organization: spatially, from the cell to the tissue to whole animal behavior, and temporally, from the embryo to the adult. The physiological function of chimeric tissue can be studied *in vivo* in the intact organism or cultured *in vitro*.

It has become clear that most, if not all, cells in mammals are competent circadian oscillators. It is not yet well understood, however, how multi-oscillator networks are set up as tissues and organs that sustain and adapt their timing to each other and to the environment. In particular, in the suprachiasmatic nucleus (SCN), which functions as the mammalian central circadian pacemaker, the oscillations of thousands of individual cells are transformed into a coherent signal that controls the circadian rhythm of locomotor activity. Incorporating mutant cells into chimeras is a means to genetically dissect the system *in vivo* by effectively creating genetic lesions at the level of individual cells. By combining mutant with wild-type cells in the SCN, the experimenter alters the components of the pacemaking ensemble, potentially altering the properties of how the cells interact to produce circadian behavior. At the same time, because the genetically different cells are combined early—at embryonic day 2.5—a practically "seamless" coupling of cellular pacemakers is achieved in chimeric tissue. The intricate intercellular connectivity is preserved, and this is critical to the extraordinary complexity of this neural structure. In short, the study of a series of mutant chimeras, which we refer to as chimera analysis, is a versatile tool for studying the nature of network properties in integrated biological systems such as the circadian system.

Properties of Aggregation Chimeric Mice

Aggregation chimeric mice were first produced by Tarkowski (1961) and by Mintz (1962) as a means to study cell regulation during mammalian development. Each chimeric mouse must be generated experimentally— chimeras cannot be produced through breeding alone. When a normal embryo is combined with one carrying a particular mutation, the chimeric individual that results will contain a unique proportion and distribution of cells from its two component cell genotypes in all of its tissues. That is, across a population of chimeric mice, no two are the same. Distribution of the component cell types in chimeras is essentially random. Furthermore, an experimental population, or series, of aggregation chimeras will cover a 0–100% range of relative cell genotypic proportions (or, put another way, the dose of mutant cells varies), with all proportions being equally represented (Falconer and Avery, 1978). By using a reliable cell marker, one ensures that the genotype of each cell can be unambiguously identified in all tissues. A single set of chimeras can be used to study the effects of chimerism in tissues throughout the body. The proportions of the two cellular genotypes among different tissues and organs are highly correlated within each chimeric individual (Musci and Mullen, 1992). This characteristic, though, tends to interfere with efforts to determine a tissue focus of the mutation based on correlations between mutant phenotypes and composition of various tissues, a strategy used in studies of mosaic fruitflies, for example (Benzer, 1973). The two cellular genotypes are finely mixed in chimeric tissue in general (Dewey *et al.*, 1976; Oster-Granite and Gearhart, 1981), including the SCN (Low-Zeddies and Takahashi, 2001). The close apposition of the two cellular genotypes in chimeric tissue enhances the potential for their functional interaction.

Chimeras can be made using any mutant mouse strain, whether spontaneous, induced, gene targeted, or transgenic. Studying an experimental series of chimeric mice can extend the utility of any given mutant line: in contrast to studying animals that are wholly homozygous mutant, heterozygote, or wild type, each chimeric mouse is a new experimental system. The variety of configurations and proportions of mutant versus normal cells represented in a series of chimeric mice potentially reveals all of the possible structural and functional outcomes of the interactions between these different cellular genotypes.

A Study of Clock Chimeras Principles for Circadian Studies

We used a series of mouse chimeras to study the effects of allowing *Clock* mutant cells to interact with wild-type cells on the circadian rhythm of locomotor behavior (Low-Zeddies and Takahashi, 2001). The *Clock* gene regulates the intrinsic circadian period and the persistence of circadian rhythmicity (or circadian amplitude) in constant conditions (Vitaterna *et al.*, 1994). These two phenotypic traits are fundamental properties of the circadian clock system. We generated a population of 130 homozygous *Clock* aggregation chimeric mice that carried from 0 to 100% of *Clock* mutant versus wild-type cells. For each of these chimeric individuals, we analyzed the cellular composition of SCN tissue and assayed their circadian behavior (Low-Zeddies and Takahashi, 2001).

Across the varied population of *Clock* chimeric mice, we observed patterns of circadian behavior that spanned a range from normal to *Clock* mutant like. In general, the comprehensive range in relative mutant contribution from 0 to 100% across a series of chimeras is advantageous in that the widest potential range of phenotypes may be seen. Furthermore, this continuous distribution is advantageous for correlational analyses. In some cases, we observed phenotypic traits that were intermediate between those characteristic of wild type and mutant. For example, we documented stable circadian period lengths that fell between the 23.5- and the 28-h periods characteristic of the wild-type and *Clock* mutant strains (Vitaterna *et al.*, 1994). In general, we found that intermediate and novel behavioral profiles that were observed in some *Clock* mutant chimeras served as valuable analytical tools. The phenotypic gradient across the chimeric series tracked a progression of mutant severity, revealing how the circadian pacemaker mechanism can be broken down incrementally as the mutant cell dose is increased. Unexpectedly, we found that some of these *Clock* chimeras that contained only homozygous *Clock* mutant and wild-type cells could behave as though they were *Clock* heterozygotes. This result indicated that the interaction between wild-type and *Clock* mutant alleles at an intercellular level in chimeras can resemble allelic interaction at an intracellular level. Finally, our study of *Clock* chimeras demonstrated that the effect of the *Clock* mutation on circadian period and its effect on circadian amplitude are likely to arise from physiological processes that involve different subsets of cells because they did not always covary in individual chimeric mice. As a general principle, how mutant versus wild-type phenotypic traits covary across a series of chimeras is informative as to their underlying mechanism and can be used to discriminate separable phenotypic traits of a mutation. The use of *Clock* chimeras, then, allowed us to reveal new aspects of how the *Clock* gene affects circadian function at the cellular level and how individual cellular oscillators within the SCN functionally interact with one another to produce an integrated circadian rhythm of behavior.

## Using Chimeras to Study Intercellular Interactions in Circadian Systems

We have found that chimera analysis can be a generally useful approach to manipulate and then probe intercellular interaction in circadian physiology. Chimeras permit investigation into the interactions between genetically different cell populations and the effect of these interactions on organismal behavior. Chimeric mice also have the potential to be a powerful model system for the *in vitro* study of cell and tissue organization of the mammalian circadian system. Numerous lines of mice already exist in which genes implicated in circadian rhythms [e.g., the mouse *Cryptochrome* genes 1 and 2 (van der Horst *et al.*, 1999; Vitaterna *et al.*, 1999), the mouse *Period* genes 1, 2, and 3 (Bae *et al.*, 2001; Zheng *et al.*, 2001), BMAL 1/Mop3 (Bunger *et al.*, 2000), and *Clock* (Vitaterna *et al.*, 1994)] have been altered or knocked out and that show robust circadian behavioral phenotypes. All of these clock gene mutations, and those that continue to be produced, can be incorporated into experimentally informative chimeras. Our results outlined previously demonstrate how subtle aspects of the functional organization of the SCN can become apparent in chimeras.

### *Chimeras Can Be Used to Study Cell Interaction in the SCN*

Chimeric mice can be used to study the intercellular interactions that mediate the coupling among cellular oscillators and confer a specialized pacemaking function to the SCN. It is now well established that circadian periodicity is an intrinsic property of individual SCN cells (Herzog *et al.*, 1998; Liu *et al.*, 1997). The chimeric SCN is a potentially useful system for recording the electrical activity of single cells in slice culture preparations, as a direct method of determining whether the rhythmicity of wild-type SCN cells can be disrupted by the presence of mutant cells and whether the two can synchronize one another. If desired, chimeric combinations of two different circadian mutations could be used to study interactions between cells with different oscillatory properties, e.g., short and long circadian period combinations. By studying the interaction between mutant and wild-type cells in chimeric tissue, one might ask which genes, when mutated, affect coupling properties among SCN neurons or their entrainment by light stimuli. For example, using chimeras, one might test the role of genes that encode neurochemicals or receptors that are thought to mediate signaling by light and/or interneuronal communication within the SCN, such as the $VPAC_2$ receptor, or gastrin-releasing peptide (Aida *et al.*,

2002; Harmar *et al.*, 2002). An alternative assay to electrical recording of cellular activity in chimeric SCN preparations is to incorporate cells from one of the lines of mice now available that carry circadian reporter transgenes to visually observe the dynamic interactions between cells of contrasting genotype. Mice have been genetically engineered to produce green fluorescent protein (GFP) or luciferase (*luc*) that are driven to oscillate with the circadian rhythm of the molecular clockwork in individual cells (Kuhlman *et al.*, 2000; Wilsbacher *et al.*, 2002; Yoo *et al.*, 2004). These constructs that utilize mouse *Period* gene promoters are ubiquitously expressed and can be imaged in living tissue in real time.

### Chimeras Can Be Used to Study the Effects of Cell Interaction on Behavior

An important question in circadian biology is, what are the mechanisms by which the output signals from a multiplicity of individual oscillators interact to form a functional pacemaker at the tissue and whole-animal levels? The properties of the multiple levels of cell and tissue organization that underlie circadian rhythms in behavior are not likely to be resolved by studying the behavior of single-cell oscillators. Using chimeras, one can manipulate the genetic composition of the network of oscillators in the SCN and assess how different cellular circadian phenotypes are ultimately integrated into a coherent behavioral output (Low-Zeddies and Takahashi, 2001). The expression of a complex behavior such as circadian locomotor activity relies on an intricate network of connections and feedback; by using chimeras, the behavior can be studied in the intact nervous system.

### Chimeras Can Be Used to Study the Control of Circadian Rhythms in Peripheral Tissues

Similarly, chimeras may be used to study intercellular processes involved in conveying circadian timing signals from the SCN to other tissues and organs in the body (the periphery). The routes of communication from the SCN to the sites that generate locomotor activity behavior are thought to be complex and multimodal. Both neural mechanisms and diffusible signaling molecules have been implicated. Chimeras may be used to test the role of candidate molecules involved in output signaling from the SCN. For example, transforming growth factor-$\alpha$ has been identified as a candidate factor that mediates SCN output controlling the timing of locomotor behavior (Kramer *et al.*, 2001). Chimeric models carrying various proportions of cells in which molecules such as this are deficient may help define their role in the circadian control of behavior.

Finally, chimera analysis can help explore how cellular circadian phenotypes are integrated in peripheral tissues. It is known that fundamental properties of circadian rhythms in tissues throughout the body differ from those of the circadian oscillators in the master pacemaker in the SCN. For example, the phases of the liver, lung, and skeletal muscle tissues lag the SCN rhythm by 7–11 h and respond differently from the SCN to shifts in the light cycle (Yamazaki *et al.*, 2000). Isolated explants from peripheral tissues express a circadian rhythm in *Period2*-driven bioluminescence, suggesting that the cellular oscillators in these peripheral tissues are also coupled, although by what mechanism is not known (Yoo *et al.*, 2004). Chimeras can be used to study how the interactions between individual cellular oscillators in nonneural tissues in the periphery compared with intercellular interactions in the neural pacemaker in the SCN. Can network properties be identified that differentiate central from peripheral tissues and account for the uniqueness of the SCN master pacemaker?

## General Applications for Chimeras in Circadian Studies

### *Chimeras Can Reveal Gene Function at a Cellular Level*

Chimeras have been conventionally employed to determine whether a mutant cell can affect the phenotype of neighboring wild-type cells, i.e., whether a cellular phenotype is cell autonomous. This remains an important use of chimeras that can help define in which cells the expression of a circadian gene is functionally important, particularly when the gene is widely expressed. Moreover, effects of mutations can be mediated directly or indirectly. For example, molecules such as receptors, signal transducers, or transcription factors, although expressed within a cell, may perform regulatory functions that affect extracellular molecules or neural communication, which in turn influence fates or functions of other cells. Thus, in the postgenomic age, even though function is often predicted from gene sequence, chimera analysis is valuable for assessing the role of specific genes in circadian physiology at the cellular level. Chimeras have also traditionally been a tool for tracing cell lineages during development and for addressing how various mutations affect the ability of cells to participate in normal development.

### *Chimeras Can Rescue Cells Carrying Mutations that Cause Early Developmental Lethality*

Chimera analysis has taken on a new relevance with the popularity of gene knockout mice. In chimeras, the effects of phenomena that interfere with interpretation of the phenotype in a pure knockout animal may be alleviated. For example, the products of many genes are essential such that inactivating them results in lethality early in development, preventing further study of discrete behavioral effects. Given the pervasiveness of circadian rhythms in physiology and the importance of timing to developmental processes, one might expect that some proportion of genes that affect the circadian timing system will also have functions during development that are essential for viability. Cells carrying genetic alterations that cause early organismal lethality can often be rescued in a chimeric setting (e.g., Barsh *et al.*, 1990; Campbell and Peterson, 1992). That is, disabling gene function in only a subset of cells within the organism may permit survival to a later developmental stage or adulthood, allowing examination of later roles for the gene in question. It can also be determined what proportion of cells and where they must be located to rescue normal function.

From a practical standpoint, when a mutation causes early lethality, the embryos used to produce homozygous mutant chimeras must come from heterozygous mutant crosses. It then becomes necessary to adopt a strategy to be able to determine the genotype of the mutant portion of the resulting chimeras because both homozygous mutant and heterozygous chimeric tissues will contain both mutant and wild-type alleles. For example, one may incorporate either two different wild-type alleles or two distinct mutant alleles in the component heterozygous embryos that can be distinguished by DNA genotyping of the resulting chimeras. Alternatively, it is possible, though challenging, to microdissect out single mutant cells from tissue in the chimera (identified by cell marker) to determine whether the cells are homozygous or heterozygous for the mutant allele. Another route would be to generate embryonic stem (ES) cell chimeras using known homozygous ES cells.

## Chimeras Can Clarify the Mechanisms behind Pleiotropic Effects of a Mutation

In some cases, knocking out certain genes in mice results in no detectable phenotype due to gene redundancy mechanisms, or the genetic manipulation produces multiple phenotypic effects that make it difficult to discriminate separate roles for the gene product. Chimera analysis can provide valuable perspective to the study of a mutant phenotype. It is now well known that observed effects of a targeted mutation may not be directly related to the function of the gene *per se* due to redundancy of function among members of a gene family, alterations in the regulation of other genes, and other indirect effects. Combining mutant cells with wild-type cells in chimeras may help tease out phenotypic effects resulting from compensatory processes that obscure interpretation of the normal function of the gene. Chimera analysis can potentially clarify and separate the effects of a given genetic manipulation. If based on different underlying cell physiological processes, the pleiotropic effects of the mutation may be dissociated across a series of mutant-normal chimeras. Conversely, by assessing their covariance across a series of chimeras, it can be determined whether different traits share a common controlling cell population.

### Chimeras Can Serve as Models for Circadian Dysfunction

Finally, chimeras that incorporate clock gene mutations may be used to model circadian dysfunction and how the loss of circadian control contributes to disease states. A gradient of chimeras may be able to mimic the breakdown in circadian organization that occurs with aging, for example.

## Technique

### Chimera Production

The production of mouse chimeras is a straightforward technique that reliably yields viable animals. Methods of aggregating two eight-cell embryos (morulae) are standard (Hogan *et al.*, 1994) (Fig. 1). Superovulation

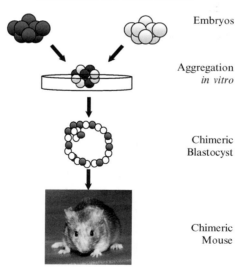

FIG. 1. A schematic diagram of the method used to create a chimeric mouse. Two genetically different 2.5-day embryos are denuded and then aggregated together in a dish. The embryos fuse together in culture overnight. The resulting chimeric blastocyst is transferred into a foster mother, where it completes development into a chimeric mouse.

is induced hormonally in female mice of both wild-type and mutant strains. Each female is then mated with a stud male of a matching genotype. On embryonic day 2.5 (vaginal plug = embryonic day 0.5), morulae are flushed from dissected oviducts into M2 mouse medium and then incubated briefly in acidic Tyrode's solution to remove the zonae pellucidae. Pairs of embryos are then pushed together and cocultured overnight in CZB$^+$ or KSOM medium at 37°, 5% $CO_2$. The following day, the aggregated embryos are transferred surgically into the uterine homs of 2.5-day pseudopregnant foster mothers where they finish development. Pseudopregnant recipient females are produced by mating naturally cycling females with vasectomized males. Chimeric pups are identifiable by the presence of variegated coat and eye pigmentation when pigmentation is used as a marker to differentiate wild-type from mutant parental strains. Coat color is a reliable indicator of the presence of chimerism in the central nervous system (CNS) because both CNS and coat pigment cells arise from the neural crest (Rawles, 1947). Nonchimeric littermates serve as useful controls to verify the lack of effect of the marker strain and to control for embryo manipulation.

## Cell Markers

To analyze the tissues of chimeric mice, it is essential to have a reliable, independent means of distinguishing mutant from wild-type cells *in situ*—a cell marker. The applicability of chimera analysis has been enhanced by the advent of strains of transgenic mice that express genes such as *LacZ* or GFP that are detected easily at the single-cell level and can serve as markers for identifying cell genotypes (Hadjantonakis and Nagy, 2001; Zambrowicz *et al.*, 1997). For example, the ROSA26 strain is ideal for chimera analysis studies. ROSA26 constitutively expresses the $\beta$-galactosidase enzyme in all of its cells (Zambrowicz *et al.*, 1997) so that ROSA26 cells can be used as the marked wild-type component or ROSA26 can be crossed with the mutant strain to produce marked mutant cells. For *LacZ* detection in ROSA26 brain tissue, sections are incubated for 24 h at 37° in an X-gal staining solution containing 1 mg/ml X-gal (5-bromo-4-chloro-3-indolyl-$\beta$-D-galactoside dissolved in dimethyl sulfoxide), 5 m$M$ $K_3Fe(CN)_6$, and 5 m$M$ $K_4Fe(CN)_6$ in a phosphate-buffered saline-based wash buffer. Alternatively, living chimeric tissues can be imaged that incorporate strains genetically engineered to produce fluorescent proteins or luciferase (Hadjantonakis and Nagy, 2001; Kuhlman *et al.*, 2000; Wilsbacher *et al.*, 2002; Yoo *et al.*, 2004). Using powerful software applications now available, it is possible to reconstruct serial sections into three-dimensional images that can help in visualizing the distribution of mutant versus wild-type cells in chimeric tissue.

## Statistical Analysis of Chimera Data

Analysis of a series of chimeric mice yields data amenable to statistical analyses. With the potential for a phenotypic gradient across a population of chimeras, it is important to use a sensitive, quantitative scoring method, such as period length or a Fourier analysis measure of circadian amplitude. For a chimera analysis to be informative requires studying a sufficient number of chimeras to cover a comprehensive phenotypic range. Statistically significant numbers of animals necessary for a given analysis will depend on the complexity of the phenotype and its genetic control, which should be determined after analyzing an initial number of chimeras. In the analysis of complex behavioral data, statistical power will increase with the number of chimeric individuals studied.

Complex biological phenomena such as circadian behavior are described most effectively by multiple quantitative measures. Multivariate statistical analyses can simplify, organize, and reveal structure in large behavioral and anatomical data sets, as are generated by large populations of chimeras. These tools have allowed us to test hypotheses about how the relationships between variables reveal SCN functional organization (Low-Zeddies and Takahashi, 2001). We found principal components and cluster analyses useful for evaluating relationships among the period, amplitude, and SCN scores in *Clock* chimeras and to facilitate comparison of the multidimensional behavior of chimeras with that of the control genotypic groups. A principal components analysis yields a unique solution of weighted linear composites of the observed variables that can reduce a multivariate data set to fewer components, making data easier to visualize and understand. Cluster analysis is a procedure for detecting natural groupings in data. The method is based on measures of dissimilarity between objects, expressed as distances in a multidimensional space (defined by the number of variables taken into account). We have found that clustering algorithms can define inherent structure in complex behavioral data and be of heuristic value for comparing multidimensional behavioral profiles.

## Other Chimeric or Mosaic Mouse Models

There are other methods of generating chimeric (carrying cells that differ in genomic content) or mosaic (all cells carry the same genomic content, but genes are activated differentially in different cell populations) mice. Technically, the preimplantation stage is the only feasible time to routinely manipulate mammalian embryos. Preimplantation chimeric mice can also be generated by injecting cultured ES cells into blastocysts. Unlike aggregation chimeras, however, which show a comprehensive range of mutant contribution to all tissues, the composition of blastocyst injection chimeras may be biased systematically according to characteristics of the ES cell line (Berger *et al.*, 1995; Ioffe *et al.*, 1995). The outcome of producing chimeras by morula aggregation compared with ES cell injection is that, across a series of chimeras, those made by aggregation show a more variable mutant contribution to all tissues. For chimeras made using ES cells, the hard work is primarily in the production and maintenance of the ES cell cultures. For morula aggregation chimeras, because the embryo "culture" occurs in the living mouse, the bulk of the effort goes toward mouse husbandry.

It is possible to create genetically composite mice by delivering genes to a limited proportion of cells *in vivo* using retroviruses (Soriano and Jaenisch, 1986), stem cell transplantation (Brustle *et al.*, 1995), or even tissue transplantation (Ralph and Lehman, 1991). These methods have appropriate applications, but also disrupt the normal development of tissues, which does not occur (after the eight-cell stage) in aggregation chimeras. Methods of generating various kinds of mosaic mice include X inactivation (Nesbitt, 1971), *laacZ/lacZ* (Bonnerot and Nicolas, 1993), and Cre-*loxP*-mediated mosaicking strains (Betz *et al.*, 1996; Dietrich *et al.*, 2000; Guo *et al.*, 2002). In contrast to aggregation chimeras, these methods do not have general applicability to all genes, they result in different patterns of juxtaposition between the component cell genotypes compared with aggregation chimeras, and they produce systematic biases in the proportions and distributions of the two contrasting cell genotypes. Finally, conditionally mutant lines of mice can be created in which particular genes are altered or deleted with spatial specificity. To create a conditional mutant requires a suitable gene promoter that reliably confers the required cell-type specificity. Production of these lines is technically demanding, requiring a substantial amount of work up front, and risks being subject to the regulatory peculiarities of each construct. The use of aggregation chimeras is a straightforward, generally applicable approach that complements more high-investment conditional genetic engineering manipulations. Indeed, a chimera analysis can be used as a first-line tool to indicate whether undertaking a more precise targeting strategy will be worthwhile and informative. We suggest that aggregation chimeras are the most efficient and straightforward chimera/mosaic model to use. With a viable, reasonably breeding mutant in hand, morula aggregation can be performed immediately and is guaranteed to produce viable experimental chimeras.

## Summary

An important goal for the future in circadian rhythms research is to determine how individual cellular oscillators are integrated into higher-order structures to produce this complex behavior. Chimera analysis using clock mutants can help achieve this goal and builds on the progress that has been made in the dissection of circadian molecular components. Any clock mutant or knockout strain can be used to produce chimeric mice, each of which is a unique and novel experimental system that can potentially exhibit new circadian biological properties.

The analysis of chimeric mice can help reveal how multicellular circadian oscillatory structures are assembled from a functional standpoint. At the cellular level, chimera experiments with genetically differing oscillator populations demonstrate how multiple oscillators interact with one another and are a potentially rich source of data for mathematical modeling studies. At the tissue level, chimeras provide a way to probe the fundamental mechanisms by which signals from the individual cellular clocks in the SCN are integrated to produce a coherent timing signal that is broadcast to the rest of the body. At the organismal level, analyzing the wheel-running activity rhythms of chimeras allows the investigator to probe the functional organization of the intact circadian system.

Chimera analysis should be considered an essential part of a growing set of tools for manipulating gene activity *in vivo*, the results of which will be used to converge upon an understanding of complex biological processes such as circadian behavior.

## References

Aida, R., Moriya, T., Araki, M., Akiyama, M., Wada, K., Wada, E., and Shibata, S. (2002). Gastrin-releasing peptide mediates photic entrainable signals to dorsal subsets of suprachiasmatic nucleus via induction of *Period* gene in mice. *Mol. Pharmacol.* **61,** 26–34.

Bae, K., Jin, X., Maywood, E. S., Hastings, M. H., Reppert, S. M., and Weaver, D. R. (2001). Differential functions of *mPer1*, *mPer2*, and *mPer3* in the SCN circadian clock. *Neuron* **30,** 525–536.

Barsh, G. S., Lovett, M., and Epstein, C. J. (1990). Effects of the lethal yellow (*Ay*) mutation in mouse aggregation chimeras. *Development* **109,** 683–690.

Benzer, S. (1973). Genetic dissection of behavior. *Sci. Am.* **229,** 24–37.

Berger, C. N., Tam, P. P., and Sturm, K. S. (1995). The development of haematopoietic cells is biased in embryonic stem cell chimaeras. *Dev. Biol.* **170,** 651–663.

Betz, U. A., Vosshenrich, C. A., Rajewsky, K., and Muller, W. (1996). Bypass of lethality with mosaic mice generated by Cre-loxP-mediated recombination. *Curr. Biol.* **6,** 1307–1316.

Bonnerot, C., and Nicolas, J. F. (1993). Clonal analysis in the intact mouse embryo by intragenic homologous recombination. *C. R. Acad. Sci. III* **316,** 1207–1217.

Brustle, O., Maskos, U., and McKay, R. D. G. (1995). Host-guided migration allows targeted introduction of neurons into the embryonic brain. *Neuron* **15,** 1275–1285.

Bunger, M. K., Wilsbacher, L. D., Moran, S. M., Clendenin, C., Radcliffe, L. A., Hoganesch, J. B., Simon, M. C., Takahashi, J. S., and Bradfield, C. A. (2000). Mop3 is an essential component of the master circadian pacemaker in mammals. *Cell* **103,** 1009–1017.

Campbell, R. M., and Peterson, A. (1992). An intrinsic neuronal defect operates in *dystonia musculorum*: A study of *dt/dt↔+/+* chimeras. *Neuron* **9,** 693–703.

Dewey, M. J., Gervais, A. G., and Mintz, B. (1976). Brain and ganglion development from two genotypic classes of cells in allophenic mice. *Dev. Biol.* **50,** 68–81.

Dietrich, P., Dragatsis, I., Xuan, S., Zeitlin, S., and Efstratiadis, A. (2000). Conditional mutagenesis in mice with heat shock promoter-driven *cre* transgenes. *Mamm. Genome* **11,** 196–205.

Falconer, D. S., and Avery, P. J. (1978). Variability of chimeras and mosaics. *J. Emb. Exp. Morph.* **43,** 195–219.

Guo, C., Yang, W., and Lobe, C. G. (2002). A Cre recombinase transgene with mosaic, widespread tamoxifen-inducible action. *Genesis* **32,** 8–18.

Hadjantonakis, A.-K., and Nagy, A. (2001). The color of mice: In the light of GFP-variant reporters. *Histochem. Cell Biol.* **115,** 49–58.

Harmar, A. J., Marston, H. M., Shen, S., Spratt, C., West, K. M., Sheward, W. J., Morrison, C. F., Dorin, J. R., Piggins, H. D., Reubi, J. C., Kelly, J. S., Maywood, E. S., and Hastings, M. H. (2002). The VPAC$_2$ receptor is essential for circadian function in the mouse suprachiasmatic nucleus. *Cell* **109,** 497–508.

Herzog, E. D., Takahashi, J. S., and Block, G. D. (1998). *Clock* controls circadian period in isolated suprachiasmatic nucleus neurons. *Nature Neurosci.* **1,** 708–713.

Hogan, B., Beddington, R., Costantini, F., and Lacy, E., eds. (1994). "Manipulating the Mouse Embryo: A Laboratory Manual," 2nd Ed., Cold Spring Harbor Laboratory Press, Cold Spring Harbor, NY.

Ioffe, E., Liu, Y., Bhaumik, M., Poirier, F., Factor, S. M., and Stanley, P. (1995). WW6: An embryonic stem cell line with an inert genetic marker that can be traced in chimeras. *Proc. Natl. Acad. Sci. USA* **92,** 7357–7361.

Kramer, A., Yang, F. C., Snodgrass, P., Li, X., Scammell, T. E., Davis, F. C., and Weitz, C. J. (2001). Regulation of daily locomotor activity and sleep by hypothalamic EGF receptor signaling. *Science* **294,** 2511–2515.

Kuhlman, S. J., Quintero, J. E., and McMahon, D. G. (2000). GFP fluorescence reports *Period1* circadian gene regulation in the mammalian biological clock. *Neuroreport* **11,** 1479–1482.

Liu, C., Weaver, D. R., Strogatz, S. H., and Reppert, S. M. (1997). Cellular construction of a circadian clock: Period determination in the suprachiasmatic nucleus. *Cell* **91,** 855–860.

Low-Zeddies, S. S., and Takahashi, J. S. (2001). Chimera analysis of the *Clock* mutation in mice shows that complex cellular integration determines circadian behavior. *Cell* **105,** 25–42.

Mintz, B. (1962). Formation of genotypically mosaic mouse embryos. *Am. J. Zool.* **4,** 432.

Musci, T. S., and Mullen, R. J. (1992). Cell mixing in the spinal cords of mouse chimeras. *Dev. Biol.* **152,** 133–144.

Nesbitt, M. N. (1971). X chromosome inactivation mosaicism in the mouse. *Dev. Biol.* **26,** 252–263.

Oster-Granite, M. L., and Gearhart, J. (1981). Cell lineage analysis of cerebellar Purkinje cells in mouse chimeras. *Dev. Biol.* **85,** 199–208.

Ralph, M. R., and Lehman, M. N. (1991). Transplantation: A new tool in the analysis of the mammalian hypothalamic circadian pacemaker. *Trends Neurosci.* **14,** 362–366.

Rawles, M. E. (1947). Origin of pignment cells from the neural crest in the mouse embryo. *Physiol. Zool.* **20,** 248–266.

Soriano, P., and Jaenisch, R. (1986). Retroviruses as probes for mammalian development: Allocation of cells to the somatic and germ cell lineages. *Cell* **46,** 19–29.

Tarkowski, A. K. (1961). Mouse chimaeras developed from fused eggs. *Nature* **190,** 857–860.

van der Horst, G. T., Muijtjens, M., Kobayashi, K., Takano, R., Kanno, S., Takao, M., de Wit, J., Verkerk, A., Eker, A. P., van Leenen, D., Buijs, R., Bootsma, D., Hoeijmakers, J. H., and Yasui, A. (1999). Mammalian *Cry1* and *Cry2* are essential for maintenance of circadian rhythms. *Nature* **398,** 627–630.

Vitaterna, M. H., King, D. P., Chang, A.-M., Komhauser, J. M., Lowrey, P. L., McDonald, J. D., Dove, W. F., Pinto, L. H., Turek, F. W., and Takahashi, J. S. (1994). Mutagenesis and mapping of a mouse gene, *Clock,* essential for circadian behavior. *Science* **264,** 719–725.

Vitaterna, M. H., Selby, C. P., Todo, T., Niwa, H., Thompson, C., Fruechte, E. M., Hitomi, K., Thresher, R. J., Ishikawa, T., Miyazaki, J., Takahashi, J. S., and Sancar, A. (1999). Differential regulation of mammalian *period* genes and circadian rhythmicity by *cryptochromes 1* and *2. Proc. Natl. Acad. Sci. USA* **96,** 12114–12119.

Wilsbacher, L. D., Yamazaki, S., Herzog, E. D., Song, E. J., Radcliffe, L. A., Abe, M., Block, G., Spitznagel, E., Menaker, M., and Takahashi, J. S. (2002). Photic and circadian expression of luciferase in *mPeriod1-luc* transgenic mice *in vivo. Proc. Natl. Acad. Sci. USA* **99,** 489–494.

Yamazaki, S., Numano, R., Abe, M., Hida, A., Takahashi, R. I., Ueda, M., Block, G. D., Sakaki, Y., Menaker, M., and Tei, H. (2000). Resetting central and peripheral circadian oscillators in transgenic rats. *Science* **288,** 682–685.

Yoo, S. H., Yamazaki, S., Lowrey, P. L., Shimomura, K., Ko, C. H., Buhr, E. D., Siepka, S. M., Hong, H. K., Oh, W. J., Yoo, O. J., Menaker, M., and Takahashi, J. S. (2004). PERIOD2::LUCIFERASE real-time reporting of circadian dynamics reveals persistent circadian oscillations in mouse peripheral tissues. *Proc. Natl. Acad. Sci. USA* **101,** 5339–5346.

Zambrowicz, B. P., Imamoto, A., Fiering, S., Herzenberg, L. A., Kerr, W. G., and Soriano, P. (1997). Disruption of overlapping transcripts in the ROSA ßgeo 26 gene trap strain leads to widespread expression of ß-galactosidase in mouse embryos and hematopoietic cells. *Dev. Biol.* **94,** 3789–3794.

Zheng, B., Albrecht, U., Kaasik, K., Sage, M., Lu, W., Vaishnav, S., Li, Q., Sun, Z. S., Eichele, G., Bradley, A., and Lee, C. C. (2001). Nonredundant roles of the *mPer1* and *mPer2* genes in the mammalian circadian clock. *Cell* **105,** 683–694.

# Section VI

# Peripheral Circadian Clocks

# [25]  Measuring Circadian Rhythms in Olfaction Using Electroantennograms

*By* Parthasarathy Krishnan,
Stuart E. Dryer, and Paul E. Hardin

## Abstract

Circadian clocks control daily rhythms in many behavioral, physiological, and metabolic processes. Despite remarkable advances in our understanding of the circadian timekeeping mechanism and how it responds to environmental cycles, relatively little is known about how the timekeeping mechanism regulates behavior, physiology, and metabolism. One of the most extensively characterized timekeeping mechanisms is that of *Drosophila melanogaster*. In this species, autonomous circadian clocks are found in many neuronal and nonneuronal tissues, including essentially all sensory structures. We have shown that sensory neurons in the antenna mediate a robust rhythm in electrophysiological responses to the food odorant ethyl acetate. This article describes how rhythms in olfactory responses are measured and provides a perspective on the generality of these rhythms and their regulation by the clock.

## Introduction

Daily rhythms in behavior, physiology, and metabolism have been observed in a diverse array of animals, plants, and microbes. These rhythms are controlled by circadian clocks, which are innate timekeeping systems that are set by daily environmental cycles (e.g., light, temperature), but continue to operate even in the absence of environmental cues. Considerable progress has been made in defining the core circadian timekeeping mechanism and how it is entrained, but comparatively little is known about how these molecular oscillators control rhythmic outputs. This is especially true in the fruit fly, *Drosophila melanogaster*, despite the fact that this animal has one of the best characterized circadian timekeeping mechanisms of any species.

Many different neuronal and nonneuronal tissues in *Drosophila* harbor autonomous circadian oscillators that are directly entrainable by light (Hall, 2003). The rhythmic outputs that these oscillators control are, however, largely a mystery. The most extensively characterized clock output is a rhythm in locomotor activity behavior, which is controlled by circadian oscillators in a group of small ventral lateral neurons (sLN$_v$s) in the brain (Frisch et al., 1994; Helfrich-Forster, 1998; Renn et al., 1999). The sLN$_v$ oscillators control this output through the rhythmic release of the neuro-peptide pigment-dispersing factor (PDF) from projections into the dorsal brain (Park et al., 2000; Renn et al., 1999). Although cellular and molecular targets of PDF have not been identified within this output pathway, down-stream signaling via ras/MAPK is required for locomotor activity rhythms (Williams et al., 2001). Another behavioral rhythm is seen in the emergence of adults from their pupal cases, a process referred to as eclosion (Konopka and Benzer, 1971). During the pupal stage, circadian oscillators in the sLN$_v$s, along with those in the prothoracic gland, are required for eclosion rhythms (Myers et al., 2003). Although neuropeptides that initiate eclosion and an RNA-binding protein that alters eclosion rhythms have been iden-tified (Hall, 2003; McNeil et al., 1998; Mesce and Fahrbach, 2002), how they coordinate to control eclosion rhythms is not known.

The presence of circadian oscillators in different *Drosophila* sensory organs suggested that, as in other animals (Barlow, 2001; Green and Besharse, 2004; Natesan et al., 2002), sensory physiology might also be under circadian control. Indeed, odor-induced electrophysiological responses in *Drosophila* antennae are robustly rhythmic. In wild-type flies, responses to the food odorant ethyl acetate are low during the day and increase to peak levels in the middle of the night during a 12-h light:12-h dark (LD) cycle (Krishnan et al., 1999). This rhythm persists during constant darkness (DD) but is abolished in *per*[01], *tim*[01], *cyc*[0], and *cry*[b] mutants, indicating that rhythms in olfactory physiology are controlled by a clock that is similar, but not identical, to that which controls locomotor activity rhythms (Krishnan et al., 1999, 2001; Tanoue et al., 2004). Circadian oscillators in antennal neurons are both necessary and sufficient for olfaction rhythms (Tanoue et al., 2004), which implies that olfactory signal transduction components (or regulators thereof) are clock output targets.

Electrophysiological responses to odors are measured in the third an-tennal segment using a technique called an electroantennogram (EAG). EAGs measure the field potential produced by neuronal activity within a localized group of sensillae in response to an odor (Ayer and Carlson, 1992). There are three major morphological classes of olfactory sensillae in *Drosophila*: basiconic, trichoid, and coeloconic (Stocker, 1994). Each sensillum has up to four sensory neurons, called odorant receptor neurons (ORNs), that express at least one of the ∼60 odorant receptor (Or) genes (Stocker, 2001). These odorant receptors are expressed in regional subsets of olfactory sensillae and are thought to define the odor specificities of these sensillae (de Bruyne et al., 2001; Gao et al., 2000; Vosshall et al., 2000). Measurements of circadian EAG responses have focused on the dorsomedial region of the third antennal segment, which is rich in basiconic sensillae and shows a robust response to the food odorant ethyl acetate (Ayer and Carlson, 1992; de Bruyne *et al.*, 2001). This article describes the apparatus and methods used to record EAG responses over circadian time in *Drosophila* and concludes with a perspective of how such analyses can be used to define the cellular and molecular mechanisms by which the clock controls olfaction.

Electroantennogram Apparatus Setup

Because EAG recordings are made from a localized region on the *Drosophila* antenna (Ayer and Carlson, 1992), the recording apparatus should possess good optics to enable reproducible electrode placement, mechanical stability to avoid vibration-related problems, and specialized electronics to amplify and record the EAG trace. The following section describes the components of the EAG recording apparatus.

*Optics, Micromanipulators, and Vibration Isolation*

We use an Olympus SZ 6045 dissecting microscope with 1–6.3× objectives and an eyepiece that provides an additional 10× magnification. The large working distance (<10 cm) allows placement of all the different manipulators in relation to the fly. Light is provided by a fiber-optic source. Any high-quality micromanipulator can be used to position electrodes. We use Huxley style manual micromanipulators (SD Instruments) to position the ground and recording electrodes. These manipulators allow both coarse and fine controlled (i.e., micrometer) movement in all three axes. Micromanipulators with the ground and recording electrodes are positioned adjacent to each other (Fig. 1). Typically, the ground electrode is positioned at a steeper angle and the recording electrode is oriented at a more oblique angle. A third coarse manipulator (Narishige, MN151) controls a glass micropipette with a long shank. This micropipette is used to lift the antenna so that the recording electrode can be placed on the anterior surface of the antenna (see under Recording Electroantennograms). A fourth coarse manipulator controls the position of the airflow tube relative to the fly (Fig. 1). This tube delivers the odor stimulus from the olfactometer (see *Olfactometer Construction*).

Vibration isolation improves the quality and duration of recordings. Any sort of system that reduces vibration can be used; indeed, an inexpensive approach is to place a heavy marble balance table on partially inflated tire inner tubes. We use a TMC Micro-g air table that is designed for electrophysiological or optical experiments. A grounded Faraday cage is

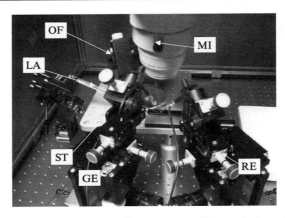

FIG. 1. The electroantennogram recording apparatus. GE, manipulator that controls the ground electrode; RE, manipulator that controls the recording electrode; OF, manipulator that holds the capillary that directs airflow from the olfactometer; LA, manipulator that holds the glass micropipette used to lift the antenna; MI, dissecting microscope.

an integral component of this table, which serves to reduce electrical noise from external sources. We use a microscope platform that is screwed directly into the surface of the vibration isolation table. This stage has a moveable frame to position the fly holder that contains immobilized flies (see *Fly Immobilization*) relative to the recording and ground electrodes.

*Electronics and Data Aquisition*

Any high-quality high-impedance amplifier suitable for intracellular or extracellular recordings can be used. We use a differential amplifier (World Precision Instruments DP301), which receives input from both the recording and the ground electrode. The signal is low pass filtered at 3 kHz, and the output is passed onto a digitizer (Axon Instruments 1322A) and from there into a computer for storage and off-line analysis of data. The amplifier output is also sent to an audio monitor (Grass, AM9). The sound is low pass filtered at 300 Hz and high pass filtered at 3 kHz. This audio monitor emits a burst of high-frequency sound when the recording electrode comes in contact with the antennal surface, thus aiding electrode position in the dark—a necessity for measuring EAGs over circadian time (see *Measuring EAG Responses over Circadian Time*). A schematic diagram shows the arrangement of the electronic components (Fig. 2). We use Axoscope software (Axon Instruments) for data analysis. Typical EAG signals evoked by ethyl acetate are 6–14 mV in the dorsomedial region of the antenna (see Fig. 4), depending on the time of day (Krishnan *et al.*, 1999).

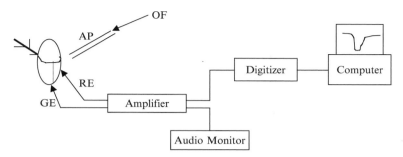

FIG. 2. Electronic components of the EAG recording apparatus. An antenna is depicted on the left as an oval. The dorsomedial surface is shown, where the dashed line represents the division between the anterior and the posterior surfaces on the third segment. The arista is shown protruding from the antenna. OF, olfactometer; GE, ground electrode; RE, recording electrode; AP, port from the air pump. Arrows show the approximate positions of the GE and RE when recording circadian EAG responses to ethyl acetate. See text for further details.

*Olfactometer Construction*

To achieve controlled and repeatable stimulation of the fly preparation with the desired odorant, an olfactometer is constructed as shown in Fig. 3. Briefly, a flowmeter regulates the airflow generated by an aquarium-type aerator. From the flowmeter, a T junction separates the airflow into a vial containing 10 ml of mineral oil and into a second vial containing 10 ml of mineral containing the odorant. A second T junction receives the output from both the vials. All connections are made using Tygon tubing. Two-way valves are connected to the incoming and outgoing ports of both the vials. This allows the control of airflow from either the vial containing the mineral oil or the vial containing the odorant. The air flowing out of the second T junction is directed to the fly preparation by means of a glass tube (3-mm inner diameter). This tube is secured to a coarse manipulator to direct the airflow onto the fly preparation. The volume of odorant imping-ing on the fly is calculated as $\Delta V = T \, dV/dt$ where $\Delta V$ is the displaced air volume, $T$ is the duration of valve opening, and $dV/dt$ is the airflow rate. Typically, the volume ranges from 10 to 15 ml of the odorant in the vapor phase. Note that with this system, the fly is always exposed to flowing air. What changes is whether the air contains odorants. This is essential to prevent mechanical stimulation from contributing to the EAG signal.

*Housing the EAG Apparatus*

The room in which the apparatus is housed is provided with a HEPA filter-based air purifier and is located in an area of the laboratory away from experimental stations using strong odoriferous chemicals such as

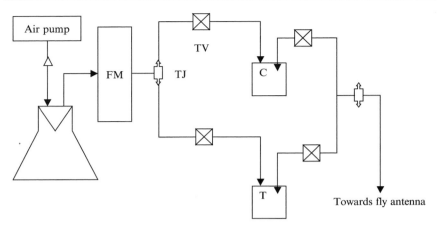

FIG. 3. Schematic diagram of an olfactometer. FM, flowmeter; TV, two-way valves; C, vial containing mineral oil; T, vial containing the odorant diluted in mineral oil; TJ, T junction. See text for further details.

acetone, methanol, and acetic acid. The room must also be able to be made completely dark to facilitate circadian measurements (see *Measuring EAG Responses over Circadian Time*).

### Preparing to Record EAG Responses

Once the equipment is in place for recording EAG responses, animals and consumables must be assembled. Animals produced for EAG measurements must be large, healthy, and unaffected by external odor sources before they are immobilized for recording. Likewise, odorants and electrodes must be freshly prepared to ensure that EAG response amplitudes are robust and reproducible. This section describes the rearing and immobilization of animals, as well as the preparation of odorants and electrodes for EAG recordings.

#### Rearing Animals

Adult female flies are larger than males and offer the advantage of increased antennal surface area, which facilitates electrophysiological measurements. We record from flies that are less than a week old, as older flies typically develop a thickened antennal cuticle, which makes it difficult to place the recording electrode. Flies are grown on standard "cornmeal, yeast, molasses" media (http://fly.bio.indiana.edu/media-recipes.htm). Care is taken so as to not overcrowd the culture vial, which will reduce the size of the flies (and their antennae) due to competition for food. Typically, 5–10 flies (80% females) are placed in vial containing about 20 ml of solidified food. Flies are maintained at 25°. A day before the recordings are made, flies are transferred to a new food vial to ensure that the animals are not exposed to the strong smell of fermenting media just before recording EAGs.

*Immobilizing Flies*

Flies are transferred to an empty glass vial, which is immersed in a bucket of ice for about a minute to produce anesthesia. The anaesthetized flies are placed on a Whatman filter paper on top of an aluminum block, which is cooled from below by ice. The flies are then gently picked up by the wings using forceps and are placed on a "L"-shaped plexiglass "fly holder" (Fig. 4), which rests on top of the aluminum block. This fly holder has 1-mm-diameter holes drilled every 10 mm along its length on the wide portion of the "L." A fly is inserted into each hole such that the head protrudes out and the rest of the body lies flat, with the wings against the surface of the short portion of the "L" (Fig. 4). A small amount of low-melting-point wax (myristic acid, 58.5° melting point) is melted using a fine silver wire heated by an adjustable power source of 10–15 V (Staco Energy, 3PN1010V). The wax is melted just until it can be spread, and care is taken not to overheat the wax. This wax is then gently applied to opposite sides of the head, making sure that the silver wire never comes into contact with the fly preparation, but is sufficiently close that the wax spreads by capillary action alone. This effectively glues the fly to the fly holder. The wings and the legs are also immobilized with melted wax.

Three to five animals are fixed in a single fly holder spaced in such a way that there is an empty hole between two flies. This is done to prevent the odor stimulus applied to one fly from affecting the neighboring fly. This immobilization procedure is done as quickly as possible to avoid dehydration of flies. With practice, about 10 flies can be immobilized in 15–20 minutes. After fixing the flies, the fly holder is moved to a moist chamber, which is composed of a 14-cm petri dish lined with a sheet of moistened Whatman filter paper. The immobilized flies remain in this chamber for up to 1 h before EAG recordings are made.

*Fabrication of Micropipettes*

Pipettes for EAG recording are made from borosilicate glass capillary tubes (Drummond #2-200-210). Capillaries are pulled to an internal tip diameter of 1–2 $\mu$m using a two-stage vertical micropipette puller

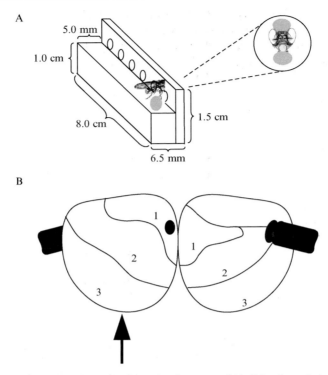

Fig. 4. Fly immobilization and electrode placement. (A) Side view of the fly holder showing an immobilized fly in the last position. The top view shows the head of the immobilized fly protruding from the hole. Wax placed on opposite sides of the head and on the fly body is depicted as gray ovals. (B) Diagram of a fly antenna showing regions rich in different types of sensillae. Left side, anterior face; right side, posterior face; top, dorsal; bottom, ventral. 1 denotes the subregion rich in large basiconic sensillae, 2 denotes the subregion rich in small basiconic sensillae, and 3 denotes the subregion rich in trichoid sensillae. Coeloconic sensillae are present at low density in subregions 2 and 3. The black oval denotes the site from which Krishnan et al. (1999) made EAG recordings. The arrow denotes where the ground electrode penetrates the antenna. The base of the arista (black shape) is shown protruding from the antenna. The diagram is based on de Bruyne et al. (2001).

(Narshige, PP830). Other electrode pullers capable of pulling patch clamp microelectrodes would also be suitable. The pipettes are back-filled with 0.17 $M$ NaCl solution. Bubbles near the tip are removed by tapping the pipette gently. The pulled pipettes can be used up to 2 h after fabrication. A glass micropipette used for lifting the antenna is made from thin capillary tubes (Drummond #1-000-0500 Microcaps) pulled using a vertical one-stage puller (Kopf 720).

*Preparation of Odor Dilutions*

Odorants are diluted in light mineral oil. Most of our studies have focused on ethyl acetate, which is serially diluted starting from 1 ml of neat ethyl acetate and 9 ml of mineral oil (both from Sigma). After vortexing, the same procedure is repeated three times to reach a dilution of $10^{-4}$ that is used for EAG measurements. Each dilution is vortexed for 20 s to ensure thorough miscibility of the odorant in mineral oil. Glass pipettes or one-time-use disposable pipettes are used to make serial dilutions of the odorant, and a fresh dilution of the odorant is made just before each recording session. Care is taken to ensure that the vials are airtight and sealed properly.

*Chloridization of Electrodes*

The silver electrode wires of both ground and recording electrodes are immersed in concentrated bleach for about 15 min until the surface of the wires are oxidized and have a dark brown or dull gray appearance. Chloridization by this method ensures a stable baseline in the EAG trace. Care should be taken while inserting the filled micropipettes into the holder so as to not scratch the surface of the chlorided silver wire.

Recording Electroantennograms

EAG recordings are done at a temperature of $\sim25°$ with a relative humidity of 50–80%. Recording is possible from various sites on the third antennal segment that are rich in different sensillae classes (Stocker, 2001), but keep in mind that this technique is not designed to measure exclusively from an individual sensillum or even one class of sensillae. The only way to ensure that recordings are made from a single sensillum is to perform single unit recordings on an individual ORN (Clyne *et al.*, 1997; de Bruyne *et al.*, 2001). Nevertheless, by taking into account the population of sensillae and their odor specificities, recordings can be made that greatly favor one class or subclass of sensillae. Ethyl acetate evokes robust single-unit responses from basiconic sensillae and EAG responses from regions rich in basiconic sensillae (Ayer and Carlson, 1992; de Bruyne *et al.*, 2001; Dobritsa *et al.*, 2003; Krishnan *et al.*, 1999). One of the highest densities of basiconic sensillae is on the dorsomedial portion of the third antennal segment (de Bruyne *et al.*, 2001). To record from this region, the antenna is lifted to expose a high density of basiconic sensillae on the anteromedial face (Krishnan *et al.*, 1999). The following describes the steps used in recording an EAG.

## Positioning the Electrodes

Because the recording is done from the anterior region of the antenna, which is not immediately accessible, the left antenna is gently lifted with the help of a micropipette with a very long shank length. The dorsomedial region of the anterior surface of the antenna contains primarily basiconic sensilla, many of which respond robustly to ethyl acetate (Ayer and Carlson, 1992; de Bruyne et al., 2001). The antenna is lifted such that the dorsomedial surface is facing the manipulator with the recording electrode (Fig. 4). The manipulator with the ground electrode is advanced until the tip of the electrode is in contact with the distal tip of the antenna. The manipulator is then tapped gently so that the electrode tip is inserted into the lumen of the antenna. Once the ground electrode is in place, the micropipette used to lift the antenna is gently moved away and the antenna is supported solely by the ground electrode in the lumen (Fig. 4). The recording electrode is then placed onto the dorsomedial surface of the third antennal segment such that the tip just impinges on the surface of the antenna (Fig. 4). When the recording electrode is resting on the surface of the antenna, a positive deflection is seen in the Axosope window and a high-frequency burst of sound is emitted from the audio monitor.

## Stimulus Delivery and Response

The airflow tube from the olfactometer is placed about 5 mm from the fly preparation. The Axoscope program is started, and an active oscilloscope window is opened and small changes in electrode positions are made until a stable baseline is achieved. After obtaining a stable baseline reading, the baseline is adjusted to zero using the DC shift knob on the amplifier. There is a constant flow of air on the preparation through the olfactometer, which can deliver odor stimuli by opening and closing of the two-way valve connected to the vial containing the odorant with the simultaneous closing and opening of the valve connected to the vial with the mineral oil. Odorant delivery evokes a change in potential, which is stored for analysis. The traces for individual flies are highly reproducible (Krishnan et al., 1999), thus only one trace is recorded from each fly. Recordings are made from at least eight independent flies, and the resulting amplitude values in millivolts are used to compute the mean and standard error of EAG responses. Note that with sustained delivery of odorant, the responses desensitize. Responses to ethyl acetate typically recover from desensitization in 30 s (Krishnan et al., 1999).

## Measuring EAG Responses over Circadian Time

Adult female flies are entrained in incubators set to 12-h light:12-h dark (LD) cycles for 4 days. Time during LD conditions is referred to as Zeitgeber time (ZT), where ZT0 is defined as lights on and ZT12 is defined as lights off. Measurements made every 4 h during an LD cycle are sufficient to reliably detect a rhythm. EAG responses measured during the light phase (i.e., ZT1, ZT5, and ZT9) are done using the protocol described earlier for recording EAGs. Typically the EAG amplitude during the light phase is close to the trough level of ~8 mV (Krishnan et al., 1999, 2001; Tanoue et al., 2004).

To record responses during the dark phase (i.e., ZT13, ZT17, and ZT21), flies are transferred from the incubators to the EAG room in a "light-tight" box that is typically used to transport negative photo-films prior to developing them. Flies are insensitive to wavelengths of light above 600 nm (Klemm and Ninnemann, 1976; Suri et al., 1998). All light sources in the EAG room are shielded with filters that emitted light in the near-infrared region (>600 nm). These filters are red translucent cellophane sheets that are spectrophotometrically tested to transmit light above 600 nm. In addition, walls of the EAG room are painted black to ensure that flies are not affected by any incidental light source. Recordings conducted during the dark are tedious at first, but with practice they become more manageable. EAG responses during the dark phase increase to a peak of ~14 mV at ZT17 and then decrease to trough levels as lights turn on (Fig. 5).

Recordings can also be made during constant dark (DD) conditions. Time during DD conditions is referred to as circadian time (CT), where CT0 is defined as the time when lights would have turned on (i.e., subjective lights on) and CT12 is defined as the time when lights would have turned off (i.e., subjective lights off). To record EAG responses under these conditions, flies are first entrained to LD cycles for 4 days as described earlier. Flies are then transferred to an incubator set to constant darkness and collected every 4 h (i.e., CT1, CT5, CT9, CT13, CT17, and CT21). Flies are transferred to the EAG room in a light-tight box, immobilized, and recorded as described previously. Under DD conditions, the phase and amplitude of the EAG rhythm are similar to that seen under LD conditions (Krishnan et al., 1999, 2001; Tanoue et al., 2004).

## Conclusion

The rhythm in EAG responses is the only clock-controlled physiological output identified in *Drosophila* to date. Along with the behavioral rhythms in locomotor activity and eclosion, the rhythm in olfaction

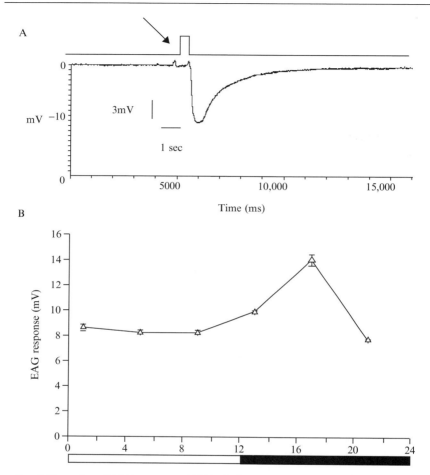

FIG. 5. Examples of EAG responses. (A) Typical EAG response to a $10^{-4}$ dilution of ethyl acetate in wild-type flies. The arrow denotes the onset of a 1-s stimulus. (B) Circadian control of EAG response to ethyl acetate in wild-type flies. The white bar denotes subjective day, and the black bar denotes subjective night.

provides an opportunity to define an entire output pathway from the timekeeping mechanism to the output gene that confers circadian control. Rhythms in EAG responses are controlled autonomously by clocks in antennal neurons (Tanoue *et al.*, 2004), which implies that the clock controls some component of the olfaction signal transduction pathway (or regulator thereof). Measuring EAG responses in mutants that alter the expression or activity of these components should provide significant insight into the molecular mechanisms by which the circadian clock controls olfaction in *Drosophila*.

Studies of olfaction rhythms have focused almost completely on the response of basiconic sensillae to ethyl acetate (Krishnan *et al.*, 1999, 2001; Tanoue *et al.*, 2004). Although a similar rhythm in EAG responses has been demonstrated for a structurally different odorant, benzaldehyde (Krishnan *et al.*, 1999), the generality of such rhythms is not known, but could be tested by determining whether other chemical classes of odorants and pheromones show a rhythm in EAG responses. Similar questions exist concerning the types of sensillae that are capable of mediating rhythmic EAG responses. While basiconic sensillae show EAG rhythms (Tanoue *et al.*, 2004), similar analyses have not been conducted on other classes of sensillae. The ability to eliminate or rescue clock function in specific subsets of olfactory neurons should, given an appropriate olfactory stimulus, provide a means to determine which classes and subclasses of sensillae support EAG rhythms (Tanoue *et al.*, 2004). Defining the odorants and sensillae that are capable of producing rhythmic EAG responses is critically important given that the olfactory system dictates a host of behaviors in *Drosophila* and other species (Beauchamp and Yamazaki, 2003; Devaud, 2003).

# References

Ayer, R. K., Jr., and Carlson, J. (1992). Olfactory physiology in the *Drosophila* antenna and maxillary palp: *acj6* distinguishes two classes of odorant pathways. *J. Neurobiol.* **23,** 965–982.

Barlow, R. (2001). Circadian and efferent modulation of visual sensitivity. *Prog. Brain Res.* **131,** 487–503.

Beauchamp, G. K., and Yamazaki, K. (2003). Chemical signalling in mice. *Biochem. Soc. Trans.* **31,** 147–151.

Clyne, P., Grant, A., O'Connell, R., and Carlson, J. R. (1997). Odorant response of individual sensilla on the *Drosophila* antenna. *Invert. Neurosci.* **3,** 127–135.

de Bruyne, M., Foster, K., and Carlson, J. R. (2001). Odor coding in the *Drosophila* antenna. *Neuron* **30,** 537–552.

Devaud, J. M. (2003). Experimental studies of adult *Drosophila* chemosensory behaviour. *Behav. Processes* **64,** 177–196.

Dobritsa, A. A., van der Goes van Naters, W., Warr, C. G., Steinbrecht, R. A., and Carlson, J. R. (2003). Integrating the molecular and cellular basis of odor coding in the *Drosophila* antenna. *Neuron* **37,** 827–841.

Frisch, B., Hardin, P. E., Hamblen-Coyle, M. J., Rosbash, M., and Hall, J. C. (1994). A promoterless *period* gene mediates behavioral rhythmicity and cyclical *per* expression in a restricted subset of the *Drosophila* nervous system. *Neuron* **12,** 555–570.

Gao, Q., Yuan, B., and Chess, A. (2000). Convergent projections of *Drosophila* olfactory neurons to specific glomeruli in the antennal lobe. *Nature Neurosci.* **3,** 780–785.

Green, C. B., and Besharse, J. C. (2004). Retinal circadian clocks and control of retinal physiology. *J. Biol. Rhythms* **19**, 91–102.

Hall, J. C. (2003). Genetics and molecular biology of rhythms in *Drosophila* and other insects. *Adv. Genet.* **48**, 1–280.

Helfrich-Forster, C. (1998). Robust circadian rhythmicity of *Drosophila melanogaster* requires the presence of lateral neurons: A brain-behavioral study of *disconnected* mutants. *J. Comp. Physiol. A.* **182**, 435–453.

Klemm, E., and Ninnemann, H. (1976). Detailed action spectrum for the delay shift of pupal emergence of *Drosophila pseudoobscura. Photochem. Photobiol.* **24**, 369–371.

Konopka, R. J., and Benzer, S. (1971). Clock mutants of *Drosophila melanogaster. Proc. Natl. Acad. Sci. USA* **68**, 2112–2116.

Krishnan, B., Dryer, S. E., and Hardin, P. E. (1999). Circadian rhythms in olfactory responses of *Drosophila melanogaster. Nature* **400**, 375–378.

Krishnan, B., Levine, J. D., Lynch, M. K., Dowse, H. B., Funes, P., Hall, J. C., Hardin, P. E., and Dryer, S. E. (2001). A new role for *cryptochrome* in a *Drosophila* circadian oscillator. *Nature* **411**, 313–317.

McNeil, G. P., Zhang, X., Genova, G., and Jackson, F. R. (1998). A molecular rhythm mediating circadian clock output in *Drosophila. Neuron* **20**, 297–303.

Mesce, K. A., and Fahrbach, S. E. (2002). Integration of endocrine signals that regulate insect ecdysis. *Front. Neuroendocrinol.* **23**, 179–199.

Myers, E. M., Yu, J., and Sehgal, A. (2003). Circadian control of eclosion: Interaction between a central and peripheral clock in *Drosophila melanogaster. Curr. Biol.* **13**, 526–533.

Natesan, A., Geetha, L., and Zatz, M. (2002). Rhythm and soul in the avian pineal. *Cell Tissue Res.* **309**, 35–45.

Park, J. H., Helfrich-Forster, C., Lee, G., Liu, L., Rosbash, M., and Hall, J. C. (2000). Differential regulation of circadian pacemaker output by separate clock genes in *Drosophila. Proc. Natl. Acad. Sci. USA* **97**, 3608–3613.

Renn, S. C., Park, J. H., Rosbash, M., Hall, J. C., and Taghert, P. H. (1999). A pdf neuropeptide gene mutation and ablation of PDF neurons each cause severe abnormalities of behavioral circadian rhythms in *Drosophila. Cell* **99**, 791–802.

Stocker, R. F. (1994). The organization of the chemosensory system in *Drosophila melanogaster:* A review. *Cell Tissue Res.* **275**, 3–26.

Stocker, R. F. (2001). *Drosophila* as a focus in olfactory research: Mapping of olfactory sensilla by fine structure, odor specificity, odorant receptor expression, and central connectivity. *Microsc. Res. Tech.* **55**, 284–296.

Suri, V., Qian, Z., Hall, J. C., and Rosbash, M. (1998). Evidence that the TIM light response is relevant to light-induced phase shifts in *Drosophila melanogaster. Neuron* **21**, 225–234.

Tanoue, S., Krishnan, P., Krishnan, B., Dryer, S. E., and Hardin, P. E. (2004). Circadian clocks in antennal neurons are necessary and sufficient for olfaction rhythms in *Drosophila. Curr. Biol.* **14**, 638–649.

Vosshall, L. B., Wong, A. M., and Axel, R. (2000). An olfactory sensory map in the fly brain. *Cell* **102**, 147–159.

Williams, J. A., Su, H. S., Bernards, A., Field, J., and Sehgal, A. (2001). A circadian output in *Drosophila* mediated by neurofibromatosis-1 and Ras/MAPK. *Science* **293**, 2251–2256.

## [26]   Circadian Effects of Timed Meals (and Other Rewards)

*By* ALEC J. DAVIDSON, ÖZGUR TATAROGLU, and MICHAEL MENAKER

Abstract

Mammals organize many of their activities around rhythmic events in their environments. Primary among these events is the daily light–dark cycle. However, for many animals, food availability is rhythmic or quasi-rhythmic and is therefore a potential synchronizing cue. While circadian rhythms in both behavior and physiological activity can be entrained in animals via meal-feeding schedules, the mechanism by which this occurs remains poorly understood. Similarities between the circadian effects of restricted feeding and the effects of chronic methamphetamine treatment may be indicative of a common mechanism. This article argues that reward (or the arousal that accompanies it) may be the final common pathway for such nonphotic circadian inputs.

## Introduction

Mammals organize many of their activities around rhythmic events in their environments. Primary among these events is the daily light–dark cycle, which acts through specialized retinal photoreceptors to synchronize (entrain) an autonomous circadian pacemaker in the suprachiasmatic nucleus (SCN) of the hypothalamus. The SCN regulates rhythmic locomotor behavior, rhythms of body temperature, and some other physiological variables and influences rhythmicity in many other autonomous circadian oscillators distributed throughout the brain and body. At present we know very little about how this complex multioscillator system is integrated: Which oscillators influence which others? Are they coupled through neural, humoral, or behavioral links? What is the adaptive significance of their phase relationships, and what are the consequences of the loss of internal synchrony?

Although the day–night cycle is the dominant environmental synchronizer for most organisms, many other aspects of the environment are rhythmic: temperature, humidity, wind velocity, and the activities of other organisms, to name a few. For many animals, food availability is rhythmic or quasi-rhythmic and is therefore a potential synchronizing cue.

METHODS IN ENZYMOLOGY, VOL. 393

Entrainment of Locomotor Activity by Meal Feeding

Animals use endogenous circadian oscillators to predict and prepare for changes in the environment, such as the rhythmic availability of food. Rodents synthesize digestive enzymes in the gut prior to the delivery of a timed daily meal (Scheving, 2000; Scheving et al., 1989). Under such experimental conditions food is only digested for a restricted time each day, and because it requires significant energy and time to prepare the gut for this undertaking, it is most efficient to have the machinery required for digestion available at the time food is expected.

Not only do cells and tissues utilize circadian clocks to organize their energy-processing activities, animals use clocks to organize food behaviors. For predators, returning to hunt at a location and time when prey were available on previous occasions increases the efficiency of the hunt. For foraging omnivores such as rodents, appropriate timing can reduce the risk of predation while foraging. While there are clearly learned components involved in the repetition of such complex behaviors, the circadian system provides one of several important time bases.

Food-anticipatory activity (FAA) is the increase in locomotor activity that occurs before a daily timed meal (Richter, 1922). The behavior has been described in a wide variety of species, underscoring its value in nature: Bees, birds, fish, reptiles, and mammals all express it (for review, see Stephan, 2001). It has been suggested by Stephan (2002) that this phenomenon is the laboratory manifestation of an internal foraging signal.

In rodents, FAA is characterized by a reorganization of wheel running or locomotor activity such that a large percentage of daily activity occurs during the 3–4 h prior to the delivery of a daily meal. This meal can be delivered as an *ad libitum* period lasting a fixed time or as a fixed amount of food delivered at the same time each day. It can be normal rat chow on a background of deprivation or a highly palatable meal on an *ad libitum* background (Mistlberger and Rusak, 1987).

While FAA was initially thought of as the result of a simple hunger signal operating like an hourglass, careful behavioral analysis has revealed that it is the output of an unidentified circadian clock separate and distinct from the SCN (the so-called "master oscillator" in mammals). The experiments that support this argument have been reviewed in detail elsewhere (Stephan, 2001). The following several examples illustrate the techniques that have been used to study FAA.

Circadian clocks have several defining features (Dunlap et al., 2004). Most importantly, circadian rhythms persist in constant conditions with periods of about 24 h and are entrainable by periodic stimuli within limits near 24 h. While persistence in constant conditions is easy to demonstrate for circadian activity patterns normally entrained by light, it is more difficult to demonstrate a free-running rhythm of FAA. While FAA readily emerges in restricted-fed (RF) animals, even those lacking an SCN (Davidson and Stephan, 1999; Krieger *et al.*, 1977; Stephan *et al.*, 1979a,b), subsequent *ad libitum* feeding results in disappearance of the behavior (Fig. 1). While this might be interpreted as a failure to persist in constant

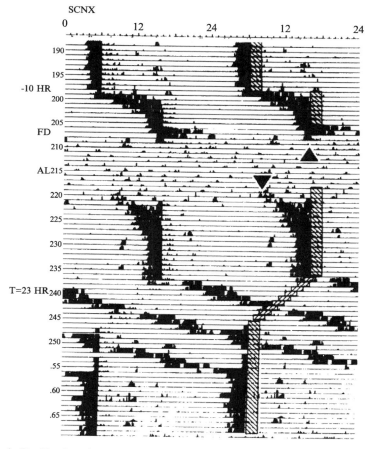

FIG. 1. Double-plotted event record of an SCNX rat on restricted feeding. Hatched rectangles indicate mealtime. Triangles indicate the beginning and end of *ad libitum* (AL) feeding. Note the transients following a 10-h phase delay of food access and the persistence of FAA during food deprivation (FD). Food delivery was scheduled at 23-h intervals starting from day 236 to 245. The free-running rhythm of FAA confirms that a self-sustained circadian pacemaker drives FAA and that this pacemaker has limits of entrainment near 24 h. Reprinted with permission from Stephan (1992).

conditions, key findings (see later) have suggested that a more likely explanation is that the output of the clock that underlies this behavior, the feeding-entrainable oscillator (FEO), becomes uncoupled from the behavioral output when the animal has constantly available food. It is clear why this might be adaptive if FAA is the result of a drive to forage. The need for timed foraging disappears if food is always present.

Although FAA disappears during *ad libitum* feeding, it readily re-emerges during subsequent food deprivation (Coleman *et al.*, 1982; Davidson *et al.*, 2003a; Rosenwasser *et al.*, 1984; Stephan, 1992) (Fig. 1). This can occur days or weeks after the cessation of RF. The onset of FAA in these instances is at roughly the same phase as during the previous RF, suggesting that the FEO was free running with a period close to 24 h. In addition, a free-running rhythm of FAA does occasionally occur. Interestingly, it is seen during manipulations designed to test another defining circadian characteristic: limits of entrainment in the circadian range. Rats will anticipate meals presented at circadian intervals only (Boulos *et al.*, 1980; Stephan, 1981). An RF schedule with a period outside this range (e.g., 23 or 31 h) will produce either no anticipation or, occasionally, a free-running rhythm of FAA (Fig. 1), even in SCN-lesioned rats.

## Entrainment of Peripheral Organs by Meal Feeding

The notion that peripheral organs contain independent circadian oscillators is not new. Andrews (1968a,b) described rhythms in isolated adrenal glands as early as 1968. In 1960, Aschoff characterized the circadian system as involving multiple oscillators that were maintained in synchrony (Aschoff, 1960). Pioneering work by Scheving described rhythmicity of the alimentary tract (Scheving, 2000; Scheving *et al.*, 1978, 1983, 1989).

However, many of these early experiments did not address the critical question of whether these organ and tissue rhythms were locally generated or were driven by other structures (e.g., SCN) (for review, see Davidson *et al.*, 2004). Most involved sampling of blood or tissue at different times of day from populations of animals, with little concern for the masking effects of light or the role of other pacemakers in regulating organ function.

More recently, isolated cell and organ preparations have served as valuable models for the study of peripheral rhythmicity. Balsalobre *et al.* (1998) reported rhythmicity of gene expression in cultured Rat-1 fibroblasts following a serum shock. This approach showed that cells previously not known to contain circadian clocks had the capacity for endogenous oscillation, at least with respect to gene expression. Shortly thereafter,

Yamazaki and colleagues (2000), using a real-time reporter of Period1 gene expression, showed that many isolated organs had this capacity for circadian oscillation. This paper and several that followed described endogenous oscillations in a variety of brain tissues (Abe *et al.*, 2002), liver (Davidson *et al.*, 2002, 2003a; Stokkan *et al.*, 2001; Yamazaki *et al.*, 2000), lung, muscle (Yamazaki *et al.*, 2000), gastrointestinal tract (Davidson *et al.*, 2003a), pineal gland, pituitary gland (Yamazaki *et al.*, 2002), blood vessels, ovaries, and a variety of tumors (unpublished observations). A common feature of isolated mammalian oscillators (except the SCN) is that they damp after varying periods of isolation. This has led to the assumption that the oscillators were slaves to the SCN and required its rhythmic input to sustain their oscillations. The hypothesis that the SCN is required to sustain these peripheral oscillations has been addressed by a large number of flawed experiments (e.g., Furukawa *et al.*, 1999; Iijima *et al.*, 2002; Sakamoto *et al.*, 1998; Terazono *et al.*, 2003). These studies measured gene expression or organ function at different times of day by sampling from populations of SCN-lesioned, behaviorally arrhythmic animals; reported that rhythmicity in peripheral structures was absent; and concluded that the hypothesis was confirmed. However, using this approach it was not possible to address the alternative interpretation—that the SCN is required not to maintain rhythmicity, but rather to maintain phase control. Perhaps peripheral oscillators are still rhythmic in SCN-lesioned animals, but are no longer synchronized among animals at each time point sampled. Under those circumstances, no rhythm could be detected by sampling populations of unsynchronized individuals, which is the result obtained in each of these experiments (for further discussion, see Davidson *et al.*, 2003b). Indeed, Yoo and colleagues suggested that peripheral clocks continue to oscillate *in vivo* in mice lacking an SCN. Instead of damping, it seems that the myriad peripheral oscillators have lost internal synchronization in the absence of the "conductor of the circadian orchestra" (Davidson *et al.*, 2003b; Yoo *et al.*, 2004).

Restricted feeding quickly resets clocks in peripheral organs (Damiola *et al.*, 2000; Kita *et al.*, 2002; Stokkan *et al.*, 2001). Indeed, rhythmic food intake may be the dominant mechanism linking the SCN to some peripheral clocks, such as those in the liver. During *ad libitum* feeding in the presence of a light cycle, food intake in nocturnal rodents occurs mostly in the early night (Davidson *et al.*, 2003a), resulting in a nocturnal phase for the rhythm of Period1 gene expression in the liver. Restricting food to the daytime shifts this rhythm by 180° while the rhythm of gene expression in the SCN remains locked to the light cycle. Whether there is additional regulation of the liver by SCN-derived signals remains undetermined, however. It seems that food itself is the stronger zeitgeber.

The Search for the Feeding-Entrainable Oscillator

After 82 years of sporadic research, neither the anatomical nor the molecular substrate for the FEO has been identified. The search for the elusive FEO, while reviewed in detail elsewhere (Mistlberger, 1994; Stephan, 2001), serves to highlight a few methodological approaches. Description of the formalisms of FAA induction and entrainment have utilized straightforward behavioral assays (e.g., wheel running, feeder approach behavior, tilt cage activity). Molecular and surgical approaches used in attempts to identify candidate structures for FEO location have been less successful.

The behavior ablation technique has been used extensively in these attempts with little success; the literature underlines the risks of overinterpretation of the results of such experiments. The ventromedial hypothalamic nucleus, known to be involved in food intake and located near another important biological clock (SCN), was an early target for lesion studies. Inouye (1982) showed that lesions of this site abolished FAA, and Krieger et al. (1977) described a similar effect on meal entrainment of corticosterone (another output of the FEO). However, subsequent studies showed that FAA recovered about 18 weeks after the lesions (Mistlberger and Rechtschaffen, 1984) and even sooner when the food access time was shortened enough to prevent the obesity that these lesions cause (Honma et al., 1987b). Interestingly, rats made obese with diet also showed a significant reduction in FAA (Persons et al., 1993). Behavior ablation results can lead to other misinterpretations as well; negative results can falsely eliminate a region, although the region may be involved in a subtle or redundant way, and positive results may be due to third variable effects, as in the example given earlier, or to damage of fibers of passage. Other papers have been more reserved in their interpretations of "successful" behavior ablation studies (e.g., Davidson et al., 2000).

Care must also be taken in the selection of activity measures used to study this phenomenon. Food-directed behaviors may be regulated very differently than wheel running or general activity. This problem was nicely illustrated in a study by Mistlberger and Rusak (1988) addressing the effects of lesions of hypothalamic paraventricular nuclei (PVN) on FAA in rats. Three subjects with complete (2) or partial (1) bilateral PVN ablation showed no anticipation of mealtime when tilt cage activity was measured. However, the authors wisely also measured food bin approach behavior in these rats and found that all three retained the ability to anticipate according to this food-directed behavior.

Molecular approaches have also been used in attempts to identify functional components of the FEO. For example, rhythmicity of the circadian clock genes Per1 and Per2 phase shifts in response to RF in the hippocampus and frontal cortex (Wakamatsu et al., 2001). However, such experiments are a first step and only provide candidate structures. Altered gene expression patterns in specific brain areas measured in vivo may reflect feedback onto those regions or output from the FEO located elsewhere. Furthermore, these studies assume that the FEO employs the same set of cycling genes as have been described for the SCN (for review, see Reppert and Weaver, 2002). This may not be the case. Indeed, studies by Pitts et al. (2003) and Rutter et al. (2001) indicate that differences may exist between the molecular mechanisms responsible for SCN rhythmicity and those involved in the generation of FAA.

Localization of the FEO will require, among other criteria, evidence of phase control of persistent rhythmicity by RF, which is maintained during subsequent ad libitum feeding and food deprivation. Utilizing these manipulations, we have ruled out a number of peripheral organs as sites of FEOs (Davidson et al., 2003a).

## Might Entrainment of the FEO Be Entrainment by Reward?

There is significant literature that documents a variety of reward-based inputs affecting or affected by circadian rhythmicity. The efficacy of electrical intracranial self-stimulation varies with circadian time (Terman and Terman, 1970, 1975, 1980); cocaine reward is also under the influence of the circadian rhythm (Abarca et al., 2002). Phase response curves (PRCs) obtained from brain stimulation reward are similar to PRCs obtained with other nonphotic cues, such as limited access to running wheels, cage changes, or social encounter with estrous females (Mrosovsky, 1996). In particular, access to a novel wheel may cause phase shifts if it is used heavily.

Running wheel access is a rewarding stimulus for rodents (Lett et al., 2000, 2002) and, like drugs of abuse [methamphetamine (MAP), morphine, heroin], increases dopamine (DA) in the nucleus accumbens (NAc) (Werme et al., 2002a). Immediate early genes such as fosB are upregulated in NAc during both wheel running and treatment with drugs of abuse. Rats given free access to wheels have increased levels of fosB in the dynorphin-containing subpopulation of the core of NAc (Werme et al., 2002b). These data suggest that wheel running and drug-induced reward trigger common pathways.

Psychoactive drugs such as amphetamine, imipramine, clorgyline, lithium, and triazolam can affect circadian systems (discussed in Ozaki et al., 1991), but the effects of methamphetamine have been characterized most fully. MAP has a striking and unique effect on locomotor activity rhythms of rats when provided chronically in drinking water. After 2–8 weeks of chronic exposure to MAP in the presence of a light–dark (LD) cycle, entrainment to the light cycle becomes abnormal (Honma *et al.*, 1986), $\alpha$ lengthens, and, in some cases, the activity rhythm splits, with one of the components remaining entrained to the LD cycle while the other free runs with a period that is longer than 24 h. In constant darkness (DD), MAP-exposed animals often simultaneously express two free-running activity components with different periods. The effects of MAP on circadian parameters are dose dependent and reversible (Honma *et al.*, 1986, 1987a). Because MAP-treated rats can simultaneously display two different circadian periods, multiple oscillators are likely to be involved. In support of this, suprachiasmatic-lesioned rats arrhythmic in constant conditions become rhythmic when given chronic MAP (Honma *et al.*, 1987a, 1989). MAP induces feeding, drinking, and body temperature rhythms, as well as rhythmic wheel running in these animals (Honma *et al.*, 1988). Furthermore, when intact rats on MAP display splitting in the presence of an LD cycle, lesions of the SCN abolish the light-entrainable component of the split behavior, but do not abolish the longer period MAP-induced component (Honma *et al.*, 1987a, 1989). These observations strongly suggest the existence of a MAP-inducible oscillator that is independent of the SCN (Hiroshige *et al.*, 1991; Honma *et al.*, 1987a). Mice and golden hamsters show similar effects of MAP (Masubuchi *et al.*, 2001; Omata and Kawamura, 1988). While the current literature describes the behavioral output of the presumptive MAP-inducible oscillator in detail, there is very little information on its anatomical, physiological, or molecular basis.

Like most rewarding drugs, MAP affects both serotonergic and (more profoundly) dopaminergic systems, and its circadian effects might well be related to alteration of these systems (Honma and Honma, 1995). In support of this idea, injections of haloperidol (a nonselective dopamine receptor antagonist, which also has minor effects on serotonergic and noradrenergic systems) cause dose-dependent phase shifts in the locomotor rhythms of SCNX rats on MAP (Honma *et al.*, 1995). However, the toxic effects of MAP on dopaminergic neurons are long lasting (Kita *et al.*, 2003), but the circadian effects are reversed rapidly after withdrawal of the drug (Honma *et al.*, 1986, 1987a), suggesting the possibility of a nondopaminergic mechanism for the circadian effects of MAP.

As in the search for the FEO and subject to the same reservations, several studies have used clock gene expression in different brain areas to identify candidate sites for the MAP-inducible oscillator. Per1 rhythms in SCN, pineal melatonin (MEL), and plasma MEL rhythms are not affected by MAP treatment in rats (Masubuchi *et al.*, 2000). However, the expression profiles of Per1, Per2, and Bmal1 in the caudate putamen (CPu) and parietal cortex were phase reversed in MAP-treated rats when compared with controls. This kind of information is difficult to interpret because the CPu and parietal cortex (as well as other areas/organs) may be on output pathway(s) of an oscillator located elsewhere, and their clock gene expression profiles may be driven rather than self-generated. MAP also induces free-running locomotor rhythms in homozygous arrhythmic CLOCK mutant arrhythmic mice (Masubuchi *et al.*, 2001), suggesting that the underlying molecular mechanism of MAP-induced rhythms may depend on NPAS2 (Rutter *et al.*, 2002) rather than or in addition to Clock or may involve an entirely different set of genes (Masubuchi *et al.*, 2001).

### Are the FEO and Methamphetamine-Inducible Oscillator the Same Clock?

MAP and circadian FAA share a number of key features. Both are circadian locomotor activity rhythms that persist in the absence of the SCN. When FAA is induced to free run (see Fig. 1), its long period and the short duration of $\alpha$ are reminiscent of MAP-induced rhythmicity. Aging seems to attenuate both rhythms (Mistlberger *et al.*, 1990; Shibata *et al.*, 1994a,b).

MAP-induced locomotor rhythms in SCNX rats can be entrained by RF regimes (Fig. 2), although the behavior does not anticipate meal time (rather it follows it) (Honma *et al.*, 1989, 1992). Therefore, food can, at the very least, act as an input to the MAP-inducible oscillator, as well as to the FEO.

Daily MAP injections, but not saline injections, produce weak anticipatory activity similar to that produced by RF (Kosobud *et al.*, 1998). Importantly, in those experiments, care was taken to prevent a dramatic alteration in the food intake pattern by the injected rats. Therefore, the anticipation must have been to the MAP injections rather than the indirect effects of these injections on food intake.

Experiments by Timberlake and colleagues suggested another kind of overlap between MAP and food effects on circadian behavior. They showed that meal feeding or MAP injections given at 31-h intervals produce so-called "circadian ensuing activity" (Pecoraro *et al.*, 2000; White and Timberlake, 1999). Rats become active approximately 24 h after the stimulus is delivered. At least in the case of the feeding experiments, the behavior appears to persist during a 2-day fast. Whether this phenomenon is due to the daily resetting of a circadian oscillator or to an interval timing mechanism is still debatable, but the similarities between effects of the feeding signal and MAP injections are intriguing.

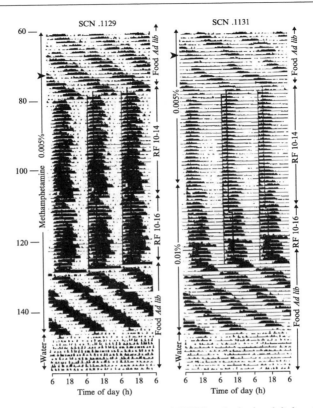

FIG. 2. Triple-plotted actograms of enucleated SCNX arrhythmic adult female rats during chronic MAP treatment. Arrow indicates the timing of enucleation. Animals were arrhythmic before the start of MAP administration. MAP-induced locomotor rhythm is clearly observed between days 60 and 75. A restricted feeding (RF) regimen entrains this rhythm robustly, and after the RF regimen, termination rhythm starts to free run again with a period length that is more than 24 h.

## Is It "Reward" After All?

If food and MAP are indeed acting on the same (non-SCN) circadian oscillator, it is reasonable to ask whether they are acting through well-known reward pathways. Cain *et al.* (2004) addressed a related question by comparing the phase-shifting effects of brain stimulation reward with those of an aversive stimulus, foot shock. Interestingly, these two stimuli, with opposite valence, produce almost identical phase response curves (Fig. 3). What they share is the ability to cause "arousal," and the authors suggest that arousing stimuli act similarly on the clock regardless of their

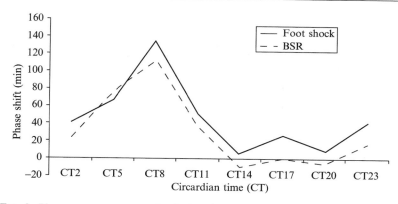

FIG. 3. Phase response curves for brain stimulation reward (BSR) and foot shock. Reprinted from Cain *et al.* (2004), with permission.

"emotional valence." They then argued that these and other nonphotic signals are mediated by cholinergic input to the IGL.

It is tempting to speculate that (with the exception of light) the entire panoply of signals affecting the mammalian circadian system—food, MAP, novel wheel, brain stimulation, foot shock, and others—may act on a single arousal-sensitive oscillator or, more accurately, pacemaker (ASP). This hypothetical ASP might be coupled to the SCN and could also affect behavior directly. With the further assumption that both the sensitivity of the ASP to specific signals and the strength of its interaction with the SCN vary between and perhaps within species, many of the inconsistencies in the literature become explicable. As far as we are aware, nothing in the existing literature rules out either this possibility or its many more complex alternatives. Further progress in unraveling this intriguing aspect of mammalian circadian organization depends heavily on localizing the non-SCN circadian pacemaker(s).

## Acknowledgments

The authors acknowledge the financial support of NSBRI Grant NCC9-58-167 to MM and NIA Grant F32 AG22741-01 to AJD.

## References

Abarca, C., Albrecht, U., and Spanagel, R. (2002). Cocaine sensitization and reward are under the influence of circadian genes and rhythm. *Proc. Natl. Acad. Sci. USA* **99**(13), 9026–9030.

Abe, M., Herzog, E. D., Yamazaki, S., Straume, M., Tei, H., Sakaki, Y., Menaker, M., and Block, G. D. (2002). Circadian rhythms in isolated brain regions. *J. Neurosci.* **22**(1), 350–356.

Andrews, R. V. (1968a). Daily variation in membrane flux of cultured hamster adrenals. *Comp. Biochem. Physiol.* **26**(2), 479–488.

Andrews, R. V. (1968b). Temporal secretory responses of cultured hamster adrenals. *Comp. Biochem. Physiol.* **26**(1), 179–193.

Aschoff, J. (1960). Exogenous and endogenous components in circadian rhythms. *Cold Spring Harb. Symp. Quant. Biol.* **25,** 11–28.

Balsalobre, A., Damiola, F., and Schibler, U. (1998). A serum shock induces circadian gene expression in mammalian tissue culture cells. *Cell* **93**, 929–937.

Boulos, Z., Rosenwasser, A. M., and Terman, M. (1980). Feeding schedules and the circadian organization of behavior in the rat. *Behav. Brain Res.* **1**(1), 39–65.

Cain, S. W., Verwey, M., Hood, S., Leknickas, P., Karatsoreos, I., Yeomans, J. S., and Ralph, M. R. (2004). Reward and aversive stimuli produce similar nonphotic phase shifts. *Behav. Neurosci.* **118**(1), 131–137.

Coleman, G. J., Harper, S., Clarke, J. D., and Armstrong, S. (1982). Evidence for a separate meal-associated oscillator in the rat. *Physiol. Behav.* **29**(1), 107–115.

Damiola, F., Le Minh, N., Preitner, N., Kornmann, B., Fleury-Olela, F., and Schibler, U. (2000). Restricted feeding uncouples circadian oscillators in peripheral tissues from the central pacemaker in the suprachiasmatic nucleus. *Genes Dev.* **14**(23), 2950–2961.

Davidson, A. J., Cantenon-Cervantes, O., and Stephan, F. K. (2004). Daily oscillations in liver function: Diurnal versus circadian rhythmicity. *Liver Int.* **24**, 179–186.

Davidson, A. J., Cappendijk, S. L., and Stephan, F. K. (2000). Feeding-entrained circadian rhythms are attenuated by lesions of the parabrachial region in rats. *Am. J. Physiol.* **278**(5), R1296–R1304.

Davidson, A. J., Poole, A., Yamazaki, S., and Menaker, M. (2003a). Is the food-entrainable oscillator in the digestive system? *Genes Brain Behav.* **2**(1), 1–8.

Davidson, A. J., and Stephan, F. K. (1999). Feeding-entrained circadian rhythms in hypophysectomized rats with suprachiasmatic nucleus lesions. *Am. J. Physiol.* **46,** R1376–R1384.

Davidson, A. J., Stokkan, K.-A., Yamazaki, S., and Menaker, M. (2002). Food-anticipatory activity and liver per1-luc activity in diabetic transgenic rats. *Phys. Behav.* **76**(1), 21–26.

Davidson, A. J., Yamazaki, S., and Menaker, M. (2003b). SCN: Ringmaster of the circadian circus or conductor of the circadian orchestra? *Novartis Found Symp.* **253**, 110–121; discussion 121–115, 281–114.

Dunlap, J. C., Loros, J. J., and Decoursey, P. J., eds. (2004). "Chronobiology: Biological Timekeeping." Sinauer, Sunderland, MA.

Furukawa, T., Manabe, S., Watanabe, T., Sehata, S., Sharyo, S., Okada, T., and Mori, Y. (1999). Daily fluctuation of hepatic P450 monooxygenase activities in male rats is controlled by the suprachiasmatic nucleus but remains unaffected by adrenal hormones. *Arch. Toxicol.* **73**(7), 367–372.

Hiroshige, T., Honma, K., and Honma, S. (1991). SCN-independent circadian oscillators in the rat. *Brain Res. Bull.* **27**(3–4), 441–445.

Honma, K., Honma, S., and Hiroshige, T. (1986). Disorganization of the rat activity rhythm by chronic treatment with methamphetamine. *Physiol. Behav.* **38**(5), 687–695.

Honma, K., Honma, S., and Hiroshige, T. (1987a). Activity rhythms in the circadian domain appear in suprachiasmatic nuclei lesioned rats given methamphetamine. *Physiol. Behav.* **40**(6), 767–774.

Honma, S., and Honma, K. (1995). Phase-dependent phase shift of methamphetamine-induced circadian rhythm by haloperidol in SCN-lesioned rats. *Brain Res.* **674**(2), 283–290.

Honma, S., Honma, K., and Hiroshige, T. (1989). Methamphetamine induced locomotor rhythm entrains to restricted daily feeding in SCN lesioned rats. *Physiol. Behav.* **45**(5), 1057–1065.

Honma, S., Honma, K., Nagasaka, T., and Hiroshige, T. (1987b). The ventromedial hypothalamic nucleus is not essential for the prefeeding corticosterone peak in rats under restricted daily feeding. *Physiol. Behav.* **39**(2), 211–215.

Honma, S., Honma, K., Shirakawa, T., and Hiroshige, T. (1988). Rhythms in behaviors, body temperature and plasma corticosterone in SCN lesioned rats given methamphetamine. *Physiol. Behav.* **44**(2), 247–255.

Honma, S., Kanematsu, N., and Honma, K. (1992). Entrainment of methamphetamine-induced locomotor activity rhythm to feeding cycles in SCN-lesioned rats. *Physiol. Behav.* **52**(5), 843–850.

Iijima, M., Nikaido, T., Akiyama, M., Moriya, T., and Shibata, S. (2002). Methamphetamine-induced, suprachiasmatic nucleus-independent circadian rhythms of activity and mPer gene expression in the striatum of the mouse. *Eur. J. Neurosci.* **16**(5), 921–929.

Inouye, S. T. (1982). Ventromedial hypothalamic lesions eliminate anticipatory activities of restricted daily feeding schedules in the rat. *Brain Res.* **250**(1), 183–187.

Kita, T., Wagner, G. C., and Nakashima, T. (2003). Current research on methamphetamine-induced neurotoxicity: Animal models of monoamine disruption. *J. Pharmacol. Sci.* **92**(3), 178–195.

Kita, Y., Shiozawa, M., Jin, W., Majewski, R. R., Besharse, J. C., Greene, A. S., and Jacob, H. J. (2002). Implications of circadian gene expression in kidney, liver and the effects of fasting on pharmacogenomic studies. *Pharmacogenetics.* **12**(1), 55–65.

Kosobud, A. E., Pecoraro, N. C., Rebec, G. V., and Timberlake, W. (1998). Circadian activity precedes daily methamphetamine injections in the rat. *Neurosci. Lett.* **250**(2), 99–102.

Krieger, D. T., Hauser, H., and Krey, L. C. (1977). Suprachiasmatic nuclear lesions do not abolish food-shifted circadian adrenal and temperature rhythmicity. *Science* **197,** 398–399.

Lett, B. T., Grant, V. L., Byrne, M. J., and Koh, M. T. (2000). Pairings of a distinctive chamber with the aftereffect of wheel running produce conditioned place preference. *Appetite* **34**(1), 87–94.

Lett, B. T., Grant, V. L., Koh, M. T., and Flynn, G. (2002). Prior experience with wheel running produces cross-tolerance to the rewarding effect of morphine. *Pharmacol. Biochem. Behav.* **72**(1–2), 101–105.

Masubuchi, S., Honma, S., Abe, H., Ishizaki, K., Namihira, M., Ikeda, M., and Honma, K. (2000). Clock genes outside the suprachiasmatic nucleus involved in manifestation of locomotor activity rhythm in rats. *Eur. J. Neurosci.* **12**(12), 4206–4214.

Masubuchi, S., Honma, S., Abe, H., Nakamura, W., and Honma, K. (2001). Circadian activity rhythm in methamphetamine-treated Clock mutant mice. *Eur. J. Neurosci.* **14**(7), 1177–1180.

Mistlberger, R. E. (1994). Circadian food-anticipatory activity: Formal models and physiological mechanisms. *Neurosci. Biobeh. Rev.* **18**(2), 171–195.

Mistlberger, R. E., Houpt, T. A., and Moore-Ede, M. C. (1990). Effects of aging on food-entrained circadian rhythms in the rat. *Neurobiol. Aging* **11**(6), 619–624.

Mistlberger, R. E., and Rechtschaffen, A. (1984). Recovery of anticipatory activity to restricted feeding in rats with ventromedial hypothalamic lesions. *Physiol. Behav.* **33**(2), 227–235.

Mistlberger, R. E., and Rusak, B. (1987). Palatable daily meals entrain anticipatory activity rhythms in free-feeding rats: Dependence on meal size and nutrient content. *Physiol. Behav.* **41**(3), 219–226.

Mistlberger, R. E., and Rusak, B. (1988). Food-anticipatory circadian rhythms in rats with paraventricular and lateral hypothalamic ablations. *J. Biol. Rhythms* **3**(3), 277–291.

Mrosovsky, N. (1996). Locomotor activity and non-photic influences on circadian clocks. *Biol. Rev. Camb. Philos. Soc.* **71**(3), 343–372.

Omata, K., and Kawamura, H. (1988). Effects of methamphetamine upon circadian rhythms in multiple unit activity inside and outside the suprachiasmatic nucleus in the golden hamster (*Mesocricetus auratus*). *Neurosci. Lett.* **95**(1–3), 218–222.

Ozaki, N., Nakahara, D., Kasahara, Y., and Nagatsu, T. (1991). The effect of methamphetamine on serotonin and its metabolite in the suprachiasmatic nucleus: A microdialysis study. *J. Neural Transm. Gen. Sect.* **86**(3), 175–179.

Pecoraro, N., Kosobud, A. E., Rebec, G. V., and Timberlake, W. (2000). Long T methamphetamine schedules produce circadian ensuing drug activity in rats. *Physiol. Behav.* **71**(1–2), 95–106.

Persons, J. E., Stephan, F. K., and Bays, M. E. (1993). Diet-induced obesity attenuates anticipation of food access in rats. *Physiol. Behav.* **54**(1), 55–64.

Pitts, S., Perone, E., and Silver, R. (2003). Food-entrained circadian rhythms are sustained in arrhythmic Clk/Clk mutant mice. *Am. J. Physiol. Regul. Integr. Comp. Physiol.* **285**(1), R57–R67.

Reppert, S. M., and Weaver, D. R. (2002). Coordination of circadian timing in mammals. *Nature* **418**(6901), 935–941.

Richter, C. P. (1922). A behavioristic study of the rat. *Comp. Psychol. Mono.* **1**, 1–55.

Rosenwasser, A. M., Pelchat, R. J., and Adler, N. T. (1984). Memory for feeding time: Possible dependence on coupled circadian oscillators. *Physiol. Behav.* **32**(1), 25–30.

Rutter, J., Reick, M., and McKnight, S. L. (2002). Metabolism and the control of circadian rhythms. *Annu. Rev. Biochem.* **71**, 307–331.

Rutter, J., Reick, M., Wu, L. C., and McKnight, S. L. (2001). Regulation of clock and NPAS2 DNA binding by the redox state of NAD cofactors. *Science* **293**(5529), 510–514.

Sakamoto, K., Nagase, T., Fukui, H., Horikawa, K., Okada, T., Tanaka, H., Sato, K., Miyake, Y., Ohara, O., Kako, K., and Ishida, N. (1998). Multitissue circadian expression of rat period homolog (rPer2) mRNA is governed by the mammalian circadian clock, the suprachiasmatic nucleus in the brain. *J. Biol. Chem.* **273**(42), 27039–27042.

Scheving, L. A. (2000). Biological clocks and the digestive system. *Gastroenterology* **119**(2), 536–549.

Scheving, L. E., Burns, E. R., Pauly, J. E., and Tsai, T. H. (1978). Circadian variation in cell division of the mouse alimentary tract, bone marrow and corneal epithelium. *Anat. Rec.* **191**(4), 479–486.

Scheving, L. E., Tsai, T. H., and Scheving, L. A. (1983). Chronobiology of the intestinal tract of the mouse. *Am. J. Anat.* **168**(4), 433–465.

Scheving, L. E., Tsai, T. H., and Scheving, L. A. (1989). Rhythmic behavior in the gastrointestinal tract. *In* "Ulcer Disease: New Aspects of Pathogenesis and Pharmacology" (S. Szabo and C. J. Pfeiffer, eds.), pp. 239–270. CRC Press, Boca Raton, FL.

Shibata, S., Minamoto, Y., Ono, M., and Watanabe, S. (1994a). Age-related impairment of food anticipatory locomotor activity in rats. *Physiol. Behav.* **55**(5), 875–878.

Shibata, S., Minamoto, Y., Ono, M., and Watanabe, S. (1994b). Aging impairs methamphetamine-induced free-running and anticipatory locomotor activity rhythms in rats. *Neurosci. Lett.* **172**(1–2), 107–110.

Stephan, F. K. (1981). Limits of entrainment to periodic feeding in rats with suprachiasmatic lesions. *J. Comp. Physiol.* **143**, 401–410.

Stephan, F. K. (1992). Resetting of a feeding-entrainable circadian clock in the rat. *Physiol. Behav.* **52**(5), 985–995.

Stephan, F. K. (2001). Food entrainable oscillators in mammals. "Handbook of Behavioral Neurobiology" (F. W. Turek and R. Y. Moore, eds.), vol. 12, pp. 223–241. Kluwer Academic, New York.

Stephan, F. K. (2002). The "other" circadian system: Food as a Zeitgeber. *J. Biol. Rhythms* **17**(4), 284–292.

Stephan, F. K., Swann, J. M., and Sisk, C. L. (1979a). Anticipation of 24-hr feeding schedules in rats with lesions of the suprachiasmatic nucleus. *Behav. Neural Biol.* **25**(3), 346–363.

Stephan, F. K., Swann, J. M., and Sisk, C. L. (1979b). Entrainment of circadian rhythms by feeding schedules in rats with suprachiasmatic lesions. *Behav. Neural Biol.* **25**(4), 545–554.

Stokkan, K. A., Yamazaki, S., Tei, H., Sakaki, Y., and Menaker, M. (2001). Entrainment of the circadian clock in the liver by feeding. *Science* **291**(5503), 490–493.

Terazono, H., Mutoh, T., Yamaguchi, S., Kobayashi, M., Akiyama, M., Udo, R., Ohdo, S., Okamura, H., and Shibata, S. (2003). Adrenergic regulation of clock gene expression in mouse liver. *Proc. Natl. Acad. Sci. USA* **100**(11), 6795–6800.

Terman, J. S., and Terman, M. (1980). Effects of illumination level on the rat's rhythmicity of brain self-stimulation behavior. *Behav. Brain Res.* **1**(6), 507–519.

Terman, M., and Terman, J. S. (1970). Circadian rhythm of brain self-stimulation behavior. *Science* **168**(936), 1242–1244.

Terman, M., and Terman, J. S. (1975). Control of the rat's circadian self-stimulation rhythm by light-dark cycles. *Physiol. Behav.* **14**(6), 781–789.

Wakamatsu, H., Yoshinobu, Y., Aida, R., Moriya, T., Akiyama, M., and Shibata, S. (2001). Restricted-feeding-induced anticipatory activity rhythm is associated with a phase-shift of the expression of mPer1 and mPer2 mRNA in the cerebral cortex and hippocampus but not in the suprachiasmatic nucleus of mice. *Eur. J. Neurosci.* **13**(6), 1190–1196.

Werme, M., Lindholm, S., Thoren, P., Franck, J., and Brene, S. (2002a). Running increases ethanol preference. *Behav. Brain Res.* **133**(2), 301–308.

Werme, M., Messer, C., Olson, L., Gilden, L., Thoren, P., Nestler, E. J., and Brene, S. (2002b). Delta FosB regulates wheel running. *J. Neurosci.* **22**(18), 8133–8138.

White, W., and Timberlake, W. (1999). Meal-engendered circadian-ensuing activity in rats. *Physiol. Behav.* **65**(4–5), 625–642.

Yamazaki, S., Numano, R., Abe, M., Hida, A., Takahashi, R., Ueda, M., Block, G. D., Sakaki, Y., Menaker, M., and Tei, H. (2000). Resetting central and peripheral circadian oscillators in transgenic rats. *Science* **288**(5466), 682–685.

Yamazaki, S., Straume, M., Tei, H., Sakaki, Y., Menaker, M., and Block, G. D. (2002). Effects of aging on central and peripheral mammalian clocks. *Proc. Natl. Acad. Sci. USA* **99**(16), 10801–10806.

Yoo, S. H., Yamazaki, S., Lowrey, P. L., Shimomura, K., Ko, C. H., Buhr, E. D., Siepka, S. M., Hong, H. K., Oh, W. J., Yoo, O. J., Menaker, M., and Takahashi, J. S. (2004). INAUGURAL ARTICLE: PERIOD2::LUCIFERASE real-time reporting of circadian dynamics reveals persistent circadian oscillations in mouse peripheral tissues. *Proc. Natl. Acad. Sci. USA* **101**(15), 5339–5346.

## [27]    Peripheral Clocks and the Regulation of Cardiovascular and Metabolic Function

By R. Daniel Rudic, Anne M. Curtis, Yan Cheng, and Garret FitzGerald

### Abstract

Circadian rhythms generated by cell autonomous biological clocks allow for the appropriate temporal synchronization of physiology and behavior, optimizing the efficiency of biological systems. Circadian oscillators and functions have been uncovered in both central and peripheral tissues. This article describes methodology, experimental design, and technical challenges pertaining to studies of circadian rhythms in the periphery. Experimental approaches are focused upon revealing the role of peripheral clocks in cardiovascular and metabolic function using *in vitro* and *in vivo* techniques.

### The Emerging Importance of Peripheral Clocks

Circadian rhythms generated by cell autonomous biological clocks allow for the appropriate temporal synchronization of physiology and behavior, optimizing the efficiency of biological systems (Panda et al., 2002; Reppert and Weaver, 2001). In mammals, the circadian timing system is hierarchical, with the master clock located in the hypothalamic suprachiasmatic nuclei (SCN) (Ralph et al., 1990). Circadian oscillators have been uncovered in both central and peripheral tissues, with the SCN presumed to coordinate cyclic gene expression in the periphery by neural and/or humoral signals (Balsalobre et al., 2000a; Cheng et al., 2002; Kramer et al., 2001a; Le Minh et al., 2001; McNamara et al., 2001). However, the autonomy of peripheral oscillators is now under debate as peripheral tissues explanted and maintained in culture demonstrate continued oscillations of Per2 for up to 20 days, and SCN lesioning does not abolish this circadian oscillation (Yoo et al., 2004). Moreover, the mechanisms of regulation of peripheral clocks and, indeed, their function remain largely obscure. Nonetheless, the SCN appears to have an important role coordinating phase in the periphery in individual animals (Yoo et al., 2004). A competitive area of research presently is to segregate dominant mechanisms of peripheral entrainment from "fine adjustment" signals, which might phase shift peripheral clocks to respond to local conditions of organ function.

METHODS IN ENZYMOLOGY, VOL. 393

While the interplay between the SCN and the periphery remains poorly understood, transcriptional regulation is central to the timing and accuracy of circadian rhythms in both cases. Pacemaker rhythms are generated and sustained by positive and negative transcriptional/translational feedback loops (Shearman *et al.*, 2000). Positive components include the bHLH-PAS proteins CLOCK and BMAL1 (also known as MOP3) driving transcription as functional heterodimers through E-box enhancer elements (CACGTG) (Gekakis *et al.*, 1998). Known targets for these effectors are the *Period* (*Per1–3*) and *Cryptochrome* (*Cry1–2*) genes (Reppert and Weaver, 2001). CLOCK and BMAL1 also drive the expression of the orphan nuclear receptor *Rev-erbα* in the SCN and liver (Preitner *et al.*, 2002). The *Bmal1* gene contains Rev-Erbα/ROR response elements in its promoter, and its expression is repressed by Rev-Erbα (Preitner *et al.*, 2002). Therefore, as *Per, Cry*, and *Rev-erbα* levels rise, *Bmal1* levels fall. The precise regulation of circadian timing systems extends beyond RNA kinetics and includes recently identified transcriptional and posttranslational modifications (Eide *et al.*, 2002; Kondratov *et al.*, 2003; Lee *et al.*, 2001, 2004).

## NPAS2 and Other Candidate Members of the Core Clock in the Periphery

NPAS2 (also known as MOP4) is a paralog of CLOCK, sharing 50% identity of the amino acid sequence. So far, it has been shown to operate in the core feedback loop in the vasculature and forebrain (Dudley *et al.*, 2003; McNamara *et al.*, 2001; Reick *et al.*, 2001). NPAS2, unlike CLOCK, is expressed predominantly in the forebrain rather than in the SCN (Reick *et al.*, 2001). We have observed a robust rhythm in NPAS2 mRNA expression in the aorta, kidney, and heart. While we found that NPAS2 can substitute for CLOCK in heterodimerizing with BMAL1 to drive E-box–dependent oscillatory gene expression, mutational analysis of the two genes reveals quite distinct phenotypes. Thus, CLOCK appears to be important in the early morning response and NPAS2 appears to have a role within locomotor activity/nighttime responses (Dudley *et al.*, 2003). NPAS2-deleted mice are unable to adapt to a daytime restricted feeding schedule (Dudley *et al.*, 2003).

BMAL2 is also a basic helix–loop–helix–Per–Arnt–Sim (bHLH-PAS) transcription factor that can form active complexes with CLOCK, NPAS2, and HIF1α (Hogenesch *et al.*, 2000). BMAL2 may also have a role in modulating the peripheral clock in the vasculature, specifically in the endothelium. BMAL2 expression coincides with BMAL1 expression in heart, liver, and kidney and in aortic endothelial and vascular smooth muscle cells (Schoenhard *et al.*, 2003). Multiple splice variants of BMAL2 have been identified (MOP9 and CLIF) that have varying transcriptional ability. Alternative splicing may allow for tissue-specific regulation of circadian rhythms by modulating interactions with selected coactivators (Schoenhard et al., 2002).

## ChIP Analysis of Promoter Occupancy

Per2 and Cry2 do not contain recognized consensus E-box promoter elements, although Per2 has several E-box sequences within its coding region. Whether these latter E boxes serve to potentiate transcriptional activity has yet to be defined. All five consensus E-box sequences within the Per1 promoter are known to contribute to transcriptional activity (Hida et al., 2000). However, the three E boxes proximal to the ATG start site of Per1 have the greatest effect on transcriptional activation (Hida et al., 2000). Chromatin immunoprecipitation is a powerful technique that can determine the sequence of events that occur on a promoter prior to transcriptional activation. We and others have used this technique to understand promoter occupancy on circadian promoters such as Per1, Per2, and Cry1 (Curtis et al., 2004; Etchegaray et al., 2003). We have successfully used ChIP analysis in HeLa cells and heart tissue using a protocol adapted from the Farnham group (http://genomecenter.ucdavis.edu/farnham/farnham/protocols/chips.html).

Tissues are harvested at circadian times, flash frozen, and pulverized. One percent formaldehyde is used to cross-link DNA to protein, and then nuclear extracts containing chromatin are prepared. Nuclear extracts are sonicated to shear DNA to lengths between 200 and 1000 bp. The time and number of pulses must be determined for each sample type. The DNA is purified by phenol/chloroform extraction and analyzed by agarose gel electrophoresis to visualize shearing efficiency. Lysates are then precleared and immunoprecipitations are performed overnight, always including a "no antibody" control or preimmune sample. Immunocomplexes are washed extensively, and 1 $\mu$l of RNase A (10 mg/ml) and 0.3 $M$ NaCl final concentration are added to samples for 4–5 h at 67° to reverse histone–DNA cross-links. DNA is extracted by phenol chloroform extraction, and polymerase chain reaction (PCR) is performed on regions bracketing the consensus sequence of interest to quantify the amount of protein or histone modification that is present on the consensus sequence. The limiting factor in this assay is the affinity of the selected antibody. In fact, many commercially available antibodies are not suitable for this method. Because the final step of this assay is a PCR reaction, nonspecific binding of the antibody to other chromatin-associated proteins may cause amplification of nonspecific DNA regions, thus reducing assay sensitivity.

Analysis of Circadian Rhythms by Serum Shock Analysis

Balsalobre *et al.* (1998) demonstrated that when cultured cells are made quiescent for 48–72 h using low serum media and then stimulated for approximately 2 h with 50% serum, circadian gene expression was induced *in vitro*. In fact, Schibler and colleagues have determined that rather than inducing circadian rhythms *de novo* within cells, serum acts to phase align existing rhythms amongst adjacent cells (Nagoshi *et al.*, 2004). Using a single cell reporter assay, they demonstrated that circadian gene oscillations do occur, even in the absence of serum shock, albeit lacking unison. Thus, in unstimulated multi-cellular system assays, circadian oscillations persist in individual cells, but are obscured due to signal interference caused by asynchrony between cells. Thus, circadian rhythms are self-sustained and cell autonomous in cultured fibroblasts.

Period lengths ($\pi$s) of genes were described to be approximately 24 h. The oscillatory expression pattern of the Period gene induced by serum shock in these cultured fibroblastic cells was remarkably similar to the circadian pattern of expression observed in the SCN of animals. In addition, the temporal sequence of expression of these circadian genes by serum shock was similar to the mRNA accumulations that occur *in vivo* (Balsalobre *et al.*, 1998; Duffield *et al.*, 2002). Thus, the serum shock model has become a popular *in vitro* approach to dissect the molecular mechanisms that comprise circadian rhythms. Many factors, such as cAMP, protein kinase C, glucocorticoid hormones, and $Ca^{2+}$ can all cause upregulation and subsequent cycling of circadian components such as Per1 and Per2 (Balsalobre *et al.*, 2000b).

Using a yeast two-hybrid screen, with RXR as the bait, in a vascular smooth muscle cell library, we identified NPAS2 as an interacting clone (McNamara *et al.*, 2001). Further analysis revealed both NPAS2 and its paralog CLOCK, but not BMAL1, interacted with the nuclear hormone receptors RXR and RAR and that the interaction was ligand dependent and involved the LxxLL motif of CLOCK and NPAS2. RXR and RAR, once activated, were able to block CLOCK/MOP4:BMAL1-induced transcription (McNamara *et al.*, 2001). Studies have identified the interaction of clock components with histone acetlytransferase enzymes and demonstrated that this interaction is important for transcriptional activity (Curtis *et al.*, 2004; Etchegaray *et al.*, 2003). Therefore, it is plausible that by blocking the LxxLL motif of CLOCK and NPAS2, which is the docking site for many of these coactivators, nuclear receptors may inhibit interaction of the clock components with coactivators, thereby repressing transcriptional activity. This interplay between nuclear receptor signaling and circadian function suggests that humoral factors, such as steroid hormones and vitamins, may serve as regulators of peripheral clock signaling.

Using the serum shock model, we examined the effect of various ligands administered at different times postserum induction upon circadian gene expression. Human vascular smooth muscle cells are grown to confluence and serum starved for 48 h, after which 50% serum is added for 2 h and then replaced with serum-free media to induce rhythms of hPer2. Retinoic acid (RA) (1 $\mu M$) is added at 6, 12, 18, and 24 h (T6, T12, T18, T24) postserum induction. The greatest shift in Per2 expression occurs when ligand is added to cells at T18, which causes a 4-h delay in the peak of Per2 from T24 to T28 and again from T48 to T52. The glucocorticoid analog dexamethasone had the opposite effect, whereby its addition at T18 caused a phase advance in Per2 from T24 to T20. In contrast, the thyroid hormone ligand T3 had no effect on Per2 oscillation.

Use of the serum shock model to study the effects of ligands does present some challenges. The "batch-to-batch" differences in commercially procured serum may produce considerable variability in the increment and phase of induction evoked. This is reflected by variability in the timing and amplitude of the circadian cycles and, in our experience, has been most prevalent when using primary cells. An additional source of potential artifact is the change of media, which alone can induce and phase shift rhythms of gene expression (Hirota et al., 2002).

Zeitgeber, Circadian Time, and Constant Darkness: Lights on, Lights off?

*Zeitgeber* refers to the most dominant environmental signal impinging on the SCN. As it relates to circadian rhythms, light is the most dominant zeitgeber. Although irrelevant to studies *in vitro*, the manner of light entrainment is an important consideration in studies involving *in vivo* model systems. A matter of confusion is that light does not generate the *periodicity* of circadian rhythms at 24 h. Light only sets the *timing* of circadian rhythms, in particular as related to gene expression, activity, and behavior and other circadian functions. A true circadian rhythm persists under constant light or dark conditions. In preparation for circadian studies, animals are first acclimated or entrained to the light–dark cycle so that each animal rhythm is aligned in unison. Light entrainment is accomplished by exposing newly procured mice for at least 2 weeks to a 12-h light/12-h dark (LD) light cycle. Time in hours under these conditions is denoted by ZT because light acts as a zeitgeber. ZT0 refers to the beginning of daylight in an entrained cycle and ZT12 is the beginning of night, under experimental conditions of 12 h in light and 12 h in dark (12:12 LD). These rhythms are diurnal rhythms. Under conditions of constant darkness (DD), one resets light timers so that after 12 h of darkness at 7 PM, mice remain in complete darkness for a minimum of 36 h before experimentation. Mice remain in constant darkness throughout the remainder of the experiment. Subjective "lights on" occurs at 7 AM and "lights off" at 7 PM. Under these conditions, time is subjective because the animals are not exposed to 12:12 L/D and is denoted as circadian time (CT). Under conditions of DD, a rhythm can be truly characterized as being circadian, although generally, diurnal rhythms in large part also resolve as circadian rhythms.

## Isolation of RNA from Murine Aorta

Truly comprehensive circadian studies can be a test of endurance, raw materials, and space. In our studies, we conducted time course experiments that spanned a minimum of 48 h, which span two circadian cycles, at 4- to 6-h time intervals. We required 4–6 mice per time point, usually 48–72 mice for thoracic aortic tissue analyses. In these studies, we took a global approach to tissue analysis, including perivascular fat. We have, by contrast, studied the distinct clocks in endothelial cells, vascular smooth muscle cells, and adipocytes *in vitro*.

At respective time points, anesthetized or euthanized (via cervical dislocation) mice are exsanguinated by cardiac puncture and infused with 10 ml of physiologic saline. Ten milliliters of diethyl pyrocarbonate (DEPC) water is than similarly infused, and aortae are dissected (Fig. 1), placed in Eppendorf tubes, and flash frozen in liquid nitrogen. Additional dissection may include removal of perivascular adipose tissue.

Processing of vascular tissue requires a rigorous method of pulverization. Arteries are elastic and resilient tissues. We have used the BioPulverizer (Biospec Products Inc.) and the Plattner mortar and pestle (Fisher Scientific) to reduce frozen aortae to fine powders. The BioPulverizer can be rapidly frozen prior to use by immersion in liquid nitrogen, whereas the Plattner system is autoclaved, cooled, and stored at $-80°$. Cross-contamination between samples is avoided by using fresh, autoclaved sheets of aluminum foil to cover the mortar, which is then flattened to a sheet using the pestle. A uniform number of hammer strikes is applied to the pestle to adequately pulverize the aorta. Using a fine, autoclaved spatula (precooled in liquid nitrogen), the "aortic pancake" is picked and immersed in a sterile and cooled Eppendorf tube containing 1 ml Trizol ($4°$, GIBCO) and further subjected to sonication (10 s, level 1–4, Sonicator 3000) or tissue tearing (1 min). Then 200 $\mu$l choloroform and 1 $\mu$l of glycogen (Boehringer) are added to the homogenate, which is then vortexed and left to sit for 3 min at room temperature. The tube is then spun at $4°$ for 15 min at 14,000$g$. The supernatant is transferred to a new tube, to which 0.5 ml isopropanol is added. The Eppendorf is inverted several times and

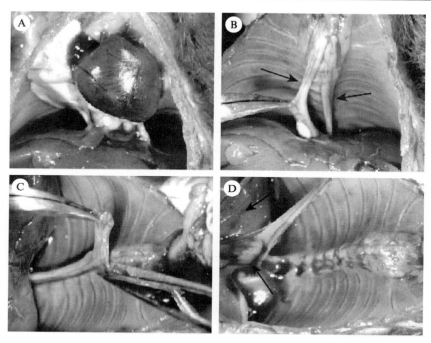

Fig. 1. Rough dissection of the murine thoracic aorta. (A) After euthanizing mice, the thoracic cavity is exposed by excising the chest wall, revealing the heart and lungs. The heart and lungs are then dissected and removed, revealing the esophagus (loosely attached to the integuments) and the aorta (aligned to the spinal cord). (B) The esophagus and aorta have some gross features that are structurally similar; however, their position within the thoracic cavity is different. The esophagus is ventral (*left arrow*) to the aorta, positioned posteriorially (*right arrow*), shown here by displacing the esophagus laterally. (C) After removal of the esophagus, the aorta is dissected by cutting at an oblique angle to the spinal cord. (D) The final segment of the thoracic aorta (*bottom arrow*) is excised at the border of the diaphragm (*top arrow*).

incubated for 30 min at room temperature. RNA is then precipitated by spinning for 20 min at 4°. The supernatant is removed carefully and the precipitated RNA is washed with 70% ethanol in DEPC water. The wash is then dispensed and the pellet left to dry for approximately 15 min. min. A single aorta may then be immersed in 50–100 $\mu$l of DEPC water, which is quantified with a spectrophotometer. We have then utilized RNA for hybridization to microarray gene chips and/or ribonuclease protection assays. Data are analyzed by COSOPT analysis, which is discussed extensively by Walker and Hogenesch (2005).

## Circadian Variation in Metabolism and Diabetes

In humans and animals, glucose and lipid homeostasis vary according to the time of day. With respect to glucoregulation, glucose, insulin, insulin release, and many metabolic enzymes have been demonstrated to oscillate with a circadian rhythm (Gagliardino and Hernandez 1971; la Fleur *et al.*, 2001; Malherbe *et al.*, 1969; Strubbe *et al.*, 1987). In fact, SCN ablation was shown to abolish rhythms in plasma levels of glucose and insulin (Strubbe *et al.*, 1987). Similarly, with respect to lipids, triglycerides and lipid-metabolizing enzymes also oscillate (Arasaradnam *et al.*, 2002; Benavides *et al.*, 1998; Castro Cabezas *et al.*, 2001; Schlierf and Dorow, 1973).

The clinical significance of rhythmic metabolic profiles is suggested by observations in patients with diabetes. Diabetics exhibit deviations from the normal circadian metabolic physiology. The Somogyi phenomenon is when certain patients with type I diabetes exhibit an exaggerated counterregulatory response to nocturnal hypoglycemia, resulting in early morning hyperglycemia (DeLawter, 1991). In animal models, insulin sensitivity also varies with time of day. Phase shifting of the circadian expression of molecular clock proteins occurs in streptozotozin-induced diabetic mice. Oishi *et al.* (2004) found that Per1 gene expression was differentially regulated after streptozotocin compared with the vehicle. Similarly, Young *et al.* (2002) also found that circadian expression of Cry and Per was altered by streptozotocin treatment. However, it is still unknown whether peripheral clocks in some way influence the development or expression of type 1 diabetes.

## Adapting Metabolic Assays to Assess Circadian Variations in Mice

### Baseline Blood Sampling

Basal levels of glucose are known to vary in blood. We have performed assays to quantify blood glucose levels at varying times of day. One important feature of performing these assays is habituating the mice to handling because initial agitation may cause variation in baseline glucose levels. Unanesthetized mice are placed in 50-ml conical tubes for restraint (Fig. 2A, left). The conical tubes are wrapped with black tape or paper in order to limit the exposure of mice to light required during dissections. We choose the saphenous vein for harvesting blood for analysis for a variety of reasons: (1) large blood volumes can be obtained, in excess of 200 $\mu$l, if needed, (2) repeat bleeds are feasible, and (3) intraorbital bleeds, which are commonly used, may result in retinal bleeds, which might potentially affect circadian rhythmicity. The hind leg is shaved in the region of the

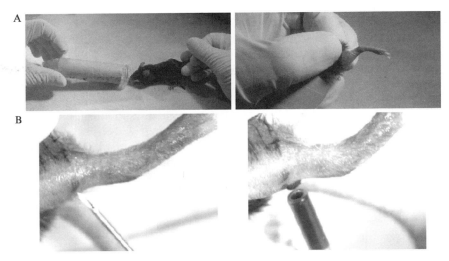

FIG. 2. Isolating whole blood via the saphenous vein. (A) Mice are restrained by gently guiding and inserting them into 50-ml conical tubes. (B) The saphenous vein is lanced using a needle and blood is collected.

gastrocnemius, revealing the saphenous vein. The left hind leg is grasped, extending the lower limb (Fig. 2A, right). A generous portion of silicon jelly (type L, Apiezon) is applied to the exposed vein, which increases surface tension, allowing the blood droplets to form beads. The saphenous vein is then lanced with a 21-gauge needle (Fig. 2B, left), and blood is collected in heparinized microcuvette collection tubes (Sarstedt) (Fig. 2B, right). Blood is placed on ice for 30 min and then centrifuged at 3000$g$ for 10 min to isolate the supernatant. We have performed these experiments with repeated sampling up to 34 h after initial sampling, at sampling frequencies of every 4–6 h, with 6–10 new mice at each time point. Repeated sampling of the same mice is also feasible, although depletion of blood volume is always a consideration. A typical ELISA/RIA assay requires 5 $\mu$l of plasma, or 10 $\mu$l of whole blood. The following guidelines may be helpful in calculating sampling frequency.

For repeated sampling daily over repeated days, where 6% is the average blood volume proportionate to body weight and 10% of this volume is the maximum that is usually permitted to be withdrawn (Hoff, 2000), a typical mouse of 25 g allows for 25 $\times$ 6% $\times$ 10% = 0.15 ml or 150 $\mu$l to be withdrawn in total, usually corresponding to three assays. This serves only as a general guide, and all experimental details are subject to review and approval in accordance with individual institutional guidelines.

## Intraperitoneal Pyruvate (PT), Glucose (GT), and Insulin Tolerance (IT) Tests

Intraperitoneal pyruvate, glucose, and insulin tolerance tests are useful *in vivo* assays that provide approximations of glucose metabolism and homeostasis. In each assay, the changes in blood glucose are measured in response to the injected substances, pyruvate, glucose, or insulin, which then provides discrete information about glucose metabolism. In the PT, injection of pyruvate results in an approximate measure of gluconeogenesis, reflected by the conversion of pyruvate into glucose over 90–180 min after a bolus injection of glucose. Intraperitoneal injection of glucose in the GT causes a gradual increase in plasma glucose. Approximately 60 min after injection, glucose levels decline due (in large part) to insulin secretion. Insulin is injected intraperitoneally for the IT, and its ability to decrease glucose levels is assessed. Although isolation of baseline bloods is completed within several minutes for a single mouse, the IT lasts 90 min for a single mouse at minimum. In these and GT studies, we measure the response 30, 60, and 90 min after the initial injection, by which time baseline glycemia (euglycemia) is usually restored. Individual mice from the different groups in the study must be staggered to obtain timely results with respect to the circadian cycle. We find that we can at most study 15 mice with two operators for this experimental paradigm. The time necessary to measure baseline glucose and inject 15 mice is about 30 min. At 30 min, which again approximates the time to the 15th injection, the second round of glucose readings ensues, now postinjection. This method of sampling is repeated at 60 and 90 min. In the PT, GT, and IT, we measure glucose levels in whole blood obtained from a scalpel blade nick in the lateral tail vein. Rapid glucose measurements are made with a handheld glucometer (LifeScan, One Touch). We have used this approach to characterize metabolic oscillations in mice, including rhythms in plasma levels of glucose, triglycerides, adiponectin, and corticosterone (Rudic *et al.*, 2004). Moreover, in mice lacking a functional molecular clock, we have demonstrated a profound hypoglycemic response to insulin and a dramatic impairment in gluconeogenesis.

## The Frequent Sampling Intravenous Glucose Tolerance Test (FSIGT)

Although the GT and IT provide a sense of the insulin response, *quantitative* assessments of insulin-mediated responses require a more rigorous test of insulin secretion and sensitivity, such as an insulin clamp or the FSIGT. The FSIGT is an extension of the GT. However, glucose is administered intravenously, and blood samplings are made with greater frequency: minimally 0, 1, 3, 5, 10, 20, and 50 min postglucose infusion. Another difference in this assay is that plasma is analyzed for both glucose and insulin. Thus, a larger sampling volume than that needed for glucometer assays is required. Approximately 20 $\mu$l/time point is necessary to obtain sufficient volume for plasma glucose and insulin analysis. Under conditions in which a single individual performs the injections, subsequent mouse infusions should occur within 5 or 10 min of the first injection. Under these conditions, 15 FSIGTs can be completed within 2 h.

Blood Pressure Studies

Radiotelemetry devices are now used to measure long-term blood pressure (BP), heart rate, and pattern of activity in small animals such as mice. This recent technology circumvents many of the problems associated with conventional BP methods such as the tail cuff device or the use of exteriorized, fluid-filled catheters (Van Vliet et al., 2000). The use of radiotelemetry has been described in more detail elsewhere (Kramer et al., 2001b; Mills et al., 2000). In brief, the implant consists of a 5-cm-long fluid-filled catheter with a thin-walled tip that refers intraarterial pressure to a sensor located in the body of the implant. The catheter can be implanted in the carotid artery or in the abdominal aorta while the transmitter body is implanted subcutaneously (carotid artery implant) or in the peritoneal cavity (abdominal aortic implant). We have successfully implanted transmitters in animals as small as 23 g using the carotid artery implant procedure. We found that anatomical placement of the transmitter on the left side of the back and leading the catheter under the skin of the left shoulder prevents necrosis from forming on the skin overlying the transmitter. The BP signal is received by a computerized receiver placed underneath each cage. This system permits BP measurements with minimal acute agitation to the animal. After 10 days postsurgery BP is restored to normal, and a circadian rhythm is evident in BP (Fig. 3).

Circadian Profiles in Cardiovascular Physiology and Disease

Certain aspects of cardiovascular physiology exhibit diurnal variation, although the contribution of central or peripheral clocks to these variations is largely conjectural. For example, diurnal variations have been identified in measures of endothelial function (Shaw et al., 2001), platelet activation (Besterman et al., 1967), fibrinolytic activity (Menon et al., 1967), arterial BP (Gupta and Scopes, 1965; Kaneko et al., 1968), glucose tolerance, and insulin sensitivity (Faiman and Moorhouse, 1967), as well as in catecholamines (Faiman and Moorhouse, 1967) and cortisol (Menon et al., 1967).

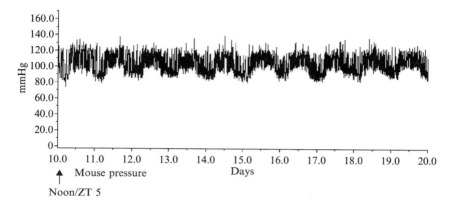

FIG. 3. Diurnal variation in blood pressure. Blood pressure was sampled at 5-min intervals by radiotelemetry in a male, C57B16 mouse. Sampling began at noon (ZT5) on the 10th day after operation.

Furthermore, cardiovascular disease has been associated with a loss of circadian rhythms. Patients with coronary heart disease lose rhythms in endothelial function (Shaw *et al.*, 2001). The incidence of acute vascular events also exhibits diurnal variation and the severity of experimentally induced stroke varies as a function of clock time (Vinall *et al.*, 2000). However, the hypothetical contribution of the molecular clock to such diurnal variations is confounded with alterations in activity and stress related to the sleep/wakefulness cycle, and genetic variants of the core clock genes remain to be related to cardiovascular events.

Cardiovascular deaths are increased in shift workers, which has been hypothesized to relate to an inversion of metabolic rhythms, such as those in plasma triglycerides and insulin (Lund *et al.*, 2001). Night shift workers in Antarctica, for example, exhibit significant elevations in integrated postprandial glucose, insulin, and triacylglycerol responses during the night compared with their responses when they are subject to normal daytime working. Perhaps abnormal metabolic responses to meals taken at night during unadapted night shifts reflect a relative resistance to insulin. This, in turn, could contribute to the documented cardiovascular morbidity associated with shift work (Lund *et al.*, 2001).

The development of ambulatory BP monitoring devices in the 1960s confirmed by direct measurement that BP varies according to the time of day (Zulch and Hossmann, 1967). In humans, BP is higher in daytime hours than at nighttime, coinciding with decreased sympathetic activity that occurs at night. In contrast, mice that exhibit nocturnal activity patterns have higher BP at night (Li *et al.*, 1999). Moreover, lesion of the SCN nearly abolishes circadian variability of BP in rodents (Janssen *et al.*, 1994; Witte *et al.*, 1998). The precision of these ablations remain to be confirmed by studies in mutant mice. Another study in mice demonstrated that light pulses increase renal nerve activity and decrease gastric vagus nerve activity, each thought, respectively, to reflect sympathetic and parasympathetic activity. These effects on sympathetic and parasympathetic activity were also abolished by ablation of the SCN (Mutoh *et al.*, 2003).

## Conclusion

Interest in peripheral clocks has blossomed. Remarkably little is known about how these clocks are entrained by the SCN, the circumstances in which they exhibit autonomy, their local regulation and the mechanisms by which it may exhibit heterogeneity, and the overall contribution of the molecular clock to peripheral functions. These studies reflect some of the practical issues related to initiating studies in this area.

## References

Arasaradnam, M. P., Morgan, L., Wright, J., and Gama, R. (2002). Diurnal variation in lipoprotein lipase activity. *Ann. Clin. Biochem.* **39,** 136–139.

Balsalobre, A., Brown, S. A., Marcacci, L., Tronche, F., Kellendonk, C., Reichardt, H. M., Schutz, G., and Schibler, U. (2000a). Resetting of circadian time in peripheral tissues by glucocorticoid signaling. *Science* **289,** 2344–2347.

Balsalobre, A., Damiola, F., and Schibler, U. (1998). A serum shock induces circadian gene expression in mammalian tissue culture cells. *Cell* **93,** 929–937.

Balsalobre, A., Marcacci, L., and Schibler, U. (2000b). Multiple signaling pathways elicit circadian gene expression in cultured Rat-1 fibroblasts. *Curr. Biol.* **10,** 1291.

Benavides, A., Siches, M., and Llobera, M. (1998). Circadian rhythms of lipoprotein lipase and hepatic lipase activities in intermediate metabolism of adult rat. *Am. J. Physiol.* **275,** R811–R817.

Besterman, E., Myat, G., and Travadi, V. (1967). Diurnal variations of platelet stickiness compared with effects produced by adrenaline. *Br. Med. J.* **1,** 597–600.

Castro Cabezas, M., Halkes, C. J., Meijssen, S., van Oostrom, A. J., and Erkelens, D. W. (2001). Diurnal triglyceride profiles: A novel approach to study triglyceride changes. *Atherosclerosis* **155,** 219–228.

Cheng, M. Y., Bullock, C. M., Li, C., Lee, A. G., Bermak, J. C., Belluzzi, J., Weaver, D. R., Leslie, F. M., and Zhou, Q. Y. (2002). Prokineticin 2 transmits the behavioural circadian rhythm of the suprachiasmatic nucleus. *Nature* **417,** 405–410.

Curtis, A. M., Seo, S. B., Westgate, E. J., Rudic, R. D., Smyth, E. M., Chakravarti, D., FitzGerald, G. A., and McNamara, P. (2004). Histone acetyltransferase-dependent chromatin remodeling and the vascular clock. *J. Biol. Chem.* **279,** 7091–7097.

DeLawter, D. E. (1991). The management of early morning hyperglycemia: Is it due to Somogyi effect or dawn phenomenon? *Md. Med. J.* **40,** 391.

Dudley, C. A., Erbel-Sieler, C., Estill, S. J., Reick, M., Franken, P., Pitts, S., and McKnight, S. L. (2003). Altered patterns of sleep and behavioral adaptability in NPAS2-deficient mice. *Science* **301,** 379–383.

Duffield, G. E., Best, J. D., Meurers, B. H., Bittner, A., Loros, J. J., and Dunlap, J. C. (2002). Circadian programs of transcriptional activation, signaling, and protein turnover revealed by microarray analysis of mammalian cells. *Curr. Biol.* **12**, 551–557.

Eide, E. J., Vielhaber, E. L., Hinz, W. A., and Virshup, D. M. (2002). The circadian regulatory proteins BMAL1 and cryptochromes are substrates of casein kinase Iepsilon. *J. Biol. Chem.* **277**, 17248–17254.

Etchegaray, J. P., Lee, C., Wade, P. A., and Reppert, S. M. (2003). Rhythmic histone acetylation underlies transcription in the mammalian circadian clock. *Nature* **421**, 177–182.

Faiman, C., and Moorhouse, J. A. (1967). Diurnal variation in the levels of glucose and related substances in healthy and diabetic subjects during starvation. *Clin. Sci.* **32**, 111–126.

Gagliardino, J. J., and Hernandez, R. E. (1971). Circadian variation of the serum glucose and immunoreactive insulin levels. *Endocrinology* **88**, 1532–1534.

Gekakis, N., Staknis, D., Nguyen, H. B., Davis, F. C., Wilsbacher, L. D., King, D. P., Takahashi, J. S., and Weitz, C. J. (1998). Role of the CLOCK protein in the mammalian circadian mechanism. *Science* **280**, 1564–1569.

Gupta, J. M., and Scopes, J. W. (1965). Observations on blood pressure in newborn infants. *Arch. Dis. Child* **40**, 637–644.

Hida, A., Koike, N., Hirose, M., Hattori, M., Sakaki, Y., and Tei, H. (2000). The human and mouse Period1 genes: Five well-conserved E-boxes additively contribute to the enhancement of mPer1 transcription. *Genomics* **65**, 224–233.

Hirota, T., Okano, T., Kokame, K., Shirotani-Ikejima, H., Miyata, T., and Fukada, Y. (2002). Glucose down-regulates Per1 and Per2 mRNA levels and induces circadian gene expression in cultured Rat-1 fibroblasts. *J. Biol. Chem.* **277**, 44244–44251.

Hoff, J. (2000). Methods of blood collection in the mouse. *Lab. Anim.* **29**, 47–53.

Hogenesch, J. B., Gu, Y. Z., Moran, S. M., Shimomura, K., Radcliffe, L. A., Takahashi, J. S., and Bradfield, C. A. (2000). The basic helix-loop-helix-PAS protein MOP9 is a brain-specific heterodimeric partner of circadian and hypoxia factors. *J. Neurosci.* **20**, RC83.

Janssen, B. J., Tyssen, C. M., Duindam, H., and Rietveld, W. J. (1994). Suprachiasmatic lesions eliminate 24-h blood pressure variability in rats. *Physiol. Behav.* **55**, 307–311.

Kaneko, M., Zechman, F. W., and Smith, R. E. (1968). Circadian variation in human peripheral blood flow levels and exercise responses. *J. Appl. Physiol.* **25**, 109–114.

Kondratov, R. V., Chernov, M. V., Kondratova, A. A., Gorbacheva, V. Y., Gudkov, A. V., and Antoch, M. P. (2003). BMAL1-dependent circadian oscillation of nuclear CLOCK: Posttranslational events induced by dimerization of transcriptional activators of the mammalian clock system. *Genes Dev.* **17**, 1921–1932.

Kramer, A., Yang, F. C., Snodgrass, P., Li, X., Scammell, T. E., Davis, F. C., and Weitz, C. J. (2001a). Regulation of daily locomotor activity and sleep by hypothalamic EGF receptor signaling. *Science* **294**, 2511–2515.

Kramer, K., Kinter, L., Brockway, B. P., Voss, H. P., Remie, R., and Van Zutphen, B. L. (2001b). The use of radiotelemetry in small laboratory animals: Recent advances. *Contemp. Top. Lab. Anim. Sci.* **40**, 8–16.

la Fleur, S. E., Kalsbeek, A., Wortel, J., Fekkes, M. L., and Buijs, R. M. (2001). A daily rhythm in glucose tolerance: A role for the suprachiasmatic nucleus. *Diabetes* **50**, 1237–1243.

Lee, C., Etchegaray, J. P., Cagampang, F. R., Loudon, A. S., and Reppert, S. M. (2001). Posttranslational mechanisms regulate the mammalian circadian clock. *Cell* **107**, 855–867.

Lee, C., Weaver, D. R., and Reppert, S. M. (2004). Direct association between mouse PERIOD and CKIepsilon is critical for a functioning circadian clock. *Mol. Cell. Biol.* **24**, 584–594.

Le Minh, N., Damiola, F., Tronche, F., Schutz, G., and Schibler, U. (2001). Glucocorticoid hormones inhibit food-induced phase-shifting of peripheral circadian oscillators. *EMBO J.* **20**, 7128–7136.

Li, P., Sur, S. H., Mistlberger, R. E., and Morris, M. (1999). Circadian blood pressure and heart rate rhythms in mice. *Am. J. Physiol.* **276,** R500–R504.

Lund, J., Arendt, J., Hampton, S. M., English, J., and Morgan, L. M. (2001). Postprandial hormone and metabolic responses amongst shift workers in Antarctica. *J. Endocrinol.* **171,** 557–564.

Malherbe, C., De Gasparo, M., De Hertogh, R., and Hoet, J. J. (1969). Circadian variations of blood sugar and plasma insulin levels in man. *Diabetologia* **5,** 397–404.

McNamara, P., Seo, S. B., Rudic, R. D., Sehgal, A., Chakravarti, D., and FitzGerald, G. A. (2001). Regulation of CLOCK and MOP4 by nucear hormone receptors in the vasculature: A humoral mechanism to reset a peripheral clock. *Cell* **105,** 877–889.

Menon, I. S., Smith, P. A., White, R. W., and Dewar, H. A. (1967). Diurnal variations of fibrinolytic activity and plasma-11-hydroxycorticosteroid levels. *Lancet* **2,** 531–533.

Mills, P. A., Huetteman, D. A., Brockway, B. P., Zwiers, L. M., Gelsema, A. J., Schwartz, R. S., and Kramer, K. (2000). A new method for measurement of blood pressure, heart rate, and activity in the mouse by radiotelemetry. *J. Appl. Physiol.* **88,** 1537–1544.

Mutoh, T., Shibata, S., Korf, H. W., and Okamura, H. (2003). Melatonin modulates the light-induced sympathoexcitation and vagal suppression with participation of the suprachiasmatic nucleus in mice. *J. Physiol.* **547,** 317–332.

Nagoshi, E., Saini, C., Bauer, C., Laroche, T., Naef, F., and Schibler, U. (2004). Circadian gene expression in individual fibroblasts: Cell-autonomous and self-sustained oscillators pass time to daughter cells. *Cell* **119,** 693–705.

Oishi, K., Kasamatsu, M., and Ishida, N. (2004). Gene- and tissue-specific alterations of circadian clock gene expression in streptozotocin-induced diabetic mice under restricted feeding. *Biochem. Biophys. Res. Commun.* **317,** 330–334.

Panda, S., Hogenesch, J. B., and Kay, S. A. (2002). Circadian rhythms from flies to human. *Nature* **417,** 329–335.

Preitner, N., Damiola, F., Lopez-Molina, L., Zakany, J., Duboule, D., Albrecht, U., and Schibler, U. (2002). The orphan nuclear receptor REV-ERB controls circadian transcription within the positive limb of the mammalian circadian oscillator. *Cell* **110,** 251–260.

Ralph, M. R., Foster, R. G., Davis, F. C., and Menaker, M. (1990). Transplanted suprachiasmatic nucleus determines circadian period. *Science* **247,** 975–978.

Reick, M., Garcia, J. A., Dudley, C., and McKnight, S. L. (2001). NPAS2: An analog of clock operative in the mammalian forebrain. *Science* **5,** 5.

Reppert, S. M., and Weaver, D. R. (2001). Molecular analysis of mammalian circadian rhythms. *Annu. Rev. Physiol.* **63,** 647–676.

Rudic, R. D., McNamara, P., Curtis, A. M., Boston, R. C., Panda, S., Hogenesch, J. B., and FitzGerald, G. A. (2004). BMAL1 and clock, two essential components of the circadian clock, are involved in glucose homeostasis. *PLoS Biol.* **2,** e377.

Schlierf, G., and Dorow, E. (1973). Diurnal patterns of triglycerides, free fatty acids, blood sugar, and insulin during carbohydrate-induction in man and their modification by nocturnal suppression of lipolysis. *J. Clin. Invest.* **52,** 732–740.

Schoenhard, J. A., Eren, M., Johnson, C. H., and Vaughan, D. E. (2002). Alternative splicing yields novel BMAL2 variants: Tissue distribution and functional characterization. *Am. J. Physiol. Cell Physiol.* **283,** C103–C114.

Schoenhard, J. A., Smith, L. H., Painter, C. A., Eren, M., Johnson, C. H., and Vaughan, D. E. (2003). Regulation of the PAI-1 promoter by circadian clock components: Differential activation by BMAL1 and BMAL2. *J. Mol. Cell Cardiol.* **35,** 473–481.

Shaw, J. A., Chin-Dusting, J. P., Kingwell, B. A., and Dart, A. M. (2001). Diurnal variation in endothelium-dependent vasodilatation is not apparent in coronary artery disease. *Circulation* **103,** 806–812.

Shearman, L. P., Sriram, S., Weaver, D. R., Maywood, E. S., Chaves, I., Zheng, B., Kume, K., Lee, C. C., van der Horst, G. T., Hastings, M. H., and Reppert, S. M. (2000). Interacting molecular loops in the mammalian circadian clock. *Science* **288,** 1013–1019.

Strubbe, J. H., Prins, A. J., Bruggink, J., and Steffens, A. B. (1987). Daily variation of food-induced changes in blood glucose and insulin in the rat and the control by the suprachiasmatic nucleus and the vagus nerve. *J. Auton. Nerv. Syst.* **20,** 113–119.

Van Vliet, B. N., Chafe, L. L., Antic, V., Schnyder-Candrian, S., and Montani, J. P. (2000). Direct and indirect methods used to study arterial blood pressure. *J. Pharmacol. Toxicol. Methods* **44,** 361–373.

Vinall, P. E., Kramer, M. S., Heinel, L. A., and Rosenwasser, R. H. (2000). Temporal changes in sensitivity of rats to cerebral ischemic insult. *J. Neurosurg.* **93,** 82–89.

Walker, J. R., and Hogenesch, J. B. (2005). RNA profiling in circadian biology. *Methods Enzymol.* **393,** Chapter 16.

Witte, K., Schnecko, A., Buijs, R. M., van der Vliet, J., Scalbert, E., Delagrange, P., Guardiola-Lemaitre, B., and Lemmer, B. (1998). Effects of SCN lesions on circadian blood pressure rhythm in normotensive and transgenic hypertensive rats. *Chronobiol. Int.* **15,** 135–145.

Yoo, S. H., Yamazaki, S., Lowrey, P. L., Shimomura, K., Ko, C. H., Buhr, E. D., Siepka, S. M., Hong, H. K., Oh, W. J., Yoo, O. J., Menaker, M., and Takahashi, J. S. (2004). LUCIFERASE::PERIOD2 real-time reporting of circadian dynamics reveals persistent circadian oscillations in mouse peripheral tissues. *Proc. Natl. Acad. Sci. USA* **101,** 5339–5346.

Young, M. E., Wilson, C. R., Razeghi, P., Guthrie, P. H., and Taegtmeyer, H. (2002). Alterations of the circadian clock in the heart by streptozotocin-induced diabetes. *J. Mol. Cell Cardiol.* **34,** 223–231.

Zulch, K. J., and Hossmann, V. (1967). 24-hour rhythm of human blood pressure. *Ger. Med. Mon.* **12,** 513–518.

# Section VII

# Cell and Tissue Culture System

## [28]   Circadian Gene Expression in Cultured Cells

*By* Emi Nagoshi, Steven A. Brown, Charna Dibner,
Benoît Kornmann, and Ueli Schibler

### Abstract

In mammals, circadian oscillators not only exist in specialized neurons of the suprachiasmatic nucleus, but in almost all peripheral cell types. These oscillators are operative even in established fibroblast cell lines, such as Rat-1 cells or NIH3T3 cells, and in primary fibroblasts from mouse embryos or adult animals. This can be demonstrated by treating such cells for a short time period with high concentrations of serum or chemicals that activate a large number of known signaling pathways. The possibility of studying circadian rhythms in cultured cells should facilitate the biochemical and genetic dissection of the circadian clockwork and should promote the discovery of new clock components.

### Introduction

Mammalian circadian oscillators were originally believed to exist only in suprachiasmatic nucleus (SCN) neurons. However, with the identification of mammalian clock and clock-controlled genes, this view changed dramatically. Robust daily oscillations in gene expression could be monitored in almost all investigated tissues (Schibler *et al.*, 2003), and in some of these tissues the amplitudes of cyclic gene expression were found to be at least as high as in the SCN. Of course, the observation of cycling gene products in peripheral tissues did not establish that these tissues contained autonomous oscillators. Indeed, such daily accumulation cycles could have been the result of cyclic humoral or neuronal signaling emanating from the SCN. Clearly, *ex vivo* experiments were required to examine whether autonomous circadian clocks govern the observed cyclic gene expression in peripheral cell types. In 1998, Balsalobre and co-workers reported that a serum shock elicited circadian oscillations in gene expression in Rat-1 fibroblasts. mRNAs encoding PER1, PER2, CRY1, REV-ERBα, BMAL1, DBP, and TEF all accumulated in a circadian fashion after a serum shock, and these mRNA accumulation cycles persisted for several days (Fig. 1A; unpublished results). Using tissue explants from transgenic rats in which firefly luciferase is expressed under the control of the *Per 1* promoter, Yamazaki *et al.* (2000) have since shown that many peripheral tissues

METHODS IN ENZYMOLOGY, VOL. 393

A

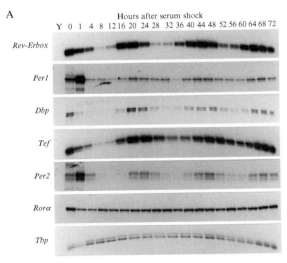

B

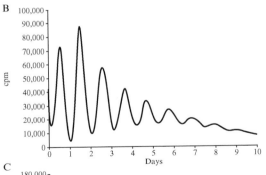

C

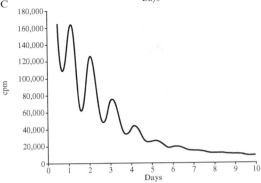

possess their own circadian clocks and that these are synchronized by the central pacemaker in the SCN.

Originally, central and peripheral oscillators were believed to be intrinsically different: the former were regarded as self-sustained and the latter as dampening rapidly without signals from the former. However, using a knockin mouse line in which luciferase expression is driven from the endogenous *mPer 2* locus, Yoo *et al.* (2004) have provided compelling evidence that circadian clocks of peripheral organs are also self-sustained. When the SCNs of these mice were ablated, circadian *mPer2-luciferase* gene expression in peripheral tissues persisted for weeks to months and it did so in an organ-autonomous fashion. In these behaviorally arrhythmic animals, all examined organs continued to oscillate with a slightly different period length, which led to a rapid desynchronization between individual tissues (Yoo *et al.*, 2004). Because the phase of clock gene expression is similar in most organs in intact animals, the SCN must synchronize the countless subsidiary clocks in the periphery. It probably does so primarily via an indirect route: namely by governing rest–activity cycles. In turn, rest–activity rhythms determine alternate cycles of absorptive and postabsorptive phases of food metabolism, which are the dominant zeitgebers (timing cues) for peripheral oscillators (Damiola *et al.*, 2000; Stokkan *et al.*, 2001).

The demonstration of circadian oscillators in peripheral cell types has not only changed our view of the mammalian circadian timing system, but it has also opened new experimental routes to investigate the mechanisms underlying circadian rhythm generation. The SCN of a rat contains about 20,000 neurons (Schibler *et al.*, 2003). These neurons together probably contain an estimated 150 ng of DNA, 600 ng of nuclear proteins, 20–40 $\mu$g

---

FIG. 1. Circadian gene expression in serum-shocked Rat-1 and mouse NIH3T3 fibroblasts. (A) Accumulation of mRNAs from various circadian genes in serum-shocked Rat-1 cells. Rat-1 cells grown to confluence were treated with 50% serum for 2 h as indicated in the text. RNA was extracted from cells snap frozen in liquid nitrogen at 4-h intervals. mRNAs specified by the indicated genes were determined by ribonuclease protection assays using antisense RNA probes. Note that TBP mRNA and ROR$\alpha$ mRNA levels, which are constant throughout the day in peripheral mouse and rat tissues, do not oscillate after serum induction. Adapted from Balsalobre *et al.* (1998), with permission. (B) Circadian bioluminescence generated by a transgenic *Bmal1-luciferase* NIH3T3 cell line after serum shock. The decrease in overall signal strength during the recorded time span is due mainly to a diminution of the cell number during the experiment (Nagoshi *et al.*, 2004). (C) Circadian bioluminescence generated by serum-shocked NIH3T3 fibroblasts transiently transfected with a *Bmal1-luciferase* expression vector. The progressive decrease in signal strength during the recorded time span is probably due to a reduction in both the cell number and the number of expression vector plasmids per cell.

of total protein, and 200–400 ng of whole cell RNA. Due to the possibility of amplifying nucleic acid, this amount of RNA is sufficient for measuring gene expression, and several groups have even performed transcriptome profiling studies with this tissue (Akhtar *et al.*, 2002; Panda *et al.*, 2002). However, even if SCN tissue from many animals were combined, protein biochemistry would hardly be feasible with such low protein amounts. Assuming that a typical regulatory protein comprises approximately one ten thousandth of the total nuclear protein mass, the purification and identification by mass spectrometry of a regulatory protein would require combined SCN tissue from about 10,000 rats, if the final recovery after purification was 10% and if 50 ng of protein was required for mass spectrometry. Making similar assumptions, a single rat liver would yield about 500 ng of this component. Likewise, fibroblasts can be grown to large numbers *in vitro*, and transgenic fibroblast cell lines expressing epitope-tagged clock proteins can be readily established. This ease of manipulation facilitates the purification of clock protein complexes (Brown *et al.*, 2005).

## Monitoring Circadian Gene Expression in Cultured Cells

### Induction of Circadian Rhythms

A short treatment of confluent, serum-starved Rat-1 fibroblasts with 50% serum (serum shock) has been shown to induce circadian mRNA accumulation for all clock genes known to be expressed in a circadian fashion in intact animals (Balsalobre *et al.*, 1998). In the original experiments, Rat-1 fibroblasts were plated at $5 \times 10^5$ cells/10-cm dish in Dulbecco's modified Eagle medium (DMEM) supplemented with 5% fetal calf serum (FCS, GIBCO) and left at 37° for 7 days. After 4 days, the cells had reached complete confluence, and after 7 days they had considerably depleted the medium for growth factors and nutrients. At this time (time zero), the medium was exchanged with prewarmed DMEM containing 50% horse serum (GIBCO), and after a 2-h incubation at 37° the medium was replaced with prewarmed serum-free DMEM. The treatment of serum-starved cells with high concentrations of serum has been known for many years to elicit the transient expression of several immediate early genes. Interestingly, the exposure of laboratory rodents kept in constant darkness to a light pulse induces the same immediate early genes in the SCN, including *cFos, FosB, JunB, Ngfi-A/Zif268*, and *Ngfi-B/Nur77*, provided that the light pulse is given during the subjective night (Kornhauser *et al.*, 1996; Morris *et al.*, 1998). As only light pulses delivered during the subjective night result in immediate early gene expression and phase shifting, it is assumed that immediate early gene expression is required for phase resetting. This speculation is corroborated by the observation that *Per1* and *Per2* also behave as immediate early genes in the SCN (Shearman *et al.*, 1997; Shigeyoshi *et al.*, 1997). Moreover, expression of both *Per1* and *Per2* mRNAs is strongly and transiently induced after a serum shock, and the levels of both of these mRNAs oscillate in a circadian manner thereafter (Fig. 1A).

As a serum shock can also serve as a mitotic stimulus to serum-starved fibroblasts, it was important to examine whether oscillations of the various clock gene transcripts were related to mitotic divisions. Because these oscillations persisted in serum-free medium and in the presence of DNA replication inhibitors (e.g., cytosine $\beta$-D arabinofuranoside), it was concluded that cell division is not required for circadian gene expression (Balsalobre *et al.*, 1998). Sera from other mammalian species, including rat, rabbit, cow, horse, and pig, were also tested and all of them had a similar potency to induce circadian gene expression (Balsalobre *et al.*, 1998). We surmise that growth factor, chemokines, or hormones present in the various sera bind to their cognate receptors, that this signaling leads to immediate early transcription, and that the immediate early transcription elicits circadian waves of clock gene expression.

Shortly after the initial discovery of serum-induced gene expression, a bewildering variety of signaling pathways were demonstrated to trigger circadian gene expression. Chemicals inducing the activation of protein kinase A (forskolin, butyryl cAMP), protein kinase C and/or MAP kinases (PMA, FGF, endothelin), G-protein-coupled receptors (endothelin), the glucocorticoid receptor (dexamethasone), and calcium ion channels (calcimycin) all elicited the immediate expression of *Per1* and the subsequent oscillation of clock gene transcripts (Akashi and Nishida, 2000; Balsalobre *et al.*, 2000a,b; Yagita and Okamura, 2000; Yagita *et al.*, 2001). Even glucose has been shown to trigger circadian rhythms of gene expression in fibroblasts, but this induction does not involve the immediate early activation of *Per* genes (Hirota *et al.*, 2002). Square wave temperature cycles applied during several days also elicit circadian gene expression that persists when cells are shifted to a constant temperature thereafter (Brown *et al.*, 2002), although once again immediate early Period gene transcription is not observed. Thus, we consider it likely that multiple pathways exist for the phase resetting of mammalian circadian oscillators. The conditions in which various chemicals and temperature have been used to induce circadian gene expression are summarized in Table I.

In principle, induction of measurable cycles in clock gene expression could be accomplished by two different mechanisms: jumpstarting of dormant oscillators or synchronization of active but dephased oscillators.

TABLE I

INDUCTION OF CIRCADIAN GENE EXPRESSION IN CELLS GROWN IN VITRO[a]

| Inducing agent | Final concentration in medium | Time of treatment | Cell line | Medium during recording | Activated pathways | Immediate early response of Per1 and Per2 | References |
|---|---|---|---|---|---|---|---|
| Horse serum in DMEM | 50% (v/v) | 2 h | Rat-1 fibroblasts | DMEM | Many growth factor, hormone, and cytokine receptors | Per1, activation Per2 activation | Balsalobre et al. (1998); Nagoshi et al. (2004) |
| | | | NIH3T3 cells | DMEM + 0.5 FCS | | | |
| Forskolin | 10 $\mu M$ | 15 min | Rat-1 cells | DMEM | Adenylate cyclase | Per1 activation | Balsalobre et al. (2000b); Yagita and Okamura (2000) |
| Calcimycin | 1 $\mu g$/ml | 15 min | Rat-1 cells | DMEM | $Ca^{2+}$ channels | Per1 activation | Balsalobre et al. (2000b) |
| Dexamethasone | 100 n$M$ | 15 min | Rat-1 cells | DMEM | Glucocorticoid receptor | Per1 activation | Balsalobre et al. (2000a,b) |
| | | | NIH3T3 cells | DMEM + 0.5–20% FCS | | | |
| Tumor promoters PMA | 1 $\mu M$ | 15 min | Rat-1 cells | DMEM | PKC MAPK | Per1 activation | Balsalobre et al. (2000b) |

| | | | | | | | |
|---|---|---|---|---|---|---|---|
| TPA | 50 nM | 2 h | NIH3T3 cells | DMEM + 1% FCS | | Per1 activation | Akashi and Nishida (2000) |
| Endothelin (ET-1) | ET-1 (final concentration 30 nM) | 2 h | Rat-1 cells, immortalized mouse embryonic fibroblastss | DMEM | PKC MAPK Phosphorylation of CREB | Per1 activation Per2 activation | Yagita et al. (2001) |
| Glucose | 5.6 mM | Left during entire experiment | Rat-1 cells | DMEM | ? | Per1 repression Per2 repression | Hirota et al. (2002) |
| Temperature | Not relevant | 12 h 33° 12 h 37° | Rat-1 cells | DMEM + 5% FCS | ? | ? | Brown et al. (2002) |

[a]Cells were grown to confluence, treated for the indicated times with the indicated substances, and kept in serum-free or serum-containing medium thereafter to record circadian gene expression.

Based on time-lapse microscopy of NIH3T3 cells expressing YFP in a circadian manner, we presently favor the second hypothesis (Nagoshi et al., 2004).

## Measurement of Immediate Early Gene Expression and mRNA Accumulation Cycles

After the treatment of cells with high serum concentrations or chemicals inducing known signaling cascades (see earlier discussion), several methods can be used to monitor oscillating transcript levels. These include Northern blot hybridization, ribonuclease protection assays, S1 nuclease protection assays, and Taqman real-time polymerase chain reaction (PCR) assays of reverse-transcribed mRNA. Ribonuclease and S1 nuclease protection assays probably yield the most precise measurements, as the hybridization follows pseudo-first-order kinetics, i.e., only the concentration of the radiolabeled probe determines the rate of hybridization. Real-time PCR is the most rapid procedure and is sufficient for most applications. As transcripts issued from immediate early genes disappear shortly after their induction, their recording necessitates short time intervals of 1 h during the first 6 h after induction (Balsalobre et al., 1998). For the monitoring of circadian cycles, 4-h intervals taken during 2 consecutive days are sufficient to assess whether circadian gene expression has been induced. Obviously, all of the assays chosen for the establishment of temporal mRNA accumulation profiles must include measurements of transcripts that do not oscillate during the examined time span. In our hands, glyceraldehyde 3'-phosphate dehydrogenase and TATA box binding protein mRNAs are convenient standard transcripts for real-time PCR experiments and nuclease protection, but others may be equally adequate. In Northern blot experiments it is advisable to label ribosomal RNA transferred to nitrocellulose or nylon membranes with methylene blue, and only membranes with even RNA amounts should be processed further.

Genome-wide profiling of circadian mRNA accumulation can be accomplished through hybridization of cRNAs or cDNAs to DNA or oligonucleotide microarrays. Using such an approach, Grundschober et al. (2001) have identified 85 transcripts that accumulate according to a daily rhythm after serum induction of Rat-1 fibroblasts. Undoubtedly, this is a minimal number, as many transcripts may have escaped this analysis. The Affymetrix biochips used in these experiments contained probes for only 9957 genes (about 25% of all genes), and the hybridization sensitivity and specificity afforded with these chips were far below what can be obtained with the latest generation of Affymetrix biochips (Gachon et al., 2004).

*Protein Analysis*

The cyclic accumulation of proteins can be measured by Western blot experiments with whole cell proteins, cytoplasmic proteins, or nuclear proteins. Because many clock and clock-controlled proteins are positively or negatively acting transcription factors, the analysis of nuclear proteins may be more informative and sensitive than that of whole cell or cytoplasmic proteins, as the concentration of transcription factors is generally higher in the nucleus than in the cytoplasm. Obviously, both cytoplasmic and nuclear proteins must be examined if the objective is to study the localization of proteins as well as their accumulation.

Various procedures for the fractionation of tissue culture cells have been published. We disrupt cells by treatment either with the nonionic detergent NP-40 or by mechanical disruption using a Dounce homogenizer. The resultant nuclei can be purified by sedimenting them through a 30% (w/v) sucrose cushion at low speed (Descombes and Schibler, 1991). Non-histone nuclear proteins can then be extracted nearly quantitatively by NUN buffer (Lavery and Schibler, 1993) or more gently via salt extraction. Regardless of the choice of procedure, the obtained extracts can be loaded directly onto SDS-containing polyacrylamide gels for subsequent Western blot analysis.

Unfortunately, detergent treatment or mechanical breakage of cells can lead to leakage of some proteins from the nucleus to the cytoplasm. It is thus essential to confirm the results obtained from cell fractionation experiments by immunohistochemical analysis in fixed cells or possibly by the microscopic analysis of GFP fusion proteins in living or fixed cells. A second prevalent problem with fractionation experiments is that most investigators use a chemiluminescence assay to reveal signals obtained in Western blots. The signals from this assay are highly nonlinear and thus will not provide reliable amplitudes of circadian protein accumulation. A convenient way to reduce this problem is to mix extracts harvested at peak and nadir expression at different ratios before gel electrophoresis and Western blotting (Fonjallaz *et al.*, 1996).

The accumulation of circadian DNA-binding proteins such as REV-ERBα, DBP, TEF, and HLF can also be examined by electrophoretic mobility shift assays (EMSA) using NUN extracts from isolated nuclei and radiolabeled double-stranded DNA fragments encompassing the binding sites for these transcription factors (Gachon *et al.*, 2004; Lopez-Molina *et al.*, 1997; Preitner *et al.*, 2002). However, EMSA experiments may not reveal the identity of individual transcription factors if multiple proteins recognize the same DNA sequence. Sometimes this problem can be overcome with more complicated two-dimensional gel shift procedures. For example, the circadian activator proteins CLOCK and BMAL1 heterodimerize to bind to a DNA sequence identical to that bound by other general and tissue-specific transcription factors such as MyoD, USF1, and USF2. To separate CLOCK/BMAL1 binding from the binding of these other factors, Ripperger and co-workers (2000) used a two-dimensional EMSA technique involving the photocross-linking of DNA-binding proteins to radiolabeled DNA probes after the first dimension.

## Automated Analysis of Circadian Gene Expression

Several authors working with a wide variety of circadian systems have successfully employed real-time recording of circadian bioluminescence in living cells (see Welsh et al., 2005; Yamazaki and Takahashi, 2005). In general, firefly luciferase is expressed under the control of cis-acting regulatory elements driving circadian transcription. The optimal way to do this is to insert a luciferase reporter gene directly into a locus using homologous recombination (Yoo et al., 2004), which leaves all surrounding regulatory sequences intact. This method should ensure that the bioluminescence recording mimics all aspects (e.g., circadian rhythms, cell type specificity, growth dependence) of the expression of the gene under study. Such knock in technology is feasible in ES cells and mice resulting from ES cells, but not in immortalized fibroblast lines. Fortunately, transgenes expressing luciferase have also been used successfully to record cyclic gene expression (see Welsh et al., 2005; Yamazaki and Takahashi, 2005). Additionally, Ueda et al. (2002) have shown that even the recording of circadian bioluminescence from cells transiently transfected with circadian reporter constructs is feasible. An example of such an experiment from our own laboratory is shown in Fig. 1C. Nevertheless, the high copy number of transiently transfected genes may result in the depletion of rare regulatory factors and could thus interfere with circadian gene expression of endogenous genes. Where possible, results obtained by transient transfections should be confirmed by other techniques, e.g., with stably transformed cell lines.

A few criteria should be respected in the construction of circadian luciferase reporter genes. First, it is advisable to use regulatory sequences from genes that drive high-amplitude circadian expression (e.g., Rev-Erbα, Per2, Bmal1, or Dbp). Second, the mRNA-encoding luciferase must be short lived. This goal can be accomplished by using 3′-untranslated sequences of mRNAs known to decay with a short half-life. We readily obtain 20-fold amplitudes in bioluminescence during the first cycle following induction using an NIH3T3 cell line expressing a luciferase reporter gene that harbors the *Bmal1* promoter, *Bmal1* trailer sequences, and the *Bmal1* polyadenylation site (see Fig. 1B; Nagoshi *et al.*, 2004).

## Choice of Cell Line

### Established Cell Lines

We have examined the robustness of circadian gene expression in various fibroblastoid cell lines (mouse LTK and NIH3T3 fibroblasts, and Rat-1 fibroblasts), the H35 hepatoma cell line, and the CL-26 colon carcinoma cell line. In general, contact-inhibited cells reveal higher amplitudes than rapidly growing cells, presumably because cell division leads to a desynchronization of circadian rhythms (Nagoshi *et al.*, 2004). This hypothesis may also explain why tumor cells are inferior to nontransformed cells when examining circadian gene expression. Even in low-serum medium, tumor cells keep proliferating once they have reached confluence, which may lead to a rapid desynchronization of circadian rhythms. In our hands, NIH3T3 cells give the highest amplitudes and desynchronize only slowly, such that rhythmic luciferase expression can still be discerned 3 weeks after a serum shock. However, these cells are more sensitive to serum depletion than Rat-1 cells so we always use a serum-containing medium when working with NIH3T3 cells. The period length of *Bmal1*-driven circadian luciferase expression in these cells is about 28 h in DMEM containing 0.5% serum and about 25 h in medium containing between 5% and 20% serum.

### Primary Cells

For some experiments it is desirable to use primary fibroblasts from mice or other organisms carrying clock gene mutations (Yagita *et al.*, 2001). We have used mouse embryonic fibroblast (MEFs) and primary fibroblasts from tail bud biopsies of adult mice. It is gratifying that primary fibroblasts show phenotypes in circadian gene expression that are qualitatively similar to those recorded for the behavior of the mice from which they were derived. Thus, no circadian rhythms can be recorded in fibroblasts from arrhythmic *mCry1/mCry2* (Yagita *et al.*, 2001) or *mPer1/mPer2* double knockout mice (Nagoshi *et al.*, 2004). Moreover, as expected on the basis of behavioral assays, MEFs or fibroblasts from adult *mPer1*$^{-/-}$ mice show a shorter period length than the corresponding cell types from wild-type mice (Pando *et al.*, 2002). Interestingly, this period difference is exacerbated in cultured cells. Thus, while the period length of wheel-running behavior of *mPer1*$^{-/-}$ mice is about 1 h shorter than that of wild-type mice, the period length of circadian gene expression is about 4 h shorter in fibroblasts derived from $mPer1^{-/-}$ mice when compared with fibroblasts prepared from wild-type mice (Pando et al., 2002).

Conclusions and Perspectives

How good a model system are fibroblasts for studying circadian gene expression and physiology? All research to date suggests that the molecular oscillators in fibroblasts have very similar properties to those operative in SCN neurons; therefore, the major differences probably lie in the phase-setting inputs. Obviously, photic entrainment cannot be studied in fibroblasts that are photoinsensitive and that do not make functional connections to photoreceptor cells. Likewise, in contrast to SCN neurons, fibroblasts probably do not entrain other cellular oscillators in the body. In fact, even when grown to confluence they do not communicate with each other in culture, and their oscillators work in an entirely cell-autonomous fashion (Nagoshi et al., 2004). Therefore, fibroblasts are inadequate as a model system to study cell–cell communications in vitro, at least under standard culture conditions. However, for many facets of circadian biology, these cells provide valuable research objects. This section lists some of the areas in which we anticipate research on fibroblast circadian rhythms to be particularly fruitful.

*Biochemical Dissection of the Core Oscillator*

The fibroblast system allows the expression of epitope-tagged proteins and should thus be useful in the affinity purification of associated proteins. This approach has already been used successfully in our laboratory for the identification of two PER1 interaction partners (Brown et al., 2005). By using RNA interference we could show that one of these is an essential component of the circadian oscillator, whereas the other modulates the amplitude of some rhythmically expressed genes. Cellular systems will also be indispensable in the analysis of kinetic parameters, such as the rate constants driving synthesis, degradation, and posttranslational modifications of clock components, as well as the rate constants determining the interaction of these components among themselves and where applicable with their cognate DNA-binding sites. Many of the techniques required for such endeavors, e.g., the metabolic labeling of clock components, fluorescence resonance energy transfer experiments in vivo, high temporal resolution of chromatin immunoprecipitation, and real-time microscopy of protein trafficking, are hardly feasible with intact animals.

## Genetics of Circadian Clock Function

Due to the development of cDNA expression vectors (Hayashizaki, 2003) and retroviral or lentiviral vectors (Berns *et al.*, 2004; Paddison *et al.*, 2004) expressing small interfering RNAs (RNAi), both genetic gain-of-function and loss-of-function experiments are now feasible on a large scale with mammalian tissue culture cells. Such genetic screens may identify essential oscillator components that do not interact directly with known clock proteins and therefore cannot be identified by the biochemical studies suggested earlier.

## Cell Biology and Physiology

Using clock proteins tagged with fluorescent proteins (e.g., with GFP, YFP, CFP, RFP), the accumulation and cellular trafficking of these proteins can be studied in real time by time-lapse microscopy. Such studies not only permit the recording of circadian rhythms in individual cells, but also the monitoring of the temporal distribution of clock components. By recording cyclic YFP accumulation in individual NIH3T3 fibroblasts, we were able to show that circadian gene expression in fibroblasts is self-sustained and that it continues in dividing cells. Moreover, using the same technology we could demonstrate that circadian oscillators appear to gate cell division (cytokinesis) to certain time windows (Nagoshi *et al.*, 2004).

Altogether, the discovery that most cells contain autonomous circadian clocks changed our conceptual overview of circadian rhythms. Now, several years later, this discovery is poised to furnish exciting new methods for the detailed analysis of circadian rhythms in mammals, and the ideas outlined here are doubtlessly only the beginning.

## Acknowledgments

We thank Nicolas Roggli for the artwork. Work from our laboratory was supported by the Swiss National Science Foundation (grant to U.S.), the State of Geneva, the NCCR program "Frontiers in Genetics," the *Bonnizzi Theler Stiftung*, and the *Louis Jeantet Foundation of Medicine*.

## References

Akashi, M., and Nishida, E. (2000). Involvement of the MAP kinase cascade in resetting of the mammalian circadian clock. *Genes Dev.* **14**, 645–649.

Akhtar, R. A., Reddy, A. B., Maywood, E. S., Clayton, J. D., King, V. M., Smith, A. G., Gant, T. W., Hastings, M. H., and Kyriacou, C. P. (2002). Circadian cycling of the mouse liver transcriptome, as revealed by cDNA microarray, is driven by the suprachiasmatic nucleus. *Curr. Biol.* **12**, 540–550.

Balsalobre, A., Brown, S. A., Marcacci, L., Tronche, F., Kellendonk, C., Reichardt, H. M., Schutz, G., and Schibler, U. (2000a). Resetting of circadian time in peripheral tissues by glucocorticoid signaling. *Science* **289**, 2344–2347.

Balsalobre, A., Damiola, F., and Schibler, U. (1998). A serum shock induces circadian gene expression in mammalian tissue culture cells. *Cell* **93**, 929–937.

Balsalobre, A., Marcacci, L., and Schibler, U. (2000b). Multiple signaling pathways elicit circadian gene expression in cultured Rat-1 fibroblasts. *Curr. Biol.* **10**, 1291–1294.

Berns, K., Hijmans, E. M., Mullenders, J., Brummelkamp, T. R., Velds, A., Heimerikx, M., Kerkhoven, R. M., Madiredjo, M., Nijkamp, W., Weigelt, B., Agami, R., Ge, W., Cavet, G., Linsley, P. S., Beijersbergen, R. L., and Bernards, R. (2004). A large-scale RNAi screen in human cells identifies new components of the p53 pathway. *Nature* **428**, 431–437.

Brown, S., Rippenger, J., Kodener, S., Fleury-Olela, F., Nagoshi, E., Vilbois, F., Rosbash, M., and Schibler, U (2005). *Science*, In press.

Brown, S. A., Zumbrunn, G., Fleury-Olela, F., Preitner, N., and Schibler, U. (2002). Rhythms of mammalian body temperature can sustain peripheral circadian clocks. *Curr. Biol.* **12**, 1574–1583.

Damiola, F., Le Minh, N., Preitner, N., Kornmann, B., Fleury-Olela, F., and Schibler, U. (2000). Restricted feeding uncouples circadian oscillators in peripheral tissues from the central pacemaker in the suprachiasmatic nucleus. *Genes Dev.* **14**, 2950–2961.

Descombes, P., and Schibler, U. (1991). A liver-enriched transcriptional activator protein, LAP, and a transcriptional inhibitory protein, LIP, are translated from the same mRNA. *Cell* **67**, 569–579.

Fonjallaz, P., Ossipow, V., Wanner, G., and Schibler, U. (1996). The two PAR leucine zipper proteins, TEF and DBP, display similar circadian and tissue-specific expression, but have different target promoter preferences. *EMBO J.* **15**, 351–362.

Gachon, F., Fonjallaz, P., Damiola, F., Gos, P., Kodama, T., Zakany, J., Duboule, D., Petit, B., Tafti, M., and Schibler, U. (2004). The loss of circadian PAR bZip transcription factors results in epilepsy. *Genes Dev.* **18**, 1397–1412.

Grundschober, C., Delaunay, F., Puhlhofer, A., Triqueneaux, G., Laudet, V., Bartfai, T., and Nef, P. (2001). Circadian regulation of diverse gene products revealed by mRNA expression profiling of synchronized fibroblasts. *J. Biol. Chem.* **276**, 46751–46758.

Hayashizaki, Y. (2003). The Riken mouse genome encyclopedia project. *C. R. Biol.* **326**, 923–929.

Hirota, T., Okano, T., Kokame, K., Shirotani-Ikejima, H., Miyata, T., and Fukada, Y. (2002). Glucose down-regulates Per1 and Per2 mRNA levels and induces circadian gene expression in cultured Rat-1 fibroblasts. *J. Biol. Chem.* **277**, 44244–44251.

Kornhauser, J. M., Mayo, K. E., and Takahashi, J. S. (1996). Light, immediate-early genes, and circadian rhythms. *Behav. Genet.* **26**, 221–240.

Lavery, D. J., and Schibler, U. (1993). Circadian transcription of the cholesterol 7 alpha hydroxylase gene may involve the liver-enriched bZIP protein DBP. *Genes Dev.* **7**, 1871–1884.

Lopez-Molina, L., Conquet, F., Dubois-Dauphin, M., and Schibler, U. (1997). The DBP gene is expressed according to a circadian rhythm in the suprachiasmatic nucleus and influences circadian behavior. *EMBO J.* **16**, 6762–6771.

Morris, M. E., Viswanathan, N., Kuhlman, S., Davis, F. C., and Weitz, C. J. (1998). A screen for genes induced in the suprachiasmatic nucleus by light. *Science* **279**, 1544–1547.

Nagoshi, E., Saini, C., Bauer, C., Laroche, T., and Schibler, U. (2004). Circadian gene expression in individual fibroblasts: Cell-autonomous and self-sustained oscillators pass time to daughter cells. *Cell* **119**, 693–705.

Paddison, P. J., Silva, J. M., Conklin, D. S., Schlabach, M., Li, M., Aruleba, S., Balija, V., O'Shaughnessy, A., Gnoj, L., Scobie, K., Chang, K., Westbrook, T., Cleary, M.,

Sachidanandam, R., McCombie, W. R., Elledge, S. J., and Hannon, G. J. (2004). A resource for large-scale RNA-interference-based screens in mammals. *Nature* **428,** 427–431.

Panda, S., Antoch, M. P., Miller, B. H., Su, A. I., Schook, A. B., Straume, M., Schultz, P. G., Kay, S. A., Takahashi, J. S., and Hogenesch, J. B. (2002). Coordinated transcription of key pathways in the mouse by the circadian clock. *Cell* **109,** 307–320.

Pando, M. P., Morse, D., Cermakian, N., and Sassone-Corsi, P. (2002). Phenotypic rescue of a peripheral clock genetic defect via SCN hierarchical dominance. *Cell* **110,** 107–117.

Preitner, N., Damiola, F., Luis Lopez, M., Zakany, J., Duboule, D., Albrecht, U., and Schibler, U. (2002). The orphan nuclear receptor REV-ERBalpha controls circadian transcription within the positive limb of the mammalian circadian oscillator. *Cell* **110,** 251–260.

Ripperger, J. A., Shearman, L. P., Reppert, S. M., and Schibler, U. (2000). CLOCK, an essential pacemaker component, controls expression of the circadian transcription factor DBP. *Genes Dev.* **14,** 679–689.

Schibler, U., Ripperger, J., and Brown, S. A. (2003). Peripheral circadian oscillators in mammals: Time and food. *J. Biol. Rhythms* **18,** 250–260.

Shearman, L. P., Zylka, M. J., Weaver, D. R., Kolakowski, L. F., Jr., and Reppert, S. M. (1997). Two period homologs: Circadian expression and photic regulation in the suprachiasmatic nuclei. *Neuron* **19,** 1261–1269.

Shigeyoshi, Y., Taguchi, K., Yamamoto, S., Takekida, S., Yan, L., Tei, H., Moriya, T., Shibata, S., Loros, J. J., Dunlap, J. C., and Okamura, H. (1997). Light-induced resetting of a mammalian circadian clock is associated with rapid induction of the mPer1 transcript. *Cell* **91,** 1043–1053.

Stokkan, K. A., Yamazaki, S., Tei, H., Sakaki, Y., and Menaker, M. (2001). Entrainment of the circadian clock in the liver by feeding. *Science* **291,** 490–493.

Ueda, H. R., Chen, W., Adachi, A., Wakamatsu, H., Hayashi, S., Takasugi, T., Nagano, M., Nakahama, K., Suzuki, Y., Sugano, S., Iino, M., Shigeyoshi, Y., and Hashimoto, S. (2002). A transcription factor response element for gene expression during circadian night. *Nature* **418,** 534–539.

Welsh, D. K., Imaizumi, T., and Kay, S. A. (2005). Real-time reporting of circadian-regulated gene expression by luciferase imaging in plants and mammalian cells. *Methods Enzymol.* **393**(11), 2005 (this volume).

Yagita, K., and Okamura, H. (2000). Forskolin induces circadian gene expression of rPer1, rPer2 and dbp in mammalian rat-1 fibroblasts. *FEBS Lett.* **465,** 79–82.

Yagita, K., Tamanini, F., van Der Horst, G. T., and Okamura, H. (2001). Molecular mechanisms of the biological clock in cultured fibroblasts. *Science* **292,** 278–281.

Yamazaki, S., Numano, R., Abe, M., Hida, A., Takahashi, R., Ueda, M., Block, G. D., Sakaki, Y., Menaker, M., and Tei, H. (2000). Resetting central and peripheral circadian oscillators in transgenic rats. *Science* **288,** 682–685.

Yamazaki, S., and Takahashi, J. S. (2005). Real-time luminescence reporting of circadian gene expression in mammals. *Methods Enzymol.* **393**(12), 2005 (this volume).

Yoo, S. H., Yamazaki, S., Lowrey, P. L., Shimomura, K., Ko, C. H., Buhr, E. D., Siepka, S. M., Hong, H. K., Oh, W. J., Yoo, O. J., Menaker, M., and Takahashi, J. S. (2004). PERIOD2::LUCIFERASE real-time reporting of circadian dynamics reveals persistent circadian oscillations in mouse peripheral tissues. *Proc. Natl. Acad. Sci. USA* **101,** 5339–5346.

## [29]  Cell Culture Models for Oscillator and Pacemaker Function: Recipes for Dishes with Circadian Clocks?

*By* DAVID J. EARNEST and VINCENT M. CASSONE

### Abstract

Primary cell cultures of avian pinealocytes and the mammalian suprachiasmatic nucleus (SCN), immortalized cell lines derived from the SCN (SCN2.2), and fibroblasts derived from mice and rats have been employed as *in vitro* models to study the cellular and molecular mechanisms underlying circadian biological clocks. This article compares and contrasts these model systems and describes methods for avian pinealocyte cultures, immortalized SCN2.2 cells, and mouse fibroblast culture. Each of these culture models has advantages and disadvantages. Avian pinealocytes are photoreceptive, contain a circadian pacemaker, and produce rhythms of an easily assayed endocrine output—melatonin. However, the molecular mechanisms underlying pinealocyte function are not understood. SCN2.2 cells express metabolic and molecular rhythms and can impose rhythmicity on cocultured cells as well as rat behavior when transplanted into the brain. Yet, the entrainment pathways are not experimentally established in these cells. Fibroblast cultures are simple to produce and express molecular clock gene rhythms, but they express neither physiological rhythmicity nor pacemaker properties. The relative merits of these culture systems, as well as their impact on understanding circadian organization *in vivo*, are also considered.

### Biological Clocks in Vertebrates

Biological rhythms are fundamental properties of nearly all living organisms studied to date. Daily, lunar, and annual cycles of biological activity have been described in cyanobacteria, algal protists, protozoa, filamentous fungi, multicellular animals, and plants (Dunlap *et al.*, 1999; Golden and Canales, 2003). Of these, the best-characterized biological rhythm is the *circadian rhythm*, which corresponds to biological rhythms with periods ($\tau$) of approximately 24 h (Pittendrigh, 1993). Almost universally, circadian rhythms are generated by endogenous *oscillators* that express $\tau$s that are rarely exactly 24 h, when organisms are placed in continuous environment light, usually constant darkness (DD), and temperature. These oscillators are in turn synchronized or *entrained* to local time via the detection of an ambient cue or *zeitgeber* such that the endogenous phase ($\phi_i$) stably corresponds to an environmental phase ($\phi_e$) (Pittendrigh, 1993). The dominant zeitgeber for most species is the light:-dark (LD) cycle, and invariably specialized photoreceptive/phototransductive mechanisms have evolved in biological clock systems. The stably entrained oscillator or population of oscillators in turn regulates multiple downstream processes, conferring to the system *pacemaker* properties. Together, the input mechanism for zeitgeber detection, the pacemakers, and their outputs compose the *biological clock*.

In vertebrates, the regulation of circadian rhythms resides in specialized neuroendocrine structures (Cassone and Menaker, 1984; Menaker *et al.*, 1978). These include the pineal gland or organ, the retina, the hypothalamic suprachiasmatic nucleus (SCN), and/or structures associated with each of the components. The relative importance of each of these components to overt circadian organization varies significantly among vertebrate taxa (Cassone, 1998). This article focuses on model systems derived from the mammalian SCN and the avian pineal gland, as these are the best-characterized pacemaker systems with cell culture applications.

The SCN is the master pacemaker for the generation and/or coordination of all molecular, biochemical, physiological, and behavioral rhythms in every mammalian species studied to date (Moore and Silver, 1998). Surgical destruction of the SCN in several rodent species (rats, mice, and hamsters), cats, and squirrel monkeys abolishes the expression of a wide array of circadian rhythms and renders these species' biological clocks insensitive to LD cycles (Moore, 1995; Moore and Silver, 1998), as the SCN of all species studied receive a specialized retinohypothalamic pathway (RHT) that is required for entrainment (Cassone *et al.*, 1988; Moore, 1995). SCN tissue explants, organotypic explants, dispersed SCN cells, and, as described later, immortalized cells derived from embryonic rat SCN express circadian patterns of gene expression, metabolic activity, peptide secretion, and electrical activity *in vitro* (Earnest and Sladek, 1987; Gillette and Reppert, 1987; Green and Gillette, 1982; Newman and Hospod, 1986). Finally, transplantation of embryonic SCN tissue, dispersed SCN cells, or immortalized SCN cells into the third ventricle confers circadian rhythms of behavioral activity to arrhythmic, SCN-lesioned rodents (Drucker-Colin *et al.*, 1984; Earnest *et al.*, 1999b; Sawaki *et al.*, 1984; Silver *et al.*, 1990, 1996). Thus, the SCN represents a tissue that comprises all features of the biological clock: a specialized input pathway and circadian oscillators with pacemaker properties capable of driving and/or entraining downstream rhythms.

Similarly, the avian pineal gland expresses all the characteristics of a biological clock, although the relative importance of the pineal gland in overt avian circadian organization is not as universal as it is for the SCN.

In oscine passeriform birds, such as the house sparrow, *Passer domesticus*, pinealectomy abolishes overt behavioral rhythms in DD (Gaston and Menaker, 1968). Further, transplantation of pineal glands into the anterior chamber of the eye of arrhythmic pinealectomized sparrows confers circadian patterns of locomotor activity to the pinealectomized birds (Zimmerman and Menaker, 1979). This effect is probably due to the rhythmic release of melatonin, as the pineal glands of all species studied produce circadian rhythms of the indoleamine hormone melatonin, such that melatonin is synthesized during the night in LD and subjective night in DD (Binkley *et al.*, 1978; Deguchi, 1979a; Kasal *et al.*, 1979; Wainwright, 1977). This rhythm is directly entrained to LD cycles via photoreceptors residing in the gland itself (Deguchi, 1979b). Further, rhythmic administration of melatonin restores the circadian patterns of activity and brain metabolism in arrhythmic, pinealectomized birds (Cassone *et al.*, 1992; Gwinner and Benzinger, 1978; Lu and Cassone, 1993). It is important to note that other avian taxa appear to be less dependent upon the pineal per se in that pinealectomy may not completely abolish rhythms in starlings, pigeons, and quail (Ebihara *et al.*, 1984; Gwinner, 1978; Underwood and Siopes, 1984), and the surgery has absolutely no behavioral effect in chickens (McGoogan and Cassone, 1999). It is likely that in these species, the rhythmic secretion of melatonin by the retina also contributes to circadian organization. Thus the avian pineal gland contains the photic input, circadian oscillators, and a molecular output, melatonin, to constitute a circadian pacemaker as well.

## Molecular Clockworks

Since the early 1990s, the mRNA and protein products of many "clock genes" have been shown to oscillate with a genotype-specific period, and their protein products have been shown to interact *in vitro*. These rhythmic profiles have suggested a generalized model for circadian rhythm generation that involves the transcription, translation, and feedback of these clock gene products on their own transcription. The proposed interlocking transcriptional/translational feedback loops are composed of "negative elements" and "positive elements" whose interactions are believed to drive overt rhythmicity (Dunlap *et al.*, 1999). Briefly, the negative elements in mammals comprise the period genes (*per1–3*) and cryptochromes (*cry1, 2*). These are transcribed in response to the dimerization of the positive elements, which include clock (*clock*) and the Bmals (*Bmal1, 2*), and their subsequent binding to E-boxes in the promoter regions of several genes. The negative elements in turn are translated, oligomerize, and reenter the nucleus where they inhibit their and other genes' transcription, closing the feedback loop.

One unexpected observation in the analysis of mammalian clock genes was that even though the SCN were clearly critical for the expression of overt rhythms, clock gene rhythms were expressed in nearly all peripheral tissues, ranging from the liver to endocrine tissues to the heart and skeletal muscles (Zylka et al., 1998). Even more surprising was the finding that serum shock of fibroblasts induced similar in vitro rhythms of clock gene expression. This raised the suggestion by many that rather than a hierarchy in which the SCN (or pineal) drove downstream rhythms, organisms may be composed of multiple peripheral oscillators that are only synchronized by upstream pacemakers.

The purpose of this article is to review well-known cell culture systems that are available to examine both the pacemaker and the oscillator properties of the avian pineal gland, the mammalian suprachiasmatic nucleus, and serum- or chemically shocked fibroblasts. After their introduction and a brief description of techniques for their study, the relative merits of each approach are addressed.

## Avian Pineal Gland

The chick pineal gland has been one of the leading models for the cellular analysis of biological clocks since the late 1970s, when several groups showed that in static organ culture systems, the gland expressed daily and circadian rhythms in the activity of arylalkylamine (serotonin)-N-acetyltransferase (AANAT), the rate-limiting step in melatonin biosynthesis (Binkley et al., 1978; Deguchi, 1979a; Kasal et al., 1979; Wainwright, 1977; Wainwright and Wainwright, 1979). In these systems, glands from many chicks ranging in age from 1 to 14 days (depending on the study) were placed in culture media for several days under LD cycles or DD. At intervals of 2–4 h, four to five glands were removed and frozen rapidly. Once the entire experiment was completed, all glands were processed for radioenzymatic assays for AANAT activity. These studies established that cultured chick pineal glands expressed a daily pattern of AANAT activity such that activity was high during the night and low during the day. Second, they revealed that the rhythms in AANAT activity persisted for several days even if the glands were placed in DD. Finally, a short pulse of light or a shift in the LD cycle was found to phase shift pineal rhythms in AANAT activity. Together, data from these studies established that the chick pineal gland contained a circadian oscillator, photoreceptors capable of entraining the circadian oscillators to environmental light, and a molecular output, activity of AANAT.

This model system was modified by Deguchi (1979b) to further establish that dispersed chick pineal cells expressed similar AANAT rhythms that were responsive to light. These studies suggested that the individual cells responsible for producing the AANAT rhythm also contained oscillators and photoreceptors because the rhythm in AANAT activity was not diminished by cell dispersal. Further, Deguchi (1981) showed that the light sensitivity of the gland expressed a rhodopsin-like action spectrum, suggesting that an opsin was the likely photopigment in these cells.

The flow-through pineal culture system developed by Takahashi and co-workers (1980) was a significant technical advance in that it reduced the significant animal usage required in the static culture preparations and allowed much finer temporal resolution of the rhythms expressed by the gland. In this system, pineal glands are removed from 2-week-old chicks and placed in a stainless-steel basket suspended within a Pasteur pipette. A peristaltic pump establishes a constant flow of oxygenated culture medium over the gland and into a fraction collector. Once collected, all fractions are assayed for melatonin content by radioimmunoassay. Using this approach, Takahashi and colleagues (1989) further demonstrated that the chick pineal clock resides in multiple pieces of the gland and that the clock within the gland is temperature compensated. While this approach provided greater temporal resolution than static culture approaches, it was not amenable to pharmacological manipulation, as the volume of perfusate was relatively great, making the cost of pharmacological and/or molecular manipulations exorbitant.

Zatz and colleagues (Natesan et al., 2002; Zatz, 1989; Zatz and colleagues, 1988; Zatz and Mullen, 1988a–d) further developed the dispersed pineal cell culture devised by Deguchi with the significant modification that melatonin production would be assayed in the culture media rather than AANAT activity, which required harvesting cultured cells. This technique is described later. A dispersed culture of pineal cells provided a preparation that was less animal intensive than the formerly described static cultures, could approach the temporal resolution of the flow-through system, and was amenable to pharmacological manipulation. This work demonstrated the role of calcium and cyclic nucleotide metabolism in the regulation of pineal melatonin output and the fact that these second messenger systems were not involved in the regulation of circadian phase in this system. Still, the approach is amenable now to molecular manipulation, and this approach is described later.

## Methods in Pinealocyte Culture

White Leghorn chicks (approximately 100 glands/experimental analysis) are sacrificed on the day of hatching and their pineal glands are removed, dispersed in trypsin, and plated in 24-well microtiter plates at approximately $10^6$ cells/well. Modified McCoy's 5A medium (GIBCO) containing 25 m$M$ HEPES, L-glutamine, penicillin, streptomycin, 10% fetal bovine serum, and 10% chicken serum are exchanged at least daily according to procedures worked out by Zatz and Mullen (1988b,d). Cells are fed by exchange of medium on this daily basis or, in experimental situations, more frequently. Cells are then maintained with this culture medium for 4 days (Fig. 1A). On day 4, cells are switched to the same medium without serum and an additional 10 m$M$ KCl. This culture procedure seems to us the least invasive culture procedure with a concomitant easy accessibility for pharmacological agents, such as those previously (and

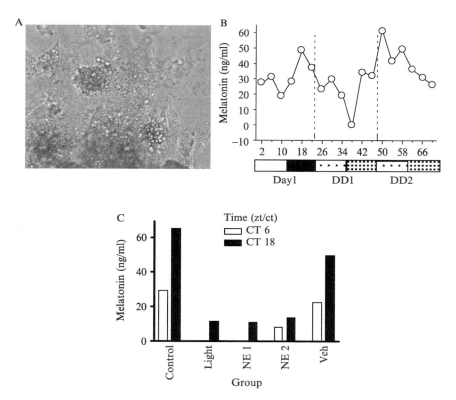

FIG. 1. (A) Interference phase-contrast photomicrograph of cultured pinealocytes. (B) Melatonin concentrations in culture media are rhythmic such that levels are high at night and persist in DD. (C) Exposure to light or norepinephrine in culture reduces the levels of melatonin in the culture media if delivered during the subjective day (CT6) or night (CT18) (Karaganis, Kumar, and Cassone, unpublished results).

elegantly) employed by Zatz and Mullen (1988b,d), for antisense oligonucleotides (ASO) and small interfering RNA (RNAi).

Because chick pinealocytes are photoreceptive, special considerations
for the configuration of $CO_2$ incubators should be considered in order to
take advantage of this model system. Zatz and colleagues have had their
incubators modified to contain two sets of fluorescent lights. One light
fixture is maintained with a red fluorescent tube and is constantly on,
whereas the other fixture, with a white tube, is clock controlled. In this
configuration, the continuous red serves as the "dark" phase in an LD
cycle, whereas the constant red corresponds to DD. This configuration has
the advantage that the constant red is visible for experimenters without the
aid of infrared viewers or other devices. In addition, the heat generated by
the ballast of the continuous red light blunts temperature fluctuations
within the chamber, which are accommodated by the thermostat. We have
modified this configuration for our genomics analysis of chick pinealocytes
so that we can place our cells in constant darkness DD without a red
background. In this case, both sets of lights are clock controlled such that
during the light phase of the experiment, one ballast with a white fluorescent tube is turned on, while the other ballast is off, and during the "dark"
phase, the white light is turned off, and the other ballast, with a fluorescent
tube that is completely blacked with electrician's tape, is turned on. In
this case, there is always only one ballast activated, and no temperature
fluctuation is present.

Cell culture rhythmicity can be assayed by a simple dilution assay of
melatonin release devised by Zatz and Mullen (1988b,d). L-[side chain-
3-$^{14}$C]tryptophan is added to the McCoy's 5A medium (normally contains
0.03 m$M$ tryptophan) at 0.5 $\mu$Ci/ml for 24 h before the start of timed
collections. Then, at the time when rhythmicity is assessed, medium is
collected into polypropylene tubes, extracted with 5 ml chloroform, and
backwashed with acid and base; and 3 ml of the final chloroform extract,
which will contain lipid-soluble melatonin but very little, if any, tryptophan, serotonin, or $N$-acetylserotonin (Zatz and Mullen, 1988b), is dried in
scintillation vials and counted in a scintillant. By collecting medium every
2–4 h, rhythms of melatonin release in LD and DD are easily assayable.

We have employed a modification of the Fraser et al. (1983) radioimmunoassay, later validated in our laboratory using sheep antimelatonin
antiserum (G/S/704-8483; Stockgrand, Guilford, Surrey, UK) and $^3$H-
labeled melatonin from Perkin-Elmer, USA. Briefly, 5–10 $\mu$l of medium
collected from the pineal cell culture (equaled in volume with medium) is
incubated with antiserum (1:750, one can use 1:1000 also) at room temperature for 45 min and then together with $^3$H-labeled melatonin (8000–10,000
cpm) at 4° for 18 h. Finally, the dextran-charcoal solution (suspension)

made in Tricine buffer (TBS) is used to separate free from bound
[3]H-labeled melatonin. All samples collected from one experiment are
assayed in one assay. The lower detection limit of this assay for the pineal
cell culture medium is ~50 pg/ml.

With this assay, we have been able to show a robust rhythm of melatonin in LD and DD (Fig. 1B), largely confirming the extensive work of Zatz
and colleagues, which is inhibited by both norepinephrine superfusion and
acute illumination (Fig. 1C). This experimental system is currently being
employed in an extensive transcriptional profiling of the pineal clock
*in vivo* (Bailey *et al.*, 2003) and *in vitro* (Karaganis, Kumar, and Cassone,
unpublished results).

## Mammalian Suprachiasmatic Nucleus and Primary Cell Culture

A variety of cell culture models have been applied to study the circadian clock mechanism in the mammalian SCN. Most *in vitro* studies have
used organotypic or dissociated primary cultures to isolate SCN neurons
and analyze their temporal patterns of endogenous cellular activity. Hypothalamic slices derived from coronal or horizontal sections through the
SCN of neonatal or young adult rodents were the first *in vitro* preparations
developed for this purpose. Early as well as recent studies using the
hypothalamic slice model established that many endogenous properties of
SCN cells, including neuronal firing rate, glucose utilization, and vasopressin secretion, oscillate independent of external input for several cycles
*in vitro* (Gillette and Reppert, 1987; Newman and Hospod, 1986; Wray
*et al.*, 1993). Organotypic cultures consisting of explanted blocks of SCN
tissue have been similarly used to demonstrate that the SCN is capable of
generating circadian rhythms in vasopressin release for multiple cycles
*in vitro* (Earnest and Sladek, 1987). Even when indigenous neural connections are presumably disrupted by trypsinization of primary cells obtained
from neonatal rats, dissociated cultures of SCN neurons continue to express circadian rhythms of vasopressin release (Watanabe *et al.*, 1993).
Primary cultures of dissociated SCN cells have also yielded important
observations indicating that individual neurons contain the cellular machinery necessary for the generation of self-sustained circadian oscillations.
When maintained on microelectrode plates, individual neurons in dissociated cultures of the rat SCN express independent oscillations in firing rate
that differ widely with regard to circadian phase and period, despite the
presence of synaptic connections (Welsh *et al.*, 1995). The effects of targeted genetic manipulations on circadian clock function in primary cultures
of SCN cells have been assessed to a limited extent by virtue of using
knockout mice. In dissociated cultures of SCN cells from *Clock* mutant

mice, multielectrode recordings indicate that neurons from heterozygotes display an increase in the period of the rhythm in the firing rate and that neurons from homozygous animals are arrhythmic, similar to the effects of the mutation on circadian wheel-running behavior (Herzog *et al.*, 1998).

## Immortalized SCN2.2 Cell Line

Despite their widespread application to study the SCN clock mechanism, primary cell cultures have some limitations that affect this analysis, such as the postmitotic nature of neuronal elements and variability in their survival *in vitro*, cultivation of SCN cells or explants/slices from multiple animals, and uniformity of these preparations with regard to their cellular composition. Because development of mitotic, immortal cell lines derived from SCN progenitors would circumvent some of the problems associated with primary cultures, we instituted this strategy using gene transfer techniques to mediate the introduction of an oncogene and expression of its immortalizing protein product in primary SCN cells (Earnest *et al.*, 1999a). The adenoviral early region 1A (E1A) gene was used to immortalize progenitors of the rat SCN because the 12S protein product of this oncogene induces DNA synthesis and extends the growth potential of primary cells derived from a variety of rat tissues (Cone *et al.*, 1988; Quinlan and Grodzicker, 1987). As established previously (Earnest *et al.*, 1989), SCN cells were specifically obtained from Long Evans rat fetuses (E15 and E16) using topographical landmarks described by Altman and Bayer (1978a,b) to delineate the SCN from adjacent regions of the developing hypothalamus that contain the anlagen of the paraventricular and supraoptic nucleus. SCN tissue from 12–18 fetuses was pooled, disaggregated with 0.125% trypsin, and plated (density = $5 \times 10^5$ cells/well) onto poly-D-lysine- or mouse laminin-coated 24-well plates (16 mm; Costar) containing Dulbecco's modified Eagle's medium (DMEM) supplemented with 10% fetal bovine serum (FBS), 2 $\mu$g/ml glucose, 292 $\mu$g/ml L-glutamine, 2.5 $\mu$g/ml fungizone, and 100 U/ml penicillin and 100 $\mu$g/ml streptomycin. These SCN cell cultures were incubated with filtered (0.45 $\mu$m pore), conditioned medium from confluent cultures of $\varphi$2 cells producing defective retroviruses that encode the adenovirus 2-adenovirus 5 hybrid E1A 12S sequence and the neomycin phosphotransferase gene (*neo*) (Cone *et al.*, 1988). After two successive incubations with fresh virus-containing medium ($\approx 10^4$–$10^5$ CFU/ml) in the presence of 2 $\mu$g/ml polybrene (Aldrich) for 1 h, retrovirus-infected cells were propagated for 7–10 days in regular growth medium, seeded onto 35-mm dishes (Corning), and selected by treatment with medium containing genetecin (G418; 250 $\mu$g/ml) for 14–18 days. The resulting neural cell lines (SCN1.4 and

SCN2.2) were characterized by (1) immunological expression of the immortalizing oncogene (E1A protein) in all G418-resistant cells after numerous passages; (2) extended growth potential without evidence of transformed phenotype (i.e., tumorigenic properties); (3) anchorage-dependent and contact-inhibited proliferation; and (4) heterogeneous cell types in various stages of differentiation. Because morphological criteria were often equivocal in distinguishing glia and neurons in the SCN2.2 line, the phenotypes of immortalized cells were defined further via immunocytochemical screening with antibodies for glial cell markers, neuron-specific antigens, and SCN-like neuropeptides. During initial passages, a few ($\approx$1–5%) of the large cells with flat morphologies displayed diffuse expression of glial fibrillary acidic protein (GFAP) within the cell cytoplasm. At all stages, the SCN2.2 line was characterized by a number of large, polygonal cells with cytoplasmic immunoreactivity for the microglial cell surface proteins (MRC OX-42), but were devoid of cells expressing the astroglial calcium-binding protein S100. Immunostaining for neuron-specific enolase (NSE), protein gene product 9.5 (PGP), microtubule-associated protein 2 (MAP-2), arginine vasopressin (AVP), gastrin-releasing peptide (GRP), somatostatin (SMT), or vasoactive intestinal polypeptide (VIP) was consistently observed in the SCN2.2 line within the perikarya of small, round cells. The SCN-like nature of immortalized cells with peptidergic immunophenotypes was further specified by the finding that the SCN2.2 line was devoid of oxytocin-expressing cells, which are indigenous to paraventricular and supraoptic regions of the hypothalamus. Treatment with succinylated concanavalin A (200 $\mu$g/ml) for 2 days enhanced expression of all SCN peptidergic cell types in SCN2.2 cell cultures. Immunocytochemical analyses were corroborated by radioimmunoassay and RNase protection assays, indicating that SCN2.2 cells are characterized by content and release of AVP, SMT, and VIP in conjunction with expression of their mRNAs. In addition, the SCN2.2 line exhibited mRNA expression for the *Per1*, *Per2*, *Per3*, *Cry1*, *Cry2*, *Clock*, and *Bmal1* genes, indicating that these immortalized cells retain the full complement of molecular elements comprising the SCN clockworks.

Because immortalized SCN cells were characterized by the conservation of many biochemical and molecular signals that distinguish mature parental cell types, including the expression of neuropeptides and core clock genes, it was critical to next determine whether the cell line also retains the distinctive functional properties of the SCN *in vivo*. A fundamental feature of the SCN *in situ* is that many of its molecular, biochemical, and physiological activities oscillate in the absence of environmental or external input. Consequently, immortalized SCN progenitors were first screened for evidence of circadian function by determining whether their uptake of 2-deoxyglucose (2-DG), a well-documented marker of circadian metabolic activity (Newman and Hospod, 1986; Schwartz, 1991), and expression of the neurotrophins, brain-derived neurotrophic factor (BDNF) and NT-3, fluctuate rhythmically *in vitro*. A single passage of cells was expanded to obtain experimental cultures on dishes (60 mm) coated with mouse laminin and maintained in MEM medium containing 10% FBS, 2 $\mu$g/ml glucose, and 292 $\mu$g/ml L-glutamine under constant temperature (37°) and 5% $CO_2$. At 4-h intervals for 2 days, confluent cultures were incubated for 1 h with $^{14}$C-labeled 2-DG (0.2 mCi/ml; American Radiological Co.) and then harvested by lysis in either 0.6 $M$ perchloric acid or TRIzol reagent (Invitrogen). Fractional products of 2-DG metabolism were measured using the methods described by Newman and colleagues (1990). Aliquots (25 $\mu$l) of recovered protein from these samples were also assayed in triplicate for BDNF and NT-3 content by enzyme-linked immunosorbent assay. Determinations of $^{14}$C-labeled 2-DG uptake and neurotrophin content were normalized for sample protein content as measured by the bicinchoninic acid method (Pierce). SCN2.2 cells exhibited robust circadian rhythms in both the uptake of 2-DG and phosphorylation to 2-DG-6-phosphate (2-DG-6P) for two cycles *in vitro* (Earnest *et al.*, 1999b). Levels of 2-DG and 2-DG-6P in immortalized cells oscillated contemporaneously and peak values were two- to threefold greater than the corresponding minimum for both rhythms. Concurrent analysis of BDNF and NT-3 revealed that SCN2.2 cells also exhibited circadian fluctuations in their content of these neurotrophins. The circadian peak in NT-3 content recurred 8 h in advance of the rhythmic maxima in BDNF levels. The rhythms of glucose utilization and BDNF content in SCN2.2 cells oscillated in a 12-h antiphase relationship similar to that observed in the SCN *in vivo* (Liang *et al.*, 1998; Schwartz, 1991). It is unlikely that the cell cycle drives the rhythms in glucose utilization and neurotrophin rhythms because (1) the circadian period of these rhythms is distinctly shorter than the 28-h generation time for SCN2.2 cells and (2) proliferative activity and DNA synthesis in this cell line are arrested upon contact between neighboring cells in confluent cultures (Earnest *et al.*, 1999a).

Functionality of the circadian clock mechanism in SCN2.2 cells was evaluated further by determining whether the clock genes, *Per1*, *Per2*, *Cry1*, and *Bmal1*, are rhythmically regulated similar to the SCN *in vivo*. Using ribonuclease protection assays or quantitative polymerase chain reaction (qt-PCR) to compare mRNA levels in cultures collected at 4-h intervals for 2 days, circadian rhythms in *Per2*, *Cry1*, and *Bmal1* expression were observed in SCN2.2 cells with 2- to 10-fold differences between peak and minimum levels. Importantly, the circadian peak of *Per2* expression occurred 12 h in advance of the rhythmic maxima for *Bmal1* mRNA levels. Collectively, these findings indicate that molecular elements of the clock mechanism are appropriately regulated and configured in SCN2.2 cells.

Circadian pacemaker properties of SCN2.2 cells were initially examined using the neural transplantation technique. In these experiments, SCN-lesioned rats exhibiting arrhythmicity or ultradian rhythms in their wheel-running behavior for 6–8 weeks received transplants of SCN2.2 cells, E1A-immortalized mesencephalic progenitors, or NIH/3T3 fibroblasts placed near the ablation site. All viable grafts of SCN2.2 cells that were later found to contain VIP-, GRP-, or AVP-immunopositive perikarya and fibers restored the circadian activity rhythm in SCN-lesioned hosts within 1–4 days after transplantation (Fig. 2). In these animals, the free-running period of the activity rhythm was shorter than that observed prior to destruction of the SCN. Complete lesions of the host SCN were validated by the absence of VIP, GRP, and AVP immunostaining. In the remaining SCN-lesioned rats that failed to show recovery of circadian rhythmicity during the posttransplantation period, SCN2.2 cells grafts were either characterized by low cell survival or devoid of SCN-like elements. Importantly, the capacity to function as a circadian pacemaker was distinctive to SCN2.2 cells because viable transplants of either E1A-immortalized mesencephalic cells or NIH3T3 fibroblasts did not restore circadian wheel-running behavior in arrhythmic, SCN-lesioned rats.

Development of a coculture model provided further opportunity to distinguish the circadian pacemaker properties of SCN2.2 cells from the oscillatory behavior of cells that are inherently nonrhythmic but show serum shock- or forskolin-induced molecular oscillations. This model was used to determine whether SCN2.2 cells drive both molecular and physiological rhythms in cocultured NIH/3T3 fibroblasts. In these studies (Allen et al., 2001), colonies of NIH/3T3 fibroblasts were established on cell-impermeable inserts (23 mm; pore size = 1 $\mu$m) and cocultured at confluence with companion wells (35 mm) containing either SCN2.2 cells or similar NIH/3T3 cultures. Beginning 24 h after coculture, cells were harvested at 4-h intervals for 52 h to determine 2-DG uptake and *Per2* mRNA expression in each coculture compartment. Because cell lines are optimal for applications of reporter and expression constructs, *Per1* transcription was analyzed in SCN2.2 and NIH/3T3 cells transfected via liposome-mediated introduction (Lipofectamine Plus; Life Technologies) with a reporter vector (pGL3, Promega) containing a 3-kb fragment of the *Per1* promoter fused to the firefly luciferase (*luc*) gene (provided by Dr. Charles Weitz, Harvard University) and a downstream cassette encoding the blasticidin resistance gene (Invitrogen). Stable integrants were selected by growing cells in the presence of blasticidin (3–10 $\mu$g/ml) for 5 days. In

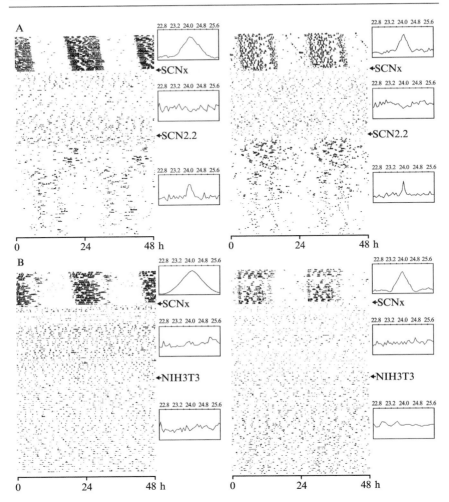

FIG. 2. Representative records of wheel-running activity in rats that received SCN lesions and transplants containing either SCN2.2 cells (A) or NIH/3T3 fibroblasts (B). All intact hosts exhibited free-running activity rhythms with periods of 24.0–24.1 h in constant dim illumination. On the days indicated, the activity patterns were first rendered arrhythmic by bilateral SCN ablation (SCNx). Circadian rhythmicity (period = 24.0 h) was later restored in both rats by transplants of SCN2.2 cells, whereas arrhythmicity persisted in animals receiving grafts of NIH/3T3 cells. Periodogram analyses of data during the last 15–25 days of intact, lesioned, and grafted intervals are shown on the right of each record (from Earnest et al., 1999a).

cocultures containing NIH/3T3 fibroblasts, SCN2.2 cells expressed circadian rhythms in 2-DG uptake, *Per2* mRNA levels (Fig. 3), and *Per1*-driven luciferase bioluminescence for two cycles. In SCN2.2 cells, the rhythm in *Per2* expression reached maximal levels about 8 h after the peak in 2-DG

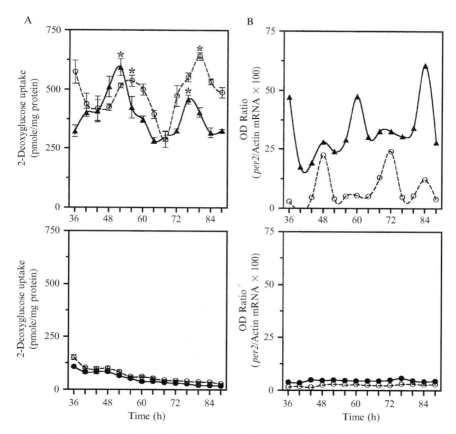

FIG. 3. SCN2.2 cells exhibit circadian pacemaker properties by driving metabolic and molecular rhythmicity in cocultured NIH/3T3 fibroblasts. Temporal patterns of 2-deoxyglucose (2-DG) uptake (A) and *Per2* expression (B) in cocultures (top) containing SCN2.2 cells on wells ($N = 4$; solid line) and NIH/3T3 fibroblasts on inserts ($N = 4$; dashed line) and in cocultures (bottom) consisting of NIH/3T3 cells on both wells ($N = 6$; solid line) and inserts ($N = 6$; dashed line). Symbols denote determinations of 2-DG uptake (mean ± SEM) and optical density (OD) ratios of *Per2*/$\beta$-actin mRNA signal at 4-h intervals. These values are plotted as a function of time such that time 0 denotes when cells located on companion wells and inserts were first cocultured together. Asterisks indicate sampling intervals, during which peak values for 2-DG uptake were significantly greater ($p < 0.05$) than those observed during the preceding or succeeding minima (from Allen *et al.*, 2001).

uptake. NIH/3T3 fibroblasts in these cocultures with SCN2.2 cells displayed comparable rhythms of 2-DG uptake, *Per2* mRNA expression, and *Per1/luc* transgene activity. However, the phase of these conferred rhythms in NIH/3T3 cells was delayed by 4–12 h relative to the circadian patterns observed in cocultured SCN2.2 cells. For the oscillations in *Per* gene expression, the 4-h phase difference between SCN2.2 and NIH/3T3 rhythms is similar to that reported between SCN and peripheral tissues *in vivo* (Lopez-Molina *et al.*, 1997; Shearman *et al.*, 1997; Zylka *et al.*, 1998).

The molecular components responsible for the distinctive oscillatory and pacemaking properties of the SCN were examined by determining whether experimental inhibition of a core clock element alters metabolic and molecular rhythmicity in SCN2.2 cells and their capacity to drive these rhythms in NIH/3T3 fibroblasts. Because ASO treatment is typically more effective on culture models, especially cell lines (Brysch and Schlingensiepen, 1994), morpholino oligonucleotides (Gene Tools, LCC, Philomath, OR) directed against rat-specific *Clock* (*rClock*) mRNA were used to knock down CLOCK protein levels. The *rClock* ASO (25-mer) was designed for the region of the mRNA containing the initiation ATG codon (Kozak, 1978) with the base composition (5'-TAC TAC AGC TTA CGG TAA ACA ACA T-3'). Invert ASO for *rClock* was designed so as to have the same length and base composition (5'-TAC AAC AAA TGG CAT TCG ACA TCA T-3'). SCN2.2 cells were plated on multiple six-well companion plates and were treated 24 h later with *rClock* morpholino ASO or invert of ASO. Delivery of morpholino ASO (1 $\mu M$) to SCN2.2 cultures was performed using a weakly basic delivery reagent, ethoxylated polyethylenimine (EPEI). Twenty-four hours after ASO treatment, SCN2.2 cells were cocultured with parallel colonies of untreated NIH/3T3 cells that were established separately on cell-impermeable inserts, and then 24 h later samples were collected at 4-h intervals to determine 2-DG uptake and clock gene mRNA expression in each coculture compartment. Inhibition of CLOCK was observed in ASO-treated SCN2.2 cells such that CLOCK protein levels were reduced (80–100 h posttransfection) to 28–59% of those found in invert-treated control cells. Similar to the effects of the *Clock* mutation on molecular and physiological rhythmicity in the SCN *in vivo* (Oishi *et al.*, 2000; Vitaterna *et al.*, 1994), this inhibition of CLOCK disrupted the rhythmic pattern of *Per2* expression and increased the period of metabolic rhythmicity in ASO-treated SCN2.2 cells. In untreated NIH/3T3 fibroblasts cocultured with ASO-treated SCN2.2 cells, metabolic rhythms showed comparable increases in period and decreases in rhythm amplitude, whereas *Per2* expression was similarly marked by non-24-h or irregular peak-to-peak intervals. This analysis indicates that ASO methods can be used to knock down the expression of specific genes in SCN2.2 cells and thus determine their involvement in the generation of endogenous molecular and metabolic rhythmicity and in the regulation of SCN2.2 output signals that control rhythmic processes in other cells.

## Shocked Fibroblast Model

Fibroblasts are a cell line model that has been frequently applied to study the molecular workings of the mammalian circadian clock because rhythmic expression of various clock and clock-controlled genes can be induced in cultures of the rat-1 and NIH/3T3 fibroblast lines by serum shock treatment (Akashi and Nishida, 2000; Balsalobre *et al.*, 1998). These findings raise an important question: what distinguishes the circadian pacemaker properties of SCN2.2 cells from the induced oscillatory behavior of serum-shocked fibroblasts? Experiments were first conducted to determine whether serum-shocked NIH/3T3 fibroblasts express molecular and physiological rhythms similar to those imposed in cocultures by SCN2.2-specific outputs. Using the methods described by Balsalobre and colleagues (1998), confluent cultures of NIH/3T3 cells were exposed to medium containing 50% adult horse serum for 2 h and then maintained under serum-free conditions. Cells were harvested at 4-h intervals for 52 h with TriZol reagent to analyze 2-DG uptake and expression of mouse-specific clock gene mRNAs. Serum shock induced rhythms of *Per1*, *Per2*, *Cry1*, and *Cry2* expression in NIH/3T3 cells. The *Per* and *Cry* gene oscillations in serum-shocked NIH/3T3 cells showed contemporary cycles in paralog comparisons, but the circadian expression profiles for the *Cry* genes were phase delayed by 4 h relative to those for the *Per* genes. This phase relationship between the rhythms in *Per* and *Cry* gene expression for serum-shocked NIH/3T3 cells is comparable to that reported for Rat-1 fibroblasts *in vitro* (Yagita *et al.*, 2001) and for the SCN *in vivo* (Shearman *et al.*, 2000). Despite these molecular oscillations, metabolic activity in serum-shocked NIH/3T3 fibroblasts was arrhythmic, with 2-DG uptake remaining consistently at levels similar to those observed in solitary NIH/3T3 cultures.

To determine whether this molecular machinery found in SCN and SCN2.2 cells propagates similar pacemaker properties in fibroblasts, serum-shocked NIH/3T3 cells were examined in a coculture environment for the capacity to confer molecular rhythmicity to untreated fibroblasts. NIH/3T3 cells derived from a single passage were separately established on Transwell inserts (75 mm) and companion wells (100 mm). At confluence, only NIH/3T3 cultures on companion wells were exposed to medium containing 50% adult horse serum for 2 h and thereafter cocultured in serum-free DMEM (supplemented with glucose and L-glutamine) with inserts containing untreated NIH/3T3 cells. Using ribonuclease protection assays, *Per1* and *Per2* mRNA levels in serum-shocked NIH/3T3 fibroblasts maintained on companion wells displayed synchronous oscillations similar to the preceding experiment. In contrast to the serum-shocked fibroblasts in these cocultures, untreated NIH/3T3 cells displayed no evidence of circadian fluctuations in *Per1* or *Per2* mRNA expression.

## Relative Merits of Different Cell Culture Approaches

All three experimental systems have merits and problems for ascribing system level function. The avian pinealocyte culture system retains its photoreceptive apparatus, which enables analyses of entrainment via a "natural" pathway, a molecular mechanism capable of generating an endogenous circadian patter, and an easily measured, physiologically relevant output, melatonin, that is known to have system level effects. The molecular biology of the melatonin-generating system has been characterized, all of the avian orthologs to mammalian clock genes have been isolated and cloned, and the temporal dynamics of their rhythmicity has been described. However, an efficient method for transforming pinealocyte cultures has not as yet been described, so hypothesis-driven molecular physiology is not possible at this time. It is hoped that problems associated with ASO and RNAi in this system will be solved.

The SCN2.2 system retains the oscillatory and pacemaker properties of the SCN *in vivo* and in coculture experiments *in vitro*. Further, the molecular dynamics of clock gene expression has been described, and experimental manipulation of clock gene levels by ASO has circadian effects on both circadian patterns of metabolism and of clock gene expression. However, because the RHT is no longer in place and the cells are not directly photoreceptive, "natural" entrainment is not possible. Further, although melatonin and glutamate have phase-dependent effects on SCN2.2 cells (Hurst *et al.*, 2002; Rivera-Bermúdez *et al.*, 2003), experimental entrainment of these cells has not been established. It is hoped that future research will identify a physiologically relevant zeitgeber for these cells *in vitro*.

Finally, serum-shocked fibroblasts express a dynamic circadian pattern of clock genes and of clock-controlled genes. The procedures for inducing these rhythms are simple, and if one is only interested in the dynamics of clock gene expression, this is an adequate model. However, the clock gene rhythm that is induced in these cells is not sufficient to generate a physiological output such as 2DG uptake. Further, because serum-shocked cells are not able to confer their own rhythmicity to cocultured cells, fibroblasts do not appear to acquire pacemaker function with serum shock.

Thus, each of these cell culture model systems has its benefits and detractions. Ultimately, however, the principles unveiled by these cell culture systems must be verified *in vivo* and *in situ*. Without this integrated approach, circadian organization cannot be fully understood.

## Acknowledgments

The authors recognize Gregg Allen, Steven Karaganis, Barbara Earnest, and Vinod Kumar for their invaluable contributions to the development of the culture models described in this review. Our studies were supported by NIH Program Project Grant P01 NS39546 (D.J.E. and V.M.C.).

## References

Akashi, M., and Nishida, E. (2000). Involvement of the MAP kinase cascade in resetting of the mammalian circadian clock. *Genes Dev.* **14,** 645–649.

Allen, G., Rappe, J., Earnest, D. J., and Cassone, V. M. (2001). Oscillating on borrowed time: Diffusible signals from immortalized suprachiasmatic nucleus cells regulate circadian rhythmicity in cultured fibroblasts. *J. Neurosci.* **21,** 7937–7943.

Altman, J., and Bayer, S. A. (1978a). Development of the diencephalon in the rat. I. Autoradiographic study of the time of origin and the settling patterns of neurons of the hypothalamus. *J. Comp. Neurol.* **182,** 945–972.

Altman, J., and Bayer, S. A. (1978b). Development of the diencephalon in the rat. II. Correlation of the embryonic development of the hypothalamus with the time of origin of its neurons. *J. Comp. Neurol.* **182,** 972–994.

Bailey, M. J., Beremand, P. D., Hammer, R., Bell-Pedersen, D., Thomas, T. L., and Cassone, V. M. (2003). Transcriptional profiling of the chick pineal gland, a photoreceptive circadian oscillator and pacemaker. *Mol. Endocrinol.* **17,** 2084–2095.

Balsalobre, A., Damiola, F., and Schibler, U. (1998). A serum shock induces circadian gene expression in mammalian tissue culture cells. *Cell* **93,** 929–937.

Binkley, S. A., Riebman, J. B., and Reilly, K. B. (1978). The pineal gland: A biological clock *in vitro. Science* **202,** 1198–1201.

Brysch, W., and Schlingensiepen, K.-H. (1994). Design and application of antisense oligonucleotides in cell culture, *in vivo*, and as therapeutic agents. *Cell. Mol. Neurobiol.* **14,** 557–568.

Cassone, V. M. (1998). Melatonin's role in vertebrate circadian rhythms. *Chronobiol. Int.* **15,** 457–473.

Cassone, V. M., Brooks, D. S., Hodges, D. B., Kelm, T. A., Lu, J., and Warren, W. S. (1992). Integration of circadian and visual function in mammals and birds: Brain imaging and the role of melatonin in biological clock regulation. *In* "Advances in Metabolic Mapping Techniques for Brain Imaging of Behavioral and Learning Functions" (F. Gonzalez-Lima, T. Finkenstaedt, and H. Scheich, eds.), pp. 299–318. Kluwer Academic Publishers, Dordrecht/Boston/London.

Cassone, V. M., and Menaker, M. (1984). Is the avian circadian system a neuroendocrine loop? *J. Exp. Zool.* **232,** 539–549.

Cassone, V. M., Speh, J. C., Card, J. P., and Moore, R. Y. (1988). Comparative anatomy of the mammalian hypothalamic suprachiasmatic nucleus. *J. Biol. Rhythms* **3,** 71–91.

Cone, R. D., Grodzicker, T., and Jaramillo, M. (1988). A retrovirus expressing the 12S adenoviral E1A gene product can immortalize epithelial cells from a broad range of rat tissues. *Mol. Cell. Biol.* **8,** 1036–1044.

Deguchi, T. (1979a). Circadian rhythms of indoleamines and serotonin N-acetyltransferase activity in the pineal gland. *Mol. Cell. Biol.* **27,** 57–66.

Deguchi, T. (1979b). A circadian oscillator in cultured cells of chicken pineal gland. *Nature* **282,** 94–96.

Deguchi, T. (1981). Rhodopsin-like photosensitivity of isolated chicken pineal gland. *Nature* **290,** 706–707.

Drucker-Colin, R., Aguilar-Roblero, R., Garcia-Hernandez, F., Fernandez-Cancino, F., and Bermudez Rattoni, F. (1984). Fetal suprachiasmatic nucleus transplants: Diurnal rhythm recovery of lesioned rats. *Brain Res.* **311,** 353–357.

Dunlap, J. C., Loros, J. J., Liu, Y., and Crosthwaite, S. K. (1999). Eukaryotic circadian systems: Cycles in common. *Genes Cells* **4,** 1–10.

Earnest, D. J., Liang, F.-Q., DiGiorgio, S. M., Gallagher, M. J., Harvey, B., Earnest, B. J., and Seigel, G. M. (1999a). Establishment and characterization of adenoviral E1A immortalized cell lines derived from the rat suprachiasmatic nucleus. *J. Neurobiol.* **39,** 1–13.

Earnest, D. J., Liang, F.-Q., Ratcliff, M., and Cassone, V. M. (1999b). Immortal time: Circadian clock properties of rat suprachiasmatic cell lines. *Science* **283,** 693–695.

Earnest, D. J., and Sladek, C. D. (1987). Circadian rhythms of vasopressin release from perifused rat suprachiasmatic explants *in vitro*: Effects of acute stimulation. *Brain Res.* **422,** 398–402.

Earnest, D. J., Sladek, C. D., Gash, D. M., and Wiegand, S. J. (1989). Specificity of circadian function in transplants of the fetal suprachiasmatic nucleus. *J. Neurosci.* **9,** 2671–2677.

Ebihara, S., Uchiyama, K., and Oshima, I. (1984). Circadian organization in the pigeon *Columbia livia*: The role of the pineal organ and the eye. *J. Comp. Physiol. A* **154,** 59–69.

Fraser, S., Cowen, P., Franklin, M., Franey, C., and Arendt, J. (1983). Direct radioimmunoassay for melatonin in plasma. *Clin. Chem.* **29,** 396–397.

Gaston, S., and Menaker, M. (1968). Pineal function: The biological clock in the sparrow? *Science* **160,** 1125–1127.

Gillette, M. U., and Reppert, S. M. (1987). The hypothalamic suprachiasmatic nuclei: Circadian patterns of vasopressin secretion and neuronal activity *in vitro*. *Brain Res. Bull.* **19,** 135–139.

Green, D. J., and Gillette, R. (1982). Circadian rhythm of firing rate recorded from single cells in the rat suprachiasmatic brain slice. *Brain Res.* **245,** 198–200.

Golden, S. S., and Canales, S. R. (2003). Cyanobacterial circadian clocks—timing is everything. *Nature Rev. Microbiol.* **1,** 191–199.

Gwinner, E. (1978). Effects of pinealectomy on circadian locomotor activity rhythms in European starlings, *Sturnus vulgaris*. *J. Comp. Physiol.* **126,** 123–129.

Gwinner, E., and Benzinger, I. (1978). Synchronization of a circadian rhytm in pinealectomized European starlings *Sturnus vulgaris*. *J. Comp. Physiol. A* **127,** 209–213.

Herzog, E. D., Takahashi, J. S., and Block, G. D. (1998). *Clock* controls circadian period in isolated suprachiasmatic nucleus neurons. *Nature Neurosci.* **8,** 708–713.

Hurst, W. J., Mitchell, J. W., and Gillette, M. U. (2002). Synchronization and phase-resetting by glutamate of an immortalized SCN cell line. *Biochem. Biophys. Res. Commun.* **298,** 133–143.

Kasal, C. A., Menaker, M., and Perez-Polo, J. R. (1979). Circadian clock in culture: N-acetyltransferase activity of chick pineal glands oscillates *in vitro*. *Science* **203,** 656–658.

Kozak, M. (1978). How do eucaryotic ribosomes select initiation regions in messenger RNA? *Cell* **15,** 1109–1123.

Liang, F.-Q., Walline, R., and Earnest, D. (1998). Circadian rhythm of brain-derived neurotrophic factor expression in the rat suprachiasmatic nucleus. *Neurosci. Lett.* **242,** 89–92.

Lopez-Molina, L., Conquet, F., Dubois-Dauphin, M., and Schibler, U. (1997). The DBP gene is expressed according to a circadian rhythm in the suprachiasmatic nucleus and influences circadian behavior. *EMBO J.* **16,** 6762–6771.

Lu, J., and Cassone, V. M. (1993). Daily melatonin administration synchronizes circadian patterns of brain metabolism and behavior in pinealectomized house sparrows, *Passer domesticus. J. Comp. Physiol. A* **173,** 775–782.

McGoogan, J. M., and Cassone, V. M. (1999). Circadian clock regulation of the electroretinogram of the chick: Effects of pinealectomy and exogenous melatonin. *Amer. J. Physiol.* **277,** R1418–R1427.

Menaker, M., Takahashi, J. S., and Eskin, A. (1978). The physiology of circadian pacemakers. *Annu. Rev. Physiol.* **40,** 501–526.

Moore, R. Y. (1995). Organization of the mammalian circadian system. *Ciba Found. Symp.* **183,** 88–106.

Moore, R. Y., and Silver, R. (1998). Suprachiasmatic nucleus organization. *Chronobiol. Int.* **15,** 475–487.

Natesan, A., Geetha, L., and Zatz, M. (2002). Rhythm and soul in the avian pineal. *Cell Tissue Res.* **309,** 35–45.

Newman, G. C., and Hospod, F. E. (1986). Rhythm of suprachiasmatic nucleus 2-deoxyglucose uptake *in vitro. Brain Res.* **381,** 345–350.

Newman, G. C., Hospod, F. E., and Patlak, C. S. (1990). Kinetic model of 2-deoxyglucose metabolism using brain slices. *J. Cereb. Blood Flow Metab.* **10,** 510–526.

Oishi, K., Fukui, H., and Ishida, N. (2000). Rhythmic expression of *Bmal1* mRNA is altered in *Clock* mutant mice: Differential regulation in the suprachiasmatic nucleus and peripheral tissues. *Biochem. Biophys. Res. Commun.* **268,** 164–171.

Pittendrigh, C. S. (1993). Temporal organization: Reflections of a Darwinian clock-watcher. *Annu. Rev. Physiol.* **55,** 16–54.

Quinlan, M. P., and Grodzicker, T. (1987). Adenovirus E1A 12S protein induces DNA synthesis and proliferation in primary epithelial cells in both the presence and absence of serum. *J. Virol.* **61,** 673–682.

Rivera-Bermúdez, M. A., Gerdin, M. J., Earnest, D. J., and Dubocovich, M. L. (2003). Melatonin regulation of basal rhythm in protein kinase C activity generated in immortalized rat suprachiasmatic nucleus cells. *Neurosci. Lett.* **346,** 37–40.

Sawaki, Y., Nihonmatsu, I., and Kawamura, H. (1984). Transplantation of the neonatal suprachiasmatic nuclei into rats with complete bilateral suprachiasmatic lesions. *Neurosci. Res.* **1,** 67–72.

Schwartz, W. J. (1991). SCN metabolic activity *in vivo. In* "Suprachiasmatic Nucleus: The Mind's Clock" (D. C. Klein, R. Y. Moore, and S. M. Reppert, eds.), pp. 144–156. Oxford, New York.

Shearman, L. P., Jin, X., Lee, C., Reppert, S. M., and Weaver, D. R. (2000). Targeted disruption of the *mPer3* gene: Subtle effects on circadian clock function. *Mol. Cell. Biol.* **20,** 6269–6275.

Shearman, L. P., Zylka, M. J., Weaver, D. R., Kolakowski, L. F., Jr., and Reppert, S. M. (1997). Two *period* homologs: Circadian expression and photic regulation in the suprachiasmatic nuclei. *Neuron* **19,** 1261–1269.

Silver, R., Lehman, M. N., Gibson, M., Gladstone, W. R., and Bittman, E. L. (1990). Dispersed cell suspensions of fetal SCN restore circadian rhythmicity in SCN-lesioned adult hamsters. *Brain Res.* **525,** 45–58.

Silver, R., LeSauter, J., Tresco, P. A., and Lehman, M. N. (1996). A diffusible coupling signal from the transplanted suprachiasmatic nucleus controlling circadian locomotor rhythms. *Nature* **382,** 810–813.

Takahashi, J. S., Hamm, H., and Menaker, M. (1980). Circadian rhythms of melatonin release from individual superfused chicken pineal glands *in vitro*. *Proc. Natl. Acad. Sci. USA* **77,** 2319–2322.

Takahashi, J. S., Murakami, N., Nikaido, S. S., Pratt, B. L., and Robertson, L. M. (1989). The avian pineal, a vertebrate model system of the circadian oscillator: Cellular regulation of circadian rhythms by light, second messengers, and macromolecular synthesis. *Recent Prog. Horm. Res.* **45,** 279–348.

Underwood, H., and Siopes, T. (1984). Circadian organization in Japanese quail. *J. Exp. Zool.* **232,** 557–566.

Vitaterna, M. H., King, D. P., Chang, A.-M., Kornhauser, J. M., Lowrey, P. L., McDonald, J. D., Dove, W. F., Pinto, L. H., Turek, F. W., and Takahashi, J. S. (1994). Mutagenesis and mapping of a mouse gene, *Clock*, essential for circadian behavior. *Science* **264,** 719–725.

Wainwright, S. D. (1977). Metabolism of tryptopan and serotonin by the chick pineal gland in organ culture. *Can. J. Biochem.* **55,** 415–423.

Wainwright, S. D., and Wainwright, L. K. (1979). Chick pineal serotonin acetyltransferase: A diurnal cycle maintained in vitro and its regulation by light. *Can. J. Biochem.* **57,** 700–709.

Watanabe, K., Koibuchi, N., Ohtake, H., and Yamaoka, S. (1993). Circadian rhythms of vasopressin release in primary cultures of rat suprachiasmatic nucleus. *Brain Res.* **624,** 115–120.

Welsh, D. K., Logothetis, D. E., Meister, M., and Reppert, S. M. (1995). Individual neurons dissociated from rat suprachiasmatic nucleus express independently phased circadian firing patterns. *Neuron* **14,** 697–706.

Wray, S., Castel, M., and Gainer, H. (1993). Characterization of the suprachiasmatic nucleus in organotypic slice explant cultures. *Microsc. Res. Tech.* **25,** 46–60.

Yagita, K., Tamanini, F., van der Horst, G. T. J., and Okamura, H. (2001). Molecular mechanisms of the biological clock in cultured fibroblasts. *Science* **292,** 278–281.

Zatz, M. (1989). Relationship between light, calcium influx and cAMP in the acute regulation of melatonin production by cultured chick pineal cells. *Brain Res.* **477,** 14–18.

Zatz, M., and Mullen, D. A. (1988a). Norepinephrine, acting via adenylate cyclase, inhibits melatonin output but does not phase-shift the pacemaker in cultured chick pineal cells. *Brain Res.* **450,** 137–143.

Zatz, M., and Mullen, D. A. (1988b). Photoendocrine transduction in cultured chick pineal cells. II. Effects of forskolin, 8-bromocyclic AMP, and 8-bromocyclic GMP on the melatonin rhythm. *Brain Res.* **453,** 51–62.

Zatz, M., and Mullen, D. A. (1988c). Two mechanisms of photoendocrine transduction in cultured chick pineal cells: Pertussis toxin blocks the acute but not the phase-shifting effects of light on the melatonin rhythm. *Brain Res.* **453,** 63–71.

Zatz, M., and Mullen, D. A. (1988d). Does calcium influx regulate melatonin production through the circadian pacemaker in chick pineal cells? Effects of nitrendipine, Bay K 8644, Co2+, Mn2+, and low external Ca2+. *Brain Res.* **463,** 305–316.

Zatz, M., Mullen, D. A., and Moskal, J. R. (1988). Photoendocrine transduction in cultured chick pineal cells: Effects of light, dark, and potassium on the melatonin rhythm. *Brain Res.* **438,** 199–215.

Zimmerman, N. H., and Menaker, M. (1979). The pineal gland: A pacemaker within the circadian system of the house sparrow. *Proc. Natl. Acad. Sci. USA* **76,** 999–1003.

Zylka, M. J., Shearman, L. P., Weaver, D. R., and Reppert, S. M. (1998). Three *period* homologs in mammals: Differential light responses in the suprachiasmatic circadian clock and oscillating transcripts outside of brain. *Neuron* **20,** 1103–1110.

# [30] Analysis of Circadian Mechanisms in the Suprachiasmatic Nucleus by Transgenesis and Biolistic Transfection

*By* Michael H. Hastings, Akhilesh B. Reddy,
Douglas G. McMahon, and Elisabeth S. Maywood

## Abstract

Analysis of the cellular and molecular mechanisms that underlie the circadian pacemaker of the suprachiasmatic nuclei (SCN) requires *in vitro* preparations amenable to genetic manipulation that can provide dynamic measures of circadian activity in real time over multiple circadian cycles. This article focuses on the value of the SCN organotypic slice for such studies. Specifically, it describes the use of tissues from genetically modified mice in which the circadian promoter of the *mPer1* gene is used to drive the expression of either firefly luciferase or destabilized green fluorescent protein optical reporters. Furthermore, we describe a procedure for biolistic (particle-mediated) transfection of SCN organotypic slices with fluorescent reporters that can be used to explore the *cis*-acting elements and *trans*-acting factors that control circadian patterning, and also the interactions between subpopulations of neuronal oscillators within the SCN assemblage.

Suprachiasmatic nuclei (SCN) of the hypothalamus are the principal circadian pacemaker driving the sleep/wake cycle and the myriad behavioral and neurophysiological rhythms consequential to it (Reppert and Weaver, 2002). Moreover, SCN orchestrate the activity of local circadian oscillators in peripheral tissues and thereby maintain the internal temporal order that underpins physiological adaptation to the solar cycle (Hastings *et al.*, 2003). Current molecular and cellular models of the circadian mechanism of the SCN are based heavily on inferences drawn from biochemical analysis of circadian factors in peripheral tissues and the properties of recombinant proteins expressed in cell lines. While very successful, this approach alone will not lead to an understanding of how the neurons of the SCN intrinsically generate a circadian signal, then synchronize as a population, and finally relay that signal to targets in the brain that ultimately maintain the sleep/wake cycle and control circadian activity in other brain regions. Therefore, more direct analysis of circadian molecular events in SCN is required.

Until recently, most studies of the SCN oscillator used electrophysiological measures to make inferences about the clockwork, usually in acute

METHODS IN ENZYMOLOGY, VOL. 393

slice preparations. Long-term organotypic slice cultures (House *et al.*, 1998) offer considerable advantages over these acute preparations and, in several cases, have yielded important insights into cellular events underlying the circadian cycle of electrical firing (Herzog *et al.*, 1997; Ikeda *et al.*, 2003; Nakamura *et al.*, 2001). Moreover, advances in real-time optical imaging techniques now make it possible to observe not only electrical functions in the SCN, which are downstream of the core clockwork, but also key molecular events at the heart of the circadian oscillator. Critical to the success of these approaches has been the development of genetically modified mice and rats bearing optical reporter genes (Kuhlman *et al.*, 2000; Yamaguchi *et al.*, 2001; Yamazaki *et al.*, 2000; Yoo *et al.*, 2004). This approach will be extended further by the refinement of biolistic transfection techniques to introduce novel reporter constructs into SCN neurons in slice culture to achieve acute genetic modification of such neurons (Ikeda *et al.*, 2003; O'Brien *et al.*, 2001).

## Organotypic Slices of Suprachiasmatic Nuclei

Organotypic slices of SCN can be made from animals of any age (Yamazaki *et al.*, 2002), although for longer-term studies, mouse or rat pups of up to 1 week of age are preferable. Following sacrifice by cervical dislocation, the brain is dissected free, taking care not to damage the optic nerves or chiasm, and placed into 1 ml ice-cold dissection medium in a petri dish. The temporal lobes are removed by lateral cuts using a razor blade, the brain is placed onto one side, and the dorsal tissues above the hypothalamus are cut away. The tissue block is then placed with the hypothalamus facing upward onto a McIlwain "tissue chopper" and sectioned at 300 $\mu$m. The sliced block is then washed using 1 ml of dissection medium into a clean petri dish for sorting. Slices containing SCN are separated free of the tissue block under low power ($\times 6$–10 magnification) using fine-drawn, glass Pasteur pipettes for dexterity. The slices are then drawn up in a small volume of medium and placed onto a filter membrane (Millipore "Millicell" PIC-MORG50). Once slices are positioned, the medium is drawn off, the membrane insert is placed into a clean petri dish, and 1 ml of culture medium plus glutamate receptor blockers (to prevent neurotoxic cell death) is pipetted under the insert. The slices are then left in a 37° $CO_2$ incubator for 3 h to stabilize, at which point they are transferred to a six-well plate with 1.1 ml of culture medium. Over the first week, the slices will exhibit cell death, especially at the periphery, as evidenced by their darkened, necrotic profiles. The slices will thin down as tissue sloughs off. The SCN and surrounding medial hypothalamus do not succumb, however, and after 7–10 days the stabilized slice takes on a silvery, translucent sheen with very clear morphological landmarks equivalent to its *in vivo* condition. The health of the tissue is evident from strong vacillatory activity of the ependymal cells lining the third ventricle, which is obvious from the whirling of detached cell clusters within the fluid-filled ventricle. The slice can be maintained in culture for several months, with periodic medium changes every 1–2 weeks.

## Dissection Medium

Gey's balanced salt solution (Sigma, G9779)
5 mg/ml glucose
100 n$M$ MK-801 (Sigma M107)
3 m$M$ MgCl$_2$
50 $\mu M$ D-APV (DL-2-amino-5-phosphonovaleric acid, Sigma A5282)

## Culture Medium

50% Eagle's basal medium (Sigma, B1522)
25% Earle's balanced salt solution (Sigma, E2888)
25% heat-inactivated horse serum (Gibco Invitrogen 26050-070)
5 mg/ml glucose
1% Glutamax (Gibco Invitrogen 35050-038)
Penicillin/streptomycin (25 $\mu$g/ml each)
pH 7.2, osmolarity of 315–320 mOsm

## Immunofluorescence

Morphology of the slice can be analyzed by immunofluorescence. After two washes in phosphate-buffered saline (PBS), slices attached to the filter are fixed in 4% paraformadehyde in 0.1 $M$ phosphate buffer for 1 h. They are immersed sideways to avoid disturbance to the tissue from fluid pushing up through the membrane. After two further washes in PBS, they are blocked with normal serum (1:200) in day 1 buffer (PBS with 1% bovine serum albumen and 0.3% Triton-X detergent) and incubated for 1 h at room temperature. Primary antibody is then added to day 1 buffer this added to the slices, which are then incubated overnight at 4°. After removal of primary and three washes in day 2 buffer (day 1 buffer diluted 1 in 3 with PBS), fluorescent-tagged secondary (1:600, Alexafluor-488 goat antirabbit) is added to day 2 buffer for a 1-h incubation. After three washes with day 2 buffer and then two washes with PBS, the membrane with the attached slice is punched free of its plastic support, washed briefly in distilled water, and coverslipped in stabilizing mountant ("Vectashield," Vecta Labs). The membrane does not compromise optical imaging of the tissue. Examples of organotypic slices immunostained after several weeks in culture for either

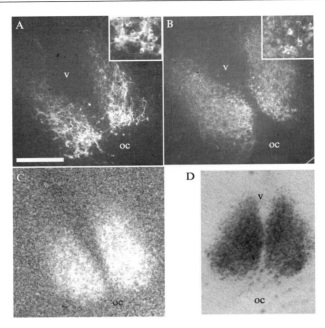

FIG. 1. Organotypic SCN cultures imaged several weeks after preparation show typical morphology and cellular composition. Immunofluorescence staining for neuropeptides PHI (A) and AVP-neurophysin (B) reveals characteristic core and shell neuronal subpopulations within organotypic SCN cultures. (Insets) Details of neurons. v, third ventricle; oc, remains of optic chiasm. Scale bar: 500 μm. (C) GFP emission from a slice prepared from a *mPer1::dsGFP* mouse and imaged at the peak of the circadian emission cycle (Kuhlman *et al.*, 2000). (D) β-Galactosidase staining of a slice prepared from a *hVipr2* transgenic mouse, illustrating expression of the hVPAC2 neuropeptide receptor in both core and shell (King *et al.*, 2003).

PHI or AVP/neurophysin (neuropeptides that identify the core and shell subdivisions of the SCN, respectively), are presented in Fig. 1A, B. Integrity of the slice can also be confirmed if it carries a transgene, e.g., a fluorescent protein driven by a circadian promoter (Fig. 1C) or β-galactosidase driven by the *Vipr2* gene that encodes a neuropeptide receptor widely expressed in SCN (King *et al.*, 2003) (Fig. 1D).

## Real-Time Recording of Circadian Activity in Organotypic SCN Slice

The molecular oscillator of the SCN is based on an autoregulatory negative feedback loop in which the protein products of the core clock genes *Period* and *Cryptochrome* suppress transcriptional activation of their cognate genes by heterodimers composed of CLOCK and BMAL

(Reppert and Weaver, 2002). Central to the clockwork, therefore, is the periodic activation and subsequent suppression of activity at the relevant promoters. This is dependent on E-box enhancer sequences in the gene (Yamaguchi *et al.*, 2000; Fig. 2A). To date, both *mPer1* and *mPer2* genes have been used for real-time imaging of SCN organotypic slices, the former as a transgene in which regulatory sequences are used to drive firefly luciferase or destabilized green fluorescent protein as optical reporters (Kuhlman *et al.*, 2000; Yamaguchi *et al.*, 2001; Yamazaki *et al.*, 2000).

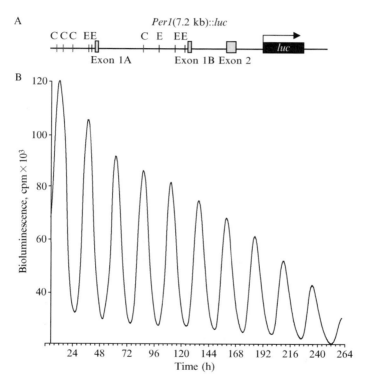

FIG. 2. Real-time recording of circadian gene expression via *mPer1::luciferase* imaging. (A) Details of a *mPer1::luc* construct (Yamaguchi *et al.*, 2000) used for acute transfection of cell lines and generation of genetically modified mice and rats. E, E-boxes, sites of circadian activation; C, $Ca^{2+}$/cAMP response elements, sites for activation by extracellular stimuli. (B) Representative trace of bioluminescence emission, detected as photon counts, from an organotypic SCN slice prepared from a *mPer1::luc* mouse and held in static culture for over 10 days. The decline in amplitude is attributed to utilization of the luciferin substrate. These biological data have a machine-like appearance, highlighting the exquisite precision of the circadian mechanism of the SCN.

*mPer2* activation has been imaged using tissue from a "knockin" mouse in which a mPER2-LUC fusion protein provides the circadian report (Yoo *et al.*, 2004). For such studies, either acute or stabilized organotypical SCN slices are cultured in medium containing luciferin substrate at an initial concentration of 0.1 $\mu M$. Emission is monitored in a light-tight incubator using photomultiplier tubes (PMT), with data stored on a PC. We have worked with two configurations. The commercially available Hamamatsu LM2400 carries two PMTs housed above a turntable that carries up to 24 petri dishes in two series of 12. The light-tight unit counts each dish for 1 min, taking data from it every 15 min. The cabinet can be housed in a $CO_2$ incubator and has a central water reservoir for humidification. Cultures dishes do not, therefore, need to be sealed to maintain humidification and buffering and so can be maintained in standard culture medium. This provides ease of access, but the acquisition system has to be switched off to open the cabinet and restarted from a new zero time once interrupted. The second configuration was designed and assembled in-house and uses PMTs incorporated into an integrated photon-counting head purchased from Hamamatsu (H9319-11) housed in a dry, heated cabinet linked to a PC for continuous data acquisition. Culture dishes are located below the PMTs in a light-tight, recessed support. They are either sealed with parafilm or the petri dish lid is replaced with a gas-permeable membrane that does not allow passage of water vapor (Potter and DeMarse, 2001). Cultures are maintained in standard culture medium or low bicarbonate, HEPES-buffered recording medium. Figure 2B shows a representative trace from an SCN organotypic slice made from a 1-week-old mouse pup and cultured for 2 weeks before recording for over 10 days in this configuration. Continuous data acquisition provides for exquisite temporal resolution of circadian output, and preparations of this type employing tissue from transgenic rats and mice have been used to uncover fundamental aspects of SCN circadian mechanisms (Herzog and Huckfeldt, 2003; Stokkan *et al.*, 2001; Yamaguchi *et al.*, 2003; Yoo *et al.*, 2004).

Recording Medium

*Stock Solution*

> Bicarbonate-free Dulbecco's modified Eagle's medium
>   (Sigma D5030)
> 0.35 mg/ml sodium bicarbonate
> 5 mg/ml glucose
> 0.01 *M* HEPES (Sigma H0887)
> Penicillin/streptomycin (25 $\mu$g/ml each)

*Working Solution*

  10 ml of stock solution
  500 μl fetal bovine serum (F7524)
  100 μl B27 supplement (Invitrogen 175044)
  100 μl Glutamax (Invitrogen)
  Syringe filter to sterilize
  pH 7.2, osmolarity of 315–320 mOsm

Fluorescent Imaging of Circadian Gene Expression in the SCN
  Organotypical Slice

We have exploited two formats for fluorescent imaging of circadian gene expression in SCN slices. First, we have used the *mPer1::dsGFP* transgenic line developed by Dr. D. G. McMahon, Vanderbilt University, in which a truncated 3.2-kbp version of the *mPer1* promoter is used to drive destabilized green fluorescent protein (Kuhlman *et al.*, 2000). When first prepared, SCN slices from these mice exhibit a high level of fluorescent emission, which has been exploited in combination with electrophysiology to examine questions pertaining to circadian entrainment and synchronization (Kuhlman *et al.*, 2003; Quintero *et al.*, 2003). With continuing time in culture, increasing levels of background autofluoresence compromise imaging. We have waited for 1 week or more until this autofluorescence subsides before imaging the slice in static culture. We use conventional epifluorescence optics (Omega Optical filter sets: ECFP XF114, EYFP XF104, EGFP XF100, supplied by Glen Spectra, Stanmore, UK) on a Leica DMIRE2 using "Open Lab" (Improvision, Warwick UK) software to automate acquisition and a Hamamatsu C4742-95 CCD camera. Post-hoc analysis is conducted off line within OpenLab and IP Lab (Scanalytics, Fairfax, VA). A representative recording from a slice prepard from a *mPer1::dsGFP* mouse is presented in Fig. 3. A robust high-amplitude oscillation in total fluorescence emission from the SCN is maintained over 5 days, here sampled every 2 h but with other recordings we have achieved 1-h resolution. Spontaneous circadian oscillations of *mPer1* activity provide cellular resolution, which allows for issues of synchronization and regional variation in rhythmic gene expressed to be addressed. For example, in Fig. 4B, recordings from four individual units, two in the dorsal shell and two in the ventral core of the SCN, are depicted. These units are putative cells that were defined statistically as areas of high variability in the recording and correspond to cellular profiles in the video image. Figure 4C shows comparable mean data for groups of shell and core units recorded for 5 days and presented as an averaged 24-h plot. It is consistently the case that higher-amplitude, more robust recordings are obtained from the shell

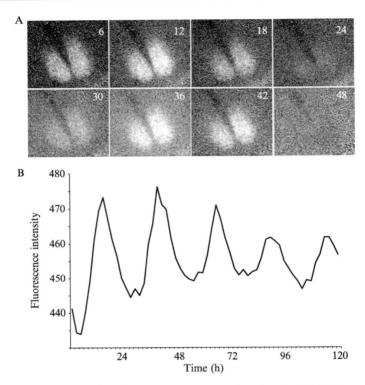

FIG. 3. Real-time recording of circadian gene expression via *mPer1::dsGFP* imaging. (A) Representative images of a single organotypical SCN slice held in static culture (over 2 weeks after initial preparation) illustrate successive peaks and nadirs over two cycles, illustrated at 6-h intervals. (B) Graphic data from the same culture recording circadian gene expression over 5 days in real time.

region of the SCN than the core, although all units appear synchronized, if not precisely simultaneous in their *mPer1* activation. Recordings from such preparations offer an unprecedented, real-time window onto the operation of SCN neurons, both individually and as an assemblage.

## Biolistic Transfection of SCN Organotypical Slices

Biolistics is a form of particle-mediated transfection in which gold microparticles coated in plasmid DNA are shot at target cells or tissues at high velocity by a blast of pressurized inert gas (Arnold *et al.*, 1994). If a particle lodges in the nucleus of a cell, the plasmid DNA may dissolve off and become transcriptionally active. This procedure has a number of

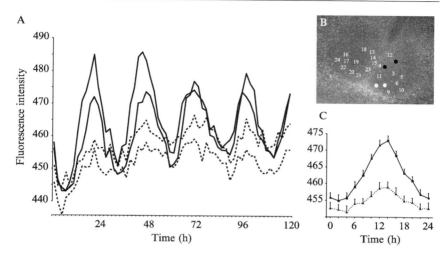

FIG. 4. Differential circadian gene expression within subdivisions of SCN. (A) Representative traces of fluorescence emission from single units within a dorsal shell (solid lines) and a ventral core (dotted line) of an organotypic SCN slice prepared from a *mPer1::dsGFP* mouse cultured for over 2 weeks and then recorded for 5 days. (B) Anatomical location of shell (black symbols) and core (white symbols) units presented in (a). (C) Mean (±SEM) fluorescence emission from shell (solid line, $n = 6$) and core (dotted line, $n = 5$) units recorded over 5 days and presented as a 24-h waveform. Note rhythmic gene expression in both subdivisions, but considerably greater amplitude in the shell unit.

advantages over other approaches, especially for postmitotic neurons that are typically refractory to chemical means of transfection. It is able to transfect neurons with large constructs (>10 kbp) and provides for multiple transfection where the microparticles are coated with a mixture of plasmids. Moreover, the preparation of microparticles is relatively rapid once the plasmid DNA is prepared. Variations on the procedure include diolistics, in which lipophilic fluorescent dyes rather than DNA are shot onto slices to reveal details of neuronal structure (Gan *et al.*, 2000). It also has the potential to be developed for gene silencing in neurons transfected with sequences encoding small hairpin RNAs (shRNA). An obvious limitation of the method is the low rate of transfection, but this is not necessarily a problem if viable experiments can be conducted on small numbers of individual neurons.

*Preparation of Bullets for Shooting*

The gold microparticles are first coated with DNA and in turn are coated onto the inner surface of plastic tubing. The tubing is cut to length, and the cartridge is fitted into the projection apparatus, in our case the Helios Gene Gun from Bio-Rad (Ikeda *et al.*, 2003; O'Brien *et al.*, 2001). The amount of gold loaded is a trade-off between increased likelihood of transfection and risk of damage to slices due to physical disruption. Routinely we use ca. 5 mg of 0.6-$\mu$m-diameter gold microspheres per cartridge. This is mixed by sonication with 50 $\mu$l of 0.05 $M$ spermidine free-base in a 1.5-ml tube. DNA is then added, between 25 and 40 $\mu$g, and mixed by gentle agitation. More than this amount causes the gold microparticles to aggregate too strongly. Calcium chloride (50 $\mu$l of 1 $M$ in Tris–HCl, pH 7.4) is then added dropwise to precipitate DNA onto the gold. This is evident by a darkening of the suspension and clear aggregation of the gold. The suspension is mixed every minute for 10 min. Meanwhile, a solution of polyvinyl pyrrolidone (PVP, 0.075 mg/ml) is prepared in dry ethanol. The PVP acts as an adhesive for coating the gold onto the plastic cartridge tubing. After a very brief spin, the aqueous supernatant is removed from the gold microparticles and they are resuspended in 3.5 ml of the PVP solution. The DNA-coated gold is mixed thoroughly and any aggregates are broken up by gentle agitation. The suspension is then drawn up into the plastic tubing to be used for the cartridges, previously cut to length, and dried under a stream of nitrogen gas. Taking care to avoid trapping any air bubbles in the tubing, it is then threaded onto the loading station supplied by Bio-Rad and left for 30 s for the gold to settle. The ethanol is then withdrawn by syringe, and the proximal end of the tubing is pushed onto the gas connection of the loading station and is then set to rotate for 1 min for the gold to coat the inner surface of the tubing. The nitrogen gas stream is then turned on with the tubing still rotating and the gold microparticles dried into place. After about 5 min, the tubing is removed from the station and cartridges are cut to length.

*Biolistic Transfection of SCN Slices*

Cartridges are loaded into the Helios Gene Gun, and helium gas pressure is set to 195 psi. The membrane insert-bearing slices are placed into a clean six-well plate and shot with the spacer frame just above the lip of the membrane support. They are then returned quickly to the culture plate. Visual inspection under 10× magnification confirms the extent of gold coverage, and continuing ependymal activity indicates the viability of the tissue. With very active promoters, e.g., CMV, a fluorescent product can be observed after overnight incubation, and both neurons and glia are evident (Fig. 5A). The intricate web of neural processes can be revealed in outstanding detail using CMV-driven enhanced yellow fluorescent protein (EYFP). With the 7.2-kb *mPer1* promoter (Yamaguchi *et al.*, 2000),

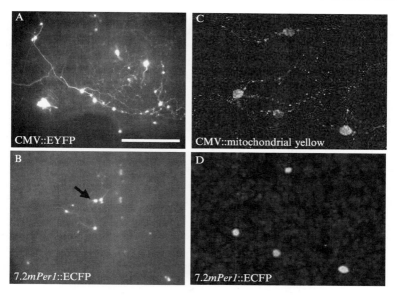

FIG. 5. Biolistic transfection of SCN organotypic slice cultures with fluorescent reporters. (A) SCN neurons and glia expressing CMV-driven EYFP. This fills up the entire cell and reveals neuronal processes that extend across the SCN and beyond it to distal target sites. Scale bar: 500 $\mu$m. (B) SCN neurons expressing *mPer1*-driven stable ECFP. The cell arrowed was cotransfected with *mPer1::dsEYFP* and imaged over several cycles (see Fig. 6). (C and D) Cotransfection of SCN neurons visualized through CMV-driven expression of mitochondrially targetted YFP and *mPer1*-driven stable ECFP. Although ECFP labeling extended into neuronal processes and was visible on epifluorescence, it was difficult to visualize without bleaching on the confocal laser settings used to generate this image.

routinely it takes about 2–3 days before the fluorescent product becomes evident, and cells expressing *mPer1* are characteristically neuronal in morphology (Fig. 5B). Double-transfected cells can be observed *in vivo* with dual wavelength epifluorescence and, after fixation, by confocal microscopy. For example, CMV-driven EYFP targeted to mitochondrial can be combined with *mPer1*-driven ECFP (Figs. 5C, D). Although the overall transfection rate is low, coming out at about 10–15 SCN neurons per slice, this procedure nevertheless has considerable potential in seeking to unravel critical features of the molecular clockwork of the SCN. For example, because of its extended half-life, the accumulation of stable fluorescent protein acts as a historical report for promoter activation in SCN neurons. By cotransfecting neurons with wild-type and mutant versions of the promoter driving cyan and yellow versions of FP, respectively, the critical elements for SCN expression can be mapped by mutagenesis. Moreover, by cotransfection of the reporter gene with cDNAs encoding potentially clock-relevant factors, their actions on the molecular clockwork within the environment of the SCN can be analyzed, complementing and extending earlier work conducted in fibroblasts and peripheral tissues.

## Imaging Circadian Gene Expression Using Biolistically Transfected Reporter Genes

Biolistic transfection can also be used to introduce dynamic reporters of circadian activity into SCN neurons where the *mPer1* promoter drives destabilized FP. Fig. 6A shows high-power views of SCN neurons cotransfected with *mPer1::ECFP* and *mPer1::dsYFP* constructs. The stable ECFP

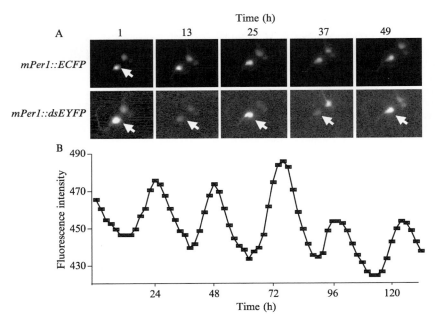

FIG. 6. Real-time imaging of circadian gene expression in biolistically cotransfected SCN neurons. (A) Representative sequential images of the cotransfected cell indicated in Fig. 5. The neuron was imaged on both the cyan channel to reveal their location by expression of stable ECFP driven by the 7.2-kbp *mPer1* promoter and the yellow channel to monitor the destabilized EYFP signal driven by the same promoter. Note circadian expression of dsYFP in arrowed neuron. Other neurons in the field were also rhythmic, but with different phases. (B) Destabilized yellow fluorescence emission driven by *mPer1* recorded for over 5 days from a cotransfected SCN neuron in a slice different from that illustrated in (A).

report reveals the individual neurons throughout the recording, whereas simultaneous monitoring on the yellow channel reports circadian activation of *mPer1* in real time. As is seen in Fig. 6B, cells can be imaged in this way through a series of stable circadian cycles. This offers considerable scope for further studies. At its simplest it can be used to screen the activity of novel constructs prior to generation of transgenic animals. Second, by transfection of mutated forms of the promoter, sequences essential for rhythmic activity can be mapped. Third, by transfection of slices from genetically modified animals, the impact of such modifications on core clock mechanisms can be assessed rapidly. Finally, by cotransfection with cDNAs encoding clock-relevant factors, their impact on rhythmic gene expression can be assessed directly in the SCN neuron. Specific neuronal populations in the SCN can be revealed using transgenic material with cell-specific reporter genes (Karatsoreos *et al.*, 2004) or by cotransfection with a suitable reporter.

## Acknowledgments

The authors are very grateful to Dr. H. Okamura (Kobe University Medical School, Japan) for provision of *Per::luc* reporter plasmids and *Per::luc* mice. The work of the authors' laboratory is supported by the Medical Research Council, UK.

## References

Arnold, D., Feng, L., Kim, J., and Heintz, N. (1994). A strategy for the analysis of gene expression during neural development. *Proc. Natl. Acad. Sci. USA* **91**, 9970–9974.

Gan, W. B., Grutzendler, J., Wong, W. T., Wong, R. O., and Lichtman, J. W. (2000). Multicolor "DiOlistic" labeling of the nervous system using lipophilic dye combinations. *Neuron* **27**, 219–225.

Hastings, M. H., Reddy, A. B., and Maywood, E. S. (2003). A clockwork web: Circadian timing in brain and periphery, in health and disease. *Nat. Rev. Neurosci.* **4**, 649–661.

Herzog, E. D., Geusz, M. E., Khalsa, S. B., Straume, M., and Block, G. D. (1997). Circadian rhythms in mouse suprachiasmatic nucleus explants on multimicroelectrode plates. *Brain Res.* **757**, 285–290.

Herzog, E. D., and Huckfeldt, R. M. (2003). Circadian entrainment to temperature, but not light, in the isolated suprachiasmatic nucleus. *J. Neurophysiol.* **90**, 763–770.

House, S. B., Thomas, A., Kusano, K., and Gainer, H. (1998). Stationary organotypic cultures of oxytocin and vasopressin magnocellular neurones from rat and mouse hypothalamus. *J. Neuroendocrinol.* **10**, 849–861.

Ikeda, M., Sugiyama, T., Wallace, C. S., Gompf, H. S., Yoshioka, T., Miyawaki, A., and Allen, C. N. (2003). Circadian dynamics of cytosolic and nuclear ca(2+) in single suprachiasmatic nucleus neurons. *Neuron* **38**, 253–263.

Karatsoreos, I. N., Yan, L., LeSauter, J., and Silver, R. (2004). Phenotype matters: Identification of light-responsive cells in the mouse suprachiasmatic nucleus. *J. Neurosci.* **24**, 68–75.

King, V. M., Chahad-Ehlers, S., Shen, S., Harmar, A. J., Maywood, E. S., and Hastings, M. H. (2003). A hVIPR transgene as a novel tool for the analysis of circadian function in the mouse suprachiasmatic nucleus. *Eur. J. Neurosci.* **17**, 822–832.

Kuhlman, S. J., Quintero, J. E., and McMahon, D. G. (2000). GFP fluorescence reports Period 1 circadian gene regulation in the mammalian biological clock. *Neuroreport* **11**, 1479–1482.

Kuhlman, S. J., Silver, R., Le Sauter, J., Bult-Ito, A., and McMahon, D. G. (2003). Phase resetting light pulses induce Per1 and persistent spike activity in a subpopulation of biological clock neurons. *J. Neurosci.* **23**, 1441–1450.

Nakamura, W., Honma, S., Shirakawa, T., and Honma, K. (2001). Regional pacemakers composed of multiple oscillator neurons in the rat suprachiasmatic nucleus. *Eur. J. Neurosci.* **14**, 666–674.

O'Brien, J. A., Holt, M., Whiteside, G., Lummis, S. C., and Hastings, M. H. (2001). Modifications to the hand-held Gene Gun: Improvements for *in vitro* biolistic transfection of organotypic neuronal tissue. *J. Neurosci. Methods* **112**, 57–64.

Potter, S. M., and DeMarse, T. B. (2001). A new approach to neural cell culture for long-term studies. *J. Neurosci. Methods* **110**, 17–24.

Quintero, J. E., Kuhlman, S. J., and McMahon, D. G. (2003). The biological clock nucleus: A multiphasic oscillator network regulated by light. *J. Neurosci.* **23**, 8070–8076.

Reppert, S. M., and Weaver, D. R. (2002). Coordination of circadian timing in mammals. *Nature* **418**, 935–941.

Stokkan, K. A., Yamazaki, S., Tei, H., Sakaki, Y., and Menaker, M. (2001). Entrainment of the circadian clock in the liver by feeding. *Science* **291**, 490–493.

Yamaguchi, S., Isejima, H., Matsuo, T., Okura, R., Yagita, K., Kobayashi, M., and Okamura, H. (2003). Synchronization of cellular clocks in the suprachiasmatic nucleus. *Science* **302**, 1408–1412.

Yamaguchi, S., Kobayashi, M., Mitsui, S., Ishida, Y., van der Horst, G. T., Suzuki, M., Shibata, S., and Okamura, H. (2001). View of a mouse clock gene ticking. *Nature* **409**, 684.

Yamaguchi, S., Mitsui, S., Miyake, S., Yan, L., Onishi, H., Yagita, K., Suzuki, M., Shibata, S., Kobayashi, M., and Okamura, H. (2000). The 5′ upstream region of mPer1 gene contains two promoters and is responsible for circadian oscillation. *Curr. Biol.* **10**, 873–876.

Yamazaki, S., Numano, R., Abe, M., Hida, A., Takahashi, R., Ueda, M., Block, G. D., Sakaki, Y., Menaker, M., and Tei, H. (2000). Resetting central and peripheral circadian oscillators in transgenic rats. *Science* **288**, 682–685.

Yamazaki, S., Straume, M., Tei, H., Sakaki, Y., Menaker, M., and Block, G. D. (2002). Effects of aging on central and peripheral mammalian clocks. *Proc. Natl. Acad. Sci USA* **99**, 10801–10806.

Yoo, S. H., Yamazaki, S., Lowrey, P. L., Shimomura, K., Ko, C. H., Buhr, E. D., Siepka, S. M., Hong, H. K., Oh, W. J., Yoo, O. J., *et al.* (2004). PERIOD2::LUCIFERASE real-time reporting of circadian dynamics reveals persistent circadian oscillations in mouse peripheral tissues. *Proc. Natl. Acad. Sci. USA* **101**, 5339–5346.

# [31] Oligodeoxynucleotide Methods for Analyzing the Circadian Clock in the Suprachiasmatic Nucleus

*By* SHELLEY A. TISCHKAU and MARTHA U. GILLETTE

## Abstract

The recent identification of specific genes responsible for the generation of endogenous circadian rhythmicity in the suprachiasmatic nucleus presents a new level of investigation into endogenous rhythmicity and mechanisms of synchronization of this circadian clock with the environmental light/dark cycle. This article describes techniques that employ antisense and decoy oligodeoxynucleotides (ODN) to determine the roles of specific molecular substrates both in endogenous rhythmicity and in regulating the effects of light on the mammalian circadian clock. Application of antisense ODN technology has revealed a role for *timeless* (*Tim*) in the core clock mechanism and established that induction of *period1* (*Per1*) is required for light responsiveness. Likewise, a decoy ODN designed to sequester activated CREB protein definitively demonstrated a requirement for CRE-mediated transcription in light signaling. Experiments designed with these molecular tools offer new insights on the interaction of cellular processes and signaling with the molecular clockworks.

## Introduction

The rotation of the earth on its axis, and the resulting daily alternation of light and darkness, imparts arguably the single-most persistent, stable environmental factor influencing the evolution of life on this planet. The importance of the ability to measure time on a daily scale is reflected by the genomic incorporation of circadian rhythmicity, or near 24-h patterns of physiology and behavior driven by gene expression, into nearly all eukaryotic organisms. This internal clock can generate and maintain rhythms in the absence of external stimuli, but it also retains the ability to respond to specific cues to allow synchronization with the solar cycle.

There is an emerging consensus that individual cells in virtually all tissues contain clocks; however, in mammals, organismic rhythmicity remains an emergent property of the suprachiasmatic nucleus (SCN). Strategically positioned in the basal hypothalamus adjacent to the optic chiasm, the SCN navigates a recurrent sequence of dynamic cellular events, driven by the coordinated function of a core group of clock genes and organized into discrete time domains. Progression through this daily cycle is characterized by waxing and waning sensitivities of the SCN to resetting stimuli, such as to light extending inappropriately into the night. This circadian oscillation in sensitive periods is underscored by clock-controlled adjustments in available molecular and biochemical substrates and activation of signal transduction pathways (Gillette and Mitchell, 2002).

The circadian clockwork is embedded in alternating transcription–translation loops of a limited number of clock genes and their products that together generate near 24-h periodicity (King and Takahashi, 2000). Clock genes, by convention, are those essential to circadian timekeeping. Several mammalian clock genes appear to be highly conserved elements of circadian clocks across phyla (Okamura *et al.*, 2002). In some cases, they have assumed additional roles, as regulators of developmental pattern formation (Li *et al.*, 2000; Xiao *et al.*, 2003) or the cell cycle (Matsuo *et al.*, 2003), which has complicated analysis of a specific gene's timekeeping role in the context of a central brain clock of mammals. Furthermore, the SCN is composed of paired clusters of ~10,000 cells. Whereas circadian clocks are properties of single cells, how timekeeping is organized within the SCN tissue is unclear. Nevertheless, insights into the positive and negative regulation of clock gene transcription and the roles of clock effector proteins and their regulation within the SCN are emerging. Studies *in vivo* and *in vitro* are defining essential elements and critical regulators. This article reviews methods for analyzing SCN rhythmicity *in vivo* and *in vitro* with new techniques that can probe the contribution of specific molecular elements to the mammalian clockwork, the regulatory pathways that intersect it, and clock-controlled outputs.

## Analysis of SCN Rhythmicity *In Vivo* and *In Vitro*

Changes in rhythmic locomotor activity in freely behaving rodents have long been considered the "gold standard" for analysis of function in mammal models of circadian timekeeping. These studies take advantage of the fact that locomotor activity cycles readily entrain to light:dark cycles and persist when the animal is placed in constant environmental lighting conditions, albeit with a somewhat altered period (*tau*, $\tau$) under these free-running conditions. Over the years, a plethora of substances have been tested by injection, either systemically or by stereotaxic direction into the ventricular system, into the SCN or other brain regions of interest. Typically substances are tested for their direct effects on the circadian timing system or for their impact on well-characterized photic or nonphotic phase shifts in behavioral rhythms. Advantages of this model system include the opportunity to examine overt effects on the entire circadian system and, more recently, the ability to directly couple changes in the expression of specific genes to alterations in animal behavior. Disadvantages include the fact that results may not reflect only what is happening in the SCN, but also within multiple layers of integration at other brain and body sites because the SCN is intricately connected to multiple input and output systems.

Direct measures of SCN activity *in vivo* have revealed an oscillatory pattern of a high activity in the daytime and low activity at night. This pattern extends from spontaneous neural activity (Inouye and Kawamura, 1982) to energy utilization (2-deoxyglucose) (Schwartz *et al.*, 1980, 1983) to mRNA expression of certain clock genes (*Per, Tim*). High daytime activity is consistent across diverse mammals, including both nocturnal (Burgoon *et al.*, 2004; Inouye and Kawamura, 1979; Kurumiya and Kawamura, 1988; Yamazaki *et al.*, 1998) and diurnal rodents (Sato and Kawamura, 1984). Other brain regions also show circadian oscillations in neural activity (Inouye and Kawamura, 1982; Szymusiak *et al.*, 1998), but only in the presence of the SCN. Thus, spontaneous oscillation in the SCN drives circadian rhythms at other brain sites. The selective pressures that have caused universal high daytime activity in the SCN are unknown; however, it follows that the signals that pattern the range of circadian behaviors, such as when sleep/arousal occur relative to the day/night cycle, must be regulated outside of the SCN.

Hypothalamic brain slices containing the paired SCN provide a mechanism for direct probing of SCN function with minimal interference from extra-SCN sources. Despite surgical deafferentation, separation of rostral and caudal components, and lack of endocrine stimulation, the cultured SCN retains its spontaneous, nearly 24-h oscillations in gene expression (Yamazaki *et al.*, 2000), neuronal firing rate (Prosser and Gillette, 1989), and peptide secretion (Shinohara *et al.*, 1995). Neuronal activity of SCN slices shows a distinct rhythm of electrical activity with a peak in midsubjective day (Green and Gillette, 1982) like those observed *in vivo* (Inouye and Kawamura, 1979). The time of the peak in electrical activity predictably displays circadian periodicity that persists up to 3 days *ex vivo* (Prosser and Gillette, 1989). Thus, this circadian rhythm in neuronal firing rate has been used extensively as a bioassay to study mechanisms of phase resetting in the SCN (Gillette and Mitchell, 2002).

The SCN expresses specific temporal domains, or windows of sensitivity, defined by continuously changing access to intrinsic signaling molecules. These sensitivities are easily probed in the SCN slice by applying stimuli to receptive regions and observing the effects on the timing of the peak oscillation of the electrical activity rhythm. Pharmacological approaches using the SCN slice preparation have identified specific cellular substrates that underlie phase resetting in response to a variety of stimuli.

Importantly, the validity of the SCN brain slice preparation has been demonstrated repeatedly by experiments performed in whole animal models using circadian wheel-running activity as a measure of rhythmicity. In most instances, results from pharmacological approaches using SCN slices *in vitro* are consistent with *in vivo* studies, particularly when the test substance is injected directly onto the SCN. For example, inhibition of nitric oxide synthase blocks light-induced phase resetting *in vivo* (Weber *et al.*, 1995a), and glutamate, the neurochemical messenger of light, stimulates the light-activated signaling pathways and phase resetting *in vitro* (Ding *et al.*, 1994). Furthermore, inhibition of protein kinase G blocks only light- or glutamate-induced phase advances, with no effect on phase delays (Ding *et al.*, 1998; Mathur *et al.*, 1996; Weber *et al.*, 1995b). The major difference between *in vivo* and *in vitro* approaches to phase resetting lies in the magnitude of the phase shift. Phase shifts *in vitro* are generally larger in amplitude than those observed *in vivo*. In the rat, light-induced phase shifts *in vivo* are commonly in the range of 1 h, whereas glutamate-induced phase shifts in this same animal model are usually around 3 h (Ding *et al.*, 1994). Similarly, we observed 3.5-h phase delays induced by PKG inhibition *in vitro*, but the same treatment into the SCN *in vivo* caused phase delays of less than 1 h (Tischkau *et al.*, 2003b).

Behavioral wheel-running activity and SCN electrical activity have been valuable tools to assess cellular mechanisms that underlie circadian rhythmicity. The discovery of circadian clock genes has provided another level of depth of inquiry into circadian clock function. Behavioral wheel running and SCN electrical activity rhythms remain essential bioassays for exploring gene function and coupling signaling pathways to circadian gene function within the clock. This article provides insight into new techniques that enable exploration of the significance of clock genes, as well as other molecular elements that activate the genome, to circadian behavior.

## Targeted Deletion of Clock Genes

The generation of animals bearing genetic deficiency in a specific protein product is often considered the definitive test for establishing the physiological role of any gene. The same is true for those genes tentatively defined as circadian clock genes. Whereas these techniques are relatively inexpensive and expeditious in several model circadian systems, such as *Drosophila* and *Neurospora*, the development of mammalian models deficient in one or more clock genes has required considerable effort. Animals are subjected to customary testing, typically wheel-running behavior under constant conditions and responsiveness to nocturnal light, to assess rhythmicity. These animals have been important for establishing the relative importance of several core clock genes, including *Clock* (King *et al.*, 1997; Vitaterna *et al.*, 1994), *bmal1* (Bunger *et al.*, 2000), *Per 1–3* (Bae *et al.*, 2001; Cermakian *et al.*, 2001; Shearman *et al.*, 2000; Zheng *et al.*, 1999, 2001), and *Cry1–2* (Okamura *et al.*, 1999; Thresher *et al.*, 1998; van der Horst *et al.*, 1999) for circadian rhythmicity. In addition, mutant animals have also established roles for an ever-expanding number of regulatory elements, such as casein kinase Iε (Lowrey *et al.*, 2000) and REV ERBα (Preitner *et al.*, 2002), and clock output molecules, such as prokineticin 2 (Cheng *et al.*, 2002), phospholipase C β4 (Park *et al.*, 2003) and D-binding protein (DBP) (Lopez-Molina *et al.*, 1997).

Experience indicates that analysis of data from these animal models is not always straightforward and must be interpreted with caution. *Bmal1*$^{-/-}$ mice exhibit the expected immediate disruption of circadian wheel-running activity under constant environmental conditions (Bunger *et al.*, 2000). In contrast, animals bearing an ENU-induced mutation of the transactivation domain of mCLK exhibit a lengthened free-running period followed by a gradual loss of rhythmicity (Antoch *et al.*, 1997; King *et al.*, 1997; Vitaterna *et al.*, 1994). Moreover, evolutionary duplication of the *Per* and *Cry* genes adds additional complexity. Deletion of *Per3* has no effect on rhythmicity (Shearman *et al.*, 2000). Whereas some discrepancies have been noted among mice with targeted deletion of either *Per1* or *Per2*, the consensus is that these animals have a partially functional clock. Most animals display significantly reduced free-running periodicity (Bae *et al.*, 2001; Cermakian *et al.*, 2001; Zheng *et al.*, 1999, 2001); animals generated by one laboratory also exhibit disrupted rhythms after prolonged exposure to constant darkness (Bae *et al.*, 2001). Likewise, mutations in either *Cry1* or *Cry2* result in altered periodicity, but retention of rhythmicity (Okamura *et al.*, 1999; Thresher *et al.*, 1998). Simultaneous disruption of *Per1* and *Per2* or of *Cry1* and *Cry2* results in animals whose rhythms are immediately disrupted when they are placed in constant darkness (van der Horst *et al.*, 1999; Zheng *et al.*, 2001).

Inconsistencies with studies in mutant animals likely stem from the developmental nature of the model, where other genes may compensate for the missing gene, or from differing genetic backgrounds of mouse strains, thereby providing misleading or false-negative results. For example, pharmacological studies have implicated a role for nitric oxide in light-induced phase resetting (Ding *et al.*, 1994; Melo *et al.*, 1997; Weber *et al.*, 1995b). However, mice deficient in either nNOS or eNOS show normal responses to nocturnal light (Kriegsfield *et al.*, 1999a,b, 2001). Furthermore, animals with a targeted deletion of protein kinase G-II display abnormal phase delays and normal phase advances, despite pharmacological data that implicate a role for PKG restricted to the late night, when light causes phase advance of rhythmicity (Ding *et al.*, 1998; Mathur *et al.*, 1996; Oster *et al.*, 2003; Weber *et al.*, 1995b). Additionally, the embryonic lethality of $mTim^{-/-}$ mice led to premature exclusion of this core gene from the mammalian clockwork (Gotter *et al.*, 2000).

### Antisense Oligodeoxynucleotides and Small Interfering RNA as Tools to Investigate Gene Function in Circadian Timekeeping

Whereas generation of tissue-specific, conditional deletion or induction of target genes will likely overcome many of the problems associated with traditional approaches to genetic deletion of clock genes, this technology remains in its infancy. Thus, antisense oligodeoxynucleotide ($\alpha$ODN) and, more recently, small interfering RNA (siRNA) approaches have been used as inexpensive models devoid of the developmental problems inherent to the whole animal targeted deletion paradigms (Estibeiro and Godfray, 2001). The distinct advantage of this approach in circadian biology is the ability to reversibly downregulate expression of the gene of interest in an adult animal in a time- and tissue-specific manner. In the brain, $\alpha$ODN and siRNA can be directed to the site of interest by stereotaxic cannula placement. Furthermore, $\alpha$ODN technology is widely adaptable for use in numerous species, whereas traditional genetic approaches are restricted primarily to mice.

Watson–Crick base pair formation dictates that $\alpha$ODN will bind to the sense stand of the RNA or DNA of interest with a high degree of specificity and thus block the formation of gene products. Typically, the $\alpha$ODN is designed to target a specific mRNA sequence. Criteria that must be considered when designing an $\alpha$ODN include (1) the uniqueness of the target sequence, (2) the length and sequence of the $\alpha$ODN, (3) modifications of the nucleotides, and (4) appropriate controls (Kashikara *et al.*, 1998). Although there is currently no way to predict the best target sequence to yield the maximal effect, two commonly successful targets include the $3'$-untranslated region and the AUG translation initiation codon. Use of the latter predicts a mechanism of action through blockage of translation by preventing binding of initiation factors. The length of the $\alpha$ODN must be considered: it must be long enough to confer specificity to the sequence of interest, yet short enough to allow uptake into the cell. In our hands, 15- to 21-mer $\alpha$ODNs are highly effective (Barnes *et al.*, 2003; Tischkau *et al.*, 2003a,b). The $\alpha$ODN sequence should always be evaluated to avoid those that will form strong secondary structures because they are self-complementary.

Modifications can be introduced to increase the half-life of the $\alpha$ODN within the cell. Common modifications include phosphorothioate, methyl phosphate, phosphoroamidate, and methyl phosphate derivatives

(Cho-Chung, 2003). The unmodified αODN is readily degraded by endo-
or exonucleases so it has a relatively short half-life of an estimated 20 min
(Kashikara *et al.*, 1998). In contrast, phosphorothioate modifications confer
nuclease resistance and are still present 8–16 h after injection into hypothalamic sites (Ogawa *et al.*, 1995). However, this same modification can
impair cellular uptake of the αODN, decrease the specificity of binding to
the RNA of interest (Kashikara *et al.*, 1998), exert nonspecific effects by
binding to proteins and small molecules (Perez *et al.*, 1994; Yakubov *et al.*,
1993) and, in some cases, can be toxic (Hebb and Robertson, 1997).

Because of the potential for nonspecific effects, design and use of
appropriate controls are critical for interpretation of data obtained in
αODN experiments. Sense ODN that is complementary to the αODN is
a commonly used control. However, sense ODN has the potential to bind
DNA for the gene of interest and thereby inhibit transcription (Kashikara
*et al.*, 1998). Therefore, a scrambled ODN with the same base composition
as the αODN but in random order is a better control. Additionally, replacing one or more nucleotides in the αODN with a different base can disrupt
complementary binding. We have found that as little as a single base pair
mismatch in a 15-mer is enough to disrupt the effectiveness of the αODN
(Barnes *et al.*, 2003; Tischkau *et al.*, 2003a,b).

## Effects of Antisense Oligodeoxynucleotides in Circadian Clock Responses to Light

αODN technology has been employed to explore the function of clock
gene proteins and other regulatory elements in the molecular events leading to phase resetting in response to nocturnal light *in vivo* or its messenger, glutamate, *in vitro*. A rapid, transient increase in *Per1* mRNA is a
hallmark of the molecular response to nocturnal light (Albrecht *et al.*, 1997;
Miyake *et al.*, 2000; Shearman *et al.*, 1997; Shigeyoshi *et al.*, 1997; Takumi
*et al.*, 1998). An unmodified αODN designed to target the region surrounding the initiation codon of *Per1* mRNA (5′-taggggaccactcatgtct-3′) blocks
glutamate-induced phase resetting of the SCN electrical activity rhythm
*in vitro* and light-induced phase delays *in vivo* (Akiyama *et al.*, 1999;
Tischkau *et al.*, 2003a); control sequences with 1- to 3-bp changes are
ineffective. These results demonstrate definitively that induction of *Per1*
mRNA is required for phase resetting in response to nocturnal light. It
follows that the PER1 protein contributes critically to molecular changes
that mediate clock resetting.

The αODN technique has also been used to examine the roles of
other proteins in the light/Glu signaling pathway. The core of the hamster
SCN contains a densely packed population of calbindin-containing cells

that receive direct retinal input and express *Fos* in response to light (Bryant *et al.*, 2000; Hamada *et al.*, 2003; Silver *et al.*, 1996). Intracerebroventricular injection of phosphothionate-modified $\alpha$ODN against calbindin (5'-aggtgcgattctgccatgg-3') significantly reduced calbindin mRNA and protein in the SCN core and attenuated both the light-induced increase in *Per1* and phase advances in response to light (Hamada *et al.*, 2003). These data implicate $Ca^{2+}$ buffering via calbindin in the light response and further the idea that $Ca^{2+}$ is critical for light-induced signaling in the SCN. Interestingly, simultaneous inhibition of *JunB* and *cFos* with $\alpha$ODN also blocked light-induced phase resetting (Schlingensiepen *et al.*, 1994).

The cGMP/protein kinase G (PKG) signal transduction cascade has also been implicated in light-induced phase shifting. Activation of cGMP/PKG occurs only in response to light that signals advance clock phase in the last half of subjective night. Pharmacological inhibition of PKG blocks light- and glutamate-induced phase advances, but not the phase-delaying effect of light/Glu in early night (Ding *et al.*, 1998; Mathur *et al.*, 1996; Weber *et al.*, 1995b). Two major isoforms of PKG have been described and a different $\alpha$ODN can be designed to the start site of each. Preliminary studies suggest that this $\alpha$ODN approach is a powerful tool for differentiating isoform-specific function of PKG in the SCN (S. A. Tischkau and M. U. Gillette, unpublished observations).

## Effects of Antisense Oligodeoxynucleotides on Circadian Clock Rhythmicity

Because short-term inhibition using $\alpha$ODN technology is effective for identifying the roles of specific proteins in light/Glu-induced phase resetting, we hypothesized that long-term inhibition might be used to identify molecular components that are required for the expression of overt rhythmicity in the SCN. Initially, we utilized this approach to examine whether mammalian *timeless (Tim)* is required for the generation of circadian rhythmicity. Traditional approaches using genetic deletion of *Tim* had been unsuccessful because the $Tim^{-/-}$ mutation yields an embryonic lethal phenotype (Gotter *et al.*, 2000). We circumvented this developmental problem by examining electrical activity rhythms in SCN-containing brain slices incubated continuously with *Tim* $\alpha$ODN (acagtccatacacc). SCN slices treated with *Tim* $\alpha$ODN expressed $\sim$ 40% of the TIM protein levels in controls and were completely arrhythmic (Barnes *et al.*, 2003). Moreover, short-term application of *Tim* $\alpha$ODN over the course of a 24-h cycle revealed a specific sensitive period where downregulation of *Tim* led to phase resetting of the circadian clock (Barnes *et al.*, 2003). Furthermore, introduction of siRNA targeting *Tim* demonstrated that *Tim* knockdown alters expression of other clock genes. Together, these data demonstrate a specific requirement for *Tim* in SCN rhythmicity and allow restoration of *Tim* as a core mammalian circadian clock element.

The success of these experiments led us to explore the role of additional elements in SCN rhythmicity. Previously, we had demonstrated a circadian rhythm of cGMP levels and PKG activity inherent to the SCN. Pharmacological disruption of this endogenous rise in PKG activity caused significant phase delay of the circadian clock *in vitro* and *in vivo* specifically at the dawn-to-dusk transition (Tischkau *et al.*, 2003b). These data led to the hypothesis that increased PKG activity at the end of subjective night is required for circadian clock progression. A corollary of this hypothesis is that continuous inhibition of PKG would lead to arrhythmicity. Utilizing isoform-specific αODNs, we can discriminate the isoform required for PKG mediation of clock function at the dawn transition (Tischkau *et al.*, 2003b). These studies demonstrate the power of αODN technology in defining a role for specific proteins in circadian clock function.

## Use of Antisense Oligodeoxynucleotides in Defining Circadian Clock Output

The αODN approach has also been used to examine the roles of proteins predicted to be components of one of the numerous SCN output programs. Vasoactive intestinal peptide (VIP), which is secreted from the SCN in a diurnal fashion (reviewed in van Esseveldt *et al.*, 2000), is an important regulator of the estrogen-induced luteinizing hormone and pro-lactin surges (van der Beek *et al.*, 1999). Intracerebroventricular or intra-SCN injection of a VIP αODN attenuates and/or phase delays rhythms of circulating corticosterone, luteinizing hormone, and prolactin (Harney *et al.*, 1996; Scarbrough *et al.*, 1996; van der Beek *et al.*, 1999). The latter is likely through an effect on neuroendocrine dopaminergic neurons in the hypothalamic arcuate nucleus and periventricular nucleus (Gerhold *et al.*, 2002). Together, these results highlight the potential of αODN technology in exploring the specific roles of the array of output networks connected to the SCN.

## Decoy Oligodeoxynucleotides as Tools to Investigate Transcriptional Control in Circadian Timekeeping

Using the same methods as for αODN, ectopic enhancer oligodeoxynucleotides sequences can be introduced in excess into cells or brain tissue. These supernumerary enhancer sequences act as *decoys*, competing with native *cis* elements for binding of specific transactivating DNA-binding proteins and thereby diminishing activity at intrinsic transcriptional regulatory

Transcriptional activation

Transcription factors

RNA Polymerase

Transcription and Expression

TGACGTCA    TGACGTCA    TGACGTCA
ACCTACTATTGATGTTGCTTTTAGC...

A

Decoy ODN Mechanism

Palindromic CRE sequences self-hybridize, forming hairpins that bind CREB.

TGACGTCA    TGACGTCA

TGACGTCA    TGAC
ACTGCACT    ACTG

TGACGTCA    TGAC
ACTGCACT    ACTG

TGACGTCA    TGAC
ACTGCACT    ACTG

TGACGTCA    TGAC
ACTGCACT    ACTG

TGACGTCA    TGAC
ACTGCACT    ACTG

TGACGTCA    TGAC
ACTGCACT    ACTG

TGACGTCA    TGAC
ACTGCACT    ACTG

No RNA Polymerase

TGACGTCA    TGACGTCA    TGACGTCA
ACCTACTATTGATGTTGCTTTTAGC...

(No Expression)

B

sites (Cho-Chung, 2003). Thus, a single-stranded 24-mer comprising three repeats of the consensus CRE sequence (trioctamer of 5′-TGACGTCA-3′) can be introduced to inhibit CRE-mediated transcription. The palindromic nature of this CRE decoy allows it to self-hybridize upon entering the cell, forming a hairpin structure that effectively binds CREB, which results in loss of the ability to activate native CRE (Park *et al.*, 1999). Previously, this ODN has been used to interfere with CRE-mediated transcription and thus potently inhibit growth in cancer cells (Park *et al.*, 1999). In primary cultures of neonatal hippocampal neurons, the CRE decoy was employed to demonstrate the importance of CRE-mediated transcription in providing protection against glutamate-induced cell death (Mabuchi *et al.*, 2001).

## Effects of Decoy Oligodeoxynucleotides in Circadian Clock Response to Light

CRE-mediated transcription has been implicated in the molecular events leading to light/glutamate-induced phase resetting of the circadian clock. Throughout the night, stimuli associated with light-induced phase resetting cause increased phosphorylation of CREB (Ding *et al.*, 1997; Ginty *et al.*, 1993) and activation of CRE-mediated transcription (Obrietan *et al.*, 1999). Furthermore, CREB may play a role in the stimulation of *Per1* (Travnickova-Bendova *et al.*, 2002; Yamaguchi *et al.*, 2000), which is required for light-induced phase resetting (Akiyama *et al.*, 1999; Tischkau *et al.*, 2003a). However, those studies fell short of establishing a requirement for CRE-mediated transcription for the molecular events leading to phase resetting in response to light.

We introduced the CRE decoy to block CRE-mediated transcription in the presence of stimuli known to mediate nocturnal light-induced phase

---

FIG. 1. Mechanism by which decoy ODN inhibits transcription. (A) Normal expression is regulated by transcription factors binding to specific sequences in the promoter region of a gene. These transcription factors provide a scaffold for RNA polymerase, which creates an RNA copy of the coding sequence and leads to expression of the gene product. (B) Decoy ODN technology is based on providing a tissue excess of short ODN recognition sequences for a specific transcription factor, such as the CRE sequence (CACGTG) in the model. Decoy ODN are constructed of the same sequence as a specific promoter element, but these ectopic-binding sites are present in excess. This causes transcription factor binding to recognition sequences in intrinsic promoter elements (e.g., CREB binding to CREs) to be outcompeted by binding to the decoy. The target transcription factors will be unavailable to bind the promoter and will fail to form the scaffold for tethering RNA polymerase at the endogenous gene target. If these transcription factors are necessary for transcription of the gene, then normal expression is interrupted.

resetting in the SCN. Our synthetic CRE decoy included one important distinction compared with previous studies in other cell systems. The CRE decoy employed in those studies was composed of phosphorothioate-modified ODN to provide nuclease resistance and increase stability within the cell (Agrawal *et al.*, 1997; Zon, 1988). Because CREB is likely to regulate many genes (Panda *et al.*, 2002), not solely those activated during light-induced transcription, we reasoned that the long-term presence of the CRE decoy could disrupt other functions and complicate interpretation of data. To restrict inhibition of CRE-mediated transcription to a narrow window, our CRE decoy was synthesized with less stable, unmodified nucleotides. This CRE decoy successfully blocked CRE-CREB binding and CRE-mediated transcription in SCN2.2 cells (Tischkau *et al.*, 2003a). When applied to SCN brain slices, the CRE decoy blocked glutamate-stimulated phase resetting of SCN electrical activity rhythms and induction of *Per1*. *In vivo*, when injected unilaterally into the SCN, the CRE decoy blocked light-induced phase resetting of behavioral wheel-running rhythms (Tischkau *et al.*, 2003a). This was the first study to combine use of this technology in cells, brain slices, and *in vivo*. In a subsequent study, intracerebroventricular injection of the CRE decoy effectively blocked CREB-DNA binding in the gerbil hippocampus, confirming the efficacy of this technique for use *in vivo* (Hara *et al.*, 2003; Fig. 1).

Conclusions

Antisense and decoy oligodeoxynucleotides provide an inexpensive, effective, and complementary alternative to molecular genetics for analyzing the function of specific gene products in the circadian timekeeping system. Whereas these approaches may be limited somewhat by cell permeability, degradation, and inability to totally knock out the RNA or protein of interest (Scarbrough, 2000), the capacity to locally antagonize the expression of a single gene product for a discrete, controlled time renders this technology highly useful. Moreover, the reversibility of $\alpha$ODN effects is particularly attractive for physiological studies. When coupled with monitoring electrical activity rhythms *in vitro* in SCN brain slices or with locomotor activity rhythms in freely behaving animals, this technology provides an efficient means to screen candidate gene products for a role in the generation or regulation of rhythmicity.

Acknowledgments

We thank the members of the Gillette laboratory who contributed conceptually and experimentally to developing these methods, P. T. Lindberg for developing the model, and S. C. Baker for manuscript preparation. Supported by Public Health Service Grants NS22155,

NS35859, and HL67007 (MUG) and by a grant from the UIUC Governor's Venture Technology Fund/Molecular and Endocrine Pharmacology Program (SAT). Any opinions, findings, and conclusions or recommendations expressed in this publication are those of the authors and do not necessarily reflect the views of the National Institutes of Neurological Diseases and Stroke, Heart, Lung, and Blood, or General Medicine.

## References

Agrawal, S., Jiang, Z., Zhao, Q., Shaw, D., Cai, Q., Roskey, A., Channavajjala, L., Saxinger, C., and Zhang, R. (1997). Mixed-backbone oligonucleotides as second generation antisense oligonucleotides: *In vitro* and *in vivo* studies. *Proc. Natl. Acad. Sci. USA* **94**, 2620–2625.

Akiyama, M., Kouzu, Y., Takahashi, S., Wakamatsu, H., Moriya, T., Maetani, M., Watanabe, S., Tei, H., Sakaki, Y., and Shibata, S. (1999). Inhibition of light- or glutamate-induced *mPer1* expression represses the phase shifts of the mouse circadian locomoter and suprachiasmatic firing rate rhythms. *J. Neurosci.* **19**, 1115–1121.

Albrecht, U., Sun, Z. S., Eichele, G., and Lee, C. C. (1997). A differential response of two putative mammalian circadian regulators, *mper1* and *mper2*, to light. *Cell* **91**, 1055–1064.

Antoch, M. P., Song, E. J., Chang, A. M., Vitaterna, M. H., Zhao, Y., Wilsbacher, L. D., Sangoram, A. M., King, D. P., Pinto, L. H., and Takahashi, J. S. (1997). Functional identification of the mouse circadian Clock gene by transgenic BAC rescue. *Cell* **89**, 655–667.

Bae, K., Jin, X., Maywood, E. S., Hastings, M. H., Reppert, S. M., and Weaver, D. R. (2001). Differential functions of *mPer1*, *mPer2*, and *mPer3* in the SCN circadian clock. *Neuron* **30**, 525–536.

Barnes, J. W., Tischkau, S. A., Barnes, J. A., Mitchell, J. W., Burgoon, P. W., Hickok, J. R., and Gillette, M. U. (2003). Requirement of mammalian *Timeless* for circadian rhythmicity. *Science* **302**, 439–442.

Bryant, D. N., LeSauter, J., Silver, R., and Romero, M. T. (2000). Retinal innervation of calbindin-D28K cells in the hamster suprachiasmatic nucleus: Ultrastructural character-ization. *J. Biol. Rhythms* **15**, 103–111.

Bunger, M. K., Wilsbacher, L. D., Moran, S. M., Clendenin, C., Radcliffe, L. A., Hogenesch, J. B., Simon, M. C., Takahashi, J. S., and Bradfield, C. A. (2000). Mop3 is an essential component of the master circadian pacemaker in mammals. *Cell* **103**, 1009–1017.

Burgoon, P. W., Lindberg, P. T., and Gillette, M. U. (2004). Different patterns of circadian oscillation in the suprachiasmatic nucleus of hamster, mouse and rat. *J. Comp. Physiol. A.* **190**, 167–171.

Cermakian, N., Monaco, L., Pando, M., Dierich, A., and Sassone-Corsi, P. (2001). Altered behavioral rhythms and clock gene expression in mice with a targeted mutation in the *period1* gene. *EMBO J.* **20**, 3967–3974.

Cheng, M. Y., Bullock, C. M., Li, C., Lee, A. G., Bermak, J. C., Belluzzi, J., Weaver, D. R., Leslie, F. M., and Zhou, Q. Y. (2002). Prokineticin 2 transmits the behavioural circadian rhythm of the suprachiasmatic nucleus. *Nature* **417**, 405–410.

Cho-Chung, Y. S. (2003). CRE-enhancer DNA decoy: A tumor target-based genetic tool. *Ann. N. Y. Acad. Sci.* **1002**, 124–133.

Ding, J. M., Buchanan, G. F., Tischkau, S. A., Chen, D., Kuriashkina, L., Faiman, L. E., Alster, J. M., McPherson, P. S., Campbell, K. P., and Gillette, M. U. (1998). A neuronal ryanodine receptor mediates light-induced phase delays of the circadian clock. *Nature* **394**, 381–384.

Ding, J. M., Chen, D., Weber, E. T., Faiman, L. E., Rea, M. A., and Gillette, M. U. (1994). Resetting the biological clock: Mediation of nocturnal circadian shifts by glutamate and NO. *Science* **266**, 1713–1717.

Ding, J. M., Faiman, L. E., Hurst, W. J., Kuriashkina, L. R., and Gillette, M. U. (1997). Resetting the biological clock: Mediation of nocturnal CREB phosphorylation via light glutamate and nitric oxide. *J. Neurosci.* **17,** 667–675.

Estibeiro, P., and Godfray, J. (2001). Antisense as a neuroscience tool and therapeutic agent. *Trends Neurosci.* **24,** S56–62.

Gerhold, L., Sellix, M. T., and Freeman, M. E. (2002). Antagonism of vasoactive intestinal peptide mRNA in the suprachiasmatic nucleus disrupts the rhythm of FRAs expression in neuroendocrine dopaminergic neurons. *J. Comp. Neurol.* **450,** 135–143.

Gillette, M. U., and Mitchell, J. W. (2002). Signaling in the suprachiasmatic nucleus: Selectively responsive and integrative. *Cell Tissue Res.* **309,** 99–107.

Ginty, D. D., Kornhauser, J. M., Thompson, M. A., Bading, H., Mayo, K. E., Takahashi, J. S., and Greenberg, M. E. (1993). Regulation of CREB phosphorylation in the suprachiasmatic nucleus by light and a circadian clock. *Science* **260,** 238–241.

Gotter, A. L., Manganaro, T., Weaver, D. R., Kolakowski, J., L. F., Possidente, B., Sriram, S., MacLaughlin, D. T., and Reppert, S. M. (2000). A time-less function for mouse timeless. *Nature Neurosci.* **3,** 755–756.

Green, D. J., and Gillette, R. (1982). Circadian rhythm of firing rate recorded from single cells in the rat suprachiasmatic brain slice. *Brain Res.* **245,** 198–200.

Hamada, T., LeSauter, J., Lokshin, M., Romero, M. T., Yan, L., Venuti, J. M., and Silver, R. (2003). Calbindin influences response to photic input in suprachiasmatic nucleus. *J. Neurosci.* **23,** 8820–8826.

Hara, T., Hamada, J., Yano, S., Morioka, M., Kai, Y., and Ushio, Y. (2003). CREB is required for acquisition of ischemic tolerance in gerbil hippocampal CA1 region. *J. Neurochem.* **86,** 805–814.

Harney, J. P., Scarbrough, K., Rosewell, K. L., and Wise, P. M. (1996). *In vivo* antisense antagonism of vasoactive intestinal peptide in the suprachiasmatic nuclei causes aging-like changes in the estradiol-induced luteinizing hormone and prolactin surges. *Endocrinology* **137,** 3696–3701.

Hebb, M. O., and Robertson, H. A. (1997). End-capped antisense oligodeoxynucleotides effectively inhibit gene expression *in vivo* and offer a low-toxicity alternative to fully modified phosphorothioate oligodeoxynucleotides. *Brain Res. Mol. Brain Res.* **47,** 223–228.

Inouye, S. T., and Kawamura, H. (1979). Persistence of circadian rhythmicity in a mammalian hypothalamic "island" containing the suprachiasmatic nucleus. *Proc. Natl. Acad. Sci. USA* **76,** 5962–5966.

Inouye, S. T., and Kawamura, H. (1982). Characteristics of a circadian pacemaker in the suprachiasmatic nucleus. *J. Comp. Physiol. A.* **146,** 153–160.

Kashikara, N., Maeshima, Y., and Makino, H. (1998). Antisense oligonucleotides. *Exp. Nephrol.* **6,** 84–88.

King, D. P., and Takahashi, J. S. (2000). Molecular genetics of circadian rhythms in mammals. *Annu. Rev. Neurosci.* **23,** 713–742.

King, D. P., Zhao, Y., Sangoram, A. M., Wilsbacher, L. D., Tanaka, M., Antoch, M. P., Steeves, T. D., Vitaterna, M. H., Kornhauser, J. M., Lowrey, P. L., Turek, F. W., and Takahashi, J. S. (1997). Positional cloning of the mouse circadian clock gene. *Cell* **89,** 641–653.

Kriegsfield, L. J., Demas, G. E., Lee, S. E., Dawson, T. M., Dawson, V. L., and Nelson, R. J. (1999a). Circadian locomotor analysis of male mice lacking the gene for neuronal nitric oxide synthase (nNOS−/−). *J. Biol. Rhythms* **14,** 20–27.

Kriegsfield, L. J., Drazen, D. L., and Nelson, R. J. (2001). Circadian organization in male mice lacking the gene for endothelial nitric oxide synthase (eNOS−/−). *J. Biol. Rhythms* **16,** 142–148.

Kriegsfield, L. J., Eliasson, M. J. L., Demas, G. E., Blackshaw, S., Dawson, T. M., Nelson, R. J., and Snyder, S. J. (1999b). Nocturnal motor coordination deficits in neuronal nitric oxide synthase knock-out mice. *Neuroscience* **89**, 311–315.

Kurumiya, S., and Kawamura, H. (1988). Circadian oscillation of the multiple unit activity in the guinea pig suprachiasmatic nucleus. *J. Comp. Physiol. A.* **162**, 301–308.

Li, Z., Stuart, R. O., Qiao, J., Pavlova, A., Bush, K. T., Pohl, M., Sakurai, H., and Nigam, S. K. (2000). A role for Timeless in epithelial morphogenesis during kidney development. *Proc. Natl. Acad. Sci. USA* **97**, 10038–10043.

Lopez-Molina, L., Conquet, F., Dubois-Dauphin, M., and Schibler, U. (1997). The DBP gene is expressed according to a circadian rhythm in the suprachiasmatic nucleus and influences circadian behavior. *EMBO J.* **16**, 6762–6771.

Lowrey, P. L., Shimomura, K., Antoch, M. P., Yamazaki, S., Zemenides, P. D., Ralph, M. R., Menaker, M., and Takahashi, J. S. (2000). Positional systemic cloning and functional characterization of the mammlain circadian mutation tau. *Science* **288**, 483–492.

Mabuchi, T., Kitagawa, K., Kuwabara, K., Takasawa, K., Ohtsuki, T., Xia, Z., Storm, D. R., Yanagihara, T., Hori, M., and Matsumoto, M. (2001). Phosphorylation of cAMP response element-binding protein in hippocampal neurons as a protective response after exposure to glutamate *in vitro* and ischemia *in vivo*. *J. Neurosci.* **21**, 9204–9213.

Mathur, A., Golombek, D. A., and Ralph, M. R. (1996). cGMP-dependent protein kinase inhibitors block light-induced phase advances of circadian rhythms *in vivo*. *Am. J. Physiol.* **270**, R1031–R1036.

Matsuo, T., Yamaguchi, S., Mitsui, S., Emi, A., Shimoda, F., and Okamura, H. (2003). Control mechanism of the circadian clock for timing of cell division *in vivo*. *Science* **302**, 255–259.

Melo, L., Golombek, D. A., and Ralph, M. R. (1997). Regulation of circadian photic responses by nitric oxide. *J. Biol. Rhythms* **12**, 319–326.

Miyake, S., Sumi, Y., Yan, L., Takekida, S., Fukuyama, T., Ishida, Y., Yamaguchi, S., Yagita, K., and Okamura, H. (2000). Phase-dependent responses of *Per1* and *Per2* genes to a light stimulus in the suprachiasmatic nucleus of the rat. *Neurosci. Lett.* **294**, 41–44.

Obrietan, K., Impey, S., Smith, D., Athos, J., and Storm, D. R. (1999). Circadian regulation of cAMP response element-mediated gene expression in the suprachiasmatic nuclei. *J. Biol. Chem.* **274**, 17748–17756.

Ogawa, S., Brown, S. E., Okana, H. J., and Pfaff, D. W. (1995). Cellular uptake of intracerebrally administered oligodeoxynucleotides in mouse brain. *Reg. Pep.* **59**, 143–149.

Okamura, H., Miyake, S., Sumi, Y., Yamaguchi, S., Yasui, A., Muijtjens, M., Hoeijmakers, J. H., and van der Horst, G. T. (1999). Photic induction of *mPer1* and *mPer2* in cry-deficient mice lacking a biological clock. *Science* **286**, 2531–2534.

Okamura, H., Yamaguchi, S., and Yagita, K. (2002). Molecular machinery of the circadian clock in mammals. *Cell Tissue Res.* **309**, 47–56.

Oster, H., Werner, C., Magnone, M. C., Mayser, H., Feil, R., Seeliger, M. W., Hofmann, F., and Albrecht, U. (2003). cGMP-dependent protein kinase II modulates *mPer1* and *mPer2* gene induction and influences phase shifts of the circadian clock. *Curr. Biol.* **13**, 725–733.

Panda, S., Antoch, M. P., Miller, B. H., Su, A. I., Schook, A. B., Straume, M., Schultz, P. G., Kay, S. A., Takahashi, J. S., and Hogenesch, J. B. (2002). Coordinated transcription of key pathways in the mouse by the circadian clock. *Cell* **109**, 307–320.

Park, D., Lee, S. E., Jun, K., Hong, Y.-M., Kim, D., Kim, Y. I., and Shin, H.-S. (2003). Translation of rhythmicity into neural firing in suprachiasmatic nucleus requires mGluR-PLC$\beta$4 signaling. *Nature Neurosci.* **6**, 337–338.

Park, Y. G., Nesterova, M., Agrawal, S., and Cho-Chung, Y. S. (1999). Dual blockade of cyclic AMP response element-(CRE) and AP-1-directed transcription by CRE transcription factor decoy oligodeoxynucleotide: Gene specific inhibition of tumor growth. *J. Biol. Chem.* **274,** 1573–1580.

Perez, J. R., Li, Y., Stein, C. A., Majumder, S., Oorschot, A. V., and Narayanan, R. (1994). Sequence-independent induction of Sp1 transcription factor activity by phosphorothioate oligodeoxynucleotides. *Proc. Natl. Acad. Sci. USA* **91,** 5957–5961.

Preitner, N., Damiola, F., Lopez-Molina, L., Zakany, J., Duboule, D., Albrecht, U., and Schibler, U. (2002). The orphan nuclear receptor REV-ERBalpha controls circadian transcription within the positive limb of the mammalian circadian oscillator. *Cell* **110,** 251–260.

Prosser, R. A., and Gillette, M. U. (1989). The mammalian circadian clock in the suprachiasmatic nuclei is reset *in vitro* by cAMP. *J. Neurosci.* **9,** 1073–1081.

Sato, T., and Kawamura, H. (1984). Circadian rhythms in multiple unit activity inside and outside the suprachiasmatic nucleus in the diurnal chipmunk (*Eutamais sibiricus*). *Neurosci. Res.* **1,** 45–52.

Scarbrough, K. (2000). Use of antisense oligodeoxynucleotides to study biological rhythms. *Methods* **22,** 255–260.

Scarbrough, K., Harney, J. P., Rosewell, K. L., and Wise, P. M. (1996). Acute effects of antisense antagonism of a single peptide neurotransmitter in the circadian clock. *Am. J. Physiol.* **270,** R283–288.

Schlingensiepen, K. H., Wollnik, F., Kunst, M., Schlingensiepen, R., Herdegen, T., and Brysch, W. (1994). The role of Jun transcription factor expression and phosphorylation in neuronal differentiation, neuronal cell death and plastic adaptations *in vivo*. *Cell. Mol. Neurobiol.* **14,** 487–505.

Schwartz, W. J., Davidsen, L. C., and Smith, C. B. (1980). *In vivo* metabolic activity of a putative circadian oscillator, the rat suprachiasmatic nucleus. *J. Comp. Neurol.* **189,** 157–167.

Schwartz, W. J., Reppert, S. M., Eagan, S. M., and Moore-Ede, M. C. (1983). *In vivo* metabolic activity of the suprachiasmatic nuclei: A comparative study. *Brain Res.* **274,** 184–187.

Shearman, L. P., Jin, X., Lee, C., Reppert, S. M., and Weaver, D. R. (2000). Targeted disruption of the *mPer3* gene: Subtle effects on the circadian clock function. *Mol. Cell. Biol.* **20,** 6269–6275.

Shearman, L. P., Zylka, M. J., Weaver, D. R., Kolakowski, J., L. F., and Reppert, S. M. (1997). Two *period* homologs: Circadian expression and photic regulation in the suprachiasmatic nuclei. *Neuron* **19,** 1261–1269.

Shigeyoshi, Y., Taguchi, K., Yamamoto, S., Takekida, S., Yan, L., Tei, H., Moriya, T., Shibata, S., Loros, J. J., Dunlap, J. C., and Okamura, H. (1997). Light-induced resetting of a mammalian circadian clock is associated with rapid induction of the *mPer1* transcript. *Cell* **91,** 1043–1053.

Shinohara, K., Honma, S., Katsuno, Y., Abe, H., and Honma, K. (1995). Two distinct oscillators in the rat suprachiasmatic nucleus *in vitro*. *Proc. Natl. Acad. Sci. USA* **92,** 7396–7400.

Silver, R., Romero, M. T., Besmer, H. R., Leak, R., Nunez, J. M., and LeSauter, J. (1996). Calbindin-D28K cells in the hamster SCN express light-induced Fos. *Neuroreport* **7,** 1224–1228.

Szymusiak, R., Alam, N., Steininger, T. L., and McGinty, D. (1998). Sleep-waking discharge patterns of ventrolateral preoptic/anterior hypothalamic neurons in rats. *Brain Res.* **803,** 178–188.

Takumi, T., Matsubara, C., Shigeyoshi, Y., Taguchi, K., Yagita, K., Maebayashi, Y., Sakakida, Y., Okumura, K., Takashima, N., and Okamura, H. (1998). A new mammalian period gene predominantly expressed in the suprachiasmatic nucleus. *Genes Cells* **3,** 167–176.

Thresher, R. J., Vitaterna, M. H., Miyamoto, Y., Kazantsev, A., Hsu, D. S., Petit, C., Selby, C. P., Dawut, L., Smithies, O., Takahashi, J. S., and Sancar, A. (1998). Role of mouse cryptochrome blue-light photoreceptor in circadian photoresponses. *Science* **282,** 1490–1494.

Tischkau, S. A., Mitchell, J. W., Tyan, S.-H., Buchanan, G. F., and Gillette, M. U. (2003a). $Ca^{2+}/cAMP$ response element-binding protein (CREB)-dependent activation of *Per1* is required for light-induced signaling in the suprachiasmatic nucleus circadian clock. *J. Biol. Chem.* **278,** 718–723.

Tischkau, S. A., Weber, E. T., Abbott, S. M., Mitchell, J. W., and Gillette, M. U. (2003b). Circadian clock-controlled regulation of cGMP/protein kinase G in the nocturnal domain. *J. Neurosci.* **23,** 7543–7550.

Travnickova-Bendova, Z., Cermakian, N., Reppert, S. M., and Sassone-Corsi, P. (2002). Bimodal regulation of *mPeriod* promoters by CREB-dependent signaling and CLOCK/BMAL activity. *Proc. Natl. Acad. Sci. USA* **99,** 7728–7733.

van der Beek, E. M., Swarts, H. J. M., and Weigant, V. M. (1999). Central administration of antiserum to vasoactive intestinal peptide delays and reduces luteinzing hormone and prolactin surges in ovariectomized, estrogen-treated rats. *Neuroendocrinology* **69,** 227–237.

van der Horst, G. T., Muijtjens, M., Kobayashi, K., Takano, R., Kanno, S., Takao, M., de Wit, J., Verkerk, A., Eker, A. P., van Leenan, D., Buijs, R., Bootsma, D., Hoeijmakers, J. H., and Yasui, A. (1999). Mammalian *Cry1* and *Cry2* are essential for maintenance of circadian rhythms. *Nature* **398,** 627–630.

van Esseveldt, K. E., Lehman, M. N., and Boer, G. J. (2000). The suprachiasmatic nucleus and the circadian time-keeping system revisited. *Brain Res. Brain Res. Rev.* **33,** 34–77.

Vitaterna, M. H., King, D. P., Chang, A.-M., Kornhauser, J. M., Lowrey, P. L., McDonald, J. D., Dove, W. F., Pinto, L. H., Turek, F. W., and Takahashi, J. S. (1994). Mutagenesis and mapping of a mouse gene, *Clock*, essential for circadian behavior. *Science* **264,** 719–725.

Weber, E. T., Gannon, R. L., Michel, A. M., Gillette, M. U., and Rea, M. A. (1995a). Nitric oxide synthase inhibitor blocks light-induced phase shifts of the circadian activity rhythm, but not *c-fos* expression in the suprachiasmatic nucleus of the Syrian hamster. *Brain Res.* **692,** 137–142.

Weber, E. T., Gannon, R. L., and Rea, M. A. (1995b). cGMP-dependent protein kinase inhibitor blocks light-induced phase advances of circadian rhythms *in vivo*. *Neurosci. Lett.* **197,** 227–230.

Xiao, J., Li, C., Zhu, N.-L., Borok, Z., and Minoo, P. (2003). *Timeless* in lung morphogenesis. *Dev. Dyn.* **228,** 82–94.

Yakubov, L., Khaled, Z., Zhang, L., Truneh, A., Vlassov, V., and Stein, C. A. (1993). Oligonucleotides interact with recombinant CD4 at multiple sites. *J. Biol. Chem.* **268,** 18818–18823.

Yamaguchi, S., Mitsui, S., Miyake, S., Yan, L., Onishi, H., Yagita, K., Suzuki, M., Shibata, S., Kobayashi, K., and Okamura, H. (2000). The 5' upstream region of *mPer1* gene contains two promoters and is responsible for circadian oscillation. *Curr. Biol.* **10,** 873–876.

Yamazaki, S., Kerbeshian, M. C., Hocker, C. G., Block, G. D., and Menaker, M. (1998). Rhythmic properties of the hamster suprachiasmatic nucleus *in vivo*. *J. Neurosci.* **18,** 10709–10723.

Yamazaki, S., Numano, R., Abe, M., Hida, A., Takahashi, R.-I., Ueda, M., Block, G. D., Sakaki, Y., Menaker, M., and Tei, H. (2000). Resetting central and peripheral circadian oscillators in transgenic rats. *Science* **288,** 682–685.

Zheng, B., Albrecht, U., Kaasik, K., Sage, M., Lu, W., Vaishnav, S., Li, Q., Sun, Z. S., Eichele, G., Bradley, A., and Lee, C. (2001). Nonredundant roles of the *mPer1* and *mPer2* genes in the mammalian circadian clock. *Cell* **105,** 683–694.

Zheng, B., Larkin, D. W., Albrecht, U., Sun, Z. S., Sage, M., Eichele, G., Lee, C., and Bradley, A. (1999). The *mPer2* gene encodes a functional component of the mammalian circadian clock. *Nature* **400,** 169–173.

Zon, G. (1988). Oligonucleotide analogs as potential chemotherapeutic agents. *Pharm. Res.* **5,** 539–549.

## [32] Assaying the *Drosophila* Negative Feedback Loop with RNA Interference in S2 Cells

By PIPAT NAWATHEAN, JEROME S. MENET, and MICHAEL ROSBASH

### Abstract

Transcriptional negative feedback loops play a critical role in the molecular oscillations of circadian genes and contribute to robust behavioral rhythms. In one key *Drosophila* loop, CLOCK and CYCLE (CLK/CYC) positively regulate transcription of *period* (*per*). The *period* protein (PER) then represses this transcriptional activation, giving rise to the molecular oscillations of *per* RNA and protein. There is evidence that links molecular oscillations with behavioral rhythms, suggesting that PER also regulates the expression of downstream genes, ultimately resulting in proper behavior rhythmicity. Phosphorylation of PER has also been shown to be critical for rhythms. DOUBLETIME (DBT) and casein kinase II (CKII) have been implicated in the phosphorylation of PER, which affects its stability as well as nuclear localization. We investigated the role of these kinases on PER transcriptional repression using the *Drosophila* S2 cell line in combination with RNA interference (RNAi) to knock down specific gene expression. This article describes the methods used to study PER repression activity in the S2 cell system as well as to exploit RNAi in this system. We also include protocols for immunocytochemistry and the application of leptomycin to differentiate direct effects on repression from indirect effects on subcellular localization. Finally, we discuss the generation of stable cell lines in the S2 cell system; these will be useful for experiments requiring homogeneous cell populations.

### Introduction

Many eukaryotic and some prokaryotic organisms regulate their metabolism, physiology, and behavior with a circadian (~24-h) period. These rhythms have been shown to depend on complex feedback circuits involving transcriptional/translational regulation and occur in a diverse set of organisms, including cyanobacteria, fungi, plants, insects, and higher mammals such as mice and humans (Allada *et al.*, 2001; Panda *et al.*, 2002). The current model for the circadian pacemaker of *Drosophila melanogaster* posits two basic helix–loop–helix (bHLH) transcription factors, CLOCK (CLK) and CYCLE (CYC), which bind to upstream E boxes and activate the transcription of the *period* (*per*) and *timeless* (*tim*) genes as well as other direct target genes. The PERIOD and TIMELESS proteins (PER and TIM) are synthesized, associate in the cytoplasm, and the heterodimer is imported into the nucleus. The PER–TIM heterodimer then acts on CLK–CYC to inhibit *per* and *tim* transcription (Allada *et al.*, 2001; Panda *et al.*, 2002). A second feedback loop involves the two *Clk*-dependent transcription factors VRILLE and PDP1, and this loop drives *Clk* mRNA oscillations (Cyran *et al.*, 2003; Glossop *et al.*, 2003).

The molecular mechanisms through which PER and TIM repress the transcriptional activation of CLK/CYC are not yet well understood, although some evidence points to the importance of posttranslational modification of clock proteins (for review, see Allada *et al.*, 2001). There are two kinases in particular, DOUBLETIME (DBT) and casein kinase II (CKII), that make important contributions to circadian rhythmicity (Kloss *et al.*, 1998, 2001; Lin *et al.*, 2002; Martinek *et al.*, 2001; Price *et al.*, 1998). These kinases are imagined to function by phosphorylating clock proteins in a time-dependent manner and affecting protein stability as well as temporally gated nuclear localization of both PER and TIM (Curtin *et al.*, 1995; Dembinska *et al.*, 1997; Kim *et al.*, 2002; Shafer *et al.*, 2002; So and Rosbash, 1997). To decipher the precise role of these two kinases in the repression of CLK/CYC transcriptional activation, we took a reductionist approach and studied PER phosphorylation, nuclear localization, and transcriptional repression in a noncycling system, namely in cultured S2 cells. *cyc*, *ckII*, and *dbt* are expressed endogenously in S2 cells, whereas *Clk*, *per*, and *tim* are not (Darlington *et al.*, 1998).

Several S2 cell studies previously addressed CLK/CYC transcriptional activation and repression by PER/TIM (see later). They used expression vectors containing *Clk*, *per*, and/or *tim* genes together with a luciferase-expressing reporter gene. The reporter was preceded by a *per* or *tim* promoter or preceded by a set of artificial E boxes (Darlington *et al.*, 1998; McDonald *et al.*, 2001). CLK synthesis from an expression vector combines with endogenous CYC to drive transcription from the reporter gene, and then PER and/or TIM is assayed for its effect on transcriptional activity.

We also used this reporter system and focused on examining the effects of DBT and CKII on PER repression activity. In parallel, we also studied the subcellular localization of PER. This is because the kinases have been proposed to affect subcellular localization, and this might be upstream of repression activity. We knocked down mRNA levels of these kinases by RNA interference (RNAi) and observed the effects on repression as well as nuclear localization with immunocytochemistry (ICC). To complement possible effects of RNAi on nuclear localization, we also utilized the nuclear export inhibitor leptomycin (LMB).

RNAi is a class of RNA-mediated gene silencing. In animals, double-stranded RNA (dsRNA) is detected, and a ribonuclease III (*Dicer*) cleaves the dsRNA into short pieces of 21–25 nucleotides called siRNAs (short interfering RNAs) (for review, see Agrawal *et al.*, 2003; Hannon, 2002; Novina and Sharp, 2004). siRNAs are recognized by the RNA-induced silencing complex (RISC). This ribonuclease complex unwinds the siRNA and helps anneal it to target mRNA, leading to its degradation. This mechanism is involved in viral defense mechanisms—viral dsRNA is detected and used to attack viral mRNA—as well as in development. Because RNAi exerts its effect on individual mRNAs, it has been used for the downregulation of specific gene expression. This is done by introducing siRNAs into mammalian cells with complementarity to an mRNA of interest. dsRNAs cannot be used in mammalian cells because they induce an interferon response (for review, see Mittal, 2004). This does not occur in S2 cells, so downregulation of gene expression can be achieved more simply, namely by just adding dsRNA to the cells; this was first demonstrated by Clemens *et al.* (2000). dsRNA enters S2 cells without any need for carrier or transfection agent; the effect is specific to the gene of interest and lasts for several days. This provides a convenient, effective, and long-lasting method of inhibiting gene expression.

Leptomycin was first isolated as an antifungal agent (Hamamoto *et al.*, 1983) and was later shown to be a nuclear export inhibitor (Wolff *et al.*, 1997). It has specificity for the CRM1–nuclear export receptor pathway (Ossareh-Nazari *et al.*, 1997). LMB has also been used successfully to inhibit protein export in *Drosophila* S2 cells (Abu-Shaar *et al.*, 1999), and the *Drosophila* homolog of *crm1* has been identified (Collier *et al.*, 2000; Fasken *et al.*, 2000). In the mammalian circadian system, mPER2 has been shown to be exported out of the nucleus via an LMB-sensitive pathway (Yagita *et al.*, 2002). *Drosophila* PER and TIM were subsequently shown to employ the same mechanism (Ashmore *et al.*, 2003; Nawathean and Rosbash, 2004).

As mentioned earlier, phosphorylation of PER by DBT and CKII has been proposed to be involved in PER degradation (by DBT) and nuclear localization (by both); however, the role of these kinases in regulating PER repression activity had not been investigated. We therefore made use of the

S2 cell system to study CLK/CYC/PER transcriptional regulation and added the specific downregulation of these kinases by RNAi. We also examined the effects of the specific nuclear export inhibitor LMB on PER repression as well as subcellular localization. This article uses both transient transfection and stable transfection strategies in the S2 cell system to study these problems.

## Methodology

### Repression Assay in S2 Cells

All experiments were performed on *Drosophila* Schneider 2 (S2) cells. This S2 cell line was derived from a primary culture of late-stage (20–24 h old) *D. melanogaster* embryos. Many characteristics of the S2 cell line suggest that it is derived from a macrophage-like lineage (Schneider, 1972). S2 cells are grown at 25° without additional $CO_2$ on tissue culture flasks (75 $cm^2$) in an insect cell culture medium (HyQ SFX-Insect, Hyclone) containing 10% fetal bovine serum (FBS, GIBCO) and 1% antibiotic–antimycotic solution (GIBCO).

### Vectors

Three promoters have been used to study CLK/CYC/PER transcription regulation: *per* and *tim* promoters (Darlington *et al.*, 1998) and the minimal sequence (CRS)-containing E-box promoter (Hao *et al.*, 1999; So *et al.*, 2000); all promoters were fused to *luciferase*. In general, these promoters behave similarly, i.e., they can be upregulated by CLK/CYC and downregulated by PER/TIM. There is, however, a smaller effect of PER/TIM in the case of the *per* or *tim* promoters, probably due to the presence of binding sites for other DNA-binding proteins within these natural promoters. In other words, the effect of PER is more pronounced on a promoter consisting of three repeats of the CRS sequence (69 nucleotides) of the *per* promoter and called p3 × 69-*luc* (So *et al.*, 2000); there is probably less of an effect from other transcription factors on this artificial promoter. *Clk* and *per* cDNAs were cloned into pAc V5/His6 (Invitrogen). They were in-frame with V5/His6 so that PER and CLK could be detected using the antiV5 antibody. These genes are driven by an *actin* promoter, which is constitutively active.

### Choice of Transiently Transfected S2 Cells or Stable S2 Cell Lines

Most of the experiments were initially performed on transiently transfected S2 cells (Nawathean and Rosbash, 2004). Indeed, transient transfections allow an easy characterization of the PER repression parameters, i.e., the relative amounts of the different plamids. Moreover, the use of dsRNA is particularly well suited for transient transfection. In contrast, stable cell lines are particularly useful when large amounts of material are desirable. For example, PER corepressors might be detectable by coimmunoprecipitation, and this is better done with stable lines.

### Double-Stranded RNA

Double-stranded RNA was generated by *in vitro* transcription of both sense and antisense strands. The dsRNA is then incubated with S2 cells long enough to ensure efficient mRNA degradation and (hopefully) turnover of the encoded protein. In preparing dsRNA, care should be taken to avoid RNase contamination by wearing gloves and using DEPC-treated water. The dsRNA can be prepared in large scale and stored in aliquots at $-20°$.

*Templates.* Templates were amplified from either cDNA or genomic DNA. In the latter case, introns were avoided. Both forward and reverse primers contain a sequence complementary to the gene of interest as well as a T7 promoter sequence at the 5' end of the primers for subsequent *in vitro* transcription by T7 polymerase (5' TTAATACGACTCACTA-TAGGGAGA 3'). Polymerase chain reaction (PCR) products should be generally about 500–1000 bp in length (Worby *et al.*, 2001). Before continuing to the next step, we purify the PCR product with a Qiagen PCR purification kit and adjust the concentration to 125 ng/$\mu$l.

*dsRNA Synthesis.* The DNA template with a T7 promoter on both sense and antisense strands is used for *in vitro* transcription using an Ambion Megascript kit.

1. In a 20-$\mu$l reaction, mix 1 $\mu$g of the template (8 $\mu$l), 2 $\mu$l each of NTPs, 2 $\mu$l of reaction buffer (10X), and 2 $\mu$l of enzyme mix (containing T7 polymerase).
2. Mix gently and spin.
3. Incubate at 37° for 4–6 h or overnight.
4. Precipitate reaction at $-20°$ for at least 30 min with 3 $M$ sodium acetate to a final concentration of 10% and 2.5 volumes of 100% ethanol.
5. Centrifuge at 14,000 rpm for 15 min at 4°.
6. Remove the supernatant, and let the RNA pellet air dry for 10–15 min. Take care not to over dry, as it will be difficult to dissolve/resuspend the pellet.
7. Resuspend in 40 $\mu$l of DEPC water.
8. To anneal the sense and antisense RNA strands, denature by heating at 65° for 30 min and cool slowly by turning off the heat and

letting the heating unit cool to room temperature. Measure the concentration at 260 nm (1 $A_{260} = 45$ $\mu g/\mu l$). Run a gel and check for RNA integrity and size (use ~1 $\mu g$ for the gel).

9. Make a 3-$\mu g/\mu l$ stock, aliquot, and store at $-20°$.

The normal yield of the *in vitro* transcription is approximately 150 $\mu g$ per 20-$\mu l$ reaction. The reaction volume can be scaled up.

## Transient Transfection

Transfection of S2 cells can be performed using different reagents. We only used Cellfectin reagent (Invitrogen) in these studies, which is a liposome formulation. Transfection is generally performed when the S2 cells are seeded at 80% of confluence. In general, we also cotransfected the DNA with pCopia *renilla luciferase*, which serves as an internal control for transfection efficiency. Copia is a constitutively active promoter, and renilla luciferase requires a different substrate than firefly luciferase. Renilla luciferase activity can therefore be used for normalization to reduce variability of the results due to other factors.

### Transfection

The following protocol is designed for one 35-mm well containing S2 cells.

1. In two 1.5-ml Eppendorfs prepare one tube with 100 $\mu l$ of SFM (serum-free medium) + 10 $\mu l$ Cellfectin (Invitrogen) and one tube with 100 $\mu l$ of SFM + DNA.
2. Mix the two solutions gently in a 1.5-ml Eppendorf and wait 30 min at room temperature.
3. Add 800 $\mu l$ of SFM.
4. Put this 1-ml solution in a well of a six-well plate pre-prepared with S2 cells. (Remove the culture medium in the well beforehand.)
5. Four hours later, add 1 ml of 20% FBS and 2% antibiotic–antimycotic medium. Incubate at 25° for 2–4 days.

### Application of dsRNA

1. Plate the cells in SFM.
2. Add dsRNA 15 $\mu g/2$ ml of media in a six-well plate format, i.e., 5 $\mu l$ RNA/well; mix well.
3. Incubate for 2–3 days (or more) for an efficient mRNA knockdown.

Usually, dsRNA is added 2 days before transfection, and cells are lysed 2 days after transfection. In our experience, 4 days of dsRNA is long enough for an efficient RNA and protein knockdown. In general, we leave the cells in SFM during the incubation with dsRNA. Alternatively, 1 ml of SFM with the 15 $\mu$g of dsRNA can be used, to which 1 ml of 20% FBS-containing media is added 4 h later. This is preferable for longer incubation times with dsRNA, i.e., more than 3 days.

*Use of Leptomycin*

Leptomycin (Sigma) has been used successfully in a range of 10–400 n$M$. We generally use it at 20–100 n$M$ with comparable results. To treat cells with LMB, we first remove the medium and then add new medium with an appropriate concentration of LMB and incubate for 8 h before proceeding to the next step, which is either a luciferase measurement or ICC.

*Luciferase Bioluminescence Measurement in a Transient S2 Cell Culture*

Both luciferase activities (firefly luciferase and *renilla* luciferase, which serves as an internal control; see earlier discussion) were quantified with the Dual-Luciferase reporter assay system (Promega) using a luminometer (#TD-20/20) from Turner Designs.

*Protocol*

1. Remove media.
2. Add 250 $\mu$l of lysis buffer per well (Promega). Shake for 10 min at room temperature.
3. Take liquid and centrifuge at 14,000 rpm for 10 min at 4°.
4. Take the supernatant and keep on ice.
5. In counting tubes, add 100 $\mu$l of luciferase substrate to 1–20 $\mu$l of the supernatant.
6. Quantify activity.
7. Add preprepared 100 $\mu$l Stop and Glo reagent (stock is 50×, dilute in Stop and Glo buffer).
8. Quantify the two luciferase activities.

*Comparisons with Other Papers.* The S2 cell system has been widely used to study circadian rhythm-relevant transcriptional regulation (e.g., see Ceriani *et al.*, 1999; Chang and Reppert, 2003; Darlington *et al.*, 1998; Nawathean and Rosbash, 2004; Rothenfluh *et al.*, 2000). Although results from different laboratories are largely consistent, there are some differences. For example, the role of TIM in modulating PER nuclear localization and PER repression activity is controversial and has been discussed elsewhere (Nawathean and Rosbash, 2004). Another quantitative difference between different studies is the strength of PER repression activity. For example, CLK:CYC transcriptional activity was reduced from 100% to 75% in one study (Darlington *et al.*, 1998) and reduced to less than 10% (10-fold) in another (Chang and Reppert, 2003). We believe that these differences are due to a number of factors. First, some reports used a fix amount of PER, which was probably suboptimal in those cases that reported weak repression activity. Second, a number of different reporter genes have been used. As described earlier, *per* and *tim* promoters probably contain other transcription factor-binding sequences, which increase background signal due to the activity of other S2 cell transcription factors. This has the effect of reducing potential PER repression activity. The artificial CRS (3 × 69) promoter sequence in contrast contains mainly, if not exclusively, CLK:CYC target E boxes. Third, experimental protocols have not been identical across different studies. For example, the amounts of the pAc*Clk* plasmid used range from 0.5 to 5 ng per well and, in some cases, the amount of plasmid was not even mentioned. Moreover, the amount of DNA needed for optimal transfection depends on the transfection reagent: 2 $\mu$g with Cellfectin and 20 $\mu$g with calcium phosphate per 1-ml reaction in our experience. Optimal conditions here are defined as the concentration of pAc*Clk* that gives maximal PER repression activity. In our experience, using a higher concentration of pAc*Clk* gives rise to lower PER repression activity. As a result, we suspect that some of the papers with lower PER activity simply used too high an amount of pAc*Clk*. This argument is just empirical (or a mix of empirical and guesswork), as the mechanism by which PER represses CLK/CYC transcription activation is not known. Therefore, it is not known what an optimal relationship between PER and CLK should be.

*Immunocytochemistry Protocol*

ICC is used to observe the subcellular localization of the protein of interest, in this case PER. We first seeded S2 cells in a six-well plate onto coverslips, which allows the convenient final mounting of the cells onto a slide. These cells can then be used for treatment with dsRNA, transfection, and/or incubation with LMB. Once ready for immunostaining, cells are fixed, blocked, and incubated with primary antibody. The subsequent secondary antibody is normally conjugated with fluorescence dye (for example, FITC). We use mounting medium containing DAPI, which will stain DNA, for ease of nuclear detection.

1. Seed S2 cells overnight or at least 2 h in serum-free media in six-well plates containing coverslips. This plate can then be used for the transfection and other treatments.
2. All the staining processes are also done in the six-well plate. Once the cells are ready for ICC, remove the medium and wash once briefly with 1 ml of cold phosphate-buffered saline (PBS).
3. Fix the cell with 1 ml 4% paraformaldehyde in PBS for 5 min at room temperature.

4. Wash three times for 10 min with cold PBS.
5. Remove the PBS and add 1 ml blocking solution (10% normal goat serum, 0.2% Triton X-100 in PBS). Place the plate on rocker at room temperature for 30 min.
6. Remove the blocking solution and add primary antibody (mouse antiV5—Invitrogen 1:500 in blocking solution). Incubate on the rocker at room temperature for 1–2 h.
7. Remove the primary antibody and wash the cell five times, each for 2 min, with cold PBS.
8. Add secondary antibody (goat antimouse FITC—Jackson research laboratory—1:200 in blocking solution; FITC excitation and emission wavelength—492/520 nm) and wrap the plate with aluminum foil to protect from light. Incubate on the rocker at room temperature for 1 h.
9. Remove the secondary antibody and wash 2 × 5 min with cold PBS, with aluminum foil wrapped around the plate. Mount onto a glass slide with mounting agent plus DAPI (Vector).

## Generating Stable Cell Lines

### Generality

To generate stable S2 cell lines, the plasmid(s) of interest first needs to be cotransfected with a selection vector, which confers drug resistance. Cells that have stably integrated the DNA are selected by applying the drug to the culture medium 4 days after the transfection. Duration of the selection depends on the selection vector used. We have used two different selection vectors shown to be appropriate for use with the *Drosophila* Expression system (DES, Invitrogen): pBSPuro (gift from M. Wilm, Heidelberg, Germany; Benting *et al.*, 2000) and pCoBlast (Invitrogen), which confer resistance to puromycin and blasticidin, respectively. A selection vector conferring hygromycin resistance (pCoHygro; Invitrogen) can also be used, but selection usually takes longer in this case (4 weeks after cotransfection, according to Invitrogen and McDonald *et al.*, 2001). With puromycin and blasticidin, only 2 weeks of selection are usually required to obtain a stable cell line. However, it may take somewhat longer depending on the transfection efficiency and cell density. When the selection is complete, the drug concentration in the medium can be decreased by a factor of two. Removing the drug completely should be avoided. This is because transgene silencing of multicopy repeats (what is usually the case in this kind of selection) can occur by a phenomenon still not well understood. In this case, drug reapplication after some weeks without treatment can induce death of almost all the cells.

Using stable cell lines to study the repression of PER on CLK/CYC transcriptional activation requires some care. In our limited experience, this is because the amount of PER repression is highly dependent on the characteristics of the cell line. For this reason, we recommend two consecutive selections: (1) with pAcClk and the reporter p3×69luc, to maximize CLK/CYC transcriptional activation of the reporter, and (2) with variable amounts of pAcper, to optimize repression. The decrease in luciferase activity after the second selection can then be directly correlated to the repression of PER on CLK/CYC transcriptional activation. The following protocol can be used for both types of selection (pBSPuro and pCoBlast).

*Selection Protocol*

*Preparation of Plate*

1. Culture S2 cells in 10% FBS and 1% antibiotic-antimycotic medium (call FBS medium) in a 75-cm$^2$ flask.
2. Remove the old medium and add 10 ml of fresh FBS medium.
3. Scrape the cells using a cell scraper (Corning).
4. Aspirate the 10 ml of culture medium containing the cells, count the cells using a hematocymeter, and dilute to $1 \times 10^6$ cells/ml with FBS medium.
5. Plate the cells in a six-well culture plate (Corning). Two milliliters containing $1 \times 10^6$ cells/ml is usually plated in one well.

*Transfection.* All transfections are performed using Cellfectin reagent (Invitrogen), which is well suited for insect cells. S2 cells are transfected with the plasmid(s) of interest and the selection vector. In the case of pBSPuro, we recommend a 5:1 ratio between reporter:selection vector. In the case of pCoBlast, the manufacturer recommendation ratio of 19:1 is fine. Decreasing the ratio generally causes an increase in the number of clones but a decrease in the incorporation of reporter into each clone, and increasing the ratio does the reverse. We recommend doing the transfection in a few wells (two to three) to increase the number of selected cells; two identical wells are usually sufficient.

Transfection is performed as described previously for transient transfection.

1. For one 35-mm well, mix 12 $\mu$l of Cellfectin and 88 $\mu$l of serum-free medium (SFM; Cellfectin at a final concentration of 12%) in a sterile 1.5-ml microcentrifuge tube or a culture tube.
2. In another tube, mix DNA (reporter and selection vector) in 100 $\mu$l of SFM.
3. Wait 10 min and then mix the two solutions gently.
4. Wait 30 min at room temperature.

5. Add 800 $\mu$l of SFM.
6. Remove the FBS medium from the wells.
7. Apply the 1 ml medium containing DNA and Cellfectin in a 35-mm well.
8. Four hours later, add 1 ml of 20% FBS, 2% antibiotic-antimycotic medium. Incubate at 25° for 3 days.

*Selection.* This step consists of killing S2 cells that have not stably integrated the selection vector by adding drug to the culture medium. Drug concentrations are 10 $\mu$g/ml for puromycin (http://www.mann.embl_heidelberg.de/GroupPages/PageLink/activities/iTAP/Experimental/Stable CellLine.html) and 25 $\mu$g/ml for blasticidin (Invitrogen). Both drugs are dissolved in sterile distilled water at a concentration of 10 mg/ml for puromycin and 25 mg/ml for blasticidin. The stock solutions are then sterilized by filtration through a 0.22-$\mu$m pore, aliquoted in sterile tubes, and frozen at −20° until use. Stock solutions are not subject to multiple freeze/thaw cycles. After the stable lines are selected, drug concentrations are decreased by half (5 $\mu$g/ml for puromycin and 12.5 $\mu$g/ml for blasticidin).

1. Three to 4 days after transfection, replace the old FBS medium with 1 ml of fresh FBS medium containing puromycin or blasticidin at a concentration of 10 $\mu$g/ml for puromycin and 25 $\mu$g/ml for blasticidin.
2. Scrape the cells and transfer them into a 1.5-ml microcentrifuge tube.
3. Harvest the cells by spinning at 3000 rpm for 5 min at room temperature.
4. Remove the supernatant and add for a wash 1 ml of 10% FBS + same drug concentration.
5. Harvest the cells a second time at 3000 rpm for 5 min at room temperature.
6. Remove the supernatant and resuspend the cells in 1.5 ml of 10% FBS + drug.
7. Transfer the cells to a 25-cc flask.
8. Change the medium + antibiotic every 3–4 days for 2 weeks.

*Application of dsRNA to Stable Cell Lines.* As described earlier for transient transfections, dsRNA can just be added to the culture medium to efficiently silence a target gene. Nonetheless, we observed that silencing efficiency is generally improved if the dsRNA is transfected like the DNA, i.e., accompanied by transfection reagent.

## References

Abu-Shaar, M., Ryoo, H. D., and Mann, R. S. (1999). Control of the nuclear localization of Extradenticle by competing nuclear import and export signals. *Genes Dev.* **13**, 935–945.

Agrawal, N., Dasaradhi, P. V., Mohmmed, A., Malhotra, P., Bhatnagar, R. K., and Mukherjee, S. K. (2003). RNA interference: Biology, mechanism, and applications. *Microbiol. Mol. Biol. Rev.* **67**, 657–685.

Allada, R., Emery, P., Takahashi, J. S., and Rosbash, M. (2001). Stopping time: The genetics of fly and mouse circadian clocks. *Annu. Rev. Neurosci.* **24**, 1091–1119.

Ashmore, L. J., Sathyanarayanan, S., Silvestre, D. W., Emerson, M. M., Schotland, P., and Sehgal, A. (2003). Novel insights into the regulation of the timeless protein. *J. Neurosci.* **23**, 7810–7819.

Benting, J., Lecat, S., Zacchetti, D., and Simons, K. (2000). Protein expression in Drosophila Schneider cells. *Anal. Biochem.* **278**, 59–68.

Ceriani, M. F., Darlington, T. K., Staknis, D., Mas, P., Petti, A. A., Weitz, C. J., and Kay, S. A. (1999). Light-dependent sequestration of TIMELESS by CRYPTOCHROME. *Science* **285**, 553–556.

Chang, D. C., and Reppert, S. M. (2003). A novel C-terminal domain of drosophila PERIOD inhibits dCLOCK:CYCLE-mediated transcription. *Curr. Biol.* **13**, 758–762.

Clemens, J. C., Worby, C. A., Simonson-Leff, N., Muda, M., Maehama, T., Hemmings, B. A., and Dixon, J. E. (2000). Use of double-stranded RNA interference in Drosophila cell lines to dissect signal transduction pathways. *Proc. Natl. Acad. Sci. USA* **97**, 6499–6503.

Collier, S., Chan, H. Y., Toda, T., McKimmie, C., Johnson, G., Adler, P. N., O'Kane, C., and Ashburner, M. (2000). The Drosophila embargoed gene is required for larval progression and encodes the functional homolog of schizosaccharomyces Crm1. *Genetics* **155**, 1799–807.

Curtin, K. D., Huang, Z. J., and Rosbash, M. (1995). Temporally regulated nuclear entry of the Drosophila period protein contributes to the circadian clock. *Neuron* **14**, 365–372.

Cyran, S. A., Buchsbaum, A. M., Reddy, K. L., Lin, M. C., Glossop, N. R., Hardin, P. E., Young, M. W., Storti, R. V., and Blau, J. (2003). vrille, Pdp1, and dClock form a second feedback loop in the Drosophila circadian clock. *Cell* **112**, 329–341.

Darlington, T. K., Wager-Smith, K., Ceriani, M. F., Staknis, D., Gekakis, N., Steeves, T. D., Weitz, C. J., Takahashi, J. S., and Kay, S. A. (1998). Closing the circadian loop: CLOCK-induced transcription of its own inhibitors per and tim. *Science* **280**, 1599–1603.

Dembinska, M. E., Stanewsky, R., Hall, J. C., and Rosbash, M. (1997). Circadian cycling of a PERIOD-beta-galactosidase fusion protein in Drosophila: Evidence for cyclical degradation. *J. Biol. Rhythms* **12**, 157–172.

Fasken, M. B., Saunders, R., Rosenberg, M., and Brighty, D. W. (2000). A leptomycin B-sensitive homologue of human CRM1 promotes nuclear export of nuclear export sequence-containing proteins in Drosophila cells. *J. Biol. Chem.* **275**, 1878–1886.

Glossop, N. R., Houl, J. H., Zheng, H., Ng, F. S., Dudek, S. M., and Hardin, P. E. (2003). VRILLE feeds back to control circadian transcription of Clock in the Drosophila circadian oscillator. *Neuron* **37**, 249–261.

Hamamoto, T., Seto, H., and Beppu, T. (1983). Leptomycins A and B, new antifungal antibiotics. II. Structure elucidation. *J. Antibiot. (Tokyo)* **36**, 646–650.

Hannon, G. J. (2002). RNA interference. *Nature* **418**, 244–251.

Hao, H., Glossop, N. R., Lyons, L., Qiu, J., Morrish, B., Cheng, Y., Helfrich-Forster, C., and Hardin, P. (1999). The 69 bp circadian regulatory sequence (CRS) mediates per-like developmental, spatial, and circadian expression and behavioral rescue in Drosophila. *J. Neurosci.* **19**, 987–994.

Kim, E. Y., Bae, K., Ng, F. S., Glossop, N. R., Hardin, P. E., and Edery, I. (2002). Drosophila CLOCK protein is under posttranscriptional control and influences light-induced activity. *Neuron* **34**, 69–81.

Kloss, B., Price, J. L., Saez, L., Blau, J., Rothenfluh, A., Wesley, C. S., and Young, M. W. (1998). The Drosophila clock gene double-time encodes a protein closely related to human casein kinase Iepsilon. *Cell* **94**, 97–107.

Kloss, B., Rothenfluh, A., Young, M. W., and Saez, L. (2001). Phosphorylation of period is influenced by cycling physical associations of double-time, period, and timeless in the Drosophila clock. *Neuron* **30**, 699–706.

Lin, J. M., Kilman, V. L., Keegan, K., Paddock, B., Emery-Le, M., Rosbash, M., and Allada, R. (2002). A role for casein kinase 2alpha in the Drosophila circadian clock. *Nature* **420**, 816–820.

Martinek, S., Inonog, S., Manoukian, A. S., and Young, M. W. (2001). A role for the segment polarity gene shaggy/GSK-3 in the Drosophila circadian clock. *Cell* **105**, 769–779.

McDonald, M. J., Rosbash, M., and Emery, P. (2001). Wild-type circadian rhythmicity is dependent on closely spaced E boxes in the Drosophila timeless promoter. *Mol. Cell Biol.* **21**, 1207–1217.

Mittal, V. (2004). Improving the efficiency of RNA interference in mammals. *Nat. Rev. Genet.* **5**, 355–365.

Nawathean, P., and Rosbash, M. (2004). The doubletime and CKII kinases collaborate to potentiate Drosophila PER transcriptional repressor activity. *Mol. Cell* **13**, 213–223.

Novina, C. D., and Sharp, P. A. (2004). The RNAi revolution. *Nature* **430**, 161–164.

Ossareh-Nazari, B., Bachelerie, F., and Dargemont, C. (1997). Evidence for a role of CRM1 in signal-mediated nuclear protein export. *Science* **278**, 141–144.

Panda, S., Hogenesch, J. B., and Kay, S. A. (2002). Circadian rhythms from flies to human. *Nature* **417**, 329–335.

Price, J. L., Blau, J., Rothenfluh, A., Abodeely, M., Kloss, B., and Young, M. W. (1998). double-time is a novel Drosophila clock gene that regulates PERIOD protein accumulation. *Cell* **94**, 83–95.

Rothenfluh, A., Young, M. W., and Saez, L. (2000). A TIMELESS-independent function for PERIOD proteins in the Drosophila clock. *Neuron* **26**, 505–514.

Schneider, I. (1972). Cell lines derived from late embryonic stages of *Drosophila melanogaster*. *J. Embryol. Exp. Morphol.* **27**, 353–365.

Shafer, O. T., Rosbash, M., and Truman, J. W. (2002). Sequential nuclear accumulation of the clock proteins period and timeless in the pacemaker neurons of *Drosophila melanogaster*. *J. Neurosci.* **22**, 5946–5954.

So, W. V., and Rosbash, M. (1997). Post-transcriptional regulation contributes to Drosophila clock gene mRNA cycling. *EMBO J.* **16**, 7146–7155.

So, W. V., Sarov-Blat, L., Kotarski, C. K., McDonald, M. J., Allada, R., and Rosbash, M. (2000). takeout, a novel Drosophila gene under circadian clock transcriptional regulation. *Mol. Cell. Biol.* **20**, 6935–6944.

Wolff, B., Sanglier, J. J., and Wang, Y. (1997). Leptomycin B is an inhibitor of nuclear export: Inhibition of nucleo-cytoplasmic translocation of the human immunodeficiency virus type 1 (HIV-1) Rev protein and Rev-dependent mRNA. *Chem. Biol.* **4**, 139–147.

Worby, C. A., Simonson-Leff, N., and Dixon, J. E. (2001). RNA interference of gene expression (RNAi) in cultured Drosophila cells. *Sci. STKE 2001*, PL1.

Yagita, K., Tamanini, F., Yasuda, M., Hoeijmakers, J. H., van der Horst, G. T., and Okamura, H. (2002). Nucleocytoplasmic shuttling and mCRY-dependent inhibition of ubiquitylation of the mPER2 clock protein. *EMBO J.* **21**, 1301–1314.

# [33]   Role of Neuronal Membrane Events in Circadian Rhythm Generation

*By* GABRIELLA B. LUNDKVIST and GENE D. BLOCK

## Abstract

Circadian clock systems are composed of an input or "entrainment" pathway by which synchronization to the external environment occurs, a pacemaker responsible for generating rhythmicity, and an output or "expression" pathway through which rhythmic signals act to modulate physiology and behavior. The circadian pacemaker contains molecular feedback loops of rhythmically expressed genes and their protein products, which, through interactions, generate a circa 24-h cycle of transcription and translation of clock and clock-controlled genes. Neuronal membrane events appear to play major roles in entrainment of circadian rhythms in mollusks and mammals. In mammals, the suprachiasmatic nuclei of the hypothalamus receive photic information via the retinohypothalamic tract. Retinal signals, mediated by glutamate, induce calcium release and activate a number of intracellular cascades involved in photic gating and phase shifting. Membrane events are also involved in rhythm expression. Calcium and potassium currents influence the electrical output of pacemaker neurons by altering shape and intervals of impulse prepotentials, afterhyperpolarization periods, and interspike intervals, as well as altering membrane potentials and thereby shaping the spontaneous rhythmic spiking patterns. Unlike the involvement of membrane events in circadian entrainment and expression, it is less clear whether electrical activity, postsynaptic events, and transmembrane ion fluxes also are essential elements in rhythm generation. Studies, however, suggest that neuronal membrane activity may indeed play a crucial role in circadian rhythm generation.

## Introduction

There is general agreement that circadian rhythms in both animals and plants are generated by autoregulatory transcriptional and posttranslational feedback loops involving "clock genes" and their protein products (for review, see Hastings *et al.*, 2003). Neuronal membrane events, such as electrical impulses and ionic currents, appear to play crucial roles in synchronization of circadian clocks to environmental light cycles and in output regulatory pathways through which clocks regulate tissue and organ targets. It is less clear, however, whether membrane activities are involved in the actual generation of the circadian rhythm. This article summarizes the role of membrane events in the synchronization and expression of circadian rhythms, with an emphasis on the mammalian system. It also addresses the question of whether electrical activity and underlying ionic fluxes participate likewise in rhythm generation.

## Input Pathways and Entrainment

### Entrainment in a Mollusk Model

The ionic mechanisms underlying light entrainment have been intensively examined in the sea hare, *Aplysia*, and the cloudy bubble snail, *Bulla gouldiana*. These mollusks are rhythmically active in light cycles and constant conditions, with *Aplysia* being primarily diurnal and *Bulla* nocturnal (Block and Davenport, 1982; Jacklet, 1969). Light is the primary entraining signal, and in the *Bulla* eye the basal retinal neurons (BRNs), but not the photoreceptor layer, exhibit endogenous rhythms in membrane potential, firing frequency, and membrane conductance (Block and Wallace, 1982; McMahon *et al.*, 1984; Michel *et al.*, 1993). Surprisingly, the BRNs and not the conventional photoreceptors appear to contain the photopigment and pathways for photic entrainment. Thus each BRN contains a circadian pacemaker (Block and Wallace, 1982; Michel *et al.*, 1993) and the critical elements of the entrainment pathway (Block and McMahon, 1984; Geusz *et al.*, 1997).

The *Bulla* retinal clock exhibits a conventional phase response curve (PRC) to light, with phase advances in the early subjective night, phase delays in late night, and a relative "dead zone" during the subjective day (Block and McMahon, 1984). Depolarizing agents, or allowing a Calcium ($Ca^{2+}$) influx via voltage-gated $Ca^{2+}$ channels (Geusz and Block, 1994), generate a PRC similar to light. In contrast, hyperpolarizing agents exhibit a PRC that is in antiphase with the light PRC (Khalsa and Block, 1990; McMahon and Block, 1987).

A model developed from these experiments proposes that membrane depolarization and associated $Ca^{2+}$ influx during the subjective day would have little effect on circadian phase against the expected background of persistent $Ca^{2+}$ influx brought about by the depolarized state of the neurons (Block *et al.*, 1993; McMahon, 1986). In contrast, hyperpolarization during the subjective day would be expected to cause a reduction of $Ca^{2+}$ influx, leading to a phase shift of the rhythm. During the subjective night, when BRNs are hyperpolarized and $Ca^{2+}$ influx is low, depolarizing agents or light pulses would be expected to lead to $Ca^{2+}$ influx and phase shifts of the rhythm.

*Entrainment in Mammals*

Entrainment appears to be a significantly more complicated process in mammals compared with invertebrates, involving a number of pathways, neuromodulators, and transmitters. Although several brain regions and nonneural tissues express endogenous circadian rhythms in transcriptional activity, hypothalamic suprachiasmatic nuclei (SCN) appear to play a central role in the generation and control of major physiological processes. As in most organisms, light is the primary environmental synchronizing signal for the mammalian clock (for review, see Meijer and Schwartz, 2003). In several mammalian species, photic information is conveyed principally to the ventrolateral (VL) "core" region of the SCN by glutamatergic input via the retinohypothalamic tract (RHT). As originally reported in rat by Moore and Lenn (1972) and Hendrickson *et al.* (1972), the RHT consists of a group of axons within the optic nerve that originate from a restricted group of retinal ganglion cells and project directly to the VL SCN. Photic input also reaches the VL SCN indirectly via neuropeptide Y (NPY) in the geniculohypothalamic tract (GHT) and the intergeniculate leaflet (IGL) of the lateral geniculate nucleus (LGN). Pathways containing 5-hydroxytryptamine (5-HT) from the median raphe nucleus provide the other major nonphotic signaling pathway to the SCN, which similar to the NPY phase shifts the clock in the absence of light and also modulates photic signaling (for review, see Hastings *et al.*, 1997). Photic entrainment of the SCN does not seem to involve rods and cones. Rather, a subpopulation of melanopsin-containing retinal ganglion neurons projecting to the SCN have been implicated in the circadian phototransduction cascade (Berson *et al.*, 2002; Hattar *et al.*, 2002; Warren *et al.*, 2003).

*From External Light Pulse to Intracellular Gene Expression: The Phototransduction Cascade*

Although the details of the molecular and cellular mechanisms underlying light-controlled phase regulation continue to be revealed, a satisfying picture has emerged over the last few decades that has led to a model for mammalian entrainment. According to this scheme, activation of photoreceptors in the rodent retina induce primarily glutamate release from SCN terminals of the RHT (for review, see Ebling, 1996). The pituitary adenylate cyclase-activating polypeptide (PACAP) is colocalized with glutamate in RHT terminals (Hannibal *et al.*, 2000) and may be involved in photic signal transduction (Chen *et al.*, 1999; Hannibal *et al.*, 1997; Harrington *et al.*, 1999; Kopp *et al.*, 2001). However, the precise physiological role of PACAP remains to be clarified (for review, see Hannibal,

2002). It has been suggested that retinal innervation in a diurnal rodent species, the degus, also includes the inhibitory transmitter $\gamma$-aminobutyric acid (GABA), which may indicate substantial differences in photic entrainment in diurnal versus nocturnal animals (Jiao and Rusak, 2003). Glutamate release causes depolarization of the retinorecipient SCN neurons (Bos and Mirmiran, 1993; Kim and Dudek, 1991; Meijer et al., 1993) that is mediated by N-methylaspartate (NMDA) and non-NMDA-type glutamate receptors (Jiang et al., 1997; Kim and Dudek, 1991; Lundkvist et al., 2002; Michel et al., 2002; Pennartz et al., 2001). Membrane depolarization and glutamate binding release the $Mg^{2+}$ block on the NMDA receptor channel and cause a postsynaptic calcium influx in the SCN neurons. Calcium currents after NMDA and non-NMDA activation have been demonstrated in the SCN (Colwell, 2001; Tominaga et al., 1994; van den Pol et al., 1992), and NMDA activation causes light-like behavioral phase shifts (Colwell et al., 1990; Mintz and Albers, 1997; Mintz et al., 1999). Activation of voltage-gated $Ca^{2+}$ channels and release from intracellular storage sites also lead to an elevation of intracellular $Ca^{2+}$ concentration. Calcium activates $Ca^{2+}$-dependent proteases, transcription factors, and kinases, including $Ca^{2+}$/calmodulin-dependent protein kinases (CaMK), protein kinase A (PKA), and extracellular signal-related kinase mitogen-activated protein kinase (ERK/MAPK). This leads to phosphorylation of the $Ca^{2+}$-activated cAMP response-binding element (CREB), which seems to be a crucial element in the photic resetting intracellular cascade in the SCN. Phosphorylated CREB binds to the cAMP responsive element (CRE) on target genes. For instance, promoters of the clock genes Period (Per) 1, mediated by CaMK II, and Per2 (Nomura et al., 2003; Travnickova-Bendova et al., 2002), as well as several immediate early genes such as c-fos (Dziema et al., 2003), all contain CREs. Tischkau et al. (2003a) demonstrated that light-induced phase resetting requires CREB-dependent Per1 activation. In a study by Kuhlman et al. (2003), increased neuronal firing after phase resetting light pulses was correlated with an elevated expression of Per1, demonstrating a link between light-induced electrical activity in the SCN and intracellular clock gene expression. Furthermore, in the neurons that expressed Per1 after the light pulses there was a persistent suppression of a hyperpolarizing potassium current. The authors concluded that the membrane becomes depolarized downstream of Per1 activation, affecting impulse frequency and thus leading to a resetting of the clock (Kuhlman et al., 2003).

Taken together, these data suggest that the light transduction cascade causes $Ca^{2+}$ influx, which leads to CREB-dependent activation of clock genes. This activation may result in the transcription of channel proteins, e.g., $K^+$ channels, which alter neuronal electrical activity downstream of intracellular gene activation via the membrane. Via this cascade, the endogenous SCN clock could be reset to the external solar cycle.

*Photic Gating*

Light-induced phase shifts in the SCN are highly correlated to photic activation of c-*fos*, which encodes for the nuclear phosphoprotein Fos, a well-investigated cellular marker of photic influence in the SCN (Amir and Stewart, 1998; Schwartz *et al.*, 1994). c-*fos* stimulation is phase dependent and can be correlated to phase shifts of locomotor activity rhythms (Rea, 1992). In a study by Hughes *et al.* (2004), mice lacking the vasoactive intestinal peptide (VIP)-activated VPAC$_2$ receptor did not show rhythmic variations in expression of c-FOS and another marker of photic induction, phosphorylated ERK. Moreover, light pulses given both during the subjective day and the subjective night induced c-FOS, indicating that gating of photic input is not present in VPAC$_2$-deficient mice. Thus, it appears that photic gating is dependent on VIP acting on the VPAC$_2$ receptor (Hughes *et al.*, 2004). The mechanisms underlying photic gating in the SCN are not clear, but they may involve more than a single mechanism. In a study by Ginty *et al.* (1993), the authors showed that light pulses resulted in phosphorylation of CREB only during the subjective night in the SCN of hamsters. Ding *et al.* (1997) extended this observation in rat SCN slices and suggested that photic gating occurs via phosphorylated CREB, as light, glutamate, and NO induce NMDA-dependent phosphorylated CREB only during the subjective night. Furthermore, circadian regulation of CREB expression in the SCN was demonstrated in a study by Obrietan *et al.* (1999), where levels of phosphorylated CREB peaked during the mid to late subjective night. Another mechanism has been suggested by Pennartz *et al.* (2001) and Colwell (2001). Pennartz *et al.* (2001) showed that the NMDA component following evoked optic nerve stimulation was increased significantly during the subjective night, suggesting that light sensitivity may be gated by diurnal variations in NMDA receptor activity. Reduced NMDA receptor activity during the subjective day would decrease $Ca^{2+}$ influx, thereby providing a "dead zone" for photic phase shifts (Pennartz *et al.*, 2001). In agreement with Pennartz *et al.* (2001), Colwell (2001) found that the magnitude and duration of NMDA-induced $Ca^{2+}$ transients peaked during the subjective night, and NMDA-evoked $Ca^{2+}$ currents peaked during the subjective night.

*Phase Shifting*

As in *Bulla*, light phase delays the SCN in early subjective night and phase advances in late subjective night, whereas light stimuli during the subjective day has little or no effect (Pittendrigh and Daan, 1976). The phase of the SCN rhythm can be determined *in vitro* by recording spontaneous electrical activity in SCN neurons, which is increased during the subjective day and lower during the subjective night (Bos and Mirmiran, 1990; Green and Gillette, 1982; Groos and Hendriks, 1982; Shibata *et al.*, 1982). Optic nerve stimulation in the SCN *in vitro* generates a similar PRC as light (Shibata and Moore, 1993b). Furthermore, glutamate, NMDA, and nitric oxide (NO) generators cause a light-resembling PRC *in vitro* (Ding *et al.*, 1994).

By applying phase-shifting agents and using extracellular single unit recordings to determine the SCN phase *in vitro*, several studies suggest that phase delays and advances of the SCN rhythm involve a number of intracellular cascades at different circadian phases (for review, see Gillette and Mitchell, 2002). Cyclic AMP/PKA and PACAP reset (phase advance) the SCN during the subjective day and have no effect during the subjective night (Gillette and Prosser, 1988; Hannibal *et al.*, 1997; Prosser and Gillette, 1989). Glutamate (light), NMDA, NO, acetylcholine, and cGMP generate phase advances during the subjective night (Ding *et al.*, 1994; Liu and Gillette, 1996). However, as mentioned earlier, glutamate has different phase-shifting effects during the early and late subjective night: it has been suggested by the Gillette laboratory that in early subjective night glutamate stimulation induces $Ca^{2+}$-mediated intracellular $Ca^{2+}$ release by ryanodine receptors that causes phase delays (Ding *et al.*, 1998). In late subjective night, glutamate stimulation instead would activate the GC/cGMP/PKG pathway, thereby causing phase advances (Tischkau *et al.*, 2003b). The photic responses can be shaped by other signals. For instance, NPY and 5-HT modulate photic signaling (for review, see Hastings *et al.*, 1997). Studies addressing the modulatory effects of PACAP in photic entrainment reveal that PACAP potentiates glutamate-induced phase delays during the early subjective night and blocks glutamate-induced phase advances during the late night (Chen *et al.*, 1999). The picture appears to be even more complicated because the effects of PACAP are dose dependent: low concentrations ($\sim$1 n$M$) of PACAP increase NMDA receptor channel conductance, whereas higher concentrations ($>$10 n$M$) inhibit NMDA receptors and instead activate cAMP (Harrington *et al.*, 1999). To our knowledge, it is not known at which concentration PACAP is released in RHT terminals, which makes it difficult to interpret the physiological role of PACAP *in vivo*. As demonstrated by Kopp *et al.* (2001), PACAP can have dual effects on $Ca^{2+}$ induction. Employing $Ca^{2+}$ imaging in primary cultures of the SCN, Kopp and colleagues (2001) showed that the glutamatergic $Ca^{2+}$ signal could be either increased or decreased by PACAP interacting with either AMPA/kainate receptors or metabotropic glutamate receptors, which in turn can modulate NMDA receptor signaling. The investigators suggested that via this dual effect on $Ca^{2+}$ release through different receptors, which was resistant to the sodium ($Na^+$) channel blocker tetrodoxin (TTX), PACAP might integrate information of "light" and "dark" signals to the SCN. This would be of importance during dusk and dawn, when the pacemaker switches its sensitivity to light. Furthermore, Dziema and Obrietan (2002) demonstrated that PACAP potentiates glutamate-evoked calcium transients by enhancing L-channel conductance via a MAPK pathway. Finally, melatonin is another mediator of light response at dusk and dawn and has been reported to gate sensitivity in the SCN via protein kinase C (Hunt *et al.*, 2001; McArthur *et al.*, 1991, 1997; Stehle *et al.*, 1989).

## Role of Transmembrane Ionic Fluxes in Rhythm Generation

### Extracellular $Ca^{2+}$ and Rhythm Generation

While the role of membrane events in the synchronization of mammalian circadian rhythms is unquestioned, the role of membrane-related processes in rhythm generation remains controversial. With regard to membrane ionic fluxes, most of the experimental focus has been on $Ca^{2+}$ with different studies reaching different conclusions on the importance of $Ca^{2+}$ flux in generating rhythmicity. Shibata and Moore (1987) investigated circadian rhythmicity of multiunit electrical activity in brain slices. Rats were studied from embryonic day 22 until postnatal day 14. In slices from animals obtained at postnatal day 1, the firing frequency rhythm was blocked if the slice was bathed in $Ca^{2+}$-free recording solution. Because there is very little synapse formation at this time, the authors concluded that the effect on rhythmicity was not due to blocking intercellular communication but rather was a direct effect on the intrinsic properties of clock neurons. Whether the oscillations within the slice were actually stopped by the $Ca^{2+}$-free medium or whether the impulse frequency mechanism was uncoupled from a normally running clock in the $Ca^{2+}$-free medium could not be determined in these experiments.

Similar results were obtained measuring glucose utilization in embryonic (E22) and adult rats (Shibata *et al.*, 1987). It was found that the normal rhythm of higher glucose utilization during the subjective day, which was present in both embryonic and adult rats, was absent when the slice was placed in $Ca^{2+}$-free medium. Glucose utilization was low during both the subjective night and the subjective day. As with studies looking at impulse frequency, the failure to observe a metabolic rhythm may not have been due to stopping the clock but rather to a failure in the coupling between the molecularly based clock and membrane electrical events whose metabolic demands underlie the rhythm in glucose utilization. Nonetheless, these two studies suggested a critical role for extracellular $Ca^{2+}$ in circadian clock function.

In contrast to the electrophysiological and metabolic studies, an experiment by Shinohara and colleagues (2000) led to a different conclusion. They measured rhythms of aspartate and glutamate release in organotypic explants from 6-day-old rats. The rhythms of both amino acids persisted for two cycles in the presence of a $Ca^{2+}$-free medium. The clear persistence of a circadian rhythm in $Ca^{2+}$-free medium led these investigators to suspect that the failure to observe electrical and metabolic rhythmicity in prior studies was probably an effect on output processes rather than on the clock itself.

### Are Action Potentials Involved in Rhythm Generation?

Several experiments have been designed in attempts to assess the role of action potentials in circadian rhythm generation. Most have suggested that blocking $Na^+$-dependent action potentials does not stop the clock. Schwartz and collaborators (1987) attempted to block $Na^+$-dependent action potentials *in vivo* through infusion of TTX into the SCN of rats. Such infusions blocked locomotor rhythmicity during the treatment; however, the posttreatment locomotor rhythm was restored with a phase predicted from a free-running clock during the TTX infusion. This suggested that in the apparent absence of $Na^+$-dependent action potentials, the basic clock mechanism continued to oscillate. Subsequent experiments applying TTX to explants and tissue slices likewise suggested that action potentials were not critical for rhythm generation. Circadian oscillations were found to be resistant to short-term treatment of TTX in explants and slices (Earnest *et al.*, 1991; Shibata and Moore, 1993a). In addition, Welsh and colleagues (1995) found that TTX did not affect the circadian rhythm in individual SCN neurons dispersed in cell culture.

### Intracellular $Ca^{2+}$ Rhythms and Membrane Activity

A study looking at circadian rhythms in intracellular $Ca^{2+}$ concentration, rather than electrical rhythmicity, reached a different conclusion about the role of membrane electrical activity in generation of the intracellular $Ca^{2+}$ rhythm. Colwell (2000) reported circadian rhythms in intracellular $Ca^{2+}$ concentration in brain slices from rat. Bath application of TTX blocked the intracellular $Ca^{2+}$ rhythm, suggesting that electrical activity and a transmembrane $Ca^{2+}$ flux were responsible for driving the intracellular rhythm. A different result, however, was obtained by Ikeda and colleagues (2003) employing a $Ca^{2+}$-sensitive fluorescent protein. These investigators were able to measure robust circadian rhythms in individual mouse SCN neurons. Simultaneous monitoring of electrical activity with a multiunit electrode array revealed that TTX blocked electrical activity but did not affect the circadian rhythm in intracellular $Ca^{2+}$ concentration. This suggested that intracellular circadian $Ca^{2+}$ oscillations are not dependent upon electrical impulse activity. It is uncertain why these results differ from those of Colwell (2000). Although species differences (i.e., mouse versus rat) may have been a factor in the differing results, Allen and his team believe that methodological differences (i.e., single neuron cameleon reporter versus Fura-2 tissue measurements) played a role in this discrepancy.

*Membrane Events May Play a Critical Role in Rhythm Generation*

Within the past few years, interest in a possible role for membrane electrical events in circadian rhythm generation has been reopened as a result of experiments in both *Drosophila* and mammals. In *Drosophila* (Nitabach *et al.*, 2002), targeted expression of dORK open rectifier $K^+$ channels in the pigment dispersing factor (PDF) expressing lateral neurons was employed as a strategy to silence impulse activity by stabilizing membrane potential at a hyperpolarized state. When maintained in continual darkness, the fruit flies failed to exhibit rhythmic locomotor activity and there were no clear rhythms in expression of the PER or TIM proteins. These investigators reported that overexpression of $Na^+$ channels can rescue rhythmicity in *Drosophila* containing dORK $K^+$ channels (Nitabach *et al.*, 2003). This recovery is most likely due to the restoration of normal membrane potential levels and impulse activity. The rescue provides elegant evidence that the loss of rhythmicity was not due to developmental abnormalities but rather to changes in membrane potential and the lack of impulse activity. The mechanism by which electrical silencing leads to behavioral and molecular arrhythmicity is not yet clear; however, the authors suggest that electrical silencing may result in a reduced activation of voltage-dependent $Ca^{2+}$ channels and consequently the loss of periodic $Ca^{2+}$ entry.

As with *Drosophila*, some data from mammals call into question again the prevailing view about the lack of involvement of membrane events in circadian rhythm generation. Mice missing the $VPAC_2$ receptor and PACAP were unable to maintain normal behavioral rhythmicity (Harmar *et al.*, 2002). Although weak free-running rhythms were present, the phase of the initial free run and the lack of robustness of the rhythm suggested major circadian dysfunction in animals lacking functional

$VPAC_2$ receptors. Importantly, there was no evidence for rhythmicity in several core clock genes including *mPer1*, *mPer2*, and *mCry1*. Consistent with these results, VIP/peptide histidine isoleucine (PHI)-deficient mice were found to exhibit profound abnormalities in locomotor rhythmicity (Colwell *et al.*, 2003). In an electrophysiological analysis, mice lacking the $VPAC_2$ receptor also failed to exhibit the midday peak in electrical activity that is characteristic of impulse rhythms from SCN brain slice (Cutler *et al.*, 2003). Electrical recordings on day 1 and day 2 *in vitro* failed to reveal circadian rhythms and the overall firing rate at nearly every phase was reduced when compared with brain slices from wild-type mice. Taken together, studies on $VPAC_2$ receptorless mice suggest an important requirement for this receptor in normal SCN rhythmicity. The measured reductions in spontaneous electrical activity in SCN neurons and the observation that these cells are capable of responding with high impulse frequencies to stimulation suggest that the loss of the $VPAC_2$ receptor may alter the resting potential of SCN neurons rather than directly affecting excitability. If this is the case, it may be similar to the situation in *Drosophila* where expression of a $K^+$ channel leads to a loss of behavioral and molecular rhythmicity through presumptive hyperpolarization of the membrane potential of PDF-containing lateral neurons.

A study following individual SCN neurons appears to support the view that membrane electrical activity is critical for circadian rhythmicity (Yamaguchi *et al.*, 2003). These investigators tracked *mPer1* rhythmicity in individual neurons in SCN brain slices from transgenic mice in which luciferase served as a reporter for *mPer1* activity. When SCN slices were exposed to culture medium containing TTX, *mPer1* rhythmicity damped in individual neurons. This result suggests that the loss of rhythmicity was due to an effect on the intrinsic oscillations within individual neurons rather than causing population arrhythmicity as individual neurons come out of synchrony with one another during the TTX treatment. Subsequent PCR quantifications showed decreased levels of *mPer1* and *mPer2*, and immunocytochemical measurements of PER1 and PER2 protein levels confirmed a lack of rhythmicity in the presence of TTX. These data have led the authors to suggest that electrical activity is required for both interoscillator synchronization and maintaining cell-autonomous oscillations. An interesting and unexplained aspect of this study is that molecular rhythmicity was not immediately lost following TTX treatment. In addition, the rhythms immediately resumed their normal phase relationship with each other after TTX treatment during 1 week. This could indicate that electrical activity, and presumably rhythmic $Ca^{2+}$ influx, is not essential for completing the circadian cycle. Rather, the membrane-derived signals may help keep the SCN molecular oscillations self-sustained by giving them a periodic stimulus that is not required to complete the timing cycle but is necessary for sustained rhythmicity under constant conditions.

Experiments in our laboratory appear to confirm a critical role of $Ca^{2+}$ in rhythm generation. With reporter technology, we have explored the role of extracellular $Ca^{2+}$ on the rhythmic expression of *mPer1* and mPER2 in SCN slice cultures. We have found that the *mPer1* rhythm disappears in hyperpolarizing medium (decreased $[K^+]$) and in medium with low concentrations of $Ca^{2+}$. Similarly, the rhythm in mPER2 expression is markedly blunted and disappears in 0 m$M$ $Ca^{2+}$. Furthermore, by buffering intracellular $Ca^{2+}$ with a membrane-permeable form of BAPTA, and thereby presumably stopping the rhythm in intracellular $Ca^{2+}$ flux, *mPer1* and mPER2 stop cycling. Our results clearly demonstrate that $Ca^{2+}$ is essential for rhythmic expression of *mPer1* and mPER2, and further suggest that a transmembrane $Ca^{2+}$ flux is critical for rhythm generation (Lundkvist *et al.*, 2003).

### Output and Rhythm Expression

In order for circadian pacemakers to perform useful work for the organism, the circadian oscillation must be coupled to effector systems. Such "output pathways" begin with coupling the intracellular molecularly derived timing signals to the cell membrane and then extend to the electrical and secretory events that couple rhythm-generating cells to one another and to cells, tissues, and organs that are regulated by biological clocks. This section focuses on the first stages of the output pathway, the processes that lead to the regulation of membrane function in pacemaker neurons.

Studies of circadian regulation of electrical membrane events are difficult in part because intracellular electrophysiological recordings are typically limited to a few hours in duration. Examining the intracellular events in SCN neurons is complicated further by virtue of the fact that SCN neurons are very small ($\sim$8–10 $\mu M$ in diameter; van den Pol, 1980), rendering experimental investigation laborious. Furthermore, circadian variations of membrane properties often disappear shortly after membrane rupture in whole cell patch clamp recordings (Schaap *et al.*, 1999). Nevertheless, through dedicated effort with both mammalian and nonmammalian systems, some important information about clock-regulated membrane events has been obtained and is summarized here.

### *Circadian Variations in Membrane Properties*

In *Bulla*, a circadian variation in $K^+$ conductance has been identified that apparently contributes to circadian fluctuations in membrane potential in the BRNs (Michel *et al.*, 1993). One underlying $K^+$ current that was rhythmic was identified as a delayed rectifier (Michel et al., 1999). Circadian fluctuations in membrane properties have also been demonstrated in the rat SCN. Jiang et al. (1997) performed whole cell patch recordings on SCN neurons in horizontal slices at 8 different 3-h intervals over 24 h. The investigators reported circadian variations in both input conductance and holding current at −60 mV ($I_{-60}$); however, the rhythms in conductance and $I_{-60}$ were approximately 7 h out of phase of each other. Input conductance was significantly higher during the subjective dawn and lower near subjective dusk, whereas $I_{-60}$ had a negative peak in the middle of the subjective day (more negative current required to hyperpolarize to −60 mV) and a trough during midsubjective night. In a follow-up study by de Jeu et al. (1998), these investigators used the perforated patch clamp technique to study membrane properties in coronal SCN slices during conditions with minimal rundown. In accordance to the finding by Jiang et al. (1997), who reported an increase in $I_{-60}$ during the midsubjective day, the Dutch group observed a circadian variation in membrane potential, which was more depolarized during the midsubjective day compared with the midsubjective night. The average amplitude of the action potentials was also increased during the subjective night, which could be a result of the more hyperpolarized state of the neurons at night. In contrast to Jiang et al. (1997), de Jeu et al. (1998) found that the circadian fluctuation in input resistance was in phase and consistent with the circadian variation in membrane potential. Differences in the experimental protocol may explain this discrepancy, e.g., circadian membrane properties in the study by Jiang et al. (1997) may have been slightly different, as membrane properties can be altered after whole cell rupture (Schaap et al., 1999). However, these alterations may not have been significant because Jiang et al. (1997) recorded during the first minute after membrane rupture. Another possible complicating factor is that the two groups may have recorded from different neuronal populations. de Jeu and colleagues (1998) recorded mainly from cluster I neurons in coronal slices, whereas Jiang et al. (1997) focused on four different morphologically separated groups of neurons in horizontal slices. In both studies, circadian variation in membrane properties was not decreased or blunted by TTX or bicuculline ($GABA_A$ receptor antagonist; de Jeu et al., 1998; Jiang et al., 1997) or by CNQX and AP-5 (glutamate receptor antagonists; Jiang et al., 1997). These results suggest that the circadian variation of membrane events is not dependent on intercellular synaptic communication. Despite methodological differences between the studies, they both reached the conclusion that SCN pacemaker neurons express circadian modulation in membrane conductances, which most likely contributes to the circadian rhythm in action potential generation.

*Ionic Currents Involved in Rhythm Expression*

Fast $Na^+$ channels, voltage-gated $Ca^{2+}$ channels, $Ca^{2+}$-sensitive $K^+$ channels, a delayed rectifier, and a transient outward (A) $K^+$ current have all been suggested to be activated during the generation of action potentials in the SCN (Bouskila and Dudek, 1995; Jiang *et al.*, 1997; Thomson and West, 1990; Walsh *et al.*, 1995). In addition, Kim and Dudek (1993) identified, in contrast to Thomson and West (1990), an inward rectifier current that was activated by hyperpolarization. Both low- and high-threshold $Ca^{2+}$ currents have been identified in a portion of the population of SCN neurons, and low-threshold $Ca^{2+}$ currents have been implicated in the regulation of spontaneous firing (Akasu *et al.*, 1993; Cloues and Sather, 2003; Huang, 1993). Interspike depolarizing currents that bring the membrane back to threshold for firing action potentials may be under circadian control. For instance, SCN action potentials are preceded by a typical slow depolarizing ramp interval that is caused by a slowly inactivating $Na^+$ current (Pennartz *et al.*, 1997). Importantly, using perforated patch clamp recordings performed in SCN slices, Pennartz *et al.* (2002) demonstrated a circadian modulation of an L-type $Ca^{2+}$ current, which most likely contributes to robust spiking and spike afterhyperpolarization (AHP) in SCN neurons. A cesium ($Cs^+$)-sensitive hyperpolarization-activated cation current ($I_h$ or H) is also present in SCN neurons (Akasu *et al.*, 1993; de Jeu and Pennartz, 1997). Akasu *et al.* (1993) suggested that this current regulates SCN firing frequency by shortening the AHP duration and/or the spike prepotentials, which would result in shortened interspike intervals. This hypothesis has been challenged by de Jeu and Pennartz (1997), who demonstrated that $Cs^+$ does not affect the spiking frequency, resting membrane potential, or the shape of the AHP. In addition, because the H current was activated at membrane potentials more negative than $-55$ mV (Akasu *et al.*, 1993; de Jeu and Pennartz, 1997), the activated level of this current at rest would not reach more than approximately 2% and would not be activated at all in SCN neurons that are depolarized during the subjective day (de Jeu and Pennartz, 1997). This observation questions the contribution of the H current to rhythmic spiking. Furthermore, it was not possible to demonstrate a circadian rhythm in activation kinetics, voltage dependence (de Jeu and Pennartz, 1997), or amplitude (Jiang *et al.*, 1995, 1997) of the H current.

The possibility of circadian regulation at the level of the AHP has found additional support in that cluster I SCN neurons (Pennartz *et al.*, 1998) with high firing frequency have significantly shorter AHP compared with low-frequency firing neurons (Cloues and Sather, 2003). R- and L-type $Ca^{2+}$ channels appear to trigger the SCN AHP, and circadian regulation of expression of apamin and iberiotoxin–insensitive $K_{Ca}$ channels may regulate the AHP waveform (Cloues and Sather, 2003). Indeed, circadian expression of a $Ca^{2+}$-activated $K^+$ channel, *Kcnma1*, has been demonstrated in the SCN (Panda *et al.*, 2002). *Kcnma1* is the mammalian ortholog of the $K^+$ channel *Slowpoke (Slo)* in *Drosophila*. The protein of *Slo* is bound by the $Ca^{2+}$-binding protein Slob, which increases Slo activity and voltage sensitivity and may be under circadian control (Claridge-Chang *et al.*, 2001; McDonald and Rosbash, 2001; Schopperle *et al.*, 1998).

## Concluding Remarks

Neuronal membrane activity plays a central role in both synchronization and expression of circadian rhythms in pacemaker cells, whereas it is generally accepted that circadian rhythm generation originates from intracellular molecular feedback loops of cyclic genes and their protein products. With powerful new scientific techniques, the field of circadian biology is rapidly gaining improved experimental leverage that allows one to study interactions between molecular clockworks and neuronal membrane physiology. What is emerging is an appreciation that membrane events may play new and unexpected roles in circadian systems. Questions, such as the role of membrane conductances and ionic fluxes in rhythm generation, can now be adequately addressed. The results of such inquiry should provide important new insights into the rhythmic regulation of life's processes.

## References

Akasu, T., Shoji, S., and Hasuo, H. (1993). Inward rectifier and low-threshold calcium currents contribute to the spontaneous firing mechanism in neurons of the rat suprachiasmatic nucleus. *Pflug. Arch.* **425**, 109–116.

Amir, S., and Stewart, J. (1998). Induction of Fos expression in the circadian system by unsignaled light is attenuated as a result of previous experience with signaled light: A role for Pavlovian conditioning. *Neuroscience* **83**, 657–661.

Berson, D. M., Dunn, F. A., and Takao, M. (2002). Phototransduction by retinal ganglion cells that set the circadian clock. *Science* **295**, 1070–1073.

Block, G. D., and Davenport, P. A. (1982). Circadian rhythmicity in *Bulla gouldiana*: Role of the eyes in controlling locomotor behavior. *J. Exp. Zool.* **224**, 57–63.

Block, G. D., Khalsa, S. B., McMahon, D. G., Michel, S., and Guesz, M. (1993). Biological clocks in the retina: Cellular mechanisms of biological timekeeping. *Int. Rev. Cytol.* **146**, 83–144.

Block, G. D., and McMahon, D. G. (1984). Cellular analysis of the Bulla ocular circadian pacemaker system. III. Localization of the circadian pacemaker. *J. Comp. Physiol. A.* **155**, 387–395.

Block, G. D., and Wallace, S. F. (1982). Localization of a circadian pacemaker in the eye of a mollusc. *Bulla. Science* **217,** 155–157.

Bos, N. P., and Mirmiran, M. (1990). Circadian rhythms in spontaneous neuronal discharges of the cultured suprachiasmatic nucleus. *Brain Res.* **511,** 158–162.

Bos, N. P., and Mirmiran, M. (1993). Effects of excitatory and inhibitory amino acids on neuronal discharges in the cultured suprachiasmatic nucleus. *Brain Res. Bull.* **31,** 67–72.

Bouskila, Y., and Dudek, F. E. (1995). A rapidly activating type of outward rectifier K+ current and A-current in rat suprachiasmatic nucleus neurones. *J. Physiol.* **488**(Pt 2), 339–350.

Chen, D., Buchanan, G. F., Ding, J. M., Hannibal, J., and Gillette, M. U. (1999). Pituitary adenylyl cyclase-activating peptide: A pivotal modulator of glutamatergic regulation of the suprachiasmatic circadian clock. *Proc. Natl. Acad. Sci. USA* **96,** 13468–13473.

Claridge-Chang, A., Wijnen, H., Naef, F., Boothroyd, C., Rajewsky, N., and Young, M. W. (2001). Circadian regulation of gene expression systems in the *Drosophila* head. *Neuron* **32,** 657–671.

Cloues, R. K., and Sather, W. A. (2003). Afterhyperpolarization regulates firing rate in neurons of the suprachiasmatic nucleus. *J. Neurosci.* **23,** 1593–1604.

Colwell, C. S. (2000). Circadian modulation of calcium levels in cells in the suprachiasmatic nucleus. *Eur. J. Neurosci.* **12,** 571–576.

Colwell, C. S. (2001). NMDA-evoked calcium transients and currents in the suprachiasmatic nucleus: Gating by the circadian system. *Eur. J. Neurosci.* **13,** 1420–1428.

Colwell, C. S., Michel, S., Itri, J., Rodriguez, W., Tam, J., Lelievre, V., Hu, Z., Liu, X., and Waschek, J. A. (2003). Disrupted circadian rhythms in VIP and PHI deficient mice. *Am. J. Physiol. Regul. Integr. Comp. Physiol.* **285**(5), R939–R949.

Colwell, C. S., Ralph, M. R., and Menaker, M. (1990). Do NMDA receptors mediate the effects of light on circadian behavior? *Brain Res.* **523,** 117–120.

Cutler, D. J., Haraura, M., Reed, H. E., Shen, S., Sheward, W. J., Morrison, C. F., Marston, H. M., Harmar, A. J., and Piggins, H. D. (2003). The mouse VPAC2 receptor confers suprachiasmatic nuclei cellular rhythmicity and responsiveness to vasoactive intestinal polypeptide *in vitro. Eur. J. Neurosci.* **17,** 197–204.

de Jeu, M., Hermes, M., and Pennartz, C. (1998). Circadian modulation of membrane properties in slices of rat suprachiasmatic nucleus. *Neuroreport* **9,** 3725–3729.

de Jeu, M. T., and Pennartz, C. M. (1997). Functional characterization of the H-current in SCN neurons in subjective day and night: A whole-cell patch-clamp study in acutely prepared brain slices. *Brain Res.* **767,** 72–80.

Ding, J. M., Buchanan, G. F., Tischkau, S. A., Chen, D., Kuriashkina, L., Faiman, L. E., Alster, J. M., McPherson, P. S., Campbell, K. P., and Gillette, M. U. (1998). A neuronal ryanodine receptor mediates light-induced phase delays of the circadian clock. *Nature* **394,** 381–384.

Ding, J. M., Chen, D., Weber, E. T., Faiman, L. E., Rea, M. A., and Gillette, M. U. (1994). Resetting the biological clock: Mediation of nocturnal circadian shifts by glutamate and NO. *Science* **266,** 1713–1717.

Ding, J. M., Faiman, L. E., Hurst, W. J., Kuriashkina, L. R., and Gillette, M. U. (1997). Resetting the biological clock: Mediation of nocturnal CREB phosphorylation via light, glutamate, and nitric oxide. *J. Neurosci.* **17,** 667–675.

Dziema, H., Oatis, B., Butcher, G. Q., Yates, R., Hoyt, K. R., and Obrietan, K. (2003). The ERK/MAP kinase pathway couples light to immediate-early gene expression in the suprachiasmatic nucleus. *Eur. J. Neurosci.* **17,** 1617–1627.

Dziema, H., and Obrietan, K. (2002). PACAP potentiates L-type calcium channel conductance in suprachiasmatic nucleus neurons by activating the MAPK pathway. *J. Neurophysiol.* **88,** 1374–1386.

Earnest, D. J., Digiorgio, S. M., and Sladek, C. D. (1991). Effects of tetrodotoxin on the circadian pacemaker mechanism in suprachiasmatic explants *in vitro*. *Brain Res. Bull.* **26**, 677–682.

Ebling, F. J. (1996). The role of glutamate in the photic regulation of the suprachiasmatic nucleus. *Prog. Neurobiol.* **50**, 109–132.

Geusz, M. E., and Block, G. D. (1994). Intracellular calcium in the entrainment pathway of molluscan circadian pacemakers. *Neurosci. Biobehav. Rev.* **18**, 555–561.

Geusz, M. E., Foster, R. G., DeGrip, W. J., and Block, G. D. (1997). Opsin-like immunoreactivity in the circadian pacemaker neurons and photoreceptors of the eye of the opisthobranch mollusc *Bulla gouldiana*. *Cell Tissue Res.* **287**, 203–210.

Gillette, M. U., and Mitchell, J. W. (2002). Signaling in the suprachiasmatic nucleus: Selectively responsive and integrative. *Cell Tissue Res.* **309**, 99–107.

Gillette, M. U., and Prosser, R. A. (1988). Circadian rhythm of the rat suprachiasmatic brain slice is rapidly reset by daytime application of cAMP analogs. *Brain Res.* **474**, 348–352.

Ginty, D. D., Kornhauser, J. M., Thompson, M. A., Bading, H., Mayo, K. E., Takahashi, J. S., and Greenberg, M. E. (1993). Regulation of CREB phosphorylation in the suprachiasmatic nucleus by light and a circadian clock. *Science* **260**, 238–241.

Green, D. J., and Gillette, R. (1982). Circadian rhythm of firing rate recorded from single cells in the rat suprachiasmatic brain slice. *Brain Res.* **245**, 198–200.

Groos, G., and Hendriks, J. (1982). Circadian rhythms in electrical discharge of rat suprachiasmatic neurones recorded *in vitro*. *Neurosci. Lett.* **34**, 283–288.

Hannibal, J. (2002). Neurotransmitters of the retino-hypothalamic tract. *Cell Tissue Res.* **309**, 73–88.

Hannibal, J., Ding, J. M., Chen, D., Fahrenkrug, J., Larsen, P. J., Gillette, M. U., and Mikkelsen, J. D. (1997). Pituitary adenylate cyclase-activating peptide (PACAP) in the retinohypothalamic tract: A potential daytime regulator of the biological clock. *J. Neurosci.* **17**, 2637–2644.

Hannibal, J., Moller, M., Ottersen, O. P., and Fahrenkrug, J. (2000). PACAP and glutamate are costored in the retinohypothalamic tract. *J. Comp. Neurol.* **418**, 147–155.

Harmar, A. J., Marston, H. M., Shen, S., Spratt, C., West, K. M., Sheward, W. J., Morrison, C. F., Dorin, J. R., Piggins, H. D., Reubi, J. C., Kelly, J. S., Maywood, E. S., and Hastings, M. H. (2002). The VPAC(2) receptor is essential for circadian function in the mouse suprachiasmatic nuclei. *Cell* **109**, 497–508.

Harrington, M. E., Hoque, S., Hall, A., Golombek, D., and Biello, S. (1999). Pituitary adenylate cyclase activating peptide phase shifts circadian rhythms in a manner similar to light. *J. Neurosci.* **19**, 6637–6642.

Hastings, M. H., Duffield, G. E., Ebling, F. J., Kidd, A., Maywood, E. S., and Schurov, I. (1997). Nonphotic signalling in the suprachiasmatic nucleus. *Biol. Cell* **89**, 495–503.

Hastings, M. H., Reddy, A. B., and Maywood, E. S. (2003). A clockwork web: Circadian timing in brain and periphery, in health and disease. *Nature Rev. Neurosci.* **4**, 649–661.

Hattar, S., Liao, H. W., Takao, M., Berson, D. M., and Yau, K. W. (2002). Melanopsin-containing retinal ganglion cells: Architecture, projections, and intrinsic photosensitivity. *Science* **295**, 1065–1070.

Hendrickson, A. E., Wagoner, N., and Cowan, W. M. (1972). An autoradiographic and electron microscopic study of retino-hypothalamic connections. *Z. Zellforsch. Mikrosk. Anat.* **135**, 1–26.

Huang, R. C. (1993). Sodium and calcium currents in acutely dissociated neurons from rat suprachiasmatic nucleus. *J. Neurophysiol.* **70**, 1692–1703.

Hughes, A. T., Fahey, B., Cutler, D. J., Coogan, A. N., and Piggins, H. D. (2004). Aberrant gating of photic input to the suprachiasmatic circadian pacemaker of mice lacking the VPAC2 receptor. *J. Neurosci.* **24**, 3522–3526.

Hunt, A. E., Al-Ghoul, W. M., Gillette, M. U., and Dubocovich, M. L. (2001). Activation of MT(2) melatonin receptors in rat suprachiasmatic nucleus phase advances the circadian clock. *Am. J. Physiol. Cell. Physiol.* **280**, C110–C118.

Ikeda, M., Sugiyama, T., Wallace, C. S., Gompf, H. S., Yoshioka, T., Miyawaki, A., and Allen, C. N. (2003). Circadian dynamics of cytosolic and nuclear $Ca^{2+}$ in single suprachiasmatic nucleus neurons. *Neuron* **38**, 253–263.

Jacklet, J. W. (1969). Circadian rhythm of optic nerve impulses recorded in darkness from isolated eye of Aplysia. *Science* **164**, 562–563.

Jiang, Z. G., Nelson, C. S., and Allen, C. N. (1995). Melatonin activates an outward current and inhibits Ih in rat suprachiasmatic nucleus neurons. *Brain Res.* **687**, 125–132.

Jiang, Z. G., Yang, Y., Liu, Z. P., and Allen, C. N. (1997). Membrane properties and synaptic inputs of suprachiasmatic nucleus neurons in rat brain slices. *J. Physiol.* **499**(Pt 1), 141–159.

Jiao, Y. Y., and Rusak, B. (2003). Electrophysiology of optic nerve input to suprachiasmatic nucleus neurons in rats and degus. *Brain Res.* **960**, 142–151.

Khalsa, S. B., and Block, G. D. (1990). Calcium in phase control of the Bulla circadian pacemaker. *Brain. Res.* **506**, 40–45.

Kim, Y. I., and Dudek, F. E. (1991). Intracellular electrophysiological study of suprachiasmatic nucleus neurons in rodents: Excitatory synaptic mechanisms. *J. Physiol.* **444**, 269–287.

Kim, Y. I., and Dudek, F. E. (1993). Membrane properties of rat suprachiasmatic nucleus neurons receiving optic nerve input. *J. Physiol.* **464**, 229–243.

Kopp, M. D., Meissl, H., Dehghani, F., and Korf, H. W. (2001). The pituitary adenylate cyclase-activating polypeptide modulates glutamatergic calcium signalling: Investigations on rat suprachiasmatic nucleus neurons. *J. Neurochem.* **79**, 161–171.

Kuhlman, S. J., Silver, R., Le Sauter, J., Bult-Ito, A., and McMahon, D. G. (2003). Phase resetting light pulses induce Per1 and persistent spike activity in a subpopulation of biological clock neurons. *J. Neurosci.* **23**, 1441–1450.

Liu, C., and Gillette, M. U. (1996). Cholinergic regulation of the suprachiasmatic nucleus circadian rhythm via a muscarinic mechanism at night. *J. Neurosci.* **16**, 744–751.

Lundkvist, G. B., Kristensson, K., and Hill, R. H. (2002). The suprachiasmatic nucleus exhibits diurnal variations in spontaneous excitatory postsynaptic activity. *J. Biol. Rhythms* **17**, 40–51.

Lundkvist, G. B., Kwak, Y., and Block, G. D. (2003). A transmembrane calcium flux is an essential process for circadian rhythm generation in mammalian SCN neurons. Abstract Viewer/Itinerary planner Washington, DC: Society for Neuroscience Program No. 16.5.

McArthur, A. J., Gillette, M. U., and Prosser, R. A. (1991). Melatonin directly resets the rat suprachiasmatic circadian clock *in vitro*. *Brain Res.* **565**, 158–161.

McArthur, A. J., Hunt, A. E., and Gillette, M. U. (1997). Melatonin action and signal transduction in the rat suprachiasmatic circadian clock: Activation of protein kinase C at dusk and dawn. *Endocrinology* **138**, 627–634.

McDonald, M. J., and Rosbash, M. (2001). Microarray analysis and organization of circadian gene expression in *Drosophila*. *Cell* **107**, 567–578.

McMahon, D. G. (1986). Cellular Mechanisms of Circadian Pacemaker Entrainment in the Mollusc *Bulla*. Dissertation thesis, University of Virginia, Charlottesville, VA.

McMahon, D. G., and Block, G. D. (1987). The Bulla ocular circadian pacemaker. I. Pacemaker neuron membrane potential controls phase through a calcium-dependent mechanism. *J. Comp. Physiol. A.* **161**, 335–346.

McMahon, D. G., Wallace, S. F., and Block, G. D. (1984). Cellular analysis of the Bulla ocular circadian pacemaker system. II. Neurophysiological basis of circadian rhythmicity. *J. Comp. Physiol. A.* **155,** 379–385.

Meijer, J. H., Albus, H., Weidema, F., and Ravesloot, J. H. (1993). The effects of glutamate on membrane potential and discharge rate of suprachiasmatic neurons. *Brain Res.* **603,** 284–288.

Meijer, J. H., and Schwartz, W. J. (2003). In search of the pathways for light-induced pacemaker resetting in the suprachiasmatic nucleus. *J. Biol. Rhythms* **18,** 235–249.

Michel, S., Geusz, M. E., Zaritsky, J. J., and Block, G. D. (1993). Circadian rhythm in membrane conductance expressed in isolated neurons. *Science* **259,** 239–241.

Michel, S., Itri, J., and Colwell, C. S. (2002). Excitatory mechanisms in the suprachiasmatic nucleus: The role of AMPA/KA glutamate receptors. *J. Neurophysiol.* **88,** 817–828.

Michel, S., Manivannan, K., Zaritsky, J. J., and Block, G. D. (1999). A delayed rectifier current is modulated by the circadian pacemaker in Bulla. *J. Biol. Rhythms* **14,** 141–150.

Mintz, E. M., and Albers, H. E. (1997). Microinjection of NMDA into the SCN region mimics the phase shifting effect of light in hamsters. *Brain Res.* **758,** 245–249.

Mintz, E. M., Marvel, C. L., Gillespie, C. F., Price, K. M., and Albers, H. E. (1999). Activation of NMDA receptors in the suprachiasmatic nucleus produces light-like phase shifts of the circadian clock *in vivo. J. Neurosci.* **19,** 5124–5130.

Moore, R. Y., and Lenn, N. J. (1972). A retinohypothalamic projection in the rat. *J. Comp. Neurol.* **146,** 1–14.

Nitabach, M. N., Blau, J., and Holmes, T. C. (2002). Electrical silencing of *Drosophila* pacemaker neurons stops the free-running circadian clock. *Cell* **109,** 485–495.

Nitabach, M. N., Blau, J., and Holmes, T. C. (2003). Pacemaker membrane excitability controls the period and coherence of the free running *Drosophila* circadian oscillator. Abstract Viewer/Itinerary planner Washington, DC: Society for Neuroscience Program No. 284.11.

Nomura, K., Takeuchi, Y., Yamaguchi, S., Okamura, H., and Fukunaga, K. (2003). Involvement of calcium/calmodulin-dependent protein kinase II in the induction of mPer1. *J. Neurosci. Res.* **72,** 384–392.

Obrietan, K., Impey, S., Smith, D., Athos, J., and Storm, D. R. (1999). Circadian regulation of cAMP response element-mediated gene expression in the suprachiasmatic nuclei. *J. Biol. Chem.* **274,** 17748–17756.

Panda, S., Antoch, M. P., Miller, B. H., Su, A. I., Schook, A. B., Straume, M., Schultz, P. G., Kay, S. A., Takahashi, J. S., and Hogenesch, J. B. (2002). Coordinated transcription of key pathways in the mouse by the circadian clock. *Cell* **109,** 307–320.

Pennartz, C. M., Bierlaagh, M. A., and Geurtsen, A. M. (1997). Cellular mechanisms underlying spontaneous firing in rat suprachiasmatic nucleus: Involvement of a slowly inactivating component of sodium current. *J. Neurophysiol.* **78,** 1811–1825.

Pennartz, C. M., de Jeu, M. T., Bos, N. P., Schaap, J., and Geurtsen, A. M. (2002). Diurnal modulation of pacemaker potentials and calcium current in the mammalian circadian clock. *Nature* **416,** 286–290.

Pennartz, C. M., de Jeu, M. T., Geurtsen, A. M., Sluiter, A. A., and Hermes, M. L. (1998). Electrophysiological and morphological heterogeneity of neurons in slices of rat suprachiasmatic nucleus. *J. Physiol.* **506,** 775–793.

Pennartz, C. M., Hamstra, R., and Geurtsen, A. M. (2001). Enhanced NMDA receptor activity in retinal inputs to the rat suprachiasmatic nucleus during the subjective night. *J. Physiol.* **532,** 181–194.

Pittendrigh, C. S., and Daan, S. (1976). A functional analysis of circadian pacemakers in nocturnal rodents. IV. Entrainment: Pacemaker as clock. *J. Comp. Physiol.* **106,** 291–331.

Prosser, R. A., and Gillette, M. U. (1989). The mammalian circadian clock in the suprachiasmatic nuclei is reset *in vitro* by cAMP. *J. Neurosci.* **9,** 1073–1081.

Rea, M. A. (1992). Different populations of cells in the suprachiasmatic nuclei express c-fos in association with light-induced phase delays and advances of the free-running activity rhythm in hamsters. *Brain Res.* **579,** 107–112.

Schaap, J., Bos, N. P., de Jeu, M. T., Geurtsen, A. M., Meijer, J. H., and Pennartz, C. M. (1999). Neurons of the rat suprachiasmatic nucleus show a circadian rhythm in membrane properties that is lost during prolonged whole-cell recording. *Brain Res.* **815,** 154–166.

Schopperle, W. M., Holmqvist, M. H., Zhou, Y., Wang, J., Wang, Z., Griffith, L. C., Keselman, I., Kusinitz, F., Dagan, D., and Levitan, I. B. (1998). Slob, a novel protein that interacts with the Slowpoke calcium-dependent potassium channel. *Neuron* **20,** 565–573.

Schwartz, W. J., Gross, R. A., and Morton, M. T. (1987). The suprachiasmatic nuclei contain a tetrodotoxin-resistant circadian pacemaker. *Proc. Natl. Acad. Sci. USA* **84,** 1694–1698.

Schwartz, W. J., Takeuchi, J., Shannon, W., Davis, E. M., and Aronin, N. (1994). Temporal regulation of light-induced Fos and Fos-like protein expression in the ventrolateral subdivision of the rat suprachiasmatic nucleus. *Neuroscience* **58,** 573–583.

Shibata, S., and Moore, R. Y. (1987). Development of neuronal activity in the rat suprachiasmatic nucleus. *Brain Res.* **431,** 311–315.

Shibata, S., and Moore, R. Y. (1993a). Tetrodotoxin does not affect circadian rhythms in neuronal activity and metabolism in rodent suprachiasmatic nucleus *in vitro*. *Brain Res.* **606,** 259–266.

Shibata, S., and Moore, R. Y. (1993b). Neuropeptide Y and optic chiasm stimulation affect suprachiasmatic nucleus circadian function *in vitro*. *Brain Res.* **615,** 95–100.

Shibata, S., Newman, G. C., and Moore, R. Y. (1987). Effects of calcium ions on glucose utilization in the rat suprachiasmatic nucleus *in vitro*. *Brain Res.* **426,** 332–338.

Shibata, S., Oomura, Y., Kita, H., and Hattori, K. (1982). Circadian rhythmic changes of neuronal activity in the suprachiasmatic nucleus of the rat hypothalamic slice. *Brain Res.* **247,** 154–158.

Shinohara, K., Honma, S., Katsuno, Y., and Honma, K. (2000). Circadian release of excitatory amino acids in the suprachiasmatic nucleus culture is $Ca(2+)$-independent. *Neurosci. Res.* **36,** 245–250.

Stehle, J., Vanecek, J., and Vollrath, L. (1989). Effects of melatonin on spontaneous electrical activity of neurons in rat suprachiasmatic nuclei: An *in vitro* iontophoretic study. *J. Neural. Trans.* **78,** 173–177.

Thomson, A. M., and West, D. C. (1990). Factors affecting slow regular firing in the suprachiasmatic nucleus *in vitro*. *J. Biol. Rhythms* **5,** 59–75.

Tischkau, S. A., Mitchell, J. W., Tyan, S. H., Buchanan, G. F., and Gillette, M. U. (2003a). Ca2+/cAMP response element-binding protein (CREB)-dependent activation of Per1 is required for light-induced signaling in the suprachiasmatic nucleus circadian clock. *J. Biol. Chem.* **278,** 718–723.

Tischkau, S. A., Weber, E. T., Abbott, S. M., Mitchell, J. W., and Gillette, M. U. (2003b). Circadian clock-controlled regulation of cGMP-protein kinase G in the nocturnal domain. *J. Neurosci.* **23,** 7543–7550.

Tominaga, K., Geusz, M. E., Michel, S., and Inouye, S. T. (1994). Calcium imaging in organotypic cultures of the rat suprachiasmatic nucleus. *Neuroreport* **5,** 1901–1905.

Travnickova-Bendova, Z., Cermakian, N., Reppert, S. M., and Sassone-Corsi, P. (2002). Bimodal regulation of mPeriod promoters by CREB-dependent signaling and CLOCK/BMAL1 activity. *Proc. Natl. Acad. Sci. USA* **99,** 7728–7733.

van den Pol, A. N. (1980). The hypothalamic suprachiasmatic nucleus of rat: Intrinsic anatomy. *J. Comp. Neurol.* **191,** 661–702.

van den Pol, A. N., Finkbeiner, S. M., and Cornell-Bell, A. H. (1992). Calcium excitability and oscillations in suprachiasmatic nucleus neurons and glia *in vitro*. *J. Neurosci.* **12,** 2648–2664.

Walsh, I. B., van den Berg, R. J., and Rietveld, W. J. (1995). Ionic currents in cultured rat suprachiasmatic neurons. *Neuroscience* **69,** 915–929.

Warren, E. J., Allen, C. N., Brown, R. L., and Robinson, D. W. (2003). Intrinsic light responses of retinal ganglion cells projecting to the circadian system. *Eur. J. Neurosci.* **17,** 1727–1735.

Welsh, D. K., Logothetis, D. E., Meister, M., and Reppert, S. M. (1995). Individual neurons dissociated from rat suprachiasmatic nucleus express independently phased circadian firing rhythms. *Neuron* **14,** 697–706.

Yamaguchi, S., Isejima, H., Matsuo, T., Okura, R., Yagita, K., Kobayashi, M., and Okamura, H. (2003). Synchronization of cellular clocks in the suprachiasmatic nucleus. *Science* **302,** 1408–1412.

# Section VIII

# Intercellular Signaling

# [34]   A Screen for Secreted Factors of the Suprachiasmatic Nucleus

*By* ACHIM KRAMER,[1] FU-CHIA YANG,[1] SEBASTIAN KRAVES, and CHARLES J. WEITZ

## Abstract

The circadian clock in the suprachiasmatic nucleus (SCN) drives daily locomotor activity rhythms presumably by secreting diffusible factors whose target sites are accessible from the third ventricle of the hypothalamus. This article describes the methodology of a systematic molecular and behavioral screen to identify "locomotor factors" of the SCN. To find SCN-secreted factors not previously documented, a hamster SCN cDNA library was screened in a yeast signal sequence trap. In a subsequent behavioral screen, newly identified and previously documented SCN factors were tested for an effect on locomotor activity rhythms by chronic infusion into the third ventricle of hamsters. Using this approach combined with further experiments, we identified transforming growth factor-$\alpha$ (TGF-$\alpha$) as a likely SCN inhibitor of locomotion.

## Introduction

The mammalian circadian clock residing in the suprachiasmatic nucleus (SCN) is thought to drive circadian rhythms of locomotor behavior by secreting diffusible factors that act locally within the hypothalamus. This concept is primarily derived from SCN transplant experiments. Grafting fetal SCN tissue into animals made arrhythmic by lesioning the SCN restores circadian rhythms of locomotor activity with the period of the donor tissue (Ralph *et al.*, 1990). This occurs even when the graft is encapsulated to prevent extension of axons but allow diffusion of secreted factors (Silver *et al.*, 1996). These results demonstrate that in the transplanted animal the SCN secretes "locomotor factors" that reach their targets by diffusion in a paracrine fashion. However, this does not exclude the possibility that the locomotor factors in the intact animal are secreted synaptically or that both paracrine and synaptic transmission are involved.

Although the identity of the locomotor factors was totally obscure at the time this project was initiated, there had been some clues about their function. In a study done by Vogelbaum and Menaker (1992), hamsters

[1]Achim Kramer and Fu-Chia Yang contributed equally to this chapter.

METHODS IN ENZYMOLOGY, VOL. 393

with functional SCN tissue of both wild-type and short-period mutant (*tau*) genotypes ("temporal chimeras") displayed locomotor activity rhythms influenced from the oscillators of both genotypes. While there was no evidence of coupling between the two underlying oscillators, locomotor activity was suppressed at times when the $\rho$ band (rest period) of one influence intersected with the $\alpha$ band (activity period) of the other, indicating that the SCN inhibits locomotor activity at one phase and probably promotes it at another, with inhibition dominating when the two influences coincided (Vogelbaum and Menaker, 1992).

The target sites of the locomotor factors had not yet been identified at the time this project was initiated. However, a general prediction of their location could be made based on the attachment site of functional SCN transplants, which restored locomotor rhythms most effectively when located in the vicinity of the SCN and exposed to the lumen of the third ventricle. In all likelihood, signals from the transplants could diffuse into the cerebrospinal fluid (CSF) of the ventricle and reach target sites accessible from within that cavity. Again, in the intact animal the locomotor factors might be released synaptically. Whether paracrine or synaptic, a candidate target site for receptors of secreted SCN locomotor factors is the subparaventricular zone (SPZ), a hypothalamic region between the SCN and the paraventricular nucleus (PVN), which receives major projections from the SCN. Although the function of this region is little understood, lesions of the SPZ disrupt circadian regulation of sleep–wake cycles, body temperature, and movement around the cage (Lu *et al.*, 2001).

The results and conclusions from the SCN lesion, transplant, and temporal chimera experiments can be integrated into a model by which the SCN regulates locomotor rhythms. This model consists of the SCN secreting locomotor factors by diffusion through the third ventricle, diffusion across synaptic terminals of SCN efferents, or both. The targets of these factors must be accessible from the third ventricle and are most likely near the SCN. The locomotor factors themselves include at least one inhibitor of activity, and probably at least one activator of activity, with the inhibitor being dominant over the activator. These inhibitors and activators could be secreted at different times of the circadian cycle to generate the observed locomotor rhythm. At the time that this project was initiated, the nature of the factors was unknown. The ultimate goal of this project was to identify these locomotor factors. Using the approach described here, we identified TGF-$\alpha$ as a likely SCN inhibitor of locomotion.

### General Strategy

Although the nature of the locomotor factors was unknown, what was known was that the SCN contains a paucity of amine and an abundance of peptide neurotransmitters (Klein *et al.*, 1991; Stephan and Zucker, 1972; Takahashi *et al.*, 2001). Among the classical amine neurotransmitters, only the inhibitory transmitter GABA is present in the SCN in a widespread manner. In contrast, nearly 30 peptides and growth factors have been identified in the SCN. Given this neurochemical composition of the SCN, it is reasonable to assume that at least one of the locomotor factors is a peptide or protein. Based on this assumption and because of technical considerations, we limited our search to peptides and proteins.

The first goal was to identify as many peptides secreted by the SCN as possible. Many peptides had already been previously documented in the SCN. We wanted to expand this list by performing a yeast signal sequence trap designed to identify peptides that had not been previously associated with the SCN as well as novel peptides from the SCN (see later). The strategy employed to find potential locomotor factors is summarized in Fig. 1. From a survey of the literature and the results from the yeast signal sequence trap, we generated a comprehensive list of secreted peptides and proteins likely to be made in the SCN.

To test whether these peptides and proteins had an effect on locomotor activity, we used a behavioral screen that involved the chronic infusion of the candidate factors into the third ventricle of hamsters (see later). Their wheel-running activity was recorded under circadian conditions (constant darkness) as we screened for disturbances in the circadian regulation of locomotor activity. We had a specific set of predictions for alterations due to the constitutive infusion of an activator or an inhibitor of locomotor activity (see later).

To determine whether a pharmacological effect in the behavioral screen is of any physiological relevance, follow-up studies (i) should confirm of the expression of positive factors in the SCN; (ii) might provide evidence for a circadian expression profile with a phase consistent with a function as an activating or inhibiting factor; (iii) should try to identify receptors at the predicted locations; and (iv) should aim for loss-of-function studies ultimately proving the involvement of the identified factor in regulating circadian activity rhythms.

## Signal Sequence Trap

### Principles of Yeast Signal Sequence Trap

The yeast signal sequence trap method was developed independently by Klein *et al.* (1996) and Jacobs *et al.* (1997). This method is based on the fact that proteins imported to the endoplasmic reticulum, including those destined for secretion, possess a characteristic signal peptide (or signal sequence) at their amino terminus (Kaiser and Botstein, 1986; von Heijne,

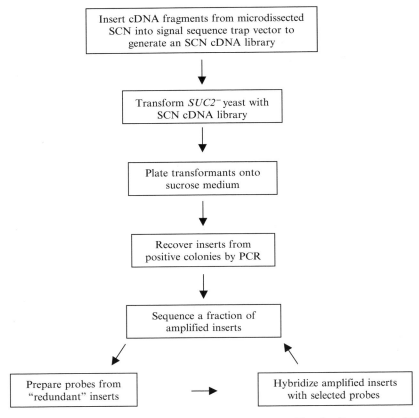

Fɪɢ. 1. Strategy for performing the signal sequence trap. For details, see text. SCN, suprachiasmatic nucleus; PCR, polymerase chain reaction.

1985). Although signal sequences consist of three critical domains—a positively charged N-terminal region, a central hydrophobic region, and a more polar C-terminal region that defines the cleavage site—there is a high level of degeneracy within any one of those elements. Therefore it is difficult to search for such sequences using bioinformatics tools.

In a functional assay, however, peptides or proteins containing a signal peptide are selected. Yeast was chosen as the model organism for the signal sequence trap because it is both easy to manipulate and its eukaryotic nature allows its secretion machinery to recognize mammalian signal peptides. This was indeed shown to be the case in the first applications of the signal sequence trap, which were able to detect secreted peptides from mammalian embryonic brain tissue (Klein *et al.*, 1996) and peripheral blood mononuclear cells (Jacobs *et al.*, 1997).

For the selection scheme of the signal sequence trap, complementation of an invertase deficiency was chosen. Invertase is an enzyme encoded by the *SUC2* gene and is necessary for yeast to grow when sucrose is the only available carbon source (Carlson *et al.*, 1983). Yeast cells secrete invertase, which breaks down sucrose into glucose and fructose monomers that can be readily metabolized in the cells. Yeast containing invertase with mutations or deletions in its signal sequence are unable to process sucrose because the invertase remains in the cytoplasm (Kaiser and Botstein, 1986). The host yeast strain that was used is missing its invertase gene and thus cannot grow on sucrose medium. A yeast expression plasmid containing a mutant invertase gene lacking its initiator methionine and signal sequence was then constructed. Transformation of this vector alone into the $SUC2^-$ yeast cannot confer growth on sucrose, as invertase can neither be made nor secreted. A library of cDNAs (in our case, derived from the SCN) was then fused in front of the mutated invertase gene. Those transformants that expressed a cDNA clone containing an initiator methionine and a signal sequence in frame with the mutated invertase gene would acquire the ability to translate an invertase fusion protein, which could then be transported through the secretory pathway. The subsequent secretion of invertase would sustain growth on sucrose. The resulting yeast colonies could then be lysed, and the cDNAs that were fused to the invertase gene amplified by polymerase chain reaction (PCR) (using universal primers from the library vector) and sequenced, thus revealing the identity of clones containing putative signal sequences.

It is known that random stretches of hydrophobic amino acids are capable of acting as signal peptides when fused to invertase (Kaiser *et al.*, 1987). By requiring an initiator methionine from the cDNA clone rather than using the native invertase methionine, we sought to increase the probability of retrieving true signal sequences (Klein *et al.*, 1996). Positives should arise only from cDNA clones encoding true signal peptides, which always lie at the amino terminus of nascent proteins near the initiator methionine; internal hydrophobic stretches of proteins should not be able to initiate translation. Additionally, we biased the library to relatively small (200–1000 bp) cDNA fragments to minimize false negatives arising from the inclusion of stop codons, which would prevent the synthesis of an invertase fusion protein. Finally, to ensure that the invertase is not hindered by the fused protein, a Kex2 cleavage site, which is recognized by native proteases in the endoplasmic reticulum (ER) of yeast, was built into the library vector so that the cDNA–library-derived peptide could be separated from the invertase protein just before secretion.

One advantage of this yeast signal sequence trap is that it is a selection rather than a screen. Thus, millions of clones can be transformed and plated, and only those yeast colonies that grow on a selective medium need to be analyzed. Another advantage is the sensitivity of the invertase system. As little as 0.6% of the wild-type invertase activity is sufficient to allow growth on sucrose medium (Kaiser *et al.*, 1987). This sensitivity theoretically permits the detection of mammalian signal peptides of suboptimal function in yeast. These advantages make the signal sequence trap suitable as a high-throughput method to try to identify a more complete set of factors secreted by the SCN.

## Construction of Signal Sequence Trap Library from SCN Tissue

*Library Vector Construction.* The library vector, pSUC2_dMSP, is illustrated in Fig. 2. It is composed of three parts. The first part is derived from pVP16 (gift from S. Hollenberg) and contains the elements necessary for replication in bacteria and in yeast. The sequence corresponding to VP16 was removed between the *Eco*RI and *Not*I sites, and the *Eco*RI site was filled in with Klenow fragment (New England BioLabs). The second part is *SUC2_dMSP*, which encodes a modified *SUC2* gene lacking the starter methionine and signal sequence. To prepare *SUC2_dMSP*, pRB58 (gift from F. Winston) was digested with *Hind*III and *Xmn*I; the resulting fragment was partially digested with *Fok*I. This generated a *SUC2* fragment lacking the starter methionine and signal sequence, and flanked by *Fok*I and *Xmn*I sites. The third part is a synthetic 102-bp linker consisting of the following sequence: 5'-ggccgcaaaagagccaaatcctccttccctggagc-aacaaacgaaactagcgatagacctttggtcca cttcacacccaacaagggctggatgaatgaccca-3'. This linker includes a sequence coding for a recognition site for the yeast protease Kex2 (Rockwell *et al.*, 2002). The ends of the linker contain *Not*I and *Fok*I overhangs. The three parts—the modified pVP16 vector, the *SUC2_dMSP* fragment, and the synthetic linker—were ligated to each other. The recircularized plasmid was cut with *Xho*I and *Not*I, and a stuffer fragment consisting of an *Eco*RI site flanked by *Xho*I and *Not*I overhangs was inserted into it.

*Library Construction.* Sixty adult male Syrian hamsters (Charles River Laboratories) are maintained on a 14:10 LD cycle for 3 weeks and then transferred to constant dim light (<1 lux), which approximates darkness, at the time of lights off. Twenty-four hours later, groups of 10 hamsters are sacrificed by decapitation every 4 h during a 24-h cycle. This circadian collection scheme is undertaken to ensure that any transcripts with a circadian oscillation would be represented in the library. Brains are removed and placed in phosphate-buffered saline (4°) for 30 s, a 1.5-mm

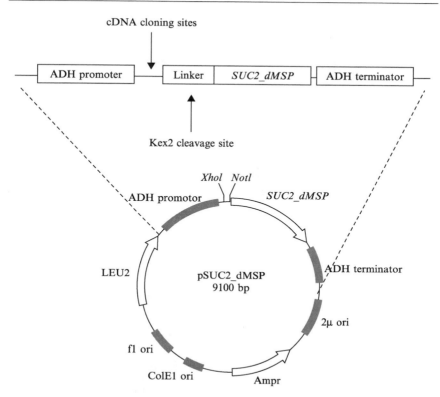

cDNA cloning sites

ADH promoter | Linker | *SUC2_dMSP* | ADH terminator

Kex2 cleavage site

*XhoI  NotI*

ADH promotor

*SUC2_dMSP*

LEU2

pSUC2_dMSP
9100 bp

ADH terminator

2μ ori

f1 ori

ColE1 ori     Ampr

FIG. 2. The yeast signal sequence trap vector, pSUC2_dMSP. The vector is a yeast expression plasmid containing a modified invertase gene lacking its starting methionine and signal sequence (*SUC2_dMSP*). A cDNA library is cloned into the *XhoI* and *NotI* sites. When a cDNA fragment containing an initiator methionine and signal sequence is cloned in frame with *SUC2_dMSP*, an invertase fusion protein is made that can then be transported through the secretory pathway. A Kex2 cleavage site was built into a linker between the inserted cDNA and *SUC2_dMSP* so that the cDNA–library-derived peptide could be separated from the invertase protein just before secretion.

coronal slice is cut, and a pyramid of anterior hypothalamus containing both SCN is dissected out, frozen on dry ice, and stored at −70°. From histological examination, it was estimated that the core of tissue is about 30–50% SCN by volume and included virtually all SCN cells. Total RNA is extracted using RNA STAT-60 reagent (Tel-Test), time-specific RNA samples are pooled, and poly(A)$^+$ RNA is selected by one round of oligo(dT) chromatography (Oligotex, Qiagen). Three micrograms of poly(A)$^+$ RNA is reverse transcribed to ultimately obtain double-stranded cDNA using the Superscript Choice System (Invitrogen) and the random primer 5′-gactagttctagatcgcgagcggccgcNNNNNNNNN-3′, which contains a *Not*I site. The cDNAs are ligated to an adapter with a *Sal*I overhang, digested with *Not*I, and size selected by agarose gel electrophoresis. Fragments between 200 and 1000 bp are ligated to *Xho*I–*Not*I-digested pSUC2_dMSP. The resulting ligation products are electoporated into *Escherichia coli* strain DH10B (Invitrogen), $2.5 \times 10^7$ transformants plated at an average of $10^6$ colonies per 15-cm dish, and plasmid DNA prepared from the bacterial colonies (Qiagen plasmid mega kit). Analysis of 24 plasmids generated from the ligation of the SCN library to the library vector indicates that 23 of 24 clones contain an excisable insert, and 22 of the inserts are between 200 and 1000 bp. Care of hamsters and all procedures are in full compliance with institutional guidelines for animal experimentation.

### Performing Signal Sequence Trap

*Yeast Transformation and Growth.* Ten micrograms of plasmid library is introduced into the yeast strain 0662 (*MATα, suc2Δ9, ura3-52, leu2-3, leu2-112, his4-519*) (gift from F. Winston) by lithium acetate transformation. Transformed yeast are plated onto complete minimal medium lacking leucine and containing 2% sugar mixture and 0.1% antimycin A. The sugar mixture consists of glucose and sucrose where the percentage of glucose ranges from 0.5 to 20% of the total sugar content. Various glucose concentrations allow the selection of different populations of positive clones presumably as a function of the effectiveness of their signal sequence in yeast. Pilot experiments reveal that the glucose might help the yeast to recover faster from the transformation procedure and thus compensate for a slightly lower efficiency of some mammalian signal peptides to function in yeast.

Yeast are allowed to grow for 4–8 days at 30°. Individual colonies are picked and streaked onto complete minimal medium lacking leucine and containing 2% sucrose and 0.1% antimycin A. This second round of selection at a higher stringency is essential to distinguish true positives from satellite colonies, which often appear due to the glucose-rich environment formed around the positive colonies. After 3 days of growth on sucrose medium, a single colony from each streak is used for lysis and analysis of the library insert. In most cases, the same colony is also grown for 3 days in complete minimal medium lacking leucine and containing 2% glucose and 15% glycerol. The liquid cultures are archived at −80°.

A total of $1.38 \times 10^8$ clones are plated (representing a 6.7-fold coverage of the SCN library), and 10,725 colonies, which grew on the selective medium, are picked for analysis.

*Recovery of Positive Clones.* Leu$^+$ transformants that grew on sucrose are lysed in 2.6 $\mu$l Lyse-N-Go$^T$ buffer (Pierce) at 95° for 5 min. Two and a half microliters is used in a 25-$\mu$l PCR to amplify the library insert (Qiagen *Taq* PCR core kit). The primers, which correspond to sequences in the library vector flanking the library insert, are 5'-gcacaatatttcaagctataccaag-ca-3' and 5'-gtgtgaagtggaccaaaggtctatc-3'. Following an initial denaturing step (94° for 5 min), amplification is carried out through 35 cycles at 94° for 1 min, 65° for 1 min, and 72° for 1 min. A random sample of PCR products is resolved in 2% agarose gels to evaluate the size and number of products. Among the ones tested, approximately 85% exhibited a single band between 200 and 1000 bp, 11% displayed multiple bands (probably due to contamination from other colonies), and 4% contained no detectable insert. PCR products are prepared for sequencing by removing excess dNTPs and unincorporated primers with exonuclease I and shrimp alkaline phosphatase (USB) and incubating the reaction at 37° for 30 min and 80° for 15 min. Sequencing is performed in a high-throughput scale using the sequencing primer 5'-atacaatcaactccaagc-3'.

*High-Throughput Colony DNA Hybridization Analysis.* To avoid sequencing redundant inserts, genes that are represented at least twice in the initial rounds of sequencing are used as probes in a high-throughput DNA hybridization analysis of the PCR products remaining to be sequenced. Probes are prepared by digesting the redundant PCR products with *MluI* and *NotI* to remove flanking sequences arising from the library vector. The resulting fragments are purified (Qiagen Qiaquick PCR purification kit) and labeled with $[\alpha$-$^{32}$P]dCTP (Stratagene Prime-It RmT random primer labeling kit). PCR products from the signal sequence trap are spotted onto Pall Biodyne B nylon membranes in arrays of 384 or 864 using the Nunc replication system. The spotted products are denatured, neutralized, and cross-linked to the membrane and then hybridized to pools of labeled probes at 65° overnight. The membranes are washed twice in 0.1X SSC, 0.1% SDS for 10 min at 65°, dried, and exposed to a phosphorimaging screen. Only PCR products that do not hybridize to any probes are then sequenced.

At the end of one round of sequencing, the redundancy of the sequences is reassessed, and new probes are designed against additional clones that are found to be represented more than two times. These probes are then included in the DNA hybridization analysis of positive clones not yet sequenced. This reiterative cross-hybridization strategy reduces the number of clones to be sequenced by 70%.

*Functional Classification of Identified Clones*

To analyze the sequences, we used National Center for Biotechnology Information BLAST search algorithms against the nonredundant database. Sequences that were homologous to genes encoding proteins of known function were then categorized according to subcellular localization, using the Swiss-Prot database as a reference. Sequences that were not homologous to genes coding for proteins of known function were classified as novel.

A total of 857 unique sequences homologous to previously characterized genes were identified (Table I). These included not only secreted factors, but also transmembrane proteins, membrane-associated proteins, and proteins targeted to intracellular compartments such as the endoplasmic reticulum, the Golgi apparatus, the lysosomes, and the endosomes.

We detected 93 mammalian genes, which encode a total of 105 peptides and proteins known to be secreted factors. This list includes 15 factors (derived from nine precursors), which had already been documented in the SCN. The majority of genes in our list ($\sim$65%) encode products with known signaling functions, including hormones, growth factors, cytokines, and extracellular matrix proteins; the rest represent secreted enzymes and carrier proteins. A significant fraction of the factors identified had been previously reported in the brain, while a yet larger number had been primarily associated with nonbrain tissues.

*False-Positive and False-Negative Clones.* It has been shown that random sequences encoding a stretch of hydrophobic residues can functionally replace the signal sequence (Kaiser *et al.*, 1987). The frequency at which we detected such clones can be estimated by the number of clones found in the signal sequence trap that encode proteins documented to be nuclear or cytoplasmic. This rate was approximately 39% and higher than false-positive rates achieved in other published signal sequence traps, which

TABLE I
SUMMARY OF cDNA CLONES ISOLATED FROM THE SCN SIGNAL SEQUENCE TRAP

| | |
|---|---|
| Matches to known proteins (unique sequences) | 857 |
| Secreted/extracellular | 93 (11%)[a] |
| Intracellular compartments | 124 (14%) |
| Transmembrane | 188 (22%) |
| Membrane associated | 77 (9%) |
| Nuclear and cytoplasmic | 332 (39%) |
| Unknown localization | 43 (5%) |
| Novel sequences | 1107[b] |

[a] Percentages refer to the number of matches to known proteins.

[b] The redundancy of the novel sequences has only partly been assessed.

ranged from 17 to 25% (Jacobs *et al.*, 1997, 1999; Taft *et al.*, 2002). The difference between our estimate and the published estimates, however, can be accounted for by our reiterative removal of redundant clones, which were mostly true positives, from each round of signal sequence trapping. Accordingly, our rate of false positives increased with each round; thus, our false-positive rate is probably reflected more accurately in the number of false positives that were detected in the initial rounds of screening, which was approximately 20% of the clones matching known sequences.

The rate of false negatives, i.e., those precursors reported to be in the SCN, but which we did not detect in the signal sequence trap, is high in that we found only 9 out of 18 known SCN precursors. There are three possible reasons for this high rate of false negatives: (1) Those transcripts we failed to detect are not present in our library. This, however, does not appear to be the case as most of them could be amplified from the signal sequence trap library by PCR using specific primers. (2) We have not performed enough signal sequence trapping. This seems to be the case, as the number of new sequences found in each round of the signal sequence trap is still not decreasing. (3) Despite the large number of mammalian signal peptides that can function in yeast, there might still be enough incompatibilities between mammals and yeast to prevent the secretion of particular mammalian factors.

*Novel Sequences.* One thousand one hundred and seven sequences were determined to be novel, i.e., they demonstrate little or no homology to sequences of known function. The redundancy of these clones, however, has not yet been fully assessed; judging by the redundancy found among the sequences that matched known proteins, the number of unique novel sequences is likely to be less. Almost half of these novel sequences displayed homology to mammalian library clones, such as bacterial artifical chromosome (BAC) clones and sequences from the human genome database. The remainder did not match any sequences in the nonredundant database. We have not yet systematically analyzed these novel sequences to determine how many contain a signal sequence, how many are expressed in the SCN, and how many are artifacts.

## Summary of Sequence Trap

The signal sequence trap is designed to identify peptides or proteins containing a characteristic signal peptide that targets them to the endoplasmic reticulum where they are then processed and sorted. Peptides and proteins destined for secretion are included in this group, and thus, we used this method and an SCN-enriched cDNA library in an effort to expand the list of factors known to be secreted by the SCN. The signal sequence trap was valuable in that we identified 90 potential SCN-secreted factors that had not been previously associated with the SCN, increasing the candidate list of locomotor factors by more than threefold. We also retrieved a large number of novel sequences. However, the fact that we were unable to retrieve more than half of the known SCN-secreted factors, despite higher than sixfold coverage of the library, suggests that there are clear limitations in this method.

While the secreted factors are the most relevant factors to this project, the transmembrane proteins that we identified are also potentially interesting. First, many transmembrane receptors have soluble isoforms that could be considered candidate locomotor factors. Second, transmembrane proteins may also be interesting with respect to the mechanism by which the SCN cells synchronize their rhythms. Little is known about this mechanism(s), although the predominant theories involve electrical coupling and synapses using both amine and peptidergic neurotransmitters (Michel and Colwell, 2001). Thus, the identities of the receptors and ion channels expressed in the SCN may provide clues to cellular communication and coupling within the SCN.

## Behavioral Screen for a Possible Role of SCN Factors in Regulating Locomotor Activity

### Principal Considerations

To test systematically the known and new SCN-secreted factors from the signal sequence trap for potential roles in activating or inhibiting locomotor activity, we chronically infused the candidate factors into the third ventricle of hamsters and looked for a reversible alteration in the circadian pattern of wheel-running activity. Any factor that met the predictions for a constitutively infused activator or inhibitor of activity was then tested further for physiological significance.

The following considerations for the behavioral screen have been made: (1) We decided to infuse the candidate factors at high concentrations with the risk of producing false positives rather than trying to block their endogenous functions (e.g., by infusing receptor antagonists or neutralizing antibodies) with the risk of missing potentially interesting effects due to functional redundancy of SCN signals. Initial results from the infusion of various peptides suggested that a large number of false positives would not be a problem as none of the peptides infused seemed to make the animals sick. (2) Because the transplant experiments indicate that the receptors for the locomotor factors are within a diffusible range of the third ventricle (LeSauter et al., 1997), we chose to infuse the candidate factors into the third ventricle just above the SCN. From there, the factors would be able to enter the cerebrospinal fluid (CSF) and diffuse through the ependymal and subependymal layers of the third ventricle to reach potential target areas. (3) As an initial screen, we chose to do a chronic infusion using an osmotic minipump (lasting 1–3 weeks) rather than an acute infusion (lasting less than a day) because there were specific patterns of activity that could be predicted for a constitutively infused activator or inhibitor (discussed later) that could be easily distinguished from the patterns expected for a factor that affected the SCN clock itself. This distinction between an output factor and a clock factor is much more subtle in a short infusion. However, because locomotor factors should be able to act within a time window of one circadian cycle, we tested the factors positive in the primary behavioral screen for their ability to be effective when acutely infused into the third ventricle. (4) Based on the general range of dissociation constants for neuropeptides and growth factors, and rough estimates of the volume and turnover rate of CSF in the third ventricle of hamsters, the concentration of candidate factors in the minipump was chosen to be 3–5 $\mu M$ to achieve complete saturation of their potential receptors.

## Chronic Infusion of SCN Factors into the Third Ventricle of the Hypothalamus

*General Strategy.* The following is protocol used for the behavioral screen. Adult male hamsters are entrained to a 14:10 LD cycle for at least 2 weeks. Their running wheel activity is then recorded under constant darkness for a period of 3–5 weeks. This period includes (1) a 1-week preoperative baseline with respect to the period, phase, and the amount of activity; (2) a 2-day postoperative recovery period during which artificial CSF (aCSF) is infused; (3) a 1- to 3-week constant infusion of the control or the test sample (the exact duration depends on the type of pump used and the specifications for that minipump lot); and (4) a 1- to 2-week postinfusion period to assess the reversibility of any observed effects. Afterward, the cannulated hamsters are sacrificed and their brains analyzed by Nissl stain to confirm cannula placement and tissue integrity.

*Hamsters and Housing Conditions.* Adult male Syrian hamsters (46–55 days old) are obtained from Charles River Laboratories, individually housed in cages equipped with running wheels, and are given *ad lib* access to standard laboratory chow and water. For the entrainment period of at least 2 weeks, the animals are kept under a 14:10 LD cycle at a room temperature of ~20°. Approximately 1 week before surgery, the animals are kept under constant darkness to assess their preoperation free-running locomotor rhythms.

*Chronic Third Ventricular Infusion.* The chronic infusion of the candidate factors is done using an osmotic minipump system. An osmotic minipump (Alzet) is filled with either aCSF [144 m$M$ NaCl, 2.7 m$M$ KCl, 1 m$M$ MgCl$_2$, 1.2 m$M$ CaCl$_2$, 2 m$M$ NaPO$_4$ (pH 7.4)] or the peptide or protein of interest. We used two types of minipumps. One minipump is designed to hold a volume of 100 $\mu$l, and the other is designed to hold a volume of 250 $\mu$l. The flow rate of both types of pumps is approximately 0.5 $\mu$l/h. Thus, the infusion period of the smaller pump is estimated to be 8–9 days, whereas the infusion period of the larger pump is estimated to be 20–21 days. The pump is connected to 6.5 cm of vinyl tubing (PVC60) (Plastics One) filled with aCSF. This length of tubing is meant to provide the animal with 2 days of aCSF to help the animal recover after the surgery. The other end of the tubing is connected to an 8-mm 28-gauge stainless steel cannula (Plastics One). The factors infused individually or as small pools of peptides (found on the same precursor) are diluted in aCSF at concentrations ranging from 0.3 to 10 $\mu M$ and filtered through a 0.22-$\mu$m filter before being injected into the pump. To test whether a possible instability of the factors could result in false-negative results, we assayed selected factors for their stability after incubation for 2 days at 37°. HPLC profiles indicate only little degradation during this period.

The surgeries are performed during subjective day, at a time in which light should not phase-shift the clock ("dead zone"). Each animal is anesthetized with sodium pentobarbital (80 mg/kg). The skull is exposed, and the cannula is implanted stereotaxically and cemented to the skull with dental acrylic. The cannula is aimed at the third ventricle just above the SCN on the midline, 0.6 mm anterior to bregma. The tubing and the minipump are inserted subcutaneously on the back of the animal. The opening on the top of the head is closed, and the animal is given the analgesic buprenorphine (0.5 mg/kg) and returned to the running wheel cage. Daily visual inspections (via infrared goggles) of the operated animals are performed to check for any ill side effects of the infusions.

*Acute Third Ventricular Infusion.* Acute cerebroventricular infusions are done using an injection needle directed to the stereotaxic coordinates of the third ventricle by a guide cannula. Surgical implantation of the guide cannula (26-gauge stainless steel cannula; Plastics One) is essentially done as described earlier for the chronic infusions. A removable cannula dummy is inserted through the guide cannula to ensure its patency until the time of the acute infusion. The opening on the top of the head is sutured, leaving the top of the guide cannula exposed, and the animal is given the analgesic buprenorphine (0.5 mg/kg) and returned to the running wheel cage, which is in constant darkness. After complete recovery from surgery, the cannula dummy is removed and an injection needle is inserted in its place, protruding 1 mm at the distal end of the guide cannula. Infusions are done in a volume of 2 $\mu$l at a rate of 2 $\mu$l/min controlled by an electronic micropump (Harvard Apparatus). The injection needle is left in place for 1 min to allow for diffusion, then gently removed from the guide cannula, and the cannula dummy is replaced. The entire infusion procedure is conducted under infrared illumination, with the aid of night-vision goggles, so as not not disturb the circadian clock, and is completed in $\sim$5 min.

*Recording and Data Analysis.* Recordings are carried out for 24 h per day from 1 week before implantation to 1–3 weeks after the infusion period. Wheel revolutions are recorded by closure of a microswitch mounted on the cage and collected and stored in 6-min bins using Clock-Lab (Actimetrics). Display and analysis of activity recordings are also performed using ClockLab.

*Histological Analysis.* At the conclusion of the behavioral analyses, the animals are anesthetized (sodium pentobarbital, 160 mg/kg) and killed by transcardiac perfusion with 50 ml phosphate buffer solution (PBS), pH 7.4, followed by 100 ml 4% formaldehyde. Immediately after the perfusion, connections among the cannula, tubing, and minipump are verified. The brains are immediately removed, postfixed for 2 h at 4°, and stored at least 24 h in cryoprotectant solution (20% sucrose in PBS). The tissue is then frozen and sectioned in the coronal plane at 40 $\mu$m using a cryostat. Cannula placement and tissue integrity are visualized by Nissl stain.

## TGF-$\alpha$ as a Likely Locomotor Inhibitory Factor

From many negative control experiments infusing artificial CSF as vehicle, the following conclusions could be drawn: (1) The animals were able to physically recover from the anesthesia and the cannulation proce-dure in a timely fashion; thus, any major effect on rhythms of locomotor activity could be ascribed to an infused factor. (2) The infusion of the vehicle did not cause noticeable illness, which affected the locomotor activity of the animal. (3) The precision of the onsets of activity during the infusion period indicated that the underlying clock was undisturbed. (4) Although the amounts of activity sometimes decreased slightly after the operation, this reduction was minimal compared with what would be pre-dicted for the infusion of an inhibitor of activity, which is a complete abolition of activity. Therefore, this system seems to be suitable for the detection of circadian inhibitors of activity. It is less clear whether this system is sensitive enough for the more subtle detection of circadian activators of activity because an infused activator of locmotor activity might not be able to override the endogenous inhibition presumably

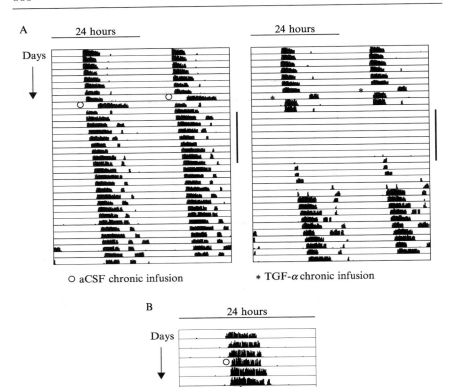

FIG. 3. Reversible inhibition of locomotor activity by TGF-$\alpha$. Wheel-running activity records from hamsters in constant darkness infused chronically (A) or acutely (B) with artificial cerebrospinal fluid (aCSF) and TGF-$\alpha$ are examples from the behavioral screen. Running wheel revolutions are given as histogram for 6-min bins. One (B) or 2 (A) days are represented horizontally; vertical lines correspond to successive days. Symbols represent the start of the infusions. (A) Chronic infusion (0.5 $\mu$l/h) of TGF-$\alpha$ (3 $\mu M$) or aCSF as a control into the third ventricle for about 8 days via an implanted cannula with an osmotic minipump. (B) Acute infusion of aCSF as a control and TGF-$\alpha$ (2 $\mu$l of a 20 $\mu M$ solution) using a previously implanted guide canula.

dictated by SCN inhibitory factors throughout the $\rho$ band or resting phase. It is still possible, however, that an exogenously infused activator would cause an increase of activity during a permissive window in which endogenous activators are low.

So far, we have infused 52 candidate locomotor factors ($n \geq 2$ for each factor). The majority of the factors infused had no effect on the animals' amount of activity or period of locomotor rhythms. In other words, they looked similar to aCSF control subjects. One peptide, TGF-$\alpha$, behaved exactly as expected for an SCN locomotor inhibitory factor (Kramer *et al.*, 2001). As seen in the wheel-running records in Fig. 3A, animals infused with TGF-$\alpha$ recuperated from the cannulation procedure, which was reflected in substantial activity during the aCSF recovery period. Immediately afterward, however, wheel-running activity ceased for the entire infusion period. The inhibitory effect was reversible in that activity returned as soon as the infusion ended with approximately the same amount of activity as preoperation levels. Furthermore, the return of activity was in phase with preoperation rhythms, indicating that only the output, and not the underlying clock, was disturbed by TGF-$\alpha$.

The fact that animals infused with TGF-$\alpha$ exhibited normal amounts of activity almost immediately after the infusion implies that TGF-$\alpha$ was not making the animals ill. This was supported by daily visual inspections and a normal body weight at the end of the observation period (approximately 2 weeks after the infusion period). Furthermore, acute infusions of TGF-$\alpha$ into the third ventricle produced a reversible cessation of locomotor activity within an hour of infusion (Fig. 3). This quick response to TGF-$\alpha$ not only argues against the inhibition of activity being a secondary effect of illness, but also is consistent with the time course of a putative circadian locomotor factor, which should act at least within the same circadian cycle of presentation to influence behavior on a daily basis.

Further evidence strengthened the case for TGF-$\alpha$ being an SCN inhibitor of locomotion (for details, see Kramer *et al.*, 2001). TGF-$\alpha$ is expressed rhythmically in the SCN and acts through the epidermal growth factor receptor (EGFR), which was found on neurons of the SPZ, a region implicated in the control of locomotor behavior. In addition, mice with a hypomorphic EGFR mutation exhibited excessive daytime locomotor activity and failed to effectively suppress activity upon light exposure.

## Acknowledgments

We thank S. Hollenberg and F. Winston for the gift of plasmids and yeast strains and F. Davis, P. Snodgrass, and X. Li for help in the initial SCN tissue collection and guidance in performing the surgery. A.K. was supported by the Deutsche Forschungsgemeinschaft,

F.-C.Y. by the National Science Foundation, S.K. by an HHMI Predoctoral Fellowship and a Quan Fellowship, and C.J.W. by the National Institutes of Health.

## References

Carlson, M., Taussig, R., Kustu, S., and Botstein, D. (1983). The secreted form of invertase in *Saccharomyces cerevisiae* is synthesized from mRNA encoding a signal sequence. *Mol. Cell. Biol.* **3**, 439–447.

Jacobs, K. A., Collins-Racie, L. A., Colbert, M., Duckett, M., Evans, C., Golden-Fleet, M., Kelleher, K., Kriz, R., La Vallie, E. R., Merberg, D., Spaulding, V., Stover, J., Williamson, M. J., and McCoy, J. M. (1999). A genetic selection for isolating cDNA clones that encode signal peptides. *Methods Enzymol.* **303**, 468–479.

Jacobs, K. A., Collins-Racie, L. A., Colbert, M., Duckett, M., Golden-Fleet, M., Kelleher, K., Kriz, R., LaVallie, E. R., Merberg, D., Spaulding, V., Stover, J., Williamson, M. J., and McCoy, J. M. (1997). A genetic selection for isolating cDNAs encoding secreted proteins. *Gene* **198**, 289–296.

Kaiser, C. A., and Botstein, D. (1986). Secretion-defective mutations in the signal sequence for Saccharomyces cerevisiae invertase. *Mol. Cell. Biol.* **6**, 2382–2391.

Kaiser, C. A., Preuss, D., Grisafi, P., and Botstein, D. (1987). Many random sequences functionally replace the secretion signal sequence of yeast invertase. *Science* **235**, 312–317.

Klein, D. C., Moore, R. Y., and Reppert, S. M. (eds.) (1991). "Suprachiasmatic Nucleus." Oxford University Press, New York.

Klein, R. D., Gu, Q., Goddard, A., and Rosenthal, A. (1996). Selection for genes encoding secreted proteins and receptors. *Proc. Natl. Acad. Sci. USA* **93**, 7108–7113.

Kramer, A., Yang, F. C., Snodgrass, P., Li, X., Scammell, T. E., Davis, F. C., and Weitz, C. J. (2001). Regulation of daily locomotor activity and sleep by hypothalamic EGF receptor signaling. *Science* **294**, 2511–2515.

LeSauter, J., Romero, P., Cascio, M., and Silver, R. (1997). Attachment site of grafted SCN influences precision of restored circadian rhythm. *J. Biol. Rhythms* **12**, 327–338.

Lu, J., Zhang, Y. H., Chou, T. C., Gaus, S. E., Elmquist, J. K., Shiromani, P., and Saper, C. B. (2001). Contrasting effects of ibotenate lesions of the paraventricular nucleus and subparaventricular zone on sleep-wake cycle and temperature regulation. *J. Neurosci.* **21**, 4864–4874.

Michel, S., and Colwell, C. S. (2001). Cellular communication coupling within the suprachiasmatic nucleus. *Chronobiol. Int.* **18**, 579–600.

Ralph, M. R., Foster, R. G., Davis, F. C., and Menaker, M. (1990). Transplanted suprachiasmatic nucleus determines circadian period. *Science* **247**, 975–978.

Rockwell, N. C., Krysan, D. J., Komiyama, T., and Fuller, R. S. (2002). Precursor processing by Kex2/furin proteases. *Chem. Rev.* **102**, 4525–4548.

Silver, R., LeSauter, J., Tresco, P. A., and Lehman, M. N. (1996). A diffusible coupling signal from the transplanted suprachiasmatic nucleus controlling circadian locomotor rhythms. *Nature* **382**, 810–813.

Stephan, F. K., and Zucker, I. (1972). Circadian rhythms in drinking behavior and locomotor activity of rats are eliminated by hypothalamic lesions. *Proc. Natl. Acad. Sci. USA* **69**, 1583–1586.

Taft, R. A., Denegre, J. M., Pendola, F. L., and Eppig, J. J. (2002). Identification of genes encoding mouse oocyte secretory and transmembrane proteins by a signal sequence trap. *Biol. Reprod.* **67**, 953–960.

Takahashi, J. S., Turek, F. W., and Moore, R. Y. (eds.) (2001). "Circadian Clocks." Kluwer Academic/Plenum, New York.

Vogelbaum, M. A., and Menaker, M. (1992). Temporal chimeras produced by hypothalamic transplants. *J. Neurosci.* **12,** 3619–3627.

von Heijne, G. (1985). Signal sequences: The limits of variation. *J. Mol. Biol.* **184,** 99–105.

# [35]   Genetic and Biochemical Strategies for Identifying *Drosophila* Genes That Function in Circadian Control

*By* F. Rob Jackson, Ginka K. Genova, Yanmei Huang, Yelena Kleyner, Joowon Suh, Mary A. Roberts, Vasudha Sundram, and Bikem Akten

## Abstract

Explicit biochemical models have been elaborated for the circadian oscillators of cyanobacterial, fungal, insect, and mammalian species. In contrast, much remains to be learned about how such circadian oscillators regulate rhythmic physiological processes. This article summarizes contemporary genetic and biochemical strategies that are useful for identifying gene products that have a role in circadian control.

## Introduction

Elucidating the mechanisms and molecules mediating the clock control of distinct physiological and behavioral processes represents a significant challenge of contemporary circadian biology. Clock control within the nervous systems of animals involves intracellular signaling mechanisms within pacemaker cells that direct the synaptic output of temporal information, as well as the anatomical pathways through which the clock directs a particular rhythmic process. It is known that both transcriptional and posttranscriptional mechanisms have a role in keeping time and conveying temporal information from the molecular oscillator to the intracellular pathways regulating synaptic output (Blau, 2001). Although quite a lot is known about components of the molecular oscillator, the intracellular signaling mechanisms that connect the oscillator to synaptic output and the regulated release of the neuropeptide pigment dispersing factor (PDF) (Park *et al.*, 2000a; Renn *et al.*, 1999) have not been well characterized. Similarly, the cellular and biochemical pathways connecting clock cells to regulated processes have not been delineated.

In both *Drosophila* and mammals, a number of transcription factors display circadian changes in abundance and participate in the regulation of clock gene transcription [by definition, core components of the circadian oscillator loop (e.g., Cyran *et al.*, 2003)]. They are also required for the regulation of other genes that more directly mediate the clock control of physiology and behavior. In *Drosophila*, at least one of these transcription factors, CLOCK, has been shown to regulate a number of different target genes within the brain that consequently exhibit circadian rhythms in expression (Ceriani *et al.*, 2003; Claridge-Chang *et al.*, 2001; Lin *et al.*, 2002; McDonald and Rosbash, 2001; Ueda *et al.*, 2002). Many of these oscillating genes contain E-box consensus motifs and are direct targets of CLOCK within the clock cell population; some may have important roles in initiating clock output. However, many neural genes exhibit circadian changes in transcriptional activity but are not direct targets of CLOCK, nor are they necessarily even expressed in cells containing the CLOCK transcription factor. This highlights the existence of intercellular (synaptic) mechanisms that communicate circadian information from neural clock cells to other elements of the nervous system. Neurobiological studies of circadian control mechanisms are thus aimed at both an understanding of the signaling pathways within clock cells that determine synaptic output and at a description of the cellular and biochemical components that connect clock cells to the rest of the organism. It is these latter elements that directly mediate circadian changes in physiological parameters.

This article highlights some of the classical genetic and biochemical methods that are being used in *Drosophila* to identify cellular and molecular elements relevant to the circadian control of behavior. We employ examples from work on several different *Drosophila* genes to illustrate genetic, biochemical, and behavioral approaches to understanding clock control mechanisms. We do not describe studies showing that PDF, a peptide released from clock neurons, is an important clock output element (Helfrich-Förster *et al.*, 2000; Park *et al.*, 2000a; Renn *et al.*, 1999), as such studies are discussed in other articles of this volume.

## Genetic Approaches to Studying Clock Control Elements

A classical genetic screen for circadian mutants remains one of the most powerful approaches for identifying genes important for the clock control of behavior. Simplistically, circadian mutations can be divided into three categories: those affecting zeitgeber input to the clock, those altering a component of the molecular oscillator, and those which perturb a factor downstream of the oscillator mechanism that is necessary for circadian control. In *Drosophila*, circadian mutants have been identified primarily by the use of two different behavioral phenotypes: locomotor activity and adult eclosion (see later). To date, all characterized clock mutations change circadian period or produce arrhythmicity. In contrast, a mutation affecting the clock control of a rhythmic process might alter circadian phase (the time at which an event occurs) or produce arrhythmicity. In particular, those mutations that alter phase or cause arrhythmicity for only one of several rhythmic behaviors (e.g., locomotor activity rhythms but not eclosion rhythms, or vice versa) are good candidates for lesions of a downstream circadian control element versus a molecular element of the clock itself. Operationally, such a screen for "rhythm-specific" mutations forms the basis for a rational genetic analysis of circadian control mechanisms.

### Behavioral Protocols for the Identification of Rhythm-Specific Mutations

While certain genetic screens for circadian mutants have made use of *in vivo* imaging methods and reporter transgenes (Stempfl *et al.*, 2002), most screens have utilized either the locomotor activity rhythm or the population eclosion rhythm (the gated emergence of adults from pupal cases). Both types of behavioral screens have identified mutations affecting the clock mechanism (Allada *et al.*, 1998; Konopka and Benzer, 1971). Importantly, these behavioral screens have also been employed for the identification of circadian control elements, and mutations have been characterized for at least four *Drosophila* genes (*lark, ebony, DC0*, and *dfmr 1*) that have differential effects on the two rhythmic behaviors. These genes are discussed in later sections. The remainder of this section provides details about the behavioral protocols employed to examine locomotor activity and eclosion rhythms in *Drosophila*.

It is of interest to note that many recent genetic screens have utilized locomotor activity as a phenotype to identify new circadian mutants (Allada *et al.*, 1998; Rutila *et al.*, 1998). While such screens may find genes important for oscillator function and the circadian control of activity, they may exclude factors that are only required for the clock control of eclosion. There are at least three studies, however, that report the use of the population eclosion rhythm as a method for identifying new circadian mutants (Jackson, 1983; Konopka and Benzer, 1971; Sehgal *et al.*, 1994). This method utilizes small pupal populations (100–200 pupae) for each strain screened and is advantageous for simultaneously screening hundreds of strains for variants. In *Drosophila melanogaster*, wild-type adult flies eclose within an approximate 6-h window of time at the beginning of the photoperiod in LD 12:12. In our current version of the eclosion screen, strains are reared in small bottle cultures (one per strain) and the emergence of adults (eclosion) is monitored twice daily to identify variants in which there is significant eclosion during the night. Because mutations that cause arrhythmic behavior or change circadian period or phase can all be associated with abnormal nighttime eclosion, this two-point screen is sufficient to identify many different types of circadian variants. Contemporary genetic screens in *Drosophila* typically utilize either chemical [e.g., ethylmethane sulfonate (EMS)] or transposon insertion mutagenesis. Of interest, there are many different collections of P transposon insertion strains available from the *Drosophila* stock center in Bloomington. A convenient source of strains carrying EMS mutagenized autosomal chromosomes is the Zuker collection (Koundakjian *et al.*, 2004). Our phenotypic screen for eclosion-rhythm mutants uses the following protocol.

1. Rear cultures representing 200 different mutagenized strains for 3–5 days at 24° and then transfer them to 18° and LD 12:12 for the remainder of development.
2. After adults have been emerging for 1–2 days, clear all cultures of adult flies a few hours before the time of lights off.
3. One hour prior to the time of lights on, examine cultures for adult eclosion using dim red light (a 2.5- to 7-W bulb behind a Kodak GBX-2 filter). In wild-type strains, there is little or no adult eclosion during the night portion of the cycle. Strains in which there is significant adult eclosion during the night (10–20 flies) are retained as candidate mutants.
4. Repeat the screen with candidates to confirm the abnormal eclosion profile.
5. Examine locomotor activity rhythms in candidate strains to identify eclosion rhythm-specific mutations.

It is known that poikilothermic (cold-blooded) animals such as *Drosophila* can entrain behavioral rhythms to both light:dark cycles and low amplitude (3–5°) temperature cycles. Thus, a variant of the eclosion screen utilizes a temperature cycle to synchronize populations (Fig. 1). An advantage of this scheme is that it may identify factors that function in the temperature entrainment of the clock. This screen also obviates the requirement to perform experiments in the dark because it uses a temperature cycle in constant light to affect behavioral entrainment. In the current paradigm, strains are reared at 25° for 3–5 days and are then transferred to constant light and a 12:12 temperature (18–24°) cycle for the remainder of development. In this paradigm, peak wild-type eclosion occurs within a several-hour window of time at the beginning of the high temperature portion of the cycle. Strains that exhibit abnormal eclosion profiles are retained as candidate mutants. The authors' laboratory is currently using

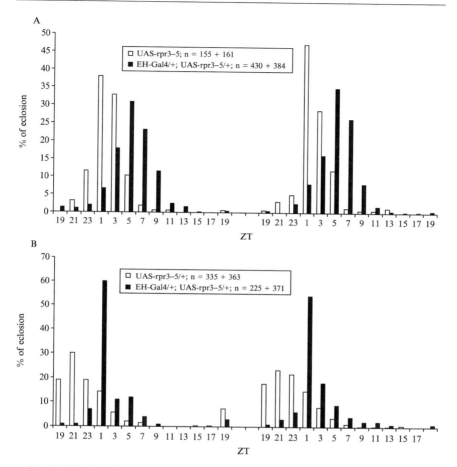

FIG. 1. Eclosion rhythms for normal flies and those lacking EH neurons. (A) Eclosion profiles for EH KO (EH-Gal4; UAS-rpr3–5) and control (UAS-rpr3–5) populations maintained in constant light (LL) and entrained to a temperature cycle consisting of 12 h of 18° and 12 h of 24°. (B) The same as A, but populations were maintained in constant darkness (DD) during entrainment to the temperature cycle. In both A and B, the control populations carried a UAS-rpr transgene and lacked the EH-Gal4 transgene. Although not shown, populations carrying only the EH-Gal4 transgene (without UAS-rpr) displayed eclosion profiles that were similar to those of UAS-rpr control flies.

this paradigm to identify *Drosophila* strains exhibiting altered entrainment of the eclosion rhythm to a temperature cycle.

A second behavioral assay for mutants is the locomotor activity rhythm, which can be used to assess the strength of rhythmicity, circadian period, and circadian phase in individuals. The details of this screen have been described in many publications (Levine *et al.*, 1994), and devices are available commercially (Trikinetics, Inc.) that utilize infrared light beams to monitor activity. In most of our current studies, we assay locomotor activity in 3- to 5-day-old flies (usually males) that are lightly anesthetized with ether and loaded into glass tubes for behavioral monitoring. Activity is monitored for at least 2 weeks and is then analyzed quantitatively for periodicity using a MatLab-based set of programs (Levine *et al.*, 2002). Combined with the eclosion assay, this screen enables the identification of rhythm-specific mutations (i.e., potential circadian control factors). Examples of such mutations are described in the next several sections.

## RNA-Binding Proteins and Posttranscriptional Mechanisms of Clock Control

*LARK and Circadian Control of Eclosion.* Genes encoding two different RNA-binding proteins have been shown to function in the *Drosophila* circadian system. One called *lark* was identified in a forward genetic screen on the basis of a gene dose-dependent effect on the daily timing of eclosion (Newby and Jackson, 1993); strains heterozygous for a *lark* null allele exhibited daily peaks of eclosion that were 2–3 h early relative to wild-type strains. In contrast, *lark*/+ heterozygotes exhibited normal locomotor activity rhythms, indicative of a rhythm-specific effect. Homozygotes have not been characterized behaviorally as *lark* is an essential gene and mutants do not survive past midembryonic stages of development (McNeil *et al.*, 1999). Moreover, this RNA-binding protein is also required for postembryonic developmental functions (McNeil *et al.*, 2001).

The *lark* gene encodes a member of the RNA recognition motif (RRM) class of RNA-binding proteins; it has a broad pattern of localization within the central nervous system (CNS) and peripheral tissues (Zhang *et al.*, 2000). In the CNS, the protein has a nuclear localization in most or all neurons, but is cytoplasmic in a population of neurosecretory cells that contain crustacean cardioactive peptide (CCAP), a peptide with an important role in the regulation of *Drosophila* ecdysis [eclosion representing the adult ecdysis (Park *et al.*, 2003)]. Interestingly, circadian changes in LARK protein abundance can be detected within CCAP cells of the ventral nervous system (McNeil *et al.*, 1998; Zhang *et al.*, 2000), and ongoing work is aimed at understanding the mechanisms that determine this protein rhythm. However, one study has indicated that the CCAP cell population is not essential for the circadian control of eclosion (Park *et al.*, 2003) and suggested that there may be functionally redundant cellular pathways for this important process. Perhaps the so-called Ap-let neuronal circuit, a group of peptidergic neurons that has been hypothesized to play a role in controlling ecdysis (Park *et al.*, 2004), contributes to the clock regulation of adult eclosion.

The broad expression pattern of LARK and a localization in CCAP cells beg the question as to whether the protein functions within other neurosecretory cell populations that are known to be important for the regulation of eclosion. To address this question, a cell-specific expression system was employed to increase LARK abundance in three different neurosecretory cell populations previously implicated in circadian control (Schroeder *et al.*, 2003). LARK overexpression in eclosion hormone (EH)-containing cells, another cell population that participates in the regulation of ecdysis (McNabb *et al.*, 1997), caused a late-eclosion phenotype wherein daily eclosion peaks were several hours later than normal. Strikingly, animals overexpressing LARK in Timeless-containing clock cells, a fraction of which make and release the PDF peptide, eclosed in an arrhythmic pattern. Interestingly, locomotor activity was also arrhythmic in most individuals with increased LARK expression in the Timeless or PDF cell populations. Finally, overexpression in CCAP cells paradoxically led to a slightly early eclosion profile. Importantly, these effects of LARK appear to be mediated by the RRM (RNA-binding) domains of the protein because expression of a mutant protein with decreased RRM activity within clock cells did not cause a behavioral phenotype (Schroeder *et al.*, 2003).

It is significant that these behavioral phenotypes are similar to those resulting from genetic ablation of the same neuronal populations (Park *et al.*, 2003; Renn *et al.*, 1999). In the case of LARK expression, however, the relevant neurosecretory cell populations appeared to differentiate and elaborate normal projections (Schroeder *et al.*, 2003), and thus did not undergo an abnormal cell death. Moreover, the overexpression of LARK in other cell populations (retinal cells and tyrosine hydroxylase-containing neurons) has no detectable effects on viability or behavior. These results suggest that LARK may function in several different neurosecretory cell types to regulate the daily timing of population eclosion. A cell-specific perturbation of *lark* mRNA expression, using RNA interference techniques, may be useful to examine the requirement for the gene in these different cell types.

The discovery of LARK mRNA targets would identify additional components of the pathway through which the protein regulates rhythmicity. Coincident with a biochemical strategy for identifying mRNA targets (see later section), it is possible to use a complementary genetic approach for the same purpose. A genetic approach is possible because of the behavioral phenotypes observed with increased LARK expression. These phenotypes permit a genetic screen for new mutations that suppress or enhance the effects of LARK overexpression. This approach is currently being employed in the laboratory, using LARK-dependent effects on the locomotor activity rhythm as a platform, to screen for such genetic modifiers (Y. Kleyner and F. R. Jackson, unpublished results). It is expected that certain suppressor mutations will identify genes encoding mRNA targets of this RNA-binding protein.

*Fragile X RNA-Binding Protein and Rhythmicity.* A gene called *dfmr1* encodes the *Drosophila* homolog of fragile-X (dFMRP), an RNA-binding protein of the KH class (Siomi *et al.*, 1993). DFMRP is expressed in most or all neurons, including PDF-containing LNv cells. It has been shown to function in the circadian system through behavioral characterization of *dfmr1* mutants (Dockendorff *et al.*, 2002; Morales *et al.*, 2002). These studies demonstrated that individual null mutants predominantly exhibit arrhythmic locomotor activity, whereas eclosion rhythms persist in mutant populations, albeit with an abnormally late phase. Thus, the mutations have differential effects on the two types of rhythms. Daily rhythms of PER and TIM abundance appear to be normal in *dfmr1* null mutants, so it appears that the effects on rhythmicity are a result of a perturbation of the circadian system that is downstream of the molecular oscillator controlling behavior. However, overexpression of *dfmr1* in TIM-containing neurons leads to long-period rhythms (Dockendorff *et al.*, 2002), consistent with a lengthening of oscillator period. Thus, the dFMRP protein can modulate molecular oscillator function even though it may normally act downstream of the clock to mediate the circadian control of behavior.

It is of interest that the analysis of *dfmr1* mutant brains has revealed subtle projection defects for the small LNv clock cell population (Dockendorff *et al.*, 2002; Morales *et al.*, 2002), and an unanswered question is whether the behavioral defects of mutants are a consequence of abnormal neuronal morphology. It might be the case that both the behavioral and the morphological phenotypes arise from misregulation of a single dFMRP target mRNA that is expressed within the LNv or another cell population. However, similar to the human fragile X protein (Brown *et al.*, 2001), *Drosophila* FMRP is likely to have many mRNA targets, and thus the two phenotypes may arise from misregulation of two different target molecules.

## Circadian Control of Locomotor Activity

*Ebony and Locomotor Activity Rhythms.* A gene called *ebony*, which is best known for a role in cuticle pigmentation (e.g., Wittkopp *et al.*, 2002), is also involved with the circadian regulation of locomotor activity. Most *ebony* mutants have arrhythmic locomotor activity or exhibit complex, multicomponent activity rhythms; however, there is no effect of *ebony* mutations on the eclosion rhythm (Newby and Jackson, 1991). These behavioral results indicate that the Ebony gene product has a function specific to the clock regulation of activity. More recently, it has been demonstrated that *ebony* mRNA shows robust circadian cycling (Claridge-Chang *et al.*, 2001; Ueda *et al.*, 2002), with peak abundance occurring near the beginning of subjective day. This result has been replicated using quantitative RT-PCR techniques (J. Suh and F. R. Jackson, unpublished studies), and those studies indicate that *ebony* mRNA abundance is highest between ZT22 and ZT4 and that there is an approximate five- to six-fold change in abundance between the day and the night portions of the cycle. Furthermore, the preliminary studies of J. Suh and F. R. Jackson (unpublished results) have also demonstrated that Ebony protein abundance changes in a circadian manner, with peak abundance during the subjective day, and importantly that such cycling occurs within the nervous system.

Surprisingly, Ebony protein does not appear to be present within neurons (Richardt *et al.*, 2002; J. Suh and F. R. Jackson, unpublished results). Rather, the protein is localized to glial cells of the larval and adult brains. Richardt *et al.* (2002) have shown that Ebony protein is localized to glia within the optic neuropil of adults; the results of J. Suh and F. R. Jackson (unpublished) demonstrated that the protein is present in a subpopulation of lateral glia within the central nervous system (CNS) of larvae and adults. Furthermore, Ebony glia within the CNS are localized near populations of biogenic amine-containing neurons, indicating that they might be required for the modulation of aminergic neuronal function. Similarly, many Ebony-containing glia are located in close proximity to PER/TIM-containing cells (both neurons and glia), and a small fraction of the Ebony cell population stains positive for nuclear PER and TIM. These results are consistent with the idea that certain Ebony-containing glia might influence pacemaker cell physiology. These intriguing findings support the hypothesis that glial cells participate directly in the orchestration of rhythmic behavior, and this idea is consistent with previous results implicating glial function in rhythmicity (Ewer *et al.*, 1992).

*cAMP/PKA Signaling Pathway.* It has been suggested that the cAMP/PKA signaling pathway plays a role in the circadian control of locomotor activity. *DCO* mutants, which have reduced PKA catalytic subunit activity, are mostly arrhythmic (Levine *et al.*, 1994; Majercak *et al.*, 1997). However, *per* RNA and PER protein cycle normally in the mutants, indicating normal molecular oscillator function, and consistent with this conclusion, the mutant populations have normal eclosion rhythms (Majercak *et al.*, 1997). These results suggest that PKA functions selectively in the pathway regulating locomotor activity. Consistent with this idea, it has also been reported that flies deficient in type II cAMP-dependent protein kinase show arrhythmic locomotor activity (Park *et al.*, 2000b). However, the targets of PKA action in this system have not been defined, and it is also not known where (in which cells) PKA is required for the circadian control of activity. It now should be possible to use cell-specific expression methods to explore the cellular requirement for PKA function.

It is of interest to note that *dunce* (*dnc*) mutants, with decreased cAMP phosphodiesterase activity and increased cAMP levels, also exhibit circadian phenotypes (Levine *et al.*, 1994). The mutants show enhanced phase resetting (i.e., larger-amplitude phase shifts) in response to light pulses that confer phase delays (phase advances are normal) and they show 23-h circadian periods in DD, roughly 1 h shorter than sibling control flies. These phenotypes suggest an effect on input or the clock mechanism in the mutants, rather than a lesion in an output pathway. At least one other study (Belvin *et al.*, 1999) implicates cAMP signaling and the cAMP response element binding (CREB) protein in rhythmicity. *Drosophila* hypomorphic *CREB* mutants exhibit an abnormally short circadian period (~23 h), similar to *dnc* flies, and it is an intriguing idea that the two phenotypes are mechanistically related. Cyclic AMP levels are known to change in a circadian manner (Levine *et al.*, 1994), with low amounts observed during the subjective night; interestingly, *Dnc* gene transcription also shows circadian changes, with peak activity near the end of subjective day (Claridge-Chang *et al.*, 2001), consistent with the observation of decreased cAMP levels in the night. dCREB activity (as monitored by a cre-luc reporter) shows a similar rhythmic pattern (Belvin *et al.*, 1999). Perhaps endogenous changes in dCREB activity mirror circadian oscillations in cAMP levels. According to this model, however, it is predicted that increased cAMP levels (as in the *dnc* mutant) would lead to increased dCREB activity. Thus, it is not apparent why hypomorphic *dCREB* and *dnc* mutations would both be associated with a short-period clock.

*ras/MAPK Signaling and Rhythmicity.* Loss-of-function alleles for the *Neurofibromatosis-1(Nf1)* gene, which encodes the *Drosophila* homolog of human neurofibromin, have also been shown to affect locomotor activity rhythms, and it has been suggested that effects of the mutations are mediated through upregulation of ras/MAPK signaling (Williams *et al.*, 2001). Whereas *Drosophila Nf1* mutants have arrhythmic locomotor activity, PER and TIM clock protein cycling seems to occur normally, indicating that the NF1 deficiency does not affect the molecular oscillator. This suggests that NF1 acts downstream of the clock mechanism for the circadian control of locomotor activity. Consistent with such a conclusion, expression of wild-type NF1 in PDF-containing LNv clock cells is not sufficient to rescue the mutant phenotype of *Nf1* mutants. It could be the case, for example, that NF1 is required in cells other than the clock neurons to promote normal rhythmicity. It is thought that cells in dorsal regions of the brain may be targets of the small LNv clock cells (Helfrich-Förster et al., 2000), and perhaps it is a deficiency for NF1 in these target neurons that results in arrhythmicity. The combination of two results from Williams et al. (2001) supports this idea. First, loss-of-function alleles affecting ras/MAPK were shown to suppress the behavioral effects of *Nf1* mutations, suggesting that there is an upregulation of MAPK signaling in *Nf1* mutant flies and that NF1 acts through MAPK signaling in a circadian output pathway regulating locomotor activity. Second, phospho-MAPK levels (indicative of increased MAPK activity) appear to be higher at night in dorsal regions of the brain close to the termini of small LNv neuronal projections. Thus, it has been postulated that NF1 might act in the dorsal targets of the small LNv cells to regulate ras/MAPK signaling activity (Williams et al., 2001). One confusing aspect of these studies is why loss-of-function ras/MAPK mutations, by themselves, do not lead to arrhythmic activity if signaling through this pathway determines the circadian control of activity. An unanswered question is whether the eclosion rhythm is modulated by NF1/MAPK signaling, as this behavioral rhythm has apparently not been examined in *Nf1* mutants.

## Identifying Cellular Output Pathways Using Flies with Cell Type-Specific Defects

An earlier section of this article mentioned that populations of flies lacking CCAP neurons continue to exhibit circadian rhythms of eclosion (albeit with slightly abnormal gating). This study (Park et al., 2003) and several other circadian studies have made use of the binary Gal4/UAS expression system (Brand and Dormand, 1995) to create flies that either lack a particular neuronal cell type or carry cell-specific gene lesions. There are currently multiple methods for generating such flies that utilize cell death genes (for "cell knockouts"), RNA interference techniques, or other kinds of cell biological perturbations. These techniques have been used by several investigators to examine the requirements for particular cell types or gene products as related to circadian rhythms (Helfrich-Förster et al., 2000; Kaneko et al., 2000; Martinek and Young, 2000; McNabb et al., 1997; Park et al., 2003; Renn et al., 1999). Several of these studies have employed the cell-specific expression of *reaper* or *hid*, apoptosis-inducing genes (White and Steller, 1995), for the elimination of particular neuronal cell types to determine their importance in the circadian system. Others have utilized the expression of toxins to perturb synaptic transmission or the expression of double-stranded RNA representing a gene of interest to achieve cell-specific RNA interference. Cell-specific RNA interference is obviously useful for determining whether specific gene products are required autonomously within particular cell types. Of interest, it is also possible to couple the Gal4/UAS method with genetic strategies that utilize mitotic recombination techniques to affect a cell-specific knockout of individual gene products (Stowers and Schwarz, 1999), although to our knowledge this approach has not been employed in circadian studies.

In the course of examining a role for LARK in EH neurons, two of us (G.K.G. and F.R.J.) employed the Gal4/UAS method and the *reaper* (*rpr*) apoptosis gene to generate flies lacking the EH cell population. Interestingly, we observed that populations of flies lacking the EH cells (EH KO flies) exhibited a late-phase eclosion phenotype (Fig. 1), whereas a previous study (McNabb *et al.*, 1997) that reported on EH KO populations did not describe such a phenotype. In our study, this phenotype was most apparent after populations had been in DD for several days following entrainment to LD or when populations were entrained to a temperature cycle rather than LD (as shown in Fig. 1). The study of McNabb *et al.* (1997) examined only the first 2 days of DD, and thus methodological differences might explain the two different results. Of note, the phasing of eclosion in control populations that were entrained to a temperature cycle was dependent on the lighting conditions; flies maintained in LL showed peak eclosion at about the time of the temperature stepup, whereas the phase of eclosion in DD was several hours earlier.

Although we did not examine locomotor activity rhythms in EH KO flies, it is expected that they will have normal activity rhythms, as it is known that EH cells undergo a programmed death after adult eclosion. Thus, the late-phase phenotype probably reflects a selective effect on the cellular output pathway regulating eclosion. Although it is not yet known how EH cells contribute to rhythmicity, the phenotype of EH cell KO populations suggests a role in circadian phase control.

## Biochemical Strategies for Identifying mRNA Targets of Clock-Regulated RNA-Binding Proteins

While genetic analysis will surely define new clock control factors, it is also possible to utilize a biochemical approach to identify additional components of certain clock control pathways. For example, it is expected that the RNA targets of LARK, dFMRP, and perhaps additional RNA-binding proteins (RBPs) will be relevant for the circadian control of behavior. Such targets would define additional components of the signaling pathways utilizing these RBPs. Some of these RNA targets might change in abundance during the circadian cycle if the relevant RBP exhibited circadian changes in activity, and thus they would be revealed by searches for clock-regulated RNAs whose temporal expression patterns were altered in RBP mutants. Other articles in this volume include sections summarizing the use of gene microarray strategies for identifying such clock-regulated mRNAs, many of which encode products important for the clock control of behavior (such as the *ebony* gene mRNA discussed in an earlier section). This section describes an alternative microarray-based "ribonomics" methodology, developed by Keene and colleagues (Tenenbaum *et al.*, 2002), that we are using to identify the RNA targets of RBPs that have a post-transcriptional role in circadian regulation. Such a strategy can be employed to augment the genetic approach used for identifying components of circadian control pathways.

Using this ribonomics methodology, we have initiated studies to identify the RNA targets of LARK and dFMRP that are bound to the proteins *in vivo*. Similar to other investigators using this approach, we begin with immunoprecipitations of LARK or dFMRP ribonucleoprotein (RNP) complexes to isolate the relevant RNAs. Rather than using cells in culture, however, we have employed tissue lysates prepared from whole animals (embryos or pupae) or hand-dissected brains as a starting point for the isolation of RNPs. We are hopeful that the use of brain preparations will enrich for targets relevant to the functions of RBPs within the nervous system. RNAs associated with these RNPs are then purified and utilized to obtain amplified quantities of cDNA or cRNA representing the targets. Finally, labeled targets are hybridized to whole genome oligonucleotide microarrays representing most predicted *Drosophila* genes. We note that the same methods, together with the cell-specific expression of an epitope-tagged RNA-binding protein [such as poly(A)-binding protein], could be used to obtain a "snapshot" of the transcriptome of identified cells.

In general, we follow a protocol similar to that described by Tenenbaum *et al.* (2002) for the isolation of *Drosophila* mRNPs, and the authors direct the reader to that review for a discussion of the procedures for optimizing RNP immunoprecipitations. In our current protocol (see Fig. 2A), we prepare tissue lysates by hand dissecting pharate adult or adult brains in *Drosophila*-SFM media (GIBCO). Once about 200 brains are collected, they are washed several times with ice-cold phosphate-buffered saline (PBS), collected at the bottom of a centrifuge tube by very gentle centrifugation, and then suspended in polysome lysis buffer. A brain cell suspension is prepared by gently dispersing brain tissue with a plastic pestle (see Appendix for a detailed protocol). Cell suspensions are frozen at $-80°$ until sufficient quantities of brains are harvested for an experiment ($\sim1000$ brains). Because our protocol is designed to isolate both nuclear and cytoplasmic RNPs, cell suspensions are thawed and sonicated to break

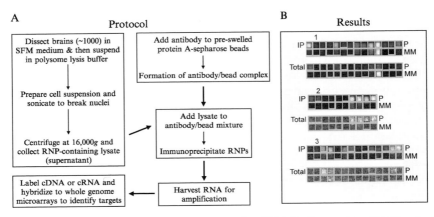

FIG. 2. Protocol for identifying *in vivo* targets of an RNA-binding protein. (A) Flowchart illustrating the procedure for harvesting and analyzing *Drosophila* brain RNPs. (B) Affymetrix chip hybridization data for three different *Drosophila* genes (1, 2, and 3) that showed enrichment after the IP procedure. Each row shows fluorescent signals after hybridization to 14 or 10 (gene 2) different oligonucleotides representing each gene. A comparison of the signals for the IP and total rows for each gene indicates enrichment. IP, immunoprecipitated sample; Total, total cellular RNA sample; PM, perfectly matched oligonucleotides; MM, mismatched oligonucleotides.

nuclear membranes. After centrifugation, supernatant fractions are collected as a source of RNPs.

Antibody-coated protein A-Sepharose beads are prepared, and immunoprecipitation reactions are carried out according to Tenenbaum *et al.* (2002) with the modifications described in the Appendix. Importantly, a portion of the lysate is saved as a sample that represents total cellular RNA to be used as an indication of the relative abundances of RNAs expressed in brain tissue. RNA is extracted from both this lysate and the immunoprecipitated (IP) sample. A higher relative abundance for a particular transcript in the IP sample (as indicated by the fluorescence hybridization signal) is an indication that it has been enriched during the procedure, which is expected of target RNAs.

A concern with the use of microarrays for target identification is obtaining sufficient quantities of target cDNA or cRNA for the hybridization procedure, and thus we have employed mRNA amplification steps to derive labeled targets. For this purpose, we have used both the Ribo-Amp-HS kit (Arcturus) and the Ovation Nanosample RNA Amplification System (NuGEN) to perform one round of RNA amplification. Other amplification kits may work equally well. We have employed the BioArray High Yield Transcript Labeling Kit (Enzo) or the Ovation Nanosample RNA Amplification System, respectively, to obtain biotin-dUTP-labeled cRNA or aminoallyl-dUTP-labeled cDNA for microarray analysis. Aminoallyl-labeled cDNA is reacted with Cy3 or Cy5 using standard reaction conditions to produce fluorescent probes, whereas biotin-labeled cRNA is detected posthybridization using streptavadin-conjugated fluorophores or an antibiotin antibody. We have used both oligonucleotide arrays made in the Tufts-NEMC Expression Array core (TEAC) and Affymetrix gene arrays in our studies. For the oligonucleotide arrays, our facility purchases sets of oligonucleotides representing most of the *Drosophila* transcriptome (Qiagen) and spots them onto glass slides. Microarray hybridizations utilize at least several micrograms of labeled cDNA or cRNA. Figure 2B shows portions of an Affymetric chip illustrating hybridization to three putative LARK target genes.

It is important to independently verify endogenous target RNAs that are identified by microarray hybridization. In addition, it is necessary to have negative controls that indicate specific association of a target to a particular RNA-binding protein. Targets can be verified by performing RT-PCR with specific primers using RNA from the immunoprecipitated RNPs. A good negative control for specificity is the use of a *Drosophila* mutant with reduced activity for the RBP of interest. If such a mutant does not exist, then use of the Sepharose bead complex in the absence of an antibody will also exclude nonspecific sticking of RNAs to the beads. Another control for specificity is the use of an unrelated antibody for RNP immunoprecipitations.

Using these procedures, we have identified 139 presumptive RNA targets for the LARK RRM protein that show from 2- to 256-fold enrichment after the immunoprecipitation procedure. These preliminary results reveal putative targets that can be grouped into several different functional categories, including synaptic transmission, translation, nuclear import/ export, and G-protein signaling. At least 15 putative targets have known or predicted functions in synaptic transmission or endocytosis. A role in the regulation of synaptic transmission is an exciting possibility for LARK given the behavioral effects of protein overexpression in peptidergic neurons (see earlier section). We are currently pursuing additional studies of these putative target genes to characterize their interactions with LARK and to define their functional roles in the circadian regulation of eclosion. A combination of molecular and genetic studies is expected to reveal new insights about the mechanisms and cellular pathways relevant for the clock control of this behavior.

Appendix

*Preparing RNP Lysate from Brain Tissue*

1. Dissect brains from pharate adults or adults in *Drosophila*-SFM media (GIBCO).

2. Keep dissected brains on ice in SFM media until ∼200 brains are collected.

3. Wash brains with ice-cold PBS several times and collect to the bottom of a 1.5-ml tube by gentle centrifugation at 700*g*.

4. Discard supernatant and add ice-cold polysome lysis buffer in a volume equal to the size of the pellet.

5. Gently disperse brain tissue using a plastic pestle and leave on ice for 30 min.

6. Freeze samples at −80° until used. Approximately five such samples (∼1000 brains) are needed for one immunoprecipitation experiment.

7. At the time of an experiment, thaw samples and pool into one tube.

8. Sonicate using an ultrasonic dismembrator (Fisher Scientific Model 100) at a setting of 2 to break nuclear membranes. Examine a small aliquot of the lysate periodically (every few seconds of sonication) to ensure that nuclei have been broken.

9. When most nuclear membranes have been lysed, centrifuge the sample at 16,000*g* for 10 min at 4°.

10. Collect the supernatant and repeat the centrifugation.

11. Collect the supernatant and store on ice until use. The protein concentration of a lysate sample generated according to this procedure is usually ∼10–15 mg/ml.

*Immunoprecipitation Procedure*

1. Prepare the protein A-Sepharose bead/antibody complex by incubating 100 ul antibody (amount may depend on the particular antibody) with 300 ul of swelled beads.

2. Rock solution gently overnight to coat beads with antibody. Prior to immunoprecipitations, save 1/10 of the lysate as a sample representing the total cellular RNA composition.

3. Carry out immunoprecipation reactions according to Tenenbaum *et al.* (2002) with the following minor modifications.

    a. Use a prehybridization/blocking step before the actual immuno-precipitation step to reduce nonspecific binding of RNAs to the antibody-coated beads.

    b. During the blocking step, tumble the antibody-coated beads for 2 h at room temperature in the immunoprecipitation buffer supplemented with 100 $\mu$g/ml yeast tRNA (Ambion). Then collect the antibody-coated beads by gentle centrifugation and use for immunoprecipitation reactions.

4. After the immunoprecipitation reaction, wash the beads eight times, 10 min each, with ice-cold NT2 buffer at 4°.

5. After washing, extract RNA from samples. At the same time, extract RNA from the nonprecipitated lysate (representing total cellular RNA).

*Note*: Certain protocols recommend using urea and increased KCl concentration in the NT2 buffer to increase specificity. We have not yet explored those options. However, different RNA-binding proteins have distinct target-binding affinities and may react differently to such stringent wash conditions.

### Detection of RNA Targets Using Microarray Hybridizations

1. Use the RiboAmp HS RNA amplification kit (Arcturus) to prepare cDNA from the RNA samples (both the nonprecipitated lysate and the immunoprecipitated sample). Other kits may work equally well for cDNA synthesis.

2. Amplify cDNA into cRNA and label with biotin using the BioArray High Yield RNA transcript labeling kit (Enzo). Hybridize equal amounts of labeled antisense RNA from precipitated and nonprecipitated samples to individual *Drosophila* GeneChip probe arrays (Affymetrix) or to other types of oligonucleotide arrays.

*Note*: Transcripts that show significant enrichment in the immunoprecipitation sample compared with nonprecipitated total cellular RNA sample are likely to represent targets of an RNA-binding protein.

### Acknowledgments

We thank Jack Keene (Duke University), Haifan Lin (Duke University), and Thoru Pederson (University of Massachusetts Medical Center) for advice about RNP precipitations. Work in the Jackson laboratory is supported by RO1 HL59873 (NIH), RO1 NS045817 (NIH), P30 NS47243 (NIH), and IBN0234724 (NSF).

### References

Allada, R., White, N. E., So, W. V., Hall, J. C., and Rosbash, M. (1998). A mutant *Drosophila* homolog of mammalian *Clock* disrupts circadian rhythms and transcription of *period* and *timeless*. *Cell* **93,** 791–804.

Belvin, M. P., Zhou, H., and Yin, J. C. P. (1999). The *Drosophila dCREB2* gene affects the circadian clock. *Neuron* **22,** 777–787.

Blau, J. (2001). The *Drosophila* circadian clock: What we know and what we don't know. *Semin. Cell Dev. Biol.* **12**, 287–293.

Brand, A. H., and Dormand, E. L. (1995). The GAL4 system as a tool for unravelling the mysteries of the *Drosophila* nervous system. *Curr. Opin. Neurobiol.* **5**, 572–578.

Brown, V., Jin, P., Ceman, S., Darnell, J. C., O'Donnell, W. T., Tenenbaum, S. A., Jin, X. K., Feng, Y., Wilkinson, K. D., Keene, J. D., Darnell, R. B., and Warren, S. T. (2001). Microarray identification of FMRP-associated brain mRNAs and altered mRNA translational profiles in fragile X syndrome. *Cell* **107**, 477–487.

Ceriani, M. F., Hogenesch, J. B., Straume, M., and Kay, S. A. (2003). Genome-wide expression analysis in *Drosophila* reveals genes controlling circadian behavior. *Cell. Mol. Neurobiol.* **23**, 223.

Claridge-Chang, A., Wijnen, H., Naef, F., Boothroyd, C., Rajewsky, N., and Young, M. W. (2001). Circadian regulation of gene expression systems in the *Drosophila* head. *Neuron* **32**, 657–671.

Cyran, S. A., Buchsbaum, A. M., Reddy, K. L., Lin, M. C., Glossop, N. R. J., Hardin, P. E., Young, M. W., Storti, R. V., and Blau, J. (2003). vrille, Pdp1, and dClock form a second feedback loop in the *Drosophila* circadian clock. *Cell* **112**, 329–341.

Dockendorff, T. C., Su, H. S., Mcbride, S. M. J., Yang, Z. H., Choi, C. H., Siwicki, K. K., Sehgal, A., and Jongens, T. A. (2002). *Drosophila* lacking dfmr1 activity show defects in circadian output and fail to maintain courtship interest. *Neuron* **34**, 973–984.

Ewer, J., Frisch, B., Hamblen-Coyle, M. J., Rosbash, M., and Hall, J. C. (1992). Expression of the *period* clock gene within different cell types in the brain of *Drosophila* adults and mosaic analysis of these cells influence on circadian behavioral rhythms. *J. Neurosci.* **12**, 3321–3349.

Helfrich-Förster, C., Tauber, M., Park, J. H., Muhlig-Versen, M., Schneuwly, S., and Hofbauer, A. (2000). Ectopic expression of the neuropeptide pigment-dispersing factor alters behavioral rhythms in *Drosophila melanogaster*. *J. Neurosci.* **20**, 3339–3353.

Jackson, F. R. (1983). The isolation of biological rhythm mutations on the autosomes of *Drosophila melanogaster*. *J. Neurogenet.* **1**, 3–15.

Kaneko, M., Park, J. H., Cheng, Y. Z., Hardin, P. E., and Hall, J. C. (2000). Disruption of synaptic transmission or clock-gene-product oscillations in circadian pacemaker cells of *Drosophila* cause abnormal behavioral rhythms. *J. Neurobiol.* **43**, 207–233.

Konopka, R. J., and Benzer, S. (1971). Clock mutants of *Drosophila melanogaster*. *Proc. Natl. Acad. Sci. USA* **68**, 2112–2116.

Koundakjian, E. J., Cowan, D. M., Hardy, R. W., and Becker, A. H. (2004). The Zuker collection: A resource for the analysis of autosomal gene function in *Drosophila melanogaster*. *Genetics* **167**, 203–206.

Levine, J. D., Casey, C. I., Kalderon, D. D., and Jackson, F. R. (1994). Altered circadian pacemaker functions and cyclic AMP rhythms in the *Drosophila* learning mutant *dunce*. *Neuron* **13**, 967–974.

Levine, J. D., Funes, P., Dowse, H. B., and Hall, J. C. (2002). Signal analysis of behavioral and molecular cycles. *BMC Neurosci.* **3**(1), 1.

Lin, Y., Han, M., Shimada, B., Wang, L., Gibler, T. M., Amarakone, A., Awad, T. A., Stormo, G. D., Van Gelder, R. N., and Taghert, P. H. (2002). Influence of the period-dependent circadian clock on diurnal, circadian, and aperiodic gene expression in *Drosophila melanogaster*. *Proc. Natl. Acad. Sci. USA* **99**, 9562–9567.

Majercak, J., Kalderon, D., and Edery, I. (1997). *Drosophila melanogaster* deficient in protein kinase a manifests behavior-specific arrhythmia but normal clock function. *Mol. Cell. Biol.* **17**, 5915–5922.

Martinek, S., and Young, M. W. (2000). Specific genetic interference with behavioral rhythms in *Drosophila* by expression of inverted repeats. *Genetics* **156**, 1717–1725.

McDonald, M. J., and Rosbash, M. (2001). Microarray analysis and organization of circadian gene expression in *Drosophila*. *Cell* **107**, 567–578.

McNabb, S. L., Baker, J. D., Agapite, J., Steller, H., Riddiford, L. M., and Truman, J. W. (1997). Disruption of a behavioral sequence by targeted death of peptidergic neurons in *Drosophila*. *Neuron* **19**, 813–823.

McNeil, G. P., Schroeder, A. J., Roberts, M. A., and Jackson, F. R. (2001). Genetic analysis of functional domains within the *Drosophila* LARK RNA-binding protein. *Genetics* **159**, 229–240.

McNeil, G. P., Zhang, X. L., Genova, G., and Jackson, F. R. (1998). A molecular rhythm mediating circadian clock output in *Drosophila*. *Neuron* **20**, 297–303.

McNeil, G. P., Zhang, X., Roberts, M., and Jackson, F. R. (1999). Maternal function of a retroviral-type zinc-finger protein is essential for *Drosophila* development. *Dev. Genet.* **25**, 387–396.

Morales, J., Hiesinger, P. R., Schroeder, A. J., Kume, K., Verstreken, P., Jackson, F. R., Nelson, D. L., and Hassan, B. A. (2002). *Drosophila* fragile X protein, DFXR, regulates neuronal morphology and function in the brain. *Neuron* **34**, 961–972.

Newby, L. M., and Jackson, F. R. (1993). A new biological rhythm mutant of *Drosophila melanogaster* that identifies a gene with an essential embryonic function. *Genetics* **135**, 1077–1090.

Newby, L. M., and Jackson, F. R. (1991). *Drosophila ebony* mutants have altered circadian activity rhythms but normal eclosion rhythms. *J. Neurogenet.* **7**, 85–101.

Park, D. K., Han, M., Kim, Y. C., Han, K. A., and Taghert, P. H. (2004). Ap-let neurons: A peptidergic circuit potentially controlling ecdysial behavior in *Drosophila*. *Dev. Biol.* **269**, 95–108.

Park, J. H., Helfrich-Förster, C., Lee, G., Liu, L., Rosbash, M., and Hall, J. C. (2000a). Differential regulation of circadian pacemaker output by separate clock genes in *Drosophila*. *Proc. Natl. Acad. Sci. USA* **97**, 3608–3613.

Park, S. K., Sedore, S. A., Cronmiller, C., and Hirsh, J. (2000b). Type II cAMP-dependent protein kinase-deficient *Drosophila* are viable but show developmental, circadian, and drug response phenotypes. *J. Biol. Chem.* **275**, 20588–20596.

Park, J. H., Schroeder, A. J., Helfrich-Forster, C., Jackson, F. R., and Ewer, J. (2003). Targeted ablation of CCAP neuropeptide-containing neurons of *Drosophila* causes specific defects in execution and circadian timing of ecdysis behavior. *Development* **130**, 2645–2656.

Renn, S. C. P., Park, J. H., Rosbash, M., Hall, J. C., and Taghert, P. H. (1999). A *pdf* neuropeptide gene mutation and ablation of PDF neurons each cause severe abnormalities of behavioral circadian rhythms in *Drosophila*. *Cell* **99**, 791–802.

Richardt, A., Rybak, A., Strortkuhl, K. F., Meinertzhagen, L. A., and Hovemann, B. (2002). Ebony protein in the *Drosophila* nervous system: Optic neuropile expression in glial cells. *J. Comp. Neurol.* **452**, 93–102.

Rutila, J. E., Suri, V., Le, M., So, W. V., Rosbash, M., and Hall, J. C. (1998). CYCLE is a second bHLH-PAS clock protein essential for circadian rhythmicity and transcription of *Drosophila period* and *timeless*. *Cell* **93**, 805–814.

Schroeder, A. J., Genova, K., Roberts, M. A., Kleyner, Y., Suh, J., and Jackson, F. R. (2003). Cell-specific expression of the lark RNA-binding protein in *Drosophila* results in morphological and circadian behavioral phenotypes. *J. Neurogenet.* **17**, 139–169.

Sehgal, A., Price, J. L., Man, B., and Young, M. W. (1994). Loss of circadian behavioral rhythms and *per* RNA oscillations in the *Drosophila* mutant *timeless*. *Science* **263**, 1603–1606.

Siomi, H., Siomi, M. C., Nussbaum, R. L., and Dreyfuss, G. (1993). The protein product of the fragile-X gene, Fmr1, has characteristics of an RNA-binding protein. *Cell* **74**, 291–298.

Stempfl, T., Vogel, M., Szabo, G., Wulbeck, C., Liu, J., Hall, J. C., and Stanewsky, R. (2002). Identification of circadian-clock-regulated enhancers and genes of *Drosophila melanogaster* by transposon mobilization and luciferase reporting of cyclical gene expression. *Genetics* **160**, 571–593.

Stowers, R. S., and Schwarz, T. L. (1999). A genetic method for generating *Drosophila* eyes composed exclusively of mitotic clones of a single genotype. *Genetics* **152,** 1631–1639.

Tenenbaum, S. A., Lager, P. J., Carson, C. C., and Keene, J. D. (2002). Ribonomics: Identifying mRNA subsets in mRNP complexes using antibodies to RNA-binding proteins and genomic arrays. *Methods* **26,** 191–198.

Ueda, H. R., Matsumoto, A., Kawamura, M., Iino, M., Tanimura, T., and Hashimoto, S. (2002). Genome-wide transcriptional orchestration of circadian rhythms in *Drosophila*. *J. Biol. Chem.* **277,** 14048–14052.

White, K., and Steller, H. (1995). The control of apoptosis in *Drosophila*. *Trends Cell Biol.* **5,** 74–78.

Williams, J. A., Su, H. S., Bernards, A., Field, J., and Sehgal, A. (2001). A circadian output in *Drosophila* mediated by neurofibromatosis-1 and Ras/MAPK. *Science* **293,** 2251–2256.

Wittkopp, P. J., True, J. R., and Carroll, S. B. (2002). Reciprocal functions of the *Drosophila* yellow and ebony proteins in the development and evolution of pigment patterns. *Development* **129,** 1849–1858.

Zhang, X., McNeil, G. P., Hilderbrand-Chae, M. J., Franklin, T. M., Shroeder, A. J., and Jackson, F. R. (2000). Circadian regulation of the LARK RNA-binding protein within identifiable neurosecretory cells. *J. Neurobiol.* **45,** 14–29.

# [36] Membranes, Ions, and Clocks: Testing the Njus–Sulzman–Hastings Model of the Circadian Oscillator

*By* Michael N. Nitabach, Todd C. Holmes, and Justin Blau

## Abstract

Current circadian clock models based on interlocking autoregulatory transcriptional/translational negative feedback loops have arisen out of an explosion of molecular genetic data obtained over the last decade (for review, see Stanewsky, 2003; Young and Kay, 2001). An earlier model of circadian oscillation was based on feedback interactions between membrane ion transport systems and ion concentration gradients (Njus *et al.*, 1974, 1976). This membrane model was posited as a more plausible alternative at the time to the even earlier "chronon" model, which was based on autoregulatory genetic feedback loops (Ehret and Trucco, 1967). The membrane model has been tested in a number of experimental systems by pharmacologically manipulating either ionic gradients across the plasma membrane or ion transport systems, but with inconsistent results. In the meantime, the scope and explanatory power of the genetic models overshadowed inquiries into the role of membrane ion fluxes in clock function. However, several recently developed techniques described in this article have provided a new glimpse into the essential role that membrane ion fluxes play in the mechanism of the core circadian oscillator and indicate that a complete understanding of the clock must include both genetic and membrane-based feedback loops.

## Membrane Model for the Circadian Clock

The heading for this section is the title of a seminal paper by Njus, Sulzman, and Hastings (Njus *et al.*, 1974). In this paper, the authors described a new model for the core circadian oscillator based on interactions between transmembrane ion gradients and the activities of ion transport structures: transporters and/or channels. According to the Njus–Sulzman–Hastings model, the primary mechanism of circadian oscillation is a feedback loop in which transmembrane ion gradients affect the activities of ion transport structures, which in turn affect ion distribution across the membrane. The time delays necessary for oscillation of the system are provided by the kinetics of activation and inactivation of membrane protein function (Njus *et al.*, 1974).

As pointed out by its originators, this model has a number of experimentally testable implications: there must be circadian changes in the activity of transmembrane ion transport pathways, there must be circadian changes in transmembrane ionic gradients, and manipulation of ion transport pathways or ionic gradients must influence clock oscillation (Njus *et al.*, 1974, 1976). The next sections describe pharmacological approaches to testing these predictions of the Njus–Sulzman–Hastings model.

## Testing the Membrane Model in the Molluscan Eye

The marine molluscs *Aplysia californica* and *Bulla gouldiana* exhibit a circadian rhythm of action potential firing in the optic nerves, which originates in a small group of nonphotoreceptor neurons at the base of the retina and can be recorded both *in vivo* and in explanted eyes in culture (Block, 1981; Block and Wallace, 1982; Blumenthal *et al.*, 2001; Jacklet, 1969; Roberts and Block, 1983). The explanted culture system has provided the opportunity to test various aspects of the membrane model: through measurements of transmembrane potential and ionic currents, through manipulation of the ionic composition of the artificial seawater culture medium, and through the application of pharmacological agents that act on ion channels.

Measurements of transmembrane conductance in *Bulla* ocular pacemaker neurons in explanted semi-intact eyes reveal a circadian rhythm. Membrane conductance in free-running conditions is highest during subjective night (when the cells are also most hyperpolarized) and decreases through subjective dawn (when the cells begin to depolarize) reaching a nadir during subjective day (Michel *et al.*, 1993; Ralph and Block, 1990). This conductance seems to be based partly on an increase in a tetraethylammonium (TEA)-sensitive potassium conductance that peaks just before subjective dawn (Michel *et al.*, 1993). *Bulla* ocular pacemaker neurons continue to exhibit a circadian rhythm of membrane conductance even when completely dissociated from the explanted eye and cultured as single isolated cells (Michel *et al.*, 1993). This suggests that the *Bulla* clock functions cell autonomously and is not based on neuronal circuit properties (Michel *et al.*, 1993). Whole-cell perforated-patch voltage-clamp recordings demonstrated that circadian variation in the magnitude of a delayed rectifier potassium current is responsible for the rhythm in TEA-sensitive conductance and that there are also rapidly inactivating and calcium-sensitive potassium currents in *Bulla* ocular pacemaker neurons that do not vary with circadian phase (Michel *et al.*, 1999).

While some treatments affecting ionic conductances or gradients influenced period or phase of the *Bulla* circadian oscillator, none of these treatments stopped circadian oscillations altogether. Both chronic depolarization—induced by raising extracellular potassium—and chronic hyperpolarization—induced by lowering extracellular sodium—lengthened the free-running period (McMahon and Block, 1987). Inhibition of chloride conductance—induced by substituting sulfate, isethionate, or glutamate for chloride in the artificial seawater or by applying a chloride channel blocker—significantly shortened the circadian period of optic nerve impulses (Khalsa *et al.*, 1990). Long-duration treatment with low-calcium artificial seawater containing no added calcium and 10 m$M$ EGTA did not interfere with circadian oscillation during the treatment, although it did induce phase shifts when applied at certain circadian phases (Khalsa *et al.*, 1993). Finally, incubation in sodium-free solutions did not invariably stop circadian oscillation, and when it did, it was attributable to an indirect effect on intracellular pH (Khalsa *et al.*, 1997).

In summary, these results in molluscan ocular pacemaker neurons are consistent with a role for transmembrane ion currents and gradients in entraining input pathways and in output pathways. However, the inability of these treatments to stop the clock did not provide strong support for the Njus–Sulzman–Hastings membrane model of the core circadian oscillator.

Testing the Membrane Model in the Mammalian
   Suprachiasmatic Nucleus

Neurons of the mammalian hypothalamic suprachiasmatic nucleus (SCN) exhibit circadian rhythms of spontaneous action potential firing rate *in vivo*, in explanted brain slices, in organotypic culture, and in dissociated culture (Bos and Mirmiran, 1990; Gillette and Reppert, 1987; Green and Gillette, 1982; Groos and Hendriks, 1982; Herzog *et al.*, 1997; Quintero *et al.*, 2003; Shibata *et al.*, 1982; Welsh *et al.*, 1995). The persistence of independently phased circadian rhythms of firing rate in dissociated SCN neurons suggests that the mammalian circadian oscillator, as in *Bulla*, is cell autonomous (Welsh *et al.*, 1995). The SCN thus provides another model system in which to test the predictions of the Njus–Sulzman–Hastings membrane model in the context of a cell autonomous circadian pacemaker.

A number of studies have examined the nature of the transmembrane ionic currents in SCN neurons and to what extent these currents are regulated by the circadian oscillator. The results reveal that SCN neurons possess a tetrodotoxin (TTX)-sensitive transient inward sodium current, a slowly inactivating persistent sodium current, three distinct outward potassium currents—transient, delayed rectifier, and calcium dependent—a hyperpolarization-activated nonspecific cation current, a nimodipine-sensitive L-type calcium current, and an inward rectifier potassium current (Akasu *et al.*, 1993; Cloues and Sather, 2003; de Jeu and Pennartz, 1997; Kononenko *et al.*, 2004; Pennartz *et al.*, 2002; Teshima *et al.*, 2003; Walsh *et al.*, 1995). Input resistance and spontaneous firing rate are higher in the day than at night (de Jeu *et al.*, 1998; Pennartz *et al.*, 2002), indicating that at least one of the currents that would be active at a typical neuronal resting potential is modulated by the circadian clock. Because the hyperpolarization-activated cation current is not modulated by the circadian clock (de Jeu and Pennartz, 1997), at least one of the outward potassium currents is likely to be subject to clock modulation (de Jeu *et al.*, 1998). Furthermore, since input resistance is higher, and resting membrane potential is more depolarized during the day (de Jeu *et al.*, 1998; Pennartz *et al.*, 2002), it is likely that one or more potassium currents is downregulated during the day. It should be noted that there is substantial anatomical, neurochemical, and functional heterogeneity in the SCN (Jobst and Allen, 2002; Lee *et al.*, 2003; Silver *et al.*, 1999), and it is likely that different classes of SCN neurons will have distinct biophysical properties.

The nimodipine-sensitive L-type calcium current is greater during the day than at night and underlies a 2- to 7-Hz oscillation in membrane potential that is present only during the day (Pennartz *et al.*, 2002). However, nimodipine or TTX treatment does not abolish the day–night differences in input resistance and average resting membrane potential (Pennartz *et al.*, 2002). These results are consistent with a model for an output pathway in which the circadian oscillator influences the spontaneous firing rate of SCN neurons by independently modulating at least two currents: an outward potassium current that is greater during the night than the day, which underlies the day–night difference in resting membrane potential and input resistence, and an L-type calcium current that is greater during the day than the night, which underlies a 2- to 7-Hz oscillation in membrane potential.

In addition to predicting that transmembrane ion fluxes vary in a circadian manner, the Njus–Sulzman–Hastings membrane model predicts that ion concentration gradients vary with circadian phase. Given the numerous intracellular signaling pathways sensitive to calcium, this ion would be a natural candidate for closing the feedback loop between oscillating ion concentrations and oscillating ion fluxes. TTX-resistant ultradian rhythms of intracellular calcium with periods of 5–8 s have been observed in cultured SCN neurons using calcium-sensitive fluorescent indicator dyes (van den Pol et al., 1992; Yamazaki et al., 1995). Circadian rhythms in intracellular calcium have also been observed, but there has been some inconsistency in the pharmacological sensitivity profiles of these rhythms. In one study that measured intracellular calcium using the calcium indicator dye Fura-2, circadian rhythms of intracellular calcium were abolished either by TTX or by the calcium channel blocker methoxyverapamil (Colwell, 2000). This suggests that circadian rhythms in intracellular calcium are a consequence of the circadian rhythm in action potential firing rate and the resulting circadian rhythm in calcium influx through voltage-gated calcium channels. However, in another study in which calcium was measured by transfection with a cDNA encoding the calcium-sensitive fluorescent protein "cameleon," circadian rhythms of intracellular calcium were reduced significantly by ryanodine but were not affected by TTX or nimodipine (Ikeda et al., 2003). Ryanodine blocks release from a subset of internal calcium stores, and its application also stops circadian rhythms in the action potential firing rate (Ikeda et al., 2003). This suggests that circadian rhythms in action potential firing rate may be a consequence— and not a cause—of circadian rhythms in intracellular calcium. Furthermore, it also suggests that intracellular calcium oscillation is a direct consequence of clock modulation of release from ryanodine-sensitive intracellular stores and is not mediated by circadian modulation of calcium influx through voltage-sensitive calcium channels in the plasma membrane. Interestingly, there has also been a report of a circadian rhythm of intracellular chloride concentration in acutely dissociated SCN neurons (Shimura et al., 2002).

In addition to descriptive studies examining circadian variation in transmembrane ion currents and intracellular ion concentration, several studies have assayed the effects on circadian clock function of manipulations of transmembrane ion fluxes. TTX infusion into the SCN of unanesthetized and unrestrained rats blocked the expression of a circadian rhythm in water drinking, but when the TTX infusion was stopped, the drinking rhythm returned with an unaltered phase (Schwartz *et al.*, 1987). This indicates that the TTX blockade of action potential firing in SCN neurons severs the circadian output pathway of the SCN, but that TTX-sensitive currents are not required for the internal circadian pacemaker mechanism of the SCN. This *in vivo* finding has been extended to an examination of the effects of TTX on both explanted SCN and dissociated SCN neurons. In several studies, the TTX blockade of action potential firing was not observed to interfere with the continued oscillation of the circadian timekeeping mechanism (Earnest *et al.*, 1991; Shibata and Moore, 1993; Welsh *et al.*, 1995), suggesting that action potentials mediated by voltage-gated, TTX-sensitive sodium channels are not an essential component of the core circadian clock mechanism. However, in a more recent study in which organotypic slice cultures of SCN were treated with TTX for 7 days, the amplitude of oscillatory activity of the *Period1* promoter within single neurons was observed to run down over the course of the treatment period, suggesting that sodium-dependent action potentials may play some role in maintaining cell-autonomous oscillations (Yamaguchi *et al.*, 2003). These results indicate that sodium-mediated action potentials are not absolutely required for cell autonomous circadian oscillations, but they do not address the role that TTX-insensitive active membrane conductances might play.

## Testing the Membrane Model in *Drosophila melanogaster*

The *Drosophila* molecular clock resides in a set of about 100 clock neurons in the adult *Drosophila* brain that control circadian rhythms of locomotor activity (Blanchardon *et al.*, 2001; Kaneko *et al.*, 2000; Renn *et al.*, 1999). A ventral–lateral subset of these clock neurons ($LN_v$s) produces a neuropeptide, PIGMENT DISPERSING FACTOR (PDF), which may signal the phase of the circadian oscillator to downstream neural circuits that directly control locomotor activity (Park *et al.*, 2000; Renn *et al.*, 1999). Because either ablation of the $LN_v$s or a *Pdf* null mutation abolishes free-running locomotor rhythms (Renn *et al.*, 1999) and because cycling clock gene expression solely in the $LN_v$s is sufficient to drive behavioral rhythms (Frisch *et al.*, 1994; Stanewsky *et al.*, 1998), the $LN_v$s are considered the pacemakers of the *Drosophila* circadian system.

A reverse genetic approach was adopted to address the Njus–Sulzman–Hastings membrane model of clock function in *Drosophila*. The electrical excitability of the $LN_v$ pacemaker neuronal membrane was decreased through a genetically targeted expression of either of two distinct $K^+$ channels: Kir2.1, a mammalian inward rectifier $K^+$ channel (also known as KCNJ2) (Baines *et al.*, 2001), and a C-terminal-truncated form of dORK, a *Drosophila* open rectifier $K^+$ channel (also known as KCNK0)

(Goldstein *et al.*, 1996; Nitabach *et al.*, 2002). These channels exhibit no voltage or time dependence of the open state and behave as $K^+$-selective holes in the cell membrane, similar to the neuronal "leak" conductance. While channels assembled from full-length dORK subunits are highly suppressed in the absence of serine phosphorylation of the C-terminal cytoplasmic domain, channels assembled from engineered truncated subunits (dORK$\Delta$) are not suppressed and thus remain constitutively open (Zilberberg *et al.*, 2000). dORK$\Delta$ and Kir2.1 channels have two effects on neuronal membrane properties that behave synergistically to silence electrical activity. First, their expression decreases the input resistance of a neuron at rest, which shunts synaptic currents and reduces their depolarizing effect. Second, their expression hyperpolarizes a neuron toward the $K^+$ equilibrium potential, thereby increasing the depolarization required to activate voltage-gated $Na^+$ and $Ca^{2+}$ channels.

The dORK$\Delta$ and Kir2.1 cDNAs were cloned into pUAST, a P element transformation vector containing the upstream activating sequence (UAS) that drives transcription in the presence of the GAL4 transcriptional activator, and transgenic flies were generated using standard *Drosophila* embryo injection techniques (Baines *et al.*, 2001; Brand and Perrimon, 1993; Nitabach *et al.*, 2002). Transgenic dORK$\Delta$ and Kir2.1 lines were then crossed to readily available lines that express the GAL4 protein in various cell-specific patterns.

When dORK$\Delta$-C or Kir2.1 was expressed in the PDF-expressing $LN_v$ pacemaker neurons using a *Pdf-GAL4* driver (Renn *et al.*, 1999), they caused severe deficits in free-running circadian locomotor rhythms, with few flies exhibiting any statistically significant rhythms (Nitabach *et al.*, 2002). This indicates that dORK$\Delta$-C or Kir2.1 expression effectively silences the $LN_v$ membrane and prevents communication with downstream neuronal targets. In this way, dORK$\Delta$-C or Kir2.1 expression is analogous to the TTX treatment of mammalian clock neurons described earlier. However, in contrast to TTX treatment of SCN neurons, electrical silencing of the $LN_v$ membrane causes the free-running intracellular molecular oscillator to run down and stop, as assayed using immunocytochemistry for the clock proteins PERIOD and TIMELESS (Nitabach *et al.*, 2002). Unlike free-running molecular oscillations in constant darkness, molecular oscillations in diurnal light–dark conditions were not prevented by electrical silencing of the $LN_v$ membrane (Nitabach *et al.*, 2002). These results can also be contrasted with *Drosophila* clock neuron expression of tetanus toxin light chain—an enzyme that cleaves synaptobrevin and prevents chemical synaptic transmission—which abolishes behavioral rhythms, but does not affect molecular clock oscillations (Kaneko *et al.*, 2000; Nitabach *et al.*, 2005). Taken together, these results suggest that free-running clock oscillations depend on the ability of the neuronal membrane to depolarize and activate voltage-gated ionic conductances, but do not depend on action potential-mediated neuronal outputs. Thus, these experiments strongly support the Njus–Sulzman–Hastings membrane model.

### Detailed Procedure for Electrical Silencing of Drosophila Pacemaker Neurons with dORKΔ Potassium Channel

dORKΔ-C cDNA was generated from full-length dORK by introducing a stop codon after amino acid number 298 by site-directed mutagenesis. dORKΔ-NC cDNA was generated by introducing point mutations by site-directed mutagenesis into each of the two "GY(or F)G" potassium-selective pore motifs present in each of the two dORKΔ-C subunits that assemble into a functional homodimer. These point mutations changed each pore motif into "AAA," which renders the resulting channel completely nonconducting and thus appropriate as a negative control.

The dORKΔ-C and dORKΔ-NC cDNAs were then fused in frame at their 3′ ends to cDNA encoding enhanced GFP (eGFP), thus yielding cDNAs encoding C-terminal eGFP fusion proteins. These fusion proteins were cloned into the multiple cloning site of the pUAST P element transformation vector (Brand and Perrimon, 1993). The pUAST multiple cloning site is downstream of five tandem repeats of the UAS binding site for GAL4 protein and an hsp70 basal promoter, and upstream of an SV40 polyadenylation signal. pUAST also contains the mini-white gene, which drives expression of an enzyme required for synthesis of the red pigment of Drosophila eyes, which allows easy identification of transgenic flies. All of these DNA regions lie in between 5′ P and 3′ P transposition sites.

The two resulting pUAST-dORKΔ plasmids were injected into fertilized syncytial Drosophila embryos at their posterior ends along with a helper plasmid that encodes the P transposase enzyme. In a fraction of the injected embryos, the transposase excised the region of the pUAST-dORKΔ plasmid between the 5′ and the 3′ P transposition sites and inserted it at a random site in one of the four Drosophila chromosomes in a nucleus that will ultimately give rise to the germ line. The adult flies that developed from the injected embryos were then individually crossed with flies carrying a null mutation in the white gene, which have white eyes. Progeny from these individual crosses that exhibited eyes with red pigment carried independent chromosomal insertions of the pUAST-dORKΔ plasmids, which were mapped and balanced using standard genetic methods.

Flies carrying various independent chromosomal insertions of the UAS-dORKΔ-C and UAS-dORKΔ-NC P element transgenes were then mated to flies carrying a chromosomal insertion of a Pdf-GAL4 P element transgene. This transgene contains cDNA encoding the yeast GAL4 transcriptional activator protein downstream of a fragment of the *Pdf* gene promoter that drives GAL4 expression only in the PDF-containing $LN_v$ subset of clock neurons. The progeny possess both a *UAS-dORKΔ* transgene and the *Pdf-GAL4* transgene. The GAL4 protein activates transcription downstream of the UAS element of the *UAS-dORKΔ* transgene, and thus induces expression of either dORKΔ-C or dORKΔ-NC solely in the PDF-containing $LN_v$s. In the case of dORKΔ-C, this hyperpolarized and decreased the input resistance of the $LN_v$s in the background of an unaltered nervous system. dORKΔ-NC served as a negative control.

This procedure can also be used in an analogous manner to ectopically express other ion channel subunits in *Drosophila* clock neurons, thereby manipulating their membrane properties in other ways to probe other aspects of the role of transmembrane ion fluxes in the function of the circadian oscillator. For example, expression of "NaChBac," a bacterial sodium channel with slow kinetics of activation and inactivation, has been used to render *Drosophila* clock neurons *hyper*excitable (Nitabach *et al.*, 2003).

## Conclusions

The Njus–Sulzman–Hastings membrane model posits that the core mechanism of the cellular circadian oscillator relies on feedback between varying transmembrane ion fluxes and varying transmembrane ion gradients. Studies in the molluscan eye and mammalian SCN in which particular ionic conductances or particular ionic gradients were manipulated failed to provide strong support for the membrane model. However, studies in *Drosophila melanogaster* in which the ability of the membrane as a whole to depolarize—and thus to activate any voltage-gated conductances—was impaired to reveal an essential role for active ionic conductances in the core mechanism of the cellular circadian oscillator. Because interlocking autoregulatory transcriptional/translational negative feedback loops are also essential for clock oscillation, a hybrid model is warranted for the core circadian oscillator incorporating aspects of both the Njus–Sulzman–Hastings membrane model and the transcriptional/translational feedback model. Further work is required to elucidate the specific nature of the required ionic conductances, as well as how the ionic and genetic feedback loops are coupled.

## Acknowledgments

The authors thank Gene Block for his valuable comments on a draft version of this chapter.

## References

Akasu, T., Shoji, S., and Hasuo, H. (1993). Inward rectifier and low-threshold calcium currents contribute to the spontaneous firing mechanism in neurons of the rat suprachiasmatic nucleus. *Pflug. Arch.* **425,** 109–116.

Baines, R. A., Uhler, J. P., Thompson, A., Sweeney, S. T., and Bate, M. (2001). Altered electrical properties in *Drosophila* neurons developing without synaptic transmission. *J. Neurosci.* **21,** 1523–1531.

Blanchardon, E., Grima, B., Klarsfeld, A., Chelot, E., Hardin, P. E., Preat, T., and Rouyer, F. (2001). Defining the role of *Drosophila* lateral neurons in the control of circadian rhythms in motor activity and eclosion by targeted genetic ablation and PERIOD protein overexpression. *Eur. J. Neurosci.* **13,** 871–888.

Block, G. D. (1981). *In vivo* recording of the ocular circadian rhythm in *Aplysia. Brain Res.* **222,** 138–143.

Block, G. D., and Wallace, S. F. (1982). Localization of a circadian pacemaker in the eye of a mollusc. *Bulla. Science* **217,** 155–157.

Blumenthal, E. M., Block, G. D., and Eskin, A. (2001). Cellular and molecular analysis of molluscan circadian pacemakers. *In* "Handbook of Behavioral Neurobiology: Circadian Clocks." Plenum, New York.

Bos, N. P., and Mirmiran, M. (1990). Circadian rhythms in spontaneous neuronal discharges of the cultured suprachiasmatic nucleus. *Brain Res.* **511,** 158–162.

Brand, A. H., and Perrimon, N. (1993). Targeted gene expression as a means of altering cell fates and generating dominant phenotypes. *Development* **118,** 401–415.

Cloues, R. K., and Sather, W. A. (2003). Afterhyperpolarization regulates firing rate in neurons of the suprachiasmatic nucleus. *J. Neurosci.* **23,** 1593–1604.

Colwell, C. S. (2000). Circadian modulation of calcium levels in cells in the suprachiasmatic nucleus. *Eur. J. Neurosci.* **12,** 571–576.

de Jeu, M., Hermes, M., and Pennartz, C. (1998). Circadian modulation of membrane properties in slices of rat suprachiasmatic nucleus. *Neuroreport* **9,** 3725–3729.

de Jeu, M. T., and Pennartz, C. M. (1997). Functional characterization of the H-current in SCN neurons in subjective day and night: A whole-cell patch-clamp study in acutely prepared brain slices. *Brain Res.* **767,** 72–80.

Earnest, D. J., Digiorgio, S. M., and Sladek, C. D. (1991). Effects of tetrodotoxin on the circadian pacemaker mechanism in suprachiasmatic explants *in vitro. Brain Res. Bull.* **26,** 677–682.

Ehret, C. F., and Trucco, E. (1967). Molecular models for the circadian clock. I. The chronon concept. *J. Theor. Biol.* **15,** 240–262.

Frisch, B., Hardin, P. E., Hamblen-Coyle, M. J., Rosbash, M., and Hall, J. C. (1994). A promoterless period gene mediates behavioral rhythmicity and cyclical per expression in a restricted subset of the *Drosophila* nervous system. *Neuron* **12,** 555–570.

Gillette, M. U., and Reppert, S. M. (1987). The hypothalamic suprachiasmatic nuclei: Circadian patterns of vasopressin secretion and neuronal activity *in vitro. Brain Res. Bull.* **19,** 135–139.

Goldstein, S. A., Price, L. A., Rosenthal, D. N., and Pausch, M. H. (1996). ORK1, a potassium-selective leak channel with two pore domains cloned from *Drosophila melanogaster* by expression in *Saccharomyces cerevisiae. Proc. Natl. Acad. Sci. USA* **93,** 13256–13261.

Green, D. J., and Gillette, R. (1982). Circadian rhythm of firing rate recorded from single cells in the rat suprachiasmatic brain slice. *Brain Res.* **245,** 198–200.

Groos, G., and Hendriks, J. (1982). Circadian rhythms in electrical discharge of rat suprachiasmatic neurones recorded *in vitro. Neurosci. Lett.* **34,** 283–288.

Herzog, E. D., Geusz, M. E., Khalsa, S. B., Straume, M., and Block, G. D. (1997). Circadian rhythms in mouse suprachiasmatic nucleus explants on multimicroelectrode plates. *Brain Res.* **757**, 285–290.

Ikeda, M., Sugiyama, T., Wallace, C. S., Gompf, H. S., Yoshioka, T., Miyawaki, A., and Allen, C. N. (2003). Circadian dynamics of cytosolic and nuclear $Ca^{2+}$ in single suprachiasmatic nucleus neurons. *Neuron* **38**, 253–263.

Jacklet, J. W. (1969). Circadian rhythm of optic nerve impulses recorded in darkness from isolated eye of *Aplysia*. *Science* **164**, 562–563.

Jobst, E. E., and Allen, C. N. (2002). Calbindin neurons in the hamster suprachiasmatic nucleus do not exhibit a circadian variation in spontaneous firing rate. *Eur. J. Neurosci.* **16**, 2469–2474.

Kaneko, M., Park, J. H., Cheng, Y., Hardin, P. E., and Hall, J. C. (2000). Disruption of synaptic transmission or clock-gene-product oscillations in circadian pacemaker cells of *Drosophila* cause abnormal behavioral rhythms. *J. Neurobiol.* **43**, 207–233.

Khalsa, S. B., Michel, S., and Block, G. D. (1997). The role of extracellular sodium in the mechanism of a neuronal *in vitro* circadian pacemaker. *Chronobiol. Int.* **14**, 1–8.

Khalsa, S. B., Ralph, M. R., and Block, G. D. (1990). Chloride conductance contributes to period determination of a neuronal circadian pacemaker. *Brain Res.* **520**, 166–169.

Khalsa, S. B., Ralph, M. R., and Block, G. D. (1993). The role of extracellular calcium in generating and in phase-shifting the *Bulla* ocular circadian rhythm. *J. Biol. Rhythms* **8**, 125–139.

Kononenko, N. I., Shao, L. R., and Dudek, F. E. (2004). Riluzole-sensitive slowly inactivating sodium current in rat suprachiasmatic nucleus neurons. *J. Neurophysiol.* **91**, 710–718.

Lee, H. S., Billings, H. J., and Lehman, M. N. (2003). The suprachiasmatic nucleus: A clock of multiple components. *J. Biol. Rhythms* **18**, 435–449.

McMahon, D. G., and Block, G. D. (1987). The *Bulla* ocular circadian pacemaker. II. Chronic changes in membrane potential lengthen free running period. *J. Comp. Physiol. A* **161**, 347–354.

Michel, S., Geusz, M. E., Zaritsky, J. J., and Block, G. D. (1993). Circadian rhythm in membrane conductance expressed in isolated neurons. *Science* **259**, 239–241.

Michel, S., Manivannan, K., Zaritsky, J. J., and Block, G. D. (1999). A delayed rectifier current is modulated by the circadian pacemaker in *Bulla*. *J. Biol. Rhythms* **14**, 141–150.

Nitabach, M. N., Blau, J., and Holmes, T. C. (2002). Electrical silencing of *Drosophila* pacemaker neurons stops the free-running circadian clock. *Cell* **109**, 485–495.

Nitabach, M. N., Blau, J., and Holmes, T. C. (2003). Pacemaker membrane excitability controls the period and coherence of the free running *Drosophila* circadian oscillator. Program No. 284.11. 2003 Abstract Viewer/Itinerary Planner. Washington, DC: Society for Neuroscience, 2003.

Nitabach, M. N., Sheeba, V., Vera, D. A., Blau, J., and Holmes, T. C. (2005). Membrane electrical excitability is necessary for the free-running larval *Drosophila* circadian clock. *J. Neurobiol.* **6**, 1–13.

Njus, D., Gooch, V. D., Mergenhagen, D., Sulzman, F., and Hastings, J. W. (1976). Membranes and molecules in circadian systems. *Fed. Proc.* **35**, 2353–2357.

Njus, D., Sulzman, F. M., and Hastings, J. W. (1974). Membrane model for the circadian clock. *Nature* **248**, 116–120.

Park, J. H., Helfrich-Forster, C., Lee, G., Liu, L., Rosbash, M., and Hall, J. C. (2000). Differential regulation of circadian pacemaker output by separate clock genes in *Drosophila*. *Proc. Natl. Acad. Sci. USA* **97**, 3608–3613.

Pennartz, C. M., de Jeu, M. T., Bos, N. P., Schaap, J., and Geurtsen, A. M. (2002). Diurnal modulation of pacemaker potentials and calcium current in the mammalian circadian clock. *Nature* **416,** 286–290.

Quintero, J. E., Kuhlman, S. J., and McMahon, D. G. (2003). The biological clock nucleus: A multiphasic oscillator network regulated by light. *J. Neurosci.* **23,** 8070–8076.

Ralph, M. R., and Block, G. D. (1990). Circadian and light-induced conductance changes in putative pacemaker cells of *Bulla gouldiana*. *J. Comp. Physiol. A* **166,** 589–595.

Renn, S. C., Park, J. H., Rosbash, M., Hall, J. C., and Taghert, P. H. (1999). A pdf neuropeptide gene mutation and ablation of PDF neurons each cause severe abnormalities of behavioral circadian rhythms in *Drosophila*. *Cell* **99,** 791–802.

Roberts, M. H., and Block, G. D. (1983). Mutual coupling between the ocular circadian pacemakers of *Bulla gouldiana*. *Science* **221,** 87–89.

Schwartz, W. J., Gross, R. A., and Morton, M. T. (1987). The suprachiasmatic nuclei contain a tetrodotoxin-resistant circadian pacemaker. *Proc. Natl. Acad. Sci. USA* **84,** 1694–1698.

Shibata, S., and Moore, R. Y. (1993). Tetrodotoxin does not affect circadian rhythms in neuronal activity and metabolism in rodent suprachiasmatic nucleus *in vitro*. *Brain Res.* **606,** 259–266.

Shibata, S., Oomura, Y., Kita, H., and Hattori, K. (1982). Circadian rhythmic changes of neuronal activity in the suprachiasmatic nucleus of the rat hypothalamic slice. *Brain Res.* **247,** 154–158.

Shimura, M., Akaike, N., and Harata, N. (2002). Circadian rhythm in intracellular Cl(-) activity of acutely dissociated neurons of suprachiasmatic nucleus. *Am. J. Physiol. Cell. Physiol.* **282,** C366–C373.

Silver, R., Sookhoo, A. I., LeSauter, J., Stevens, P., Jansen, H. T., and Lehman, M. N. (1999). Multiple regulatory elements result in regional specificity in circadian rhythms of neuropeptide expression in mouse SCN. *Neuroreport* **10,** 3165–3174.

Stanewsky, R. (2003). Genetic analysis of the circadian system in *Drosophila melanogaster* and mammals. *J. Neurobiol.* **54,** 111–147.

Stanewsky, R., Kaneko, M., Emery, P., Beretta, B., Wager-Smith, K., Kay, S. A., Rosbash, M., and Hall, J. C. (1998). The cryb mutation identifies cryptochrome as a circadian photoreceptor in *Drosophila*. *Cell* **95,** 681–692.

Teshima, K., Kim, S. H., and Allen, C. N. (2003). Characterization of an apamin-sensitive potassium current in suprachiasmatic nucleus neurons. *Neuroscience* **120,** 65–73.

van den Pol, A. N., Finkbeiner, S. M., and Cornell-Bell, A. H. (1992). Calcium excitability and oscillations in suprachiasmatic nucleus neurons and glia *in vitro*. *J. Neurosci.* **12,** 2648–2664.

Walsh, I. B., van den Berg, R. J., and Rietveld, W. J. (1995). Ionic currents in cultured rat suprachiasmatic neurons. *Neuroscience* **69,** 915–929.

Welsh, D. K., Logothetis, D. E., Meister, M., and Reppert, S. M. (1995). Individual neurons dissociated from rat suprachiasmatic nucleus express independently phased circadian firing rhythms. *Neuron* **14,** 697–706.

Yamaguchi, S., Isejima, H., Matsuo, T., Okura, R., Yagita, K., Kobayashi, M., and Okamura, H. (2003). Synchronization of cellular clocks in the suprachiasmatic nucleus. *Science* **302,** 1408–1412.

Yamazaki, S., Inouye, S. T., and Kuroda, Y. (1995). TTX-resistant $Ca^{2+}$ oscillation in cultured hypothalamus: Similarity to the mammalian circadian pacemaker. *Neuroreport* **6,** 1306–1308.

Young, M. W., and Kay, S. A. (2001). Time zones: A comparative genetics of circadian clocks. *Nature Rev. Genet.* **2,** 702–715.

Zilberberg, N., Ilan, N., Gonzalez-Colaso, R., and Goldstein, S. A. (2000). Opening and closing of KCNK0 potassium leak channels is tightly regulated. *J. Gen. Physiol.* **116,** 721–734.

# Section IX

# Photoresponsive Clocks

# [37]  Mammalian Photoentrainment: Results, Methods, and Approaches

*By* Stuart N. Peirson, Stewart Thompson,
Mark W. Hankins, and Russell G. Foster

## Abstract

Research on circadian biology over the past decade has paid increasing attention to the photoreceptor mechanisms that align the molecular clock to the 24-h light/dark cycle, and some of the results to emerge are surprising. For example, the rods and cones within the mammalian eye are not required for entrainment. A population of directly light-sensitive ganglion cells exists within the retina and acts as brightness detectors. This article provides a brief history of the discovery of these novel ocular photoreceptors and then describes the methods that have been used to study the photopigments mediating these responses to light. Photopigment characterization has traditionally been based on a number of complementary approaches, but one of the most useful techniques has been action spectroscopy. A photopigment has a discrete absorbance spectrum, which describes the probability of photons being absorbed as a function of wavelength, and the magnitude of any light-dependent response depends on the number of photons absorbed by the photopigment. Thus, a description of the spectral sensitivity profile (action spectrum) of any light-dependent response must, by necessity, match absorbance spectra of the photopigment mediating the response. We provide a step-by-step approach to conducting action spectra, including the construction of irradiance response curves, the calculation of relative spectral sensitivities, and photopigment template fitting, and discuss the underlying assumptions behind this approach. We then illustrate action spectrum methodologies by an in-depth analysis of action spectra obtained from rodless/coneless (*rd/rd cl*) mice and discuss, for the first time, the full implications of these findings.

## Introduction

Our aims in this review are to introduce the topic of circadian entrainment and describe how light has been used as a stimulus to deduce the photopigments that mediate the effects of light on the clock. We provide a brief history of the discovery of novel ocular photoreceptors in mammals and then describe in detail action spectrum techniques and the statistical methods of data analysis. Finally, we illustrate action spectrum methodologies by an in-depth analysis of the action spectra obtained from rodless/coneless (*rd/rd cl*) mice and discuss, for the first time, the full statistical significance of these findings. Although this article concentrates on mammals, the methods described can be applied to the characterization of photopigments in any organism.

*Photoentrainment*

A biological clock is only of any use if internal clock time is adjusted to environmental time. The classic example of a mismatch between biological and environmental time is "jet lag." In the natural environment, many factors could set or entrain circadian rhythms to a specific phase of the 24-h rotation of the earth. Light, temperature, food availability, or even social contact could indicate the time of day. However, only those factors that provide the most reliable indicator will be selected by evolution to entrain circadian clocks, and for most species, the 24-h change in the quantity and quality of light at dawn and dusk provides the entraining signal for photoentrainment (Foster and Helfrich-Forster, 2001; Roenneberg and Foster, 1997). In the absence of light, other signals, such as temperature or food, can act as zeitgebers (time givers), but if animals are exposed to conflicting signals, such as light and temperature, light will invariably be selected as the zeitgeber (Aschoff, 1981).

Colin Pittendrigh was one of the first to make a systematic study of photoentrainment. He explored the effects of short pulses of light on free-running rhythms in a variety of animals, including *Drosophila*, maintained in constant darkness. He showed that light had different effects on the clock at different phases of the circadian cycle. As the animals were kept under constant darkness, he used the phase of activity of the animal to determine the position of the clock. In a nocturnal species, activity onset was considered to indicate the beginning of the night and was designated circadian time (CT) 12 (the subjective night would span CT 12–24), whereas in a diurnal species the start of activity was considered to signal dawn and was designated CT 0 (the subjective day would span CT 0–12). Pittendrigh observed that light pulses given to free-running animals during their subjective day would have no marked effect on the clock, whereas light during the first half of the subjective night (CT 12–18) would phase delay the clock—the animal would start activity later the next day. In contrast, light given during the second part of the subjective night (CT 18–24) would phase advance the clock—activity would start earlier the next day. Pittendrigh studied a range of both nocturnal and diurnal species, and remarkably, all showed the same basic phase response curve (PRC) to light, although the precise shape would vary between species. For a more detailed discussion, see Johnson *et al.* (2003), Pittendrigh (1981), and Pittendrigh and Daan (1976).

In all organisms studied to date, light around dusk will delay the clock and light around dawn will advance the clock, and in this way activity is broadly aligned to the expanding and contracting natural light/dark cycle throughout the year. The sensory demands that bring about photoentrainment appear to have imposed a unique set of selection pressures, which have led to the evolution of specialized photoreceptor systems. In non-mammalian vertebrates, pineal and deep brain photoreceptors are known to play an important, but ill-defined, role in circadian organization (Ekstrom and Meissl, 2003; Shand and Foster, 1999). In contrast, photoentrainment in the mammals relies exclusively on ocular photoreceptors (Foster, 1998).

## Questioning Assumptions about the Eye

Studies on the classical image-forming functions of the eye have attracted the attention of such distinguished figures as Sir Isaac Newton, Thomas Young, John Dalton, Jan Evangelista Purkinje, and Ramon e Cajal. Indeed, the eye has been the subject of serious study for more than 200 years, and in broad terms its functions were thought to be well understood. The rods and cones of the outer retina detect light, and the cells of the inner retina provide the initial stages of image construction before topographically mapped signals travel down the optic nerve to specific sites in the brain for advanced visual processing (Rodieck, 1998). However, studies in the last decade have demonstrated that image formation is not the sole function of the mammalian eye and that the rods and cones are not the only photoreceptive cells within the retina. At first this seemed inconceivable. How could something as important as another class of light-responsive cell within the eye have been missed? Indeed, when the first results suggesting the existence of these novel receptors were presented, they generated a hostile response. There appeared to be a perfectly good set of photoreceptors in the outer retina and so there was need to propose something new. William Ockham's (1285–1349) law of parsimony, *Entia non sunt multiplicanda praeter necessitatem* (Entities should not be multiplied without necessity), was quoted with devastating effect by referees and granting bodies alike.

The discovery of nonrod, noncone ocular photoreceptors resulted from studies on how biological clocks are regulated by light. Vision scientists could account for the responses they were measuring whereas circadian biologists could not. The response of the circadian system to light is not like a visual response. The circadian system needs relatively bright and long exposure to light to achieve photoentrainment. For example, the visual system of the hamster is some 200 times more sensitive to light than the circadian system, and hamsters will not entrain to a light stimulus shorter than 30 seconds (Nelson and Takahashi, 1991). The human circadian system is more than a thousand times less responsive to light than our visual system and, again, needs a long-duration signal to entrain (Foster and Hankins, 2002). Further, the circadian system is sensitive to gross changes in environmental light rather than just the specific patterns of light (Foster and Hankins, 2002). These differences in the response of the circadian system to light are mirrored by very different projections from the eye to the clock centers in the brain. The master circadian pacemaker in mammals, the suprachiasmatic nuclei (SCN), receives its retinal projections from the retinohypothalamic tract (RHT), which is formed from a small number of morphologically distinct retinal ganglion cells (RGCs). These RGCs (around 1% of the total) tend to be distributed evenly over the entire retina and send an unmapped or random projection to the SCN (Provencio et al., 1998). In contrast, the ganglion cells of the visual system send a highly mapped projection to the visual centers of the brain, such that a point on the retina maps precisely to a group of cells in the visual cortex. The visual system is thus able to deduce both how much light and where it occurs in specific regions of the environment, whereas the SCN receives only information about the general brightness of environmental light (Rodieck, 1998). Thus the mammalian eye has parallel outputs, providing both image and brightness information (Foster and Hankins, 2002).

These features led circadian biologists to question the assumption that the rods and cones are the only photosensory cells of the retina. Part of the reason for this more open-minded attitude can be attributed to the fact that the researchers asking these questions were not classically trained visual neuroscientists, but came from areas such as reproductive physiology and animal behavior. Combined with this naiveté, however, was a first-hand knowledge that the vertebrate central nervous system has many enigmatic photoreceptors that help entrain circadian rhythms to the local light/dark cycle (Ekstrom and Meissl, 2003; Shand and Foster, 1999). Knowing that birds, reptiles, amphibians, and fish utilize specialized photosensory cells in the basal brain and pineal to regulate their circadian rhythms made it much easier to ask whether there might be dedicated photoreceptors in the retina for the same task. It did not seem absurd to ask whether rods, cones, uncharacterized retinal photoreceptors, or a combination thereof might regulate the circadian rhythms of mammals.

*A New View of the Mammalian Eye*

Eye loss (like constant darkness) results in free-running rhythms in all mammals studied, including humans. However, claims have been made from time to time that mammals have nonocular photoreceptors. It has even been suggested that humans have photoreceptors behind the knee, as bright light shone in this region apparently shifted human circadian rhythms (Campbell and Murphy, 1998). This was an amusing idea but nothing more, as nobody could replicate these findings (Wright and Czeisler, 2002). All the experimental evidence shows that the circadian system of mammals is entrained exclusively via photoreceptors within the eye (Foster, 1998; Foster *et al.*, 2003b). Although ocular, the rods and cones appear not to be needed for this task. The first detailed experiments that led to this finding were in mutant mice that lacked all rods and most of their cone photoreceptors. This mutant strain of mouse, known as the retinal degeneration or *rd/rd* mouse, is visually blind (Carter-Dawson *et al.*, 1978; Provencio *et al.*, 1994). Despite the massive, but not complete, loss of their rods and cones, these animals showed apparently normal circadian responses to light. Enucleation confirmed that the photoreceptors mediating these effects were indeed located within the eye (Foster *et al.*, 1991). At the very least, these studies in *rd/rd* mice and later studies in other rodents (David-Gray *et al.*, 1998) and humans (Zeitzer *et al.*, 2000) showed that the processing of light by the eye for vision was very different from the way the eye processes light information for the clock. However, suggestive as these early studies were, they did not conclusively demonstrate the existence of a new ocular photoreceptor. Although the rods and cones were massively reduced in these mammals, they were never completely eliminated. There was the possibility that even a small number of photoreceptors may still be sufficient to maintain normal circadian responses to light. As a result, a new model, the *rd/rd cl* mouse, was developed that lacked all functional rods and cones. Remarkably, these mice still showed normal circadian responses to light (Freedman *et al.*, 1999; Lucas *et al.*, 1999). Again, by blocking light reaching the eye, the effects of light on the circadian system were abolished so there had to be a novel ocular photoreceptor. Because these mice lacked an outer retina, the new photoreceptor cells were assumed to be located within the inner retina, and RGCs were strong candidates (Provencio *et al.*, 1998). The search was on for these novel sensory cells using a range of different approaches.

As discussed in detail later, the spectral response or action spectrum for adjusting the circadian clock of *rd/rd cl* mice demonstrates the involvement of an novel opsin/vitamin A-based photopigment with a wavelength of maximum sensitivity ($\lambda_{max}$) of ~480 nm ($OP^{480}$) (Hattar *et al.*, 2003; Lucas *et al.*, 2001). The known mouse photopigments peak at ~360 nm (UV cone) (Jacobs et al., 1991), ~498 nm (rod) (Bridges, 1959), and ~508 nm (green cone) (Sun et al., 1997) and do not show any significant fit to the 480-nm action spectrum in rd/rd cl mice. These results provided overwhelming evidence that a novel photopigment contributes to the regulation of circadian responses to light in the mouse. Studies in both rats (Berson et al., 2002) and human subjects have also suggested a photopigment with a similar spectral sensitivity (Brainard et al., 2001; Hankins and Lucas, 2002; Thapan et al., 2001). A cellular analysis of these photoreceptors in rodents showed that at least some of the RGCs that project to the SCN are intrinsically light sensitive. This was demonstrated by recording from single RGCs from the normal rat retina isolated either surgically or pharmacologically from the rods and cones (Berson et al., 2002) or by monitoring light-induced changes in calcium concentration in the RGCs of the whole retina from rd/rd cl mice (Sekaran et al., 2003). Significantly, these light-sensitive RGCs express the candidate photopigment melanopsin (Hattar et al., 2002; Provencio et al., 2000). Evidence that melanopsin plays an important role in the transduction of light information in the intrinsically photosensitive RGCs comes from several sets of experiments. The first showed that melanopsin knockout mice have attenuated phase-shifting responses to light and that the photosensitive RGCs fail to respond to light in these melanopsin knockout animals (Lucas et al., 2003; Panda et al., 2003; Ruby et al., 2002). Most recently, mice in which rods, cones, and melanopsin-based photoreceptors have all been ablated fail to show any responses to light, arguing that these three classes of photoreceptor account for all light detection within the mouse eye (Hattar et al., 2003; Panda et al., 2003).

Although highly suggestive, melanopsin knockout data do not confirm that melanopsin is the photopigment of the photosensitive RGCs (Foster and Bellingham, 2002; Peirson et al., 2004). Gene ablations studies alone can only indicate that a gene is critical; biochemistry on the protein product is required to define its function. In an attempt to address the biochemical role of melanopsin, studies have expressed the protein in COS cells and found that after reconstitution with 11-cis-retinal the pigment showed a maximal absorbance between 420 and 440 nm (Newman et al., 2003). This $\lambda_{max}$ is in marked contrast to action spectra for nonrod, noncone photoreception in rd/rd cl mice, which defined a photopigment close to 480 nm ($OP^{480}$). If ~430 nm is the true $\lambda_{max}$ of melanopsin, then it is difficult to see how melanopsin could be the photopigment responsible.

## Action Spectra

Photoreceptor characterization has traditionally been based on a number of complementary approaches, but one of the most useful has been action spectroscopy. A photopigment has a discrete absorbance spectrum, which describes the probability of photons being absorbed as a function of wavelength, and the magnitude of any light-dependent response depends on the number of photons absorbed by the photopigment. Thus, a description of the spectral sensitivity profile (action spectrum) of any light-dependent response must, by necessity, match absorbance spectra of the photopigment mediating the response, provided that any confounding factors, such as screening pigments or absorption of ocular media, are taken into account (Lythgoe, 1979).

Despite the fact that light can be used in a variety of different ways (as an energy source, as a sensory stimulus, or as a regulatory signal), relatively few photopigments appear to have evolved (Wolken, 1995). This may relate to the demanding task of a photopigment molecule. It must be able to absorb a photon with high probability and, again with high probability (high quantum efficiency), pass on this information to a transduction mechanism (Wolken, 1995). In nature, a wide variety of different light responses are regulated by only a few types of photopigment, and these have highly conserved absorbance spectra. For example, both action spectra for phototaxis in the algae *Chlamydomonas* and visual photosensitivity in humans are described by the same standard absorbance spectrum for an opsin/vitamin A photopigment (Wolken, 1995). In the same way, flavoprotein-based photoresponses can also be described by conserved absorbance spectra (Smyth *et al.*, 1988). Because action spectra reflect photopigment absorbance spectra and because individual photopigment families have characteristic absorbance spectra (akin to a spectral fingerprint), the photopigments mediating a response can be deduced using action spectroscopy. It is worth stressing that when an action spectrum is conducted correctly, it provides one of the only approaches that can be used to both identify a class of protein and simultaneously link that protein to a particular cellular or behavioral function.

Considerable progress has been made in identifying the different photoreceptor organs and photopigments of animal circadian systems, and action spectroscopy has played a critical role in this endeavor. For example, the identification $OP^{480}$ in rodents and many other photopigments associated with extraretinal responses to light have been identified by using this technique (Table I). This article provides a detailed overview of action spectroscopy and the potential use of this powerful technique in understanding the photopigments of the circadian system. We address why action spectra have been used, consider the procedures of how to undertake action spectra, and detail the statistical analysis of data. Finally, we illustrate action spectrum methodologies by an analysis of action spectra obtained from *rd/rd cl* mice and discuss, for the first time, the full statistical significance of these findings.

TABLE I
ACTION SPECTROSCOPY OF NONIMAGE-FORMING RESPONSES TO LIGHT

| Biological response | $\lambda_{max}$ | Species | Reference |
|---|---|---|---|
| Frontal organ response | 500–520 | Clawed toad | Korf et al. (1981) |
| Pineal NAT* suppression | 495 | Rat | Bronstein et al. (1987) |
| Pineal melatonin suppression | 500 | Atlantic salmon | Max and Menaker (1992) |
| Pineal NAT* suppression | 500 | Chicken | Deguchi (1981) |
| LH† secretion | 492 | Japanese quail | Foster et al. (1985) |
| Melanin aggregation | 461 | Clawed toad | Lythgoe and Thompson (1984) |
| Contraction of isolated iris | 500 | Frog | Barr and Alpern (1963) |
| Phase-shifting rd/rd | 511 | Mouse | Provencio and Foster (1995) |
| | 480 | Mouse | Yoshimura et al. (1994) |
| Pupillary light response | 479 | Mouse | Lucas et al. (2001) |
| Melatonin suppression | 446–477 | Human | Brainard et al. (2001) |
| | 459 | Human | Thapan et al. (2001) |
| Cone ERG‡ | 483 | Human | Hankins and Lucas (2002) |
| Intrinsic horizontal cell responses | 477 | Roach | Jenkins et al. (2003) |
| Melanopsin retinal ganglion cells | 484 | Rat | Berson et al. (2002) |
| Phase-shifting rd/rd cl | 481 | Mouse | Hattar et al. (2003) |

*NAT, N-acetyltransferase.
†LH, Luteinizing hormone.
‡ERG, Electroretinogram.

## Investigating Photopigments

The first law of photochemistry states that only the radiation absorbed by a molecule can produce a photochemical effect. As a result, the most direct method of investigating and characterizing a photopigment molecule is by absorption spectroscopy. This section outlines these methods and discusses why this technique has rarely been used to study the circadian photopigments.

### Absorption Spectroscopy

The light-sensitive pigment rhodopsin was first described in 1877 by Franz Boll, who noted that the "visual purple" of the retina bleached upon exposure to light. George Wald and co-workers subsequently determined that this molecule was composed of an opsin protein combined with a chromophore formed from the aldehyde of vitamin A, which was ultimately termed *retinal*. The primary event in light detection is the absorption of a photon by 11-*cis*-retinal and its photoisomerization to the all-*trans* state. This conformational change in the retinal chromophore causes a change in the opsin protein, which in turn activates elements of the phototransduction cascade (Pepe, 2001). As well as determining the involvement of vitamin A in vision, Wald (1968) also studied the stages of visual excitation, along with the intermediates of the bleaching and regeneration of rhodopsin, winning the 1967 Nobel Prize in Medicine for his work on the primary visual processes.

The analysis of photopigments in solution by absorption spectroscopy has provided an ideal means of characterizing the visual pigments of rod photoreceptors. Typically, this involves solubilizing the membrane fragment of the retina in detergent and then purifying the pigment in an organic solvent (De Grip, 1982). All visual pigments have similarly shaped spectra, with small structural differences in the opsin protein determining the $\lambda_{max}$ (Bowmaker and Hunt, 1999; Knowles and Dartnall, 1977). In 1953, H. J. A. Dartnall demonstrated that vitamin A-based photopigments have a characteristic absorption profile, and when expressed on a frequency scale $(1/\lambda)$, the shape of this profile is the same irrespective of the $\lambda_{max}$. Thus although the $\lambda_{max}$ may vary, the shape of the absorption spectrum remains the same. In short, a single curve defined by the $\lambda_{max}$ will provide the entire absorption profile of a vitamin A-based photopigment. Additional templates have been provided by Lamb (1995), Partridge and DeGrip (1991), and Govardovskii *et al.* (2000) based on an improved fit to empirical data.

Wald described a variant form of the vitamin A chromophore in the visual pigments of freshwater fish, which use 3-dehydroretinal as an alternative form of retinal. Photopigments based on 3-dehydroretinal (also called vitamin $A_2$) form a subgroup of visual pigments called porphyropsins. Those based on the "normal" retinal or vitamin $A_1$ are called rhodopsins (Lythgoe, 1979). Spectrophotometric studies showed that the precise shape of the absorption spectrum is determined by the type of chromophore, either vitamin $A_1$ or vitamin $A_2$ (Bridges, 1965; Bridges *et al.*, 1979; Knowles and Dartnall, 1977).

It is often difficult to get a pure photopigment extract. One way to deal with the absorption of impurities within a preparation (e.g., hemoglobin) is to conduct difference spectra. Two spectra are measured: one before and a second after light exposure. The bleached spectrum is then subtracted from the initial absorption spectrum to deduce the contribution of the light-sensitive pigment alone (Knowles and Dartnall, 1977). However, it is often difficult to extract a large amount of photopigment in a sufficiently pure form. As a result, absorption spectroscopy is only suitable if the photopigment naturally occurs in high abundance, as with rhodopsin in the bovine eye, or can be functionally expressed at high concentrations (DeGrip *et al.*, 1999).

*Microspectrophotometry*

Microspectrophotometry (MSP) provides a partial solution to the problems of isolating a sufficient amount and purity of photopigment in solution. MSP utilizes a narrow monochromatic beam of light, which is passed through the photopigment-containing outer segment of a photoreceptor cell to obtain the absorbance of the pigment. The advantage of this approach is that it enables photopigments to be characterized within their native photoreceptors. For further details of MSP, see Bowmaker (1984). MSP of human rod photoreceptors has shown a remarkable match with the spectral sensitivity of human scotopic vision—as derived from action spectroscopy. Similarly, MSP of the human cone photopigments clearly matches the spectral sensitivity of human photopic vision (Brown and Wald, 1963; Wald, 1958).

*Why Use Action Spectra?*

Unfortunately, approaches such as absorption spectroscopy and MSP cannot be applied to the study of photopigments where the location of the photoreceptor itself may be unknown or where the photopigment concentration may be too low for detection. In such cases, action spectra provide an indirect method of analyzing the photopigments. Action spectra determine the spectral sensitivity of a biological response at different wavelengths. To recap, if a relatively low photon dose at a particular wavelength elicits a marked biological response, then this suggests that the photopigment is relatively sensitive at this wavelength. Conversely, if a higher photon dose at another wavelength is needed to cause the same level of response, then this indicates reduced sensitivity. If the reciprocal of the light response is plotted against wavelength ($\lambda$), then the action spectrum should mirror the absorption spectrum of the pigment mediating the response (Lythgoe, 1979).

Methods for Action Spectroscopy

An overview of the methods of action spectroscopy is provided. We consider the selection of an appropriate biological response and discuss experimental design, the generation of irradiance response curves (IRCs), the construction of action spectra, and template fitting. Finally, we outline some of the pitfalls that may be encountered with these approaches.

*The Biological Response*

The first consideration of any action spectrum is the choice of a suitable biological response. Ideally, the effect should be graded in response to light, with increased stimulus intensity (irradiance; see later) causing a larger effect over a fairly broad dynamic range. Furthermore, the response needs to be fairly rapid and show little variation in response to the same stimulus. All this is critical given that repeated measures will need to be collected over an extended period of time. Sometimes these experiments can take years (Foster *et al.*, 1985).

As discussed later, the primary objective in constructing an action spectrum is to generate a set of IRCs at different wavelengths of light. The precise number of wavelengths, irradiance levels, and replicates per irradiance level will be a compromise based on such factors as the variability in response and the length of time required to undertake the experiments. The most important consideration, however, is resolution of the IRCs. Ultimately it is not sensible to generate a large number of IRCs at different wavelengths if each is poorly resolved. It is far better to use the same resources to generate a smaller number of well-resolved IRCs. The resolution of an IRC is reflected by the $R^2$ value obtained when fitting a curve to a representative number of data points (see later).

Another consideration relates to the use of short-wavelength stimuli. Many artificial light sources produce relatively little short-wavelength light. Organic molecules and biological pigments absorb in the UV region of the spectrum, and in addition, short-wavelength light is scattered to a much greater extent than longer wavelengths (Rayleigh scatter) (Lythgoe, 1979). Because responses to short-wavelength stimuli are modified by these factors, the responses measured often provide less reliable measures of relative sensitivity. As a result, it is often beneficial to bias the resources to an analysis of the longer wavelengths within the dynamic range of the response.

*Measures of Light*

The measurement of light can be confusing. Table II defines the main units that are used to measure light (defined as that part of the electromagnetic spectrum between ~350 and 750 nm). For most purposes, the radiometric measure of irradiance provides the most useful unit of light measurement for action spectroscopy. In contrast to radiometric measures, photometric measures of light attempt to mimic the spectral sensitivity of the human visual system and do not measure all wavelengths of the electromagnetic spectrum with equal efficiency. As a result, photometric detectors/units should never be used to measure the flux of a monochromatic light source for action spectroscopy.

In all interactions of light and matter, energy is transferred in discrete packets or quanta of energy, termed *photons*. Photopigments essentially act as photon counters, and therefore it is critical to define the photon flux

TABLE II
UNITS OF RADIOMETRY AND PHOTOMETRY

**A. Radiometric measures**

Measurement of electromagnetic energy within the optical spectrum, which includes ultraviolet radiation, visible light, and infrared radiation. An ideal radiometric detector has a "flat" spectral response

▼                                    ▼

| **Irradiance** | **Radiance** |
|---|---|
| Measure of radiant energy from all directions over a 180° field of view | Measure of radiant energy viewed from a specific direction or region in space |
| Common units: | Common units: |
| erg/s/cm$^2$ | erg/s/cm$^2$/sr |
| $\mu$W/cm$^2$ | $\mu$W/cm$^2$/sr |
| photons/cm$^2$/s | photons/cm$^2$/s/sr |

**B. Photometric measures**

Measuring human visual responses to radiant energy. A measurement of visible light that falls between the wavelengths of 400 and 700 nm. The spectral response of a photometric detector is not flat but attempts to reproduce that of the average human eye. Two average human eye responses are used: a photopic response (maximum sensitivity at a wavelength of 555 nm) and a scotopic response (maximum sensitivity at a wavelength of 507 nm). By convention, photometric measurements are considered photopic unless otherwise stated

▼                                    ▼

| **Illuminance (illumination)** | **Luminance** |
|---|---|
| Measure of light from all directions over a 180° field of view | Measure of light viewed from a specific direction or region in space |
| Common units: | Common units: |
| lux (lx) | candela (cd) |
| lumen (lm)/m$^2$ | lumen (lm)/sr |
| phot (ph) = lm/cm$^2$ | cd/m$^2$ = lm/m$^2$/sr |
| foot candles (fcd) = lm/ft$^2$ | lm/m$^2$/sr = lx/sr |
|  | footlamberts (fL) = fcd/sr |

of a light stimulus. As the energy per photon is proportional to $1/\lambda$, short wavelengths possess relatively greater photon energies than longer wavelengths. Consequently, the use of stimuli of the same energy at different wavelengths will not present a comparable stimulus to a photopigment. IRCs must be constructed using measures of photon flux, which can be calculated as described later. Irradiance is measured most commonly using a power meter, typically in $\mu$W/cm$^2$. The energy per photon ($N_p$) can be calculated for a specific wavelength $\lambda$ (in nm) as

$$\text{Energy per photon}(N_p) = hc/\lambda \qquad (1)$$

where $h$ is Planck's constant ($6.625 \times 10^{-34}$ W/s$^2$) and $c$ equals the speed of light in a vacuum ($3.00 \times 10^{17}$ nm/s), giving $hc$ a value of $1.99 \times 10^{-16}$ W/s or

$1.99 \times 10^{-10}$ in $\mu$W/s. The photon flux can then be calculated by dividing the irradiance (in $\mu$W/cm$^2$) by the energy per photon:

$$\text{Photon flux}(\Phi_p) = \frac{irradiance}{N_p} \tag{2}$$

For example, for monochromatic light at 500 nm, the energy per photon is $3.98 \times 10^{-19}$ W/s or $3.98 \times 10^{-13}$ $\mu$W/s. For an irradiance of 0.03 $\mu$W/cm$^2$, this would give a photon flux of $7.44 \times 10^{10}$ photons/cm$^2$/s.

## Constructing Irradiance Response Curves (IRCs)

For each wavelength, the IRC is plotted as log photon flux ($x$ axis) versus response magnitude ($y$ axis) as shown in Fig. 1A. A family of IRCs is then used to derive an equivalent response at different wavelengths and is usually set at half-maximum, often referred to as an $EC_{50}$. The $EC_{50}$ is then calculated from an IRC by fitting a sigmoid function. A sigmoid curve can be defined by four parameters:

$$\text{Response} = \frac{(top - bottom)}{1 + 10^{[(EC_{50}-n)xk]}} \tag{3}$$

Where *top* and *bottom* correspond to the maximum saturating response and baseline response, respectively; the $EC_{50}$ is the photon flux necessary to elicit a half-maximum response (around which the curve rotates); and $k$ is the slope of the curve, also known as the Hill slope or slope factor (Motulsky and Christapoulos, 2003).

Ideally, every IRC at each wavelength should have an empirically derived saturating response, and a four-parameter model should be used [Eq. (3)]. However, in practice this may not be achieved, as it may not be possible to produce a sufficient photon flux to obtain a maximal response for every wavelength from the light source, particularly at shorter wavelengths. If the saturating response is known for most of the IRCs, then this value may be applied as the maximum for all IRCs. Most biological responses will produce no detectable response at a low photon flux, and thus the baseline can be set to 0. This leaves just two remaining parameters to fit: the $EC_{50}$ and the Hill slope [Eq. (3)].

Fitting a sigmoidal model to the IRCs can be performed using any good statistical package, and the resulting $EC_{50}$ and slope will be given. However, data analysis using a standard statistical package can be very time-consuming. To address this issue, we have assembled an Excel-based workbook for the automated analysis of IRCs, the construction of action spectra, and the fitting of rhodopsin (vitamin $A_1$) and porphyropsin (vitamin $A_2$) photopigment templates. This system enables analysis of up to 10

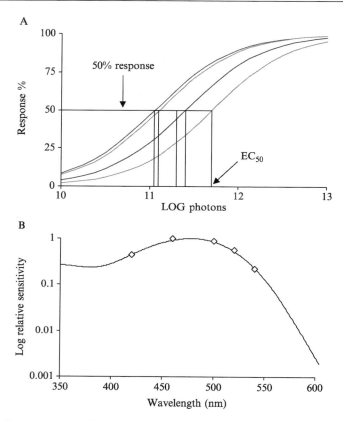

Fig. 1. Basic principles of action spectroscopy. (A) Irradiance response curves (IRCs) of response magnitude versus log photon dose are plotted for each wavelength. Five schematic IRCs are shown. The photon dose required to elicit a half-maximum response is then calculated ($EC_{50}$). (B) The relative sensitivity is derived from the $EC_{50}$, and this value is plotted against wavelength to give the action spectrum ($\diamond$). A photopigment template may then be fitted to determine the $\lambda_{max}$ of the photopigment mediating the biological response under analysis (solid line). (See color insert.)

independent wavelengths, with 10 irradiance levels per wavelength and 10 samples per irradiance. Tests for univariance are included (see later) and, additionally, EC values can be calculated at any level. This workbook is available on request or as a free download from our Web site (s.peirson@imperial.ac.uk; http://wwwfom.sk.med.ic.ac.uk/medicine/about/divisions/neuro/npmdepts/visualneuro/default.html).

*Nonlinear Regression*

The mathematical approaches used to fit IRCs, as well as photopigment templates, fall under the broad title of nonlinear regression. For those using a statistical package to analyze data or with a severe phobia of numbers, the details of these methods can be overlooked. However, understanding the way in which these models work can be very useful in determining whether the results obtained are accurate, for diagnosing problems associated with curve fitting, and for confirming that the software is actually doing what it is supposed to do. No matter how good the software, if it is used incorrectly, the output is meaningless. Rubbish in, rubbish out!

Unlike linear regression, the best fit of a nonlinear model cannot be calculated simply and requires an iterative approach. These methods are dependent on computationally intensive algorithms that minimize the sum of squares of the regression ($SS_{regression}$)—in other words, that the model accurately describes the data. The $SS_{regression}$ is derived from the total sum of the squared differences between data and model (Fig. 2A). For an IRC, this involves determining the difference between data points and model at each irradiance level. This difference is then squared so as to become sign independent, as an equal distribution of points above and below the model would inaccurately result in a $SS_{regression}$ of 0. Thus

$$SS_{regression} = \sum_{n=1}^{i} \left[ \left( (d_1 - r_1)^2 + (d_2 - r_2)^2 + \ldots (d_i - r_i)^2 \right) \right] \qquad (4)$$

where $d_n$ represents the data value at irradiance $n$ and $r_n$ is the model value at that point. The $SS_{regression}$ is then compared with the total sum of squares ($SS_{total}$), which assumes a straight line through the mean ($\bar{x}$), as shown in Fig. 2B. Again the values are summated:

$$SS_{total} = \sum_{n=1}^{i} \left[ \left( (d_1 - \bar{x})^2 + (d_2 - \bar{x})^2 + \ldots (d_i - \bar{x}) \right)^2 \right] \qquad (5)$$

The measure of fit ($R^2$) is a value between 0 and 1 and may be thought of as the fraction of the total variance that is explained by the model. When $R^2 = 0$, then the model is no better than a horizontal line drawn through the mean, and when $R^2 = 1$, the model describes data perfectly, fitting every point. $R^2$ is calculated from

$$R^2 = 1 - \frac{SS_{regression}}{SS_{total}} \qquad (6)$$

Due to the iterative nature of nonlinear regression, the starting point for the parameters to be fitted is also important. A reasonable approximation must

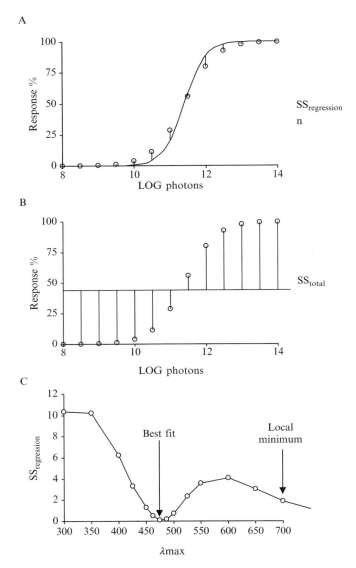

Fig. 2. Fundamental concepts of nonlinear regression. (A) A mathematical model is fitted to data with a number of variable parameters. The squared difference is calculated at each data point to give the sum of squares of the regression ($SS_{regression}$). The parameters are changed systematically to minimize the $SS_{regression}$ to provide an optimum fit. (B) A straight line is plotted through the mean of data, and the squared difference at each data point is summated to give the total sum of squares ($SS_{total}$). The fit of the model ($R^2$) is the proportion

be used, or a local minimum may be achieved, at which point the iterative procedure will stop. This is visualized easily by thinking of the best fit as the lowest point of a landscape formed by the sum of squares, with the curve-fitting algorithm changing the parameters to try and reach the lowest point of this landscape. However, regions may exist within this landscape with a low SS, but not the lowest possible SS. These local minima will lead to the search for the minimal SS being terminated—often with a resulting poor fit (see Fig. 2C). Nonlinear regression cannot therefore be conducted from an arbitrary starting point—a reasonable approximation must be provided for each parameter. As a final note, $R^2$ is normally used to denote the fit of nonlinear regression, whereas $r^2$ is used to denote the fit of linear regression (Motulsky and Christapoulos, 2003).

### Calculating Relative Sensitivity

Once IRCs have been constructed and the $EC_{50}$ calculated for each wavelength, the relative sensitivity can be derived. As sensitivity is based on the photon flux required to elicit the same response; a low photon flux is indicative of high sensitivity to this wavelength, and *vice versa*. To calculate relative sensitivity, divide the lowest photon flux obtained (i.e., the wave-length to which the response is most sensitive) by the $EC_{50}$ photon flux at each wavelength (on a linear scale, not in log units). This will produce a value between 0 and 1, with the wavelength of maximum sensitivity having a value of 1 and higher photon fluxes possessing values below 1. The action spectrum is simply the plot of relative sensitivity ($y$ axis) against wavelength ($x$ axis), as shown in Fig. 1B.

Instead of using the $EC_{50}$ for construction of an action spectrum, it is also possible to use the photon flux required to elicit any level of response. This may be necessary if it is difficult to produce a half-maximum response with the light source available, as the $EC_{50}$ would be an extrapolation beyond available data. If the slopes of the IRCs are comparable across all wavelengths, then the action spectrum will be the same using an $EC_{20}$, $EC_{80}$, or any other level of response. Use of a lower level of response (e.g., $EC_{20}$) may be useful when the biological response itself screens the light available to the photopigment. For example, the light-induced change in melanin distribution within a photosensitive melanophore will screen the amount of light reaching the photopigment. A lower EC value, with

---

of the variance explained by the model. A value of 1 indicates a perfect description of data, whereas a value of 0 indicates the fit is no better than a straight line through the mean. (C) An unsuitable starting point for a parameter may lead to a poor fit being obtained due to the iterative procedure terminating when reaching a local minimum $SS_{regression}$ rather than the true best fit.

minimal screening, may provide a more accurate representation of the unscreened photopigment response. However, the danger here is that the level of noise will be much higher when the EC value is low and the response close to threshold. Differences in IRC slope are suggestive of differing response mechanisms, as described later, and so care must be taken when conclusions are drawn from sliding EC values.

## Fitting a Template

As discussed earlier, photopigments have characteristic absorption spectra, and the shape of an action spectrum can therefore be used to characterize the type of photopigment mediating a response. Given that the most common photopigments in the vertebrates are vitamin A based, the use of these pigment templates provides a good starting point for the analysis of circadian responses to light. If a vitamin A-based template provides a poor match, then the use of alternative templates is of course an option (Wolken, 1995).

The fitting of a photopigment template is best achieved when relative sensitivity ($y$ axis) is expressed as a logarithm. The reason for this becomes clear when one considers the calculations for $R^2$. As the $SS_{regression}$ is calculated from the squared difference between data and model, a difference of 10% near the wavelength of maximum sensitivity will contribute proportionally more to the $SS_{regression}$ than a 10% difference at any other wavelength. When calculated on a logarithmic scale, a 10% difference would give the same contribution to the $SS_{regression}$ at all wavelengths. This is important in that it ensures that all data points contribute equally to the fit of the photopigment template. Fitting a photopigment template relies on the same nonlinear protocol as for fitting an IRC. The only variable here is the $\lambda_{max}$, and the $SS_{regression}$ is minimized by changing this parameter to provide the best possible fit, with a resulting $R^2$ value as described earlier.

## Conditions Necessary for Reliable Action Spectra

The construction of action spectra is dependent on several assumptions, and if these are violated then the resulting action spectra may be distorted and not truly reflect the absorption spectrum of the photopigment driving the response. Conclusions based on action spectroscopy may not be reliable if there is (1) failure to show univariance, (2) lack of reciprocity, and (3) screening effects of additional pigments that modify the light reaching the photopigment.

1. Univariance is evident when the driving mechanism is the same at all wavelengths. This is apparent in a family of IRCs when they all show a similar

Hill slope ($k$ value). If IRCs demonstrate markedly different slopes, then this suggests a different response mechanism at different wavelengths, as might occur if different photopigments contribute to the same biological response. The law of photochemical equivalence states that every atom, ion, or molecule undergoing a reaction must absorb a photon. The proportion of molecules that undergo a specific reaction following absorption of a photon is termed the *quantum efficiency* of the reaction. For example, the absorption of light by retinal in an opsin-based photopigment has been measured at 0.67, i.e., two-thirds of the photons absorbed lead to an isomerization from the 11-*cis* to all-*trans* form (Dartnall, 1968). To ensure that the action spectrum is derived from a single photopigment, univariance must hold. Testing for univariance can be accomplished by determining the proportion of the variance attributable to the slope in each IRC. By fitting IRCs with a fixed slope and subsequently freefitting with independently variable slopes, the improvement in fit may be quantified statistically against the reduction in degrees of freedom (produced by fitting an additional parameter). Based on an $F$ distribution, one can then determine whether the reduction in variance achieved is significantly greater than that expected to occur by chance. These calculations are provided within the action spectra analysis sheet (see earlier) and can be used for any specified slope.

2. The law of reciprocity states that photochemical action is dependent on the product of light intensity and duration of exposure. Thus, exposure to low irradiances for longer durations should yield the same effect as higher irradiances for shorter durations. Circadian responses to light have long integration times, as such that reciprocity holds for durations up to 45 min in some rodents (Nelson and Takahashi, 1991). In contrast, visual responses integrate light information rapidly, and reciprocity holds for durations of usually less than 3 s. However, reciprocity will break down when saturating stimuli are applied, as the addition of extra photons cannot initiate any further response. It therefore becomes critical to know the dynamic range of the response to light, hence the required use of an IRC, and the recommended use of an $EC_{50}$.

3. The modification of light by pigments overlaying a photopigment may modify the apparent spectral sensitivity of the biological response greatly. Several factors may affect the wavelength of incident light, including increased scattering of short wavelengths by Rayleigh scatter (Lythgoe, 1979) and the absorption by pigments such as hemoglobin and melanin in overlying tissues (Foster and Follett, 1985; Hartwig and van Veen, 1979). A common, and often overlooked, problem is if the biological response itself changes the light exposure to the photopigment, as can occur with nonconsensual pupil constriction and the light-induced dispersion of melanin in photosensitive melanophores.

Phase Shifting in the *rd/rd cl* Mouse

To illustrate the methodologies discussed earlier, we have conducted a more detailed analysis of the circadian action spectra published from the *rd/rd cl* mouse (Hattar *et al.*, 2003).

*Materials and Methods*

*Phase-Shifting Experiments.* Singly housed C3H/He mice lacking rod and cone photoreceptors (*rd/rd cl*) aged between 80 and 250 days are maintained in constant darkness. A single monochromatic light pulse (half bandwidth 10 nm) of defined irradiance is applied after 7–10 days at circadian time 16 (4 h after activity onset), at a point in the PRC that produces maximum phase delays. Seven wavelengths were investigated: 420, 460, 471, 506, 540, 560, and 580 nm, with between four and seven animals used for each irradiance level. For full methodological details, see Foster *et al.* (1991).

Irradiance data are corrected for the percentage lens transmission specific to the wavelength and genotype. To describe lens transmission in each genotype, lenses are extracted from age-matched wild-type ($n = 4$) and *rd/rd cl* ($n = 4$) mice and scanned in air using a Shimadzu UV-2101 (PC) S spectrophotometer fitted with an integrating sphere (Douglas, 1989). Percentage transmission relative to transmission at 600 nm is determined at 1-nm intervals between 305 and 700 nm (Hattar *et al.*, 2003).

*Action Spectrum Construction.* The size of the delaying phase shift in free-running locomotor activity induced by the different light treatments is plotted against irradiance and fitted to a sigmoid function. The maximum observed phase shift is at 506 nm (197 min), and all IRCs are plotted with this value as 100% (Fig. 3). Irradiances that produce half-maximum phase shifts ($EC_{50}$) are used to calculate relative sensitivity and are plotted against wavelength to produce the action spectrum. This spectrum is fitted to the absorbance spectrum of a vitamin $A_1$-based pigment with a $\lambda_{max}$ determined by nonlinear regression (Fig. 4). The use of a four-parameter model for IRCs is not appropriate, as at long (580 nm) and short (420 nm) wavelengths, saturating responses are not obtained. A global model is therefore used with a common maximum and minimum (see earlier discussion). The vitamin $A_1$-based visual pigment template used is that described by Govardovskii *et al.* (2000). No significant improvement is observed in the IRC fit when enabling a variable slope, suggesting that there is no significant difference in the slope of the IRCs (i.e., univariance holds, $p = 0.383$). As univariance is demonstrated, the mean slope across all wavelengths is used (Hill slope = 1.82).

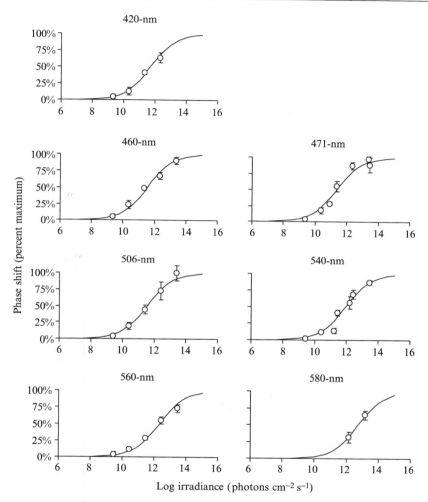

FIG. 3. Irradiance response curves for *rd/rd cl* phase-shifting behavior. Phase shifting was analyzed from wheel-running behavior using monochromatic light of 420-, 460-, 471-, 506-, 540-, 560-, and 580-nm wavelengths (four to seven animals per wavelength). Data were fitted using a sigmoid function, using a fixed slope as described in the text.

*Accuracy of Action Spectrum.* As each data point on an IRC is based on several biological replicates, it has an associated variance. To determine how much the $\lambda_{max}$ of the resulting action spectrum is affected by errors associated with IRC variability, a Monte Carlo approach is used to

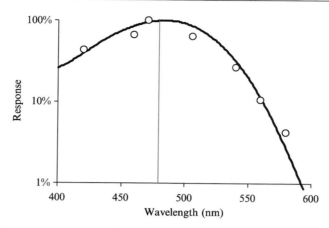

Fig. 4. Action spectrum for *rd/rd cl* phase shifting, as derived from relative sensitivity to monochromatic light pulses from 420 to 580 nm. The logarithm of relative sensitivity is plotted against wavelength (○), and a vitamin $A_1$-based visual pigment template was then fitted to data (solid line). The optimum fit was obtained with an $A_1$ Govardovskii template with a $\lambda_{max}$ of 484 nm ($R^2 = 0.951$).

generate confidence limits for the $\lambda_{max}$ (using PopTools add-in for Excel, see http://www.cse.csiro.au/poptools). Each irradiance level is reiterated with the addition of random, normally distributed noise, based on the observed variance at that irradiance. After each reiteration, all seven IRCs are curve fitted to calculate $EC_{50}$ values, and a visual pigment template is fitted to provide a $\lambda_{max}$. Ninety-five percent confidence limits are established based on 1000 such iterations. This procedure is also performed using bootstrapped data, using resampling of actual data values rather than an assumed normal distribution, at each irradiance level, again using 1000 iterations.

*Results*

The irradiance response curves for circadian phase shifting are shown in Fig. 3. The relative sensitivity was determined as the number of photons required to elicit a 50% response, and these values were plotted against wavelength as an action spectrum (Fig. 4). A vitamin $A_1$-based visual pigment template was fitted to these data as described earlier, with a variable Hill slope giving an optimum fit at 483 nm and a fixed Hill slope of 1.82 producing a $\lambda_{max}$ of 484 nm ($R^2 = 0.951$). The small difference (2–3 nm) in the action spectrum $\lambda_{max}$ from that of our previously published result (Hattar *et al.*, 2003) is most probably due to the use of the

Govardovskii template, which is based on an empirical fit to a wide range of rod and cone pigments from $\lambda_{max}$ 357–620 nm as determined by MSP rather than rod opsin extracts in solution (Dartnall, 1953; Govardovskii et al., 2000).

The application of Monte Carlo analysis to calculate the accuracy of the $\lambda_{max}$ derived from this action spectrum resulted in 95% confidence limits within the 477- to 494-nm range. Use of bootstrapped data produced similar results, with the 95% confidence limits ranging from 479 to 492 nm.

## Discussion

Analysis of photopigments in solution by absorption spectroscopy or MSP has provided an ideal means of characterizing the visual pigments of rod and cone photoreceptors.

The action spectrum for phase shifting the circadian rhythm in locomotor behavior in the *rd/rd cl* mouse is best fitted by the absorption spectrum of a vitamin $A_1$-based photopigment with a $\lambda_{max}$ of 484 nm. This $\lambda_{max}$ is remarkably similar to the sensitivity observed for the pupillary light response in this genotype ($\lambda_{max} = 479$) (Lucas et al., 2001). This $\lambda_{max}$ is also comparable to the spectral sensitivity of the intrinsically photosensitive, melanopsin-expressing RGCs of the rat ($\lambda_{max} = 484$) (Berson et al., 2002). Collectively, data provide overwhelming evidence for the existence of a novel opsin/vitamin $A_1$-based photopigment within the mammalian eye that is capable of regulating the circadian system and other irradiance detection tasks (Foster and Hankins, 2002).

An alternative hypothesis to the involvement of opsin-based photopigments is that the flavoprotein-based cryptochromes (CRYs) act as photopigments within the inner retina. The origins of this hypothesis have, to a large extent, arisen from the presumed photoreceptive role for the CRYs in other organisms rather than any hard data in mammals (Foster et al., 2003a). Nevertheless, proponents of the crytochrome photopigment hypothesis have suggested that data from action spectroscopy do not provide a sufficiently reliable approach to make any firm conclusions (Cashmore, 2003; Sancar, 2000). These two primary criticisms are addressed.

*Action Spectrum Results from* rd/rd cl *Mice Could Equally Well Describe a Cryptochrome-Based Photopigment (Cashmore, 2003).* The shape of the *rd/rd cl* action spectrum fits closely a vitamin $A_1$-based photopigment template ($R^2 = 0.951$). Although this is a remarkably close fit, we have until now not shown that a vitamin $A_1$-based template provides a demonstrably better fit than any other photopigment template, such as a flavoprotein (Smyth et al., 1988; Wolken, 1995) or the expressed human cryptochrome (hCRY) (Hsu et al., 1996). Here several photopigment

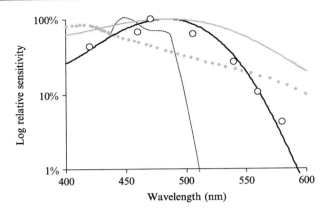

FIG. 5. Action spectra for *rd/rd cl* phase shifting (○) plotted against absorption spectra for vitamin $A_1$ (solid black line), vitamin $A_2$ (solid gray line), flavin (dashed black line), and hCRY1 (dashed gray line). The optimum fit to *rd/rd cl* phase-shifting data is provided by a vitamin $A_1$-based photopigment ($R^2 = 0.951$).

templates have been fit to the *rd/rd cl* action spectrum, and the results are shown in Fig. 5. The hCRY spectrum provides $R^2 = 0.553$ and the $A_2$ template $R^2 = 0.058$, whereas a flavoprotein spectrum provides no better fit than a straight line through the mean ($R^2 < 0$). Thus a vitamin $A_1$-based photopigment template with an $R^2 = 0.951$ provides by far the best fit to data. Cashmore (2003) has argued that the cryptochromes may use different chromophores and, as a result, it is impossible to know their precise absorption spectrum. If this were the case, one would have to argue that CRY photopigments describe precisely the same absorption spectrum as an opsin/vitamin $A_1$-based photopigment. This view is not only unlikely, but difficult to reconcile with a recent action spectrum for cryptochrome in plants that provides an excellent fit to a flavoprotein photopigment template (Ahmad *et al.*, 2002; Foster *et al.*, 2003a).

*Action Spectra Do Not Provide Sufficient Accuracy to Link Responses to Specific Photopigments.* Cashmore (2003) argued that there is only a poor match between action spectroscopy and absorption spectra of photopigments. To investigate this point statistically, we used a Monte Carlo algorithm to generate confidence limits for *rd/rd cl* action spectrum data. This analysis represents the first study to apply confidence limits to action spectrum data, and the results are shown in Fig. 6. Results demonstrate that the $\lambda_{\max}$ of the photopigment-mediating circadian responses to light lies within the 477- to 494-nm range (95% confidence limits). Action spectra can, therefore, describe a photopigment with considerable accuracy, and

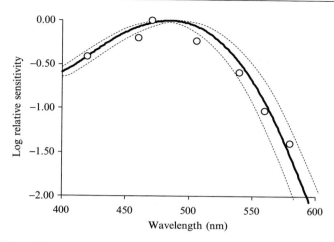

FIG. 6. Ninety-five percent confidence limits for *rd/rd cl* phase-shifting action spectrum, derived using a Monte Carlo algorithm with 1000 iterations with the addition of normally distributed noise. Solid black line indicates optimum fit at 484 nm, and dashed lines indicate 95% confidence limits from 477 to 494 nm. See text for further details.

our statistical analysis would not support the conjecture that action spectroscopy does not provide sufficient accuracy to link biological responses to specific photopigments. This general point has also been demonstrated in other studies where action spectra and absorption spectra have been compared directly. For example, an action spectrum based on electrophysiological recordings from macaque rods shows the same $\lambda_{max}$ as the $\lambda_{max}$ derived from MSP analysis of the rod photopigments in this species (Baylor *et al.*, 1984; Bowmaker and Dartnall, 1980).

Collectively, data from action spectroscopy and results showing that ablation of rods, cones, and melanopsin-based photoreceptors can account for all light detection within the mouse eye (see earlier) leave little room for the involvement of a cryptochrome photopigment in the regulation of circadian responses to light in mammals. Indeed, one of the primary advocates for a photopigment role for the cryptochromes stated, "It would thus appear that the primary photopigment in nonvisual photoreception is melanopsin dependent, but not cryptochrome dependent" (Panda *et al.*, 2003).

Conclusions

Photopigment characterization based on absorbance spectroscopy provides an ideal means of studying rod and cone photoreceptors. However, such approaches cannot be utilized when the location of the photoreceptor is unknown or where the photopigment concentration is too low for reliable detection. This is often the case with the light-detecting systems that regulate plant and animal photoentrainment. As a result, action spectroscopy has provided the primary means of characterizing these photopigments (e.g., Table I). It is worth stressing that when an action spectrum is conducted correctly, it provides one of the only approaches that can be used to both identify a class of molecule, and simultaneously link that molecule to a particular cellular or behavioral function.

The techniques of action spectroscopy can be both complicated and labor intensive; as with all methods, if these techniques are not used correctly, misleading results will be generated. Because most circadian biologists are unfamiliar with action spectrum methodologies and because the validity of these techniques has been brought into question (Cashmore, 2003), we have taken this opportunity to detail how action spectra should be undertaken and validated. It is hoped that we have made this technique more accessible and, in so doing, stimulated the readership to use this powerful methodology in understanding the nature of circadian photopigments in a broad range of organisms.

## References

Ahmad, M., Grancher, N., Heil, M., Black, R. C., Giovani, B., Galland, P., and Lardemer, D. (2002). Action spectrum for cryptochrome-dependent hypocotyl growth inhibition in Arabidopsis. *Plant Physiol.* **129**, 774–785.

Aschoff, J. (1981). Biological rhythms. *In* "Handbook of Behavioural Neurobiology." Plenum Press, New York.

Barr, L., and Alpern, M. (1963). Photosensitivity of the frog iris. *J. Gen. Physiol.* **46**, 1249–1265.

Baylor, D. A., Nunn, B. J., and Schnapf, J. L. (1984). The photocurrent, noise and spectral sensitivity of rods of the monkey *Macaca fascicularis*. *J. Physiol.* **357**, 575–607.

Berson, D. M., Dunn, F. A., and Takao, M. (2002). Phototransduction by retinal ganglion cells that set the circadian clock. *Science* **295**, 1070–1073.

Bowmaker, J., and Hunt, D. M. (1999). Molecular biology of photoreceptor spectral sensitivity. *In* "Adaptive Mechanisms in the Ecology of Vision" (S. N. Archer, M. B. A. Djamgoz, E. R. Loew, J. C. Partridge, and S. Vallerga, eds.), pp. 439–462. Kluwer, A. Dordrecht, Netherlands.

Bowmaker, J. K. (1984). Microspectrophotometry of vertebrate photoreceptors. *Vision Res.* **24**, 1641–1650.

Bowmaker, J. K., and Dartnall, H. J. (1980). Visual pigments of rods and cones in a human retina. *J. Physiol.* **298**, 501–511.

Brainard, G. C., Hanifin, J. P., Greeson, J. M., Byrne, B., Glickman, G., Gerner, E., and Rollag, M. D. (2001). Action spectrum for melatonin regulation in humans: Evidence for a novel circadian photoreceptor. *J. Neurosci.* **21**, 6405–6412.

Bridges, C. (1959). The visual pigments of some common laboratory animals. *Nature* **184**, 727–728.

Bridges, C. D. B. (1965). The grouping of visual pigments about preferred positions in the spectrum. *Vision Res.* **5,** 223–238.

Bridges, C. D. B., Fong, S. L., and Alvarez, R. A. (1979). Separation by programmed-gradient high-pressure liquid chromatography of vitamin A isomers, their esters, aldehydes, oximes and vitamin $A_2$: Presence of retinyl ester in dark-adapted goldfish pigment epithelium. *Vision Res.* **20,** 355–360.

Bronstein, D. M., Jacobs, G. H., Haak, K. A., Neitz, J., and Lytle, L. D. (1987). Action spectrum of the retinal mechanism mediating nocturnal light-induced suppression of rat pineal gland N-acetyltransferase. *Brain Res.* **406,** 352–356.

Brown, P. K., and Wald, G. (1963). Visual pigments in human and monkey retinas. *Nature* **200,** 37–43.

Campbell, S. S., and Murphy, P. J. (1998). Extraocular circadian phototransduction in humans. *Science* **279,** 396–399.

Carter-Dawson, L. D., La Vail, M. M., and Sidman, R. L. (1978). Differential effect of the rd mutation on rods and cones in the mouse retina. *Invest. Ophthalmol. Vis. Sci.* **17,** 489–498.

Cashmore, A. R. (2003). Cryptochromes: Enabling plants and animals to determine circadian time. *Cell* **114,** 537–543.

Dartnall, H. (1953). The interpretation of spectral sensitivity curves. *Br. Med. Bull.* **9,** 24–30.

Dartnall, H. J. (1968). The photosensitivities of visual pigments in the presence of hydroxylamine. *Vision Res.* **8,** 339–358.

David-Gray, Z. K., Janssen, J. W., DeGrip, W. J., Nevo, E., and Foster, R. G. (1998). Light detection in a 'blind' mammal. *Nature Neurosci.* **1,** 655–656.

De Grip, W. J. (1982). Purification of bovine rhodopsin over concanavalin A-Sepharose. *Methods Enzymol.* **81,** 197–207.

DeGrip, W. J., Klaassen, C. H., and Bovee-Geurts, P. H. (1999). Large-scale functional expression of visual pigments: Towards high-resolution structural and mechanistic insight. *Biochem. Soc. Trans.* **27,** 937–944.

Deguchi, T. (1981). Rhodopsin-like photosensitivity of isolated chicken pineal gland. *Nature* **290,** 702–704.

Douglas, R. H. (1989). The spectral transmission of the lens and cornea of the brown trout (*Salmo trutta*) and goldfish (*Carassius auratus*): Effect of age and implications for ultraviolet vision. *Vision Res.* **29,** 861.

Ekstrom, P., and Meissl, H. (2003). Evolution of photosensory pineal organs in new light: The fate of neuroendocrine photoreceptors. *Philos. Trans. R. Soc. Lond. B. Biol. Sci.* **358,** 1679–1700.

Foster, R. G. (1998). Shedding light on the biological clock. *Neuron* **20,** 829–832.

Foster, R. G., and Bellingham, J. (2002). Opsins and melanopsins. *Curr. Biol.* **12,** 543–544.

Foster, R. G., and Follett, B. K. (1985). The involvement of a rhodopsin-like photo-pigment in the photoperiodic response of the Japanese quail. *J. Comp. Physiol. A* **157,** 519–528.

Foster, R. G., Follett, B. K., and Lythgoe, J. N. (1985). Rhodopsin-like sensitivity of extra-retinal photoreceptors mediating the photoperiodic response in quail. *Nature* **313,** 50–52.

Foster, R. G., Hankins, M., Lucas, R. J., Jenkins, A., Munoz, M., Thompson, S., Appleford, J. M., and Bellingham, J. (2003a). Non-rod, non-cone photoreception in rodents and teleost fish. *Novartis Found Symp.* **253,** 3–23; discussion 23–30, 52–55, 102–109.

Foster, R. G., and Hankins, M. W. (2002). Non-rod, non-cone photoreception in the vertebrates. *Progr. Retinal Eye Res.* **21,** 507–527.

Foster, R. G., and Helfrich-Forster, C. (2001). The regulation of circadian clocks by light in fruitflies and mice. *Philos. Trans. R. Soc. Lond. B Biol. Sci.* **356,** 1779–1789.

Foster, R. G., Provencio, I., Bovee-Geurts, P. H., and DeGrip, W. J. (2003b). The photoreceptive capacity of the developing pineal gland and eye of the golden hamster (*Mesocricetus auratus*). *J. Neuroendocrinol.* **15,** 355–363.

Foster, R. G., Provencio, I., Hudson, D., Fiske, S., DeGrip, W., and Menaker, M. (1991). Circadian photoreception in the retinally degenerate mouse (*rd/rd*). *J. Comp. Physiol. A* **169,** 39–50.

Freedman, M. S., Lucas, R. J., Soni, B., von Schantz, M., Munoz, M., David-Gray, Z. K., and Foster, R. G. (1999). Regulation of mammalian circadian behavior by non-rod, non-cone, ocular photoreceptors. *Science* **284,** 502–504.

Govardovskii, V. I., Fyhrquist, N., Reuter, T., Kuzmin, D. G., and Donner, K. (2000). In search of the visual pigment template. *Vis. Neurosci.* **17,** 509–528.

Hankins, M. W., and Lucas, R. J. (2002). The primary visual pathway in humans is regulated according to long-term light exposure through the action of a non-classical photopigment. *Curr. Biol.* **12,** 191–198.

Hartwig, H.-G., and van Veen, T. (1979). Spectral characteristics of visible radiations penetrating into the brain and stimulating extra-retinal photoreceptors. *J. Comp. Physiol. A* **120,** 277–282.

Hattar, S., Liao, H. W., Takao, M., Berson, D. M., and Yau, K. W. (2002). Melanopsin-containing retinal ganglion cells: Architecture, projections, and intrinsic photosensitivity. *Science* **295,** 1065–1070.

Hattar, S., Lucas, R. J., Mrosovsky, N., Thompson, S., Douglas, R. H., Hankins, M. W., Lem, J., Biel, M., Hofmann, F., Foster, R. G., and Yau, K. W. (2003). Melanopsin and rod-cone photoreceptive systems account for all major accessory visual functions in mice. *Nature* **424,** 75–81.

Hsu, D. S., Zhao, X., Zhao, S., Kazantsev, A., Wang, R. P., Todo, T., Wei, Y. F., and Sancar, A. (1996). Putative human blue-light photoreceptors hCRY1 and hCRY2 are flavoproteins. *Biochemistry* **35,** 13871–13877.

Jacobs, G. H., Neitz, J., and Deegan, J. F. (1991). Retinal receptors in rodents maximally sensitive to ultraviolet light. *Nature* **353,** 655–656.

Jenkins, A., Munoz, M., Tarttelin, E. E., Bellingham, J., Foster, R. G., and Hankins, M. W. (2003). VA opsin, melanopsin, and an inherent light response within retinal interneurons. *Curr. Biol.* **13,** 1269–1278.

Johnson, C. H., Elliott, J. A., and Foster, R. (2003). Entrainment of circadian programs. *Chronobiol. Int.* **20,** 741–774.

Knowles, A., and Dartnall, H. J. A. (1977). "The Eye." Academic Press, London.

Korf, H. W., Liesner, R., Meissl, H., and Kirk, A. (1981). Pineal complex of the clawed toad, *Xenopus laevis* Daud.: Structure and function. *Cell Tissue Res.* **216,** 113–130.

Lamb, T. D. (1995). Photoreceptor spectral sensitivities: Common shape in the long-wavelength region. *Vision Res.* **35,** 3083–3091.

Lucas, R. J., Douglas, R. H., and Foster, R. G. (2001). Characterization of an ocular photopigment capable of driving pupillary constriction in mice. *Nature Neurosci.* **4,** 621–626.

Lucas, R. J., Freedman, M. S., Munoz, M., Garcia-Fernandez, J. M., and Foster, R. G. (1999). Regulation of the mammalian pineal by non-rod, non-cone, ocular photoreceptors. *Science* **284,** 505–507.

Lucas, R. J., Hattar, S., Takao, M., Berson, D. M., Foster, R. G., and Yau, K. W. (2003). Diminished pupillary light reflex at high irradiances in melanopsin-knockout mice. *Science* **299,** 245–247.

Lythgoe, J. N. (1979). "The Ecology of Vision." Clarendon Press, Oxford.

Lythgoe, J. N., and Thompson, M. (1984). A porphyropsin-like action spectrum from *Xenopus* melanophores. *Photochem. Photobiol.* **40,** 411–412.

Max, M., and Menaker, M. (1992). Regulation of melatonin production by light, darkness, and temperature in the trout pineal. *J. Comp. Physiol. A* **170,** 479–489.

Motulsky, H., and Christapoulos, A. (2003). "Fitting Models to Biological Data Using Linear and Nonlinear Regression." GraphPad Software, Inc.

Nelson, D., and Takahashi, J. (1991). Sensitivity and integration in a visual pathway for circadian entrainment in the hamster (*Mesocricetus auratus*). *J. Physiol.* **439,** 115–145.

Newman, L. A., Walker, M. T., Brown, R. L., Cronin, T. W., and Robinson, P. R. (2003). Melanopsin forms a functional short-wavelength photopigment. *Biochemistry* **42,** 12734–12738.

Panda, S., Provencio, I., Tu, D. C., Pires, S. S., Rollag, M. D., Castrucci, A. M., Pletcher, M. T., Sato, T. K., Wiltshire, T., Andahazy, M., Kay, S. A., Van Gelder, R. N., and Hogenesch, J. B. (2003). Melanopsin is required for non-image-forming photic responses in blind mice. *Science* **301**(5632), 525–527.

Partridge, J. C., and De Grip, W. J. (1991). A new template for rhodopsin (vitamin A1 based) visual pigments. *Vision Res.* **31,** 619–630.

Peirson, S., Bovee-Geurts, P. H., Lupi, D., Jeffery, G., DeGrip, W. J., and Foster, R. G. (2004). Expression of the candidate circadian photopigment melanopsin (*Opn4*) in the mouse retinal pigment epithelium. *Mol. Brain Res.* **123,** 132–135.

Pepe, I. M. (2001). Recent advances in our understanding of rhodopsin and phototransduction. *Prog. Retin. Eye Res.* **20,** 733–759.

Pittendrigh, C. S. (1981). Circadian systems: Entrainment. *In* "Handbook of Behavioral Neurobiology: Biological Rhythms" (J. Aschoff, ed.), Vol. 4, pp. 95–124. Plenum Press, New York.

Pittendrigh, C. S., and Daan, S. (1976). A functional analysis of circadian pacemakers in nocturnal rodents. IV, Entrainment: Pacemaker as clock. *J. Comp. Physiol. A* **106,** 333–355.

Provencio, I., Cooper, H. M., and Foster, R. G. (1998). Retinal projections in mice with inherited retinal degeneration: Implications for circadian photoentrainment. *J. Comp. Neurol.* **395,** 417–439.

Provencio, I., and Foster, R. G. (1995). Circadian rhythms in mice can be regulated by photoreceptors with cone-like characteristics. *Brain Res.* **694,** 183–190.

Provencio, I., Rodriguez, I. R., Jiang, G., Hayes, W. P., Moreira, E. F., and Rollag, M. D. (2000). A novel human opsin in the inner retina. *J. Neurosci.* **20,** 600–605.

Provencio, I., Wong, S., Lederman, A. B., Argamaso, S. M., and Foster, R. G. (1994). Visual and circadian responses to light in aged retinally degenerate mice. *Vision Res.* **34,** 1799–1806.

Rodieck, R. W. (1998). "The First Steps in Seeing." Sinauer, Sunderland, MA.

Roenneberg, T., and Foster, R. G. (1997). Twilight times: Light and the circadian system. *Photochem. Photobiol.* **66,** 549–561.

Ruby, N. F., Brennan, T. J., Xie, X., Cao, V., Franken, P., Heller, H. C., and O'Hara, B. F. (2002). Role of melanopsin in circadian responses to light. *Science* **298,** 2211–2213.

Sancar, A. (2000). Cryptochrome: The second photoactive pigment in the eye and its role in circadian photoreception. *Annu. Rev. Biochem.* **69,** 31–67.

Sekaran, S., Foster, R. G., Lucas, R. J., and Hankins, M. W. (2003). Calcium imaging reveals a network of intrinsically light-sensitive inner-retinal neurons. *Curr Biol.* **13,** 1290–1298.

Shand, J., and Foster, R. G. (1999). The extraretinal photoreceptors of non-mammalian vertebrates. *In* "Adaptive Mechanisms in the Ecology of Vision" (S. N. Archer, M. B. A. Djamgoz, E. R. Loew, J. C. Partridge, and S. Vallerga, eds.), pp. 197–222. Kluwer, A Dordrecht, Netherlands.

Smyth, R. D., Saranak, J., and Foster, K. W. (1988). Algal visual systems and their photoreceptor pigments. *Prog. Phycol. Res.* **6**, 255–286.

Sun, H., Macke, J. P., and Nathans, J. (1997). Mechanisms of spectral tuning in the mouse green cone pigment. *Proc Natl. Acad. Sci. USA* **94**, 8860–8865.

Thapan, K., Arendt, J., and Skene, D. J. (2001). An action spectrum for melatonin suppression: Evidence for a novel non-rod, non-cone photoreceptor system in humans. *J. Physiol.* **535**, 261–267.

Wald, G. (1958). Photochemical aspects of visual excitation. *Exp. Cell. Res.* **14**, 389–410.

Wald, G. (1968). Molecular basis of visual excitiation. *Science* **162**, 230–239.

Wolken, J. J. (1995). "Light Detectors, Photoreceptors, and Imaging Systems in Nature." Oxford Univ. Press, New York.

Wright, K. P., Jr., and Czeisler, C. A. (2002). Absence of circadian phase resetting in response to bright light behind the knees. *Science* **297**, 571.

Yoshimura, T., Nishio, M., Goto, M., and Ebihara, S. (1994). Differences in circadian photosensitivity between retinally degenerate CBA/J mice (*rd/rd*) and normal CBA/N mice (+/+). *J. Biol. Rhythms* **9**, 51–60.

Zeitzer, J. M., Dijk, D. J., Kronauer, R., Brown, E., and Czeisler, C. A. (2000). Sensitivity of the human circadian pacemaker to nocturnal light: Melatonin phase resetting and suppression. *J. Physiol.* **526**, 695–702.

# [38]  Cryptochromes and Circadian Photoreception in Animals

*By* Carrie L. Partch and Aziz Sancar

## Abstract

Cryptochromes are flavin- and folate-containing blue-light photoreceptors with a high degree of similarity to DNA photolyase, which repairs ultraviolet-induced DNA damage using blue light to initiate the repair reaction. Cryptochromes play essential roles in the maintenance of circadian rhythms in mice and *Drosophila*, and genetic data indicate that cryptochromes function as circadian photoreceptors in these and other animals. However, the photochemical reactions carried out by cryptochromes are not known at present.

## Introduction

In animals, synchronization of the circadian clock with the environmental light/dark cycle requires contribution from multiple photoreceptor systems. Genetic studies in mice have revealed functional redundancy between retinaldehyde-based opsins and flavin-based cryptochromes in circadian photoreception. These studies have revealed the role of three photoreceptor systems in this process: (1) visual opsins, (2) the nonvisual opsin melanopsin, and (3) cryptochromes (Sancar, 2003; Van Gelder and Sancar, 2003). This article reviews the experiments used thus far to elucidate the role of cryptochromes in circadian photoreception in animals. In addition to their putative photoreceptor function, cryptochromes also constitute an integral component of the transcriptional feedback loop that generates the circadian clock (Thresher et al., 1998; van der Horst et al., 1999; Vitaterna et al., 1999). However, this light-independent function of cryptochromes is not covered in any detail.

Cryptochromes were initially identified as putative photoreceptors by their high degree of homology to the light-activated DNA repair enzyme photolyase (Ahmad and Cashmore, 1993; Hsu et al., 1996). Animal cryptochromes are 60- to 70-kDa proteins that share 30–50% sequence identity with photolyase along the first 500 amino acids and contain the same two chromophore/cofactors: methenyltetrahydrofolate (MTHF) and a flavin in the form of FAD. A small number of photolyases contain 8-hydroxy-5-deazariboflavin instead of folate as the second chromophore. Photolyase family members have no apparent sequence homology to other classes of flavoproteins, perhaps because photolyase utilizes flavin in its two electron-reduced and photochemically excited state $1(FADH^-)^*$ as opposed to most other flavoproteins, which operate from the oxidized, ground state of flavin ($FAD_{ox}$). Both cryptochromes and photolyases have a positively charged groove along one face of the protein that binds the phosphodiester backbone of DNA, with a hole in the middle that, in the case of photolyase, allows entry of an ultraviolet (UV)-induced cyclobutane pyrimidine dimer or pyrimidine-pyridimidone (6–4) photoproduct into the active site cavity close to the flavin for repair. The significance of the conservation of this groove and the hole in cryptochromes is not yet understood.

Despite strong genetic evidence in plants and animals for its role as a photoreceptor, the mechanism of action, or photocycle, of cryptochrome is currently not known. However, detailed mechanistic studies have been carried out on photolyase and its mechanism of action is well understood (Sancar, 2003). The enzyme binds its substrate independently of light, and catalysis is initiated by light (Fig. 1). The photoantenna chromophore MTHF ($\lambda_{max} = 380$– $420$ nm) absorbs a photon of blue light (350–450 nm) and transfers the excitation energy to the catalytic cofactor $FADH^-$ by Förster resonance energy transfer. Alternatively, the $FADH^-$ ($\lambda_{max} = 360$ nm) may become excited by absorbing a photon directly. The excited $1(FADH^-)^*$ singlet state transfers an electron to the pyrimidine dimer, generating an $FADH°$ blue neutral radical and a pyrimidine dimer radical. The latter undergoes bond rearrangements to generate two canonical pyrimidines and restores $FADH°$ to its catalytically competent form ($FADH^-$)

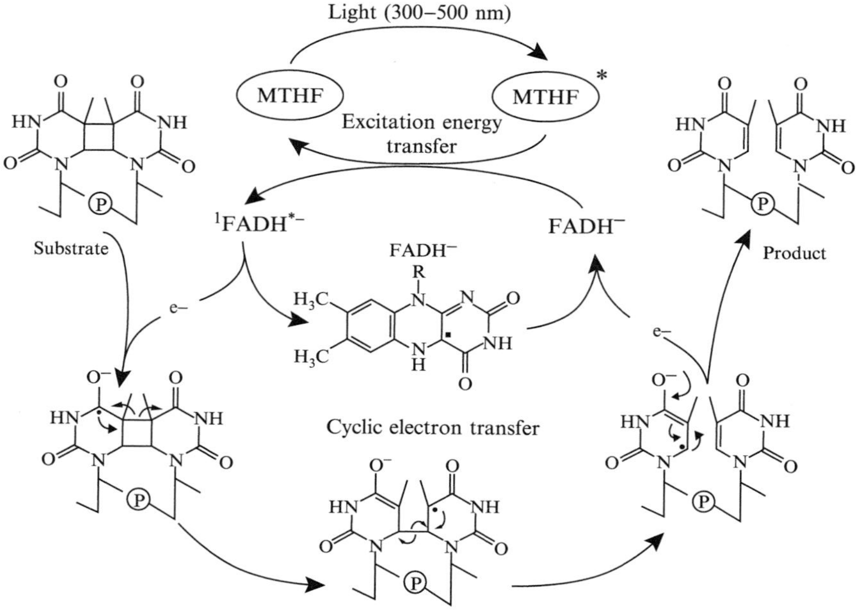

FIG. 1. Reaction mechanism of *Escherichia coli* cyclobutane pyrimidine dimer photolyase. MTHF absorbs a 300- to 500-nm photon and transfers the excitation energy to $FADH^-$ by resonance energy transfer. The $1(FADH^-)^*$ transfers an electron to the pyrimidine dimer to generate a pyrimidine and pyrimidine$^{\circ-}$; back electron transfer to $FADH^\circ$ restores the catalytic cofactor to the active reduced form and the dimer is converted to canonical bases (Sancar, 2003).

by back electron transfer to complete the photocycle; the repaired DNA subsequently dissociates from the enzyme. Although cryptochromes by definition lack DNA repair activity, it is hypothesized that they utilize a similar photocycle to regulate light-dependent signaling. Structurally, cryptochromes are also defined by the presence of extended C-terminal domains ranging from 40 to 220 amino acids that are not homologous to any known protein. Studies of cryptochromes from two different organisms (*Arabidopsis thaliana* and *Drosophila melanogaster*) indicate that these unique C-terminal domains are involved in regulating light-dependent signaling by cryptochromes (Rosato *et al.*, 2001; Yang *et al.*, 2000). All mammalian and bird species analyzed so far, including humans and mice, have two cryptochrome isoforms (Cry1, Cry2), and some amphibians possess up to seven cryptochromes. The variable C-terminal domain sequences are the only predominant difference between most cryptochrome isoforms.

## Mammalian Cryptochromes

### Biochemical Characterization

*Structural Aspects.* Although cryptochrome/photolyase family members contain two noncovalently bound chromophores, only the FAD is absolutely required for activity. The crystal structure of *Escherichia coli* photolyase, the prototype of this family of proteins, is shown in Fig. 2A. The enzyme consists of an N-terminal $\alpha/\beta$ domain and a C-terminal $\alpha$-helical domain connected by a long interdomain loop (Park *et al.*, 1995). The photolyase-like domain of human cryptochrome 2 (hCRY2) was homology modeled on the *E. coli* crystal structure and is predicted to have a similar tertiary structure (Fig. 2B) (Ozgur and Sancar, 2003). The FAD is deeply buried within the C-terminal $\alpha$-helical domain, held tightly in place by contact with 14 amino acids in photolyase, most of which are conserved in cryptochromes (Park *et al.*, 1995). The second chromophore, MTHF, is loosely bound in a shallow cleft between the two domains and is easily lost

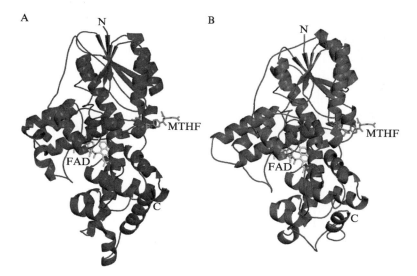

FIG. 2. Crystal structure of *E. coli* photolyase and homology-modeled human cryptochrome 2. (A) Ribbon diagram representation of *E. coli* photolyase showing the N-terminal $\alpha/\beta$ domain, the C-terminal $\alpha$-helical domain, and the positions of the two cofactors (Park *et al.* 1995). (B) The model of the hCRY2 tertiary structure was generated using the experimentally determined structures of *E. coli* and *A. nidulans* photolyases as templates, excluding the N-terminal 22 and C-terminal 80 amino acids of hCRY2 that have no homology to photolyase (Ozgur and Sancar, 2003). (See color insert.)

during purification. It acts as a photoantenna, increasing the efficiency of DNA repair by photolyase five- to 10-fold and dominates the absorption spectrum with a peak ranging from 377 to 410 nm depending on the source of the enzyme. In the absence of folate, the absorption spectrum of purified cryptochrome/photolyase family members is characteristic of the FAD and its oxidation state. During purification, the FADH$^-$ cofactor of photolyases becomes oxidized in the majority of photolyases to yield either the flavin neutral radical (FADH$^\circ$) or FAD$_{ox}$. Enzyme preparations that are blue in color contain the neutral radical form of flavin, due to its strong absorbance at long wavelengths, from 380 to 625 nm, and preparations with oxidized flavin are yellow, due its absorbance at 370 and 430 nm. The catalytically inactive FADH$^\circ$ can be reduced *in vitro* to FADH$^-$ either chemically or by photoreduction, in which a tryptophan residue in the apoenzyme transfers an electron to the excited state FADH$^\circ$ (Payne *et al.*, 1987). The oxidation state of the catalytic flavin in cryptochromes is not known at present, although an action spectrum of hypocotyl elongation performed in *Arabidopsis*, a cryptochrome-dependent response, suggests that the flavin may be active in the one (FADH$^\circ$)- or two-electron (FAD$_{ox}$) oxidized form in plants, which would suggest a radically different photochemistry from photolyase (Ahmad *et al.*, 2002). In contrast, a *Vibrio cholerae* cryptochrome purified from *E. coli* contains the flavin in the FADH-form, suggesting a photolyase-like reaction mechanism (Worthington *et al.*, 2003).

*Purification and Spectroscopic Properties.* With current protocols, purification of most animal cryptochromes from heterologous sources does not yield protein with stoichiometric amounts of chromophores in sufficient quantities for biochemical studies. In contrast to *Arabidopsis* cryptochrome 1 (AtCry1), which can be purified as a recombinant protein from *E. coli* with stoichiometric amounts of FAD, expression and purification of human CRY1 and CRY2 as MBP fusion proteins in *E. coli* yielded moderate quantities of protein with grossly substoichiometric amounts (1–5%) of FAD and even less folate; efforts to supplement the apoprotein with FAD and folate were unsuccessful (Hsu *et al.*, 1996). Absorption spectra of recombinant hCRY1 and hCRY2 expressed in *E. coli* (Fig. 3A) show the characteristic absorbance of oxidized flavin at 420 nm, with residual absorbance extending all the way to 700 nm. However, it is doubtful that this represents the active form of cryptochrome, as many photolyases known to be active only when the flavin is in the FADH$^-$ form exhibit similar spectra when overexpressed and purified from heterologous sources (Sancar, 2003).

Attempts to purify animal cryptochromes from native sources have been difficult because of the lack of a biochemical assay for cryptochrome

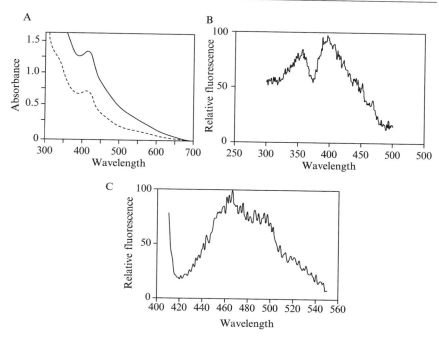

FIG. 3. Spectroscopic properties of mammalian cryptochromes. (A) Dashed and solid lines represent the absorbance spectra of hCRY1 and hCRY2, respectively, purified from *E. coli* (Hsu *et al.*, 1996). (B) Uncorrected fluorescence emission spectrum of hCRY2 purified from HeLa cells with λ emission at 520 nm, indicative of FAD. (C) Uncorrected fluorescence emission spectrum of hCRY2 purified from HeLa cells with λ excitation set at 400 nm reveals a peak at 460 nm and a shoulder at 510 nm, indicative of the presence of both FAD and MTHF, respectively (Ozgur and Sancar, 2003).

function. However, affinity purification of recombinant, FLAG-tagged hCRY2 from a stably transfected HeLa cell line yielded small quantities of protein (5–15 μg hCRY2 from 10-liter HeLa suspension cultures) with an estimated chromophore stoichiometry of 30% (Ozgur and Sancar, 2003). Chromophore stoichiometry was estimated by fluorescence spectroscopy; the fluorescence excitation spectrum of purified hCRY2 with emission set at 520 nm (Fig. 3B) is characteristic of FAD with maxima at 370 and 430 nm, and the fluorescence emission spectrum with excitation set at 400 nm (Fig. 3C) is indicative of the presence of both MTHF (major peak, 460 nm) and FAD (shoulder, 505 nm) (Sancar *et al.*, 1984). Finally, expressing hCRY2 in insect cells using the baculovirus system yielded abundant protein with no detectable chromophore.

*Enzymatic Activities.* Several *in vitro* activities associated with animal cryptochromes, such as DNA binding and autophosphorylation, have been described (Bouly *et al.*, 2003; Ozgur and Sancar, 2003; Shalitin *et al.*, 2003). Because mammalian cryptochromes have dual roles as light-independent regulators of the molecular clock and as circadian photoreceptors in the eye, it is unclear whether these *in vitro* activities are physiologically relevant for cryptochrome in the photocycle, the molecular clock, or both.

Purified hCRY2 binds to single-stranded DNA with high affinity ($K_D \sim 5 \times 10^{-9}$ $M$) and double-stranded DNA weakly ($K_D \sim 10^{-7}$ $M$), as measured by electrophoretic mobility shift assay (Ozgur and Sancar, 2003). This is in contrast to photolyase, which binds to damage in single- and double-stranded DNA with comparable affinities ($K_D \sim 10^{-9}$ $M$) (Sancar *et al.*, 1985). hCRY2 also bound with slightly higher affinity to UV-damaged DNA, although the magnitude of increase in affinity for damaged over undamaged DNA is significantly less than that of photolyase. Unlike photolyase, DNA binding by hCRY2 was not affected by light, and no repair by cryptochrome has been detected *in vivo* or *in vitro*.

It has been reported that plant and human cryptochrome 1 have autophosphorylating kinase activities (Bouly *et al.*, 2003; Shalitin *et al.*, 2003). The kinase activity of purified AtCry1 was also tested on a variety of classic kinase substrates such as histones, casein, and myelin-binding protein and it appears that the kinase activity is limited to autophosphorylation. *In vitro* autophosphorylation of AtCry1 occurred only on serine residues, depended on the presence of flavin in a reducing environment, and was stimulated by blue light. Because both AtCry1 and AtCry2 have previously been shown to be phosphorylated rapidly *in vivo* in response to blue light, this autophosphorylation may be involved in regulating signal transduction *in vivo* (Shalitin *et al.*, 2002, 2003). hCRY1 purified from insect cells was shown to bind to ATP cellulose and autophosphorylate in solution (Bouly *et al.*, 2003).

## Expression of Cryptochrome in the Retina

The retina is the exclusive site of circadian photoreception in mammals (Wright and Czeisler, 2002). While visual pigments in rods and cones unquestionably contribute to circadian photoreception, they are not essential for circadian phototransduction. Mice and humans with certain retinal degeneration diseases lose complete function of the visual photoreceptors in the outer retina and retain circadian photoreception. Therefore, the inner retina must contain photoreceptors capable of sensing and transmitting light information in the absence of the visual photoreceptors. Currently, two candidate photoreceptive pigments are known to be expressed in the inner retina: melanopsin and the two mammalian cryptochromes (Miyamoto and Sancar, 1998; Provencio *et al.*, 2000).

*Cryptochrome Expression in Mouse Retina.* To examine the expression of cryptochromes in the retina by bright-field microscopy, polymerase chain reaction fragments of *mCry1* (nucleotides 1074–1793) and *mCry2* (nucleotides 1040–1649) are subcloned into the pBluescipt SK+ plasmid, and $^{35}$S-UTP-labeled sense and antisense RNA probes are generated *in vitro* with T3 and T7 RNA polymerase. Frozen sections of retina (20 $\mu$m thick) are fixed for 20 min in 4% formaldehyde in phosphate buffer, treated with proteinase K (10 $\mu$g/ml) for 10 minutes, acetylated with acetic anhydride in 0.1 $M$ triethanolamine, and dehydrated with sequential ethanol dehydration. $^{35}$S-labeled sense and antisense probes diluted in hybridization buffer (50% formamide, 10% dextran sulfate, 20 m$M$ Tris–HCl, pH 8.0, 0.3 $M$ NaCl, 0.2% Sarcosyl, 0.02% salmon sperm DNA, and 1X Denhardt's solution) are placed on the sections and incubated at 55° overnight. The sections are washed at 65° (50% formamide, 2X SSC, 0.1 $M$ dithiothreitol) for 30 min and then treated with RNase A (1 $\mu$g/ml) for 30 min at 37°. Sections are washed again for 30 min at 65°, dipped in nuclear emulsion (Kodak NTB-2), and exposed to X-ray film for 2 weeks at 4°. Slides are stained after the emulsion autoradiography for 1 min with hematoxylin, washed with dH$_2$O, dehydrated with ethanol, and then treated with xylene and mounted. Examination of *Cry1* and *Cry2* mRNA levels in the mouse retina by *in situ* hybridization reveals moderate expression of *mCry1* and a high level of *mCry2* mRNA in both the inner nuclear layer and the ganglion cell layer of the retina (Fig. 4A) (Miyamoto and Sancar, 1998). The arrows in Fig. 4 indicate clusters of ganglion cells that express *Cry1* and *Cry2*.

*Cryptochrome Expression in Human Retina.* CRY2 protein levels are measured in the human retina by immunohistochemistry (Fig. 4B) as follows: 5-mm trephine punches of preserved human donor eyes are cryosectioned (10 $\mu$m), pretreated in 0.15% H$_2$O$_2$, and washed thoroughly in phosphate-buffered saline (PBS) before incubation in 0.02 mg/ml affinity-purified CRY2 antibody (Alpha Diagnostics, Inc.) on 0.1 $M$ PBS, 0.5% Triton X-100, and 10% normal goat serum for 12–26 h at 4° (Thompson *et al.*, 2003). Sections are then washed in PBS three times and incubated in a goat antirabbit biotinylated secondary antibody (1:50; Jackson ImmunoResearch) for 2 h. After washing with PBS, sections are incubated in an avidin–biotin–peroxidase mixture (Vectastain ABC Kit; Vector Laboratories) for 1 h and in 3,3'-diaminobenzidine tetrahydrochloride (DAB, Sigma-Aldrich) for 10 min followed by a brief treatment with DAB and 0.03% H$_2$O$_2$. Slides are washed in PBS, mounted in a glycerin–PBS mixture, and analyzed with a microscope equipped with either epifluorescence or differential interference contrast optics. Antibody specificity is

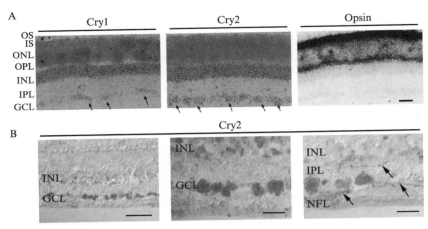

FIG. 4. Expression of cryptochrome in the mammalian retina. (A) Bright-field micrographs of *in situ* hybridizations comparing the expression of *mCry1*, *mCry2*, and opsins in the mouse retina. Bar: 30 $\mu$m (Miyamoto and Sancar, 1998). Arrows indicate clusters of ganglion cells expressing cryptochromes. (B) hCRY2 immunoreactivity in the human retina. Arrows indicate cryptochrome expression in ganglion cell extensions. Bar: 50 $\mu$m in the lower left box and 25 $\mu$m in the lower right two boxes (Thompson *et al.*, 2003). Retinal histology: OS, outer segment; IS, inner segment; ONL, outer nuclear layer; OPL, outer plexiform layer; INL, inner nuclear layer; IPL, inner plexiform layer; GCL, ganglion cell layer. (See color insert.)

determined either by eliminating primary antibody or by preadsorption of the primary antibody with 0.1 mg/ml CRY2 peptide (Alpha Diagnostics, Inc.) overnight at 4° before incubation with tissue. Expression of hCRY2 is detected in approximately 70% of retinal ganglion cells in both the macula and the peripheral retina with some staining also present in the inner nuclear layer. Interestingly, hCRY2 in the retina was found to be mostly cytoplasmic using 4′,6-Diamidino-2-phenylindole (DAPI) labeling of nuclei and anti-CRY2 immunofluorescence. This was confirmed by subcellular fractionation and Western analysis of retinal extracts. Intriguingly, hCRY2-reactive immunostaining was visible in some axonal processes extending into the inner plexiform layer and nerve fiber layer, as indicated by the arrows in Fig. 4B.

Retinal ganglion cells that directly innervate the site of the molecular clock in the brain (the suprachiasmatic nucleus, SCN) represent approximately 1% of total ganglion cells in the mouse. The majority of these cells are directly photosensitive by whole cell current clamp recordings, depolarizing in response to blue/green light with a maximum response at 480 nm (Berson *et al.*, 2002; Hattar *et al.*, 2002). This response was attributed to melanopsin; however, reconstituted melanopsin has an absorption peak at 420 nm (Newman *et al.*, 2003).

## Genetic Analysis

Genetic studies have been carried out on mice with mutations inactivating each of the various candidate circadian photopigments to quantitatively assess the contribution of each candidate gene to circadian photoreception. These studies have highlighted the contributions of three classes of photopigments in this process: the visual opsins, melanopsin, and cryptochromes. There are two common assay end points used to quantify photoreceptor input to the suprachiasmatic nucleus (SCN): behavioral analysis, which measures the synchronization of circadian behavior with a given light/dark cycle, and quantification of gene induction in the SCN in response to light. The use of behavioral analysis to analyze the effects of the loss of cryptochromes on photoreception is complicated by the essential, light-independent role of cryptochromes in the molecular clock mechanism (Griffin *et al.*, 1999; Kume *et al.*, 1999; Thresher *et al.*, 1998; van der Horst *et al.*, 1999; Vitaterna *et al.*, 1999). Both $Cry1^{-/-}$ and $Cry2^{-/-}$ mice exhibit abnormalities in the lengths of their intrinsic circadian rhythms, and $Cry1^{-/-}$ $Cry2^{-/-}$ mice are arrhythmic in constant darkness, indicative of total loss of the molecular clock. The apparently normal behavioral response of $Cry1^{-/-}$ $Cry2^{-/-}$ mice in light/dark cycles can be attributed to masking, which is the acute behavioral response to light with no lasting effect on the phase and period of the rhythm. Given that several processes, circadian and noncircadian, govern an animal's behavioral response to light, molecular analysis of phototransduction by gene induction in the SCN in response to light is the most quantitative and reliable assessment of the contribution of a photoreceptor to circadian phototransduction.

A molecular readout of light signaling to the SCN is typically measured by irradiating mice in the middle of the dark period of their circadian cycle (ZT18-20) with a range of white light doses to generate a dose–response curve. Light given at this point in the circadian cycle rapidly induces robust expression of mRNA of the immediate early gene c-*fos* and the clock genes *Per1* and *Per2*. The level of gene induction is measured quantitatively by *in situ* hybridization of 20-$\mu$m slices of the SCN using $^{35}$S-labeled probes against a specific gene. The c-*fos* gene is used as the molecular target in assays involving cryptochrome knockout mice, as the disruption of both cryptochrome genes causes constitutively high expression of *Per1* and *Per2*, making direct comparisons of *Per* levels between cryptochrome knockout mice and other genotypes impractical (Selby *et al.*, 2000; Vitaterna *et al.*, 1999). Although Fos protein is not necessary for light-induced phase shifting, induction of c-*fos* transcription in the SCN serves as a robust marker of photic input to the circadian clock (Honrado *et al.*, 1996). Examination of c-*fos* induction by a variety of chemical agents in immortalized fibroblast lines generated from wild-type and $Cry1^{-/-}$ $Cry2^{-/-}$ mice indicates that there are no gross alterations of the well-established signal transduction pathways involved in c-*fos* induction in $Cry1^{-/-}$ $Cry2^{-/-}$ mice and that c-*fos* is a suitable target for use in comparing photoresponses in wild-type and $Cry1^{-/-}$ $Cry2^{-/-}$ mice (Thompson *et al.*, 2004).

To deconvolute the contributions of various pigments in the retina, genetic approaches were used to eliminate one or more of the candidate pigments (rod and cone opsins, cryptochromes, all opsins, or cryptochromes plus all opsins) and then c-*fos* induction was tested in these animals. *rd/rd* mice were used to eliminate pigments from the outer retina, as the *rd* mutation causes retinal degeneration, resulting in complete histological destruction of the outer retina and a near complete loss of visual pigments by 12 weeks of age and has been used in many studies investigating the role of nonvisual pigments in circadian photoreception. A second approach has been to utilize $rbp^{-/-}$ mice whose retinas are histologically normal but when placed on a vitamin A-free diet lack all opsin photoreception due to depletion of the opsin chromophore retinaldehyde (Quadro *et al.*, 1999). These mice lack plasma retinol-binding protein (RBP), the only known serum transport protein for mobilizing hepatic retinol stores to other tissues, including the retina where retinol is converted to retinaldehyde for use as the opsin chromophore. In $rbp^{-/-}$ animals maintained on a vitamin A-free diet, the animals become progressively blind; after 6–10 months on a vitamin A-free diet, no electroretinogram signal can be detected and HPLC measurements show that retinal is below sensitive detection limits (0.5 ng per pair of eyecups), reduced 500-fold from wild-type levels (Thompson *et al.*, 2001).

rd/rd *and* rd/rd Cry1$^{-/-}$ Cry2$^{-/-}$ *Mice.* To assess the role of cryptochromes in circadian phototransduction, *rd/rd* and *rd/rd* $Cry1^{-/-}$ $Cry2^{-/-}$ mice on a 12-h light/dark cycle are irradiated at ZT18 with various doses of white light for a total of 30 min, sacrificed immediately, and the brains frozen under yellow light. Coronal sections of frozen brain (18 $\mu$m) are fixed and hybridized with a [35]S-labeled c-*fos* antisense RNA probe (nucleolides 855–1577) using the *in situ* hybridization protocol described earlier with standard autoradiography and quantified using a density-calibrated Leica M420 macroscope. Representative SCN slices and quantification of gene induction in wild-type, *rd*, $Cry1^{-/-}$ $Cry2^{-/-}$, and *rd/rd* $Cry1^{-/-}$ $Cry2^{-/-}$ mice are shown in Fig. 5 (Selby *et al.*, 2000). Under low irradiance ($10^4$ $\mu$mol/ m$^2$ or less photons), c-*fos* induction was severely attenuated in $Cry1^{-/-}$ $Cry2^{-/-}$ and *rd/rd* $Cry1^{-/-}$ $Cry2^{-/-}$ mice, virtually indistinguishable from

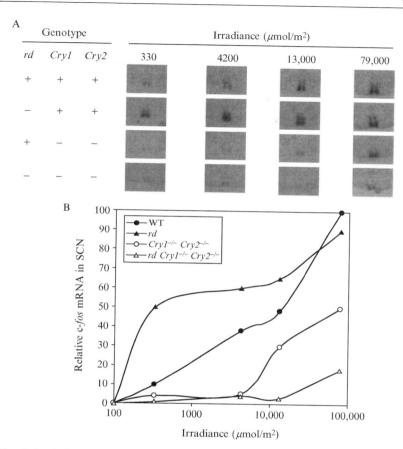

Fig. 5. Analysis of the role of cryptochromes and visual opsins in the photoinduction of c-*fos* in the SCN by *in situ* hybridization. (A) Representative slices exhibiting the strongest signal at each light dose in the SCN are shown for each of the four genotypes. (B) Dose–response plot of c-*fos* induction in the SCN of wild-type and mutant mice. Levels of c-*fos* are plotted relative to the wild type at the highest dose used (79,000 $\mu$mol/m$^2$ photons), which is taken as 100% (Selby *et al.*, 2000).

the uninduced, background level. From these induction curves, it is estimated that the loss of cryptochromes reduces photosensitivity approximately 10- to 20-fold in animals with intact rods and cones and 3000-fold in *rd/rd* animals. Thus, it appears that even in the presence of the visual opsins, the lack of cryptochromes seriously compromises photoinduction of c-*fos*, which is then reduced drastically in their absence. The residual gene induction measured in the *rd/rd Cry1$^{-/-}$ Cry2$^{-/-}$* is attributed to melanopsin (*Opn4*). Data from *Opn4*$^{-/-}$ mice indicate that there are only minor effects of the loss of melanopsin on circadian phototransduction in the presence of the visual opsins; however, all photoresponses are lost in the *rd/rd Opn4*$^{-/-}$ mice (Hattar *et al.*, 2003; Panda *et al.*, 2002, 2003; Ruby *et al.*, 2002). These data considered in their entirety suggest that photo-transduction to the SCN by cryptochrome requires melanopsin or the outer retina—how this is accomplished in mechanistic terms is not known at present.

rbp$^{-/-}$ and rbp$^{-/-}$ Cry1$^{-/-}$ Cry2$^{-/-}$ *Mice.* Studies with mice of the *rbp*$^{-/-}$ genotype were conducted to assess the relative contributions of opsins and cryptochromes to circadian photoreception. Mice of this background raised on a vitamin A-free diet for 6–10 months have less than 0.2% of the ocular retinal of wild-type mice and yet induction of both *Per1* and *Per2* mRNA in the SCN of *rbp*$^{-/-}$ mice is normal (Thompson *et al.*, 2001). In order to address the role of cryptochromes in the remaining photorespon-siveness in *rbp*$^{-/-}$ mice, *rbp*$^{-/-}$ Cry1$^{-/-}$ Cry2$^{-/-}$ mice are generated and depleted of ocular retinal on a vitamin A-free diet (Thompson *et al.*, 2004). Gene induction in response to light is performed essentially as before; mice are exposed to 80,000 $\mu$mol/m$^2$ photons of white light at ZT18-20 and killed 30–45 min after initiation of the light pulse. As seen in Fig. 6A, triple mutant mice raised on a vitamin A-free diet have virtually no c-*fos* induc-tion compared with *rbp*$^{-/-}$ controls, indicating that cryptochromes are required for photoreception in animals depleted of ocular retinal. Accord-ingly, the majority of triple mutant mice have lost all behavioral responses to light/dark cycles, as shown in Fig. 6B. Moreover, the sensitivity of pupillary photoresponse in these animals was reduced three logs relative to wild-type mice and one log relative to *rd/rd* Cry1$^{-/-}$ Cry2$^{-/-}$ animals, indicating that retinal in both the outer and the inner retina had indeed been depleted. These data strongly indicate a photoreceptive role for mouse cryptochromes, although their light-dependent mechanism of signaling to the SCN remains to be determined.

Zebrafish Cryptochromes

Unlike mammals, which rely strictly on their eyes for all photorecep-tion, some animal species receive extraocular photoreceptive input into the circadian clock. Among the most well studied are avian species such as the Japanese quail and chicken, where the pineal gland in the brain has demon-strated activity as a photoreceptive organ for the circadian clock, and zebrafish (*Danio rerio*), where peripheral clocks in internal organs such as heart and liver are locally entrained by light (Whitmore *et al.*, 2000). Several cell lines (PAC1, Z3) have been derived from zebrafish embryos and retain

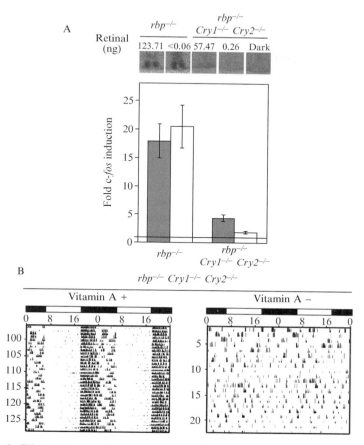

Fig. 6. Elimination of photoresponses in the absence of functional opsins and cryptochromes. (A) Photoinduction of c-*fos* in the SCN analyzed by *in situ* hybridization. Representative slices of the SCN are shown for each genotype. The bar graph represents fold-induction over unirradiated controls. Genotypes are indicated along the *x* axis. Gray bars represent mice on a vitamin A-supplemented diet, and open bars indicate mice on a vitamin A-deficient diet (ocular retinal <10 ng). Error bars represent SEM. (B) Behavioral analyses of $rbp^{-/-}$ $Cry1^{-/-}$ $Cry2^{-/-}$ mice on control and vitamin A-free diets. Actograms of mice from each diet regimen are shown. Ocular retinaldehyde levels in these two mice were 98.7 and 0.5 ng, respectively (Thompson *et al.*, 2004).

photoreceptive input into the clock. These are attractive model systems for studying light input into circadian clocks (Pando *et al.*, 2001; Whitmore *et al.*, 2000). In particular, the Z3 cell line has been rather useful in studying the circadian clock and photoentrainment mechanism in zebrafish.

The zebrafish Z3 cell line expresses all the animal core circadian clock components: Period, Clock, Bmal, and Cryptochrome. Most notably, the cell line undergoes circadian rhythms of clock gene expression that conform to the given light/dark cycle, indicating that the cell line contains the requisite photoreceptors for entraining the clock to light (Pando *et al.*, 2001). The molecular readout of circadian photoreception in the Z3 cell line is the light-dependent gene induction of *zPer2* mRNA. Induction of *zPer2* mRNA is rapid and robust, reaching levels 10- to 15-fold over dark controls within 2 h and is easily measured quantitatively by the RNase protection assay (Cermakian *et al.*, 2002). Unlike the mammalian clock, in which *Per2* expression is rhythmically regulated as a key component of the molecular clock, *zPer2* expression is strictly regulated by light and is therefore thought to be the mechanism by which the zebrafish clock synchronizes to changes in light cycles.

Zebrafish express seven cryptochrome isoforms; four of the zebrafish cryptochromes (zCry1a,b and zCry2a,b) share significant homology with mammalian cryptochromes and can act as inhibitors of the Clock/Bmal heterodimer that acts as the core transcriptional regulator of the clock, similar to the light-independent function of cryptochromes in the mammalian clock (Kobayashi *et al.*, 2000). Two cryptochromes (zCry3, zCry4) that have lost the ability to act as transcriptional repressors in reporter gene assays *in vitro* and an additional gene related to bacterial cryptochromes all have unknown function. Expression of six zebrafish cryptochromes (zCry1a,b-4) has been measured in the Z3 cell line; each cryptochrome displays a distinct expression profile with several of the cryptochrome mRNAs expressed abundantly in naïve, dark-grown Z3 cells and nearly all are strongly induced by exposure to a light/dark cycle (Cermakian *et al.*, 2002).

*Action Spectrum of* zPer2 *Induction by Light*

The isolation of the Z3 cell line as a model system for studying circadian photoreception has, for the first time, facilitated the identification of dedicated circadian photoreceptors in a simple, well-defined system. An action spectrum is a measurement of the efficiency of the output response (*zPer2* mRNA induction) as a function of the wavelength of light used. Various doses of monochromatic light at wavelengths ranging from 320 to 580 nm were used to induce *zPer2* expression, and the slope of the dose response of each wavelength was calculated as the relative efficiency of that wavelength to elicit the response (Fig. 7A) (Cermakian *et al.*, 2002). The action spectrum of *zPer2* induction in Z3 cells, shown in Fig. 7B, reveals a peak located at 380–400 nm and minimal induction over 450 nm. These data are

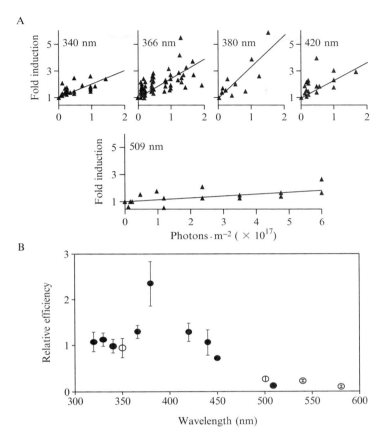

FIG. 7. Action spectrum of *zPer2* induction in the Z3 cell line. (A) Dose-dependent induction of *zPer2* by five monochromatic wavelengths of light. (B) Action spectrum showing the relative efficiency of *zPer2* induction by each wavelength of light used. Relative efficiency is the slope of the linear regression of data in A ($\times 10^{-17}$). Open circles represent single experiments. Error bars represent standard errors for the slope of the regressions (Cermakian *et al.*, 2002).

consistent with either a cryptochrome or UV/blue opsin absorbance spectrum, and the shape of the action spectrum is remarkably similar to the absorption spectrum of the *V. cholerae* cryptochrome VcCry1 (Worthington *et al.*, 2003). The zebrafish Z3 cell line provides an attractive model system for the investigation of the signal transduction of circadian photoreception and cryptochrome function *in vivo*.

*Drosophila* Cryptochrome

As in mammals, circadian photoreception in *Drosophila* consists of multiple photoreceptive input pathways utilizing both compound eyes and extraocular photoreception in the Hofbauer–Buchner eyelet and pacemaker cells. *Drosophila* has one cryptochrome (dCRY) that acts as a cell autonomous photoreceptor and is sufficient for most aspects of circadian light sensitivity and entrainment to light/dark cycles (Hall, 2000). There are currently no null mutations in *dCry*; the sole mutant available for study of cryptochrome function (*cry^b*) has a single amino acid substitution in the highly conserved flavin-binding domain (D542N), which presumably cannot interact stably with the catalytic flavin chromophore (Stanewsky *et al.*, 1998). *cry^b* mutant flies retain behavioral rhythmicity in light/dark conditions but are unable to shift the phase of their behavior in response to pulses of white light, indicating functional redundancy with rhodopsin and other opsins. Interestingly, wild-type flies exhibit two peaks in the action spectrum for phase shifting: a 420- and a 480-nm peak (Helfrich-Forster *et al.*, 2002). *glass^{60j}; so^1* double mutants, which lack all known external and internal eye structures, lost only the 480-nm peak, suggesting that the remaining 420-nm peak was contributed by cryptochrome. Combination of the *cry^b* mutation with the *glass^{60j}* mutation generated flies that lacked all known external and internal eye structures in addition to cryptochromes and resulted in flies that were visually and circadian blind (Helfrich-Forster *et al.*, 2001).

Functionally, dCRY is thought to signal light information to the clock through light-dependent interactions with the integral clock proteins dTIM and dPER, regulating the ability of the dPER-dTIM complex to inhibit CLOCK-mediated transcription (Ceriani *et al.*, 1999; Rosato *et al.*, 2001). Interestingly, in yeast two-hybrid assays, the C-terminal extension of dCRY was required to modulate the light dependence of these interactions; in its absence, all interactions became light independent, suggesting a light-dependent conformational change in dCRY involving the C-terminal domain. In addition, dTIM and dCRY protein levels are sensitive to blue light, undergoing rapid proteolytic digestion in response to light, whereas protein levels in mutant *cry^b* flies appear to lack light sensitivity (Lin *et al.*, 2001; Naidoo *et al.*, 1999). The signal transduction mechanisms utilized by dCRY, involving light-dependent protein–protein interactions and light-mediated protein degradation, are in agreement with its proposed role as a cell autonomous circadian photoreceptor.

Conclusions

Cryptochromes are flavin and folate-containing blue-light photoreceptors. Their role in regulating the circadian clock in mice and *Drosophila* has been shown unambiguously. Genetic data strongly indicate that cryptochromes function as circadian photoreceptors in these and other animals. However, direct photochemical evidence for their photoreceptive function remains to be determined.

## Acknowledgments

This work was supported by NIH Grant GM31082 to A.S. C.L.P. is supported by NIMH predoctoral National Research Service Award MH70151-01.

## References

Ahmad, M., and Cashmore, A. R. (1993). HY4 gene of *A. thaliana* encodes a protein with characteristics of a blue-light photoreceptor. *Nature* **366,** 162–166.

Ahmad, M., Grancher, N., Heil, M., Black, R. C., Giovani, B., Galland, P., and Lardemer, D. (2002). Action spectrum for cryptochrome-dependent hypocotyl growth inhibition in Arabidopsis. *Plant Physiol.* **129,** 774–785.

Berson, D. M., Dunn, F. A., and Takao, M. (2002). Phototransduction by retinal ganglion cells that set the circadian clock. *Science* **295,** 1070–1073.

Bouly, J. P., Giovani, B., Djamei, A., Mueller, M., Zeugner, A., Dudkin, E. A., Batschauer, A., and Ahmad, M. (2003). Novel ATP-binding and autophosphorylation activity associated with Arabidopsis and human cryptochrome-1. *Eur. J. Biochem.* **270,** 2921–2928.

Ceriani, M. F., Darlington, T. K., Staknis, D., Mas, P., Petti, A. A., Weitz, C. J., and Kay, S. A. (1999). Light-dependent sequestration of TIMELESS by CRYPTOCHROME. *Science* **285,** 553–556.

Cermakian, N., Pando, M. P., Thompson, C. L., Pinchak, A. B., Selby, C. P., Gutierrez, L., Wells, D. E., Cahill, G. M., Sancar, A., and Sassone-Corsi, P. (2002). Light induction of a vertebrate clock gene involves signaling through blue-light receptors and MAP kinases. *Curr. Biol.* **12,** 844–848.

Griffin, E. A., Jr., Staknis, D., and Weitz, C. J. (1999). Light-independent role of CRY1 and CRY2 in the mammalian circadian clock. *Science* **286,** 768–771.

Hall, J. C. (2000). Cryptochromes: Sensory reception, transduction, and clock functions subserving circadian systems. *Curr. Opin. Neurobiol.* **10,** 456–466.

Hattar, S., Liao, H. W., Takao, M., Berson, D. M., and Yau, K. W. (2002). Melanopsin-containing retinal ganglion cells: Architecture, projections, and intrinsic photosensitivity. *Science* **295,** 1065–1070.

Hattar, S., Lucas, R. J., Mrosovsky, N., Thompson, S., Douglas, R. H., Hankins, M. W., Lem, J., Biel, M., Hofmann, F., Foster, R. G., and Yau, K. W. (2003). Melanopsin and rod-cone photoreceptive systems account for all major accessory visual functions in mice. *Nature* **424,** 76–81.

Helfrich-Forster, C., Edwards, T., Yasuyama, K., Wisotzki, B., Schneuwly, S., Stanewsky, R., Meinertzhagen, I. A., and Hofbauer, A. (2002). The extraretinal eyelet of Drosophila: Development, ultrastructure, and putative circadian function. *J. Neurosci.* **22,** 9255–9266.

Helfrich-Forster, C., Winter, C., Hofbauer, A., Hall, J. C., and Stanewsky, R. (2001). The circadian clock of fruit flies is blind after elimination of all known photoreceptors. *Neuron* **30,** 249–261.

Honrado, G. I., Johnson, R. S., Golombek, D. A., Spiegelman, B. M., Papaioannou, V. E., and Ralph, M. R. (1996). The circadian system of c-fos deficient mice. *J. Comp. Physiol. A.* **178,** 563–570.

Hsu, D. S., Zhao, X., Zhao, S., Kazantsev, A., Wang, R. P., Todo, T., Wei, Y. F., and Sancar, A. (1996). Putative human blue-light photoreceptors hCRY1 and hCRY2 are flavoproteins. *Biochemistry* **35,** 13871–13877.

Kobayashi, Y., Ishikawa, T., Hirayama, J., Daiyasu, H., Kanai, S., Toh, H., Fukuda, I., Tsujimura, T., Terada, N., Kamei, Y., Yuba, S., Iwai, S., and Todo, T. (2000). Molecular analysis of zebrafish photolyase/cryptochrome family: Two types of cryptochromes present in zebrafish. *Genes Cells* **5,** 725–738.

Kume, K., Zylka, M. J., Sriram, S., Shearman, L. P., Weaver, D. R., Jin, X., Maywood, E. S., Hastings, M. H., and Reppert, S. M. (1999). mCRY1 and mCRY2 are essential components of the negative limb of the circadian clock feedback loop. *Cell* **98,** 193–205.

Lin, F. J., Song, W., Meye-Bernstein, E., Naidoo, N., and Sehgal, A. (2001). Photic signaling by cryptochrome in the *Drosophila* circadian system. *Mol. Cell. Biol.* **21,** 7287–7294.

Miyamoto, Y., and Sancar, A. (1998). Vitamin B2-based blue-light photoreceptors in the retinohypothalamic tract as the photoactive pigments for setting the circadian clock in mammals. *Proc. Natl. Acad. Sci. USA* **95,** 6097–6102.

Naidoo, N., Song, W., Hunter-Ensor, M., and Sehgal, A. (1999). A role for the proteasome in the light response of the timeless clock protein. *Science* **285,** 1737–1741.

Newman, L. A., Walker, M. T., Brown, R. L., Cronin, T. W., and Robinson, P. R. (2003). Melanopsin forms a functional short-wavelength photopigment. *Biochemistry* **42,** 12734–12738.

Ozgur, S., and Sancar, A. (2003). Purification and properties of human blue-light photoreceptor cryptochrome 2. *Biochemistry* **42,** 2926–2932.

Panda, S., Provencio, I., Tu, D. C., Pires, S. S., Rollag, M. D., Castrucci, A. M., Pletcher, M. T., Sato, T. K., Wiltshire, T., Andahazy, M., Kay, S. A., Van Gelder, R. N., and Hogenesch, J. B. (2003). Melanopsin is required for non-image-forming photic responses in blind mice. *Science* **301,** 525–527.

Panda, S., Sato, T. K., Castrucci, A. M., Rollag, M. D., DeGrip, W. J., Hogenesch, J. B., Provencio, I., and Kay, S. A. (2002). Melanopsin (Opn4) requirement for normal light-induced circadian phase shifting. *Science* **298,** 2213–2216.

Pando, M. P., Pinchak, A. B., Cermakian, N., and Sassone-Corsi, P. (2001). A cell-based system that recapitulates the dynamic light-dependent regulation of the vertebrate clock. *Proc. Natl. Acad. Sci. USA* **98,** 10178–10183.

Park, H. W., Kim, S. T., Sancar, A., and Deisenhofer, J. (1995). Crystal structure of DNA photolyase from *Escherichia coli*. *Science* **268,** 1866–1872.

Payne, G., Heelis, P. F., Rohrs, B. R., and Sancar, A. (1987). The active form of *Escherichia coli* DNA photolyase contains a fully reduced flavin and not a flavin radical, both *in vivo* and *in vitro*. *Biochemistry* **26,** 7121–7127.

Provencio, I., Rodriguez, I. R., Jiang, G., Hayes, W. P., Moreira, E. F., and Rollag, M. D. (2000). A novel human opsin in the inner retina. *J. Neurosci.* **20,** 600–605.

Quadro, L., Blaner, W. S., Salchow, D. J., Vogel, S., Piantedosi, R., Gouras, P., Freeman, S., Cosma, M. P., Colantuoni, V., and Gottesman, M. E. (1999). Impaired retinal function and vitamin A availability in mice lacking retinol-binding protein. *EMBO J.* **18,** 4633–4644.

Rosato, E., Codd, V., Mazzotta, G., Piccin, A., Zordan, M., Costa, R., and Kyriacou, C. P. (2001). Light-dependent interaction between *Drosophila* CRY and the clock protein PER mediated by the carboxy terminus of CRY. *Curr. Biol.* **11,** 909–917.

Ruby, N. F., Brennan, T. J., Xie, X., Cao, V., Franken, P., Heller, H. C., and O'Hara, B. F. (2002). Role of melanopsin in circadian responses to light. *Science* **298,** 2211–2213.

Sancar, A. (2003). Structure and function of DNA photolyase and cryptochrome blue-light photoreceptors. *Chem. Rev.* **103,** 2203–2237.

Sancar, A., Smith, F. W., and Sancar, G. B. (1984). Purification of *Escherichia coli* DNA photolyase. *J. Biol. Chem.* **259,** 6028–6032.

Sancar, G. B., Smith, F. W., and Sancar, A. (1985). Binding of *Escherichia coli* DNA photolyase to UV-irradiated DNA. *Biochemistry* **24**, 1849–1855.

Selby, C. P., Thompson, C., Schmitz, T. M., Van Gelder, R. N., and Sancar, A. (2000). Functional redundancy of cryptochromes and classical photoreceptors for nonvisual ocular photoreception in mice. *Proc. Natl. Acad. Sci. USA* **97**, 14697–14702.

Shalitin, D., Yang, H., Mockler, T. C., Maymon, M., Guo, H., Whitelam, G. C., and Lin, C. (2002). Regulation of Arabidopsis cryptochrome 2 by blue-light-dependent phosphorylation. *Nature* **417**, 763–767.

Shalitin, D., Yu, X., Maymon, M., Mockler, T., and Lin, C. (2003). Blue light-dependent *in vivo* and *in vitro* phosphorylation of Arabidopsis cryptochrome 1. *Plant Cell* **15**, 2421–2429.

Stanewsky, R., Kaneko, M., Emery, P., Beretta, B., Wager-Smith, K., Kay, S. A., Rosbash, M., and Hall, J. C. (1998). The cryb mutation identifies cryptochrome as a circadian photoreceptor in *Drosophila*. *Cell* **95**, 681–692.

Thompson, C. L., Selby, C. P., Van Gelder, R. N., Blaner, W. S., Lee, J., Quadro, L., Lai, K., Gottesman, M. E., and Sancar, A. (2004). Effect of vitamin A depletion on nonvisual phototransduction pathways in cryptochromeless mice. *J. Biol. Rhythms* **19**, 504–517.

Thompson, C. L., Blaner, W. S., Van Gelder, R. N., Lai, K., Quadro, L., Colantuoni, V., Gottesman, M. E., and Sancar, A. (2001). Preservation of light signaling to the suprachiasmatic nucleus in vitamin A-deficient mice. *Proc. Natl. Acad. Sci. USA* **98**, 11708–11713.

Thompson, C. L., Rickman, C. B., Shaw, S. J., Ebright, J. N., Kelly, U., Sancar, A., and Rickman, D. W. (2003). Expression of the blue-light receptor cryptochrome in the human retina. *Invest. Ophthalmol. Vis. Sci.* **44**, 4515–4521.

Thompson, C. L., Selby, C. P., Partch, C. L., Plante, D. T., Thresher, R. J., Araujo, F., and Sancar, A. (2004). Further evidence for the role of cryptochromes in retinohypothalamic photoreception/phototransduction. *Brain Res. Mol. Brain Res.* **122**, 158–166.

Thresher, R. J., Vitaterna, M. H., Miyamoto, Y., Kazantsev, A., Hsu, D. S., Petit, C., Selby, C. P., Dawut, L., Smithies, O., Takahashi, J. S., and Sancar, A. (1998). Role of mouse cryptochrome blue-light photoreceptor in circadian photoresponses. *Science* **282**, 1490–1494.

van der Horst, G. T., Muijtjens, M., Kobayashi, K., Takano, R., Kanno, S., Takao, M., de Wit, J., Verkerk, A., Eker, A. P., van Leenen, D., Buijs, R., Bootsma, D., Hoeijmakers, J. H., and Yasui, A. (1999). Mammalian Cry1 and Cry2 are essential for maintenance of circadian rhythms. *Nature* **398**, 627–630.

Van Gelder, R. N., and Sancar, A. (2003). Cryptochromes and inner retinal non-visual irradiance detection. *Novartis Found. Symp.* **253**, 31–42; discussion 42–55, 102–109, 281–284.

Vitaterna, M. H., Selby, C. P., Todo, T., Niwa, H., Thompson, C., Fruechte, E. M., Hitomi, K., Thresher, R. J., Ishikawa, T., Miyazaki, J., Takahashi, J. S., and Sancar, A. (1999). Differential regulation of mammalian period genes and circadian rhythmicity by cryptochromes 1 and 2. *Proc. Natl. Acad. Sci. USA* **96**, 12114–12119.

Whitmore, D., Foulkes, N. S., and Sassone-Corsi, P. (2000). Light acts directly on organs and cells in culture to set the vertebrate circadian clock. *Nature* **404**, 87–91.

Worthington, E. N., Kavakli, I. H., Berrocal-Tito, G., Bondo, B. E., and Sancar, A. (2003). Purification and characterization of three members of the photolyase/cryptochrome family glue-light photoreceptors from *Vibrio cholerae*. *J. Biol. Chem.* **278**, 39143–39154.

Wright, K. P., Jr., and Czeisler, C. A. (2002). Absence of circadian phase resetting in response to bright light behind the knees. *Science* **297**, 571.

Yang, H. Q., Wu, Y. J., Tang, R. H., Liu, D., Liu, Y., and Cashmore, A. R. (2000). The C termini of Arabidopsis cryptochromes mediate a constitutive light response. *Cell* **103**, 815–827.

# [39] Nonvisual Ocular Photoreception in the Mammal

*By* Russell N. Van Gelder

## Abstract

Rodents blind from outer retinal (rod and cone) degeneration still retain several light-dependent phenomena, including entrainment of the circadian clock and pupillary light responsiveness. This paradox is explained by the presence of intrinsically photosensitive retinal ganglion cells in the inner retina. These cells have unique properties, including a novel action spectrum, resistance to bleaching and adaptation under continuous light, and resistance to vitamin A depletion. Two candidate classes of photopigment have been proposed: melanopsin and cryptochromes. Physiologic analysis of circadian entrainment and pupillary light responsiveness in mice lacking these proteins leads to three conclusions: (1) outer and inner retinal photoreceptors provide partially redundant information to the inner retina, (2) melanopsin is required for inner retinal phototransduction in the absence of rod and cone signaling, and (3) cryptochromes contribute to the amplitude of inner retinal phototransduction but are not strictly required.

## Historical Introduction

Although the eye is primarily the organ subserving form vision, it has been recognized for at least a century and a half that the vertebrate eye may also harbor nonvisual photoreceptors. Brown-Sequard (1847) for example, noted that the isolated iris of the eel continues to contract in response to light. The existence of a mammalian nonvisual photoreceptor was first posited in 1927 by Keeler, the discoverer of the rodless mouse, who noted that these mice had persistent pupillary light responses despite an anatomical absence of the rods. This led to Keeler's prescient prediction:

> The controversy as to whether or not the rodless eye serves as an organ of vision has placed at stake some of our fundamental conceptions of retinal structure and correlated function.... If the vetebrate rod is necessary as an end-organ in which may be set up a neural impulse resulting in vision when it meets the brain, and, if visual purple [rhodopsin] is required in the chemical process initiating this impulse, then we may suppose that a rodless mouse will not see in the ordinary sense. Nevertheless, *we can imagine the possibility of other forms of stimulation by light, such as through absorption by pigment*

*cells, the contraction of the iris, or direct stimulation of the internal nuclear or ganglionic cells in the case of absence or faulty development of the external nuclear layer or of the rods.*

Subsequent work demonstrated that Keeler's rodless mouse (which was lost in the first half of the last century) is allelic to the modern *rd1* mutation (Pittler *et al.*, 1993).

The nonvisual photoreceptive systems of mammals elicited little additional interest for many years until Ebihara's observation that *rd/rd* mice retain the ability to entrain and phase shift their circadian rhythms in response to light (Ebihara and Tsuji, 1980). These observations were greatly extended by Foster and colleagues (reviewed in Foster, 2002), who demonstrated that entrainment persists even in mice devoid of all rods and cones [compounding the *rd* mutation or the transgenic *rdta* mutation with *cl*, a transgenic line driving diphtheria toxin under the control of a cone-specific promoter (Freedman *et al.*, 1999)]. Light suppression of pineal melatonin (Lucas *et al.*, 1999) and pupillary light responses (Lucas *et al.*, 2001) are also retained in these animals. In the latter case, an action spectrum could be derived, detailing the wavelength sensitivity of the preserved photoreceptor underlying these phenomena. With a peak sensitivity of ~480 nm, the responsible pigment obeyed univariance (suggesting a single pigment) and had a spectrum that could be reasonably well fit by a Dartnell curve of a typical opsin. A similar system appears to be operant in humans, based on action spectra that have been derived for suppression of pineal melatonin by light at night. Two groups independently derived similar spectra, which are consistent with a photopigment with peak absorption ~460 nm (Brainard *et al.*, 2001; Thapan *et al.*, 2001).

The Melanopsin Hypothesis

A remarkable confluence of experimental lines suggested a strong candidate photopigment for these phenomena. Provencio *et al.* (1998) first identified melanopsin in the dermis of *Xenopus laevis* as a potential opsin mediating the light-dependent aggregation of melanophores in this tissue. Subsequent work demonstrated a mammalian homolog for melanopsin, whose expression appeared to be limited to the inner retina (the portion preserved in outer retinal degeneration) (Provencio *et al.*, 2000). At about the same time, Berson and colleagues (2002) used a retrograde dye tracing method to identify cells projecting from the retina directly to the suprachiasmatic nucleus (SCN; the retinohypothalamic tract). Patch clamp recordings of these cells (but not other retinal ganglion cells) revealed intrinsic photosensitivity (Berson *et al.*, 2002). A rough action spectrum of these cells appeared to match that of the pigment characterized by Lucas *et al.* (2001; Berson *et al.*, 2002) as responsible for the preserved pupillary light constriction in *rd/rd;cl* mice. The photoreceptive cells had unique properties, including slow latency and sustained firing in response to stimulus. Antibody staining suggested that melanopsin was expressed in a minority of retinal ganglion cells (Provencio *et al.*, 2002). A knockin of lacZ to the melanopsin locus revealed that melanopsin was specifically expressed in those cells projecting to the nonvisual processing centers in the brain, including the SCN, the olivary pretectal nucleus, and the intergeniculate leaflet of the thalamus (Hattar *et al.*, 2002). Subsequent work has suggested that most of the retinohypothalamic tract retinal ganglion cells (but not all) are melanopsin expressing.

Knockout mice for melanopsin were generated by several laboratories (Lucas *et al.*, 2003; Panda *et al.*, 2002; Ruby *et al.*, 2002). Remarkably, these mice had minimal phenotype, showing slightly reduced magnitude of circadian phase shifts following bright light exposure (Panda *et al.*, 2002; Ruby *et al.*, 2002) and subtly reduced pupillary responses to very bright light (Lucas *et al.*, 2003). However, when the melanopsin knockout was compounded with outer retinal degeneration or dysfunction mutations (either *rd/rd* or the combination of *cgna3-* and *gnat-*), total loss of nonvisual photoreception (including circadian entrainment, pupillary light response, and melatonin suppression) was observed (Hattar *et al.*, 2003; Panda *et al.*, 2003). The minimal phenotype seen in melanopsin mutations is thought to be due to redundancy between the classical photoreceptive systems and the inner retinal photoreceptors, which may have an anatomical basis in connections between the intrinsically photosensitive retinohypothalamic-projecting neurons and the outer, visual retina (Hattar *et al.*, 2002).

These experiments would suggest a simple model for nonvisual mammalian photoreception: photic irradiance information may be transmitted by either the classical rods and cones or by the intrinsically photosensitive retinal ganglion cells, with melanopsin serving as the inner retinal photoreceptor. However, several lines of evidence are not easily reconciled with this model.

1. Vitamin A depletion fails to reduce the sensitivity of the inner retinal photoreceptor for retinohypothalamic signaling. Mice carrying the *rbp* mutation (retinol-binding protein) are unable to mobilize retinol—the precursor for all retinaldehyde-based photopigments, including all known opsins—from the liver to the eye. These animals become visually blind following 4 months of vitamin A depletion (Quadro *et al.*, 1999). However, after 10 months of depletion, these animals have unattenuated induction of the immediate early gene c-*fos* in the SCN following light exposure (Thompson *et al.*, 2001). Subsequent work has demonstrated that these

animals also retain pupillary light responses and circadian entrainment to light–dark cycles (Thompson *et al.*, 2004). These data suggest either that melanopsin is extraordinarily resistant to vitamin A depletion or that a non–vitamin-A-based photopigment is involved in inner retinal photoreception, and melanopsin is serving an essential nonphotopigment role.

2. *In vitro* expression of melanopsin yields a photopigment with properties varying from the physiological properties of the presumed pigment. When melanopsin was overexpressed in COS cells, the resulting pigment had a difference spectrum with a peak absorption of 424 nm (Newman *et al.*, 2003), quite different from the ~479-nm peak absorption seen for the maximum pupillary (Lucas *et al.*, 2001) and circadian phase shifting (Hattar *et al.*, 2003) responses in murine studies. Additionally, the pigment activated transducin only weakly. These results could be the result of abnormal posttranslational modifications in the COS cells, use of an alternate chromophore to 11-*cis* retinaldehyde by melanopsin, or misfolding of the protein in this system, but it remains to be explained.

3. Failure of inner retinal photoreceptors to bleach or adapt. As described by Berson *et al.* (2002), the inner retinal photoreceptors maintain a sustained firing rate in response to continuous irradiation without evidence of bleaching or adaptation. Such sustained responses have not been reported for opsin-based photoreceptive systems and are difficult to reconcile with a photopigment requiring *cis–trans* isomerization of retinaldehyde.

### The Cryptochrome Hypothesis

Do other pigments play a role in mammalian nonvisual photoreception? Cryptochromes are members of the photolyase family of photopigments (reviewed in Cashmore, 2003; Sancar, 2000) and are potential flavin-based photopigments. In the plant *Arabidopsis thaliana* (where cryptochromes were first discovered), this class of proteins functions critically in blue-light developmental photoresponses (Ahmad and Cashmore, 1993, 1996) as well as contributing to circadian entrainment (Somers *et al.*, 1998). The fruit fly *Drosophila melanogaster* has a single cryptochrome gene; mutants lose the ability to phase shift their circadian clock in response to light (Stanewsky *et al.*, 1998). The *Drosophila* cryptochrome behaves as a photopigment in heterologous expression systems, binding Timeless and Period in a light-dependent fashion in yeast (Ceriani *et al.*, 1999; Rosato *et al.*, 2001) and showing light-dependent degradation in *Drosophila* tissue culture cells (Busza *et al.*, 2004). Mammals express two cryptochrome homologs (Hsu *et al.*, 1996) and both are expressed in the inner retina as well as the SCN (Miyamoto and Sancar, 1998; Thompson *et al.*, 2003). Mice lacking either cryptochrome 1 or 2 show abnormal free-running periods, whereas mice lacking both are arrhythmic under constant dark conditions (Thresher *et al.*, 1998; van der Horst *et al.*, 1999; Van Gelder *et al.*, 2002; Vitaterna *et al.*, 1999). The arrhythmic phenotype is caused by the cryptochromes and functions in the central circadian clockworks as a major negative feedback regulator within the core clock transcription–translation feedback loop (reviewed in Reppert, 2000; Van Gelder and Herzog, 2003).

However, cryptochrome-deficient mice maintain rhythmicity under LD conditions, due to the direct, "masking" effect of light on locomotor activity in rodents (Mrosovsky, 2001; Mrosovsky *et al.*, 2001). When the two cryptochrome mutations were compounded with *rd/rd* retinal degeneration, an additive phenotype was seen in masking/circadian entrainment: two-thirds of triply mutant (i.e., *rd/rd;mCry1−/−;mCry2−/−*, hereafter *rd;cry-*) mice were arrhythmic in light–dark conditions (Selby *et al.*, 2000; Van Gelder *et al.*, 2002). A second assay—immediate early gene induction of c-*fos* in the SCN following light exposure—was also additively affected by mutations in *rd* and the cryptochromes. While *rd/rd* mice, if anything, have supranormal induction of c-*fos* following light exposure in the subjective night (as initially reported by Foster *et al.*, 1993) and cryptochrome mice have ~50% of normal c-*fos* induction, *rd;cry-* mice demonstrate <20% of wild-type c-*fos* induction. Two models could potentially account for these findings. In the first, cryptochrome could be participating as a photopigment or accessory protein to nonvisual (i.e., inner retinal) photoreception in the retina, in addition to its function as a core clock component. In the second, the observed additivity of phenotype could be due to the conjoint phenotype of loss-of-clock function with loss of outer retinal function.

To distinguish these possibilities, the pupillary light responses of *rd/rd* mice with and without cryptochromes were measured. The experimental setup employed is shown in Fig. 1. An infrared-sensitive CCD video camera with close-up and macro lenses is tripod mounted. Light sources (halogen and/or xenon) with narrow bandpass glass filters are used for illumination (the camera has an internal infrared light source). Irradiance is measured with a calibrated photometer. Unanesthetized mice are tolerized to handling by daily practice for at least 1 week. [We have found that ketamine anesthetic induces reversible cataracts that interfere with transmission of light to the retina (Calderone *et al.*, 1986) and chloral hydrate interferes with pupillary dilation.] Mice are held by the scruff in the calibrated light beam for the 30-s to 3-min recording period (depending on the experiment; Fig. 2). Subsequent to the experiment, the video images are enlarged on a display screen and pupillary diameter measured (Fig. 3);

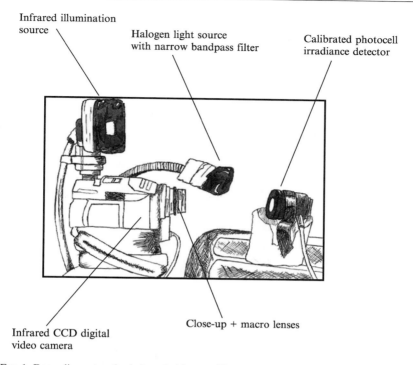

FIG. 1. Recording setup for infrared video pupillometry. Drawing by Theresa Jauregui.

the ratio of pupillary diameter to corneal diameter is typically used. Fluence–response curves can be generated by exposing mice to variable fluences of narrow bandpass-filtered light. As first noted by Lucas *et al.* (2001), outer retinal degenerate mice have fluence response curves about one log less sensitive than wild-type mice (suggesting the outer retina is responsible for ∼90% of the pupillary light response drive). When compounded with knockout alleles of both cryptochromes 1 and 2, *rd/rd* mice lose another log of sensitivity (Van Gelder *et al.*, 2003b) and are thus <1% as sensitive as wild type. As one would not expect a circadian rhythm phenotype to manifest as an alteration in pupillary light sensitivity, these results would suggest that cryptochromes do participate in inner retinal photoresponses independent of their role in the central clock mechanism. [However, one should note that many mice lacking Bmal1/Mop3 show behavioral arrhythmicity even in an LD 12:12 cycle (Bunger *et al.*,

Fig. 2. Scruff immobilization of mouse for pupillary light recording. Drawing by Theresa Jauregui.

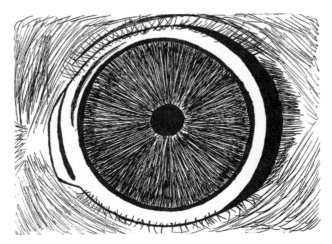

Fig. 3. Schematic of digital infrared pupil image on monitor viewing. Pupil diameter relative to corneal diameter is measured and correlated to camera video time to derive data. Drawing by Theresa Jauregui.

2000)—a result that is complicated by Mop3, a transcription factor required for cryptochrome transcription (Van Gelder *et al.*, 2003a).] The definitive role of cryptochromes in inner visual photoreception awaits physiological recordings on cryptochrome-deficient animals.

## References

Ahmad, M., and Cashmore, A. R. (1993). HY4 gene of *A. thaliana* encodes a protein with characteristics of a blue-light photoreceptor. *Nature* **366,** 162–166.

Ahmad, M., and Cashmore, A. R. (1996). Seeing blue: The discovery of cryptochrome. *Plant Mol. Biol.* **30,** 851–861.

Berson, D. M., Dunn, F. A., and Takao, M. (2002). Phototransduction by retinal ganglion cells that set the circadian clock. *Science* **295,** 1070–1073.

Brainard, G. C., Hanifin, J. P., Greeson, J. M., Byrne, B., Glickman, G., Gerner, E., and Rollag, M. D. (2001). Action spectrum for melatonin regulation in humans: Evidence for a novel circadian photoreceptor. *J. Neurosci.* **21,** 6405–6412.

Brown-Sequard, C. E. (1847). Recherches experimentales sur l'action de la lumiere et de celle d'un changement de temperature sur l'iris, dans les cinq classes d'animaux vertebres. *C. R. Acad. Sci.* **25,** 482–483.

Bunger, M. K., Wilsbacher, L. D., Moran, S. M., Clendenin, C., Radcliffe, L. A., Hogenesch, J. B., Simon, M. C., Takahashi, J. S., and Bradfield, C. A. (2000). Mop3 is an essential component of the master circadian pacemaker in mammals. *Cell* **103,** 1009–1017.

Busza, A., Emery-Le, M., Rosbash, M., and Emery, P. (2004). Roles of the two *Drosophila* CRYPTOCHROME structural domains in circadian photoreception. *Science* **304,** 1503–1506.

Calderone, L., Grimes, P., and Shalev, M. (1986). Acute reversible cataract induced by xylazine and by ketamine-xylazine anesthesia in rats and mice. *Exp. Eye. Res.* **42,** 331–337.

Cashmore, A. R. (2003). Cryptochromes: Enabling plants and animals to determine circadian time. *Cell* **114,** 537–543.

Ceriani, M. F., Darlington, T. K., Staknis, D., Mas, P., Petti, A. A., Weitz, C. J., and Kay, S. A. (1999). Light-dependent sequestration of TIMELESS by CRYPTOCHROME. *Science* **285,** 553–556.

Ebihara, S., and Tsuji, K. (1980). Entrainment of the circadian activity rhythm to the light cycle: Effective light intensity for a Zeitgeber in the retinal degenerate C3H mouse and the normal C57BL mouse. *Physiol. Behav.* **24,** 523–527.

Foster, R. G. (2002). Keeping an eye on the time: The Cogan Lecture. *Invest. Ophthalmol. Vis. Sci.* **43,** 1286–1298.

Foster, R. G., Argamaso, S., Coleman, S., Colwell, C. S., Lederman, A., and Provencio, I. (1993). Photoreceptors regulating circadian behavior: A mouse model. *J. Biol. Rhythms* **8,** S17–S23.

Freedman, M. S., Lucas, R. J., Soni, B., von Schantz, M., Munoz, M., David-Gray, Z., and Foster, R. (1999). Regulation of mammalian circadian behavior by non-rod, non-cone, ocular photoreceptors. *Science* **284,** 502–504.

Hattar, S., Liao, H. W., Takao, M., Berson, D. M., and Yau, K. W. (2002). Melanopsin-containing retinal ganglion cells: Architecture, projections, and intrinsic photosensitivity. *Science* **295,** 1065–1070.

Hattar, S., Lucas, R. J., Mrosovsky, N., Thompson, S., Douglas, R. H., Hankins, M. W., Lem, J., Biel, M., Hofmann, F., Foster, R. G., and Yau, K. W. (2003). Melanopsin and rod-cone photoreceptive systems account for all major accessory visual functions in mice. *Nature* **424,** 75–81.

Hsu, D. S., Zhao, X., Zhao, S., Kazantsev, A., Wang, R. P., Todo, T., Wei, Y. F., and Sancar, A. (1996). Putative human blue-light photoreceptors hCRY1 and hCRY2 are flavoproteins. *Biochemistry* **35,** 13871–13877.

Keeler, C. E. (1927). Iris movements in blind mice. *Am. J. Physiol.* **81,** 107–112.

Lucas, R. J., Douglas, R. H., and Foster, R. G. (2001). Characterization of an ocular photopigment capable of driving pupillary constriction in mice. *Nature Neurosci.* **4,** 621–626.

Lucas, R. J., Freedman, M. S., Munoz, M., Garcia-Fernandez, J. M., and Foster, R. G. (1999). Regulation of the mammalian pineal by non-rod, non-cone, ocular photoreceptors. *Science* **284,** 505–507.

Lucas, R. J., Hattar, S., Takao, M., Berson, D. M., Foster, R. G., and Yau, K. W. (2003). Diminished pupillary light reflex at high irradiances in melanopsin-knockout mice. *Science* **299,** 245–247.

Miyamoto, Y., and Sancar, A. (1998). Vitamin B2-based blue-light photoreceptors in the retinohypothalamic tract as the photoactive pigments for setting the circadian clock in mammals. *Proc. Natl. Acad. Sci. USA* **95,** 6097–6102.

Mrosovsky, N. (2001). Further characterization of the phenotype of mCry1/mCry2-deficient mice. *Chronobiol. Int.* **18,** 613–625.

Mrosovsky, N., Lucas, R. J., and Foster, R. G. (2001). Persistence of masking responses to light in mice lacking rods and cones. *J. Biol. Rhythms* **16,** 585–588.

Newman, L. A., Walker, M. T., Brown, R. L., Cronin, T. W., and Robinson, P. R. (2003). Melanopsin forms a functional short-wavelength photopigment. *Biochemistry* **42,** 12734–12738.

Panda, S., Provencio, I., Tu, D. C., Pires, S. S., Rollag, M. D., Castrucci, A. M., Pletcher, M. T., Sato, T. K., Wiltshire, T., Andahazy, M., Kay, S. A., Van Gelder, R. N., and Hogenesch, J. B. (2003). Melanopsin is required for non-image-forming photic responses in blind mice. *Science* **301,** 525–527.

Panda, S., Sato, T. K., Castrucci, A. M., Rollag, M. D., DeGrip, W. J., Hogenesch, J. B., Provencio, I., and Kay, S. A. (2002). Melanopsin (Opn4) requirement for normal light-induced circadian phase shifting. *Science* **298,** 2213–2216.

Pittler, S. J., Keeler, C. E., Sidman, R. L., and Baehr, W. (1993). PCR analysis of DNA from 70-year-old sections of rodless retina demonstrates identity with the mouse rd defect. *Proc. Natl. Acad. Sci. USA* **90,** 9616–9619.

Provencio, I., Jiang, G., De Grip, W. J., Hayes, W. P., and Rollag, M. D. (1998). Melanopsin: An opsin in melanophores, brain, and eye. *Proc. Natl. Acad. Sci. USA* **95,** 340–345.

Provencio, I., Rodriguez, I. R., Jiang, G., Hayes, W. P., Moreira, E. F., and Rollag, M. D. (2000). A novel human opsin in the inner retina. *J. Neurosci.* **20,** 600–605.

Provencio, I., Rollag, M. D., and Castrucci, A. M. (2002). Photoreceptive net in the mammalian retina. *Nature* **415,** 493.

Quadro, L., Blaner, W. S., Salchow, D. J., Vogel, S., Piantedosi, R., Gouras, P., Freeman, S., Cosma, M. P., Colantuoni, V., and Gottesman, M. E. (1999). Impaired retinal function and vitamin A availability in mice lacking retinol-binding protein. *EMBO J.* **18,** 4633–4644.

Reppert, S. M. (2000). Cellular and molecular basis of circadian timing in mammals. *Semin. Perinatol.* **24,** 243–246.

Rosato, E., Codd, V., Mazzotta, G., Piccin, A., Zordan, M., Costa, R., and Kyriacou, C. P. (2001). Light-dependent interaction between Drosophila CRY and the clock protein PER mediated by the carboxy terminus of CRY. *Curr. Biol.* **11,** 909–917.

Ruby, N. F., Brennan, T. J., Xie, X., Cao, V., Franken, P., Heller, H. C., and O'Hara, B. F. (2002). Role of melanopsin in circadian responses to light. *Science* **298,** 2211–2213.

Sancar, A. (2000). Cryptochrome: The second photoactive pigment in the eye and its role in circadian photoreception. *Annu. Rev. Biochem.* **69,** 31–67.

Selby, C. P., Thompson, C., Schmitz, T. M., Van Gelder, R. N., and Sancar, A. (2000). Functional redundancy of cryptochromes and classical photoreceptors for nonvisual ocular photoreception in mice. *Proc. Natl. Acad. Sci. USA* **97,** 14697–14702.

Somers, D. E., Devlin, P. F., and Kay, S. A. (1998). Phytochromes and cryptochromes in the entrainment of the Arabidopsis circadian clock. *Science* **282,** 1488–1490.

Stanewsky, R., Kaneko, M., Emery, P., Beretta, B., Wager-Smith, K., Kay, S. A., Rosbash, M., and Hall, J. C. (1998). The cryb mutation identifies cryptochrome as a circadian photoreceptor in *Drosophila*. *Cell* **95,** 681–692.

Thapan, K., Arendt, J., and Skene, D. J. (2001). An action spectrum for melatonin suppression: Evidence for a novel non-rod, non-cone photoreceptor system in humans. *J. Physiol. (Lond)* **535,** 261–267.

Thompson, C. L., Blaner, W. S., Van Gelder, R. N., Lai, K., Quadro, L., Colantuoni, V., Gottesman, M. E., and Sancar, A. (2001). Preservation of light signaling to the suprachiasmatic nucleus in vitamin A-deficient mice. *Proc. Natl. Acad. Sci. USA* **98,** 11708–11713.

Thompson, C. L., Rickman, C. B., Shaw, S. J., Ebright, J. N., Kelly, U., Sancar, A., and Rickman, D. W. (2003). Expression of the blue-light receptor cryptochrome in the human retina. *Invest. Ophthalmol. Vis. Sci.* **44,** 4515–4521.

Thompson, C. L., Selby, C. P., Van Gelder, R. N., Blaner, W. S., Lee, J., Quadro, L., Lai, K., Gottesman, M. E., and Sancar, A. (2004). Effect of vitamin A depletion on nonvisual phototransduction pathways in cryptochromeless mice. *J. Biol. Rhythms* **19,** 504–517.

Thresher, R. J., Vitaterna, M. H., Miyamoto, Y., Kazantsev, A., Hsu, D. S., Petit, C., Selby, C. P., Dawut, L., Smithies, O., Takahashi, J. S., and Sancar, A. (1998). Role of mouse cryptochrome blue-light photoreceptor in circadian photoresponses. *Science* **282,** 1490–1494.

van der Horst, G. T., Muijtjens, M., Kobayashi, K., Takano, R., Kanno, S., Takao, M., de Wit, J., Verkerk, A., Eker, A. P., van Leenen, D., Buijs, R., Bootsma, D., Hoeijmakers, J. H., and Yasui, A. (1999). Mammalian Cry1 and Cry2 are essential for maintenance of circadian rhythms. *Nature* **398,** 627–630.

Van Gelder, R. N., Gibler, T. M., Tu, D., Embry, K., Selby, C. P., Thompson, C. L., and Sancar, A. (2002). Pleiotropic effects of cryptochromes 1 and 2 on free-running and light-entrained murine circadian rhythms. *J. Neurogenet.* **16,** 181–203.

Van Gelder, R. N., and Herzog, E. D. (2003). Oscillatory mechanisms underlying the murine circadian clock. *Sci. STKE* **209,** tr7.

Van Gelder, R. N., Herzog, E. D., Schwartz, W. J., and Taghert, P. H. (2003a). Circadian rhythms: In the loop at last. *Science* **300,** 1534–1535.

Van Gelder, R. N., Wee, R., Lee, J. A., and Tu, D. C. (2003b). Reduced pupillary light responses in mice lacking cryptochromes. *Science* **299,** 222.

Vitaterna, M. H., Selby, C. P., Todo, T., Niwa, H., Thompson, C., Fruechte, E. M., Hitomi, K., Thresher, R. J., Ishikawa, T., Miyazaki, J., Takahashi, J. S., and Sancar, A. (1999). Differential regulation of mammalian period genes and circadian rhythmicity by cryptochromes 1 and 2. *Proc. Natl. Acad. Sci. USA* **96,** 12114–12119.

# Section X

# Sleeping Flies

# [40]  Essentials of Sleep Recordings in *Drosophila:* Moving Beyond Sleep Time

*By* Rozi Andretic and Paul J. Shaw

## Abstract

The power of *Drosophila* genetics can be used to facilitate the molecular dissection of sleep regulatory mechanisms. While evaluating total sleep time and homeostatic processes provides valuable information, other variables, such as sleep latency, sleep bout duration, sleep cycle length, and the time of day when the longest sleep bout is initiated, should also be used to explore the nature of a genetic lesion on sleep regulatory processes. Each of these variables requires that the recording interval used to identify periods of sleep and waking be determined accurately and empirically. This article describes the procedures for recording sleep in *Drosophila* and associated methodological constraints. In addition, it provides results from a normative data set of 1037 *Canton-S* female flies and 639 male flies to illustrate the nature and variability of sleep variables that one can extract from 24 h of data collection in *Drosophila*.

## Introduction

Sleep is defined by periods of quiescence that are associated with a reduced responsivity to the external environment and are homeostatically regulated (Campbell and Tobler, 1984; Carskadon and Dement, 2000; Hendricks *et al.*, 2000; Shaw *et al.*, 2000). In *Drosophila melanogaster*, quiescent episodes can be measured easily and reliably using an apparatus originally designed for investigating circadian rhythms in locomotor behavior (Hamblen *et al.*, 1998). Whereas the circadian periodicity of locomotor activity is tightly regulated, average sleep time and other sleep-related parameters can display a high degree of variability. Thus, while *Drosophila* shows great promise for quickly identifying sleep regulatory process, care must be taken to accurately describe sleep phenotypes. This article describes the procedures for recording sleep in *Drosophila* and associated methodological constraints. In addition, it provides results from a normative data set of 1037 *Canton-S* (*Cs*) female flies and 639 male flies to illustrate the nature and variability of sleep phenotypes that one can expect to find when initiating sleep studies.

Procedure

We measure quiescence in *Drosophila* utilizing the activity monitoring system (AMS) developed by TriKinetics (Waltham, MA). Analysis of sleep and wake values is done on a population of flies of the same age, genotype, and sex exposed to similar environmental conditions (e.g., food, lighting, temperature, humidity). The minimum number of flies for a given genotype that need to be tested to get a reliable estimate of sleep parameters is usually 16. Repeated measurements across several generations are preferred as sleep parameters in some genotypes can show variability over time in apparently similar conditions. Application of standardized procedures in fly culturing, collecting, and recording is crucial to decrease variability between repeated experiments. The same conditions used for maintaining fly cultures should also be used when recording sleep (e.g., type of food, temperature, humidity, and lighting). Flies are collected using $CO_2$ anesthesia when they are 0–24 h old; males are separated from females, aged in same sex vials for 3 days, and used for experiment not earlier than day 4, as the adult pattern of sleep stabilizes by that time. Exposure to $CO_2$ anesthesia should be kept to a minimum held constant for all fly collections and avoided at least 2 days before recording. Flies are loaded into TriKinetics glass tubes using a mouth aspirator. Tubes are 65 mm long, 6.5 mm in diameter, and have fresh food on one end, enabling the continuous recording over several days. The monitor holds 32 tubes and collects infrared beam interruptions caused when a fly walks across the midline of the tube. The first full day of recording, the baseline day, is the first 24 h of recording, from the time of lights on to lights off. When flies are monitored for more than 3 consecutive days they should be transferred into clean tubes with fresh food. This is particularly important if mated females are used, as larvae will start to develop and will lead to faulty activity and sleep values. Data collection can be accomplished in time bins ranging from 30 s to 1 h, and raw data consist of the number of beam interruptions (activity) or lack of beam interruptions (quiescence) in a given time window. In our work we routinely use a 5-min time window; advantages of this time window compared with shorter or longer ones are discussed in the following sections (Hendricks *et al.*, 2000; Shaw *et al.*, 2000, 2002).

From raw data, which consist of the total number of infrared crossing in each of 288 5-min bins in a 24-h recording period for each fly, we routinely extract several basic numerical values, including sleep time, sleep bout duration, the number of sleep bouts, the interval from the start of one sleep episode to the beginning of the subsequent episode (interbout interval or sleep cycle), the latency to the first sleep episode of the night, and the time of day in which the longest sleep bout occurs. Because quiescent episodes lasting 5 min or more are characterized by increased arousal thresholds and are regulated homeostatically (see later), they can be used as a surrogate measure for sleep (Shaw et al., 2000). Thus, bins without any beam interruptions are used to calculate minutes of sleep for each hour of the 24-h day (Fig. 1). Sleep values can then be used for within- or between-group analysis; either on the basis of an hour-by-hour difference in sleep amount or difference in average sleep amount for larger time periods (during the day, during the night, or during 24 h).

In addition to parameters relating to sleep, we evaluate both the amount and the intensity of locomotor activity. We regularly calculate both the total number of beam crossings/hour and the average number of beam crossings for each waking minute (counts per waking minute). The average total number of counts per hour yields an estimate of total activity without taking into account minutes of sleep within a given hour. This measure cannot distinguish between a poorly active but awake fly and a very active fly that sleeps for a large portion of a given hour. In contrast, counts/waking minute distinguishes between hypoactive and hyperactive flies irrespective of sleep time. Moreover, counts/waking minute provides a much better indicator of the health of a fly, where a reduction in the intensity of activity is suggestive of a sick or otherwise impaired fly. This is particularly true when an increase in the amount of time a fly is quiescent is accompanied by a significant decrease in counts/waking minute. Under these circumstances, the decrease in activity might indicate that other factors, such as health status, are possibly confounding data. To exclude this possibility, flies must be observed visually to determine whether they show locomotor anomalies and are awake but not crossing the midline of the tube (see later).

## Constraints

The recording interval used to quantify sleep and waking parameters is constrained by the behavior of the fly within the apparatus. We have determined that 5 min is a very robust interval that can be used to accurately identify periods of sleep and waking in a wide range of experimental conditions (Shaw et al., 2000). Nonetheless, we would like to emphasize that the appropriate interval must be determined empirically and can be influenced by age, sex, and genotype. Thus, while a 5-min interval is very accurate, it should not be viewed as the absolute standard. However, when shorter or longer intervals are required, their accuracy should be justified and accompanied by empirical data. When considering alternate time intervals, the following arguments should be considered.

First, the accuracy of the apparatus to detect movement in a given interval will define the shortest recording time that can be used to identify waking. Waking will be miscoded as sleep if the fly is awake and does not

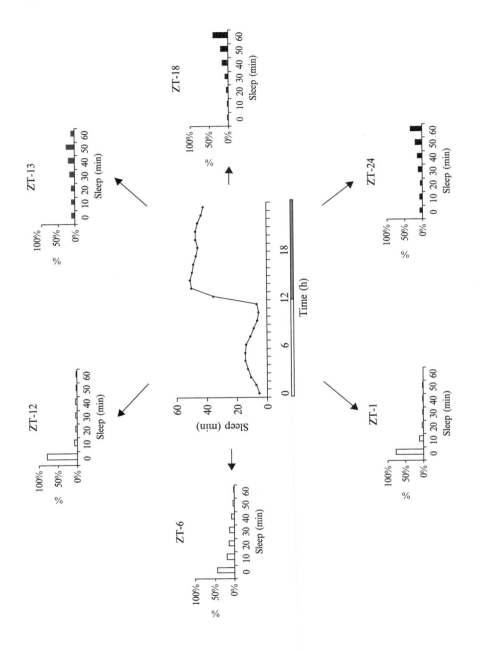

cross the midline within the specified interval. As the time interval is lengthened, the probability that the fly will cross the midline increases. As the time interval is shortened, the probability that an awake fly will cross the midline will decrease. To assess the accuracy of the defined interval, flies are placed into TriKinetics tubes and monitored visually for goal-directed behavior (e.g., eating, walking, grooming). An error is counted if the fly is awake for any portion of the interval and does not cross the midline of the tube by the end of that interval. We have found that young wild-type male and female flies will cross the infrared beam in 95–98% of all 5-min intervals that exhibit goal-directed behavior (Shaw et al., 2000). However, when the interval is shortened to 1 min, the error rate can be as high as 20–30%. Although one might expect that the error rate will be lower in hyperactive flies, this need not be the case, i.e., the error is determined by the temporal spacing of activity, not its absolute levels. Indeed, we have found that the error rate in some strains of hyperactive flies can be substantially higher than in Cs flies when using 1-min intervals.

Second, the minimum duration of quiescent episodes that result in an elevated arousal threshold defines the shortest recording time that can be used to identify sleep. Sleep episodes will not be detected when periods of sleep and waking occur within the same time interval. As the time interval is lengthened, there is an increased probability that periods of sleep will be interspersed with waking. However, as the time interval is shortened, there is an increased probability that a quiescent fly will not display elevated arousal thresholds. Such episodes fail to meet an important criterion for sleep and thus cannot be used to quantify sleep parameters. We have found that an awake fly will respond to 90% of environmental perturbations, whereas flies that are quiescent for 5 min or more respond to only 10% of these stimuli (Shaw et al., 2000). To evaluate arousal thresholds, we videotape flies that have been placed into TriKinetics tubes and are exposed to a vibratory or visual stimulus each hour. Because we are interested in identifying the shortest recording time that can be used to identify sleep, we have focused our attention on the time of day when sleep episodes are short (see later). An observer that is blind to the animals' prior sleep history scores the responses of the flies for 30 s. A response is

---

Fig. 1. Circadian distribution of sleep. (Center) The minutes of sleep (mean ± SEM) for each hour of the 24-h day in 1037 female Cs flies maintained on a 12:12 LD schedule. Female flies obtain the majority of their sleep during the dark period (gray bar). (Surround) Frequency distribution of sleep times during representative 1-h periods expressed in 10-min bins. Zero minutes of sleep indicate that the flies have been awake the entire hour; 60 min of sleep indicate that the fly has been sleeping the entire hour. Note that during the dark period a substantial percentage of flies have been awake for 30 min or more (white bar indicates lights on, gray bar indicates lights off).

defined as an abrupt change in ongoing behavior. After the responses are recorded, sleep and wake behaviors in the preceding 10-min interval are determined by visual observation. We have found that when *Cs* flies are quiescent for only 1 min they are much more responsive to environmental perturbation (50% responses versus 10% for 5-min episodes). To explore this result further, we have visually identified quiescent episodes and tabulated their number and duration. Interestingly, we found that at this time of day, 50% of all sleep episodes were shorter than 5 min, i.e., flies become quiescent for a short period of time (<5 min), after which they resumed their activity. These data suggest that quiescent periods whose duration is less than 5 min exhibit arousal thresholds similar to that seen in an awake behaving animal. Thus, based on the empirical results outlined earlier, a 5-min interval seems best suited to accurately record both sleep and waking behavior in young *Cs* flies.

Basic Characteristics of Sleep

We have thoroughly evaluated sleep in a normative data set of 1037 *Cs* flies maintained on a 12:12 LD schedule. We present these data as an example of the number and kind of sleep and wake parameters that can be extracted from 24 h of data collection. As described previously, flies are diurnal and sleep is largely confined to the dark period (Fig. 1, center). On average, female flies sleep between 40 and 50 min per hour during the night. Closer inspection of data, however, reveals that there is a high degree of diversity in the amount of sleep that individual flies obtain during any given hour of the circadian day (Fig. 1, surround). For example, at zeitgeber time 18 (ZT-18), fewer than half of the flies sleep for the entire hour and 30% of all flies have been awake for 30 min or more. This observation has important implications for studies investigating the biochemical and molecular basis of sleep. That is, collecting flies without identifying those individuals that are asleep or are awake for a specified amount of time can obliterate molecular or biochemical differences that might exist between different conditions.

In addition to measuring total sleep time, we routinely evaluate several other parameters to gain insight into sleep regulatory process. These parameters are summarized in Table I and include sleep latency, average sleep bout duration, sleep cycle length, and the time of day when the longest sleep bout is initiated. The latency to the first sleep bout of the night provides a measure on how readily a fly can get to sleep. Prolonged sleep latency may indicate a deficit in mechanisms underlying sleep initiation. The duration of sleep episodes or bouts during the night indicates how well a fly can stay asleep. Decreased sleep bout duration may be indicative of a

TABLE I

QUANTIFICATION OF SLEEP AND WAKE VARIABLES DURING LIGHT AND DARK PERIODS

| Sleep | n | | Sleep time | Bout number | Bout duration | Bout interval | Longest bout | Latency to sleep |
|---|---|---|---|---|---|---|---|---|
| Females | 1037 | Light | 147.5 ± 4.2 | 11.3 ± 0.2 | 12.3 ± 0.8 | 79.5 ± 2.8 | 31.5 ± 1.3 | |
| | | Dark | 571.4 ± 3.4 | 10.4 ± 0.2 | 99.7 ± 4 | 92.5 ± 2.8 | 257.4 ± 5.1 | 14 ± 0.7 |
| Males | 639 | Light | 369.8 ± 4 | 10.2 ± 0.2 | 46.2 ± 1.7 | 70.7 ± 1.8 | 162.8 ± 4.2 | |
| | | Dark | 563.9 ± 3.8 | 9 ± 0.2 | 88.9 ± 3.1 | 112.2 ± 3.8 | 236.1 ± 5.6 | 10 ± 1.1 |
| Wake | | | Wake time | Bout number | Bout duration | Bout interval | Longest bout | |
| Females | | Light | 573 ± 4.2 | 11.2 ± 0.2 | 122.3 ± 5.5 | 83.9 ± 2.9 | 238.6 ± 5.4 | |
| | | Dark | 149.9 ± 3.4 | 10.6 ± 0.2 | 15 ± 0.8 | 90.4 ± 2.7 | 41.4 ± 1.4 | |
| Males | | Light | 350.8 ± 4 | 10.4 ± 0.2 | 39.1 ± 0.8 | 73.8 ± 1.7 | 154 ± 2.4 | |
| | | Dark | 156.1 ± 3.8 | 9.4 ± 0.2 | 17.7 ± 0.4 | 110.5 ± 3.6 | 47.1 ± 1.4 | |

deficit in mechanisms underlying sleep maintenance. The length of the sleep cycle and the number of sleep bouts provide insight into sleep timing mechanisms. As seen in Table I, the number of sleep bouts and cycle length are very stable across the circadian day even though flies sleep substantially more at night. Thus, as in mammals, the timing of sleep is precisely controlled (Kleitman, 1963). Disruptions in cycle length and sleep bout number may be indicative of deficits in mechanisms regulating sleep timing. Sleep is believed to be regulated by both circadian and homeostatic processes. In mammals, and presumably flies, the clock alternately promotes and maintains both sleep and waking at specific times of the day while the homeostatic process monitors sleep need, which accumulates during waking and is dissipated during sleep (Borbely, 1982; Daan *et al.*, 1984; Edgar *et al.*, 1993; Franken *et al.*, 1991). The interaction between these two processes determines the timing, duration, and quality of both sleep and wakefulness. If the sleep bout duration reflects homeostatic drive, then the time of day when the longest sleep bout is initiated may provide a rough estimate of homeostatic drive and/or the interaction between homeostatic and circadian processes. Displacement of the longest sleep episode to later or earlier circadian times could suggest either that homeostatic mechanisms have been altered or that sleep and/or wake promoting signals from the clock have been modified.

We routinely describe our results using average data for a population of flies (Table I). However, it is often helpful to identify important trends within a data set by examining the frequency distributions of sleep parameters. The importance of this approach and its implications for experimental design was discussed earlier (Fig. 1). We believe that the variability found in sleep parameters within a genotype is an asset that can be exploited to further elucidate fundamental sleep mechanisms. However, we understand that those not expecting this degree of variability may find it disconcerting. To prepare those who wish to use *Drosophila* as a model system to study sleep, we have plotted the frequency distributions for several sleep parameters in Fig. 2. Although these data represent all 1037 flies, we have found that approximately 160 age- and sex-matched animals provide a good representation for a particular genotype. As can be seen in Fig. 2A, total daily sleep time is normally distributed. Nonetheless, subsets of these flies sleep less than 360 min/day and more than 1200 min/day. The range of sleep times found in wild-type *Cs* flies has implications for determining criteria that can be used to identify both short- and long-sleeping mutant flies. This determination is further complicated by the observation that the mean sleep time in a population of *Cs* flies can vary across generations. For example, the mean total sleep times in 10 independent groups of 16 *Cs* females flies evaluated during a 3-month period ranged

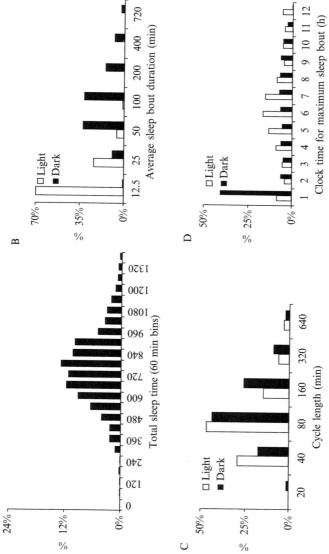

FIG. 2. Frequency distribution of sleep parameters in 1037 female Cs flies. (A) Total sleep time. (B) Average sleep bout duration during the day (white bars) and night (dark bars). (C) Interval from the beginning of one sleep episode to the beginning of the subsequent sleep episode. (D) Time of day when the longest sleep bout is initiated. The first hour indicates either the first hour of the light period (white bars) or the first hour of the dark period (dark bars).

from $540 \pm 50$ to $820 \pm 45$ min/day. Thus, both the magnitude of the difference and its reproducibility over generations should be considered along with statistical considerations when identifying "short" and "long" sleeping mutants.

During the light period, sleep episodes in female $Cs$ flies are uniformly short and, as expected, increase in length during the dark period (Fig. 2B). Nonetheless, subsets of flies obtain their sleep in short bouts ($<25$ min), whereas others sleep in long bouts ($>400$ min). In our experience, sleep time and average sleep bout duration are frequently dissociable and likely reflect separate regulatory processes. In contrast to sleep bout duration, the frequency distribution for sleep cycle length (the interval from the start of one sleep episode to the beginning of the subsequent episode) is quite similar during both the day and the night (Fig. 2C). Thus, initiation of a sleep episode does not appear to be strongly influenced by circadian signals. Nonetheless, the average sleep bout duration and the time of day when the longest sleep bout is initiated are likely to reflect an interaction between circadian and homeostatic processes as they do in mammals (Borbely, 1982; Daan *et al.*, 1984; Edgar *et al.*, 1993; Franken *et al.*, 1991). Thus, sleep time is stably increased during the midafternoon and during this time flies tend to initiate longer sleep bouts (Fig. 2D). At the beginning of the circadian night, when homeostatic drive is presumed to be high and circadian signals are conducive for sleep, 40% of female $Cs$ flies initiate their longest sleep of the night. This result could not have been readily predicted by evaluating data presented in Fig. 1 alone or by examining mean data for this parameter.

Sleep Deprivation

Although the quantification of basic sleep parameters such as total sleep time and average bout duration can be quite informative, the regulation of sleep is eveluted more precisely using sleep deprivation and evaluating the subsequent homeostatic response (Shaw and Franken, 2003). The size of the homeostatic response is evaluated by constructing cumulative difference plots (Edgar and Seidel, 1997; Shaw *et al.*, 2002).

This is accomplished for each individual fly, first by subtracting the minutes of sleep during deprivation and recovery from the corresponding baseline value and summing the difference score with the preceding hour. A negative slope indicates that sleep is being lost; a positive slope indicates sleep gained and a slope of zero indicates that recovery is complete. Sleep rebound is calculated as a ratio of the amount of sleep recovered divided by that lost, i.e., [(maximum value when the slope was zero–minimum value)/ minimum value]. Nonetheless, it is important to note that sleep deprivation experiments are inherently problematic in that the deprivation stimulus itself may be the cause of a given outcome (Rechtschaffen, 1998). We discuss some of these limitations and their potential solutions briefly.

Currently we utilize two sleep deprivation methods. The first method is to keep flies awake by gently tapping on their tubes when they become immobile (Shaw et al., 2000). This method is superior to other deprivation methods because (1) the stimulus is administered to an individual fly based on its behavior; (2) the experimenter controls the complexity of the stimulus and utilizes the least intense intervention required to induce waking in each individual; and (3) one can monitor the response of the fly to the stimulus in order to detect aberrant behavior. A stimulus that we found to be particularly effective when conducting these experiments was to suddenly lower or snap one end of the tube by 1–2 mm. This stimulus is effective presumably because it initiates a geotactic response.

Using this principle, we created an automated system: the sleep nullifying apparatus (SNAP) (for details, see Shaw et al., 2002). We have determined that the stimulus provided by this apparatus is benign (see later). Nonetheless, if results obtained from the SNAP indicate that a fly is resistant or sensitive to sleep loss, we immediately conduct manual sleep deprivation experiments to ensure that the outcome cannot be explained by artifacts of either the SNAP apparatus or with the genotype under investigation. In addition, when flies show an exaggerated homeostatic response, we evaluate their arousal thresholds as described earlier to ensure that flies are indeed sleeping and not merely awake but inactive.

Although manual sleep deprivation methods have many advantages, they are not feasible for screening large numbers of flies. Thus, we have conducted a series of experiments to validate the SNAP apparatus (Shaw et al., 2000, 2002). Each of these experiments can be adapted for use in any automated system. To determine whether the stimulus perceived by the fly in the SNAP apparatus was as mild as we believed, we evaluated the behavior of a bang-sensitive paralytic mutant that is extremely sensitive to mechanical shock. Not only did we find that flies carrying this mutation did not become paralyzed by the deprivation stimulus, but we also demonstrated that their activity levels did not differ from wild-type flies during the 12-h course of deprivation (Shaw et al., 2002). These data confirmed that the stimulus used to keep flies awake is quite benign. An alternative strategy used to determine whether a particular outcome was the result of the stimulus intensity rather than sleep loss per se was to administer the same number of stimuli without producing continuous wakefulness. Using a modified protocol, we subjected flies to the same total number of stimuli that accrued during 12 h of sleep deprivation without producing 12 h of continuous wakefulness and found no long-term consequences of the mechanical stimulus (Shaw et al., 2002). To determine whether the deprivation stimulus activates nonspecific stress pathways, we evaluated genes that respond to qualitatively different stressors, including metabolic stress (*SNF4a, Hif1a*), chemical stress (*mpk2*), and humoral stress (*turandot*). Results demonstrated that these genes were not transcriptionally upregulated by sleep deprivation (Shaw *et al.*, 2002). Finally, to test the assumption that the mutation exerts its effects on sleep independently of its effects on health and fitness, we evaluate a particular mutant and determine whether it is hyperresponsive to nonspecific environmental challenges by exposing it to heat stress, oxidation, starvation, and desiccation. Based on our experience, mutants that are less fit can still be used to evaluate the effects of short-term sleep loss but, due to confounding factors, should be used cautiously when testing hypotheses requiring long-term sleep deprivation.

## Sexual Dimorphism

To simplify the discussion, this article has focused exclusively on female flies. However, it is important to note that there is a pronounced sexual dimorphism in average sleep time in *Cs* flies (Shaw *et al.*, 2000). The difference in total sleep time is accounted for by significantly more sleep during the light period. Thus, while both male and female *Cs* flies tend to obtain similar amounts of sleep at night, males sleep more during the day and their sleep bouts are longer (Table I). This phenomenon is prevalent in many wild-type strains as well as in many mutant lines (data not shown). Interestingly, even flies mutant for circadian genes that do not display noon-time siesta when housed under constant conditions show pronounced sexual dimorphism in average sleep time during 24 h (data not shown). The sexual dimorphism in sleep amounts may also extend to other sleep parameters, including sleep homeostasis (Hendricks *et al.*, 2003; Shaw *et al.*, 2002). Thus, male flies mutant for the clock gene *cycle* ($cyc^{01}$) show little if any homeostatic response following sleep loss whereas their sisters exhibit a vigorous compensatory sleep rebound that is six times larger than wild-type flies and is not complete until they have slept an additional 3 min over baseline for each minute of sleep lost (Hendricks *et al.*, 2003; Shaw *et al.*, 2002). Taken together, these data emphasize the importance of evaluating sleep in both sexes.

## Conclusions

A number of sleep parameters can be identified readily in *D. melanogaster* using standard off-the-shelf equipment (Hendricks *et al.*, 2000; Shaw *et al.*, 2000). The data collection interval used to accurately record sleep and waking episodes can be influenced by age, sex, and genotype and must be determined empirically. Total sleep time constitutes only one of several variables that can be used to evaluate the genetic basis of sleep regulation. Other variables, such as sleep latency, sleep bout duration, sleep cycle length, and the time of day when the longest sleep bout is initiated, should also be used to explore the nature of the genetic lesion on sleep regulatory processes. Results for a given genotype should be replicated over several generations due to an inherent variability in many of these parameters. Sleep deprivation experiments provide invaluable information but should be accompanied by the appropriate controls to ensure that the outcome is not due to confounding variables. Because many sleep parameters are sexually dimorphic, both male and female flies should be tested.

## Acknowledgment

Research conducted at The Neurosciences Institute was supported by Neurosciences Research Foundation.

## References

Borbely, A. A. (1982). A two process model of sleep regulation. *Hum. Neurobiol.* **1**(3), 195–204.

Campbell, S. S., and Tobler, I. (1984). Animal sleep: A review of sleep duration across phylogeny. *Neurosci. Biobehav. Rev.* **8**(3), 269–300.

Carskadon, M., and Dement, W. C. (2000). Normal human sleep: An overview. *In* "Principles and Practice of Sleep Medicine" (M. Kryger, T. Roth, and W. C. Dement, eds.), pp. 15–25. Saunders, Philadelphia.

Daan, S., Beersma, D. G., and Borbely, A. A. (1984). Timing of human sleep: Recovery process gated by a circadian pacemaker. *Am. J. Physiol.* **246**(2 Pt. 2), R161–R183.

Edgar, D. M., Dement, W. C., and Fuller, C. A. (1993). Effect of SCN lesions on sleep in squirrel monkeys: Evidence for opponent processes in sleep-wake regulation. *J. Neurosci.* **13**(3), 1065–1079.

Edgar, D. M., and Seidel, W. F. (1997). Modafinil induces wakefulness without intensifying motor activity or subsequent rebound hypersomnolence in the rat. *J. Pharmacol. Exp. Ther.* **283**(2), 757–769.

Franken, P., Tobler, I., and Borbely, A. A. (1991). Sleep homeostasis in the rat: Simulation of the time course of EEG slow-wave activity. *Neurosci. Lett.* **130**(2), 141–144.

Hamblen, M. J., White, N. E., Emery, P. T., Kaiser, K., and Hall, J. C. (1998). Molecular and behavioral analysis of four period mutants in *Drosophila melanogaster* encompassing extreme short, novel long, and unorthodox arrhythmic types. *Genetics* **149**(1), 165–178.

Hendricks, J. C., Finn, S. M., Panckeri, K. A., Chavkin, J., and Williams, J. A. (2000). Rest in Drosophila is a sleep-like state. *Neuron* **25**(1), 129–138.

Hendricks, J. C., Kirk, D., Panckeri, K., Miller, M. S., and Pack, A. I. (2003). Gender dimorphism in the role of cycle (BMAL1) in rest, rest regulation, and longevity in *Drosophila melanogaster*. *J. Biol. Rhythms* **18**(1), 12–25.

Kleitman, N. (1963). "Sleep and Wakefluness." University of Chicago Press, Chicago.

Rechtschaffen, A. (1998). Current perspectives on the function of sleep. *Perspect. Biol. Med.* **41**(3), 359–390.

Shaw, P. J., Cirelli, C., Greenspan, R. J., and Tononi, G. (2000). Correlates of sleep and
    waking in *Drosophila melanogaster*. *Science* **287**(5459), 1834–1837.
Shaw, P. J., and Franken, P. (2003). Perchance to dream: Solving the mystery of sleep through
    genetic analysis. *J. Neurobiol.* **54**(1), 179–202.
Shaw, P. J., Tononi, G., Greenspan, R. J., and Robinson, D. F. (2002). Stress response genes
    protect against lethal effects of sleep deprivation in Drosophila. *Nature* **417**(6886), 287–291.

## [41]  *Drosophila melanogaster:* An Insect Model for Fundamental Studies of Sleep

*By* Karen S. Ho and Amita Sehgal

### Abstract

In 2000, *Drosophila melanogaster* joined the ranks of vertebrates and invertebrates with a defined behavioral sleep state. The characterization of this sleep state revealed striking similarities to sleep in humans: sleep in flies has both circadian and homeostatic components, it is influenced by sex and age, and it is affected by pharmacological agents such as caffeine and antihistamines. As in mammals, arousal thresholds in flies increase with sleep deprivation. Furthermore, changes in brain electrical activity accompany the change from wake to sleep states. Not only do flies and vertebrates share these behavioral and physiological traits of sleep, but they are likely to share at least some genetic mechanisms underlying the regulation of sleep as well. This article reviews the methods currently used to identify and characterize the *Drosophila* sleep state. As these methods become more refined and our understanding of *Drosophila* sleep more detailed, the powerful techniques afforded by this organism are likely to unveil deep insights into the function(s) and regulatory mechanisms of sleep.

### Introduction

Sleep is a tightly regulated, reversible state of quiescence with defined behavioral, homeostatic, and electrophysiological traits. It is important for both survival and proper brain function: rats die after 3–4 weeks of complete sleep deprivation (Rechtschaffen and Bergmann, 2002; Rechtschaffen *et al.*, 1989), and human performance in cognitive tasks declines with increasing amounts of deprivation (Rogers *et al.*, 2003). Furthermore, animals throughout the phyla sleep (Hartse, 1994), suggesting that, despite the perils of adopting this quiescent state each day, sleep confers a significant adaptive advantage.

Sleep is under the control of both a circadian clock and a homeostatic mechanism. The circadian clock maintains sleep with 24-h periodicity. In addition, sleep is homeostatically controlled: Loss of sleep causes the animal to subsequently attempt to regain it, exhibiting what is known as a sleep rebound. The two-process model of sleep regulation takes both these influences into account: In this model, circadian and homeostatic factors determine when sleep will occur (Borbely and Achermann, 2000). Although a great deal is now known about the molecular mechanism of the circadian clock (Stanewsky, 2003), very little is known about the homeostatic mechanism. Thus, the homeostatic mechanism is an important focus of sleep research today.

Sleep accomplishes a restorative function, but the exact nature of this function remains a mystery. Several theories have been proposed (Rechtschaffen, 1998), among them the following ideas:

1. Sleep replenishes energy stores depleted during wakefulness (Benington and Heller, 1995).

2. Sleep aids in the removal of harmful by-products that accumulate in the brain as a result of high levels of monoaminergic neural firing during waking (Hartmann, 1973).

3. Sleep is involved in neural plasticity, and thus in memory consolidation (Benington and Frank, 2003).

4. Sleep is involved in hormonal and immunologic processes (Krueger and Majde, 2003).

5. Sleep is involved in thermoregulation (Rechtschaffen *et al.*, 1989).

Unfortunately, these and other current theories for the functions of sleep lack evidence of a causative, rather than correlative, relationship between sleep and any of these vital processes. To gain such evidence, knowledge of the molecular mechanism of sleep, and the consequences of having that mechanism go awry, is necessary.

Poised to make significant contributions to this vein of discovery is the fruit fly, *Drosophila melanogaster. Drosophila* researchers originally identified the sleep state (Hendricks *et al.*, 2000; Shaw *et al.*, 2000) on the basis of behavioral criteria used to identify sleep in other organisms (Campbell and Tobler, 1984), and we continue to use behavioral measures to identify mutant phenotypes. The characteristics are (1) a consolidated state of immobility regulated in a circadian fashion, (2) a species-specific posture and/or preferred resting site, (3) an increased arousal threshold that is reversible, and (4) a homeostatic component in which the animal attempts to regain lost sleep after deprivation. By adopting these behavioral criteria for sleep, it has been possible to identify sleep states in fish (Tobler and Borbely, 1985; Zhdanova *et al.*, 2001), scorpions (Tobler and Stalder, 1988), cockroaches (Tobler, 1983), bees (Kaiser and Steiner-Kaiser, 1983; Sauer et al., 2003), and the mollusk *Aplysia* (Strumwasser, 1971), among many other organisms (Campbell and Tobler, 1984), in addition to *Drosophila*.

Lack of a sleep-like electroencephalogram (EEG) signature has plagued the characterization of many nonmammalian sleep states, drawing criticism that those animals do not actually sleep. However, not all animals have the requisite neuroanatomy to generate cortical waves in the same manner as mammals, and thus nonmammalian sleep architecture is likely to be significantly different than that of mammals (Hartse, 1994). Thus, it is reasonable to expect that changes in brain electrical activity accompany sleep, but those changes may not be exactly the same from organism to organism across phyla. Neurophysiological approaches to studying *Drosophila* sleep and salience have been initiated and indicate that *Drosophila*, too, have electrophysiological correlates of sleep and waking (Nitz et al., 2002; van Swinderen and Greenspan, 2003; van Swinderen et al., 2004).

*Drosophila* share a significant number of genes important for neural development and function with mammals (Hewes and Taghert, 2001; Nassel, 2002; Yoshihara et al., 2001). These similarities suggest that vertebrate and invertebrates may share the same basic genetic mechanisms for sleep as they do for circadian rhythms, learning and memory, and aging (Finch and Ruvkun, 2001; Paulsen and Morris, 2002; Stanewsky, 2003). A vast and powerful array of genetic resources is at the fly researchers' disposal, and this arsenal of techniques has been used successfully to identify genes involved in several complex behaviors (Carlson, 1993; Cirelli, 2003; Greenspan and Ferveur, 2000). Because many aspects of sleep appear to be controlled genetically, *Drosophila* presents a useful system for analysis of this complex behavioral state (Hendricks, 2003; Shaw and Franken, 2003; Tully, 1996). In fact, in the time since *Drosophila* sleep was first described (Hendricks et al., 2000; Shaw et al., 2000), significant progress has been made in identifying candidate genes involved in sleep regulation. This article describes the methods used to study sleep and characterize mutant phenotypes in *Drosophila*.

## Baseline Sleep: Locomotor Assay and Videography

To measure baseline sleep, the rest and activity behaviors of individual flies are tracked over the course of at least a few days. From this information, the average number of minutes of sleep and waking, the length of each sleep bout, and amount of sleep consolidation can be gleaned. A widely used automated system for scoring waking activity is the *Drosophila* Activity Monitoring System (TriKinetics). Each fly is placed in a glass tube containing standard cornmeal:molasses food at one end. If nonvirgin females are to be used, the food may alternatively consist of 5% sucrose/ 2% agar, which prevents eggs from nonvirgin females from developing and causing erroneous activity counts. The tube is then placed in a monitor that holds 32 tubes, with an infrared beam bisecting each tube. The number of infrared beam breaks caused by each fly walking back and forth in the tube is counted by the TriKinetics system and summed in bins of 5 min or less. Baseline sleep is measured in flies as the total amount of sleep they attain per 24-hr day when left undisturbed on a 12-h light:12-h dark (LD) schedule, which mimics natural conditions and allows flies lacking an endogenous clock to consolidate sleep during the dark period. Baseline sleep can also be measured in constant darkness (DD), which is useful for assessing sleep in free-running conditions. Fly stocks are generally maintained and previously entrained at 25°, 50–60% humidity, in incubators and on standard food (Roberts, 1986).

Videotape analyses of Canton-S flies in the monitoring system showed that 97% of rest bouts last 5 min or more, with the majority of these bouts lasting 30 min or more (Hendricks *et al.*, 2000). When the beam remains unbroken for more than 5 min, the fly exhibits a relaxed posture and quiescent behavior associated with sleep. In contrast, immobile periods less than a minute long were infrequent and often found interspersed with high activity, and thus may signify brief rest rather than sleep (Hendricks *et al.*, 2000). Using an ultrasound system to detect fine movements of the fly, Shaw *et al.* (2000) independently corroborated these observations. Furthermore, a fly at rest for 5 min or longer exhibits increased arousal thresholds (Huber *et al.*, 2004). Thus, a 5-min or longer interval with zero activity counts is scored as "sleep." Using this 5-min definition of sleep, we have graphed the average baseline sleep of wild-type Canton-S flies throughout the course of the day (Fig. 1). The validity of using a 5-min designation for sleep, however, may change with different genotypes. If a 5-min definition for sleep is used, data from *Drosophila* activity monitors should ideally be collected in bin sizes smaller than 5 min. As bin size decreases, however, computer memory and processing speed become limiting factors.

Baseline sleep will vary with the age and sex of the flies in the study. Young 1- to 3-day-old flies sleep more than older ones (Shaw *et al.*, 2000). While females spend little time resting during the day, males spend some time sleeping during the day and sleep about the same amount at night (Huber *et al.*, 2004) (Fig. 1). Interestingly, sleep of both sexes during the day seems to be qualitatively different than sleep at night: arousal thresholds are lower during the day than during the night (Huber *et al.*, 2004).

Because sleep is characterized by many behavioral changes, not all of which can be reported by activity monitoring systems, videotape analyses

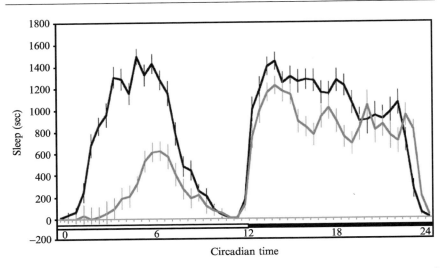

FIG. 1. Baseline sleep in an isogenic line of wild-type Canton-S flies. Average amounts of sleep (in seconds) per 30-min bin for 3- to 6-day-old males (black, $n = 32$) and 3- to 6-day-old nonvirgin females (gray, $n = 29$). The corresponding light:dark (LD) cycle is represented by the white (12-h light period) and black (12-h dark period) bars below the graph. Highly active times for both sexes are at dawn (CT 0) and dusk (CT 12), hence sleep is lowest at these times. Note that females sleep less than males during the day and that males sleep almost as much during the day as during the night. Periods of immobility lasting 5 min or longer in the TriKinetics activity monitors were scored as sleep. Data were collected in 5-min bins. Bars represent standard error.

of each new sleep mutant under study is a valuable means of phenotypic characterization (Hendricks *et al.*, 2000; Shaw *et al.*, 2000). This type of analysis also serves to distinguish a putative sleep mutant from one that has altered activity for other reasons. For example, 5 min of inactivity may not be sleep for a behavioral mutant that sits still while grooming excessively for over 5 min, and thus such a mutant may be misscored as a sleep mutant. Videography is thus a valuable addition to activity monitor data in sleep analysis. However, because analysis of videographic data is time-consuming, it is more feasible to use videography as a secondary screen, once an initial pool of new sleep mutants has been identified.

## Sleep Deprivation and Rebound

Sleep rebound is the homeostatic response of an organism to sleep deprivation and constitutes one of the hallmarks of sleep. Sleep deprivation has been achieved using manual or automated methods [manually tapping on tubes (Hendricks *et al.*, 2000) or mechanically stimulating flies via an automated system (Hendricks *et al.*, 2001; Huber *et al.*, 2004; Shaw *et al.*, 2000)] to provide gentle, randomized stimulation to the flies within the monitor tubes. Sleep deprivations have been performed for 3, 6, 12, and 24 h (Hendricks *et al.*, 2000; Huber *et al.*, 2004; Shaw *et al.*, 2000). Shorter (less than 12 h) deprivations are usually restricted to the dark period of the circadian cycle, when flies exhibit the most consolidated sleep. Several studies have now shown that flies deprived in this fashion display a rebound, sleeping more during the period following the deprivation (Fig. 2). In these studies, an important control is a group of flies that are handled like the deprived flies but are not subjected to the stimulation; these flies do not show a rebound (Hendricks *et al.*, 2000). Another important control is

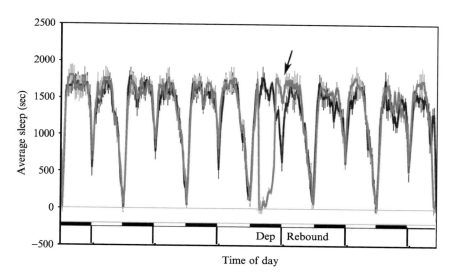

FIG. 2. Baseline sleep, deprivation, and sleep rebound in wild-type male Canton-S flies. Average sleep (in seconds) of males during 2.5 baseline days, a 12-h nighttime deprivation, and 2.5 recovery days. Flies (shown in gray) were deprived in TriKinetics monitors using a custom-made apparatus that provides a randomized, mechanical stimulus, allowing activity to be monitored continuously throughout the deprivation (B. Joiner, unpublished work). Handled control flies (black) were kept in the same incubator to control for environmental effects but not subjected to mechanical stimulation during the deprivation period. Note that during this deprivation period rest decreases dramatically in stimulated flies, but not in controls. In the recovery period following the deprivation, deprived flies exhibit a marked increase in rest known as rebound. Data were collected in 30-s bins and scored for sleep using 5 min or more of immobility as the sleep criterion. Bars represent standard error. Experimental group, $n = 15$; handled control group, $n = 16$. (Data kindly provided by B. Joiner.)

stimulation of a similar group of flies during their active phase. Such a deprivation carried out during the 12-h light cycle does not generate a significant rebound in Canton-S flies, demonstrating that the stimulation itself is not overly harmful to the locomotor activity of the flies (Hendricks et al., 2000; Huber et al., 2004; Shaw et al., 2000). An additional control demonstrated that an automated sleep deprivation system was not harmful to *stress-sensitive B* (*sesB*) flies, a mutant strain for which excessive mechanical stimulation is lethal (Shaw et al., 2000).

The length of time a deprivation is carried out can vary. For many lines, a 6-h deprivation seems to suffice to elicit a measurable rebound (Hendricks et al., 2001). However, this is not the case for male Canton-S flies, which seem to exhibit significant rebounds only after a 12-h deprivation (Huber et al., 2004). A 24-h deprivation has the additional advantage in that the extended deprivation serves to control for differences in sleep distribution caused by circadian and other influences. For instance, if a fly line tends to sleep a great deal during daytime hours, it may not show as large a rebound after nighttime deprivation as a line that confines its major rest period to the nighttime hours. The former line may be classified as a sleep mutant, even though the real defect lies in its circadian phase. Twenty-four-hour deprivation protocols can ameliorate this problem.

The amount of sleep lost, relative to baseline, can be calculated if flies are monitored continuously during the deprivation. The ability to monitor flies as they are deprived, and thus calculate exactly how much sleep they lose due to the intervention, is a new development that was not possible in previous sleep studies. Using this technique of applying mechanical stimulus to flies while they are in the monitors, an elegant study has shown that, like mammals, flies increase sleep intensity in response to increasing sleep debt: the more sleep is lost, the more time is spent in recovery sleep and the more consolidated the sleep becomes (Huber et al., 2004).

Arousal Threshold

In mammals, increased sleep need is associated with increased arousal thresholds (Roth et al., 1982), and the same is true in the fly. An initial study of arousal threshold showed that as flies lose more and more sleep, they require increasingly more intense stimulation, i.e., an increased number of taps on a petri dish in which they are housed, in order to keep the entire population active (Hendricks et al., 2000). Another group showed that while active flies responded to low levels of vibratory stimulation (0.05 and 0.1$g$), sleeping flies rarely responded; they responded only when the vibratory stimulation was increased to 6$g$ (Shaw et al., 2000). In a more recent study, two new methods to measure arousal threshold were devised (Huber *et al.*, 2004). In one method, a mechanized flap was pushed against the sides of the activity tubes housing the sleep-deprived flies to create vibration and noise. The escape response of the fly was measured as beam breaks following the stimulus. The arousal threshold was monitored once an hour every hour for both a baseline day and the day after sleep deprivation. As expected, the latency of the escape response was dependent on whether the fly was awake or asleep at the time the stimulus was administered. Active flies responded to the stimulus much more frequently than flies that had been immobile >5 min prior to the stimulus. A surprising result from this study showed that the arousal threshold for immobile flies at night was greater than that for immobile flies during the day (Huber *et al.*, 2004). This interesting observation may indicate a circadian or environmental influence on sleep throughout the day. Furthermore, perhaps some immobile periods during the day represent quiet wakefulness rather than sleep. In any case, this result suggests that, in addition to reporting the total amount of sleep per 24 h, sleep should probably also be reported as daytime and nighttime totals.

The second method devised to measure arousal threshold consisted of heating the flies using a Peltier thermoelectric device. The fly was placed in the device before or after sleep deprivation, and its avoidance response to heat was measured in terms of latency to move away from the heat source. The group found, again, that sleep-deprived flies showed a slower response time than rested flies (Huber *et al.*, 2004).

Sleep Intensity

*Drosophila* sleep research is limited in that sophisticated electrophysiological studies for measuring brain activity have not yet been developed, although progress has been made even on this front (see later). Instead, currently we must rely on clever methods of extracting data from the activity monitoring system. Huber *et al.* (2004) have done just that: they show that sleep episodes are longer following deprivation and that recovery sleep is more consolidated, with fewer brief awakenings interrupting each episode compared with baseline. These are important observations because they represent a method by which sleep intensity may be measured in flies. In mammals, deep slow wave sleep is reflected in high delta power on the EEG, and this measure reflects sleep drive: the longer and more complete the deprivation, the higher and more extensive the delta power sleep (Franken *et al.*, 1991; Tobler and Borbely, 1986). Recovery slow wave sleep in mammals is also more consolidated, with fewer brief awakenings (Franken *et al.*, 1991). The number of brief awakenings, as defined by a short burst of activity preceded and followed by sleep, can be obtained readily from data collected from *Drosophila* activity monitors. Important- ly, brief awakenings decrease while the duration and continuity of sleep episodes increase in response to increasing sleep deprivation of Canton-S flies (Huber *et al.*, 2004). This study thus demonstrates that sleep homeo- stasis in flies shares many of the same behavioral features described in mammals.

### Pharmacological Studies

Pharmacological studies of sleep in *Drosophila* further demonstrate the similarities between vertebrate and invertebrate sleep. Caffeine is an an- tagonist of the mammalian adenosine receptor and is a popular stimulant, reducing sleep and promoting wakefulness in a dose-dependent manner (Fredholm *et al.*, 1999). Similarly, feeding flies caffeine increases wakeful- ness (Fig. 3), indicating that the molecular mechanism for caffeine response and promotion of wakefulness in mammals is conserved in flies (Hendricks *et al.*, 2000; Shaw *et al.*, 2000). Conversely, a specific agonist of the mammalian adenosine receptor A1 cyclohexyladenosine (CHA) has the opposite effect when microdialyzed to the basal forebrain area of mam- mals: it increases NREM sleep and decreases waking (Strecker *et al.*, 2000). Feeding flies CHA has a similar effect: flies fed increasing doses of CHA sleep more compared with controls fed diluent alone (Hendricks *et al.*, 2000).

The antihistamine hydroxyzine also affects sleep in flies, in much the same way it does in mammals (Shaw *et al.*, 2000). While adenosine acts on a sleep-promoting system (Basheer *et al.*, 2000), antihistamines work by antagonizing a wake-promoting system (Pace-Schott and Hobson, 2002). Together, these results suggest that both wake- and sleep-promoting sys- tems are present in the fly, and these systems are at least pharmacologically similar to those in mammals.

*Drosophila* also serves as a suitable animal model to study the mechan- isms of novel sleep- and wake-promoting drugs. Modafinil (diphenylmethl- sulfinyl-2 acetamide) is one such example. It draws a great deal of interest because some studies show that it promotes vigilance without the negative side effects of other wake-promoting drugs (Buguet *et al.*, 1995; Edgar and Seidel, 1997; Pigeau *et al.*, 1995). Modafinil sustains wakefulness in *Drosophila* as well as mammals, and the rebound response to modafanil- induced wakefulness is significantly smaller than rebound resulting from deprivation by conventional means (Hendricks *et al.*, 2003a). Thus, *Dro- sophila* could serve as a useful animal model for determining the mode of drug reception and response to hypnotics and stimulants.

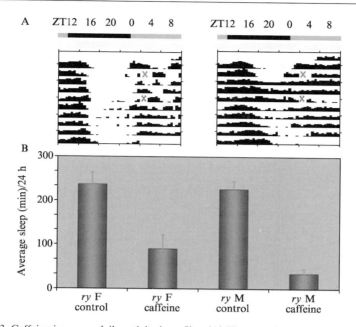

FIG. 3. Caffeine increases daily activity in *ry* flies. (A) Three- to 6-day-old *rosy* (*ry*) female flies previously entrained on a 12:12 LD cycle were monitored in constant darkness (DD) for 1.5 days and then flipped into new tubes containing standard sucrose (5%):agar (2%) food (left actogram) or sucrose:agar food with 0.5 mg/ml caffeine added (at the first X, right actogram). Flies were then monitored continuously in DD for the next 3 days and were then flipped into new tubes containing fresh sucrose:agar food (second X). Note that caffeine-treated animals become very active, even in the dark period when they would normally sleep, and that immediately after caffeine is withdrawn they return to their original circadian rhythm. Handled controls were flipped into new tubes at the same time as the experimental group, but not fed caffeine. Flipping flies into tubes did not significantly alter sleep–wake behavior (left actogram). (B) Average amount of sleep (in minutes) per day for female controls (*n* = 19), females fed caffeine (*n* = 30), male controls (*n* = 22), and males fed caffeine (*n* = 32). Note that caffeine treatment significantly reduced the amount of daily sleep of experimental groups compared with controls. Data were collected in 30-min bins, and a 30-min sleep criterion was used to score sleep. Similar results are obtained by collecting data in 5-min bins and using the 5-min sleep criterion (K. Ho, unpublished results). Bars represent standard error.

## Electrophysiology Studies

A remarkable set of experiments spanning three papers from Ralph Greenspan's group demonstrated spike-like potentials in the medial *Drosophila* brain that correlate with sleep/wake state (Nitz *et al.*, 2002; van Swinderen and Greenspan, 2003; van Swinderen *et al.*, 2004). The work further showed that local field potentials (LFPs) in the 20- to 30-Hz range, induced by a visual stimulus, decrease during sleep (van Swinderen and Greenspan, 2003). A control probe measuring slow potentials in the optic lobe in response to this stimulus did not show a dependence on activity state. Careful controls were also done to exclude the possibility that the spikes were the result of movement of the fly, and the sleep state of the fly being recorded was confirmed using an arousal threshold assay (Nitz et al., 2002; van Swinderen and Greenspan, 2003; van Swinderen et al., 2004). Thus the authors concluded that the sleep state of Drosophila is accompanied by lowered brain activity as measured by LFPs. These experiments mark the beginning of the exploration of Drosophila sleep using electrophysiology.

Electrophysiological correlates of the sleep–wake state in the invertebrate Aplysia, as well as recordings from the optomotor interneuron during the sleep–wake state in the honeybee, have also been made (Kaiser and Steiner-Kaiser, 1983; Strumwasser, 1971). Taken together, these experiments suggest that the sleep state in Drosophila and other invertebrates is accompanied by changes in the electrical activity of the brain. As techniques for recording from the brain of Drosophila improve, a characterization of sleep stages, if they exist in Drosophila, and identification of sleep-active brain regions may also become possible.

## Sleep-Relevant Genes

The aforementioned work lays the groundwork for identifying Drosophila mutants with aberrant sleep patterns. Mammalian sleep is certainly influenced by a number of genetic factors (Franken et al., 1999), but to carry out a systematic forward genetic screen for sleep mutants in mammals is a serious undertaking in terms of time and cost (Kapfhamer et al., 2002; Tafti et al., 1999; Veasey et al., 2000). The rapid discovery of genes involved in sleep state control is the area in which Drosophila can make a great contribution. In the field of circadian rhythms, Drosophila has played the leading role in the discovery of many of the genes that encode proteins of the core clock (Hendricks, 2003). The human homolog of one such gene, period, is mutated in people afflicted with Advanced Phase Sleep Syndrome, which affects the circadian rhythm of sleep onset, rather than sleep architecture (Toh et al., 2001). The pharmacological studies described earlier corroborate the idea that many sleep-promoting and wake-promoting systems are also shared between flies and vertebrates. Thus, it is highly likely that Drosophila will aid in the discovery of genes involved in the regulation of sleep and that this knowledge will be applicable to human sleep.

## Analysis of Existing Genetic Mutants to Identify Candidate Sleep Genes

The circadian mutants were among the first candidates tested for sleep phenotypes, as they modify the distribution of sleep and wake activities by virtue of their roles in the circadian clock. Two of the core clock mutants in *Drosophila*—*period* ($per^{01}$) and *timeless* ($tim^{01}$)—exhibit normal sleep-rebound phenotypes, indicating that circadian control is separable from the genetic mechanism of sleep (Hendricks *et al.*, 2003b). Although these circadian mutants have an altered distribution of sleep, the overall levels of sleep are not significantly altered (Fig. 4). Similarly, *mPeriod1*(*mPer1*) and *mPeriod2*(*mPer2*) mutant mice, in which the mammalian *period* orthologs *mPer1* and *mPer2* are disrupted, display normal sleep rebounds in response to deprivation (Kopp *et al.*, 2002; Shiromani *et al.*, 2004).

As exception to the rule, a strain of flies mutant for the circadian gene *cycle* does appear to have a sleep phenotype (*ry cyc^{01}*). Females of this mutant strain exhibit rebounds that are much longer in duration than those in *ry* flies, whereas males do not exhibit much of a rebound at all

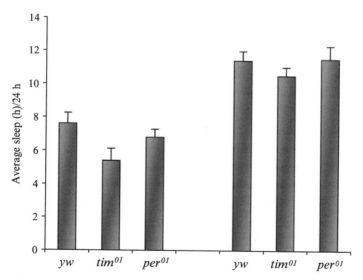

FIG. 4. Circadian rhythm mutants *per* and *tim* have normal amounts of daily sleep. Average amounts of daily sleep for the background *yw* strain, *yw tim^{01}*, and *yw per^{01}* flies monitored in DD (left group) and LD (right group). Data were collected in 5-min bins and analyzed for sleep using a 5-min sleep criterion. Note that *yw tim^{01}* and *yw per^{01}* flies are driven by the LD cycle to exhibit a rhythm in locomotor activity in LD due to masking by light, while they are completely arrhythmic in DD. Despite the arrhythmicity of their behavior, they nevertheless display normal amounts of sleep. Bars represent standard error (Modified from Hendricks *et al.*, 2003b).

(Hendricks *et al.*, 2003b; Shaw *et al.*, 2002). However, because this phenotype appears to be extremely background sensitive, it has been difficult to
ascertain whether the sleep phenotype is due to a lesion in *cyc* itself or is
the result of a more complex genetic interaction in the background of
*ry cyc$^{01}$* flies (K.Ho, unpublished observations).

Another candidate gene with a potential role in sleep is *CREB*, encoding the cAMP response element-binding protein (CREB). *CREB* plays
important roles in neural plasticity, learning and memory, and circadian
rhythm in mammals and in flies. Its phosphorylated, activated form is
found at high levels in the mammalian cortex during normal waking and
during waking due to deprivation, suggesting that CREB may regulate, or
be regulated by, the sleep–wake cycle (Cirelli and Tononi, 2000). To
determine the role of CREB in *Drosophila* sleep, Hendricks *et al.* (2001)
examined the daily sleep durations of three classical mutants in which
CREB activity is predicted to be either higher or lower than normal.
Furthermore, heat shock-inducible transgenes designed to express activating (dCREB2a) and dominant-negative (dCREB2b) mutant forms of
CREB were also used to manipulate CREB activity in flies. Results from
analyzing sleep in all these lines are remarkably consistent: Lowering
CREB activity increases amounts of daily sleep, whereas raising CREB
activity lowers the amounts of sleep.

To determine whether *CREB* affects sleep homeostasis in flies, transgenic *CREB2a* and *2b*, along with their background line, were sleep deprived for 6 h and then baseline and recovery sleep were analyzed. Flies in
which the dominant-negative *dCREB2b* was overexpressed showed a significant increase in rebound sleep compared with background, and this
result was confirmed using the *dCREB2* classical mutant. Overexpression
of activated *CREB* had little effect on rebound sleep, but the lack of a
statistically significant effect may simply indicate that activated CREB is
not the limiting factor in waking or in rebound. Taken together, these
studies showed that *Drosophila* CREB plays a role in maintaining wakefulness, as mammalian expression data for CREB had suggested (Cirelli
and Tononi, 2000).

A satisfying indication of the congruence of mammalian and insect
sleep came when Graves *et al.* (2003) showed that CREB also plays a role
in maintaining wakefulness in mice. This study took advantage of a line of
mutant mice that lack $\alpha$ and $\Delta$ isoforms of the CREB protein, reducing
overall CREB protein levels throughout the brain to 15% of wild-type
levels. The study showed that *CREB* $\alpha\Delta$ mutant mice spend an average of
100 min more per day in NREM sleep than wild-type sibling controls. EEG
spectral analysis of *CREB* $\alpha\Delta$ mutant mice suggested that they have a
deficit in maintaining cortical arousal, as reflected in decreased theta

activity during wake and REM stages. This study suggests that, like flies, mice rely on a CREB-mediated pathway to maintain wakefulness.

### Differential Display and Microarray Experiments to Identify Candidate Sleep Genes

A differential display screen for genes expressed preferentially during sleep or waking, independent of circadian time, identified *Drosophila* genes with human counterparts known or postulated to play a role in sleep. Examples of some of the genes identified encode the neuromodulator fatty acid synthase, the metabolic protein cytochrome P450, and dopamine *N*-acetyltransferase (Dat) (Shaw *et al.*, 2000). Dat is an enzyme involved in the catabolism of monoamines, including dopamine, tryptamine, and serotonin (Hintermann *et al.*, 1995). A mutant allele of $Dat(Dat^{lo})$, in which enzymatic activity is low due to disruption of one of the two normal isoforms of this gene (Hintermann *et al.*, 1996), exhibits normal baseline activity behavior, but a rest rebound that is greater than that of wild-type controls (Shaw *et al.*, 2000). In animals where the $Dat^{lo}$ allele was placed over a deficiency for the *Dat* region, these effects were even more pronounced (Shaw *et al.*, 2000). This phenotype fits a model for sleep in mammals in which the high monoaminergic activity that occurs during waking must be counteracted by restorative sleep. An animal with decreased ability to break down monoamines, as is the case with the $Dat^{lo}$ fly, may have an increased sleep need as a result. This increased sleep need is reflected in its extended rebound.

Whole genome profiling of sleep has been done for mammals (e.g., Cirelli and Tononi, 1999; Mackiewicz and Pack, 2003; Tafti *et al.*, 2003), and it is merely a matter of time before the same studies are conducted in *Drosophila*. Microarray technologies will allow the comparison of enriched and depressed transcripts during wakefulness, sleep, and sleep rebound in fly heads. It will also be valuable to compare expression profiles of flies under hypoxic stress and immune stress to determine whether any of these states share similarities with sleep deprivation, as many theories about sleep function center around these phenomena. Processes involved in recovery sleep may be indicated by coordinate gene regulation. For example, membrane trafficking and maintenance have been proposed to have roles in mammalian sleep because genes with these functions were identified in a mammalian chip screen for sleep-related genes (Cirelli *et al.*, 2004). To distinguish between circadian and sleep genes, expression profiles of sleep-deprived flies and those undergoing rebound sleep can be compared with the expression profile of handled controls sacrificed at the same time of day. Candidates affected specifically by sleep state, as opposed to circadian time of day, can be characterized further by examining mutants of the corresponding genes for sleep behavior phenotypes.

## Forward Genetic Screens for Sleep Genes

The handful of *Drosophila* mutants that have been characterized for sleep phenotypes to date (discussed earlier) demonstrate the promise and power of *Drosophila* in sleep research. The fly sleep assays for baseline rest and rebound, assays for arousal thresholds and sleep intensity, and the pharmacological assays described earlier can be used to discover and characterize new sleep mutants. Screens can be conducted using ethyl-methane sulfonate (EMS) to generate point mutations and small deletions or by creating fly lines of independent P-element transposon insertions, interrupting endogenous genes to create possible phenotypes. In fact, a large-scale endeavor has been initiated to generate at least one P-element insertion into every identified gene in the fly genome and to make all the generated lines available for public use (Spradling *et al.*, 1999). Furthermore, publicly available deficiency kits can be used not only for initial screening, but also for mapping mutations once they are identified. Although loss-of-function adult phenotypes cannot be obtained for many genes because the homozygous mutant is lethal, overexpression screens may reveal a role for these genes in adult behavior. Some false positives may be identified in this manner, but researchers have met with success using this method to identify not only developmental genes, but also the gene *shaggy*, which encodes a glycogen synthase kinase ortholog involved in regulating circadian rhythms (Martinek *et al.*, 2001; Rorth *et al.*, 1998). The main difficulty underlying all these screening methods is the design of the screen itself: unlike classical screens for developmental genes in which often a single fly could be relied upon for phenotypic analysis, the variability of the sleep phenotype necessitates screening more than one member of a mutant line for a sleep phenotype. In some cases, many flies are required to generate statistical confidence. In a screen of isofemale lines, discussed later, eight flies were screened per line at a time (Huber *et al.*, 2004). Currently, monitors hold 32 flies, so only four lines can be screened at once in a monitor. Monitors are costly in terms of price and space, so large-scale screens are not undertaken lightly.

In all screens, the importance of obtaining or making a control line that shares a near-identical genetic background with the mutant under study cannot be overemphasized. As in all behavioral assays, the variability of the sleep phenotype between lines and among flies of the same line can be great. To identify a sleep phenotype accurately, mutant flies must be compared with the appropriate background strain.

Progress on the genetic front has been made by Huber *et al.* (2004), who identified a "short sleeper" strain when they examined 117 wild-type isofemale (derived from one female) lines whose baseline rest pattern fell more than two standard deviations from the mean. Using the same criteria, they also identified isofemale lines with long sleeper phenotypes. It is now possible to carry out quantitative trait loci (QTL) analysis by generating recombinants from the short and long sleeper lines. This would allow one to estimate the number of genes contributing to the short sleeper phenotype as well as their approximate chromosomal locations. Coupled with further mapping techniques, such as the use of publicly available deficiency kits, candidate short-sleeper and long-sleeper genes can be identified.

## Shortcomings and Future Prospects

*Drosophila* holds exciting prospects as a model organism for sleep, yet every model organism has both strengths and shortcomings. At present a number of challenges are posed to fly sleep researchers. First and foremost is the inherent variability in behavioral data, necessitating the characterization of large numbers of individual flies to obtain statistically significant results. This can be a potentially serious drawback for a genetic screen, when the ease and efficiency of the assay determine how rapidly new mutants can be identified and subsequently mapped. However, the relatively large number of genes discovered in screens for *Drosophila* behavioral mutants eloquently attests that such screens have been, and can be, done effectively (Carlson, 1993; Greenspan and Ferveur, 2000; Konopka and Benzer, 1971; Toma *et al.*, 2002; Tully, 1996).

A second challenge *Drosophila* sleep researchers face lies in the quantification of sleep. The popular method of using TriKinetics monitors, while convenient, does not allow positive scoring of sleep. Instead, the *absence* of beam breaks over a specific period of time constitutes sleep using this method. When the beam is not broken the fly could, in fact, be doing many things, e.g., running back and forth in half the tube, eating, grooming, or digging at the cotton (an escape behavior documented by Hendricks *et al.*, 2000). Each new mutant may spend a different proportion of its time engaged in these activities, and thus the proportion of time spent in each of these activities should be documented by videography. This can be quite a time-consuming undertaking, but currently it affords the greatest resolution for monitoring fly sleep.

Pharmacological studies have the disadvantage that many drugs have pleiotropic effects, and if a drug having nothing to do with sleep compromises the locomotor, coordination, or other system of the fly, the fly may appear to be sleeping more in response to a drug when in fact it is simply unable or unwilling to move. Proper controls for each new drug must be devised and may include dose–response curves and geotaxic or other ambulatory assays for locomotor control. In addition, receptor specificities for drugs, based on work done in mammals, may not be the same in *Diptera* (e.g., Magazanik and Fedorova, 2003).

Sleep rebound in *Drosophila* has now been measured in many ways, including duration of rebound, arousal threshold during rebound, and degree of consolidation by bout length and number of brief awakenings. The characterization of the wild-type Canton-S *Drosophila* strain represents the foundation upon which an understanding of sleep homeostasis can be built. At the present time, it is unclear which of these measures provides the best quantitative means to reflect sleep drive and sleep need, and thus a number of these measures should be taken in order to completely characterize new fly sleep mutants. Indeed, in mammals, some sleep deprivation protocols do not result in a sleep rebound of longer sleep duration, but rather one of greater sleep intensity, as reflected in the delta frequency power of the EEG (Franken *et al.*, 1991; Tobler and Borbely, 1986). Thus, a fly sleep mutant with a rebound quantitatively similar to wild-type flies may still in fact sleep less or more intensely than its wild-type siblings. Other measures of sleep intensity should be devised in order to uncover this phenotype. Similarly, arousal threshold assays based on using heat or vibratory stimuli might falsely identify a mutant as a sleep mutant simply because it lacks the appropriate thermosensation, proprioception, or other abilities that allow it to sense the stimulus. Nevertheless, the design of properly controlled baseline and rebound assays can likely distinguish *bona fide* sleep mutants from those with general neurological problems. Finally, the measure of consolidation, whether by average bout length, brief awakenings, or other methods, still begs the question as to the importance of consolidation for fly sleep. Of the mammals, humans exhibit well-consolidated sleep, whereas rats and mice do not display such large amounts of consolidation. Thus, across-species variation in consolidation with no apparent deleterious consequences on the animals' fecundity or life span suggests we do not yet understand the importance of consolidation to sleep quality (Endo *et al.*, 1998). However, the fact that consolidation increases during rebound in rats (Franken *et al.*, 1991), as well as in flies (Huber *et al.*, 2004), suggests that consolidation is important.

In characterizing sleep mutants, a rescue experiment in which a transgene encoding the gene of interest is introduced into the mutant strain to provide a "rescue" of the mutant phenotype is of primary importance. This is the acid test that can associate the mutant phenotypes observed with a single gene, and without this assay, one cannot soundly conclude that the phenotypes are all due to the loss of that particular gene. Given the variable nature of sleep and its current measures, a rescue experiment for every candidate mutant should be done.

Despite these challenges, the genetic amenability of *Drosophila* and its utility for behavioral studies hold great promise for the sleep researcher. More than 100 years of knowledge and resources can be brought to bear toward understanding the function of the 250,000 neurons in its nervous system. *Drosophila* are complex enough to model sophisticated human behaviors, but simple enough, with only four chromosomes and roughly 13,600 genes in the genome (Adams *et al.*, 2000), for discoveries to be made quickly. With *Drosophila*, it is possible to ask deep questions about the nature of sleep, such as what important restorative functions does sleep perform? What is the molecular nature of the sleep homeostat? How does sleep quantity and quality affect learning and memory, longevity, fecundity, and overall fitness? What are the molecular triggers of sleep, how is sleep regulated, and what is the molecular interaction between sleep- and wake-promoting systems? These are the questions that entice and beckon. *Drosophila* sleep, and in sleeping, urge us on.

## Acknowledgments

The authors heartily thank J. D. Alvarez, Susan Harbison, Julie Williams, and Quan Yuan for helpful comments on the manuscript. We thank Joan Hendricks and Allan Pack for many engaging discussions, Bill Joiner for providing data shown in Fig. 2, and Suresh Venkatasubramanian for expert help with data analysis. K.H. is supported by a National Sleep Foundation Pickwick Fellowship and A.S. is an Associate Investigator of the Howard Hughes Medical Institute.

## References

Adams, M. D., Celniker, S. E., Holt, R. A., Evans, C. A., Gocayne, J. D., Amanatides, P. G., Scherer, S. E., Li, P. W., Hoskins, R. A., Galle, R. F., *et al.* (2000). The genome sequence of *Drosophila melanogaster*. *Science* **287**, 2185–2195.

Basheer, R., Porkka-Heiskanen, T., Strecker, R. E., Thakkar, M. M., and McCarley, R. W. (2000). Adenosine as a biological signal mediating sleepiness following prolonged wakefulness. *Biol. Signals Recept.* **9**, 319–327.

Benington, J. H., and Frank, M. G. (2003). Cellular and molecular connections between sleep and synaptic plasticity. *Prog. Neurobiol.* **69**, 71–101.

Benington, J. H., and Heller, H. C. (1995). Restoration of brain energy metabolism as the function of sleep. *Prog. Neurobiol.* **45**, 347–360.

Borbely, A. A., and Achermann, P. (2000). Sleep homeostasis and models of sleep regulation. *In* "Principles and Practice of Sleep Medicine" (M. Kryger, T. Roth, and W. Dement, eds.), pp. 377–399. Saunders, Philadelphia.

Buguet, A., Montmayeur, A., Pigeau, R., and Naitoh, P. (1995). Modafinil, d-amphetamine and placebo during 64 hours of sustained mental work. II. Effects on two nights of recovery sleep. *J. Sleep Res.* **4**, 229–241.

Campbell, S. S., and Tobler, I. (1984). Animal sleep: A review of sleep duration across phylogeny. *Neurosci. Biobehav. Rev.* **8,** 269–300.

Carlson, J. (1993). Molecular genetics of *Drosophila* olfaction. *Ciba Found. Symp.* **179,** 150–161; discussion 162–166.

Cirelli, C. (2003). Searching for sleep mutants of *Drosophila melanogaster*. *Bioessays* **25,** 940–949.

Cirelli, C., Gutierrez, C. M., and Tononi, G. (2004). Extensive and divergent effects of sleep and wakefulness on brain gene expression. *Neuron* **41,** 35–43.

Cirelli, C., and Tononi, G. (1999). Differences in brain gene expression between sleep and waking as revealed by mRNA differential display and cDNA microarray technology. *J. Sleep Res.* **8**(Suppl. 1), 44–52.

Cirelli, C., and Tononi, G. (2000). Differential expression of plasticity-related genes in waking and sleep and their regulation by the noradrenergic system. *J. Neurosci.* **20,** 9187–9194.

Edgar, D. M., and Seidel, W. F. (1997). Modafinil induces wakefulness without intensifying motor activity or subsequent rebound hypersomnolence in the rat. *J. Pharmacol. Exp. Ther.* **283,** 757–769.

Endo, T., Roth, C., Landolt, H. P., Werth, E., Aeschbach, D., Achermann, P., and Borbely, A. A. (1998). Effect of frequent brief awakenings from nonREM sleep on the nonREM-REM sleep cycle. *Psychiat. Clin. Neurosci.* **52,** 129–130.

Finch, C. E., and Ruvkun, G. (2001). The genetics of aging. *Annu. Rev. Genom. Hum. Genet.* **2,** 435–462.

Franken, P., Dijk, D. J., Tobler, I., and Borbely, A. A. (1991). Sleep deprivation in rats: Effects on EEG power spectra, vigilance states, and cortical temperature. *Am. J. Physiol.* **261,** R198–R208.

Franken, P., Malafosse, A., and Tafti, M. (1999). Genetic determinants of sleep regulation in inbred mice. *Sleep* **22,** 155–169.

Fredholm, B. B., Battig, K., Holmen, J., Nehlig, A., and Zvartau, E. E. (1999). Actions of caffeine in the brain with special reference to factors that contribute to its widespread use. *Pharmacol. Rev.* **51,** 83–133.

Graves, L. A., Hellman, K., Veasey, S., Blendy, J. A., Pack, A. I., and Abel, T. (2003). Genetic evidence for a role of CREB in sustained cortical arousal. *J. Neurophysiol.* **90,** 1152–1159.

Greenspan, R. J., and Ferveur, J. F. (2000). Courtship in Drosophila. *Annu. Rev. Genet.* **34,** 205–232.

Hartmann, E. (1973). "Functions of Sleep." Yale Univ. Press, New Haven, CT.

Hartse, K. M. (1994). Sleep in insects and nonmammalian vertebrates. *In* "Principles and Practice of Sleep Medicine" (M. Kryger, T. Roth, and W. C. Dement, eds.), pp. 95–104. Saunders, Philadelphia.

Hendricks, J. C. (2003). Invited review: Sleeping flies don't lie: The use of *Drosophila melanogaster* to study sleep and circadian rhythms. *J. Appl. Physiol.* **94,** 1660–1673.

Hendricks, J. C., Finn, S. M., Panckeri, K. A., Chavkin, J., Williams, J. A., Sehgal, A., and Pack, A. I. (2000). Rest in *Drosophila* is a sleep-like state. *Neuron* **25,** 129–138.

Hendricks, J. C., Kirk, D., Panckeri, K., Miller, M. S., and Pack, A. I. (2003a). Modafinil maintains waking in the fruit fly *Drosophila melanogaster*. *Sleep* **26,** 139–146.

Hendricks, J. C., Lu, S., Kume, K., Yin, J. C., Yang, Z., and Sehgal, A. (2003b). Gender dimorphism in the role of cycle (BMAL1) in rest, rest regulation, and longevity in *Drosophila melanogaster*. *J. Biol. Rhythms* **18,** 12–25.

Hendricks, J. C., Williams, J. A., Panckeri, K., Kirk, D., Tello, M., Yin, J. C., and Sehgal, A. (2001). A non-circadian role for cAMP signaling and CREB activity in *Drosophila* rest homeostasis. *Nature Neurosci.* **4,** 1108–1115.

Hewes, R. S., and Taghert, P. H. (2001). Neuropeptides and neuropeptide receptors in the *Drosophila melanogaster* genome. *Genome Res.* **11,** 1126–1142.

Hintermann, E., Grieder, N. C., Amherd, R., Brodbeck, D., and Meyer, U. A. (1996). Cloning of an arylalkylamine N-acetyltransferase (aaNAT1) from *Drosophila melanogaster* expressed in the nervous system and the gut. *Proc. Natl. Acad. Sci. USA* **93,** 12315–12320.

Hintermann, E., Jeno, P., and Meyer, U. A. (1995). Isolation and characterization of an arylalkylamine N-acetyltransferase from *Drosophila melanogaster*. *FEBS Lett.* **375,** 148–150.

Huber, R., Hill, S. L., Holladay, C., Biesiadecki, M., Tononi, G., and Cirelli, C. (2004). Sleep homeostasis in *Drosophila melanogaster*. *Sleep* **27,** 628–639.

Kaiser, W., and Steiner-Kaiser, J. (1983). Neuronal correlates of sleep, wakefulness and arousal in a diurnal insect. *Nature* **301,** 707–709.

Kapfhamer, D., Valladares, O., Sun, Y., Nolan, P. M., Rux, J. J., Arnold, S. E., Veasey, S. C., and Bucan, M. (2002). Mutations in Rab3a alter circadian period and homeostatic response to sleep loss in the mouse. *Nature Genet.* **32,** 290–295.

Konopka, R. J., and Benzer, S. (1971). Clock mutants of *Drosophila melanogaster*. *Proc. Natl. Acad. Sci. USA* **68,** 2112–2116.

Kopp, C., Albrecht, U., Zheng, B., and Tobler, I. (2002). Homeostatic sleep regulation is preserved in mPer1 and mPer2 mutant mice. *Eur. J. Neurosci.* **16,** 1099–1106.

Krueger, J. M., and Majde, J. A. (2003). Humoral links between sleep and the immune system: Research issues. *Ann. NY. Acad. Sci.* **992,** 9–20.

Mackiewicz, M., and Pack, A. I. (2003). Functional genomics of sleep. *Respir. Physiol. Neurobiol.* **135,** 207–220.

Magazanik, L. G., and Fedorova, I. M. (2003). Modulatory role of adenosine receptors in insect motor nerve terminals. *Neurochem. Res.* **28,** 617–624.

Martinek, S., Inonog, S., Manoukian, A. S., and Young, M. W. (2001). A role for the segment polarity gene shaggy/GSK-3 in the *Drosophila* circadian clock. *Cell* **105,** 769–779.

Nassel, D. R. (2002). Neuropeptides in the nervous system of *Drosophila* and other insects: Multiple roles as neuromodulators and neurohormones. *Prog. Neurobiol.* **68,** 1–84.

Nitz, D. A., van Swinderen, B., Tononi, G., and Greenspan, R. J. (2002). Electrophysiological correlates of rest and activity in *Drosophila melanogaster*. *Curr. Biol.* **12,** 1934–1940.

Pace-Schott, E. F., and Hobson, J. A. (2002). The neurobiology of sleep: Genetics, cellular physiology and subcortical networks. *Nature Rev. Neurosci.* **3,** 591–605.

Paulsen, O., and Morris, R. G. (2002). Flies put the buzz back into long-term-potentiation. *Nature Neurosci.* **5,** 289–290.

Pigeau, R., Naitoh, P., Buguet, A., McCann, C., Baranski, J., Taylor, M., Thompson, M., and Mac, K. I. I. (1995). Modafinil, d-amphetamine and placebo during 64 hours of sustained mental work. I. Effects on mood, fatigue, cognitive performance and body temperature. *J. Sleep Res.* **4,** 212–228.

Rechtschaffen, A. (1998). Current perspectives on the function of sleep. *Perspect. Biol. Med.* **41,** 359–390.

Rechtschaffen, A., and Bergmann, B. M. (2002). Sleep deprivation in the rat: An update of the 1989 paper. *Sleep* **25,** 18–24.

Rechtschaffen, A., Bergmann, B. M., Everson, C. A., Kushida, C. A., and Gilliland, M. A. (1989). Sleep deprivation in the rat. X. Integration and discussion of the findings. *Sleep* **12,** 68–87.

Roberts, D. B. (1986). Basic *Drosophila* care and techniques. In "*Drosophila:* A Practical Approach" (D. B. Roberts, ed.), pp. 1–19. IRL Press, Washington, DC.

Rogers, N. L., Dorrian, J., and Dinges, D. F. (2003). Sleep, waking and neurobehavioural performance. *Front. Biosci.* **8,** s1056–s1067.

Rorth, P., Szabo, K., Bailey, A., Laverty, T., Rehm, J., Rubin, G. M., Weigmann, K., Milan, M., Benes, V., Ansorge, W., and Cohen, S. M. (1998). Systematic gain-of-function genetics in Drosophila. *Development* **125,** 1049–1057.

Roth, T., Roehrs, T., and Zorick, F. (1982). Sleepiness: Its measurement and determinants. *Sleep* **5**(Suppl. 2), S128–S134.

Sauer, S., Kinkelin, M., Herrmann, E., and Kaiser, W. (2003). The dynamics of sleep-like behaviour in honey bees. *J. Comp. Physiol. A Neuroethol. Sens. Neural Behav. Physiol.* **189,** 599–607.

Shaw, P. J., Cirelli, C., Greenspan, R. J., and Tononi, G. (2000). Correlates of sleep and waking in *Drosophila melanogaster. Science* **287,** 1834–1837.

Shaw, P. J., and Franken, P. (2003). Perchance to dream: Solving the mystery of sleep through genetic analysis. *J. Neurobiol.* **54,** 179–202.

Shaw, P. J., Tononi, G., Greenspan, R. J., and Robinson, D. F. (2002). Stress response genes protect against lethal effects of sleep deprivation in *Drosophila. Nature* **417,** 287–291.

Shiromani, P. J., Xu, M., Winston, E. M., Shiromani, S. N., Gerashchenko, D., and Weaver, D. R. (2004). Sleep rhythmicity and homeostasis in mice with targeted disruption of mPeriod genes. *Am. J. Physiol. Regul. Integr. Comp. Physiol* **287,** R47–R57.

Spradling, A. C., Stern, D., Beaton, A., Rhem, E. J., Laverty, T., Mozden, N., Misra, S., and Rubin, G. M. (1999). The Berkeley *Drosophila* Genome Project gene disruption project: Single P-element insertions mutating 25% of vital *Drosophila* genes. *Genetics* **153,** 135–177.

Stanewsky, R. (2003). Genetic analysis of the circadian system in *Drosophila melanogaster* and mammals. *J. Neurobiol.* **54,** 111–147.

Strecker, R. E., Morairty, S., Thakkar, M. M., Porkka-Heiskanen, T., Basheer, R., Dauphin, L. J., Rainnie, D. G., Portas, C. M., Greene, R. W., and McCarley, R. W. (2000). Adenosinergic modulation of basal forebrain and preoptic/anterior hypothalamic neuronal activity in the control of behavioral state. *Behav. Brain Res.* **115,** 183–204.

Strumwasser, F. (1971). The cellular basis of behavior in Aplysia. *J. Psychiatr. Res.* **8,** 237–257.

Tafti, M., Chollet, D., Valatx, J. L., and Franken, P. (1999). Quantitative trait loci approach to the genetics of sleep in recombinant inbred mice. *J. Sleep Res.* **8**(Suppl. 1), 37–43.

Tafti, M., Petit, B., Chollet, D., Neidhart, E., de Bilbao, F., Kiss, J. Z., Wood, P. A., and Franken, P. (2003). Deficiency in short-chain fatty acid beta-oxidation affects theta oscillations during sleep. *Nature Genet.* **34,** 320–325.

Tobler, I. (1983). Effect of forced locomotion on the rest-activity cycle of the cockroach. *Behav. Brain Res.* **8,** 351–360.

Tobler, I., and Borbely, A. A. (1985). Effect of rest deprivation on motor activity of fish. *J. Comp. Physiol. A* **157,** 817–822.

Tobler, I., and Borbely, A. A. (1986). Sleep EEG in the rat as a function of prior waking. *Electroencephalogr. Clin. Neurophysiol.* **64,** 74–76.

Tobler, I., and Stalder, J. (1988). Rest in the scorpion: A sleep-like state? *J. Comp. Physiol. A* **163,** 227–235.

Toh, K. L., Jones, C. R., He, Y., Eide, E. J., Hinz, W. A., Virshup, D. M., Ptacek, L. J., and Fu, Y. H. (2001). An hPer2 phosphorylation site mutation in familial advanced sleep phase syndrome. *Science* **291**, 1040–1043.

Toma, D. P., White, K. P., Hirsch, J., and Greenspan, R. J. (2002). Identification of genes involved in *Drosophila melanogaster* geotaxis, a complex behavioral trait. *Nature Genet.* **31**, 349–353.

Tully, T. (1996). Discovery of genes involved with learning and memory: An experimental synthesis of Hirschian and Benzerian perspectives. *Proc. Natl. Acad. Sci. USA* **93**, 13460–13467.

van Swinderen, B., and Greenspan, R. J. (2003). Salience modulates 20–30 Hz brain activity in *Drosophila. Nature Neurosci.* **6**, 579–586.

van Swinderen, B., Nitz, D. A., and Greenspan, R. J. (2004). Uncoupling of brain activity from movement defines arousal states in *Drosophila. Curr. Biol.* **14**, 81–87.

Veasey, S. C., Valladares, O., Fenik, P., Kapfhamer, D., Sanford, L., Benington, J., and Bucan, M. (2000). An automated system for recording and analysis of sleep in mice. *Sleep* **23**, 1025–1040.

Yoshihara, M., Ensminger, A. W., and Littleton, J. T. (2001). Neurobiology and the *Drosophila* genome. *Funct. Integr. Genom.* **1**, 235–240.

Zhdanova, I. V., Wang, S. Y., Leclair, O. U., and Danilova, N. P. (2001). Melatonin promotes sleep-like state in zebrafish. *Brain Res.* **903**, 263–268.

# Section XI

# Circadian Biology of Populations

## [42]   Molecular Evolution and Population Genetics of Circadian Clock Genes

By Eran Tauber and Charalambos P. Kyriacou

### Abstract

This article discusses a number of common methodologies used in the field of population genetics and evolution and reviews their application within circadian rhythm research. We examine the basic principles behind phylogenetic analysis and how these can be used to illuminate clock gene evolution. We then discuss genetic variation between and within species and show how neutrality tests can reveal the signatures of selection or drift on clock genes. These tests are particularly important for moving beyond "just so" stories when discussing the evolution of clock phenotypes, and we provide relevant circadian examples. We also focus on methods that can be used to study genetic variation, such as quantitative trait loci analysis. We discuss the various bootstrapping or resampling techniques that can be applied to generate confidence intervals in the various methodologies and then examine the use of interspecific transformation studies, which can, and have, provide some useful insights, not only into clock gene evolution in particular, but "behavioral" gene evolution in general. Finally, we assess gene/protein alignments and protein structure predictions and their implicit evolutionary bases.

### Introduction

Natural genetic variation can be taken as the output of a long, ongoing experiment carried outdoors. Unfortunately, the experiment (evolution) was not well controlled, the objectives were not defined, and too many factors were involved. According to Kimura (1983), only a very small fraction of this variation is maintained by natural selection and has functional importance. The rest of the variation is effectively neutral, with no consequences for Darwinian fitness. Birth of the "neutral theory" led to heated exchanges between selectionists and neutralists over decades, but since the late 1980s, more constructive advances in the statistical analyses of DNA or protein sequences have provided an arsenal of methods to identify the signature of natural selection, starting with "neutrality" as the null hypothesis. If a positive result is obtained favoring a selective scenario of some kind, then the researcher is more comfortable and more confident of trying to dissect out the possible selective forces that are shaping this natural variation. If the result is a failure to reject the null hypothesis, then we would argue that a researcher needs to be very cautious in interpreting their data via any selective process, as it could end up as being just another superficial "just so" story that has provided such a checkered history in evolutionary thinking.

Sequence information is therefore a gold mine in which to study evolutionary process such as speciation, natural selection, migration, admixture, and other demographic changes, and this applies to any chronobiologist who wishes to pursue the evolutionary history of clock genes. However, most of us are too busy chasing new clock genes and mechanisms in our favorite organism to bother with this kind of approach (possibly because it involves maths and statistics, so it puts us off). Thus the approach has been either left to *bone fide* population geneticists that are outside the clock field and who will use variation in a clock gene sequence as cannon fodder for their statistical tests, with little or no interest in the underlying biology, or to the occasional clocky types like us who try to keep this end up. Natural molecular variation thus provides us with the opportunity to identify special adaptations of clock genes and, as we will show, can also provide some further insights into how clocks work, thereby enhancing the conventional approach, which uses mutants and transgenic lines. Rather than simply presenting a dry practical approach to the basic techniques in this area, we prefer to illustrate them liberally with chronobiological examples.

## Phylogenetic Analysis

Imagine that we have obtained sequences for a clock gene in a number of different species. The first thing we might like to do is to generate a phylogeny of these sequences in order to see whether the gene topology follows the generally accepted species order. Of course, gene trees need not follow species trees for all sorts of reasons, convergent evolution being one, in which two very unrelated species converge on a similar sequence solution that has been imposed by a similar selective agent. The gene sequences will therefore appear more similar than they should compared with the "real" species phylogeny.

There are a large number of methods for making phylogenetic trees. In fact, one need not even use gene sequences to make trees, for example, Carson (1983), perhaps most famously, used polytene bands from *Drosophila* salivary chromosomes to investigate the evolution of Hawaiian drosophilids. He found that he could move from one species inversion to another, usually in single steps, sometimes by invoking an hypothetical intermediate inversion (which he had not actually observed), by which he could jump to the next species inversion. He observed that the ancestral species appeared to live on the geologically older islands, whereas the more derived species were generally found on the newer volcanic islands. He was able to root his tree by observing that a species in the oldest island had an inversion similar to a south American species, suggesting that the founders of Hawaiian drosophilids came from the New World and that by a series of island hops, skips, and jumps they had populated the islands as they formed via volcanic activity via sequential founder effects. A "founder effect" in this context simply means that perhaps a single or a few inseminated females had populated the new island, found a new environment with a novel selective agent, and speciation had run wild. Speciation is very intense on islands, which are isolated, and therefore a captive gene pool is not subject to immigration, which will tend to slow down the speciation process by gene infiltration.

This kind of phylogeny is based on parsimony. A sequence of polytene bands in species 1, let us call them ABCDE, becomes ADCBE in species 2 so that in one step we can see that the BCD order has been inverted. Species 3 has the order AEBCD, which is one further step from species 2, and has the inversion of DCBE. We cannot in one jump go from species 1 to 3, thus the most parsimonious phylogeny has an evolution from species 1 to 2 to 3. There are other solutions to jumping from species 1 directly to 3, but they require more than one intermediate step and so are not "parsimonious." This very simple example provides the fundamental basis for parsimony analysis, where the smallest number of individual steps between the species provides the best tree.

Let us now imagine that a distantly related species, species 4, has the order ABCED, with an inversion of DE compared with species 1. We can thus "root" species 1, 2, and 3 with their distant cousin species 4, which lies closest to species 1. Thus species 1 must be the most ancestral species, and we have placed an evolutionary direction on our tree (i.e., we have gone from an unrooted tree where the evolutionary order from ancestral to derived could have been species 1 to 2 to 3, or species 3 to 2 to 1, to a rooted tree, where the species order must be species 1 to 2 to 3).

This principle, which is illustrated so beautifully with Carson's inversions, has been applied to a clock gene, *period* (*per*), in the drosophilids. Within *per* there is a repetitive minisatellite of *ACNGGN* hexamers that encode threonine–glycine (Thr-Gly) repeats in *Drosophila melanogaster*. Because the sequences of these Thr-Gly-encoding cassettes differ in the synonymous position and also in the number of cassettes, it is possible to track the evolution of these Thr-Gly haplotypes in a similar way to inversions (Fig. 1). By sheer luck, a series of natural alleles collected in Europe, containing a number of Thr-Gly length variants, from 14 to 23 pairs of

Thr-Glys, provided the perfect material to analyze the evolution of this region (Costa *et al.*, 1991).

By a series of single step deletions and duplications of cassettes (that is the predominant way in which repeated sequences evolve, not by base substitutions), it was possible to evolve all variants from the European (*Thr-Gly*)23b allele (see Fig. 1). This provides a simple example of a molecular phylogeny based on parsimony, or Occam's razor, with the minimum number of steps providing the best tree.

Of course the prerequisite for generating any kind of tree is to have an alignment of the sequences, which can be accommodated using a number of available algorithms, with CLUSTAL perhaps being the best known (Thompson *et al.*, 1994). Because most, if not all, phylogenetic algorithms will exclude sites where one or few of the sequences have gaps, any inclusion of sequences that introduce gaps to the alignment could result in a decrease in the reliability or resolution of the tree due to a decrease in the number of sites used for the calculations. Finally, although alignments are produced automatically, visual inspection and adjustment of the alignment using multiple alignment methods (including structure information where possible) are highly recommended.

Tree methods can be classified into two groups: maximum likelihood (ML) (Felsenstein, 1981) and parsimony trees versus methods based on distance such as neighbor joining (Saitou and Nei, 1987) and UPGMA

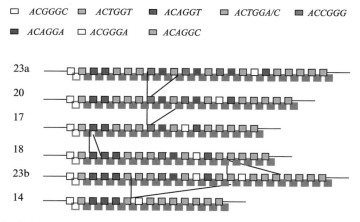

FIG. 1. Phylogeny of *period* Thr-Gly repeat alleles in *D. melanogaster*. Each colored box gives the encoding hexamer for a Thr-Gly dipeptide. Note how deletions and duplications of cassettes determine the observed variation in repeat number. All variants coalesce back to the (*Thr-Gly*)23b allele. (See color insert.)

(Sokal and Sneath, 1963). We have dealt with a simple case of a parsimony tree earlier. In ML methods, the probability of obtaining the observed data is calculated using different models, with different tree structures, branch lengths, different models of sequence evolution and so on, and a tree is chosen that provides the best probability (likelihood) for observing data. The popularity of ML trees was hindered greatly because they are very demanding on computer time, although this problem has receded more recently. Distance-based methods, in contrast, do not consider the whole sequence, but calculate a distance between pairs of aligned sequences and then use the pairwise distance matrix to fit a tree (see later). Distance methods do not perform as well as ML, but for a large data set they are the practical alternative, as they are much faster to compute.

Let us take an example of the simplest "distance" method. One can imagine that if one has a long gene sequence of 5000 bp in 10 species that the computing power required to generate every possible phylogeny, with every species in every possible position, and then counting all the possible steps in order to find the most parsimonious tree is considerable. Distance methods use a matrix whereby the number of changes between every pair of species is calculated, and the tree is constructed from this matrix. So if species A and C differ by 2 bp in a sequence, then the distance to their common ancestor is 1 bp in each direction (Fig. 2A). If species B has five changes from A and seven from C, then its average distance is $(5 + 7)/2$, or six changes from A, C, or three changes in each direction from the common ancestor. Finally, if species D has 11 changes from A, 13 from B, and 12 from C its average distance from A,B,C is $(11 + 12 + 13)/3$ or 12. In terms of the tree, this translates to six changes in each branch from the common ancestor of D from A, B, C (Fig. 2A). This way of constructing a tree is called the unweighted pair group method of averages (UPGMA), requires far less computer power, and was developed by Sokal and Sneath (1963). The method is sensitive to the order in which branches are added so successive randomization of the order is advised in order to obtain the best tree.

A more sophisticated and commonly used variant on this distance theme is the neighbor-joining method of Saitou and Nei (1987). In addition, there are many different ways of modulating the distance between two sequences by according them different weights, for example, a transversion versus a transition or, if the amino acids are being used in the phylogeny, by weighting the different types of amino acid changes or considering the number of changes required in the codon.

Phylogenies of clock genes have been used in many different ways, for example, to resolve the phylogenetic relationship in different insect groups (e.g., 8, 9). The relatively rapid amino acid evolution, often seen in

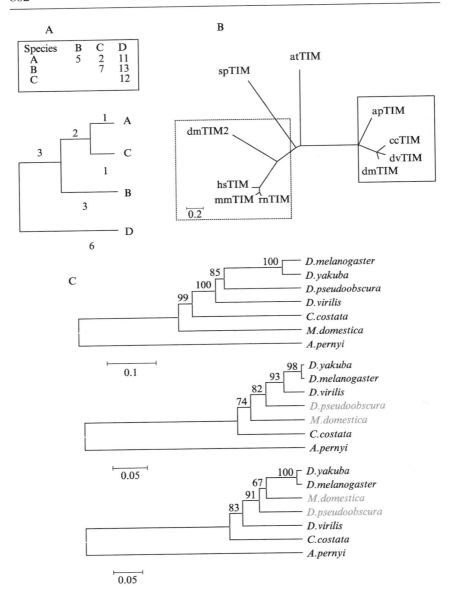

Fig. 2. (A) Construction of simple phylogeny from a distance matrix using UPGMA (see text). (B) Neighbor-joining tree for TIMELESS. Protein sequences were aligned with ClustalW (available online at http://www.ebi.ac.uk/clustalw) using default parameters. The tree was produced with MEGA2.1 using the Poisson correction model. Bootstrap values shown are for 500 replications. The scale represents substitutions per site for mouse

"behavioral" genes such as *per*, makes this gene particularly informative for resolving lower taxonomic levels, such as within the *Lepidoptera*, where 86% of the alignable amino acids undergo substitutions in 26 species surveyed (Regier *et al.*, 1998). On a more global level, a maximum likelihood method was used to examine the evolutionary relationships of the prokaryotic *kaiABC* genes in 70 different genomes and concluded that the first *kai* duplication occurred more than 3500 Mya (Dvornyk *et al.*, 2003). Phylogenies have also been used to identify the relationships among the four mammalian *per* genes (e.g., 11, 12) and, similarly, which of the two *timeless* paralogs in *Drosophila* and *Lepidoptera* were ancestral (Fig. 2B) (Benna *et al.*, 2000; Gotter and Reppert, 2001). Less conventionally perhaps, phylogeny has also been used to show that different fragments of PER show unusual phylogenetic patterns within Diptera, implicating interesting and divergent evolutionary dynamics on selected regions of the protein (Piccin *et al.*, 2000) (Fig. 2C).

Neutrality Tests

Natural selection can take different forms. *Directional* selection refers to the process where a population becomes dominated by few advantageous alleles or removal of detrimental alleles. In either case, selection causes a form of "bottleneck" that is limited to the selected site and is characterized by a local loss of genetic variation. A new neutral mutation may accumulate sporadically in different individuals, giving what is called "singletons." In contrast, *balancing* selection keeps a few alleles in intermediate frequencies, preserving polymorphism. As the preserved alleles

(mmTIM), rat (rnTIM), human (hsTIM), *C. costata* (ccTIM), *D. melanogaster* (dmTIM, dmTIM2 or TIMEOUT), *D. virilis* (dvTIM), *Antheraea pernyi* (apTIM), *Schizosaccharomyces pombe* (spTIM), and *Arabidopsis thaliana* (atTIM). Note the long branch for insect TIMs (solid box), suggesting a rapid evolution, even though TIM must have duplicated relatively recently in the common ancestors of Lepidoptera and Diptera from the ancestral TIM gene (dashed box), as TIM2, but not TIM, has been documented in the honeybee genome. (C) UPGMA trees of insect PER. A DNA tree (top) was constructed using the Kimura two-parameter model; protein trees (middle, bottom) were made using the Poisson correction model. Bootstrap values shown are for 500 replications. The scale represents substitutions per site. (*Top*) DNA tree that reflects the species phylogeny. (*Middle*) Tree based on protein alignment, excluding the PAS domain, and again largely reflects the species tree. (*Bottom*) Tree based on the PAS domain. Note how the PAS domain "switches" its topology for *Musca domestica* (100 Myr since common ancestor with *D. melanogaster*, gray) and *D. pseudoobscura* (30 Myr, gray). This correlates with the rescuing ability of the *Musca* and *D. pseudoobscura per* transgenes in *D. melanogaster* transformants, which are extremely robust for the former and extremely poor for the latter (14, 53) and may reflect a coevolution between the PAS domain of PER and a partner molecule, possibly TIM.

accumulate neutral changes over time, a peak of increased polymorphism is formed surrounding the site under selection.

Under neutrality, the amount of variation, $\theta$, is related to the rate of mutation $\mu$ (per generation, per site) and the effective population size $N_e$ as follows, $\theta = 4N_e\mu$. For a sex-linked gene, such as *period* in *Drosophila*, $\theta = 3N_e\mu$ because there are effectively only three X chromosomes (two in the female, one in the male). At the levels of the mutation rate found in nature, $\theta$ approximates the *heterozygosity* in the population or, in other words, the probability of picking at random two alleles from a population that will differ at a given site. Usually, both the population size and the mutation rate are unknown, but different aspects of the allelic variation can be used to estimate $\theta$ such as taking the number of segregating sites, $K$, in a population of sequences (corrected for sequence length). Assuming an infinite allele model (i.e., each mutation gives rise to a new allele not seen before, and every existing mutation creates a new allele), the more polymorphism that exists, the higher the number of segregating sites.

Another estimate of $\theta$ is termed $\pi$ and is based on the average number of differences between all pairs of sequences, so if we imagine we have four sequences, that gives six pairwise comparisons, and we simply calculate the mean pairwise diversity by dividing the total diversity by 6. Under neutrality, $\theta = K = \pi = 4N_e\mu$ so the two measurable estimates of $\theta$, which are $\pi$ and $K$, should be equal and this is the essence of the Tajima test (1989), which gives a test statistic $D$. Under directional selection, there will be a few individuals carrying new neutral mutations adjacent to the selected site, i.e., singletons, and these decrease the number of segregating sites $K$ compared with neutrality but also decrease $\pi$ because there are so few singletons, so one is usually comparing two identical sequences. However, under balancing selection, there is a decrease in the number of segregating sites, $K$, but a relative increase in $\pi$ as one is more often than not comparing different alleles. Thus, if we compare $\pi$ to $K$ to obtain $D$, an excess of rare variants (e.g., singletons), which is the footprint of purifying (directional) selection, will give a negative $D$ value. A positive $D$ indicates an excess of intermediate frequency variants, suggesting balancing selection. The Tajima test is very popular, although it is not very powerful (i.e., neutrality is rarely rejected), and is also very sensitive to demographic assumptions. Modifications and improvements of this general theme can be found in Fu and Li (1993).

Another family of neutrality tests uses information on both polymorphism and *divergence*. The idea here is that under neutrality, the number of segregating sites within a species (polymorphism) and the number of fixed differences between closely related species (divergence) will both be proportional to the mutation rate. The HKA test (Hudson *et al.*, 1987) compares variability within and between species in two or more loci. Selection is inferred when there is a difference between two loci in the ratio of divergence to polymorphism. Usually, one of the loci, which is considered to be neutral, serves as a reference to test neutrality in the locus under question. Like the Tajima test, the HKA test is also strongly dependent on the demographic model. The McDonald–Kreitman test (McDonald and Kreitman, 1991) uses a similar idea, but here sites are classified into different types (typically, synonymous *vs* nonsynonymous, fixed *vs* polymorphic) and the ratio of classes within species is compared with the ratio of classes that are different (fixed) between species in a 2 × 2 contingency table. In contrast to the previous neutrality test, the McDonald–Kreitman test is expected to perform well under different demographic models. A plethora of neutrality tests have been developed, including tests based on linkage disequilibrium [Kelly Zn's, (1997)] or tests based on the ratio of synonymous versus nonsynonymous substitutions (Hughes and Nei, 1988); the tests just described serve only as representative examples.

Some of these tests have been applied to clock genes, particularly *per*, in *Drosophila*. As mentioned earlier, a section near the middle of *per* gene encodes Thr-Gly (and Ser-Gly) repeats, which is highly polymorphic in length at both intra- and interspecific levels. As well as providing an early and elegant example of the evolution of a coding minisatellite by duplications and deletions (Costa *et al.*, 1991), this region has been explored in a number of population genetics and molecular evolution studies (Costa *et al.*, 1992; Peixoto *et al.*, 1992, 1993; Rosato *et al.*, 1996, 1997). These evolutionary studies led to hypotheses that subsequently revealed a function for this region in circadian temperature compensation of circadian clock in natural populations (Costa and Kyriacou, 1998; Peixoto *et al.*, 1998; Sawyer *et al.*, 1997).

Both *D. melanogaster* and *Drosophila simulans* have natural length alleles, with the predominant variants having 17, 20, or 23 repeat units in *D. melanogaster* (Costa *et al.*, 1991; Yu *et al.*, 1987) and 23, 24, and 25 repeat unit in *D. simulans* (Rosato *et al.*, 1994). The Thr-Gly region cannot be tested directly for departure from neutrality because the assumption of infinite site model (see earlier) on which the neutrality tests are based is violated (Watterson, 1975) and because of the rapid mutation rate that is associated with repeats [several orders of maginitude higher than the base substitution rate, see Rosato *et al.* (1997)]. For these reasons, initial population analyses avoided that region altogether. For example, Kliman and Hey (1993) examined a region N-terminal to the repeat and observed that it showed no characteristic signatures of selection using the tests outlined earlier. However, because of the rapid mutation rate, the Thr-Gly repeat turns out to be an "island" of dynamic changes that does not follow the rules for the upstream region because any one Thr-Gly length variant can be associated with more than one flanking haplotype (Peixoto *et al.*, 1992) and each haplotype can be associated with more than one length variant (Rosato *et al.*, 1996). Nevertheless, Rosato *et al.* (1997), using a modified analysis that used patterns of linkage disequilibrium with the repeat, were able to determine that balancing selection was maintaining the Thr-Gly length variation.

In *D. simulans*, the story is different. The three major Thr-Gly length alleles are in perfect linkage with the specific flanking amino acid variation within European populations (Rosato *et al.*, 1994). Here, the Tajima test showed significant departure from neutrality, indicating that the adjacent Thr-Gly polymorphism is maintained by balancing selection (Rosato *et al.*, 1994). Interestingly, more recent studies of natural African isofemale *D. simulans* lines have found an apparent recombinant among one of the three Thr-Gly lengths and the flanking haplotype. Remarkably, of these "recombinant" lines, one turned out to be a natural clock mutant with a very long period (Rogers *et al.*, 2004). The genetic mapping of this behavioral phenotype turned out to be complex, with a mapping close to *per*, but with significant autosomal modifiers. It was as if disturbing the linkage disequilibrium between the Thr-Gly repeat length and the adjacent flanking amino acid variation on some genetic backgrounds was also disturbing the clock.

In fact, coevolution between the Thr-Gly repeat length and adjacent amino acids had been proposed, theoretically, by comparing repeat length and amino acid substitutions in comparisons between *Drosophila* species (Nielsen *et al.*, 1994; Peixoto *et al.*, 1993). A highly significant correlation had been observed between the difference in length of the repeat between any two species, with their amino acid divergence in the immediate flanking regions. This might initially appear not surprising because the difference in repeat length in any two species may simply reflect the time elapsed since their common ancestor, and of course, the greater the amino acid divergence in the nonrepetitive flanking regions. However, given the erratic evolutionary dynamics of repetitive regions, one would not expect any direct relationship between repeat length and time. To examine this further, variation in the silent, synonymous position of these flanking regions, which is under little selection and should also reflect the time since the common ancestor between two species, was also studied. However, the correlation between the pairwise difference between repeat lengths and pairwise synonymous site variation was not significant (Peixoto *et al.*, 1993). Thus, the correlation between repeat length and flanking amino acid divergence is not an inevitable consequence of time, but instead may reflect coevolution between the repeat and its flanking regions at the protein level so that as repeat length changes between species, amino acid changes must coevolve and compensate for changes in protein structure in that region.

This hypothesis was tested experimentally by making chimeric genes between two fly species (another tool for the evolutionary chronobiologist, see later): one with a long and one with a relatively short Thr-Gly/Ser-Gly repeat and manipulating the chimeric junction so the repeat was either adjacent to its conspecific or heterospecific flanking region. Transgenic flies carrying a chimeric junction with a heterospecific combination of repeat and flanking region were almost arrhythmic (Peixoto *et al.*, 1998). Other chimeric junctions that added more (but not all) of the conspecific flanking region to the repeat gave robust rhythmicity, but remarkably temperature-sensitive periods. These results were quite startling in their magnitude and would have remained undiscovered had it not been for the statistical tests on the sequences (Peixoto *et al.*, 1998). Thus it appears that changes in repeat length drive sequence variation in the flanks, or *vice versa*, in order to stabilize the PER protein.

So what happens if you have polymorphism within a species, with no amino acid changes in the flanking region, as in *D. melanogaster*, where there are four major natural variants carrying 14, 17, 20, and 23 Thr-Gly (Costa *et al.*, 1992)? The answer is that a linear relationship exists between Thr-Gly length and temperature compensation in this series of alleles, with the (*Thr-Gly*) 20 allele being the best compensated (Sawyer *et al.*, 1997). This almost perfect linear relationship between the 14-17-20 and 23 series of variants breaks down when very rare variants with repeat length such as 15, 21, and 24 are considered. Also, their temperature compensation is much worse. One reason for this may be that (Thr-Gly)3 is the conformational monomer that generates a turn (Castiglione-Morelli *et al.*, 1995). Thus, moving from one repeat length to another in multiples of (Thr-Gly)3 provides additional stability by maintaining complete -turn structures. The final fillip to this story is that the two predominant alleles in Europe, the (*Thr-Gly*)$_{20}$ and (*Thr-Gly*)$_{17}$, which contribute equally to 90% of the natural length variation, show a latitudinal cline in Europe, with high levels of the (*Thr-Gly*)$_{17}$ allele in southern Europe and *vice versa* in the north (Costa *et al.*, 1992). At hot temperatures, the (*Thr-Gly*)$_{17}$ has a circadian period that is almost exactly 24 h, although this period shortens as the temperature is reduced (Sawyer *et al.*, 1997). The (*Thr-Gly*)$_{20}$, however, has a period shorter than 24 h, but the period is stable at all temperatures. Thus, we might imagine that (*Thr-Gly*)$_{17}$ is better adapted to generally warmer environments (Mediterranean) and that the (*Thr-Gly*)$_{20}$ is better adjusted to harsher northern environments with bigger fluctuations and generally colder annual temperatures. In other words, we appear to have a case of balancing selection. This interpretation is saved from being another a "just so" story because balancing selection had left its signature in the DNA sequences of these variants (Rosato *et al.*, 1997).

The fact that allelic variation in any organism can show latitudinal clines is often taken as an indication of some kind of balancing selection. However, it should always be remembered that clines can arise from many processes, e.g., drift, population admixture, or historical processes. Clock genes such as *period* (see earlier) and *timeless* (M. Zordan, unpublished results) show clines, as do clock phenotypes (e.g., Michael *et al.*, 2003). However, when possible, it is always wise to use neutrality tests before any conclusion is drawn about the selective process, if any, that has given rise to the spatial patterning.

### Polymorphism in Human Clock Genes

Molecular polymorphism in human clock genes provides a special case where a considerable amount of data was obtained not by molecular evolutionists, but by clinicians who were interested in associating different clock alleles with sleep and psychological disorders for diagnostic purposes. A polymorphism in *hClock* in the 3′-UTR region (Katzenberg *et al.*, 1998) was correlated with morning–evening tendencies in a large sample of normal subjects, whereas the *hPer1* polymorphism (Katzenberg *et al.*, 1999) or *hTim* (Pedrazzoli *et al.*, 2000) did not have any effect. The most astonishing example of a human clock polymorphism, however, is the case of "advanced sleep phase syndrome" where a mutation in a serine codon of *hPer2* shortens the free-running circadian period, thereby generating the behavioral syndrome (Toh *et al.*, 2001). This serine corresponds to the same serine residue that is mutated in the classic fly mutant, *per$^s$* (Baylies *et al.*, 1987; Yu *et al.*, 1987).

The large body of data of human clock polymorphism might be used in the future to explore the evolution of these genes and the nature of these polymorphisms using the methodology described here. One problem, however, working with human genes is that the human population size has expanded dramatically in the past 10–100,000 years and therefore is not at stationary equilibrium (Kreitman, 2000), making the use of neutrality tests and the detection of positive selection difficult and challenging with the existing statistical methods.

### Quantitative Trait Loci Analysis

The approach of quantitative trait loci (QTL) is another demonstration for the use of naturally occurring genetic variation and has been adopted by clock biologists, recognized as a natural alternative for identifying clock genes (as opposed to forward genetic screens). Briefly, the method is based on crossing two pure lines that differ substantially in a quantitative (e.g., clock) trait. Different crossing designs are used to interbreed these lines, eventually producing many descendant lines, each having a different segment from the parental line. The phenotype of these lines is measured and specific segments with a phenotypic effect can be identified. Mapping of these segments (QTL) is carried using molecular markers such as restriction length polymorphism (RFLP) and simple repeat polymorphism (SSLP). These markers are typically highly polymorphic, phenotypically neutral, and spread through the whole genome in high abundance.

QTL analysis was used extensively to study the circadian system of the mouse (e.g., Hofstetter *et al.*, 2003; Shimomura *et al.*, 2001) and led to the identification of loci significantly affecting different aspects (including the free-running period $\tau$) of circadian rhythmicity. Notably, all QTLs do not map to the same regions of canonical clock genes, demonstrating the advantage of this approach in discovering new clock components. A similar approach has also been used to explore the clock of the flowering plant, *Arabidopsis taliana* (e.g., Michael *et al.*, 2003; Swarup *et al.*, 1999), using natural lines. Here, a few QTLs coincide with QTLs for timing of flowering, demonstrating a link between the circadian clock and seasonal timing. Light sensitivity of the circadian clock also been explored using QTL in mice (Yoshimura *et al.*, 2002), suggesting polygenic architecture. In *Arabidopsis*, a similar approach led to the identification of new photoreceptor alleles (El-Din El-Assal *et al.*, 2001; Maloof *et al.*, 2001).

## Statistical Resampling Methods in Molecular Evolution

Statistical resampling methods are a family of techniques (such as Monte-Carlo simulation and bootstrapping) that derive distributions, test statistics, and other statistical properties by running computer simulations rather than formulating mathematical solutions. Typically, raw data are manipulated (randomly shuffled, permutated, resampled, etc.) to produce many pseudo-samples that are used in the simulation and effectively generate the values that will be considered as "significant." It is this property that earned the bootstrap technique its name (Efron, 1982), referring to someone who pulls himself up by his own bootstraps. The resampling approach became popular in molecular evolution as it is particularly suitable for evolutionary problems, which are often complicated and are not data rich (bear in mind that any natural evolutionary experiment was carried out only once).

One application of resampling is in calculating statistical confidence levels for neutrality tests. The theory behind many neutrality tests is based on genealogies. In coalescence simulation, the computer randomly generates many genealogies (typically 500–1000) that resemble the original data set in terms of number and length of sequences, the number of segregating sites, or other aspects of molecular diversity that the user defines. Each set of sequences is produced by generating a genealogy, usually assuming a neutral infinite sites model with a large constant population size (Hudson, 1990). The distribution of different statistics (e.g., Tajima D) can be calculated from the pseudo-data sets, and this in turn can provide the 95% confidence limits, which is then compared with the observed statistic from real data to accept or reject the null hypothesis. Most programs (see later) will allow introducing recombination into the simulation, which usually has the effect of narrowing the confidence limits and increasing the statistical power of the tests.

Another application of resampling is the bootstrap analysis used in phlyogenetics, first introduced by Felsenstein (1981). The purpose of this analysis is to provide some confidence estimation to the different groups of sequences (clades) that are formed in a phylogenetic tree. Imagine we have 10 sequences each of 15 bp so that each nucleotide position represents a column in the alignment. Buttons numbered 1–15 are put into a hat, and 15 are drawn at random, but each button is replaced in the bag after each draw. Columns or positions from original data may therefore be represented more than once in the bootstrap data set or not at all in the new shuffled sequence. For each new shuffled data set, a tree is made using the same method that has been used with original data and the procedure is repeated (typically to generate 1000 trees). If the original topography of the tree based on real data is largely preserved, then that suggests the tree is robust. At each node of the tree will be a bootstrap value that describes the percentage of trials in which shuffled data also supported that clade. Typically, values over 70% are taken as a reliable estimate for a group. However, one should remember that bootstrap data merely reflect the variation in the original data set and the model used to generate the tree in the first place. Thus, bootstrap values cannot be used to assess the correctness of the whole tree.

Interspecific Transformations

Another method in the molecular evolutionists' toolbox is interspecific transformation. This was nicely demonstrated with the *frequency* gene in *Neurospora crassa*, when the ortholog from the distantly related *Sordaria fimicola* was shown to rescue the rhythmic conidiation of a *frq*-null allele, even though *Sordaria* does not possess this circadian developmental program, thereby demonstrating the central role of *frq* in the clock, irrespective of different types of outputs (Merrow and Dunlap, 1994). Interspecific transformations are therefore important to reveal functional equivalence of clock genes from widely divergent taxa. A very modest rescue of fly *per-null* mutants was even obtained with the murine *mPer1* and *mPer2* genes (Shigeyoshi *et al.*, 2002).

With more closely related species, interspecific transformations have somewhat revolutionized the view, commonly held by many in evolution, that adaptive characters evolve by small changes in many genes. This was first demonstrated by Petersen *et al.* (1988) when they transformed the *Drosophila pseudoobscura per* gene into *D. melanogaster* and "converted" the hosts circadian locomotor behavior into a pattern that mimicked that of the donor species. This extraordinary and completely unexpected result revealed that a single gene, *per*, had the potential to encode a species-specific instructive role for behavior and was soon followed by similar work showing that species-specific ultradian courtship song cycles are also *per* controlled by their respective species orthologs (Wheeler *et al.*, 1991). More recently, *D. pseudoobscura per* transformants have revealed that the characteristic circadian mating rhythm of *D. pseudoobscura* is also transferred to the host species (Tauber *et al.*, 2003). When male and female *D. melanogaster* transformants carrying either the conspecific *D. melanogaster per* transgene or the *D. pseudoobscura* were mixed together, a strong assortative mating was observed based on the conspecificity of the transgene. What appeared to be happening was that when one *per* transformant class was feeling sexy, the other one was not, and *vice versa*. These results provided an additional evolutionary perspective on initial work that had revealed the importance of *per* and photoperiods on mating rhythms in both *Drosophila* and melon flies (Miyatake *et al.*, 2002; Sakai and Ishida, 2001). If one extrapolates, it is not difficult to see that *per* could conceivably initiate a speciation event by altering temporal mating patterns and provide an example of incipient speciation in the laboratory.

## Protein Alignments and Predictions

Implicit in any alignment is that the protein-coding regions under inspection are homologous and related by decent. The finding that PER, which was essentially a "pioneer" protein until the mid-1990s, was related by sequence to the transcription factors ARNT and SIM gave rise to the PAS family (Huang *et al.*, 1993). Since then, it has become clear that the PAS domain is represented in a large number of clock proteins, spanning both animals and fungi (Ponting *et al.*, 1997). Sequence similarity outside the immediate PAS domain has also been detected between the two PAS containing proteins *Neurospora* WC1 and BMAL1, suggesting a possible common ancestor for these two functionally related clock proteins (Lee *et al.*, 2000). PAS domains are also found in bacterial photoactive yellow protein, PYP (Borgstahl *et al.*, 1995), revealing a light-signaling function for this structural unit, with obvious implications for the evolution of ancient clock proteins as a response to light–dark cycles (Crosthwaite *et al.*, 1997). More recently, another "pioneer" clock protein, TIM, has been shown, using structure prediction algorithms, to contain sequences that bear some similarity to ARM domains (Vodovar *et al.*, 2002). ARM domains were discovered initially in ARMADILLO (-CATENIN), a *Drosophila* developmental protein, whose stability and nuclear/cytoplasmic partition are a focus for the proteins of the wingless pathway (Huber *et al.*, 1997; van Noort *et al.*, 2002). Thus, the putative ARM domains of TIM may have been the stimulus for recruiting proteins from the wingless pathway during evolution of the clock, and several further members of this pathway are known to play important roles in the clock mechanism of the fly (Akten *et al.*, 2003; Grima *et al.*, 2002; Kloss *et al.*, 1998; Ko *et al.*, 2002; Lin *et al.*, 2002; Martinek *et al.*, 2001; Sathyanarayanan *et al.*, 2004). Thus, in the case of PER PAS and TIM, simple alignments in the former, and more sophisticated conformational predictions in the latter, have provided interesting global perspectives on the evolution of the clock.

### Conclusions

This brief introduction into the types of evolutionary analyses one can perform on clock genes is not comprehensive. There are many variations on the general themes, in terms of both phylogenetic and population analyses. For example, a technique called "evolutionary trace" uses trees to discover functionally important residues within clades and could be used to examine the important surfaces of any two clock proteins that are believed to physically associate (Lichtarge *et al.*, 1996). In the population genetics literature, it seems that almost every issue of the front-line journal *Genetics* has yet another statistical test developed to identify selection on sequence variation. We have focused only on those techniques that have been used so far in the clock field, and we apologize for the somewhat partisan nature of the literature presented, but that is what we do. The following useful computer programs are for those interested in this area.

### Computer Program for Phylogeny and Molecular Evolution

DnaSP (http://www.ub.es/dnasp/DnaSP32Inf.html) is a Windows software package that performs extensive population genetics and sequence polymorphism analyses of DNA sequence data (Rozas and Rozas, 1999).

It includes analysis of polymorphism and divergence, neutrality tests, recombination, and much more. Statistical significance is calculated by coalescence simulation. The interface is friendly and the graphic output is nice.

ProSeq (http://helios.bto.ed.ac.uk/evolgen/filatov/proseq.html) is another Windows program created by Filatov (2002). It will carry polymorphism analysis and many neutrality tests. PreSeq also carries coalescence simulation. There is a large overlap between DnaSP and ProSeq, but there are some tests unique to each of the programs. Unlike DnaSP, ProSeq also functions as a sequence editor (including chromotogram viewer and contig assembly).

MEGA (http://www.megasoftware.net/), which stands for *molecular evolution genetic analysis* (Kumar *et al.*, 2001), also performs tests of selection but its main power is in producing trees. The user can choose from a variety of distance estimation models and tree-making methods (UPGMA, neighbor joining, minimum evolution, and maximum parsimony). The program provides a bootstrap test of the phylogeny, tests of the molecular clock, and much more. It handles both DNA and protein sequences and is very flexible in defining groups of sequences, domains, etc. The program is very friendly and produces high-quality graphics.

TREE-PUZZLE (http://www.tree-puzzle.de/) uses maximum likelihood to reconstruct phylogenetic (Schmidt *et al.*, 2002). It implements quartet puzzling, which is a fast tree search algorithm, allowing the analysis of large data sets. It automatically assigns estimations of support to each internal branch. Windows, UNIX, and Mac versions are available. Finally, the collection of 193 phylogeny packages maintained by Joe Felsenstein at the University of Washington (http://evolution.genetics.washington.edu/phylip/software.html) is helpful.

## Acknowledgments

C.P.K. is pleased to acknowledge a research grant from NERC and a Royal Society Wolfson Research Merit Award.

## References

Akten, B., Jauch, E., Genova, G. K., Kim, E. Y., Edery, I., Raabe, T., and Jackson, F. R. (2003). A role for CK2 in the *Drosophila* circadian oscillator. *Nature Neurosci.* **6,** 251–257.

Baylies, M. K., Bargiello, T. A., Jackson, F. R., and Young, M. W. (1987). Changes in abundance or structure of the *per* gene product can alter periodicity of the *Drosophila* clock. *Nature* **326,** 390–392.

Benna, C., Scannapieco, P., Piccin, A., Sandrelli, F., Zordan, M., Rosato, E., Kyriacou, C. P., Valle, G., and Costa, R. (2000). A second *timeless* gene in *Drosophila* shares greater sequence similarity with mammalian tim. *Curr. Biol.* **10,** R512–R513.

Borgstahl, G. E., Williams, D. R., and Getzoff, E. D. (1995). A structure of photoactive yellow protein, a cytosolic photoreceptor: Unusual fold, active site, and chromophore. *Biochemistry* **34,** 6278–6287.

Carson, H. (1983). Chromosomal sequences and inter-island colonisations in Hawaiian *Drosophila. Genetics* **103,** 465–482.

Castiglione-Morelli, M. A., Guantieri, V., Villani, V., Kyriacou, C. P., Costa, R., and Tamburro, A. M. (1995). Conformational study of the Thr-Gly repeat in the *Drosophila* clock protein, PERIOD. *Proc. R. Soc. Lond. B Biol. Sci.* **260,** 155–163.

Costa, R., and Kyriacou, C. P. (1998). Functional and evolutionary implications of natural variation in clock genes. *Curr. Opin. Neurobiol.* **8,** 659–664.

Costa, R., Peixoto, A. A., Barbujani, G., and Kyriacou, C. P. (1992). A latitudinal cline in a *Drosophila* clock gene. *Proc. R. Soc. Lond. B Biol. Sci.* **250,** 43–49.

Costa, R., Peixoto, A. A., Thackeray, J. R., Dalgleish, R., and Kyriacou, C. P. (1991). Length polymorphism in the threonine-glycine-encoding repeat region of the *period* gene in *Drosophila. J. Mol. Evol.* **32,** 238–246.

Crosthwaite, S. K., Dunlap, J. C., and Loros, J. J. (1997). *Neurospora wc-1* and *wc-2*: Transcription, photoresponses, and the origins of circadian rhythmicity. *Science* **276,** 763–769.

Dvornyk, V., Vinogradova, O., and Nevo, E. (2003). Origin and evolution of circadian clock genes in prokaryotes. *Proc. Natl. Acad. Sci. USA* **100,** 2495–2500.

Efron, B. (1982). "The Jackknife, Bootstrap, and Other Resampling Plans." Philadelphia.

El-Din El-Assal, S., Alonso-Blanco, C., Peeters, A. J., Raz, V., and Koornneef, M. (2001). A QTL for flowering time in *Arabidopsis* reveals a novel allele of CRY2. *Nature Genet.* **29,** 435–440.

Felsenstein, J. (1981). Evolutionary trees from DNA sequences: A maximum likelihood approach. *J. Mol. Evol.* **17,** 368–376.

Filatov, D. (2002). ProSeq: A software for preparation and evolutionary analysis of DNA sequence data sets. *Mol. Ecol. Notes* **2,** 621–624.

Fu, Y. X., and Li, W. H. (1993). Statistical tests of neutrality of mutations. *Genetics* **133,** 693–709.

Gotter, A. L., and Reppert, S. M. (2001). Analysis of human *Per4. Brain Res. Mol. Brain Res.* **92,** 19–26.

Grima, B., Lamouroux, A., Chelot, E., Papin, C., Limbourg-Bouchon, B., and Rouyer, F. (2002). The F-box protein slimb controls the levels of clock proteins *period* and *timeless. Nature* **420,** 178–182.

Hofstetter, J. R., Trofatter, J. A., Kernek, K. L., Nurnberger, J. I., and Mayeda, A. R. (2003). New quantitative trait loci for the genetic variance in circadian period of locomotor activity between inbred strains of mice. *J. Biol. Rhythms* **18,** 450–462.

Huang, Z. J., Edery, I., and Rosbash, M. (1993). PAS is a dimerization domain common to *Drosophila period* and several transcription factors. *Nature* **364,** 259–262.

Huber, A. H., Nelson, W. J., and Weis, W. I. (1997). Three-dimensional structure of the armadillo repeat region of beta-catenin. *Cell* **90,** 871–882.

Hudson, R. (1990). Gene genealogies and the coalescent process. *Oxf. Surv. Evol. Biol.* **7,** 1–44.

Hudson, R. R., Kreitman, M., and Aguade, M. (1987). A test of neutral molecular evolution based on nucleotide data. *Genetics* **116,** 153–159.

Hughes, A. L., and Nei, M. (1988). Pattern of nucleotide substitution at major histocompatibility complex class I loci reveals overdominant selection. *Nature* **335,** 167–170.

Katzenberg, D., Young, T., Finn, L., Lin, L., King, D. P., Takahashi, J. S., and Mignot, E. (1998). A CLOCK polymorphism associated with human diurnal preference. *Sleep* **21,** 569–576.

Katzenberg, D., Young, T., Lin, L., Finn, L., and Mignot, E. (1999). A human *period* gene (HPER1) polymorphism is not associated with diurnal preference in normal adults. *Psychiatr. Genet.* **9,** 107–109.

Kelly, J. K. (1997). A test of neutrality based on interlocus associations. *Genetics* **146,** 1197–1206.

Kimura, M. (1983). "The Neutral Theory of Molecular Evolution." Cambridge Univ. Press, Cambridge, MA.

Kliman, R. M., and Hey, J. (1993). DNA sequence variation at the *period* locus within and among species of the *Drosophila melanogaster* complex. *Genetics* **133,** 375–387.

Kloss, B., Price, J. L., Saez, L., Blau, J., Rothenfluh, A., Wesley, C. S., and Young, M. W. (1998). The *Drosophila* clock gene *double-time* encodes a protein closely related to human *casein kinase Iepsilon. Cell* **94,** 97–107.

Ko, H. W., Jiang, J., and Edery, I. (2002). Role for *Slimb* in the degradation of *Drosophila period* protein phosphorylated by *doubletime. Nature* **420,** 673–678.

Kreitman, M. (2000). Methods to detect selection in populations with applications to the human. *Annu. Rev. Genomics Hum. Genet.* **1,** 539–559.

Kumar, S., Tamura, K., Jakobsen, I. B., and Nei, M. (2001). MEGA2: Molecular evolutionary genetics analysis software. *Bioinformatics* **17,** 1244–1245.

Lee, K., Loros, J. J., and Dunlap, J. C. (2000). Interconnected feedback loops in the Neurospora circadian system. *Science* **289,** 107–110.

Lichtarge, O., Bourne, H. R., and Cohen, F. E. (1996). An evolutionary trace method defines binding surfaces common to protein families. *J. Mol. Biol.* **257,** 342–358.

Lin, J. M., Kilman, V. L., Keegan, K., Paddock, B., Emery-Le, M., Rosbash, M., and Allada, R. (2002). A role for *casein kinase 2alpha* in the *Drosophila* circadian clock. *Nature* **420,** 816–820.

Maloof, J. N., Borevitz, J. O., Dabi, T., Lutes, J., Nehring, R. B., Redfern, J. L., Trainer, G. T., Wilson, J. M., Asami, T., Berry, C. C., Weigel, D., and Chory, J. (2001). Natural variation in light sensitivity of *Arabidopsis. Nature Genet.* **29,** 441–446.

Martinek, S., Inonog, S., Manoukian, A. S., and Young, M. W. (2001). A role for the segment polarity gene *shaggy/GSK-3* in the *Drosophila* circadian clock. *Cell* **105,** 769–779.

McDonald, J. H., and Kreitman, M. (1991). Adaptive protein evolution at the *Adh* locus in *Drosophila. Nature* **351,** 652–654.

Merrow, M. W., and Dunlap, J. C. (1994). Intergeneric complementation of a circadian rhythmicity defect: Phylogenetic conservation of structure and function of the clock gene *frequency. EMBO J.* **13,** 2257–2266.

Michael, T. P., Salome, P. A., Yu, H. J., Spencer, T. R., Sharp, E. L., McPeek, M. A., Alonso, J. M., Ecker, J. R., and McClung, C. R. (2003). Enhanced fitness conferred by naturally occurring variation in the circadian clock. *Science* **302,** 1049–1053.

Miyatake, T., Matsumoto, A., Matsuyama, T., Ueda, H. R., Toyosato, T., and Tanimura, T. (2002). The *period* gene and allochronic reproductive isolation in *Bactrocera cucurbitae. Proc. R. Soc. Lond. B Biol. Sci.* **269,** 2467–2472.

Nielsen, J., Peixoto, A. A., Piccin, A., Costa, R., Kyriacou, C. P., and Chalmers, D. (1994). Big flies, small repeats: The "Thr-Gly" region of the *period* gene in Diptera. *Mol. Biol. Evol.* **11,** 839–853.

Pedrazzoli, M., Ling, L., Finn, L., Kubin, L., Young, T., Katzenberg, D., and Mignot, E. (2000). A polymorphism in the human *timeless* gene is not associated with diurnal preferences in normal adults. *Sleep. Res. Online* **3,** 73–76.

Peixoto, A. A., Campesan, S., Costa, R., and Kyriacou, C. P. (1993). Molecular evolution of a repetitive region within the *per* gene of *Drosophila*. *Mol. Biol. Evol.* **10**, 127–139.

Peixoto, A. A., Costa, R., Wheeler, D. A., Hall, J. C., and Kyriacou, C. P. (1992). Evolution of the threonine-glycine repeat region of the period gene in the *melanogaster* species subgroup of *Drosophila*. *J. Mol. Evol.* **35**, 411–419.

Peixoto, A. A., Hennessy, J. M., Townson, I., Hasan, G., Rosbash, M., Costa, R., and Kyriacou, C. P. (1998). Molecular coevolution within a *Drosophila* clock gene. *Proc. Natl. Acad. Sci. USA* **95**, 4475–4480.

Petersen, G., Hall, J. C., and Rosbash, M. (1988). The *period* gene of *Drosophila* carries species-specific behavioral instructions. *EMBO J.* **7**, 3939–3947.

Piccin, A., Couchman, M., Clayton, J. D., Chalmers, D., Costa, R., and Kyriacou, C. P. (2000). The clock gene *period* of the housefly, *Musca domestica*, rescues behavioral rhythmicity in *Drosophila melanogaster*: Evidence for intermolecular coevolution? *Genetics* **154**, 747–758.

Ponting, C. P., and Aravind, L. (1997). PAS: A multifunctional domain family comes to light. *Curr. Biol.* **7**, R674–R677.

Regier, J. C., Fang, Q. Q., Mitter, C., Peigler, R. S., Friedlander, T. P., and Solis, M. A. (1998). Evolution and phylogenetic utility of the *period* gene in *Lepidoptera*. *Mol. Biol. Evol.* **15**, 1172–1182.

Rogers, A. S., Escher, S. A., Pasetto, C., Rosato, E., Costa, R., and Kyriacou, C. P. (2004). A mutation in *Drosophila simulans* that lengthens the circadian period of locomotor activity. *Genetica* **120**, 223–232.

Rosato, E., Peixoto, A. A., Barbujani, G., Costa, R., and Kyriacou, C. P. (1994). Molecular polymorphism in the period gene of *Drosophila simulans*. *Genetics* **138**, 693–707.

Rosato, E., Peixoto, A. A., Costa, R., and Kyriacou, C. P. (1997). Linkage disequilibrium, mutational analysis and natural selection in the repetitive region of the clock gene, *period*, in *Drosophila melanogaster*. *Genet. Res.* **69**, 89–99.

Rosato, E., Peixoto, A. A., Gallippi, A., Kyriacou, C. P., and Costa, R. (1996). Mutational mechanisms, phylogeny, and evolution of a repetitive region within a clock gene of *Drosophila melanogaster*. *J. Mol. Evol.* **42**, 392–408.

Rozas, J., and Rozas, R. (1999). DnaSP version 3: An integrated program for molecular population genetics and molecular evolution analysis. *Bioinformatics* **15**, 174–175.

Saitou, N., and Nei, M. (1987). The neighbor-joining method: A new method for reconstructing phylogenetic trees. *Mol. Biol. Evol.* **4**, 406–425.

Sakai, T., and Ishida, N. (2001). Circadian rhythms of female mating activity governed by clock genes in *Drosophila*. *Proc. Natl. Acad. Sci. USA* **98**, 9221–9225.

Sathyanarayanan, S., Zheng, X., Xiao, R., and Sehgal, A. (2004). Posttranslational regulation of *Drosophila* PERIOD protein by protein phosphatase 2A. *Cell* **116**, 603–615.

Sawyer, L. A., Hennessy, J. M., Peixoto, A. A., Rosato, E., Parkinson, H., Costa, R., and Kyriacou, C. P. (1997). Natural variation in a *Drosophila* clock gene and temperature compensation. *Science* **278**, 2117–2120.

Schmidt, H. A., Strimmer, K., Vingron, M., and von Haeseler, A. (2002). TREE-PUZZLE: Maximum likelihood phylogenetic analysis using quartets and parallel computing. *Bioinformatics* **18**, 502–504.

Shigeyoshi, Y., Meyer-Bernstein, E., Yagita, K., Fu, W., Chen, Y., Takumi, T., Schotland, P., Sehgal, A., and Okamura, H. (2002). Restoration of circadian behavioural rhythms in a period null *Drosophila* mutant (*per01*) by mammalian period homologues *mPer1* and *mPer2*. *Genes Cells* **7**, 163–171.

Shimomura, K., Low-Zeddies, S. S., King, D. P., Steeves, T. D., Whiteley, A., Kushla, J., Zemenides, P. D., Lin, A., Vitaterna, M. H., Churchill, G. A., and Takahashi, J. S. (2001).

Genome-wide epistatic interaction analysis reveals complex genetic determinants of circadian behavior in mice. *Genome Res.* **11,** 959–980.

Sokal, R., and Sneath, P. (1963). "Principles of Numerical Taxonomy." Freeman, San Francisco.

Swarup, K., Alonso-Blanco, C., Lynn, J. R., Michaels, S. D., Amasino, R. M., Koornneef, M., and Millar, A. J. (1999). Natural allelic variation identifies new genes in the *Arabidopsis* circadian system. *Plant J.* **20,** 67–77.

Tajima, F. (1989). Statistical method for testing the neutral mutation hypothesis by DNA polymorphism. *Genetics* **123,** 585–595.

Tauber, E., Roe, H., Costa, R., Hennessy, J. M., and Kyriacou, C. P. (2003). Temporal mating isolation driven by a behavioral gene in *Drosophila*. *Curr. Biol.* **13,** 140–145.

Thompson, J. D., Higgins, D. G., and Gibson, T. J. (1994). CLUSTAL W: Improving the sensitivity of progressive multiple sequence alignment through sequence weighting, position-specific gap penalties and weight matrix choice. *Nucleic Acids Res.* **22,** 4673–4680.

Toh, K. L., Jones, C. R., He, Y., Eide, E. J., Hinz, W. A., Virshup, D. M., Ptacek, L. J., and Fu, Y. H. (2001). An *hPer2* phosphorylation site mutation in familial advanced sleep phase syndrome. *Science* **291,** 1040–1043.

van Noort, M., Meeldijk, J., van der Zee, R., Destree, O., and Clevers, H. (2002). Wnt signaling controls the phosphorylation status of beta-catenin. *J. Biol. Chem.* **277,** 17901–17905.

Vodovar, N., Clayton, J. D., Costa, R., Odell, M., and Kyriacou, C. P. (2002). The *Drosophila* clock protein *timeless* is a member of the Arm/HEAT family. *Curr. Biol.* **12,** R610–R611.

Watterson, G. A. (1975). On the number of segregating sites in genetical models without recombination. *Theor. Popul. Biol.* **7,** 256–276.

Wheeler, D. A., Kyriacou, C. P., Greenacre, M. L., Yu, Q., Rutila, J. E., Rosbash, M., and Hall, J. C. (1991). Molecular transfer of a species-specific behavior from *Drosophila simulans* to *Drosophila melanogaster*. *Science* **251,** 1082–1085.

Yoshimura, T., Yokota, Y., Ishikawa, A., Yasuo, S., Hayashi, N., Suzuki, T., Okabayashi, N., Namikawa, T., and Ebihara, S. (2002). Mapping quantitative trait loci affecting circadian photosensitivity in retinally degenerate mice. *J. Biol. Rhythms* **17,** 512–519.

Yu, Q., Jacquier, A. C., Citri, Y., Hamblen, M., Hall, J. C., and Rosbash, M. (1987). Molecular mapping of point mutations in the *period* gene that stop or speed up biological clocks in *Drosophila melanogaster*. *Proc. Natl. Acad. Sci. USA* **84,** 784–788.

## [43]  Testing the Adaptive Value of Circadian Systems

*By* CARL HIRSCHIE JOHNSON

Abstract

Circadian clocks are thought to enhance reproductive fitness. However, most of the evidence that supports the adaptiveness of clocks is not rigorous and falls into the category of "adaptive storytelling." Approaches that an evolutionary biologist would consider appropriate to address this issue are described along with an analysis of the evidence—past and present—that has been evoked to demonstrate the adaptive value of circadian systems.

Background

*Have We Demonstrated that Circadian Clocks Enhance Fitness?*

We who dedicate our lives toward understanding circadian clocks obviously believe that it is an important phenomenon. For a biologist, "an important phenomenon" is usually one that enhances reproductive fitness and/or has medical/agricultural significance. The medical significance of circadian clocks is a debated subject, with some recent suggestions of a key role for circadian systems in, for example, depressive disorders or cancer. However, for most circadian biologists, the idea that circadian clock systems enhance fitness is a credo that is beyond criticism. There is a vast literature on the daily phasing of behaviors and metabolic events that are interpreted to enhance fitness, e.g., the hypothesis that an internal clock allows the anticipation of regular daily events such as dawn or dusk (Daan, 1981; Dunlap *et al.*, 2004; Horton, 2001; Sharma, 2003).

However, for the bulk of circadian-regulated events, we should consider whether a rigorous, critical evolutionary biologist would agree that we circadian biologists have demonstrated that circadian clocks enhance fitness. Until the 1970s, it was common for evolutionary biologists to interpret the phenomenon they observed according to a line of reasoning that might be paraphrased, "if a biological phenomenon is present, it must have been selected by evolution and therefore of adaptive significance." This dogma was ridiculed and largely discredited in a seminal paper by Gould and Lewontin (1979) entitled "The Spandrels of San Marco." Gould and Lewontin labeled this line of thinking an "adaptationist program" that provided "just-so" explanations of biological phenomena that are intellectually satisfying but might have little basis in the reality of the history of evolution. For example, many biological phenomena might have evolved (1) as a random trait that was neither adaptive nor nonadaptive, (2) as a trait physiologically linked to another trait where the linkage is either not currently present or not obvious, (3) as a trait that was once adaptive, but is no longer (e.g., the appendix of humans), (4) as a trait that evolved for one purpose but later was recruited to another task, and so on. To illustrate the concept, consider the case of human noses and spectacles—because noses provide such an excellent platform for mounting spectacles, it would be easy to assume in the absence of knowledge of the history of noses and spectacles that noses evolved so as to provide a place for spectacles to reside. These issues of adaptive storytelling in the context of circadian systems have been admirably discussed by DeCoursey (2004).

An example from the circadian literature of a "just-so" story that the author has personally promulgated is that of "temporal separation" of photosynthesis and nitrogen fixation in cyanobacteria. In nitrogen-fixing unicellular bacteria, nitrogen fixation is often phased to occur at night. Nitrogen ($N_2$) fixation is inhibited by low levels of oxygen, which poses a dilemma for photosynthetic bacteria because photosynthesis generates oxygen. Mitsui et al. (1986) proposed that the nocturnal phasing of nitrogen fixation was an adaptation to permit $N_2$ fixation to occur when photosynthesis was not evolving oxygen, and the author has repeated this hypothesis in several publications (Johnson et al., 1996, 1998). This hypothesis would predict that cyanobacterial growth in constant light would be slower than in a light/dark (LD) cycle because nitrogen fixation would be inhibited under these conditions and therefore the growing cells might rapidly become starved for metabolically available nitrogen. The problem is that cyanobacteria grow perfectly well in constant light—in fact, they grow faster in constant light than in LD cycles, presumably because of the extra energy they derive from the additional photosynthesis. This result is inconsistent with the "temporal separation" hypothesis. It does not mean that the "temporal separation" hypothesis is incorrect—in fact, the author believes that under appropriate (but as yet untested) conditions of medium, light, and carbon dioxide, the "temporal separation" hypothesis will emerge triumphant. Nevertheless, the point here is that "temporal separation in cyanobacteria" is an example of a "just-so" circadian story that we like to tell without its being rigorously supported by appropriate data. This was the conclusion of Gould and Lewontin (1979) for many investigations in the field of population biology, and this criticism is on target for most of the fitness arguments in the field of circadian biology.

*Definitions*

Before further verbal perambulations, two key terms should be defined. The fitness of a genotype is the average per capita lifetime contribution of individuals of that genotype to the population after one or more generations (Futuyma, 1998). Fitness is a measure of reproductive success and the passing on of genes. Fitness may be influenced by longevity, survival, growth, development, and other factors, but these ancillary factors are not direct measures of fitness. For example, consider the case of a mighty but sterile lion (an example told to the author by Michael Menaker in an introductory biology class). This lion might dominate a pride and survive to a ripe old age, but it is the wimpy but fertile lion that skulks in the shadows, surreptitiously inseminating the lionesses, who will pass his genes to the next generation. The skulking lion might die an early death (due, no doubt, to being caught *flagrante delicto* by the mighty lion), but his reproductive fitness is greater than that of the long-lived mighty lion. We will come back to this point in the context of appropriate measures of fitness.

The second key term is *adaptation*. This term is used in two different ways by evolutionary biologists. An adaptation is an aspect of the phenotype that is the product of evolution by natural selection in a particular environmental context and represents a solution to some challenge presented by the environment. In this sense, an adaptation is a feature of an organism that enhances its reproductive success relative to other possible features (Futuyma, 1998). However, the process of adaptation refers to ongoing phenotypic/genetic evolutionary change driven by natural selection in a given environmental context. So an adaptation is the result of the process of adaptation.

Strictly speaking, an adaptation can only be assumed to be adaptive when it first appears. As time goes on, the feature may persist for any of three reasons. First, the feature might still be adaptive for the original reason (selective pressure remains). Second, the selective pressure has relaxed, but in the absence of selection against the feature, it may persist passively (no longer adaptive). Third, since its original appearance, other features may have become linked to the original feature so that even if the original selective pressure is relaxed, the feature persists because so many other processes depend upon it (no longer adaptive for the original reason). Many rigorous evolutionary biologists do not accept the use of the term *adaptation* for a feature that falls into either of the latter two categories. Probably most evolutionary biologists would evaluate the adaptive significance of a trait in both the context of its phylogenetic history and the context of the environment in which the organism naturally lives.

## Approaches Used by Evolutionary Biologists

In general, evolutionary biologists would like to know about a putative adaptation (1) how it evolved and (2) whether it enhances fitness now. One approach to understanding adaptation is by understanding the history of the relevant adaptation. For example, a comparative phylogenetic approach would ask when did clocks originate and in which groups? What do those groups have in common? This approach might not be very successful in the case of circadian clocks in eukaryotes because all eukaryotic groups appear to have clocks. However, it might be very successful among prokaryotes, as cyanobacteria are the only prokaryotes with well-documented circadian systems. As other prokaryotes join the circadian fold, we can ask what selective pressure the "in groups" have in common that might have favored the evolution of circadian clocks. Moreover, even though all eukaryotic groups have circadian systems, the actual genes involved are different among fungi, plants, and animals, implying that circadian clocks are a product of convergent evolution. Therefore, phylogenetic comparisons of what is similar and what is dissimilar among these groups could yield important clues to their evolutionary history.

To test the current fitness advantage of a circadian clock, an evolutionary biologist might use a number of approaches, each of which have been applied to the analysis of clock-enhanced fitness. For example, we could manipulate phenotype and assess its effect, as in the study of DeCoursey et al. (2000; Fig. 3). We could compete genotypes, as in the study of Ouyang et al. (1998; Figs. 1 and 2), or we could compare natural variation in the environment with variation in clock properties in the organism, as in the studies of Costa et al. (1992) and Michael et al. (2003). Each of these approaches as applied to clocks are discussed in greater detail later.

## Adaptive Significance of Clocks

### Extrinsic versus Intrinsic Value of Circadian Clocks

In the particular case of circadian oscillators, we might also make a distinction between "extrinsic" and "intrinsic" value. The original adaptation of circadian clocks was presumably to enhance reproductive fitness in natural environments, which are cyclic (24 h) conditions. We can refer to this situation as an adaptation to extrinsic conditions. However, some researchers have proposed that circadian clocks may additionally provide an "intrinsic" adaptive value (Klarsfeld and Rouyer, 1998; Paranjpe et al., 2003; Pittendrigh, 1993). That is, circadian pacemakers may have evolved to become an intrinsic part of internal temporal organization and, as such, may have become intertwined with other traits that influence reproductive fitness in addition to their original role for adaptation to environmental cycles. Note that a rigorous evolutionary biologist would no longer consider an intrinsic value for clocks to be an adaptation if their original extrinsic value has been lost. However, if clocks retain extrinsic value and additionally accrue intrinsic value, then they would still be considered an adaptation. If it is true that circadian clocks have acquired intrinsic value for internal temporal programming, they would be expected to be of adaptive value to an organism in constant environments where the original selective pressure has relaxed. In support of this hypothesis, populations of *Drosophila melanogaster* raised for hundreds of generations in constant conditions retain rhythmicity and the ability to entrain to various LD cycles, indicating that even in the absence of environmental selection, the components of the circadian system are maintained (Paranjpe et al., 2003). However, it is possible that this experiment was not of sufficient duration to answer the question—a counter example is that of cave animals that frequently lose robust behavioral rhythmicity in the constant environment of caverns (Blume et al., 1962). A test of extrinsic versus intrinsic value for cyanobacteria is described later.

*Selected Examples from Pre-1980 Literature*

The adaptive significance of circadian clocks was a topic of great interest to the previous generation of circadian biologists. The literature is so vast that only a few illustrative cases in insects and plants will be discussed, namely where the researchers attempted to go beyond adaptive storytelling and devise experimental tests. The most famous example is probably that of Pittendrigh and Minis (1972). These workers tested the longevity of *Drosophila* adults maintained in either constant light or LD cycles of 21, 24, and 27 h. They reported that flies lived significantly longer on the LD cycle of 24 h, implying an optimal "resonance" of the internal clock's period with the period of the environmental cycle. This result was touted as supporting the adaptive value of having an endogenous clock whose period is similar to that of the environment. However, a "fly in the ointment" of this investigation is that Colin Pittendrigh's laboratory was unable to repeat the result after Pittendrigh moved his laboratory from Princeton University to Stanford University (T. Page, personal communication). At about the same time, Jurgen Aschoff's laboratory reported a similar approach to test adaptive significance in blowflies, which died sooner on non-24-h light/dark cycles as compared with 24 h cycles (von Saint Paul and Aschoff, 1978) or when exposed to repeated shifting of the zeitgeber (Aschoff et al., 1971). Another study on insects comparing 24-h cycles with non-24-h cycles tested the larval developmental rate of flesh flies on different combinations of photoperiods and cycle time; long photoperiods with 24-h cycle times allowed the most rapid development (Saunders, 1972).

In plants, several studies from the 1950s addressed the adaptive significance of circadian oscillators. For example, tomatoes were found to grow optimally when maintained on LD cycles that were similar to those encountered in nature; in other words, tomatoes on LD 12:12 outgrew those on LD 6:6 h or LD 24:24 (Highkin and Hanson, 1954; Hillman, 1956; Withrow and Withrow, 1949). Remarkably, tomato plants on an LD 12:12 cycle grew even faster than those in continuous light, even though the plants in constant light were receiving twice as much photonic energy (Hillman, 1956). In addition, there was an interdependence between the temperature and the optimal light cycle; at colder temperatures (when the clock might be expected to run slower), the optimal LD cycle was longer than at higher temperatures (Went, 1960). The $Q_{10}$ for the effect was about 1.2, which is within the range that would be expected for the temperature dependence of the period of a circadian oscillator. Those data indicated that tomato plants were optimally adapted to growth in light/dark cycles that were similar to those found in nature and implied that a circadian timekeeper was responsible for the adaptation.

A defect of these pre-1980 experiments was that the measures of reproductive fitness were indirect, either longevity, growth rate, developmental rate, and so on. As mentioned earlier in the definition of fitness, these parameters may influence fitness, but these ancillary factors are not direct measures of fitness. For example, if a slowly growing plant produces more seeds that successfully germinate than a rapidly growing plant, the slow grower may be more fit. Moreover, in the case of the plant studies from the 1950s, many species of plants other than tomato do not exhibit reduced growth in non-24-h LD cycles. Therefore, the results with tomatoes mentioned in the preceding paragraph were not generalizable to many other plant species. One such species is *Arabidopsis*, which is commonly used as a research material. *Arabidopsis* plants grow perfectly well in constant light; in fact they often grow faster in constant light than in LD 12:12. A more direct test of fitness in plants would measure fecundity, e.g., by measuring the number and germinating ability of the seeds. One report used that type of assay in *Arabidopsis* and found that clock-disrupted mutant plants produce fewer viable seeds than wild-type plants on LD 4:20 (Green *et al.*, 2002). However, LD 4:20 is not a photoperiod that *Arabidopsis* is likely to encounter in nature, and more reasonable LD 8:16 and LD 16:8 cycles did not show a fecundity difference between wild-type and clock-disrupted mutants. Therefore, it is difficult to assess the significance of these data.

The next generation of research into the adaptive value of circadian systems must apply more direct measures of fitness than was true of the pre-1980 studies. More recently, progress toward those goals has been made as described next.

## Tests of Adaptive Significance

### Laboratory Studies of Circadian Clocks and Reproductive Fitness

From 1975 to 1995, most circadian researchers focused on studies of circadian physiology and molecular mechanism—investigations of adaptive significance were rare during this time. Perhaps circadian biologists thought the question had been answered. Or perhaps the isolation of circadian clock mutants in *Drosophila* and *Neurospora* that exhibited dramatically different free-running periods (or even arrhythmicity) but which appeared to grow and reproduce as well as wild type in the laboratory might have discouraged such studies? More recent studies have found some mutations that affect longevity in the laboratory (Hurd and Ralph, 1998; Klarsfeld and Rouyer, 1998), but it is not clear that these are clock-specific effects ("clock genes" may affect more processes than just clocks). Since 1995, however, a number of researchers have returned to this topic, using measures that more closely addressed reproductive fitness *per se*.

The authors' laboratory tested the adaptive significance of circadian programs by using competition experiments between different strains of the cyanobacterium *Synechococcus elongatus* (Ouyang *et al.*, 1998; Woelfle *et al.*, 2004). For asexual microbes such as *S. elongatus*, differential growth of one strain under competition with other strains is a good measure of reproductive fitness. In pure culture, because the strains grew at about the same rate in constant light and in LD cycles, there did not appear to be a significant advantage or disadvantage in having different circadian periods when the strains were grown individually. The fitness test was to mix different strains together and to grow them in competition to determine whether the composition of the population changes as a function of time. The cultures were diluted at intervals to allow growth to continue.

Different period mutants were used to answer the question, "Does having a period that is similar to the period of the environmental cycle enhance fitness?'' The circadian phenotypes of the strains used had free-running periods of about 22 h (B22a, C22a) and 30 h (A30a, C28a). These strains were determined by point mutations in three different clock genes: *kaiA* (A30a), *kaiB* (B22a), and *kaiC* (C22a, C28a). Wild type has a period of about 25 h under these conditions. When each of the strains was mixed with another strain and grown together in competition, a pattern emerged that depended on the frequency of the LD cycle and the circadian period. When grown on a 22-h cycle (LD 11:11), the 22 h-period mutants could overtake wild type in the mixed cultures. On a 30-h cycle (LD 15:15), the 30 h-period mutants could defeat either wild type or the 22 h-period mutants. On a "normal" 24-h cycle (LD 12:12), the wild-type strain could overgrow either mutant (Ouyang *et al.*, 1998). Note that over many cycles, each of these LD conditions have equal amounts of light and dark (which is important, as photosynthetic cyanobacteria derive their energy from light); it is only the frequency of light versus dark that differs among the LD cycles. Figure 1 shows results from the competition between wild type and the mutant strains (Ouyang *et al.*, 1998). Clearly, the strain whose period most closely matched that of the LD cycle eliminated the competitor. Under a nonselective condition (in this case, constant light), each strain was able to maintain itself in the mixed cultures. Because the mutant strains could defeat the wild-type strain in LD cycles in which the periods are similar to their endogenous periods, the differential effects that were observed are likely to result from the differences in the circadian clock. A genetic test was also performed to demonstrate that the clock gene mutation was specifically responsible for the differential effects in the competition experiment (Ouyang *et al.*, 1998). Because the growth rate of the various cyanobacterial strains in pure culture is not detectably different, these results are most likely an example of "soft selection" where the reduced fitness of one genotype is seen only under competition (Futuyma, 1998).

In a test of the extrinsic versus intrinsic value of the clock system of cyanobacteria, wild type was competed with an apparently arrhythmic strain (CLAb). As shown in Fig. 2, the arrhythmic strain was defeated rapidly by wild type in LD 12:12, but under competition in constant light, the arrhythmic strain grew slightly better than wild type (Woelfle *et al.*, 2004). Taken together, results show that an intact clock system whose free-running period is consonant with the environment significantly enhances the reproductive fitness of cyanobacteria in rhythmic environments; how-ever, this same clock system provides no adaptive advantage in constant environments and may even be slightly detrimental to this organism. Therefore, the clock system does not appear to confer an intrinsic value for cyanobacteria in constant conditions.

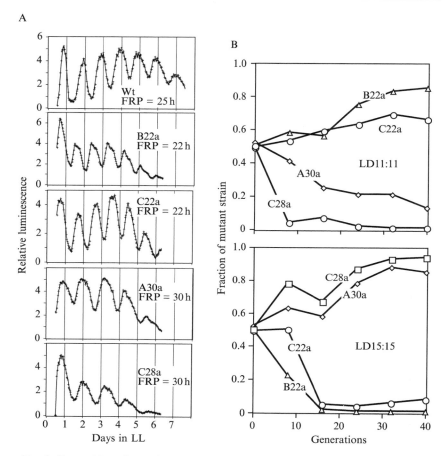

FIG. 1. Competition of cyanobacterial cultures in LD 11:11 and LD 15:15 cycles. (A) The circadian phenotypes of wild type (AMC343, FRP ~25 h) and mutants [mutations in *kaiB* (B22a, FRP ~22 h), *kaiA* (A30a, FRP ~30 h), and *kaiC* (C22a, FRP ~22 h; C28a, FRP ~30 h)]. (B) Kinetics of competition between wild-type and mutant strains in mixed cultures exposed to LD 11:11 (*upper*) or LD 15:15 (*lower*). Data are plotted as the fraction of the mutant strain in the mixed culture versus the estimated number of generations.

Attempts to apply measures of reproductive fitness to clock function in other organisms in the laboratory have been partially successful, especially in the fruit fly *Drosophila*. One investigation studied male fecundity as sperm production in singly mated flies and found that clock mutations that disrupt circadian rhythmicity also decrease sperm production (Beaver *et al.*, 2002). Those authors speculate that the significant declines in fertility

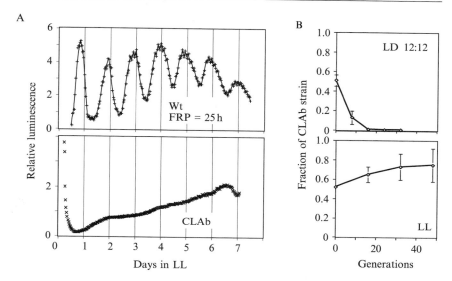

FIG. 2. Competition of a clock-disrupted strain of cyanobacteria with wild type. (A) The circadian phenotypes of luminescence emission from wild type and the clock-disrupted strain CLAb. (B) Competition between wild type and CLAb in LD12:12 (*top*) and LL (*bottom*) plotted as the fraction of mutant in the mixed culture versus the estimated number of generations (mean ± SD).

observed with singly mated flies are not readily detected in the usual laboratory populations where males and females are housed together for several days and therefore have the opportunity to mate multiple times (Beaver *et al.*, 2002). However, while clock mutations also affect oogenesis in *Drosophila* females, this effect appears to be pleiotropic and does not involve the circadian clock (Beaver *et al.*, 2003). Another group that studied the "trade-off" between fecundity and longevity in *Drosophila* reported the unexpected result that lifetime egg production was higher in LL (when the flies are arhythmic) than in LD (Sheeba *et al.*, 2000).

A different tack toward the issue of the adaptive significance of circadian systems is the question of whether these clocks could be involved in a reproductive isolation that could drive speciation and/or prevent interspecific hybridization. This "temporal mating isolation" was addressed in a study in *Drosophila* (Tauber *et al.*, 2003). The authors found that species-specific timing of mating behavior significantly affects mate choice. This phenomenon could help prevent interspecies mating and thereby decrease the overall fitness of any particular species; moreover, it could play (or has played in the past) a role in providing permissive conditions for speciation (Tauber *et al.*, 2003).

A reasonable criticism of laboratory studies in general is that these settings are artificial and cannot mimic the selective pressures found in nature. This criticism is likely to be more potent for organisms that live in complex environments and/or exhibit complex behaviors. This may partially explain why the competition experiment appears to have been successful for the relatively "simple" cyanobacteria (Ouyang *et al.*, 1998; Woelfle *et al.*, 2004), but not for *Drosophila* (Klarsfeld and Rouyer, 1998). Clearly, studies in natural settings are most appropriate to answer questions of adaptive significance. Such studies are discussed in the next section.

### Studies of Organisms from Natural Environments

As mentioned previously, there is a large body of literature on the daily phasing of behaviors and metabolic events in nature that can be interpreted to enhance fitness (Daan, 1981; Dunlap *et al.*, 2004; Horton, 2001; Sharma, 2003). This section focuses on that subset of those studies that are experimental and/or that use new molecular tools to address the issue of adaptive significance of circadian clocks in populations in nature.

One way to assess the adaptive value of circadian clocks would be to search for evidence of natural selection acting upon circadian parameters in nature. If clocks have adaptive value, then natural selection should be acting upon their properties. One type of evidence could be variation in circadian rhythm properties where selective strength varies. For example, where natural selection may be presumed to have relaxed, we would predict that the robustness of circadian expression will decline. In support of this prediction, the expression of circadian rhythms has often been found to be absent or reduced in animals that have evolved for a long time in cave environments in which the conditions are no longer cyclic (Blume *et al.*, 1962; Koilraj *et al.*, 2000).

Similarly, over a gradient of a relevant environmental condition—what we might call a selective gradient—we might expect to observe gradation of a responding clock property. An excellent example of this type of environmental condition is the latitudinal gradation in annual day length and temperature. Day length and temperature are both highly relevant to daily clocks and associated property of photoperiodism. In support of the prediction that these gradients influence clock properties, there is a positive correlation between the circadian period and the latitude from which samples of the plant *Arabidopsis* have been isolated from nature (Michael *et al.*, 2003). Another latitudinal cline of interest is that of polymorphism in the *Drosophila period* gene. There are differing lengths of the threonine–glycine-encoding repeat region of the *period* gene that vary over the latitudes of Europe (Costa *et al.*, 1992). This polymorphism may be related to the temperature compensation of the *Drosophila* clock (Sawyer *et al.*, 1997), and therefore a result of a selective pressure of temperature ranges that vary with latitude. Another study found polymorphism in the *kaiABC* clock genes of cyanobacteria over a restricted geographical continuum of divergent environmental conditions (Dvornyk *et al.*, 2002). These observations support the conclusion that natural selection is still operating upon "clock genes" in natural populations, and therefore that circadian clocks continue to confer a fitness advantage (however, this statement makes the crucial and almost certainly false assumption that "clock genes" only act in clock-related processes).

Another type of data that suggests the continuing action of natural selection comes from analyses of quantitative trait loci (QTL). In the plant *Arabidopsis*, two studies have used recombinant inbred lines (RILs) of different isolates from nature (from northern Europe versus the Cape Verde Islands). These studies found a much larger variation in circadian period, phase angle, and amplitude among crosses between different isolates (RILs) than was found in the parental lines (Michael *et al.*, 2003; Swarup *et al.*, 1999). Therefore, the hybridizations unmasked a large amount of genetic variation that was not obvious in the parental lines. This result implies that natural selection had favored combinations of alleles that counterbalance each other's effect on clock properties so as to attain similar emergent phenotypes. It was only when disparate lines were crossed so as to create RILs that contained allele combinations that had not coevolved was the underlying genetic variation uncovered. Again, these QTL results indicate that natural selection is still acting upon genetic variation in *Arabidopsis* to optimize allelic combinations, implying that the clock system itself remains of adaptive value (Michael *et al.*, 2003).

In a heroic effort, DeCoursey *et al.* (2000) combined an experimental approach with studies of day-active chipmunks in nature. DeCoursey removed chipmunks from the wild and lesioned the site of the master circadian clock [the suprachiasmatic nuclei (SCN)] in the brain of some of the animals (= SCN-X). Lesioning of the SCN results in a highly disrupted and often arhythmic activity pattern for animals monitored in the laboratory. A control group was sham operated (= sham). After these operations, the chipmunks were repatriated to the study site. A second control group was animals that were neither operated upon nor removed from their burrows (= control). Animals were fitted with radio collars and monitored individually for activity and survival. Figure 3A depicts the overall results. In the first 10–15 days after repatriation, there were significant losses to predation of chipmunks in the SCN-X and sham groups, and

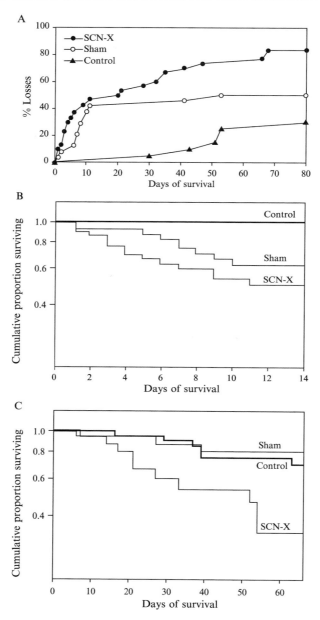

FIG. 3. Mortality of chipmunks during the 80 days following operations and/or collaring and repatriation. (A) Mortality relative to treatment group. (B) Survival curves for the three treatment groups for days 1–14. (C) Survival curves for days 15–80. SCN-X, suprachiasmatic nuclei lesioned; Sham, sham operated; Control, unoperated and undisrupted (DeCoursey *et al.*, 2000; reprinted here courtesy of Pat DeCoursey).

these data are shown in greater detail in Fig. 3B. This effect is almost certainly due to difficulties that the operated animals experienced in resettling to a burrow. Subsequent to day 15, however, the sham group and the control group stabilized and exhibited minor predation (Fig. 3C). However, the SCN-X group continued to experience consistent predation after day 15 (Fig. 3C).

Although predation was never directly observed, several lines of evidence suggested that the major predators of the chipmunks were weasels (DeCoursey et al., 2000). The major behavioral difference between SCN-X chipmunks and sham/control groups observed in this study was an increase of the nighttime activity of the animals in their dens (DeCoursey et al., 2000). DeCoursey et al. (2000) concluded that this nocturnal restlessness allowed the weasel predator to detect the chipmunks (by vibration or other cues) and descend into the burrows to capture and kill the prey. For circadian biologists, these results were both a delight and a surprise. The delight was that the results confirmed the importance of the SCN and, by implication, the clock (perhaps the SCN has important nonclock functions, however?). The surprise was that predation seemed to cue on nocturnal restlessness rather than the phasing of surface activity. One might imagine that a simple system could adequately prevent nocturnal restlessness without the necessity to elaborate a circadian clock with entrainment capability. Is it really true that the function of the clock regulating chipmunk activity is merely to suppress nighttime activity? If so, does this mean that the oft-hypothesized purpose of the circadian clock to anticipate daily environmental events is another "just-so" story? Of course, natural selection builds on preexisting characteristics, and it is possible that natural selection enlisted a circadian mechanism that originally evolved to satisfy more stringent specifications in order to accomplish a plebian task in chipmunks—to suppress nocturnal activity.

The DeCoursey study was inspired and heroic, but it should also be noted that it measured survival, not reproductive fitness per se. Would the value of a complex circadian system be vindicated if a similar study could be undertaken for more years in which reproductive capability was also measured? Such an endeavor would be Herculean, but to be consistent with the major thesis of this article, it would be necessary to ultimately test the adaptive value of the clock in chipmunks.

Natural Selection and the Evolution of Clocks: "Escape from Light?"

The question of whether circadian clocks are adaptive is linked with identifying the selective forces that encouraged the original evolution of these timers. In view of the presence of circadian oscillators in cyanobacteria, whose ancestors appear in the fossil record at least 3.5 billion years ago, we might infer that circadian clocks were an ancient invention of evolution (although it remains possible that circadian timers may have been absent in ancient cyanobacteria and evolved relatively recently). Perhaps a strong initial driving force for the early evolution of circadian clocks could have been the advantage inherent in phasing cellular events that are sensitive to the daily bombardment of sunlight so that they occur in the night. This idea has been called the "escape from light" hypothesis (Pittendrigh, 1965, 1993). Cells exhibit many light-sensitive processes. For example, there are many pigments in cells that probably do not act as sensory "photo-receptors" but nevertheless absorb light because of obligatory associated cofactors (e.g., cytochromes) and whose activity is thereby modulated by light and dark. For most photosynthetic organisms, light is the sole provid-er of energy, and they must suffer the damaging side effects of the sunlight to obtain their sustenance. Moreover, DNA can be mutated by exposure to ultraviolet (UV) light, and the genome may be more sensitive to UV irradiation at some phases of the cell division cycle (e.g., during S phase when DNA is partially unwrapped from histones to allow replication). In fact, there are numerous examples of microorganisms with 24-h cell divi-sion cycles in which DNA replication and cell division occur during the night (Edmunds, 1988). Therefore, cellular metabolism will be sensitive to sunlight, especially to the high-energy blue and UV light.

If this hypothesis about the early evolution of circadian programs is correct, then present-day organisms might retain a restriction of light-sensitive processes to the night. Such a test comes from the eukaryotic alga *Chlamydomonas* (Nikaido and Johnson, 2000). As shown in Fig. 4, these algae are more sensitive to UV light near sunset and into the early night. The rhythmic sensitivity persists in constant conditions, albeit with a re-duced amplitude. Even though there is some sensitivity in the late daytime, UV light is strongly scattered at twilight and is therefore nearly absent from sunlight around the time of sunset, so the UV sensitivity of *Chlamy-domonas* near dusk would not be expected to pose a significant problem in the natural environment. In *Chlamydomonas*, the circadian clock regulates the timing of the cell division cycle (Goto and Johnson, 1995), and the UV-sensitive phases (Fig. 4) correspond with the times in which S/G2 would be expected to occur. These data are consistent with the "escape from light" hypothesis that the daily cycle of UV radiation may have created a selective pressure favoring the evolution of circadian clocks (Pittendrigh, 1993).

The hypothesis dovetails with the discovery of a role for cryptochromes in circadian systems. Cryptochromes are pigmented photoreceptors involved

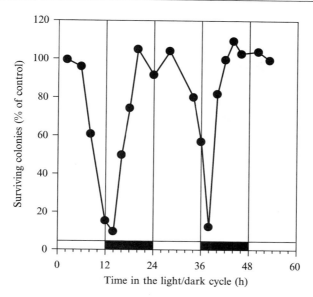

FIG. 4. Survival of *Chlamydomonas* cells after irradiation by UV light as a function of time in an LD cycle. *Chlamydomonas* cultures were plated onto agar medium and treated with equal amounts of UV light at different phases of a 12-h light:12-h dark cycle. Survival was measured as the colony-forming ability of cells following treatment as compared with that of cells not irradiated with UV light (modified from Nikaido and Johnson, 2000).

in blue-light-mediated entrainment and photoperiodism. Cryptochromes share sequence homology to another blue-light-activated protein from which they probably evolved: DNA photolyase, which uses blue-light energy to repair UV-induced damage of DNA. Possibly, an ancestral photolyase that repaired DNA damage inflicted by the daily cycle of UV light may have been enlisted for duties in biological timing mechanisms and evolved into cryptochromes (Gehring and Rosbash, 2003; Nikaido and Johnson, 2000). Indeed, based on the "escape from light" hypothesis, a clock-related role for a DNA photolyase-type enzyme was anticipated and predicted by Pittendrigh in 1965.

## For Future Studies

It is difficult to imagine that circadian clocks are not adaptive. The new generation of studies on the adaptive significance of clock systems encourages the view that circadian clocks were adaptive in the past and retain their value to the present day. Nevertheless, from the perspective of a rigorous evolutionary biologist, we have not yet conclusively demonstrated that some common manifestations of the circadian system (e.g., anticipation, locomotor activity, temporal separation, leaf movements) have adaptive value.

To counter this criticism, we must embrace the methods and criteria of evolutionary biologists. In particular, for experimental studies the end points we measure must be reproductive fitness, not longevity, survival, growth, etc. In the case of plants, these measures could include fecundity as measured by (1) seed production and germinating ability as in Green *et al.* (2002) and/or (2) pollen production or pollination success. In the case of animals, these measures can include reproductive capacity, e.g., sperm or egg production as in Beaver *et al.* (2002) and Sheeba *et al.* (2000). Whenever possible, reproductive success measured by the generation of viable and fertile progeny is the ultimate measure. When we use mutants, it is not enough to show that a clock mutant decreases fitness—this might not be mediated by a clock effect, but via a pleiotropic effect (Beaver *et al.*, 2003). To demonstrate a clock-specific effect, mutants should be used whenever possible in ways that allow them to outcompete wild-type organisms under rhythm-specific conditions that are optimal for the mutants (as in Figs. 1 and 2; Ouyang *et al.*, 1998; Woelfle *et al.*, 2004). Another experimental approach is to use artificial selection under new rhythm-specific environments to determine whether clock systems "evolve." Examples of this approach in *Drosophila* include the selection of strains with different clock phase angles (Pittendrigh, 1967) or exposure for many generations to constant conditions (Paranjpe *et al.*, 2003). Such an evolutionary approach could identify selective pressures to which circadian systems can respond.

For studies in nature, measurements of genetic variation are a favorite assay of evolutionary biologists; for example, the comparison of genetic variation among various ecotypes can be assayed by QTL analyses (as in Michael *et al.*, 2003; Swarup *et al.*, 1999). Assays of polymorphisms in clock properties within the *same* species are particularly valuable if made among populations isolated from different environments in nature where the putative selective pressures on the circadian clock are different, such as latitudinal clines (affecting temperature and photoperiod, as in Costa *et al.*, 1992; Michael *et al.*, 2003), altitudinal clines (affecting temperature but not photoperiod), caves (Blume *et al.*, 1962; Koilraj *et al.*, 2000), and so on. An especially exciting approach enabled by new tools would be to release genetically traceable strains with different clock genotypes/phenotypes into the wild and assess their relative reproductive success in nature. With new batteries of genetic markers available, we can now sample a population and determine who were the most successful parents (Avise, 1994). In the case of the mighty/sterile versus wimpy/fertile lions, we could check the cubs in the pride and confirm who was the most successful father. Do circadian programs enhance fitness? Which properties? Under which conditions? Let us find out!

## Acknowledgments

I am grateful to Dave McCauley for his sage advice and assistance with this paper and our studies of fitness in cyanobacteria. I thank Pat DeCoursey for permission to reprint Fig. 3 and Terry Page for his ever insightful comments on this and many other manuscripts. I also apologize to the many scientists whose important contributions to these issues were not included in this review; the literature on this topic is vast and I could not include all the relevant studies.

## References

Aschoff, J., von Saint Paul, U., and Wever, R. (1971). Die Lebensdauer von Fliegen unter dem Einfluss von Zeit-Verschiebungen. *Naturwissenschaffen* **58**, 574.

Avise, J. C. (1994). "Molecular Markers, Natural History and Evolution." Chapman & Hall, New York.

Beaver, L. M., Gvakharia, B. O., Vollintine, T. S., Hege, D. M., Stanewsky, R., and Giebultowicz, J. M. (2002). Loss of circadian clock function decreases reproductive fitness in males of *Drosophila melanogaster*. *Proc. Natl. Acad. Sci. USA* **99**, 2134–2139.

Beaver, L. M., Rush, B. L., Gvakharia, B. O., and Giebultowicz, J. M. (2003). Noncircadian regulation and function of clock genes *period* and *timeless* in oogenesis of *Drosophila melanogaster*. *J. Biol. Rhythms* **18**, 463–472.

Blume, J., Bünning, E., and Gunzler, E. (1962). Zur Aktivitätsperiodik bei Höhlentieren. *Naturwissenschaften* **49**, 525.

Costa, R., Peixoto, A. A., Barbujani, G., and Kyriacou, C. P. (1992). A latitudinal cline in a *Drosophila* clock gene. *Proc. R. Soc. Lond. B.* **250**, 43–49.

Daan, S. (1981). Adaptive daily strategies in behavior. *In* "Handbook of Behavioral Neurobiology; Biological Rhythms," (J. Aschoff, ed.), Vol. 4, pp. 275–298. Plenum Press, New York.

DeCoursey, P. J. (2004). Chapter 2 in "Chronobiology: Biological Timekeeping" (J. C. Dunlap, J. J. Lords, and P. J. DeCoursey, eds.), pp. 27–65. Sinauer, Sunderland, MA.

DeCoursey, P. J., Walker, J. K., and Smith, S. A. (2000). A circadian pacemaker in free-living chipmunks: Essential for survival? *J. Comp. Physiol. A.* **186**, 169–180.

Dunlap, J. C., Loros, J. J., and DeCoursey, P. J. (2004). *In* "Chronobiology: Biological Timekeeping," 406 pages. Sinauer, Sunderland, MA.

Dvornyk, V., Vinogradova, O., and Nevo, E. (2002). Long-term microclimatic stress causes rapid adaptive radiation of *kaiABC* clock gene family in a cyanobacterium, *Nostoc linckia*, from "Evolution Canyons" I and II, Israel. *Proc. Natl. Acad. Sci. USA* **99**, 2082–2087.

Edmunds, L. N. (1984). Circadian oscillators and cell cycle controls in algae. *In* "The Microbial Cell Cycle" (P. Nurse and E. Streiblová, eds.), pp. 209–230. CRC Press, Boca Raton, FL.

Futuyma, D. J. (1998). "Evolutionary Biology," 3rd Ed., Sinauer, Sunderland, MA.

Gehring, W., and Rosbash, M. (2003). The coevolution of blue-light photoreception and circadian rhythms. *J. mol. Evol.* **57**, S286–S289.

Goto, K., and Johnson, C. H. (1995). Is the cell division cycle gated by a circadian clock? The case of *Chlamydomonas reinhardtii*. *J. Cell Bio.* **129,** 1061–1069.

Gould, S. J., and Lewontin, R. C. (1979). The spandrels of San Marco and the Panglossian paradigm: A critique of the adaptationist programme. *Proc. R. Soc. Lond. B.* **205,** 581–598.

Green, R. M., Tingay, S., Wang, Z.-Y., and Tobin, E. M. (2002). Circadian rhythms confer a higher level of fitness to *Arabidopsis* plants. *Plant Physiol.* **129,** 576–584.

Highkin, H. R., and Hanson, J. B. (1954). Possible interaction between light–dark cycles and endogenous daily rhythms on the growth of tomato plants. *Plant Physiol.* **29,** 301–302.

Hillman, W. S. (1956). Injury of tomato plants by continuous light and unfavorable photoperiodic cycles. *Am. J. Bot.* **43,** 89–96.

Horton, T. H. (2001). Conceptual issues in the ecology and evolution of circadian rhythms. *In* "Handbook of Behavioral Neurobiology; Circadian Clocks," (J. S. Takahashi, F. W. Turek, and R. Y. Moore, eds.), Vol. 12, pp. 45–57. Plenum Press, New York.

Hurd, M. W., and Ralph, M. R. (1998). The significance of circadian organization for longevity in the golden hamster. *J. Biol. Rhythms* **13,** 430–436.

Johnson, C. H., Golden, S. S., Ishiura, M., and Kondo, T. (1996). Circadian clocks in prokaryotes. *Mol. Microbiol.* **21,** 5–11.

Johnson, C. H., Golden, S. S., and Kondo, T. (1998). Adaptive significance of circadian programs in cyanobacteria. *Trends Microbiol.* **6,** 407–410.

Klarsfeld, A., and Rouyer, F. (1998). Effects of circadian mutations and LD periodicity on the life span of *Drosophila melanogaster*. *J. Biol. Rhythms* **13,** 471–478.

Koilraj, A. J., Sharma, V. K., Marimuthu, G., and Chandrashekaran, M. K. (2000). Presence of circadian rhythms in the locomotor activity of a cave-dwelling millipede. *Glyphiulus cavernicolus sulu Chronobiol. Int.* **17,** 757–765.

Michael, T. P., Salomé, P. A., Yu, H. J., Spencer, T. R., Sharp, E. L., McPeek, M. A., Alonso, J. M., Ecker, J. R., and McClung, C. R. (2003). Enhanced fitness conferred by naturally occurring variation in the circadian clock. *Science* **302,** 1049–1053.

Mitsui, A., Kumazawa, S., Takahashi, A., Ikemoto, H., and Arai, T. (1986). Strategy by which nitrogen-fixing unicellular cyanobacteria grow photoautotrophically. *Nature* **323,** 720–722.

Nikaido, S. S., and Johnson, C. H. (2000). Daily and circadian variation in survival from ultraviolet radiation in *Chlamydomonas reinhardtii*. *Photochem. Photobiol.* **71,** 758–765.

Ouyang, Y., Andersson, C. R., Kondo, T., Golden, S. S., and Johnson, C. H. (1998). Resonating circadian clocks enhance fitness in cyanobacteria. *Proc. Natl. Acad. Sci. USA* **95,** 8660–8664.

Paranjpe, D. A., Anitha, D., Kumar, S., Kumar, D., Verkhedkar, K., Chandrashekaran, M. K., Joshi, A., and Sharma, V. K. (2003). Entrainment of eclosion rhythm in *Drosophila melanogaster* populations reared for more than 700 generations in constant light environment. *Chronobiol. Int.* **20,** 977–987.

Pittendrigh, C. S. (1965). Biological clocks: The functions, ancient and modern, of circadian oscillations. *In* "Science and the Sixties. Proc. Cloudcraft Symp.," pp. 96–111. Air Force Office of Scientific Research.

Pittendrigh, C. S. (1967). Circadian systems. I. The driving oscillation and its assay in *Drosophila pseudoobscura*. *Proc. Natl. Acad. Sci. USA* **58,** 1762–1767.

Pittendrigh, C. S. (1993). Temporal organization: Reflections of a Darwinian clock-watcher. *Annu. Rev. Physiol.* **55,** 16–54.

Pittendrigh, C. S., and Minis, D. H. (1972). Circadian systems: Longevity as a function of circadian resonance in *Drosophila melanogaster*. *Proc. Natl. Acad. Sci. USA* **69,** 1537–1539.

Saunders, D. S. (1972). Circadian control of larval growth rate in *Sarcophaga argyrostoma*. *Proc. Natl. Acad. Sci. USA* **69,** 2738–2740.

Sawyer, L. A., Hennessy, J. M., Peixoto, A. A., Rosato, E., Parkinson, H., Costa, R., and Kyriacou, C. P. (1997). Natural variation in a *Drosophila* clock gene and temperature compensation. *Science* **278,** 2117–2120.

Sharma, V. K. (2003). Adaptive significance of circadian clocks. *Chronobiol. Int.* **20,** 901–919.

Sheeba, V., Sharma, V. K., Shubha, K., Chandrashekaran, M. K., and Joshi, A. (2000). The effect of different light regimes on adult life span in *Drosophila melanogaster* is partly mediated through reproductive output. *J. Biol. Rhythms* **15,** 380–392.

Swarup, K., Alonso-Blanco, C., Lynn, J. R., Michaels, S. D., Amasino, R. M., Koornneef, M., and Millar, A. J. (1999). Natural allelic variation identifies new genes in the *Arabidopsis* circadian system. *Plant J.* **20,** 67–77.

Tauber, E., Roe, H., Costa, R., Hennessy, J. M., and Kyriacou, C. P. (2003). Temporal mating isolation driven by a behavioral gene in *Drosophila*. *Curr. Biol.* **13,** 140–145.

von Saint Paul, U., and Aschoff, J. (1978). Longevity among blowflies *Phormia terraenovae* R.D. kept in non-24-hour light-dark cycles. *J. Comp. Physiol.* **127,** 191–195.

Went, F. W. (1960). Photo- and thermoperiodic effects in plant growth. *In* "Cold Spring Harbor Symposia on Quantitative Biology, Biological Clocks," Vol. 25, pp. 221–230. Cold Spring Harbor Press, Cold Spring Harbor, NY.

Withrow, A. P., and Withrow, R. B. (1949). Photoperiodic chlorosis in tomato. *Plant Physiol.* **24,** 657–663.

Woelfle, M. A., Ouyang, Y., Phanvijhitsiri, K., and Johnson, C. H. (2004). The adaptive value of circadian clocks: An experimental assessment in cyanobacteria. *Curr. Biol.* **14,** 1481–1486.

# Section XII

# Circadian Clocks Affecting Noncircadian Biology

# [44]    A "Bottom-Counting" Video System for Measuring Cocaine-Induced Behaviors in Drosophila

*By* REBECCA GEORGE, KEVIN LEASE, JAMES BURNETTE, and JAY HIRSH

## Abstract

Cocaine exposure elicits a set of stereotypic behaviors in *Drosophila* that are strikingly similar to the cocaine-induced behaviors observed in vertebrates. This provides a valuable model for the study of cocaine abuse and has led to the discovery of a connection between the cocaine response pathway and the circadian system. This article describes a simplified assessment of cocaine-induced behavior combined with an image acquisition system, which allows the assay to be semiautomated. With this new system, cocaine response can be evaluated in a fraction of the time required by the previous assay, and subjectivity in scoring is reduced dramatically.

## Introduction

*Drosophila melanogaster* has become established as a valuable model in which to study various pharmacological agents, including drugs of abuse. Studies using the fruit fly as a model for cocaine abuse revealed an unexpected connection between cocaine response and the circadian gene pathways, demonstrating that many circadian gene products are required for normal responses to cocaine (Andretic *et al.*, 1999).

In *Drosophila*, cocaine exposure elicits a set of stereotypic behaviors that are strikingly similar to the cocaine-induced behaviors observed in vertebrates. When exposed to cocaine, flies respond with behaviors that vary as a function of dose, ranging from increased grooming and locomotor circling to severe whole body tremors and death (McClung and Hirsh, 1998). Further studies show that the molecules and mechanisms underlying the cocaine response are remarkably conserved as well, including the neurotransmitters dopamine and serotonin (Bainton *et al.*, 2000; Li *et al.*, 2000), the protein kinase A (Park *et al.*, 2000) and the G-protein-signaling pathways (Li *et al.*, 2000). *Drosophila* also develop sensitization (McClung and Hirsh, 1998), an enhanced motor responsiveness to repeated doses of cocaine (Epstein and Altshuler, 1978; Kalivas *et al.*, 1998; Shuster *et al.*, 1977; Vanderschuren and Kalivas, 2000). In vertebrate studies, sensitization appears to be correlated with persistent and heightened craving for drugs of abuse, and thus it is thought to be involved in the addictive process in humans (Kalivas *et al.*, 1998; Robinson *et al.*, 1993).

Beyond equipping researchers to analyze the pathways suspected to be involved in cocaine response, *Drosophila* also provides a powerful investigative tool for uncovering new cocaine-response pathways, such as those involving circadian gene products (Andretic and Hirsh, 2000; Andretic *et al.*, 1999) and the trace amine tyramine (McClung and Hirsh, 1999). The involvement of the circadian system and of tyramine was unexpected from previous vertebrate studies of cocaine, but circadian gene participation has now been confirmed in mice. Flies with mutations in the circadian genes *period, clock, cycle*, and *doubletime* show ablated or reduced sensitization to repeated cocaine exposures (Andretic *et al.*, 1999). Abarca *et al.* (2002) showed that mice with a knockout of one of the three mouse *period* genes, *mper1*, were defective in sensitization to repeated cocaine exposures. These mice were also deficient in a conditioned place preference assay that measures the rewarding properties of cocaine. Conversely, *mper2* knockout mice are somewhat enhanced in sensitization and conditioned place preference (Abarca *et al.*, 2002). Because *mper2* has a more important role in circadian rhythmicity than *mper1* (Albrecht *et al.*, 2001; Bae *et al.*, 2001; Zheng *et al.*, 1999), these results support the idea that there may be noncircadian roles for circadian genes.

Because these results indicate that *Drosophila* can be used to discover novel cocaine-response pathways with relevance to vertebrates, it is advantageous to make the assessment of cocaine-induced behaviors in flies as efficient as possible. This article describes a simplified method for measuring cocaine-induced behaviors in flies that facilitates both forward and reverse genetic studies.

Behavioral Scoring

Previous studies of *Drosophila* cocaine responses from our laboratory used a video assessment of behaviors (Hirsh, 2000; McClung and Hirsh, 1998). Flies were exposed to volatilized free-base cocaine (NIDA Drug Supply Program, NIH) boiled off a hot nichrome wire and then transferred to a ~2.5 × 4-cm by 4-mm-high glass viewing chamber that was observed from a macro-video camera mounted above. Cocaine-induced behaviors were assessed using a behavioral scoring system, as shown in Table I [adapted from (Hirsh and McClung, 1998)]. Sensitization, an enhanced responsiveness to repeated doses of cocaine, was measured in this system as an enhanced fraction of flies showing more severe behaviors of 5 and above with repeated exposures.

Although this system is effective for rating the severity of the effect of a cocaine exposure, behavior scoring is time-consuming and subject to

TABLE I
BEHAVIORAL SCORING

| | |
|---|---|
| Capable of standing/walking | 0. Normal behavior |
| | 1. Intense nearly continuous grooming |
| | 2. Stereotyped locomotion, extended proboscis; some locomotion with simultaneous grooming |
| | 3. Slow stereotypic locomotion in a circular pattern, extended proboscis |
| | 4. Rapid twirling, sideways or backwards locomotion |
| Incapable of standing/walking | 5. Hyperkinetic behaviors, including bouts of rapid rotation, wing buzzing, erratic activity with flies often bouncing off the wall of the container |
| | 6. Severe whole body tremor, no locomotion, usually overturned with legs contracted to body |
| | 7. Total akinesia or dead |

examiner variation. Scoring a group of flies requires watching each 5-min recording repeatedly and following each fly individually. With higher numbers of affected flies, scoring becomes increasingly difficult. Flies showing hyperkinetic behaviors often bounce off the walls of the viewing chamber, but they also careen into other flies, making it difficult to distinguish between flies affected by the cocaine and flies that have been knocked off their feet by the erratically behaving flies.

## Bottom-Counting Assay

In our new cocaine behavioral assay, we take advantage of the normal negative geotaxis and positive phototaxis of flies by allowing the flies that are unimpaired or weakly affected by cocaine to remove themselves from the field of view. This assay is similar in principle to that used by Bainton *et al.* (2000), who gauged the severity of cocaine responses by counting the number of flies present in within 1 cm of the bottom of a 1-foot tube after stimulating the flies by manually tapping the vial; however, it is considerably different in practice and more amenable to semiautomation. Our modified recording apparatus consists of a glass platform over an upward-pointing video camera (Fig. 1). When a glass vial is positioned on the platform over the camera and the camera is focused on the floor of the vial, the camera records an image of only flies on or near the bottom of the vial. The camera has a very narrow focal plane such that as soon as the flies crawl up the side of the vial they are enough out of focus to disappear (Fig. 2A–C). A paper strip around the bottom two-thirds of the vial darkens the lower portion of the vial; this strengthens the tendency

Fig. 1. Camera and platform Setup. (A) Side view. (B) Top view. The 15-cm$^2$ platform is constructed of 1/4-in. lucite. Six 1-in.-diameter holes were cut in a hexagonal pattern, and shortened microscope slides were cemented with epoxy to the bottoms of these holes to form transparent mounts for the fly vials. The fly vials were further steadied using 1-in. i.d. lucite tubes glued to the platform. A 3/4-in. o.d. B/W 1/3-in. CCD bullet camera (CCTV Imports, Covington, LA, Model #3747; URL: http://www.spycameras4less.com) is placed in a ring stand mount focusing on the bottom of the vials. The camera focus is adjusted by unscrewing the protective lens cover and unscrewing the focusing lens in front the CCD chip to a focus of ~1 cm. The analog signal from this camera is digitized using a Canopus (San Jose, CA, Model ADVC-100) analog-to-digital firewire converter. The platform is mounted to a stepper motor (MCG, Inc, part #IS23005) with screws into a gear affixed to the shaft.

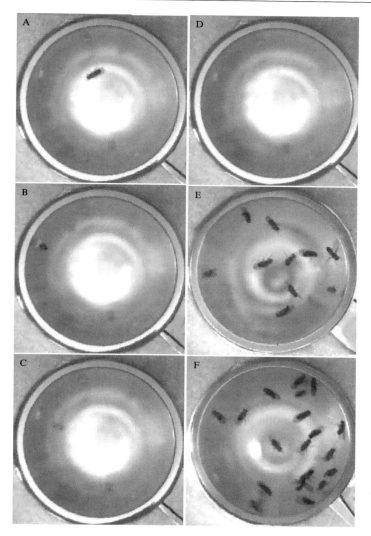

FIG. 2. The camera's view of the bottom surface of the vial. All vials contain 20 flies. (A–C) The progress of a fly in a control vial at 10-s intervals as it leaves the bottom surface of the vial and proceeds up the side away from the camera. (D) Control, no cocaine. (E) Low dose of cocaine. (F) High dose of cocaine.

of the flies to move toward the brighter upper portion. In a glass vial containing wild-type flies, the view recorded by the camera will be empty for the majority of the time because all the flies will be at the top of the vial, far enough out of focus to be invisible (Fig. 2D).

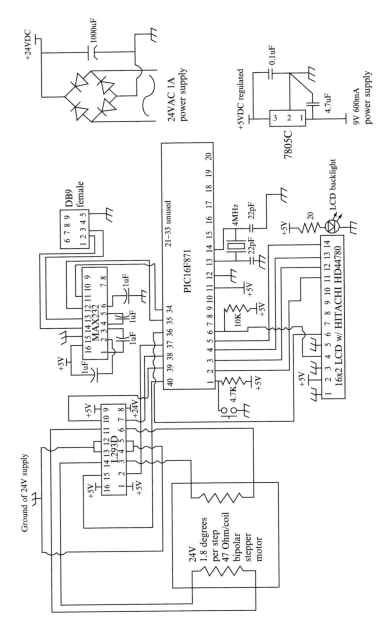

FIG. 3. Circuit diagram of the stepper motor controller. The microcontroller is a PIC16F871 chip, programmed using the USB flash microchip PIC flash programmer kit 149 (Carl's Electronics, Sterling, MA; URL: http://www.electronickits.com/). The program was written in the PicBasic programming language, compiled to assembly language using the PicBasic Pro Compiler (MeLabs, Inc.). The image-acquisition signal is sent to a Power G4 Mac desktop computer through a MAX232 integrated circuit, which converts the TTL

In a vial of flies exposed to the smoke of free-base cocaine, the flies generally show a range of responses to cocaine. Unaffected and weakly affected flies continue to demonstrate normal phototaxis and negative geotaxis, and these flies congregate in the top one-third of the vial. Flies demonstrating mild cocaine behaviors 1, 2, and 3 are generally also in the top one-third of the vial. Flies at behavior level 4 are starting to show more severe effects of cocaine and can be found in the top or the bottom of the vial. By definition, flies demonstrating the most severe behaviors (5, 6, and 7) are unable to stand upright and therefore fall down to the floor of the vial (Fig. 2E and F). With this new recording arrangement, severely affected flies can now simply be counted as the number of flies that are visible on the floor of the vial. Generally, this number will be equal to the number of flies showing behaviors 5, 6, and 7, plus some flies demonstrating behavior 4.

To track the onset and progression of cocaine behavior, we record video from flies in the same vial in which they were exposed to cocaine. We begin recording data after 1 min of cocaine exposure, counting the number of flies present on the bottom of the vial at 1-min intervals for at least 5 min. Occasionally, flies fall from the top of the vial, and if they are affected only weakly by cocaine, they usually right themselves and climb back up the walls again immediately. Because counts of the flies on the bottom are taken at 1-min intervals, the occasional unaffected fly that may be counted among the flies on the bottom of the vial will be averaged out over the 5-min recording time.

## Automated Carousel and Image Capture

To further increase the efficiency of this assay, the upward-pointing camera has been integrated into a six position "fly carousel" that allows six vials of flies to be assayed simultaneously (Fig. 1). The carousel consists of

logic pulses used by the microcontroller into the RS-232 signals required for serial communication with the software. To advance the stepper motor, the microcontroller sends TTL logic pulses to an L293D integrated circuit push–pull dual H bridge to energize the stepper motor coils. The motor is advanced in half-step mode ($0.9°$ steps) to allow precise positioning of the platform, and the number of steps between vials was determined empirically. The coils are energized for 20 ms per step, long enough to provide the torque necessary for starting and stopping the stage after the required number of steps, yet short enough to smoothly turn the stage and avoid dislodging the flies from the sides of the assay vials. To keep the user apprised of the progress of the behavioral assay, the microcontroller communicates the vial number and time of the assay to an LCD panel through a built-in Hitachi HD44780 interface. Basic and assembled code used for programming the chip is available at http://www.virginia.edu/biology/Fac/hirsh_fly_carousel/Fly%20Carousel%20 Documentation/index.html.

a revolving platform powered by a stepper motor; the motor rotates the platform at 3-s intervals to consecutively position each vial precisely over the camera. Each time the platform stops moving, the camera records a still image and 3 s of video from the bottom of the current vial. The still images are recorded as JPEG files on an attached desktop computer, and the video is saved to VHS tape.

Image acquisition and platform rotation are both directed from the desktop computer by a custom-written ImageJ plugin. ImageJ is a Java-based image processing package written and maintained by Wayne Rasband and the National Institutes of Health. The program is in the public domain and is available at http://rsb.info.nih.gov/ij/index.html. The code for the plugin is available at http://www.virginia.edu/biology/Fac/hirsh_fly_carousel/Fly%20Carousel%20Documentation/index.html. These processes are coordinated through a programmable microcontroller, which allows communication between the computer and the stepper motor (Fig. 3). Quicktime for Java (http://developer.apple.com/quicktime/qtjava/index.html) provides access to the image stream from the camera and allows the ImageJ plugin to capture still images from the stream.

The flowchart for the subsequent software operations is sketched in Fig. 4. To begin image capture, the user selects "start grab" on the ImageJ window and presses the reset button on the microcontroller. The microcontroller then signals the plugin software to capture an image. The plugin records the beginning time, grabs an image from the video image stream, and places it in memory labeled with the vial number. After capturing an image, the plugin signals to the microcontroller directing the motor to advance one position: the microcontroller sends TTL logic pulses to an integrated circuit to energize the stepper motor coils, and the motor advances, rotating the platform to position the next vial over the camera. The microcontroller then sends another image-capture signal to the plugin software.

When the platform has completed one full revolution, the six images are saved. The plugin then calculates the length of time to pause the platform movement to ensure a 1-min interval between successive pictures of each vial. After the appropriate pause, the image acquisition and platform rotation start again. This cycle continues until the required number of revolutions has been completed.

Proof of Principle

To test the reliability of using this counting method compared with scoring the behavior of individual flies, vials of flies were videotaped from below for 5 min, examining either wild-type or $per^0$ mutant flies (Fig. 5).

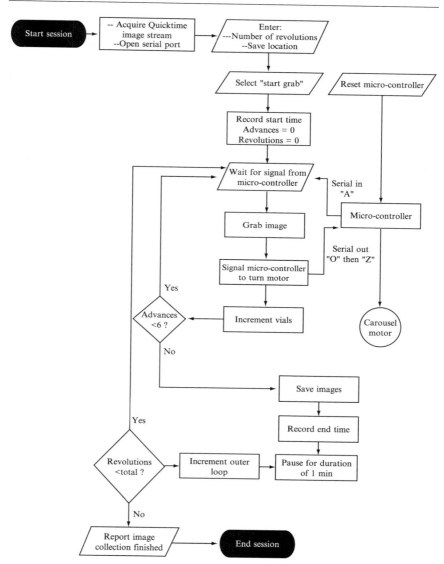

Fig. 4. Flowchart of the image acquisition software.

Each video was then scored in two different ways. First, we watched the flies on the bottom of the vial and scored behaviors using our established seven-level system. The tape was then reanalyzed, pausing every 30 s and counting the number of flies on the bottom, without regard to what

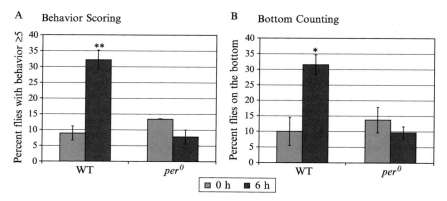

FIG. 5. Circadian mutants show altered cocaine responses, as indicated by both behavioral scoring (A) and the semiautomated bottom-counting assay (B). For each strain, $n = 90$. Flies were exposed to 85 $\mu$g free-base cocaine at 0, 3, and 6 h. Graphs compare the severity of the cocaine response on the first and third exposures. Significant sensitization, determined by $\chi^2$ analysis, is indicated by asterisks: * $P < 0.05$ and ** $P < 0.001$.

behavior they were exhibiting. Because the counting method included some behavior level 4 flies, the numbers collected by simply counting were slightly higher than the numbers collected by behavioral analysis. However, the responses followed the same patterns both within each exposure and between consecutive exposures (Fig. 4), demonstrating a high degree of consistency between the methods. The bottom-counting method shows that $per^0$ flies are defective in sensitization to repeated cocaine exposures, as shown previously by scoring individual flies (Andretic et al., 1999).

In addition to providing a much faster method for processing behavioral experiments, we observe less variance between replicate vials when using this method. This may be due to recording the fly behavior in the same vial in which they are exposed instead of transferring them to a viewing chamber, thus minimizing variability in handling. It also increases the amount of time during which the flies are able to move around in the cocaine smoke, potentially allowing a more even cocaine exposure among all flies within a vial. Most importantly, though, scoring can be done in a fraction of the amount of time required by the previous assay, and subjectivity in scoring is reduced dramatically.

# References

Abarca, C., Albrecht, U., and Spanagel, R. (2002). Cocaine sensitization and reward are under the influence of circadian genes and rhythm. *Proc. Natl. Acad. Sci. USA* **99,** 9026–9030.

Albrecht, U., Zheng, B., Larkin, D., Sun, Z. S., and Lee, C. C. (2001). *MPer1* and *mper2* are essential for normal resetting of the circadian clock. *J. Biol. Rhythms* **16,** 100–104.

Andretic, R., Chaney, S., and Hirsh, J. (1999). Circadian genes are required for cocaine sensitization in *Drosophila*. *Science* **285,** 1066–1068.

Andretic, R., and Hirsh, J. (2000). Circadian modulation of dopamine receptor responsiveness in *Drosophila melanogaster*. *Proc. Natl. Acad. Sci. USA* **97,** 1873–1878.

Bae, K., Jin, X., Maywood, E. S., Hastings, M. H., Reppert, S. M., and Weaver, D. R. (2001). Differential functions of mPer1, mPer2, and mPer3 in the SCN circadian clock. *Neuron* **30,** 525–536.

Bainton, R. J., Tsai, L. T., Singh, C. M., Moore, M. S., Neckameyer, W. S., and Heberlein, U. (2000). Dopamine modulates acute responses to cocaine, nicotine and ethanol in *Drosophila*. *Curr. Biol.* **10,** 187–194.

Epstein, P. N., and Altshuler, H. L. (1978). Changes in the effects of cocaine during chronic treatment. *Res. Commun. Chem. Pathol. Pharmacol.* **22,** 93–105.

Hirsh, J. (2000). *In* "*Drosophila* Protocols" (W. Sullivan, M. Ashburner, and R. S. Hawley, eds.), pp. 617–623. Cold Spring Harbor Laboratory Press, Cold Spring Harbor, NY.

Kalivas, P. W., Pierce, R. C., Cornish, J., and Sorg, B. A. (1998). A role for sensitization in craving and relapse in cocaine addiction. *J. Psychopharmacol.* **12,** 49–53.

Li, H., Chaney, S., Forte, M., and Hirsh, J. (2000). Ectopic G-protein expression in dopamine and serotonin neurons blocks cocaine sensitization in *Drosophila melanogaster*. *Curr. Biol.* **10,** 211–214.

McClung, C., and Hirsh, J. (1998). Stereotypic behavioral responses to free-base cocaine and the development of behavioral sensitization in *Drosophila melanogaster*. *Curr. Biol.* **8,** 109–112.

McClung, C., and Hirsh, J. (1999). The trace amine tyramine is essential for sensitization to cocaine in *Drosophila*. *Curr. Biol.* **9,** 853–860.

Park, S., Sedore, S., Cronmiller, C., and Hirsh, J. (2000). PKAII-deficient *Drosophila* are viable but show developmental, circadian and drug response phenotypes. *J. Biol. Chem.* **275,** 20588–20596.

Robinson, T. E., and Berridge, K. C. (1993). The neural basis of drug craving: An incentive-sensitization theory of addiction. *Brain Res. Brain Res. Rev.* **18,** 247–291.

Shuster, L., Yu, G., and Bates, A. (1977). Sensitization to cocaine stimulation in mice. *Psychopharmacology (Berl.)* **52,** 185–190.

Vanderschuren, L. J., and Kalivas, P. W. (2000). Alterations in dopaminergic and glutamatergic transmission in the induction and expression of behavioral sensitization: A critical review of preclinical studies. *Psychopharmacology (Berl.)* **151,** 99–120.

Zheng, B., Larkin, D. W., Albrecht, U., Sun, Z. S., Sage, M., Eichele, G., Lee, C. C., and Bradley, A. (1999). The mPer2 gene encodes a functional component of the mammalian circadian clock. *Nature* **400,** 169–173.

## [45]   The Circadian Clock and Tumor Suppression by Mammalian *Period* Genes

*By* CHENG CHI LEE

## Abstract

*Period* (*Per*) genes are key circadian rhythm regulators in mammals. Expression of mouse *Per* (*mPer*) genes has a diurnal pattern in the suprachiasmatic nucleus and in peripheral tissues. Genetic ablation mPER1 and mPER2 function results in a complete loss of circadian rhythm control based on wheel-running activity in mice. In addition, these animals also display apparent premature aging and a significant increase in neoplastic and hyperplastic phenotypes. When challenged by $\gamma$ radiation, *mPer2*-deficient mice respond by rapid hair graying, are deficient in p53-mediated apoptosis in thymocytes, and have robust tumor occurrences. Studies have demonstrated that the circadian clock function is very important for cell cycle, DNA damage response, and tumor suppression *in vivo*. The temporal expression of genes involved in cell cycle regulation and tumor suppression, such as *c-Myc*, *Cyclin D1*, *Cyclin A*, *Mdm-2*, and *Gadd45α*, is deregulated in *mPer2* mutant mice. Genetic studies have demonstrated that many key regulators of cell cycle and growth control are also important circadian clock regulators, confirming the critical role of circadian function in organismal homeostasis.

## Introduction

While identifying genes that mapped on human chromosome 17, a cDNA encoding a polypeptide with about 20% amino acid homology to the Drosophila *Period* (dPer) protein was isolated. This gene, originally named Rigui, is the human homolog of the mouse *Period1* (*mPer1*) gene (Sun *et al.*, 1997). A BLAST search of GeneBank identified two additional genes with homology to mPER1. One, *mPer2*, encoded a polypeptide highly homologous to mPER1 (Albrecht *et al.*, 1997). The other gene, *mPer3*, exhibited less homology to *mPer1* (Zylka *et al.*, 1998). All three mPERs are PAS domain protein containing two stretches of polypeptide repeats known as PASA and PASB, which is also found in several other clock proteins, including CLOCK, NPAS2, and BMAL1.

*In situ* hybridization studies revealed that all three *mPer* genes are expressed in the suprachiasmatic nucleus (SCN), the clock structure important for circadian rhythm control in rodents (Albrecht *et al.*, 1997; Sun *et al.*, 1997; Zheng *et al.*, 1999). The expression of *mPer* genes in the SCN displays circadian control, with peak levels at about CT4–CT6 for *mPer1* and *mPer3* and CT10–CT14 for *mPer2*. In addition to the SCN, other structures of the brain also expressed *mPer* genes in a temporal or constitutive fashion. However, the timing of peak expression in these structures, such as the pars-tuberalis and purkinje cells, differs from the SCN in that peak expression of *mPer1* occurs at about CT12 and CT24, respectively. Constitutive expression was also observed in other regions of the brain, such as the hippocampus (Albrecth *et al.*, 1997; Sun *et al.*, 1997). In addition to neural structures, all three *mPer* are expressed in a circadian manner in peripheral tissues. Fibroblast and lymphoblast cells expressed *mPer* genes in a temporal fashion, and in culture cells, circadian regulation of clock gene expression can be reset by serum shock (Balsalobre *et al.*, 1998). These observations suggested that mPERs have a role outside of the SCN.

## Genetic Studies Demonstrated that mPER1 and mPER2 are Circadian Regulators

Gene knockout studies in mice revealed differential roles for the three mPER proteins in the mammalian circadian rhythm (Zheng *et al.*, 1999, 2001). The circadian phenotype of mPER1 deficiency revealed a defect in the precision control of circadian period length. Period length can be highly variable in an individual animal, but the loss of circadian rhythm under free-running conditions is rarely observed in mice without mPER1 function (see Fig. 1A). Mice without mPER2 function have transient rhythm with a shortened period length of 22 h. A majority of mutant mice lose the persistence of circadian rhythm when placed in free-running conditions (see Fig. 1B). Thus, the loss of mPER2 function had a greater impact on mammalian clock function than that observed in mice lacking mPER1. That PER2 is a key circadian gene is further supported by its role in humans with familial advance sleep phase syndrome (Toh *et al.*, 2001). The loss of mPER3 has no apparent impact on central clock function and appears to be expendable (Lee *et al.*, 2004). These studies suggest that *mPer3* is under the control of the clock rather than a central clock regulator. Animals without mPER1 and mPER2 function have no circadian rhythm. These mice are completely compliant to cyclic external cues such as very short light–dark (LD) cycles (Zheng *et al.*, 2001). The loss of clock function is also manifest at the physiological and molecular level as arrhythmicity in body temperature and as deregulated expression of clock-controlled genes *in vivo*, demonstrating that mPER1 and mPER2 are indeed endogenous clock regulators (Shiromani *et al.*, 2004; Zheng *et al.*, 2001).

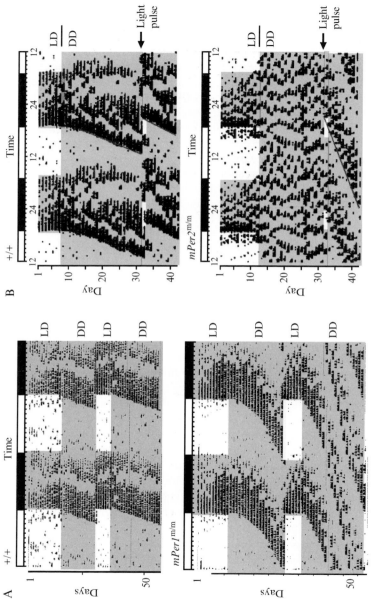

Fɪɢ. 1. (A) Double plots of actograms from wild-type and *mPer1* mutant mice. Note the change in period length during the free-running condition of the same *mPer1* mouse after LD entrainment. (B) Double plots of actograms of wild-type and *mPer2* mutant mice. Note the loss of rhythmicity during the free-running condition and the transient rhythm after the light pulse.

Noncircadian Phenotypes of *mPer1* and *mPer2* Mutant Mice

Both *mPer1* and *mPer2* mutant mice are morphologically indistinguishable from wild-type mice, indicating that these genes are not essential for cell differentiation and development. However, animals without mPER1 and mPER2 function display a morphological phenotype that mimics premature aging. The penetration of this phenotype is very high. Compared with wild-type mice, mice without mPER1 and mPER2 function display a more rapid decline in fertility and subsequent litter size from each successive breeding. Physical phenotypes are manifest by general loss of soft tissues and development of kyphosis in animals at about 12–14 months of age (see Fig. 2). It is unclear whether the pathological changes that mimic premature aging in fact reflect greater susceptibility to environmental stress, including infection. In addition, a low but significant number of animals develop tumors, suggesting that animals without circadian function have compromised cell proliferation control *in vivo* (Fu *et al.*, 2002). While the relative contribution of both mPER1 and mPER2 to the phenotype is unclear, we have focused on *mPer2*, as the circadian phenotype of this gene mutation is more severe. Careful examination of organs from older *mPer2* mutant mice revealed that the most consistent abnormality is hyperplasia of the salivary glands (Fig. 3). Animals without mPER2 function have enlarged salivary glands starting from about 5 months of age. By 8–12 months of age, salivary glands from *mPer2*-deficient mice will have enlarged by about twofold compared with wild-type mice (see Fig. 3). Both

FIG. 2. Physical phenotype of an *mPer1/mPer2* mutant mouse aged about 14 months. Note the prominent kyphosis and a general decline in physical appearance. (See color insert.)

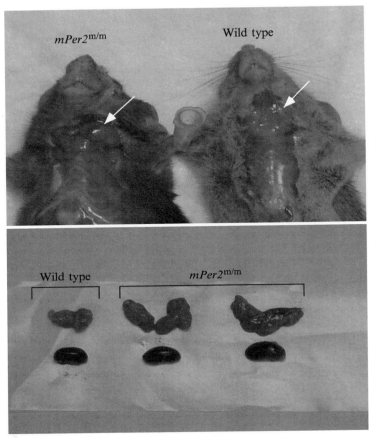

FIG. 3. Abnormal cell growth control in *mPer2* mutant mice. (*Top*) *mPer2* and wild-type mice of about 8 months of age. The white arrow points to the salivary gland region. Note the enlargement of the salivary gland in the *mPer2* mutant mouse. (*Bottom*) The relative size of kidneys and salivary glands from wild-type and *mPer2* mutant mice. (See color insert.)

$mPer2^{m/m}$ and $mPer2^{-/-}$ animals display the same phenotype, indicating that the loss of mPER2 function is the underlying cause (Fu *et al.*, 2002).

## Tumor Suppression Function of mPER2

Hyperplasia is often associated with a lower rate of cell apoptosis, a greater cell proliferation rate, or both. $\gamma$ irradiation, a potent genotoxic agent, was used to analyze cellular changes associated with the apoptotic response and the long-term survival of the animals. Studies revealed that *mPer2* mutant mice respond abnormally to a sublethal dose (4 Gy) of $\gamma$ radiation (Fu *et al.*, 2002). A very pronounced graying of the coat was observed in *mPer2* mutant but not wild-type mice starting at about 2 months after irradiation. From 5 months after irradiation, *mPer2* mutant mice also developed lymphoma at a robust rate compared with wild-type controls. These observations are consistent with Knudson's "two-hit" hypothesis for a tumor suppressor function and implicate *mPer2* as a tumor suppressor gene. Interestingly, a similar coat graying response to a sublethal level of $\gamma$ radiation has been observed previously with mice heterozygous for ataxia telangiectasia mutation (ATM) (Barlow *et al.*, 1999). The ATM kinase is a tumor suppressor and key regulator of the cellular response to ionizing radiation (IR). It remains unclear whether there is a shared pathway between mPER2 and ATM kinase action in cancer prevention and the response to IR. Molecular analysis of the cellular response to $\gamma$ radiation revealed that thymocytes from *mPer2* mutant mice are either less sensitive or have a slower apoptotic response to IR. Clock genes, including *mPer1* and *mPer2*, are induced by $\gamma$ irradiation and these inductions are deregulated in *mPer2* mutant mice *in vivo* (Fu *et al.*, 2002). The induction of clock genes by ultraviolet (UV) light radiation suggests that these proteins are early responsive factors to genotoxic challenge (Kawara *et al.*, 2002). Thymocytes from *mPer2* mutant mice apparently survive genotoxic stress by undergoing G2/M checkpoint arrest. At the molecular level, after IR, *mPer2*-deficient thymocytes exhibit lower p53 protein accumulation and slower release of cytochrome c from mitochondria compared with those from wild-type mice (Fu *et al.*, 2002). Thus, *mPer2*-deficient cells display less sensitivity in detecting genotoxic damage induced by IR. Conceptually, this defect in the surveillance mechanism for DNA damage will result in a higher cancer rate as more cells with DNA mutations escape the cellular apoptotic mechanism.

## Circadian Clock Regulates Cell Cycle Genes *In Vivo*

While there have been many observations reported on the relationship between circadian rhythm and cancer, the key to understanding the circadian role in cancer is to identify the genetic relationship between them. Given the abnormal cell growth phenotype of the salivary gland and the abnormal response to DNA damage observed in *mPer2* mutant mice, we have focused on genes predicted to be important for cell proliferation and DNA damage control. One key regulator of the cell division cycle and modulator of DNA damage control is *c-Myc* (Evan and Vousden, 2001). The protooncogene *c-Myc* regulates the G0 to G1 transition of cell cycle.

In many human tumors, *c-Myc* expression is highly elevated, and over-expression *c-Myc* in transgenic mice results in neoplasia (Jamerson *et al.*, 2004). Studies revealed that *c-Myc* is under circadian regulation and that loss of mPER2 function was associated with a deregulated expression level of *c-Myc* (Fig. 4). Reporter assay studies revealed that the NPAS2/BMAL1 circadian transcription complex repressed *c-Myc* promoter expression, indicating that *c-Myc* is a first-order clock-controlled gene (Fu *et al.*, 2002). The identification of *c-Myc* as a circadian-regulated gene directly links circadian function to a key regulator of the cell cycles. Because *c-Myc* is a major transcription factor, *c-Myc* target genes such as *Gadd45α* were also deregulated when mPER2 function is lost (Fig. 4). GADD45α is involved in cell growth arrest, and its expression is induced in response to DNA damage mediated by p53 (Evan and Vousden, 2001). The expression of *Gadd45α* is regulated negatively by c-MYC (Tao and Umek, 1999). Figure 4 demonstrates an inverse relationship between *c-Myc* and *Gadd45α* expression levels during the diurnal cycle in both wild-type and *mPer2* mutant animals. This observation establishes the importance of circadian function in the cell cycle and in the DNA damage response. Studies with *Cry* mutant mice revealed that the circadian clock also controls the G2/M phase via *Cyclin B/Cdc2* expression and phosphoryation through WEE1 (Matsuo *et al.*, 2003). Thus, genetic and molecular evidence points to circadian regulation of multiple stages of the cell cycle pathway.

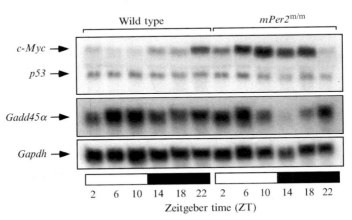

FIG. 4. Northern analysis of circadian clock-controlled genes *c-Myc* and *Gadd45α* in wild-type and *mPer2* mutant mice. The corresponding level of *p53* and *Gapdh* mRNA is used as an internal control. The blot was sequentially hybridized with the indicated radiolabeled cDNA probes. Note the inverse relationship between *c-Myc* and *Gadd45α* mRNA expression levels and the deregulation of their expression in *mPer2* mutant mice.

While cancer is a multistep mechanism often linked to deregulate cell cycle control, the involvement of circadian-regulated components important for cell cycle opens new insight into the mechanism of neoplasia.

## New Links Between Growth Regulators and Circadian Clock

Studies have shown that several other genes implicated in tumorgenic pathways are also key circadian regulators. For example, casein kinase II (CKII) is a key regulator of the phosphorylation of major proteins important for preventing cancer, including *p53*, *c-Myc*, and adenomatous polyposis coli (APC) (Vogelstein *et al.*, 2000). CK II is frequently upregulated in cancer cells and its overexpression in transgenic mice results in lymphoma (Xu *et al.*, 1999). Genetic studies demonstrate that a phenotype of CK II mutation in plants, neurospora, and flies is circadian abnormality (Lin *et al.*, 2002; Yang *et al.*, 2002). It is unclear whether CKII will also regulate circadian function in mammals. Because mutation of the $\beta$ subunit of CKII is embryonic lethal in mice, its role in circadian regulation cannot be ascertained (Buchou *et al.*, 2003). Another example that suggests a common pathways between circadian regulation and cell growth regulation is the casein kinase 1ε (CK1ε). CK1ε mutation results in circadian abnormalities, including the *tau* mutant of hamster and *doubletime* mutant in flies (Harms *et al.*, 2003; Lowrey *et al.*, 2000). CK1ε is also an important positive regulator of $\beta$-catenin. $\beta$-Catenin is a key player in the *wnt* signaling pathway (Gao *et al.*, 2002) and is a potent cell proliferation promoter. Overaccumulation of $\beta$-catenin as a result of mutant APC protein is the underlying basis for familial colorectal cancer (Clevers, 2004). Two other examples include glycogen synthase kinase 3 (GSK3) and protein phosphatase 2A (PP2A). The *Drosophila* GSK3 mutant (*shaggy*) has circadian abnormality (Harms *et al.*, 2003). Its mammalian homolog GSK-3$\beta$ is known as a negative regulator of $\beta$-catenin. It has been shown that protein phosphatase 2A is a key circadian gene in *Neurospora* and flies (Sathyanarayanan *et al.*, 2004; Yang *et al.*, 2004). In mammalian cells, PP2A is a key regulator of *c-Myc* during the oncogenic transformation of human cells (Yeh *et al.*, 2004). Other studies showed that GADD45$\alpha$ is directly associated with the two components of the APC complex, namely phosphatase 2A and GSK-3$\beta$. GADD45$\alpha$ promotes the dephosphorylation of GSK-3$\beta$ and the degradation of $\beta$-catenin after UV radiation (Hildesheim *et al.*, 2004). Together, these observations point to a common pathway shared by regulators of circadian rhythm and cell proliferation. Perhaps the circadian mechanism that evolved to sustain an organism is also the same mechanism that protects (apoptosis) and nurtures (proliferation) the cells through evolution. Therefore, malfunction of the circadian clock would lead to increased tumor rates during the life span of the animal, as the mechanism that protects and nurtures the cells would also be deficient. One predicted outcome of this hypothesis is that genes identified to be important for neoplasia are also likely to be good candidates for circadian regulation and *vice versa*.

## Acknowledgments

I thank Dr. J. Lever for comments on this manuscript. This work is supported in part by funding from NIH.

## References

Albrecht, U., Sun, Z. S., Eichele, G., and Lee, C. C. (1997). A differential response of two putative mammalian circadian regulators, mper1 and mper2, to light. *Cell* **91,** 1055–1064.

Balsalobre, A., Damiola, F., and Schibler, U. (1998). A serum shock induces circadian gene expression in mammalian tissue culture cells. *Cell* **93,** 929–937.

Barlow, C., Eckhaus, M. A., Schaffer, A. A., and Wynshaw-Boris, A. (1999). Atm haploinsufficiency results in increased sensitivity to sublethal doses of ionizing radiation in mice. *Nature Genet.* **21,** 359–360.

Buchou, T., Vernet, M., Blond, O., Jensen, H. H., Pointu, H., Olsen, B. B., Cochet, C., Issinger, O. G., and Boldyreff, B. (2003). Disruption of the regulatory beta subunit of protein kinase CK2 in mice leads to a cell-autonomous defect and early embryonic lethality. *Mol. Cell. Biol.* **23,** 908–915.

Clevers, H. (2004). Wnt breakers in colon cancer. *Cancer Cell* **5,** 5–6.

Evan, G. I., and Vousden, K. H. (2001). Proliferation, cell cycle and apoptosis in cancer. *Nature* **411,** 342–348.

Fu, L., Pelicano, H., Liu, J., Huang, P., and Lee, C. C. (2002). The circadian gene Period2 plays an important role in tumor suppression and DNA damage response *in vivo. Cell* **111,** 41–50.

Gao, Z. H., Seeling, J. M., Hill, V., Yochum, A., and Virshup, D. M. (2002). Casein kinase I phosphorylates and destabilizes the beta-catenin degradation complex. *Proc. Natl. Acad. Sci. USA* **99,** 1182–1187.

Harms, E., Young, M. W., and Saez, L. (2003). CK1 and GSK3 in the *Drosophila* and mammalian circadian clock. *Novartis Found. Symp.* **253,** 267–277.

Hildesheim, J., Belova, G. I., Tyner, S. D., Zhou, X., Vardanian, L., and Fornace, A. J., Jr. (2004). Gadd45a regulates matrix metalloproteinases by suppressing DeltaNp63alpha and beta-catenin via p38 MAP kinase and APC complex activation. *Oncogene* **23**(10), 1829–1837.

Jamerson, M. H., Johnson, M. D., and Dickson, R. B. (2004). Of mice and myc: c-Myc and mammary tumorigenesis. *J. Mammary Gland Biol. Neoplasia* **9,** 27–37.

Kawara, S., Mydlarski, R., Mamelak, A. J., Freed, I., Wang, B., Watanabe, H., Shivji, G., Tavadia, S. K., Suzuki, H., Bjarnason, G. A., Jordan, R. C., and Sauder, D. N. (2002). Low-dose ultraviolet B rays alter the mRNA expression of the circadian clock genes in cultured human keratinocytes. *J. Invest. Dermatol.* **119,** 1220–1223.

Lee, C., Weaver, D. R., and Reppert, S. M. (2004). Direct association between mouse PERIOD and CKIepsilon is critical for a functioning circadian clock. *Mol. Cell. Biol.* **24,** 584–594.

Lin, J. M., Kilman, V. L., Keegan, K., Paddock, B., Emery-Le, M., Rosbash, M., and Allada, R. (2002). A role for casein kinase 2alpha in the *Drosophila* circadian clock. *Nature* **420,** 816–820.

Lowrey, P. L., Shimomura, K., Antoch, M. P., Yamazaki, S., Zemenides, P. D., Ralph, M. R., Menaker, M., and Takahashi, J. S. (2000). Positional syntenic cloning and functional characterization of the mammalian circadian mutation tau. *Science* **288,** 483–492.

Matsuo, T., Yamaguchi, S., Mitsui, S., Emi, A., Shimoda, F., and Okamura, H. (2003). Control mechanism of the circadian clock for timing of cell division *in vivo. Science* **302,** 255–259.

Sathyanarayanan, S., Zheng, X., Xiao, R., and Sehgal, A. (2004). Posttranslational regulation of *Drosophila* PERIOD protein by protein phosphatase 2A. *Cell* **116,** 603–615.

Shiromani, P. J., Xu, M., Winston, E. M., Shiromani, S. N., Gerashchenko, D., and Weaver, D. R. (2004). Sleep rhythmicity and homeostasis in mice with targeted disruption of mPeriod genes. *Am. J. Physiol. Regul. Integr. Comp. Physiol.* **287,** R47–R57

Sun, Z. S., Albrecht, U., Zhuchenko, O., Bailey, J., Eichele, G., and Lee, C. C. (1997). RIGUI, a putative mammalian ortholog of the *Drosophila* period gene. *Cell* **90,** 1003–1011.

Tao, H., and Umek, R. M. (1999). Reciprocal regulation of gadd45 by C/EBP alpha and c-Myc. *DNA Cell Biol.* **18,** 75–84.

Toh, K. L., Jones, C. R., He, Y., Eide, E. J., Hinz, W. A., Virshup, D. M., Ptacek, L. J., and Fu, Y. H. (2001). An hPer2 phosphorylation site mutation in familial advanced sleep phase syndrome. *Science* **291,** 1040–1043.

Vogelstein, B., Lane, D., and Levine, A. J. (2000). Surfing the p53 network. *Nature* **408,** 307–310.

Xu, X., Landesman-Bollag, E., Channavajhala, P. L., and Seldin, D. C. (1999). Murine protein kinase CK2: Gene and oncogene. *Mol. Cell. Biochem.* **191,** 65–74.

Yang, Y., Cheng, P., and Liu, Y. (2002). Regulation of the Neurospora circadian clock by casein kinase II. *Genes Dev.* **16,** 994–1006.

Yang, Y., He, Q., Cheng, P., Wrage, P., Yarden, O., and Liu, Y. (2004). Distinct roles for PP1 and PP2A in the Neurospora circadian clock. *Genes Dev.* **18,** 255–260.

Yeh, E., Cunningham, M., Arnold, H., Chasse, D., Monteith, T., Ivaldi, G., Hahn, W. C., Stukenberg, P. T., Shenolikar, S., Uchida, T., Counter, C. M., Nevins, J. R., Means, A. R., and Sears, R. (2004). A signalling pathway controlling c-Myc degradation that impacts oncogenic transformation of human cells. *Nature Cell Biol.* **6,** 308–318.

Zheng, B., Albrecht, U., Kaasik, K., Sage, M., Lu, W., Vaishnav, S., Li, Q., Sun, Z. S., Eichele, G., Bradley, A., and Lee, C. C. (2001). Nonredundant roles of the mPer1 and mPer2 genes in the mammalian circadian clock. *Cell* **105,** 683–694.

Zheng, B., Larkin, D. W., Albrecht, U., Sun, Z. S., Sage, M., Eichele, G., Lee, C. C., and Bradley, A. (1999). The mPer2 gene encodes a functional component of the mammalian circadian clock. *Nature* **400,** 169–173.

Zylka, M. J., Shearman, L. P., Weaver, D. R., and Reppert, S. M. (1998). Three period homologs in mammals: Differential light responses in the suprachiasmatic circadian clock and oscillating transcripts outside of brain. *Neuron* **20,** 1103–1110.

# Author Index

## A

Abarca, C., 515, 842
Abbott, S. M., 596, 598, 599, 601, 628
Abe, H., 454, 516, 517, 595
Abe, M., 270, 272, 273, 288, 289, 297, 483, 484, 487, 513, 543, 580, 583, 595
Abel, T., 303, 784
Abodeely, M., 37, 39, 42, 43, 44, 50, 56, 67, 76, 77, 78, 79, 130, 310, 379, 395, 409
Abovich, N., 78, 96
Abu-Shaar, M., 612
Achermann, P., 773, 788
Acosta, D., 401
Adachi, A., 288, 290, 334, 336, 347, 367, 370, 374, 458, 552
Adachi, M., 424
Adams, C. A., 255
Adams, M. D., 789
Adler, N. T., 512
Aeschbach, D., 788
Agami, R., 555
Agapite, J., 66, 137, 669, 673, 674
Agosto, J., 84, 85, 87, 97, 99, 105, 106, 107, 108, 109, 110, 123, 125
Agrawal, N., 612
Agrawal, S., 603, 604
Aguade, M., 804
Aguilar-Henonin, L., 34
Aguilar-Roblero, R., 559
Ahmad, M., 720, 727, 730, 732, 749
Aida, R., 457, 482, 515
Akaike, N., 686
Akashi, M., 290, 328, 334, 379, 390, 395, 547, 549, 573
Akasu, T., 635, 685
Akhtar, R. A., 367, 373, 546
Akiyama, M., 457, 482, 513, 515, 599, 603
Akten, B., 48, 72, 78, 79, 80, 99, 310, 379, 395, 812
Al-Ghoul, W. M., 629
Alam, N., 595

Albers, H. E., 457, 460, 626
Alberts, A. S., 303
Albrecht, U., 273, 290, 419, 456, 482, 515, 525, 551, 597, 598, 599, 783, 840, 850, 851
Albus, H., 456, 459, 460, 626
Aldape, K. D., 371
Allada, R., 43, 44, 46, 50, 52, 65, 67, 72, 73, 75, 81, 82, 83, 87, 96, 99, 100, 111, 131, 157, 159, 160, 208, 310, 344, 379, 380, 395, 611, 665, 812, 859
Allen, C. N., 269, 276, 457, 580, 588, 625, 626, 630, 631, 634, 635, 685, 686
Allen, G., 569, 571
Allenbach, L., 33
Allis, C. D., 477
Alonso, J. M., 34, 808, 809, 821, 828, 829, 834
Alonso-Blanco, C., 809, 829, 834
Alpern, M., 704
Alster, J. M., 596, 598, 600, 628
Alt, S., 155
Altman, J., 566
Altshuler, H. L., 841
Alvarez, C. E., 158
Alvarez, M. E., 17
Alvarez, R. A., 705
Amado, D. A., 80
Amanatides, P. G., 789
Amarakone, A., 52, 344, 347, 367, 370, 664
Amasino, R. M., 27, 809, 829, 834
Amaya, E., 206, 207, 209, 214
Amherd, R., 785
Amir, S., 147, 627
Amsterdam, A., 225
Andahazy, M., 702, 721, 748
Anderson, E. R., 457
Anderson, F. E., 206
Andersson, C. R., 821, 825, 828, 834
Andretic, R., 145, 148, 149, 841, 842, 850
Andrews, R. V., 512
Andronis, C., 379

863

# I

# J

# Index

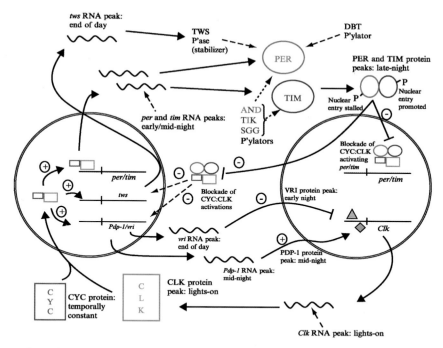

HALL, CHAPTER 4, FIG. 1. Expansions of the basic pacemaker mechanism subserving daily rhythmicity in *Drosophila*: two interlocked feedback loops and their posttranslational regulators. This diagram is based loosely on one presented by Glossop *et al.* (1999), the investigators whose experiments prompted them to invoke "extra" gene functions and feedback phenomenettes. These components act beyond the elementary feedback events that are rooted in the negative effects of PER and TIM on their own genes' transcribability, as diagramed here within parts of the left half of the diagram (which has been modified twice from the Glossop *et al.* scheme, i.e., as it was tweaked within a figure appearing in Hall, 2003a). Genes functioning within the schemes ($n = 2$, really, as described later) are represented by italicized symbols, their protein products by the corresponding abbreviations in all uppercase; and activating/stimulating *vs* inhibitory/repressive functions and events are designated by "plus" and "minus" symbols, respectively. The left half of this picture and the right-hand portion each represent the nucleus within one circadian pacemaker cell. The mechanistic components implied to be operating within and hovering around the two circles are roughly separated in terms of gene product dynamics that occur for *per*, *tim*, *Clk*, and *cyc* (left) and occurring in a pretty much temporally distinct manner with regard to *Clk* product dynamics and those of their gene expression regulators (*vri*, *Pdp-1*) (right). Thus factors entering the nucleus on the right and acting within it are oscillating in a largely out-of-phase fashion as compared with the "interlocked" loop components functioning within the left-hand circle and feeding into that nucleus (in terms of the *per*, *tim*, *Pdp-1*, *vri*, and *tws* genes chosen to be depicted within it). As to how these genes and their products are currently believed to act and interact: In brief, and focusing for the moment only on the transcriptional regulators PER:TIM binding to CYC:CLK, the former two proteins (at least PER) poison the actions of CYC:CLK, which results in not only in depression of *per/tim* activation, but also releases

CYC:CLK-dependent repression of *Clk*. The negative function just implied is mediated by a transcription factor encoded by the *vri* gene, but *Clk* begins to be transcriptionally activated at a later cycle time by the action of the transcription factor of *Pdp-1* [specified as PDP-1ε by Cyran *et al.* (2003) and thus symbolically abbreviated here]. Therefore, CLK production eventually goes into its rising phase, resulting from action of the PDP-1 activator. As CLK accumulates, it stimulates increases in *per* and *tim* transcription rates, but inasmuch as the cooperative action of CYC:CLK has also promoted VRI production, whose result is to repress *Clk* transcription, PDP-1-regulated activation of *Clk* is stalled in terms of the latter's primary and secondary products beginning to rise toward their lights-on maximum (*Clk* mRNA and CLK protein rise and fall in concert within an LD cycle). The posttranslationally acting elements of these mechanisms involve an array of catalytic activities, the specific actions and eventual effects of which are verbally noted within portions of the bipartite diagrams. These schemes include symbols for kinase subunits encoded by the genes *dbt*, *And*, *Tik*, and *sgg* (color codings for the latter three gene products are meant to indicate the blue DBT enzyme as using TIM for its rhythm-relate substrate, whereas the bicolored AND and TIK designations stand for the fact that these two enzyme subunits participate in phosphorylating both PER and TIM). Another kind of enzymatic function schematized here involves the *tws* gene, which specifies a protein phosphatase subunit; it is believed to be activated transcriptionally by CLK:CYC, because the cycling of *tws* mRNA shown is flattened by the effects of a *cyc*-null mutation. However, the contribution of TWS/phosphatase to clock gene dynamics—apparently meaning in the main stabilization effects on PER levels, including those within the nucleus—is more complicated, because this polypeptide forms only one (regulatory) subunit of a phosphatase holoenzyme called PP2A. Another such subunit emanates from the *widerborst* (*wdb*) gene whose mRNA seems to cycle, but the relevant peak was observed at the end of the night, less than two times above the apparent transcript level 12 h earlier (or later), whereas *tws* mRNA (albeit for only one of the transcript isoforms emanating from this PP2A subunit gene) maxes out at the end of the day compared with the trough level that is *ca.* five times lower at the end of the night. For simplicity, and because this recently emerged phosphatase contribution to the overall mechanism is difficult to appreciate fully, only TWS is included with the diagram. (For example, particulars of the chrono roles that seem to be played by not only *wdb*-encoded polypeptide, but also by the *catalytic* PP2A subunit encoded by a gene known as *mutagenic star* are left out.)

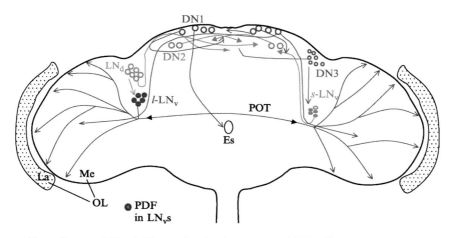

HALL, CHAPTER 4, FIG. 2. Neuronal projection patterns of CNS cells expressing clock genes in the brain of a *Drosophila* adult. The three clusters of lateral neurons (LNs) and the three groups of dorsal ones (DNs) are depicted in somewhat of a cartoon fashion—not so much in terms of positions of these cell groups, but, for example, because space limitation forced many fewer of the dorsally located DN3s to be diagrammed here than the actual number: *ca.* 40 bilaterally symmetrical pairs of such cells. For the other cell groups, numbers of colored dots are more representative, in that a given hemibrain (left or right side) contains approximately 4–5 small (*s*)-LN$_v$ cell bodies ("small" referring to *relative* sizes of such perikarya). Five for the large (*l*)-LN$_v$ group nearby, six for the more dorsally located LN$_d$s, and for the two additional groups of DN cells: 15 for DN1s (such that the real number of these cells, as for the DN3s, is diagrammatically underdepicted) and two for DN2s. Totals (including the DN3 number noted earlier) are approximately 70–75 pairs of brain neurons expressing the *period* gene, the *timeless* one, or both (implying that not all the neurons indicated could be observed to coexpress both of the clock gene products, as discussed briefly within the text). These cell counts were originally presented by Ewer *et al.* (1992) and Frisch *et al.* (1994); updated by Kaneko and Hall (2000); and have appeared in various reports of Helfrich-Förster, as summarized in 2003. Within that review is a diagram that includes a large number of small circles (compared with the colored dots within the present picture), representing the *ca.* 1800 *per*-expressing glial cells that are located in several cortical and neuropile regions of the central brain proper and those of the optic lobes (see Fig. 3 for what these glia may be about). The neurites projecting in various directions and potentially to various target neurons are depicted by thin colored lines, accompanied by arrowheads when the so-called targets (referring to distal extents of certain axon terminals) could not be observed in definitive fashion. This intrabrain "wiring diagram" is based on stainings for the PDH immunoreactivity that is present in all of the *l*-LN$_v$s and in of most *s*-LN$_v$s (e.g., Helfrich-Förster, 2003) as well as by application of transgene-driven neurite markers (Kaneko and Hall, 2000; Park *et al.*, 2000a; Renn *et al.*, 1999; Stoleru *et al.*, 2004; Veleri *et al.*, 2003). Focusing on elements of these projection patterns, *l*-LN$_v$ perikarya send axons across the brain midline [via the posterior optic track (POT)] to contralateral LN regions and also project into the optic lobes (OL), but only as far as the medulla (Me), which is underneath the distal-most lamina (La); it is unknown whether a given *l*-LN$_v$ cell body might elaborate both POT neurites as well as centrifugal fibers. *s*-LN$_v$ cells project fibers into a dorsomedial brain region, with these nerve terminals being near the mushroom body MB calyces (see Fig. 3). The integrity of the MB structure may be necessary for thorough normality of the behavioral rhythmicity of *Drosophila* (Helfrich-Förster *et al.*, 2002b). The PDF-nonexpressing LN$_d$ perikarya send their (transgenically marked) axons mainly into the dorsal brain (Kaneko and Hall, 2000), but there is also a minor projection that courses toward the LN region (Stoleru *et al.*, 2004). Many of the DN cells also, and in the main, project locally to regions relatively near the locations of s-LN$_v$ and LN$_d$ axon terminals. However, certain DN1 cells also (or instead) project fibers to the vicinity of LN$_v$ perikarya and to that of the esophagus (Es). An analogous situation pertains to DN3 neurites, some of which project rather "dorsolocally" (Kaneko and Hall, 2000), with others coursing toward putative LN$_v$ targets (Veleri *et al.*, 2003). This diagram is based largely on one that appeared in Hall (2003a), as augmented by the LN$_d$-to-LN$_v$ and by the DNs-to-LN-region projections that were uncovered more recently. These six clusters worth of clock gene-expressing neurons (including the two groups that are also PDF-containing) reappear with less anatomical detail in Figs. 3 and 4 in terms of functional meanings associated with several of the cell groups.

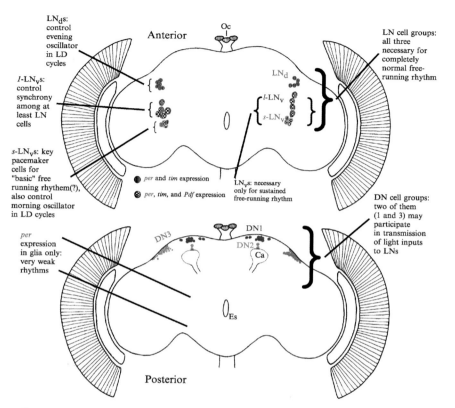

LN_ds: control evening oscillator in LD cycles

*l*-LN_vs: control synchrony among at least LN cells

*s*-LN_vs: key pacemaker cells for "basic" free running rhythm(?), also control morning oscillator in LD cycles

*per* expression in glia only: very weak rhythms

Anterior

Oc

LN_d

*l*-LN_v

*s*-LN_v

● *per* and *tim* expression

⊛ *per*, *tim*, and *Pdf* expression

LN_vs: necessary only for sustained free-running rhythm

LN cell groups: all three necessary for completely normal free-running rhythm

DN cell groups: two of them (1 and 3) may participate in transmission of light inputs to LNs

DN3

DN1

DN2

Ca

O_Es

Posterior

HALL, CHAPTER 4, FIG. 3. Neural substrates for the daily locomotor rhythms of *Drosophila*. The basic diagram is from Hall (1998), as augmented by Hall (2003a), and as then modified substantially here. The brain-behavioral responsibilities of the various cell types and neuronal cluster locations diagrammed within the current version of this scheme have been deduced from various points summarized later, with the caveat that this recitation of relevant functions is not comprehensive and that certain features of these illustrative conclusions must be thought of as tentative. In any case, essentially all LN cells are "removed" in a given *disco* mutant individual (Helfrich-Förster, 1998; Zerr *et al.*, 1990); such neurons also have their synaptic-output chrono-functioning ruined by combining the pan-*per* or pan-*tim*-(*gal4*) drivers with UAS-*tetanus-toxin* (Kaneko *et al.*, 2000b). The three kinds of variants just noted largely genocopy the effects of *per*-or *tim*-null mutations on locomotor rhythmicity in DD (Dowse *et al.*, 1989; Dushay *et al.*, 1989; Hardin *et al.*, 1992; Helfrich-Förster, 1998; Kaneko *et al.*, 2000b; Renn *et al.*, 1999, but see Blanchardon *et al.*, 2001). Proviso: *tim*[0] mutations are from time to time noted as allowing for a bit more locomotor rhythmicity (expressed as the proportion of animals tested that slip into the significantly periodic range), compared with the behavior that is observed routinely for *per*-null mutants (see, e.g., Sehgal *et al.*, 1994; Yang and Sehgal, 2001). PER presence within only a subset of the LNs is mediated by one of the "promoterless" *per* transgenics, discussed in several sections of the text (Frisch *et al.*, 1994), focused attention on LN/rhythmic behavior connections from the opposite perspective (quasi-restoration of normal behavior as opposed to disruption of it). Loosely similar inferences resulted from applying a transgenic type carrying *glass*-gene regulatory sequences fused to PER-coding ones (Vosshall and Young, 1995). Specific elimination of $s$-LN$_v$ and $l$-LN$_v$ cells, mediated by *Pdf*-promoter or *per*-like enhancer-trap driving of cell-killing factors (Blanchardon *et al.*, 2001; Renn *et al.*, 1999), resulted in degrees of arrhythmicity that were not as bad as those caused by the clock-null mutations noted earlier; moreover, LN$_v$-ablated brains caused the flies containing them progressively to lose rhythmicity in DD as opposed to becoming immediately arrhythmic upon proceeding from LD-cycling conditions into constant ones. The behavior of several *disco*-mutant individuals is similar to that of the doubly transgenic type just implied: quasi-rhythmic locomotion early in DD→thoroughgoing arrhythmicity by day 2 or 3 (Wheeler *et al.*, 1993), although the overall penetrance of rhythmic behavior is considerably worse when both LN$_d$s and LN$_v$s are gone (i.e., as caused by either of the two *disco* mutations tested for their effects on locomotor rhythmicity). With respect to the small subset of the latter cluster pair, $s$-LN$_v$s keep getting harped on as "the SCN of the fly" (*cf*. Tischkau and Gillette, 2005), in part because these neurons *may* be able to sustain oscillations of clock gene products in DD better than do the other LN cell types (see text). Concentrating on diel behavior in LD cycles, transgenic ablations of the LN$_d$s or the LN$_v$s, dovetailing with the LN$_v$-specific effects of a *Pdf*-null mutation, led to the supposition that there is differential (neuronal) regulation of "evening" *vs* "morning" oscillators that support locomotor anticipations of lights off *vs* lights on, respectively (Grima *et al.*, 2004; Stoleru *et al.*, 2004). The diagram at the bottom deals with chrono-behavioral functions of DN cell groups, as are mainly explicated within Fig. 4. Also suggested (within this figure) is the possibility of "PER-glial" involvement in behavioral rhythm regulation due to rare $per^+$//$per$-null genetic-mosaic individuals (Ewer *et al.*, 1992) in which no LN expression of the normal allele was detectable (also see Method 12); but these flies exhibited weak, long-period rhythms, correlated with the presence of $per^+$ in various glia only (depending on the brain region where that allele was retained within the mosaic individual in question). However, *disco* flies exhibit robust and normally cyclical expression of PER in brain glia (Zerr *et al.*, 1990), which therefore is insufficient for routine behavioral rhythmicity in the almost-certain *anatomical* absence of the LN's in this mutant (see earlier discussion).

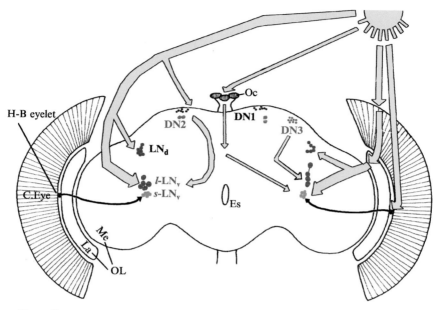

HALL, CHAPTER 4, FIG. 4. Input routes for light-mediated resetting of the behavioral clock of *Drosophila* and molecules functioning in these anatomical pathways. This diagram was adapted from Helfrich-Förster *et al.* (2001) as presented in modified form within the review of Hall (2003a); also see two reviews by Helfrich-Förster (2002, 2003). With respect to photic influences on the rhythm system of the fly, these three articles mainly summarize results reported by Ohata *et al.* (1998), Stanewsky *et al.* (1998), Yasuyama and Meinertzhagen (1999), Emery *et al.* (2000a,b), Helfrich-Förster *et al.* (2001, 2002a), and Malpel *et al.* (2002); also see the "postreview" papers of Mealey-Ferrara *et al.* (2003), Rieger *et al.* (2003), Veleri *et al.* (2003), and Klarsfeld *et al.* (2004). Essentially all of the individual studies just cited were based on applications of mutations, transgenics, or dietary treatments (see later) and stemmed from earlier and more basic experiments in which only the external eyes were eliminated or inactivated, which nevertheless allowed *D. melanogaster* adults to entrain to light:dark (LD) cycles (e.g., Dushay *et al.*, 1989; Helfrich, 1986). The current picture shows the input routes inferred to be knocked out or remaining under the influence of a given "treatment" (usually a genetic one). One element of the scheme is based on the entrainability to LD cycles, and reentrainability to shifted ones, of *Drosophila* that express a *norpA*-null mutation; it causes physiological blindness of photoreceptor cells (PRs) within the compound eyes (C.eye), the ocelli (Oc), and probably the H-B eyelet as well; or of flies that express an *eyes-absent* (*eya*) mutation in combination with an *ocelliless* one. The afferents of C.eye PRs may transmit results of their light inputs to dendrites of *l*-LN$_v$ cells (Helfrich-Förster, 2003; Helfrich-Förster *et al.*, 2002a); axon terminals projecting from H-B PRs have been more definitively observed to overlap with those particular LN arborizations as well as with those of the *s*-LN$_v$s (Helfrich-Förster *et al.*, 2002a); however, it is unknown whether Oc afferents connect with LN cells (or with any of the DNs for that matter), either directly or via dorsal-brain interneurons (Helfrich-Förster, 2003). Additional visual system variants have come into play as follows: (A) A *sine oculis* (*so*) mutant, which is missing C.eyes and Oc, although not all flies in this strain are thoroughly devoid of the former structures (e.g., Helfrich-Förster *et al.*, 2002a) so mutant *Drosophila* retain their H-B eyelets. (B) A *glass* (*gl*-null) mutation that wipes out all external PRs and the eyelet as well (Helfrich-Förster *et al.*, 2001). (C) A cryptochrome mutation (*cry^b*), which leads to little or no function of CRY, a blue-light absorbing protein present in many *Drosophila* tissues, including various neurons (see later). (D) A *histidine decarboxylase* (*hdc*) mutant, which lacks histamine in both the external eyes and the H-B eyelet such that these structures are unable to mediate neurotransmission to CNS neurons [Rieger *et al.* (2003) and references therein]. Usage of this *hdc* mutant can be regarded as a valuable adjunct to applications of pleiotropic PR-anatomical mutants such as *so* or *gl*, which are known or suspected to suffer some brain damage beyond their external PR and "circadian CNS" problems (see Helfrich-Förster and Homberg, 1993; Klarsfeld *et al.*, 2004). In any case, one can draw on these genetic tools to do things like conceptually subtracting the circadian-PR functions of external structures (*so*) from those of external eyes + H-B eyelet (*gl*), or one can cause flies to retain only the eyelet function (*so/cry^b*). Nutritional rearing provides an additional tool to affect functioning of the (eventual) animals' visual system, in particular by growing flies on vitamin A-deficient medium (Ohata *et al.*, 1998; *cf.* Zimmerman and Goldsmith, 1971); this results in severe knockdowns of the rhodopsins that are naturally contained within the main eyes, the simple ones atop the head, and the eyelet underneath the retina. Thus, with regard to all external photoreceptors and the H-B eyelet (summing to seven such PR structures), sensitivity for LD entrainment is sharply reduced in the mutant types that are, for example, missing their C.eyes or them plus Oc (Helfrich-Förster *et al.*, 2002a; Stanewsky *et al.*, 1998). These results are consistent with the effects of rearing *Drosophila* on vitamin A-less food (Ohata *et al.*, 1998). Because LD entrainability was not eliminated in these situations, a light-to-clock input role of extraocular photoreception was implied (e.g., in the aforementioned studies from the 1980s). Reentraining flies to either advanced or delayed LD cycles (in separate experiments) implied that the eyelet, cooperating with deep brain photoreception (CRY), is involved mainly in mediating behavioral phase delays (Helfrich-Förster *et al.*, 2002a). A further study of this sort applied the mutants and genetic combinations just noted; also *Drosophila* devoid of their C.eyes only (*eya*); or of them plus CRY (*eya/cry^b*); and even flies that were missing all relevant external structures (*n* = 5 eyes) plus the H-B eyelet, along with communication between these PR organs and second-stage interneurons of the visual system (*so/gl/hdc*). Most of these experiments, in turn, were based on entrainability of animals partially or massively gutted of putative circadian photoreceptive functions in non-12:12 LD cycles (Rieger *et al.*, 2003). Thus it was deduced, from analysis of behavior in 16:8 or (especially) 20:4 LD cycles, that all overtly eye-like PR types (C.eye, Oc, and H-B) donate functions pertinent to entrainment in such heavily L-enriched conditions; also that the C.eyes and Oc both contribute to (non entrained) period lengthening that can occur for certain mutants (notably *hdc* expressed by itself) in "long photoperiods" (again, as encompassed by LD cycles totaling 24 h in duration). However, most of this period-lengthening effect was deduced to "go through" the H-B eyelet and the light absorption CRY (Rieger *et al.*, 2003). Focusing on deep brain circadian photoreception, the *cry^b* mutation eliminates behavioral phase shifts that are normally induced by short light pulses in the early or late subjective night (in DD). Within the brain, CRY is found in LN pacemaker neurons and DN cells as well. Transgenically mediated restoration of *cry^+* function to certain LNs only (in flies homozygous for *cry^b*) allowed for solid light pulse responsiveness, as if extraocular photoreception is mediated partly by pacemaker cells themselves. That *cry^+* is expressed within DNs is interesting in light of the DN1- and DN3-to-LN axonal pathways shown in Fig. 3 and that the aforementioned *glass* mutation eliminates at least a substantial subset of the DN1s as well as the H-B eyelet. Anatomical studies of that internally located, eye-like entity revealed an eyelet-to-LN region axonal projection (as originally specified in empirical detail by Yasuyama and Meinertzhangen, 1999), designated by the two mildly curving horizontal lines. The external-PR-less and DN-disrupted $gl$ mutant could be readily reentrained in LD tests (as implied earlier), albeit with reduced sensitivity. This is similar to the effects of a $norpA$-null mutation (acting by itself) or of carotenoid depletion (see earlier discussion), keeping in mind that those two effects should not alter light reception by any of the relevant CNS neurons. Such resynchronization responses exhibited even lower sensitivity when a $norpA$ mutation was combined with $cry^b$, and the responsiveness was worse still (perhaps zero) in $gl$ $cry^b$ double mutants. Both of these doubly mutant types should be anatomically and/or physiologically nonfunctional for external + eyelet + deep-brain photoreception. Inasmuch as the $gl$-null $cry^b$ combination leads to extremely poor nonentrainability, the $glass$ mutation (but not $norpA$) is presumed to affect a "fourth pathway" by which photic stimuli get to the deep brain clock. Further mutationally based findings are on point (Rieger $et$ $al.$, 2003): With the external PRs gone and the DNs rendered nonfunctional in $so/cry^b$ flies, only half of these doubly mutant individuals—despite their retention of H-B eyelet functioning— could entrain to 12:12 LD cycles, fewer to 8:16 or 16:8, and none to 4:20 or 20:4. When the $hdc$ neurochemical mutation was combined with $cry^b$, presumably to eliminate all five of the separate input routes (C.eyes, Oc, H-B, LNs, and DNs) in a similar manner to the effects of $gl$ $cry^b$, a minority of $hdc/cry^b$ double mutants entrained to 12:12 and none in the other (> or < 12-h) photoperiods. This genetic combination can be inferred to be less severe than that of $gl$ $cry^b$ (Helfrich-Förster $et$ $al.$, 2001) or perhaps roughly equivalent ($cf.$ Mealey-Ferrara $et$ $al.$, 2003). In any case the effects of these maximally pleiotropic mutant combinations reinforce the reality of pathway #4 and that it would involve the DNs-to-LNs route noted earlier.

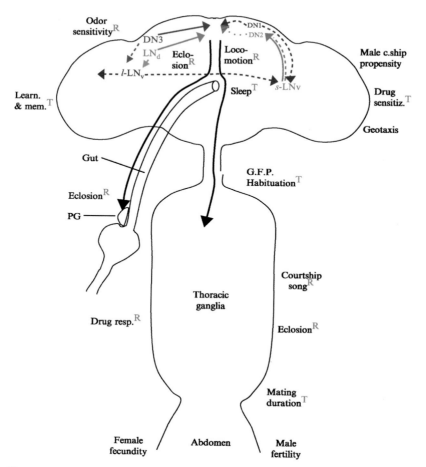

HALL, CHAPTER 4, FIG. 5. *Drosophila* phenotypes caused by rhythm mutations and tissue expressions of the corresponding genes. This diagram summarizes many of the biological correlates of such genetic effects and gene product locations. The former are indicated by uppercase words or phrases. These are placed within or next to a body region from which one or more rhythm-related mutations are known or presumed to regulate the phenotype or at least in some way to influence it. Abbreviations associated with certain such attributes of the behavior or physiology of the fly: C.SHIP, courtship; G.F.P., giant fiber pathway. Most of the phenotypes referred to have a temporal component, e.g., daily rhythmicity (**R**) of locomotion, eclosion, odor sensitivity, drug resp. (responsiveness), or *ca.* 1-min rhythmicity of courtship song. Other biological parameters, such as Sensitiz. (sensitization) to DRUG exposures, are associated with some sort of "time base" (**T**) not explicitly connected to rhythmicity; but certain additional phenotypes, notably female fecundity and male fertility, are not. The expression patterns of genes connected to *Drosophila* chronobiology are only alluded to for most of the relevant tissues, as opposed to being diagrammed in detail with regard to tissues and organs within a given body region and within which particular genes are known to make their products in these locations. Inside the head, colored brain neurons that contain "clock proteins" are shown along with most features of their neurite "wiring diagram" (*cf.* Fig. 2). Additional substances are colocalized within certain such cells, axons, or both; e.g., the PDF neuropeptide within ventro(v)-Lateral Neurons of the brain; cryptochrome in most of the six neuronal (N) cell groups. Clock genes are also expressed in many peripherally located sensory structures, as exemplified by the pictorial allusion to an antennal rhythm of odor sensitivity. However, certain rhythm-related genes are probably not neurally expressed within the head; e.g., *takeout*, which makes its products in fat-body tissue juxtaposed to the brain and is surmised to influence male c.ship thereby (via unknown molecular and cellular actions). An eclosion-regulating endocrine structure within the anterior throrax—the prothoracic gland (PG)—is pulled off to the side for diagrammatic purposes (same for the gut, within which certain clock genes make their products but for unidentified biological reasons). Other thoracic structures are not specifically depicted. In this regard it should be pointed out that the *period* clock gene, the one that has been assessed most extensively for tissue expression patterns, makes its thoracic ganglia products only within glial cells. These portions of the *Drosophila* CNS are the three anterior-most such ganglia within the ventral nerve cord (VNC). Elements of this figure depict eclosion control as originating (if you will) in anterior structures (LNs, PG). This regulation is ultimately "read out" via operation of the VNC and the locomotor events associated with emergence of an adult from the pupal case. The same kind of anterior → posterior pathway naturally underpins overall features of the daily behavioral cycles of the adult. However, the locomotion label is not reiterated with regard to this VNC involvement, because no gene-defined chronobiological factors are known to regulate the adult activity rhythm of the fly from within this body region. In contrast, an eclosion indicator is tacitly placed in association with the VNC because of expression of a specific gene in this portion of the CNS; its products are coexpressed with an eclosion-involved neuropeptide called CCAP within the VNC, in context of the relevant *lark* genetic variants influencing the timing of periodic adult emergence. (LARK protein is also contained within CCAP neurons in the brain, which project axons through the neck into the VNC.) The abdominal ganglion (not designated) is actually within the thorax and is the most posteriorly located portion of the VNC. As the diagram vaguely suggests, control of mating duration may occur via expression of clock genes in abdominal ganglionic cells (again, these are mostly or exclusively glia). The abdomen per se is barely included within the diagram, even though three clock genes are known to be expressed in various posterior organs. Among them are the female and male reproductive systems, whose local functions could be abnormal in the clock mutants that exhibit the aforementioned deficits in fecundity or fertility.

10 μm

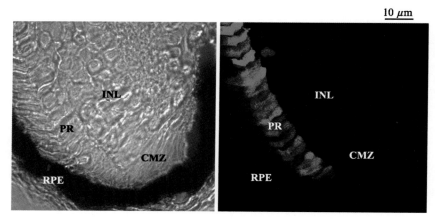

HAYASAKA *ET AL.*, CHAPTER 6, FIG. 2. A small portion of the nocturnin promoter is sufficient to drive GFP reporter gene expression specifically to differentiated photoreceptor cells. A piece of the *Xenopus nocturnin* promoter (−108/+20) was cloned upstream of a GFP reporter and used to generate transgenic *Xenopus*. Phase-contrast (left) and fluorescent (right) images of a portion of a transgenic tadpole eye show GFP expression in the cell bodies of the rod and cone photoreceptors (PR), but not in the other layers of the retina nor in the ciliarly marginal zone (CMZ) where undifferentiated retinal stem cells are located. RPE, retinal pigment epithelium; INL, inner nuclear layer. Reprinted with permission from Liu and Green (2001).

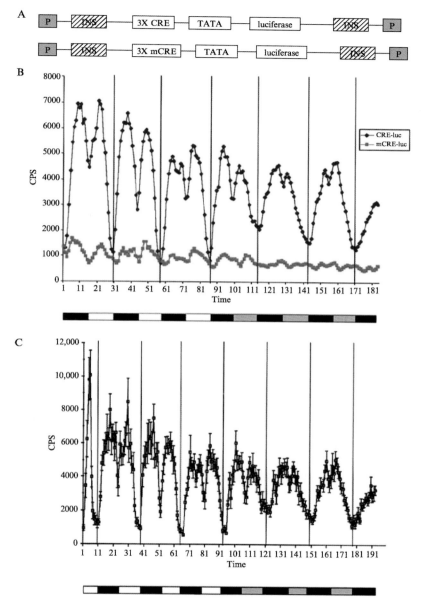

IIJIMA-ANDO AND YIN, CHAPTER 13, FIG. 1. Cycling of a dCREB2-responsive reporter. (A) Constructs used to generate CRE-luc and mCRE-luc transgenic flies. P, P transposable element inverted repeats; INS, SCS, and SCS' insulator elements; TATA, TATA box sequence from hsp 70 promoter. These constructs were cloned into the pCaSpeR transformation vector. (B) *In vivo* cycling of CRE-luc and mCRE-luc reporter expression as measured in a Packard TopCount luminometer. All time points represent an average of data points from 30 flies. The bar below the graph indicates light/dark conditions. White box, light period; black box, dark period; gray box, dark period during former light hours (subjective day). Vertical bars in the graph represent lights out and are spaced 24 h apart. (C) Similar graph as in B but with standard error bars added.

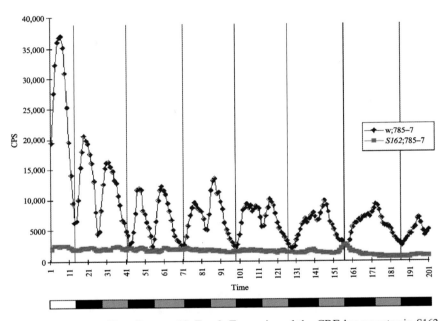

IIJIMA-ANDO AND YIN, CHAPTER 13, FIG. 2. Expression of the CRE-luc reporter in S162 mutant flies. *S162*/FM7 females were mated to males homozygous for the CRE-luc reporter transgene. Escaper males of the genotype *S162*/Y;CRE-luc/+ were assayed in the luminometer. Traces represent an average of data from 15 flies (*S162* mutants) or 30 flies (wild type). See Fig. 1 legend for a description of the graph.

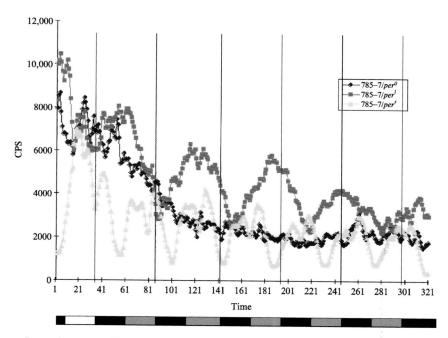

IIJIMA-ANDO AND YIN, CHAPTER 13, FIG. 3. Expression of the CRE-luc reporter in *per* mutant backgrounds. Females homozygous for one of three *per* mutations (described in the text) were crossed to males homozygous for the CRE-luc reporter, and male progeny of the genotype *per*/Y;CRE-luc/+ were assayed in the luminometer. *per⁰* or *per* null; *per¹* or *per* long; *perˢ* or *per* short. Flies were entrained on a 12-h light:12-h dark cycle for 4 days before the start of the experiment. Flies were then switched to constant darkness for the duration of the experiment. Each trace represents the average of data from 40 flies.

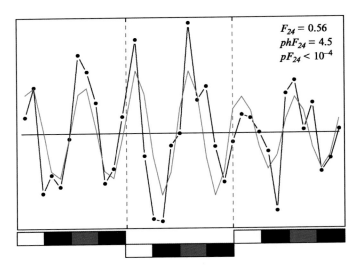

$$F_{24} = 0.56$$
$$phF_{24} = 4.5$$
$$pF_{24} < 10^{-4}$$

WIJNEN *ET AL.*, CHAPTER 15, FIG. 2. Extraction of the 24-h Fourier component to detect circadian periodicity. Black dots and lines indicate expression data from three 2-day LD/DD time course experiments for the *Ugt35b* gene. Data are presented in the form of $\log_2$ expression ratios normalized per experiment. The 24-h Fourier component shown in red was extracted from these normalized $\log_2$ expression ratios after correction for variations in amplitude between experimental repetitions (see text). The $F_{24}$ score (expressed in range 0–1) indicates the relative strength of the extracted circadian component, and $pF_{24}$ represents its peak phase relative to the onset of light at ZT0. $pF_{24}$ represents the probability of observing an $F_{24}$ score from randomly permuted data that is of equal or greater strength than the extracted Fourier component. Parts of this figure are reprinted from Claridge-Chang *et al.* (2001), with permission from Elsevier.

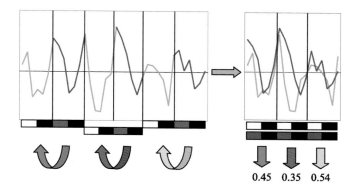

0.45   0.35   0.54

WIJNEN *ET AL.*, CHAPTER 15, FIG. 3. Calculation of 24-h autocorrelations to detect circadian periodicity. Normalized $\log_2$ expression ratio data for the gene *Ugt35b* (same as in Fig. 2) are used as an example. The 24-h autocorrelation for each 2-day experimental repetition is calculated by fitting the six time points collected during the first day of the time course in the presence of a light/dark cycle (orange lines) to those collected on the second day under conditions of constant darkness (blue lines). For details, see text.

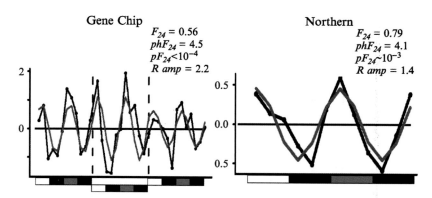

WIJNEN *ET AL.*, CHAPTER 15, FIG. 4. Independent verification of circadian periodicity. (Left) The circadian pattern detected by microarray analysis for the *Ugt35b* gene (black pattern) along with its extracted 24-h Fourier component (red pattern; see Fig. 2). (Right) Data from a 2-day LD/DD time course Northern analysis for the same gene. The Northern signals for *Ugt35b* were normalized to a loading control and expressed as $\log_2$ ratios relative to the experimental average. The extracted 24-h Fourier component is shown in red. Note the high $F_{24}$ scores and significant $pF_{24}$ values, as well as the similarity in the peak phases ($phF_{24}$) for microarray and Northern data. The relative amplitude (R amp) indicates the estimated fold change at the peak under the assumption that the patterns are perfect sine waves. For a detailed description of R amp calculation, see the on-line supplement for Claridge-Chang *et al.* (2001). Parts of this figure are reprinted from Claridge-Chang *et al.* (2001), with permission from Elsevier.

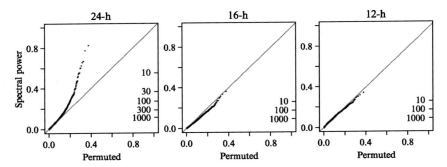

WIJNEN *ET AL.*, CHAPTER 15, FIG. 5. Global analysis of circadian periodicity using quantile–quantile plots of spectral power. Global trends of rhythmic expression in the heads of wild-type fruit flies are visualized for a microarray data set consisting of four LD/DD time course experiments. Quantile–quantile plots of spectral power (24-h, 16-h, 12-h Fourier components) are shown at circadian (24-h) and noncircadian (16-h, 12-h) periods. The real Fourier score for each probe set is indicated on the $y$ axis, whereas quantiles on $x$ values represent results from 1000 permutations of the time ordering for all probe sets. Quantile–quantile pairs connect identical ranks in the value-ordered sets for real and permuted data. Tick marks on the right show the number of probe sets selected by the indicated lower threshold values. The upward deviation from the diagonal (brown) seen at the 24-h period for highly ranked quantile–quantile pairs indicates enrichment in time traces with a strong circadian component. Fourier components with 16- and 12-h periods do not show similar deviations.

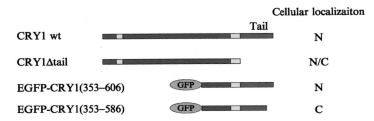

TAMANINI *ET AL.*, CHAPTER 20, FIG. 1. Examples of deletion, chimeric, and combined deletion/chimeric constructs used to delineate the C-terminal NLS in the mCRY1 protein. Proteins were expressed transiently in COS7 cells, and the subcellular localization of mCRY1 was determined by (immuno)fluorescence. The subcellular localization of the expressed protein is indicated on the right (N, nuclear; N/C, both nuclear and cytoplasmic; C, cytoplasmic).

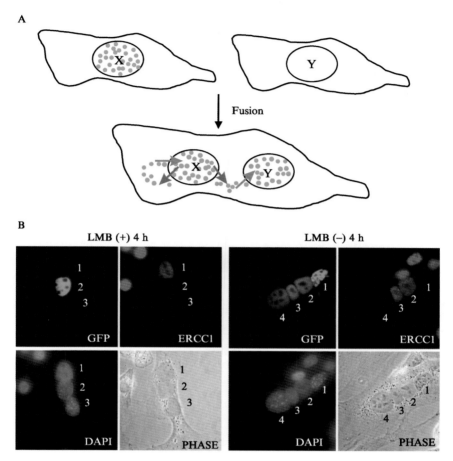

TAMANINI *ET AL.*, CHAPTER 20, FIG. 2. Analysis of protein shuttling using the heterokaryon assay. (A) Schematic representation of the principle of the heterokaryon assay. For an explanation, see text. (B) Example of the heterokaryon shuttling assay in which transfected NIH3T3 cells that transiently express nuclear mPER2(1–916)-GFP were fused to non-transfected HeLa cells in the presence of cycloheximide (CHX; used to block *de novo* protein synthesis) and in the presence or absence of leptomycin B (LMB; used to inhibit NES-mediated nuclear export). Shown are the subcellular localization of mPER2-GFP(1–916) (top left), as well as phase-contrast photographs of the heterokaryon (bottom right) and their DAPI-stained nuclei (bottom left). HeLa cell-derived nuclei were identified by immunofluorescence using antibodies recognizing the human ERCC1 (top right). Reprinted from Yagita *et al.* (2002), with permission.

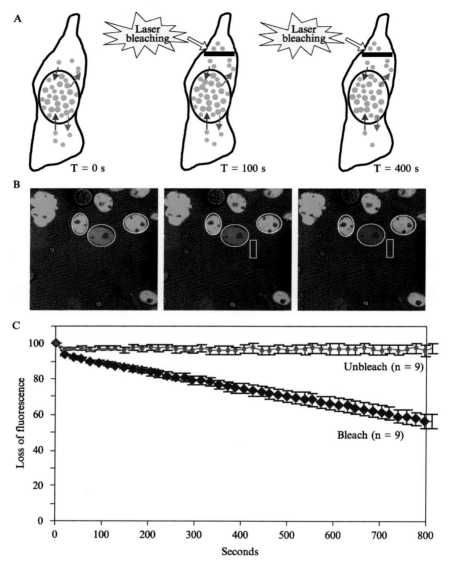

Tamanini *et al.*, Chapter 20, Fig. 3. Analysis of protein shuttling using the fluorescence loss in photobleaching (FLIP) technique. (A) Schematic representation of the principle of FLIP: GFP-tagged proteins are photobleached by laser pulses when crossing a defined area of the cytoplasm. Fluorescence is detected in the nucleus and the loss of signal is calculated. (B) Confocal images of COS7 cells transiently expressing the shuttling mPER2(1–916)-GFP protein taken at various time points after the start of the bleaching experiment. Note the loss of fluorescence in nuclei of cells subject to photobleaching. (C) Graphic representation of the loss of nuclear fluorescence in time in mPER2(1–916)-GFP expressing COS7 cells that received photobleaching light pulses and their neighboring (unbleached) cells.

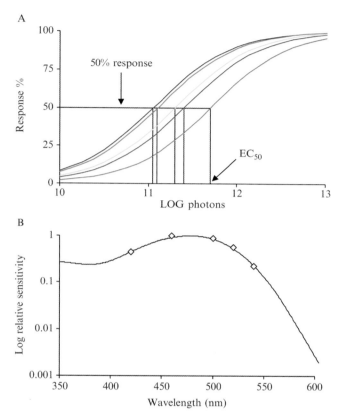

PEIRSON *ET AL.*, CHAPTER 37, FIG. 1. Basic principles of action spectroscopy. (A) Irradiance response curves (IRCs) of response magnitude versus log photon dose are plotted for each wavelength. Five schematic IRCs are shown. The photon dose required to elicit a half-maximum response is then calculated ($EC_{50}$). (B) The relative sensitivity is derived from the $EC_{50}$, and this value is plotted against wavelength to give the action spectrum ($\diamond$). A photopigment template may then be fitted to determine the $\lambda_{max}$ of the photopigment mediating the biological response under analysis (solid line).

A 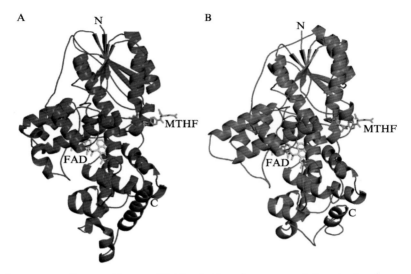 B

PARTCH AND SANCAR, CHAPTER 38, FIG. 2. Crystal structure of *E. coli* photolyase and homology-modeled human cryptochrome 2. (A) Ribbon diagram representation of *E. coli* photolyase showing the N-terminal $\alpha/\beta$ domain, the C-terminal $\alpha$-helical domain, and the positions of the two cofactors (Park *et al.* 1995). (B) The model of the hCRY2 tertiary structure was generated using the experimentally determined structures of *E. coli* and *A. nidulans* photolyases as templates, excluding the N-terminal 22 and C-terminal 80 amino acids of hCRY2 that have no homology to photolyase (Ozgur and Sancar, 2003).

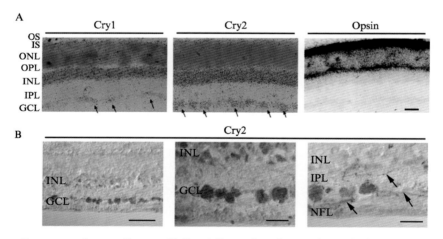

PARTCH AND SANCAR, CHAPTER 38, FIG. 4. Expression of cryptochrome in the mammalian retina. (A) Bright-field micrograms of *in situ* hybridizations comparing the expression of *mCry1*, *mCry2*, and opsins in the mouse retina. Bar: 30 $\mu$m (Miyamoto and Sancar, 1998). Arrows indicate clusters of ganglion cells expressing cryptochromes. (B) hCRY2 immunore-activity in the human retina. Arrows indicate cryptochrome expression in ganglion cell extensions. Bar: 50 $\mu$m in the lower left box and 25 $\mu$m in the lower right two boxes (Thompson *et al.*, 2003). Retinal histology: OS, outer segment; IS, inner segment; ONL, outer nuclear layer; OPL, outer plexiform layer; INL, inner nuclear layer; IPL, inner plexiform layer; GCL, ganglion cell layer.

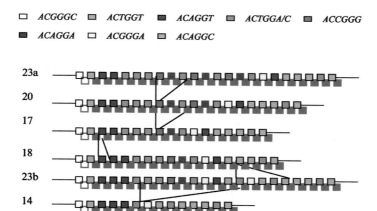

23a

20

17

18

23b

14

TAUBER AND KYRIACOU, CHAPTER 42, FIG. 1. Phylogeny of *period* Thr-Gly repeat alleles in *D. melanogaster*. Each colored box gives the encoding hexamer for a Thr-Gly dipeptide. Note how deletions and duplications of cassettes determine the observed variation in repeat number. All variants coalesce back to the (*Thr-Gly*)*23b* allele.

LEE, CHAPTER 45, FIG. 2. Physical phenotype of an *mPer1/mPer2* mutant mouse aged about 14 months. Note the prominent kyphosis and a general decline in physical appearance.

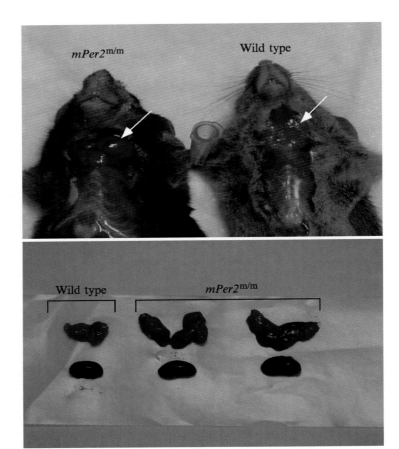

LEE, CHAPTER 45, FIG. 3. Abnormal cell growth control in *mPer2* mutant mice. (*Top*) *mPer2* and wild-type mice of about 8 months of age. The white arrow points to the salivary gland region. Note the enlargement of the salivary gland in the *mPer2* mutant mouse. (*Bottom*) The relative size of kidneys and salivary glands from wild-type and *mPer2* mutant mice.